RISK ASSESSMENT OF CHEMICALS

RISK ASSESSMENT OF CHEMICALS

An Introduction

2nd edition

Edited by

C.J. van Leeuwen

European Commission, Joint Research Centre, Ispra, Italy
Netherlands Organization for Applied Scientific Research TNO,
Zeist, The Netherlands

and

T.G. Vermeire

National Institute for Public Health and the Environment, Bilthoven, The Netherlands

A C.I.P. Catalogue record for this book is available from the Library of Congress.

ISBN 978-1-4020-6101-1 (HB)
ISBN 978-1-4020-6102-8 (e-book)

Published by Springer,
P.O. Box 17, 3300 AA Dordrecht, The Netherlands.
www.springer.com

Printed on acid-free paper

The first publication of this book (1995) was made possible with financial support provided by The Netherlands Ministry of Housing, Spatial Planning and Environment (VROM). The second edition of this book was made possible with the support of the National Institute for Public Health and the Environment, the European Commission and many colleagues working at different research institutes, the chemical industry and the following organizations: the U.S. Environmental Protection Agency, the Ministry of the Environment in Japan, Health Canada and Environment Canada, and the Organization for Economic Co-operation and Development.

PREFACE

Chemicals are used to make virtually every man-made product and play an important role in the everyday life of people around the world. The chemical industry is the third largest industrial sector in the world and employs millions of people. Since 1930, global production of chemicals has risen from 1 million tonnes to over 400 million tonnes annually. In 2004 the global sales were estimated at € 1776 billion. The EU accounts for approximately 33% of global sales. This gradual increase in the production and widespread use of chemicals was not without "cost". While chemicals play an important role in products for health and well-being, they may also pose risks to human health and the environment.

In 1992, at the United Nations Conference on Environment and Development (UNCED) in Rio de Janeiro, agreement was reached on an action plan for sustainable development in a number of policy areas. "Agenda 21" was born. The management of chemicals features prominently in Agenda 21, including the need to expand and accelerate the international assessment of chemical risks and strengthen national capacities for the management of chemicals. In the light of all of this, it is no coincidence that chemicals were again high on the agenda of the World Summit on Sustainable Development in Johannesburg in 2002. In South Africa our heads of state and governments undertook to minimize all adverse effects of chemicals within one generation, by the year 2020.

With the new legislative framework for industrial chemicals, i.e. REACH, Europe has moved from words to deeds in meeting the Johannesburg goal. REACH stands for Registration, Evaluation, Authorization and Restriction of CHemicals. The Regulation creates one system for the evaluation of all industrial chemicals with regard to their production, formulation, use and disposal. It will provide a high level of protection of human health and the environment and, at the same time, enhance the competitiveness of the EU chemicals industry.

Successful implementation of REACH will be a challenge. It will involve 30,000 chemicals, 30,000 companies, a newly created European Chemicals Agency and many other stakeholders. REACH will also be a scientific challenge. It will boost further scientific research into sustainable chemistry. It will also make us aware of the scarce human resources currently available to meet these challenges. Therefore I hope that the scientific community will shoulder its responsibility for training students in chemistry, technology, biology, toxicology and other sciences related to the development, assessment and management of chemicals.

The present volume is the 2nd edition of a book published in 1995. It is an introduction to the risk assessment of chemicals and contains basic background information on sources, emissions, distribution and fate processes for the estimation of exposure of plant and animal species in the environment and humans exposed via the environment, consumer products and in the workplace. It includes chapters on environmental chemistry, toxicology and ecotoxicology, as well as information on estimation methods and intelligent testing strategies. It describes the basic principles and methods of risk assessment in their legislative frameworks (EU, USA, Japan and Canada). The book is intended to be used by students in technology, health and environmental sciences. It also provides background material for those who are currently involved in the risk assessment of chemicals. I hope that this book will contribute to meeting the challenges we are currently facing throughout the world.

Janez Potočnik
Commissioner for Science and Research
European Commission

EDITORS

Cornelis Johannes (Kees) Van Leeuwen (1955) studied biology at the University of Utrecht (UU), where he received his masters' degree in 1980 *(cum laude)* and obtained his PhD further to a thesis about the ecotoxicological effects of pesticides in 1986. He began his career in 1980 as a plant ecologist at the University of Groningen and, shortly thereafter, became head of the Laboratory of Ecotoxicology at the Ministry of Transport and Public Works. He served in a research and advisory role in the implementation of the Netherlands' Pollution of Surface Waters Act. In 1987 he joined the Chemicals Division of the Ministry of Housing, Spatial Planning and Environment (VROM). In 1991 he became head of the Risk Assessment and Environmental Quality Division. He held a part-time professorship in biological toxicology at the Institute for Risk Assessment Sciences (IRAS) at the University of Utrecht. From 1997-2002 he was deputy director of the Centre for Substances and Risk assessment (CSR) at the National Institute for Public Health and the Environment (RIVM). He has been member of various policy and expert groups in the European Union, the International Rhine Committee, the Organization for Economic Co-operation and Development, the European and Mediterranean Plant Protection Organization and the Council of Europe. He was chairman of the OECD Hazard Assessment Advisory Body, vice-chairman of the OECD Joint Committee on Chemicals and member of the Scientific Committee on Toxicology, Ecotoxicology and the Environment (CSTEE) of the European Commission and external advisor to the Long-range Research Initiative (LRI) of CEFIC. In 2002 he became director of the Institute for Health and Consumer Protection of the Joint Research Centre of the European Commission. In this role he was responsible for the European Chemicals Bureau (ECB), the European Centre for the Validation of Alternative Methods (ECVAM), the Biotechnology and GMO unit, the Physical and Chemical Exposure unit and Biomedical Materials and Systems. In 2007 he retired from the European Commission and was appointed as principal scientist at TNO Quality of Life in the Netherlands.

Theodorus Gabriël (Theo) Vermeire (1953) studied chemistry and toxicology at the University of Utrecht. He received his MSc and teaching qualifications in 1978. After a 3-year teaching period as a volunteer in Zambia, he joined the Dutch Directorate-General of Environmental Protection in 1982 and started his career in risk assessment as a toxicologist contributing to projects of the WHO International Programme on Chemical Safety (IPCS) and UNEP International Register of Potentially Toxic Chemicals (currently: UNEP Chemicals). In 1987, he joined the National Institute for Public Health and the Environment (RIVM) in Bilthoven, the Netherlands and has served in several scientific and managerial functions up to this day. As a project leader, he was involved in many projects in the area of toxicological standard setting, human and environmental exposure assessment, human toxicological dose-response assessment and the development of risk assessment methodologies and tools. Major projects include the development of the Netherlands' Uniform System for the Evaluation of Substances (industrial chemicals, plant protection products and biocides) and the European Union System for the Evaluation of Substances (industrial chemicals and biocides). His present position at RIVM is deputy head of the RIVM Expertise Centre for Substances. As an expert with a wide knowledge of toxicology and risk assessment, he has been involved in many expert groups developing risk assessment guidance for IPCS, the European Union and the Organization for Economic Co-operation and Development. Further to his interest in capacity building and teaching he has been involved in the organization of international risk assessment courses and EU twinning projects and taken part in them. He is a member of the Scientific Committee of the European Environment Agency and editor of the journal Human and Ecological Risk Assessment.

CONTRIBUTORS AND REVIEWERS

CONTRIBUTORS

Dr. J. Alter
U.S. Environmental Protection Agency
Office of Pollution Prevention & Toxics
Washington, DC, USA

Dr. V. C. Armstrong
Environment Health Consultant
Ottawa, Canada

Dr. C. Auer
U.S. Environmental Protection Agency
Office of Pollution Prevention & Toxics
Washington, DC, USA

Dr. A.J. Baars
National Institute for Public Health and the Environment
Bilthoven, The Netherlands

Dr. J.G.M. Bessems
National Institute for Public Health and the Environment
Bilthoven, The Netherlands

Dr. B.J. Blaauboer
Utrecht University
Institute for Risk Assessment Sciences
Utrecht, The Netherlands

Dr. D.N. Brooke
Centre for Environment Risk Management
Garston, Watford, United Kingdom

Prof. J.C. Dearden
Liverpool John Moores University
School of Pharmacy and Chemistry
Liverpool, United Kingdom

Dr. J.H.M. De Bruijn
European Commission
Joint Research Centre
European Chemicals Bureau
Ispra, Italy

Dr. R. Diderich
Organization for Economic Co-operation and Development
Health and Safety Division
Paris, France

Dr. R.-U. Ebert
Helmholtz Centre for Environmental Research-UFZ
Department of Ecological Chemistry
Leipzig, Germany

Dr. P.J. Hakkinen
Gradient Corporation
20 University Road
Cambridge, MA, USA

Dr. B.G. Hansen
European Commission
DG Environment
Brussels, Belgium

Dr. R. Kühne
Helmholtz Centre for Environmental Research-UFZ
Department of Ecological Chemistry
Leipzig, Germany

Prof. Dr. M.S. McLachlan
Stockholm University
Department of Applied Environmental Science
Stockholm, Sweden

Dr. M.E. Meek
Health Canada
Existing Substances Division
Ottawa, Canada

Dr. C. Money
ExxonMobil
Machelen, Belgium

Dr. J.J.A. Muller
National Institute for Public Health and the Environment
Bilthoven, The Netherlands

Dr. M. Nendza
Analytical Laboratory
Luhnstedt, Germany

Dr. T.I. Netzeva
European Commission
Joint Research Centre
European Chemicals Bureau
Ispra, Italy

Dr. A. Paschke
Helmholtz Centre for Environmental Research-UFZ
Department of Ecological Chemistry
Leipzig, Germany

Dr. G. Patlewicz
European Commission
Joint Research Centre
European Chemicals Bureau
Ispra, Italy

Dr. Ir. W.J.G.M. Peijnenburg
National Institute for Public Health and the Environment
Bilthoven, The Netherlands

Drs. M.G.J. Rikken
National Institute for Public Health and the Environment
Bilthoven, The Netherlands

Drs. E. Rorije
National Institute for Public Health and the Environment
Bilthoven, The Netherlands

Prof. Dr. G. Schüürmann
Helmholtz Centre for Environmental Research-UFZ/
Technical University Bergakademie Freiberg
Leipzig/Freiberg, Germany

Dr. D.T.H.M. Sijm
National Institute for Public Health and the Environment
Bilthoven, The Netherlands

Prof. Dr. W. Slob
National Institute for Public Health and the Environment
Bilthoven, The Netherlands

Dr. E. Toda
Ministry of the Environment
Environmental Health and Safety Division
Tokyo, Japan

Dr. T.P. Traas
National Institute for Public Health and the Environment
Bilthoven, The Netherlands

Prof. Dr. Ir. D. Van De Meent
National Institute for Public Health and the Environment
Bilthoven, The Netherlands

Ing. P. Van Der Poel
National Institute for Public Health and Environmental Protection
Bilthoven, The Netherlands

Dr. Ir. J.G.M. Van Engelen
National Institute for Public Health and the Environment
Bilthoven, The Netherlands

Prof. Dr. C.J. Van Leeuwen
European Commission/TNO Quality of Life
Ispra, Italy / Zeist, The Netherlands

Drs. T. Vermeire
National Institute for Public Health and the Environment
Bilthoven, The Netherlands

Dr. A.P. Worth
European Commission
Joint Research Centre
European Chemicals Bureau
Ispra, Italy

Drs. P.G.P.C. Zweers
National Institute for Public Health and the Environment
Bilthoven, The Netherlands

REVIEWERS

Dr. S. Bradbury
U.S. Environmental Protection Agency
Office of Pesticide Programs
Washington, DC, USA

Prof. Dr. J. Bridges
University of Surrey
Robens Institute
Surrey , United Kingdom

Dr. F. Christensen
European Commission
Joint Research Centre
European Chemicals Bureau
Ispra, Italy

Dr. M. Comber
Exxon Mobil Petroleum & Chemical
Machelen, Belgium

Prof. Dr. M. Cronin
Liverpool John Moores University
School of Pharmacy and Chemistry
Liverpool, England

Drs. C. De Rooij
Solvay SA
Brussels, Belgium

Dr. J. Doe
Syngenta Crop Protection
Greensboro, USA

Drs. R. Donkers
European Commission
DG RELEX
Washington DC, USA

Dr. W. De Wolf
DuPont
Mechelen, Belgium

Dr. S. Eisenreich
European Commission
Joint Research Centre
European Chemicals Bureau
Ispra, Italy

Dr. J. Fentem
Unilever Colworth
Safety & Environmental Assurance Centre
Bedford, United Kingdom

Prof. Dr. H. Greim
Technical University of Munich
Institute of Toxicology and Environmental Hygiene
Munich, Germany

Dr. J. Hermens
Utrecht University
Institute for Risk Assessment Sciences
Utrecht, The Netherlands

Prof. Dr. B. Jansson
Stockholm University
ITM Department of Applied Environmental Science
Stockholm, Sweden

Dr. J. Jaworska
Procter & Gamble
Central Product Safety
Strombeek–Bever, Belgium

Dr. E. Karhu
European Commission
DG Enterprise
Brussels, Belgium

Dr. D. Kotzias
European Commission
Joint Research Centre
Physical and Chemical Exposure Unit
Ispra, Italy

Prof. Dr. D. Krewski
University of Ottawa
Ottawa, Canada

Prof. Dr. H. Kromhout
Utrecht University
Institute for Risk Assessment Sciences,
Utrecht, The Netherlands

Prof. Dr. M. Matthies
Universität Osnabrück
Institut für Umweltsystemforschung
Osnabrück, Germany

Prof. Dr. E. Millstone
University of Sussex
SPRU Science & Technology
Brighton, United Kingdom

Dr. S. Morishita,
Chemicals Evaluation Office,
Ministry of the Environment
Tokyo, Japan.

Dr. L. Musset
Organization for Economic Co-operation and Development
Environment, Health and Safety Division
Paris, France

Dr. D. Owen
Shell Chemicals
London, United Kingdom

Dr. F. Pedersen
European Commission
DG Environment
Brussels, Belgium

Dr. M. Scheringer
Eidgenössische Technische Hochschule Hönggerberg
Zürich, Switzerland

Prof. Dr. W. Seinen
Utrecht University
Institute for Risk Assessment Sciences,
Utrecht, The Netherlands

Dr. H. Shiraishi
National Institute for Environmental Studies,
Research Centre for Environmental Risk
Tsukuba, Japan

Dr. M. Sobanska
European Commission
Joint Research Centre
European Chemicals Bureau
Ispra, Italy

Dr. P. Thomas
Akzo Nobel Technology & Engineering
Environmental Chemistry and Regulatory Affairs
Arnhem, The Netherlands

Dr. J. Tolls
Henkel KGaA
Düsseldorf, Germany

Dr. D. Turnheim
Organization for Economic Co-operation and Development
Environment, Health and Safety Division
Paris, France

Dr. H. Tyle
Environmental Protection Agency
Chemicals Division
København, Denmark

Prof. Dr. M. Van Den Berg
Utrecht University
Institute for Risk Assessment Sciences,
Utrecht, The Netherlands

Ir. P. Van Der Zandt
European Commission
DG Environment
Brussels, Belgium

Dr. C.A.M. Van Gestel
Vrije Universiteit
Institute of Ecological Science
Amsterdam, The Netherlands

Prof. Dr. N. M. Van Straalen
Vrije Universiteit
Institute of Ecological Science
Amsterdam, The Netherlands

Dr. G. Veith
International QSAR Foundation
Two Harbors, MI, USA

Dr. R. Visser
Organization for Economic Co-operation and Development
Environment, Health and Safety Division
Paris, France

ACKNOWLEDGEMENTS

The second edition of this book is the result of a joint effort between the National Institute for Public Health and the Environment (RIVM) in the Netherlands, the European Commission and many experts in the field of risk assessment and the management of chemicals throughout Europe, the USA, Japan and Canada. The aspects of the first edition that have made it useful as a reference to students, scientists and risk managers have been retained. The second edition has been further expanded to include an update on current testing procedures, data evaluation and interpretation, fate, effects, legislation and terminology. New authors have been added to broaden the input and provide expanded coverage of these fields.

The editors would like to thank the Commissioner, Janez Potočnik, and the Director-General of RIVM, Marc Sprenger, for their support and final approval of this project. The early contacts that were established with Reinout Woittiez and Jan Roels at RIVM were crucial in catalyzing and managing this process. Thanks!

We would also like to thank the authors who made important contributions to the first edition of this book but were not able to contribute to the second edition: Joop Hermens, Tjalling Jager, Robert Kroes (†), Frank De Leeuw, Ton De Nijs, Jan Ros, Jaap Struijs, Wieke Tas, Martin Van Den Berg and Peter Van Der Zandt. We appreciate that you allowed us to update your previous contributions. We would also like to thank Christina Cowan, Mark Blainey and Ad Ragas for their contributions. In a period of particular change in the assessment and management of industrial chemicals we were very pleased that so many colleagues around the world volunteered to contribute to this book. We really appreciate your efforts on top of all the other obligations you face!

Special thanks are due to those who reviewed one or more chapters of this second edition. We would like to thank the members of the European Centre for Ecotoxicology and Toxicology of Chemicals (ECETOC) for their support. The involvement of industry experts in the drafting and review of the chapters has made it a real multi-stakeholder effort. We would like to thank Laura Bosatelli and Katrien Kouwenberg for their technical assistance and secretarial support. For the English correction we would like to thank Susan Hunt. Last, but not least, we would like to thank the Studio of RIVM (Wout Niezen, Jan de Bie and André Berends) for their creative inspiration. They turned our contributions (manuscripts, draft figures, tables and boxes) into camera-ready copy for the publisher.

EXPLANATORY NOTES

Prefixes to the names of units

M	mega (10^6)
k	kilo (10^3)
d	deci (10^1)
c	centi (10^{-2})
m	milli (10^{-3})
μ	micro (10^{-6})
n	nano (10^{-9})
p	pico (10^{-12})
f	femto (10^{-15})

Chemical prefixes

o	ortho
m	meta
p	para
n	normal
sec	secondary
tert	tertiary

Units

Å	Ångstrom (0.1 nm)
atm	atmosphere
°C	degree Celsius or centigrade
cal	calorie
d	day
g	gram
h	hour
ha	hectare
J	Joule
K	degree absolute (Kelvin)
kg	kilogram
L	litre
m	metre
M	molar (mol/litre)
min	minute
Pa	Pascal (unit of pressure; 100kPa = 1 bar)
s	second
V	Volt
W	Watt
y	year

Abbreviations

ACD	Allergic Contact Dermatitis
ADI	Acceptable Daily Intake
ADME	Absorption, Distribution, Metabolism and Excretion
AEC	Anion Exchange Capacity
AF	Assessment Factor or Application Factor
a.i.	Active ingredient
AIM	Analog Identification Methodology, USEPA
AIST	National Institute of Advanced Industrial Science and Technology, Japan
ALARA	As Low As Reasonably Achievable
ANN	Artificial Neural Network
ANOVA	ANalysis Of VAriance
APHA	American Public Health Association
ASTM	American Society for Testing and Materials
ATP	Adaptation to Technical Progress
AUC	Area Under the blood/plasma concentration vs. time Curve, representing the total amount of substance reaching the plasma
AVS	acid volatile sulphide
B	Bioaccumulation
BAF	Bioaccumulation Factor
BBA	Biologische Bundesanstalt für Land- und Forstwirtschaft
BCF	Bioconcentration Factor
BfR	German federal Institute for Risk Assessment
BIAC	Business and Industry Advisory Committee
BLM	Biotic Ligand Model
BMD	Benchmark Dose
BMF	Biomagnification Factor
BOD	Biological Oxygen Demand
b.p.	Boiling point
bw	body weight
CA	Competent Authority
CAS	Chemical Abstract Services
CBA	Cost-Benefit Analysis
CBB	Critical Body Burden
CBI	Confidential Business Information
CBR	Critical Body Residue
CCPs	Capacity Controlling Properties
CDC	Centre for Disease Control (and prevention)
CEC	Cation Exchange Capacity
CED	Critical Effect Dose
CEN	European Standardization Organization
CES	Critical Effect Size
CEPA	Canadian Environmental Protection Act

CFCs	Chlorofluorocarbons
ChemRTK	Chemical Right-to-Know initiative, USEPA
CICAD	Concise International Chemical Assessment Document, IPCS
C&L	Classification and Labelling
CMR	Carcinogenic, Mutagenic and toxic to Reproduction
CNS	Central Nervous System
COD	Chemical Oxygen Demand
ComET	Complex Exposure Tool, Canada
ComHaz	Complex Hazard tool, Canada
CSA	Chemical Safety Assessment
CSCL	Chemical Substances Control Law, Japan
CSR	Chemical Safety Report
CTV	Critical Toxicity Value
CT50	Clearance Time, elimination or depuration expressed as half-life
C.V.	coefficient of variation
Cyt	cytochrome
DfE	Design for the Environment program, OPPT
dfi	daily food intake
DIN	Deutsche Industrie Norm (German norm)
DNA	DeoxyriboNucleic Acid
DNEL	Derived No Effect Level
DOC	Dissolved Organic Carbon
DOM	Dissolved Organic Matter
DSL	Domestic Substances List, Canada
DT50	Degradation half-life or period required for 50 percent dissipation / degradation
DU	Downstream User
EASE	Estimation and Assessment of Substance Exposure [Model]
EbC50	Effect Concentration measured as 50% reduction in biomass growth in algae tests
EC	European Communities
EC10	Effect Concentration measured as 10% effect
EC50	median Effect Concentration
ECA	(1) Environmental Contaminants Act, Canada; (2) Enforceable Consent Agreement, USEPA
ECB	European Chemicals Bureau
ECHA	European CHemicals Agency
ECETOC	European Centre for Ecotoxicology and Toxicology of Chemicals
ECVAM	European Centre for the Validation of Alternative Methods
ED50	median Effective Dose
EEB	European Environment Bureau
EEC	European Economic Community
EEM	Emission Estimation Model
EEV	Estimated Exposure Value
Eh	Electrode potential
EHPV	Extended HPV chemicals programme, USEPA
EINECS	European Inventory of Existing Commercial Chemical Substances
ELS	Early Life Stage
EN	European Norm
ENEV	Estimated No Effects Value
EP	(1) European Parliament; (2) Equilibrium Partitioning
EPA	Environmental Protection Agency
ErC50	Effect Concentration measured as 50% reduction in growth rate in algae tests
EQO	Environmental Quality Objective
EQS	Environmental Quality Standard
ERA	Environmental Risk Assessment
ES	Exposure Scenario
ESD	Emission Scenario Document
ESIS	European chemical Substances Information System
EST	Embryonic Stem cell Test
EU	European Union
EUSES	EU System for the Evaluation of Substances [software tool in support of the TGD]
F	Variance ratio
FACA	Federal Advisory Committee Act, USA
FAO	Food and Agriculture Organization, UN
FCV	Final Chronic Value
FDA	Food and Drug Administration
FELS	Fish Early Life Stage
FFRP	Furniture Flame Retardancy Partnership, USEPA
FIFRA	Federal Insecticide, Fungicide and Rodenticide Act, USA
FYI	For Your Information submissions under TSCA
GAP	Good Agricultural Practice
GC	Gas Chromatography
GC-MS	Gas Chromatography-Mass Spectrometry
GHS	Globally Harmonised System of classification and labelling, UN
GLC	Gas-Liquid chromatography
GLP	Good Laboratory Practice, OECD
GPE	Greatest Potential for Exposure (of the general population), Canada
H	Henry coefficient
HC5	Hazardous Concentration for 5% of the

	species
H2E	Hospitals for a Healthy Environment program, USEPA
HEDSET	EC/OECD Harmonised Electronic Data Set (for data collection of existing substances)
HELCOM	Helsinki Commission - Baltic Marine Environment Protection Commission
HOMO	Highest Occupied Molecular Orbital
HPLC	High Pressure Liquid Chromatography
HPV	High Production Volume
HPVC	High Production Volume Chemical (> 1000 t/y)
HPVIS	HPV Information System, USEPA
HRA	Health Risk Assessment
IBT	Inherent Biodegradability Test
IC	Industrial Category
ICAPO	International Council on Animal Protection in OECD Programmes
IC50	median Immobilization Concentration or median Inhibitory Concentration
ICCA	International Council of Chemical Associations
ICHC	International Conference on Harmonization of Technical Requirements for Registration of Pharmaceuticals for Human Use
ILSI	International Life Science Institute
IOMC	Inter-Organization Programme for the Sound Management of Chemicals,
IPCS	International Programme on Chemical Safety
IPE	Intermediate Potential for Exposure (of the general population), Canada
ISO	International Organization for Standardization
ISO/DIS	International Organization for Standardization/Draft International Standard
ITC	Interagency Testing Committee, USEPA
ITS	Intelligent Testing Strategies
IUCLID	International Uniform Chemical Information Database
IUR	Inventory Update Rule under TSCA
i.v.	intravenous
JMPR	Joint Meeting of Experts on Pesticide Residues, WHO/FAO
k	Rate constant
K	Partition coefficient or equilibrium constant or distribution ratio or carrying capacity
K_{oa}	*n*-octanol-air partition coefficient
K_{oc}	organic carbon normalised solids-water partition coefficient
K_{ow}	*n*-octanol-water partition coefficient
K_p	solids-water partition coefficient
log	Logarithm (common, base 10)
ln	Logarithm (natural, base e)
L(E)C50	median Lethal (Effect) Concentration
LAEL	Lowest Adverse Effect Level
LC50	median Lethal Concentration
LD50	median Lethal Dose
LEV	Local Exhaust Ventilation
LFER	Linear Free Energy Relationship
LLNA	Local Lymph Node Assay
LOAEL	Lowest Observed Adverse Effect Level
LOEC	Lowest Observed Effect Concentration
LOED	Lowest Observed Effect Dose
LOQ	Limit Of Quantitation
LPE	Lowest Potential for Exposure (of the general population), Canada
LUMO	Lowest Unoccupied Molecular Orbital
m	Mean of population
MAC	Maximum Allowable Concentration
MAD	Mutual Acceptance of Data
MATC	Maximum Acceptable Toxic Concentration
MC	Main Category
MDS	Minimum Data Set
MFO	Mixed Function Oxidase
MIC	Minimum Inhibitory Concentration
M/I	Manufacturer / Importer
MITI	Ministry of International Trade and Industry, Japan
MM	Micromass (test)
MOA	Mode Of Action
MOE	Margin Of Exposure
MOS	Margin Of Safety
m.p.	Melting point
MRL	Maximum Residue Limit
MS	Mass spectrometry
MS-test	Multi-species test
MSDS	Material Safety Data Sheet
MW	Molecular Weight
NAEL	No Adverse Effect Level
n or N	Total number of individuals or variates
N-DSL	Non-Domestic Substances List, Canada
NF	Norme Française
NGO	Non-Governmental Organization
NIMBY	Not In My BackYard
NITE	National Institute for Technology Evaluation, Japan

No.	Number (in tables and parentheses)
NOAEL	No Observed Adverse Effect Level
NOEC(L)	No Observed Effect Concentration (Level)
NPPTAC	National Pollution Prevention and Toxics Advisory Committee, USEPA
NSN	New Substances Notification, Canada
NTP	National Toxicology Program, US
OCT	OECD Confirmatory Test (biodegradation)
OECD	Organization for Economic Cooperation and Development
OM	Organic Matter
OPPT	Office of Pollution Prevention and Toxics, USEPA
OSPAR	Oslo and Paris Convention for the protection of the marine environment of the Northeast Atlantic
p	Level of significance (probability of wrongfully rejecting the null hypothesis)
P	Persistent
P2	Pollution Prevention framework, USEPA
PAH	Polycyclic Aromatic Hydrocarbon
PBDE	PolyBrominated Diphenyl Ether
PBPK	Physiologically-Based PharmacoKinetic modelling
PBT	Persistent, Bioaccumulative and Toxic
PBTK	Physiologically-Based ToxicoKinetic modelling
PCBs	PolyChlorinated Biphenyls
PCDD	PolyChlorinated Dibenzo Dioxin
PCDF	PolyChlorinated Dibenzo Furan
PCRM	Physicians Committee for Responsible Medicine, USA
PEC	Predicted Environmental Concentration
PETA	People for the Ethical Treatment of Animals, USA
PFOA	Perfluorooctanoic acid
PFOS	Perfluorooctyl sulfonate
PLS	Partial Least Square
PMN	Premanufacture Notification under TSCA
PNC	Pre-Notification Consultation, OECD
PNEC	Predicted No Effect Concentration
p.o.	per os
POC	Particulate Organic Carbon
PoD	Point of Departure
POM	Particulate Organic Matter
POP	Persistent Organic Pollutant
PPA	Pollution Prevention Act, USA
ppb	Parts per billion
PPE	Personal Protective Equipment
ppm	Parts per million
PPORD	Product and Process Oriented Research and Development
PRTR	Pollutant Release and Transfer Register
PSI	Predetermined Set of Information, OECD
PSL	Priority Substances List, Canada
QA	Quality Assurance
QAAR	Quantitative Activity-Activity Relationship
QC	Quality Control
QSAAR	Quantitative Structure-Activity-Activity Relationship
QSAR	Quantitative Structure-Activity Relationship
r^2	Squared correlation coefficient or Coefficient of (multiple) determination
R (phrases)	Risk phrases according to Annex III of Directive 67/548/EEC
RAR	Risk Assessment Report
RBC	Red Blood Cell
RBT	Ready Biodegradability Test
RCF	Root Concentration Factor
REACH	Registration, Evaluation, Authorization and restriction of CHemicals
RIVM	National Institute for Public Health and the Environment, the Netherlands
RMM	Risk Management Measure
RNA	RiboNucleic Acid
RP	Reference Point
RRM	Risk Reduction Measure
RWC	Reasonable Worst Case
S (phrases)	Safety phrases according to Annex III of Directive 67/548/EEC
s^2	sample variance
σ	standard deviation of population
SAB	USEPA's Science Advisory Board
SAICM	Strategic Approach to International Chemicals Management
SAM	Standardized Aquatic Microcosm
SAR	Structure-Activity Relationships
SARA	Superfund Amendment and Reauthorization Act, USA
SCE	Sister Chromatic Exchange
SD	Standard deviation of series
SDS	Safety Data Sheet
SE	Standard error of mean
SETAC	Society of Environmental Toxicology And Chemistry
SF	Sustainable Futures initiative, OPPT
SIDS	Screening Information Data Set, OECD
SIEF	Substance Information Exchange Forum
SimHaz	Simple Hazard tool, Canada

SME	Small and Medium Enterprise
SimET	Simple Exposure Tool, Canada
SNac	Significant New Activity, Canada
SNAN	Significant New Activity Notice, Canada
SNIF	Summary Notification Interchange Format (new substances)
SNUR	Significant New Use Rule under TSCA
sp.	Species (when part of a bionomial)
SQO	Sediment Quality Objective
SSD	Species Sensitivity Distribution
SS-test	Single Species test
STP	Sewage Treatment Plant
$t_{1/2}$	Half-life
TCDD	2,3,7,8-tetrachloro-dibenzo-*p*-dioxin
TDI	Tolerable Daily Intake
TEER	Trans-Epithelial Electrical Resistance
TEF	Toxicity Equivalency Factor
TGD	Technical Guidance Document on risk assessment, EU
TIE	Toxicity Identification Evaluation
TLC	Thin Layer Chromatography
TLV	Threshold Limit Value
TNsG	Technical Notes for Guidance (for Biocides)
TNO	The Netherlands Organization for Applied Scientific Research
TSCA	Toxic Substances Control Act, USA
TSCF	Transpiration Stream Concentration Factor
TTC	Threshold of Toxicological Concern
TUAC	Trade Union Advisory Committee
TWA	Time-Weighted Average
UC	Use Category
UDS	Unscheduled DNA Synthesis
UN	United Nations
UNCED	UN Conference on Environment and Development
UNEP	United Nations Environment Programme
USEPA	Environmental Protection Agency, USA
vB	very Bioaccumulative
VOC	Volatile Organic Compound
vP	very Persistent
vPvB	very Persistent and very Bioaccumulative
VSD	Virtually Safe Dose
WEC	Whole Embryo Culture
WHO	World Health Organization
UV	Ultraviolet
v/v	volume/volume (concentration)
WoE	Weight of Evidence
w/v	weight/volume (concentration)
ww	wet weight
WWTP	Waste Water Treatment Plant

CONTENTS

PART III. EFFECTS ASSESSMENT AND RISK CHARACTERIZATION

PART IV. DATA

1. GENERAL INTRODUCTION

C.J. VAN LEEUWEN

1.1 INTRODUCTION

Over the last few decades there has been considerable activity in the field of risk assessment. This has mainly taken place in international bodies such as the Organization for Economic Co-operation and Development (OECD), the World Health Organization (WHO) - especially in the context of its International Programme on Chemical Safety (IPCS) - the European and Mediterranean Plant Protection Organization (EPPO), the Council of Europe and the European Centre for Ecotoxicology and Toxicology of Chemicals (ECETOC) [1-10]. Various directives and regulations in which risk assessment plays a crucial part have been issued by the European Community [11-14] and similar activities are taking place in other parts of the world, e.g., the U.S., Canada and Japan. Most of these developments would not have taken place without the contributions of many expert advisory bodies and individual scientists.

Historically, risk assessments have primarily focused on risks to human beings. It has gradually become apparent, however, that the ecological implications of large-scale environmental pollution should also receive attention. A situation has now been reached whereby detrimental ecological effects, caused e.g., by deforestation, food production (agriculture), excessive energy consumption, as well as the production and use of chemicals, have begun to threaten biological diversity and ecosystem integrity, and thus humanity's very existence. Accidents such as that at Chernobyl, the Sandoz disaster on the river Rhine, and recent cases of massive river pollution in China with benzene and cadmium, have increased awareness of the ecological and economic consequences inherent in such disasters.

Risk assessment is a central theme in the control of chemicals. Despite the role of risk assessment as the scientific foundation for many national and international regulatory guidelines, the phrase "risk assessment" means different things to different people and is often surrounded by misunderstandings and controversy. Some points of controversy involve the interpretation of scientific studies. Others have to do with science policy issues. Still others centre on definitions and on the distinctions between risk assessment and risk management. Some important definitions are given in Table 1.1.

The scope and nature of risk assessments range widely, from broadly based scientific analyses of air pollutants affecting a nation as a whole, to site-specific studies concerning chemicals in a local water supply. Some assessments are retrospective, focusing on the effects of a pollution incident, for example, the risks posed by a particular chemical dump site. Others seek to anticipate or predict possible future harm to human health or the environment, for example of a newly developed pesticide approved for use on food crops. In short, risk assessment takes many different forms, depending on its intended scope and purpose, the available data and resources, and other factors [15].

Risk management decisions may have local, regional or national consequences, but measures taken by a single country may also have world-wide consequences. Pollution does not recognize national borders. That is why the risk management of chemicals has become an important issue on the international agenda.

The development and international harmonization of risk assessment methodologies is recognized to be a great challenge. In Agenda 21 of the United Nations Conference on Environment and Development (UNCED), chapter 19 was entirely devoted to the management of chemicals [16]. The first recommendation of UNCED was to expand and accelerate the international assessment of chemical risks (Table 1.2), which requires mutual acceptance of hazard and risk assessment methodologies. Mutual acceptance of hazard and risk assessment methodologies (Figure 1.1) is considered to be the second essential step in the risk management process of chemicals, after international agreement was reached on the mutual acceptance of data by the member countries of the OECD [17]. The implementation of Agenda 21 is a long-term commitment. Therefore, it is no coincidence that chemicals were again high on the agenda of the World Summit on Sustainable Development in Johannesburg in 2002.

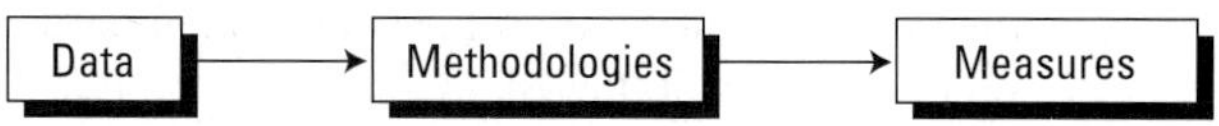

Figure 1.1. Mutual acceptance of data (cf. words) and hazard or risk assessment methodologies (cf. grammar) is essential to arrive at mutually accepted risk reduction measures (cf. language).

C.J. van Leeuwen and T.G. Vermeire (eds.), Risk Assessment of Chemicals: An Introduction, 1–36.

Table 1.1. Definitions of terms commonly used in the field of risk assessment and management.

Hazard is the inherent capacity of a chemical or mixture to cause adverse effects in man or the environment under the conditions of exposure

Risk is the probability of an adverse effect on man or the environment occurring as a result of a given exposure to a chemical or mixture

Risk assessment is a process which entails some or all of the following elements: hazard identification, effects assessment, exposure assessment and risk characterization

Hazard identification is the identification of the adverse effects which a substance has an inherent capacity to cause, or in certain cases, the assessment of a particular effect

Effects assessment, or more precisely, dose-response assessment is the estimation of the relationship between dose or level of exposure to a substance, and the incidence and severity of an effect

Exposure assessment is the determination of the emissions, pathways and rates of movement of a substance and its transformation or degradation in order to estimate the concentrations/doses to which human populations or environmental compartments are or may be exposed

Risk characterization is an estimate of the incidence and severity of the adverse effects likely to occur in a human population or environmental compartment due to actual or predicted exposure to a substance, and may include "risk estimation", i.e., the quantification of that likelihood

Risk management is a decision-making process that entails weighing political, social, economic, and engineering information against risk-related information to develop, analyse and compare regulatory options and select the appropriate regulatory response to a potential health or environmental hazard

Risk reduction is taking measures to protect man and/or the environment from the risks identified

Safety is defined as the strong probability that adverse effects will not result from the use of a substance under specific conditions, depending on quantity and manner of use

In this chapter a description is given about the risk management process in general. In Section 1.2 the 8 different steps are described. They reflect the current regulatory practice in most countries, where the work is mainly done by the public authorities. In Section 1.3 a number of changes are described that reflect recent developments such as the focus on risk reduction and responsible care *(reversal of the burden of proof)*, risk communication, the importance of stakeholder participation in all stages of the risk management process, risk assessment policy and integration in risk assessment. In Section 1.4 disciplines, roles and responsibilities in the risk management process are described. How risks are expressed is explained in Section 1.5 and risk perception is described in Section 1.6. Section 1.7 focuses on uncertainty, variability and precaution and Section 1.8 provides some concluding remarks. Finally, Section 1.9 gives a more detailed overview of the different chapters of the entire book.

1.2 THE RISK MANAGEMENT PROCESS

Risk encompasses impacts on public health and on the environment, and arises from exposure and hazard. Risk does not exist if exposure to a harmful substance or situation does not or will not occur. Hazard is determined by whether a particular substance or situation has the potential to cause harmful effects. The risk management process is triggered by concerns about the risks of particular uses of a chemicals or particular situations.

Table 1.2. Environmentally-sound management of toxic chemicals as recommended by UNCED [16].

a. Expanding and accelerating the international assessment of chemical risks
b. Harmonization of classification and labelling of chemicals
c. Information exchange on toxic chemicals and chemical risks
d. Establishment of risk reduction programmes
e. Strengthening of national capabilities and capacities for management of chemicals
f. Prevention of illegal traffic in toxic and dangerous products

Figure 1.2. The conventional wisdom is that risk management should not influence the processes and assumptions made in risk assessment. Regulatory practice, however, shows that the two elements depend on each other like Yin and Yang.

Risk assessment and risk management are closely related but different processes, with the nature of the risk management decision often influencing the scope and depth of a risk assessment [15]. In simple terms, risk assessors ask "How risky is this situation?" and risk managers then ask "What are we willing to accept?" and "What shall we do about it?" Risk assessment is usually seen as the objective/scientific part of the process and risk management as the subjective/political part. The distinction between these two components is important, though controversial. The conventional wisdom - which needs rethinking (Figure 1.2) - is that risk management should not influence the processes and assumptions made in risk assessment: the two functions should be kept conceptually and administratively separate [18]. Risk assessment provides *information* based on the analysis of scientific data which describe the form, magnitude, and characteristics of a risk, i.e. the likelihood of harm to humans or the environment. Although risk assessment is mainly a scientific task, political decisions are required on matters such as: "What are we trying to protect and to what extent should it be protected?" Endpoints, unacceptable effects, magnitude of uncertainty factors are controversial topics and based on implicit political choices. Questions about risk often have no scientific answers or the answers are multiple and contestable.

Risk management is about taking *measures* based on risk assessments and considerations of a legal, political, social, economic, and engineering nature. It is mainly a political process, although science is involved in the gathering of technical, social or economic information. The entire risk management process consists of eight steps (Figure 1.3), in which steps 1-4 belong to the risk assessment phase, while steps 5-8 are in the domain of risk management.

1.2.1 Hazard identification (step 1)

Hazard identification is the identification of the adverse effects that a substance has an inherent capacity to cause. It is the likelihood of harm due to exposure that distinguishes risk from hazard. Hazard identification involves gathering and evaluating data on the types of health effects or disease that may be produced by a chemical and exposure conditions under which environmental damage, injury or disease will be produced. For example, a toxic chemical that is hazardous to human health does not constitute a risk unless humans are exposed to it. The observed effects in humans may include reproductive defects, neurological defects or cancer. Ecological hazards include lethal effects, such

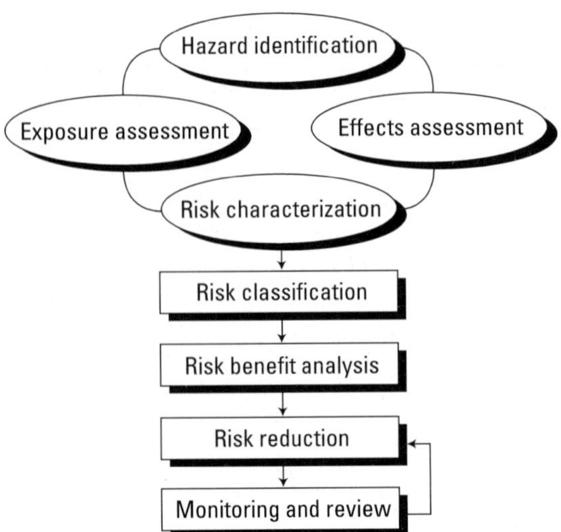

Figure 1.3. Steps in the risk management process.

as fish or bird mortality and sub-lethal effects on the growth and reproduction of various populations. This information may come from experimental laboratory studies, accidents or from other sources such as measured residues in fish or high concentrations detected at the workplace.

Hazard identification may also involve characterization of the behaviour of a chemical within the body and its interactions with organs, cells, or genetic material. The principal question is whether data from populations in which toxic effects and exposure occur suggest a potential problem for other populations under similar exposure conditions. Once a hazard (potential risk) has been identified, a number of other steps become important.

1.2.2 Exposure assessment (step 2)

Exposure can be assessed by measuring exposure concentrations, once chemicals are produced, used and emitted. With new chemicals, exposure assessments can only be predictions. This involves estimating emissions, pathways and rates of movement of a substance and its transformation or degradation in order to obtain concentrations or doses to which human populations or environmental compartments are or may be exposed. It involves describing the nature and size of the populations or compartments exposed to a substance, and the magnitude and duration of their exposure. The evaluation may concern past or current exposures, or anticipated future exposures. Multimedia exposure models are often used, especially in environmental exposure assessment (Chapter 4). Exposure assessment is also an uncertain part of risk assessment because of the lack of information on emission factors during the production of chemicals *(point-source pollution)*, and about the use of chemicals in various products and their emissions *(diffuse sources of pollution)*. The enormous geographic variability caused by differences in abiotic conditions, such as climate (e.g. temperature, humidity, wind speed, and precipitation), hydrology (e.g. different dilution factors in streams, lakes and rivers), geology (e.g. soil type) and biotic conditions (differences in ecosystem structures and functions) also contribute to this uncertainty. Exposure varies with time and depends on process-technology and the safety measures taken. It is therefore not surprising that measured environmental concentrations often differ by several orders of magnitude [19]. The same applies to occupational exposure and direct exposure to consumer products. It may be concluded that measurements of actual concentrations can help to reduce uncertainties in exposure assessment, but only for existing chemicals, not for new ones!

In health risk assessment (HRA) the various exposure routes are often combined in order to determine a total daily intake, expressed as mg per kg body weight per day. In ecological risk assessment (ERA) there is no single PEC or total daily intake, in fact, there are many PECs. This complexity is often simplified by deriving PECs for single environmental compartments: water, sediment, soil and air.

1.2.3 Effects assessment (step 3)

Effects assessment or, more precisely, dose-response assessment, is the estimation of the relationship between dose or level of exposure to a substance, and the incidence and severity of an effect. It sometimes involves the description of the quantitative relationship between the degree of exposure to a substance and the extent of a toxic effect or disease, but reliable quantitative precision cannot always be achieved. Data are generally obtained from (quantitative) structure-activity relationships (Chapters 9 and 10), read-across and *in vitro* studies or from experimental plant and animal laboratory studies or, less frequently, from experimental field studies with plants or animals, or epidemiologic studies of ecosystems and human populations (Chapters 6 and 7) or combinations of these (Chapter 11). Different dose-response relationships may be found if a substance produces different toxic effects. For instance, short-term exposure to high concentrations of benzene may produce lethal effects (acute toxic effects), whereas cancer may be induced as a result of long-term exposure to relatively low concentrations (chronic carcinogenic effects).

For most chemicals, no effect levels (NELs) derived from studies in laboratory animals are converted into predicted or estimated NELs (PNELs or DNELs) for humans or the environment by applying *assessment factors* usually in the range of 10-10,000 [2,20-22]. Assessment factors are numbers reflecting the estimated degree or amount of uncertainty when experimental data from model systems are extrapolated to humans or ecosystems. The rationale for assessment factors is that if no assessment factors are applied large groups of the human population or large parts of ecosystems will remain unprotected. This is because laboratory tests cover only a small part of the variety of responses that may occur in ecosystems and in human populations [2,20-22]. Experiments can yield both "false positives" and "false negatives". Extrapolation involves numerous scientific uncertainties and assumptions, which in turn involve policy choices.

In HRA, risk assessment focuses on one single species. Uncertainty is restricted to differences in sensitivity between laboratory mammals and humans, variations in exposure routes and differences in sensitivity between individuals *(intraspecies variation)*. In ERA millions of species may be exposed via a variety of routes (see Chapter 7). Therefore, many NELs can be determined. Differences in sensitivities between species *(interspecies variation)* play an important part in ERA. This complexity in ERA is often simplified by deriving predicted no effect concentrations (PNECs) for different environmental compartments: water, sediment, soil and air.

Please note that E stands for Effects in the acronym DNEL, PNEL and PNEC and for Exposure in the acronym PEC (predicted environmental concentration).

1.2.4 Risk characterization (step 4)

Risk characterization is the estimation of the incidence and severity of the adverse effects likely to occur in a human population or environmental compartment due to actual or predicted exposure to a substance, and may include risk estimation, i.e. the quantification of that likelihood. It generally involves the integration of the previous three steps [23]:

1. Hazard identification.
2. Effects assessment, i.e. the determination of the DNEL or PNEC.
3. Exposure assessment, i.e. the determination of the PEC or human intake or exposure.

A framework to define the significance of the risk is developed, and all the assumptions, uncertainties, and scientific judgements from the preceding three steps are considered. In many international regulatory frameworks environmental risks are often expressed as PEC/PNEC ratios, i.e. as *risk quotients* (Figure 1.4). For human risks a similar comparison between exposure and the NEL is usually made. It should be noted that these ratios or comparisons provide no absolute measure of risks. Nobody knows the real risks of chemicals where the exposure exceeds the PNEC or NEL. We only know that the likelihood of adverse effects increases as the exposure/effect level ratios increase. Thus, exposure/ effect ratios are internationally accepted substitutes for risks. It should also be noted that there is no such thing as precise risk assessments and scientists will always differ in the conclusions they draw from the same set of data, particularly if they contain some implicit value judgements.

At the present level of understanding we cannot adequately predict adverse effects on ecosystems, nor can we predict what part of the human population will be affected. We are only able to assess risks in a very general and simplified manner. In fact, the best we can do is provide a *relative risk ranking*. Risk ranking enables us to compare single chemicals or groups of chemicals once the risks of the respective chemicals have been assessed in a consistent "simplified" manner. Nevertheless, relative risk ranking allows us to replace

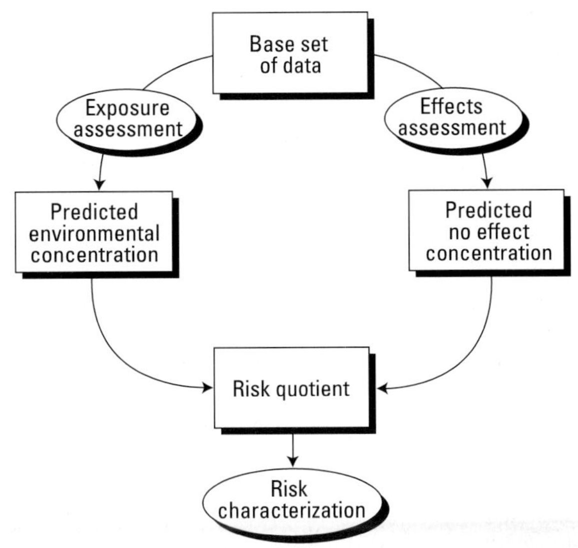

Figure 1.4. Risk characterization: a systematic procedure through estimation of exposure and effects.

dangerous processes, techniques or chemicals with safer alternatives in the risk management phase, without knowing the precise risks.

1.2.5 Risk classification (step 5)

Once a risk characterization has been made the focus turns to risk management. The first step in the risk management phase is the classification, i.e., the *valuation of risks* in order to decide if risk reduction is required. It is obvious that risks cannot be evaluated solely on the basis of scientific considerations, but who can decide what is acceptable? Decisions about risk classification are related to *risk acceptance* and must always be taken in a situation of some residual uncertainty. This is the field of policy-makers. According to Bro-Rasmussen [24] the term "acceptability" has become a crucial new element to be considered as a constituent part of the risk management process. The problem of defining operational criteria for "acceptable" and "unacceptable" risks is especially important in relation to the environment. Defining acceptable risk cannot be reduced to a mechanical exercise. It requires scientific knowledge as well as an appreciation of the limits of that knowledge. It requires a good understanding of the context of the risk and it requires willingness, by regulatory agencies as well as by their critics, to deal openly with these difficult, value-laden issues. Acceptability varies with time and place. What was acceptable in the past may not be acceptable in the future, and vice versa (Table 1.3). What may be acceptable in one country may be totally unacceptable in another. Cultural influences on risk management in legal and institutional frameworks are significant. It is important to realize that discussions on acceptability go back to our roots: to our youth, education and culture. In conclusion: risk classification is related to risk acceptability, which in turn is a risk-related, technical, social, cultural, political, educational and economic (conjuncture-dependent) phenomenon.

Over the past decade there has been growing support for defining two risk levels that may help to avoid lengthy debates about acceptability, because the area under discussion is restricted. These risk levels are known as:

- The upper limit, i.e. the maximum permissible level (MPL).
- The lower limit, i.e. the negligible level (NL).

These two risk limits create three zones: a black (high risk) zone, a grey (medium risk) zone and a white (low risk) zone. Actual risks in the black zone above the MPL are unacceptable and further risk management measures (RMMs) are necessary. Actual risks in the white zone below the NL (the *de minimus* level) are negligible (Figure 1.5) and further RMMs are not strictly required [25,26]. In the Netherlands, the lower limit for chemicals has generally been defined as 1% of the upper limit (Table 1.4). This approach has been adopted to take into account factors such as:

- Multiple exposure (additivity of risks and synergistic effects).
- Uncertainties in the estimates (limited testing and specific sensitivity).
- To leave a sufficient margin to distinguish between MPL and NL.

In the grey zone between the upper and lower limits, risk reduction is required based on the *ALARA* principle (as low as reasonably achievable). This is a powerful risk management principle. Managers are expected to do everything possible to reduce risks up to a limit they can justify to their organization and justify to the regulatory authorities. In general, the aim is to reduce risks until the cost of doing so is disproportionate to the benefit.

1.2.6 Identification and risk-benefit analysis of risk reduction options (step 6)

Once risk classification has been completed and risk reduction is thought necessary, the next consideration is the identification and analysis of options for risk reduction, and eventually selection of the most appropriate risk reduction option(s). The options for the risk reduction of chemicals range from slight adaptation of the production process or the intended use of the chemical to a complete ban on the production or use of a chemical. To that end a risk-benefit analysis *sensu lato* is carried out by drawing up of a balance sheet of the respective risks and benefits of a proposed risk-reducing intervention as compared to the baseline, i.e. the situation of not imposing risk reduction.

It is essential to remember that the result of risk classification is only one of the many aspects involved in the selection of regulatory options for risk reduction. This is the most difficult step in the risk management process, because it is a multifactorial task in which the risk manager has to consider not only the risk assessment but also other important aspects (Figure 1.6), such as:

- *Technical feasibility:* are measures technically feasible?
- *Social and economic factors:* e.g. what are the costs, do the measures affect employment or, in the case of extremely high risks, do we need to remove people from their homes?

Table 1.3. Changes in the perception of health and environmental risks and their solutions.

1970	1990
• Sectoral (air or surface water)	• Multiple media (including soil, sediment and groundwater)
• Localized	• Diffuse pollution
• Human health and well-being	• Ecosystem health, production functions and goods
• Local/regional	• National/international
• Limited economic damage	• Great economic damage
• End-of-pipe solutions	• Integrated approaches

Table 1.4. Risk limits for chemicals. From [23].

	Maximum permissible level	Negligible level
Man: *individual risk*		
chemicals with threshold	10^{-6}/y	10^{-8}/y
chemicals without threshold	PNEL	1% of PNEL
Man: *cumulative risk*		
chemicals without threshold	10^{-5}/y	10^{-7}/y
Ecosystems	PNEC[a]	1% of PNEC

[a] The PNEC is determined by using fixed assessment factors (little data) or variable assessment factors (adequate data set) calculated by means of a statistical extrapolation model with an arbitrary cut-off value set at a protection level of 95% of the species [24].

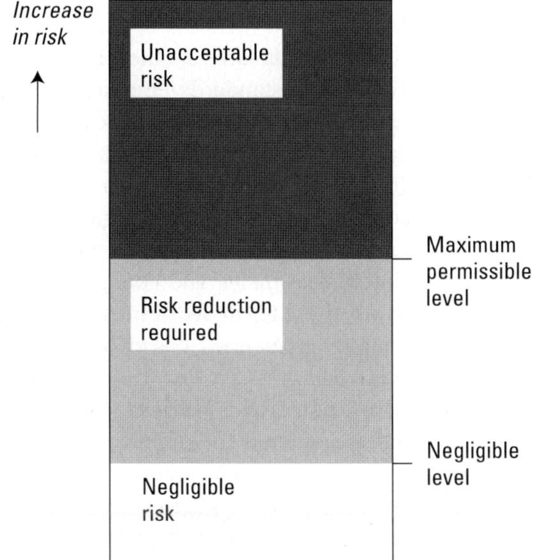

Figure 1.5. Risk limits and risk reduction.

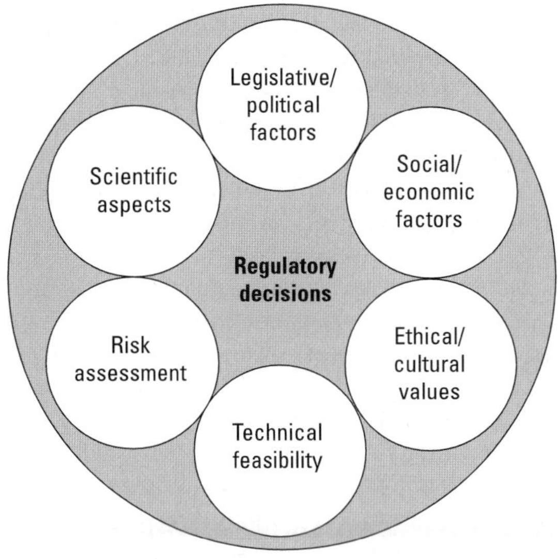

Figure 1.6. Elements in risk management. Modified from the U.S. Congress, Office of Technology Assessment [27].

- *Ethical and cultural values:* e.g. will a potential measure discriminate against specific groups in our society?
- *Legislative/political factors:* legal, regulatory, policy, and litigative constraints or risks, i.e. do we have appropriate regulatory, monitoring and enforcement tools?
- *Scientific aspects:* the limits of science are manifest at different levels; how great are the uncertainties in methodologies, measurements and other observations, extrapolations; do the risks affect mortality, morbidity or both and what assumptions have been made?

Selecting risk-reducing options will trigger "acceptability" discussions, not only about the predicted risks themselves but also about the anticipated consequences of risk reduction measures. This requires *risk communication*: a process by which stakeholders discuss risks and consequences with one another. Because the perception of risks (see Section 1.6) often differs widely, risk communication typically requires a sensitive approach and should involve genuine dialogue. The role of risk communication will be discussed in more detail in Section 1.3.2.

The use of a *cost-benefit analysis*, where the risks reduced by a proposed intervention are juxtaposed to estimate the net benefits (or net costs) to society and thus cover all major changes that will occur as a consequence of imposing a restriction compared with the baseline, is sometimes, but not always, a useful tool in risk management. To gauge benefit in an absolute sense, it is necessary to assign a value to the risk avoided (e.g. lives saved, lifetime extended). In general, the philosophy is that the greater the risk, the greater the incentive to reduce it. Estimated values of saving one additional "statistical life" can vary by at least six orders of magnitude [29,30]. Another relevant term used in this context is *cost-effectiveness* (determination of that action which maximizes the level of risk reduction per unit cost). Environmental risks are also difficult to quantify, although clean-up costs for polluted soil or sediment, as well as loss of fish stocks can be quantified. Cost-benefit analyses are useful in many contexts, certainly in ranking investments in some order of priority and effectiveness. However, this approach can only be a guideline, another input into a decision.

In conclusion, selecting the options for risk reduction using risk-benefit analysis is a multifaceted task centering on discussions about acceptability. Acceptability revolves around facts, value judgements and communication. It is this part of the risk management process in particular, where the lines between science, science policy and policy become fuzzy, that much conflict arises over where the boundary should be drawn [18]. Some of the forces at work in policymaking regarding human health and the environment are shown in Figure 1.7.

1.2.7 Risk reduction (step 7)

Risk reduction is taking measures to protect humans and/or the environment against the risks identified. Apart from the factors explained above, a number of additional factors should be taken into account before a risk management decision is taken, including those related to the implementation of RMMs. These considerations include: effectiveness, practicality, monitorability, equity, administrative simplicity, consistency, public acceptability, time, and the nature of the legislative mandate. There are different approaches to risk management (Chapter 2). In this Section only a brief summary will be given.

1. Classification and labelling

Notifiers of chemicals are required to provisionally classify and label dangerous substances on the basis of the intrinsic properties of the chemical. The decision on how to classify and label a substance is based on a series of criteria which themselves are based on the results of standard laboratory tests. The classification and labelling includes assigning a symbol (Figure 1.8), a risk phrase and a safety phrase [31,32]. Classification and labelling can be considered to be the first risk management tool for chemicals.

2. Safety standards

Safety or quality standards are another approach to chemicals control. Such standards are set with the intention of protecting human health and the environment. The terms criteria, guidelines, objectives, and standards, are often used. In this sequence the nature of the values moves from recommendations towards legally binding provisions. The use and interpretation of these terms varies between different agencies and countries. For the purposes of this book, these terms are defined as follows:

- *Criteria* are quality guidelines based on the evaluation of scientific data.
- *Guidelines* are numerical limits or narrative statements that are applied to support and maintain designated uses of the environment or to protect human health.
- *Objectives* are numerical limits or narrative statements that have been established to protect and maintain human health or designated uses of the environment at a particular site.

- *Standards* are fixed upper limits of exposure for certain chemicals that are laid down in enforceable laws or regulations by one or more levels of government.

Well-known examples of standards are the air, water and soil quality standards as well as the threshold limit values (TLVs) for airborne concentrations of industrial chemicals at the workplace. Environmental quality standards and TLVs are the control levels at which exposure is currently considered acceptable. They do not provide assurance of safety. Guidelines, objectives and standards for chemicals are derived from criteria, often by applying safety factors. Another example is the acceptable daily intake (ADI). The ADI is derived by applying a safety factor to no observed effect levels (NOELs) obtained from toxicological studies. An ADI is an estimate of the daily exposure dose that is unlikely to have any detrimental effects even if exposure occurs over a lifetime.

Absolute safety is a special case in safety standards. The most obvious example is the so-called Delaney clause, enacted by the U.S. Congress in 1958 as an amendment to the Food and Drug Act. This requires that no (food) additive shall be deemed to be safe if it is found to induce cancer when ingested by man or animal. The introduction of this amendment posed significant problems for the US authorities. In practice, the US authorities abandoned this approach in the mid-1980s.

3. Risk reduction measures *(sensu proprio)*
RMMs may comprise [33]:

- Technical measures such as redesign of production and use processes, closed systems, separation of man

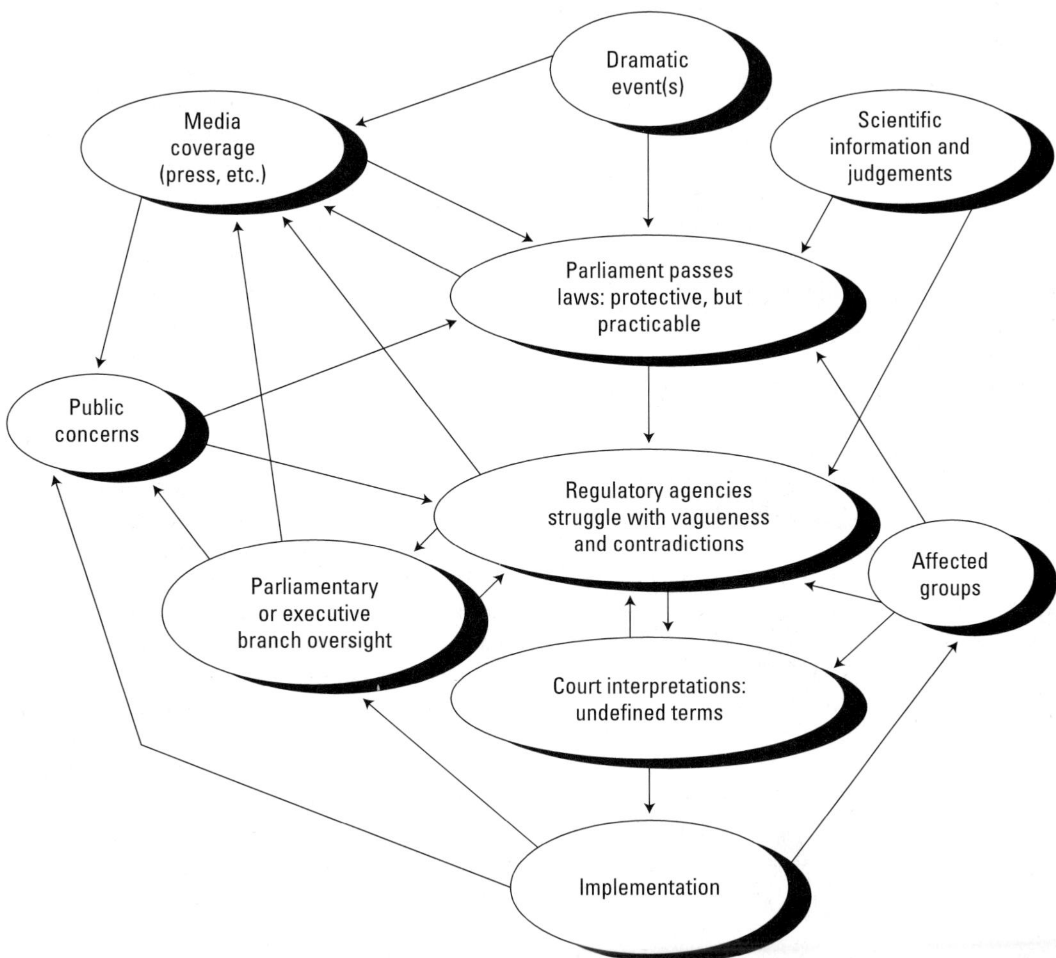

Figure 1.7. Forces in health and environmental policy-making. Modified from Lave and Malès [28]. With permission. Copyright 1989 American Chemical Society.

and sources (by construction measures), exhaustion, ventilation, separation and clarification techniques, physical, chemical and biological treatment.
- Organizational measures such as restriction to certain specific workplaces, limiting time of operation or work activities, training, monitoring and surveillance, prohibiting eating, drinking and smoking at the site.
- Instructions, information and warnings regarding normal use or safe use. This may include classification and labelling as described above.
- Personal protection measures such as gas and dust filter masks, independent air equipment, goggles, gloves and protective clothing.
- Product-substance related measures. Examples include limiting the concentration of a substance in a preparation or article.

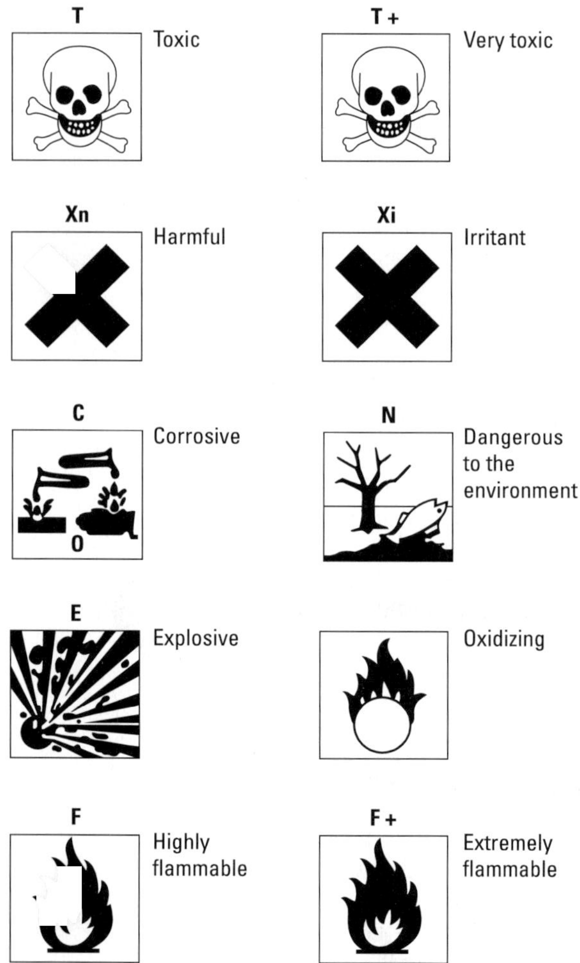

Figure 1.8. Classification and labelling of chemicals is the most frequently applied management tool in the control of chemicals.

- Instructions to limit the use of a substance or product. This can be implemented by limiting certain applications and uses and the restriction of uses with releases, etc.

1.2.8 Monitoring and review (step 8)

Monitoring and review is the last step in the risk management process. Monitoring is the process of repetitive observation for defined purposes of one or more chemical or biological elements according to a pre-arranged schedule over space and time, and using comparable and preferably standardized methods. Monitoring is undertaken to ensure that previously formulated standards are being met. In this sense monitoring serves an important function in enforcement (Figure 1.3), i.e. control. Monitoring serves a number of purposes [34]:
- The *control function* to verify the effectiveness of risk reduction (control) strategies and check for compliance.
- The *signal or alarm function* to be able to detect sudden (adverse) changes in human health and the environment. Ideally, the monitoring system should be designed such that the causes can be traced immediately.
- The *trend (recognition) function* to enable the prediction of future developments based on time-series analysis.
- The *instrument function* to help in the recognition and clarification of underlying processes.

Monitoring plays an important role in both environmental and health risk management. In health risk management biomonitoring is a part of the exposure-disease continuum as depicted in Figure 1.9. It can be used for consumer and occupational safety. Both biological monitoring and biochemical effect monitoring are crucial methods to help better understand the complex relationships between external and internal exposure and consequently, the potential adverse health effects that may result from exposure. Just like ambient monitoring, biological monitoring and biochemical effect monitoring should be regarded as exposure monitoring methods with high specificity for the substance being measured. Both methods give a measure of the total actual exposure regardless of the route of exposure [35]. Typical examples of biological monitoring are the determination of metals in blood or urine, unchanged substances (e.g. PCBs) in e.g. adipose tissue or blood, specific metabolites of a chemical in urine or volatile compounds in exhaled

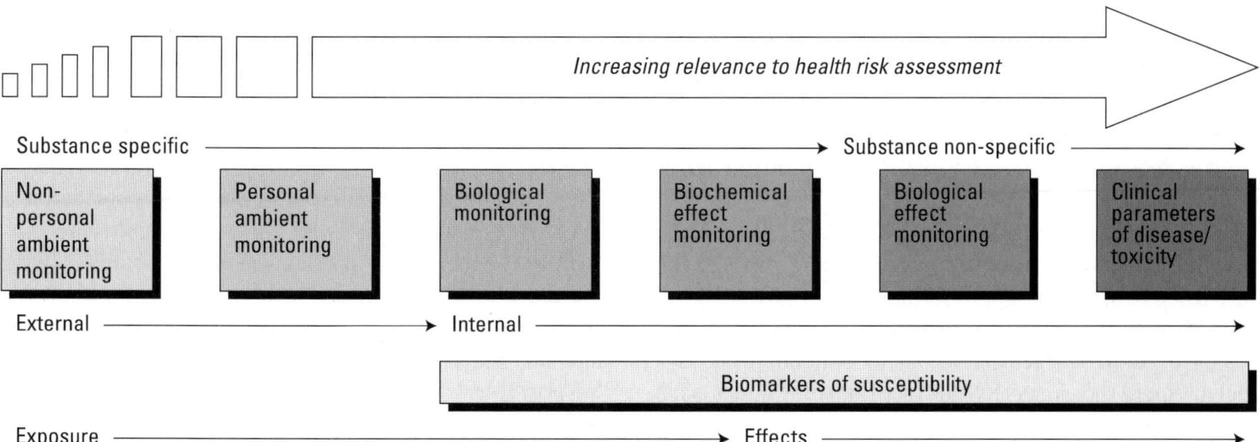

Figure 1.9. Monitoring techniques as part of the exposure-disease continuum according to ECETOC [35]. Non-personal external (ambient) monitoring includes static air monitoring, monitoring of soil, drinking, ground or surface water monitoring, and "food basket" monitoring. Personal external monitoring includes personal air monitoring and dermal exposure monitoring. With permission.

breath. Biochemical effect monitoring includes the determination of adducts of a specific chemical to DNA or a protein, or increased or decreased levels of specific enzyme activities.

Besides monitoring there are many other ways to review environmental and health management measures such as: audits and inspections, voluntary agreements and programmes, reporting (e.g. in case of voluntary agreements), market investigations, economic instruments, product registers, technology assessments, performance measurements and indicators for human health and sustainable development. These are equally important tools used to arrive at sustainable patterns of production, use and disposal of chemicals [36-39].

1.3 FURTHER DEVELOPMENTS IN THE RISK MANAGEMENT PROCESS

Section 1.2 described the different steps in the risk management process. These steps reflect the current practices of risk assessment and risk management. Compared to the first version of our book published in 1995, many developments have taken place and a number of them need to be highlighted in this new edition. They are crucial elements in the REACH legislation [40]. The major changes are shown in Table 1.5.

First of all, in the context of the REACH legislation [40], the focus has shifted from risk assessment to risk management, i.e. the implementation of RMMs, and from the principle of the authorities identifying and regulating the risks to industry taking its own responsibility for doing the assessments and implementing the necessary control measures to adequately control the risks.

Secondly, the risk management process has been put in a much wider context. The planning of risk assessments, the problem formulation phase, risk communication and stakeholder participation have all become more important. Communication with all the stakeholders at all stages in the process is crucial. In this respect the report of the US Presidential/Congressional Commission on risk assessment and risk management [41] and the guidelines for ecological risk assessment [42] have had a substantial impact.

The third major development is the inclusion of risk assessment policy as a specific component of risk management, as advocated by the Codex Alimentarius Commission [43]. This particular inclusion can help us to understand disagreements arising from differences in up-stream framing assumptions [43].

The fourth relevant development is the need for further integration between human health and

Table 1.5. Major changes in the risk management process during the last decade.

1. Focus on risk reduction and responsible care
2. Risk communication and stakeholder participation
3. Risk assessment policy and the role of science
4. Integration in risk assessment

environmental risk assessment [44-48]. These major trends and paradigm shifts will be discussed here in Section 1.3.

1.3.1 Focus on risk reduction and responsible care

Under the REACH legislation [40] emphasis will be placed on industry taking its own responsibility for the safe use of chemicals. This will take the form of a formal requirement to draw up *exposure scenarios*. These scenarios will be used as a tool to indicate what risk management measures (RMMs) will be used under what operational conditions to ensure that risks are adequatly controlled during the manufacture and use of chemicals. According to REACH, exposure scenarios will be developed for manufacturing processes and for identified uses of the substance on its own or in a preparation and for all life-cycle stages resulting from these uses.

Exposure scenarios are essential for risk management at the various life-cycle stages to ensure safe handling and adequate control of risks related to human health (workers, consumers and the general population exposed via the environment) and the environment. To be able to make realistic estimates of the exposures, it is important, as a first step, to determine which RMMs are already in place. These measures are an integral part of the overall process of developing exposure scenarios for identified uses of a substance on its own or in a preparation and the life-cycle stages resulting from these uses. How to arrive at appropriate exposure scenarios is an iterative process described in more detail in Chapters 2 and 12. Although a manufacturer or importer is not required to be proactive in seeking information on the uses of their chemicals, it will be beneficial to be so. It would allow them to develop a Chemicals Safety Assessment (CSA) covering all identified uses. Thus, already early in the process of developing the CSA, the manufacturer or importer should identify the uses of their chemicals and obtain sufficient information to develop exposure scenarios, e.g. by approaching customers, to be able to adequately control risks. Relevant RMMs should therefore be taken as a starting point for the development of exposure scenarios under the assumption that the described and recommended measures are implemented.

While in the past the entire risk management process, as described in Figure 1.3, was the responsibility of authorities (except for the implementation of risk reduction), the responsibility has now shifted to industry (manufacturers and importers in collaboration with their downstream users). Furthermore the focus has changed from risk assessment to risk management. These are fundamental changes by which the main policy objectives of REACH are achieved, i.e., the *reversal of the burden of proof* from the authorities to industry for testing and risk assessment and a shift in the focus on identification and implementation of RMMs to controlling the risks of chemicals. In this way REACH could be considered a legal instrument for implementing “responsible care”.

Little experience has so far been obtained with exposure scenarios and these new iterative approaches to reduce risks. This redesign of the risk management and the practical tools approach needed to implement “responsible care” or “risk reduction first” will be developed further over the next few years following a stakeholder participation process that will be described in more detail in the next section.

1.3.2 Risk communication and stakeholder participation

Risk communication is an essential interactive process among the stakeholders, i.e. risk managers, risk assessors, and those who may directly or indirectly be affected by the risk management decision. The general principles of risk management decision-making are given in Box 1.1. Risk communication is the link between risk assessment and risk management. Stakeholders who could potentially be included in any particular risk assessment are representatives of industry, public and occupational health professionals, public pressure groups, academic experts, specific consumer groups and private citizens. These stakeholders can participate in a number of ways, including assisting in the development of management goals, proposing assessment endpoints, providing valuable insight and information, and reviewing assessment results. Timely engagement of all the stakeholders will help to ensure that different technical perspectives, public values, perceptions, and ethics are considered [33,41,42].

Although the circumstances of stakeholder involvement will vary widely between risk assessments (depending on the regulatory and management context of the assessment), active stakeholder participation helps to ensure understanding and acceptance of assessment results and management actions.

Stakeholder participation and risk communication are key elements in a broad framework for risk management that was developed by the US Presidential/Congressional Commission [41]. This general framework was designed to help all types of risk managers - government officials, private sector business, and individual members of

Box 1.1. General principles for risk management decision-making [41]

A good risk management decision:

- Addresses a clearly articulated problem in its public health and ecological context.
- Emerges from a decision-making process that elicits the views of those affected by the decision, so that differing technical assessments, public values, knowledge, and perceptions are considered.
- Is based on a careful analysis of the weight of evidence that supports conclusions about a problem's potential risks to human health and the environment.
- Is made after examining a range of regulatory and non-regulatory risk management options.
- Reduces or eliminates risks in ways that:
  - Are based on the best available scientific, economic, and other technical information.
  - Account for their multisource, multimedia, multichemical, and multirisk contexts.
  - Are feasible, with benefits reasonably related to their costs.
  - Give priority to preventing risks, not just controlling them.
  - Use alternatives to command-and-control regulation, where applicable.
  - Are sensitive to political, social, legal and cultural considerations.
  - Include incentives for innovation, evaluation, and research.
- Can be implemented effectively, expeditiously, flexibly, and with stakeholder support.
- Can be shown to have a significant impact on the risks of concern.
- Can be revised and changed when significant new information becomes available, while avoiding "paralysis by analysis."

the public - make good risk management decisions. This Commission also broadened the definition of risk management. In their view risk management is "the process of identifying, evaluating, selecting, and implementing actions to reduce risks to human health and to ecosystems", whereas the goal of risk management was defined as "scientifically sound, cost-effective, integrated actions that reduce or prevent risks while taking into account social, cultural, ethical, political, and legal considerations". The framework consists of six consecutive stages:

1. Define the problem and put it in context.
2. Analyze the risks associated with the problem in context.
3. Examine the options for addressing the risks.
4. Make decisions about which option to implement.
5. Take actions to implement the decision.
6. Conduct an evaluation of the actions.

Every stage of this framework (Figure 1.10) relies on defining risks in a broader context, involving stakeholders, and repeating the process, or part of it, when needed. The problem formulation phase is the most important step. It establishes the goals, breadth, and focus of the assessment. It is a systematic planning step that identifies the major factors to be considered, linked to the regulatory and policy context of the assessment [42]. This step requires an intensive dialogue between all stakeholders to define the goals of the assessment (Box 1.2). The importance of problem formulation was also highlighted in the USEPA guidelines for ecological risk assessment [42]. Shortcomings consistently identified were: (1) absence of clearly defined goals, (2) endpoints

Box 1.2. Defining the problems and putting them in a context [41]

1. Identifying and characterizing an environmental health problem, or a potential problem, caused by chemicals or other hazardous agents or situations.
2. Putting the problem into its public health and ecological context.
3. Determining risk management goals.
4. Identifying risk managers with the authority or responsibility to take the necessary actions.
5. Implementing a process for engaging stakeholders.

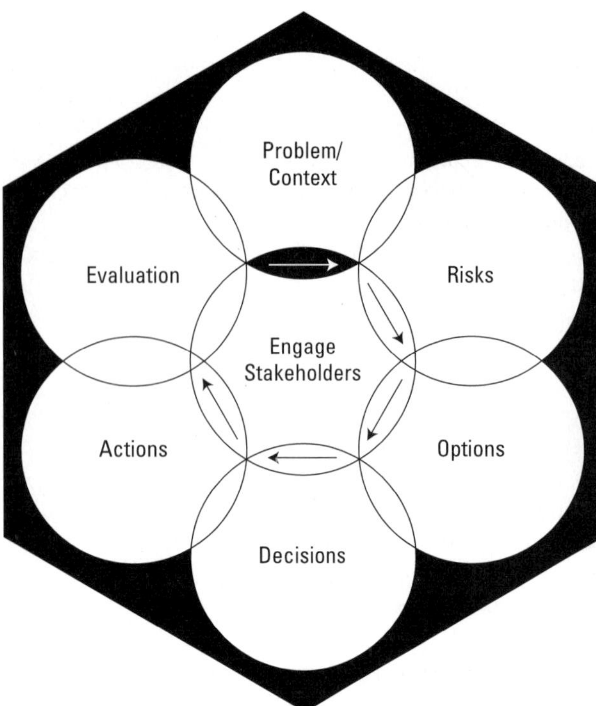

Figure 1.10. Framework for risk management according to the US Presidential/Congressional Commission [41].

that were ambiguous and difficult to define and measure, and (3) failure to identify important risks. These and other shortcomings can be avoided through rigorous development of the products of problem formulation as described by the USEPA [42].

The identification of "what" needs to be assessed is also known as the determination of the *assessment endpoints*. Further dialogue and interaction between risk managers and all other stakeholders will subsequently lead to a scientific and technical way of "how" to measure risk. This is the identification of the *measurement endpoints*. Essentially, this second step is a translation from higher-level decision criteria from the manager to the assessors by formulating evaluation questions with specific assessment and measurement endpoints and a testable hypothesis [49]. It is also during the problem formulation phase that the nature and extent of integration must be defined [45-47]. The involvement of stakeholders at all stages in the risk management process has great advantages (Box 1.3).

1.3.3 Risk assessment policy and the role of science

Risk assessment policy

A Dutch State-Secretary of science once said that science does not play a decisive role in important political decisions. Whether you agree or not, there is a growing recognition that science, on its own, cannot settle policy questions (see also Sections 1.2 and 1.7), and consequently that policy-makers need to take both scientific considerations and other legitimate factors into account (Figure 1.6). A common approach on the part of the policy-makers and their advisors [50] was to represent these deliberations and policy-making processes in terms of a model that did not acknowledge prior framing judgements. An important part of these framing assumptions concern what the Codex Alimentarius Commission (CAC) calls "risk assessment policy" [43]. According to the CAC *risk assessment policy* comprises documented guidelines on the scope of the assessment, the range of options (and associated judgements for their application) at appropriate decision points in the risk assessment such that the scientific integrity of the process is maintained [43]. Very often, the key difficulty facing risk managers, expert advisors and policy analysts has been to understand how, within the policy-making process, scientific considerations and other relevant factors can be distinguished and separated from each other and yet ultimately brought together to arrive at informed, systematic, complete, unbiased and

Box 1.3. Seven benefits of engaging stakeholders [41]

1. Supports democratic decision-making.
2. Ensures that public values are considered.
3. Develops the understanding needed to make better decisions.
4. Improves the knowledge base for decision-making.
5. Can reduce the overall time and expense involved in decision-making.
6. May improve the credibility of agencies responsible for managing risks.
7. Should generate better accepted, more readily implemented risk management decisions.

transparent decisions. The relevance of risk assessment policy has been demonstrated in a critical analysis of trade disputes [50]. The study showed that:

1. Different judgements were made about what the breadth and scope of scientific risk assessments should be.
2. Different judgements were made about the ways uncertainties should be handled by risk assessors, and the significance that should be ascribed to them.
3. Different judgements were made about the benchmarks by reference to which the available evidence is interpreted.
4. Different judgements were made about the "chosen level of protection" i.e. the extent to which risks and uncertainties are socially acceptable.

These *prior framing assumptions* [50], may have to do with very practical questions related to the management context, identification of the assessment and measurement endpoints, or to specific questions related to the data and the risk assessment methodologies [49,50]. Just to mention a few:

1. What management decisions will the risk assessment support?
2. What are the time constraints on performing the risk assessment?
3. What is the budget for the risk assessment, including the collection and generation of additional data and/ or modelling?
4. Is there going to be more than 1 assessment (i.e. more than 1 alternative to be examined)?
5. What is the maximum level of uncertainty that will still allow for a decision to be made, and how should uncertainties be handled?
6. What are the reference conditions against which possible adverse effects or risks will be compared?
7. Which impacts are deemed to be within the scope of the assessment and which are outside it?
8. What kind of evidence can be included and what can be discounted?
9. How should the available evidence be interpreted?
10. How much of different kinds of evidence would be necessary or sufficient to justify different judgements?

In conclusion, risk assessment policy judgements have routinely played a key role in risk policy-making processes, but they have often remained implicit, unacknowledged and unexamined [50]. As a result, the CAC [43] concluded that: "the determination of risk assessment policy should be included as a specific component of risk management. Risk assessment policy should be established by risk managers in advance of risk assessment, in consultation with risk assessors and all other interested parties. The procedure aims at ensuring that the risk assessment process is systematic, complete, unbiased and transparent...Where necessary, risk managers should ask risk assessors to evaluate the potential changes in risk resulting from different risk management options."

The role of science

Further to the increased emphasis on risk communication and stakeholder participation different levels of scientific involvement can be distinguished. The first type of approach articulated by policy-officials can be encapsulated in what is termed a *technocratic model.* A technocratic model assumes that risk policy can and should be decided solely by reference to scientific considerations and expert advice. In short it is "on, and only on, the basis of sound science" [50]. This reflects the thinking of the 1980s, but has its roots in the 1890s. [51] The technocratic model is incapable of explaining how to make decisions in conditions of acknowledged scientific uncertainties and neglects many other relevant factors, as given in Figure 1.6.

In response to the inadequacies of the technocratic model, an increasingly large portion of public policy-makers and their advisors now represent the processes in which they participate as a *decisionist model* [50]. This closely corresponds to the model described in Figure 1.3. It assumes that risk policy is, and should be, the product of a two-stage process, the first of which is purely scientific (risk assessment) and a second one that includes economic, social, technical, political and other considerations, often called risk management. This model reflects the thinking of the early 1990s. The decisionist model assumes that the risk assessment phase is entirely independent on any and all risk management considerations and judgements which, of course, it is not (Figure 1.2). For instance, every mandatory risk assessment of a chemical starts with an explicit political decision about the core set of data - the basis - on which a risk assessment will be performed. These discussions have been dominated by politicians, not only decades ago, e.g. in discussions about minimum data requirements (pre-marketing set of data or base set in the OECD and the EU respectively), but also more recently in the context of the political discussion about REACH.

The results of the study of trade disputes [50] can effectively be incorporated in a third model on how science and governance should interact. This third model

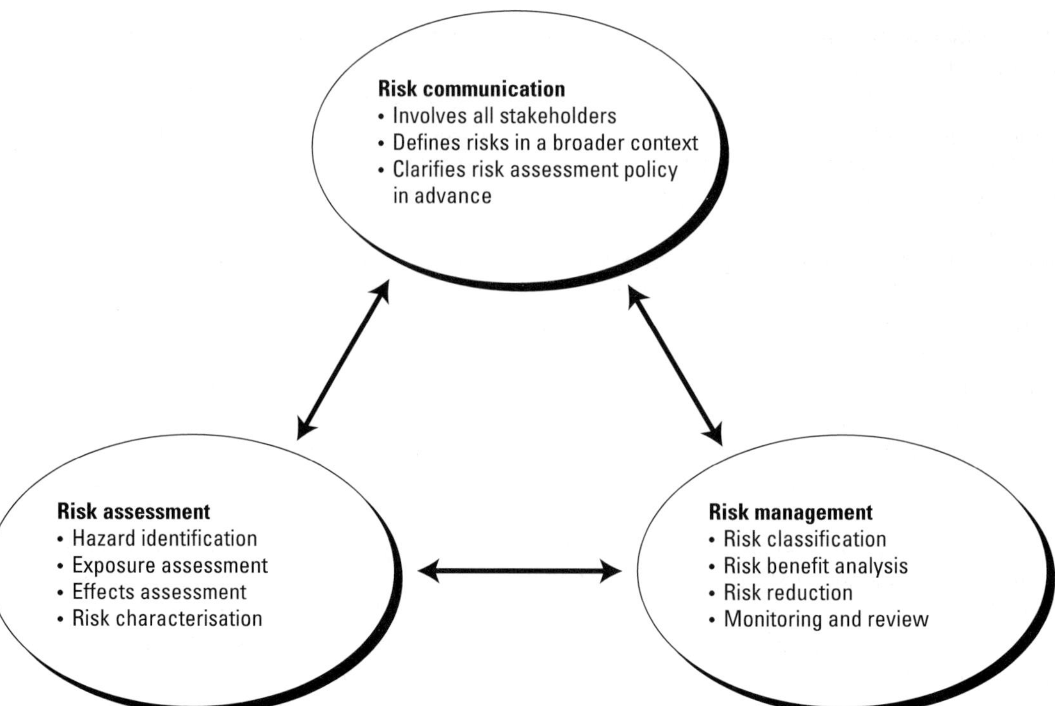

Figure 1.11. Risk analysis. Future processes of risk management should focus on risk communication in an interactive dialogue with all stakeholders and clarification of risk assessment policy at all stages [41-43].

emphasises the importance of risk assessment policy as given by the CAC [43]: "risk assessment policy should be established by risk managers in advance of risk assessment, in consultation with risk assessors and all other interested parties." This third model, the *transparent model* [50], assumes that not just science-based risk assessments play a role in policy-making processes, but that the risk assessments are also influenced by the socio-economic, cultural and political contexts in which they are developed. The transparent model assumes that non-scientific considerations play a distinctive up-stream role in setting the framing assumptions that shape the ways in which risk assessments are performed. It implies that rather than leaving those assumptions implicit, and leaving risk assessors to take responsibility for non-scientific judgements, risk managers should provide their risk assessors with explicit upstream framing guidance. In this way the transparent model can be considered as a three-stage iterative process (Figure 1.11) encompassing:

1. Risk communication over the entire process embedded in an iterative dialogue engaging all stakeholders at all phases (Figure 1.10) with a focus on risk assessment policy in advance of the actual risk assessment [41-43,50]. This phase is dominated by legislative, socio-economic and political considerations, particularly in relation to public and occupational health and environmental protection.
2. Risk assessment, (steps 1-4 in Figure 1.3) dominated by scientific considerations.
3. Risk management decision-making (steps 5-8 in Figure 1.3) based on technical, economic and social information.

Does this mean that the role of science in the overall risk management process has decreased over time? First of all, the process of risk assessment has not changed fundamentally in the past 25 years [41,52]. Secondly, scientists will continue to play a crucial role in the problem formulation and dialogue with risk managers in the development of risk assessment methodologies, in making explicit what we want to protect and at what levels, and in providing clarity about the uncertainties in assessments that were made [41,42,47,49]. The real change is the increased awareness that scientists are part of an overall risk management process in which the role of other stakeholders (Box 1.3) has increased. This new way of thinking is gradually implicitly, and sometimes explicitly, finding its way into current practice. Examples include the development of the REACH legislation [40], the development of Technical Guidance Documents [22] and other REACH implementation projects [33,53]. It is obvious that further work is necessary to provide clarity on the quality of the input (data), testing/assessment strategies, risk assessment models and their assumptions, simple tools to quantify uncertainties, guidance about acceptable levels of uncertainty and further practical guidance concerning the risk characterisation/risk management interface.

The debate on how to use and implement scientific expertise continues! The role of scientific expertise in EU policy making remains under discussion. In multiple scientific committees, experts provide guidance to regulators and decision-makers about the potential risks to human health and the environment. Proponents of a "technocratic" approach claim the credibility problem of supranational regulation caused by extensive politicization. They want to provide far-reaching delegation of powers to independent experts. Representatives of a "democratic" approach argue for a more socially inclusive use of expertise by providing a broader participatory mechanism for a variety of stakeholders as described above [54,55]. These views are conflicting. Gaps remain in the scope of the operational guidelines for the inclusion of scientific evidence in the legislative process, and in relation to information quality, the interpretation of evidence and the reporting of results. There is a lack of institutional mechanisms to ensure the integrity, quality, and effective operation of the scientific advisory system. In a recent review it was concluded that there are weaknesses in the effective use of scientific evidence in policy-making and regulatory decision-making processes by the European Union and a structured programme of reform has been proposed [56,57].

A further focus on risk assessment policy as advocated by the CAC [43] will allow us to arrive at considerable improvements regarding informed, systematic, complete, unbiased and transparent decision-making. A recent bulletin of the US Office of Management and Budget [58] contains clear proposals to make the risk assessment process better understood, more transparent and more objective. It also broadens the set of circumstances in which risk assessment needs to be done. The purpose of this proposed risk assessment bulletin is to enhance the technical quality and objectivity of risk assessments prepared by federal agencies in the USA by establishing uniform, minimum standards. Under the REACH legislation [40] emphasis will be placed on industry's own responsibility for the safe use of chemicals. Once uniform, minimum standards have been formulated and internationally agreed, correct implementation by regulatory agencies, as well as industries, will enhance the scientific and technical quality of the risks assessments. It will foster international collaboration on risk assessments of chemicals *(sharing the burden)*. It will facilitate communication about risks and it will embed risk assessment more deeply in the decision-making process.

1.3.4 Integration in risk assessment

As indicated above, it is important to deal with both human health risks and environmental risks. Let us first look at the effects part of the risk equation, i.e., the toxicological and ecotoxicological effects. From a scientific viewpoint, studies into the mechanism of toxicity should be a central element of risk. This is because the mechanism of toxicity is often similar across a wide range of species, even though the observed endpoints may vary [44,46,59]. The most obvious benefits of integrated assessment come from the sharing of information and even collaboration in the generation of hazard information by health and ecological risk assessors. Successful integration of human health and environmental (ecological) risk assessment must begin with the recognition that, for pragmatic rather than scientific reasons, the strategies for these areas have developed independently of one another. This is despite the fact that in many situations human health risk and environmental risks are interdependent. For many uses of chemicals there is a legal requirement that an assessment be made for the risks to both human and ecological health, but commonly these assessments are conducted separately [44].

What has been said about the integration of toxicological effects also applies to the exposure part of risk assessment. First of all, the same process may cause exposure to both workers and the environment and controlling worker exposure to exhaust ventilations, for example, may cause an environmental problem. Human and environmental exposure assessment are therefore linked by the same "determinants of exposure". Secondly, when developing exposure scenarios these must be based on integrated thinking in order to avoid or reduce problem shifting.

This integration of human and ecological risk assessment, both effects assessment and exposure assessment, can provide better input for decision-making. The move towards integration to achieve more fully informed decisions must come from the realization that decisions are currently not always fully informed [46] and often are not made in a cost effective manner [59]. Integration has been one of the major

Table 1.6. Types of integration in the risk management process and why they are needed [45-48].

Exposure and effects – This is the most fundamental type of integration in risk assessment, i.e. the interaction of exposure estimates with estimates of the relationship between exposure and effects to estimate risks.

Multiple agents – Assessments should integrate risks to humans and the environment from all agents that are relevant to the decision.

Multiple routes - Assessments should integrate risks to human health and the environment from all routes of exposure relevant to the decision.

Multiple endpoints – Assessments should consider all potentially significant endpoints for both human health and ecological receptors that are relevant to the decision.

Multiple receptors – Assessments should consider all classes of human and ecological receptors that are relevant to the decision.

Multiple scales in dimensions – Extrapolations in risk assessment can occur in various dimensions including time (short to long term), place (one site to other sites), space (local to regional), biological scale (small species to larger ones), or mechanisms (molecular processes to physiology and responses at individual to population level).

Life cycle - Assessments may need to integrate the risks from the entire life cycle of the chemical or product.

Normal use, accidents and incidents. Risk assessments tend to focus on normal uses and permitted discharges of e.g. waste-water as a result of production. Extreme events such as peak discharges, accidents and uses that are not permitted or illegal may dominate ecological and health risks and should be integrated into health and ecological risk assessments.

Management alternatives – When decisions are based on comparison of alternatives, assessments should consider the risks from relevant alternatives in an integrated manner.

Socio-economics and risks – The actual level of protection provided depends on the relative acceptability to society of the costs and benefits. Currently, the integration of social sciences into risk assessments is largely limited to weighing the costs of a regulated party against the benefits to the public.

Stakeholder participation – Integration of active stakeholder participation will help to improve the risk assessment and acceptance of management actions (see Box 1.3).

trends in environmental risk assessment. This increase in integration is predictable, as narrowly focused assessments often have failed to provide adequate answers in the past [45-47]. Apart from the integration of human health and environmental effects and exposure assessments, there are a variety of other types of integration that should be explored further to improve risk-based decision-making. This integrative thinking will also help to focus efforts and resources in the risk assessment and risk management process. These types of integration are given in Table 1.6.

1.4 DISCIPLINES, ROLES AND RESPONSIBILITIES IN RISK MANAGEMENT

The assessment of risks associated with the production, use and disposal of chemicals is a task that cannot be undertaken without adequate knowledge of chemistry (including process technology), toxicology and biology (Figure 1.12). Yet, the complexity of the subject requires the involvement of other disciplines: mathematics, statistics and informatics. These disciplines play an essential part in disentangling, analyzing and quantifying the complex interactions between substances, species

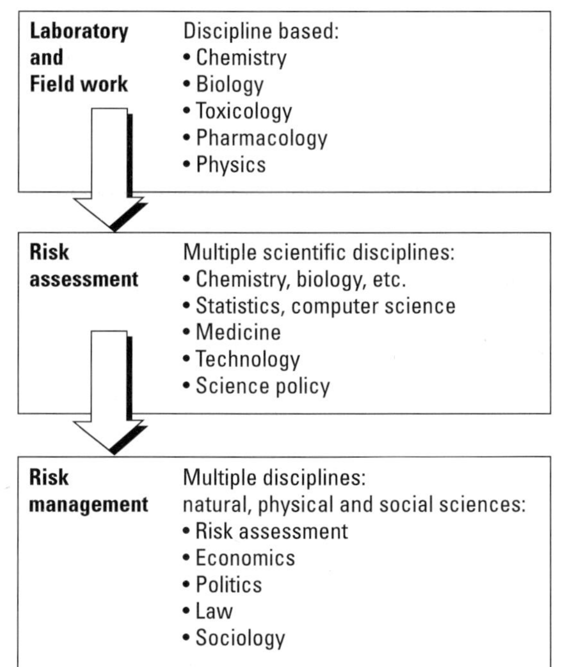

Laboratory and Field work
Discipline based:
• Chemistry
• Biology
• Toxicology
• Pharmacology
• Physics
Risk assessment
Multiple scientific disciplines:
• Chemistry, biology, etc.
• Statistics, computer science
• Medicine
• Technology
• Science policy
Risk management
Multiple disciplines:
natural, physical and social sciences:
• Risk assessment
• Economics
• Politics
• Law
• Sociology

Figure 1.12. Disciplines involved in the risk management process. Modified from Patton [15].

and systems, often using models. These complex systems may be either ecosystems with numerous species and functions to be protected, or "human environments" or "technospheres", in which attention is focused on only one species: man. Human populations may be exposed directly, i.e. at the workplace (occupational exposure) and through consumer goods such as detergents, or indirectly via the environment. Of course, other disciplines are involved as well, including physics, medicine, geology, hydrology, and epidemiology.

The feature distinguishing risk assessment from the underlying sciences is this: after evaluating standard practices within the discipline, the most relevant information from each of these areas is brought together to describe the risk. This means that individual studies, or even groups of studies, from a single discipline, may be used to develop risk assessments, although they are not, in themselves, generally regarded as risk assess-ments nor can they alone generate risk assessments [15]. In other words, risk assessment is multi-disciplinary team work (Figures 1.12). Risk management is also a multi-disciplinary process. It draws on data, information, and principles from many other disciplines and specialists with different kinds and levels of expertise representing many different organizations and interested parties (Figures 1.11 and 1.12). Risk communication is a vital part of the process of involving, informing and advising people about how they can manage potential risks. Risk communication requires an understanding by manufacturers and importers of the information needs of users. The development of appropriate risk-based communication systems, including the provision of complementary information using, for example, websites and educational campaigns, should be pursued. Roles and responsibilities of the different stakeholders are different. In conclusion:

- Scientific experts need independence so that they are willing to speak "truth to power". However, they should not be involved in decision-making and used only as providers of input to regulatory decision-making. Persson [55] stated it very clearly: "it is essential to distinguish the role of the expert from the role of the decision-maker." Scientific experts inform and decision-makers/regulators should: (a) manage the overall risk management process, (b) decide on the risk management options, and (c) take their responsibility as they are accountable.
- Involving stakeholders and incorporating their recommendations where possible, re-orients the decision-making process from one dominated by regulators to one that includes those who must live with the consequences of the decision. This not only fosters successful implementation, but can also promote greater trust in government institutions [41].

1.5 HOW RISKS ARE EXPRESSED

Risk can be defined as the probability of an adverse effect in an organism, system or (sub)population caused under specified circumstances by exposure to an agent. Risk has three characteristic variables: the type, magnitude and probability of the hazard. In quantitative terms, risks are often expressed in terms of probability estimates ranging from zero (harm will not occur) to one (absolute certainty that harm will occur). A distinction is made between *chemicals with and without threshold levels*. In the case of *chemicals without threshold levels*, e.g. many carcinogens, often a linear relationship is assumed between exposure (dose) and effect (incidence of cancer). This means that, in a statistical sense, it is always possible that an effect will occur. In such cases, the risk number represents the *probability* of additional cancer cases occurring. For example, an estimate for chemical X might be expressed as 1×10^{-6}, or simply 10^{-6}. This figure can also be written as 0.000001, or one in a million, which means that one additional case of cancer is projected in a population of one million people exposed to a certain level of chemical X over their lifetimes. Similarly, smoking 1 packet of cigarettes a day produces a potential risk of lung cancer of 5 x 10^{-3} per year (Table 1.7) or 1 in 200 per year. These risks signify additional cases to the background incidence of cancer in the general population. The American Cancer Society has published statistics that indicate that the background incidence of cancer in the general population is 1 in 3 over a lifetime [15]. It should be noted that not all carcinogens cause non-threshold effects and that non-carcinogenic effects may also be non-threshold.

Not all chemicals present non-threshold cancer risks, but they may affect developmental, reproductive, neurobehavioural, and other body functions. Such effects are often associated *with a threshold level* and a non-linear S-shaped relationship between dose and effect. There is a threshold level below which there is no effect, albeit that the precise level of this threshold will vary between individuals. In other words, unlike chemicals with non-threshold effects, risk is not assumed to be present in all doses or concentrations. Typically, such substances are regulated by determining NELs in test species by applying a predetermined or calculated assessment factor (AF) to arrive at an ADI or DNEL for man [21,45] or PNEC for ecosystems [2,45,60] :

$$\text{ADI or DNEL or PNEC} = \text{NEL} / \text{AF} \qquad (1.1)$$

An ADI or DNEL is a rough estimate of the daily exposure to which human populations (including sensitive subgroups) may be subjected that is not likely to cause harm during a lifetime. For chemicals with threshold levels, values are not typically given as probability of occurrence, but rather as *levels of exposure* estimated to be without harm. These values are typically expressed in mg (of the chemical) per kg of body weight per day. A PNEC is a rough estimate of the exposure level at which ecosystems will suffer no harm. PNECs are typically expressed as mg/L air or water or as mg/kg soil or sediment.

The uncertainty in a DNEL, PNEC or ADI may be one of several orders of magnitude (i.e. powers of 10). As exposures typically vary over time and space, and plant and animal species vary widely in their susceptibility to toxicants, the question may be asked: can effects, exposures and risks be expressed by a single figure or do we need to provide ranges of concentrations? Let's look at a carcinogenic compound. The statement for a carcinogenic compound that the risk of a specified exposure concentration is A x 10^{-B} is actually shorthand for the general truth that "we are Y% sure that the risk is no more than A x 10^{-B} for Z% of the population". Real risks cannot be given for chemicals which pose a threat at certain thresholds (Section 1.2.4). Where the PEC/PNEC ratio is less than 1, we are V% sure that the exposure concentration does not exceed the NEC for W species which were tested for ecosystem X, comprising a total of Y species at time Z. Where the PEC/PNEC ratio is greater than 1, it is not at all clear what the risks is. Absolute certainty in risk assessment is impossible.

Patton [15] stresses a number of other important points. Firstly, the numbers themselves do not tell the whole story. For instance, even though the numbers are identical, a cancer risk value of 10^{-6} for the "average exposed person" (perhaps exposed through the food supply) is not the same thing as a cancer risk of 10^{-6} for a "most exposed individual" (perhaps someone exposed because he lives or works in a highly contaminated area). It is important to know the difference. By omitting the qualifier "average" or "most exposed" the risk is incompletely described, which would mean a failure in risk communication.

Secondly, numerical estimates are only as good as the data is they are based on ("garbage in, garbage out"). Just as important as the quantitative aspect of risk characterization (the risk numbers) then, are the qualitative aspects. How extensive is the database supporting the risk assessment? Does it include human epidemiological data as well as experimental data? Does the laboratory data base include test data on more than one species? If multiple species are tested, do they all respond similarly to the test substance? Are extrapolations being made from more or less sensitive varieties, species and endpoints? What are the data gaps, the missing pieces of the puzzle? What are the scientific uncertainties? What science-policy decisions are made to address these uncertainties? What working-assumptions underlie the risk assessment? What is the overall confidence level in the risk assessment? All of these qualitative considerations are essential to deciding what reliance to place on a number and to determining potential risk.

1.6 PERCEPTION OF RISKS

The perception of risks (and benefits) varies between individuals and public, business, labour, and other stakeholders. Moreover, they change with time (Table 1.3) and across cultures. People continually assess situations and decide whether the risks associated with a particular action can be justified. In certain circumstances, harmful effects are clearly attributable

Table 1.7. Annual mortality rate associated with certain occurrences and activities in the Netherlands [25].

Activity/occurrence	Annual mortality rate	
Drowning as a result of dike collapse	10^{-7}	1 in 10 million
Bee sting	2×10^{-7}	1 in 5 million
Struck by lightning	5×10^{-7}	1 in 2 million
Flying	1.23×10^{-6}	1 in 814,000
Walking	1.85×10^{-5}	1 in 54,000
Cycling	3.85×10^{-5}	1 in 26,000
Driving a car	1.75×10^{-4}	1 in 5,700
Riding a motorbike	2×10^{-4}	1 in 1,000
Smoking cigarettes (1 packet a day)	5×10^{-3}	1 in 200

to a particular course of action. However, in other cases, the impact of such effects may be uncertain and need not be immediately obvious. People use different methods to evaluate their own individual risks and environmental risks. In some cases the perception of a group of people may alter the priorities assigned to reducing competing risks. Risks that are involuntary or "novel" seem to arouse more concern than those that are voluntary or "routine", i.e. accepted. Environmental risks are largely of an involuntary nature. "Natural" contaminants and toxins in food may be considered acceptable even though they may cause illness, while food additives whose introduction (or identification) in foodstuffs is to assist in preservation may not be acceptable to some people [61].

Hazards that are delayed in their effect, such as extinction of populations or species caused by long-term accumulation of persistent pollutants in food webs, are usually difficult to observe, assess and control. As a result, hazards of this type are often regarded as being more serious than those that happen immediately. Others, such as Lovell [53], state that "outcomes which are rare, unpredictable, and catastrophic, such as chemical plant explosions, are viewed as more disturbing than those that are common, regular, and small in size, such as road accidents, even if the overall cost in human life and suffering may be similar. There seems to be a "dread" component to people's perception of certain types of risks".

Table 1.7 gives examples of the various risks to which man can be exposed. The risks inherent in these activities give some indication of the magnitude of the risk added to natural circumstances due to human interference. Some are *voluntary* risks, e.g. smoking, others are of an *involuntary* nature, e.g. being struck by lightning. Although the risks of, say, smoking and driving a car are comparatively great, they are widely accepted. On the other hand, even the presence of minute quantities of (natural) carcinogenic substances in food is not readily accepted by the public at large. Although the risk-benefit equation should be a major determinant, both the risk and the benefit are frequently not fully understood and people develop irrational fears [62].

1.7 UNCERTAINTY, VARIABILITY AND PRECAUTION

1.7.1 Uncertainty and variability

Risk assessment controversies often revolve around disagreements regarding the nature, interpretation, and justification of methods and models used to evaluate incomplete and uncertain data. When science is used for regulatory purposes, decision-makers need to be informed not only of the available scientific knowledge but also of relevant uncertainties and lacunae in the knowledge base. We need to distinguish between uncertainty and variability. Uncertainty can often be reduced by obtaining or generating more information. This is one of the reasons why we apply tiered testing and assessment strategies. Variability is a natural phenomenon and cannot be reduced (Box 1.4). The aim of uncertainty analysis is to identify major sources of uncertainty in either hazard or exposure assessment. Any risk assessment carries uncertainty with it. An evaluation of uncertainty therefore should assist in communicating these uncertainties to improve decision-making in the light of the uncertainty associated with the outcome of the risk assessment [33,63]. The probability that any given chemical presents a hazard to man and/or the environment can be difficult to determine, but it is essential that rigorous scientific methods be used in any such assessment. Mathematical approaches to risk

Box 1.4. Risk assessment according to Aristotle

"It is the mark of an instructed mind to rest easy with the degree of precision which the nature of the subject permits and not to seek an exactness where only an approximation of the truth is possible."

assessment help to expose a problem to logical analysis, and to identify areas of uncertainty. This type of analysis provides an intellectual basis for decision-making or determining further research needs. In other words: risk assessment is driven by doubt, not by certainty.

Mathematical analysis can, unfortunately, be used for hiding inconvenient information or muddled thinking behind a façade of apparent technical and scientific expertise [64]. It is important to realize that mathematical assumptions are still assumptions and require estimates of the errors implicit in them. Using ranges of values rather than only a central estimate is a necessary adjunct to risk assessment and forms the basis of a sensitivity analysis, which tests how general the findings of an assessment may be. Risk assessment in practice is far from ideal and is hampered by four types of uncertainty:

1. Lack of information.

Very often basic data are lacking or inadequate to make precise predictions. Where essential data are lacking the use of expert judgement, estimation methodologies or even default values becomes necessary. This lack of basic data [65,66] applies to toxicological data (Table 1.8) and is likely to be even greater for data on emissions, fate and exposure concentrations. The data situation for lower volume chemicals is even worse! This general lack of data applies to the 100,000 chemicals on the European Inventory of Existing Commercial Chemical Substances (EINECS). For new chemicals, plant protection products and biocides, the actual situation may be slightly better because basic information is required for notification and registration.

Table 1.8. Estimation of available toxicological data (%) for about 2500 High Production Volume Chemicals [66].

Acute oral toxicity	77
Repeated dose toxicity	58
Genetic toxicity *in vivo*	38
Genetic toxicity *in vitro*	67
Reproductive toxicity	26
Teratogenicity	32
Acute ecotoxicity (fish and daphnids)	68
Short-term toxicity (green algae)	45
Effects on soil organisms	30

2. Measurement uncertainties.

Measurement uncertainties include low statistical power due to insufficient observations, difficulties in making measurements, inappropriateness of measurements, and human error (incorrect measurements, misidentifications, data recording errors and computational errors).

3. Observation conditions.

Uncertainties related to conditions of observation include spatiotemporal variability in climate, soil type, sensitivity, ecosystem structure, differences between natural and laboratory conditions, and differences between tested or observed species and species of interest for risk assessment.

4. Inadequacies of models.

Inadequacies of models include a fundamental lack of knowledge concerning underlying mechanisms, failure to consider multiple stresses, responses of all species, extrapolation beyond the range of observations, and instability of parameter estimates. In fact two related types of uncertainties can be distinguished: quantifiable uncertainties (the "known unknowns") and undefined uncertainties that cannot be described or quantified (the "unknown unknowns"). The PEC/PNEC approach is an example of such "unknown unknowns" (Sections 1.2.4 and 1.5). The same is true for laboratory-based soil quality criteria because there is a fundamental lack of knowledge about the differences in the bioavailability of the chemical between the laboratory and the field.

Suter [64] distinguishes between three types of uncertainty, i.e. stochasticity, error and ignorance, whereas Ricci et al. [67] identify six elements:

1. Scientific judgements and defaults that are imposed on stakeholders by regulators when scientific evidence is contradictory and causation is unknown.
2. Misspecified models that exclude key variables or wrongly or incompletely formulate the relationships between them
3. Statistical uncertainties that combine aspects of model misspecification and choice of model with estimation and inference, heterogeneities of statistical

and physiological parameters, confounders, effect modifiers, measurement errors and missing or censored data.
4. Deterministic representations where the formal description of physical processes gives an illusion of complete and certain knowledge of future outcomes and their magnitude.
5. Probabilistic representations where the analysis concerns assessing events that have not yet occurred.
6. Statistical representations where the analysis concerns inference about a population's parameters from observed exposures and outcomes determined from experimental or observational study.

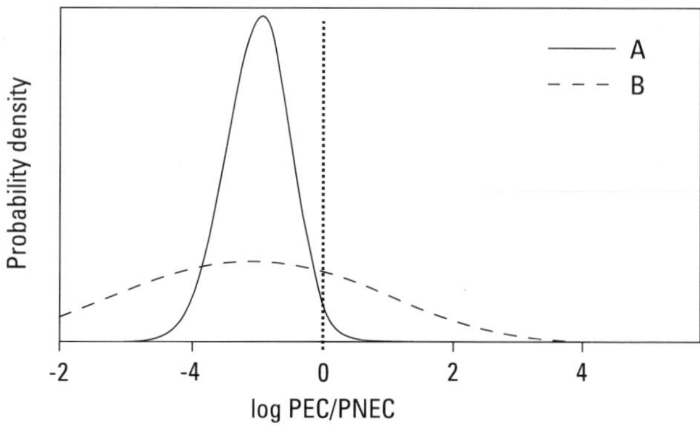

Figure 1.13. Probability distributions of two hypothetical chemicals with the same median PEC/PNEC ratio. Redrawn from Jager et al [68]. With permission. Copyright Elsevier.

In risk assessment reports of chemicals, generally two ways of dealing with uncertainty can be seen: a *deterministic* approach and a *probabilistic* approach [68]. In the deterministic approach, uncertainty is not explicitly addressed by the application of "reasonable worst-case assumptions" in hazard and exposure assessments. The advantage of the simple deterministic approach is that it is quick and easy to apply and takes uncertainties into account without having to specify uncertainty about elements in the assessment that are difficult to estimate. It also avoids the problem of communicating risks in terms of probability and statistics that are often difficult to follow for non-experts. Therefore it has proven to be very efficient in taking regulatory decisions. The disadvantage of the deterministic approach to uncertainty is that several reasonable worst-case assumptions can be combined leading to unrealistic assessment outcomes and outcomes that are not transparent [33,68,69]. The deterministic approach gives a false sense of accuracy and ignores variability in the population [68].

Uncertainties occur throughout the different steps of the risk assessment and should thus be addressed as an integral part of the work during the assessment and not as an add-on in the reporting at the end of the assessment.

1.7.2 Quantifying uncertainty and validation

Quantifying uncertainty

Quantifying estimates of uncertainty, as is sometimes done in *probabilistic risk assessments* (PRA), may help in making more rational decisions on the risk of toxic substances and can help to achieve a better balance between assessment, uncertainty and safety [33,68,70]. The advantage of a quantitative treatment of uncertainty is that assumptions about variability and uncertainty must be backed up by explicit information [71]. The application of uncertainty analysis to decision-making is far from routine as virtually all decisions are still based on point estimates of exposure and effects. As sources and magnitude of uncertainties will differ between chemicals, it means that some substances can be assessed with greater confidence than others, as illustrated in Figure 1.13. The disadvantages of PRA are obvious. Additional information is needed to estimate the uncertainty in hazard and exposure assessment that may be difficult, time-consuming or expensive to generate [72,73]. Examples of PRA [74-76] show that it is also more complex and more difficult to communicate. Furthermore, no scientific consensus exists about standard methods and their validation. There seems to be a guarded interest in uncertainty analysis, but currently it does not receive high priority. There seems to be a gap between the scientist and the risk manager [68]. Perhaps this is related to the fact that there are many sources of uncertainty. Some of these uncertainties can be quantified, whereas sometimes major sources of uncertainty are simply non-quantifiable. This may create a false sense of security and certainty. An additional disadvantage of PRA is that regulators need to decide on an acceptable risk level for the outcome of the risk assessment [33].

The current state of science is that some elements of uncertainty analysis and some probabilistic approaches are already part of the guidance on effects assessment [22]. Similar approaches are not routinely applied in the areas of exposure assessment and risk characterisation. Jager et al. [68] have tried to list the options that are currently available to revise risk assessments in order to deal with uncertainties in the risk characterization stage (Figure 1.14). They arrived at three options in order of preference:

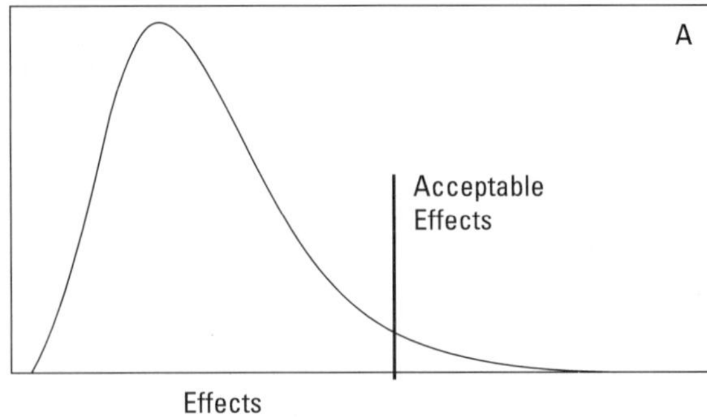

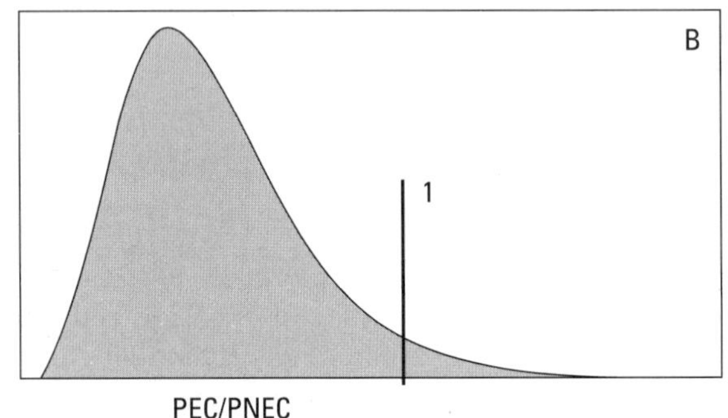

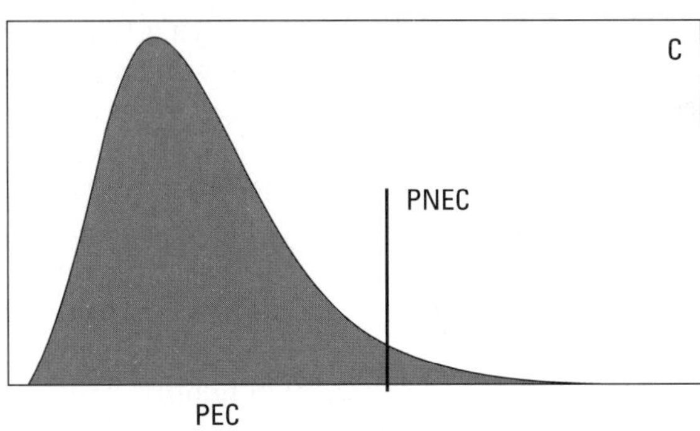

Figure 1.14. Options for uncertainty in risk characterization. Redrawn from Jager et al [68]. With permission. Copyright Elsevier.

A. Establish a dose-effect relationship for human populations and ecosystems. The result of the risk characterization stage will be a probability distribution of effects. Decisions can be based on an acceptable level of effects.
B. Revise the assessment factors in the effects assessment to yield a median, or most likely, PNEC Instead of a conservative estimate and attaching uncertainty to these factors (e.g. instead of a factor of 1000 use an assessment factor of 100 with a factor of 10 uncertainty)
C. Leave the effects assessment as it is now. In that case, only uncertainty in the exposure estimate needs to be quantified. The result of the risk characterization will be a probability that the PEC exceeds a fixed, worst-case PNEC.

Further discussions and developments are needed to improve transparency and to address variability and uncertainty [76-78]. Recently, pragmatic proposals have been made to use three ways of getting to grips with uncertainty in risk assessments that have different levels of complexity, resource intensity (time and money) and data needs [33]. With a view to developing maximum workability, a tiered strategy has been developed:

Tier 1: Qualitative uncertainty assessment. Uncertainty assessment using a deterministic approach linked to scenario analysis.

Tier 2: Simple (semi-quantitative) analysis. This is a simple semi-quantitative probabilistic analysis providing insight into the influence of uncertainty on the risk quotient or risk characterization ratio (RCR).

Tier 3: Full quantitative PRA.

Such a system can be applied in a pragmatic manner following the principle of: "as simple as possible and as complex as needed."

Validation

Risk assessments are generally performed by applying risk assessment models, such as EUSES [79]. These models are crucial as they increase transparency and predictability. They play an important role in risk communication and risk acceptance. The more specific models are, the more data are needed and the more difficult they are to generalize. On whatever scale we use models - biological models such as the zebra fish or rat, local or multi-compartment exposure models, or risk assessment models - they remain a distortion of the truth and their output is input-dependent. Users must be confident that their models actually correspond with the systems being studied. The process of obtaining this confidence has been referred to as model validation. Model validation can be defined as any process that is designed to assess the correspondence between the model and the system. The purpose of validation is to improve the credibility and reliability of predictive methods. Validation must be viewed as an iterative process in which predictions are tested, models are refined, and then new predictions are tested [80].

Risk assessment is a broad term that encompasses a variety of analytical techniques that are used in different situations, depending upon the nature of the hazard, the available data, and needs of the decision-makers.

As a consequence it is essential to realize that (model) validation is also context-specific, i.e. validation needs to be placed in the context of the risk management decision to be made. Models for priority setting of chemicals for future evaluation or a safety evaluation of a specific chemical require different data to be assessed using different models with different input and output qualities. Too much focus on model validation without putting the whole exercise in a broader regulatory context may lead to "validation paralysis" or "paralysis by analysis". At least three kinds of studies can contribute to validation:

- Improved measurements of specific quantities and testing of assumptions.
- Experimental testing of models under reasonably realistic conditions.
- Monitoring of effects or other investigations to determine the level of agreement between predictions and the actual observations.

Using a model to quantify risks and their uncertainties would, in principle, permit more useful risk assessments, but if the model itself is a poor representation of reality, the results might be totally meaningless. Furthermore, it should be stressed that although uncertainties in effects data, exposure data and methodologies in risk assessment are important, uncertainties and prior framing assumptions in risk policy may be even more important. Transparency in this area is of the utmost importance in order to improve decision-making (see Section 1.3 and Figures 1.6, 1.7 and 1.11).

1.7.3 Precaution

Taking regulatory action on the basis of *the precautionary principle* is sometimes interpreted as an alternative to taking action based on an assessment of risks [81]. In practice, however, many references in international law to the precautionary principle refer to the use of this approach when there are threats of serious irreversible damage, but there is a lack of conclusive scientific evidence. For instance, in the Rio Declaration of 1992 principle 15 states: "in order to protect the environment, the precautionary approach shall be widely applied by States according to their capability. Where there are threats of serious or irreversible damage, lack of full scientific certainty shall not be used as a reason for postponing cost-effective measures to prevent environmental degradation." Some people have argued that this definition is legally unenforceable [67], whereas others say it is enforceable and have defended the precautionary principle against five common charges, namely that it is: (1) ill-defined, (2) absolutist, (3) leads to increased risk-taking, (4) is a value-judgement or an ideology and (5) is unscientific or marginalizes the role of science. Those who defend the precautionary principle argue that, in principle, it is no more vague or ill-defined than other decision principles and that it can be made precise through elaboration and practice [81-83]. According to Kriebel et al. [83] the precautionary principle has four central components:

1. Taking preventive action in the face of uncertainty.
2. Shifting the burden of proof to the proponents of an activity.
3. Exploring a wide range of alternatives to possibly harmful actions.
4. Increasing public participation in decision-making.

They argue that a shift to more precautionary policies creates opportunities and challenges for scientists to think differently about the ways they conduct health and environmental studies and communicate results. According to these authors the precautionary principle highlights this tight, challenging link between science and policy, which is in line with the observations made in Sections 1.3 and 1.4.

In 2000, the European Commission published a Communication on the precautionary principle [84] providing a general framework for its use in EU policy (see Box 1.5). The aim was to outline the Commission's approach to using the precautionary principle, to establish guidelines for it, to build a common understanding of how to assess, manage and communicate risk that science is not yet able to evaluate fully, and to avoid unwarranted recourse to the precautionary principle as a disguised form of protectionism [84]. The precautionary principle is a cornerstone of the REACH legislation [40] and of EU health and environmental management in general [84,85]. In the Communication of the Commission it is clearly stated that the precautionary principle should be considered as part of a structured approach to the analysis of risk which comprises three elements: risk assessment, risk management and risk communication, as shown in Figure 1.11. The precautionary principle is particularly relevant to the management of risk. In the Commission Communication four elements need to be highlighted in the context of this book.

1. The precautionary principle is based on the assumption that a thorough scientific evaluation of the risks is performed which is as objective and complete as possible prior to decision-making: "the implementation of an approach based on the precautionary principle should start with a scientific evaluation, as complete as possible, and where possible, identifying at each stage the degree of scientific uncertainty."

Box 1.5. The precautionary principle according to the European Commission [84]

Where action is deemed necessary, measures based on the precautionary principle should be, *inter alia:*

1. **Proportional** to the chosen level of protection.
2. **Non-discriminatory** in their application.
3. **Consistent** with similar measures already taken.
4. **Based on an examination of the potential benefits and costs** of action or lack of action (including, where appropriate and feasible, an economic cost/benefit analysis).
5. **Subject to review,** in the light of new scientific data.
6. **Capable of assigning responsibility for producing the scientific evidence** necessary for a more comprehensive risk assessment.

Proportionality means tailoring measures to the chosen level of protection. Risk can rarely be reduced to zero, but incomplete risk assessments may greatly reduce the range of options to risk managers. A total ban may not be a proportional response to a potential risk in all cases. However, in certain cases, it is the sole response to a given risk.

Non-discrimination means that comparable situations should be treated differently, and that different situations should be treated in the same way, unless there are objective grounds for doing so.

Consistency means that measures should be of comparable scope and nature to those already taken in equivalent areas in which all scientific data are available.

Examining costs and benefits entails comparing the overall cost to the Community of action and lack of action, in both the short and long term. This is not simply an economic cost-benefit analysis: its scope is much broader, and includes non-economic considerations, such as the efficacy of possible options and their acceptability to the public. In the conduct of such an examination, account should be taken of the general principle and the case of law of the Court such that the protection of health takes precedence over economic considerations.

Subject to review in the light of new scientific data, means measures based on the precautionary principle should be maintained as long as scientific information is incomplete or inconclusive, and the risk is still considered too high to be imposed on society, in view of the chosen level of protection. Measures should be periodically reviewed in the light of scientific progress, and amended as necessary.

Assigning responsibility for producing scientific evidence is already a common consequence of these measures. Countries that impose a prior approval (marketing authorization) requirement on products that they deem dangerous *a priori* reverse the burden of proving injury, by treating them as dangerous unless and until businesses do the scientific work necessary to demonstrate that they are safe.

Where there is no prior authorization procedure, it may be up to the user or to public authorities to demonstrate the nature of a danger and the level of risk of a product or process. In such cases, a specific precautionary measure might be taken to place the burden of proof on the producer, manufacturer or importer, but this cannot be made a general rule.

2. The second relevant element in the Communication of the Commission is the separation of roles and responsibilities between scientist and decision-makers: "decision-makers need to be aware of the degree of uncertainty attached to the results of the evaluation of the available scientific information. Judging what is an "acceptable" level of risk for society is an eminent political responsibility. Decision-makers faced with an unacceptable risk, scientific uncertainty and public concerns have a duty to find answers. Therefore, all these factors have to be taken into account".
3. The third relevant element is related to risk communication: "the decision-making procedure

should be transparent and should involve as early as possible and to the extent reasonably possible all interested parties".

4. Actions taken under the precautionary principle must in certain cases include a clause reversing the burden of proof and placing it on the producer, manufacturer or importer: "measures based on the precautionary principle may assign responsibility for producing the scientific evidence necessary for a comprehensive risk evaluation."

These views are comparable to the views expressed in the policy paper "Premises for Risk Management" published by the Dutch Ministry of the Environment in 1989 [25] and the US Presidential/Congressional Committee in 1997 [41] as discussed in Section 1.3.2. The Commission has added a number of important elements and conditions to make the precautionary principle more operational [84]. These are given in Box 1.5. In reviewing the role of the precautionary principle in the EU risk assessment process on industrial chemicals [81] it was concluded that the main reason for doing so was the uncertainties in the risk assessment (or the underlying effects or exposure data), which were, according to the scientific experts, so high that a "normal" level of certainty could not be obtained. In the next decade it will be a scientific, legal, and political challenge to make the precautionary principle a legally enforceable and practical tool in health and environmental management. As stated above, the precautionary principle is a cornerstone of REACH. Further general guidance or rules need to be developed that will support policy-makers in their decisions as to whether this uncertainty is so large that action is warranted or whether it is acceptable to wait until further information becomes available.

1.8 CONCLUDING REMARKS

It is not uncommon that during the selection of options for risk reduction, fundamental questions about the principles of risk assessment are raised. It is also not uncommon that discussions on risk reduction re-open discussions on data needs and risk assessment methodologies; zinc is an example [73]. Risk assessment is an important tool, but not if it is used to postpone decision-making *("paralysis by analysis")* or as a cover for a deregulatory agenda [86-88]. Risk assessors need to know:

- Their science.
- The multiple uncertainties in risk assessment.
- The multiple media and many spatial levels of risk assessment (Table 1.3).
- The limited relevance of science, i.e., the science underlying most risk assessments is inconclusive [88].
- The limited information content of effects and exposure data, including monitoring data [89-90].
- The limitations of risk assessment in general (Box 1.4).
- The difference between data, information and knowledge (Figure 1.15).
- That information must be credible and verifiable.
- That information for decision-making should be timely and affordable to those who need it and should be communicated in a manner which is understandable, efficient, and transparent [91].
- The limitations on how risks can be expressed (Section 1.5).
- The importance of risk communication in general (Section 1.3).
- The stakeholders and the context in which they are working (Figure 1.6).
- The forces in health and environmental decision-making (Figure 1.7).
- The consequences assessments may have in terms of follow-up risk reduction measures.
- The different roles and responsibilities in risk management (Section 1.4).
- The different perceptions of risk (Section 1.6).

It has become common to distinguish between: (1) data, (2) information, and (3) knowledge and wisdom. *Data* can be defined as basic observations or measurements. Data can be transmitted, combined and analyzed using a variety of tools such as EUSES [79]. *Information* refers to the products of analysis and interpretation, such as a risk assessment report. Knowledge is created by accumulating information by e.g. interacting, aggregation, filtering and transmission to the risk manager for example. *Knowledge* is internalized; it is information in the mind, in a context, based on personal perceptions and experiences allowing it to be transformed into action. It is familiarity gained by experience. *Wisdom* concerns interaction with stakeholders, management of the bigger picture (Figure 1.6), re-applied knowledge and experience from lessons learned, prudence, good judgement and reflection.

A few concluding remarks will be made using this distinction between data, information, knowledge and wisdom as shown in Figure 1.15.

1. Data

In fact, roughly three tiers of data (testing and measurements) can be distinguished, from initial to comprehensive (Table 1.9). Very few chemicals are in the

Table 1.9. Stages in risk assessment and required effects and exposure information.

Tiers	Stages	Effects data	Exposure data
Tier-1	initial or preliminary	short-term toxicity	basic physicochemical data, equilibrium partitioning
Tier-2	refined	chronic toxicity	steady-state model predictions
Tier-3	comprehensive	more chronic, epidemiological and field data	measurements and (non)steady-state model predictions

data-rich category [65,66]. Risk management decisions can be postponed where tier 3 testing and measurement is seen as the decisive level (Figure 1.16). Tier 3 data (Figure 1.17) is costly, time-consuming and not always necessary. Rather than taking the defensive approach by generating more data about a chemical to prove that its risks are acceptable, a proactive approach may be taken by looking for harmless substitutes, for which tier 1 data may suffice. For example, the replacement of a persistent toxic chemical by a readily biodegradable toxic compound can take place on the basis of tier 1 data. That is why classification and labelling (Figure 1.8) is such an important risk management tool.

2. Information

Risk assessments tend to be uncertain and highly variable and their quality varies considerably [58]. This is also true of socio-economic, financial and technical projections about the consequences of risk reduction. Risk assessment can be a very a time-consuming activity [73], complex and difficult to communicate despite the availability of international guidance [22,33] and risk assessment models [79]. "Paralysis by analysis" may be a realistic threat to the future risk assessment and management of chemicals. Our current ability to generate new data often exceeds our ability to evaluate it [89]. In my view, further simplifications of information and information-flows are necessary in order to manage chemicals in the near future. Simple methodologies such as relative risk ranking on the basis of tier 1 data will not lose their relevance [93]. The implementation of REACH [40] will require a pragmatic and target-oriented approach to manage risks. Government agencies, the regulated community, and stakeholders face the challenges of generating and interpreting data for risk assessments in a cost-effective and efficient manner [94]. Chapman [95] has recently made a plea for simpler approaches to regulating chemicals. She suggested moving away from risks and assessing the riskiness of chemicals, i.e., (1) their capacity to cause harm, (2) their novelty (a matter of the degree to which something is different from what we know), (3) their persistence and (4) their mobility.

3. Wisdom and knowledge

It is difficult to predict which methodologies will be implemented in the near future to speed-up the assessment and management process, but the key will be further simplification based on Aristotle's knowledge and wisdom about risk assessment (Box 1.4). In the regulatory process, risk assessment will never provide "the correct answer" and risk management will never provide "absolute" solutions. To assume otherwise would

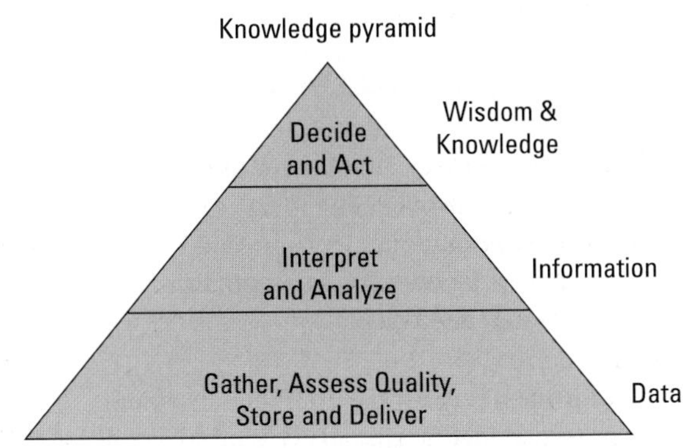

Figure 1.15. The knowledge pyramid. Modified from [91].

be to accept that there will be no further changes in the knowledge, views, values, rights and duties accepted by society and its individual members over time [61]. The multiple uncertainties in risk assessment mean that it is possible for its conclusions to be attacked from both sides. Arguments over whether or not assumptions in risk assessments are scientifically valid often amounts to debate about whether it is better to err on the side of "false positives" (if there is an error, it is more likely to be a false indication of danger) or "false negatives" (if there is an error, it is more likely to be a false indication of safety). Those who might be harmed by the substance being assessed will generally favour false positives; those who would gain from the substance will generally favour false negatives [18]. Different groups often interpret the results from the same study in different ways [95].

Risk assessment can be most useful when those who rely on it to inform the risk management process understand the context, its nature and its limitations, and use it accordingly. This means that decision-makers must at least understand that the process is assumption and value laden; that they are aware what assumptions were used in the assessment in question, and what values they reflect. They must also be aware that the risk estimate is expressed as a range, with a given certainty that the true average lies within that range; that variability is expressed to the degree that it is known; and that uncertainties can be reduced but often at high cost. Managing risks implies *management of the simplicity-complexity dilemma* (Figure 1.17). Risk managers must take all these factors into account when making a decision, along with political and economic factors which are not related to the risk assessment (Figure 1.6). Wisdom and knowledge are a prerequisite for *informed decision-making*.

Risk management of chemicals is an international challenge. Frameworks differ in scope and depth and continue to undergo dramatic changes [96]. New challenges will continue to arrive [40,97].

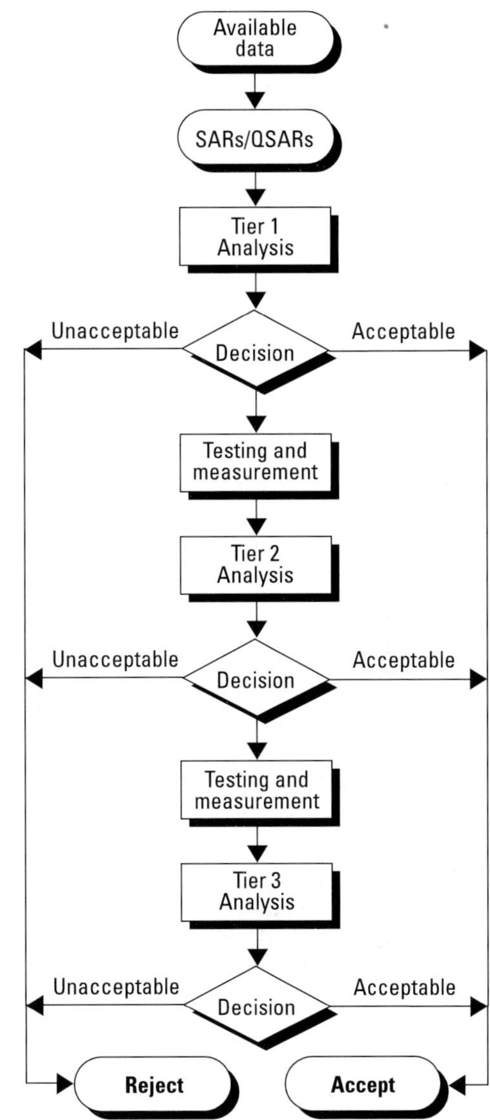

Figure 1.16. Diagram of the risk assessment process. Modified from Cairns Jr., Dickson and Maki [92].

1.9 CONTENTS OF THE BOOK

Applying risk assessment techniques to analyse the risks of chemicals to man and the environment is the subject of this book (Figure 1.18). It provides basic information to understand the process of risk assessment of chemicals arising from normal production, their use and disposal. Risk assessment for major accidental releases is not dealt with. The same applies to the various monitoring techniques that can be used for the enforcement of risk reduction measures. The contents of the book fall into 5 main sections.

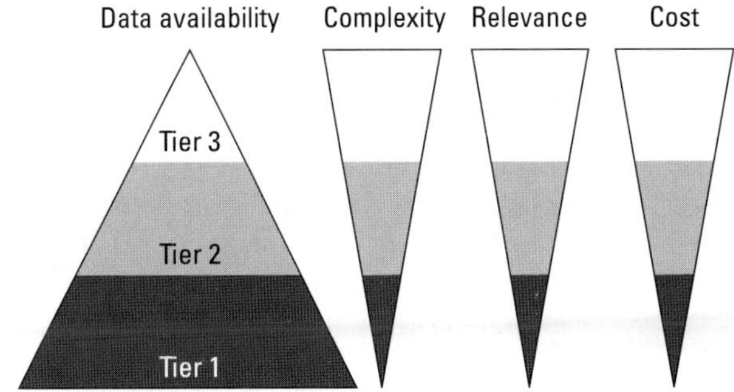

Figure 1.17. The simplicity-complexity dilemma.

- Part I deals with general issues in the risk management of industrial chemicals (Chapter 1).
- Part II is about exposure assessment. It starts with sources and emissions (Chapter 2), transport, accumulation and transformation processes (Chapter 3) and two chapters on exposure assessment, i.e., environmental and human exposure assessment (Chapters 4 and 5).
- Part III is related to human health and ecological effects assessment and risk characterization (Chapters 6 and 7).
- Part IV is about data and data estimation. It describes aspects of data needs, sources and quality evaluation (Chapter 8), the prediction of physicochemical properties and fate (Chapter 9), and the prediction of endpoints of toxicity and ecotoxicity (Chapter 10). Chapter 11 is devoted to so-called "Intelligent Testing Strategies".
- Part V is about risk assessment and management of industrial chemicals in the EU (REACH), USA, Japan and Canada (Chapters 12-15), whereas the OECD chemicals programme to support international cooperation on the assessment and management of chemicals is presented in Chapter 16. Most chapters, where relevant, include a section on further reading and a list of references for those who want more information about data, methodologies or processes.
- In addition, the book contains a glossary of the major key issues and terminology. Risk terminology is difficult and may cause confusion as risk assessors may disagree on terminology [98,99]. We have tried to be consistent with the risk terminology because without a common set of definitions, a meaningful discussion of this complex subject area is impossible.

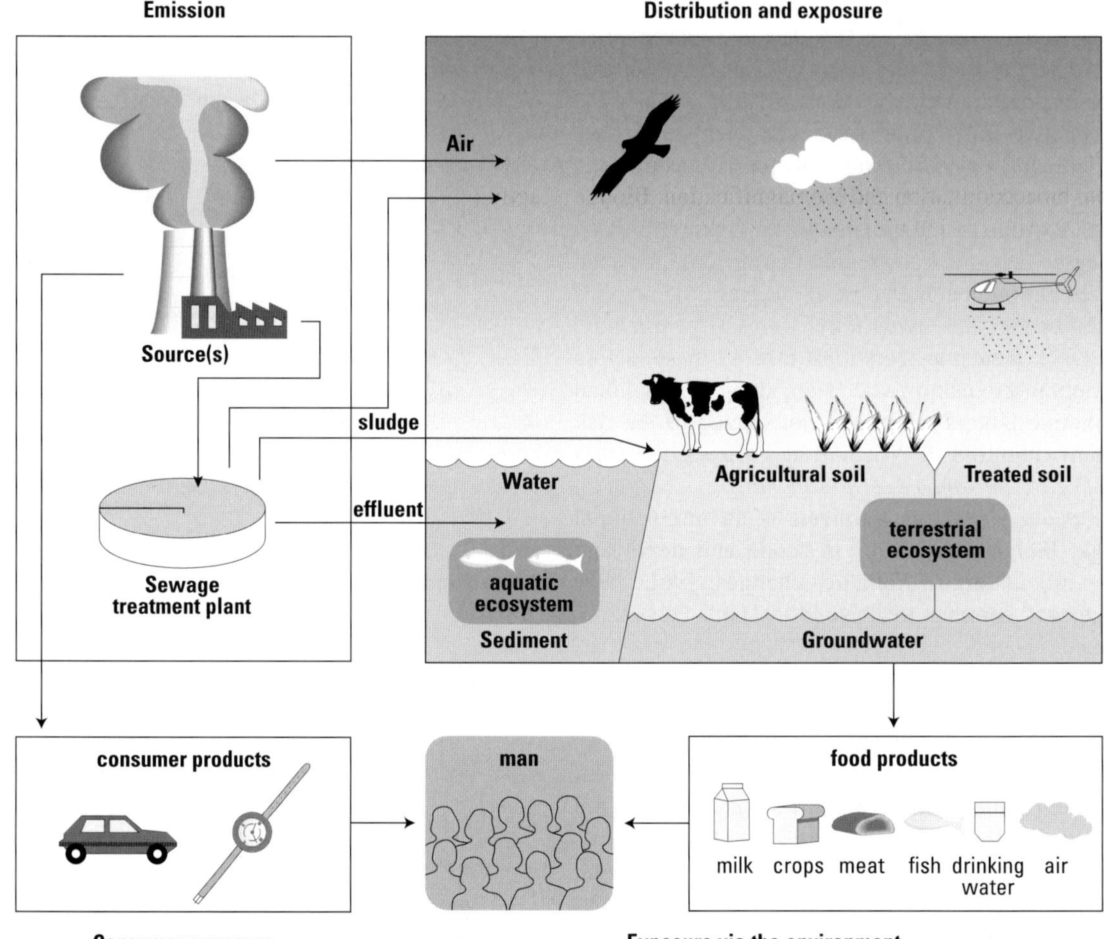

Figure 1.18. Elements of the European Union System for the Evaluation of Substances covered in this book [79]. The evaluation system comprises man, ecosystems (both aquatic and terrestial) and micro-organisms in sewage treatment plants.

The 16 chapters are summarized below.

Chapter 1. General introduction.
This chapter covers the general principles of risk assessment and risk management. It describes the role of risk assessment and other socio-economic and policy factors which contribute to the overall process of risk management of chemicals. Important definitions are given which are used in this field.

Chapter 2. Emissions of chemicals to the environment.
This chapter deals with the sources and emissions of chemicals into the environment, the life cycles of chemicals, point and diffuse sources of pollution and the classification of chemicals into main, industrial and use categories, as well as the development of "exposure scenarios". This provides important information for estimating emissions.

Chapter 3. Transport, accumulation and transformation processes.
This chapter highlights the transport, transformation and accumulation processes, e.g. advection, dispersion, volatilization, sorption, sediment transport, wet and dry deposition, bioaccumulation and biomagnification. Biotic and abiotic transformation processes are also included.

Chapter 4. Environmental exposure assessment.
The central theme of this chapter is environmental exposure assessment, i.e. the determination of exposure concentrations. It reviews compartmental models for surface water, groundwater, soil and air, as well as multimedia approaches.

Chapter 5. Human exposure assessment.
The central theme of this chapter is (external) human exposure assessment, i.e. the determination of exposure concentrations as a result of (a) exposure through the environment. It also highlights (b) consumer exposure assessment and (c) occupational exposure.

Chapter 6. Toxicity testing for human health risk assessment.
The main theme of this chapter is the assessment of health effects in man. It describes short and long-term toxicity, reproductive toxicity, mutagenicity, carcinogenicity, sensitization and irritation. Extrapolation methodologies and assessment factors are given which are used for the determination of DNELs for man.

Chapter 7. Ecotoxicological effects.
This chapter deals with ecotoxicological effects assessment for the aquatic and terrestrial environments. It describes single-species tests with aquatic and terrestrial species as well as multi-species studies. Extrapolation methodologies and safety factors are given which can be used to derive PNECs for ecosystems. It also examines the issue of mixture toxicity and the assessment of PBT and vPvB chemicals.

Chapter 8. Data: needs, availability, sources and evaluation.
This chapter addresses the input of any risk assessment, i.e. the data related to releases of chemicals, fate, exposure and effects. The focus of this chapter is on effects data.

Chapter 9. Predicting fate-related physicochemical properties.
This chapter describes basic physicochemical properties such as water solubility, melting point, boiling point, Henry's law constant, vapour pressure (P_v), the octanol-water partition coefficient (K_{ow}). Structure-activity relationships (SARs) and quantitative structure-activity relationships (QSARs) are given for various physicochemical parameters, (bio)accumulation and (bio)degradation.

Chapter 10. Predicting toxicological and ecotoxicological endpoints.
This chapter is about SARs and QSARs for basic toxicological and ecotoxicological properties. The application of SARs and QSARs can help to overcome the problem of data gaps and reduce animal testing.

Chapter 11. Intelligent Testing Strategies.
This chapter brings together the previous chapters on exposure and effects assessment. It describes testing strategies combining use and exposure information and effects information obtained from QSARs, read-across methods, thresholds of toxicological concern (TTCs), and *in vitro* tests prior to *in vivo* testing, as this is a more rapid, efficient, and cost-effective way of performing a risk assessment of chemicals.

Chapter 12. The management of industrial chemicals in the EU.
This chapter is about REACH. It summarizes the main features and requirements of the REACH legislation.

Chapter 13. The management of industrial chemicals in the USA.
This chapter is about the Toxic Substances Control Act in the USA It summarizes the main features and requirements of the legislation of industrial chemicals in the USA, including voluntary initiatives such as the Challenge Programme on High Production Volume Chemicals.

Chapter 14. The management of industrial chemicals in Japan.
This chapter is about chemicals management in Japan. It summarizes the main features and requirements of the legislation of industrial chemicals in Japan. It describes how risk assessment is applied in this regulatory context.

Chapter 15. The assessment and management of industrial chemicals in Canada.
This chapter summarizes the main features and requirements of the legislation of industrial chemicals in Canada, including the methodology of selecting priority chemicals. The relevant elements of how risk assessments are performed in Canada are included as well.

Chapter 16. The OECD chemicals programme.
This chapter describes the OECD activities relevant for the testing, assessment and management of industrial chemicals.

REFERENCES

1. Organization for Economic Co-operation and Development. 1992. Report of the OECD workshop on quantitative structure-activity relationships (QSARs) in aquatic effects assessment. OECD Environment Monographs 58. OECD, Paris, France.
2. Organization for Economic Co-operation and Development. 1992. Report of the OECD workshop on the extra-polation of laboratory aquatic toxicity data to the real environment. OECD Environment Monographs 59. OECD, Paris, France.
3. Organization for Economic Co-operation and Development. 1992. Report of the OECD workshop on the application of simple models for environmental exposure assessment. OECD Environment Monographs 69. OECD, Paris, France.
4. Organization for Economic Co-operation and Development. 1993. Occupational and consumer exposure assessment. OECD Environment Monographs 69. OECD, Paris, France.
5. Mercier M. 1988. Risk assessment of chemicals: a global approach. In Richardson ML, ed, *Risk Assessment of Chemicals in the Environment.* Royal Society of Chemistry, Cambridge, UK, pp 73-91.
6. European and Mediterranean Plant Protection Organization and Council of Europe. 1993. Decision-making scheme for the environmental risk assessment of plant protection products. EPPO Bull 23:1-165.
7. European Centre for Ecotoxicology and Toxicology of Chemicals. 1992. Estimating environmental concentrations of chemicals using fate and exposure models. ECETOC Technical Report 50. Brussels, Belgium.
8. European Centre for Ecotoxicology and Toxicology of Chemicals. 1993. Environmental hazard assessment of substances. ECETOC Technical Report 51. Brussels, Belgium.
9. European Centre for Ecotoxicology and Toxicology of Chemicals. 2004. Soil and sediment risk assessment of organic chemicals. ECETOC Technical Report 92. Brussels, Belgium.
10. European Centre for Ecotoxicology and Toxicology of Chemicals. 2004. Targeted Risk Assessment. ECETOC Technical Report 93. Brussels, Belgium.
11. Commission of the European Communities. 1967. Council Directive 67/548/EEC of 18 August 1967 on the approximation of the laws, regulations and administrative provisions relating to the classification, packaging and labelling of dangerous substances. *Off J Eur Communities,* L96/1.
12. Commission of the European Communities. 1993. Council Regulation 793/93/EEC of 23 March 1993 on the evaluation and control of the risks of existing substances. *Off J Eur Communities*, L84/1.
13. Commission of the European Communities. 1991. Council Directive 91/414/EEC of 15 July 1991 concerning the placing of plant protection products on the market. *Off J Eur Communities*, L230/1.
14. Commission of the European Communities. 1993. Council Directive for biocides. 1993. Proposal of the Commission of the European Communities, COM(93) 351 def/SYN 465. Brussels, Belgium.
15. Patton DE. 1993. The ABCs of risk assessment. *EPA Journal* 19:10-15.
16. United Nations. 1992. Environmentally sound management of toxic chemicals including prevention of illegal international traffic in toxic and dangerous products. Agenda 21, chapter 19. United Nations Conference on Environment and Development, Rio de Janeiro, Brazil.
17. Organization for Economic Co-operation and Development. 1981. Decision of the Council concerning mutual acceptance of data in the assessment of chemicals, Annex 2. OECD Principles of good laboratory practice.

OECD, Paris, France.
18. Carnegie Commission on Science, Technology, and Government. 1993. Risk and the environment. Improving regulatory decisionmaking. Carnegy Commission, New York, NY.
19. European Centre for Ecotoxicology and Toxicology of Chemicals. 1988. Concentrations of industrial chemicals measured in the environment: The influence of physico-chemical properties, tonnage and use pattern. ECETOC Technical Report 29. Brussels, Belgium.
20. McColl RS. 1990. Biological safety factors in toxicological risk assessment. Health and Welfare Canada, Ontario, Canada.
21. Vermeire T, Stevenson H, Pieters MN, Rennen M, Slob W, Hackert BC. 1999. Assessment factors for human health risk assessment: a discussion paper. *Crit Rev Toxicol* 29:439-490.
22. Commission of the European Communities. 2003. Technical Guidance Document in support of Commission Directive 93/67/EEC on risk assessment for new notified substances, Commission Regulation (EC) No 1488/94 on risk assessment for existing substances and Directive 98/8/EC of the European Parliament and of the Council concerning the placing of biocidal products on the market. Joint Research Centre, European Chemicals Bureau, Brussels, Belgium.
23. Van De Meent D. 1988. Environmental chemistry: instrument in ecological risk assessment. In De Kruijff HAM, De Zwart D, Ray PK, Viswanathan PN, eds, *Manual on Aquatic Ecotoxicology*. Kluwer Academic Press, Dordrecht, The Netherlands, pp 31-35.
24. Bro-Rasmussen F. 1988. Hazard and risk assessment and the acceptability of chemicals in the environment. In Richardson ML, ed, *Risk Assessment of Chemicals in the Environment*. Royal Society of Chemistry, Cambridge, UK, pp 437-450.
25. Premises for Risk Management. 1989. Risk limits in the context of environmental policy. Annex to the Dutch National Environmental Policy Plan (to Choose or to Lose) 1990-1994. Second Chamber of the States General, session 1988-1989, 21137, No 5. The Hague, The Netherlands.
26. Van Leeuwen CJ. 1990. Ecotoxicological effects assessment in the Netherlands: recent developments. *Environ Management* 14:779-792.
27. US Congress, Office of Technology Assessment. 1993. Researching health risks. OTA-BBS-571, Washington, DC.
28. Lave LB, Malès EH. 1989. At risk: the framework for regulating toxic substances. *Environ Sci Technol* 23:386-391.
29. Roberts LEJ. 1988. Risk assessment and risk accep-tance. In Richardson ML, ed, *Risk Assessment of Chemicals in the Environment*. Royal Society of Chemistry, Cambridge, UK, pp 7-32.
30. Tengs TO, Adams ME, Pliskin JS, Gelb Safran D, Siegel JE, Weinstein MC, Graham JD. 1995. Five-hundred lifesaving interventions and their cost-effectiveness. *Risk Analysis* 15:369-390.
31. Commission of the European Communities. 1992. Council Directive 92/32/EEC of 30 April 1992 amending for the seventh time Directive 67/548/EEC on the approximation of the laws, regulations and administrative provisions relating to the classification, packaging and labelling of dangerous substances. *Off J Eur Communities*, L154/1.
32. United Nations. 2005. Globally harmonized system of classification and labelling of chemicals (GHS). First revised edition. United Nations, Geneva, Switzerland.
33. European Chemicals Bureau. 2005. Scoping study on the technical guidance document on preparing the chemical safety report under REACH. Reach Implementation Project 3.2. Report prepared by CEFIC, ECETOC, RIVM, the Federal Institute for Risk Assessment (BfR), Federal Institute for Occupational Safety and Health (BAuA), Okopol, DHI Water & Environment and TNO. European Commission, Joint Research Centre, Ispra, Italy.
34. De Zwart D. 1994. Monitoring water quality in the future. Part B. biomonitoring. National Institute of Public Health and Environmental Protection, Bilthoven, The Netherlands.
35. European Centre for Ecotoxicology and Toxicology of Chemicals. 2005. Guidance for the interpretation of biomonitoring data. ECETOC Document No 44. ECETOC, Brussels, Belgium.
36. European Environment Agency. 1999. Environment in the European Union at the turn of the century. Environmental assessment report No 2. European Environment Agency, Copenhagen, Denmark.
37. United Nations Environment Programme. 1995. Environmental management tools. *Industry and Environment* 18:1-131.
38. World Health Organization. 2004. *Environmental Health Indicators for Europe. A Pilot Indicator-based Report*. WHO Europe, Copenhagen, Denmark.
39. US Environmental Protection Agency. 2004. Status and future directions of the High Production Volume Challenge programme. US Environmental Protection Agency. Office of Pollution Prevention and Toxics, Washington, DC.
40. Commission of the European Communities. 2006.

Regulation (EC) No 1907/2006 of the European Parliament and of the Council of 18 December 2006 concerning the Registration, Evaluation, Authorisation and Restriction of Chemicals (REACH), establishing a European Chemicals Agency, amending Directive 1999/45/EC and repealing Council Regulation (EEC) No 793/93 and Commission Regulation (EC) No 1488/94 as well as Council Directive 76/769/EEC and Commission Directives 91/155/EEC, 93/67/EEC, 93/105/EC and 2000/21/EC. *Off J Eur Union*, L 396/1 of 30.12.2006.

41. US Presidential/Congressional Commission on Risk Assessment and Risk Management. 1997. Final Report, Volumes 1 and 2, Washington, DC.
42. US Environmental Protection Agency. 1998. Guidelines for Ecological Risk Assessment. Report EPA/630/ R-95/002F. Fed Reg 63(93):26846-26924. USEPA, Washington, DC.
43. Codex Alimentarius Commission, 2003. Procedural Manual 14th edition. World Health Organization/Food and Agricultural Organization of the United Nations, Rome, Italy.
44. Bridges J. 2003. Human health and environmental risk assessment: the need for a more harmonised approach. *Chemosphere* 52:1347-1351.
45. Suter II GW, Vermeire T, Munns WR Jr, Sekizawa J. 2003. A framework for the integration of health and ecological risk assessment. *Human Ecol Risk Assess* 9: 281-301.
46. Suter II GW, Vermeire T, Munns WR, Sekizawa J. 2005. An integrated framework for health and ecological risk assessment. *Toxicol Applied Pharmacol* Issue 2, (Suppl 1) 207:611-616.
47. Suter GW II, Munns WR Jr, Sekizawa J. 2003. Types of integration in risk assessment and management, and why they are needed. *Human Ecol Risk Assess* 9:273-279.
48. World Health Organization and International Programme on Chemical Safety. 2001. Report on Integrated Risk Assessment. WHO/IPCS/IRA/01/12. Geneva, Switzerland.
49. van Leeuwen C, Biddinger G, Gess D, Moore D, Natan T, Winkelmann D. 1998. Problem formulation. In Reinert KH, Bartell SM, Biddinger GR, eds, *Ecological Risk Assessment Decision-Support System: A Conceptual Design*. Proceedings from SETAC Ecological Risk Assessment Modelling Workshop 1994. Society of Environmental Toxicology and Chemistry (SETAC), Pensacola, Fl, pp 7-15.
50. Commission of the European Communities. 2004. Science in trade disputes related to potential risks: comparative case studies. Technical Report Series, EUR 21301 EN, Joint Research Centre, Institute for Prospective Technology Studies, Seville, Spain.
51. Van Zwanenberg P, Millstone E. 2005. BSE: *Risk, Science, and Governance*. Oxford University Press, Oxford, UK.
52. National Research Council. 1983. *Risk assessment in the Federal Government: managing the process*. National Academy Press, Washington, DC.
53. European Chemicals Bureau. 2005. Scoping study on the development of a technical guidance document on information requirements on intrinsic properties of substances (RIP 3.3-1). Report prepared by CEFIC, DK-EPA, Environmental Agency of Wales and England, ECETOC, INERIS, KemI and TNO. European Commission, Joint Research Centre, Ispra, Italy.
54. Majone G. 2000. The credibility crisis of community regulation. *J Common Market Studies* 38:273-303.
55. Persson T. 2005. The role of scientific expertise in EU policy-making. In Hansen O, Ruden C, eds, *Science for a safe chemical environment*. Royal Institute of Technology, Stockholm, Sweden, pp 179-189.
56. Allio L, Ballantine B, Meads R. 2006. Enhancing the role of science in the decision-making of the European Union. *Regul Toxicol Pharmacol* 44:4-13.
57. Ballantine B. 2005. Enhancing the role of science in the decision-making of the European Union. EPC Working Paper 17. European Policy Centre, Brussels, Belgium.
58. US Office of Management and Budget. 2006. Proposed Risk Assessment Bulletin. Draft being released for peer review and public comment before 15-06-06. Washington, DC.
59. Bradbury S, Feijtel T, van Leeuwen K. 2004. Meeting the scientific needs of ecological risk assessment in a regulatory context. *Environ Sci Technol* 38/23, 463a-470a.
60. Posthuma L, Suter II GW, Traas TP. eds, 2002. *Species Sensitivity Distributions in Ecotoxicology,* Lewis Publishers. Boca Raton, Fl.
61. Lovell DP. 1986. Risk assessment - general principles. In Richardson ML, ed, *Toxic Hazard Assessment of Chemicals*. Royal Society of Chemistry, London, UK, pp 207-222.
62. Gow JS. 1990. Introduction. In Richardson ML, ed, *Risk Assessment of Chemicals in the Environment*. Royal Society of Chemistry, Cambridge, UK, pp 3-7.
63. Warren-Hicks W, Moore D, eds, *Uncertainty Analysis in Ecological Risk Assessment*. A Special Publication of SETAC. Society of Environmental Toxicology and Chemistry, Pensacola, Fl.
64. Suter GW. 1993. *Ecological Risk Assessment*. Lewis Publishers, Chelsea, MI.
65. Van Leeuwen CJ, Bro-Rasmussen F, Feijtel TCJ, Arndt

R, Bussian BM, Calamari D, Glynn P, Grandy NJ, Hansen B, Van Hemmen JJ, Hurst P, King N, Koch R, Müller M, Solbé JF, Speijers GAB, Vermeire T. 1996. Risk assessment and management of new and existing chemicals. *Environ Toxicol Pharmacol* 2:243-299.

66. Allanou R, Hansen BG, Van Der Bilt Y. 1999. Public availability of data on EU high production volume chemicals. Report EUR 18996 EN, European Commission, Joint Research Centre, Ispra, Italy.
67. Ricci PF, Rice D, Ziagos J, Cox Jr LA. 2002. Precaution, uncertainty and causation in environmental decisions. *Environ Internat* 29:1-19.
68. Jager T, Vermeire TG, Rikken MGJ, van der Poel P. 2001. Opportunities for a probabilistic risk assessment of chemicals in the European Union. *Chemosphere* 43:257-64.
69. Scheringer M, Steinbach D, Escher B, Hungerbuhler K. 2002. Probabilistic approaches in the effect assessment of toxic chemicals. What are the benefits and limitations? *Environ Sci Pollut Res* 9:307-314.
70. MacLeod, M, Fraser, AJ, MacKay, D. 2002. Evaluating and expressing the propagation of uncertainty in chemical fate and bioaccumulation models. *Environ Toxicol Chem* 4:700-709.
71. Slob W, Pieters M. 1998. A probabilistic approach for deriving acceptable human intake limits and human health risks from toxicological studies: a general framework. *Risk Analysis* 18:787-798.
72. Bodar CWM, Berthault F, De Bruijn JHM, Van Leeuwen CJ, Pronk MEJ, Vermeire TG. 2003. Evaluation of EU risk assessments of existing chemicals (EC Regulation 793/93). *Chemosphere* 53:1039-1047.
73. Bodar WM, Pronk MEJ, Sijm DTHM. 2005. The European Union risk assessment on zinc and zinc compounds: the process and the facts. Integrated. *Environ Assessm Managem* 1:301-319.
74. Bosgra S, Bos PMJ, Vermeire TG, Luit RJ, Slob W. 2005. Probabilistic risk characterisation: an example with di(2-ethylhexyl)phthalate. *Regul Toxicol Pharmacol* 43:104-113.
75. Jager T, Den Hollander HA, Van der Poel P, Rikken MGJ. 2001. Probabilistic environmental risk assessment for dibutylphtalate (DBP). *Human Ecol Risk* Ass 7 (6):1681-1697.
76. Verdonck AM, Aldenberg T, Jaworska J, Vanrolleghem PA. 2003. Limitations of current risk characterization methods in probabilistic environmental risk assessment. *Environ Toxicol Chem* 22:2209-2213.
77. Weed DL. 2005. Weight of evidence: a review of concept and methods. *Risk Analysis* 25:1545-1557.
78. Verdonck AM, Van Sprang PA, Jaworska J, Vanrolleghem PA. 2005. Uncertainty and precaution in European environmental risk assessment of chemicals. *Water Sci Technol* 52:227-234.
79. Commission of the European Communities. 2005. EUSES 2.0, the European Union System for the Evaluation of Substances. European Chemicals Bureau (ECB) and National Institute of Public Health and the Environment (RIVM), Bilthoven, The Netherlands. Available from the ECB. European Chemicals Bureau, Ispra, Italy. (http://ecb.jrc.it/REACH/).
80. National Research Council. 1993. Issues in risk assessment. National Academy Press, Washington, DC.
81. De Bruijn J, Hansen B, Munn S. 2003. Role of the precautionary principle in the EU risk assessment process on industrial chemicals. *Pure Appl Chem* 75:2523-2541.
82. Sandin P, Petersen M, Hanson SO, Ruden C, Juthe A. 2002. Five charges against the precautionary principle. *J Risk Res* 5:287-299.
83. Kriebel D, Tickner J, Epstein P, Lemons J, Levins R, Loechler EL, Quinn M, Rudel R, Schettler T, Stoto M. 2001. The precautionary principle in environmental science. *Environ Health Persp* 109:871-876.
84. Commission of the European Communities. 2000. Communication from the Commission on the precautionary principle. COM (2000) 1 final. Brussels, Belgium.
85. European Environment Agency. 2001. Late lessons from early warnings: the precautionary principle 1896-2000. Environmental issue report No 22. European Environment Agency, Copenhagen, Denmark.
86. Anonymous. 2006. Safety catch. Risk assessment is a useful environmental tool, but not if it is used as a cover for a deregulatory agenda. *Nature* 442:223-224.
87. Macilwain C. 2006. Safe and sound? *Nature* 442:242-243.
88. Hart JW, Jensen NJ. 1990. The myth of the final hazard assessment. *Regul Toxicol Pharmacol* 11:123-131.
89. National Research Council. 2006. *Human Biomonitoring for Environmental Toxicants*. National Academic Press, Washington, DC.
90. Paustenbach D, Galbraith D. 2006. Biomonitoring and biomarkers: exposure assessment will never be the same. *Environ Health Perspect* 114:1143-1149.
91. Taskforce on a Canadian information system for the environment. 2002. Informing environmental decisions. Interim report. Environment Canada. Ottawa, Canada.
92. Cairns Jr J, Dickson KL, Maki AW. 1987. *Estimating Hazards of Chemicals to Aquatic Life*. STP 675. American Society for Testing and Materials, Philadelphia, PA.
93. Hansen BG, van Haelst AG, van Leeuwen K, van der

Zandt P. 1999. Priority setting for existing chemicals: the European Union risk assessment method. *Environ Toxicol Chem* 8:772-779.

94. Bradbury S, Feijtel T, Van Leeuwen K. 2004. Meeting the scientific needs of ecological risk assessment in a regulatory context. *Environ Sci Technol* 38/23:63a-470a.
95. Chapman A. 2006. Regulating chemicals. From risks to riskiness. *Risk Analysis* 26:603-616.
96. Power M, McCarty LS. 1998. A comparative analysis of environmental risk assessment/risk management frameworks. *Env Sci Technol* May:224A-231A.
97. European Commission. 2005. Opinion on the appropriateness of existing methodologies to assess the risks associated with engineered and adventitious products of nanotechnologies. Scientific Committee on Emerging and Newly Identified Health Risks (SCENIHR), Brussels, Belgium.
98. Organization for Economic Co-operation and Development. 2003. Description of selected key generic terms used in chemical hazard/risk assessment. Joint project with IPCS on the harmonization of hazard/risk assessment terminology. OECD Environment, Health and Safety Publications. Series on Testing and Assessment 44. OECD, Paris, France.
99. Christensen FM, Andersen O, Duijm NJ, Harremoes P. 2003. Risk terminology - a platform for common understanding and better communication. *J Hazardous Materials* A103:181-203.

2. EMISSIONS OF CHEMICALS TO THE ENVIRONMENT

P. van der Poel, D.N. Brooke and C.J. van Leeuwen

2.1 INTRODUCTION

The essence of risk assessment of chemicals (Chapter 1) is the comparison of *exposure* (the concentration of the substance to which organisms are exposed) with *effects* (the highest concentration at which no effects are expected on organisms or ecological systems).

This chapter deals with the release (emission) of substances into the environment, eventually leading to exposure of organisms and humans via the environment. Exposure of workers is not covered. Section 2.2 describes the origin of all substances but the subsequent sections (as in other chapters) will focus on xenobiotic and natural substances produced or released due to human activities. Section 2.3 deals with the entry of substances into the environment and types of emissions in relation to variations in level, time and place. Emission prevention and reduction is also considered. Section 2.4 focuses on data availability, measurements and methods for the estimation of emissions and their feasibility. The composition of mass balances and the use of emission factors are discussed. Techniques for emission estimation based on a minimum of data are presented and three examples for the estimation of emissions are also given. Section 2.5 provides a short overview of the development of emission estimates and the development of the computer program EUSES. Section 2.6 is an introduction to the estimation of emissions under the new EU chemicals legislation REACH. It highlights the development of exposure scenarios and the challenges ahead. Finally, selected references for further reading are provided in Section 2.7.

2.2 CHEMICAL SUBSTANCES

2.2.1 Origin of chemical substances

Chemical substances emitted into the environment can originate from (a) *lifeless materials*, such as rock from the earth's crust, (b) *fossil fuels*, created out of dead organisms, (c) *organisms* (biomass) present on earth, or (d) *chemical synthesis*. Figure 2.1 presents a simplified diagram showing the pathways of all types of substances into the environment, i.e. the original substances from sources, man-made substances (including their inevitable by-products) and the combustion/degradation products. The types of substances and their origins are briefly described below.

a. Substances from lifeless materials
Lifeless materials contain many substances which are, in effect, generally the essence of the material, for example, metals in rock, salts in seawater, nitrogen and the inert gases, such as helium and argon, in air. Many of these natural resources (minerals, ores, etc.) are exploited and utilized by man and in many cases transformed into other substances. For example, some metals may be present in the earth's crust in a pure form (gold or silver) or they may be extracted or released chemically from compounds present in ores (iron and zinc). Nitrogen from the air is transformed into ammonia and nitric acid in nitrogen fixating plants, while bromine and chlorine are produced electrochemically from brine (e.g. water from the Dead Sea for bromine, and rock salt for chlorine).

The elements of the periodic table provide the essential basis for all possible substances. They are all widely distributed and to some extent present in minerals and ores. The heavy metals are of special interest because of their potentially adverse effects.

b. Substances present in fossil fuels
Fossil fuels contain mainly organic compounds such as hydrocarbons. These substances may be emitted, as shown in Figure 2.1. Examples include emissions of benzene from petrol, benzo[a]pyrene from creosote and tar, and both from oil and coal. Fossil fuels are used for heating (natural gas, oil, coal, peat), electricity generation by power plants (natural gas, coal, oil and peat), in combustion engines for transport (petroleum distillates and chemically modified compounds) and for the production of a wide range of synthetic organic substances (e.g. plastics, dyes and pesticides). Combustion of fuels leads to combustion products.

c. Substances from organisms
All organisms consist of substances formed by biochemical reactions in the organism itself or taken up in their food. Plants are the source of a wide range of many different types of compounds and produce the elementary building blocks, such as carbohydrates, glycerides (natural oils and fats) and proteins (amino acids).

C.J. van Leeuwen and T.G. Vermeire (eds.), Risk Assessment of Chemicals: An Introduction, 37–71.

Many substances originating from plants and animals (including microorganisms) are utilized directly by man as food, medicines, construction materials, dyes, clothing materials, etc., or are chemically modified (e.g. amoxillins synthesized from penicillin produced by fungi). Usually, substances produced by organisms are easily decomposed into other substances and hence, when released into the environment, readily biodegradable. This means that they are transformed and completely broken down by plants or animals. In Figure 2.1 organisms (biomass) are incorporated in the environment. Biomass and fossil fuels may be combusted with or without human interference. This pathway is not shown in Figure 2.1 to provide a clearer overview.

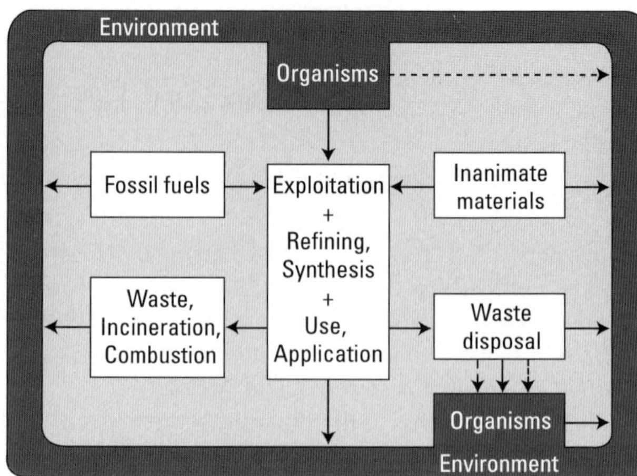

Figure 2.1 Simplified diagram showing possible origin of substances and pathways into the environment. Original substances from biomass, fossil fuels or lifeless materials, by-products in chemical synthesis, synthesized substances and combustion or degradation products.

d. Substances from chemical synthesis

There are many tens of thousands of man-made substances, many of which do not - or rather did not - occur in the natural environment (xenobiotics). They are used for various purposes. In chemical synthesis by-products may also be formed for which there is no use and which are therefore considered as waste. Man-made substances will not be discussed further here, as they will be considered in detail below.

Please note that the classification of chemicals into the four categories above is pragmatic and arbitrary. Cadmium for example is a chemical element widely found (in the form of several different compounds) in the abiotic environment. Cadmium is released and emitted as a result of the exploitation and use of ores (e.g. phosphate and zinc ores). Cadmium is also present in fossil fuels, accumulates in organisms (e.g. in mussels) and is used in chemical synthesis.

2.2.2 Desirable and undesirable substances

Substances (formed or present) may be either *desirable* or *undesirable*. Substances produced by man intentionally are, of course, desirable, unlike by-products and many substances formed unintentionally due to human activities. The dioxin compound 2,3,7,8-TCDD, for example, is an undesirable substance present as a by-product in many compounds such as chlorophenols and 2,4,5-trichlorophenoxyacetic acid (2,4,5-T); another example is PCB 153 which is formed in some pigments. Dioxins like 2,3,7,8-TCDD and PCBs like PCB 153 are also formed unintentionally in waste incineration under certain conditions. In principle, all products of incomplete combustion (e.g. aldehydes and polycyclic aromatic hydrocarbons) are undesirable. Complete combustion gives rise to the formation of water, carbon dioxide (a greenhouse gas), and noxious compounds like NO_x. Some unaltered compounds may be emitted as well, like heavy metals.

A substance may be desirable from one point of view as a useful substance for a certain purpose but be undesirable due to the effect it has on organisms and/or humans via the environment. The use of quite a number of substances, therefore, has been abandoned or restricted in many countries. Examples of substances abandoned in many countries are PCBs and persistent chlorinated pesticides, such as dieldrin. So, a substance can sometimes be both desirable and undesirable. Cadmium produced from a zinc ore and used is desirable. If more cadmium than needed is produced in this way, or if it is released during the processing of phosphate ores, it is undesirable. It should also be noted that "new" substances may be formed from emitted compounds due to successive reactions. A well-known example is the reaction of volatile organic compounds (VOCs) with NO_x. These processes are considered in Chapter 3, and therefore they have not been included in Figure 2.1.

2.3 EMISSIONS AND SOURCES

2.3.1 Entry into the environment

In the risk assessment of substances the best results will be obtained if data are available on concentrations in all environmental compartments. However, this kind of monitoring is carried out only for a few priority substances and in a limited number of situations, mainly because of high costs. The next best option is risk assessment based on real emission data. As these data are also not often available, modelling has to be applied in most cases, with subsequent distribution modelling. In addition future risks, namely the accumulation of persistent toxic chemicals, can be predicted on the basis of emission scenarios/models.

From here on, "emission" will be understood as the result of human activities leading to the release of substances from the technosphere into the environment. Therefore, emissions are directly related to the way man handles resources. Substances of natural origin are already present in the environment, although those present in rock and deep soil layers as part of a matrix are not in direct contact with organisms. It is only after release due to processes such as erosion and abrasion that they enter the environment. These processes are caused by climate factors such as rain, temperature and wind, and volcanic activity.

As indicated in the introduction to this chapter, the focus is on xenobiotic and natural substances produced or released due to human activities. With substances produced intentionally - and their by-products - emissions can take place at any stage of the life cycle of the substance (Figure 2.2). The life of any substance starts with the production or formation stage. A distinction can be made between substances produced as a raw material for the synthesis of other substances (intermediates) and all other substances. For both categories, the life cycle starts in the chemical or petrochemical industry, except for mining and refining of ores and other minerals, which will not be considered here as our focus is on organic substances.

For intermediates three situations can be considered. Firstly, an intermediate may be processed directly in the same reactor without isolation. Secondly, an intermediate may be isolated and processed at the same site (on-site treatment). And finally, an intermediate may be isolated and transported to another factory for processing (off-site treatment). The reason for this distinction is the differences in the level of process releases (emissions). The lowest emissions occur when the intermediate is converted without being isolated in the same vessel. The highest emissions occur when the intermediate is produced and isolated at one site, and processed at another site.

The following potential stage is the formulation process, where substances are mixed and blended to obtain preparations. The production of paint is an example of formulation. It is possible for a substance to go through more than one formulation process. This is the case, for example, when a substance is formulated first into a pigment and than added to a paint formulation.

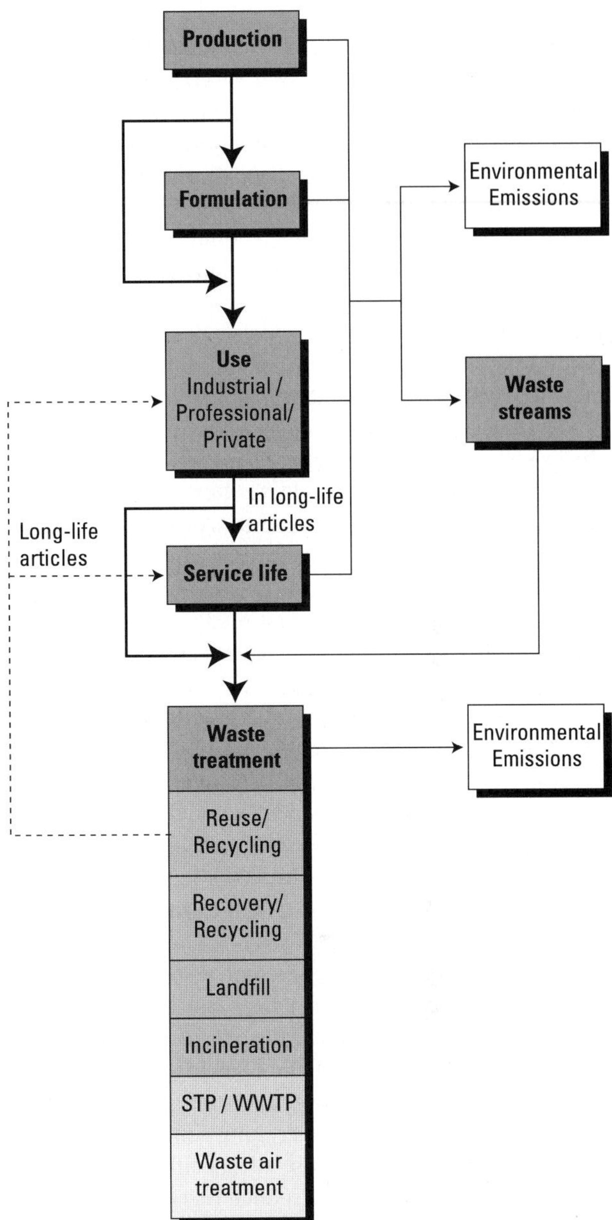

Figure 2.2. Possible stages in the life cycle of anthropogenic substances.

Next comes the use stage, where the substance (as such or in a preparation) is used or applied for a certain purpose. As can be seen in Figure 2.2, a distinction has been made between private use, professional use, and industrial use. To illustrate this distinction, consider the use of a certain paint. An automotive paint might be used by the public at large as do-it-yourself paint (private use), in car refinishing at a body repair shop (professional use) and at a car manufacturer (industrial use). In general, private use is more diffuse than industrial use. There are also differences in the use of emission-reducing techniques. Such techniques are likely to be found in industrial use, where they may be imposed by regulation as well as adopted for commercial reasons. Private use is not usually subject to the imposition of such techniques. Professional use consists of small to medium point sources where to some extent emission-reducing techniques may be imposed or adopted.

Another distinction that has been made is between the use of a substance as a processing aid and its function in articles (industrial products like plastic articles, finished coating layers, etc.). As a processing aid the substance itself is not needed after completion of the process, for example, catalysts in chemical reactions. In other cases, the substance intentionally becomes part of a product or closed system in which it fulfils a certain function, for example, a pigment for polymers used to make toys, and substances used as a dielectric medium in transformers. Processing aids are emitted during processing and/or released into air streams or wastewater and become (hazardous) waste materials. It should be noted that sometimes processing aids become enclosed in a matrix. This is the case, for example, when a catalyst for a polymerisation process becomes incorporated in a plastic.

Substances in articles can enter a stage in their life cycle (the service life) which may last for a long time. Releases can then occur in a diffuse way, due to processes such as migration, abrasion and leaching or evaporation. Estimating emissions from products or articles is further complicated because the exact composition of products and knowledge about the use of specific chemicals in products is often lacking. Product registers, which are available in some countries, may be used to overcome this problem.

The last stage in the life cycle is waste treatment. Releases into waste streams can occur at all stages in the life cycle of a substance, in addition to direct emissions into the environment. Waste streams can be wastewater streams, liquid and solid waste streams, and waste gas streams.

More and more wastewater from households and industry is treated in sewage treatment plants (STPs) and waste water treatment plants (WWTPs). STPs collect diffuse emissions from households, professional activities and often also some industrial activities and form, as such, a kind of point source for the aquatic environment. WWTPs are industrial point sources for substances released during processes in the industry or industries discharging to these plants. The level of connection to municipal STPs varies from country to country, as can be seen in Table 2.1. In the table a distinction is made between primary treatment, secondary treatment and tertiary treatment. Primary treatment refers to treatment by a physical and/or chemical process involving settlement of suspended solids, or other processes in which the biological oxygen demand (BOD) of the incoming waste water is reduced by at least 20% and the total suspended solids of the incoming waste water are reduced by at least 50%. Secondary treatment refers to treatment by a process generally involving biological treatment and a secondary settlement or other process, resulting in removal of at least 70% of BOD and 75% of chemical oxygen demand (COD). Tertiary treatment refers to any additional treatment beyond primary and secondary treatment, intended to reduce the level of BOD or COD or to remove other pollutants. Waste water collected in tankers from houses with no direct connection to the sewage system is considered to be connected to urban waste water collection system. Household waste water not collected by a waste water collection system is generally discharged directly into the environment (onto land or into a river, lake or the sea), though sometimes the householder may chemically treat the waste before discharge [1].

Waste streams may be treated to recover substances or materials for further use in a similar or different form. The fate of an individual substance will depend, among other things, on whether it is a target for recovery or not. Examples of substances recovered for re-use for the same purpose include catalysts and extraction solvents. These may require some treatment to make them suitable for re-use. Another example is the recovery of silver from photographic baths - other substances present in the baths are not considered to be sufficiently valuable to be recovered, and so may be disposed of in waste water.

Lubricating oil is an example of a product that could be recovered and used again after treatment and possible reformulation to clean it up. These steps may involve the removal of some substances present in the used oil to waste streams, as well as the addition of other substances. The recovery of paper fibres for recycling into new paper can result in the removal of substances such as pigments and dyes from the fibres and their presence in waste water or sludges. However, depending on the treatment used (which will depend on how the recovered fibres are to be used) substances may remain on the fibres and so be incorporated into the new product. The shredding of used

Table 2.1. Level of connection to sewage treatment plants (%) of populations in EU member states in 2002 [1]. In those cases where data are based on another year, the year is given in parentheses. A distinction is made between primary treatment, secondary treatment and tertiary treatment. Further explanation is given in the text. No data have been provided for Belgium, Denmark, Greece, Italy and Portugal.

Country	Population not connected to waste water collection systems (%)	Population connected to waste water treatment by type (%)			
		Without treatment	Primary treatment	Secondary treatment	Tertiary treatment
Czech Republic	20	8	-	-	-
Germany (2001)	5	2	0	5	88
Estonia	28	1	1	24	46
Spain	0	11	1	62	26
France (2001)	18	2	2	51	27
Ireland (2001)	7	23	41	21	8
Cyprus (2000)	65	0	0	0	35
Latvia (2003)	-	-	2	35	33
Lithuania (2003)	27	11	32	7	21
Luxembourg (2003)	0	5	7	66	22
Hungary	38	5	22	25	11
Malta (2001)	0	87	-	-	-
Netherlands	1	0	0	14	85
Austria	14	0	0	-	-
Poland (2003)	-	-	3	25	31
Slovenia	37	30	10	18	5
Slovakia (2003)	45	3	-	-	-
Finland	19	0	0	0	81
Sweden	15	0	0	5	80
United Kingdom[a]	2	0	1	59	38

[a] Data for waste water in the UK refer to England and Wales only

tyres to give a material for use in construction will leave behind any additive substances present in the material. The inclusion of substances in new products, whether by accident or by design, will lead to a new service life phase, however this may well be in a different area to the original use.

The treatment of waste materials which have no further use and for which recycling or composting is not a valid option, basically consists of incineration and dumping (land filling). Non-hazardous waste, such as domestic waste containing a mixture of biodegradable, combustible and inert materials, is often land filled. Table 2.2 presents an overview of the percentages of municipal waste land filled and incinerated in the 15 "old" EU member states.

Hazardous wastes are often incinerated and as a result the organic and inorganic substances are destroyed or decomposed to a certain extent, depending on the temperature and residence time. Non-combustible hazardous wastes are land filled at special sites, where they are isolated from contact with the environment. Other possibilities are physical/chemical treatment and biological treatment. Table 2.3 gives an overview for some EU member states of the distribution of the treatment of hazardous waste over the various kinds of treatment, based on statistical data [3].

From the above it can be seen that a number of new waste streams can be generated through the treatment of waste and substances can move partially or completely into these streams, depending on the processes used

Table 2.2. Fraction (%) of municipal waste land filled and incinerated in 15 EU member states. Other methods of waste treatment, e.g., recycling and composting, are not included. These figures are based on statistical data for the amounts of municipal waste land filled and incinerated according to [2].

Country	% land filled			% incinerated		
	1995	2000	2004	1995	2000	2004
EU (15 countries)	78	74	69	22	26	31
Belgium	57	34	23	43	66	77
Denmark	25	16	8	75	84	92
Germany	72	55	42	28	45	58
Greece	100	100	100	0	0	0
Spain	93	90	90	7	10	10
Cyprus (2000)	54	57	54	46	43	46
Ireland	100	100	100	0	0	0
Italy	95	90	83	5	10	17
Luxembourg	34	33	31	66	67	69
Netherlands	53	23	7	47	77	93
Austria	79	75	48	21	25	53
Portugal	100	78	77	0	22	23
Finland	100	85	86	0	15	14
Sweden	48	37	16	53	63	84
United Kingdom	90	92	90	10	8	10

Table 2.3. Distribution of treatment (% of total) of hazardous waste (in kilotonnes) in European countries for which adequate data sets were available [3].

Country	Total amount	Physico/chemical treatment (%)	Biological treatment (%)	Incinerated without energy recovery (%)	Land filled and deposited into or onto land (%)
Czech Republic	464	40	25	5	29
Estonia	6004	1	-	0	99
Finland	998	-	-	8	91
Germany	14580	20	-	13	38
Netherlands	1605	5	-	13	37
Romania	2228	-	-	3	92
Slovakia	1148	39	41	5	12
Slovenia	31	19	-	42	35
Switzerland	926	26	-	45	29
United Kingdom	3896	29	-	2	51

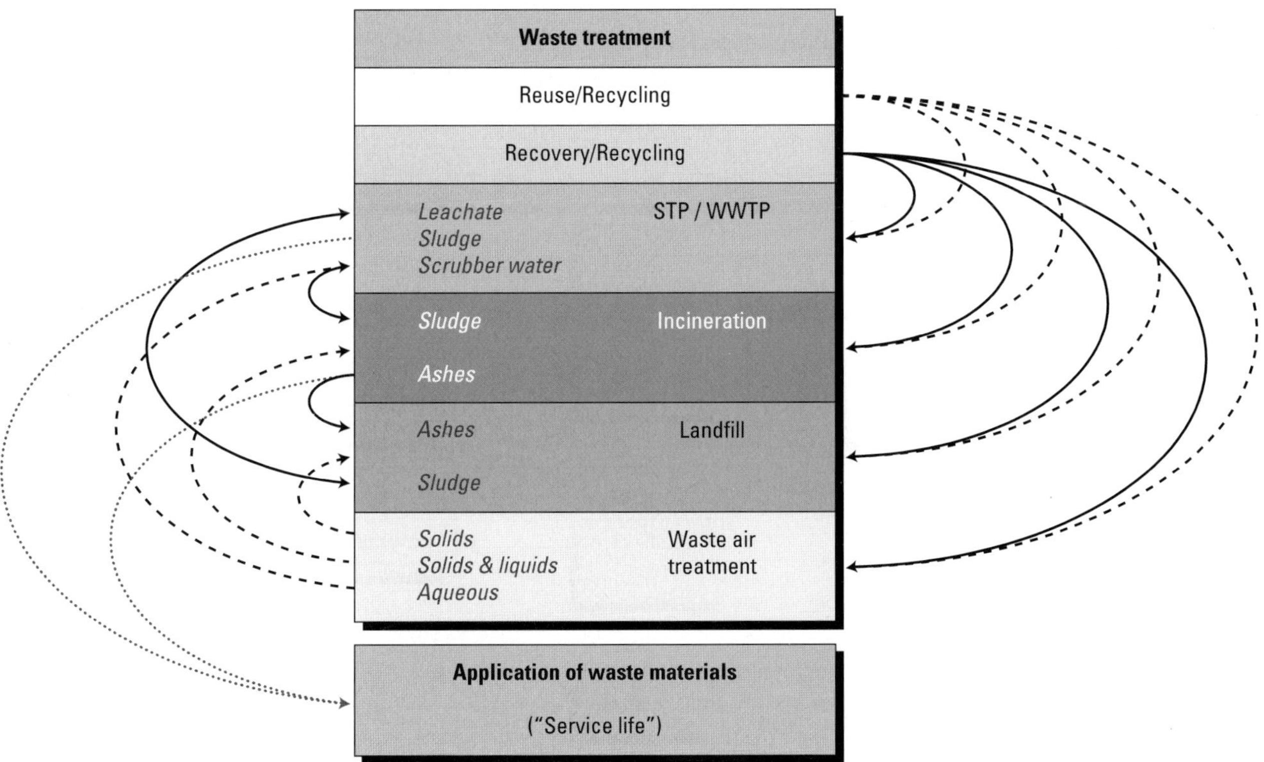

Figure 2.3. Possible flows of a substance in waste streams from previous stages in life cycles at the waste treatment stage, where STP is a municipal or other sewage treatment plant and WWTP is an industrial wastewater treatment plant. Waste treatment will always lead to (minor) emissions to the environment as recovery is generally less than 100%.

and the properties of the substance. Figure 2.3 presents the waste treatment box from Figure 2.2, showing the possible flows of a substance into waste streams. Where substances or materials in waste streams are reused, recovered and recycled, other waste streams may be generated and treated in a particular way (solid and broken arrows on the right hand side of Figure 2.3). These streams may be aqueous with treatment occurring either in an STP or a WWTP. Solid and liquid waste streams may be incinerated and solids may be landfilled. Waste air streams may be treated in several ways, for example, with cyclones (which may give solid waste), scrubbers (giving aqueous waste) and condensers (which may give liquid waste).

Treatment of water in an STP/WWTP produces sludge (Figure 2.3), which can be either incinerated or applied to agricultural soil, through which a substance may unintentionally end up in soil. Incineration yields ashes, which are either landfilled or applied, e.g. as an underlayer in road construction (Figure 2.3). Some substances may completely or partly remain present in ashes (certain metal compounds). At landfills, leachate (rainwater percolating through the landfill) may contain substances which could be treated in an STP (Figure 2.3). Solid waste resulting from waste air treatment may be landfilled or incinerated, like liquid waste, and aqueous waste may be treated in an STP/WWTP.

Figure 2.4 shows a chain of human activities related to phosphate ore from which fertilizer and fodder phosphate are produced. Phosphate ore contains a certain amount of cadmium. As cadmium cannot be completely separated from the process during wet chemical processing, cadmium is emitted into the environment where it may eventually cause toxic effects if no measures are taken. The cadmium released is distributed over the products and the gypsum which is released to surface waters.

2.3.2 Types of emissions and sources

Emissions occur in many different ways: for short periods or continuously for many years, at fixed levels or with wide fluctuations. Depending on the type of risk assessment, it is interesting to know the maximum emissions leading to peak concentrations for acute

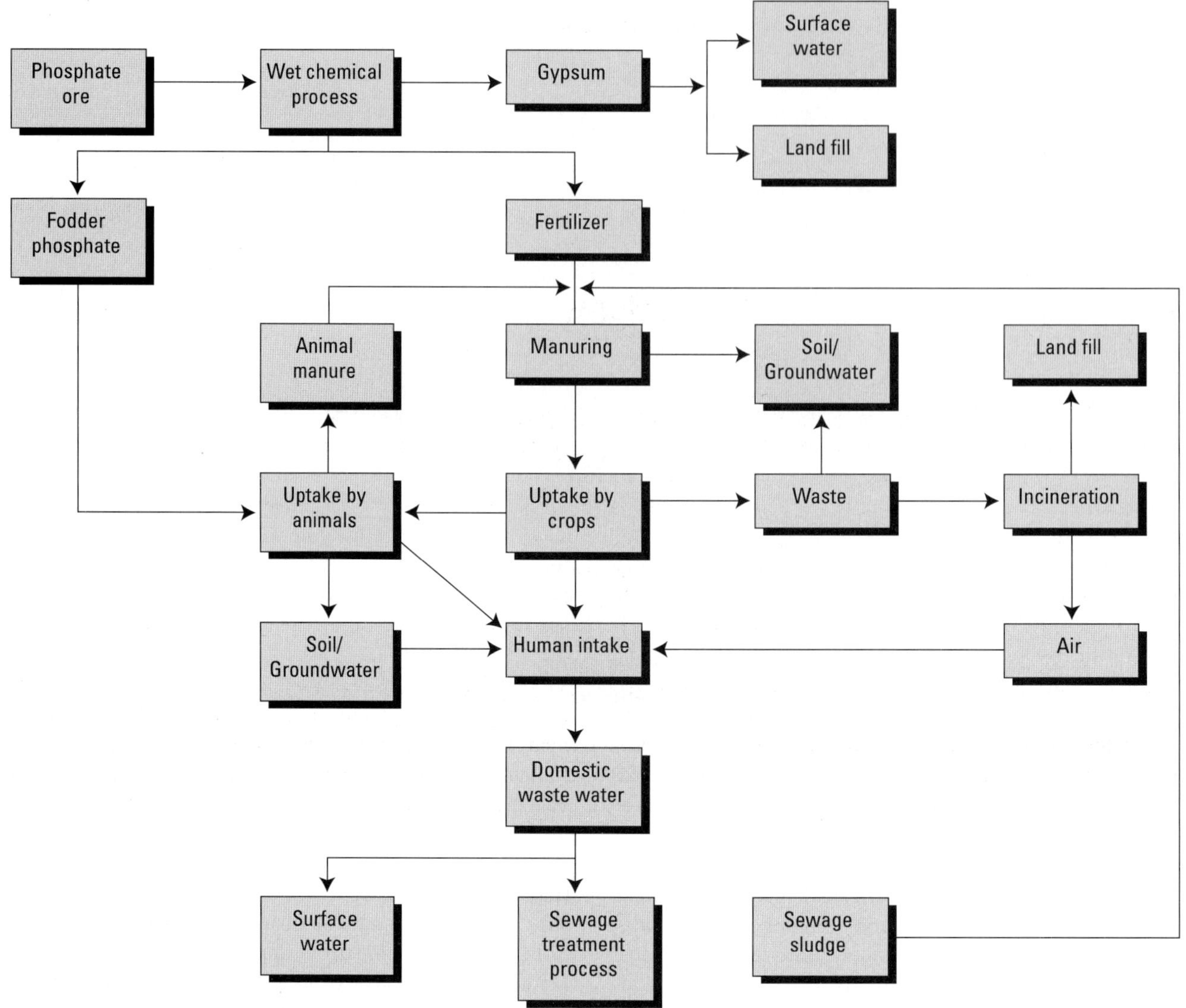

Figure 2.4. Pathways of cadmium released from phosphate ore. Modified from [4].

effects, or the total emissions leading to average concentrations (a "background" concentration) for chronic exposure effects. In general, a distinction can be made between (a) continuous emissions, (b) block emissions, and (c) peak emissions, as shown in Figure 2.5. In practice, the magnitude of both continuous emissions and block emissions will vary widely in time and place. For example, releases of chemicals in effluents from STPs into surface water will vary greatly. Emissions may come from large point sources or they may be diffuse. Additionally, a distinction can be made between stationary and mobile sources.

Table 2.4 gives some definitions of types of emissions and sources.

2.3.3 Emissions in relation to risk assessment

Emissions (direct releases) from sources (processes) into the air, surface waters and soil may take place at any stage in the life cycle of a substance. It is important to distinguish between emissions and the load to an environmental compartment where the actual risk may occur. The load to air, water and soil is the result not only of direct emissions but also of transport and distribution processes in the environment. Figure 2.6 shows the different pathways to surface waters from sources (processes), via transport, and via the distribution processes.

Emissions to the air, for example, take place due to

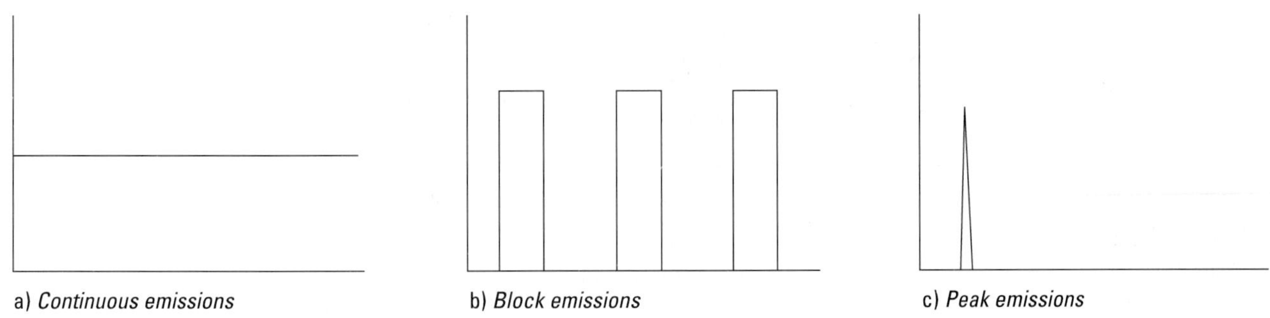

Figure 2.5. Types of emission.

evaporation and the dusting of solids (e.g. in handling and transhipment) and through releases from chimneys. Indirectly, there may be a load to the air when volatile substances such as solvents evaporate from STPs or surface water.

Many substances reach the soil due to wet and dry deposition from the air. This load is the indirect result of emissions into the air. Direct releases to soil will occur as a result of leakages from industrial sites or storage tanks and handling, and when chemicals are applied onto the soil directly as occurs with pesticides (and substances other than biocides used in formulations as solvents or extenders, etc.) and fertilizers on agricultural land. Indirect exposure of the soil also takes place when waste water pollutants enter the soil due to the application of sewage sludge as a fertilizer in agricultural areas.

As stated above, the essence of hazard and risk assessment is the comparison of exposure with effects. In many cases, this comparison has to be made for a specific location and period. In other words, "where" and "when" are crucial. In the case of water a general distinction can be made between surface water (fresh, brackish and salt) and groundwater. Small rivers with a point source discharge are highly exposed. With groundwater flows near agricultural activities the load of pesticides and their degradation products can be considerable. In this situation the local soil (i.e., the soil directly exposed to

Table 2.4. Definitions of types of emissions and sources.

Type	Definition and example
Continuous emission	emissions with an almost constant emission flow rate over a prolonged period. *Example:* the emission of a substance from a continuous production process such as an oil refinery.
Block emissions	emissions with a flow rate which is reasonably constant over certain time periods with regular intervals with a low or even zero background emission. *Example:* the emissions from traffic; emissions are high during rush hours and low at night.
Peak emissions	emissions where a relatively large amount is discharged in a short time; the time intervals between peaks and the peak height can vary greatly. *Example:* the discharge of spent liquid (reaction mixture) after isolation of a synthesized substance in a batch process, or a discharge after a process failure.
Point sources	sources, either single or multiple, which can be quantified by means of location and the amount of substance emitted per source and emission unit (e.g. amount per time unit). *Example:* a chemical plant or a power plant (usually a factory with several plants is considered a single point source).
Diffuse sources	large numbers of small point sources of the same type. *Example:* emission of solvents from painted objects (maintenance of buildings, boats, vehicle, fences, etc.).

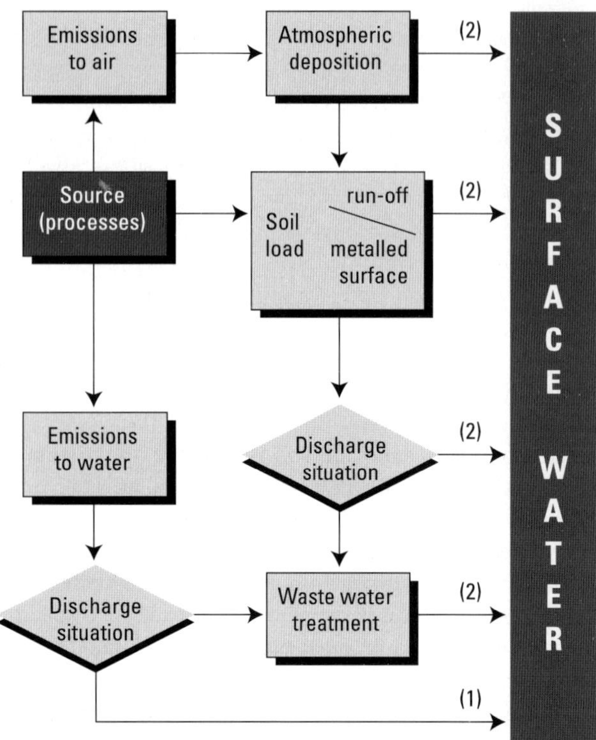

Figure 2.6. Direct (1) and indirect (2) emissions to surface water from processes.

Box 2.1. Examples of risk reduction measures (RRMs)

Substance flow measures:

- recycling of waste
- substitution of substances in products
- quality of raw materials and products

Process optimization:

- good housekeeping
- process-internal recycling
- substitution of processing aids
- process optimization

End-of-pipe treatment:

- waste-water treatment
- gas-flow treatment
- waste destruction and disposal

the emission) should be included in the risk assessment. In general, we can distinguish between agricultural soil, industrial soil and other soil (e.g. urban and rural). Although air is renewed quickly, local pollution may give rise to risks. In a busy street, traffic emissions may lead to effects on humans during rush hours. This is an example of time dependency as well. Another aspect to be considered is air quality in the neighbourhood of large point sources, such as incinerators or industrial plants. It is clear that there is not only a general need for complete emission data, but also a specification of time and place (Section 2.4).

2.3.4 Prevention and risk reduction measures

If the outcome of a risk assessment is critical, risk reduction is the next step. Risk reduction means emission reduction. Box 2.1 presents various examples of risk reduction measures (RRMs).

End-of-pipe treatment is the traditional way of solving these problems. It has been effective in many cases. However, quite often the result is that the problem is merely shifted to another environmental compartment. Prevention, i.e. process optimization, is increasingly regarded as the best alternative and recycling as the next best. But these options, too, may just shift the problem, especially when alternative materials and energy use are involved in the evaluation. There is no objective method by which to decide which option is best. Therefore, the decision should be a pragmatic one, based on an integrated approach to process management. Table 2.5 gives various options for some example substances. Under the REACH system, the consideration of risk management measures will become an integral part of the assessment process (see Section 2.6).

If no emission-reducing measures are taken, emissions of persistent substances eventually lead to accumulation in the environment. A classic example is the accumulation of PCBs in sediments. Figure 2.7 illustrates simulations carried out on the concentration of PCB 153 in the upper layer of Lake Ketelmeer with and without remediation [6]. It appears that whether emission-reducing measures are taken or not, the concentration will return to the same level after 5 to 6 years (emissions of PCB 153 are diffuse and often emanate from unknown sources). Remediation of contaminated sites is therefore not always appropriate and the problem needs to be solved at the source (prevent or reduce emissions).

2.4 DATA AVAILABILITY AND GENERATION

2.4.1 Measurements

The most direct way of gathering information about emissions is to carry out measurements in effluents and emitted gas flows. However, a measurement is just one

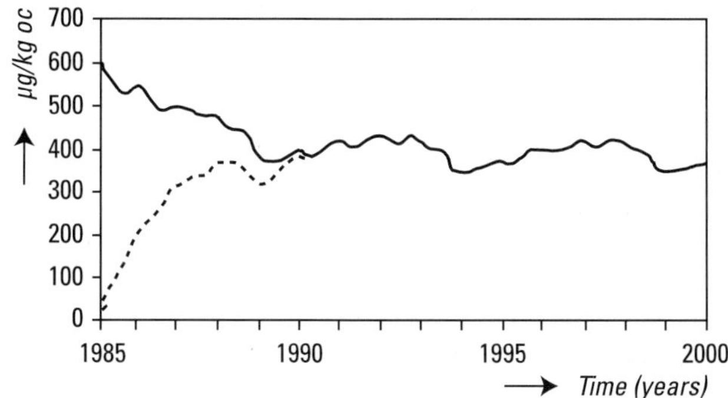

Figure 2.7. Simulation of PCB 153 concentration in the top layer of aquatic sediments in Lake Ketelmeer in 1985, without emission reduction in the period 1985-2000, without removal of polluted sediments (——) and with removal of polluted sediments in 1985 (- - - -). From [6].

sample, taken in a particular period from a stream with flows and concentrations varying in time. Therefore, the result of a single measurement must be converted into more generally applicable emission data, based on knowledge of the process or activity during the sampling period and in general over time, especially in the event of block or peak emissions. These data should include information on production and process conditions, which are often more difficult to obtain (especially afterwards) than the sample itself. A combination of this information and the analytical results can provide insight into emissions from the process at times other than the sampling time, and for similar processes elsewhere.

Other measured data can also be used for emission calculations. These include measurements in daily practice of the quantity or quality of raw materials, like ores or concentrations in waste or products, as

Table 2.5. Some options for reducing or preventing emissions for a number of example substances.

Substance	Process (chain leakage)	Substance flow measures	Process optimization	End-of-pipe measures
Cadmium	waste incineration	substitution of Cd in products		electrostatic filtration
	agriculture	reduction of Cd-content in phosphate products		
	metal plating		electrolysis	precipitation
2,3,7,8-TCDD	waste incineration	avoidance of strong variation in waste composition	temperature control in process	scrubbing and adsorption
Dieldrin	agriculture	substitution by less harmful pesticides		
Chloroform	pharmaceutical industry	solvent substitution	improvement process control (closed equipment, vapour return)	adsorption, incineration
2-Propanol	chemical industry			scrubbing and treatment in STP
PCB 153	(all processes)	substitution by other substances		

well as measurements from pilot scale or laboratory processes. Some priority pollutants, like cadmium and dioxins, have received a great deal of attention and specific measurements can provide a basis for their risk assessment. Emissions are usually measured when a problem is suspected (risk definition). The effects of RMMs are also monitored in the control phase. However, for thousands of substances there is little or no measured data available.

2.4.2 Specific calculations

The following types of calculations can be carried out to convert the results of various measurements into information about emissions:

1. Mass-balance calculations.
2. Calculations based on the process characteristics and properties of the substance.
3. A combination of 1 and 2.

1. Mass-balance calculations

If data from measurements are available for all flows except one, the missing flow can also be quantified. The basic formula for the mass balance of a process or activity over a certain period is:

$$I = E_w + E_a + E_s + W + P + dS + D \tag{2.1}$$

where

I = input (amount produced, purchased, etc.)
E_w = amount discharged with wastewater
E_a = amount emitted into the air
E_s = amount released to the soil
W = amount in outgoing waste
P = amount in outgoing product
dS = difference in amount in storage at start and end of period
D = amount degraded (thermally, biologically and chemically).

An example of this is the determination of chloroform emissions into the air from a pharmaceutical plant using chloroform as a solvent. In principle, the input (the amount bought by the company) should be equal to the output, i.e. the amounts released in the effluent and in the waste (which can be measured), plus the emissions to air and soil (which are very difficult to measure). Figure 2.8 presents the method used to determine mass balances in a study on emissions from pharmaceutical plants [7].

2. Calculations based on the process characteristics and properties of the substance

These calculations can be used to estimate the release of a substance into only one environmental compartment. For example, they can be used to calculate the evaporation of a solvent from open tanks, or polycyclic hydrocarbons from wood treated with creosote. In these cases the physicochemical properties of the substance play an important role. Other examples include discharges to wastewater of dyes from the textile industry, or chemicals from the photographic industry, which are mainly determined by the process characteristics.

3. A combination of 1 and 2

Sewage treatment is an example of a process in which many substances occur somewhere, at sometime. Other examples are landfills and waste incineration; both processes at the end of a substance chain. The behaviour of substances in STPs has been studied extensively. Elements like heavy metals have been measured mostly by monitoring in practice, while the behaviour of organics has also been examined by process simulation modelling [7 – 9]. The influence of properties like water solubility, vapour pressure, and octanol-water partition coefficient, together with reasonably well-defined process characteristics, can be estimated with the help of exposure models (Chapter 4). More difficult, however, is the incorporation of biodegradation (Chapter 3), because of the uncertainties in processes like microbial adaptation and biokinetics. In fact, the presence of specific micro-organisms in sludge is never constant and varies in every STP. Table 2.6 gives ranges of results for several calculations carried out with different models [8]. It should be noted that these calculations were made for pharmaceutical manufacturers in The Netherlands, where the STPs usually have a low pollution load. In other circumstances very different values may be found.

The same kinds of activities and behaviour play a role in landfills, although the process characteristics (time scale and/or anaerobic conditions) are quite different. Much less is known about the behaviour of organic substances in landfills. Waste incineration presents very specific problems in emission assessment. For heavy metals mass balances can be determined, but organics are assumed to decompose completely. In older and badly controlled incinerators, fractions of the more stable organic substances are evaporated, and there may be significant formation of other toxic substances like dioxins.

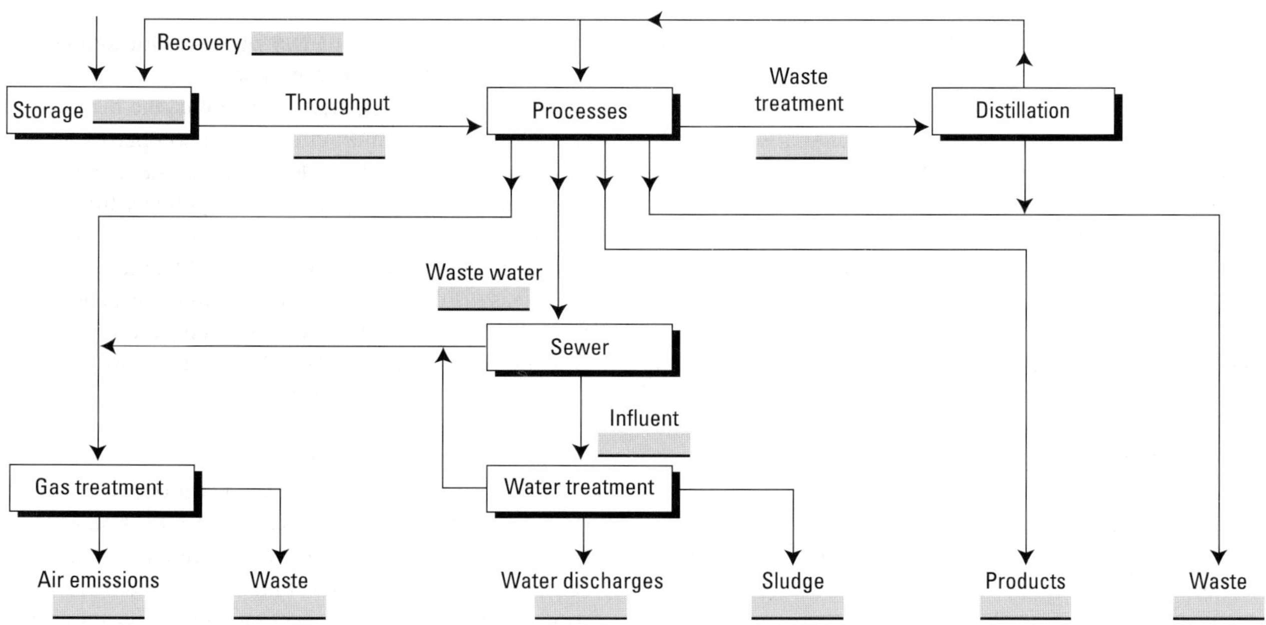

Figure 2.8. Simplified diagram of the processing and composition of mass balances of solvents. Each grey block contributes to the mass balance of the solvents [7].

2.4.3 The application of emission factors

Measurements and specific calculations, for example in the form of mass balances, can only be carried out for a limited number of actual sources and substances. The results of such measurements and calculations can be used to derive emission factors, provided they can be related to the size of the activity (e.g. the production volume of a process). These emission factors can be used to calculate the emissions of other substances handled in a similar way for which on-site measurements are not feasible or possible.

Emission factors can be used at different levels: (1) apparatus or plant, (2) industrial sector, and (3) national. It is important to make this distinction because the emission factors may be very different. If they are determined for well-defined technical situations, like specific types of pumps, gas burners or cars used under defined circumstances, or for specific industrial processes where various specific environmental measures are taken, there will be no misunderstanding about them. However, there are larger scale emission factors, for example, an average for a whole country. These relate to the penetration of technologies and form an average of all

Table 2.6. Fate and behaviour of some solvents in active-sludge plants with a low load, based on model calculations[a] [8].

Substance	Water	Air	Sludge	Degradation
Toluene	1- 2	31 - 69	0-1	32-67
Methanol	<<1	<<1	<<1	> 99
Acetone	1 - 2	<<1	<<1	98 - 99
Dichloromethane	2- 3	20 - 56	<<1	40 - 77
Tetrachloromethane	1- 2	94 - 99	0-2	0 - 2
1,2-Dichloroethane	19 - 30	30 - 50	0 - 1	20 - 50
Trichloroethylene	1- 2	84 - 95	0-1	3-12
Monochlorobenzene	2- 4	63 - 85	1-8	6-25

[a] Numbers represent percentages.

individual installations. These emission factors are much less precise. Such emission factors can be estimated, for example, for cars in France or Germany in 2000, or for the ceramic industry in Europe in 2002, where each year and scale has its own value. Even on a global scale emission factors may be useful, for example, for emission calculations of PCBs based on world production. What type of emission factor should be used depends on the purpose and precision of the calculation; these, of course, are related to the specific hazard and risk assessment goal, which in turn may depend on the scale of the problem and the distribution of the substance in particular.

There are many databases with emission factors and they are growing in number [11–16]. However, they cover a limited number of substances. Most of the emission factors listed are directly related to the size of the activity and, in the case of industrial processes, to the production or the amount (or surface) of the material treated. In many cases, however, loss percentages related to the input, based on the concept of mass balances or other calculations are useful, especially if only information on the total amount of the substance is available. Some examples of this approach are given in the next section.

2.4.4 Generic approach to emission estimation

From the previous sections it will be clear that emissions can be calculated on the basis of monitoring data (measurements) or mass balances in only a few cases. For new substances such data are normally not available at all. Due to this lack of information it is necessary to use generalized approximations, for example, for:

1. Emission factors for processes.
2. Substance properties and process conditions (capacity of processes, etc.).
3. Data from practice (type of process and formulations used).

In this section the approaches set out in the Technical Guidance Document (TGD) will be described [17] but first some comments need to be made about the spatial scales and time scales for which these emission calculations are performed.

Spatial scales

A wide variety of spatial scales can be distinguished: the global scale, the continental scale, the fluvial scale, the regional scale and the local scale (Figure 7.2). The most appropriate scale depends on the nature of the substance to be assessed, including the effect it has on the environment and its emission pattern. Substances causing ozone depletion have a continental or even global impact, while exposure to certain substances used in specialized industrial processes is highly controlled and therefore limited. Sometimes operative policy dictates the choices to be made. When "environmental concentrations are not allowed to exceed the maximum permissible level beyond the border of a site" (e.g. a chemical plant), a local scale of 100 or 1000 m, for example, can be chosen. Whatever scenario is chosen, it should be remembered that, after their release, substances do not stop at borders; they may be transported over great distances. Persistent organic pollutants have been measured in arctic ice, deposited there from the air after emissions in far away civilized areas.

For an initial exposure assessment of a substance produced or marketed in small quantities (e.g. <100 tonnes/y), a local scale is generally selected (Table 2.7). On this scale, emissions are considered to come from a main point source. Protection targets are assumed to be exposed near this source. In the case of pesticide use (also referred to as an emission in this context), this may be a piece of agricultural land. Larger quantities of substances may have significant effects on a larger scale and a regional scale could be added to the assessment. At the regional scale, emissions are regarded as diffuse and continuous and multi-compartment steady-state models are often used for the estimation of environmental transport and transformation (Chapter 4). Estimated continental concentrations can be used as background data to estimate concentrations on a regional scale.

If a point source can be identified, any effects will predominantly occur locally. If the sources are diffuse, effects may become apparent on a regional scale. However, where a persistent substance is emitted in small amounts locally, it is possible that regional concentrations could ultimately become as significant as local concentrations.

Time scales

The choice of the time scale to be considered in the emission and exposure assessment depends on factors such as the frequency and duration of emissions, the generation times of the organisms being considered in the effects assessment (Chapter 7) or the level of refinement of the assessment [17]. For organisms with a short generation time, such as micro-organisms (e.g. in a STP) and most aquatic organisms, an emission episode can cover a considerable part of their life cycle. Thus, the exposure concentrations to these organisms during such episodes can be considered to be "continuous" and compared with no effect levels derived from long-term (chronic) toxicity data, even though environmental

Table 2.7. Examples of spatial scales for the calculation of environmental concentrations used as input for indirect exposure calculations according to the TGD [17].

Compartment	Local assessment	Regional assessment
Surface water	annual average concentration after complete mixing of STP-effluent	steady-state concentration in surface water
Air	annual average concentration at 100 m from source or STP (maximum)	steady-state concentration in air
Agricultural soil	concentration averaged over 180 days after 10 years of sludge application and aerial deposition	steady-state concentration in agricultural soil
Porewater	concentration in porewater of agricultural soil as defined above	steady-state concentration in porewater of agricultural soil
Groundwater	concentration in porewater of agricultural soil as defined above	steady-state concentration in porewater of agricultural soil

concentrations averaged over a year may be much lower. The exposure of terrestrial organisms can be assumed to be only marginally influenced by temporal fluctuations in emission rates, as most substances are not emitted directly to soil. This does not always apply to pesticides, however, which are usually applied over short periods of time, leading to peak concentrations in soil, water and air. These concentrations are generally compared with short-term toxicity data for non-target organisms. With long-term exposure to chemicals, for example due to the persistence of the substance or frequent or widescale emissions, long-term toxicity data should be used. For human beings, episodes of emissions to the environment cover a short period of their lives. It can therefore be assumed that they are exposed to environmental concentrations averaged over a longer period (e.g. 1 to 10 years), which are derived from average emission rates (Table 2.7). Consumers and workers, however, may be exposed to episodic concentrations once or repeatedly, even extending throughout their lifetime.

Emission calculations according to the TGD

The TGD [17] considers three types of categories: 4 *main categories* (MCs); 16 *industrial categories* (ICs); and 55 *use or function categories* (UCs). These are listed in Table 2.8.

The main categories are intended to classify the exposure relevance of the use(s) of a substance in four broad categories: (1) use in closed systems; (2) use resulting in inclusion into or onto a matrix; (3) non-dispersive use; and (4) wide dispersive use. In the context of environmental risk assessment these categories are also used to characterize emission scenarios for the estimation of emissions to the environment during specific stages of the life cycle of the substance.

The industrial categories specify the branch of industry (including personal and domestic use, and use in the public domain) in which the substance is used. Relevant emissions are likely to occur in the application of the substance as such or in the application and use of preparations and products containing the substance. The industrial categories are rather broad in many cases and many important emission sources (processes) are not specifically included and hence have to be allocated to IC 16 “Others”.

The use category or function category represents the specific function of the substance. There are 55 categories with a varying level of detail. For substances used in photography, for example, there is UC 42 “Photochemicals”. Several specific types of photochemicals have a specific UC, such as fixing agents (UC 21). For chemicals applied in polymers, however, there is no specific UC for plastics additives. More significant is the lack of specific use categories for particular functions substances may have in fields like photography and polymer processing.

For emission estimation the TGD [17] contains so-called A and B Tables. These generic tables are based on expert judgement with additional in-depth knowledge based on use category documents (UCDs) on textile dyes [18], photochemicals [19], metalworking fluids [20], and paper chemicals [21]. There are A and B Tables for each of the 16 industrial categories.

Table 2.8. Categories used by the EU [17].

Main categories			
I	Use in closed system - non-isolated intermediates - isolated intermediates stored on-site - isolated intermediates with controlled transport	II III IV	Use resulting in inclusion into or onto a matrix Non-dispersive use Wide dispersive use

Industrial categories			
1	Agricultural industries	9	Mineral oil and fuel industry
2	Chemical industry: basic chemicals	10	Photographic industry
3	Chemical industry: chemicals used in synthesis	11	Polymers industry
4	Electrical/electronic industry	12	Pulp, paper and board industry
5	Personal/domestic	13	Textile processing industry
6	Public domain	14	Paints, lacquers and varnishes industry
7	Leather processing industry	15	Engineering industries; civil and mechanical
8	Metal extraction, refining and processing industry	16	Others

Use categories			
1	Absorbents and adsorbents	31	Impregnation agents
2	Adhesive, binding agents	32	Insulating materials
3	Aerosol propellants	33	Intermediates (monomers; pre-polymers)
4	Anti-condensation agents	34	Laboratory chemicals
5	Anti-freezing agents	35	Lubricants and additives
6	Anti-set-off and anti-adhesive agents	36	Odour agents
7	Anti-static agents	37	Oxidizing agents
8	Bleaching agents	38	Plant protection products; agricultural
9	Cleaning/washing agents and additives (detergents; soaps; dry cleaning solvents; optical brighteners in detergents)	39	Biocides, non-agricultural (disinfectants; preservative products; pest control products; specialist biocides)
10	Colouring agents (dyestuffs; pigments; colour forming agents; fluorescent brighteners)	40	pH-regulating agents
		41	Pharmaceuticals (veterinary medicines)
11	Complexing agents	42	Photochemicals (desensitisers; developers; fixing agents; photosensitive agents; sensitisers; anti-fogging agents; light stabilisers; intensifiers)
12	Conductive agents (electrolytes; electrode materials)		
13	Construction materials and additives		
14	Corrosion inhibitors	43	Process regulators (accelerators; activators; catalysts; inhibitors; siccatives; anti-siccatives; cross-linking agents; initiators; photoinitiators; etc.)
15	Cosmetics		
16	Dust binding agents		
17	Electroplating agents	44	Reducing agents
18	Explosives (blasting agents; detonators; incendiaries)	45	Reprographic agents (toners for photocopying machines; toner additives)
19	Fertilisers		
20	Fillers	46	Semiconductors (photovoltaic agents)
21	Fixing agents	47	Softeners (coalescing agents; bates in leather technology devulcanising agents; emollients; swelling agents; water softeners; plasticisers)
22	Flame retardants and fire preventing agents		
23	Flotation agents		
24	Flux agents for casting	48	Solvents
25	Foaming agents (chemical/physical blowing agents; frothers)	49	Stabilisers
26	Food/feedstuff additives	50	Surface-active agents
27	Fuels (gasoline; kerosine ; gas oil; fuel oil; petroleum gas; non-mineral oil)	51	Tanning agents
		52	Viscosity adjustors (pour-point depressants; thickeners; thixotropic agents; turbulence supressors; viscosity index improvers)
28	Fuel additives (anti-fouling agents; anti-knock agents; deposit modifiers; fuel oxidizers)		
29	Heat transferring agents (cooling agents; heating agents)	53	Vulcanising agents
30	Hydraulic fluids and additives	54	Welding and soldering agents
		55	Others

The A Tables provide the estimated total release fractions (emission factors) of the production volume to air, (waste) water and industrial soil during production, formulation, industrial/professional use, private use, and recovery, according to their industrial category. The TGD distinguishes between the *production volume* (the amount produced in the EU, to be used in calculations relating to production) and the *tonnage* or *market volume* (production volume plus the total amount imported into the EU minus the total amount exported from the EU, to be used for calculations for other life cycle steps). The total amount of the substance released is averaged over the year and used for the calculation of the regional predicted environmental concentration (PEC).

The B Tables are used to determine releases from point sources for the calculation of the local PEC. This means the estimation of the daily capacity of the representative largest point source for the particular stage of the life cycle and process to be expected (reasonable worst-case situation). Based on the volume of the substance the B Tables provide estimates for the *fraction of the main source* and the *number of days* that releases occur. So, the main source is the emission source where the largest fraction of the production volume or market volume of the substance is handled (i.e., produced, formulated, etc., depending on the stage of the life cycle considered). The number of days is the number of emission days per year that activities take place during which the substance will be released.

The A and B Tables cover both low production volume chemicals (LPVCs) and high production volume chemicals (HPVCs). In the EU, HPVCs are defined as being chemicals produced or imported in quantities of at least 1000 tonnes per year by at least one company. During the development of the A and B Tables a different approach was taken. Individual threshold tonnages for considering a substance as high production volume were set for each industry category and for some specific use categories.

The TGD also contains *emission scenario documents* (ESDs). An ESD is defined as a set of conditions for sources, pathways, production processes and use patterns that quantify the emissions (or releases) of a chemical [22]. An ESD should ideally include all the following stages: (1) production, (2) formulation, (3) industrial use, (4) professional use, (5) private and consumer use, (6) service life of product/article, (7) recovery, and (8) waste disposal (incineration, landfill), but in many cases existing ESDs may be limited to only one or a few stages. ESDs provide methods for calculating emissions which often allow the use of specific data related to the substance or use and provide default values for use when specific information is not available.

Chapter 7 in Part IV of the TGD [17] contains various ESDs (see Table 2.9), which were available in suitable form at the time of publication. These ESDs cover one or more stages of the life cycle for a specific industrial branch and/or a specific group of chemicals or chemical products. Some cover only biocidal products. In many cases the time aspect is not covered.

Further ESDs have been developed by EU countries, the United States and Canada. The OECD task force on environmental exposure assessment (see also Chapter 16) develops ESDs at the OECD level, in order to be able to compare differences in conditions of production, use etc., between OECD member countries, and to avoid duplication of effort as far as possible. Draft ESDs produced by lead countries are circulated to member countries for comments, amended by the lead countries and published by the OECD. The documents published by the OECD as at summer 2006 are listed in Table 2.10.

The OECD has also published guidelines on the production of ESDs [22]. All OECD ESDs are available from the OECD website (see also Chapter 16).

Both the A and B Tables and ESDs may have specific emission factors for functions that substances may have which are not covered by one of the use categories of the TGD. For emission estimation the TGD states that releases of a substance at different stages of its life cycle should be estimated from (in order of preference):

- Specific information for the given substance (e.g., from producers, product registers or open literature).
- Specific information from the emission scenario documents (use category documents) for several industrial categories as well as for some of the 23 biocidal product types as given in Part IV, Chapter 7 (see Table 2.10).
- Emission factors as included in the release tables of Appendix I (A and B Tables, see below).

The A and B Tables serve as a kind of safety net, enabling emission estimation with a minimum of data if ESDs are not available. Both the A and B Tables and emission scenarios of the ESDs have been implemented in the latest version of EUSES (see Section 2.5). Usually the emission scenarios for the local situation are based on one of two possibilities:

1. Tonnages (production volume / market volume).
2. Consumption and use.

1. Emission scenarios based on production/market volume

In general no regional tonnages will be known for

Table 2.9. Emission scenario documents of the TGD [17]. The following abbreviations are used: IC is the industrial category, BPT is the biocidal product type and ww is wastewater. For the life cycle stage: 1 stands for production, 2 for formulation, 3 for industrial use, 4 for professional use, 5 for private use, 6 for service life, and 7 for waste treatment.

Area	Emission scenario document	Environmental compartments	Time aspect	Stage of the life cycle
IC-3	Chemical industry: chemicals used in synthesis[1]	ww	no	1, 3
IC-5	Personal/domestic[2]	ww	no	1, 2
IC-6	Public domain		yes	5
IC-7	Leather processing industry	ww	yes	3
BPT 9	Biocides used as preservatives			
IC-8	Metal extraction, refining and processing industry[3]	ww	yes	3, 7
IC-10	Photographic industry	ww	yes	3, 7
IC-12	Pulp, paper and board industry	ww	yes	3, 7[4]
IC-13	Textile processing industry	ww	yes	3, 6
BPT 9	Biocides used as preservatives in textile wet processing			
IC-14	Paints, lacquers and varnishes industry[5]	ww, air, soil	no	2, 3, 7
IC-15	Others: releases of additives in the rubber industry[6]	ww	yes	3
BPT 2	Private and public health area disinfectants and other biocidal products			
BPT 6, 7 and 9	Biocides used as preservatives in various applications[7]	ww	yes	3, 7
BPT 22	Embalming and taxidermy fluids	ww, soil	yes	3

[1] Only for substances with UC 33 "Intermediates" at HPVC level.
[2] Only for use categories 9 "Cleaning/washing agents" and 15 "Cosmetics".
[3] Only for use categories 29 "Heat transferring agents" and 35 "Lubricants and additives" applied in metalworking fluids.
[4] Waste treatment: paper recycling.
[5] For 20 particular applications / coating types.
[6] The production of synthetic "raw rubber" is counted in the industrial category IC 11 (polymers industry).
[7] IC 12 "Pulp, paper and board industry".

any given substance. In which event the EU tonnage according to the TGD should be used in the calculations, except for private use where the 10 % rule applies (page 32 of Part II of the TGD [17]). For diffuse emissions from households the standard scenario of the TGD is based on an average waste water flow of 200 L per capita per day for a population of 10,000 inhabitants. If the use of a substance were to be evenly distributed over the population (consumers) and STPs in the region and over the week, the fraction of this substance reaching the standard STP of the TGD (EUSES) would be the *number of inhabitants connected to the STP / number of inhabitants in the region.* This means a fraction of 10,000 / 20 x 10^6 = 0.0005 with the defaults provided in the TGD. As the use of (formulations containing) substances is never distributed evenly over the population and from day to day, a safety factor of four was assumed at the time. This means that the fraction of the main source is 0.002. This value is used in the emission tables of the TGD. In this case the number of emission days is equal to 365. There may be other applications where a point source is considered. For the ESD on private and public health area disinfectants and other biocidal products, for example, the fraction of the main source of the model hospital has been estimated to be 0.007. This fraction was calculated from the average number of beds per hospital in a region and the total number of hospital beds in that region.

2. Emission scenarios based on the consumption and use

This type of emission scenario applies either the average consumption per inhabitant or the (estimated) use in a process. An ESD might also use consumption data

Table 2.10. Emission scenario documents published by the OECD. The following abbreviations are used: IC is industrial category, BPT is biocidal product type, ww is wastewater, and sw is surface water. For the life cycle stage: 1 stands for production, 2 for formulation, 3 for industrial use, 4 for professional use, 5 for private use, 6 for service life, and 7 for waste treatment.

Area	Emission scenario document (ESD)	Environmental compartments	Time aspect	Stage of the life cycle
BPT 8	Wood preservatives	ww. sw	yes	3-6
IC-11	Plastic additives	ww. air	yes	2, 3, 6
IC-6, 12	Water Treatment Chemicals	ww	yes	4, 6
IC-10	Photographic Industry[1]	ww	yes	3, 4, 7
IC-11	Rubber additives[1]	ww, air, soil	yes	2, 3, 6[2]
IC-13	Textile Finishing	ww, air	yes	3, 4. 6
IC-7	Leather Processing	ww	yes	3
IC-4	Photoresist use in semiconductor manufacturing[3]	ww	yes	3
IC-8, 9	Lubricants and lubricant additives	ww, air, sw	yes	2-4, 6[4]
IC-14	Automotive spray application[5]	ww	yes	3, 4
IC-8	Metal finishing	ww	yes	3
BPT 21	Antifoulants	ww, sw	yes	3, 4
BPT 18	Insecticides for stables and manure storage systems	sw, soil	yes	3
IC-12	Kraft pulp mills	ww, air	yes	3
IC-12	Non-integrated paper mills	ww, air	yes	3
IC-12	Recovered paper mills	ww, air	yes	3, 7

[1] Revised from version in TGD [17].
[2] Tyre abrasion.
[3] Only for non-volatile substances used in photoresists.
[4] Waste treatment: metal working fluids only.
[5] Only for non-volatile components.

(expressed as g/cap/d or L/cap/d) for products such as cleaning products, soaps, etc. Such an emission scenario applies: (1) an emission factor, (2) the concentration of the substance in the product, and (3) a penetration factor. The penetration factor is the fraction of the product on the market containing the specific substance. Some examples of the use of ESDs and the A and B Tables to estimate chemical emissions are presented in the next section.

2.4.5 Three examples of estimating emissions

In this section three examples are presented on the use of ESDs and the use of the A and B Tables according to the procedure as explained in Section 2.4.4. The examples were selected to cover a wide range of applications of the different approaches. As companies develop many new chemicals, especially intermediates, the first example concerns the production of an intermediate. The second example deals with the use of a chemical in an industrial branch for which a comprehensive ESD exists (IC 10 Photographic Industry). The third example deals with the application of a biocidal product for which a specific ESD was developed and which offers a choice between scenarios based on tonnage and consumption.

Example 1. Production of an intermediate

Question: What are the emission factors for wastewater and air, the fraction of the main source, and the number of emission days for an intermediate, which is produced at a level of 15,000 tonnes per year? The substance has a vapour pressure of 550 Pa and is stored on-site. An intermediate is a substance that is produced during a chemical process before the desired product is obtained. So, it is synthesized as a raw material for the manufacture of a certain end product. According to Table 2.8 there are two industrial categories for the chemical industry:

Table 2.11. Emission factors (as fractions) to wastewater for intermediates (UC 33) according to the ESD in the TGD [17] during production.

Process	Emission factor
Wet	0.003
Dry	0

IC 2 "Chemical industry: basic chemicals"
IC 3 "Chemical industry: chemicals used in synthesis".

Logically, intermediates belong in IC 3. According to Table 2.8 UC 33 "Intermediates" applies. As can be seen from Table 2.9 there is an ESD for IC 3 covering intermediates at the HPVC level (footnote Table 2.9). The emission factors according to this ESD are presented in Table 2.11. It should be noted that only the environmental compartment wastewater is covered in this ESD. For the compartments air and (industrial) soil the same generic A Table should be used as for LPVCs. The A Table for the life cycle stage production of intermediates is presented here as Table 2.12.

No information is supplied in the ESD for the capacity of the process. Hence, the same B Tables as for LPVCs have to be used. The relevant B Tables for LPVCs and HPVCs are presented in Tables 2.13 and 2.14, respectively. It should be noted that in this case a substance is only recognized as an HPVC if the production volume exceeds 7000 tonnes/year.

Answer: The emission factors are found in Tables 2.11 and 2.12. Air: the intermediate is stored on-site, which means that the main category is 1b. Because of the vapour pressure of 550 Pa, the emission factor for air is 0.0001. Wastewater: as it is not clear from the question whether it is a dry or wet process the worst-case situation is considered by default. So, the emission factor for

Table 2.12. Emission factors (as fractions) to air, wastewater, and soil for intermediates (UC 33) according to the appropriate A Table of the TGD [17] at production.

Air	Vapour pressure (Pa)		Main category[a]		
			1a	1b	1c
	<1		0	0	0
	1 – 10		0	0	0.00001
	10 – 100		0	0.00001	0.0001
	100 – 1000		0.00001	0.0001	0.001
	1000 – 10.000		0.0001	0.001	0.01
	≥10.000		0.001	0.01	0.025
Wastewater	Process	Production volume (tonnes/y)			
	Wet[b]	<1000	0.02		
		≥1000	0.007		
	Dry[c]		0		
Soil	-	-	0.0001		

[a] Main category 1a concerns non-isolated intermediates, main category 1b concerns isolated intermediates stored on-site, and main category 1c concerns intermediates stored off-site.

[b] Wet means a "wet process", where water is used either during reaction, work up or vessel cleaning.

[c] Dry means a "dry process', where water is used in none of the stages.

wastewater is 0.003. Number of emission days: despite the fact that the question does not make clear that we are dealing with an HPVC, the default of 7000 tonnes implies that we need to consider the intermediate as an HPVC anyway. This means that Table 2.14 has to be used. The tonnage of 15,000 then results in a fraction of the main source of 0.75 and 300 emission days.

Example 2. Photochemicals from photographic materials at processing (industrial use)
Question: What is the emission to wastewater of a substance that is used as a sensitizer in colour negative films? The ESD for the photographic industry (IC 10) should be used. For the calculation of the emission to wastewater, expressed as kg/d ($Elocal_{water}$), the following equation is used:

$$Elocal_{water} = C \cdot W \cdot S \cdot (1 - R) \qquad (2.2)$$

where

C = Content of the substance in photographic material (kg/m^2)
W = Surface of photographic material processed per day (m^2/d)
S = Fraction that dissolves from the emulsion layer in the bath solution during processing (-)
R = Fraction removed or converted during processing (-).

It should be noted that in, principle, the parameter S represents the emission factor. The notifiers may supply specific data for the values of the parameters, or the defaults presented in Table 2.15 – 2.17 should be used.

Answer: as no content of the substance in the photographic material is known the default value for sensitizers in Table 2.15 of 25 mg/m^2 is used, which means that $C = 2.5 \cdot 10^{-5}$ kg/m^2. The surface of photographic material treated per day is also not specified, so the default for colour film in Table 2.15 is used: $W = 680$ m^2/d. As no data are known for the fractions dissolved and removed or converted during processing either, the defaults in Table 2.17 are used: $S = 1$ and $R = 0$. The emission to wastewater is calculated according to equation 2.2: $Elocal_{water} = 2.5 \cdot 10^{-5} \times 680 \times 1 \times (1 - 0) = 0.017$ kg/d.

Table 2.13. Estimates for the fraction of the main source and the number of days for emission of intermediates being LPVCs.

Production volume T (tonnes/yr)	fraction of the main source (fms)	number of emission days (nds)
<10	1	fms · T
10 – 50	0.9	fms · T
50 – 100	0.8	0.6667 fms · T
100 – 1000	0.75	0.4 fms · T
1000 – 2500	0.6	0.2 fms · T
≥ 2500	0.6	300

Table 2.14. Estimates for the fraction of the main source and the number of days for emission of intermediates being HPVCs (default ≥ 7000).

Production volume T (tonnes/yr)	fraction of the main source (fms)	number of emission days (nds)
<10 000	1	300
10 000 – 50 000	0.75	300
50 000 – 250 000	0.6	300
≥ 250 000	0.5	300

Table 2.15. Defaults for the parameter *C* "content of the substance in photographic material" for the emission scenario for substances (Photochemicals, UC 42) used in photographic materials at the life cycle stage of industrial use. Data obtained from the ESD in the TGD [17].

Type of substance	Content (mg/m^2) in paper	Content (mg/m^2) in film
Sensitizers	1	25
Stabilizers (UC 49)	5	100
Fungicides (UC 39)	30	150
Silver (as Ag)	500	12,000
Halides (Cl^-, Br^-, I^-)	300	7000
Split of products:		
- masking compounds in colour negative films	40	80
- remaining groups of colour couplers	80	800
- stabilizers (UC 49)	0	80
Wetting agents (UC 50)	10	300
Filter dyestuffs (IC 10)	50	250

Table 2.16. Defaults for the parameter W "Surface of photographic material processed per day" for the emission scenario for substances (Photochemicals, UC 42) used in photographic materials at the life cycle stage of industrial use. Data obtained from the ESD in the TGD [17].

Photographic process		W (m^2/d)
C41	Colour negative film	680
RA-4	Colour paper	4950
E-6	Colour reversal film	120
R-3	Colour reversal paper	350
BW-N	Black and white negative	40
BW-N	Black and white positive	270
BW-X	X-Ray	110
BW-R	Black and white	80
ECN-2	Cine and Television film negative	35
ECP-2	Cine and Television film positive	350
VNF-1	Cine and Television film reversal	35

Example 3. Use of disinfectants for sanitary purposes in hospitals

Question: what is the emission to wastewater of an active substance in a biocidal product used in hospitals for sanitary purposes and to disinfect brushes, when the concentration for sanitary purposes is 3500 mg/L and 5000 mg/L for the disinfection of brushes? The emission scenario for this application is covered in the ESD for biocidal product type 2 "Private and public health disinfectants and other biocidal products" of the TGD. All types of biocides have UC 39 "biocides, non-agricultural". This ESD presents two emission scenarios, one based on the tonnage and one based on the consumption. For the emission scenario based on tonnage the equation below is used to calculate the releases to wastewater discharged to an STP:

see next page (2.3)

where

TONNAGEreg	=	Relevant tonnage in the region for this application (tonnes/yr)
$F_{hospital}$	=	Fraction for the hospital (connected to STP) (-)
F_{water}	=	Emission factor for wastewater (-)
Temission	=	Number of emission days (d/y).

The defaults for the emission scenario parameters are presented in Table 2.18. It should be noted that according to the ESD the regional tonnage is derived from the EU tonnage by applying a multiplication factor for the region, with 0.1 as a default. As the tonnage is input provided by the producer, importer, etc., there is no default value for it.

The emission scenario based on consumption makes a distinction between the use of disinfectants for sanitary purposes (floors, furniture, objects) and for the disinfection of brushes. In this emission scenario the

Table 2.17. Defaults for the parameters S "fraction dissolved during processing from emulsion layer in the bath solution" (emission factor) and R "fraction removed or converted during processing" for the emission scenario for substances (Photochemicals, UC 42) used in photographic materials at the life cycle stage of industrial use. Data obtained from the ESD in the TGD [17].

Parameter	Unit	Default
S	-	1
R	-	0

equation below is used to calculate of the emission to wastewater discharged to an STP:

see below (2.4)

where

Q_{water_san} = Amount of water with active substance for sanitary purposes (L/d)
Q_{water_obj} = Amount of water with active substance for disinfection of brushes (L/d)
C_{san} = Concentration at which active substance is used for sanitary purposes (kg/L)
C_{san} = Concentration at which active substance is used for the disinfection of brushes (kg/L)
$Fsan_{water}$ = Emission factor for wastewater for sanitary purposes (-)
$Fobj_{water}$ = Emission factor for wastewater for disinfection of brushes (-)

The defaults for the emission scenario parameters are given in Table 2.19. It should be noted that there are no defaults for the concentrations at which the active substance should be used. This concentration follows from the prescribed use of the disinfectant by the company bringing the product on the market.

In the ESD the subscript 3 in the symbols for several parameters refers to the life cycle stage processing which was used at the time of development. In the latest version of the TGD life cycle stage 3 "processing" has been split into industrial use, professional use and private use. So, it would be better to take 'professional use' for this application.

Answer: As tonnage is not known the scenario is based on the consumption defaults as presented in Table 2.19 for the amounts of water with active substances for sanitary purposes and the emission factors for wastewater for sanitary purposes and the disinfection of brushes. With the concentrations given (conversion from mg/L to kg/L) the emission to wastewater can be calculated using equation 2.4: $Elocal_{water} = 25 \times 3500 \cdot 10^{-6} \times 0.55 + 25 \times 5000 \cdot 10^{-6} \times 0.95 = 0.167$ *kg/L.*

2.5 DEVELOPMENT OF EMISSION ESTIMATION AND RISK ASSESSMENT TOOLS

In the early 1980s several countries began to develop assessment systems for substances based on the PEC/NEC approach [26-29]. Generally, emission estimates have to be made in order to establish a PEC. Because of the lack of data on many substances, the only way to achieve this is through considering known emissions of existing substances used for the same purposes and in the same processes. In several countries use category studies were carried out for several substance applications to supply the necessary information. Some use category documents (UCDs) that were developed covered textile dyes [18,30], photochemicals [19,31], metalworking fluids and hydraulic fluids [20], paper chemicals [21, 32], intermediates [33], paint production [34], and plastic additives [35]. On the basis of these use category results, emission scenarios can be made with emission estimates for the relevant stages of the life cycle covered in the document. Total emission data alone are not sufficient for a risk assessment, as explained in Section 2.3.4. Location and time must be included. In the case of textile dyes, for instance, the estimates for the dyeing process itself could be based on the known capacities of dye houses. If such parameters are not known for the country or region being studied, general assumptions have to be

$$Elocal_{water} = \frac{TONNAGEreg \cdot 10^3 \cdot F_{hospital} \cdot F_{water}}{Temission} \quad (2.3)$$

$$Elocal_{water} = Q_{water\_san} \cdot C_{san} \cdot Fsan_{water} + Q_{water\_obj} \cdot C_{obj} \cdot Fobj_{water} \quad (2.4)$$

Table 2.18. Emission scenario with defaults to calculate the release of disinfectants used for sanitary purposes in hospitals, based on the tonnage applied.

Variable/parameter	Symbol	Unit	Default
Relevant tonnage in the region for this application	TONNAGEreg	tonnes/yr	
Fraction for the hospital (STP)	$F_{hospital}$	-	0.007
Emission factor for wastewater	F_{water}	-	0.75
Number of emission days	Temission	d/yr	260

Table 2.19. Emission scenario with defaults to calculate the release of disinfectants used for sanitary purposes in hospitals based on the consumption.

Variable/parameter	Symbol	Unit	Default
Amount of water with active substance for sanitary purposes	Q_{water_san}	L/d	25
Amount of water with active substance for disinfection of brushes	Q_{water_obj}	L/d	25
Concentration at which active substance is used for sanitary purposes	C_{san}	kg/L	
Concentration at which active substance is used for the disinfection of brushes	C_{san}	kg/L	
Emission factor for wastewater for sanitary purposes	$Fsan_{water}$	-	0.55
Emission factor for wastewater for the disinfection of brushes	$Fobj_{water}$-	-	0.95

made for the number of point sources, the maximum fraction of the main source, and the number of days in production use (by expert judgement). So, gradually ESDs were developed instead of UCDs. Most of the early ESDs do not cover the time aspect and often only one environmental compartment is considered.

During the development of the TGD the importance of emission estimation for the existing substances became evident and the A and B Tables, which were initially intended for LPVCs, were extended to cover HPVCs as well. In the Netherlands, emission scenarios for pesticides were already available in 1992 [36]. Between 1993 and 1996 the first emission scenarios for biocides were developed [37–39]. More work covering various biocide applications was later carried out in various countries. Examples include the Finnish calculation models for wood preservatives for wood in service and for slimicides in the paper industry and the Danish guidelines for assessing the environmental risks associated with industrial wood preservatives. For wood preservatives the OECD started projects to produce ESDs for all aspects of wood preservation [40–43] and for antifouling agents [44, 45]. The European Union Biocides Environmental Emission Scenarios (EUBEES) working group developed environmental emission scenarios for biocides. At present, the work on the development of ESDs for new and existing substances and biocides is being harmonised by different task forces of the OECD.

As indicated above, in the 1980s risk assessment techniques began to be developed in various places. Examples in the Netherlands include a system for the evaluation of new substances [46], followed by a system for setting priorities for existing substances [47]. In 1994, the first version of the uniform system for the evaluation of substances (USES) was launched [48]. USES was developed with the aim of integrating existing assessment systems in the Netherlands. It also incorporated the emission scenarios for pesticides and biocides from the evaluation systems for pesticides [36, 37]. As more emission scenarios from ESDs became available, these scenarios were introduced in newer versions (USES versions 2 and 3). USES was developed for use on personal computers and also contained the A and B Tables of the TGD. Further to modification (modules for pesticides and biocides were left out, for example) the European Union system for the evaluation of substances (EUSES) was developed [49]. It was published in 1996. In 2004, EUSES 2.0 was launched [50]. This version was based on the new TGD [17]. It also comprises the ESDs for new and existing substances and for biocides. As such, EUSES can be seen as a useful decision-support tool. It provides a platform for assessing the risks of chemicals according to the TGD. In the next decade further work will be needed to implement the REACH legislation [51]. This will be the subject of Section 2.6.

2.6 EMISSION ESTIMATION AND REACH

The emission estimation methods described in the previous sections were developed, at least in part, in response to regulations requiring the risk assessment of chemical substances in the EU, i.e., Council Regulation (EEC) 793/93 (Existing Substances Regulation), the 7th Amendment to Directive 67/548/EEC (Directive 92/32/EEC for the risk assessment of new substances), and Directive 98/6/EC (the Biocidal Products Directive). The evaluation of this legislation prompted the EU to develop a new regulatory framework for chemicals, i.e. Registration, Evaluation, Authorisation and Restriction of Chemicals or REACH [51]. REACH is explained in more detail in Chapter 12. In this section we will describe the development of exposure scenarios under REACH and the challenges ahead.

2.6.1 Development of exposure scenarios

REACH [51] has modified the approach to the estimation of emissions and more specifically the development of *exposure scenarios* (ESs). What is described in this section goes beyond emission estimation as discussed in the earlier sections of this chapter. It also includes aspects related to worker and consumer exposure, in addition to environmental emissions which are the main focus of the rest of this chapter. Exposure scenarios under REACH provide an integrated approach to controlling risks, as shown in Box 2.2, and form an integral part of the *chemical safety assessment* (CSA) and *chemical safety report* (CSR). The elements in the ES/CSA process under REACH are presented in Figure 2.9.

The importance of exposure scenarios and how they are applied differently under REACH can be explained by five questions: when, who, why, what and how?

When?
Under REACH the development of exposure assessments, including the generation of ESs and exposure estimation, is required for those who have to register under REACH and who manufacture or import classified chemicals in quantities of 10 tonnes or more per year. In this case classified means that the substance meets the criteria to be classified as dangerous under Directive 67/548/EEC or is assessed to be a persistent, bioaccumulative and toxic (PBT) or a very persistent and very bioaccumulative (vPvB) chemical (REACH Article 14; see also Chapter 12).

> **Box 2.2 Definition of exposure scenario**
>
> *Exposure scenario* means the set of conditions, including operational conditions and risk management measures, that describe how the substance is manufactured or used during its life-cycle and how the manufacturer or importer controls, or recommends downstream users to control, exposures of humans and the environment. These exposure scenarios may cover one specific process or use or several processes or uses as appropriate.

Who?
With regard to "who?" it is important to note that REACH differs from previous regulatory systems in that the risk assessments, including the development of exposure scenarios, are carried out by the manufacturer/importer and/or users of the substances, rather than by the regulatory authorities. This is also known as "reversal of the burden of proof". To start with the manufacturer or importer of the chemical substance has to develop the ESs needed for controlling risks throughout the life cycle (as part of the CSR which forms part of the registration dossier). As can be seen in Figure 2.9, these also have to be "translated" into a language which can be understood by the downstream users of the substances and attached to the *safety data sheets* (SDSs) communicated downstream. The downstream user (DU) in turn has to check whether his use is covered by an ES and ensure that he is using the substance in a way which is at least as well controlled as set out in the ES. If his use is not covered by the ES he can notify his supplier of his use, who in turn can prepare an ES covering the need of the DU (he can also choose not to support that DU). The DU can also decide not to inform his supplier of the use, but in that event he assumes responsibility for assessing that use and developing an appropriate ES.

Why?
The question "why?" can be answered easily. The assessment is required to demonstrate that the substance can be used safely, and to describe how this can be achieved. The key part of this new policy is implemented in the definition of exposure scenario as given in Box 2.2.

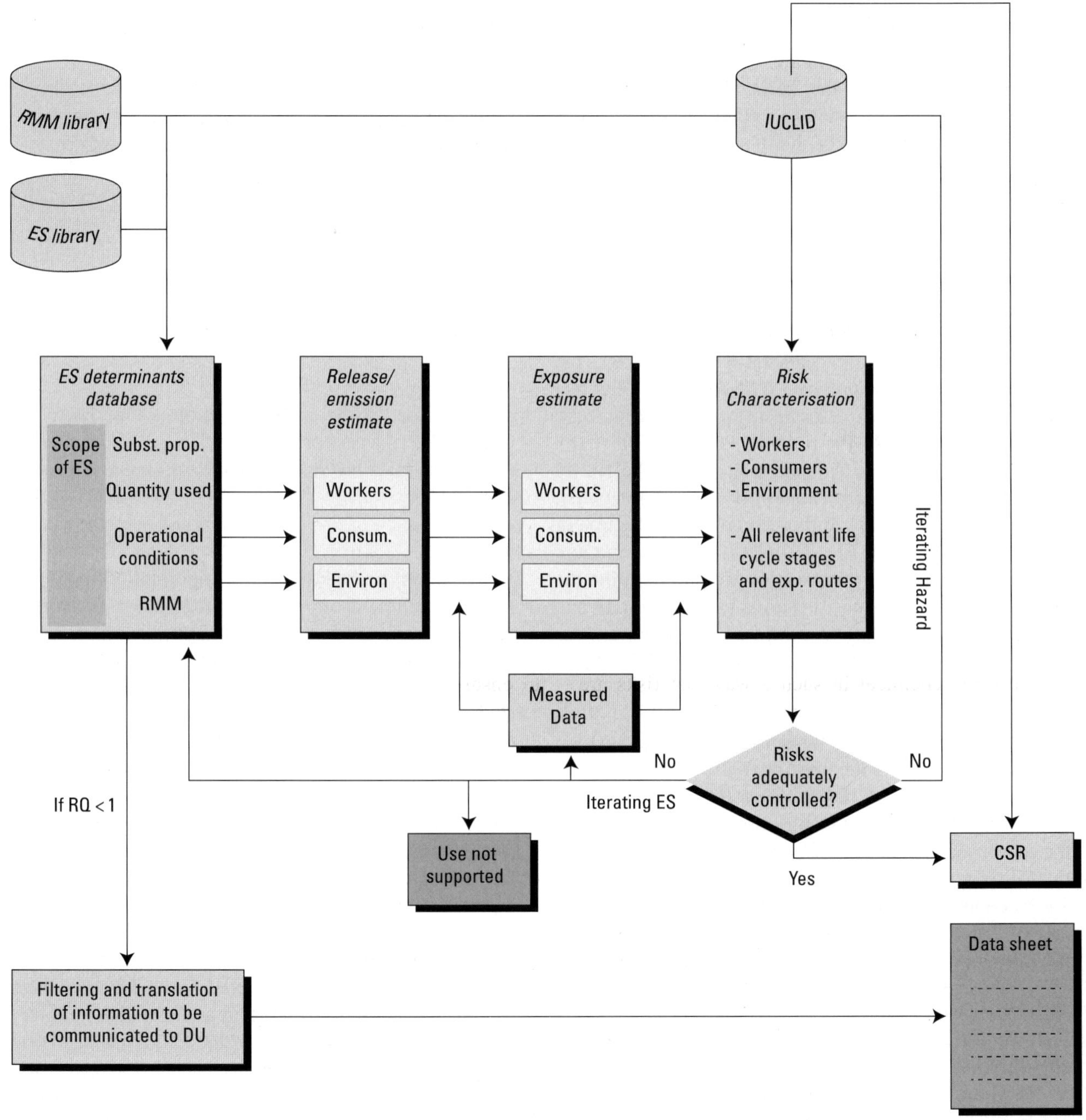

Figure 2.9. Elements of the exposure estimation, exposure scenario (ES) development and chemical safety assessment (CSA) process under REACH [51,52]. RQ = risk quotient, RMM = risk management measure.

What?

The question "what?" is related to the definition as given above. It is about "risk management first". Evidence needs to be provided by manufacturers and/or importers showing how risks throughout the life cycle of the chemical can be adequately controlled. As such, the focus of the ES and subsequent exposure and risk assessment (called safety assessment under REACH) is on risk management. Exposure scenarios under REACH have a *dual role*. One role is to provide the *basis for exposure estimation* (in preparing the CSA). They enable a quantitative release and exposure estimation

Box 2.3 Exposure scenarios under REACH

An **exposure scenario** (ES) sets out for a given use how the substance can be used in a way that risks are adequately controlled, by describing the conditions for use:

- **Process descriptions** (incl. quantity used)
- **Operational conditions** (incl. frequency and duration of specified operations)
- **Risk management measures** (process and emission control, personal protective equipment, good hygiene, etc.)
- **Other information** relevant to the safe use of the chemical

Exposure Scenarios are developed as part of the Chemical Safety Assessment (CSA)

by describing the *determinants of exposure*, i.e. the parameters that affect the exposure level. Their other role is to be *a communication tool* to the user, showing how to use the chemical in such a way that risks are controlled. They will become part of the SDSs. The sets of conditions or determinants of exposure, also called the *drivers of exposure*, need to be known and changed, if necessary, if iterations in the risk assessment show that risks are not adequately controlled.

How?

The remainder of this section is devoted to the question "how?" A short overview is provided in Box 2.3. It is important to note that exposure scenarios as defined in Box 2.2 are a completely new concept. Experience with the implementation of this concept will need to be gained over the next few years. Data, methodology and communication tools are still in development at this stage. Only preliminary guidance is currently available [53], but as the concept of ES and its implementation in the context of REACH is crucial, we will try to describe it in some detail.

Six steps to developing an exposure scenario

The proposed procedure for the development of an exposure scenario can be described in six consecutive steps as shown below in Table 2.20. It is important to determine what risk management measures (RMMs) are already in place. RMMs form an integral part of the overall process of developing ESs for identified uses of a substance on its own or in a preparation. An ES is a description of a control strategy for substances, giving realistic operational conditions for manufacture of a substance or identified (downstream) use(s) of a substance, a group of substances or a preparation. It prescribes appropriate RMMs that should be in place during the manufacture or use of a substance, including a manufacturer's own use, downstream uses, the service life of articles and the waste phase, under a given set of operational conditions (Figure 2.2). The ES is intended for risk management at the various life cycle stages to ensure safe handling and adequate control of risks related to human health (workers and consumers) and the environment [51-53]. In this section a number of general principles for the development of exposure scenarios are described including a description of work processes that may be conducted to identify uses and to assess the exposure of and risks to workers, consumers and the environment. Key terminology related to the development of ESs under REACH is given in Table 2.21.

Table 2.20. Six steps to develop an exposure scenario under REACH.

Step 1	Identification of uses and use processes
Step 2	Description of manufacturing or use process
Step 3	Development of a "tentative" ES
Step 4	Exposure estimation and risk characterisation
Step 5	Defining the "final" ES
Step 6	Developing the annex to the SDS

Step 1: Identification of uses and use processes

According to REACH [51], exposure scenarios should be developed for the manufacturing processes, for identified uses and for life cycle steps resulting from the identified uses of the substance on its own or in a preparation. The manufacturer or importer needs to build up a picture of the life cycle of the substance which they produce or import. While a manufacturer or importer registering a chemical will have information on his own manufacturing process(es) and use(s) of the chemical, information on uses further down the chemical supply chain may be more sparse. It is expected that significant information exchange will be needed with downstream users, particularly in the case of extended supply chains or where substances are included in preparations with many uses.

REACH provides any downstream user with the

Table 2.21. Key terminology related to the development of exposure scenarios under REACH [51].

Term	Definition
Exposure scenario	the set of conditions, including operational conditions and risk management measures, that describe how the substance is manufactured or used during its life cycle and how the manufacturer or importer controls, or recommends downstream users to control exposures of humans and the environment. These exposure scenarios may cover one specific process or use or several processes or uses as appropriate.
Use and exposure category	an exposure scenario covering a wide range of processes or uses, where the processes or uses are communicated, as a minimum, in terms of the brief general description of use.
Use	any processing, formulation, consumption, storage, keeping, treatment, filling into containers, transfer from one container to another, mixing, production of an article or any other utilisation.
Manufacturer	any natural or legal person established within the Community who manufactures a substance within the Community.
Downstream user	any natural or legal person established within the Community, other than the manufacturer or the importer, who uses a substance, either on its own or in a preparation, in the course of his industrial or professional activities. A distributor or a consumer is not a downstream user.
Supplier of a substance or a preparation	any manufacturer, importer, downstream user or distributor placing on the market a substance, on its own or in a preparation, or a preparation.
Recipient of a substance or a preparation	a downstream user or a distributor being supplied with a substance or a preparation.
Importer	any natural or legal person established within the Community who is responsible for import.
Registrant	the manufacturer or the importer or the producer or importer of an article submitting a registration for a substance.
Distributor	any natural or legal person established within the Community, including a retailer, who only stores and places on the market a substance, on its own or in a preparation, for third parties.
Substance	a chemical element and its compounds in the natural state or obtained by any manufacturing process, including any additive necessary to preserve its stability and any impurity deriving from the process used, but excluding any solvent which may be separated without affecting the stability of the substance or changing its composition.
Preparation	a mixture or solution composed of two or more substances.
Article	an object which during production is given a special shape, surface or design which determines its function to a greater degree than does its chemical composition.
Intermediate	a substance that is manufactured for and consumed in or used for chemical processing in order to be transformed into another substance.

possibility to make a use known to his supplier for the purpose of making his use an "identified use" for which an ES should be developed to the extent that the supplier accepts and supports this use. Although a manufacturer or importer is not obliged to be pro-active in seeking information on uses of his chemicals, it will be beneficial for the manufacturer or importer to do so. This will allow him to develop the CSA covering all identified uses. Thus, already early in the process of developing the CSA, the manufacturer or importer should identify the uses of his chemical and obtain sufficient information for developing an ES, e.g. by approaching his customers.

The information required should be adequate and sufficient to develop the "tentative" or provisional ES (see step 4; Table 2.20), i.e. it should include general information on a particular use that facilitates eventual grouping of information from various downstream users, more specific information on the use processes including relevant exposure determinants, and information on RMMs already in place and their possible efficiency, if available. A step-by-step strategy for identifying uses of chemicals and more specific details of use processes may be followed: (1) use in-house information to define identified uses, (2) use publicly available information, and (3) communicate with downstream users. Preliminary guidance is provided for this [53].

Step 2: Description of manufacturing or use process
This description of the processes forms the basis for developing a "tentative" or provisional ES (see step 3). This process description should be centred on typical operational conditions and typical risk management measures already implemented, based on the assumption that these are sufficient for ensuring adequate control of risks. A general identification of emission pathways at the different life-cycle steps needs to be considered in the description of the processes, i.e., manufacturing, formulation, industrial use, professional use, consumer use, service-life (the use of an article containing the substance, generally for over more than a year, resulting in the emission of the substance) and the waste phase (Figure 2.2).

In the description of the manufacturing process or identified use, a number of individual activities may be identified. The possible contribution of each of these activities to the overall exposure should be considered. To make sure that the ES covers relevant exposures, the manufacturing or use process(es) should be analysed using a list of main determinants for exposure. Determinants of exposure are needed to describe in more detail how these processes and activities lead to exposure and how this exposure is quantified.

Some determinants of exposure will form part of the ES process characteristics, operational conditions, quantities used, risk management measures within the control of the manufacturer or downstream user. Other determinants are external to the ES. These may relate to the substance, for example, its physicochemical properties, or biodegradation. Other determinants relate to the surroundings, for example, the properties of the receiving environment (for workers, consumers and the environment, as appropriate), risk management measures under the control of others (e.g. municipal waste water treatment), exposure factors such as inhalation rates, market penetration and others. The list of determinants acts as a link between the ES and the exposure estimate which is the quantitative part of the CSA.

The process descriptions should include the registrant's own manufacture and use, for which sufficient information will normally be available for a thorough description of the operational conditions and a subsequent assessment of exposure and risks. For downstream industrial uses there may be wide variations in the amount of information available to a registrant. Consumer uses of substances (mainly as ingredients of preparations and articles) will be largely similar for comparable types of products. Some information may be found in publicly available surveys on consumer use of various types of chemicals, including the duration of each use and the frequency of use (how often, how much), which may form the basis for the development of ESs.

The further a chemical travels down the supply chain, the more likely exposure to it will occur as a preparation (and, as a consequence, heterogeneous exposure). Whilst the primary manufacturer of the chemical (and REACH registrant) may have a notion of the circumstances of use, they are most unlikely to be privileged to information on proprietary preparations that is necessary to describe subsequent exposures and risks. This "differentiation of knowledge" within the supply chain must be accounted for within REACH information flows if risks are to be managed to equivalent levels throughout the chain.

The product use categories approach, categorising products in relation to their uses (e.g. paints, cleaners, lubricants, adhesives, detergents, etc.) can provide enough information to allow rough estimates of exposure to be made. It should be noted, however, that such categories can only be used as ESs if they meet the basic requirements of providing a basis for exposure estimation and the user with sufficient information about what he should do in order to use the chemical safely. The categories may be related to basic use information, for example, the frequency and duration of contact, and the amount of the substance used [53].

Step 3: Development of a "tentative" ES
A "tentative" or provisional ES is developed for the process for which an exposure estimate and a risk characterization are required. The purpose of the "tentative" ES is to assess whether risks are adequately controlled. Typical characteristics which may be included in an ES are given in Table 2.22.

For the development of a "tentative" exposure scenario for a process, determinants of importance for defining the process and the subsequent exposure assessment should be extracted from the process description. This also applies to the assumptions about which activities are conducted and which risk management measures (including their efficiency) are in place. The description will normally be relatively simple. Taken together with the process description, the "tentative" ES will allow a subsequent assessment of exposure and risks through the risk assessment process by using suitable risk assessment tools. The "tentative" ES could either be developed by the manufacturer or importer or by the downstream user. Another option is to use or modify an ES already developed for a similar process [53].

The "tentative" ES should contain the main determinants of exposure, but the level of detail depends

Table 2.22. Typical characteristics of an exposure scenario [53].

ES characteristics	Examples of parameters (not exhaustive)	Remarks
Life cycle of substance or product to which the ES refers	Manufacture or import, synthesis, compounding, formulation, use, service life, waste phase	Identify relevant exposures for all target groups, supports selection of suitable broad ES
Process characteristics	Industrial category, use category	Manufacture or use activity
Operational conditions	Type of activity/use Duration of activity/use Frequency of activity/use Temperature, pH, etc. Containment of process [open/closed]	Determines type of exposure (short term vs. long term) and choice of PNEC or DNEL
Preparation characteristics	Weight fraction of substance Migration rate	Determines exposure of humans and environment for preparations or products
Used quantity	Use rate [tonnes/year] Amount handled [kg/day, etc]	Determines the exposure potential per time
Risk Management Measures (within control)	Local exhaust ventilation On-site waste (water) treatment Personal Protective Equipment	RMMs as part of process or under direct control by DU

on the available data. In some cases, only minimal exposure information may be available while in other cases, an extensive data set on exposure conditions may exist. The tentative ES can be used to explore whether the available data is sufficient to reach a conclusion on the adequate control of risks, or whether further detailing is necessary. The minimum level of detail in a "tentative" ES is logically linked to the Tier-1 exposure model data requirements (see Chapters 4 and 5).

Although the ES will form part of the registration of a substance, its initial development will be largely driven by the process of manufacture or use and the specific operational conditions and RMMs that apply to the process, rather than by the substance and its properties. Many processes used for the manufacture (synthesis) of substances or the formulation of preparations are more or less standardized and the same set of RMMs are used; hence the tentative ES developed may apply to more than one substance. The same applies to the industrial use of substances as auxiliary chemicals in industrial processes. Often ESs for such processes will be relatively broad and applicable to a range of specific activities or uses where only generally applicable RMMs are needed. In general there are three different types of RMMs:

1. Process or product integrated measures (for example, substances marketed in matrices, or contained processing) which result in reduced emissions to the immediate receiving environment (internal or external);
2. Process or product external measures under the control of the user of the substance (for example, good housekeeping, on-site pollution abatement, PPE) which mitigate exposure or release;
3. Measures outside the direct control of the user (for example, public sewage treatment plant).

Step 4: Exposure estimation and risk characterisation

The information in the "tentative" ES will be fed into the risk estimation tool. In order to complete the initial or preliminary risk assessment the following information is needed: ES information, substance information and the characteristics of the surroundings. The resulting exposure levels are compared with available effect levels to determine whether the risks are adequately controlled (see Chapters 4 and 5 for further details on exposure assessment, Chapters 6 and 7 for the determination of (no) effect levels and Chapter 12 for a general introduction on REACH). The preliminary risk assessment is carried out for all relevant target groups and compartments. In cases where adequate control of risks for all exposed groups or compartments cannot be demonstrated based on the "tentative" ES, iterations are needed on one (or more) of the ES (process, including operational conditions and RMMs), the exposure modelling, or the hazard data. For individual uses, RMMs that have been developed specifically for such cases may be applicable, or special use instructions may be introduced. Thus, isolating specific activities from a broad array of processes offers an option for solving specific problems efficiently. If adequate control of risk during manufacture and use can be demonstrated with the tentative ES, the tentative ES will become the final ES for the substance(s) and process(as) considered.

Step 5: Defining the "final" ES

When adequate control of risks has been demonstrated, the operational conditions and the required RMMs are brought together in the "final" ES for the process of manufacture or use of the substance. Note, however, that when substances are formulated into preparations, the same ES, including operational conditions and RMMs, will, of course, need to apply to all of the individual substances in this preparation. A broad ES may be applicable for a range of substances falling within the boundaries of the ES in relation to substance properties, for example.

The procedure described for developing a suitable ES for a process or activity as part of the CSA for a substance is considered an efficient way forward in cases where there is no ES. However, it is assumed that over time, as more and more ESs are developed, such ESs may be standardized and collated into a library, which should be made available to subsequent registrants. A classification system may be useful for labelling ESs in a way which allows subsequent registrants to identify whether an ES already available fits their process. This is seen as a beneficial option in relation to the sequential registration proposed under REACH, where the high production volume chemicals will be registered first.

Step 6: Developing the annex to the SDS

The "final" ES or a summary consisting of relevant extracts of the "final" ES will be supplied to downstream users as an annex to the safety data sheet. A structured format for the ES should be used with standard headings to facilitate a proper communication to the downstream users. When more than one ES has been developed for a substance (e.g. due to different uses requiring significantly different RMMs), separate annexes are required. If the supplier of a chemical is aware of his customer's intended use, he may then provide the safety data sheet and the specific annex with the ES for this use. Alternatively, the safety data sheet and all available annexes to ESs for all the identified uses may be provided. The information given in the annex must be sufficient to be able to identify precisely the use process and RMMs required under the specified operational conditions.

2.6.2 Challenges ahead

In order to implement the ES concept, new or revised approaches are needed. Estimates of the release of chemicals from processes are still required, like those included in the ESDs or A Tables. However, such estimates need to be accompanied by information on the risk management measures in place which result in the estimate. Note that the emission factors themselves do not form part of the exposure scenario; rather it is the measures described in the scenario leading to the emission estimates which are used in the calculation of exposure in the accompanying exposure assessment.

The ideal form of emission factor would be one which included only the emission reduction measures which are integral to a process through, for example, the design of the equipment. This provides a baseline for the emissions and would allow further external risk management measures to be added in a modular fashion, should these be necessary to demonstrate safe use.

It has not been common practice to date to describe risk management measures explicitly in relation to emission factors. The factors in the A Tables were based on a degree of experience, and so incorporate the effects of measures which could be expected to be seen in most or all cases. However, these are not explicitly described in relation to the factors.

The same is true for emission factors in most ESDs. There are one or two exceptions. The ESD on plastics additives [54] presents emission factors based on the presence of abatement equipment for air emissions. Such equipment was assumed to be present in a certain percentage of companies at the time of producing the document, and to be present at all larger sites. A lower tonnage cut-off was suggested, below which the presence of such equipment should not be assumed and the emission factors should be increased. The revised OECD guidance for producing ESDs [22] includes the identification of risk management measures in place, related to the emission factors as an important component of new documents.

There is a need to develop a new library or database of emission factors for use in developing emission scenarios. These should be related as far as possible to specific types of equipment and their intrinsic emission control measures. These can then be combined with factors for the effectiveness of risk management measures to give the resulting emissions.

It is likely that ESDs and the A Tables will be useful at least in the initial development of scenarios for REACH. A wider group of actors have obligations to meet under REACH than under previous legislation. They will need a tool to help them find appropriate sources of information to support the development of exposure scenarios and the selection of emission factors. The Matrix project [55] is developing such a tool. This tool uses a series of identifiers to locate the correct emission module/factors. The identifiers may depend on the particular life cycle stage under consideration. Examples of identifiers include industrial category, use or function of the substance, method of application, etc. The identifiers are provided by the user in a sequence which guides them to the appropriate information.

This work has shown that the current system of identifiers, comprising the main industry and use categories, has limitations when used in this type of approach. These categories do not always provide sufficient detail on the actual use of the substance and are sometimes rather broad. They also do not include branches of industry using chemicals which have developed significantly over recent years, for example, the semi-conductor industry. Thus alongside the need to develop further emission scenario documents to cover the range of chemical uses, there is also a need for a better classification or identification system to facilitate the proper description of these uses. Such a system will also help in categorizing emission scenarios as they are developed, and hence in locating suitable ESs to use in developing registrations.

The Matrix project has also begun to analyse existing information sources, such as the OECD ESDs, to present the information in a form suitable for the searching tool above [56]. This involves dividing them into emission estimation modules (EEMs), each of which refers to a specific emission situation – one life cycle step and one emission pathway (to one receiving compartment). These EEMs form the content of the matrix. In addition, tools (software and manuals) are being developed to support the development of branch-specific Exposure Scenarios and emission estimates as part of the chemical safety assessment. A pilot study on additives in the plastics industry has been carried out [57]. This provides a basis for initial estimates of emissions from basic physicochemical information and tonnage for relevant areas of the life cycle, which could be equivalent to the "tentative" ES. If needed, further levels of iteration can be included in which more detailed or specific information (such as the specific type or purpose of the additive, specific emission factors, efficiency of emission reduction measures, etc.) can be incorporated to refine the emission estimation. The tools can be used by manufacturers, importers, and downstream users.

2.7 FURTHER READING

1. Commission of the European Communities. 2003. Technical Guidance Document in support of Commission Directive 93/67/EEC on risk assessment for new notified substances, Commission Regulation (EC) No1488/94 on risk assessment for existing substances and Directive 98/8/EC of the European Parliament and of the Council concerning the placing of biocidal products on the market. Joint Research Centre, European Chemicals Bureau, Ispra, Italy.
2. Commission of the European Communities. 2004. European Union System for the Evaluation of Substances 2.0 (EUSES 2.0). Prepared for the European Chemicals Bureau by the National Institute of Public Health and the Environment (RIVM), Bilthoven, The Netherlands (RIVM Report 601900005). Available via the European Chemicals Bureau, Ispra, Italy (http://ecb.jrc.it).
3. European Chemicals Bureau. 2005. Technical guidance document on preparing the chemical safety report under REACH. REACH Implementation Project 3.2. Report prepared by CEFIC, RIVM, the Federal Institute for Risk Assessment (BfR), Federal Institute for Occupational Safety and Health (BAuA), Ökopol, DHI Water & Environment and TNO Chemistry. European Commission, Joint Research Centre, Ispra, Italy (http://ecb.jrc.it/REACH/).
4. European Chemicals Bureau. 2007. Technical guidance document on preparing the chemical safety report under REACH. REACH Implementation Project 3.2. Report prepared by CEFIC, RIVM, the Federal Institute for Risk Assessment (BfR), Federal Institute for Occupational Safety and Health (BAuA), Ökopol, DHI Water & Environment and TNO Chemistry. European Commission, Joint Research Centre, Ispra, Italy (http://ecb.jrc.it/REACH/). (Under preparation).
5. Organization for Economic Co-operation and Development. 2004. OECD Series on Emission Scenario Documents. Number 3. Emission Scenario Document on plastics additives. ENV/JM/MONO(2004)8. OECD, Paris, France.

REFERENCES

1. Commission of the European Communities. 2006. Eurostat news release 37/2006 of 22 March 2006 for the

world water day. Eurostat, Luxembourg.

2. Commission of the European Communities. Eurostat. 2006. Tables: municipal waste land filled and municipal waste incinerated. Eurostat, Luxembourg.
3. Commission of the European Communities. Eurostat. 2005. Tables: disposal of hazardous waste. Eurostat, Luxembourg,
4. Ros JPM, Slooff W. 1988. Integrated criteria document cadmium. Report 758476004. National Institute for Public Health and the Environment (RIVM), Bilthoven, The Netherlands.
5. National Environmental Policy Plan. 1989. Second Chamber of the States General, session 1988-1989, 21 137, Nos. 1-2. The Hague, The Netherlands.
6. Ministry of Transport and Public Works. 1990. Aquatic sediments. Report for the third national water policy document of The Netherlands. Report 90.038. Institute for Inland Water Management and Waste Water Treatment (RIZA), Lelystad, The Netherlands.
7. Ros JPM, Van Der Poel P, Slootweg J. 1990. Target group study pharmaceutical industry. Report 73301002. National Institute for Public Health and the Environment (RIVM), Bilthoven, The Netherlands.
8. Struijs J, Duvoort-Van Engers LE, Gerlofsma A, Ros JPM. 1987. Risk assessment system new substances: behaviour of a substance in an STP. Report 738622001. National Institute for Public Health and the Environment (RIVM), Bilthoven, The Netherlands.
9. Struijs J. 1991. SimpleTreat 3.0: a model to predict the distribution and elimination of chemicals by sewage treatment plants. Report 719101025. National Institute for Public Health and the Environment (RIVM), Bilthoven, The Netherlands.
10. Ros JPM, Van Der Poel P. 1989. Target group study pharmaceutical industry, Part 1: sources and emissions. Report 738509002. National Institute for Public Health and the Environment (RIVM), Bilthoven, The Netherlands [in Dutch].
11. Van Der Most PFJ, Veldt C. 1992. Emission factors manual PARCOM-ATMOS. Emission factors for air pollutants 1992. TNO Environmental and Energy Research, Reference 92-235. Apeldoorn, The Netherlands.
12. Mulder W, Verbeek A, Eggels PG. 1993. Emission factors: Losses at fugitive emission sources in the storage and handling of liquids. Publication series emission registration. Environmental Protection Inspection, The Hague, The Netherlands [in Dutch].
13. Veldt C. 1993. Emission factors: Micropollutants from combustion processes. Publication series emission registration. Environmental Protection Inspection, The Hague, The Netherlands [in Dutch].
14. Veldt C, Van Der Most PFJ. 1993. Emission factors: volatile organic compounds from mobile sources. Publication series emission registration. Environmental Protection Inspection, The Hague, The Netherlands [in Dutch].
15. Groenewegen RJJ, De Groot JLB, Veldt C. 1993. Emission factors: plastics and rubber processing industry. Publication series emission registration. Environmental Protection Inspection, The Hague, The Netherlands [in Dutch].
16. US Environmental Protection Agency. Compilation of air pollutant emission factors, (AP-42). Washington, DC. See http://www.epa.gov/ttn/chief/ap42/index.html. Office of air and radiation and Office of air quality planning and standards, Washington, DC.
17. Commission of the European Communities. 2003. Technical guidance document in support of Commission Directive 93/67/EEC on risk assessment for new notified substances, Commission Regulation (EC) No. 1488/94 on risk assessment for existing substances and Directive 98/8/EC of the European Parliament and of the Council concerning the placing of biocidal products on the market. Joint Research Centre, European Chemicals Bureau, Ispra, Italy
18. Ros JPM. 1985. Risk assessment system new substances: expected releases of textile dyes. Report 851502001. National Institute of Public Health and the Environment (RIVM), Bilthoven, The Netherlands [in Dutch].
19. Ros JPM, Bogte JJ. 1985. Risk assessment system new substances, subject: Expected releases of photochemicals. Report 851502002. National Institute for Public Health and the Environment (RIVM), Bilthoven, The Netherlands [in Dutch].
20. Van Der Poel P, Ros JPM. 1987. Risk assessment system new substances: expected releases of metalworking fluids and hydraulic fluids. Report 738620001. National Institute for Public Health and the Environment (RIVM), Bilthoven, The Netherlands [in Dutch].
21. Ros JPM, Berns JAS. 1988. Risk assessment system new substances: Expected releases of paper chemicals. Report 738620002. National Institute for Public Health and Environmental Protection (RIVM), Bilthoven, The Netherlands [in Dutch].
22. Organization for Economic Co-operation and Development. 2000. OECD Series on Emission Scenario Documents. Number 1. Guidance Document on Emission Scenario Documents. Unclassified. ENV/JM/MONO(2000)12. OECD, Paris, France.
23. The Council of the European Communities. 1986. Fourth Community Action Programme on the Environment

(1987-1992). Office of official publications of the European Communities, Luxembourg.
24. Organisation for Economic Co-operation and Development. 1987. Decision-Recommendation of the Council on the Systematic Investigation of Existing Chemicals. OECD Publications, C(87)90 Final. Paris, France.
25. Organization for Economic Co-operation and Development. 2004. Manual for Investigation of HPV Chemicals. OECD Secretariat. http://www.oecd.org/document/7/0,2340,en_2649_34379_1947463_1_1_1_1,00.html. OECD, Paris, France
26. Callahan M. 1982. Assessment of potential exposure from low volume chemicals and site-limited intermediates. Final report. U.S. Environmental Protection Agency, Washington, DC.
27. Rodriguez VE. 1987. Generic engineering assessment spray coating: Occupational and environmental releases. US Environmental Protection Agency, Washington, DC.
28. Office fédéral de l'environnement, des forêts et du paysage. 1989. Ordonnance sur les substances: instructions pour le contrôle autonome. Bern, Switzerland.
29. Swedish National Chemicals Inspectorate. 1989. Systems for testing and hazard evaluation of chemicals in the aquatic environment. KEMI Report 4/89. Solna, Sweden.
30. Department of the Environment, Environmental Research. 1988. Industry category document: UK dye production and use in the textile industry. Report CR36/88. Department of the Environment, London, UK.
31. Department of the Environment, Environmental Research. 1988. Industry category document: The UK photographic industry. Report CR39/88. Department of the Environment, London, UK.
32. Cathie K, Staves J, Kirkpatrick N. 1991. Consultancy report, paper recycling industry: Review of processes and effluent. Composition PIRA International, 33/TENQ/012/588. Department of the Environment, London, UK.
33. Ros JPM, Van Der Poel P. 1990. Risk assessment system new substances: expected releases of intermediates. National Institute for Public Health and the Environment (RIVM), Bilthoven, The Netherlands [in Dutch].
34. Prats A. 1992. Study on the identification and quantification of the principal discharges of paint and varnish additives in the environment. INERIS-Environment Ministry Contract, Ministère de l'environnement, Paris, France.
35. Jolly AM, Willoughby B, Karas GC, Hobbs SJ. 1994. Use category document plastic additives. Department of the Environment, Building Research Establishment, Watford, UK.
36. Emans HJB, Beek MA, Linders BJH. 1992. Evaluation system for pesticides (ESPE) 1. Agricultural pesticides. Report 679101004. National Institute for Public Health and the Environment (RIVM), Bilthoven, The Netherlands.
37. Luttik R, Emans HJB, Van Der Poel P, Linders JBHJ. 1993. Evaluation system for pesticides (ESPE) 2. Non-agricultural pesticides. Report 679102021. National Institute for Public Health and the Environment (RIVM), Bilthoven, The Netherlands.
38. Luttik R, Van Der Poel P, van den Hoop MAGT. 1995. Supplement to the methodology for risk evaluation of non-agricultural pesticides (ESPE) 2, incorporated in the Uniform System for the Evaluation of Substances (USES). 679102028. National Institute for Public Health and the Environment (RIVM), Bilthoven, The Netherlands.
39. Montfoort JA, Van Der Poel P, Luttik R. 1996. The use of disinfectants in livestock farming. Supplement to the evaluation method of non-agricultural pesticides for the Uniform System for the Evaluation of Substances (USES). 679102033. National Institute for Public Health and the Environment (RIVM), Bilthoven, The Netherlands.
40. Organization for Economic Co-operation and Development. 2003. OECD Series on Emission Scenario Documents. Number 1. Emission Scenario Document for Wood Preservatives. PART 1. Unclassified. Environment Directorate. OECD, Paris, France.
41. Organization for Economic Co-operation and Development. 2003. OECD Series on Emission Scenario Documents. Number 2. Emission Scenario Document for Wood Preservatives. PART 2. Unclassified. Environment Directorate. OECD, Paris, France.
42. Organization for Economic Co-operation and Development. 2003. OECD Series on Emission Scenario Documents. Number 3. Emission Scenario Document for Wood Preservatives. PART 3. Unclassified. Environment Directorate. OECD, Paris, France.
43. Organization for Economic Co-operation and Development. 2003. OECD Series on Emission Scenario Documents. Number 4. Emission Scenario Document for Wood Preservatives. PART 4. Unclassified. Environment Directorate. OECD, Paris, France.
44. Organization for Economic Co-operation and Development. 2005. OECD Series on Emission Scenario Documents. Number 13. Emission Scenario Document on Antifouling Products. ENV/JM/MONO(2005)8. Unclassified. OECD Environment Directorate, Paris, France.

45. Organization for Economic Co-operation and Development. 2005. OECD Series on Emission Scenario Documents. Annex to Emission Scenario Document on Antifouling Products No. 13. OECD Environment Directorate. Paris, France.
46. De Nijs ACM, Toet C, Vermeire TG, Van Der Poel P, Tuinstra J. 1993. Dutch risk assessment system for new chemicals: DRANC. *Sci Total Environ*, Suppl 1993:1729-1747.
47. Van De Meent D, Toet C. eds 1992. Dutch priority setting system for existing chemicals: A systematic procedure for ranking chemicals according to increasing estimated hazards. Report 679120001. National Institute for Public Health and the Environment (RIVM), Bilthoven, The Netherlands.
48. USES. 1994. Uniform system for the evaluation of substances, version 1.0. National Institute of Public Health and the Environment, Ministry of Housing, Spatial Planning and Environment, Ministry of Welfare, Health and Cultural Affairs. Distribution 1114/ 150, Ministry of Housing, Spatial Planning and the Environment. The Hague, The Netherlands.
49. Commission of the European Communities. 1996. EUSES, the European Union System for the Evaluation of Substances. National Institute for Public Health and the Environment (RIVM), Bilthoven, The Netherlands.
50. Commission of the European Communities. 2004. European Union System for the Evaluation of Substances 2.0 (EUSES 2.0). Prepared for the European Chemicals Bureau by the National Institute for Public Health and the Environment (RIVM), Bilthoven, The Netherlands (RIVM Report 601900005). Available via the European Chemicals Bureau, Ispra, Italy (http://ecb.jrc.it).
51. Commission of the European Communities. 2006. Regulation (EC) No 1907/2006 of the European Parliament and of the Council of 18 December 2006 concerning the Registration, Evaluation, Authorisation and Restriction of Chemicals (REACH), establishing a European Chemicals Agency, amending Directive 1999/45/EC and repealing Council Regulation (EEC) No 793/93 and Commission Regulation (EC) No 1488/94 as well as Council Directive 76/769/EEC and Commission Directives 91/155/EEC, 93/67/EEC, 93/105/EC and 2000/21/EC. *Off J Eur Union*, L 396/1 of 30.12.2006.
52. Vermeire, T and Torslov J. 2006. Preparatory document for the development of IT tools supporting REACH CSA/CSR assessment (RIP 3.2-2 Task III). European Chemicals Bureau, Ispra, Italy.
53. European Chemicals Bureau. 2005. Technical guidance document on preparing the chemical safety report under REACH. REACH Implementation Project 3.2. Report prepared by CEFIC, RIVM, the Federal Institute for Risk Assessment (BfR), Federal Institute for Occupational Safety and Health (BAuA), Ökopol, DHI Water & Environment and TNO Chemistry. European Commission, Joint Research Centre, Ispra, Italy (http://ecb.jrc.it/REACH/).
54. Organization for Economic Co-operation and Development. 2004. OECD Series on Emission Scenario Documents. Number 3. Emission Scenario Document on plastics additives. ENV/JM/MONO(2004)8. OECD, Paris, France.
55. Van Der Poel P, Bakker J, Hogendoorn EA, Vermeire TG. 2005. Branch and product related emission estimation tool for manufacturers, importers and downstream users within the REACH system. Project A. Technical guidance for identifying the appropriate emission scenarios. Report 601200006. National Institute for Public Health and the Environment (RIVM), Bilthoven, The Netherlands.
56. Bunke D, Ahrens A, Müller S. 2005. Branch and product related emission estimation tool for manufacturers, importers and downstream users within the REACH system. OECD Matrix Project, Results of Project B1. R+D Project FKZ 204 67 456/02. Paris, France.
57. Reihlen A, Bunke D, Müller S. 2005. Branch and product related emission estimation tool for manufacturers, importers and downstream users within the REACH system. Manual for environmental emission estimation of plastic additives. R+D Project FKZ 204 67 456/02. Dessau, Germany.

3. TRANSPORT, ACCUMULATION AND TRANSFORMATION PROCESSES

D.T.H.M. Sijm, M.G.J. Rikken, E. Rorije, T.P. Traas, M.S. McLachlan and W.J.G.M. Peijnenburg

3.1 INTRODUCTION

This chapter will deal with the phenomena which determine the concentration of substances in the environment as well as within organisms. Knowing the concentrations of chemicals in different compartments as well as their further fate within these compartments is one of the key issues in a chemicals' risk assessment procedure. Once the concentrations in the various relevant environmental compartments are known or estimated, they can be compared with information on the hazards of a substance in that compartment. The relevant environmental compartments may be water, sediment, soil, air, or biota.

After entering the environment, chemicals are transported, distributed over the various environmental compartments and may be transformed into other chemicals. Transport can occur within a compartment, such as in air or in soil, or between compartments (e.g., between air and water, air and soil or water and soil). Transformation processes in the environment involve chemical degradation (e.g., hydrolysis) or microbial degradation, i.e., biodegradation. Chemicals may also be transformed within organisms, i.e., biotransformation. In most cases, degradation is beneficial because less hazardous substances are formed. However, some examples are known in which more hazardous compounds are formed in the degradation process. Usually, toxic effects only occur when chemicals are inside organisms. Therefore, understanding the uptake of chemicals, which relates to bioaccumulation, is of the utmost importance in risk assessment. Often, however, there is no direct link between the extent of bioaccumulation or toxicity, and the concentration of a chemical in the environment. It is the aspect of bioavailability that determines whether a chemical is actually taken up and able to exert toxicity. Bioavailability may, therefore, be briefly defined as the fraction of a chemical present in an environmental compartment that, within a given timeframe, is available for uptake by an organism. At the end of this chapter, the characteristic processes underlying bioavailability will be discussed.

In the following sections several processes are described:

- Transport processes: describe and are helpful in understanding and predicting how chemicals distribute in the environment after being released into it. This information is of paramount importance for risk assessment to find out where in the environment organisms are likely to be exposed to the substance or where in the environment the substance could do any harm. The actual concentration in each environmental compartment will depend on many parameters and it this concentration that should be compared with the hazardous properties of a substance to assess the risk of the substance.
- Bioaccumulation processes: describe how concentrations in biota are sometimes higher than those in the surrounding environment or in the prey or food of organisms.
- Abiotic transformation processes: describe how chemicals can be chemically altered by abiotic processes in the environment and thus affect the fate and reduce the concentration of the substance in the environment.
- Biodegradation processes: describe how micro-organisms in the aquatic environment transform substances and thus affect the fate and reduce the concentration of the substance in the environment.
- Biotransformation processes: describe how organisms, after having taken up the substance, can transform it and thereby reduce the concentration of the substance in the organism.
- Bioavailability processes: describe which fraction of the concentrations in an environmental compartment is really relevant for organisms and which parameters affect this fraction.

3.2 TRANSPORT PROCESSES

3.2.1 Transport mechanisms

In this section, some relevant transport mechanisms will be described. Two kinds of transport mechanisms are distinguished: (1) *intramedia transport,* which is transport away from a source in one environmental medium, and (2) *intermedia transport*, which is transport from one environmental medium to another. Intramedia transport is important in relation to the mobile environmental

C.J. van Leeuwen and T.G. Vermeire (eds.), *Risk Assessment of Chemicals: An Introduction*, 73–158.

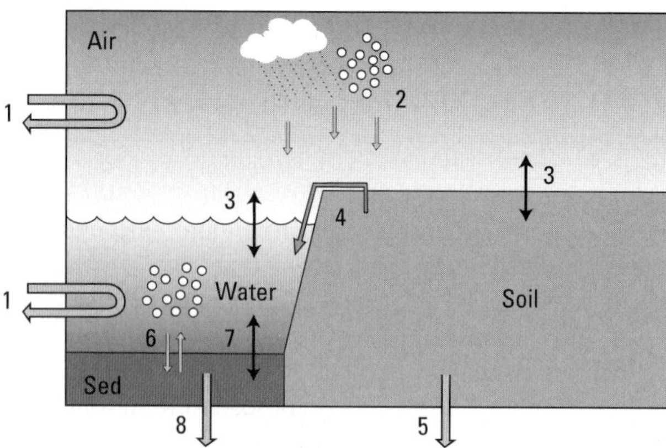

Figure 3.1. Intramedia and intermedia transport processes. 1, 5, 8: advective and dispersive intramedia transport, 2, 3, 4, 6, 7: advective and dispersive intermedia transport.

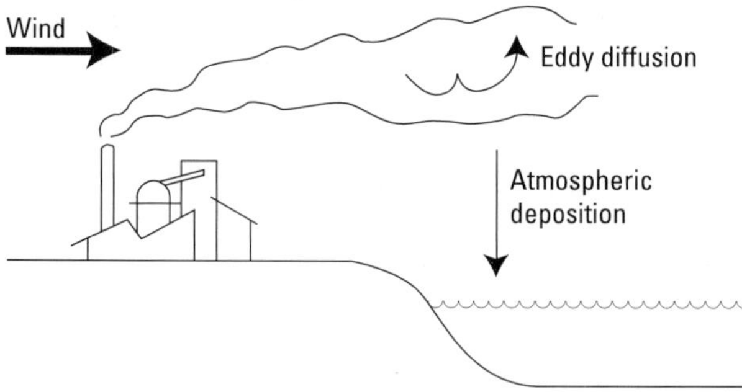

Figure 3.2. After release into air, a chemical is carried downwind and diluted *(intramedia transport)*; atmospheric deposition carries the chemical from air to water and soil *(intermedia transport)*.

media: air, water and groundwater; intermedia exchange takes place between all media, but is most important for transport of chemicals to the stationary media: sediment and soil (Figures 3.1 and 3.2).

Intramedia transport takes place through the mechanisms of advection and dispersion. Advection causes a chemical to travel from one place to another as a result of the flow of the medium in which it occurs; locally emitted packages or "puffs" of a chemical are carried as far as the wind or water current can take it during the residence time in that medium. Dispersion mechanisms (molecular diffusion, eddy diffusion) make the chemical move down concentration gradients until the concentration gradients disappear. The residence time of the chemical in the medium is an important factor since besides intramedia transport other removal processes occur at the same time. If, for example, a chemical is emitted into air and its degradation in air is rapid, the effective residence time of the chemical in air is short. Consequently, there is little time for the advective and dispersive processes to take place. In one medium, advection and dispersion always operate together. If a chemical is emitted continuously into air or water, the combined operation of advection and dispersion results in the formation of a plume. At short distances from emission sources, concentrations are usually affected most by intramedia transport. The result is observed as dilution.

Intermedia transport (air-water, water-sediment, etc.) also takes place by advective and dispersive mechanisms. Advective intermedia transport takes place if a chemical is transported from one environmental compartment to another by a physical carrier. Examples are deposition of fog, raindrops and aerosol particles from air to water or soil, sedimentation and resuspension of particulate matter across the watersediment interface, and percolation of water through soil. Advective transport is a one-way phenomenon: the chemical is carried by the medium in which it resides in the direction in which the medium flows. Intermedia dispersion, like intramedia dispersion, is diffusive in nature and follows concentration gradients. Examples are volatilization and gas absorption (air-water and air-soil), the direction depending on the concentration difference between the media, and diffusive exchange of chemicals between sediment and water. The driving force of intermedia transport is the tendency of chemicals to seek equilibrium between different phases.

3.2.2 Equilibrium partitioning between phases

In systems that consist of more than one phase, chemicals tend to migrate from one phase to another if the phases are not in equilibrium. The third law of thermodynamics states that systems spontaneously seek a minimum value for the Gibbs free energy, G. As a result, migration in multi-phase systems continues until this minimum has been reached. At this point of minimum G the system has reached a state of equilibrium. Equilibrium has traditionally been characterized as the point where the chemical potential, μ (the change in Gibbs free energy of a phase with a change in the amount of chemical), has the same value in the different phases. An alternative way of stating the same is to say that at equilibrium the phases have the same fugacity. This way of expressing the equilibrium condition has been promoted by Mackay as a useful method of describing multi-compartmental environmental systems. For an overview of this subject, the reader is referred to the book Multimedia Models,

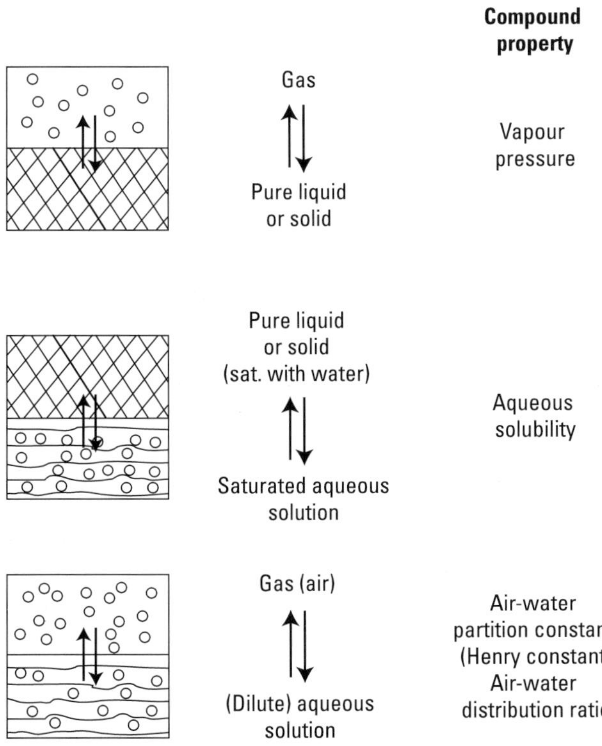

Figure 3.3. Important compound properties showing the equilibrium partitioning between two phases. From Schwarzenbach [2]. With permission.

by this author [1]. Fugacity measures the tendency of a chemical to escape from the phase it is in. Fugacity is sometimes called "escape tendency" or "escape pressure". Since fugacity is the change in Gibbs free energy (J/mol or Pa · m³/mol) with concentration (mol/m³) it is clear that the fugacity is expressed in units of pressure (Pa). Often the term fugacity capacity (Pa · m³/mol) is used [1] that provides a measure of the capacity a medium has to store the chemical, or in other words, is a measure of the ability of the medium to prevent the chemical from escaping that medium.

For practical purposes, it is important to note that it is often observed experimentally that the equilibrium ratio of concentrations in two phases is constant if the concentrations are sufficiently low. If the partition coefficient (K_{12}) is known, the general Equation 3.1 can be used to derive the concentration in one phase from the concentration in the other phase if both are at equilibrium.

$$C_1 / C_2 = \text{constant} = K_{12} \tag{3.1}$$

where

C_1 = concentration in phase 1 (mol/m^3)
C_2 = concentration in phase 2 (mol/m^3)
K_{12} = partition coefficient

For two immiscible liquids this is known as the Nernst distribution law and the constant concentration ratio is called the Nernst constant. For air-water systems, the equilibrium equation is known as Henry's law. For solids-water systems, the equilibrium constant is known as the partition coefficient, K_p (common for aquatic systems), or distribution constant, K_d (more common for terrestrial systems). This is shown in Figures 3.3, 3.4 and 3.5 and is also explained in Chapter 9. Partition coefficients for many chemicals are available from laboratory or field measurements. However, for many chemicals experimental data are not available and estimation methods must be used (Chapter 9). In general, the applicability of these estimation methods is limited to those classes of (organic) chemicals for which empirical relationships have been derived. Extrapolation beyond these limits may lead to errors of several orders of magnitude. For metals, no generally applicable estimation methods are known. This is because values of K_p depend strongly on the composition of the solid and aqueous phases among which the metal is distributed. Especially pH is an important parameter in this respect, and K_p values usually decrease proportionally with decreasing pH.

Sediment-water, suspended matter-water and soil-water equilibria

Equilibrium partitioning between water and solids is the result of adsorption of the chemical onto the surface of particles. For low concentrations of the chemical in water, the equilibrium ratio is usually a constant, as in Equation

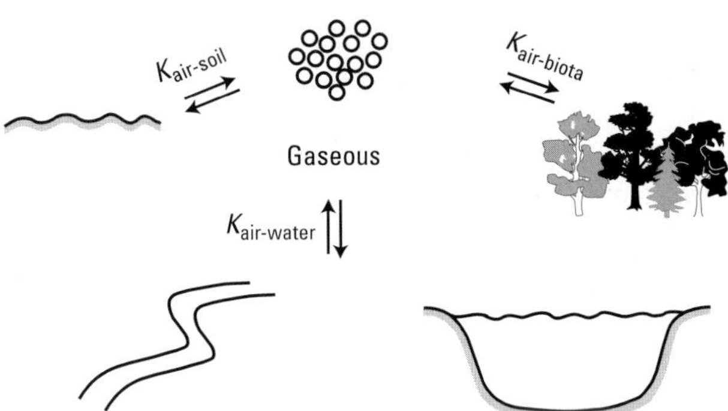

Figure 3.4. Gas exchange between the atmosphere and the earth's surface. From Schwarzenbach [2]. With permission.

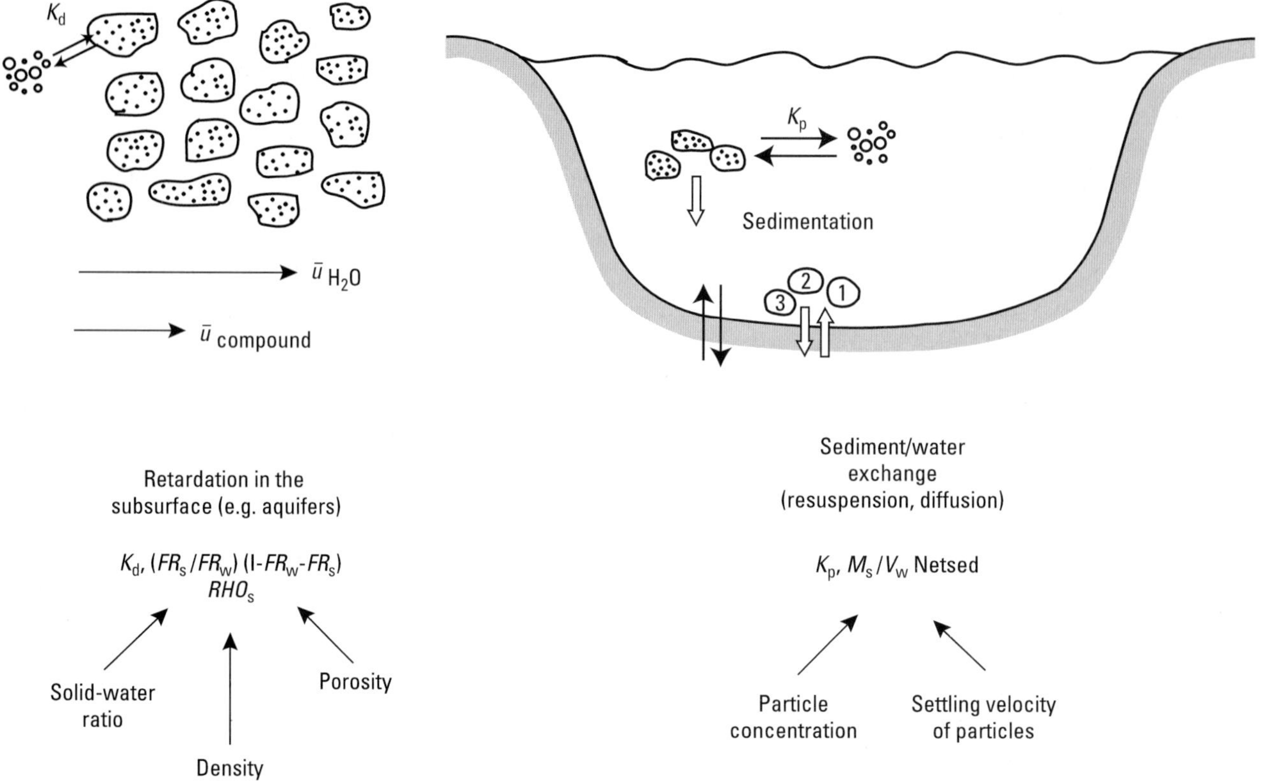

Figure 3.5. Solids-water exchange in natural waters. See text for the explanation of abbreviations. From Schwarzenbach [2]. With permission.

3.1. For higher concentrations, it is often observed experimentally that the equilibrium ratio depends on the concentration. In such cases, the equilibrium relationship between the concentrations is given by a non-linear sorption isotherm. Different mathematical expressions, reflecting different theoretical approaches to the sorption mechanisms, may be used to describe the non-linearity of the sorption isotherm. The Freundlich-isotherm equation is often used (without making assumptions about the nature of the underlying mechanism) to fit experimentally observed non-linear sorption (Figure 3.6). Commonly used estimation methods for partition coefficients are based on the assumption that there is a "hydrophobic sorption" mechanism. This mechanism is generally modelled based on the organic carbon content of the soil, sediment or suspended solids and the octanol-water partition coefficient of the chemical, using simple regression equations:

$$\log K_p = \log (K_{oc} \cdot f_{oc}) = a \log K_{ow} + b + \log f_{oc} \quad (3.2)$$

where

K_p = solids-water partition coefficient (L/kg)
K_{oc} = organic carbon referenced solids-water partition coefficient (L/kg)
f_{oc} = organic carbon content of the solid (kg/kg)
K_{ow} = *n*-octanol-water partition coefficient of the chemical.

Normalization to the organic carbon content of particulate matter has become standard procedure in this field of research. This procedure is based on the experimental observation that the K_p of organic chemicals is often proportional to the organic matter content of the solid phase. It can be inferred from this that interaction with organic matter plays a prominent role in sorption of organic substances to sediment and soil. Instead of the organic carbon content, f_{oc}, the organic matter content, f_{om}, is sometimes used. Since the organic carbon content of organic matter in different solids has similar values, the ratio of f_{om} to f_{oc} is taken as a fixed value (approximately 1.7) for most purposes. This estimation

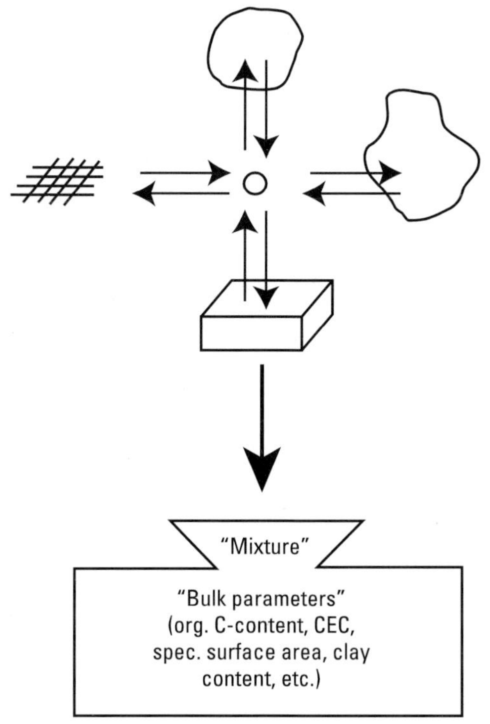

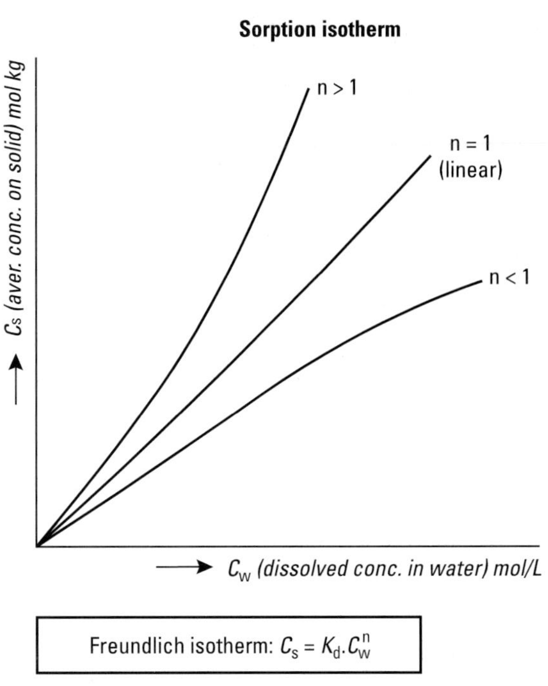

Figure 3.6. Sorption equilibrium between solids and water. From Schwarzenbach [2]. With permission.

method is valid only for non-ionic organic chemicals and cannot be applied to:

- Acidic or basic chemicals that occur to some extent in an ionic form.
- Anionic and cationic surfactants.
- Metals.

Solids-water partition coefficients are expressed in the dimension "unit volume of water per unit mass of solid". The commonly reported format is L/kg, as experimentally observed solids-water concentration ratios are conveniently expressed as, e.g., mol/kg or mol/L. The physical meaning of this dimension can be understood by reading it as "the volume of water (in litres) which contains that amount of the chemical which is equal to the amount present in one kg of solid material". For many purposes, however, we are not just interested in the concentration ratio, but also in the distribution of the chemical over the phases. Obviously, this distribution depends on both the partition coefficient and the relative volumes of the phases. In surface water, the solids-water ratio is much smaller than in sediment and soil systems. As a result, the extent of partitioning of a certain chemical into the particle phase of sediment or soil is much greater than in surface water. Partitioning is fully described by the intermedia equilibrium constant and the mass-balance equation. In the case of a system containing water and suspended matter, the mass balance equation becomes:

$$V_w \cdot C_{tot} = V_w \cdot C_w + M_s \cdot C_s \quad (3.3)$$

where

C_s	=	concentration of the chemical in the solid phase (mol/kg)
C_w	=	concentration of the chemical dissolved in the water phase (mol/L)
C_{tot}	=	total concentration of the chemical in the aqueous system (mol/L)
V_w	=	volume of the aqueous system (L)
M_s	=	mass of the solid in the aqueous system (kg).

The fraction of the chemical dissolved in water is derived by combining Equations 3.1 and 3.3:

$$FR_{water} = \frac{C_w}{C_{tot}} = \frac{1}{1 + K_p \cdot M_s / V_w} \quad (3.4)$$

More generally, in heterogeneous aqueous systems, the

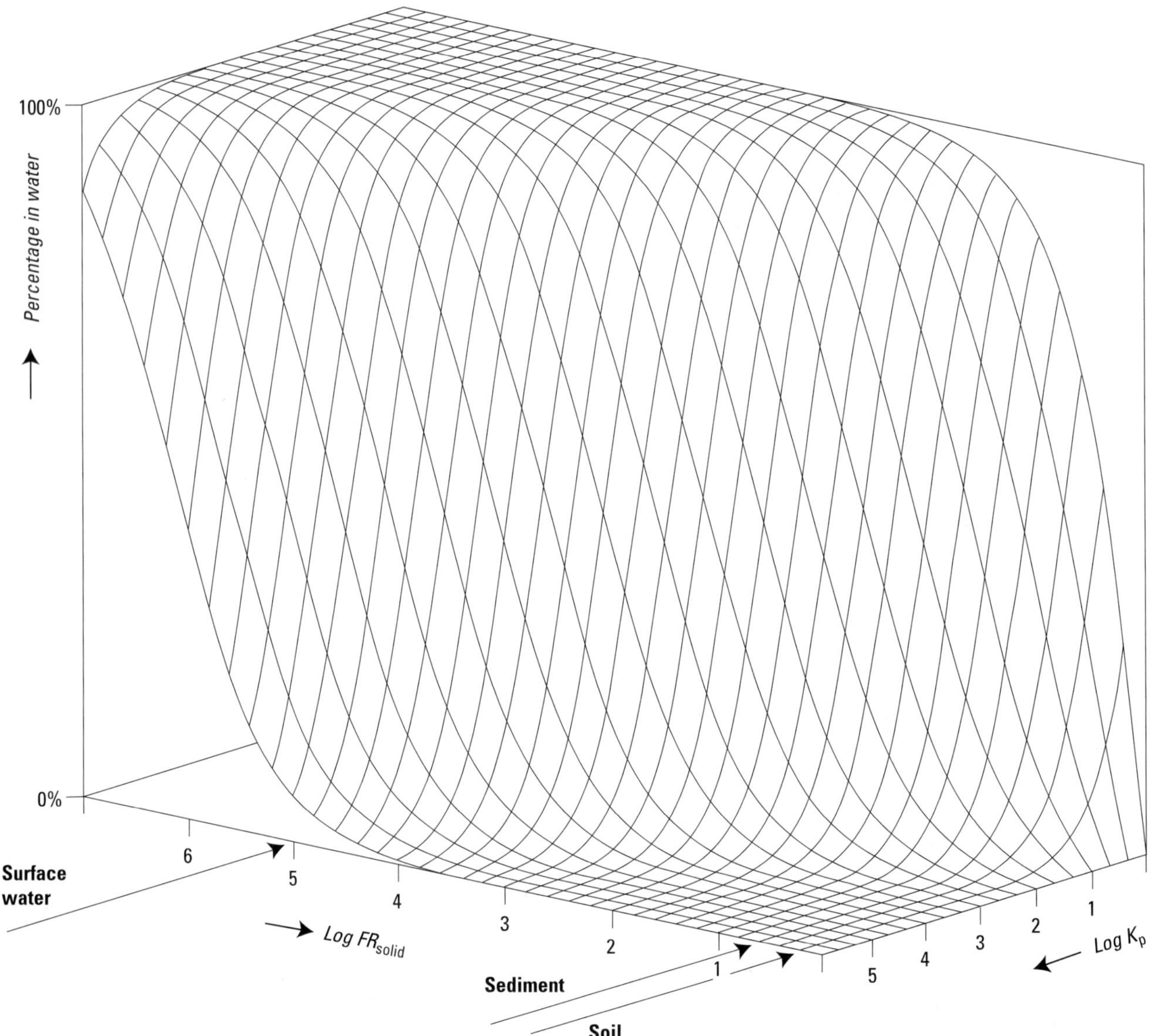

Figure 3.7. Fraction of a chemical in the water phase of a solid-water system as a function of the solid-water ratio (FR_{solid}) and the partition coefficient (K_p).

fraction of the chemical present in the water phase can be calculated according to Equation 3.5:

$$FR_{water} = \frac{FR_w}{FR_w + FR_s \cdot K_p \cdot RHO_s} \tag{3.5}$$

where

FR_{water} = fraction of the chemical present in the water phase of the heterogeneous system

FR_w = volume fraction of the water phase in the system

FR_s = volume fraction of the solid phase in the system

K_p = partition coefficient (L/kg)

RHO_s = density of the solid phase (kg/L).

In Figure 3.7 the results of Equation 3.5 are plotted for different solids-water ratios and different partition coefficients. It is evident that with an increasing solids-water ratio and partition coefficient, the fraction of the chemical in the solid phase of the system also increases. For a chemical with a K_p value of 10^5 L/kg, only some 10% would be associated with the particles (typically

10 mg/L on a dry weight basis, FR_w = small) in surface water. In a typical soil system where $FR_w = FR_s = 40\%$, only as little as 10^{-6} % of the same chemical would be present in the water phase.

Air-water and air-soil equilibrium

Henry's law constant can be derived from the ratio of the vapour pressure (P_v) and solubility of the pure compound (Equation 3.6). This is only correct if vapour pressure and solubility refer to the same state of the compound (liquid or solid) and to the same temperature. The air-water concentration ratio can be derived from Henry's law constant by reworking it into a "dimensionless" partition coefficient (Equation 3.7). Dimensionless air-soil concentration ratios can be obtained in the same way (Equation 3.8 and Figure 3.4):

$$H = \frac{P^S_{L,S}}{S_{L,S}} \tag{3.6}$$

$$K_{\text{air-water}} = \frac{C_{\text{air}}}{C_{\text{water}}} = \frac{H}{R \cdot T} = \frac{P^S_{L,S} / S_{L,S}}{R \cdot T} \tag{3.7}$$

see below (3.8)

where

- H = Henry's law constant (Pa·m³/mol)
- $P^s_{L,S}$ = vapour pressure of the pure liquid or solid (Pa)
- $S_{L,S}$ = solubility of the pure liquid or solid in water (mol/L)
- $K_{\text{air-water}}$ = "dimensionless" air-water distribution ratio
- R = gas constant (8.314 Pa·m³/(mol·K))
- T = temperature at the air-water interface (K)
- $K_{\text{air-soil}}$ = "dimensionless" air-soil distribution ratio
- K_p = soil-water partition coefficient (L/kg)
- M_s = mass of the solids in the aqueous system (kg)
- V_w = volume of the aqueous system (L)
- C = concentration (mol/L).

Air-aerosol equilibrium

Air-aerosol partition coefficients are usually not reported in the literature. It is more common to report the fraction of the chemical that occurs in association with the aerosol phase. Often an inverse proportionality between the fraction associated with aerosol and the chemical's P_v is observed. The fraction associated with the aerosol phase can be estimated according to Junge's equation (Equation 3.9) [3]:

$$FR_{\text{airosol}} = \frac{c\,\Theta}{P^s_L + c\,\Theta} \tag{3.9}$$

where

- FR_{aerosol} = fraction of the chemical in air associated with aerosol
- Θ = aerosol surface area per volume unit (m²/m³)
- P^s_L = vapour pressure of the pure compound in the liquid state (Pa)
- c = constant (Pa·m).

The constant c depends on the heat of condensation and molecular weight for many organics. It is assumed to be 0.17 Pa·m. The local pollution climate determines the aerosol surface density. A typical value for aerosol surface area under rural conditions is 3.5 x 10^{-4} m²/m³. For more polluted urban/industrialized areas Θ is estimated to be 1.1 x 10^{-3} m²/m³. Substitution of these values in Equation 3.9 shows that gas-particle partitioning is important for organic compounds with a P_v lower than approximately 10^{-3} Pa. Since P_v is strongly temperature dependent, the fraction of a substance absorbed to particles will also be temperature dependent. For certain organics this may imply that in tropical regions the pollutant will be in the gas phase, while in arctic regions it will be in the particle phase.

New insights into the partitioning of organic chemicals between air and aerosols indicate that this process may well be an absorption phenomenon of chemicals on the aerosol. Regression equations based upon the octanol-air partition coefficient (K_{oa}) were developed to quantify the air-aerosol equilibrium. The octanol-air partition coefficient has been shown to have a linear correlation with the compound vapour pressure,

$$K_{\text{air-soil}} = \frac{C_{\text{air}}}{C_{\text{soil}}} = \frac{C_{\text{air}}}{C_{\text{water}}} \cdot \frac{C_{\text{water}}}{C_{\text{soil}}} = \frac{K_{\text{air-water}}}{K_p \cdot M_s / V_w} \tag{3.8}$$

Table 3.1. Correspondence between spatial and temporal scales of atmospheric transport.

Horizontal transport		Time	Vertical transport	
Local	0-10 km	seconds		
	0-30 km	hours	boundary layer	0-3 km
Mesoscale	< 1000 km	days		
Continental	< 3000 km	days	troposphere	< 12 km
Hemisphere		months		
Global		years	stratosphere	< 50 km

indicating that the vapour pressure of a compound can be used to examine the influence of organic carbon on the gas-particle partitioning [4].

3.2.3 Intramedia transport in air

Transport, transformation and removal (by deposition) are mainly confined to a thin layer of the atmosphere, approximately 2 to 3 km, usually called the planetary boundary layer. Advective transport is caused by a horizontal wind which is driven by gradients in atmospheric pressure. Close to the earth's surface the wind changes both in velocity and direction as it experiences friction due to the roughness of the terrain. These fluctuations in speed and direction are referred to as mechanical turbulence and affect the dilution rate of air pollutants considerably. Another type of turbulence is caused by the upward movement of air heated on the earth's surface by solar radiation. As a result, cold air replaces the rising hot air. The vertical sizes of these turbulences can range over several orders of magnitude (10^{-3} - 10^{2} m). Turbulence is a very effective mixing process and is generally far more important than molecular diffusion. Some examples of large-scale meteorological processes that affect the advective transport and dispersion of pollutants are:

- Wind shear, which is the gradual change in direction and velocity of the advective flow with height, caused by friction at the earth's surface.
- Large-scale vertical atmospheric motions due to high or low pressure systems in clouds or introduced by terrain effects (e.g., mountains, etc.).

The scales of time and space are closely linked in atmospheric transport (Table 3.1). Therefore, the atmospheric residence time of a pollutant determines how far the pollutant will be transported away from its source. In the direct vicinity (< 30 km) of a source, concentrations are mainly controlled by advection and dispersion.

3.2.4 Intramedia transport in water

Before developing and/or applying water models the basics of transport in surface water systems should be understood. Then, depending on the purpose of the model, a specific model type may be chosen to estimate surface water concentrations. A distinction between different types of water models can be made by looking at a number of different aspects, such as:

- Complexity with respect to the modelling of dilution.
- Complexity with respect to the modelling of the fate of the chemical after discharge.
- Generic versus site-specific models.
- Steady-state versus (quasi)dynamic computations.

Choosing the right model for a specific application very much depends on whether or not we are interested in the mixing process in the receiving water body. It is obvious that the discharge of effluents in surface water will not result in an instantaneous mixing. The turbulence of the receiving water will cause dispersion of the chemicals in the discharge in all directions until a homogeneous concentration is achieved. When deciding whether or not a certain discharge may cause adverse effects in the environment, it is important to know the range and the degree of mixing. There are three successive stages in the mixing process of an effluent in a river:

- *Near field:* vertical mixing of the discharged effluent over the depth of the water layer. The mixing is determined by the initial momentum and the buoyancy of the effluent jet.
- *Mixing zone:* transverse mixing over the width of the river, determined by the turbulence and flow of the receiving water. For continuous discharges a gradual spread over the cross-section is observed.
- *Far field:* as the cross-sectional mixing is completed, longitudinal dispersion will determine the concentration distribution of the discharge.

These different mixing stages are shown in Figure 3.8, for both a continuous discharge and a chemical spill in a

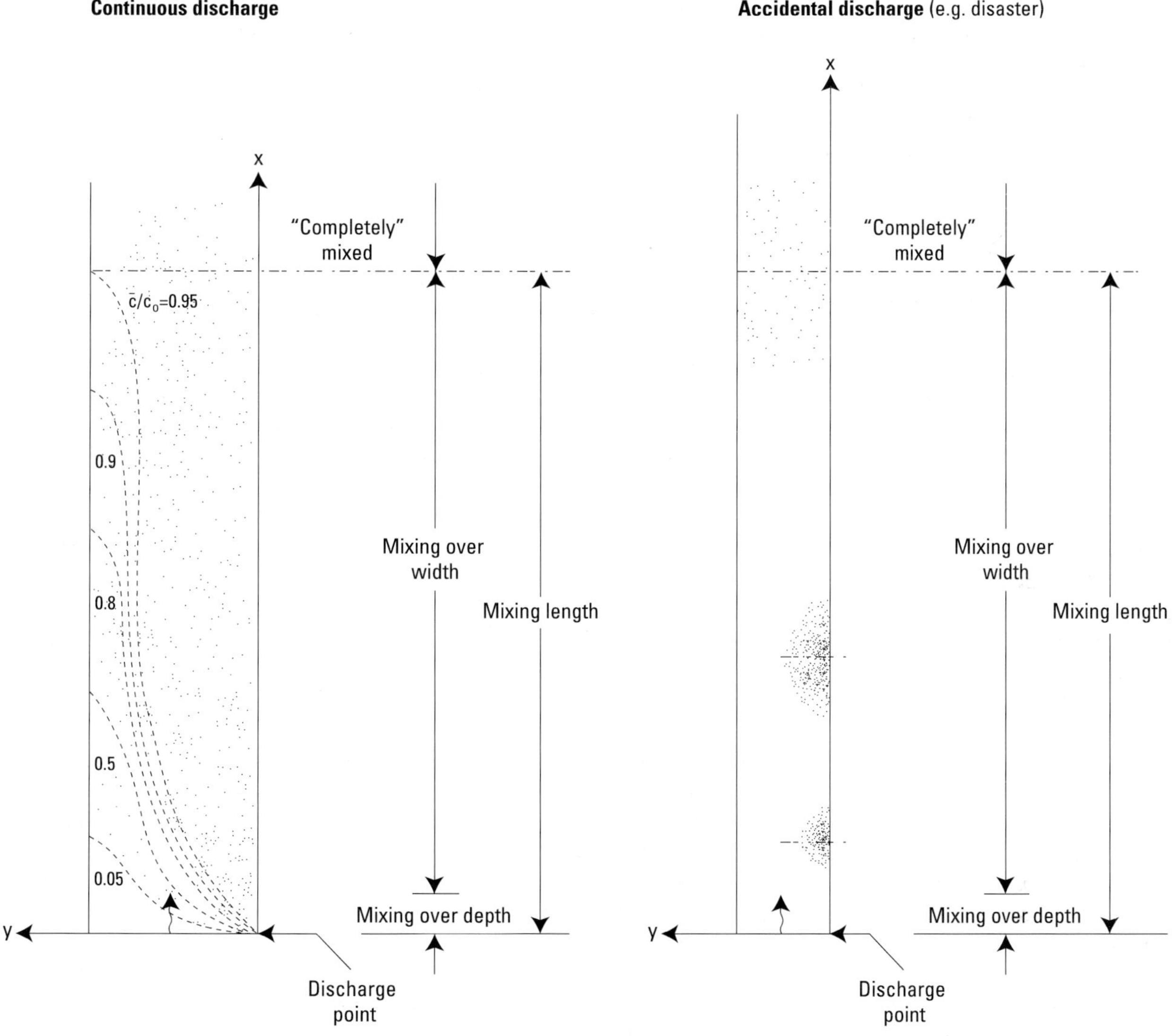

Figure 3.8. Stages in the mixing of effluent in river water. The dashed lines represent the relative (c/c_0) iso-concentration lines. From Van Mazijk and Veldkamp [5]. With permission.

river. The description of the mixing processes is restricted to river systems, for reviews of transport processes in non-river systems, see [6]. In most cases mixing over depth is achieved much faster than over the width of the river system, because of the initial momentum and buoyancy of the discharged effluent and due to the width to depth ratio of most systems. As the mixing over depth is a local or near-field phenomenon, the distribution of a compound in a river is usually described by a two-dimensional model over the width and length of the system being considered, although the z-component in many systems is also often important where there is turbulence. In the third stage complete transverse mixing is accomplished and a one-dimensional model will suffice. Dispersion in lakes and seas differs in that, as might be imagined, the third and last phase of complete mixing may never be reached. In rivers or canals the dispersion is "bounded" by the borders of the system. In lakes and seas "unbounded" dispersion takes place.

These stages of the mixing process, as well as the type of water flow are of major importance in the development or choice of the water models to be used. To study the concentration distribution within the mixing zone, a two-dimensional model should be used. An accidental chemical spill coming down the river Rhine could be modelled with a one-dimensional model. All aquatic dispersion models assume that the compound is fully dissolved. Indirectly (i.e., by means of equilibration of exchange between the solid phase and the aquatic phase, the dispersion models take into account sedimentation and successive resuspension processes. Only first-order degradation or transformation processes can be incorporated. If the local distribution of a compound is not the major topic of interest or when sedimentation/resuspension, sorption or complex degradation processes are involved, a box or compartment model may be more appropriate. Most models do not take stratification into account, although stratification may be very important in lakes, estuaries, and the marine environment.

The subject of dispersion and mixing of solutes and suspended materials in turbulent natural streams has been described extensively [7,8]. Whether or not a one or two-dimensional model should be used is determined by the length of the mixing zone. When assuming that, on average, the depth of a river is 0.4 times the width of a river, this length can be estimated from [9]:

$$L_{mix} = \frac{0.4\,\bar{u} \cdot w^2}{D_y} \tag{3.10}$$

where

- L_{mix} = length of the mixing zone (m)
- D_y = transverse dispersion coefficient (m^2/s)
- w = width of the water system (m)
- $\bar{u}$ = average flow velocity over the cross-section of the river (m/s).

Depending on the width, flow and turbulence of the system, the mixing zone can range from 500 m for relatively narrow, highly turbulent systems up to 10-100 km for large, wide rivers like the Rhine or the Meuse.

3.2.5 Intermedia transport

Intermedia transport is the result of two fundamental processes, namely intermedia diffusion and intermedia advection. In Figure 4.11 of Chapter 4, a diagram is given of the most important diffusion and advection processes that take place in the environment. A detailed theoretical description of these processes may be found in [10]. The most important interfaces and the corresponding intermedia processes are described below.

Soil leaching and sediment burial

Quality management of soil and sediment focuses on the health of the ecosystem and usually involves only the upper layers of these compartments. For this purpose, transport of chemicals from the upper layer downward is regarded as a removal process similar to advective and dispersive transport of a chemical away from the source in air and water.

Transport from the upper layer of the soil to the groundwater takes place through leaching with percolating water. If we choose to treat groundwater as part of the soil system, it should be considered as an intramedia transport phenomenon (from upper soil to lower soil). However, if we choose to treat groundwater as a separate medium, soil leaching should be regarded as intermedia transport (soil to groundwater). Background information on transport in porous media can be found in Spitz and Moreno [11] and will not be considered in detail here.

In most multimedia models (Chapter 4), the process of soil leaching is simplified by assuming equilibrium between the solid phase and pore water phase at all times and in all places. Leaching of the chemical from the upper soil layer can then be treated as a first-order removal process:

see below (3.11)

where

- $LEACH$ = removal of the chemical from the upper soil layer (mol/s)
- $RAIN$ = rate of wet precipitation (m/s)
- FR_{inf} = fraction of rain water that infiltrates into the soil
- $AREA_{soil}$ = soil area (m^2)
- FR_w = volume fraction of the water phase of soil
- FR_s = volume fraction of the solid phase of soil

$$LEACH = \frac{RAIN \cdot FR_{inf}}{FR_w + FR_s \cdot K_p \cdot RHO_s} \cdot AREA_{soil} \cdot C_{soil} \tag{3.11}$$

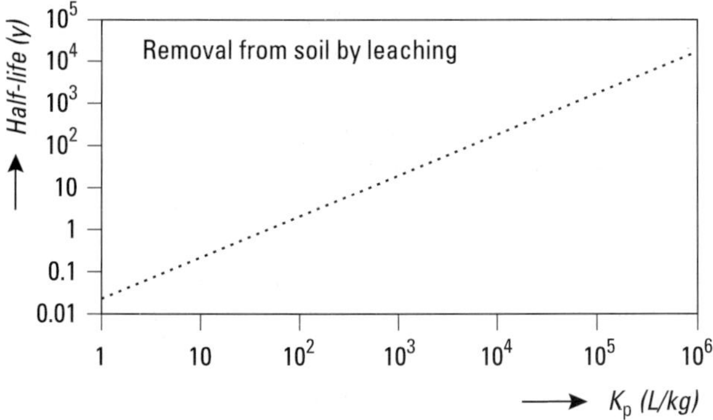

Figure 3.9. Half-lives for the removal of substances by leaching from a top layer of soil at different values of the soil-water partition coefficient K_p. Mixing depth=5 cm; $FR_w=FR_s=0.4$; RAIN=760 mm/y; $FR_{int}=0.4$.

K_p	=	soil-water partition coefficient (L/kg)
RHO_s	=	density of the solid phase of soil (kg/L)
C_{soil}	=	concentration in soil (mol/m^3).

It is clear that leaching is an important factor for chemicals with a small K_p value. This is illustrated in Figure 3.9, where calculated half-lives for leaching from soil are plotted for different values of the soil-water partition coefficient. Analogous transport phenomena take place in sediment. Surface water may seep into the sediment, thereby carrying the chemical from the upper sediment layer down and *vice versa*. The mass flows resulting from this can be derived by analogy with Equation 3.11. An additional phenomenon occurs in areas where there is continuous sedimentation. In this situation sediment is continuously buried under freshly deposited material. If only the upper layer of the sediment is considered in quality management, the contaminated upper layer is, in fact, transported to the deeper sediment. This "transport" process from the upper sediment layer by burial can be described by a first-order removal process with Equation 3.12:

$$BURIAL = NETSED \cdot AREA_{sed} \cdot C_{sed} \qquad (3.12)$$

where

$BURIAL$	=	apparent burial mass flow from the sediment compartment (mol/s)
$NETSED$	=	net sedimentation rate (m/s)
$AREA_{sed}$	=	area of the sediment-water interface (m^2)
C_{sed}	=	bulk concentration in sediment (mol/m^3).

Wet and dry atmospheric deposition

Chemicals are transported from the atmosphere to water and soil by atmospheric deposition (Figure 3.10). In atmospheric chemistry it is customary to present these different mechanisms as being composed of wet (precipitation-mediated) deposition mechanisms and dry deposition mechanisms. Wet deposition is the sum of rain-out (in-cloud processes) and wash-out (below-cloud processes). Dry deposition is the sum of

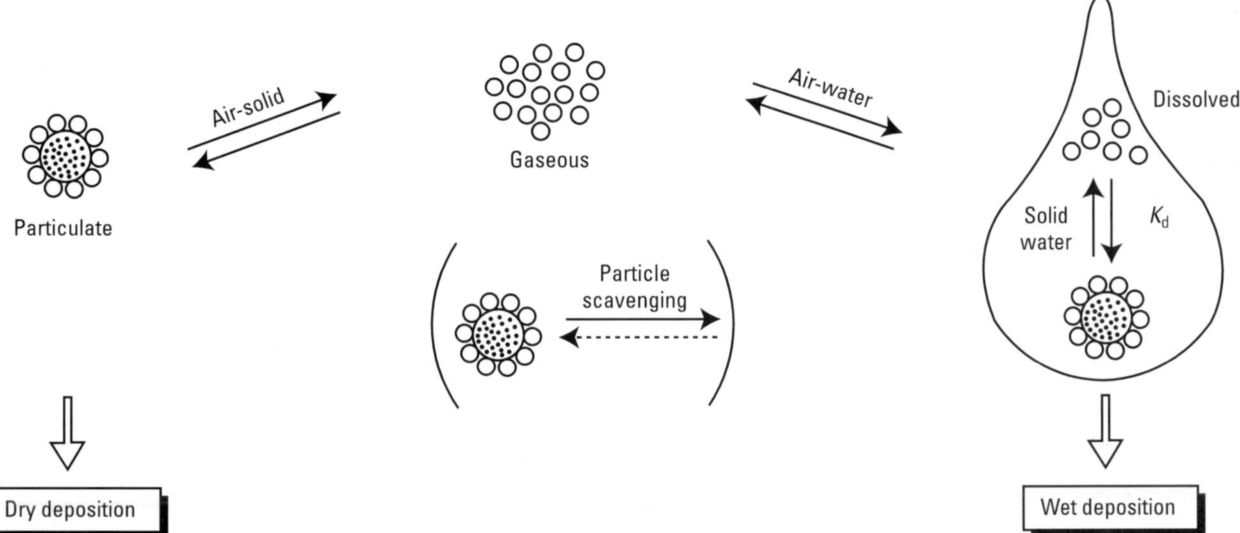

Figure 3.10. Mechanisms of atmospheric deposition. From Schwarzenbach [2]. With permission.

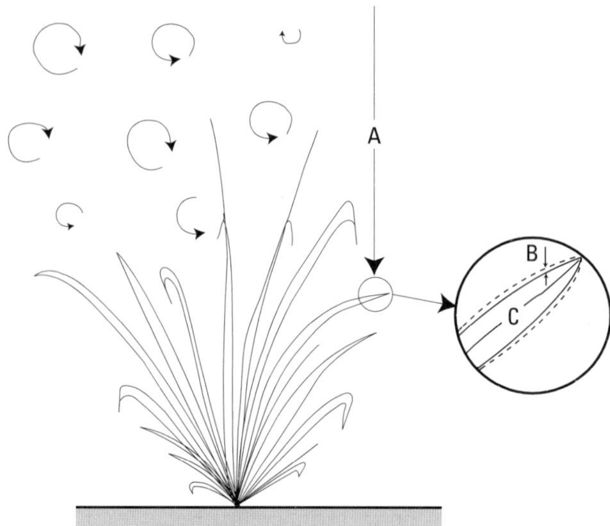

Figure 3.11. Three-step mechanism of dry deposition. A. Transport from the mixed layer to the laminar sublayer in the immediate vicinity of the surface. This transport is controlled by turbulent diffusion in the mixed layer. B. Transport through the laminar sublayer is typically in the order of 0.1-1 mm. For gases this process is controlled by molecular diffusion, for aerosols by Brownian diffusion. C. Absorption to the surface. The chemical nature and biological reactivity of both the receiving surface and depositing material determines how much material is actually removed at the surface. From Fowler [12]. With permission.

aerosol deposition and gas absorption. In multimedia environmental chemistry, the latter mechanism is usually treated as one part of a bi-directional exchange mechanism. Rain-out, wash-out and aerosol deposition are one-way advective transport processes: the chemical is carried from the atmosphere to water and soil. This is true even if the chemical has a greater fugacity in water or soil. Gas absorption is a diffusive mechanism. There is only net absorption of chemicals from the gas phase by water or soil if the fugacity in air is greater than the fugacity in water or soil. If the fugacity in water or soil is greater, the result will be the reverse: net volatilization. This will generally be the case if a chemical is emitted to water or soil. In such cases the fugacity in water or soil will be greater than in air, resulting in continuous volatilization into the atmosphere, although at the same time deposition will occur. Gas absorption and volatilization are discussed in a separate section below, but it should be noted that in this case absorption and volatilization occur simultaneously and it is the net difference that accounts for the effective intermedia transport.

Dry deposition

Transport of chemicals from air to water and soil by dry deposition (Figure 3.11) can be viewed by analogy with an electric current passing through a series of resistances. In the case of dry deposition, the main resistances occur at the air-surface interface: transport of the chemical from the air to the interface, diffusion across the interface and transport from the interface to the solid surface. Thus, the deposition velocity v_d is dependent on the atmospheric turbulence, the chemical composition and the physical structure of both the receiving surface and the depositing material. For highly soluble or chemically reactive gases (e.g., nitric acid, HNO_3) the surface resistance is small, especially when the surface is wet. For fatty materials like many organic compounds, the canopy resistance of trees and plants will be small because the resistance at the vegetation surface is low (i.e. the cuticle is a good and easily accessible sink), resulting in high deposition velocities.

For chemicals for which dry deposition is an important fate process, the pollutant can be either scavenged from the atmosphere in its gaseous form by soil or vegetation, or attached to a carrier particle for which removal rates can be described as a function of the physical parameters of the particle, of which the size is most important. Small particles tend to behave like gases; larger particles (> 2 μm) are efficiently removed from the atmosphere by deposition under the influence of gravity. Inertial impaction is important for particles with a diameter of between 0.1 and 10 μm. This effect greatly depends on the velocity of the air and the intensity of the turbulence, which varies with the properties of the landscape. Since the lifetime of atmospheric particles is a function of particle size, it is important to know the sizes of the particles as they leave the source. Removal of a chemical from air by dry deposition of aerosols is proportional to the concentration of the chemical in aerosol particles and the deposition velocity of these particles; larger particles (> 10 μm) are deposited primarily by sedimentation and chemicals associated with larger particles will, in general, be deposited close to the source. The rate of deposition to water or soil can be expressed according to Equation 3.13:

see next page (3.13)

where

$DRYDEP_{aerosol}$ = rate of removal of the chemical from the atmosphere by dry deposition of aerosol particles (mol/s)

$vd_{aerosol}$ = deposition velocity of aerosol particles (m/s)

$AREA_{water\ or\ soil}$ = area of the air-water or air-soil interface (m^2)

C_{air} = bulk concentration in air (mol/m^3)

$FR_{aerosol}$ = fraction of the chemical associated with aerosol (Equation 3.9).

A similar equation can be written for dry deposition by gas absorption. This mechanism is explained in connection with volatilization below.

Wet deposition

Wet deposition includes the following processes:

a. Wash-out or below-cloud scavenging, a process which occurs *below* the clouds and by which gases or particles are absorbed by falling raindrops.
b. Rain-out or in-cloud scavenging, a process which occurs *in* the clouds: the gases or particles are scavenged by the cloud droplets and the chemical is removed during the next rainfall.

The efficiency of the wet deposition process varies greatly. It depends on meteorological factors such as the duration, intensity and type of precipitation (snow, rain, hail), as well as on the size and the number of droplets. Other specific parameters, like solubility in rain and snow, are important too. Wash-out is an efficient removal mechanism for soluble gases (low Henry's law constant) and for aerosols with a diameter greater than 1 μm. For less soluble gases (higher Henry's law constants) the falling droplet will absorb only a very small amount of the compounds below the cloud. Wash-out plays an important role when concentrations *below* the cloud are much higher than the concentrations in the cloud, e.g., for plumes close to the source. *In* clouds the uptake of aerosols by cloud droplets is a very efficient process. In many cases wash-out is the most important removal mechanism for aerosols. In general, the removal rate by wet deposition can be described by a first-order process defined by a scavenging coefficient Λ, consisting of a gas and aerosol scavenging component (Equation 3.14):

see below (3.14)

where

$WET\text{-}DEP$ = rate of removal of the chemical from the atmosphere by wet deposition (mol/s)

Λ = overall scavenging coefficient (1/s)

Λ_{gas} = gas scavenging coefficient (1/s)

$\Lambda_{aerosol}$ = aerosol scavenging coefficient (1/s)

$AREA$ = total (water and soil) interfacial area (m^2)

z_{air} = height of the mixed air layer (m)

C_{air} = concentration in air (mol/m^3).

For most purposes, it is sufficient to assume that the rain phase is in equilibrium with the gas phase. The gas scavenging coefficient Λ_{gas} can then be estimated from the dimensionless air-water distribution ratio $K_{air\text{-}water}$, the rain intensity and the height of the air layer:

see below (3.15)

where

Λ_{gas} = gas phase scavenging coefficient (1/s)

$RAIN$ = rain intensity (m/s)

z_{air} = height of the mixed air layer (m)

FR_{gas} = fraction of the chemical in the gas phase

$K_{air\text{-}water}$ = dimensionless air-water distribution constant (m^3/m^3)

$FR_{aerosol}$ = fraction of the chemical in the aerosol phase.

As a practical approach to estimating the aerosol scavenging coefficient $\Lambda_{aerosol}$, Mackay [1] has suggested that during rainfall in the atmosphere, each drop sweeps through a volume of air about 200,000 times its own volume (Equation 3.16):

$$\Lambda_{aerosol} = \frac{RAIN}{z_{air}} \cdot 2 \cdot 10^5 \cdot FR_{aerosol} \tag{3.16}$$

where

$$DRYDEP_{aerosol} = vd_{aerosol} \cdot AREA_{water\ or\ soil} \cdot C_{air} \cdot FR_{aerosol} \tag{3.13}$$

$$WETDEP = \Lambda \cdot AREA \cdot z_{air} \cdot C_{air} = (\Lambda_{gas} + \Lambda_{aerosol}) \cdot AREA \cdot z_{air} \cdot C_{air} \tag{3.14}$$

$$\Lambda_{gas} = \frac{RAIN}{z_{air}} \cdot \frac{FR_{gas}}{K_{air\text{-}water}} = \frac{RAIN}{z_{air}} \cdot \frac{1 - FR_{aerosol}}{K_{air\text{-}water}} \tag{3.15}$$

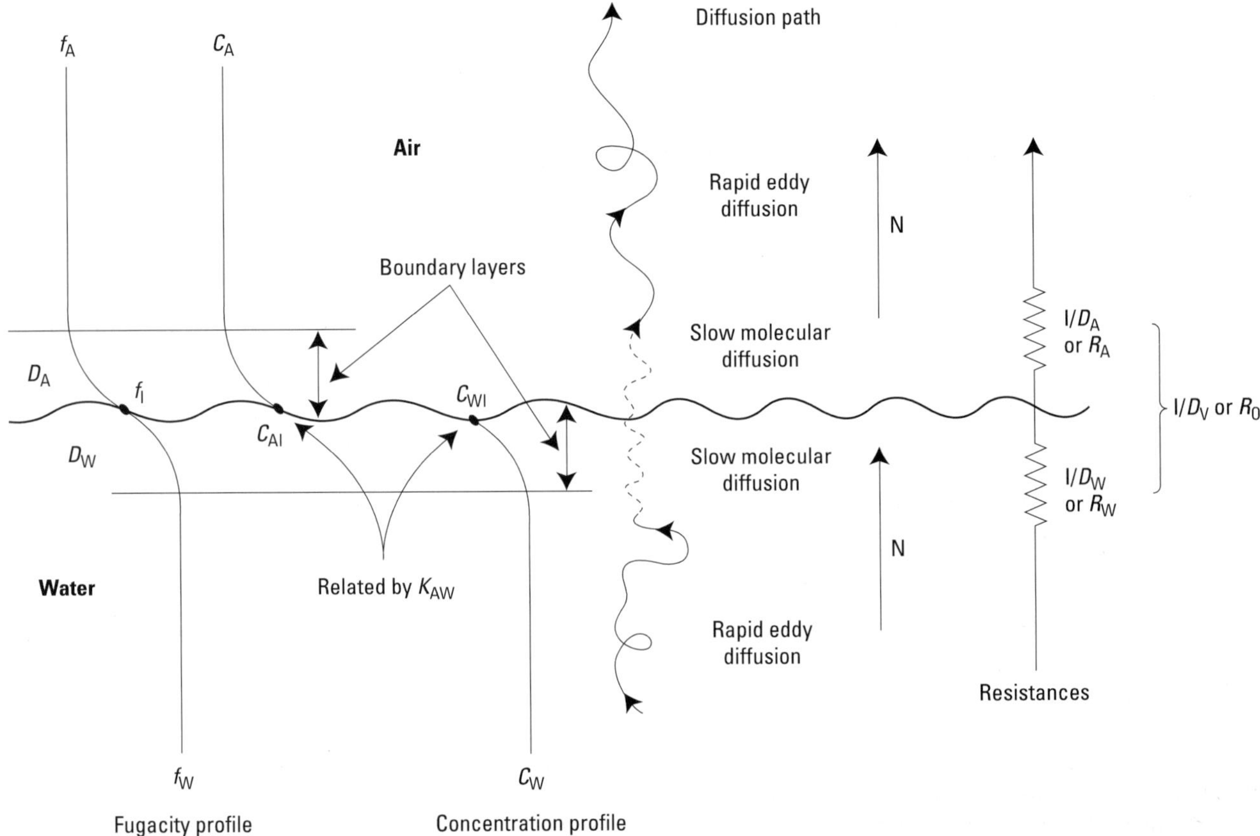

Figure 3.12. Mass transfer of a chemical between two phases, air and water, according to the two resistances concept described by Mackay [1]. With permission.

$RAIN$ = rain intensity (m/s)
z_{air} = height of the mixed air layer (m)
$FR_{aerosol}$ = fraction of the chemical in the aerosol phase.

It is important to note that the tendency to associate with aerosol particles is different for different chemicals; and different chemicals are associated with different particle-size fractions in the aerosol. Therefore, both the aerosol deposition velocity, $vd_{aerosol}$ in Equation 3.13, and the aerosol scavenging coefficient in Equation 3.14 are greatly chemical-dependent.

Volatilization and gas absorption

Transport of a chemical from water and soil to the gas phase of air and *vice versa* is commonly described with the two-resistance approach, as originally introduced almost a century ago by Whitman [13]. In this concept, the resistance to intermedia transfer is considered to be concentrated in two thin films on either side of the interface. Transport through this interfacial double layer has to take place by molecular diffusion and is, therefore, slow in comparison with transport to and from the interface. This concept was used by Liss and Slater [14] as a basis for modelling the transfer of gases across the air-sea interface. This is shown in Figure 3.12 for exchange between air and water. The direction of transport depends on the concentrations in air and water. If the actual concentration of the chemical in water is higher than the equilibrium concentration in water, the chemical will volatilize from the water phase into the gas phase. If the actual concentration in air is higher than the equilibrium concentration in air, the water phase will absorb the chemical from the gas phase. In fugacity terminology: the net diffusion is from the phase in which the highest fugacity exists to the phase with the lowest fugacity. At the interface, the air and water concentrations are in equilibrium and the fugacities are equal. The rate masstransfer (volatilization or gas absorption) is usually quantified by means of an "overall" mass-transfer coefficient. The mass-transfer coefficient is expressed in the dimension of velocity (m/s). This process can be looked upon as if the chemical is pushed through the interface by a piston that moves with a velocity equal to the overall mass-transfer coefficient. The mass flux across the interface is given by Equation 3.17:

see below (3.17)

where

$VOLAT$ = rate of removal from water by volatilization (mol/s)
$ABSORB$ = rate of absorption to water from air (mol/s)
$AREA_{water}$ = area of the air-water interface (m^2)
K_{water} = water-based overall mass-transfer coefficient (m/s)
K_{air} = air-based overall mass-transfer coefficient (m/s)
$K_{air\text{-}water}$ = dimensionless air-water distribution constant (m^3/m^3)
C_{water} = concentration in water (mol/m^3)
C_{air} = concentration in air (mol/m^3).

As indicated in Equation 3.17, the flux can be expressed on the basis of either one of the phases. The piston velocities in the two phases are different. However, the same amount of chemical is transported towards and away from the interface, but the concentrations in the two phases differ! In the usually much "thinner" air, the piston has to move faster than in water. The water and air-based overall mass-transfer coefficients are derived by Equations 3.18 and 3.19:

$$K_{water} = \frac{kaw_{air} \cdot kaw_{water}}{kaw_{air} + kaw_{water} / K_{air\text{-}water}} \quad (3.18)$$

and

$$K_{air} = \frac{kaw_{air} \cdot kaw_{water}}{kaw_{air} \cdot K_{air\text{-}water} + kaw_{water}} \quad (3.19)$$

where

K_{water} = water-based overall mass-transfer coefficient (m/s)
K_{air} = air-based overall mass-transfer coefficient (m/s)
kaw_{air} = partial mass-transfer coefficient for the air side of the air-water interface (m/s)
kaw_{water} = partial mass-transfer coefficient for the water side of the air-water interface (m/s)
$K_{air\text{-}water}$ = dimensionless air-water distribution constant (m^3/m^3).

Note that the ratio of the air and water-based mass-transfer coefficients is equal to the dimensionless intermedia partition coefficient. Transport through the air and water films takes place by molecular diffusion. The partial mass-transfer coefficients are, therefore, proportional to the diffusion coefficients of the chemical in air and water, and inversely proportional to the thickness of the films. Since the molecular diffusion coefficients of different chemicals do not differ much, the partial mass-transfer coefficients have nearly the same values for all chemicals. The values depend on the turbulence of the interface. Typical values are 10^{-3} and 10^{-5} m/s for kaw_{air} and kaw_{water}, respectively. If the concentration in air is negligible, only volatilization occurs. Volatilization can then be treated as a first-order removal process from water:

see below (3.20)

The rate constant for volatilization is:

see below (3.21)

$$\begin{aligned} VOLAT \text{ or } ABSORB &= AREA_{water} \cdot K_{water} \cdot (C_{water} - C_{air} / K_{air\text{-}water}) \\ &= AREA_{water} \cdot K_{air} \cdot (C_{air} - C_{water} \cdot K_{air\text{-}water}) \end{aligned} \quad (3.17)$$

$$VOLAT = K_{water} \cdot AREA_{water} \cdot C_{water} = \frac{kaw_{air} \cdot kaw_{water}}{kaw_{air} + kaw_{water} / K_{air\text{-}water}} \cdot AREA_{water} \cdot C_{water} \quad (3.20)$$

$$k_{volat} = K_{water} \cdot AREA_{water} / VOLUME_{water} = K_{water} / DEPTH_{water} \quad (3.21)$$

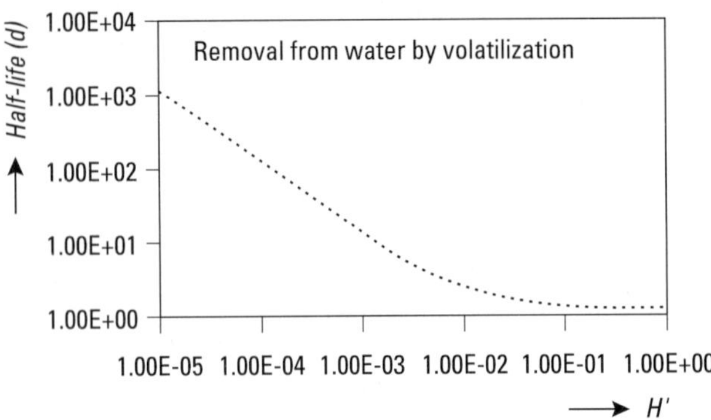

Figure 3.13. Half-lives for the removal of a substance by volatilization from a body of water (depth 2 m), plotted for different values of the dimensionless Henry's law constant.

where

k_{volat}	=	pseudo first-order rate constant for volatilization from water (1/s)
K_{water}	=	water-based overall mass-transfer coefficient (m/s)
$AREA_{water}$	=	area of the air-water interface (m^2)
$VOLUME_{water}$	=	volume of the water compartment (m^3)
$DEPTH_{water}$	=	depth of the water column (m).

As can be seen from Equations 3.20 and 3.21, different chemicals with different Henry's law constants volatilize at different rates. This is illustrated in Figure 3.13, where for a typical water, two metres deep, volatilization half-lives are plotted against the dimensionless air-water distribution ratio $K_{air\text{-}water}$. For small values of $K_{air\text{-}water}$, the half-life is inversely proportional to $K_{air\text{-}water}$. For greater $K_{air\text{-}water}$ values, chemicals volatilize at maximum speed and the half-life becomes small and independent of $K_{air\text{-}water}$. Similar equations can be derived for volatilization from soil or vegetation and gas absorption to soil or vegetation. Advanced readers are referred to specialized textbooks [15,16].

Soil run-off

Part of the rainwater that reaches the soil runs off to surface water. In urban areas, where most of the surface is paved, nearly all the precipitation is collected in sewerage systems, from where it may either be redirected to a waste water treatment facility or discharged into surface water. In rural areas the rainwater runs off directly into the surface waters. With the run-off, soil particles are washed away (eroded). Chemicals dissolved in water or associated with the soil particles, are transported by this mechanism from soil to water. If we assume that the water which runs off from soil is in equilibrium with the soil, the mass flow of a chemical resulting from run-off can be quantified according to Equation 3.22 (see below).

$$RUN\text{-}OFF = \left[\frac{RAIN \cdot FR_{run}}{FR_w + FR_s \cdot K_p \cdot RHO_s} + EROSION_{soil\ i}\right] \cdot AREA_{soil} \cdot C_{soil} \quad (3.22)$$

where

$RUN\text{-}OFF$	=	mass flow of chemical due to run-off from soil to water (mol/s)
$RAIN$	=	rate of wet precipitation (m/s)
FR_{run}	=	fraction of rainwater that infiltrates into soil i
FR_w	=	volume fraction of the water phase of soil
FR_s	=	volume fraction of the solid phase of soil
K_p	=	soil-water partition coefficient (L/kg)
RHO_s	=	density of the solid phase of soil (kg/L)
$EROSION_{soil\ i}$	=	rate at which soil is washed from soil i into surface water (m/s)
$AREA_{soil}$	=	soil area (m^2)
C_{soil}	=	concentration in soil (mol/m^3).

Sediment-water exchange

The transport of chemicals across the sediment-water interface can be treated in the same manner as air-water and air-soil exchanges. In this case there is an advective transport component: i.e., sedimentation (and resuspension); and a diffusive transport component: i.e., direct adsorption onto (and desorption from) the sediment. To estimate the rate of advective transport from water to sediment by sedimentation of suspended particles, we need to know the concentration of the chemical on the particles. For most purposes it is sufficient to assume equilibrium between the suspended particles and water phase. The removal from water by sedimentation can then be obtained from Equation 3.23:

see below (3.23)

where

SED = removal of the chemical from water by sedimentation (mol/s)

$SETTL_{vel}$ = gross settling velocity of suspended particles (m/s)

$AREA$ = area of the sediment-water interface (m^2)

$SUSP$ = concentration of suspended particles in the water column (kg/m^3)

C_{susp} = concentration in suspended particles (mol/kg)

K_p = suspended matter-water partition coefficient (m^3/kg)

C_{water} = concentration in water (mol/m^3).

Taking resuspension into account, the equation for net removal from the water column due to sedimentation *(NETSED)* becomes as follows:

see below (3.24)

where

$RESUSP_{rate}$ = resuspension rate (m/s)

C_{sed} = concentration in sediment matter (mol/m^3).

Diffusive transport between sediment and water, by direct adsorption and desorption across the sediment-water interface, is analogous to diffusive transport across the air-water and air-soil interfaces and can be described with a two-film resistance model:

see below (3.25)

where

$ADSORB_{sed}$ = removal of the chemical from water by direct adsorption onto the sediment (mol/s)

kws_{water} = partial mass-transfer coefficient on the water side of the sediment-water interface (m/s)

kws_{sed} = partial mass-transfer coefficient on the pore water side of the sediment-water interface (m/s)

$AREA_{sed}$ = total area of the system (air-water and air-soil interfaces in m^2)

C_{water} = concentration in water (mol/m^3).

Since the quotient of the mass-transfer coefficients for adsorption and desorption is equal to the volume-based sediment-water partition coefficient, removal of a chemical from sediment can be calculated with Equation 3.26:

see below (3.26)

where

$DESORB_{sed}$ = removal of the chemical from sediment by direct desorption to water (mol/s)

$K_{sed\text{-}water}$ = dimensionless sediment-water partition coefficient

C_{sed} = concentration in sediment (mol/m^3).

A value of 0.01 m/h [17] may be taken for the mass-transfer coefficient on the water-side of the sediment-water interface, kws_{water}. According to Mackay [17], mass-transfer on the pore water side of the sediment-water interface is treated as molecular diffusion in the aqueous phase of a porous solid material, characterized by an effective diffusivity of 2×10^{-6} m^2/h and a diffusion

$$SED = SETTL_{vel} \cdot AREA \cdot SUSP \cdot C_{susp} = SETTL_{vel} \cdot AREA \cdot SUSP \cdot K_p \cdot C_{water} \quad (3.23)$$

$$NETSED = AREA \cdot (SETTL_{vel} \cdot SUSP \cdot K_p \cdot C_{water} \cdot RESUSP_{rate} \cdot C_{sed}) \quad (3.24)$$

$$ADSORB_{sed} = \frac{kws_{water} \cdot kws_{sed}}{kws_{water} + kws_{sed}} \cdot AREA_{sed} \cdot C_{water} \quad (3.25)$$

$$DESORB_{sed} = [\frac{kws_{water} \cdot kws_{sed}}{kws_{water} + kws_{sed}} / K_{sed\text{-}water}] \cdot AREA_{sed} \cdot C_{sed} \quad (3.26)$$

path length of 2 cm. This gives kws_{sed} a value of 0.0001 m/h. It should be noted, however, that additional processes that are typically of a non-equilibrium nature, may greatly affect the net mass-transfer of all kinds of chemicals. For instance bioturbation can play a key role in the sediment side resistance, essentially eliminating it in some cases. As the extent of bioturbation is not governed by thermodynamic principles, and as, in general, very limited information is available on this and similar topics, it will not be extensively discussed here. Instead, the reader is referred to the textbook of Thibodeaux [15].

With this, we conclude the section on transport processes in and between media. Intramedia and intermedia transport processes result in different concentrations in environmental compartments. Species living in these compartments are exposed to these environmental concentrations. This may result in bioaccumulation, which is the subject of Section 3.3.

3.3 BIOACCUMULATION

3.3.1 Introduction

Many xenobiotics are released into the environment. Consequently, most aquatic and terrestrial organisms, as well as plants, are exposed to these chemicals. Some xenobiotics are taken up and bioaccumulate in high concentrations. *Bioaccumulation* produces higher concentrations of a chemical in an organism than in its immediate environment, including food. Particularly in aquatic organisms, *bioconcentration* describes the process which leads to higher concentrations of xenobiotics in the organisms than in water. For aquatic as well as higher organisms, *biomagnification* describes the process which occurs when food is the major source of bioaccumulation. Biomagnification refers to those cases where concentrations in an organism (on a lipid-wt basis for organic contaminants) exceed concentrations in the consumed prey. The extent to which compounds accumulate and the routes by which they are taken up and excreted may differ between species. Concentrations in organisms can also be lower than those in their prey if substances are biotransformed easily, thereby leading to *trophic dilution* [18,19].

Chemicals are taken up by biota via different routes, from air, water, soil and sediment, and each process depends on environmental and physiological factors. Mammals breathe air and will therefore take up chemicals which occur in air. Fish ventilate water for their oxygen supply and therefore take up chemicals which occur in the aqueous phase. Fish may be temporarily exposed to accidental spills of pollutants in water, or continuously to ubiquitously occurring xenobiotics. Terrestrial organisms in soil may be exposed to pesticide sprays or to chemicals present in dump sites. Plants are usually found in soil and air, or sediment and water, and therefore take up chemicals from several compartments. All organisms, except most plants and some other primary producers, may be exposed to chemicals via food.

Different models are used to describe and predict bioaccumulation, bioconcentration and biomagnification. Each type of bioaccumulation is measured differently and depends on the type of organism and chemical involved.

This section will discuss bioaccumulation processes in aquatic and terrestrial organisms. Both uptake and elimination processes will be addressed, as well as the models used to describe and predict bioaccumulation. Methods used for measuring bioaccumulation will also be described.

3.3.2 Aquatic bioaccumulation processes

Most of the knowledge gained on aquatic bioaccumulation processes stems from studies on fish, although some (but less) is known on other aquatic organisms, from phytoplankton, zooplankton, oysters, mussels to marine mammals. Since risk assessment models usually take into account bioconcentration in fish and environmental classification and PBT-assessment is based on bioconcentration factors in fish, the section mainly focuses on bioaccumulation in fish. PBT stands for Persistent, Bioaccumulative, and Toxic chemicals.

For many aquatic organisms, the major route of uptake of xenobiotics is from water and the route of elimination is to water. Bioconcentration, therefore, is the net result of uptake, distribution and elimination processes of a substance due to aqueous exposure. The bioconcentration factor BCF is defined as the ratio of the concentration in an organism (C_o) and that in the surrounding water (C_w) at steady-state:

$$\mathrm{BCF} = C_o / C_w \tag{3.27}$$

Bioaccumulation is similar to bioconcentration, but relates to all routes of exposure. The bioaccumulation factor BAF is defined as the ratio of the concentration in an organism (C_o) and that in the surrounding water (C_w) at steady-state, where uptake may occur via all routes of exposure:

$$\mathrm{BAF} = C_o / C_w \tag{3.28}$$

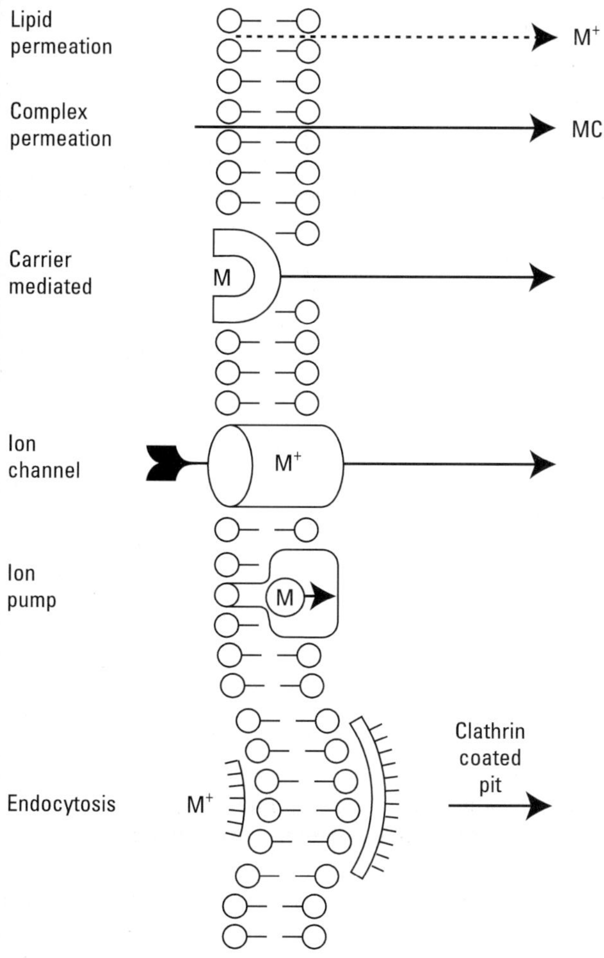

Figure 3.14. Different mechanisms of membrane passage for xenobiotic contaminants. M can be either a metal, an organometal or an organic chemical. From Phillips [20]. With permission.

Biomagnification describes the process which occurs when food is the major source of bioaccumulation. The biomagnification factor BMF is defined as the ratio of the concentration in an organism (C_o) and that in its food (C_{food}) at steady-state:

$$BMF = C_o / C_{food} \tag{3.29}$$

In the following paragraphs, certain aspects of bioaccumulation will be described, e.g., uptake processes, elimination processes, bioconcentration, bioconcentration models and methods for measuring bioconcentration.

Uptake processes

There are several processes leading to the uptake of chemicals by organisms. Each process involves the passage of compounds across a biological membrane, mediated by a carrier or as a single solute (Figure 3.14). Passive diffusion is the major uptake process for many organic chemicals as well as some metals and organometals. The driving force for uptake by passive diffusion is a fugacity difference between water and the organism (Chapter 4). Usually, passive diffusion is described as being driven by a concentration gradient.

However, with bioaccumulation processes, a concentration gradient will never lead to higher concentrations of xenobiotics in organisms compared with the surrounding medium. Bioaccumulation, therefore, is better described by the concept of fugacity. Organisms usually have a much higher capacity to store xenobiotics per unit of volume than water. For example, some metals bind to proteins, such as metallothionein, and may therefore be stored in relatively high concentrations within an organism. Organic chemicals are usually stored in lipids, and may thus reach high concentrations in an organism on a volume basis. Organometals can be stored by either lipids or proteins.

The fugacity of a chemical is the ratio of concentration to storage or fugacity capacity. The concentration of xenobiotics in water is usually small, but since its storage capacity (solubility) is also small, the fugacity is relatively large. The concentration in the organism is small initially and may reach higher concentrations than in water during the course of uptake, but due to its high storage capacity the fugacity of the chemical in the organism is relatively low. Thus, chemicals are transported from high to low fugacity by passive diffusion. For the sake of clarity, however, all equations will use concentrations rather than fugacities.
In addition to passive diffusion, other uptake processes may play a role in the uptake of contaminants (Figure 3.14). Metals, particularly, can be taken up by complex permeation, by carrier mediated processes, by ion channel, or by ATPases. For example, cadmium (Cd^{2+}) may be taken up either by Ca^{2+}-ATPases or as a cadmium-xanthate complex in fish [21].

Although there is no regulation in the uptake of chemicals by passive diffusion, organisms are able to regulate the uptake of chemicals by other, active uptake processes.

Elimination processes

Different processes lead to a reduction in the concentration of chemicals in an organism (Figure 3.15). Again, analogous to uptake processes, passive and active mechanisms are responsible for the elimination of

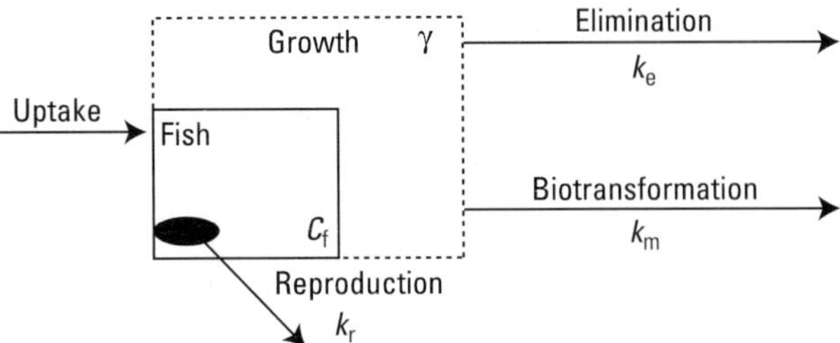

Figure 3.15. Different processes which reduce the concentration of xenobiotic contaminants in an organism (C_f): physico-chemical elimination (k_e), biotransformation (k_m), growth (γ) and reproduction (k_r). From [22]. With permission. Copyright 1992 American Chemical Society.

chemicals. Most hydrophobic chemicals are excreted by passive diffusion, either to water or via faeces. Growth is another way of diluting chemicals: the same number of moles of a compound in a small organism results in a higher concentration than in a bigger organism. Reproductive transfer of chemicals either via lactation (milk production) in mammals or via the mother to the egg can significantly reduce the concentration of chemicals in the organism. Biotransformation processes can also convert some chemicals into other, usually more hydrophilic ones, and thus reduce the concentration of the parent compound. Finally, some aquatic organisms are able to regulate elimination and consequently reduce the concentration of some metals.

Bioconcentration

Bioconcentration is the net result of the uptake, distribution and elimination processes of a substance due to aqueous exposure. The magnitude of bioconcentration depends on a variety of physicochemical and physiological factors.

For organic chemicals which bioconcentrate in lipid tissues mainly by passive exchange processes from and to water, the magnitude of bioconcentration largely depends on the hydrophobicity expressed via the *n*-octanol-water partition coefficient (K_{ow}; Chapter 9), and the lipid content of the organism.

For metals, bioconcentration depends more on physiological processes. The presence of active uptake and elimination processes, as well as the capacity of an organism to induce the synthesis of a metal storage protein, metallothionein, are manifestations of physiological processes which may differ greatly between organisms. An important physicochemical property of some heavy metals which influences bioconcentration is the similarity of these metals to essential ions, like that of cadmium to calcium. For metals there is no clear relationship between a physicochemical parameter and either the uptake rate constant, the elimination rate constant, or the BCF. However, inverse relationships occur between BCF or BAF and metal exposure concentration for essential and non-essential metals [23]. This not only complicates the theoretical aspect of using BCF/BAF values as an intrinsic property of a substance, but also results in elevated variability when data are compiled. BCFs determined from natural conditions, which are characterized by low exposure concentrations, can be as high as 300,000, but are generally meaningless in the context of evaluating potential for toxicity in relation to environmental hazard [23]. In addition, many aquatic organisms are able to regulate internal metal concentrations through active regulation, storage or combinations thereof [23,24]. Factors that influence metal uptake and bioaccumulation act at almost every level of abiotic and biotic complexity, including: water geochemistry; membrane function; vascular and intercellular transfer mechanisms; and intracellular matrices. In addition, physiological processes (usually renal, biliary or branchial) generally control elimination and detoxification processes. Storage adds additional controls on steady-state concentrations within the organism. When metal bioaccumulation is predominantly via mechanisms that demonstrate saturable uptake kinetics, BCFs will thus decline at higher exposure concentrations.

Bioconcentration models

Models are used to describe and predict bioconcentration. They serve to mathematically describe the increase or decrease in the concentration of xenobiotics in an organism. Simple models regard an organism as one homogenous compartment and the surrounding medium as another: the two-compartment model. In addition, rate constants are assumed to be first-order rate constants, independent of the concentration of the chemicals. More complicated models may regard the surrounding medium and the organism as different compartments, and involve different order rate constants.

The one-compartment model

For organic chemicals, bioconcentration is usually described by the exchange of a chemical from water to the organism and vice versa. Therefore, in theory, a two-compartment model with first-order kinetics can be applied. However, since the concentration of the chemical in the water is not influenced by the organism the use of a one-compartment model can be justified from a mathematical point of view [25]. In this model,

Table 3.2. Uptake rate constants of xenobiotics in various aquatic organisms [25,26].

Compound	Species	Uptake rate constant (L/(kg·d))
Metals		
Chromium	trout	0.12 - 0.5
Cadmium	trout	0.003 - 0.12
Cadmium + 0.1 mM EDTA	trout	< 0.015
Cadmium + 1 mM Citrate	trout	3
Cadmium + 0.1 mM Potassiumethylxanthate	trout	0.3
Organic chemicals		
Phenol	trout	20-50
Halogenated phenols	trout	200-450
Polychlorinated biphenyls	trout	200-450
Polychlorinated benzenes	trout	200-450
Organometals		
Triphenyltin	trout	0.1 - 5
Tributyltin	trout	4 - 30
Tributyltin	oyster	75 - 1000
Tributyltin	mussel	70 - 17,290
Tributyltin	clam	250
Tributyltin	amphipod	70 - 1230
Tributyltin	snail	1.8 - 9.5
Tributyltin	crab	0.11 - 1000

the exchange of a compound thus takes place between water and the organism:

$$\text{water} \xrightarrow[k_w]{\text{uptake}} \text{organism} \xrightarrow[k_e]{\text{elimination}} \text{surrounding medium}$$

The increase or decrease in the concentration of a xenobiotic in an aquatic organism over time is described by Equation 3.30:

$$dC_o / dt = k_w C_w - k_e C_o \qquad (3.30)$$

where

C_o = the concentration of the chemical in the organism (mol/kg)

C_w = the concentration of the chemical in water (mol/L)

k_w = the uptake rate constant from water (L/(kg·d))

k_e = the overall elimination rate constant (1/d).

Table 3.2 shows the uptake rate constants of different chemicals. While the uptake rate constant of hydrophobic chemicals, such as halogenated benzenes, biphenyls and phenols, is approximately constant within one species, those of metals and organometals may differ widely, and depend on environmental conditions, such as the presence of hydrophilic (citrate) or hydrophobic (xanthate) ligands. The uptake rate constants of metals and other hydrophilic contaminants are usually much lower than those of hydrophobic compounds (Table 3.2). Furthermore, the uptake rate constants of metals may differ by several orders of magnitude under different environmental conditions.

Substances can be eliminated from the organism via different routes, where k_r is elimination via the respiratory surface (gills, skin or lungs for terrestrial organisms) k_f excretion via egested faeces, k_m for metabolic transformation, k_g for pseudo elimination via growth dilution and k_p for elimination via reproductive cells or offspring. The overall elimination rate k_e is therefore the sum of rate constants (1/d) for all major elimination routes:

$$k_e = k_r + k_f + k_m + k_g + k_p \qquad (3.31)$$

A number of elimination rate constants are given in Table 3.3.

Table 3.3. Elimination rate constants of xenobiotics in various aquatic organisms [21,26].

Compound	Species	Elimination rate constant (1/d)
Metals		
Chromium	trout	0.03 - 0.7
Cadmium	trout	0.003
Nickel	trout	0.01
Organic chemicals		
DDT	trout	0.01
Lindane	trout	0.06
Phenol	trout	> 0.06
Chlorophenols	trout	> 0.7
Polychlorinated biphenyls	trout	< 0.0001 - 0.3
Polychlorinated benzenes	trout	< 0.003 - 0.7
Organometals		
Methylmercury	trout	0
Triphenyltin	guppy	0.005 - 0.014

The rate constants k_w and k_e are independent of the concentrations in water and the organism. However, they may be dependent on the organism and on the properties of the compound.

When an organism is continuously exposed to a chemical (C_w = constant), Equation 3.30 is integrated to:

$$C_o\,(t) = (C_w k_w/)k_e\,[1-e^{-k_e t}] \tag{3.32}$$

If the exposure concentration in water varies with time, numerical solutions may be applied to solve Equation 3.30.

The uptake rate constant can be derived from the initial uptake of the chemical by the organism, when elimination is assumed to be negligible:

$$C_o = k_w C_w t \tag{3.33}$$

After long exposure times (t $\rightarrow \infty$), the term $e^{-k_e t}$ in Equation 3.32 approaches zero, and a steady-state will be achieved ($dC_o/dt = 0$). Subsequently, the bioconcentration factor (BCF) can be determined:

$$\text{BCF} = C_o\,/\,C_w = k_w\,/\,k_e \tag{3.34}$$

The ratio of the concentration of the chemical in fish and water $C_o\,/\,C_w$ only represents the bioconcentration at a steady-state. If the concentrations in fish (C_o) and water (C_w) are determined before the steady-state has been attained, the ratio $C_o\,/\,C_w$ will underestimate the BCF. However, when the ratio is determined in a situation where the concentration in water has decreased faster than the concentration in the organism, the ratio will overestimate the BCF.

In the environment, organisms may be exposed to chemicals only for short periods of time. When exposure stops and the concentration of the chemical in water decreases or reaches zero, the chemical will be eliminated from the organism. The rate of elimination is usually determined under laboratory conditions. Mathematically, elimination rate constants are determined with Equation 3.30 provided that $C_w = 0$, which will result in a decrease in the concentration in the organism. Integration of Equation 3.30, gives Equation 3.35:

$$C_o\,(t) = C_o\,(t{=}0)\,e^{-k_e t} \tag{3.35}$$

where C_o (t=0) is the concentration in the organism at the start of the elimination period (mol/kg).

The biological half-life ($t_{1/2}$) of a compound can be derived from the elimination rate constant, and is the time required to reduce the concentration of a compound in the organism to half its original value. Hence, when C_o ($t_{1/2}$) = 1/2 C_o (t=0) is substituted in Equation 3.35, this leads to:

$$t_{1/2} = (\ln 2)/k_e \tag{3.36}$$

Figure 3.16 shows how uptake and elimination rate constants are derived. Figure 3.17 shows the relationships

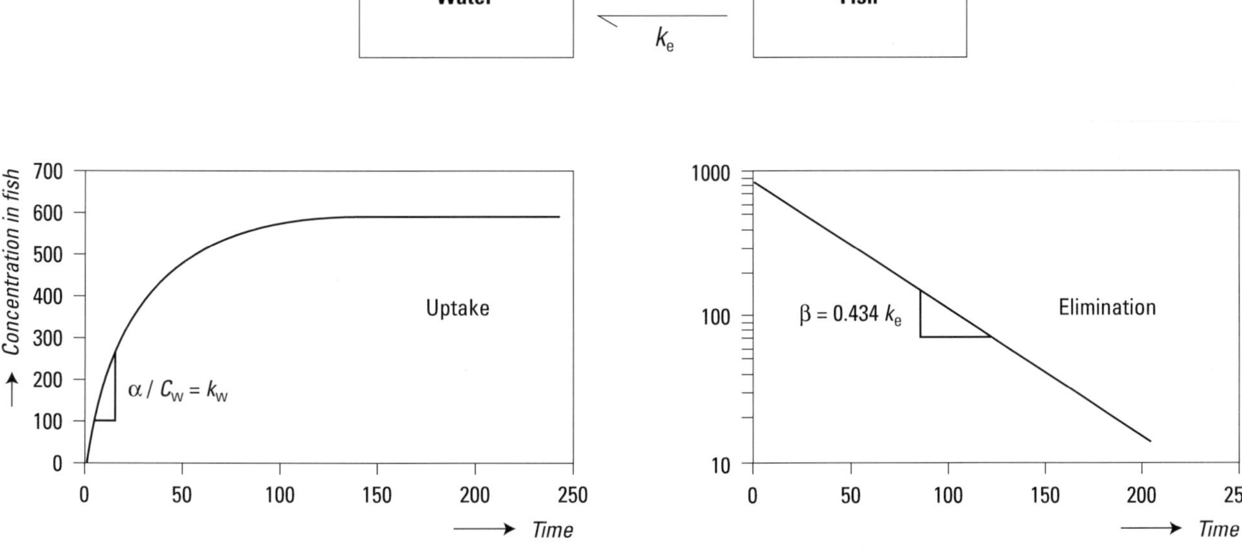

Figure 3.16. Hypothetical curves for a first-order one-compartment bioaccumulation model, in which k_w is the uptake rate constant and k_e is the elimination rate constant. α is the slope from which k_w is determined, β is the slope from which k_e is determined. From [25]. With permission.

of k_w, k_e, and BCF with hydrophobicity (K_{ow}) for organic chemicals. Uptake rate constants increase with K_{ow}, and become constant for hydrophobic chemicals with log K_{ow} > 3-4. Elimination rate constants are constant for hydrophilic chemicals and decrease with K_{ow} for chemicals with log K_{ow} > 3-4. Since the BCF is the ratio of the uptake and elimination rate constants, BCF increases with K_{ow} for all hydrophobic chemicals.

The uptake of chemicals by aquatic organisms from water usually occurs via the respiratory surfaces. Since larger organisms usually have a relatively smaller respiratory surface than smaller organisms, it has been shown for fish that the uptake rate constant for different weight classes depends largely on the size of the fish. The rationale for this is that larger organisms usually require less oxygen per unit of volume for metabolic processes. Since the exchange of chemicals is related to the exchange surface according to Fick's law, this implies that small organisms will both take up and eliminate chemicals faster than large aquatic organisms. Uptake rate constants for hydrophobic chemicals in guppy (0.1 g) are usually around 1000 L/(kg·d), while those in large rainbow trout (750 g) are around 50 L/(kg·d). The following allometric relationship between fish weight (W in g) and uptake rate constant was derived for hydrophobic organic chemicals with a log K_{ow} > 3 [28]:

$$k_w = (550 \pm 16) W^{-0.27 \pm 0.05} \qquad (3.37)$$

Both uptake and elimination rate constants are thus (allometric) functions of the weight of an organism [26,28-31].

For metals there is no clear relationship between a physicochemical parameter and either the uptake rate constant, the elimination rate constant, or the BCF.

Although the accumulation of metals does not necessarily take place by passive diffusion, the first-order kinetic model can be successfully applied to describe the uptake and elimination kinetics of metals. However, a steady-state is not always observed for metals. Due to a very high storage capacity of metallothionein for instance, continuous uptake of metals may occur, resulting in ever increasing concentrations in aquatic organisms.

The chemical speciation of metals greatly affects bioconcentration and largely depends on environmental properties such as pH, salinity, oxygen concentration, and dissolved organic carbon, among other things (Figure 3.18). The aqueous concentration of the free ion can be predicted from these properties. Complex ligands, such as hydroxyl and carbonate ions, play a prominent role in regulating speciation (Figure 3.18). Bioaccumulation can thus be predicted based on the free-ion concentration. Complexation of metals with natural humic and fulvic

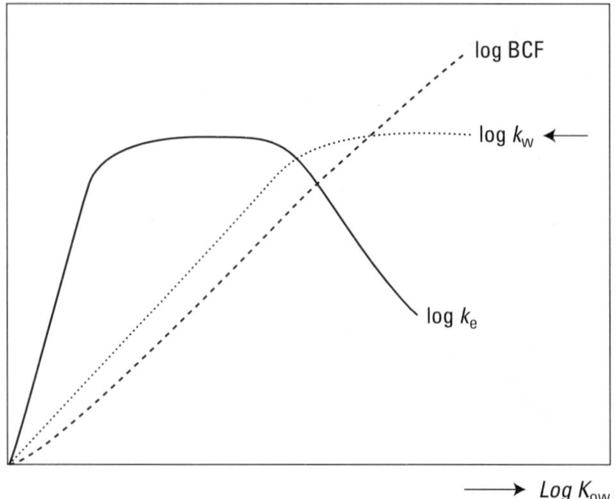

Figure 3.17. The relationship of k_w, k_e and BCF with hydrophobicity (K_{ow}) for organic chemicals. From [27]. Copyright ©1986. Reprinted by permission of Alliance Communications Group, Allen Press, Inc.

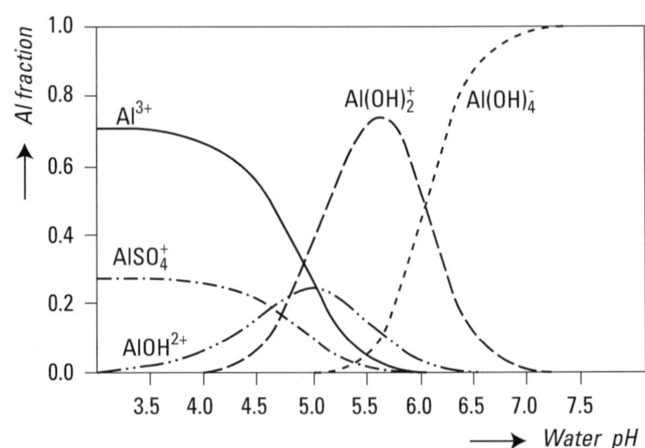

Figure 3.18. The chemical speciation of aluminium (Al) is influenced by salinity, pH and ligand. The pH-dependent activities of the different Al forms in the surrounding water have been plotted. From [32]. With permission.

substances generally reduces the uptake of the metal [33]. In some cases, however, the uptake rate constant of the metal complex may be higher than the free ion, for instance, when the complex is more hydrophobic than the metal.

Multiple-compartment models

In some cases, a one-compartment model cannot sufficiently describe bioconcentration. Usually, this occurs when there are two or more stages in which elimination rates differ (Figure 3.19). The simplest form of a multiple-compartment model is when the organism occupies not one, but two-compartments, each with its own bioconcentration kinetics. The result is an initially fast and later slow elimination rate of the chemical from the entire organism (Figure 3.19). The rationale for a two-compartment organism (two-compartment model) is, for example, that one compartment quickly releases the xenobiotics when in contact with a clean surrounding medium, while the second compartment only slowly releases the chemicals to the first compartment, which in turn quickly eliminates them to the medium.

An example of the mathematical description of the elimination kinetics of a two-compartment model is given in Equation 3.38:

$$C_o = Ae^{-\alpha t} + Be^{-\beta t} \qquad (3.38)$$

where A and B are constants (mol/kg) and α and β represent kinetic rate constants (1/d).

More complicated modelling approaches exist, such as the physiologically-based pharmacokinetic (PBPK) models (Section 6.3.2), which make use of the blood to organ distribution coefficients of chemicals, the size and constitution of organs, and blood perfusion rates through organs, are much more complicated.

Methods for measuring bioconcentration

While the literature on the bioconcentration of xenobiotics in aquatic organisms is extensive, few standard methods have been developed. The Organization for Economic Cooperation and Development (OECD) has produced standard protocols for measuring bioconcentration in fish [34]. These test guidelines, which employ different species and test conditions, are summarized in Table 3.4. No aquatic organisms other than fish are used in existing OECD protocols. This should be particularly useful for metals, since the bioconcentration of these chemicals in invertebrates and molluscs is usually greater than in fish.

The American Society for Testing and Materials (ASTM) has also published a procedure for identifying bioconcentration in fish and marine molluscs [35], which is very similar to that of the OECD. The main difference is that the ASTM stipulates that exposure should continue until an apparent steady-state has been reached. If a steady-state is not obtained, the observed 28-d BCF may be taken as an apparent BCF, while the OECD procedure derives the k_w/k_e ratio (Equation 3.34).

The US Environmental Protection Agency has adopted procedures for identifying bioconcentration in

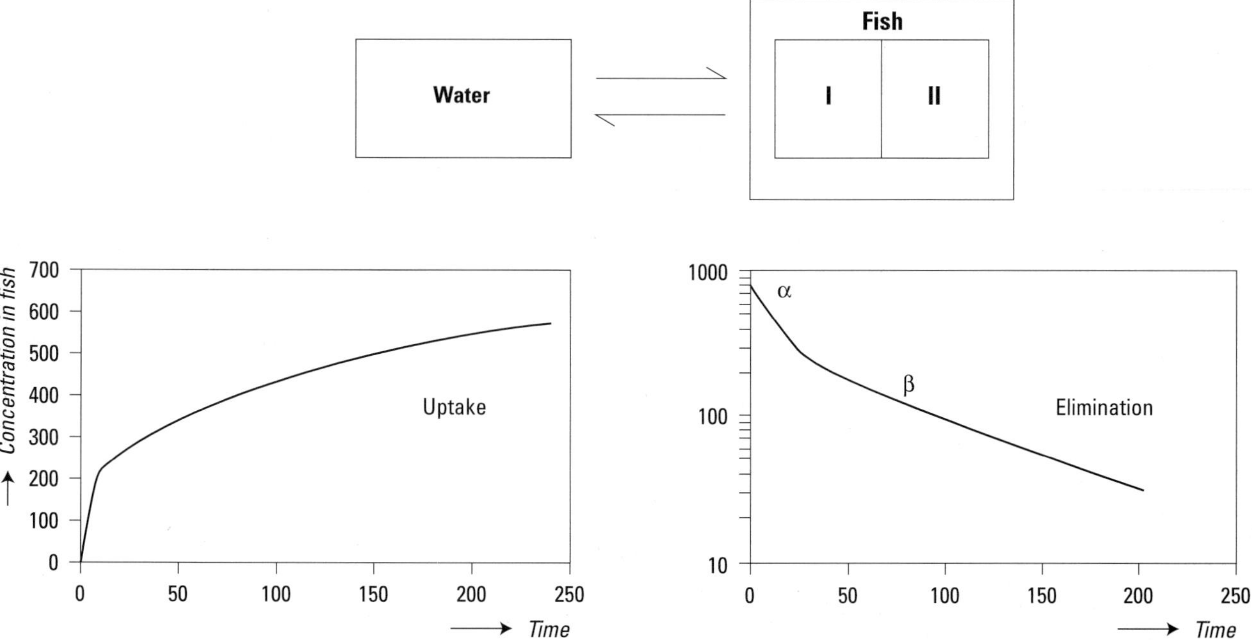

Figure 3.19. Biphasic uptake and elimination as an example of a two-compartment bioaccumulation model; α is the slope of the initial, fast elimination period, β is the slope of the slower elimination period. From [25]. With permission.

fish [36] and oysters [37]. These procedures include a flow-through technique and are suitable for both organic and inorganic compounds.

3.3.3 Factors affecting bioconcentration

Modelling bioconcentration is regarded as a relatively simple process. In well-defined examples, a simple first-order one-compartment model can be applied to describe and predict BCFs and bioconcentration kinetics. Many chemicals, however, do not follow these simple rules; moreover, bioconcentration (kinetics) may be species dependent. Chemical and biological aspects may thus modify bioconcentration. The following factors are important: molecular weight, molecular size, molecular charge, speciation, surface/volume ratios, morphology, and biotransformation. These factors will be discussed separately below.

Chemical aspects may influence bioconcentration by affecting the membrane passage properties of the chemical and its bioavailability, i.e., the freely dissolved chemical in the aqueous phase. The major biological aspects influencing bioconcentration are: bioconcentration kinetics (surface/volume ratios and morphology), and the rate and extent to which chemicals are biotransformed.

Molecular weight

Several values have been suggested for the molecular weight (MW) cut-off value above which absorption across fish tissues becomes negligible. The EU TGD [38] indicates that molecules with a MW greater than 700 g/mol are less likely to be absorbed and bioconcentrate, whereas the US EPA exempts chemicals with a MW of above 1100 g/mol in the PBT assessment conducted under the Toxic Substances Control Act (USEPA 1999). Anliker et al. [40] suggested that a pigment could be excluded from a fish bioaccumulation test if it has both a MW of greater than 450 and a cross-sectional diameter of over 1.05 nm (as the second smallest van der Waals diameter). Rekker et al. [41] suggested that a calculated log K_{ow} of > 8 can be used on its own, or in combination with a molecular weight of > 700-1000 to conclude (with confidence) that the compound is unlikely to bioaccumulate. While there has been limited experimental evidence for a MW cut-off, Burreau et al. [42] did demonstrate reduced bioconcentration and no biomagnification for high MW polybrominated diphenyl ethers, with 6 or more bromines, their MW ranging from 644 to 959 g/mol. Considering that molecular size and shape can vary considerably for substances with similar MW, molecular weight alone is insufficient to allow absorption predictions. However, it does suggest that once the MW is in the region of 700-1100, depending on

Table 3.4. OECD test guidelines for measuring bioconcentration in aquatic organisms [34].

OECD Guideline	305A Sequential static fish test	305B Semi-static fish test	305C Test for the degree of bioconcentration in fish	305D Static fish test	305E Flow-through fish test
Recommended species	catfish, zebrafish, carp	zebrafish	yearling carp	guppy, zebrafish	rainbow trout, sheepshead minnow, bluegill, fathead minnow, spot, silverside, shiner, perch, English sole, staghorn, sculpin, three-spined stickleback
Supply of test water	static	semi-static	flow-through	static	flow-through
Concentration of test water	< 0.1 LC50 > 3 levels	< 0.02 LC50 > 1 level	< 0.01 and < 0.001 LC50, 2 levels	< 0.01 and < 0.001 LC50 2 levels	< 0.02 LC50
Carrier of test substance	ethanol or acetone (< 0.5 ml/L)	acetone (25 ml/L)	recommended solvents and surfactants	dimethyl-sulfoxide *t*-butanol (< 0.1 ml/L)	recommended solvents (< 0.1 ml/L)
Test period					
- uptake	± 2 weeks	2 or 4 weeks	8 weeks	8 d	8 h - 90 d
- steady-state	mandatory	optional	mandatory	mandatory	mandatory
- elimination	mandatory	mandatory	mandatory	mandatory	optional
Dilution water	artificial	artificial	well water or city water pretreated with activated carbon	well water or artificial	test organisms can live in it
Biomass (g/L)	< 1	< 0.8	< 8	< 0.4	< 15
Sampling frequency					
- water	1 L	7 levels	> 16 levels	> 12	28
- fish	19	7 levels	8 levels	> 12	9
Measurement of lipid content	mandatory	optional	optional	mandatory	optional
BCF	C_{fish}/C_w at steady-state	C_{fish}/C_w at steady-state	C_{fish}/C_w at steady-state	C_{fish}/C_w at steady-state	k_w/k_e at 80% steady-state

other factors, a reduced BCF may be expected. Hence, while recognizing the uncertainties in the interpretation of experimental results, de Wolf et al. [43] recommended that to demonstrate a reduced BCF a substance should have either:

- a MW in excess of 1100 g/mol,
- or a MW of 700 – 1100 g/mol with other indicators (see later discussion).

Molecular size

Molecular size deals with the dimensional properties of chemicals together with their potential transport across biological membranes. Since bioconcentration starts with the transport from the bulk water to the respiratory surface and subsequently follows uptake of chemicals across a bilipid membrane (Figure 3.14), the molecular size of a chemical is very important in determining whether it will be able to be transported across this membrane.

Molecular size may be considered as a more refined approach, specifically taking into account molecular shape and flexibility, rather than relying on MW alone. For some hydrophobic chemicals, such as hexabromobenzene, octachloronaphthalene,

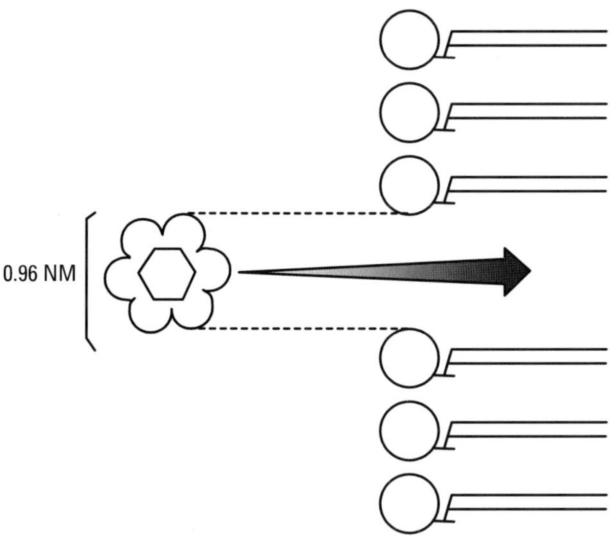

Figure 3.20. Diagram showing the transfer of a hydrophobic molecule across the polar heads of a bilipid membrane in relation to the effective cross-section of the membrane's cavity for neutral organic chemicals. Reprinted from [27]. Copyright ©1986. Reprinted by permission of Alliance Communications Group, Allen Press, Inc.

octachlorodibenzo-*p*-dioxin, decabromobiphenyl, disperse dyestuffs, organic pigments, a fluorescent whitening agent and azopigments, no bioconcentration in guppy was observed when exposed in water [27,40,44,45,46]. This absence of bioconcentration was assumed to be due to the size of the molecules, which may have prevented them from penetrating the gill membrane, and for some chemicals, due to their limited solubility in n-octanol (see below). To permeate the polar surface of the membrane the molecule must be small enough to pass through "holes" in the lipid membrane (Figure 3.20). In guppy, the critical cross-sectional diameter is 0.95 nm, above which little or no uptake occurs. In other fish, however, such as rainbow trout and goldfish, the uptake of some bigger molecules has been observed. Hence, species differences may influence the uptake of big molecules due to the composition of the membrane [47,48], and a simple parameter may not be sufficient to explain when reduced uptake occurs. Dimitrov et al. [49] have tried to develop a more mechanistic approach to address this concept, using MW, size, and flexibility in their BCF estimates. They found that for compounds with a log $K_{ow} > 5.0$, a threshold value of 1.5 nm for the maximum cross-sectional diameter (i.e., molecular length) could discriminate between chemicals with BCF > 2,000 from those with BCF < 2,000. This critical value was found to be comparable with the architecture of the cell membrane, i.e., half the thickness of the lipid bilayer of a cell membrane. This is consistent with a possible switch in uptake mechanism from passive diffusion through the bilayer to facilitated diffusion or active transport. Dimitrov et al. [50] and Dimitrov et al. [51] later used this parameter to assess experimental data on a wide range of chemicals. The conclusion was that a chemical with maximum cross-sectional diameter over 1.74 nm would not have a BCF > 5,000, and would not meet the European Union PBT criteria for vB (very Bioaccumulative) chemicals [38].

In other studies, accumulation has been shown not to occur with hydrophobic chemicals whose length exceeds 4.3 nm. This has been found for linear polydimethylsiloxanes in fish from water [52] and *n*-alkanes in rats from food. Limited bioaccumulation was observed for alkanes larger than $C_{27}H_{56}$. This critical length of 4.3 nm corresponds to the average distance between the polar heads in a bilipid layer of a cell membrane. The length of the polydimethylsiloxanes is also very close to the length of the bilipid layer (Figure 3.21). Molecular weight did not explain reduced uptake, since one of the silicone substances with a molecular weight of 1,050 was detected in fish. Tolls et al. [53] did observe uptake in fish of some non-ionic surfactants with an apparent equal length to long chain alkanes, which seems contradictory to the earlier proposed cut-off molecular length by Opperhuizen et al. [47]. However, the uptake of the long non-ionic surfactants may be explained by internal molecular flexibility reducing the effective molecular length below 4.3 nm. In conclusion, there would appear to be no clear cut-off value for molecular size beyond which no absorption will take place. While recognizing the uncertainties in the interpretation of experimental results, de Wolf et al. [43] recommended:

- A maximum effective molecular length of 4.3 nm indicates no uptake and indicates that a chemical is not bioconcentrating.
- A maximum cross-sectional diameter of 1.74 nm indicates that a chemical would not have a BCF > 5,000.
- A maximum cross-sectional diameter of 1.74 nm plus a MW of 700 – 1,100 would suggest that a chemical will not have a BCF > 2,000.

Steric factors thus seem to influence the transport of large chemicals across membranes.

Lipinksi's rule of 5

Lipinski et al. [54] identified five physical chemical characteristics that influence solubility and absorption

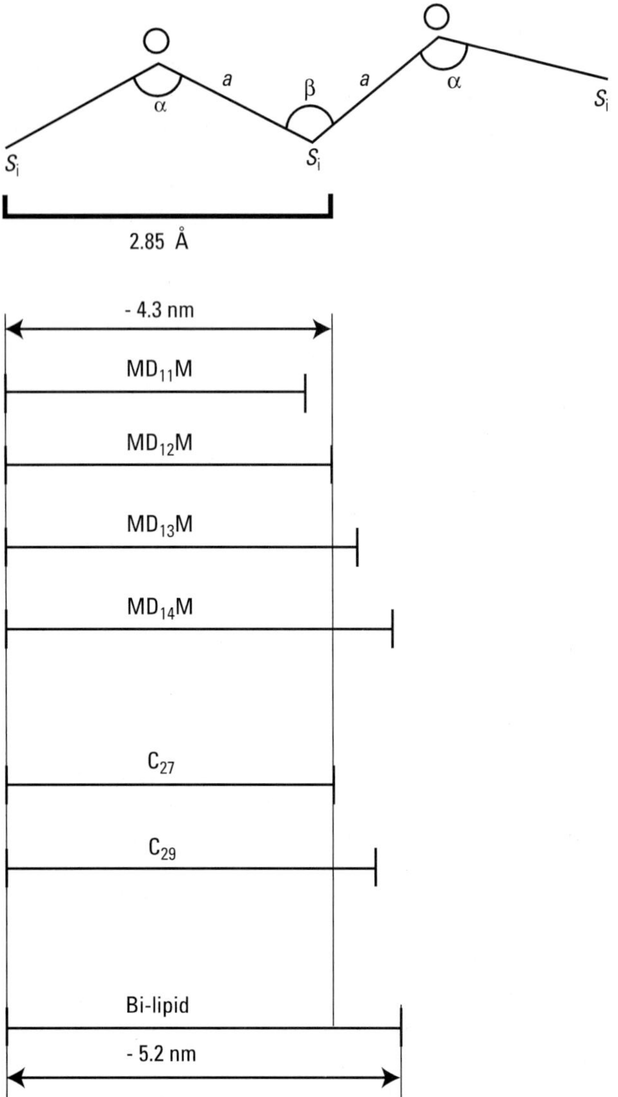

Figure 3.21. Relationship between the molecular length of n-alkanes and polydimethylsiloxanes (PDMS) for the membrane permeation of neutral organic chemicals. MD_nM refers to polydimethylsiloxanes, where n refers to the number of dimethylsiloxanes units. C_m refers to linear alkanes where m refers to the number of methylene units. Top: length of a Si-O-Si fragment of PDMS oligomers. Bottom: lengths of linear PDMS, linear alkanes, and the thickness of a bilipid membrane. From [47]. Copyright ©1987. Reprinted by permission of Alliance Communications Group, Allen Press, Inc.

across the mammalian intestinal lumen using more than 2,200 drug development tests. These characteristics have been rigorously reviewed [55,56] and are used to develop commercial models to estimate absorption in mammals. They are also commonly used by the human and veterinary pharmaceutical industry. Organic chemical absorption is, however, not similar in all vertebrates, since mammalian dietary absorption rates for the same chemical may differ by more than two orders of magnitude between humans and ruminants. This is likely not due to the membrane properties, but rather other characteristics of the digestive process controlling the absorption. "Lipinksi's Rule of 5" may be extrapolated from the mammalian intestinal membrane to fish gills, which in turn would account for the prediction of poor solubility and poor absorption from chemical structure. A chemical is then not likely to cross a biological membrane in quantities sufficient to exert a pharmacological or toxic response when it has more than 5 Hydrogen (H-)bond donors, 10 H-bond acceptors, a MW greater than 500, and a log K_{ow} value greater than 5 [54]. Wenlock et al. [55] studied about 600 additional chemicals and found that 90% of the absorbed compounds had fewer than 4 Hydrogen (H)-bond donors, < 7 H-bond acceptors, MW less than 473, and a log K_{ow} value less than 4.3. More recent work by Vieth et al. [57] and Proudfoot [56] supports these lower numbers. Molecular charge and the number of rotational bonds will also affect absorption by passive diffusion across a membrane or diffusion between cells. The "leakiness" of a tissue, or its ability to allow a chemical to passively diffuse through it, is measured using trans-epithelial electrical resistance (TEER) and can be used to compare tissue capabilities. A low TEER value indicates the tissue has greater absorption potential. Although the studies by Lipinski et al. [54], Wenlock et al. [55], Vieth et al. [57] and Proudfoot [56] focused on absorption across the intestinal lumen, the more restrictive TEER for fish gills [43] implies that the equations and concepts can be re-applied to conservatively estimated absorption in fish.

Other indicators for low uptake

There are other indicators for low uptake that could also be used to suggest that a chemical, despite having a log K_{ow} in excess of 4.5, has a low bioconcentration potential such as lack of experimentally observed gill or skin permeability, and low or reduced uptake in mammalian studies. Cell culture models or perfused gill preparations offer many advantageous features for the analysis of chemical transport across membranes and can be used to expedite identification of compounds with less favourable uptake properties, and to evaluate structure-absorption relationships [e.g., 28,58-62]. Both these systems show relatively high variation, however, this can be significantly reduced and the uptake rate constants determined once they are normalized with a reference chemical.

Limited fat or octanol solubility

The concept of having a value relating a chemicals' solubility in fat or octanol to reduced uptake is derived from two considerations. Firstly, that octanol is a reasonable surrogate for fish lipids, and secondly, that if a substance has a reduced solubility in octanol this may result in a reduced uptake. The former forms the basis of the majority of models for predicting BCF using log K_{ow}. In general, a hydrophobic substance has a low aqueous solubility (S_w), a high solubility in octanol (S_{oct}) and a high BCF. However, when a hydrophobic substance has a low solubility in fat or octanol, the resulting ratio S_{oct}/S_w could range from very low to very high, with no clear idea of how this would affect the magnitude of the BCF. Still, it could be argued that a very low solubility in octanol could be used as an indication that only low body burdens build up in an aquatic organism. Chessells et al. [63] demonstrated a decrease in lipid solubility with increasing K_{ow} values for highly hydrophobic compounds (log $K_{ow} > 6$). It was suggested that this led to reduced BCFs. Banerjee and Baughman [64] demonstrated that by introducing a term for lowered octanol/lipid solubility into the calculated log K_{ow} BCF relationship, they could significantly improve the prediction of bioconcentration for highly hydrophobic chemicals. Experimental K_{ow} values already reflect the lower octanol solubility.

Morphology

The rates of uptake and elimination of xenobiotic compounds are also affected by the morphology of organisms. For instance, in their larval stages midge loose their skin several times during growth. Contaminants attached to the skin will thus be actively removed from the organism.

Uptake of chemicals, as well as oxygen, occurs through the skin of aquatic organisms, such as fish. Both the composition and thickness of the skin as well as the surface area of the skin, compared with the gills, explain the low uptake rates of xenobiotic compounds through the skin compared with the gills.

Biotransformation

Biotransformation (Section 3.6) is one of the processes which decreases the concentration of a parent chemical in an organism. In general, it transforms the chemical to more polar products [65]. In bioaccumulation models, biotransformation is treated as an elimination process, alongside elimination through physicochemical processes, growth dilution, excretion by lactation and reproduction.

Biotransformation only takes place after the chemical has been transported to a site where it can be transformed through enzymatic catalytic action. In this process, the compound must reach the enzyme and then bind with it. Consequently, both transport rate or internal distribution and the capacity of the enzyme to bind and biotransform the chemical will determine the biotransformation rate. In addition, the enzyme requires cofactors to enable the transformation.

Species differ widely in their capacity for biotransformation, which largely depends on the presence or absence and specific activity of enzymes (Section 3.6).

3.3.4 Biomagnification

When the concentration of a chemical becomes higher in the organism than in its food (and the major uptake route is food) this is called biomagnification. Biomagnification is usually important only for chemicals reaching relatively high concentrations in food compared to very low concentrations in other surrounding media, such as water for aquatic organisms, air for terrestrial organisms and soil and sediment for soil and benthic organisms. In this section the uptake from food, sediment and multiple media will be discussed, together with methods for measuring biomagnification.

Uptake from food

Uptake from food occurs in the gastrointestinal tract (GIT). After release of the contaminants in the GIT lumen, the chemicals may cross the lipid membranes by the same mechanisms as described above (Figure 3.14). Food digestion is the key process that leads to a positive thermodynamic gradient between the gut content and the organism [66], which is responsible for biomagnification.

Biomagnification, uptake from food and elimination to the surrounding medium can be modelled in a similar way to bioconcentration (Equation 3.30):

$$\text{food} \xrightarrow{k_f} \text{organism} \xrightarrow{k_e} \text{surrounding medium}$$

where k_f is the uptake rate constant from food (kg/kg$_{bw}$·d), which can be expressed as the product of the uptake efficiency from food, E_f, and the feeding rate f (kg$_{food}$/kg$_{bw}$·d). Biomagnification can thus be described mathematically as:

$$dC_o / dt = E_f \cdot f \cdot C_{food} - k_e C_o \quad (3.40)$$

Table 3.5. Dietary uptake efficiencies (Ef) of PCBs in fish [68]. With permission. Copyright Elsevier.

Compound	C_{food} (μg/g)	Species	E_f (%)
Biphenyl:			
Dichloro-	10	guppy	56
Trichloro-	10	guppy	49-60
Tetrachloro-	1-51	guppy, Coho salmon	10-77
Pentachloro-	1-12	coho salmon	30-73
Hexachloro-	1-50	guppy, Coho salmon	44-81
Octachloro-	50	guppy	31-40
Decachloro-	50	guppy	19-26
Aroclor 1242a	20	channel catfish	73
Aroclor 1254a	15	rainbow trout	68

[a] Aroclor is an industrial PCB mixture, in which 12 refers to the biphenyl molecule, and 42 and 54 refer to the percentage of chlorination.

where C_{food} is the concentration in the food (mol/ kg_{food}). It must be recognized that k_e is again an overall elimination rate constant (see Equation 3.31). When the contaminant concentration is constant (C_{food} = constant) and the feeding rate is also constant, Equation 3.40 can be solved. However, f may depend on the biological species and life stage. Poikilothermic organisms, in general, have lower feeding rates than homoeothermic organisms.

When f is known and constant, Equation 3.40 can be solved:

$$C_o(t) = (E_f \cdot f \cdot C_{food}) / k_e \cdot [1 - e^{-k_e t}] \tag{3.41}$$

which is similar to Equation 3.32 for uptake from water [26,67]. Some dietary uptake efficiencies, E_f, for a number of individual PCB congeners and commercial PCB mixtures in fish are given in Table 3.5. The feeding rate constant f is approximately 0.02 - 0.05 $kg_{food}/kg_{bw} \cdot d$ for fish.

Consequently, for exposure via food a biomagnification factor (BMF) can be derived for steady-state conditions, as shown in Equation 3.42:

$$BMF = E_f \cdot f / k_e = C_o / C_{food} \tag{3.42}$$

Uptake from sediment

Some aquatic organisms, such as many aquatic invertebrates, are sediment-dwelling organisms or deposit feeders. They are able to digest sediment or detritus, which serves as a food source. Uptake from sediment may be significant for these organisms.

Deposit feeders show a wide variety of feeding types. Surface deposit feeders, such as the clam *Macoma*, feed primarily on the upper few millimetres of sediment. "Conveyor belt" species ingest particles as deep as 20 to 30 cm below the surface (Figure 3.22). In sediments with a distinct vertical concentration gradient, these organisms would be exposed to substantially different pollutant concentrations than surface feeders.

The concentration of contaminants measured in sediment does not always reflect the exposure of the organisms to xenobiotics. Most deposit feeders selectively ingest the finer particles which contain higher amounts of organic carbon, while they discard the larger particles. This behaviour can concentrate the organic content of the ingested sediment by more than one order of magnitude compared to that of the original sediment. As a result of selective feeding, the pollutant concentration measured in the original sediment may underestimate the actual dose ingested by selective deposit feeders.

In addition, concentrations of contaminants in the interstitial water may differ from concentrations in the overlying water. Surface deposit-feeding bivalves, such as the clam *Macoma* sp., ventilate an insignificant amount of interstitial water, but ventilate large amounts of overlying water. Free-burrowing amphipods and polychaetes, however, ventilate interstitial water almost exclusively while buried in the sediment. Many bivalves are filter feeders, unlike the clam *Macoma* which is a deposit feeder. Filter feeders use their gills to ventilate large amounts of water. The organic carbon is filtered from the water and used as a food source. As the organic carbon often contains large amounts of pollutants, this can provide an important route of uptake for these species.

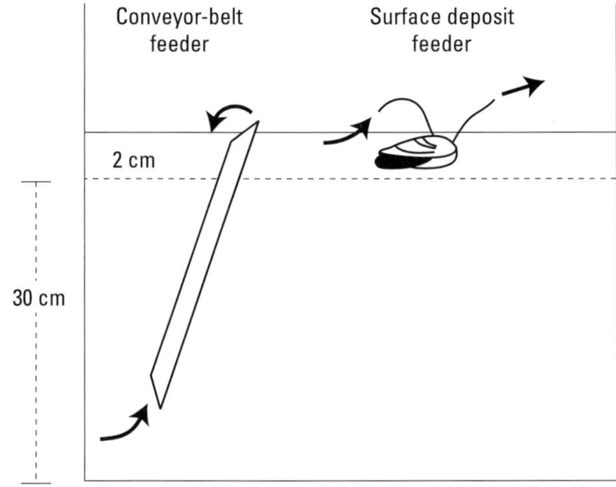

Figure 3.22. The effect of feeding depth on pollutant exposure. The conveyor-belt and surface deposit feeding modes illustrate the range in feeding depth by deposit feeders. From [69]. With permission.

Species differences thus result in the uptake of contaminants from different sources: surface and deeper sediment, interstitial and overlying water.

Other benthic organisms often studied are the larvae of the midge (*Chironomus* sp.). Midges go through several larval stages in the development from egg to adult. The larval stages last from a few days to several months in sediment. Midges connect the aqueous and terrestrial food web, since the larvae are a food source for invertebrates and fish, while the adults provide a food source for birds. Midge larvae feed on organic material in the sediment (see Chapter 7, Sections 7.4 and 7.5). The uptake of xenobiotics takes place predominantly via interstitial water, as in the case of worms.

Multimedia uptake from water, food and sediment

Xenobiotic compounds can be taken up by aquatic organisms from water, food or sediment. The most important route of uptake depends on the physicochemical properties of the compound as well as on the habitat, the diet and the physiological properties of the organism.

To be able to address the most significant contribution of each of the three routes, information on the mechanisms and kinetics of the various uptake processes is required. The first-order bioaccumulation model provides a helpful tool for this. The three uptake routes are shown in Figure 3.23.

In Figure 3.23 k_s is the uptake rate constant for chemicals from sediment ($kg_{sediment}/kg_{bw}\cdot d$), which is derived in a similar manner to the uptake from food. Each uptake rate constant k_w, k_f and k_s can be substituted by the product of uptake efficiency (E_w, E_f and E_s) and the flows of water (V_w) passing through the gills, food through the GIT (*f*) and sediment through the GIT (*S*) of the organism [70]:

$$k_w = V_w E_w \tag{3.43}$$

$$k_f = f \cdot E_f \tag{3.44}$$

$$k_s = S \cdot E_s \tag{3.45}$$

Hence, the change in concentration of the chemical in the organism can be described by Equation 3.46:

see below (3.46)

Kinetic rate constants have been derived for several aquatic organisms, such as guppy, rainbow trout and clams.

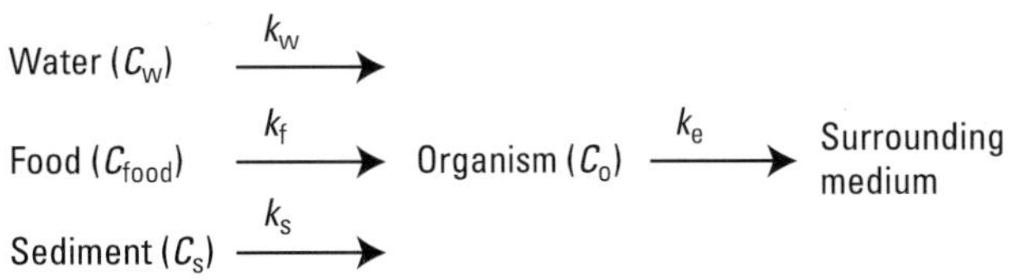

Water (C_w) $\xrightarrow{k_w}$
Food (C_{food}) $\xrightarrow{k_f}$ Organism (C_o) $\xrightarrow{k_e}$ Surrounding medium
Sediment (C_s) $\xrightarrow{k_s}$

Figure 3.23. Comparison of three uptake routes, water, sediment and food, and elimination to the surrounding medium.

Uptake from water

For several classes of organic compounds, uptake rate constants in small fish (< 1 g) are approximately 1000 L/(kg·d). The flow of water ventilated across the gills is approximately 2000 L/(kg·d). Hence, the uptake efficiency of chemicals from water (E_w) is approximately 50%. For larger fish almost the same extraction efficiency has been reported. Larger fish, however, have lower uptake rate constants due to lower ventilation rates. For example, fish of 100 g have uptake rate constants of approximately 100 L/(kg·d).

$$dC_o/dt = (V_w E_w C_w + f \cdot E_f C_{food} + S E_s C_s) - k_e C_o \tag{3.46}$$

Uptake from food

For many aquatic organisms the feeding rate (f), is approximately 0.01-0.05 $kg_{food}/(kg_{bw} \cdot d)$. Table 3.5 shows that uptake efficiencies of hydrophobic chemicals are approximately 50%, which corresponds with the average digestibility of food components.

Uptake from water and food

When comparing the contribution of uptake from water and food for small fish, Equation 3.46 shows that the concentration of the compound in food has to be five orders of magnitude higher than in water, before uptake from food makes a significant contribution to the concentration in small fish. If the food is assumed to be a smaller aquatic organism, the bioconcentration factor of this prey would have to exceed 10^5. Consequently, only extremely hydrophobic chemicals, such as chlorinated biphenyls, naphthalenes or dibenzo-*p*-dioxins with more than three chlorine atoms, will be taken up primarily via the food chain. For chemicals with BCFs lower than 10^5, overall bioaccumulation will therefore mainly take place through uptake from water.

For larger fish, uptake from food contributes more significantly to the total bioaccumulation for chemicals with lower hydrophobicity. Larger fish have significantly lower ventilation volumes than smaller fish, while feeding rates are almost equal, thus favouring the uptake from food for less hydrophobic chemicals. Larger fish also have lower growth rates than smaller fish with corresponding lower overall elimination rates (Equation 3.32).

Uptake from sediment

Information on the uptake of organic chemicals from sediment by aquatic organisms is relatively scarce. It is assumed that the possible significance of uptake from sediment occurs in two extreme situations:

1. Organisms are able to digest sediment, in which event sediment acts as food and the total amount of chemical which is taken up from sediment depends on both the rate of sediment ingestion and the uptake efficiency.
2. Organisms are completely unable to digest sediment, in which event the uptake of the chemical by the organism from sediment will be determined by both the desorption rate constant of the chemical from the sediment and the sediment residence time in the gastrointestinal tract. If it is assumed that the sediment ingestion rate is very high, the desorption rate will probably determine the uptake efficiency.

Uptake from water, food and sediment

Table 3.6 illustrates the flow rates and uptake efficiencies of organic chemicals for the uptake routes water, food and sediment in guppy, rainbow trout and clam. The relative contribution to the uptake of hydrophobic chemicals from food and sediment can be calculated with Equations 3.47 and 3.48 (see below).

where

f	=	feeding rate $(kg_{food} / (kg_{bw} \cdot d))$
E	=	uptake efficiency from food (f), water (w) or sediment (s)
C	=	concentration in food (f; mol/kg), water (w; mol/L) or sediment (s; mol/kg)
V_w	=	flow of water passing through gills $(L/(kg_{bw} \cdot d))$
S	=	amount of sediment passing the GIT $(kg_{sediment} / (kg_{bw} \cdot d))$
BCF	=	bioconcentration factor (L/kg)
K_p	=	soil-water partition coefficient (L/kg).

When the data from Table 3.6 are entered in Equations 3.47 and 3.48, the relative contribution of uptake from food and sediment compared with water can be determined. For guppy, uptake from food becomes important for hydrophobic chemicals with a BCF greater than 100,000; the same applies to rainbow trout for chemicals with a BCF greater than 12,000. Uptake from sediment may be important for chemicals with a K_p greater than 1,500 in the case of guppy, and greater than 1,700 in the case of clam.

Methods for measuring biomagnification

Biomagnification can only be measured when sufficient information is available on the type of food, the amount of food ingested, the uptake efficiency from food as well as excretion processes. A simple model as described

$$(\text{food}) \div (\text{water}) = (f \cdot E_f C_f) \div (V_w E_w C_w) = (f \cdot E_f) \div (V_w E_w / \text{BCF}) \tag{3.47}$$

$$(\text{sediment}) \div (\text{water}) = (S \cdot E_s C_s) \div (V_w E_w C_w) = (S \cdot E_s) \div (V_w E_w / K_p) \tag{3.48}$$

Table 3.6. Flow rates and uptake efficiencies of organic chemicals from water, food and sediment in three aquatic species [69,71][a].

		Guppy	Rainbow trout	Clam
Water	E_w (-)	0.5	0.5	0.65
	V_w (L/(kg$_{bw}$·d))	2000	240	100
Food	E_f (-)	0.5	0.5	
	f (kg$_{food}$/(kg$_{bw}$·d))	0.02	0.02	
Sediment	E_s (-)	0.5[b]		0.38
	S (kg$_{sediment}$/(kg$_{bw}$·d))	1.3		0.1

[a] E is the uptake efficiency from water (w), food (f), or sediment (s), V_w is the rate of water across the gills, f is the rate of food across the gut and S is the rate of sediment across the gut.

[b] Value is assumed to be equal to E_f.

above can be used, but usually organisms have multiple food sources, each having its own specific contaminant concentrations, uptake rates and efficiencies. Recently, a dietary test [73,74] has been developed that exposes fish to chemicals via the diet, and measures both the uptake rate during exposure as well as depuration when fish are transferred to clean food. This approach allows for determination of the elimination half-life, dietary assimilation efficiency and biomagnification factor. It also allows for the determination of the bioconcentration factor when it is assumed that the uptake rate can be derived from allometric relationships. Dietary bioaccumulation tests are practically much easier to conduct for poorly water-soluble substances than the OECD 305 guideline [34], because a higher and more constant exposure to the substance can be administered via the diet than via water. A further advantage is that multiple substances, including mixtures, can be investigated in a single test.

Invertebrate accumulation studies generally involve sediment-dwelling species (such as annelids (oligochaetes) and insects), although molluscs may also be tested. Like the fish dietary test, the spiking of sediment circumvents exposure problems for poorly soluble substances. Several standardized guidelines exist or are in development:

- ASTM E1022-94 describes a method for measuring bioconcentration in saltwater bivalve molluscs using a flow-through technique [35]. A similar test is described in OPPTS 850.1710 [37].
- Proposed OECD test guideline for a bioaccumulation test with benthic oligochaetes [74]. Worms are exposed to the substance by means of spiked (artificial) sediment. The worms are then transferred to clean sediment and allowed to depurate. Results may be expressed as a ratio of the concentration in worms and sediment at steady-state (either as a BAF or BSAF), although the kinetic value is generally preferred. A similar test is described by the ASTM [75].
- The ASTM [75] describes several bioaccumulation tests with spiked sediment using a variety of organisms.

Many of these are based on techniques used in successful studies and expert opinion rather than a specific standard method. Non-standard tests may also be encountered in the scientific literature, involving many species.

Bioaccumulation models

Food chain or food web models can be used to predict bioaccumulation in aquatic and terrestrial organisms [30] and humans [66] These models integrate uptake from water, air and dietary sources such as detritus (water or sediment), plants or animals.

Concentrations in organisms in a food chain can be modelled by linking a set of Equations 3.46 to describe uptake from water and consecutive food sources. The following equations describe concentrations in a food chain consisting of algae (C_A), daphnids (C_D) and fish (C_F). Algae are exposed to water, daphnids are exposed to water and algae, and fish (e.g., carp or bream) are exposed to water, sediment and daphnids:

see next page (3.49)

Rates for ventilation, food uptake, growth and reproductive effort can be estimated based on allometric equations (Equation 3.39). Bearing in mind that the BCF and the BMF can be expressed as ratios of rate constants (Equation 3.34 and 3.42), the steady-state concentrations can be calculated as:

see below (3.50)

where

d_X = proportion of diet item X in the diet of a species [-] with $0 \le d_X \le 1$.

Where species have several dietary sources, a more complex *food web* exists where fluxes between different species can occur simultaneously. Such a model is mathematically very similar to multimedia models as described in Section 4.5. The great advantage of these models is that food webs of any dimension can be described with as many food sources as required, and concentrations in all species can be calculated simultaneously [76]. In general, food web models successfully predict steady-state concentrations of persistent halogenated organic pollutants which are slowly metabolized [77,78]. However, these models are still relatively difficult to use for screening a large number of chemicals.

A different, simpler approach can be taken by estimating the BAF of species at different trophic levels that account for both water and food uptake with empirical regressions [79] or a semi-empirical BAF model [19]. These are calibrated on measured BAF data and calculate a maximum BAF for persistent organic chemicals in selected generic trophic levels (algae, invertebrates and fish). The only required input is the K_{ow} value for the chemical. The main discrepancies between model predictions and measured BAF values often are due to biotransformation of a chemical by the organism (Section 3.6) and to an overestimation of bioavailable concentrations in the water column and in sediment (Section 3.7).

3.3.5 Accumulation in terrestrial plants

In terrestrial ecosystems, plants have the greatest biomass. Understanding bioaccumulation in these primary producers is therefore important. Metal uptake from soil in vegetables has been studied [80], but in the following we will focus on organic contaminants. When exploring xenobiotic uptake in plants, it is useful to distinguish between uptake into roots and uptake into foliage.

Uptake into roots

Chemicals in soil can be transferred to the root surface via the soil water, via the gas phase in soil pores, or via direct contact with soil particles. From the surface, the chemicals may pass through the epidermis into the cortex (outer tissue of the root). The cortex is separated from the vascular tissue of the root by the endodermis. Passage through the endodermal pores depends on chemical polarity and the molecular configuration of the xenobiotics. Once across the endodermis, xenobiotics can migrate via bulk transport with the sap in the xylem. Xylem is the principal transport system for conducting water and minerals upward from the roots. Transport through the xylem is induced by evapotranspiration of water vapour from the foliage to air. If the chemical overcomes the endodermal barrier and is effectively transported in the xylem, it may leave the root system via the stem and eventually be released from the foliage to the atmosphere. At every stage along this journey, transport of the chemical can be retarded due to partitioning into plant tissues or chemical transformation (which can lead to bound residues or mobile transformation products).

Organic contaminants are generally taken up into roots passively, i.e., the plant does not expend energy to regulate the level of the chemical in the roots. Thus the maximum capacity of the roots to store a chemical is defined by the equilibrium partition coefficient of the chemical between the root and the surrounding medium. Similarly to Equation 3.54 below, the root-soil water partition coefficient $K_{\text{root-water}}$ can be approximated by:

see next page (3.51)

$$dC_A / dt = (V_{w,A} E_{w,A} C_w) - k_e C_A$$
$$dC_D / dt = (V_{w,D} E_{w,D} C_w + f \cdot E_f C_A) - k_e C_D$$
$$dC_F / dt = (V_{w,F} E_{w,F} C_w + f \cdot E_f C_D + S E_s C_s) - k_e C_F \tag{3.49}$$

$$C_A = \text{BCF}_A \cdot C_w$$
$$C_D = \text{BCF}_D \cdot C_w + d_A \cdot \text{BMF}_D \cdot C_A$$
$$C_F = \text{BCF}_F \cdot C_w + d_S \cdot \text{BMF}_D \cdot C_S + d_D \cdot \text{BMF}_D \cdot C_D \tag{3.50}$$

where

$K_{\text{root-water}}$ = dimensionless root-water partition coefficient (m^3/m^3)

$v_{\text{a-root}}$ = volume fraction of air in the root (m^3/m^3)

$v_{\text{w-root}}$ = volume fraction of water in the root (m^3/m^3)

$v_{\text{l-root}}$ = volume fraction of lipid equivalents in the root (m^3/m^3)

K_{oa} = octanol-air partition coefficient of the chemical (m^3/m^3).

Note that $K_{\text{root-water}}$ is defined for the chemical concentration in soil solution. When assessing root uptake with respect to the bulk soil concentration, soil properties also play a role. A strong tendency to partition to soil solids (i.e., a high organic carbon content of the soil or a high K_{ow} of the chemical, Equation 3.52) will reduce the chemical concentration in the soil solution, and hence reduce the uptake into roots. Therefore, the root-soil partition coefficient is largely independent of K_{ow}

Briggs et al. [81] measured a bioconcentration factor for roots (BCF_{root}), defined as the quotient of the chemical concentration in the roots and the chemical concentration in the aqueous solution surrounding the roots. They developed an empirical equation to predict $BCF_{root.}$

$$BCF_{root} = 10^{(0.77 \log K_{ow} - 1.52)} + 0.82 \qquad (3.52)$$

where

BCF_{root} = root-water bioconcentration factor (L/kg wet root).

For hydrophilic chemicals with $\log K_{ow} < 1$, BCF_{root} is constant at 0.82. This can be explained by partitioning of the chemical into the water in the root (i.e., 0.82 in Equation 3.52 is equivalent to $v_{\text{w-root}}$ in Equation 3.51). For more lipophilic chemicals with a higher K_{ow}, sorption of the chemical to the root solids becomes important, and BCF_{root} increases with increasing K_{ow}. This increase is less pronounced than predicted by Equation 3.51, but this may be due to the higher K_{ow} chemicals not reaching equilibrium with the root tissue (see below). Equation 3.52 was demonstrated up to a $\log K_{ow}$ of ~4.

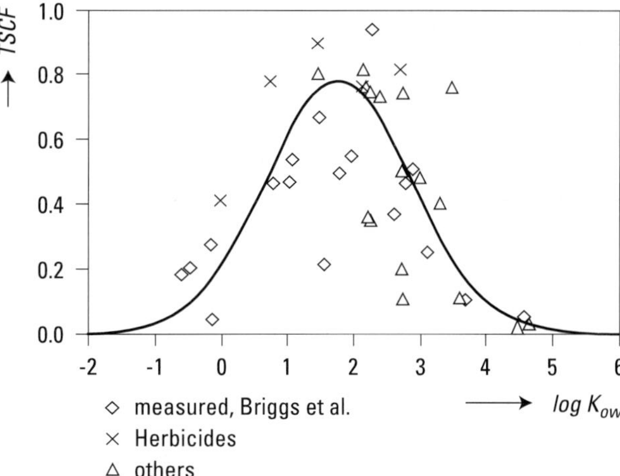

Figure 3.24. Transport of organic chemicals within plants as a function of K_{ow}; adapted from Briggs et al. [81]. From: Trapp and Matthies [88]. With permission.

Compounds with higher K_{ow} have been observed to be present primarily on the surface of roots [82]. It would appear that their partitioning into root solids is limited by their rate of transport into the root [83]. In addition, the fraction of the chemical in soil that is dissolved in soil water may be so low that transfer of the chemical from soil particles directly to the root surface and/or incorporation of soil particles in the (analyzed) root tissue may be the dominant uptake mechanisms.

Food safety is one issue where organic contaminant uptake in roots may be particularly relevant. In this context it is important to differentiate between root vegetables (e.g., carrots) and tubers (e.g., potatoes). A tuber is not a root morphologically, but rather a modified stem. It does not serve to supply water to the foliage; but rather receives its water and nutrients via the phloem from the foliage. Hence chemical uptake from soil is limited to diffusion through the tuber, and uptake can be expected to be considerably slower than for roots, at least for more polar chemicals [84].

The transport of soil contaminants to foliage is another potential consequence of root uptake. The analogy of a chromatographic column has been used to describe this process, with the aqueous xylem sap as the mobile phase and the hydrophobic root tissue as the stationary phase [85]. The greater the partitioning

$$K_{\text{root-water}} = v_{\text{a-root}} \cdot K_{\text{air-water}} + v_{\text{w-root}} + v_{\text{l-root}} \cdot K_{oa} \qquad (3.51)$$

coefficient between the stationary and mobile phases, the greater the retention time of the chemical on the column, i.e., the more slowly the chemical will move through the root. Briggs et al. [81] quantified this process using the transpiration stream concentration factor (TSCF), the quotient of the chemical concentration in the xylem sap sampled in the stem and the chemical concentration in the aqueous solution surrounding the roots. They found TSCF to be related to K_{ow} (see Figure 3.24), and derived the following equation:

$$\mathrm{TSCF} = 0.784\, e^{-\frac{(\log K_{ow} - 1.78)^2}{2.44}} \tag{3.53}$$

where

TSCF = transpiration stream concentration factor (dimensionless)

Although Figure 3.24 shows considerable variability, a clear decrease in TSCF can be seen for chemicals with $\log K_{ow} > 2$. This can be attributed to the increased retention of these chemicals on the root "chromatographic column". The decrease in TSCF for chemicals with $\log K_{ow} < 2$ may be attributable to dissociation of these more polar chemicals coupled with the reduced permeability of the endodermis to ionic organic molecules [89]. One family of plants, namely *Cucurbitaceae*, deviates markedly from the behaviour illustrated in Figure 3.24. Members of this family, e.g., zucchini (courgette) and squash, can take up large quantities of very hydrophobic chemicals from soil [90]. The mechanism by which this occurs has not yet been elucidated.

To estimate the flux of chemicals from the roots to the foliage, a simple approach is to multiply the TSCF by the plant transpiration rate and the chemical concentration in soil pore water [88]. However, this approach neglects the temporal dimension of the root chromatography and other processes, such as root growth, that can be important for chemicals with a higher K_{ow}. For an example of a more sophisticated modelling approach, see Trapp [90].

Uptake into foliage

Contaminants can enter the foliage either via the stem, as discussed above, or from the atmosphere. The aerial parts of the plant, including the foliage, are covered by a cuticle which acts as a barrier to reduce water loss from the plant, and prevents penetration of airborne particles. The cuticle is covered in cuticular waxes. The leaf surface also contains small pores, or stomata, which open and close according to environmental conditions. The stomata play an important role in regulating gas exchange processes and in transpiration. Chemicals from the atmosphere can enter the foliage through the cuticle or the stomata. Once incorporated in cuticular waxes, a chemical may diffuse through the cuticle, the epidermal cells, the mesophyll cells, and eventually reach the phloem (the stream that carries assimilation products from the leaves to the stem and the roots). Chemicals entering the cells through the stomata bypass the cuticular barrier. The permeation rates of chemicals through the cuticle vary widely for different species and environmental conditions. Permeation is related to the hydrophobicity and molar volume of the chemical, as well as to the cuticle structure and composition [91].

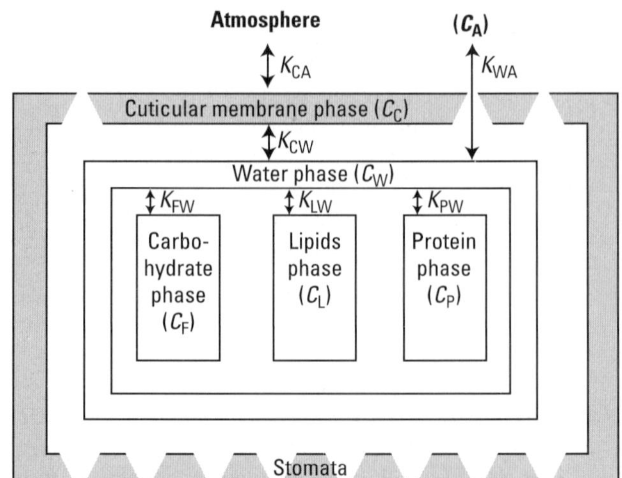

Figure 3.25. Multiple-compartment model for bioaccumulation of organic chemicals from the atmosphere in plant leaves. C is concentration. K is distribution coefficient, subscripts c, w, a, f, l and p refer to cuticle, water, air, carbohydrate, lipid and protein. Reprinted from [92]. Copyright Elsevier.

Organic contaminants are generally taken up into and released from foliage passively, i.e., the plant does not expend energy to regulate the level of the chemical in the foliage. Thus the equilibrium partition coefficient of the chemical between the foliage and the surrounding air ($K_{foliage\text{-}air}$) plays an important role in regulating exchange of the chemical with the atmosphere. To estimate $K_{foliage\text{-}air}$, the foliage can be treated as a mixture of several different phases, such as air, water, lipids, carbohydrate, cuticle, and protein (Figure 3.25). The protein and carbohydrate compartments are often neglected and $K_{foliage\text{-}air}$ approximated by [93]:

$$K_{foliage\text{-}air} = v_{a\text{-}fol} + v_{w\text{-}fol} / K_{aw} + v_{l\text{-}fol} \cdot K_{oa} \tag{3.54}$$

where

$K_{foliage\text{-}air}$ = foliage-air partition coefficient (m^3/m^3)

$v_{a\text{-}fol}$ = volume fraction of air in the foliage (m^3/m^3)

$v_{w\text{-}fol}$ = volume fraction of water in the foliage (m^3/m^3)

$v_{l\text{-}fol}$ = volume fraction of lipid equivalents in the foliage (m^3/m^3)

K_{oa} = octanol-air partition coefficient of the chemical (m^3/m^3).

Values of 0.19, 0.7, and 0.01 have been used as generic values for $v_{a\text{-}fol}$, $v_{w\text{-}fol}$, and $v_{l\text{-}fol}$, respectively. Equation 3.54 yields only a rough approximation of $K_{foliage\text{-}air}$. For instance, the $K_{foliage\text{-}air}$ of a given chemical has been shown to vary by more than one order of magnitude between different plant species with similar lipid contents [94].

One factor that distinguishes foliage-air partitioning from root-soil partitioning or biota-water partitioning is its strong temperature dependence. A change in the foliage temperature of 25°C, a not atypical diurnal variation, can change $K_{foliage\text{-}air}$ by more than one order of magnitude [95]. Consequently, the foliage-air exchange of a chemical can be a considerably more dynamic process.

Another feature that distinguishes foliage uptake from root uptake is that transport to the plant surface is typically the limiting step for foliage uptake, whereas it is usually transport within the plant for root uptake. Therefore, it is necessary to examine the different atmospheric deposition processes in some more detail. These include:

1. Deposition of gaseous chemical.
2. Dry deposition of chemical sorbed to dust or atmospheric particulate matter.
3. Wet deposition of contaminants dissolved in water droplets or sorbed to particulate matter.
4. Deposition of resuspended soil particles.
5. Direct application, as for example in the use of pesticides.

Furthermore, it is necessary to consider the manner by which chemicals can be eliminated from foliage. These include:

1. Volatilization.
2. Shedding of the leaves or leaf parts.
3. Transport with the phloem.
4. Chemical transformation.

Deposition of gaseous chemical is a diffusive process governed by a diffusion gradient between the atmosphere and the foliage. This gradient is defined by the difference between the gaseous chemical concentration in the atmosphere ($C_{air\text{-}gas}$), and the gaseous chemical concentration that the atmosphere would have if it were in equilibrium with the foliage. The latter is equal to the concentration in the foliage ($C_{foliage}$) divided by $K_{foliage\text{-}air}$. If $C_{air\text{-}gas} > C_{foliage} / K_{foliage\text{-}air}$, then there will be net diffusion from the atmosphere to the foliage, i.e., deposition. If $C_{air\text{-}gas} < C_{foliage} / K_{foliage\text{-}air}$, then there will be net volatilization. In analogy to Equation 3.17, the mass flux N_{gas} is equal to:

see below (3.55)

where

$N_{foliage\text{-}gas}$ = air to foliage chemical flux by gaseous deposition (mol/h, a positive value means net deposition, a negative value net volatilization)

$k_{foliage\text{-}gas}$ = deposition velocity for gas transfer to foliage (m/h)

$A_{foliage}$ = surface area of the foliage (m^2)

$C_{air\text{-}gas}$ = gaseous chemical concentration in the air (mol/m^3) where $C_{air\text{-}gas} = (1 - FR_{aerosol}) \cdot C_{air}$

$C_{foliage}$ = chemical concentration in the foliage (mol/m^3).

$k_{foliage\text{-}gas}$, the deposition velocity for gas exchange between the atmosphere and the foliage, depends on the shape and surface properties of the leaf, the surface roughness of the foliage canopy, and meteorological conditions. In general, the more exposed the foliage and the more turbulent the atmosphere, the larger $k_{foliage\text{-}gas}$ will be. $A_{foliage}$, the foliage surface area, must be defined using the same reference planes as for $k_{foliage\text{-}gas}$ (the alternatives include the plane of the leaf or the (horizontal) plane of the canopy). Values of $k_{foliage\text{-}gas}$ that have been measured using the plane of the canopy as the frame of reference are 8 m/h for grassland [96], 28 m/h for a mature coniferous forest, and 130 m/h for a mature deciduous forest in temperate latitudes [97].

$$N_{foliage\text{-}gas} = k_{foliage\text{-}gas} \cdot A_{foliage} \cdot (C_{air\text{-}gas} - C_{foliage} / K_{foliage\text{-}air}) \qquad (3.55)$$

Dry deposition of xenobiotics associated with atmospheric particles is a complex process. The flux to the surface of the foliage depends on the size of the particles that the xenobiotic is associated with, the meteorological conditions, the orientation of the leaf surfaces, and the aerodynamic "roughness" of the canopy, while the retention of the xenobiotics on the leaf depends on the stickiness of the particles and the leaf surface, as well as the rate of removal of the particles from the surface. The latter can be greatly influenced by precipitation, but precipitation is also a vector of particle associated xenobiotics to the leaf surface as a result of particle washout and scavenging in the atmosphere. Rather than trying to resolve the complexities of this process, a more pragmatic approach is generally taken in which an average net deposition velocity for a particle-associated chemical is employed:

$$N_{foliage\text{-}part} = vd_{foliage\text{-}part} \cdot A_{foliage} \cdot C_{air\text{-}part} \tag{3.56}$$

where

$N_{foliage\text{-}part}$	=	air to foliage chemical flux by deposition of aerosol associated chemical (mol/h)
$vd_{foliage\text{-}part}$	=	net deposition velocity for aerosol associated chemical to foliage (m/h)
$A_{foliage}$	=	surface area of the foliage (m^2)
$C_{air\text{-}part}$	=	particle-associated chemical concentration in the air (mol/m^3 air) ($C_{air\text{-}part} = FR_{aerosol} \cdot C_{air}$).

The same care must be taken to define a plane of reference for $A_{foliage}$ and $k_{foliage\text{-}part}$ as for gaseous deposition. $k_{foliage\text{-}part}$ is typically lower than $k_{foliage\text{-}gas}$. For instance, for the grassland mentioned above, the corresponding $k_{foliage\text{-}particle}$ was 3 m/h for a range of xenobiotics of pyrogenic origin (PCDD/Fs and PAHs) [96], For an extensive review of particle-bound deposition of xenobiotics to foliage, see Smith and Jones [98].

As outlined in Section 3.2.5, chemicals can also be deposited dissolved in precipitation. This is the major form of atmospheric deposition for chemicals that preferentially partition into water. These tend to be the same chemicals that are taken up efficiently from soil into roots.

Soil particle resuspension, for instance by rain splash, can be an important source of contaminants in foliage. Adhesion to and retention of soil particles on the leaves are important determinants of the extent of contaminant uptake in foliage. Like the deposition of atmospheric particles, the adhesion and retention of soil particles will depend on the properties of the particles and the plant surface, as well as the meteorological conditions. This pathway is most likely to be important for chemicals with high K_{oa} [99]. The mass loading of resuspended soil particles on the foliage of agricultural crops, such as grass and corn, ranges from 0.2 to > 20% of soil per dry weight plant [98]. Values at the lower end of the range quoted above can be expected under most conditions. This pathway can be particularly important for indirect human exposure, where harvesting practices can result in additional incorporation of soil into fodder fed to livestock.

Direct application of chemicals to plants, e.g., via biocide application, typically leads to a situation where the levels of chemical in the foliage are high compared to the surrounding air and soil. Chemical uptake will not continue; rather, the levels of the chemical in the foliage will then be determined by the elimination processes. Elimination may occur via volatilization as described in Equation 3.5, or it may occur via transformation of the chemical. In addition, some plants shed portions of their cuticular waxes under certain conditions, and this shedding can be a meaningful loss mechanism for contaminants associated with these waxes. Shedding of leaves is the most important elimination mechanism for persistent chemicals with high $K_{foliage\text{-}air}$ values. Weak acids can also be transported via the phloem in the plant's vascular system. In this way, some pesticides can be transported from the foliage to the roots and other plant organs [100]. Finally, plant growth can be viewed as a form of elimination, as the production of new biomass dilutes the xenobiotics in the foliage, giving lower concentrations.

Elimination processes do not influence xenobiotic levels following direct application only. Xenobiotic levels in foliage are always determined by the balance between uptake and elimination rates. Volatilization, transformation, shedding, and export via the phloem tend to lower the concentrations in the foliage and, in general, these processes ensure that the levels in the foliage stay at or below the concentration that would be expected if the plants were in equilibrium with the atmosphere. There is, however, an interesting exception. Persistent chemicals with a low K_{ow} and a high $K_{foliage\text{-}air}$ value can accumulate in foliage to much higher levels than could be obtained by equilibration with the air or the soil. This is because these chemicals are readily taken up by the roots and translocated to the foliage, while their volatilization from the foliage is comparatively slow. Hence the contaminants are "pumped" via the translocation stream into a foliar "trap" [88]. This is the plant equivalent of biomagnification.

Factors influencing bioaccumulation in plants
From the above discussion it is clear that a multitude of factors influence bioaccumulation in plants. These include the properties of the chemical (e.g., K_{ow}, K_{oa}, size, dissociation), properties of the plant (e.g., lipid content, leaf orientation, leaf and canopy roughness, shedding of foliage, transpiration rates), properties of the soil (e.g., organic carbon content, properties affecting transfer of soil to foliage), and properties of the atmosphere (e.g., temperature, wind speed, particle size distribution, precipitation). It is the interaction of these variables that determines the plant accumulation in a given situation. The reader is referred to McLachlan [99] for an analysis of the factors affecting bioaccumulation in foliage.

Models of accumulation in terrestrial plants
The approaches used to model bioaccumulation in aquatic organisms can also be applied to plants. The basis of all models is a mass balance equation describing the rate of change in chemical inventory in a given compartment as the difference between the rate of chemical uptake and the rate of chemical elimination.

$$d(V_{plant} \cdot C_{plant}) / dt = N_{plant\ uptake} - N_{plant\ elim} \quad (3.57)$$

where

V_{plant} = plant volume (m^3)
C_{plant} = concentration in the plant compartment modelled (mol/m^3)
$N_{plant\ uptake}$ = chemical flux to the plant compartment modelled by all pathways (mol/h)
$N_{plant\ elim}$ = chemical flux from the plant compartment modelled by all pathways (mol/h).

The uptake and elimination fluxes can be calculated on the basis of the equations and information given above. A number of different models have been developed with varying degrees of sophistication depending on the intended application. One-compartment models of just the foliage or just the roots have been utilized, as have multi-compartment models that include both roots and foliage, as well as other plant organs [e.g., 84, 101].

3.3.6 Accumulation in terrestrial invertebrates

Uptake from soil pore water often constitutes an important source of uptake for animals which are in continuous contact with soil, such as earthworms, nematodes and other small or soft-bodied invertebrates. For many other terrestrial invertebrates, ingested food is the primary route of uptake of contaminants. Animals may feed on living plant material (phytophagous), dead organic matter (saprophagous), or on living animal material (predacious). For many invertebrates, prediction of bioaccumulation depends on the properties of the substance, the soil, and the species in question.

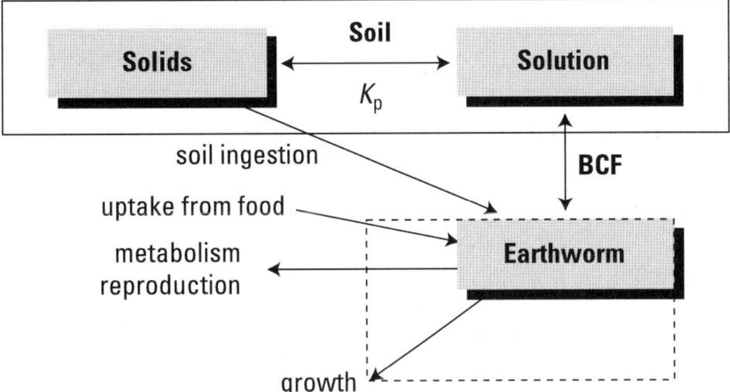

Figure 3.26. Processes affecting the concentration of xenobiotics in earthworms. Thick lines represent the equilibrium partitioning theory; thin lines represent processes that may influence the validity of this theory. From Jager [102]. Copyright ©1998. Reprinted by permission of Alliance Communications Group, Allen Press, Inc.

A significant part of the invertebrate biomass in soil consists of earthworms (Oligochaeta). These feed on organic material in the soil, and respiration takes place via the skin. Bioaccumulation of organic chemicals in earthworms can be described based on partitioning of the chemicals between the soil, pore water and the internal phases of the worm, lipid and water.

The first process that determines uptake is sorption of the chemical to soil, as determined by the solids-water partition coefficient K_p (Figure 3.26), and is used to calculate the pore water concentration from the total concentration in soil. The second process is bioconcentration in the earthworm from the pore water. The BCF was modelled by Jager [102] with the following equation:

$$BCF_{earthworm} = \frac{0.84 + 0.012 \cdot K_{ow}}{RHO_{earthworm}} \quad (3.58)$$

where

$BCF_{earthworm}$ = the earthworm bioconcentration factor (L/(kg wet wt.)).
K_{ow} = the *n*-octanol-water partition coefficient
$RHO_{earthworm}$ = the density of the earthworm (set at 1 kg wet wt./L)

Table 3.7. Dietary uptake efficiencies for cadmium (Cd) in terrestrial invertebrates [109]. With kind permission of Springer Science and Business Media.

Species	Food	Cd concentration in food (μmol/g)	Uptake efficiency (%)
Snail	agar	1.48	55-92
Isopod	poplar leaves	0.03-0.37	10-60
Centipedes	isopod hepatopancreas	1.21-10.2	0-7
Millipedes	maple leaves	-	8-40
Pseudoscorpion	collembolans	0.2	59
Mites	green algae	0.15	17
Insects	green algae	0.09-0.15	9
	collembolans	0.23	35

Several processes can lead to concentrations in earthworms that deviate from this prediction. Biotransformation of substances in the worm could lead to a lower BCF than predicted, but is generally not expected to play a major role, due to the more limited metabolic capacity of invertebrates. There are many examples where biotransformation in invertebrates plays a minor role, e.g., PAHs [103].

The data collected by Jager [102,104] and Jager et al. [105] do not indicate that the food exposure route actually leads to much higher body residues than expected on the basis of bioconcentration alone, in contrast to previous findings [106]. The uptake of chemicals from soil particles in the gut of the earthworm is limited, most likely due to the limited capacity for digestion of food in the earthworm gut, thereby causing only a small increase in chemical potential in the gut.

Field data indicate that the worm model overpredicted the earthworm accumulation, when estimating the soil pore water with a partition coefficient based on the K_{oc} [107]. Reliable estimation of soil pore water concentrations strongly depends on the QSAR used to calculate the pore water concentration. Many factors are responsible for an overestimation of soil pore water concentrations, most notably depletion of the substance in the pore water (due to e.g., slow desorption kinetics), biodegradation etc. In cases where the soil pore water concentration can be reliably determined, the BCF earthworm model can be applied [108], including for substances with a high K_{ow}. In general, the BCF model is correct but pore water concentrations are often overestimated using standard QSARs to estimate the K_p, leading to a conservative estimate of worm concentrations.

Uptake of metals by invertebrates is determined by different factors than for organic chemicals. Species differences are observed that may be related to differences in feeding physiology and trace element requirements. For instance, large differences between almost zero and approximately 90% in the uptake efficiency of cadmium (Table 3.7) have been observed for different species [109]. The uptake efficiency of cadmium corresponds well with food uptake efficiency. This suggests that the mechanisms determining the amount of assimilated metals are related to those regulating food uptake and the assimilation of nutrients. The general differences found between detritivores, herbivores, and carnivores are related to food digestibility and the distribution of elements between egested and digested fractions. In addition, absorption and elimination rates generally decease with species weight [31]. The exposure concentration generally influences uptake rates due to some form of saturating uptake kinetics.

The distribution of metals over the various invertebrate organs and cell fractions is usually far from uniform. In earthworms, metals accumulate mainly in the chloragogenous tissue that lines the gut, while in snails, metals accumulate in the midgut gland, the gut and the foot. Related species have a characteristic internal sequestration over cell fractions and organs, but the relative accumulation potential differs between species [110]. The distribution of metals over the various binding sites inside the cell is affected by the binding affinity of the endogenous ligands, the number of binding sites and the presence of competing metals [111,112]. Metals specifically bind to metal-rich granules in the cell, and to the inducible metal-binding proteins. Metal-binding proteins similar to mammalian metallothionein have been identified in various terrestrial invertebrates, such as slugs, midges, freshflies, cockroaches and earthworms.

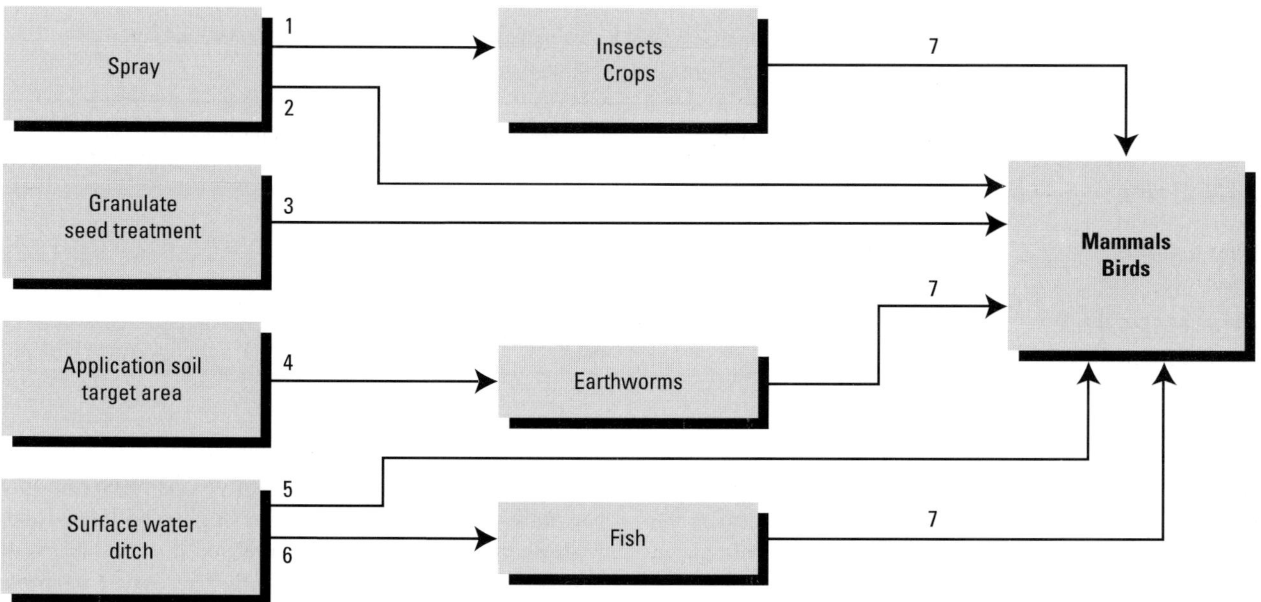

Figure 3.27. Food as a major source of contaminants for mammals and birds in a simplified food web. 1=Application of spray, 2=Drinking from leaves/crop, 3=Ingestion of granules/treated seeds, 4=Bioconcentration soil-worm, 5=Drinking from surface water, 6=Bioconcentration water-fish, 7=Consumption. From USES [113].

The rate of synthesis of this protein is considered a key factor in metal regulation, where the protein acts as a first scavenger of metal in the cytosol. Metal binding to metallothionein may diminish its binding to other molecules, including those which are the targets for metal toxicity. The inducibility of metal proteins differs between biological species. It seems that the metal binding to the inducible metallothionein offers protection against short-term metal exposure, while the granules are a sink for metals, to protect against long-term exposure [110].

Metal elimination processes in invertebrates are also species-dependent. The springtail *Orchesella cincta* is an invertebrate with a unique excretion mechanism. Excretion of metals occurs through exfoliation of the mid-gut epithelium at every moult, which is very regular. This excretion mechanism is an important component in cadmium tolerance [111]. Excretion of metals in isopods depends on the storage in the hepatopancreas of metals which are not available for elimination.

Bioaccumulation of organic chemicals in terrestrial invertebrates can be understood using the principles of equilibrium partitioning for reasonable worst-case predictions. Partitioning of the chemical over the different phases in soil can lead to lower estimates of bioavailability, e.g., due to slow desorption of chemicals and complexation to organic and organic ligands in pore water. For metals, speciation in soil and metal partitioning over cell fractions and organs complicate the prediction of metal accumulation. This will be discussed in more detail in Section 3.7.

3.3.7 Accumulation in mammals and birds

Higher organisms, such as mammals and birds, are toppredators, and form the end point of biological pathways along which contaminants may accumulate in increasing concentrations. Thus they may be subject to adverse effects.

Food is the major route of uptake for mammals and birds (Figure 3.27). Essential to bioaccumulation is the choice of food. Plants and lower organisms are the prey of mammals and birds. Since the concentrations of contaminants vary significantly between the prey, the choice of food largely determines the concentration of contaminants in higher organisms. Polar bears as well as the Inuit people contain high concentrations of PCBs due to the fact that fish is their major food source: fish that has accumulated high concentrations of PCBs. Herbivores contain fewer hydrophobic chemicals, but may consume more metals, due to deposition on leaf surfaces.

Significant differences may also occur within the same region where animals have different feeding

Table 3.8. Geometric mean cadmium and lead concentrations in liver and kidney of three small mammals in De Kempen, a highly polluted area in The Netherlands [114,115].

Species	Organ	Cadmium (μg/g)	Lead (μg/g)
Talpa europaea	kidney	180	48
	liver	152	13
Sorex araneus	kidney	127	36
	liver	155	3.1
Microtus agrestis	kidney	1.8	4.2
	liver	0.33	1.2

strategies. For example, the mole *Talpa europaea* predominantly feeds on worms and insect larvae, the mouse *Sorex araneus* preys on worms as well as on small insects and snails, while the mouse *Microtus agrestis* pre-dominantly eats grass stems, fruit and seed. Worms and insects accumulate heavy metals to a large extent, while plants do not. The result is that the kidneys of the mole and the mouse *Sorex araneus* may contain high concentrations of cadmium and lead, while the herbivorous mouse *Microtus agrestis* contains very low concentrations of these metals (Table 3.8).

When the exact composition of the diet, the concentration of the contaminant in the diet items and the uptake efficiencies are available, uptake from food can be modelled using Equation 3.39, for example. Other models take a different approach, focusing on bioenergetics. These models try to relate the amount of energy a higher organism requires for growth, reproduction, warmth, migration, etc., to the amount, caloric content and digestive efficiency of the food they ingest [116,117]. For example, warm-blooded animals such as mammals have low growth efficiencies compared to cold-blooded animals, such as fish, due to homeothermy and the associated high activity level. This means that relative food intake in Equation 3.39 will be higher for mammals than for a fish, and relative growth will be slower thus leading to less growth dilution in mammals than in fish, a lower overall elimination rate (Equation 3.32) and a higher BMF. An essential difference between birds/mammals and fish/plankton is elimination rates. The elimination rate drives biomagnification: if it is high it does not matter how much an organism eats, and no biomagnification will occur.

Differences in the diet composition, energy content and digestibility of the diet of herbivorous and carnivorous birds are responsible for the larger BMF often observed for carnivorous birds when compared to BMFs for herbivorous birds. Carnivorous birds not only eat more food, but also eat more highly polluted food, both phenomena leading to increased concentrations of contaminants in carnivorous birds. The earlier example of the herbivorous mouse and the carnivorous mole and mouse, illustrates that the same applies to herbivorous and carnivorous mammals.

Thus, biomagnification of hydrophobic chemicals which are very slowly excreted will be most pronounced in birds and mammals that prey on organisms which are already relatively highly contaminated. Fish-eating birds, such as the herring gull and the cormorant, as well as higher animals, such as seals and polar bears, therefore accumulate high concentrations of such chemicals.

BCFs for meat and milk from cattle have been shown to be directly proportional to the K_{ow}, on a logarithmic scale [118].

3.3.8 Methods for measuring terrestrial bioaccumulation

Hitherto, no standard protocols have been developed for measuring bioaccumulation in terrestrial ecosystems, plants and animals. Therefore, the results of biomagnification studies vary greatly, making it very difficult to predict or describe biomagnification with any confidence.

3.4 ABIOTIC TRANSFORMATION PROCESSES

3.4.1 Introduction

Following its release into the environment, a chemical may undergo various biotic and abiotic processes which modify its chemical structure. Degradation or transformation of a compound refers to the disappearance of the parent compound from the environment by a change in its chemical structure. When this change is brought about by microorganisms, the degradation process is called primary biodegradation or biotransformation. In this process fractions of the chemical structure are incorporated into cellular material or used as an energy source by the organism. Often micro-organisms are capable of converting the chemical to simple molecules and ions, such as carbon dioxide, methane, water and chloride. This process is referred to as mineralization.

Transformation of chemicals in the environment can also occur by abiotic processes. The most important abiotic transformation processes can be divided into four separate categories:

- *Hydrolysis:* alteration of the chemical structure by direct reaction with water.
- *Oxidation:* a transformation process in which electrons are transferred from the chemical to a species accepting the electrons; the oxidant.
- *Reduction:* the reverse of oxidation; electron transfer takes place *from* a reductant to the chemical to be reduced.
- *Photochemical degradation:* transformation due *to* interaction with sunlight.

Transformation and mineralization processes can alter the physicochemical and toxicological properties and reduce exposure concentrations of chemicals released in the environment. Where biotransformation is carried out by higher organisms, the formation of polar transformation products (metabolites) can also provide an important method of detoxification (Section 3.6), albeit that metabolism in some cases can cause toxification.

The rate of degradation of a specific chemical will depend on its availability for reaction, its intrinsic reactivity, the availability of the reactant and the reactivity of the reactant. Generally, the availability and reactivity of both the chemical and the reactant depend to a large extent on environmental conditions like pH, temperature, light intensity and redox conditions. This section deals with the most important abiotic transformation processes and the main environmental conditions affecting kinetics and product formation. It will be shown that the quantification of transformation processes requires careful consideration of the intrinsic chemical properties due to the variable nature of the environmental system.

3.4.2 Hydrolysis

The chemical reaction of organic compounds with water is called hydrolysis. In a typical hydrolysis reaction hydroxide replaces another chemical group. Figure 3.28 shows a number of hydrolytically unstable compound families and the products formed by hydrolysis. However, certain functional groups, including alkanes, alkenes, benzenes, biphenyls, (halogenated) polycyclic aromatics (e.g., PAHs and PCBs), alcohols, esters and ketones, are often inert to hydrolysis.

The importance of hydrolysis stems from the fact that upon introduction of a hydroxyl group, additional polar products are formed which are more water soluble and are generally less lipophilic than the parent compound. Hydrolysis reactions are commonly catalyzed by hydrogen or hydroxide ions. Because the concentrations of hydrogen ion $[H^+]$ and hydroxide ion $[OH^-]$ change by definition with the pH of the water, the rate of hydrolysis directly depends on the pH. It is generally observed that hydrolysis reactions proceed according to a pseudo first-order reaction:

$$-dC/dt = k_h \cdot C \tag{3.59}$$

where

dC/dt	=	the decay of the concentration of the chemical undergoing hydrolysis is as a function of time
C	=	the chemical concentration
k_h	=	the pseudo first-order rate constant for hydrolysis at constant pH.

The constant k_h contains the contributions of the acid and base-catalyzed processes and the contribution from hydrolysis due to water attack. Because water is always present in excess, its concentration is not affected by the course of the hydrolysis process taking place. Consequently k_h can be rewritten as:

$$k_h = k_a \cdot [H^+] + k_b \cdot [OH^-] + k_n \tag{3.60}$$

where

k_a	=	second-order reaction rate constant for the acidcatalyzed process (L/(mol·s))
k_b	=	second-order reaction rate constant for the base-catalyzed process (L/(mol·s))
k_n	=	second-order reaction rate constant for the neutral hydrolysis process (1/s).

Experimentally, a known quantity of the compound is introduced into a solution of fixed pH and the disappearance of the compound is followed over time. By integrating Equation 3.59, the concentration of the chemical typically declines exponentially with increasing

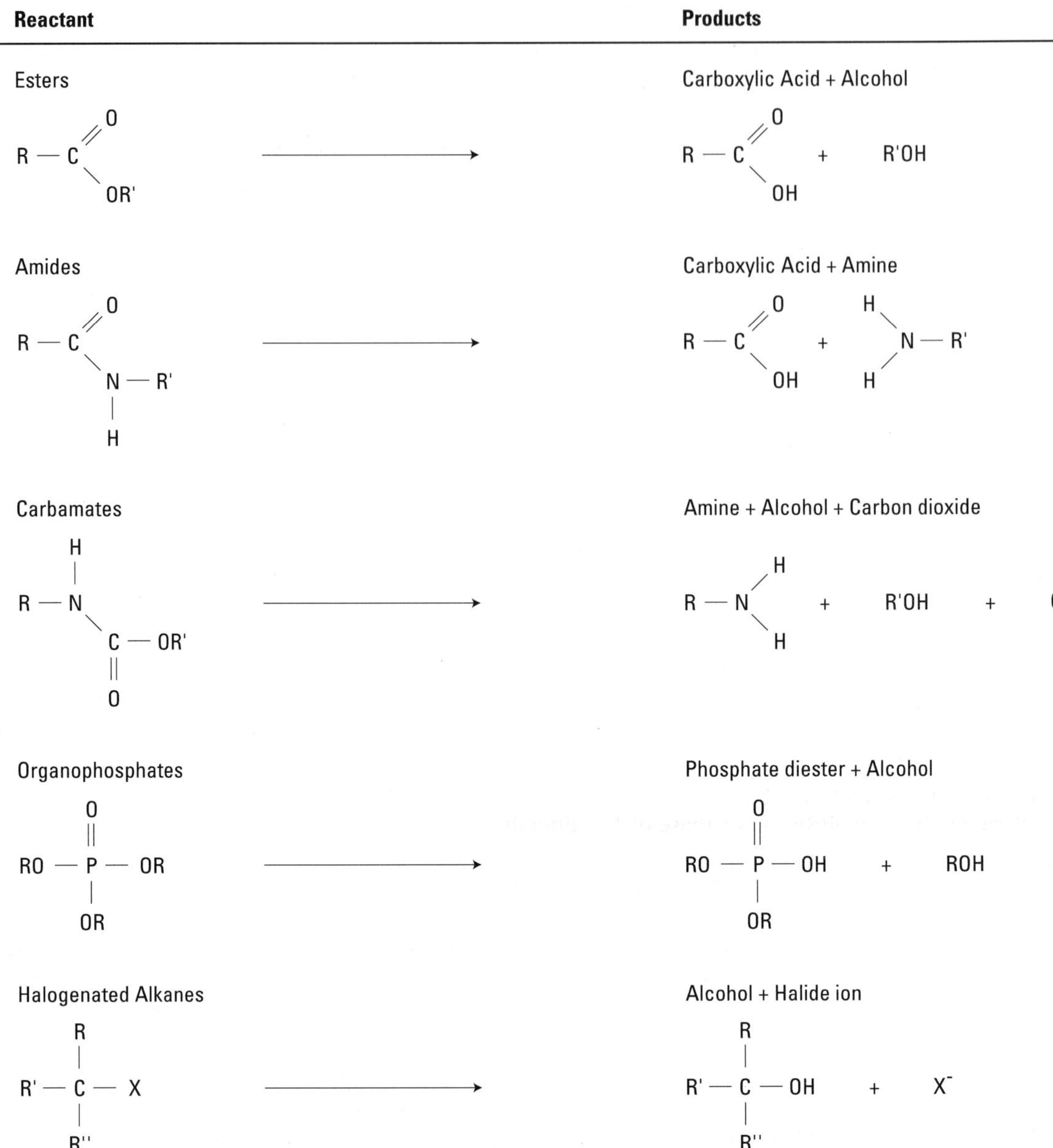

Figure 3.28. Some examples of hydrolytically unstable chemicals and the products formed by hydrolysis. (R, R', R'' represents an aromatic ring or aliphatic chain and X is a halogen atrom).

time:

$$\ln C_t = \ln C_0 - k_{obs} \cdot t \tag{3.61}$$

where

C_t = the concentration at time t

C_0 = the concentration of the chemical at the beginning of the experiment

k_{obs} = the observed pseudo first-order rate constant (1/s).

From the results of a series of such experiments at different pH levels, a pH rate profile can be constructed by plotting the base 10 logarithms of the observed rate constants as a function of the pH of the experimental solutions. Figure 3.29 shows the pH rate profile of the hydrolytic transformation of phenyl acetate to yield acetic acid and phenol. Under acid conditions (pH < 3), specific acid catalysis is the predominant mechanism. In this pH region, the logarithm of k_{obs} decreases by a

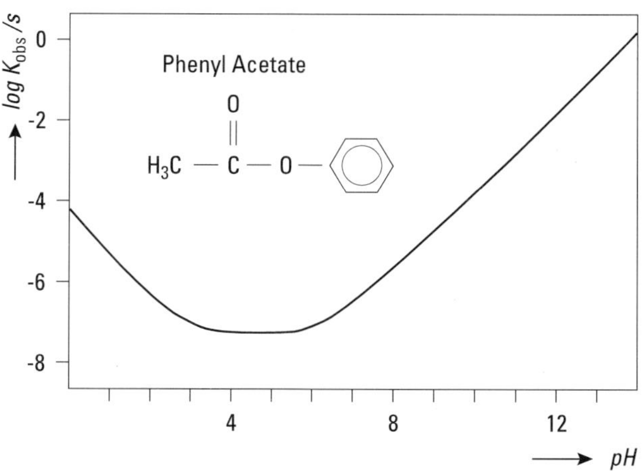

Figure 3.29. Hydrolysis pH rate profile of phenyl acetate at 25° C. Profile developed by Burns and Baughman [119] from rate constant data summarized by Mabey and Mill [120]. With permission.

unit slope -1 with increasing pH. At less acidic pH (pH > 4), the hydrogen ion concentration is so small that the specific acid catalyzed hydrolytic reaction is too slow to be seen in the profile. Between pH 4 and 6, the neutral mechanism (independent of pH) predominates. Finally, at pH > 8, due to base catalysis, an increase of k_{obs} directly proportional with increasing OH concentrations, becomes visible. The numerical values of the second-order rate constants k_a and k_b can be calculated by dividing k_{obs} by the molar concentration of either H^+ or OH^-, in the relevant section of the pH rate profile.

3.4.3 Oxidation

Oxidation is the chemical process in which an electron-deficient particle (the oxidant) accepts electrons from the compound to be oxidized. Examples of oxidants that occur under environmental conditions in sufficiently high concentrations and also react quite quickly with organic compounds are:

- Alkoxy radicals (RO·)
- Peroxy radicals (RO_2·)
- Hydroxyl radicals (HO·)
- Singlet oxygen ($^1O_2$)
- Ozone (O_3)

Most of these oxidants are directly or indirectly generated from chemicals that interact with solar radiation, forming an "excited state" of the molecule. Compounds in this photo-chemically excited state either react directly with oxygen or cleave to form radicals which subsequently react with oxygen. Oxidations are the main

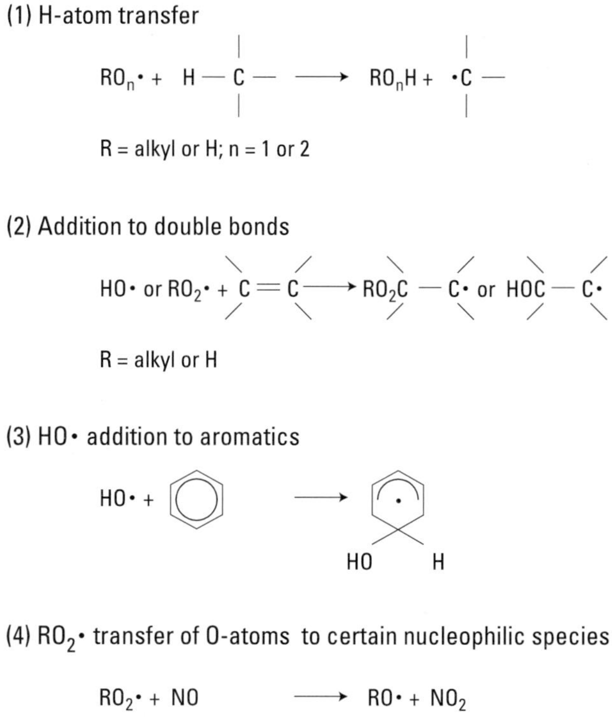

(1) H-atom transfer

$RO_n\cdot + H{-}C{-} \longrightarrow RO_nH + \cdot C{-}$

R = alkyl or H; n = 1 or 2

(2) Addition to double bonds

$HO\cdot$ or $RO_2\cdot + C{=}C \longrightarrow RO_2C{-}C\cdot$ or $HOC{-}C\cdot$

R = alkyl or H

(3) HO• addition to aromatics

$HO\cdot$ + (benzene) $\longrightarrow$ (hydroxycyclohexadienyl radical, HO H)

(4) RO_2• transfer of O-atoms to certain nucleophilic species

$RO_2\cdot + NO \longrightarrow RO\cdot + NO_2$

Figure 3.30. The general reaction pathways for environmental oxidation. From Mill [122]. With kind permission of Springer Science and Business Media.

transformation routes for most organic compounds in the troposphere and also transform various micropollutants in surface waters [121]. Most radical oxidants exhibit similar chemistry for aliphatic and aromatic structures. Four common processes are known:

1. H-atom transfer.
2. Addition to double bonds.
3. HO· addition to aromatics.
4. RO_2· transfer of O atoms to nucleophilic species.

These general reaction pathways are given in Figure 3.30. If the rate of oxidation of a specific chemical in aquatic and atmospheric systems must be predicted, three kinds of information and data are required:

a. The identities and concentrations of the oxidants in the environmental compartment.
b. The rate constant for oxidation by each oxidant at a specific site in a molecule.
c. The kinetic rate law for each process.

The simplest form of the oxidation rate law can be written as follows:

$$R_{ox} = k_{ox} \cdot [C] \cdot [OX] \tag{3.62}$$

where

Table 3.9. Half-lives (d) for tropospheric oxidation of various classes of organic compounds in the northern hemisphere.

Alkanes	1 - 10
Alcohols	1 - 3
Aromatics	1 - 10
Olefins	0.06 - 1
Halomethanes	100 - 47,000

R_{ox} = rate of oxidation of a chemical C (mol/(L·s))

k_{ox} = the specific second-order rate constant for oxidation at a specific temperature (L/(mol·s))

$[C]$ = molar concentration of the chemical C

$[OX]$ = molar concentration of the oxidant.

The parameter k_{ox} contains contributions from each of the four common oxidation processes listed above. Although many different kinds of RO_2· or RO· radicals may be present in a natural system, the simplifying assumption can be made that the structure of R has little effect on its reactivity [122]. Rate constants for reactions of most radical oxidants are known for a large number of organic molecules. The concentrations of the major oxidants in less heavily polluted aquatic and atmospheric systems are also known. By combining these data it can be derived that, in general, the hydroxyl radical is the only oxidant of importance in atmospheric systems. In aquatic systems the concentration of ·OH is so low that its contribution is negligible compared with RO2· or RO·. To illustrate the differences in reactivity of the hydroxyl radical to various organic chemicals, the half-lives for gas-phase oxidation of various classes of chemicals in the northern hemisphere are given in Table 3.9.

The half-life is defined as the time required to reduce the concentration of a chemical by 50%. From this table it is clear that chloro-fluoro-hydrocarbons (CFCs or halomethanes), in particular, may remain in the troposphere for prolonged periods of time. This enables them to reach the stratosphere, where they pose a threat to the ozone layer.

3.4.4 Reduction

Reduction is the chemical process by which electrons are transferred from an electron donor (reductant) to the compound to be reduced. The redox half-reactions leading to reduction of a 1,2-substituted alkane are shown as a diagram in Figure 3.31. In this example, Fe^{2+} is used as the reductant. Following the transfer of 2 electrons from 2 molecules of Fe^{2+} to the halogenated compound, Fe^{3+}, the free halide ion and the product of reduction (in this case ethene) are formed.

It has been shown that reductive reaction pathways can contribute significantly to the removal of several micropollutants. Nitroaromatics, azo-compounds, halogenated aliphatic and aromatic compounds (including PCBs and even dioxins) can be reduced under certain environmental conditions [123]. Reduction can take place in a variety of reducing (non-oxic) systems, including sewage sludge, anaerobic biological systems, saturated soil systems, anoxic sediments, reducing iron porphyrin systems, solutions of various chemical reagents, as well as in the gastronomic tract of invertebrate species. It has also been shown that the reduction rate of specific halogen compounds depends on environmental factors, such as the prevailing redox potential, temperature, pH and the physical and chemical properties of the micropollutant to be reduced.

As in hydrolytic transformation, usually more polar products are formed from the parent compound

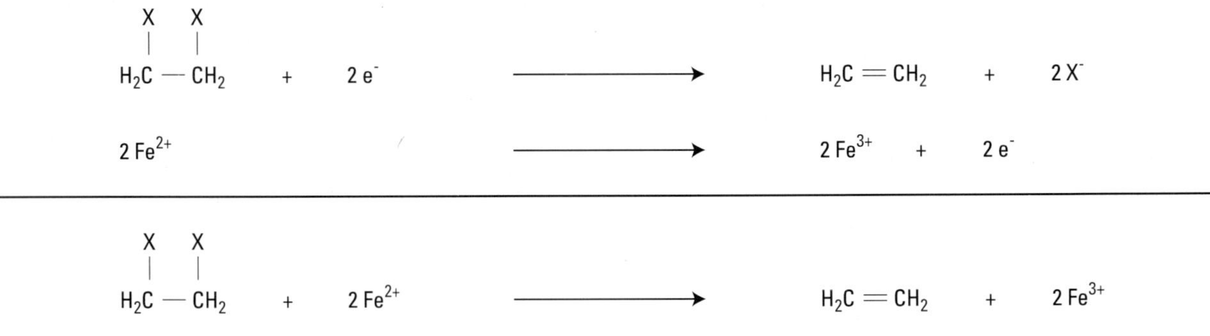

Figure 3.31. Example of a reductive transformation: electron transfer from Fe^{2+} to 1,2-dihalogen substituted ethane (X denotes a halogen atom).

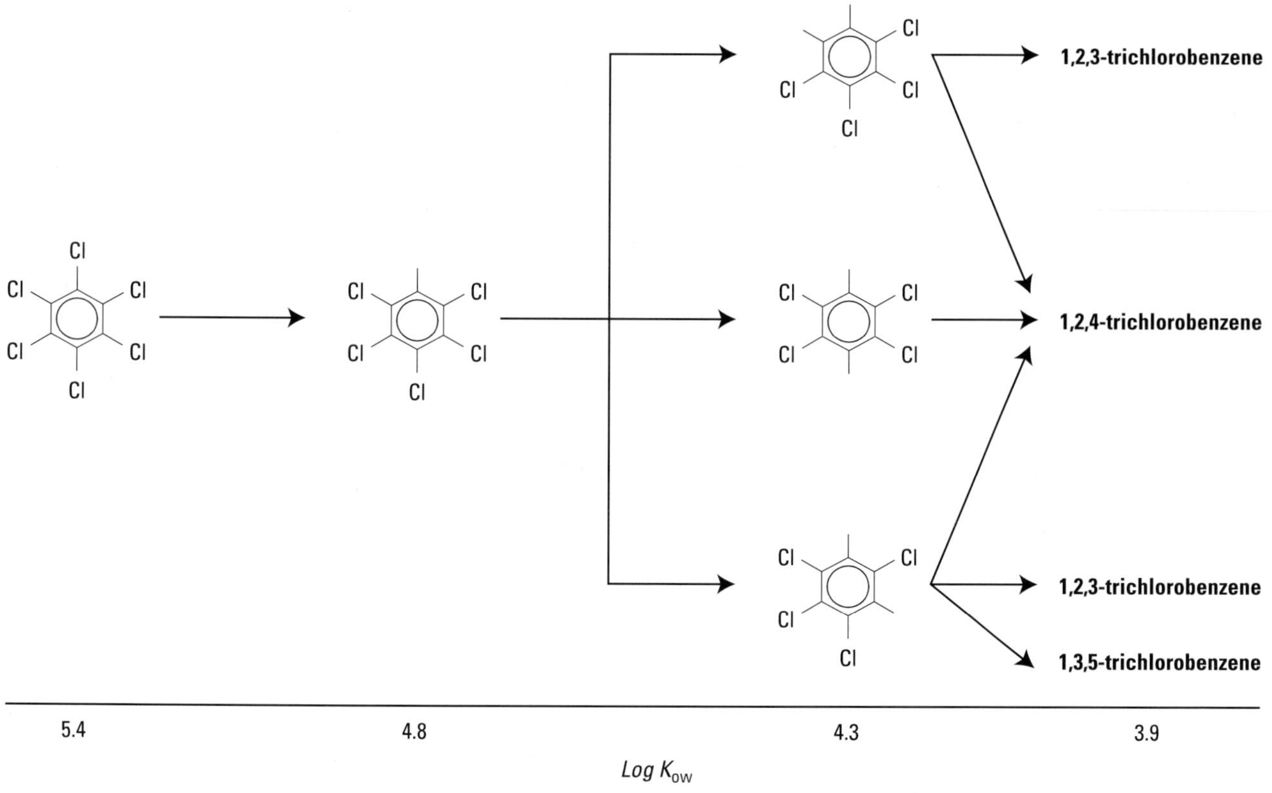

Figure 3.32. Products formed by reductive dehalogenation of hexachlorobenzene and the corresponding values of log K_{ow}.

by reduction, which makes them more susceptible to further chemical attack and less likely to accumulate. The products formed by reductive dehalogenation of hexachlorobenzene, for example, are shown in Figure 3.32, together with the corresponding values of log K_{ow}. At present, insufficient information is available on the nature of the reductants responsible for the main reductive transformations in natural systems. Nevertheless, it has been shown in most studies that reductive transformations generally follow pseudo first-order reaction kinetics (Equation 3.59). Values for the rate constant k (1/s) have been reported for various chemicals under varying environmental conditions.

3.4.5 Photochemical degradation

Figure 3.33 gives a few typical examples of photo-chemical transformation processes. As can be derived from this figure, interaction with sunlight can initiate a wide variety of photolytic processes. The primary requirement for photo-chemical processes is the penetration of radiation (light, including UV light) in aqueous and atmospheric environments. Following absorption of a photon by a compound, the photon energy either needs to be transferred to the reactive site within the molecule or transferred to another molecule, which may subsequently undergo a photo-chemical transformation. Although all photochemical reactions are initiated by the absorption of a photon, not every photon induces a chemical reaction. Besides chemical reactions, possible processes which excited molecules may undergo include the reemission of light through fluorescence and phosphorescence, the internal conversion of the photon's energy into heat and the excitation of other molecules. The fraction of absorbed photons which causes the desired reaction is termed the quantum yield (Φ) and is given in Equation 3.63 (see below).

The quantum yield is always less than or equal to 1.

$$\Phi = \frac{\text{moles of a given species formed or transformed}}{\text{moles of photons absorbed by the system}} \tag{3.63}$$

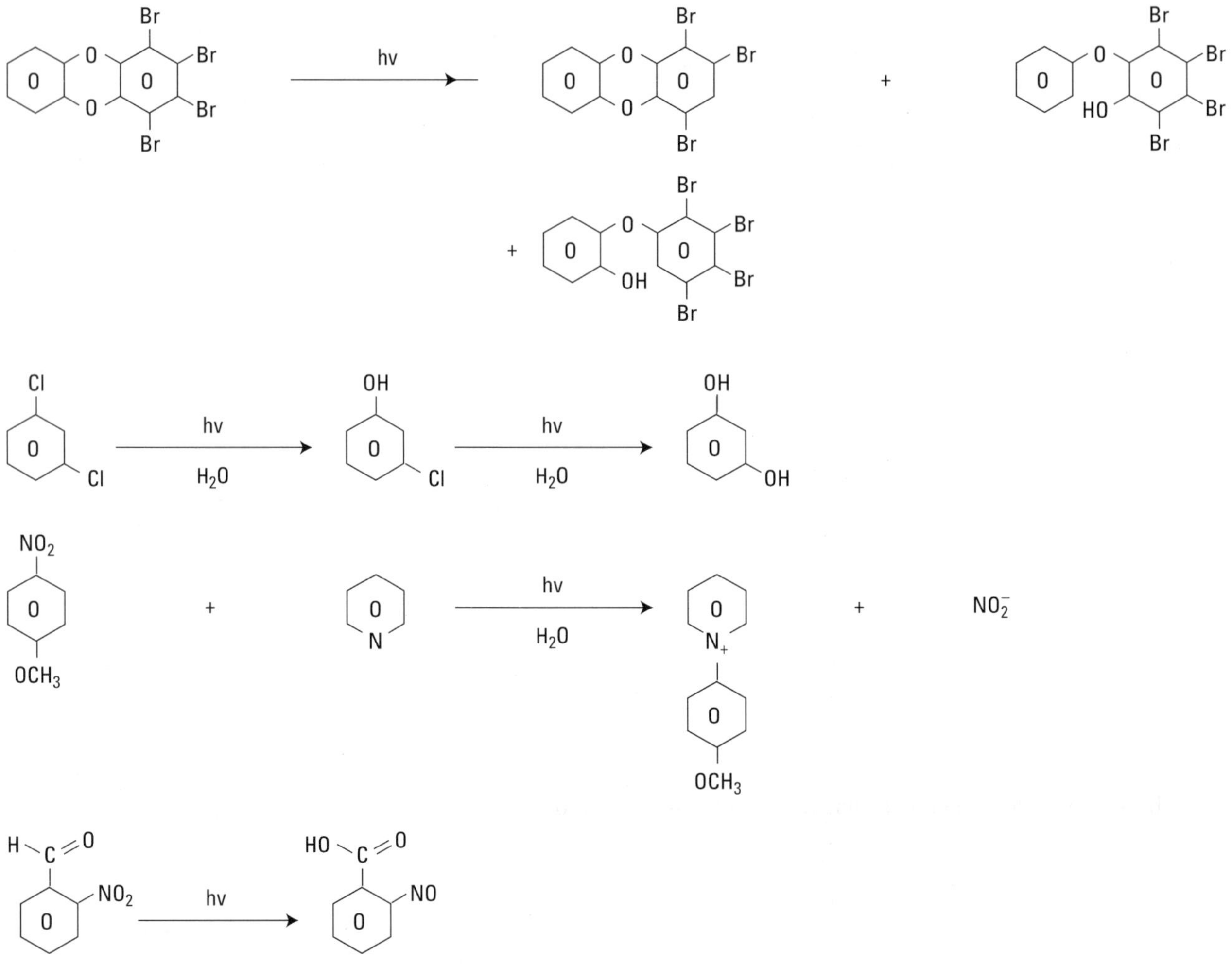

Figure 3.33. Some typical examples of photochemical transformation processes.

Quantum yields may vary over several orders of magnitude depending on the nature of the molecule which absorbs light and the reactions it can undergo. Two types of photochemical conversions are generally distinguished:

a. Direct photoreactions, in which the reacting molecule itself directly absorbs light.
b. Indirect or sensitized photolysis, in which a light-absorbing molecule transfers its excess energy to an acceptor molecule causing the acceptor to react.

The direct photoreaction rate of chemicals is proportional to the absorption of light at a specific wave length and the quantum yield. The absorption rate constant is directly related to the light intensity and extinction coefficient of the compound at a specific wave length. The molar absorption coefficient and the quantum yield are both molecular properties. Therefore, in principle the direct photolysis of environmental chemicals can be described as a second-order process:

$$-dC/dt = k_p \cdot I \cdot C \qquad (3.64)$$

where

k_p	=	the second-order photochemical reaction rate constant
C	=	the concentration of the parent compound
I	=	the light intensity.

Since the rates of all photochemical reactions are proportional to light intensity, it is evident that the significance of the phototransformation of a certain chemical will change with time and place. In this process factors such as time of the day or year, location (climate) and weather (cloud cover) play a major role.

In the aquatic environment, an important fraction of sunlight is absorbed by dissolved and particulate matter. This clearly reduces the rates of direct phototransformation, and changes the solar spectrum in deeper water layers. However, this dissolved and particulate matter is also capable of initiating indirect photoconversions. Given the complexity of these indirect conversions, and the many variables that influence the rate of indirect photolysis, it has so far only been possible, to a limited extent, to derive general, mathematical equations for rate constants in natural water systems.

Given the various direct and indirect transformations that can take place due to interaction with solar radiation, a variety of primary and secondary photoproducts is often observed. Since penetration of light is usually only possible in oxic systems, most photoproducts formed are in an oxidized state, compared with the parent compound.

3.4.6 Methods for measuring abiotic degradation

Standard methods for measuring abiotic degradation are available only for hydrolysis as a function of pH [124]. In this method, the aqueous concentration of the test substance is determined as a function of time, at a specific temperature and a specific pH. The experiments are carried out for at least three pH values, enabling the calculation of the contributions of acid and base-catalyzed, as well as neutral hydrolysis processes. Basically the same procedure is generally used of following the decay of a chemical as a function of time, given the action of an abiotic reagent to measure the contribution of additional abiotic transformation processes.

In photochemical transformation processes the amount of light absorbed by the chemical concerned and the intensity of the light source as a function of wave length are the main factors determining rates of reaction. In all the methods for measuring rates of photolysis described in the literature, these factors are taken into careful consideration. Often use is made of a reference compound with well-known absorption characteristics for which the quantum yield has also been assessed as a function of wavelength (actinometer) [125].

3.5 BIODEGRADATION

3.5.1 Introduction

Microbial degradation plays a key role in the removal of synthetic chemicals from the aquatic and terrestrial environment. Initially it was considered an undesirable process associated with the diminished durability of man-made products. However, if biodegradation of a chemical is too slow it may accumulate in environmental compartments and organisms and eventually result in primary and secondary poisoning in the food web. In addition, it may reduce the quality of drinking water and affect the various functions of surface waters. The persistence of chemicals due to resistance to microbial -attack has been found to be objectionable for several reasons: aesthetically (plastics, foaming surfactants in the past), due to the ecological risk (surfactants, PCBs, DDT, aldrin, etc.) and even as a hazard to human health (dioxins and pesticides in food and drinking water). By contrast with non-biological elimination processes such as hydrolysis or photochemical degradation, biodegradation in the oxygen-containing biosphere is, generally, equivalent to conversion into inorganic end-products, such as carbon dioxide and water. This phenomenon has been named *ultimate* biodegradation or *mineralization* and may be regarded as a true sink in aerobic compartments. In the anaerobic environment, however, microbial degradation processes are generally much slower and may not always result in complete mineralization. Transformation of the parent compound into another organic structure (product) is referred to as *primary degradation or biodegradation*.

The organization of heterotrophic micro-organisms is characterized by catabolic versatility. In order to survive, more metabolic changes are possible than for higher organisms. The ability of the organism to make such changes is called *adaptation* or acclimatization. Mixed microfloras, rather than monocultures, are responsible for the elimination of substances from the biosphere, and because adaptation of the microbial ecosystem to a xenobiotic compound is so important, a more operational definition would be useful. Adaptation can be described as a change in the microbial community that increases the rate of biodegradation of a chemical as a result of prior exposure to that compound. This definition does not distinguish between mechanisms such as gene transfer or mutation, enzyme induction and population changes. The enzymatic machinery of micro-organisms consists of constitutive enzymes, which are involved in fundamental metabolic cycles (e.g., hydrolysis), and adaptive or induced enzymes. These enzymes enable bacteria to utilize organic compounds which are not appropriate for immediate use.

Environmental factors affect the population distribution and biochemistry of bacteria. Sediment and soil are more or less aerobic unless the oxygen consumption

Table 3.10. Free energy of redox reactions in the saturated zone of soil. Calculations are based on data from [126,127].

Environment (electron acceptor)	-DG (kJ)	Relative to oxygen %	Reaction equation
Oxygen	472.5	100	$O_2 + CH_2O \rightarrow CO_2 + H_2O$
Nitrate	462.8	97.9	$4/5\ NO_3^- + 4/5\ H^+ + CH_2O \rightarrow 2/5\ N_2 + CO_2 + 7/5\ H_2O$
Nitrate/nitrite	332.8	70.4	$2\ NO_3^- + CH_2O \rightarrow 2\ NO_2- + CO_2 + H_2O$
Pyrolusite	364.2	77.1	$2\ MnO_2 + 4\ H^+ + CH_2O \rightarrow 2\ Mn^{2+}\ CO_2 + 3\ H_2O$
Manganite	320.9	67.9	$4\ MnOOH + 8\ H^+ + CH_2O \rightarrow 4\ Mn^{2+} + CO_2 + 7\ H_2O$
Hausmannite	330.6	70.0	$2\ Mn_3O_4 + 12\ H^+ + CH_2O \rightarrow 6\ Mn^{2+} + CO_2 + 7\ H_2O$
Hematite	60.0	12.7	$2\ Fe_2O_3 + 8\ H^+ + CH_2O \rightarrow 4\ Fe^{2+} + CO_2 + 5\ H_2O$
Magnetite	27.1	5.7	$2\ Fe_3O_4 + 12\ H^+ + CH_2O \rightarrow 6\ Fe^{2+} + CO_2 + 7\ H_2O$
Sulphate	98.1	20.8	$1/2\ SO_4^{2-} + H^+ + CH_2O \rightarrow 1/2\ H_2S + CO_2 + H_2O$
H2 production	26.0	5.5	$H_2O + CH_2O \rightarrow 2\ H_2 + CO_2$
Methanogenic	91.4	19.3	$CH_2O \rightarrow 1/2\ CH_4 + 1/2\ CO_2$

by micro-organisms, due to an abundance of substrate, is higher than the oxygen supply by diffusion. Aerobic bacteria use oxygen both as a reactant for the oxidation of organic compounds, and as a terminal electron acceptor. The latter is necessary for the conversion of the organic compound, as an energy source, into carbon dioxide. This reaction, also known as *dissimilation*, produces the energy required during the formation of biomass from the organic compound *(assimilation)*. Facultative anaerobic bacteria use oxygen but have the capability to change to another electron acceptor if their environment turns anaerobic. Other electron acceptors are nitrate, utilized by denitrifying bacteria and (particularly in marine environments) sulfate, used by sulfate-reducing bacteria. Oxygen is very toxic to the obligate anaerobic bacteria, which can only use alternative electron acceptors. The methanogens or methane-producing bacteria derive energy from the conversion of hydrogen and carbon dioxide (electron acceptor) into methane. The considerable decrease in energy supply by the different electron acceptors from oxygen to the organic compound itself explains why microbial processes are faster in the aerobic world (Table 3.10).

Biodegradation of synthetic chemicals does not always result in bacterial growth. When (exponential) growth does not occur the degradation process is called *cometabolism*, in which micro-organisms while growing on another, widely available, substrate also have the capacity to transform other compounds (xenobiotics) without deriving any benefit from that transformation [128].

3.5.2 Aerobic biodegradation and metabolic pathways

A wide variety of chemicals can serve as nutrients for bacteria, which are used for their growth and energy requirements. However, the variety of biochemical mechanisms needed for these processes is much narrower, since one mechanism can often be used by the organism for a whole array of related compounds and many of the degradation intermediates are similar. If a substance is completely mineralized, its Theoretical Oxygen Demand (ThOD) and Theoretical Carbon dioxide production ($ThCO_2$) can be calculated from the elemental composition of a substance. The final oxidation products are given in chemical Equations 3.65 and 3.66, without and with nitrification respectively (see next page), where X = any halogen.

Subsequently the ThOD expressed in mg O_2 consumed per mg substance is subsequently deducted from the above equations:

(without nitrification)
see next page (3.67)

(with nitrification)
see next page (3.68)

where *MW Oxygen* is the molecular weight of oxygen (15.9994 au) and *MW subst.* is the molecular weight of the test substance.

The $ThCO_2$ in mg CO_2 produced per mg substance follows easily for mineralization both with and without nitrification:

see below (3.69)

with c being the number of carbon atoms in substance $C_cH_hO_o$.

The biodegradability of a substance, when measured in a laboratory test, is often reported as a percentage. This percentage is calculated based on the theoretically maximum mineralization, i.e., ThOD when oxygen uptake has been used as a biodegradation parameter in the test, or $ThCO_2$ when the production of carbon dioxide has been measured as a mineralization parameter.

Microbial metabolism databases

A large number of more specialized microbial degradation pathways can be found in the University of Minnesota Biocatalysis/Biodegradation Database [129]. This free internet database started collecting microbial metabolism information from literature sources in 1995 and provides detailed information on microbial biocatalytic reactions and biodegradation pathways for primarily xenobiotic, chemical compounds. Mostly aerobic microbial pathways are covered, but some coverage of anaerobic biodegradation pathways is also present. In general, all microbes try to transform chemicals in the direction of a limited number of "central metabolites".

A very useful reference when dealing with the degradation processes of these central metabolites, is the Kyoto Encyclopedia of Genes and Genomes (KEGG) Pathway database [130]. A number of the more important xenobiotic degradation pathways from the UMBBD have also been copied into the KEGG pathway database.

Three of the major oxidative mechanisms may illustrate how bacteria can break down hydrocarbons. However, both environmental conditions and chemical structures may hinder or impede these reactions in specific cases:

ω-oxidation

ω-oxidation is the initial attack on an aliphatic chain at the terminal methyl group which is oxidized to yield a fatty acid. The pathway leads through the primary alcohol and the corresponding aldehyde to a carboxylic acid which, for example, is illustrated by the *n*-octanol degradation pathway in the UMBBD, http://umbbd.ahc.umn.edu/oct/oct_map.html. This reaction requires oxygen in the first step when addition of molecular oxygen to the hydrocarbon takes place, catalyzed by an oxygenase enzyme [131]. Adaptive enzymes are likely to be involved in the initial attack of unsubstituted aliphatics, and they certainly are involved if the chain is branched or has functional groups. w-oxidation is normally followed by ß-oxidation when dealing with alkane chains.

ß -oxidation

β-oxidation is a sequential oxidation two carbons at a time of a fatty acid chain, catalyzed by enzymes. First, thio-ester formation of the carboxyl group with Coenzyme A (CoA) takes place; two hydrogens are removed to give the α,β-unsaturated derivative. Hydration gives the β-hydroxy and dehydrogenation the β-keto derivative. CoA is added between the α and β carbons, acetyl CoA is split off (Figure 3.34) yielding a fatty acid CoA ester which is two carbons shorter. This reaction takes place in all living cellular organisms and does not need molecular oxygen. β-oxidation is used in fatty acid metabolism, and can be found in the KEGG Pathway database with an indication of the enzymes involved in all single steps in the metabolism; http://www.genome.jp/kegg/pathway/map/map00071.html. The reaction is hindered by the presence of methyl groups in the

$$C_cH_hO_oN_nNa_{na}P_pS_sX_x + (c + 1/4\,(h - x - 3n) + na/4 + 5p/4 + 3s/2)\,O_2 + (3p/2 + s)\,H_2O \rightarrow cCO_2 + 1/2\,(h - x - 3n)\,H_2O + nNH_3 + na/2\,Na_2O + pH_3PO_4 + sH_2SO_4 + xHX \quad (3.65)$$

$$C_cH_hO_oN_nNa_{na}P_pS_sX_x + (c + 1/4\,(h - x) + 5n/4 + na/4 + 5p/4 + 3s/2)\,O_2 + (n/2 + 3p/2 + s)\,H_2O \rightarrow cCO_2 + 1/2\,(h - x)\,H_2O + nHNO_3 + na/2\,Na_2O + pH_3PO_4 + sH_2SO_4 + xHX \quad (3.66)$$

$$ThOD\ (mg\ O_2 / mg\ subst.) = (MW\ Oxygen / MW\ subst.) \times (c + 1/4\,(h - x - 3n) + na/4 + 5p/4 + 3s/2) \quad (3.67)$$

$$ThOD\ (mg\ O_2 / mg\ subst.) = (MW\ Oxygen / MW\ subst.) \times (c + 1/4\,(h - x) + 5n/4 + na/4 + 5p/4 + 3s/2) \quad (3.68)$$

$$ThCO_2\ (mg\ CO_2 / mg\ subst.) = (MW\ Carbondioxide / MW\ subst.) \times c \quad (3.69)$$

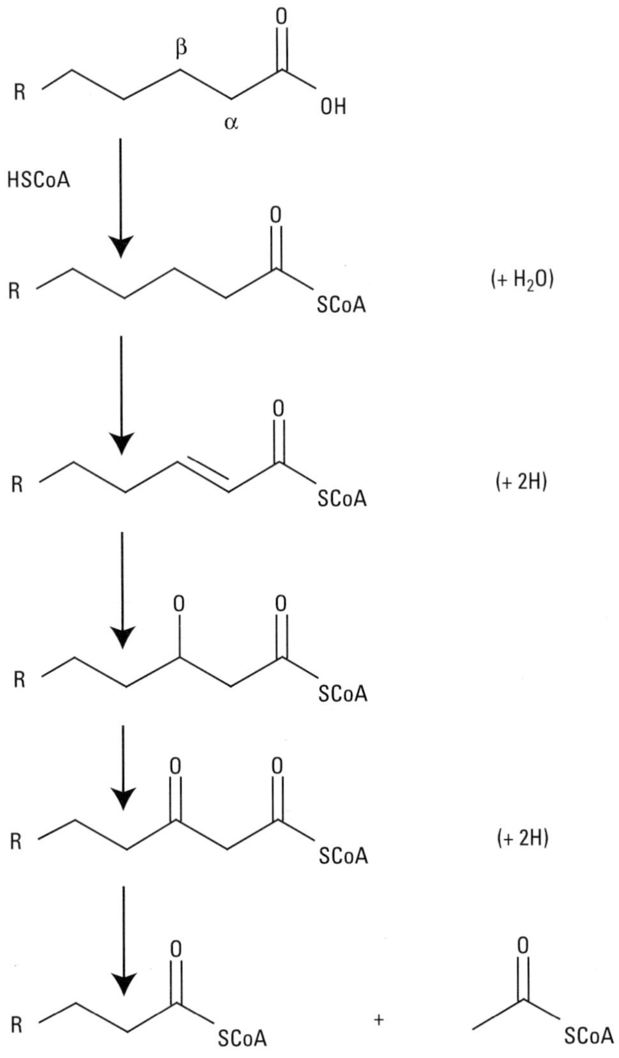

Figure 3.34. β-oxidation of aliphatic hydrocarbons by bacteria. (H)SCoA = Coenzyme A.

β-position: the formation of a keto group would then require a (highly unstable) pentavalent carbon atom. Although an alternative reaction, α-oxidation, may take over, followed by β-oxidation, etc., tertiary carbons in an alkyl chain generally considerably reduce the ease of biodegradability. Studies on surfactants have shown that quarternary carbon atoms [131], especially at the end of the hydrophobic chain, may even completely impede biodegradation.

Aromatic ring oxidation

Aromatic ring oxidation starts with the formation of catechol from benzene or benzene derivatives such as benzoate, phenol and others, by means of enzymecatalyzed oxidation with molecular oxygen, for example, by means of cytochrome P450 enzymes. Recent literature suggests that cytochrome P450 enzymes have two subunits clenching O_2 in such a way that at the catalytic site of the enzyme an oxygen-atom is available for reaction with the aromatic substrate. [132]. In the case of aromatic ring oxidation, electron withdrawing substituents reduce the electron density of the aromatic ring functioning as a substrate for the enzyme-catalyzed oxidation with molecular oxygen. This makes the ring a less suitable target for electrophilic attack by catalytic enzymes performing oxidation with molecular oxygen, like cytochrome P450. After the first oxidation, the ring is cleaved between or adjacent to the two hydroxylated carbons (Figure 3.35). An example of aromatic ring oxidation in the UMBBD database would be the toluene pathway, http://umbbd.ahc.umn.edu/tol/tol_map.html. Although there are several different possible pathways for the breakdown of toluene, the ring opening is constantly performed by catechol formation.

3.5.3 Anaerobic biodegradation

Anaerobic microbial activity is carried out in the absence of O_2 as a terminal electron acceptor (TEA) in respiration. Anaerobic compartments are usually characterized according to the alternative for oxygen that is used by the microbes as the terminal electron acceptor. Alternatives to oxygen in respiration may be (in order of energetic favourability) $N0_3^-$ (nitrate reducing environment), FeIII (iron reducing), MnIV (manganese reducing, SO_4^{2-} (sulfate reducing), and even CO_2 in the case of methanogenesis. In the methanogenic compartment, mineralization is defined as conversion into monocarbon end-products such as methane and carbon dioxide. The role of anaerobic biodegradation in anoxic sediment, soil and groundwater has attracted attention more recently. This is of particular relevance to biodegradation since research has shown that different organic contaminants, such as benzene, toluene, and chlorinated compounds, will have different microbial degradation rates depending upon these TEA conditions. Rates under nitrate reducing conditions are often faster than under methanogenic conditions as nitrate reduction is more energetically favourable. Similarly, aromatic compounds may biodegrade more readily under nitrate reducing conditions than under sulfate reducing conditions. Some compounds, such as chlorinated compounds and MTBE, may actually biodegrade at higher rates under proper anaerobic conditions, than under aerobic conditions.

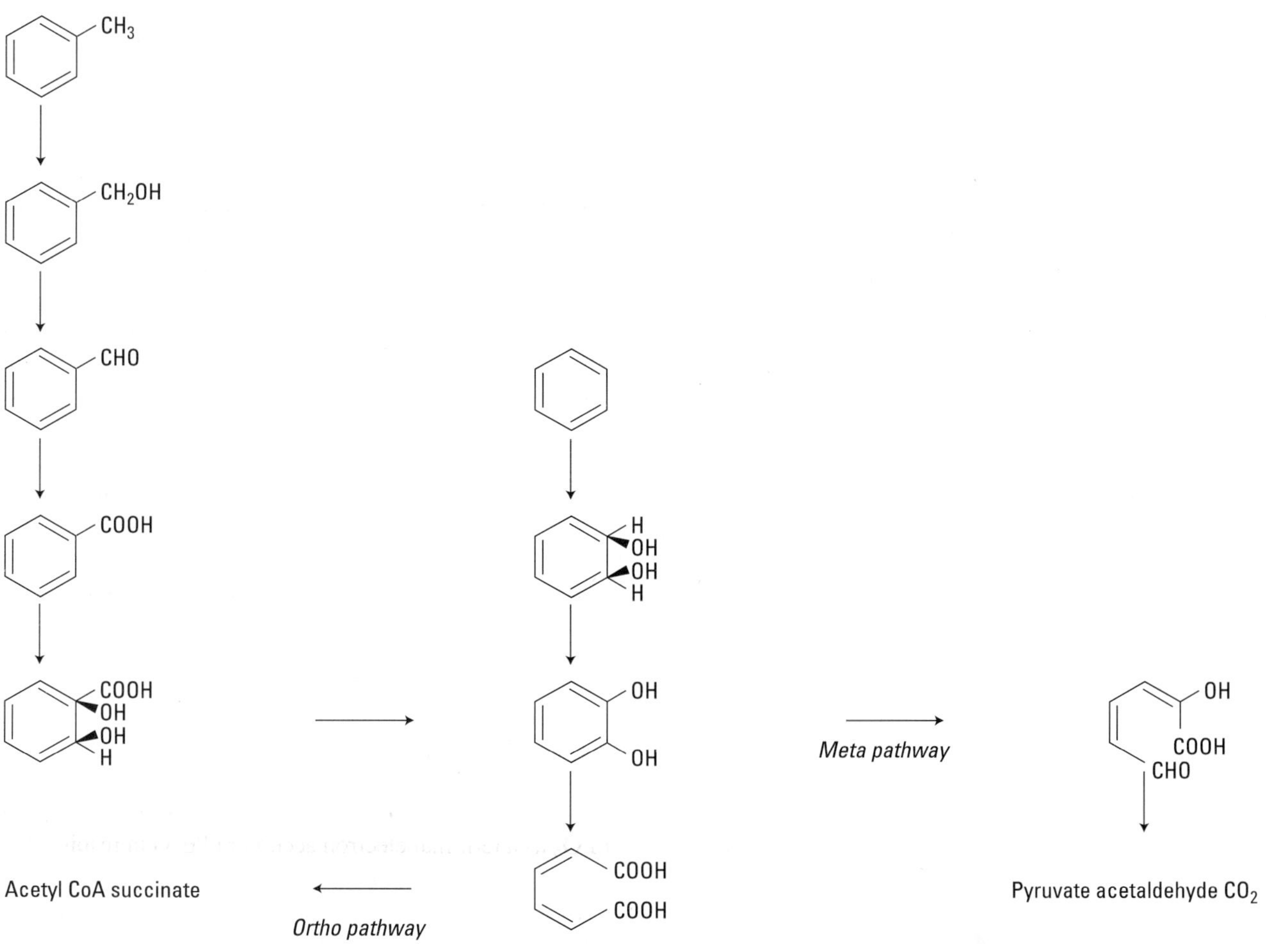

Figure 3.35. Aromatic oxidation by bacteria after catechol formation. Left *ortho* and right, *meta* fission.

Formerly, it was believed that anaerobic biotic processes could be neglected as they were considered rather slow in general, compared to aerobic biodegradation. In addition, experimental studies with anaerobic bacteria are difficult to conduct. Only two decades ago a standard method for ultimate biodegradation of chemicals under methanogenic circumstances was investigated and became eligible for adoption by the OECD [133]. In this method the degree of mineralization is calculated from the measured amount of carbon dioxide and methane produced due to digestion of the tested compound relative to the theoretical amount, which can be calculated according to a stoichiometry as given in Equation 3.70 (see below).

Compared to oxidation mechanisms, the number of catabolic routes is restricted. Primary, rather than ultimate, degradation is more common and transformation rates are slower. Nevertheless, for environmental exposure and risk assessment it is necessary to consider anaerobic biodegradation. The appearance of xenobiotic substances in drinking water is an environmental problem that has received increased attention lately, and shows the importance of the degradability (or the lack thereof) of aerobically persistent metabolites in the anaerobic compartment. Cases have been observed where pharmaceutically active compounds and their metabolites are found in drinking water. In these cases the intake of drinking water depends on the groundwater, and the

$$C_cH_hO_o + (c - h/4 - o/2)\, H_2O \rightarrow (c/2 + h/8 + o/4)\, CH_4 + (c/2 - h/8 + o/4)\, CO_2 \qquad (3.70)$$

degree of artificial recharge of the groundwater aquifers. In highly urbanized areas, such as the city of Berlin, underground water aquifers used for the drinking water supply are partly recharged using aerobically treated wastewater from the municipal STPs. This implies that aerobically persistent substances or metabolites will enter this anaerobic compartment. Their anaerobic biodegradability subsequently determines whether or not such metabolites are found again in the drinking water intake [134].

The importance of the potential anaerobic biodegradability of substances (or their aerobically persistent metabolites) is increasing since anaerobic treatment of sludge in communal waste water treatment plants is becoming more common in order to save energy and reduce the volume of sludge produced. After the digesting process in the anaerobic reactor the sludge will re-enter the aerobic world, carrying the chemicals and reaction products with it, because it is anticipated that the sludge will be applied to agricultural soil. Chemicals or products formed due to primary degradation in the anaerobic reactor may be very stable under such conditions, however, in the sludge they may enter the soil compartment, where they may be susceptible to mineralization by aerobic micro-organisms or subject to other transformation processes. The importance of primary biodegradation is not overestimated as reaction products are usually more polar than the parent compound. Thus, the distribution of products favours the aqueous phase where exposure to aerobic micro-organisms may lead to further and, probably, ultimate degradation.

A special type of primary degradation, known as reductive dehalogenation, may illustrate this. If a compound has undergone a reductive dehalogenation reaction, a halogen atom has been replaced with a hydrogen, rendering a product which is less hydrophobic. It has been found that this transformation process may occur under reducing conditions and that anaerobic micro-organisms are involved, although the dehalogenation reaction at the alkyl carbons seems non-specific and not exclusively mediated by biological processes. Several other chlorinated aromatic chemicals, which have proved rather persistent in the aerobic hydrosphere, have also been shown to undergo reductive dechlorination, yielding products which are generally less problematic in the aquatic environment. These studies provide enough evidence to formulate the following general rules [135]:

- Reductive dehalogenation at the aromatic ring requires strictly anaerobic conditions.
- Specific microbial enzyme systems are involved.
- Higher halogenated aromatic molecules are less persistent than their lower halogenated congeners.

Although in soil and sediment the last rule may be counteracted by reduced bioavailability, it has important consequences. The opposite generally applies for aerobic degradation because cleavage of the aromatic ring is increasingly hindered by the number of halogen substituents. Thus, biodegradation depends on both the chemical structure and the environmental conditions. In addition, structure-biodegradability relationships for aerobic biodegradation principally differ from those observed for anaerobic transformation processes [136,137].

3.5.4 Reasons for the environmental persistence of chemicals

Microbial communities in the natural environment seem catabolically versatile, in the sense that a specific habitat may contain one or more species or populations which are capable of degrading every synthetic chemical. As shown above, for some persistent chemicals' habitats at different redox levels are complementary. This raises the question: why do some manmade chemicals persist in the environment for such a long time? Several mechanisms are responsible for slow biodegradation of chemicals. Generally, the rate and extent of biodegradation of a chemical depends on both its chemical structure and the prevailing environmental conditions. However, slow microbial degradation of a substance in some environmental compartments should also be considered in relation to slow transport of that chemical to environmental subcompartments where crucial transformation reactions can be carried out by micro-organisms. The following properties or conditions have a significant influence on the biodegradation of synthetic chemicals:

1. Chemical structure.

Type, number and position of substituents on aliphatic or aromatic structures may cause "violation of comparative biochemistry and enzyme specificity", as described by Alexander [138]. Effects of substitution have already been discussed in the three examples of major metabolic pathways for biochemical oxidation. The influence of the molecular structure on its biodegradability in the *aerobic* environment is shown in Table 3.11. It should be emphasized, however, that there are many exceptions to these general rules. The following example may illustrate the application of these general rules: 2,3,7,8-TCDD, popularly known as dioxin, has at least three distinctive

Table 3.11. Influence of molecular structure on the biodegradability of chemicals in the aerobic environment.

Type of compounds or substituents	More biodegradable	Less biodegradable
Hydrocarbons	linear alkanes > C_{12}	linear alkanes < C_{12}
	alkanes with not too high molecular weight	high molecular weight alkanes
	linear chain	branched chain
	-C-C-C-	-C-O-C-
	aliphatic	aromatic
	mono- and bicyclic aromatic	polycyclic aromatic
Aliphatic chlorine	Cl more than 6 carbons from terminal carbon	Cl at 6 or less carbon atoms from terminal C
Substituents to an aromatic ring	-OH	-F
	$-CO_2H$	-Cl
	$-NH_2$	$-NO_2$
	$-OCH_3$	$-CF_3$

aspects that contribute to its high environmental persistence. In this molecule *ether groups* link together *aromatic* moieties that are substituted with chlorine (Figure 3.36). A more extensive description of structure-biodegradability relationships is given in Chapter 9.

2. Environmental conditions.
Temperature is an important factor and especially around and below 4°C, microbial processes become very slow. The optimum temperature for *psychrophilic (cold-loving)* bacteria is between 0 and 20°C and for *mesophyllic (moderate temperature loving)* bacteria it is between 20 and 40°C. In seawater 15°C is the borderline between different microbial ecosystems. The inorganic nutrient status of the surface water affects the biodegradation rate and in some coastal waters may even exceed the temperature effect. The presence of auxiliary organic nutrients may also play a role, and the occurrence of cometabolism has already been mentioned. Failure of biodegradation may be due to the presence of other, more easily degradable compounds used in preference to the specific xenobiotic compound. This phenomenon is known as *diauxism*. Unlike seawater, which is a well-buffered system of pH 8, inland waters can vary up to 5 pH units in acidity, thereby determining the form in which some chemicals exist. The availability of some natural organic substrates may also facilitate cometabolism of the pollutant. However, even if it were possible to find two aquatic ecosystems characterized by similar environmental parameters, the outcome of a biodegradability experiment might be quite different for the same chemical. The presence and influence of high population densities of "specialized" degraders is evident. Some aquatic ecosystems may have been previously exposed to a chemical or another pollutant which shares a common enzyme system of such a specific degrader. The presence and density of specific degraders is often highly decisive for biodegradation to occur within a limited period of time.

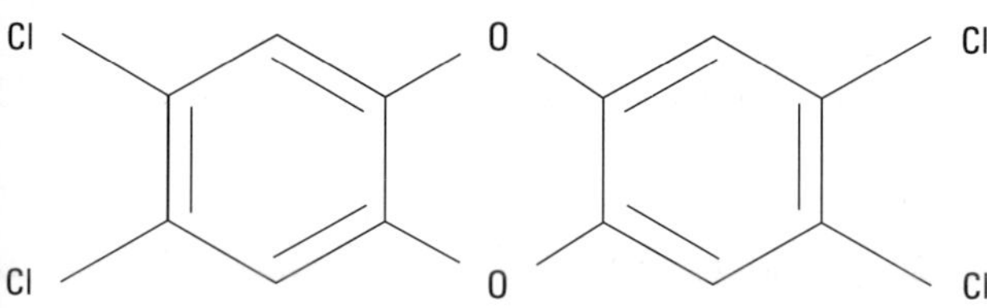

Figure 3.36. Molecular structure of 2,3,7,8-tetrachloro-*p*-dioxin (TCDD).

3. Bioavailability.
If a chemical is trapped in microsites, e.g., in inorganic material such as clay minerals or the organic matrix of sediment or soil, interaction with micro-organisms may be physically impossible, which impedes biodegradation.

3.5.5 Kinetics of biodegradation

In most kinetic models the chemical is considered a "substrate" and, as such, limiting to bacterial growth.

What these models all have in common is that they combine mass-transfer (from substrate to biomass) with saturation phenomena that are similar to non-linear Michaelis-Menten kinetics for biotransformation (Section 3.6.5). A popular expression for simulating biological processes (e.g., biodegradation) is the Monod-function:

$$\mu = \mu_{max} \cdot C / (K_c + C) \qquad (3.71)$$

where

μ = the growth rate of biomass (1/d)
μ_{max} = the maximum growth rate (1/d)
C = the concentration of growth-limiting substrate (mg/L)
K_c = the half-saturation coefficient (mg/L).

K_c is the concentration which allows the micro-organism to grow at half the maximum growth rate. Monod kinetics are different from, but still based upon Michaelis-Menten kinetics for enzymes. Monod kinetics can be thought of as describing a chain of enzymatically mediated reactions with a limiting step described by Michaelis-Menten kinetics. This is why the equations for both kinetic models are identical. In water the concentration of a xenobiotic is very low, usually much lower than the saturation constant, and as a consequence the non-linear rate equation is simplified to first-order kinetics.

Prior to the degradation of many organic compounds, a period is observed in which no degradation of the chemical is evident. This time interval is designated an *acclimation period*, or alternatively an adaptation or lag period. It is defined as the length of time between the addition or entry of the chemical into an environment and evidence of its detectable loss. During this interval, no significant change in concentration is noted, but then its disappearance becomes evident and the rate of degradation often becomes rapid due to the exponential growth of the micro-organisms. This rapid elimination phase is often termed the *log* (for logarithmic) *phase*, and is described by first-order kinetics, as referred to above. In biodegradation tests, as well as in the environment, the elimination percentage reaches a plateau. Concentration of the substance has then become so low that (exponential) growth of the micro-organism is no longer possible using the substance as the main substrate. Generally, the plateau phase never reaches 100% elimination when expressed in terms of the mineralization parameters DOC (Dissolved Organic Carbon) removal, O_2 uptake and/or CO_2 production. This is due to the fact that part of the organic carbon is used for growth of the bacterial mass (and is thus not mineralized), and the fact that the bacterial mass will be producing dissolved organic matter itself. These considerations also form the basis of the "pass levels" chosen for biodegradation test results; substances are considered "completely" biodegraded or mineralized when > 60% of the ThOD or $ThCO_2$, or > 70% DOC removal is reached within a certain time span (normally 28 days). An example of a degradation curve resulting from a laboratory degradation test showing the lag, log and plateau phase is given in Figure 3.37.

Biodegradation in sediment or soil is often described in terms of biological half-lives. If the half-life of the chemical is indeed independent of its concentration, the degradation rate equation is first-order for the chemical concentration:

$$dC / dt = -k \times C = -\frac{\ln 2}{t_{1/2}} \times C \qquad (3.72)$$

where

C = the concentration of chemical in wet sediment or soil (mg/L)
$t_{1/2}$ = the biological half-life of the chemical (d)
k = the biodegradation rate constant (1/d) for wet soil or sediment.

It is not impossible that biodegradation occurs in the particulate phase. However, several studies have provided evidence that a chemical associated with sediment or soil particles is not available for biodegradation because micro-organisms only utilize dissolved chemicals [139]. The rate of biodegradation in a solids-water system is adequately described by first-order kinetics of disappearance from the aqueous phase. In this model sorption may diminish the overall degradation rate, and depending on differences in process rates, two extreme scenarios can be distinguished:

1. Partitioning of a compound between the particle and the aqueous phase is governed by a thermodynamic equilibrium occurring at a fast rate with respect to degradation processes. The rate of elimination will then become strongly dependent on the organic carbon-water partitioning constant (K_{oc}) of a substance (see Equation 3.2 and accompanying text). With increasing K_{oc} the concentration of a substance in the pore water subsequently becomes very low, and hence the elimination rate due to biodegradation becomes proportionally low (compare Equations 3.3.-3.5 of this chapter).
2. Biodegradation in the aqueous phase is relatively fast but overall elimination (and hence the biodegradation

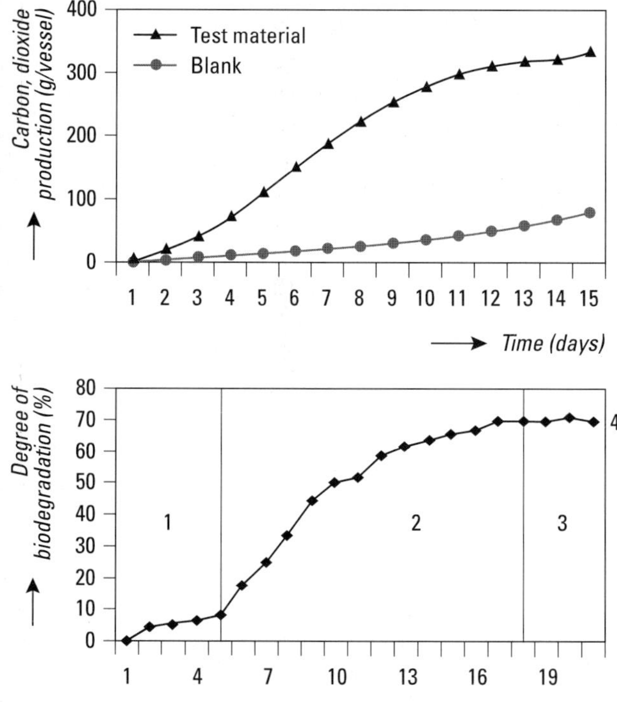

Figure 3.37. Above: CO_2 production curve in a OECD 301B Modified Sturm test. By subtracting the CO_2 blank production from the test material production and dividing by the Theoretical CO_2 Production ($ThCO_2$) the corresponding biodegradation curve is calculated (below), showing lag phase (1), exponential growth or log phase (2), plateau phase (3) and mean degree of biodegradation (4) of a readily biodegradable substance.

kinetics) from the solids-water system is controlled by slow desorption.

This again illustrates the difficulties associated with extrapolation from a standard biodegradation test to an environmental half-life of a laboratory-derived degradation rate. When assessing the environmental risk of a chemical, it is important to realize that even a relatively easily biodegradable chemical can become more or less persistent when it ends up in an environmental compartment where its bioavailability becomes limited.

3.5.6 Assessing biodegradability and biodegradation rates

Estimated rate constants of degradation processes, particularly for biodegradation, generally, have larger margins of uncertainty than those of exchange processes. In principle, there are two approaches to obtaining biodegradation rate constants for a particular compartment:

1. A theoretical approach, making use of QSARs (Chapter 9).
2. An experimental approach, on the basis of standardized test results.

Despite major efforts, it has so far proved difficult to formulate generally applicable QSARs for the most relevant chemical elimination process, i.e., aerobic biodegradation in water and soil. At present, the interpretation of experimental studies is the only way to estimate rates of aerobic biodegradation.

If experimental work consisted of field studies or simulations of the "natural" environment, rate constants for microbial degradation in the relevant compartment for a number of chemicals would have been obtained. Unfortunately, the number of chemicals studied so far is limited to the category of pesticides (where such studies are obligatory for market introduction) and a few household and industrial chemicals with high production volumes. Most biodegradability data are derived from standard methods that make use of some artificial environment accommodating water, auxiliary nutrients, the test chemical and an inoculum (mixed microflora). Furthermore results from a field study in one environmental compartment cannot readily be used to estimate biodegradability in other compartments, without rigorously accounting for environmental factors.

The first biodegradability tests were the result of legislation on detergents that came into force soon after the introduction of synthetic surfactants in the early 1960s. A particular type of anionic surfactant, i.e., the slowly biodegradable branched alkyl benzene sulfonates, caused heavy foaming problems which appeared more serious than just a nuisance. Foaming had an adverse effect on water quality because it hindered the proper functioning of biological waste-water treatment plants. Therefore, elimination of surface-active properties from waste water during the short retention time in a treatment installation became a legislative requirement. Test methods were designed in such a way that elimination of surface-active properties due to microbial transformation was the test criterion.

Primary degradation was assessed by an analytical method that is specific to the whole range of certain synthetic surfactants. These test methods appeared satisfactory because the detergents that passed the test, specified in the detergent legislation, indeed did not foam during waste-water treatment and generally did not cause effluent toxicity problems. A common system was published by the OECD in 1976 [140]. It consists of two stages which differ both in principle and in the conclusions which can be drawn from them. These tests are suitable for anionic and non-ionic surfactants:

- The OECD screening test (OST) is a static flask test which is relatively quick and simple to carry out. This test should be considered as an "acceptance test", not as a "rejection test". It selects "soft" surfactants which do not have to be tested further because high biodegradability is expected in sewage treatment plants.
- The OECD confirmation test (OCT) is based on a simulation of the conditions existing in an activated sludge plant. This test should be used for any surfactant which may not have passed the OST, either to confirm or disprove the first results obtained.

In recent decades, the OST and OCT have been modified and other methods have also been added by the OECD. This was done to design a three-tier test system to evaluate the *biodegradability* of industrial chemicals as a *property* that is part of their ecotoxicity [133]. A major difference with respect to OST and OCT is the use of a non-specific analytical parameter (O_2 uptake, CO_2-development or DOC removal) to make the system suitable for any chemical, irrespective of its physicochemical properties. In principle, there is no need to develop an analytical method before a biodegradability test can be conducted. A second advantage of this approach is the fact that a non-specific parameter represents mineralization instead of primary degradation. For the very diverse group of "new chemicals" this is obviously a safer approach. The system is also known as the OECD hierarchy, which refers to the three different levels of testing, as follows.

1. Ready biodegradability.

Ready biodegradability refers to stringent tests which provide limited opportunity for biodegradation and acclimatization to occur. It may be assumed that a chemical which is regarded as "readily biodegradable" will rapidly biodegrade in the environment and may be considered as such.

2. Inherent biodegradability.

Inherent biodegradability refers to tests which allow prolonged exposure of the test compound to micro-organisms, a more favourable test compound to biomass ratio, and chemical or other conditions which favour biodegradation. A compound giving a positive result in this type of test may be classified as "inherently biodegradable". However, due to the favourable conditions employed, its rapid and reliable biodegradation in the environment should not be assumed. Inherent biodegradability tests, e.g., the Zahn-Wellens test, incorporated into OECD test guideline 302B, have their origin in industrial testing of the suitability of waste water to be treated in an industrial or municipal waste water treatment plant. They were not designed to distinguish between sorption to sludge, volatilization or biological degradation (although the shape of the elimination curve can give an indication of the process taking place), but originally sought to answer the question whether a substance would be *removed* (by any means) from the waste water stream when put through an STP. Extrapolation of test results from inherent biodegradability tests to the environment is therefore difficult.

3. Simulation.

Aerobic and anaerobic simulation tests provide data for biodegradation under specified environmentally-relevant conditions. These tests simulate the degradation in a specific environment by use of indigenous biomass, media, relevant solids (i.e., soil, sediment, activated sludge or other surfaces) to allow sorption of the chemical, and a typical temperature which represents the particular environment. A low concentration of test substance is used in tests designed to determine the biodegradation rate constant, whereas higher concentrations are normally used for identification and quantification of major transformation products for analytical reasons. Low concentrations of chemicals in these tests refer to concentrations (e.g., less than 1 μg/L to 100 μg/L) low enough to ensure that biodegradation kinetics obtained in the test reflect those expected in the environment being simulated. Biodegradation is measured either by radio-labelling techniques or by specific chemical analyses. See Table 3.12 for a number of simulation test guidelines and their analytical techniques.

Ready biodegradability tests (RBT) are designed for a quick selection of "soft" chemicals to avoid further costly and time-consuming research. Furthermore, unlike in the OST, a wide range of physicochemical and biological properties have to be determined. To meet the demands of simplicity and cost efficiency, there are six different methods in the OECD scheme, which are reasonably complementary. The methods listed in Table 3.12 are all based on the principle that biodegradation is monitored as the degree of mineralization. This is done by means of sum parameters, such as the elimination of dissolved organic carbon (DOC), oxygen uptake or carbon

Table 3.12 Ready biodegradability tests (RBT) and inherent biodegradability tests (IBT) according to the OECD. Population densities are in colony forming units (CFU) per ml. From [133,141].

OECD test guideline	Summary parameter	Population density (CFU/ml)
Ready biodegradability		
301E: Modified OECD screening test	DOC	$(0.5 - 2.5) \times 10^2$
301B: CO_2 evolution	CO_2	$(2 - 10) \times 10^5$
301F: Manometric respirometry test	O_2	$(2 - 10) \times 10^5$
301A: DOC Die-away test	DOC	$(2 - 10) \times 10^5$
301D: Closed bottle test	O_2	$(0.5 - 2.5) \times 10^3$
301C: Modified MITI(I) test	O_2	$(2 - 10) \times 10^5$
306: Biodegradability in Seawater	DOC	
Inherent biodegradability		
302B: Zahn-Wellens test	DOC	$(0.7 - 3) \times 10^7$
302A: Modified SCAS test	DOC	$(2 - 10) \times 10^7$
302C: Modified MITI(II) test	O_2	$(0.7 - 3) \times 10^6$
Simulation tests		
303A: Activated Sludge Units	DOC	
303B: Biofilms	DOC	
304A: Inherent Biodegradability in Soil	$^{14}CO_2$	
307: Aerobic and Anaerobic Transformation in Soil	$^{14}CO_2$ / CO_2	
308: Aerobic and Anaerobic Transformation in Aquatic Sediment Systems	$^{14}CO_2$ / CO_2	
309: Aerobic Mineralization in Surface Water	$^{14}CO_2$ / CO_2	

dioxide production. Without employing expensive ^{14}C-techniques, this is only possible if the test compound is the sole carbon and energy source for micro-organisms. The predictive value of a positive result in any of the RBTs is postulated as follows [142]:

- A substance will be completely removed in a biological treatment plant, even if physicochemical removal mechanisms, such as sorption on withdrawn sludge or volatilization in the aeration tank, are negligible.
- The half-life of the substance in surface water is less than 5 days.
- Biological half-lives in aerobic soils, assuming that the chemical is biodegraded only in the pore water, are dependent on the solids-water partition coefficient and may range from 0.1 (low sorption) to 300 days for sorptive chemicals due to their decreased bioavailability to the microbes.

The extrapolation of a positive result from a ready biodegradability test to an environmental half-life of 5 days is based on the results of comparison of real world data with laboratory test data [142]. This approach has also been adopted in the EU Technical Guidance Document on Risk Assessment, although a safety factor of 3 has been applied to these findings, leading to a maximum environmental half-life of 15 days for a readily biodegradable substance in EU risk assessments [38], which has also been implemented in the European Uniform System for Evaluation of Substances (EUSES).

Inherent biodegradability tests (IBT) are designed to demonstrate the potential biodegradability of a compound. Unlike in RBTs, the conditions for biodegradation to occur are more favourable as indicated, for example, by the much higher population densities for IBTs when compared to RBTs, as shown in Table 3.12. In addition, these methods have a screening function as persistent chemicals are also detected. A negative result indicates that a chemical is clearly persistent and, tentatively, that no further research on biodegradation has to be done. The MITI(II) test is an IBT and has a more favourable biomass to chemical ratio than the MITI(I) test, the latter being an RBT, as indicated in Table 3.12. The other IBTs are the Zahn-Wellens test, which has some elements of an industrial waste water treatment system, and the semi-continuous activated sludge test (SCAS), having a hydraulic residence time typical for very low-loaded biological treatment systems. Obviously, the predictive value of a negative result in an IBT is zero degradation in aerobic compartments. Nevertheless, extrapolation to the "natural" environment on the basis of only the simple RBTs and IBTs can be problematic, as most chemicals are *negative* in an RBT but *positive* in an IBT. These chemicals are probably not persistent in the environment, and may already be fully or partly mineralized in a biological waste water treatment system, for example. This has been shown in test systems which are simulations of such engineered ecosystems. Extrapolation from an IBT to an environmental half-life is hardly possible as the concentration dependence of the degradation rate, and the dependence on population densities can differ for different substances. Therefore, for a comprehensive risk assessment biodegradation rates in any compartment of concern have to be established by means of simulation tests.

Simulation tests may be subdivided according to the environment that they are designed to simulate, e.g., a) STPs, b) soil, c) aquatic sediments, and d) surface water. The activated sludge test (OECD 303A) is a method which is very similar to the OCT for detergents. However, it differs in that mineralization is analyzed (without ^{14}C-techniques), instead of primary degradation. The method is designed to determine biodegradation of water-soluble organic compounds in a continuously operated test system (where the previous test were all batch tests), simulating the activated sludge process in waste water treatment plants. The OECD 303B Biofilms test is designed to assess biodegradability in waste water treatment involving biofilms, namely, percolating or trickling filters, rotating biological contractors or fluidized beds.

Aerated soils are aerobic, whereas water-saturated or water-logged soils are frequently dominated by anaerobic conditions. The surface layer of aquatic sediments can be either aerobic or anaerobic, whereas the deeper sediment is usually anaerobic. These conditions in soil or sediment may be simulated by using aerobic or anaerobic tests described in the test guidelines (OECD 307 and OECD 308).

OECD testing guideline 309 is a laboratory shake flask batch test to determine rates of aerobic biodegradation in samples of natural surface water (fresh, brackish or marine). Very low concentrations of the test substance are used, in order to mimic environmental conditions. Often ^{14}C-labelled substances are employed in these tests to be able to accurately measure these low concentrations.

Standard biodegradation test results (like the OECD biodegradation testing battery) play an important role in the assessment of the environmental persistency of a substance, as performed e.g., in the PBT assessment described in the EU Technical Guidance Document on Risk Assessment [38]. Ready biodegradability test results are often the only available data. However, failure to meet the ready biodegradability criteria does not in itself constitute environmental persistency. Many substances are currently considered potentially PBT (Persistent, Bioaccumulative, and Toxic) or potentially vPvB (very Persistent and very Bioaccumulative) solely based on their failure to pass a ready biodegradability test (in combination with meeting the bioaccumulation and toxicity criteria). To further evaluate the potential risk of such potential PBT or vPvB chemicals, it is recommended (e.g., in the EC [38]) that the Persistency criterion be scrutinized first, as this does not require any animal testing, contrary to any further bioaccumulation and toxicity testing. A recent workshop on simulation testing of environmental persistence [143] led to several recommendations on how to improve current procedures.

Firstly, screening test data (RBTs) should be fully explored by considering test results that do not reach the threshold level to see whether or not PBT de-selection would be warranted. A simple adaptation of the test guideline which would increase the usefulness of the screening test data for the evaluation of environmental persistency would be to routinely extend the biodegradation test period to 42 or even 56 days (instead of 28 days).

Secondly, the role of inherent test results, which constitute the largest part of older test data, should be taken into account in the assessment of environmental persistency. Currently simulation tests are thought to be the best way to shed the persistency label for substances that fail the biodegradation screening test, since inherent biodegradation tests have such little relevance to environmental degradation rates (see above). However,

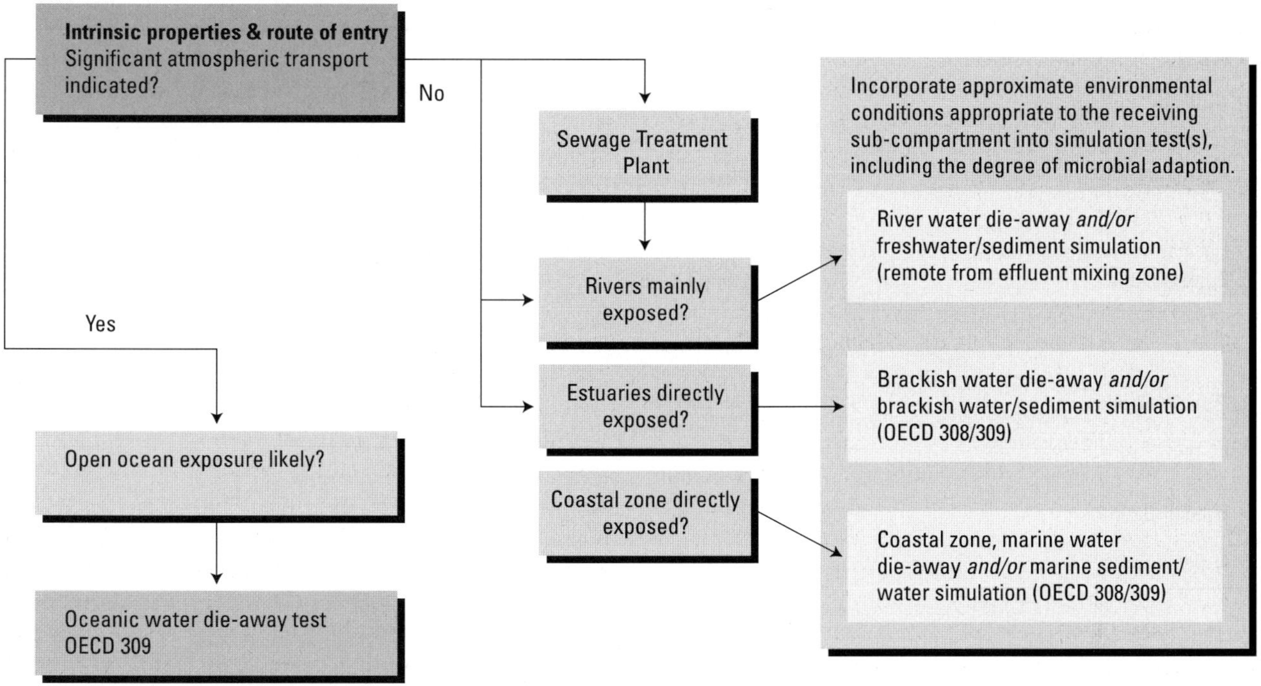

Figure 3.38. Flow diagram for selecting the appropriate environmental compartments and subsequent simulation test conditions.

existing inherent test results can, at the very least, play a role in avoiding unnecessary simulation testing (for substances that do not pass the 20% mineralization threshold in inherent tests), and could be used to prioritize those substances that are most likely to show appreciable degradation in a simulation test.

Finally, for those substances requiring further testing, a testing strategy is proposed where the environmental compartment(s) to which emissions take place, or to which significant atmospheric or other transport is expected, determine the type of simulation test and the test conditions best suited to the evaluation of persistence in the environment [143]. A flow diagram for the selection of the appropriate simulation test is given in Figure 3.38.

3.6 BIOTRANSFORMATION

3.6.1 Introduction

Organisms in the environment are surrounded by a large number of chemicals which are potentially harmful. Many of these compounds will be taken up by organisms. If the concentration of a chemical in an organism becomes too high, this affects its normal functioning. The organism has two major ways of eliminating a chemical: it is either excreted in its original form (the parent compound), or the structure is altered by the organism. When a chemical is transformed by micro-organisms it is called biodegradation (Section 3.5). When a chemical is transformed by other organisms, it is called biotransformation. Biotransformation influences the fate of a compound by decreasing its amount due to conversion into a new xenobiotic compound, the metabolite. Biotransformation can therefore be defined as an enzymecatalyzed conversion of one xenobiotic compound into another.

Biotransformation reactions involve enzymes, which act as biological catalysts. This mechanism distinguishes it from physicochemical conversions (such as photolysis) where no enzymes are involved. For biochemical reactions of lipids, proteins, carbohydrates and other normal body constituents, the term metabolism is used, while for xenobiotics the term biotransformation is more appropriate.

3.6.2 Effects of biotransformation on xenobiotics

In general, biotransformation leads to the conversion of the parent compound into a more water soluble form. As a result these more hydrophilic compounds may be more easily excreted from the body than the parent compound

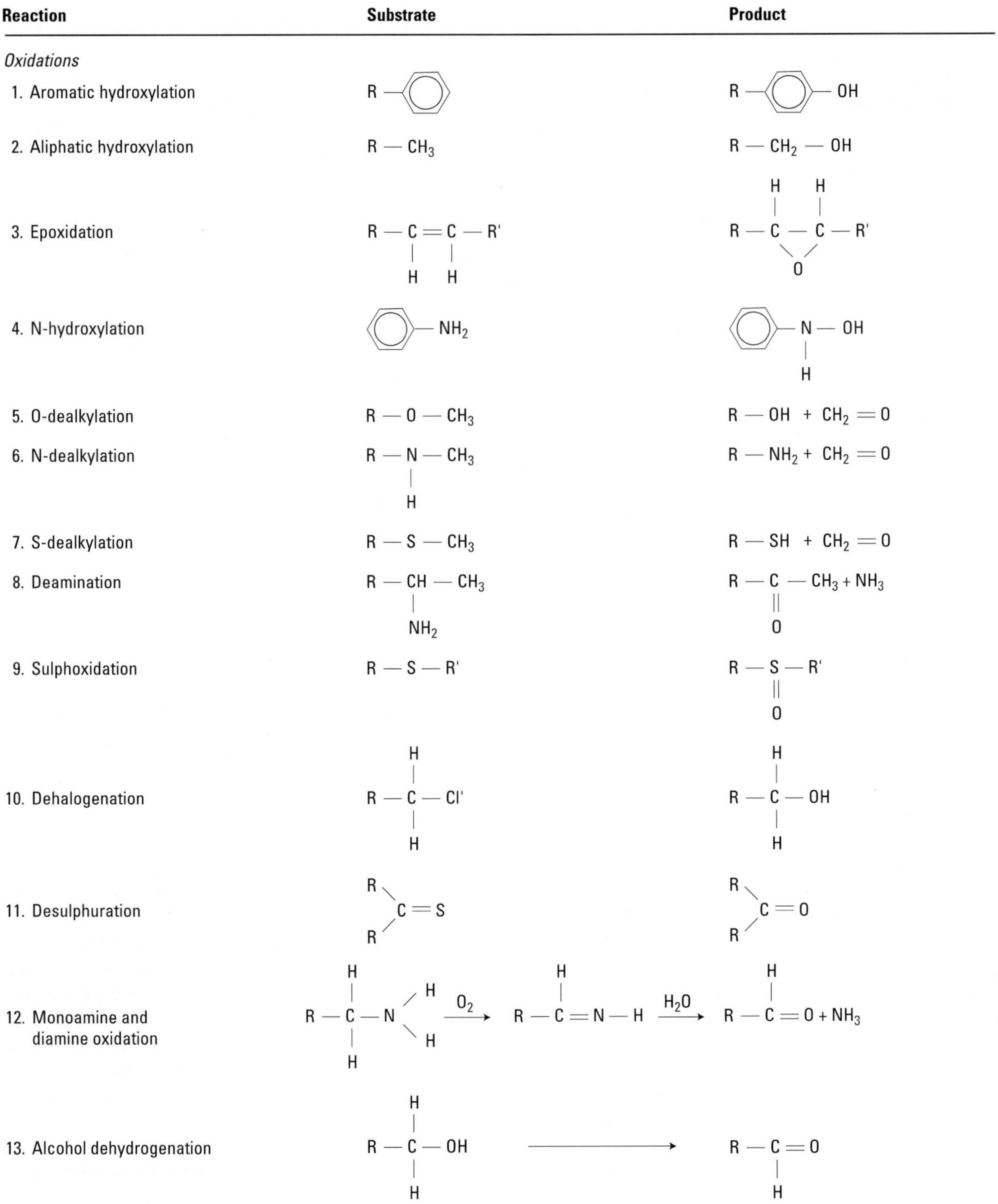

Reaction	Substrate	Product
Oxidations		
1. Aromatic hydroxylation	$R-C_6H_5$	$R-C_6H_4-OH$
2. Aliphatic hydroxylation	$R-CH_3$	$R-CH_2-OH$
3. Epoxidation	$R-CH=CH-R'$	$R-CH-CH-R'$ (epoxide, bridged by O)
4. N-hydroxylation	$C_6H_5-NH_2$	$C_6H_5-NH-OH$
5. O-dealkylation	$R-O-CH_3$	$R-OH + CH_2=O$
6. N-dealkylation	$R-NH-CH_3$	$R-NH_2 + CH_2=O$
7. S-dealkylation	$R-S-CH_3$	$R-SH + CH_2=O$
8. Deamination	$R-CH(NH_2)-CH_3$	$R-C(=O)-CH_3 + NH_3$
9. Sulphoxidation	$R-S-R'$	$R-S(=O)-R'$
10. Dehalogenation	$R-CH_2-Cl'$	$R-CH_2-OH$
11. Desulphuration	$R_2C=S$	$R_2C=O$
12. Monoamine and diamine oxidation	$R-CH_2-NH_2 \xrightarrow{O_2}$	$R-CH=N-H \xrightarrow{H_2O} R-CH=O + NH_3$
13. Alcohol dehydrogenation	$R-CH_2-OH \longrightarrow$	$R-CH=O$

Figure 3.39. The most common biotransformation reactions of xenobiotics in biota.

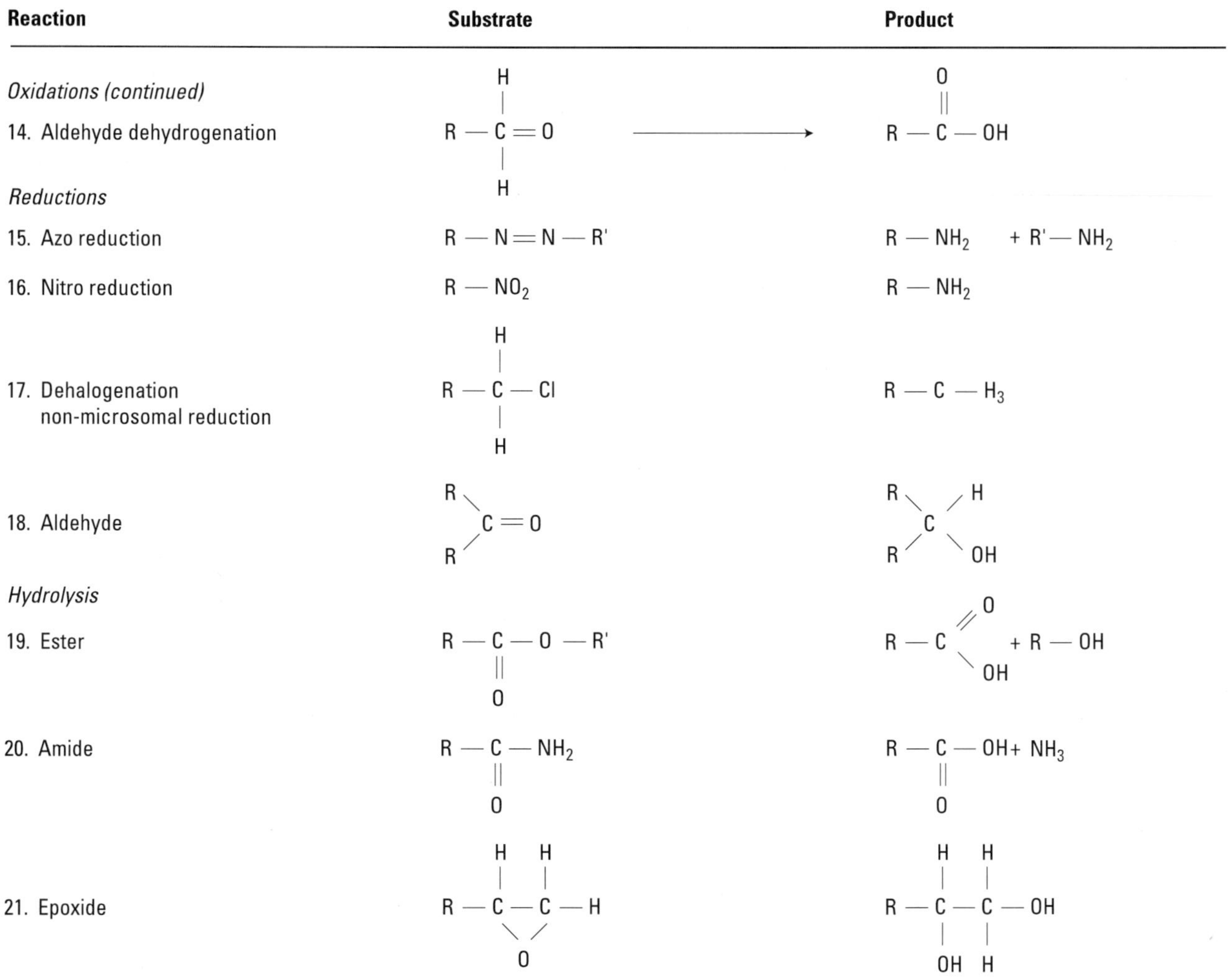

Reaction	Substrate	Product
Oxidations (continued)		
14. Aldehyde dehydrogenation	R—C(H)(H)=O	→ R—C(=O)—OH
Reductions		
15. Azo reduction	R—N=N—R'	$R—NH_2$ + $R'—NH_2$
16. Nitro reduction	$R—NO_2$	$R—NH_2$
17. Dehalogenation non-microsomal reduction	R—C(H)(H)—Cl	$R—C—H_3$
18. Aldehyde	$R_2C=O$	$R_2C(H)(OH)$
Hydrolysis		
19. Ester	R—C(=O)—O—R'	R—C(=O)(OH) + R—OH
20. Amide	$R—C(=O)—NH_2$	R—C(=O)—OH + NH_3
21. Epoxide	R—C(H)—C(H)—H (with O bridging the two C)	R—C(H)(OH)—C(H)(H)—OH

Figure 3.39. The most common biotransformation reactions of xenobiotics in biota (continued).

(Figure 3.39). When the chemical structure of a compound is altered, many properties of the compound are likely to be altered as well. Hence the biotransformation product will behave differently within the organism with respect to tissue distribution, bioaccumulation, persistence, and route and rate of excretion.

Biotransformation may also influence the toxicity of a compound. This can be either beneficial or harmful to an organism. Biotransformation may prevent the concentration in the organism from becoming so high as to produce a toxic response. However, a metabolite may be formed which is more toxic than the parent compound. Transformation into a more toxic compound is called *bioactivation*. Reduction of toxicity due to transformation to a less harmful product is called *detoxification*.

Enzymes, the catalysts of biotransformation reactions, determine the qualitative and quantitative aspects of biotransformation. Enzymes can be affected by many variables, such as age, sex, and temperature. The biotransformation of xenobiotics often involves enzymes that have a relatively low degree of substrate specificity compared with enzymes involved in the metabolism of constitutive compounds. Many organisms are able to biotransform a wide variety of chemicals that differ greatly in structure but have functional groups in common. The biotransformation of many xenobiotics is usually determined in the liver. In this organ enzyme activity is high compared with other parts of the body. However, other tissues (e.g., muscle) may contribute significantly to the total biotransformation rate. Because of the relatively large size of muscles, the total biotransformation may, in some cases, exceed that of the liver. Furthermore, at the point of entry, such as in skin or the intestinal wall, biotransformation rates may

Table 3.13 The most important enzyme systems which metabolize pesticides [20]. With permission.

Enzyme system	Location	Compounds metabolized
Phase-I reactions:		
Mixed function oxidases	Microsomes, notably from vertebrate liver and insect fat body	Many liposoluble pesticides
Phosphatases	Present in nearly all tissues and subcellular fractions of species	Organophosphorus insecticides and "nerve gases"
Carboxyesterases	In most tissues of insects and vertebrates	Malathion and malaoxon
Epoxide hydroxylase	Microsomes, particularly in the mammalian liver	Dieldrin, heptachlor and arene epoxides
DDT dehydrochlorinase	Virtually all insects and vertebrates	*p,p'*-DDT and *p,p'*-DDD
Phase-II reactions:		
Glucuronyl transferases	Mainly in microsomes; widespread in vertebrates other than fish and insects	Compounds with labile hydrogen, including hydroxylated metabolites
Glutathione-S-transferases	70,000 g supernatants of vertebrates livers and also insects	Chlorinated compounds, e.g.γ-HCH; also some epoxides

also be important in affecting the chemical structure of substances entering the organism.

3.6.3 Types of biotransformation reactions

There are two types of biotransformation reactions: Phase-I non-synthetic reactions and phase II synthetic reactions [20,144,145]. Phase-I reactions include hydrolysis, reduction, and oxidation; Phase-II reactions are usually conjugation reactions. The Phase-II reactions most studied are glucuronide, sulfate, acetyl and glutathione conjugation (Table 3.13). During Phase-I reactions the molecule is changed by the introduction of polar groups, such as hydroxy (-OH), carboxyl (-COOH) or amino (-NH_2) groups. The products of Phase-I reactions are often reactive compounds which can be easily conjugated in Phase-II reactions. The conjugated products will then be excreted. Which type of reaction will occur depends on the chemical structure of the compound. Phase-I and Phase-II reactions usually consist of several steps. In Figure 3.40, only the parent compound and the reaction product are indicated.

Phase-I

Oxidation

Oxidation of many organic compounds with a variety of functional groups is observed (Figure 3.40). Many aromatic and aliphatic compounds are hydroxylated. Other substrates for oxidation reactions are alkylated amino compounds (e.g., nicotine or morphine). N-alkyl and O-alkyl groups are de-alkylated by oxidative reactions, especially the methyl groups. The primary step in an oxidation reaction is often insertion of an oxygen atom into the compound. Subsequently, mono or dihydroxylated compounds may be formed, which could react further to ketones, with epoxides possibly being formed. Epoxides can be very reactive, and thus very harmful to the organism.

Many oxidation reactions are catalyzed by enzymes in the smooth endoplasmic reticulum (SER) of cells of many types of tissues. The oxidative enzymes are formed by a group of haemoproteins called cytochrome P-450 dependent enzymes (Figure 3.40). Cytochrome P-450 enzymes are part of an enzyme system which is commonly named mixed function oxidase (MFO). This name is derived from the fact that the major property of the system is to build one atom of molecular oxygen into a substrate, and to reduce the other oxygen atom to water. The MFO -system consists of several components, in which cytochrome P-450 has a key function (Figure 3.40). In the MFO reaction pathway both oxygen and substrate bind to the iron-haem group of cytochrome P-450. Oxidation by the MFO -system consists of the following steps:

a. The substrate SH binds to the oxidized (Fe^{3+}) cytochrome P-450.

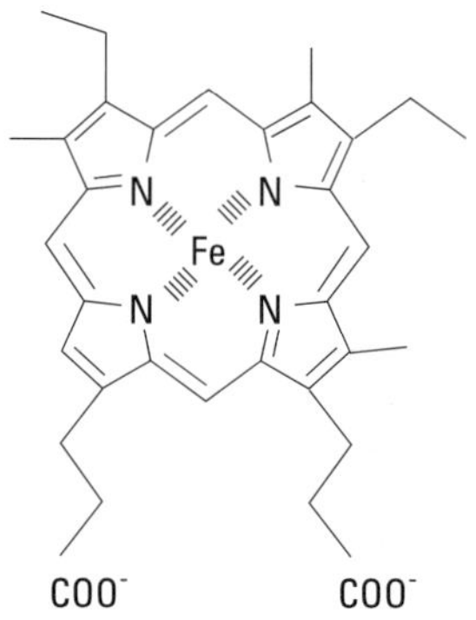

ion-protoporphyrin IX

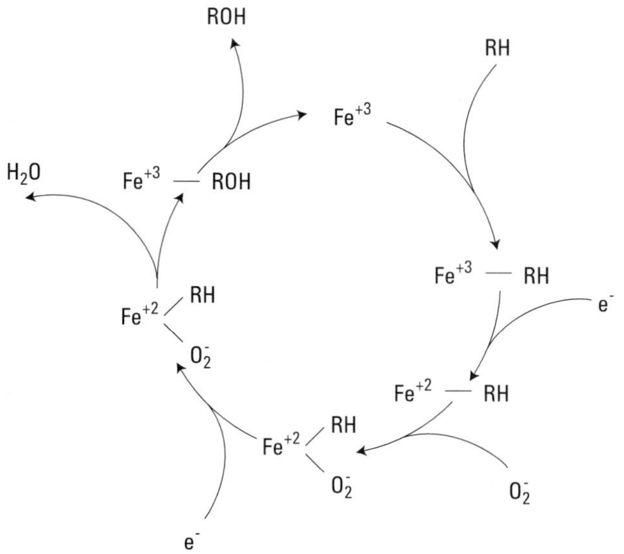

Figure 3.40. Mechanism of oxidation by cytochrome P-450. From [146]. With permission.

b. The complex formed receives an electron from NADPH by a flavoprotein.
c. The reduced (Fe^{2+}) cytochrome P-450 complex binds an oxygen molecule.
d. This complex accepts a second electron from NADH, via a second flavoprotein. This electron can also be transferred from NADPH.
e. The second reduction activates the oxygen molecule in the complex, which leads to the formation of water, the oxidized substrate and the oxidized enzyme. Hence the enzyme is ready for the next cycle.

The overall reaction is:

$$SH + NADPH + H^+ + O_2 \xrightarrow{\text{P-450}} SOH + NADP^+ + H_2O$$

This reaction is valid for a large number of xenobiotics, such as drugs, pesticides, and organic solvents. Cytochrome P-450 oxidation of constitutive substrates occurs in steroid metabolism.

Substrates can bind to cytochrome P-450 in two different ways. Some bind to the protein part, others to the haem part of cytochrome P-450. This can be seen spectro-photometrically, as binding results in a spectral change. Substrates which bind to the protein part of cytochrome P-450 cause a shift in the absorption maximum to 390 nm. They are called type I substrates. The other group of substrates binding to the haem part causes an absorption-maximum shift to 420 nm. These are called type II substrates.

At low concentrations some substrates give type I interactions and type II interactions at high concentrations. Other compounds form stable complexes with the haem iron, thus blocking the enzyme. Induction and inhibition of MFO enzymes, especially those dependent on cytochrome P-450, have been studied in detail over the last few decades.

Reduction

Compounds which undergo a reductive reaction include halogenated organic chemicals, ketones, nitro and azo compounds (Figure 3.39). The compounds to be reduced usually accept the electrons donated either by NADH or NADPH. In the cell NADH or NADPH usually donate the electrons. In mammals aromatic nitro compounds are also reduced by the micro-organisms present in the gut. It is unknown whether this process also occurs in the gut of fish. In addition, it should be noted that cytochrome P-450 enzymes are also involved in reductive reactions.

Hydrolysis

Compounds which undergo hydrolytic reactions include esters, epoxides and amides (Figure 3.39). During a hydrolytic reaction the molecule is broken down into two different molecules, for example, an ester is hydrolyzed into an acid and an alcohol. Hydrolytic reactions occur in many species. Various enzymes are involved in several types of tissues.

Phase-II

In Phase-II reactions a large polar group is introduced into the molecule. This may change it into a compound which is sufficiently hydrophilic for rapid excretion. Most compounds require such a conjugation reaction. Conjugation reactions occur with chemicals with functional groups such as -COOH, -OH and $-NH_2$ (Table 3.14). Large groups or entire compounds such as sugars and amino acids are covalently bonded to the xenobiotic.

Table 3.14. Phase-II conjugation reactions [145].

Reaction	Functional group
Glucoronic acid	-OH, -COOH, $-NH_2$, -NH, -SH, -CH
Sulphate	aromatic -OH, aromatic $-NH_2$, alcohols
Glycine	-COOH
Acetyl	aromatic $-NH_2$, aliphatic $-NH_2$, hydrazides, $-SO_2$, $-NH_2$
Methyl	aromatic -OH, $-NH_2$, -NH, -SH
Glutathion	epoxides, organic halides

In general, conjugation reactions make compounds more water soluble, thereby facilitating excretion from the body. For those substances in which the parent compound is the toxic agent, these metabolic pathways clearly represent a detoxification mechanism.

However, Phase-II reactions may also bioactivate compounds. Examples of different types of conjugation reactions are given in Figure 3.41. Phase-II biotransformation reactions require energy to drive the reaction. This is provided by activating a cofactor (or substrate) to high-energy intermediates such as PAPS, acetyl-CoA or UDPGA (see below). Since these cofactors are activated by ATP, the energy status of the organ is important in determining cofactor availability. Five major pathways for Phase-II reactions are:

- Glucuronic acid conjugation.
- Sulfate conjugation.
- Acetyl conjugation.
- Glutathione conjugation.
- Glucose conjugation.

These major Phase-II metabolic mechanisms are explained below.

Glucuronic acid conjugation

Before conjugation of glucuronic acid to the polar group of a substrate can take place, the glucuronic acid (GA) has to be activated. The activated glucuronic acid (UDPGA) is formed by enzyme reactions. The general reaction for glucuronic acid conjugation is:

$$\text{UDPGA} + \text{R-XH} \xrightarrow{\text{GT}} \text{R-X-GA} + \text{UDP}$$

where X is O, COO or NH, UDPGA is uridine diphosphoglucuronic acid and GT is glucuronyltransferase.

Glucuronide formation is one of the most common routes of conjugation for many compounds. The reaction involves condensation of the foreign compound or its (Phase-I) biotransformation product with D-glucuronic acid. The interaction of UDPGA with the acceptor compound is catalyzed by glucuronyltransferase. Several isoenzymes of this smooth endoplasmic reticulum (SER) enzyme are known. As a result, a wide range of substrates may form glucuronides in the above reaction (Table 3.14). These glucuronides are eliminated from the body in the urine or bile. The general occurrence in many species, the broad range of possible substrates, and the chemical diversity of accepted compounds, make conjugation with glucuronic acid qualitatively and quantitatively the most important conjugation reaction.

Sulphate conjugation

In this conjugation mechanism, sulfate is donated by the PAPS molecule, a reaction which is catalyzed by sulfotransferase. Sulphate has to be activated into the PAPS molecule before it can be conjugated to a substrate. The general reaction for sulphate conjugation is:

$$\text{PAPS} + \text{R-XH} \xrightarrow{\text{ST}} \text{R-X-SO}_3 + \text{PAP}$$

where X is O or NH, PAPS is 3'-phosphoadenosyl-5'-phophosulphate, ST is sulfotransferase and PAP is 3',5'-adenosine diphosphate.

Sulphate is added to the substrate through a reaction mediated by sulphotransferase, which is usually found in the cytoplasm of the cell. Again several isoenzymes of sulphotransferase are known. As with glucuronidation, a variety of substrates may form sulphate derivatives (Table 3.14).

Acetyl conjugation

The general expression for this type of reaction is:

$$\text{R-XH} + \text{acetyl-CoA} \xrightarrow{\text{AT}} \text{R-X-COCH}_3 + \text{CoA}$$

where acetyl-CoA is acetyl-coenzyme A, AT is N-acetyltransferase and X is NH.

Acetyl is added to the compound by conjugation with the amino group, with acetyl-CoA acting as a cofactor. The reaction is catalyzed by an acetyltransferase. When X = COOH, the nitrogen-containing glycine is added to the xenobiotic, also resulting in nitrogen conjugation, this is called glycine conjugation (Table 3.14). These reactions do not always result in a more water-soluble product.

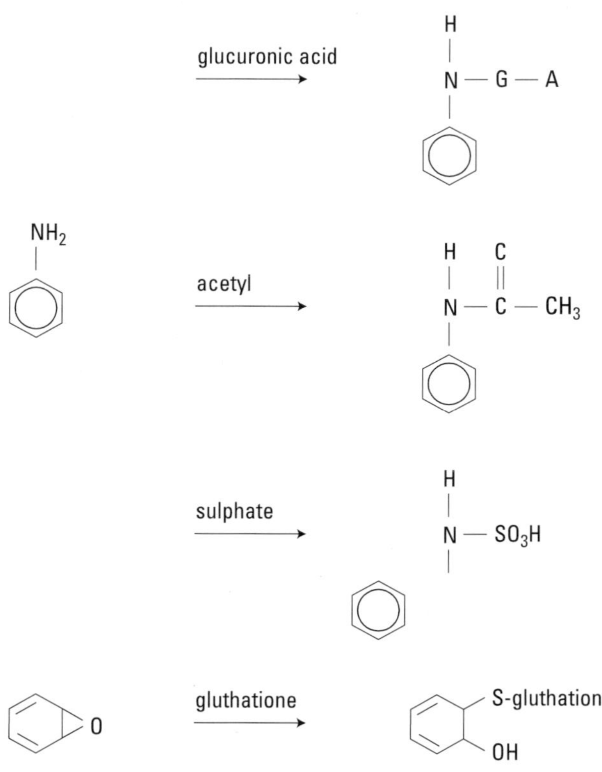

Figure 3.41. Some general phase-II biotransformation reactions involving aniline and benzene.

Glutathione conjugation
Glutathione is conjugated in the first step of mercapturic acid formation. The general expression is:

$$\text{RX + glutathione} \xrightarrow{\text{transferase}} \text{R-S-glutathione}$$

$$\text{R-S-glutathione} \xrightarrow[\text{acetylase}]{\text{peptidase}} \text{R-S-mercapturate}$$

where RX is an aromatic ring or a halide compound.

Conjugation with glutathione may reduce the toxicity of certain molecules and their metabolites. Many compounds which contain a reactive group, such as chloride, nitro or epoxides, are conjugated with glutathione. Glutathione conjugates often involve reactive (electrophilic) (intermediate) compounds, while the conjugated products are proof of exposure to compounds forming those intermediates. To determine the occupational exposure of industrial workers to these compounds, mercapturates are often analyzed in urine.

Glucose conjugation
In this conjugation mechanism, glucose is donated by UDP-glucose (UDPG: uridine diphosphoglucurose), a reaction which is catalyzed by glucosyltransferase, which is localized in the microsomal fraction.

Some examples where conjugation reactions take place at the more polar groups of a molecule resulting from Phase-I reaction are provided in Figure 3.42. Hydrophobic xenobiotics are excreted, but excretion of constitutive hydrophobic waste products also takes place. For many compounds, biotransformation reactions mediate reactive intermediates (Figure 3.43).

3.6.4 Factors influencing enzyme activity

Enzymes involved in biotransformation can be found in practically all organisms: bacteria, yeasts, plants and all classes of animals. However, large differences have been found in Phase-I and Phase-II enzyme activities between species. Quantitative (identical reactions but at different rates) as well as qualitative (different reactions) differences are known. These differences in biotransformation often complicate the extrapolation of results obtained for laboratory test species to man. In addition, there are individual variations in enzyme activity.

Animals
There are major qualitative and quantitative differences between species. Generally, terrestrial organisms have a better developed biotransformation system than those living in an aquatic environment. Fish usually have lower enzyme activity than mammals and birds. The reason suggested for this difference is that fish have less need to biotransform compounds as they can excrete compounds in water relatively easily. Some examples of qualitative differences in mammals are: dogs cannot acetylate aromatic amino compounds, while N-acetyl transferase and UDP-glucuronyl transferase are absent in cats; guinea pigs do not form mercapturic acid conjugates and pigs do not have a sulfate conjugation mechanism. Some differences in Phase-II reactions are given in Table 3.15.

The presence of cytochrome P-450 can also vary widely between species. Fish and most crustaceans have a higher cytochrome P-450 concentration (per mg microsomal protein) than *Daphnia magna* [65]. However, fish generally have a lower concentration cytochrome P-450 per mg microsomal protein than mammals such as rats and rabbits. Even between certain fish or mammalian

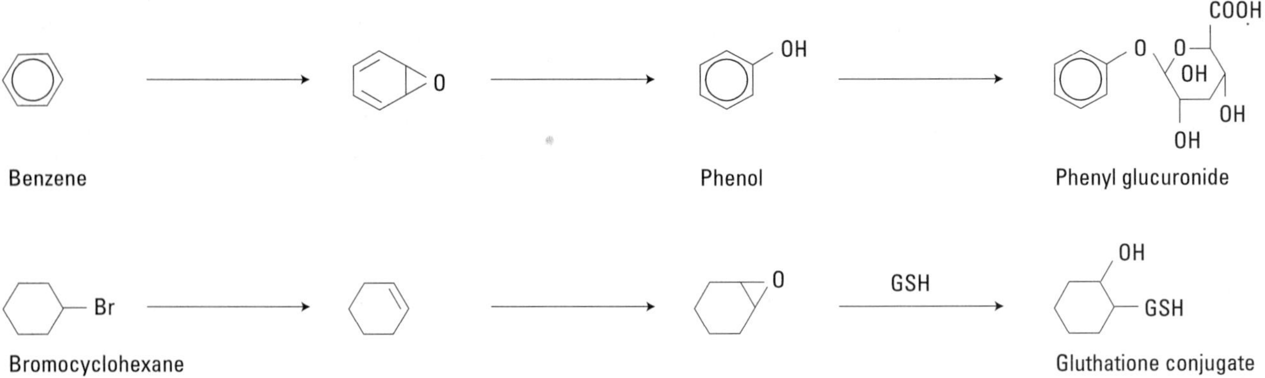

Figure 3.42. The role of phase-I and phase-II reactions in the mechanism of biotransformation of benzene and bromocyclohexane.

species there are marked differences in cytochrome P-450 concentration.

Sex, age, diet

The activity of enzymes may be influenced by hormones. For example, sex-specific forms of cytochrome P-450 are known. The age of an organism is important for the rate of biotransformation. Large differences in enzyme activity may be seen, especially between very young, adult and very old animals.

Diet has a substantial influence on enzyme activity. In general, herbivores take up a wider variety of

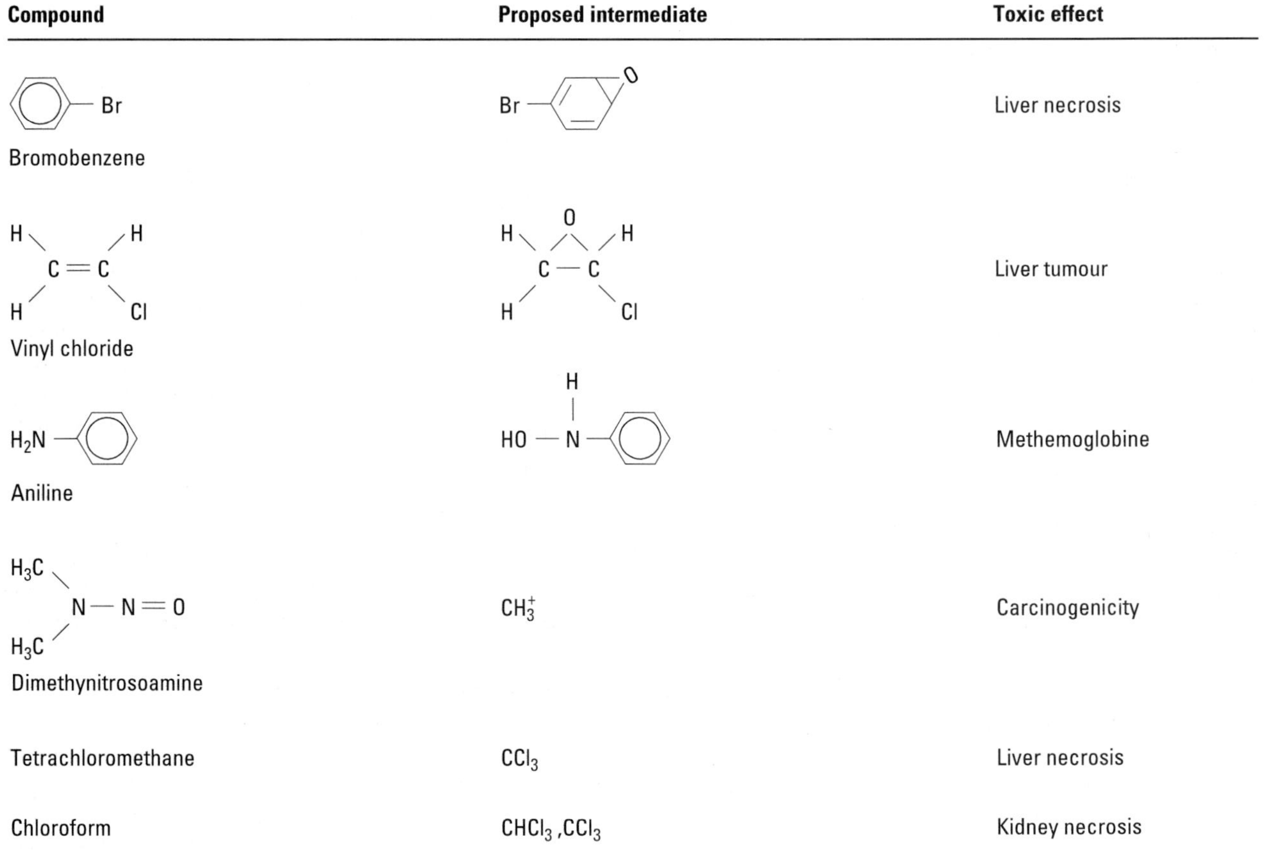

Compound	**Proposed intermediate**	**Toxic effect**
Bromobenzene		Liver necrosis
Vinyl chloride		Liver tumour
Aniline		Methemoglobine
Dimethynitrosoamine	CH_3^+	Carcinogenicity
Tetrachloromethane	CCl_3	Liver necrosis
Chloroform	$CHCl_3$, CCl_3	Kidney necrosis

Figure 3.43. The biotransformation of different xenobiotic compounds to reactive intermediates.

Table 3.15. Species variation for phenol conjugation with glucuronic acid and sulphate [144]. With permisssion.

Species	Conjugation of phenol (percentage of total excretion)	
	glucuronic acid	sulphate
Pig	100	0
Rabbit	46	45
Rat	25	68
Man	23	71
Cat	0	87

xenobiotics than carnivores and usually have a higher enzyme activity, and very specialized carnivores have lower biotransformation enzyme activity. It has been suggested that this is caused by the fact that the prey has already biotransformed many xenobiotics. The protein, carbohydrate and fat content of the diet also influence biotransformation rates. For example, higher protein content decreases some enzyme activities. For aflatoxin-B1 the type of diet influences both the route and the rate of biotransformation in mammals.

Temperature/season
It is very difficult to determine the influence of these parameters separately. With many compounds, enzyme induction in aquatic organisms is higher in summer due to higher temperatures. However, in some cases adaptation to the temperature may occur, resulting in comparable biotransformation rates at different temperatures.

Plants
Most of the available literature deals with the biotransformation of pesticides. The rate of pesticide biotransformation is generally slower in plants than in animals. In part this can be attributed to the lack of efficient circulatory and excretory systems in plants. Plants are able to perform Phase-I biotransformation reactions of oxidation, reduction and hydrolysis, as well as conjugation. However, unlike animals, conjugation usually leads to storage of the compound in the plant rather than excretion from the body.

3.6.5 Methods to measure biotransformation

Enzyme kinetics
In order to determine the rate of enzyme reactions and to obtain a better understanding of the mechanism of enzymatic reactions, an understanding of enzyme kinetics is important. Enzymes catalyze chemical reactions through the formation of an enzyme substrate complex, followed by conversion of the complex into the enzyme and a product. This process can be described by the equation:

$$E + S \underset{k_2}{\overset{k_1}{\rightleftarrows}} ES \underset{k_4}{\overset{k_1}{\rightleftarrows}} E + P$$

where
E = enzyme
S = substrate
P = product
k_1, k_2, k_3, k_4 = rate constants.

When the enzyme concentration is constant, the initial rate *(V)* of the reaction increases with the substrate concentration. Assuming the concentration of the substrate to be considerably larger than the enzyme concentration, and the concentration of the product to be negligible, the initial velocity can be described by:

$$V = V_{max}\,[S]\,/\,(K_m + [S]) \tag{3.73}$$

Where
V = the initial rate of the reaction
V_{max} = the theoretical maximum rate of the reaction
K_m = the substrate concentration at 1/2 V_{max}
$[S]$ = the substrate concentration.

This is called the Michaelis-Menten equation. At very high substrate concentrations K_m becomes negligible, and the equation simplifies to $V = V_{max}$. The Michaelis-Menten equation can also be expressed as:

$$K_m = [S]\,(V_{max}\,/\,V - 1) \tag{3.74}$$

K_m, being the substrate concentration at half the theoretical maximum rate of the reaction, is also called the Michaelis constant. Any enzymatic reaction is characterized by its value of K_m, being independent of the enzyme concentration.

For most hydrophobic environmental contaminants biotransformation may be adequately described by a first-order model (Section 3.4) [65]. Both *in vivo* and *in vitro* methods are available to measure the biotransformation rate constant.

In vitro methods

Two *in vitro* methods make use of isolated cells of the organ in which biotransformation is measured. As a rule, liver cells (hepatocytes) are used, as the liver is regarded as the principal organ responsible for the biotransformation of many xenobiotics. In all *in vitro* systems controlled exposure is an issue, since in such systems losses through evaporation from, as well as adsorption to, the test vessels can significantly reduce the nominal concentrations, often much faster than the duration of the test. Relating the test results to the nominal exposure concentration thus often overestimates the actual concentration and thus may affect the outcome of these *in vitro* tests.

a. Quantifying biotransformation products.
One *in vitro* method uses liver cells which are held under optimum conditions with regard to temperature, pH and nutrition. Xenobiotics are introduced into the medium in which the cells are held. The biotransformation products in the medium and cells have to be quantified using analytical methods. If no reference biotransformation products are available, only information on the number and nature of some of the physicochemical properties of these biotransformation products is obtained. The advantages of this method are that it is easy to conduct and small amounts of chemical are needed to discover the biotransformation pathway. However, a disadvantage is the limited exposure time of the cells. Induction of the enzymes involved in biotransformation will not be detected by this method. Moreover, extrapolation of the results to the *in vivo* situation is often unclear.

b. Quantifying enzyme activity.
The second *in vitro* method determines the rate of a specific biotransformation reaction with the help of a reference compound. Usually, cells or cell fractions of an organ are used, such as the microsomal or the cytosolic fraction, which are kept under optimum conditions. To quantify enzyme activity, the rate of formation of a biotransformation product from a reference compound is determined. By using this method enzyme activities in different organs or tissues of an organism can be compared. However, each species has its own optimum conditions for biotransformation of the reference chemical, which complicates interspecies comparison. The disadvantage is that the extrapolation to *in vivo* situations is not well-established.

In vivo methods

In vivo methods to measure biotransformation clearly have several advantages compared with *in vitro* methods. Laboratory studies with animals reflect biotransformation in a field situation more realistically. This is because kinetic and physiological factors are expressed in *in vivo* laboratory studies, but not in *in vitro* studies. Basically, four *in vivo* methods are available to measure biotransformation:

a. Quantifying biotransformation products.
The amount of biotransformation products formed in time is measured. Which products are formed and in which type of tissue needs to be known. The rate at which the products appear provides information on the biotransformation rate. An associated problem with this method is that the complete biotransformation pathway has to be elucidated to obtain information on the dominant biotransformation products. These products have to be synthesized to allow quantification. A method often used to deal with these problems is the use of radio-labelling techniques, possibly in combination with separation techniques such as GC or HPLC. Biotransformation products can be quantified by measuring the amount of radioactivity.

b. Enzyme inhibition.
When biotransformation enzymes are inhibited, the xenobiotic is eliminated from the organism only by diffusion. If biotransformation is the major pathway for excretion, the elimination of a compound from an organism with active enzymes will be faster than that of the same organism with inhibited enzymes. The difference between the two situations determines the biotransformation rate constant. To obtain this information, it is necessary to know which enzymes are responsible for the biotransformation of the xenobiotic. The agent which selectively inhibits the activity of the enzymes also has to be known. A disadvantage of this method is that it compares two different treatments of the organism. The inhibitor may influence physiological processes in the organism. In addition, the inhibitor may not completely block the biotransformation pathway, or may block only one pathway when various pathways are possible. Piperonyl butoxide is a commonly used P-450 enzyme inhibitor.

c. Mass balance.
The mass-balance analysis describes all the unexplained loss of material from the xenobiotic to biotransformation. The exact amount of xenobiotic introduced in the organism has to be known, as well as how much remains in the organism, how much has been eliminated from the

organism, and how much was lost from the total system. Losses of the xenobiotic, due to adsorption to glass or evaporation, for example, must be measured separately in a reference system.

d. Physicochemical properties.
When compounds accumulate less in organisms than expected based on their hydrophobicity, this is often related to biotransformation. The resulting lower BCF is attributed to an elevated elimination rate constant, due to biotransformation.

3.6.6 Biotransformation of some specific groups of compounds

PAHs

The toxicity of polycyclic aromatic hydrocarbons (PAHs) is mostly due to their carcinogenicity. PAHs usually have to be activated by biotransformation to become carcinogenic agents [146]. The epoxide which is formed by MFO activity can bind to DNA and initiate a carcinogenic effect. This mechanism has been studied extensively for benzo[a]pyrene (Figure 3.44). The epoxide is a suitable substrate for Phase-II conjugation, which facilitates rapid excretion. Hence, biotransformation reactions for benzo[a]pyrene result in both bioactivation and detoxification. Recent studies indicate that more polar PAHs (containing nitro, amino and hydroxy groups) are directly carcinogenic, and do not need to be activated by biotransformation.

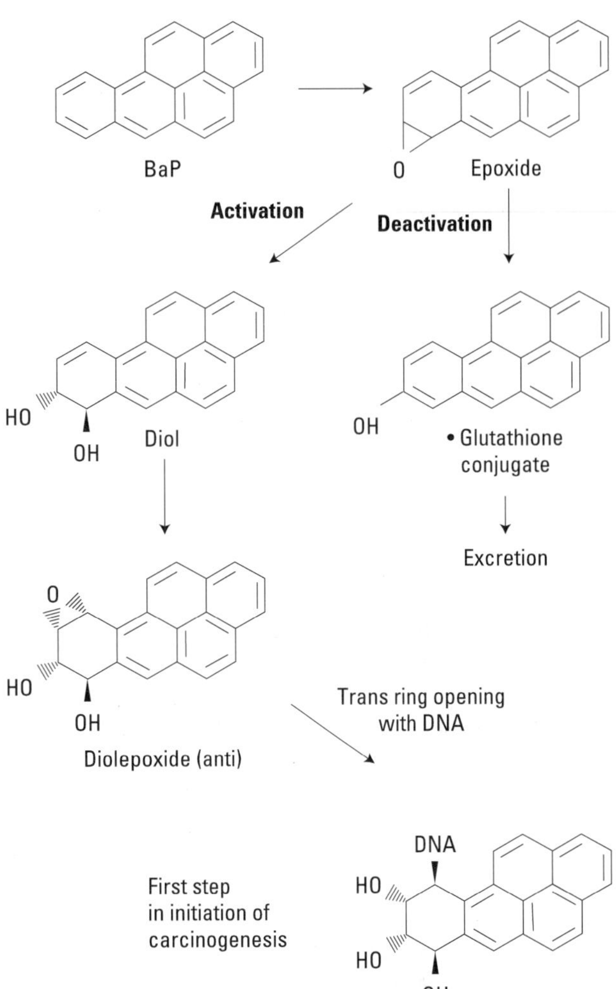

Figure 3.44. The biotransformation pathways of benzo(a)pyrene and binding to the DNA of reactive intermediates. From [146]. With permission.

PCBs

The isomer-specific composition of mixtures of polychlorinated biphenyls (PCBs) shows dramatic changes after uptake by an organism [147,148]. When the liver and adipose tissues of organisms exposed to a commercial mixture of PCBs were analyzed, the number of isomers had decreased compared with the original mixture. As the missing isomers were not found in the faeces or urine, it was concluded that biotransformation plays a key role. *In vivo* experiments revealed that the major route of biotransformation of PCBs starts via epoxide formation (Figure 3.45). However, sulfur-containing metabolites, dechlorination and re-arrangement of chlorines also constitute biotransformation pathways. The rate of biotransformation is also determined by the isomeric structure, the number of chlorines and the animal species. The chlorine-sub-stitution pattern of the molecule largely determines where epoxide formation takes place (Figure 3.46). In general, the following rules apply to the biotransformation of PCBs:

1. Hydroxylation is preferred at the *para* position (4) in the ring containing the lowest number of chlorine atoms, unless this position is sterically hindered by *m,m*-dichloro (3,5) substitution.
2. The *para* position relative to a chlorine in the ring is preferred for hydroxylation.
3. Two adjacent vicinal hydrogen atoms in the molecule may increase the rate of oxidative biotransformation, but this is not a prerequisite.
4. An increasing number of chlorines decreases the rate of biotransformation.
5. Different species may have different biotransformation pathways for the same isomer.

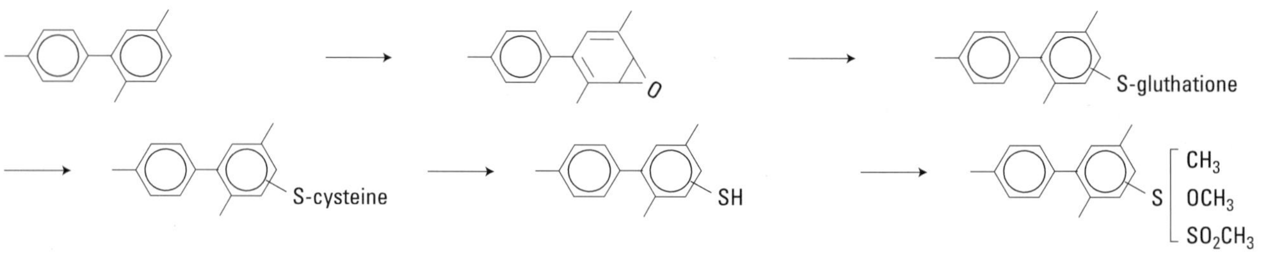

Figure 3.45. The major biotransformation route of PCBs. From Safe [149]. With kind permission of Springer Science and Business Media.

PCDDs and PCDFs

The biotransformation of polychlorinated dibenzo-p-dioxins and furans (PCDDs and PCDFs) is comparable with PCBs, and mainly influenced by the number and position of the chlorine atoms [150]. The following relationships have been determined for the biotransformation of PCDDs and PCDFs:

1. Hydroxylation on the lateral positions (2,3,7 and 8) is preferred.
2. Two vicinal hydrogen atoms, both preferably on the lateral positions, increase the biotransformation rate, but this is not a prerequisite.
3. Oxygen bridge cleavage may occur, but is not the major route for most congeners.

DDT

Most insecticides owe their toxicity to their ability to interact with the central nerve system [151]. In insects, this is well-developed, and almost comparable in organization to that in mammals. The major route of biotransformation of DDT [1,1-di-(p-chlorophenyl) 2,2,2-tri-chloroethane] is by forming DDE (Figure 3.47). DDT-resistant houseflies detoxify DDT mainly to non-insecticidal DDE. The ability to change DDT to DDE appears to be a major factor in the survival of DDT-exposed flies. The rate of biotransformation varies greatly between fly strains and individual specimens. Grasshoppers show a natural tolerance to DDT. This tolerance depends partly on biotransformation in the cuticle and gut. In addition, there is rapid passage of ingested DDT through the gut of the grasshopper without significant absorption. The combination of these factors prevents DDT from reaching its site of action in the nervous system. A small injected dose of DDT is fatal to these insects, while they can withstand large oral or dermal doses of DDT.

As a major biotransformation product of DDT, DDE still has significant hydrophobic properties. Hence, DDE also shows significant biomagnification. Higher DDE concentrations have been found in species at the top of the food chain, such as birds of prey. DDE itself is believed to inhibit the supply of calcium for egg-shell formation. The resulting egg shell thinning in birds of prey affected breeding success in the 1960's and 1970's. The impact of DDT on predators illustrates a case in which the combined effects of biotransformation and physicochemical properties eventually lead to secondary poisoning in the environment.

OP-esters

Organophosphorus compounds are neurotoxic compounds, which interact with the enzyme acetylcholinesterase (AChE) [151]. The interaction causes disturbances in the central nerve system. The neurotoxicity of organophosphorous compounds is substantially increased by biotransformation. The biotransformation reaction causes a substitution of the sulfur, bound to phosphor in the phosphorthionate, by oxygen (Figure 3.48). The biotransformation products are called oxon analogues. Oxon analogues have a higher affinity for the enzyme AChE than the original organophosphorus compounds. An oxon analogue inhibits enzyme activity. Acetylcholine-mediated neurotransmission is blocked, causing a neurotoxic effect. Hence, the Phase-I oxidation reaction, required for higher aqueous solubility, leads to bioactivation

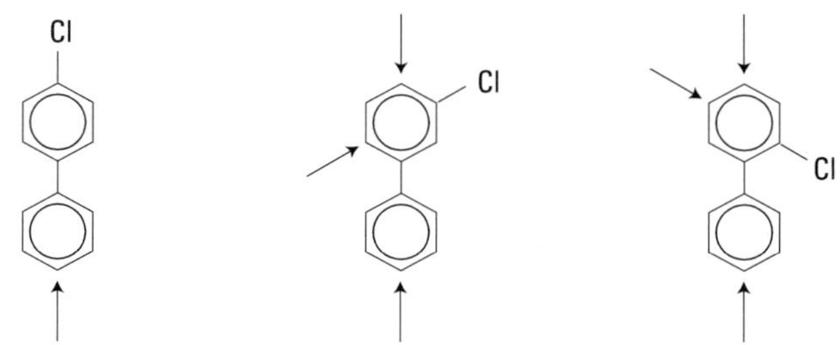

Figure 3.46. The preferred oxidation positions in a PCB molecule and the role of the chlorine position in the molecule in cytochrome P-450 catalyzed biotransformation reactions. From [147]. With permission.

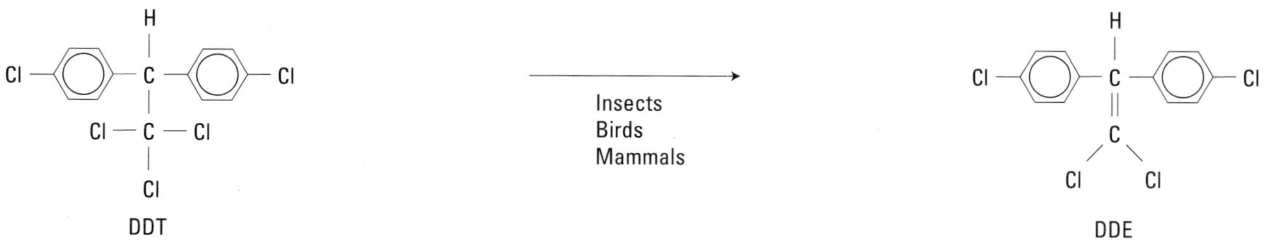

Figure 3.47. The main route of biotransformation of DDT to DDE.

of the compound. When the oxon analogues are subsequently hydrolyzed, the affinity for the enzyme acetylcholinesterase reduces.

Synthetic pyrethroids

Unlike natural pyrethroids, which degrade mainly via oxidation, hydrolytic degradation is an important route of biotransformation for synthetic pyrethroids (Figure 3.49) [151]. The two routes of biotransformation result in rapid degradation of synthetic pyrethroids in the environment. Hence, the fate of pyrethroids in the environment differs substantially from that of persistent chlorinated insecticides such as DDT or lindane.

3.6.7 Enzyme inhibition and induction

Enzyme inhibition occurs when the activity of an enzyme or enzyme system is reduced relative to control levels. Several mechanisms of inhibition are possible:

1. Competition for active sites or cofactors of enzymes.
2. Inhibition of transport components in multi-enzymatic systems.
3. Decreased biosynthesis or increased breakdown of enzymes or cofactors.
4. Changes in enzyme conformation.
5. Cell necrosis.

When enzyme induction occurs, more, or more active, enzymes are present. This usually results in an increase in the rate of metabolism and biotransformation reactions. However, it should be noted that compounds do not necessarily induce the enzyme involved in their own biotransformation. In principle enzyme induction is a reversible process. Elimination of the inducing agent results in a return to basal enzymatic activity. The duration of induction is a function of the dose and the inducing agent.

Several classes of compounds are known to induce enzymes. In many cases induction of cytochrome P-450 enzymes is studied, although in some studies induction of enzymes catalyzing hydrolytic and Phase-II reactions is also determined.

For xenobiotics two types of cytochrome P-450 induction can be distinguished: phenobarbital (PB) and 3-methylcholantrene (3-MC) type induction. These two model compounds induce different groups of cytochrome P-450 isoenzymes. The PB type of induction causes increasing protein and phospholipid synthesis, as well as induction of NADPH-cytochrome P-450 reductase and cytochrome P-450 2B and 3A isoenzymes. The net effect of these biochemical changes is enhanced biotransformation of a large number of chemicals.

The pattern of induction in the liver of 3-MC

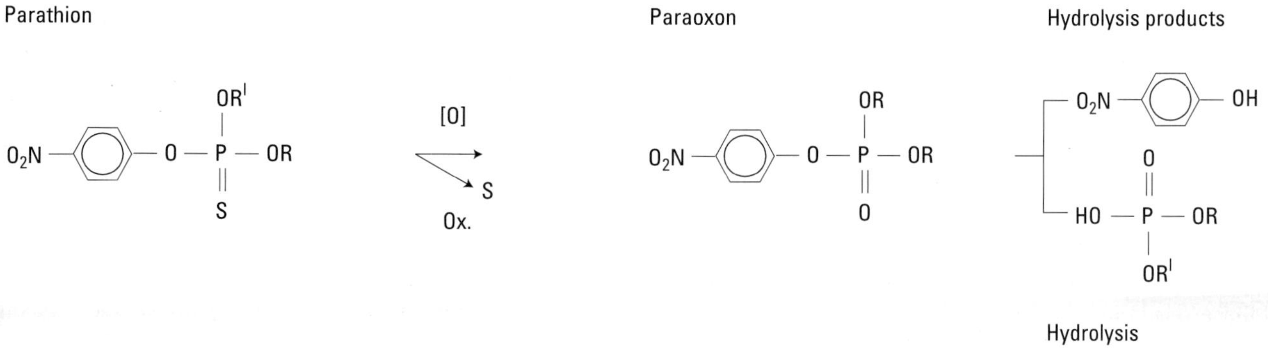

Figure 3.48. The biotransformation routes of organophosphorus compounds.

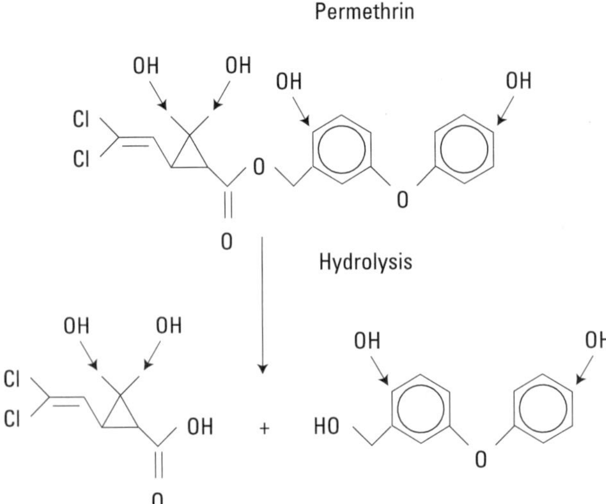

Figure 3.49. The hydrolytic degradation of synthetic pyrethroids.

(or benzo(a)pyrene) treatment is very different. The marked increase in liver weight, protein and phospholipid synthesis, and NADPH-cytochrome P-450 reductase observed for PB does not occur. Instead there is a highly selective induction of cytochrome P-450 1A1 and 1A2 isoenzymes. The 3-MC inducible isoenzymes of cytochrome P-450 are also responsible for the transformation of certain PAHs into bioactive intermediates, as occurs with benzo(a)pyrene (Figure 3.45). The differences between the two types of enzyme induction are summarized in Table 3.16. Other major classes of inducing agents include halogenated pesticides (DDT, aldrin, hexachlorobenzene, lindane, chlordane), polychlorinated and brominated biphenyls, chlorinated dioxins and furans, steroids and related compounds (e.g., testosterone), as well as metals, such as cadmium.

The chemical structure determines which type of enzyme induction occurs. Most chlorinated biphenyls and DDT induce cytochrome P-450 isoenzymes comparable with PB induction. Chlorinated dioxins, furans and some PAHs have a 3-MC type induction. For chlorinated biphenyls, chlorine substitution on the ortho position(s) influences the strength and type of induction. The strong enzyme induction properties of dioxins provide an example. Dioxin molecules with four chlorine atoms on the lateral (2,3,7,8) positions, e.g., 2,3,7,8-TCDD, exhibit slow elimination from the liver. This is primarily caused by the chlorine atoms in these positions effectively blocking a Phase-I oxidation reaction by cytochrome P-450. The persistence of TCDD in the liver cells gives a continuous receptor-mediated signal for cytochrome P-450 synthesis. As a result, strong and prolonged induction of this type of enzyme activity can be observed even after exposure to relatively small amounts of dioxins and PCBs.

Clearly, the stereospecificity of the molecule plays an important role in the type of cytochrome P-450 induction. This can be effectively illustrated with the group of PCBs as model compounds. Depending on the number of ortho chlorines in the molecule, the two aromatic rings can obtain a planar configuration towards each other. For mechanistic reasons involving a cytosolic receptor protein (the Ah receptor) this planar configuration is most easily obtained for PCBs which lack ortho chlorines. As this Ah receptor mediates the induction of cytochrome P-450 1A1 and 1A2 isoenzymes, non-orthosubstituted PCBs are the most potent inducers of the 3-MC type of induction. With an increasing number of chlorine atoms at the ortho position, the possibility of the biphenyl molecule obtaining a planar configuration strongly diminishes due to steric hindrance by the chlorines. When the number of ortho chlorines increases from one to four, the 3-MC type of cytochrome P-450 induction is gradually replaced by a PB type of induction, involving cytochrome P-450 2B1 and 2B2 iso-enzymes [152].

3.6.8 Effect of enzyme induction on toxicity

Phase-I and Phase-II enzymes either bioactivate or detoxify the xenobiotics taken up by the organism. Hence, the effect of the induction of Phase-I and Phase-II enzymes may increase or decrease the toxicity of the compounds. When the enzymes activate the compounds, the effect of enzyme induction is harmful to the organism. When the enzymes have a detoxifying effect, enzyme induction is beneficial.

It should be recognized that if induction of Phase-I enzymes only is studied, the overall biological or toxicological effect cannot be adequately ascertained. This is due to the possibility that concurrent induction of Phase-II enzymes may partly obscure the hazardous effect of Phase-I biotransformation products. The formation of reactive epoxides from aromatic or unsaturated hydrocarbons by cytochrome P-450 is an example of this. The formation of these potentially hazardous intermediate metabolites can form a direct threat to the organism due to interaction with macromolecules. If, however, the concurrent induction of glutathione conjugation occurs, the chances of detoxifying the reactive biotransformation products significantly increase. The above mechanism also applies in situations where one compound promotes the formation of carcinogenic products of a second

Table 3.16. Characteristics of the hepatic effects of PB and 3-MC [144]. With permission.

Characteristic	PB	3-MC
Onset of effects	8-12 h	3-6 h
Time of maximal effect	3-5 d	1-2 d
Persistence of induction	5-7 d	5-12 d
Liver enlargement	marked	slight
Protein synthesis	large increase	small increase
Liver blood flow	increase	no effect
Biliary flow	increase	no effect
Enzymes:		
– cytochrome P-450 1A1 + 1A2	increase	no effect
– cytochrome P-450 2B1 + 2B2	no effect	increase
– NADPH-cytochrome reductase	increase	no effect

compound. This mechanism is found in combinations of dioxins and PAHs, in which the former compound may act as a tumour promoter. The net effect eventually depends on the concurrent induction of Phase-II enzymes, which could detoxify Phase-I biotransformation products. For organophosphorous compounds, induction of the enzymes transforming parathion to paraoxon causes a greater toxic effect. When, however, Phase-II enzymes which degrade paraoxon to inactive products are also induced to a similar extent, the net effect is comparable to a situation where no enzyme induction occurs.

To summarize: determining the enzyme induction of a single enzyme will not provide sufficient information on the overall effect on the organism.

3.7 BIOAVAILABILITY

3.7.1 Introduction

General

The multitude of processes discussed above in this chapter, determine the total concentration of a chemical in the environment and its distribution over the compartments. There is ample evidence from toxicity and bioaccumulation studies in the field and in the laboratory, that the total amount of a chemical that is present in one environmental compartment is not by definition indicative of adverse effects actually occurring, nor is the extent of bioaccumulation directly related to the total amount present. Instead, biota are usually effectively exposed to only a fraction of the total chemical load. To complicate matters, the effective fraction has been shown to be dependent on the species and the time scale considered. The composition of the aquatic or the soil matrix affects this effective fraction. For fish it has been shown that uptake takes place predominantly via the gills. Due to competition with H^+ at the gill membranes, uptake of toxic metals like Zn and Cu is reduced at decreasing pH (i.e., increased concentration of H^+). In soils, the uptake of hydrophobic organic chemicals occurs mainly via the pore water. Consequently, the uptake of hydrophobic organic chemicals by earthworms is reduced in soils containing higher amounts of organic carbon while all other soil properties remain unchanged.

Bioavailable fraction and the concept of bioavailability

The fraction of a chemical's concentration that is effectively available for interaction with biota is termed the bioavailable fraction. Although other definitions apply, the bioavailable fraction is the fraction of the total amount of a chemical present in a specific environmental compartment that, within a given time span, is either available or can be made available for uptake by organisms, micro-organisms or plants. This can be from either the direct surroundings of the organism or the plant (mediated by the aqueous phase) or by ingestion of food, soil or sediment. Adverse effects are assumed to be proportional to the bioavailable fraction. Most evidence reported in the scientific literature points to the freely dissolved concentration (i.e., aqueous activity or fugacity) of a chemical as being the fraction actually bioavailable for large numbers of biota. Only freely dissolved chemicals are capable of interacting with biological

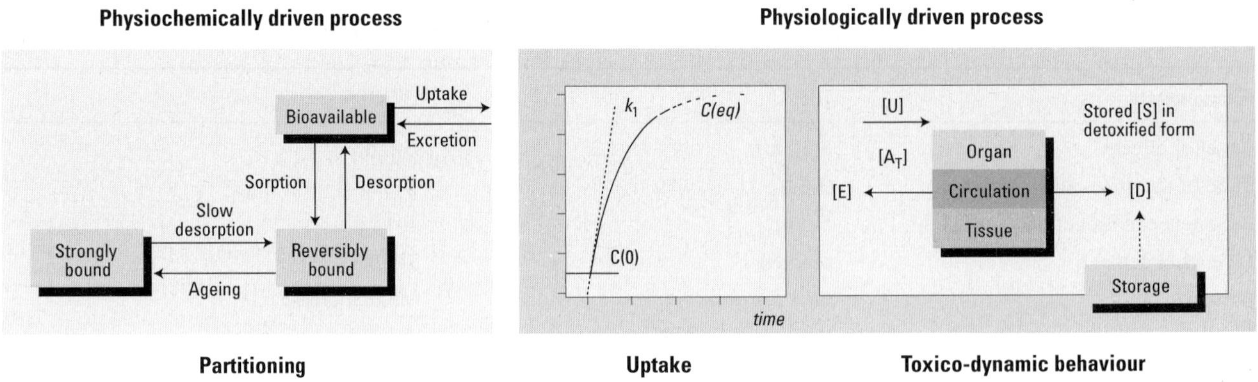

Physiochemically driven process **Physiologically driven process**

Partitioning | Uptake | Toxico-dynamic behaviour

Figure 3.50. Schematic representation of the processes underlying the bioavailability concept.

membranes and although contaminants associated with ingested food particles must cross biological membranes within the gut, digestive (catabolic) processes acting therein make this a unique route of exposure that is separate from other routes. Exposure via ingestion may contribute significantly to the overall uptake of a contaminant from the environment, although for most species its relative importance is poorly understood.

As advocated by various authors in Hamelink et al. [153], bioavailability should be treated as a dynamic process. The dynamic approach of "bioavailability" should comprise two distinct and different phases: a physicochemically driven desorption process, and a physiologically driven uptake process requiring the identification of specific biotic species as an endpoint. It should be borne in mind that the quantitative influence of solid phase constituents on toxicant binding is considerably larger in the soil and sediment compartments than in the aqueous compartment. "Toxicological bioavailability" is the third aspect that can be identified as a better-defined subdiscipline of the often vague concept of "bioavailability". Toxicological bioavailability refers to the redistribution of chemicals to targets within an individual, and thus to the dose of the chemical at the target tissue. Most organisms have developed the redistribution of chemicals and strong binding to inert granules as part of their internal detoxification strategy against excess amounts of toxic chemicals. The concept of bioavailability is presented as a diagram in Figure 3.50.

Equilibrium partitioning

In its simplified form the concept of bioavailability builds on the Equilibrium Partitioning (EP) theory. Basically, the EP approach states that organisms do not take up chemicals from soil or sediment directly (ingestion of solid material) but only from the freely dissolved phase in the pore water. It is thus assumed within the EP approach that uptake via ingestion of solid particles is either not an important exposure route, or that it may be described on the basis of the concentration of the chemical in the water phase. A chemical tends to distribute itself between the solid, water and organism phases until it is in thermodynamic equilibrium. Assuming that indeed equilibrium is obtained, this implies that the chemical residues in organisms can be predicted if we know the distribution coefficient of the chemical (partitioning between solids and water) and the bioconcentration factor (partitioning between water and the organism). In a simplified modification of the EP theory, the pore water concentration represents the bioavailable phase. Although simplified, the usefulness of using pore water concentrations as the bioavailable fraction has been demonstrated for a broad range of chemicals and a broad range of species (for an overview see, e.g., [104]). Deviations from EP are limited and relate mainly to organisms for which uptake via the pore water is not obvious. Evidence is lacking for "hard-bodied" species like insects, for example. The EP concept is illustrated in Figure 3.51. See also Chapters 7 and 11.

Speciation

As indicated above, it is the freely dissolved form of the contaminant at the interface of the biological membrane and the aquatic phase which is actually transported across the membrane. For most biota the aqueous activity or fugacity of a chemical is therefore the best indicator of bioavailability: a reduction in the aqueous activity or fugacity translates directly into reduced bioavailability of the contaminant. Reduction of the aqueous activity of organic and inorganic chemicals is primarily related to the presence of particulate and dissolved material. For

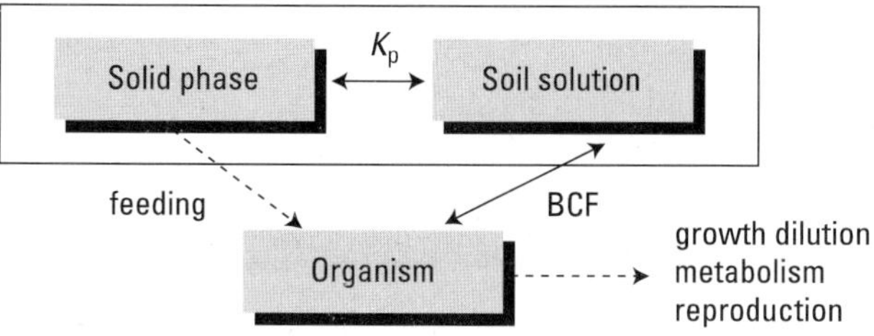

Figure 3.51. Graphical illustration of the Equilibrium Partitioning concept. K_p = partition coefficient, BCF = bioconcentration factor.

charged species like metal cations, and charged organic molecules, cationic species (like H^+, Na^+, Ca^{2+}, K^+) present in the aqueous phase will compete for binding to the sorption sites present on the particulate and dissolved material, and on the biotic membranes. This will reduce the effective uptake of these species.

The term "speciation" is widely used in this respect and covers the various forms (i.e., "chemical species") in which a molecule can exist. Common chemical species include charged and neutral organic molecules, free metal ions, complexes of metal ions with anionic ligands commonly present in water (like charged or neutral Metal-hydroxide species, metal-chloride species, and methyl-sulphate species), and whether or not the species is sorbed to macro-molecules, such as dissolved organic carbon (DOC), in water. The stronger the chemical is sorbed by particulate or dissolved material or complexed to anionic ligands, the less freely dissolved chemical is left, so that less of the chemical is usually taken up. In such cases, the apparent bioconcentration will be less, due to the smaller bioavailable fraction. On the other hand, bioconcentration may not be influenced by speciation when the bioavailable fraction is expressed as the freely dissolved fraction. Speciation calculations are used to calculate the freely dissolved fraction on the basis of the total concentration of the chemical in the water phase, the number of available abiotic ligands, the activity of competing species, and the equilibrium constants for binding of the chemical to each of the ligands. In the case of metals, the activity of the free metal ion is used as the general expression for quantifying uptake and possible adverse effects. The activity of either the charged and/or the neutral molecule is used as the expression of the bioavailable fraction of organic compounds. Present databases contain extensive information with regard to binding to inorganic ligands and many specific organic ligands, but there are major deficiencies with respect to appropriate values for binding by humic substances and heterogeneous solid phases. Furthermore, most work has focused on speciation in the bulk aqueous system, while speciation in micro-environments, such as at the surface of fish gills, needs to be defined.

Charged organic compounds

The processes determining the fate of neutral compounds have been described in detail in previous paragraphs of this chapter, as well as in Chapter 4. Various compounds are present in the environment as positively or negatively charged molecules. Surfactants may be negatively charged, positively charged or both; weak organic bases and acids are charged depending on the pH of the environment; metals are present in various forms and many organometallics can be present as cations.

Charged chemical species have different bioconcentration properties than neutral species. This will have a major impact on the uptake of chemicals. Passive diffusion through the lipid membrane is the main route of uptake for many chemicals. Charged chemicals will usually be transported across the lipid membrane at a much lower rate. The uptake of cationic surfactants and triorganotins, however, has been observed, but they may be taken up as neutral molecules. An example which illustrates this is the uptake of chlorinated phenols at different pH values by fish [154]. These weak acids are negatively charged at a pH > pK_a, and neutral at a pH < pK_a.

$$\mathrm{ROH} \rightleftarrows \mathrm{RO^-} + \mathrm{H^+}$$

$$K_a = \frac{[\mathrm{RO^-}]\,[\mathrm{H^+}]}{[\mathrm{ROH}]}$$

$$R = \mathrm{C_6H_xCl_y} \text{ with } x + y = 5$$

The uptake of these chlorinated phenols was independent of pH, when the pH < (pK_a-1). When the experimental pH was increased, the uptake of the chlorinated phenols decreased (Figure 3.52). The uptake rate of the phenols at pH > pK_a depends on their degree of ionization. However, uptake was more than one order of magnitude higher than expected on the basis of the concentration of the non-ionized form. A possible explanation for this phenomenon is that fish are to some extent able to buffer the pH in their gills. Consequently, the pH in the bulk water is not equal to the pH in the water at the gills. Ionization of the chlorinated phenols, however, was highly affected by pH, and charged molecules were taken up at a much lower rate than neutral molecules.

A specific feature that is currently receiving increased attention by scientists and risk regulators, is the observation that carbonaceous geosorbents, such as black carbon (BC), coal or kerogen, are able to bind organic pollutants like PAH or PCBs very effectively [155,156]. The sources of BC are partly natural (weathering of graphitic rocks, forest fires, condensation of organic matter) and partly anthropogenic (traffic, industry, fuel combustion, domestic fires). This effective binding is similar to that to Activated Carbon (AC) as used in many water cleaning technologies. From an ecotoxicological point of view, this binding can be regarded as advantageous, because the negative effects of organic pollutants on organisms will be reduced. It has been announced that new policies for polluted soil and sediment will be based on an evaluation of contaminant fluxes, rather than an evaluation of the total concentration in the soil or sediment. This implies that more attention will be focused on reducing the actual risk caused by bioavailable fractions of contaminants, rather than on the risks inferred from total organic pollutant concentrations.

Critical body residues

A general principle in pharmacology is that the concentration of a chemical at the receptor of toxicity determines the effect. This principle is translated in ecotoxicology into the Critical Body Residue (CBR) concept. This concept assumes that the total body concentration is proportional to the concentration at the target or receptor, and the CBR is defined as the threshold concentration of a substance in an organism that marks the transition between no effect and adverse effect. CBRs are relatively constant between groups of organic chemicals with a similar mode of action and between different organisms, and comparison of body concentrations with CBRs may be an effective tool for site-specific risk assessment of toxicants. The CBR concept integrates internal transport and metabolism processes, as well as toxicity at specific sites. The only aspect that is lacking in the cascade of processes that initiate adverse effects due to the presence of elevated concentrations of chemicals, is the external transport of the toxicants to an organism. This aspect is taken into account by means of the concept of bioavailability.

3.7.2 Underlying concepts of bioavailability

Figure 3.51 shows the three basic concepts underlying bioavailability:

1. Partitioning of the chemical between the solid and the aqueous phases, also termed *environmental availability*.
2. Physiological driven uptake, also termed *environmental bioavailability*.
3. Toxico-dynamic interactions at the site of toxicity, also termed *toxicological bioavailability*.

1. Environmental availability.

Various processes induce deviations from equilibrium partitioning in the environment. Consequently, equilibrium partitioning coefficients derived in an optimized laboratory setting cannot be used in these cases to quantify water concentrations. Sequestration or “aging” is the process by which chemicals tend to become less available in time for uptake by organisms, for partitioning into the aqueous phase, or for extraction by means of “soft” chemical extraction techniques. With increasing contact time, the chemicals appear to migrate deeper into the organic matrix, are irreversibly bound to the matrix, or tend to be precipitated at the surface of the solid phase. For metals, the latter process is strongly pH-dependent and precipitation is often observed at high pH values. The use of equations where sorption is estimated from hydrophobicity (in the case of organic chemicals) or from short-term partitioning experiments at low pH (for metals), will fail to predict the effect of sequestration. Instead, other means of obtaining good estimates of pore water concentrations or actual measurements under field conditions that do not meet the basic requirement of equilibrium, need to be used. There are also regression equations linking partition coefficients determined in field samples to the substrate-related parameters affecting partitioning (like pH and organic carbon content) available for the purpose of indirectly quantifying sequestration.

2. Environmental bioavailability.

The uptake of chemicals involves the passage of compounds across a biological membrane, mediated by a carrier or a single solute. Compounds may enter tissues through passive diffusion, facilitated diffusion and by active transport mechanisms. Passive diffusion is the major uptake process for many organic chemicals, as well as some metals and organometals. The driving force for uptake is a fugacity difference between water and the organism. Although some inorganic and organic metal complexes may be directly taken up while other ligands may compete with organisms for the metal, it is the free metal ion that is supposed to be capable of passing biological membranes. As a consequence, metal availability and toxicity are functions of water chemistry,

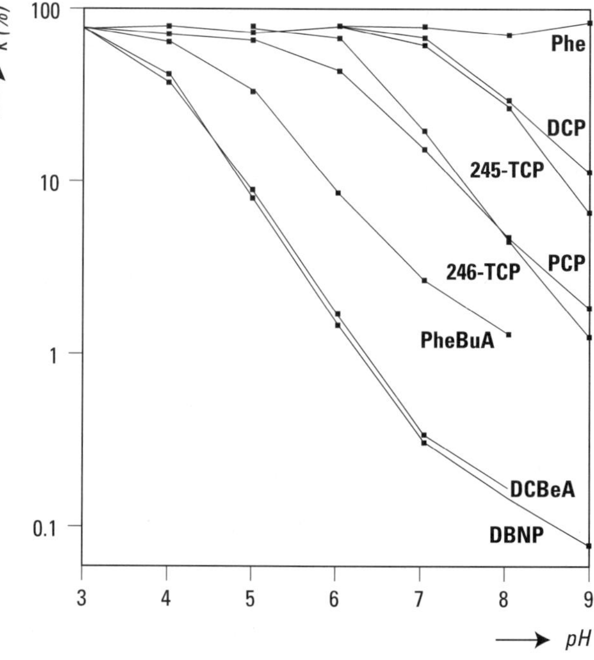

Figure 3.52. The relationship between the uptake rate (k as percentage of the uptake rate measured at the lowest experimental pH) of phenols by guppy and the pH of the water, where Phe is phenol, DCP is 2,4-dichlorophenol, 245-TCP is 2,4,5-trichlorophenol, 246-TCP is 2,4,6-trichlorophenol, PCP is pentachlorophenol, PheBuA is 4-phenylbutyric acid, DCBeA is 3,4-dichlorobenzoic acid and DBNP is 2,6-dibromo-4-dinitrophenol. From [154]. Copyright Elsevier.

as pore water chemistry (speciation) determines the free metal ion activity. Formation of inorganic and organic metal complexes and sorption of metals and organic micropollutants to particulate matter have been shown to reduce toxicity. As a result, the relationship of toxicity to total or dissolved concentrations can be highly variable, and depends on the ambient water chemistry.

3. Toxicological bioavailability.
This aspect was discussed above on the basis of the CBR concept. For metals, the activity of the free ion has been shown to correlate best to toxicity. However, competition with other cations (most notably the macro-elements Ca, Mg, Na, and protons) for uptake at specific biotic ligands also affects uptake and toxicity. The latter aspect, together with the pharmacological principle mentioned above, formed the basis for the recently developed Biotic Ligand Model (BLM). The BLM too, assumes that the effect is proportional to the concentration of metal bound to the target site, and that this target site (biotic ligand) is in direct contact with the external aquatic environment. The BLM uses chemical speciation modelling to quantify the activity of the free metal ion and seems well capable of describing the acute toxicity of metals to fish, when the gill is the target site. Although some BLMs have also been developed for other aquatic organisms, such as crustaceans and algae, it remains unclear whether the assumption that the target is in direct contact with the external environment is always valid. Chronic exposure may modify the gill-metal binding characteristics. It therefore remains unclear whether gill-metal binding constants derived in BLMs for acute toxicity may also be applicable for predicting the effects of long-term metal exposure. As only a fraction of the total metal body burden is biologically available for interaction with sites of toxic action, a better understanding of the internal compartmentalization of metals in organisms and its consequences for toxicity is required. The same is true for polar or ionic organic compounds, whereas the basic assumption of partitioning of hydrophobic organic compounds between the aqueous phase and the body tissues provides a sound basis for predicting the toxicologically bioavailable fraction. The principles behind BLMs are described in Chapter 7 and in a special issue of the journal "Comparative Biochemistry and Physiology: Part C Toxicology and Pharmacology" [157].

3.7.3 Inclusion of bioavailability in risk assessment

As discussed above and as summarized as early as in 1994 by Hamelink et al. [153], the principles underlying the bioavailability concept are well-known. What is lacking *grosso modo* however, is a well-designed database containing quantitative information on the large array of parameters that jointly determine bioavailability. This refers not only to a lack of validated expressions of the kinetics and thermodynamics of the fate processes determining the activity of the available fraction of the chemical in the water phase and in the solid matrices, but also to the kinetics of the most important uptake and elimination processes by biota, as well as the kinetics of the biological response of biotic species to elevated exposure to increased contaminant levels. This lack of knowledge is most apparent for inorganic species and hinders their proper inclusion for risk assessment purposes.

For neutral organic compounds a correction for bioavailability on the basis of standardized interactions of the chemical with the organic material present in the system, is widely accepted. Assuming that the differences in the nature of the organic matter present, and that the composition of the aqueous phase in terms of parameters potentially capable of affecting sorption of organic chemicals to organic ligands, do not affect the chemical interaction of the contaminant with the organic sorbents, most environmental risk assessment procedures include a correction of the standards by means of K_{oc}.

For metals, metalloids, and ionic organic species, no generic correction is available yet. Instead, the common procedure is to include bioavailability considerations in second-tier risk assessment procedures, taking local conditions specifically into account; general approaches are currently the subject of scientific and regulatory debate.

3.8 FURTHER READING

1. EU TGD part 3 section 4.4 PBT assessment, P criterion (as in the general references). [38]
2. Mackay D. 1991. *Multimedia Environmental Models. The Fugacity Approach.* Lewis Publisher, Chelsea, MI. [1]
3. Schwarzenbach RP, Gschwend PM, Imboden DM. 2003. *Environmental Organic Chemistry 2nd edition.* Wiley Interscience, New York, NY. [16]
4. UMBBD database on the internet (as in the general references). [129]
5. KEGG pathway database on the internet (as in the general references). [130]
6. Alexander, M. (1994) *Biodegradation and Bioremediation.* Academic Press, San Diego, CA. ISBN 0-12-049860-X [158]

REFERENCES

1. Mackay D. 1991. *Multimedia Environmental Models. The Fugacity Approach.* Lewis Publ, Chelsea, MI.
2. Schwarzenbach RP. 1992. Phase-transfer of organic pollutants in the environment. Course on environmental chemistry of organic pollutants. European Environmental Research Organization, Wageningen, The Netherlands.
3. Junge CE. 1977. Basic considerations about trace constituents in the atmosphere in relation to the fate of global pollutants. In: Suffet IH, ed, *Fate of Pollutants in the Air and Water Environment. Part I, Advances in Environmental Science and Technology, Vol 8.* Wiley Interscience, New York, NY, pp. 7-25.
4. Kauneliene V, Cicenaite A, Jegorova I, Zaliauskiene A, Bergqvist P-A. 2003. Tentative air concentrations of PAHs and PCBs in urban area of Lithuania. *Environ Res Engineering Man* 24:33-39.
5. Van Mazijk A, Veldkamp RG. 1989. Waterkwaliteits modelering oppervlaktewater. Collegenotes, Technical University, Delft, The Netherlands [in Dutch].
6. Janse J. 2005. Model studies on the eutrophication of shallow lakes and ditches. Ph.D. Thesis, Wageningen University. Wageningen, The Netherlands.
7. Neely WB. 1982. The definition and use of mixing zone. *Environ Sci Technol* 16:519A-5121A.
8. Csanady GT. 1973. Turbulent diffusion in the environment. *Geophysics and Astrophysics Monographs,* Vol 3. D Reidel Publ Co, Dordrecht, The Netherlands.
9. Fischer HB, Imberger J, List EJ, Koh RCY, Brooks RH. 1979. *Mixing in Inland and Coastal Waters.* Academic Press, New York, NY.
10. Van de Meent D. 1993. SIMPLEBOX, a generic multimedia fate evaluation Model. Report 672720001. National Institute for Public Health and Environmental Protection (RIVM), Bilthoven, The Netherlands.
11. Spitz K, Moreno J. 1996. *A practical guide to groundwater and solute transport modeling.* Wiley Interscience, New York, NY.
12. Fowler D. 1980. Removal of sulphur and nitrogen compounds from the atmosphere and by dry deposition. In: Drablos D, Tollan A, eds, *Ecological impact of Acid Precipitation.* Oslo-As, Norway, pp. 22-32.
13. Whitman WG. 1923. The two-film theory of gas absorption. *Chem Metal Eng* 29:146-150.
14. Liss PS, Slater PG. 1974. Flux of gases across the air-sea interface. *Nature* 247:181-184.
15. Thibodeaux LJ. 1996. *Environmental Chemodynamics: Movement of Chemicals in Air, Water, and Soil, 2nd ed.* John Wiley, New York, NY.
16. Schwarzenbach RP, Gschwend PM, Imboden DM. 2003. *Environmental Organic Chemistry 2nd edition.* Wiley-Interscience, New York, NY.
17. Mackay D, Paterson S, Cheung B, Neely WB. 1985. Evaluating the environmental behaviour of chemicals with a Level-III model. *Chemosphere* 14:335-374.
18. McLachlan MS. 1996. Bioaccumulation of hydrophobic chemicals in agricultural food chains. *Environ Sci Technol* 30:252-259.
19. Arnot J, Gobas FAPC. 2003. A generic QSAR for assessing the bioaccumulation potential of organic chemicals in aquatic food webs. *QSAR Comb Sci* 22: 337-345.
20. Phillips DJH. 1993. Bioaccumulation. In: Calow P, ed, *Handbook of Ecotoxicology, Vol 1.* Blackwell Sci Publ, Oxford, UK, pp. 378-396.
21. Block M. 1991. Uptake of cadmium in fish. Effects of xanthates and diethyldithio-carbamate. PhD Thesis, Uppsala University, Uppsala, Sweden.

22. Sijm DTHM, Seinen W, Opperhuizen A. 1992. Life-cycle biomagnification study in fish. *Environ Sci Technol* 26:2162-2174.
23. McGeer JC, Brix KV, Skeaff JM, DeForest DK, Brigham SI, Adams WJ, Green A. 2003. Inverse relationship between bioconcentration factor and exposure concentration for metals: implications for hazard assessment of metals in the aquatic environment. *Environ Toxicol Chem* 22:1017-1037.
24. Adams WJ, Conard B, Ethier G, Brix KV, Paquin PR, Di Toro DM. 2000. The challenges of hazard identification and classification of insoluble metals and metal substances for the aquatic environment. *Human Ecol Risk Assess* 6:1019-1038.
25. Tas JW. 1993. Fate and effects of triorganotins in the aqueous environment. Bioconcentration kinetics, lethal body burdens, sorption and physicochemical properties. PhD Thesis, University of Utrecht, The Netherlands.
26. Sijm DTHM, Pärt P, Opperhuizen A. 1993. The influence of temperature on the uptake rate constants of hydrophobic compounds determined by the isolated perfused gills of rainbow trout *(Oncorhynchus mykiss)*. *Aquat Toxicol* 25:1-14.
27. Gobas FAPC, Opperhuizen A, Hutzinger O. 1986. Bioconcentration of hydrophobic chemicals in fish: relationship with membrane permeation. *Environ Toxicol Chem* 5:637-646.
28. Sijm DTHM, Verberne ME, de Jonge WJ, Pärt P, Opperhuizen A. 1995. Allometry in the uptake of hydrophobic chemicals determined *in vivo* and in isolated perfused gills. *Toxicol Appl Pharmacol* 131:130-135.
29. Sijm DTHM, Hermens JLM. 1999. Internal effect concentrations: Link between bioaccumulation and ecotoxicity for organic chemicals. In: Beek B, ed, *The handbook of environmental chemistry, Vol 2-J. Bioaccumulation: new aspects and developments.* Springer-Verlag, Berlin, Germany, pp. 167-199.
30. Hendriks JA, van der Linde A, Cornelissen G, Sijm DTHM. 2001. The power of size. 1. Rate constants and equilibrium ratios for accumulation of organic substances related to octanol-water partition ratio and species weight. *Environ Toxicol Chem* 20:1399-1420.
31. Hendriks JA, Heikens A. 2001. The power of size. 2. Rate constants and equilibrium ratios for accumulation of inorganic substances related to species weight. *Environ Toxicol Chem* 20:1421-1437.
32. Leland HV, Kuwabara JS. 1985. Trace metals. In: Rand GM, Petrocelli SR, eds, *Fundamentals of Aquatic Toxicology.* Hemisphere, Washington DC, pp. 374-415.
33. Niimi AJ. 1987. Biological half-lives of chemicals in fishes. *Rev Environ Contam Toxicol* 99:1-46.
34. Organization for Economic Co-operation and Development. (1996). Bioaccumulation: Flow-through Fish Test. OECD Guideline for the testing of chemicals No. 305. OECD, Paris, France.
35. American Society for Testing and Materials. 2003. E1022-94. Standard guide for conducting bioconcentration tests with fishes and saltwater bivalve mollusks. ASTM International, West Conshohocken, PA, United States.
36. US Environmental Protection Agency. 1996. Ecological Effects Test Guidelines. OPPTS 850.1730 Fish BCF. Public Draft. Office of Prevention, Pesticides and Toxic Substances. Washington, DC.
37. US Environmental Protection Agency. 1996. Ecological Effects Test Guidelines. OPPTS 850.1710 Oyster BCF. Public Draft. Office of Prevention, Pesticides and Toxic Substances. Washington, DC.
38. European Commission. 2003. 2nd edition of the Technical Guidance Document in support of Commission Directive 93/67/EEC on risk assessment for new notified substances, Commission Regulation (EC) No. 1488/94 on risk assessment for existing substances and Directive 98/8/EC of the European Parliament and of the Council concerning the placing of biocidal products on the market. Office for Official Publications of the European Communities, Luxembourg.
39. US Environmental Protection Agency. 1999. Category for persistent, bioaccumulative and toxic new chemical substances. *Fed Reg* 64:60194-60204. United States Environment Protection Agency, Office of Research and Development, Washington, DC.
40. Anliker R, Moser P, Poppinger D. 1988. Bioaccumulation of dyestuffs and organic pigments in fish. Relationships to hydrophobicity and steric factors. *Chemosphere* 17:1631-1644.
41. Rekker RF, Mannhold R. 1992. *Calculation of drug lipophilicity*. Weinheim, Germany: VCH. (Cited at www.voeding.tno.nl/ProductSheet.cfm?PNR=037e)
42. Burreau S, Zebuhr Y, Broman D, Ishaq R. 2004. Biomagnification of polychlorinated biphenyls (PCBs) and polybrominated diphenyl ethers (PBDEs) studies in pike *(Esox lucius)*, perch *(Perca fluviatilis)* and roach *(Rutilus rutilus)* from the Baltic Sea. *Chemosphere* 55:1043-1052.
43. De Wolf W, Comber M, Douben P, Gimeno S, Holt H, Léonard M, Lillicrap A, Sijm D, van Egmond R, Weisbrod A, Whale G. 2007. Animal use replacement, reduction and refinement: development of an Integrated Testing Strategy for Bioconcentration of Chemicals in Fish. *IEAM* 3:3-17.
44. Opperhuizen A, Sijm DTHM. 1990. Bioaccumulation

and biotransformation of polychlorinated dibenzo-p-dioxins and dibenzofurans in fish. *Environ Toxicol Chem* 9:175-186.

45. Sijm DTHM, Wever H, Opperhuizen A. 1993. Congener-specific biotransformations and bioaccumulation of PCDDs and PCDFs from fly ash in fish. *Environ Toxicol Chem* 12:1895-1907.
46. Anliker R, Moser P. 1987. The limits of bioaccumulation of organic pigments in fish: Their relation to the partition coefficient and the solubility in water and octanol. *Ecotox Environ Saf* 13:43-52.
47. Opperhuizen A, Damen HWJ, Asyee GM, van der Steen JMD, Hutzinger O. 1987. Uptake and elimination by fish of polydimethylsiloxanes (silicones) after dietary and aqueous exposure. *Toxicol Environ Chem* 13:265-285.
48. Morris S, Allchin CR, Zegers BN, Haftka JJH, Boon JP, Belpaire C, Leonards PEG, Van Leeuwen SPJ, De Boer J. 2004. Distribution and fate of HBCD and TBBPA brominated flame retardants in North Sea estuaries and aquatic food webs. *Environ Sci Technol* 38:5497-5504.
49. Dimitrov SD, Dimitrova NC, Walker JD, Veith GD, Mekenyan OG. 2002. Predicting bioconcentration factors of highly hydrophobic chemicals. Effects of molecular size. *Pure Appl Chem* 74:1823-1830.
50. Dimitrov SD, Dimitrova NC, Walker JD, Veith GD, Mekenyan OG. 2003. Bioconcentration potential predictions based on molecular attributes - an early warning approach for chemicals found in humans, birds, fish and wildlife. *QSAR Comb Sci* 22:58-68.
51. Dimitrov SD, Dimitrova NC, Parkerton T, Comber MHI, Bonnell M, Mekenyan OG. 2006. Baseline model for identifying the bioaccumulation potential of chemicals. *SAR QSAR Environ Res* 16:531-554.
52. Opperhuizen A. 1986. Bioconcentration of hydrophobic chemicals in fish. In: Poston TM, Purdy R, eds, *Aquatic toxicology and environmental fate, Vol 9,* STP 921. American Society for Testing and Materials (ASTM), Philadelphia, USA, pp. 304-315.
53. Tolls J, Haller M, Labee E, Verweij M, Sijm DTHM. 2000. Experimental determination of bioconcentration of the nonionic surfactant alcohol ethoxylated. *Environ Toxicol Chem* 19:646-653.
54. Lipinski CA, Lombardo F, Dominy BW, Feeney PJ. 1997. Experimental and computational approaches to estimate solubility and permeability in drug discovery and development settings. *Advanced Drug Delivery Reviews* 23:3-25.
55. Wenlock MC, Austin RP, Barton P, Davis AM, Leeson PD. 2003. A comparison of physiochemical property profiles of development and marketed oral drugs. *J Med Chem* 46:1250-1256.
56. Proudfoot JR. 2005. The evolution of synthetic oral drug properties. *Bioorganic Medicinal Chemistry Letters* 15:1087-1090.
57. Vieth M, Siegel MG, Higgs RE, Watson IA, Robertson DH, Savin KA, Durst GL, Hipskind PA. 2004. Characteristic physical properties and structural fragments of marketed oral drugs. *J Med Chem* 47:224-232.
58. Wood CM, Kelly SP, Zhou B, Fletcher M, O'Donnell M, Eletti B, Pärt P. 2002. Cultured gill epithelia as models for the freshwater fish gill. *Biochim Biophys Acta - Biomembranes* 1566:72-83.
59. Hidalgo IJ, Li J. 1996. Carrier-mediated transport and efflux mechanisms in Caco-2 cells. *Advanced Drug Delivery Reviews* 22:53-66.
60. Bailey CA, Bryla P, Malick AW. 1996. The use of the intestinal epithelial cell culture model, Caco-2, in pharmaceutical development. *Advanced Drug Delivery Reviews* 22:85-103.
61. Pärt P. 1990. The perfused fish gill preparation in studies of the bioavailability of chemicals. *Ecotox Environ Saf* 19:106-115.
62. Pärt P, Saarikoski J, Tuurula H, Havaste H. 1992. The absorption of hydrophobic chemicals across perfused gills of rainbow trout. *J Exp Biol* 6:339-348.
63. Chessells M, Hawker DW, Connell DW. 1992. Influence of solubility in lipid on bioconcentration of hydrophobic compounds. *Ecotox Environ Saf* 23:260-273.
64. Banerjee S, Baughman GL. 1991. Bioconcentration factors and lipid solubility. *Environ Sci Technol* 25:536-539.
65. Sijm DTHM, Opperhuizen A. 1989. Biotransformation of organic chemicals by fish: a review of enzyme activities and reactions. In: Hutzinger O, ed, *Handbook of Environmental Chemistry, Vol 2E, Reactions and Processes.* Springer-Verlag, Heidelberg, Germany, pp. 163-235.
66. Kelly BC, Gobas FAPC, McLachlan MS. 2004. Intestinal absorption and biomagnification of organic contaminants in fish, wildlife and humans. *Environ Toxicol Chem* 23: 2324-2336.
67. Opperhuizen A. 1992. Bioconcentration and biomagnification: is a distinction necessary? In: Nagel R, Loskill R, eds, Bioaccumulation in aquatic systems. Contribution to the assessment. Proceedings of an international workshop, Berlin, VCH Publishers, Weinheim, Germany, pp. 67-80.
68. Opperhuizen A, Schrap SM. 1988. Uptake efficiencies of two polychlorobiphenyls in fish after dietary exposure to five different concentrations. *Chemosphere* 17:253-262.
69. Lee II H. 1991. A clam's eye view of the bioavailability

of sediment-associated pollutants. In: Baker R, ed, *Organic Substances and Sediments in Water, Volume III.* Lewis Publisher Inc, Chelsea, MI, pp. 73-93.

70. Opperhuizen A. 1991. Bioaccumulation kinetics: experimental data and modelling. In: Angeletti G, Bjørseth A, eds, *Organic Micropollutants in the Aquatic Environment,* Proc Sixth European Symp Lisbon, Portugal, 1990. Kluwer Acad Publ, Dordrecht, The Netherlands, pp. 61-70.
71. Schrap SM. 1991. Bioavailability of organic chemicals in the aquatic environment. *Comp Biochem Physiol* 100C:13-16.
72. Parkerton T, Letkinski D, Febbo E, Davi R, Dzamba C, Connelly M, Christensen K, Peterson D. 2001. A practical testing approach for assessing the bioaccumulation potential of poorly water soluble organic chemicals. Presentation at the SETAC Europe annual meeting, Madrid, Spain. Manuscript No. 00.7014. Annandale NJ: ExxonMobil Biomedical Sciences Inc.
73. Anonymous. 2004. Fish, dietary bioaccumulation study – Basic Protocol. Document submitted to the European Union Technical Committee for New and Existing Substances subgroup on determining of Persistent, Bioaccumulative and Toxic substances. European Chemicals Bureau, Ispra, Italy.
74. Organization for Economic Co-operation and Development. 2005. Proposal for a new guideline for the testing of chemicals - Bioaccumulation: Sediment test using benthic oligochaetes. Revised Draft. October 2005. OECD, Paris, France.
75. American Society for Testing and Materials. 2000. E1688-00a. Standard guide for determination of the bioaccumulation of sediment-associated contaminants by benthic invertebrates. ASTM International, West Conshohocken, PA.
76. Sharpe S, Mackay D. 2000.A framework for evaluating bioaccumulation in food webs. *Environ Sci Technol* 34:2373-2379.
77. Arnot J, Gobas FAPC. 2004. A food web bioaccumulation model for organic chemicals in aquatic ecosystems. *Environ Toxicol Chem* 23:2343-2355.
78. Traas TP, Van Wezel AP, Hermens JLM, Zorn M, Van Hattum AGM, Van Leeuwen CJ. 2004. Prediction of environmental quality criteria from internal effect concentrations for organic chemicals with a food web model. *Environ Toxicol Chem* 23:2518-2527.
79. Voutsas E, Magoulas K, Tassios D. 2002. Prediction of the bioaccumulation of persistent organic pollutants in aquatic food webs. *Chemosphere* 48:645-651.
80. Morrison LN, Cohen AS. 1980. Plant uptake, transport and metabolism. In: Hutzinger O, ed, *Handbook of Environmental Chemistry, 2A. Reactions and Processes.* Springer-Verlag, Heidelberg, Germany, pp. 193-219.
81. Briggs GG, Bromilow RH, Evans AA. 1982. Relationships between lipophilicity and root uptake and translocation of non-ionised chemicals by barley. *Pestic Sci* 13:495-504.
82. Iwata Y, Gunther FA, Westlake WE. 1974. Uptake of a PCB (Aroclor 1254) from soil by carrots under field conditions. *Bull Environ Contam Toxicol* 11:523-528.
83. Wild E, Dent J, Thomas GO, Jones KC. 2005. Direct observation of organic contaminant uptake, storage, and metabolism within plant roots. *Environ Sci Technol* 39:3695-3702.
84. Kulhánek A, Trapp S, Sismilich M., Janku J, Zimová M, 2005. Crop-specific human exposure assessment for polycyclic aromatic hydrocarbons in Czech soils. *Sci Tot Environ* 339:71-80.
85. McCrady JK, McFarlane C, Lindstrom FT. 1987. The transport and affinity of substituted benzenes in soybean stems. *J Experim Biol* 38:1875-1890.
86. Briggs GG, Rigitano RLO, Bromilow RH. 1987. Physico-chemical factors affecting uptake by roots and translocation to shoots of weak acids in barley. *Pestic Sci* 19:101-112.
87. Hülster A, Müller JF, Marschner H. 1994. Soil-plant transfer of polychlorinated dibenzo-p-dioxins and dibenzofurans to vegetables of the cucumber family *(Cucurbitaceae). Environ Sci Technol* 28:1110-1115.
88. Trapp S, Matthies M. 1998. *Chemodynamics and environmental modeling.* Springer, Berlin, pp. 115-127.
89. Briggs GG, Rigitano RLO, Bromilow RH. 1987. Physico-chemical factors affecting uptake by roots and translocation to shoots of weak acids in barley. Pestic Sci 19:101-112.
90. Trapp, S. 2002. Dynamic root uptake model for neutral lipophilic organics. *Environ Toxicol Chem* 2:203-206.
91. Kerstiens G. 2006. Parameterization, comparison, and validation of models quantifying relative change of cuticular permeability with physicochemical properties of diffusants. *J Exper Botany* 57:2525-2533.
92. Müller JF, Hawker DW, Connell DW. 1994. Calculation of bioconcentration factors of persistent hydrophobic compounds in the air/vegetation system. *Chemosphere* 29:623-640.
93. Paterson S, Mackay D, Bacci E, Calamari D. 1991. Correlation of the equilibrium and kinetics of leaf-air exchange of hydrophobic organic chemicals. *Environ Sci Technol* 25:866-871.
94. Kömp P, McLachlan MS. 1997. Interspecies variability

of the plant/air partitioning of polychlorinated biphenyls. *Environ. Sci. Technol.* 31:2944-2948.

95. Kömp P, McLachlan MS. 1997. The influence of temperature on the plant/air partitioning of semivolatile organic compounds. *Environ Sci Technol* 31:886-890.
96. Czub G, McLachlan MS. 2004. A food chain model to predict the levels of lipophilic organic contaminants in humans. *Environ Toxicol Chem* 23:2356-2366.
97. Horstmann M, McLachlan MS. 1998. Atmospheric deposition of semivolatile organic compounds to two forest canopies. *Atmos Environ* 32:1799-1809.
98. Smith KEC, Jones KC. 2000. Particles and vegetation: implications for the transfer of particle-bound organic contaminants to vegetation. *Sci Tot Environ* 246:207-236.
99. McLachlan MS 1999. A framework for the interpretation of measurements of SOCs in plants. *Environ Sci Technol* 33:1799-1804.
100. Rigitano FLO, Bromilow RH, Briggs GG, Chamberlain K. 1987. Phloem translocation of weak acids in *Ricinus communis. Pest Sci* 19:113-133.
101. Mackay D, Foster KL, Patwa Z, Webster E. 2006. Chemical partitioning to foliage: The contribution and legacy of David Calamari. *Environ Sci Pollut Res* 13:2-8.
102. Jager T. 1998. Mechanistic approach for estimating bioconcentration of organic chemicals in earthworms (Oligochaeta). *Environ Toxicol Chem* 17:2080-2090.
103. Stroomberg, GJ, Zappey H, Steen RJCA, Van Gestel CAM, Ariese F, Velthorst NH, Van Straalen NM. 2004. PAH biotransformation in terrestrial invertebrates - a new phase II metabolite in isopods and springtails. *Comp Biochem Physiol* Pt C 138:129-137.
104. Jager T. 2004. Modelling ingestion as an exposure route for organic chemicals in earthworms (Oligochaeta). *Ecotox Environ Saf* 57:30-38.
105. Jager T, Fleuren RHLJ ,Hogendoorn EA, De Korte G. 2003. Elucidating the routes of exposure for organic chemicals in the earthworm, *Eisenia andrei* (Oligochaeta). Environ Sci Technol 37:3399-3404.
106. Belfroid A, Meiling J, Drenth HJ, Hermens J, Seinen W, Van Gestel K. 1995. Dietary uptake of superlipophilic compounds by earthworms *(Eisenia andrei). Ecotox Environ Saf* 31:185-191.
107. Environment Agency. 2007. Verification of bioaccumulation models for use in environmental standards – Part B – Terrestrial models – Draft Report.
108. Van der Wal L, Jager T, Fleuren RHLJ, Barendregt A, Sinnige TL, Van Gestel CAM, Hermens JLM. 2004. Solid-phase microextraction to predict bioavailability and accumulation of organic micropollutants in terrestrial organisms after exposure to a field-contaminated soil. *Environ Sci Technol* 38:4842-4848.
109. Janssen MPM, Bruins A, De Vries TH, Van Straalen NM. 1991. Comparison of cadmium kinetics in four soil arthropod species. *Arch Environ Contam Toxicol* 20:305-312.
110. Vijver MG, Van Gestel CAM, Lanno RP, Van Straalen NM, Peijenburg WJGM. 2004. Internal metal sequestration and its ecotoxicological relevance: a review. *Environ Sci Technol* 38:4705-4712.
111. Posthuma L, Van Straalen NM. 1993. Heavy-metal adaptation in terrestrial invertebrates: a review of occurrence, genetics, physiology and ecological consequences. *Comp Biochem Physiol* 106C:11-38.
112. Vijver MG, Van Gestel CAM, Van Straalen NM, Lanno RP, Peijnenburg WJGM. 2006. Biological significance of metals partitioned to subcellular fractions within earthworms *(Apporrectodea caliginosa). Environ Toxicol Chem* 25:807-814.
113. USES. 1994. Uniform system for the evaluation of substances, version 1.0. National Institute of Public Health and Environmental Protection, Ministry of Housing, Spatial Planning and the Environment, Ministry of Welfare, Health and Cultural Affairs. VROM distribution No. 11144/150, The Hague, The Netherlands.
114. Ma WC. 1987. Heavy metal contamination in the mole, *Talpa europaea,* and earthworms as an indicator of metal bioavailability in terrestrial environments. *Bull Environ Contam Toxicol* 39:933-938.
115. Ma WC, Denneman W, Faber J. 1991. Hazardous exposure of groundliving small animals to cadmium and lead in contaminated terrestrial ecosystems. *Arch Environ Contam Toxicol* 20:266-270.
116. Norstrom RJ, McKinnon AE, DeFreitas AS. 1979. A bioenergetics based model for pollutant accumulation in fish: simulation of PCB and methylmercury residue levels in Ottawa river yellow perch *(Perca flavescens). J Fish Res Board Can* 33:248-267.
117. DeBruyn AMH, Gobas FAPC. 2006. A bioenergetic biomagnification model for the animal kingdom. *Environ Sci Technol* 40:1581-1587.
118. Travis CC, Arms AD. 1988. Bioconcentration of organics in beef, milk and vegetation. *Environ Sci Technol* 22:271-274.
119. Burns LA, Baughman GL. 1985. Fate modelling. In: Rand GM, Petrocelli SR, eds, *Fundamentals of Aquatic Toxicology.* Hemisphere Publ Corp, Washington DC, USA, pp. 558-584.
120. Mabey W, Mill T. 1978. Critical review of hydrolysis of organic compounds in water under environmental conditions. *J Phys Chem Ref Data* 7:383-415.

121. Haag WR, Yao CCD. 1992. Rate constants for reaction of hydroxyl radicals with several drinking water contaminants. *Environ Sci Technol* 26:1005-1013.
122. Mill T. 1980. Chemical and photo oxidation. In: Hutzinger O, ed, *The Handbook of Environmental Chemistry, Volume 2, part A: Reactions and Processes*. Springer Verlag, Berlin, Germany, pp. 77-105.
123. Wolfe NL, Macalady DL. 1992. New perspectives in aquatic redox chemistry: abiotic transformations of pollutants in groundwater and sediments. *J Contam Hydrol* 9:17-34.
124. Organization for Economic Co-operation and Development. 1986. OECD guidelines for the testing of chemicals. Hydrolysis as function of pH. Guideline 111. OECD, Paris, France.
125. Zepp RG. 1982. Experimental approaches to environmental photochemistry. In: Hutzinger O, ed, *The Handbook of Environmental Chemistry, 1st ed Vol 2/part B*. Springer-Verlag, Berlin, Germany, pp. 19-41.
126. Bolt TL, Bruggewert GM. 1976. *Soil Chemistry. Part A: Basic Elements*. Elsevier Sci Publ, Amsterdam, The Netherlands.
127. Thauer RK, Jungermann K, Decker K. 1977. Energy conservation in chemoautotrophic anaerobic bacteria. *Bact Rev* 41:100-148.
128. Horvath RS. 1972. Microbial cometabolism and the degradation of organic compounds in nature. *Bact Rev* 36:146-155.
129. UMBBD, University of Minnesota Biodegradation and Biocatalysis Database, 2006, http://umbbd.ahc.umn.edu/
130. KEGG Pathway, Kyoto Encyclopedia of Genes and Genomes Pathway Database, 2006, http://www.genome.jp/kegg/pathway.html
131. Swisher RD. 1987. *Surfactant Biodegradation*. Marcel Dekker, New York, NY.
132. Kadiyala V, Spain JC. 1998. A two-component monooxygenase catalyzes both the hydroxylation of p-nitrophenol and the oxidative release of nitrite from 4-nitrocatechol in Bacillus sphaericus JS905. *Appl Environ Microbiol* 64:2479-2484.
133. Organization for Economic Co-operation and Development. 1981 and 1993. OECD guidelines for the testing of chemicals. Degradation and Accumulation. OECD, Paris, France.
134. Khan SJ, Rorije E. 2002. Pharmaceutically active compounds in aquifer storage and recovery. In: Dillon PJ, ed, *Management of Aquifer Recharge for Sustainability*. A.A. Balkema Publishers, The Netherlands, pp. 169-174.
135. Tiedje JM, Boyd SA, Fathepure BZ. 1987. Anaerobic degradation of chlorinated aromatic hydrocarbons. *J Ind Microbiol Supp 1, Developments in Industrial Microbiology* 27:117-127.
136. Rorije E, Peijnenburg WJGM, Klopman G. 1998. Structural requirements for the anaerobic biodegradation of organic chemicals: A fragment model analysis. *Environ Toxicol Chem* 17:1943-1950.
137. Klopman G, Saiakhov R, Tu M, Pusca F, Rorije E. 1998. Computer-assisted evaluation of anaerobic biodegradation products. *Pure and Applied Chemistry* 70:1385-1394.
138. Alexander M. 1973. Nonbiodegradable and other recalcitrant molecules - Biotechnology report. *Biotechol Bioengineer* 15:611-647.
139. Klecka GM. 1985. Biodegradation. In: Neely WB, Blau GE, eds, *Environmental Exposure from Chemicals, Vol 1*. CRC Press Inc, Boca Raton, FL, pp. 109-155.
140. Organization for Economic Co-operation and Development. 1976. Proposed method for the determination of the biodegradability of surfactants used in synthetic detergents. OECD, Paris, France.
141. King EF. 1981. Biodegradability testing. Notes on water research 28. Water Research Centre, Medmenham, UK.
142. Struijs J, van den Berg R. 1995. Standardized biodegradability tests: extrapolation to aerobic environments. *Water Research* 29:255-262.
143. STEP. 2004. Simulation testing of environmental persistence (STEP): a two day workshop, Rotterdam, 4-5 October 2004.
144. Doull J, Klaassen CD, Amdur MO, eds, 1986. *Casarett and Doull's Toxicology, the Basic Science of Poisons*. Macmillan Publ Comp, New York, NY.
145. Lech JJ, Vodicnik MJ. 1985. Biotransformation. In: Rand GM, Petrocelli SR, eds, *Fundamentals of Aquatic Toxicology*. Hemisphere, Washington, DC, pp. 526-557.
146. Homburger F, Hayes JA, Pelikan EW. 1983. *A Guide to General Toxicology*. Karger/Base, New York, NY.
147. Kimbrough RD, Jensen AA, eds. 1989. *Halogenated Biphenyls, Terphenyls, Naphtalenes, Dibenzodioxins and Related Products*. Elsevier Sci Publ, Amsterdam, The Netherlands.
148. Safe SH. 1994, Polychlorinated biphenyls (PCBs): environmental impact, biochemical and toxic responses, and implications for risk assessment. *Crit Rev Toxicol* 24:87-149.
149. Safe SH, ed. 1987. *Polychlorinated biphenyls (PCBs): Mammalian and Environmental Toxicology*. Springer-Verlag, Heidelberg, Germany.
150. Van Den Berg M, De Jongh J, Poiger H, Olson JR. 1994. The toxicokinetics and metabolism of polychlorinated dibenzo-p-dioxins (PCDDs) and dibenzofurans (PCDFs), and their relevance for toxicity. *Crit Rev Toxicol* 24:1-74.

151. Matsumura F. 1985. *Toxicology of Insecticides.* Plenum Press, New York, NY.
152. Safe SH. 1990. Polychlorinated biphenyls (PCBs), dibenzo-p-dioxins (PCDDs), dibenzofurans (PCDFs), and related compounds: Environmental and mechanistic considerations which support the development of toxic equivalency factors (TEFs). *Crit Rev Toxicol* 21:51-88.
153. Hamelink JL, Landrum PF, Bergman HL, Benson WH. 1994. *Bioavailability. Physical, chemical and biological interactions.* CRC Press, Boca Raton, FL.
154. Saarikoski J, Lindström R, Tyynelä M, Viluksela M. 1986. Factors affecting the absorption of phenolics and carboxylic acids in the guppy *(Poecilia reticulata). Ecotox Environ Saf* 11:158-173.
155. Koelmans AA, Jonker MTO, Cornelissen G, Bucheli TP, Van Noort PCM, Gustafsson Ö. 2006. Black Carbon: The reverse of its dark side. *Chemosphere* 63:365-377.
156. Cornelissen G, Gustafsson Ö, Bucheli TD, Jonker MTO, Koelmans AA, van Noort PCM. 2005. Critical Review. Extensive sorption of organic compounds to Black Carbon, Coal, and Kerogen in sediments and soils: mechanisms and consequences for distribution, bioaccumulation and biodegradation. *Environ Sci Technol* 39: 6881-6895.
157. Paquin PR, Gorsuch JW, Apte S, Batley GE, Bowles KC, Campbell PG, Delos CG, Di Toro DM, Dwyer RL, Galvez F, Gensemer RW, Goss GG, Hostrand C, Janssen CR, McGeer JC, Naddy RB, Playle RC, Santore RC, Schneider U, Stubblefield WA, Wood CM, Wu KB. 2002. The biotic ligand model: a historical overview. *Comp Biochem Physiol C Toxicol Pharmacol* 133: 3-35.
158. Alexander, M. 1994. *Biodegradation and Bioremediation.* Academic Press, San Diego, CA. ISBN 0-12-049860-X.

4. ENVIRONMENTAL EXPOSURE ASSESSMENT

D. Van de Meent and J.H.M. De Bruijn

4.1 INTRODUCTION

4.1.1 Use of models in the assessment of exposure concentrations

Organisms, man included, are exposed to chemicals through environmental media. Assessment of exposure concentrations can be done by measurement or by other means of estimation, e.g. model-based computation. For the risk assessment of existing situations, both measurement and modelling can be used; to assess the risks posed by new chemicals or new situations, modelling is the only option. Although it may seem natural to assume that measurement yields more certainty, this is not necessarily so. Chemical analyses are usually carried out on samples, taken at specific locations and times. Observed concentrations reflect the variations in concentration in space and time. Unless measurement programmes are designed to yield the "typical" or "average" concentrations desired in risk assessment practice, available measurements may be biased, more often than not towards enhanced concentrations. By contrast, modelled concentrations generally do reflect the "typical" or "average" concentrations needed. Therefore, modelling may be of use in risk assessment even in existing situations where measurement would seem to be the natural option to choose. This obvious shortcoming of many risk assessment models (*viz.* their inability to accurately predict concentrations at specific times and specific locations) is thus turned into an advantage. Moreover, if we are to assess bioavailable concentrations, modelling may be preferable, since the bioavailability of many chemicals is often more adequately estimated by modelling than by analytical measurements. Ideally, both exposure levels and no-effect levels should be expressed in terms of internal concentrations at the site in the organism where the actual toxic effect occurs. Such comparisons by-pass problems of bioavailability, uptake and elimination kinetics, and metabolism. However, lack of internal exposure and internal effect/no-effect data makes this procedure impracticable for the time being. Therefore we have to base exposure assessments on external concentrations in environmental media. This introduces uncertainty into the risk assessment since the ratio of external to internal concentrations (bioavailability) may not be the same in the assessment of exposure levels and no effect levels. In order to minimize this source of uncertainty, it is desirable to express both exposure and no effect levels in terms of bioavailable concentrations.

The essential first stage in creating and using models is the conceptualization stage, i.e., deciding what kind of representation of reality is to be created. Obviously, model builders need to carry out this process of fundamental decision-making very carefully, while model users should realize that in selecting an existing model for their specific purpose, decisions of the same kind are implicitly made. During conceptualization, modellers (builders *and* users) need to reflect on the purpose of their modelling effort: what is being modelled and what for? Conceptualization involves making fundamental choices about what aspects of reality are relevant to the purpose of the specific modelling process and which aspects of reality are to be left unaccounted for. At this stage, the modeller chooses the level of sophistication required to meet the objectives of the modelling task. In general, simple models are to be preferred over sophisticated models, since the more sophisticated the model is, the more data and labour-intensive (and therefore costly) the modelling activity, and the more difficult interpretation of the results becomes. Moreover, the results of simple model calculations are easier to communicate and, therefore, may serve the purpose of decision support better. Objective criteria should be applied to make this choice (Box 4.1).

The great advantage of using models is that they allow us to conceptualize this relationship; we can use our knowledge of the processes to describe the relationship in terms of the characteristics of the environment and properties of the chemical. The utility of models is that they allow us to evaluate the results of the many processes occurring simultaneously which would not otherwise be apparent (Box 4.2). The processes affecting the concentration of a chemical are relatively well understood and may even look simple. It is the multitude of processes acting in parallel that makes the result difficult to understand.

Models are used as instruments in risk assessment and risk management to describe the relationship between emissions and concentrations and to predict the results of management measures. This use is not undisputed. Both scientists and decision-makers have often criticized the

C.J. van Leeuwen and T.G. Vermeire (eds.), Risk Assessment of Chemicals: An Introduction, 159–193.

Box 4.1. Criteria for choosing the level of model sophistication

- Purpose of the model.
  For the purpose of identification of critical environmental compartments and *a priori* estimation of risks associated with the introduction of new chemicals, a relatively simple screening with a multimedia box model may be sufficient. Prediction of the effect of emission reduction on concentrations at specific times and places may require the use of a more sophisticated dynamic two or three-dimensional air, water or groundwater quality model.
- Acceptable uncertainty.
  The required level of confidence should follow from the use of the modelling results. If simple modelling demonstrates that the margin between the calculated concentration and the predicted no effect level (PNEC) is sufficiently great for the purpose of the modelling activity, no further increase in the level of confidence is required.

Box 4.2. Purpose of models

- Provide insight.
  Models provide a way to interpret observations logically. The use of models can help us to understand certain aspects of reality. They may help to identify cause-effect relationships that are not apparent in an initial review of the data. Used in this way, models primarily serve to provide insight into "how theory operates", as Lassiter put it, rather than "how the system operates" [1]. Models are useful for quantifying the implications of our assumptions about reality: they provide a way of testing the adequacy of the current state-of-the-art of theory to describe reality. A good way to gain better understanding with a model involves systematic variation of parameters to find the parameters which the model output is most sensitive to. This sensitivity-analysis procedure helps to identify the key processes and pathways for the chemical.
- Support decision-making.
  Modelling provides a means of eliminating the vagueness inherent to decision-making. Reasoning is made more explicit when the possible results of alternative strategies for risk reduction and the uncertainties associated with it are properly quantified. A powerful way to use models in decision-making is in the "what-if scenario", which can help to identify the most effective strategy.

unquestioned confidence that modellers are thought to have in the results of their calculations. In fact, it is often believed that modellers fail to recognize the difference between the real world and the models they make of it. Scientists may question the validity of models as representative of reality, whereas decision-makers may doubt their predictive value. Model users should realize that it is impossible to make perfect predictions of real-world behaviour. Only the most dominant processes affecting the fate of a chemical can be accounted for in a model. To stress this, models have been called "cartoons of reality". Rightfully so, as models always reflect the subjective view of the modeller and different models are needed to express different aspects of reality.

Readers are referred to specialized textbooks and documents for further information on this subject [2-8].

4.1.2 Mass balance modelling

Many of the models used in risk assessment of toxic substances are compartment models, also referred to as box models. The environment is thought to be made up of homogeneous, well-mixed compartments. Compartments can represent segments of the environment, or even entire environmental media. Examples of the former are the spatially segmented air and water transport models and layered soil models. The latter is used in multimedia (air, water, soil, etc.) fate models and in physiology-based pharmaco-kinetic models (blood, tissue, etc.). Compartment models apply the principle of mass conservation: the mass of a substance in a compartment appears or disappears only as a result of mass flows of a substance into or out of the compartment. What compartment models have in common is that the mass balance equation is used as their basic instrument.

Compartment models are therefore often referred to as mass balance models. Because mass balance modelling is used so widely in the environmental risk assessment of toxic substances, its principles will be explained here. We shall first derive a mass balance equation for one compartment, then a mass balance model for more compartments.

One compartment

If a substance is added to or taken from a compartment, the mass of that substance in the compartment changes. This change can be quantitatively expressed in a mass balance equation, in which all incoming and outgoing mass flows of the substance are accounted for:

see below (4.1)

where ΔM and ΔC and are changes in mass and concentration within a time interval Δt, respectively, and V is the (constant) volume of the compartment. Note that the change is in unit mass per unit time (e.g. $kg.d^{-1}$): a sum of mass flows. If nothing is added or taken away (or if gains and losses match exactly), the mass of substance in the compartment does not change: a steady state. If ΔM, ΔC and Δt are infinitesimally small, equation (1) becomes what is mathematically known as a differential equation. Differential equations describe at what rate a variable (here: mass of a substance in a compartment) changes. If the mass at starting time (t=0) is known (the initial condition), a differential equation can be used to derive the mass at other times. The art of mass balance modelling is thus to properly quantify the mass flows of a substance going into and out of the compartments.

For the purpose of mass balance modelling it is useful to distinguish between mass flows that take place independently of what happens in the compartment and mass flows that do depend on the conditions within the compartment. Emissions and imports are examples of the first category. As explained in Chapter 2, the rate at which mass is brought into the compartment by these processes may be constant or time-dependent, and may relate to the mass of a substance *outside* the compartment, but bears no relationship to the mass of a substance *within* the compartment. These mass flows need to be specified to the model as so-called "forcings". If a constant emission of E ($kg.d^{-1}$) is forced upon a compartment, which contains M_0 kg of the substance at t=0, and nothing else happens, the mass balance equation becomes:

$$\frac{dM}{dt}\left(= V\,\frac{dC}{dt}\right) = E \qquad (4.2)$$

of which the integral form or solution is

$$M = M_0 + E \cdot t \qquad (4.3)$$

How this solution is obtained is not further explained here. Readers may want to refresh their knowledge of this mathematical calculation method by reviewing a standard text on differential calculus, e.g. Wikipedia [http://en.wikipedia.org/wiki/Differential_equation].

The result of a constant inflow of a substance is that the mass of the substance in the compartment continuously increases. Note that this occurs at the constant rate of E $kg.d^{-1}$ (Figure 4.1). This second category applies more in general. As explained in Chapter 3, loss rates generally depend on the mass of a substance in the compartment. Often, this relationship is assumed to be linear: loss is modelled as a first-order process, which means that mass flow is assumed to be directly proportional to mass in the compartment. For instance, the loss due to reaction with chemical or microbial agents (degradation) is often characterized by (pseudo) first-order kinetics:

$$\text{loss} = k \cdot M \qquad (4.4)$$

where k is a (pseudo) first-order reaction rate constant (d^{-1}). Because the reaction rate is proportional to the first power of the concentration C (C^1), the degradation mass flow is described by a first-order differential equation. This is why reaction kinetics are referred to as first-order. It should be noted that first-order reaction kinetics are the exception, rather than the rule. Zero-order kinetics, in which the reaction is independent of C (formally proportional to C^0), second-order kinetics (reaction rate proportional to C^2) and broken order kinetics (proportional to $C^{1.5}$) commonly occur. Second-order kinetics will generally apply when a substance reacts with a chemical agent: the reaction is first-order in relation to both the substance degraded and the reactant. It is only because the concentration of the reactant is

$$\frac{\Delta M}{\Delta t}\left(= V\,\frac{\Delta C}{\Delta t}\right) = \text{gains} - \text{losses} = \sum \text{mass flows} \qquad (4.1)$$

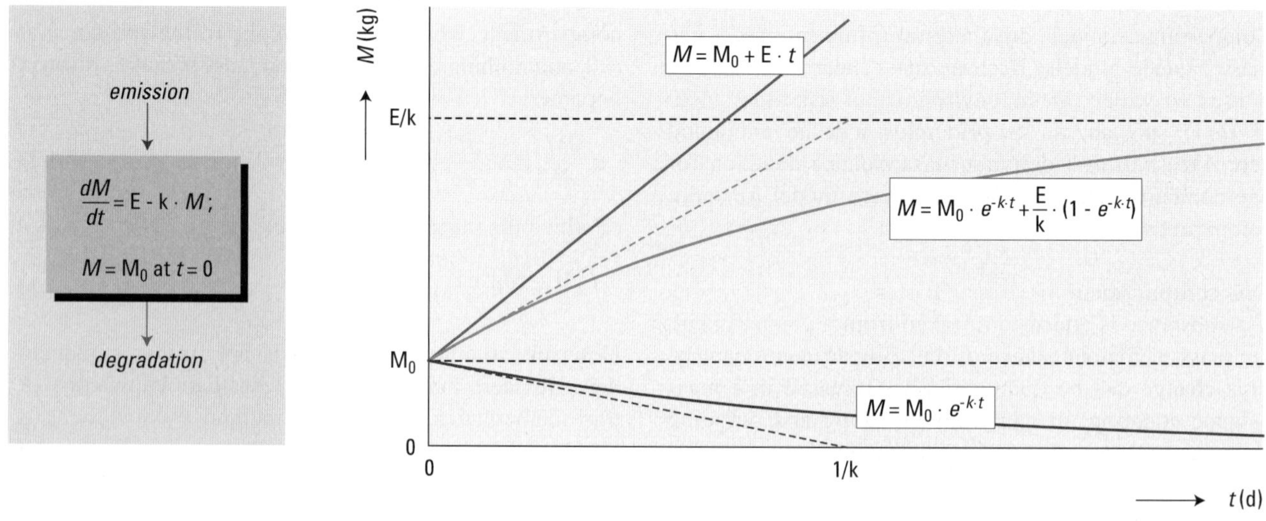

Figure 4.1. Elementary form of a one-compartment mass balance model, showing the differential mass balance equation and its solution for the cases of emission only (red), degradation only (blue) and both (green).

often approximately constant that the reaction appears proportional only to C^1. This is called pseudo first-order reaction kinetics.

If degradation is the only process, the mass balance equation becomes:

$$\frac{dM}{dt}\left(= \mathrm{V}\,\frac{dC}{dt}\right) = -\,\mathrm{k} \cdot M \tag{4.5}$$

the solution of which is:

$$M = \mathrm{M}_0 \cdot e^{-\mathrm{k} \cdot t} \tag{4.6}$$

First-order degradation results in an exponential decrease of mass in the compartment (Figure 4.1). Note that, in agreement with Equation 4.5, the rate of change decreases from its initial value of $-\mathrm{M}_0.\mathrm{k}$ to zero as t approaches infinity.

If both emission and degradation act on a compartment, the combined result will be:

$$\frac{dM}{dt}\left(= \mathrm{V}\,\frac{dC}{dt}\right) = \mathrm{E} - \mathrm{k} \cdot M;\ M = \mathrm{M}_0 \text{ at } t = 0 \tag{4.7}$$

the solution of which is:

$$M = \mathrm{M}_0 \cdot e^{-k \cdot t} + \frac{\mathrm{E}}{\mathrm{k}}\,(1 - e^{-k \cdot t}) \tag{4.8}$$

(see Figure 4.1). Equations 4.7 and 4.8 illustrate how the mathematical solution of the mass balance equation yields a mass-time profile of a substance in a compartment as a function of the initial conditions (here: mass at t=0, M_0), forcings (here: emission rate, E) and the parameters of the mass flow rate equations (here: the degradation rate constant, k). Note that eventually (at t=∞), the mass of substance in the compartment, M (kg) will reach a level at which the loss by degradation, k.M (kg.d^{-1}), exactly matches the constant emission, E (kg.d^{-1}), so that the mass of substance in the compartment is maintained at the steady-state level of E/k (kg).

There are many other loss mechanisms that need to be accounted for in the mass balance equation, such as such as advective or diffusive outflow. Because losses due to all mechanisms i are proportional to M, and can each be represented by a first-order rate constant k_i (d^{-1}), the full mass balance equation keeps the same simple format of Equation 4.7, namely:

see below (4.9)

and its solution takes the same format as Equation 4.8:

$$M = \mathrm{M}_0 \cdot e^{-\sum_i k_i \cdot t} + \frac{\mathrm{E}}{\sum_i k_i}\,(1 - e^{-\sum_i k_i \cdot t}) \tag{4.10}$$

$$\frac{dM}{dt}\left(= \mathrm{V}\,\frac{dC}{dt}\right) = \text{gains} - \text{losses} = \mathrm{E} - \sum_i k_i \cdot M_i \qquad M = \mathrm{M}_0 \text{ at } t = 0 \tag{4.9}$$

More compartments

Models usually comprise many compartments and describe the transport of a substance in and between these compartments. Such multicompartment mass balance models contain one mass balance equation for each compartment in the model. As in the above situation for one compartment, losses are all assumed to obey first-order kinetics. Where more than one compartment is involved, losses may be due to degradation or export, but losses may also represent mass flows from one compartment to another. For a set of n compartments, this leads to a set of n mass balance equations, all of which will have the same format as Equation 4.9, with n unknown masses M_i and a suite of first-order rate constants which describes the losses from the compartments. We shall work this out for three compartments, as shown in the diagram in Figure 4.2.

Each of the compartments receives an emission. For the sake of simplicity, emissions will be assumed to be constant and imports considered to be included in the emission flows. The emission flows into the compartments i are denoted by E_i ($kg.d^{-1}$). Degradation occurs in the three compartments. Again, in the interests of readability, the degradation flows will be considered to include possible exports. The resulting mass flows from the compartments i, out of the system are characterized by pseudo first-order loss rate constants k_i and denoted by $k_i.M_i$ ($kg.d^{-1}$). There are six intercompartment mass-transfer flows, each proportional to the mass in the source compartments denoted by $k_{i,j}.M_i$ ($kg.d^{-1}$). On this basis, and assuming all initial masses to be zero, the three differential mass balance equations become:

see below (4.11)

For this system of three compartments there is an equation equivalent to Equation 4.10, *i.e.* the analytical solution of the one-compartment system, which expresses the mass of the substance at all times. It is not possible to formulate precisely how the three masses in the three compartments change with time. Solutions can be approximated quite well, however, with computer-based numerical techniques which will not be described here.

As in the one-compartment system, the three-compartment system will eventually (at $t=\infty$) reach to a steady state in which emission is equally balanced by degradation ($dM_i/dt = 0$) and masses reach their constant steady state level, M^*_i:

see below (4.12)

There is an analytical solution for this system of three linear equations with three unknowns. The set of steady-state masses for which the mass balance equations become zero can be derived directly from Equation 4.12 quite easily through simple algebraic manipulation. Readers are encouraged to work this out as an exercise. Solving sets of equations algebraically becomes increasingly tedious for larger sets. Linear algebra (matrix calculus) is used to obtain solutions to large sets of linear equations. Readers may want to refresh their knowledge of this mathematical technique by reviewing a standard text on linear algebra and its application in solving systems of linear equations, e.g. in Wikipedia [http://en.wikipedia.org/wiki/System_of_linear_equations].

For this purpose, we reformulate (4.12) in vector/matrix notation and define:

$$\frac{dM_1}{dt} = E_1 - (k_1 + k_{1,2} + k_{1,3}) \cdot M_1 + k_{2,1} \cdot M_2 + k_{3,1} \cdot M_3;\ M_1 = 0 \text{ at } t = 0 \tag{4.11}$$
$$\frac{dM_2}{dt} = E_2 + k_{1,2} \cdot M_1 - (k_2 + k_{2,1} + k_{2,3}) \cdot M_2 + k_{3,2} \cdot M_3;\ M_2 = 0 \text{ at } t = 0$$
$$\frac{dM_3}{dt} = E_3 + k_{1,3} \cdot M_1 + k_{2,3} \cdot M_2 - (k_3 + k_{3,1} + k_{3,2}) \cdot M_3;\ M_3 = 0 \text{ at } t = 0$$

$$\text{balance}_1 = E_1 - (k_1 + k_{1,2} + k_{1,3}) \cdot M^*_1 + k_{2,1} \cdot M^*_2 + k_{3,1} \cdot M^*_3 = 0 \tag{4.12}$$
$$\text{balance}_2 = E_2 + k_{1,2} \cdot M^*_1 - (k_2 + k_{2,1} + k_{2,3}) \cdot M^*_2 + k_{3,2} \cdot M^*_3 = 0$$
$$\text{balance}_3 = E_3 + k_{1,3} \cdot M_1 + k_{2,3} \cdot M^*_2 - (k_3 + k_{3,1} + k_{3,2}) \cdot M^*_3 = 0$$

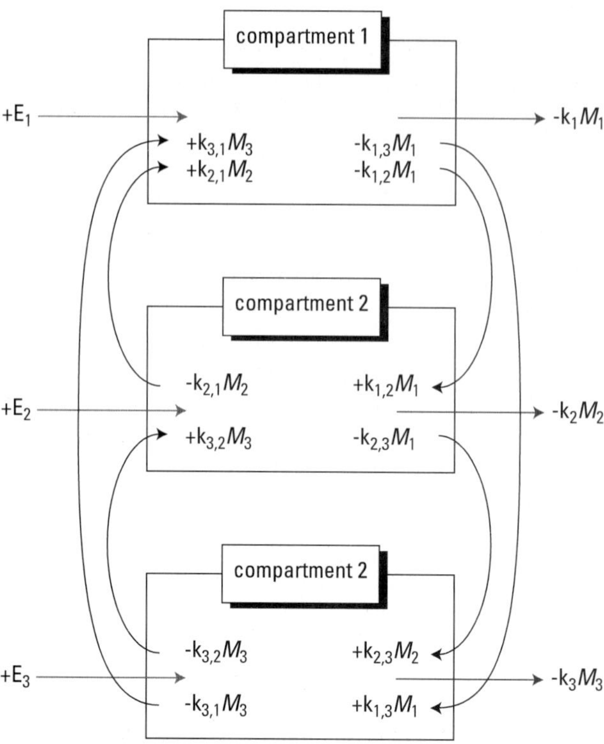

Figure 4.2. Diagram of a three-compartment mass balance model. Intercompartment mass-transfer represents a loss to the source compartment and a gain to the receiving compartment.

see below (4.13)

Using this, the three mass balance equations of 4.12 can be rewritten into a one-line linear-algebraic equation:

$$\mathbf{e} + \mathbf{A} \cdot \mathbf{m} = 0 \qquad (4.14)$$

thus

$$\mathbf{A} \cdot \mathbf{m} = -\,\mathbf{e} \qquad (4.15)$$

Multiplication of both ends of equation 4.15 by the inverse of the coefficients matrix, $\mathbf{A}^{-1}$

$$\mathbf{A}^{-1} \cdot \mathbf{A} \cdot \mathbf{m} = \mathbf{1} \cdot \mathbf{m} = \mathbf{m} = -\,\mathbf{A}^{-1} \cdot \mathbf{e} \qquad (4.16)$$

then yields the vector of steady-state masses in the compartments:

$$\mathbf{m} = -\,\mathbf{A}^{-1} \cdot \mathbf{e} \qquad (4.17)$$

Various standard software packages, such as Microsoft Excel, can be used to carry out matrix inversion.

4.1.3 Model types

The models described in this chapter represent just a few of the many different types of models that are available to serve a variety of modelling purposes. As a guide for potential model users who want to select a model for a given purpose, some of the main terms that are often used to describe and categorize models are listed and explained here:

Modelling objective

The objective of the exposure models discussed in this chapter is to describe what happens to micropollutants after their release into the environment. These kinds of models are called distribution models (Section 3.2), physiologically-based (bio)kinetic models (PB-(B)K, (Sections 6.3 and 6.4), multimedia fate models, and water-quality models, etc. They are different from population models, economic models and meteorological models, as well as statistical models or even effect models.

Basic approach

All the models dealt with in this chapter are mathematical models which are used to describe mass flows and concentrations quantitatively. This method of modelling is often contrasted with other basic approaches like descriptive modelling or physical modelling. Descriptive models generalize the phenomena to be modelled in qualitative or semi-quantitative scientific terminology. This sort of modelling is applied during the conceptualization stage of quantitative mathematical models. In physical modelling, reality is simulated by building physical, usually small-scale, models of natural situations.

$$\mathbf{m} = \begin{bmatrix} M_1 \\ M_2 \\ M_3 \end{bmatrix},\ \mathbf{e} = \begin{bmatrix} E_1 \\ E_2 \\ E_3 \end{bmatrix},\ \mathbf{A} = \begin{bmatrix} -(k_1 + k_{1,2} + k_{1,3}) & k_{2,1} & k_{3,1} \\ k_{1,2} & -(k_2 + k_{2,1} + k_{2,3}) & k_{3,2}) \\ k_{1,3} & k_{2,3} & -(k_3 + k_{3,1} + k_{3,2}) \end{bmatrix} \qquad (4.13)$$

Scientific method
Different approaches can be taken in mathematical fate modelling. The models in this chapter are deterministic and take the mechanistic or theoretical approach. The philosophy behind this approach is that fate is determined by mechanisms or processes that can be quantitatively described on a theoretical basis. The results of deterministic model calculations are always the same and do not depend on chance. Deterministic models differ in this sense from stochastic models, in which some of the factors influencing fate are allowed to have some random variation. Deterministic models may be formulated on a mechanistic or an empirical basis. Empirical model formulations makes use of relationships that are empirically found to be valid. As a consequence, they can only be applied to the conditions for which the relationship was found. Mechanistic model formulations are based on a theoretical understanding of the process; the range of applicability can be rationalized. Therefore, mechanistic model formulations are usually preferred above extrapolation models, whereas empirical formulations may be better for interpolation models.

Computational approach
Deterministic fate models may differ in the way in which the processes are represented and the solution is derived. Many simple models derived from only a few equations can be solved algebraically. The result, the analytical solution, is an equation that explicitly expresses the model output (here the exposure concentration) as a function of the influencing factors. Equation 4.8 is an example of this. More complicated models often do not have an analytical solution, they require a numerical approximation. Examples of numerical solutions are dynamic simulations of the time-dependent output of multicompartment mass balance models, as well as the linear-algebraic steady-state solution of multicompartment models as in Equation 4.17 .

Dimensionality
Fate models also differ in spatial and temporal dimensionality. With respect to space there are zero, one, two and three-dimensional models. In zero-dimensional models, there is no spatial variation in concentrations. Zero-dimensionality is used in multimedia models which describe the distribution and fate of chemicals in homogeneous environmental compartments. Fate models for one compartment usually have spatial variability in one (layered soil models) or more (air and water quality models) directions. With respect to time there are steady-state models and dynamic models. Steady-state models give the concentration in the compartments at the time when a steady-state has been reached approximately, whereas dynamic models yield concentration-time series (Figure 4.1).

4.1.4 Models versus measurements

When assessing data on exposure to chemicals, a range of concentrations may be available, e.g. from measurements in the environment. As stated in Section 4.1.1, it may appear that measurements always give more reliable results than model estimations. However, even measured exposure concentrations can have a considerable uncertainty attached to them, due to temporal and spatial variations. Therefore, when carrying out an exposure assessment it may be very useful to compare the estimated and measured concentrations in order to select the "right" data for use in the risk characterization phase. This comparison can be done in three steps [2]:

1. Selection of reliable data by evaluation of the analytical techniques used and the time scale of the measurements.

The techniques used for sampling, processing and detection have to be evaluated in the light of the physicochemical properties of the chemical. For example, filtering water samples may considerably reduce the concentrations of highly sorptive chemicals. This need not pose a problem as long as the data are compared with the bioavailable predicted no effect concentration (PNEC). Measurement of concentrations in sediment, however, may be more relevant in this case. Care should also be taken in assessing measurements at or below the analytical detection limit. Reported average values may be strongly influenced because concentrations below the detection limit are reported either as zero or as a certain fraction of this detection limit.

With regard to the time scale, information is required on whether the data were obtained from occasional sampling or from more frequent monitoring programmes. This measuring incidence has to be taken into account in the emission scenario. Monthly measurements in surface water, for instance, may very well overlook periodically high concentrations due to intermittent releases.

2. Correlating these data to the appropriate emission and modelling scenarios.

The measured data must be allocated to a certain spatial scale to enable comparison with specific modelling scenarios. Concentrations measured near point sources, e.g. the outlet of a sewage treatment plant, must be compared with model estimations set up for a similar small area. In addition, measured concentrations of chemicals that are emitted from many point sources or area sources can only be properly compared with estimates from larger scale models that take the fate of the chemical in the environment into account.

3. Comparing representative data with corresponding model estimates and undertaking a critical analysis of the differences between the two.

The results of model estimations and measured data are compared. Three different situations can occur [2]:

a. The calculated concentrations are approximately equal to the measured data, indicating that the most relevant sources have probably been taken into account and the appropriate estimation model has been selected (although sheer luck cannot be excluded, agreement may be due to balancing overestimations and underestimations!).
b. The calculated concentrations are much higher than the measured data, for which there may be several explanations: elimination of the substance under environmental conditions may be much faster than calculated in the model; emissions may have been overestimated, a different time scale may have been used; or the measured concentrations may represent "background" levels whereas at specific locations much higher concentrations may occur.
c. The calculated concentrations are much lower than the measured data, which may be due to the reverse of the reasons given under b.

In principle, data from measurements in the environment should be given more weight than model calculations, provided that they are representative of the emission scenario and have been adequately measured. Making a comparison with model estimations, however, is probably always useful since it is the only way to validate the assumptions made in models. Each time model predictions are validated by monitoring or laboratory data, confidence in the model's predictive power will increase. Hence, greater confidence can be placed in the resulting risk assessments and the conclusions drawn from them. Thus, monitoring and laboratory data have complementary roles, alongside fate models, in comprehensive risk assessments.

4.2 AIR MODELS

4.2.1 Introduction

Modelling the dispersion of trace components in the atmosphere, including their physical and chemical transformations, is an essential element in the general study of the environmental behaviour of trace components and in determining the functional relationships between emissions and concentrations or deposition levels. Measurements and models are closely interrelated. Measurements are necessary for setting parameters and the validation of models, on the one hand, while model results may provide support in the evaluation, generalization or extrapolation (in space and time) of measurements on the other.

The general structure of atmospheric models is shown in Figure 4.3. The input requirements are meteorological parameters and emission data. Terrain data (roughness, length, land-use or orography) may also be required. The output of the model consists of spatial and/or temporal information on concentration and deposition levels, i.e., the atmospheric input to soil or surface water. The inner part of the model deals with atmospheric processes (advection, dispersion, chemistry and deposition). The complexity of this part may vary, depending on the output requirements. For example, the approach taken in a model which is suitable for estimating concentration levels in the direct vicinity of a point source will be totally different from the approach taken in a model to estimate the global distribution of a persistent pollutant. Atmospheric chemistry may be treated in a complex non-linear way, e.g., to describe ozone formation or, as in the case of relatively slow reacting pollutants, as a pseudo first-order loss process.

This section first gives a short overview of different model types, followed by some examples of operational air models. Next, the use of a local air model for the risk assessment of new and existing chemicals is described. The section ends with a description of the data requirements of air models.

4.2.2 Model types

Compartment or box models with little spatial resolution, like the multimedia fate models, are perhaps the most simple tools for making a first estimate of ambient levels. In an atmospheric box model the pollutants are assumed to be mixed homogeneously. Changes in concentrations result from chemical transformation, emission, deposition and transport across the boundaries. Box models should

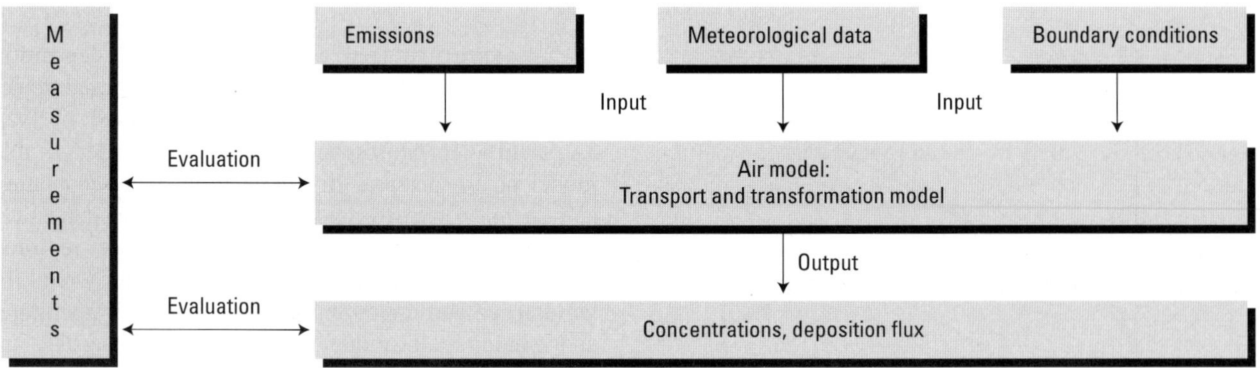

Figure 4.3 General structure of atmospheric models.

preferably be used only for indicative purposes, as the assumptions made in the model may not be met in practice. However, since the fate of pollutants in other compartments (soil, surface water, etc.) is also described in multimedia box models, they provide a valuable tool in risk assessment, as demonstrated by the widespread application of Mackay-type multimedia models (Section 4.5).

Dispersion of a chemical within compartments or boxes is not taken into account. Atmospheric dispersion of chemicals in air can be described by two different numerical approaches, the *Eulerian* or the *Lagrangian* approach. Both methods have their advantages and limitations. In the development of operational models approximations have to be made. The Eulerian approach uses a regular grid of air compartments, for which concentrations and depositions are calculated by solving mass balance equations (see Section 4.1.2). Eulerian models generally require a substantial amount of computer time. Under the Lagrangian approach the processes taking place in an air parcel travelling with the atmospheric motion are followed. Lagrangian models are either *source-oriented*, i.e., the air parcel (also called "puff") originates at a specific source and is followed on its journey downwind from the source, or *receptor-oriented*, i.e., the air parcel is followed travelling over source areas, picking up emissions until it arrives at the selected receptor area. In Lagrangian models advection is treated in a relatively simple way which makes the model computationally less demanding.

4.2.3 Some examples of operational models

Gaussian plume model

An air model commonly used is the Gaussian plume model (GPM). This Lagrangian model describes the dispersion in the direct vicinity (maximum 30 km) of a source. Assuming that turbulence is a random process, it is expected that the mean concentration of material emitted from a point source will have a two-dimensional Gaussian distribution perpendicular to the mean wind direction. Figure 4.4 shows the horizontal and vertical Gaussian distributions. In its simplest form the GPM describes concentrations at a specified location, $C_{x,y,z}$, according to the following equation:

see below (4.18)

where

Q = source strength ($kg.s^{-1}$)

u = wind speed ($m.s^{-1}$)

H = (effective) source height, i.e., the sum of stack height and plume rise (m)

σ_y = dispersion coefficient in horizontal direction (m)

σ_z = dispersion coefficient in vertical direction (m).

The values of σ_y and σ_z depend on travel distance (or travel time) and atmospheric stability. The most widely used expressions to correlate σ_y and σ_z with atmospheric variables are based on the Pasquill stability classes

$$C_{x,y,z} = \frac{Q}{2\pi u \sigma_y \sigma_x} \cdot \left[e^{-\left(\frac{y^2}{2\sigma_y^2}\right)} \right] \left\{ e^{\left(\frac{-(z-H)^2}{2\sigma_z^2}\right)} + e^{\left(\frac{-(z+H)^2}{2\sigma_z^2}\right)} \right\} \tag{4.18}$$

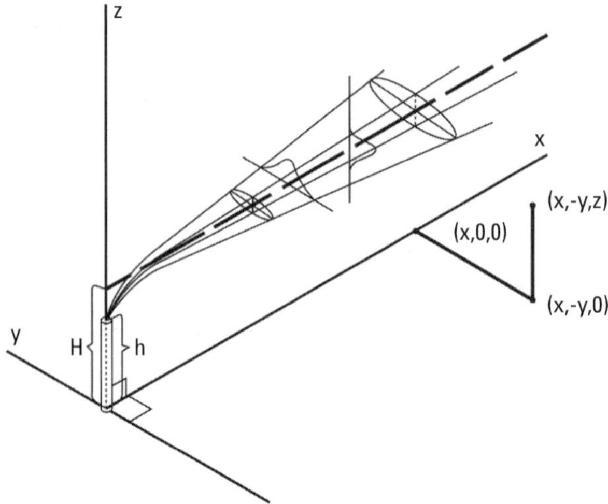

Figure 4.4. Horizontal and vertical Gaussian distributions according to a Gaussian plume model (GPM). From [10].

and were developed by Gifford [9]. The dispersion coefficients are presented in graphic and numerical form in [10] and [11]. These correlations are commonly referred to as the Pasquill-Gifford curves. For use in the GPM formula, analytical expressions are empirically determined:

$$\sigma_z = R_z x^{r_z} \quad (4.19)$$

$$\sigma_y = R_y x^{r_y} \quad (4.20)$$

R_y, R_z, r_y and r_z are empirical parameters, which depend on the stability class and averaging time. Parameter values can be found in textbooks, e.g. Seinfeld [12].

As many assumptions are made in the GPM, the model has some serious drawbacks. It can only be used if:

- There is a steady-state (constant emissions, constant wind and homogeneous turbulence).
- Deposition and chemical transformation can be neglected.
- Wind speed is over 1 $m.s^{-1}$ (the GPM can not be applied under calm weather conditions).
- Distances are less than approximately 30 km (for flat terrain, otherwise even shorter!).

A modified GPM, known as the "National model" [13] is used in The Netherlands to calculate frequency distributions of concentrations for various receptor points around a source, using statistical meteorological data.

Operational model for Priority Substances

A flexible atmospheric transport model for the calculation of long-term averaged concentrations and deposition fluxes of low-reactive pollutants is the operational model for priority substances (OPS), as described by Van Jaarsveld [14]. Atmospheric processes included in this model are dispersion, dry deposition, wet scavenging and chemical transformation. The model uses statistical meteorological data. The minimum set of required meteorological information consists of 6-hourly data for wind speed and direction, global radiation, temperature, and precipitation amount and duration. These data are pre-processed by a separate program to calculate the necessary statistics. The averaging period can range from one month to more than 10 y. The receptor points may be defined on a regular grid in a model domain ranging from the local scale (100 m around a source) up to the scale of the European continent (approximately 2000 x 2000 km), or they may be defined as exact geographical (x,y) coordinates. The last option can be used when the user wishes to compare the model results with measured values from monitoring stations, for example. Emissions can be defined as any combination of point sources and (diffuse) area sources with variable horizontal dimensions. For a more detailed description of the model structure the reader is referred to Van Jaarsveld [14].

To avoid the above shortcomings and to take into account larger spatial scales, a large number of transport models have been, and are still, developed. It is, however, beyond the scope of this book to discuss these models in detail. Various operational models have been reviewed in several papers and reports [7,15,16]. Aspects that should be considered when selecting a suitable atmospheric model are given in Box 4.3.

4.2.4 Application of a local air model in the risk assessment of industrial chemicals

Models such as the OPS model are highly flexible and can be adjusted to take into account specific information on scale, emission sources, weather conditions, etc. This type of information is generally not available for new chemicals and many existing chemicals. Hence, a generic exposure assessment is carried out based on a number of explicit assumptions with a number of fixed default parameters. How to conduct a local exposure assessment of this kind was described by Toet and De Leeuw [17]. Using the OPS model, the authors carried out a number of default calculations in order to find a relationship between the basic properties of substances, in terms of the vapour pressure and Henry's law constant, and the concentration in air and deposition flux to soil near a point source. The following assumptions/model settings were made:

Box 4.3. How to select a suitable atmospheric model

a. Spatial scale.
Is it sufficient to calculate concentrations on a local scale (less than approximately 30 km around the source) or is it necessary to include the contribution made by long-range transport of emissions on a continental scale (approximately 2000 km) or on a global scale? The relevant spatial scale relates to the atmospheric residence time of the component: transport of persistent organic pollutants occurs on a global scale; transport and deposition of heavy metals is a typical continental problem. However, in many cases it may be sufficient to include the continental or global contribution as a "background" contribution in local scale calculations.
b. Temporal scale.
Are long-term (yearly) averaged or short-term (hourly) averaged concentration or deposition values required? Episodic models are designed to predict hourly averaged concentrations during short periods of several days. For these models large amounts of meteorological input is needed as the variation in meteorological conditions in time and space has to be taken into account. In long-term models the description is generally simplified by using statistical information.
c. Components.
What are the chemical properties of the modelled component? For reactive species and for secondary pollutants, i.e., pollutants which are not directly emitted but photochemically produced in the atmosphere, atmospheric chemistry has to be included in the model. For relatively inert species, a simpler approach can be used. Special models have been developed to describe the transport of heavy gases and particle-bound pollutants.
d. Computer facilities.
The availability of computer resources may be one of the most stringent selection criteria.
e. Required accuracy.
The accuracy of the various steps in the causal chain from emission to environmental effect should be more or less the same. There is no need for a complex, detailed atmospheric model when little is known about emissions and their spatial distribution.

- Realistic, average atmospheric conditions, obtained from a 10-year data set of weather conditions for The Netherlands.
- Transport of gaseous and aerosol-bound chemicals was calculated separately; partitioning between gas and aerosol was estimated using the Junge-Pankow equation (Equation 3.9 in Chapter 3).
- Because of the short distance, losses due to deposition and atmospheric reactions were neglected.
- Assumed source characteristics:
  - Source height: 10 m, representing the height of buildings in which production, processing or use takes place.
  - Heat content of emitted gases: 0, meaning no extra plume rise caused by the excess heat of vapours compared with outdoor temperature is assumed.
  - Source area: 0 m, representing an ideal point source.
- Calculated concentrations for long-term averages.

The concentration in air at a distance of 100 m from the point source was estimated. This distance was arbitrarily chosen to represent the average size of an industrial site. The deposition flux of gaseous and aerosol-bound chemicals was estimated in the same way as the estimation of atmospheric concentrations, by means of an estimation method and with the help of the OPS model [17]. The deposition flux to soil was averaged over a circular area around the source with a radius of 1 km, to represent the local agricultural area. Deposition velocities were used for three different categories:

- Dry deposition of gas/vapour: estimated at 0.01 $cm.s^{-1}$.
- Wet deposition of gas/vapour: determined with the OPS model.
- Dry and wet deposition of aerosol particles, determined with the OPS model using an average particle size distribution.

Based on these assumptions and model settings, calculations were carried out for both gaseous and aerosol-bound substances. These calculations were carried out for a source strength of 1 $kg.s^{-1}$, as it has been shown that concentrations and deposition fluxes are proportional to the source strength. The results of the default calculations are given in Table 4.1.

The results in Table 4.1 show that local atmospheric

Table 4.1. Results of default calculations with the OPS model for a source strength of 1 kg/s. Concentrations at 100 m and deposition flux averaged over a circle with a radius of 1 km.

Log H ($Pa \cdot m^3/mol$)	Gaseous substances: Conc. at 100 m (kg/m^3)	Gaseous substances: Deposition flux 1000 m ($kg/m^2 \cdot s$)	Aerosol-bound substances: Conc. at 100 m (kg/m^3)	Aerosol-bound substances: Deposition flux 1000 m ($kg/m^2 \cdot s$)
< -2	$24 \cdot 10^{-6}$	$5 \cdot 10^{-10}$	$24 \cdot 10^{-6}$	$1 \cdot 10^{-8}$
-2 - 2		$4 \cdot 10^{-10}$		
> 2		$3 \cdot 10^{-10}$		

concentrations are independent of the physicochemical properties of the compounds. Hence, once the emission from a point source is known, the concentration at 100 m from the source can be estimated with a relatively simple relationship:

$$C_{air} = \frac{E_{air}}{Estd} \cdot Cstd_{air} \tag{4.21}$$

where

C_{air} = concentration in air (in gas phase as well as aerosol-bound) at 100 m from point source ($kg.m^{-3}$)

E_{air} = emission rate to air ($kg.s^{-1}$)

$Cstd_{air}$ = standard concentration in air at a source strength of 1 $kg.s^{-1}$ (= 24 x 10^{-6} $kg.m^{-3}$)

Estd = standard source strength (1 $kg.s^{-1}$).

The deposition flux can also be calculated relatively simply, although it is slightly more complex because of its dependence on the fraction of the chemical associated with the aerosols:

see below (4.22)

where

Dp_{total} = total deposition flux ($kg.m^{-2}.s^{-1}$)

$FR_{aerosol}$ = fraction of the chemical bound to aerosol (Chapter 3, Equation 3.11)

$Dstd_{aerosol}$ = standard deposition flux of aerosol-bound compounds (1 x 10^{-8} $kg.m-2.s^{-1}$)

$Dstd_{gas}$ = standard deposition flux of gaseous compounds as a function of the Henry coefficient ($kg.m^{-2}.s^{-1}$); see also Table 4.1.

Based on an uncertainty analysis of this model calculation Toet and De Leeuw concluded that specific information on source height, the heat content of the emitted plume and the particle distribution of the emitted aerosols would greatly improve the overall accuracy of the estimated concentrations [17]. Unfortunately, the necessary data will often not be available.

4.2.5 Input requirements for air models

Clearly, there will be a close relationship between input requirements and the complexity with which the atmospheric processes are described. Input requirements, including their temporal and spatial resolution, therefore depend on the model. However, all model applications require at least the following information [7,8]:

1. Emission data.

In addition to the pollutant emission rate, these data include information on the source itself, i.e. geographical location, stack height, volumetric exhaust rate, temperature of flue gases, etc. Emissions can be defined as point or diffuse sources. Close to a point source the maximum concentration depends on temporal variations in the emission rates. For example, the diurnal profile of concentrations in a traffic-related situation will be

$$Dp_{total} = \frac{E_{air}}{Estd} \cdot [FR_{aerosol} \cdot Dstd_{aerosol} + (1 - FR_{aerosol})Dstd_{gas}] \tag{4.22}$$

parallel to the diurnal variations in traffic intensity. During the morning and evening rush hours increased concentrations will be observed. Averaged concentrations are independent of temporal variations in emission rate.

2. Physical and chemical data.
The gas-particle partition and deposition parameters are required. First estimates can be based on vapour pressure and solubility data. A rough indication of the photochemical degradation rate must be provided. Quantitative Structure-Activity Relationships (QSARs) to estimate the reaction rate with the OH-radical, the most powerful oxidant in the atmosphere, are available (see Chapter 9).

3. Meteorological data.
Wind speed and direction are the most important meteorological parameters. Data on atmospheric stability (or atmospheric turbulence), mixing height, temperature, solar radiation or cloud cover and precipitation are also needed. Depending on the type of model, statistical data (yearly averaged values, wind roses, etc.) or short-term (e.g. 1 h averaged values) are required.

4. Terrain type.
Terrain data are generally not used in a first estimate of ambient levels. Many models assume a flat terrain; more complex models will require information on surface characteristics (terrain type, land-use, roughness, length, etc.).

4.3 WATER MODELS

4.3.1 Introduction

Besides air models, models to estimate the distribution of chemicals in surface water are generally the most frequently used models in environmental exposure assessment. Over the past few decades many different surface water models have been developed, tailored to specific needs or specific surface water systems. They range from very simple mathematical equations, where the concentration in a river is estimated from the concentration in a specific effluent divided by a specified dilution factor, to highly sophisticated models where concentrations in a whole river or an entire water system are estimated, for example. Simple models ignore the removal processes of the chemical after its discharge into a water system, whereas more sophisticated models evaluate processes such as volatilization, adsorption and settling, as well as biotic and abiotic degradation. This section describes some of the model types most frequently used in the exposure assessment of chemicals. The data requirements for water models are also considered. For basic processes such as advective and dispersive transport in water, partitioning between water and sediment and volatilization from the water body see Chapter 3, Section 3.2.

4.3.2 Simple dilution models

The simplest type of water model is a dilution model which divides the concentration of a chemical found in a domestic or industrial discharge effluent, by a specific stream dilution factor. This dilution factor may be a generic one, selected to perform a standard exposure assessment for regulatory purposes, or a site-specific value based on the volumetric flows of the discharge and the river. Seasonal differences in river flows and the time-dependence of the effluent flow may also be taken into account. Using a simple dilution model the final concentration in a river after complete mixing (C_∞) can be obtained from:

$$C_\infty = \frac{C_w Q_w + C_e Q_e}{Q_w + Q_e} \tag{4.23}$$

where C_w and C_e (mol.m^{-3}) are the chemical concentration in the river and effluent, with a flow of Q_w and Q_e (m^3.s^{-1}), respectively. For new chemicals or chemicals with only one source C_w becomes zero, resulting in the simplest dilution model:

$$C_\infty = C_e \cdot \mathrm{DF} \tag{4.24}$$

where DF is the dilution factor of the effluent (Q_e / (Q_w+Q_e)). In a generic assessment this dilution factor could be an average or median value or a 90% or 95% value of all DFs for the particular region or country under consideration. In the first step of an exposure assessment for a new chemical entering the European market a DF of 10 is applied [18]. It should be noted that these simple dilution models assume the homogeneous distribution of the chemical in river water and provide no information on the advection and dispersion of the chemical in the water system where the discharge occurs.

A more realistic approximation of exposure concentrations can be obtained by looking at the distribution of all DFs that are relevant to the emission sites of a specific chemical. For household chemicals that

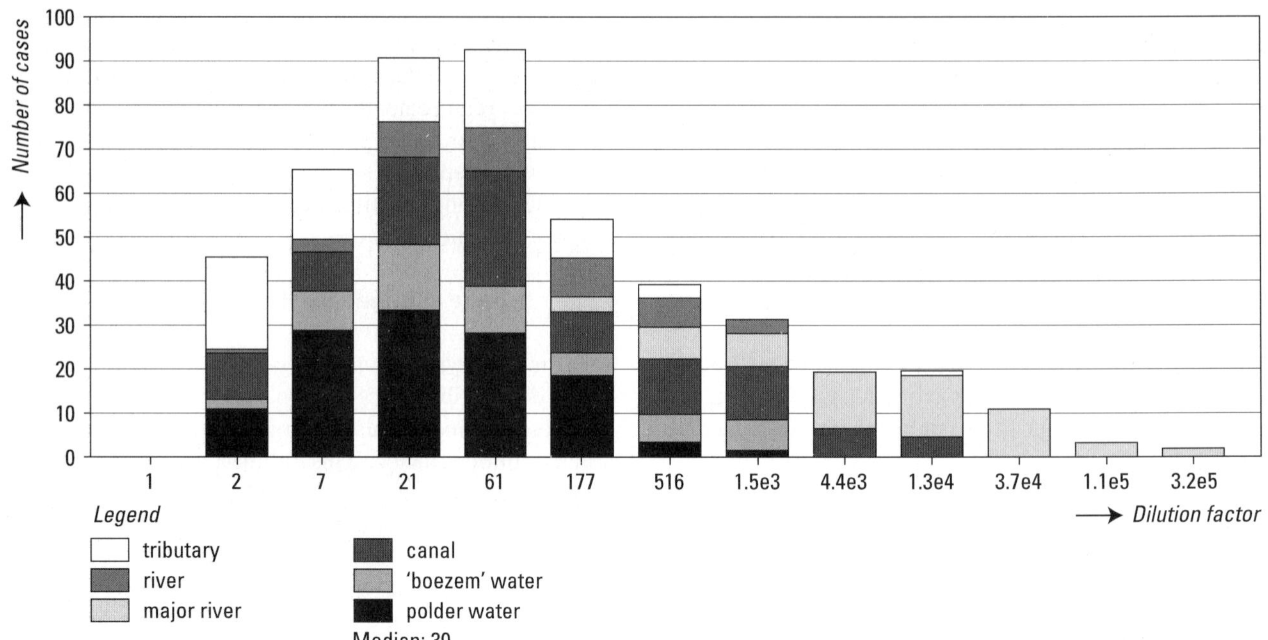

Figure 4.5. Histogram of the dilution factor (DF) at 1000 m downstream of the emission point for all waste water treatment plants in The Netherlands discharging to nearby surface waters. From [21]. With permission. Copyright Elsevier.

are typically emitted into the aquatic environment after passing through a waste water treatment facility, this can be achieved by a statistical evaluation of all waste water and river flows at all discharge locations in a specific region. This type of analysis has been incorporated in several models for the USA [19,20] and The Netherlands [21,22]. De Nijs and De Greef used a dispersion model to estimate the dilution of effluents from all waste water treatment plants (WWTPs) in The Netherlands [21]. They calculated the mixing lengths, dilution factors, and other important parameters such as the Reynolds number for every individual WWTP, and used these data to calculate the overall distribution of dilution factors at 1000 m from the outlet of a WWTP. A histogram of this distribution is given in Figure 4.5.
From these results it was concluded that the DFs show considerable variation. The median value for municipal treatment plants in The Netherlands was adopted as the dilution factor for EUSES [23].

These simple dilution models still do not take into account the fate of the chemical after discharge into the aquatic compartment. Examples of models that estimate adsorption, degradation and volatilization in the receiving water body are HAZCHEM [24], PG ROUT [25,26] and GREAT-ER [27, 28]. All these removal processes are approximated by a first-order decay rate constant k as used in Equation 4.25. Adsorbed and dissolved concentrations can be calculated if the partition coefficients are known. It should be noted, however, that within a relatively short distance of the outlet of a waste water discharge (1000 m is a "normal" value [17,22]) these removal processes have relatively little effect on the final concentration compared with dilution by river water. Only for chemicals with a short biodegradation half-life and a high sorption coefficient is it likely that any significant removal from the water body will be seen within the first few kilometres [29]. Therefore, simple dilution models may very often give satisfactory predictions.

4.3.3 Dispersion models

The subject of dispersion and mixing of solutes and suspended materials in turbulent natural streams has been extensively discussed [30-32]. Examples of dispersion models describing the concentration profiles (x,y) as a function of the location in the surface water system are the "Alarmmodel Rhine" [33], Dilmod [20] and CORMIX1 models [34]. Typical examples of dispersion models are spill models, which are used to calculate the concentration of a chemical after accidental release into a water body. Normally, these models are concerned with relatively short time scales. Advection and dispersion are generally the most important processes in a short time

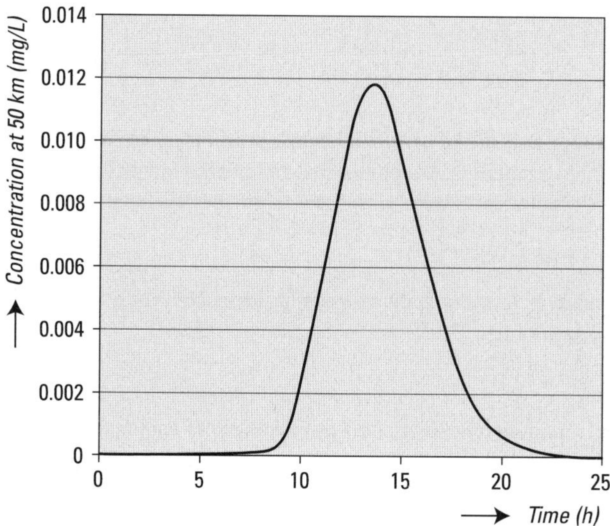

Figure 4.6. Concentration of a chemical X 50 km downstream from the spill point, according to equation 4.25.

scale. Evaporation, adsorption and degradation may also play a role but usually these processes have less effect on the local concentration than the dilution itself [7]. With an instantaneous point emission of this kind, the distribution of the concentration downstream of the mixing zone can be modelled in accordance with Fischer [32]:

$$C_{x,t} = \frac{M / A}{\sqrt{4\pi D_x t}} \, e^{-\frac{(x-ut)^2}{4D_x t} - kt} \qquad (4.25)$$

where

- $C_{x,t}$ = concentration at x metres downstream of the emission point at time t after discharge ($g.m^{-3}$)
- M = the amount of spilled chemical (g)
- A = cross-sectional area of river (m^2)
- D_x = one-dimensional longitudinal dispersion coefficient ($m^2.s^{-1}$)
- t = time (s)
- x = longitudinal distance downstream of emission point (m)
- u = average flow velocity ($m.s^{-1}$)
- k = first-order decay coefficient (s^{-1}).

Note that when applying this one-dimensional model it may be necessary to check the validity of the implicit assumptions that instantaneous transversal and vertical mixing take place.

As can be seen from this formula, the model can be applied generically or in a site-specific manner by inserting either standard or actual values for the hydrological parameters. However, to apply spill models to specific sites, such as production or storage facilities, more sophisticated two-dimensional models may be required. These models go beyond the scope of this book. However, advanced readers are referred to Fischer [32] or Trapp and Matthies [5]. A typical example of how an instantaneous point emission can be assessed based on Equation 4.25 is given in Box 4.4 and Figure 4.6.

4.3.4 Compartment models

Compartment models describe the transfer and transformation of pollutants and/or nutrients through a surface water system. In these models the surface water system is divided into a number of segments, where each segment contains a set of constituents (Figure 4.7). Thus the pollutant and nutrient transport and transformation processes can be described as fluxes between the constituents and neighbouring segments. The model was developed as a set of ordinary differential equations with one mass balance equation for each compartment. Most surface water compartment models include a water and a sediment layer. Depending on the chemical under investigation, the sorption to organic material, detritus and inorganic material (sand or clay particles) can be described and, if necessary, sedimentation of these particles and resuspension can be included. For organic compounds transformation processes (biodegradation, photolysis, hydrolysis, etc.) should be incorporated as well as volatilization to the air above. Within the sediment layer special attention should be given to reducing conditions as well as sediment burial. Most of these processes are described in Chapter 3. A compartment model for surface waters is in fact a simplified multimedia model taking only the water and sediment layers into account. The main difference is the number of water and sediment compartments and the advective and dispersive transport between the segments.

Basic textbooks dealing with compartment models have been published by Dickson [3] and Jørgensen [36,37]. Typical examples of compartment models are EXAMS [38,39] and WASP4 [40], both developed and used by the USEPA, and DELWAQ [41]. Because of their complexity, these models generally require a considerable amount of data [7]. Whole river systems can also be modelled by connected river reaches, which are regarded as well-mixed segments or compartments. Examples of these are GREAT-ER [27, 28], PG-ROUT [25, 26] or RhineBox [42].

Box 4.4. Assessment of an accidental spill in a river

a. Problem definition.

At a chemical factory located on a large river a production facility fails and 100 kg of a toxic chemical X enters the river in a very short period of time. A drinking water inlet that serves a large city is situated 50 km downstream from the factory. Should the competent authorities temporarily close this inlet?

b. Information available.

Drinking water quality guideline for chemical X: 10 $\mu g.L^{-1}$. Degradation rate constant in surface water (k): 10^{-6} s^{-1}. River characteristics: depth (d): 4 m, width (w): 100 m, average flow velocity (u): 1 $m.s^{-1}$.

c. Solution.

The first question to be addressed is what type of model to use: a one-dimensional or a two-dimensional spill model? To answer this question, as explained in section 3.2, the length of the mixing zone (L_{mix}) must be known first. This length can be estimated with Equation 3.10 (Chapter 3):

$$L_{mix} \approx 0.4\,\frac{u \cdot w^2}{D_y} \qquad (4.26)$$

in which the transverse dispersion coefficient D_y ($m^2.s^{-1}$) is estimated using Equation 4.27:

$$D_y = 0.6\,(\pm 0.3) d u_* \qquad (4.27)$$

where the shear stress velocity u_* ($m.s^{-1}$) is the velocity of the water at the sediment-water interface. This can be estimated from the average flow velocity with Equation 4.28:

$$u_* = \frac{u}{C}\sqrt{g} \qquad (4.28)$$

where g is the gravitation constant (9.81 $m.s^{-2}$) and C is the Chezy coefficient. The Chezy coefficient can be estimated from the Manning coefficient ($n_{Manning}$), which are both measures of the sediment roughness, according to Equation 4.29:

$$C = \frac{1.5\,R_h^{1/6}}{n_{Manning}} \qquad (4.29)$$

where the Manning coefficient ranges from 0.020 for normal rivers and canals, to 0.035 for highly turbulent mountain rivers [35]. The hydraulic radius R_h (m) is defined as follows:

$$R_h = \frac{wh}{w + 2h} \qquad (4.30)$$

d. Calculation.

With a width of 100 m and a depth of 4 m a hydraulic radius of 3.7 m is obtained. Using a value of 0.025 for the Manning coefficient, Equation 4.29 gives a value of 74.6 for the Chezy coefficient. Applying this value to equation 4.28 we obtain a shear stress velocity of 0.04 $m.s^{-1}$ which, when used in Equation 4.27, results in a transverse dispersion coefficient of 0.1 $m^2.s^{-1}$. With a width of 100 m and an average flow velocity of 1 m/s Equation 4.27 gives a mixing length of approximately 40 km. Hence it can be assumed that complete mixing has been achieved in the river. Therefore the one-dimensional model (Equation 4.25) can be used to predict the concentration of the chemical in time at 50 km. The one-dimensional longitudinal dispersion coefficient D_x also has to be available. An estimate of the longitudinal dispersion coefficient is provided by Fischer [32]:

$$D_x = 0.011 \frac{u^2 w^2}{d u_*} \quad (4.31)$$

From this equation a longitudinal dispersion coefficient of 655 $m^2.s^{-1}$ is calculated. On the basis of all this information Equation 4.25 can be used to calculate the concentration-time profile at the drinking water inlet. This profile is shown in Figure 4.6. Note that degradation and sediment-water exchange are not considered in this example. From Figure 4.6 it can be concluded that the maximum concentration at the water inlet (12 $\mu g.L^{-1}$) slightly exceeds the drinking water quality guideline (10 $\mu g.L^{-1}$) approximately 13 hours after the spill. Therefore, appropriate action should be taken by the regulating authorities.

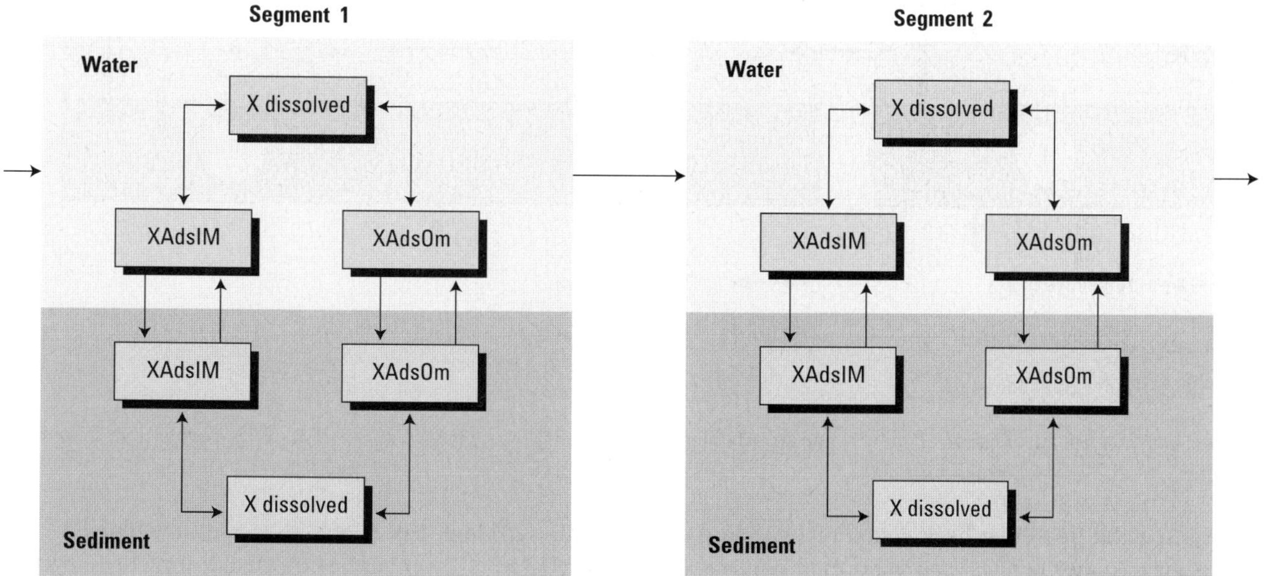

Figure 4.7. Conceptual water model including two segments, a water and a sediment compartment each with three state variables.

4.3.5 Estimation of the elimination of a chemical in a waste water treatment plant

To calculate the effluent concentrations of industrial and household chemicals it is essential to know whether the chemicals pass through a WWTP before being discharged into the aquatic environment. In the EU the average level of connection to sewage treatment facilities is approximately 80% (at least for primary purification) [17]. The actual situation, however, varies quite widely (Chapter 2). In many countries very little of the municipal sewage is treated, often in poorly run plants, while in other countries almost all sewage passes through secondary and sometimes even tertiary treatment.

In general, three different processes are involved in the removal of a chemical in a WWTP: biodegradation by micro-organisms, adsorption to sludge and volatilization. Therefore, the removal percentage depends on the physicochemical and biological properties of the chemical as well as on the operating conditions of the WWTP. If actual measured data on the degree of removal of a chemical in a WWTP are absent, this can be estimated with WWTP simulation models. An example of such a model is SIMPLETREAT, developed by Struijs et al. [43]. In this model steady-state concentrations are estimated for a WWTP consisting of a primary settler, an aeration tank and a liquid solid separator. The model, shown as a diagram in Figure 4.8, has nine compartments. The degree of removal in this model can be estimated from the octanol-water partition coefficient (K_{ow}) or, if available, the suspended solids-water partition coefficient (K_p), the Henry's law constant and the results of biodegradation tests. Depending on the outcome of standard ready biodegradation or inherent biodegradation

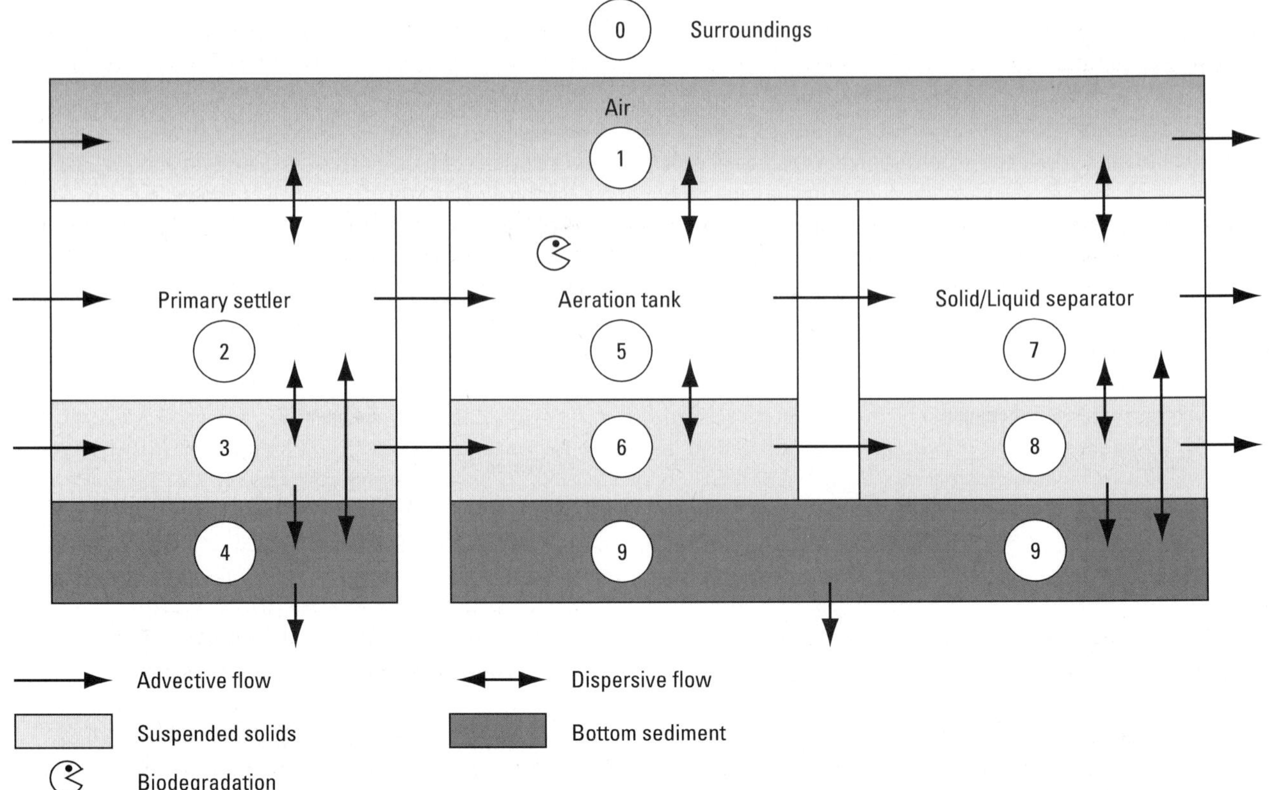

Figure 4.8. Conceptual diagram of the sewage treatment plant model SIMPLETREAT. From [43]. With permission. Copyright Elsevier.

tests, a specific first-order degradation rate constant can be assigned to the chemical and the overall removal due to degradation, adsorption and volatilization calculated. See Section 3.5.6 and [44] for details of how to derive rate constants from biodegradation tests. If no biodegradability test data are available, rate constants may be estimated using SARs (e.g. BIOWIN).

4.3.6 Data requirements for water models

Like air models, the data requirements for water models largely depend on the type of model to be used. Data on the emission scenario, chemical properties and environmental characteristics are required [6,7]:

a. Emission scenario.
The emission scenario of the chemical largely determines the choice of model. Concentrations arising from continuous discharges, e.g. of household chemicals passing through a WWTP, can be calculated with a steady-state model. The yearly production volume and average or worst-case degradation percentage in a WWTP need to be known to calculate the concentration in the effluent. For batch processes the number and length of emission periods in a year must be known in order to decide whether a steady-state calculation will suffice or a dynamic model is needed. The above spill models may be used but require information on volume and total amount (kg) of the spill.

b. Chemical properties.
For simple dilution models and dispersion models that do not model the fate of the chemical after its discharge, data on molecular weight and water solubility will normally suffice. More sophisticated models will need information on sorption properties (K_p or K_{oc} values), ionisation constants, vapour pressure or Henry's law constant in order to estimate the volatilization rate and abiotic and biotic degradation rate constants (hydrolysis, photolysis, oxidation, biodegradation in surface water and sediment).

c. Environmental characteristics.
The data requirements also largely depend on the model chosen. Simple dilution models require information on effluent and river flows to be able to estimate the dilution

factor. An average value or a worst-case value based on the seasonal variation in the river flow may be used. As described in Section 4.3.2, a statistical distribution of all DFs in a region may also be used. This, of course, entails an extensive data set on effluent and river water flow patterns.

Data on system geometry and hydrology are needed when more site-specific analyses are to be performed. These include flow volumes, river depths, areas, rainfall, entering stream and non-point source flows, and even groundwater flows. Furthermore, evaporation rates, wind speed, suspended particle and sediment loads, dissolved organic carbon content, water pH and temperature, etc. may also all be required.

4.4 SOIL MODELS

4.4.1 Introduction

There is growing concern about the possible impact that chemicals may have on organisms that live in soil and sediment. Not only because heavily polluted sites have been discovered in many countries but also because the potential risks of diffuse, long-term distribution of persistent chemicals are becoming more and more apparent. Moreover, in some countries drinking water supplies are threatened due to pollution of groundwater aquifers. Hence, exposure assessment for soil and groundwater has become an indispensable part of the risk assessment of chemical substances. Traditionally, the development and design of soil and groundwater models is closely related to the way chemicals enter the soil. Typical exposure scenarios include:

- Use of pesticides and fertilizers on agricultural land.
- Use of sludge from waste water treatment plants on agricultural land.
- Deposition of (persistent) chemicals, including pesticides, from the air on natural as well as agricultural soil.
- Contaminated sites.
- Run-off from motorways to nearby soils.

Naturally, soil models should be tailored to this variety of exposure situations. It goes beyond the scope of this section to describe all current developments in these areas. Therefore, the following section gives a short description of the relevant processes that should be taken into account in soil models, followed by a brief overview of the most frequently used models. Subsequently, use of soil models in the risk assessment of new and existing chemicals is discussed. The section ends with some comments on data requirements.

Groundwater models are not discussed in this section, although it is recognized that leaching from soil into groundwater is an important process for some chemicals. However, a description of two or three-dimensional plume migration in groundwater falls outside the scope of the generic risk assessment of a chemical. Models typically used in the risk assessment of pesticides to describe spreading and drift directly after application are not described here either.

4.4.2 Fate processes in soil modelling

Soil is the most heterogeneous of all environmental compartments. It can be viewed as a system consisting of four phases: air, water, solids and biota. The system has numerous and large gradients, temperature and moisture content are highly variable, it has a high level of reactive surfaces as well as a high biotic level, and finally, it contains aerobic as well as anaerobic layers or zones. The actual value or the presence or absence of all of these factors to a large extent determine the fate of a chemical in soil, which makes it very difficult to accurately describe the fate of chemicals in soil. Moreover, soil use plays an important part in the way a chemical is introduced into the soil.

An overview of the relevant fate-determining processes that are usually taken into consideration in soil models is given in Figure 4.9. The mobility of a chemical in soil is largely determined by its air-soil and water-soil partition coefficients, which dictate the extent to which the chemical partitions into the immobile solid phase. Soil sorption influences migration through the soil core, volatilization from the soil surface and lateral and vertical transport. In addition, bioavailability to soil organisms, including plant-uptake, as well as biodegradation by soil micro-organisms are largely dependent on the fraction of the chemical not adsorbed to the solid fraction of the soil. As described in detail in Chapter 3, for organic compounds the sorptive capacity of the soil is directly related to the organic matter or organic carbon content of the soil. Most soil models take this dependency into account. In addition to the organic matter content of the soil, a number of other important soil properties have been identified that may affect the soil buffering capacity and retention capacity for heavy metals and organic pollutants [45,46]. These capacity-controlling properties (CCPs) are summarized in Table 4.2. The qualitative explanation of the influence of these CCPs on the fate and mobility of chemicals as given in this table may look obvious. However, since most of these CCPs are highly interdependent, it will probably take a long time before

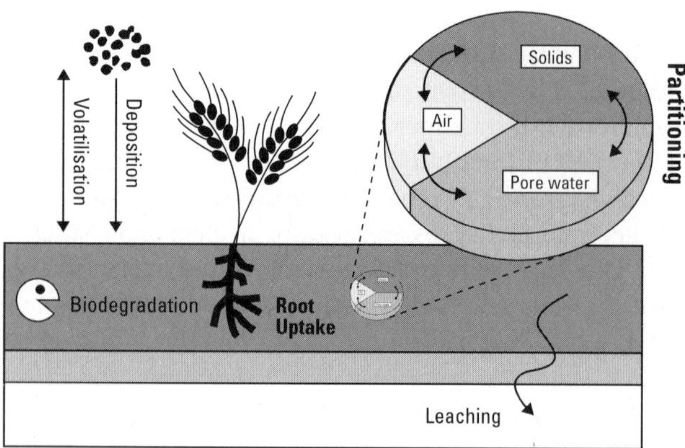

Figure 4.9. Processes determining the fate of chemicals in soil.

these relationships are quantitatively described in such a way that they can be used in general predictive soil models.

Although sorption and degradation may also occur in groundwater, one or both of these processes are often neglected in groundwater models because clay or organic matter content and microbial activity are low. In fact, they depend on the depth and origin of the soil layers. Leaching is also influenced by climate and vegetation and special conditions such as water blocking layers or fissures, which are usually not included in models.

4.4.3 Model types

More than with air and water models, the choice of a soil model (in terms of computational approach and dimensionality) depends on the modelling purpose. Models to evaluate the fate of pesticides, for instance, are often dynamic because they have to describe the remaining concentration at some point in time after a single application. However, steady-state models may well be sufficient to evaluate the long-term accumulation of persistent chemicals in natural areas due to continuous atmospheric deposition. In principle, two types of models are used to assess chemical fate in soil: those that simulate chemical fate in the unsaturated zone, and those that simulate the chemical fate in the saturated zone in the groundwater. Most unsaturated zone models are one-dimensional and simulate vertical transport only. The output of these models is often used as input for groundwater models. Groundwater models are usually two-dimensional (horizontal transport), although three-dimensional models (including vertical movement) are also available. Most models assume equilibrium conditions between the solid, pore water and air phases. The sorption constants, however, may vary according to the properties of the different layers considered.

Models which are frequently used to evaluate the fate of chemicals include the pesticide root zone model (PRZM) and the seasonal soil compartment model (SESOIL). The PRZM model simulates the vertical movement of pesticides in unsaturated soil, both within and below the plant root zone and extending to the water table [47]. Leaching, erosion, run-off, plant uptake, foliar wash-off and volatilization are taken into consideration. Degradation is incorporated by using first-order reaction rate constants. The model is validated by comparing the results with field data [48,49] and has been modified for use with Central European soils and climates [50]. The SESOIL model is designed to predict the migration of organic substances and metals through unsaturated soil zones and leaching to adjacent groundwater [51]. Vertical advection, volatilization, adsorption, cation exchange, complexation of metals, hydrolysis and first-order decay are all incorporated. The model generates average monthly concentration profiles with mass distributions in each phase and for each soil layer. It has been validated in several studies [52]. Quite similar models, differing mainly in the way in which some fate-determining processes are described, are EXSOL [53] and PESTLA [54].

4.4.4 Use of soil models in the risk assessment of industrial chemicals

Unlike the situation for pesticides, there is often little data available for use in soil fate models. Hence, in order to obtain some insight into the possible risks posed by chemicals after they have reached the soil, a number of assumptions have to be made and extrapolation steps taken. A very simple but straightforward way to calculate the concentration of a chemical in soil after direct application or application via sewage sludge is described in the guidance document for the risk assessment of new chemicals in the EU market [18]. The initial concentration in soil is obtained by assuming that the amount of chemical (directly applied or in sewage sludge) is fully mixed with the top layer of soil by ploughing (usually to a fixed depth of 20 cm). This may lead to overestimation, since it does not take into account removal processes occurring after application. Alternative approaches have recently been proposed in the EUSES program [23] for risk assessment and by ECETOC in a report on environmental exposure assessment [27]. In EUSES an adapted version of the PESTLA model is used to calculate concentrations in the upper 20 cm of the soil and in the uppermost metre of the groundwater [55]. PESTLA is a dynamic process model based on one-

Table 4.2. Important soil capacity-controlling properties (CCPs) for heavy metals and toxic organic chemicals, as described by Stigliani et al. [45,46].

CCP	Environmental effect
Cation or anion exchange	CEC and AEC depend on inorganic clay mineral content and type, organic matter (OM) content, capacity (CEC or AEC) and soil pH. Soil with a low CEC or AEC has a poor capacity to retain cations (e.g. metals) or anions (e.g. organic anions) by sorption
pH	lowering pH often increases heavy-metal solubility, decreases CEC and alters the soil microbial population
Redox potential (Eh)	decreasing redox potential (more reducing conditions) dissolves iron and manganese oxides, which mobilizes oxide-sorbed chemicals. Increasing redox potential (more oxidizing conditions) mobilizes heavy metals by dissolving metal sulphides
Organic matter (OM)	decreasing the OM content reduces CEC, soil pH buffering capacity, sorption capacity for chemicals, soil-water holding capacity and alters physical structure (e.g. increases erodibility), and decreases microbial activity
Structure	altering soil structure can reduce drainage and thereby increase redox potential, increase soil erodibility, affect the rate of chemical release to drainage water and alter pH
Salinity	increasing the salinity makes toxic chemicals soluble by altering the ion-exchange equilibrium, increasing complexation in solution and decreasing thermodynamic activities in solution. It can also decrease microbial activity
Microbial activity	altering the microbial activity and population ecology can reduce degradation of toxic organics (and increase accumulation), and alter redox potential and pH

dimensional convection/dispersion transport equations for reacting and degradable solutes in soil. The model was developed for the evaluation of pesticide leaching from soil into the water-saturated zone and can be used to support decision-making on the admission of pesticides. Because the model was developed for the evaluation of pesticides which are applied periodically, the model includes a pulse type single-dose application at the upper boundary. This type of application is similar to the sewage sludge application. Another type of input, for example, the daily dose due to atmospheric deposition, can be incorporated in the model as well. A number of features of this model were chosen to represent a reasonable worst-case:

- Sorption characteristics which reflect the chosen reference: sandy soil with relatively low organic matter content and a phreatic aquifer.
- The substance is assumed to be distributed directly after application to the upper 5 cm of the soil.
- Uptake of water and substance by plants (a culture of maize was used) results in a reduction in the substance concentration over a longer time scale.
- The accumulation in soil and the maximum concentration in deeper groundwater, in response to the substance dose rate, are both assumed to be linear.
- Precipitation data from a relatively high rainfall year are used in the calculations (75 percentile).
- However, the model is not suitable for volatile substances.

The PESTLA model was used to model the accumulation and leaching potential for various combinations of the organic matter sorption coefficient (K_{om}) and the half-life for biodegradation (DT50) for a single application of 1 kg/ha. Figure 4.10 shows the results: the percentage leached below a depth of one metre (Figure 4.10A) and the fraction remaining in the top layer (Figure 4.10B). These data, together with the actual dose rate, can then be used to calculate the soil and groundwater concentrations. The dose rate is calculated from the amount of the chemical present in sewage sludge using a sewage treatment model (Section 4.3.5), and the deposition flux resulting from emissions to air (Section 4.2.4). Figure 4.10A shows that significant leaching to groundwater occurs only for chemicals with a half-life in soil of more than 40 d and K_{om} values of less than 200 L.kg^{-1}. Figure 4.10B shows that accumulation in the top soil layer is expected to occur only for chemicals with a half-life of more than 40 d. Accumulation may become relevant for chemicals where K_{om} is greater than 20 L.kg^{-1}.

ECETOC [27] takes a similar approach. A soil module was developed to calculate the steady-state

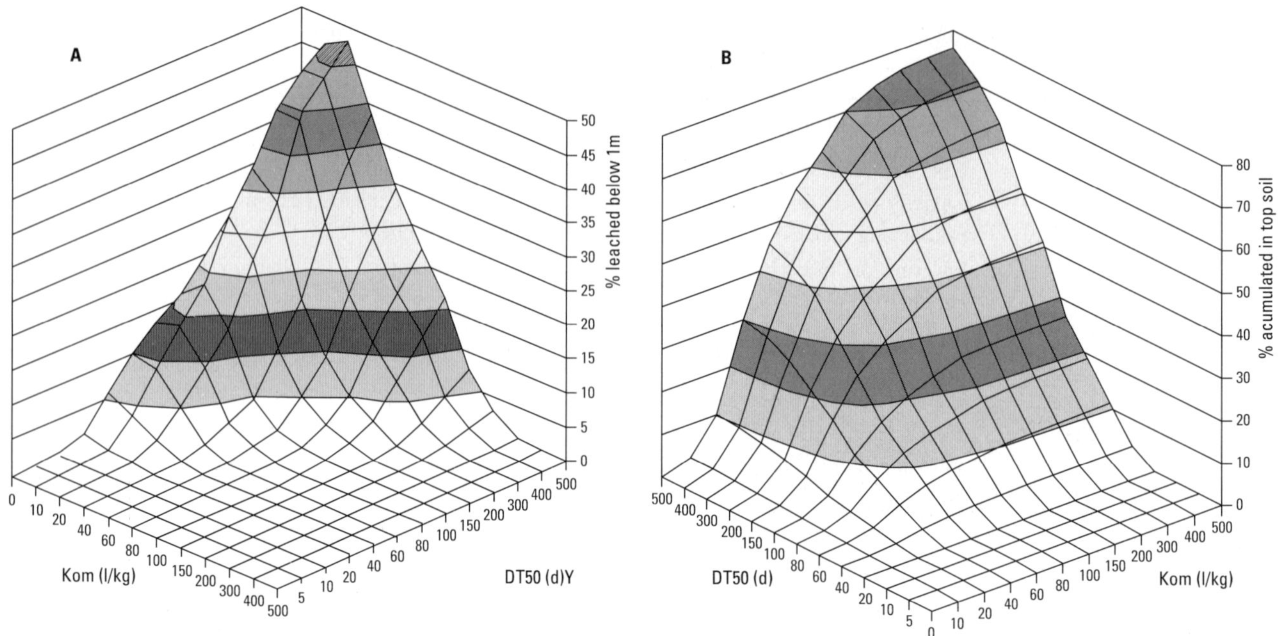

Figure 4.10. Percentage leached below a depth of 1 m (A) and percentage still present in topsoil (B) as a function of Kom (L.kg^{-1}) and DT50 (d), according to Swartjes et al. [55].

concentration in soil close to a point source after many years of exposure. Deposition from air to soil and sludge application control the input of the chemical, whereas sorption, evaporation, biodegradation and leaching ultimately determine the soil and groundwater concentrations. A distinction is made between natural soil where the chemical is acquired from deposition only, and arable soil where it is obtained from both deposition and sludge application. The steady-state concentration in arable soil can be calculated with the following equation:

$$C_{soil} = \frac{Dp_{total} + Sl_{appl}}{(k_{degr} + k_{leach} + k_{evap}) \cdot H_{soil} \cdot R_{soil}} \qquad (4.32)$$

where

C_{soil} = concentration in soil at steady-state (kg.kg^{-1})
Dp_{total} = total deposition flux (kg.m^{-2}.s^{-1}), according to equation 4.22
Sl_{appl} = substance application rate via activated sludge (kg.m^{-2}.s^{-1})
k_{degr} = (bio)degradation rate constant in soil (s^{-1})
k_{leach} = removal rate constant for leaching (s^{-1})
k_{evap} = removal rate constant for evaporation (s^{-1})
H_{soil} = soil depth (m)
R_{soil} = density of soil (kg.m^{-3}).

The values for k_{degr}, k_{leach} and k_{evap} are calculated on the basis of the physicochemical properties of the chemical, the environmental characteristics, the mass-transfer coefficients between soil, water and air, and information on abiotic and microbial biodegradation rate (for details of the calculation see [55]). To calculate the concentration in natural soil, the sludge application term is omitted from Equation 4.32.

4.4.5 Data requirements for soil models

The typical data requirements for soil and groundwater models can be divided into application data, physicochemical properties, soil characteristics and meteorological conditions [6,7]:

a. Application data.
Pesticide models typically need application rates (usually discontinuous) and duration, and where relevant, initial concentration details. Direct application information on wet and dry deposition is also needed.

b. Physicochemical properties.
Data on chemical identity, molecular weight, Henry's law constant and octanol-water partition coefficient are the minimum requirements to be able to estimate partitioning in soil. Measured solids-water distribution coefficients

and measured (pseudo) first-order (bio)degradation rate constants are essential for proper fate estimation. When considering root uptake by plants, soil-plant biotransfer factors may be required (Section 4.6.4). Biodegradation rate constants can be extrapolated from standard biodegradation tests (Section 3.5.6 and [44]). Estimation software can provide indicative parameter values in the absence of measured data.

c. Soil characteristics.
Data on soil density, porosity, moisture content, organic matter or organic carbon content are essential. Some models assume one single homogeneous layer, whereas in other models the soil is divided into several horizontal layers, each with its own properties. Biodegradation and sorption may differ considerably between these layers. Data on pH, cation or anion exchange capacity and redox potential may be required when these factors are quantitatively correlated to fate-determining properties. Groundwater models need information on aquifer depth and width and on chemical input coordinates, hydraulic conductivity and hydraulic gradient, vertical dispersivity and withdrawal rates of abstraction locations.

d. Meteorological conditions.
Annual precipitation, evapotranspiration and run-off must be known to be able to determine the water flux through the soil layers. Temperature and wind speed control evaporation, while sunlight intensity influences photodegradation.

4.5 MULTIMEDIA MODELS

4.5.1 Introduction

If a chemical is released into one medium and resides there until it is removed by degradation or advection, single-media models may be perfectly suitable for estimating the environmental concentration. If, however, a chemical:

(1) is released into several compartments simultaneously, or

(2) after release into one compartment is transported to other compartments,

it becomes necessary to account for the intermedia transport processes so that its ultimate fate in the overall environment can be assessed. Multimedia models are specifically designed to do this. This section on multimedia models starts with a short description of their features and the explicit and implicit assumptions usually made. The use of these models in exposure assessment is described together with their limitations. Subsequently, some information on data requirements and on the different models available is given, following which a number of sample calculations are presented to illustrate the use of these models.

4.5.2 Characteristics and assumptions

Multimedia fate models are typical examples of compartment mass balance models (Section 4.1.2). The total environment is represented as a set of spatially homogeneous (zero-dimensional) compartments; one compartment for each environmental medium in which the chemical is assumed to be evenly distributed (Figure 4.11). Typical compartments considered in models are: air, water, suspended solids, sediment, soil and aquatic biota. Multimedia mass balance modelling was initiated in the early 1980's by Mackay and co-workers [6,56-59]. The example was soon followed by others [59-63]. In the EU, the model SimpleBox used in The Netherlands was adopted as the basis for the risk assessment model EUSES [64,65]. While the early Mackay School models described a fixed, "unit world", which was meant to represent a global scale, later models by the Mackay School and others have enabled users to customize the environment and define smaller and more open spatial scales. More recently, the use of spatially resolved multimedia fate models has become more common [65-76].

A typical regional multimedia model describes a region between 10^4 and 10^5 km^2. In this generic form, the models can account for emissions into one or more compartments, exchange by import and export with compartments "outside" the system (air and water), degradation in all compartments and intermedia transport by various mechanisms (Figure 4.11). Mass flow kinetics, formulated slightly differently in models by different authors, are usually defined as simply as possible: mass flows are either constant (emission, import) or controlled by (pseudo) first-order rate constants (degradation, intermedia transport), as in Equation 4.2. In all the models, the user has to set parameter values for these mass flows to provide input for the model.

Using a number of criteria, such as equilibrium or non-equilibrium, steady-state or non-steady-state, and based on whether to take the degradation of the chemical into account in the calculation or not, Mackay and Paterson introduced a classification of multimedia models [57]. This classification begins with a Level I model which describes the equilibrium partitioning of a given amount of a chemical between the above media. The Level II model simulates a situation where a chemical is

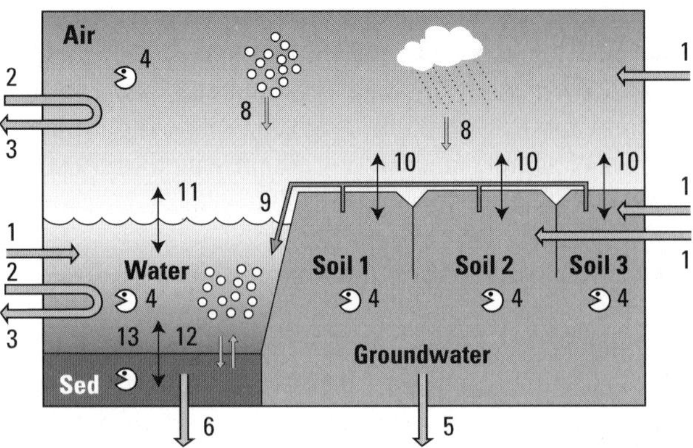

Figure 4.11. Diagram of a multimedia mass balance model concept. 1 = Emission, 2 = Import, 3 = Export, 4 = Degradation, 5 = Leaching, 6 = Burial, 7 = Wet deposition, 8 = Dry aerosol deposition, 9 = Run-off, 10, 11 = Gas absorption and volatilization, 12 = Sedimentation and resuspension, 13 = Sorption and desorption. From [61]

continuously discharged into a multimedia environment in which partitioning, advection and degradation take place. Transport between the media is infinitely rapid, so that thermodynamic equilibrium between the media is maintained. At Level III, realistic intermedia transport kinetics are assumed, so that media may not be in thermodynamic equilibrium. Level III models calculate steady-state concentrations in all compartments. Finally, Level IV models assume a non-steady-state and yield time-related chemical concentrations. An overview of these different models is given in Table 4.3.

4.5.3 Data requirements for multimedia models

The degree of accuracy of multimedia models depends, among other things, on whether all the potentially relevant phenomena have been taken into account and whether realistic data for a chemical have been used. Table 4.4 gives the physicochemical information that is typically required in order to run a multimedia model.

Level I calculation requires knowledge of intermedia partition coefficients (air-water, water-solids) only. Calculation at level II and above requires additional knowledge of degradation rate constants in air, water, sediment and soil. Unfortunately, measured partition coefficients and rate constants are not always available. In the absence of measured data, partition coefficients can be estimated from basic substance properties, using structure-activity relationships (SAR; Chapter 9). Easy to use software, e.g. EPI Suite, is also available from the USEPA: http://www.epa.gov/opptintr/exposure/pubs/episuite.htm to support such estimates. The consequence of using estimated model input data is that the accuracy of the model output will also depend on the quality of the SAR methods that have been used. Very often biodegradation rate constants are extrapolated from standard degradation tests (Chapter 3, Section 3.5.6), or even estimated using SARs (e.g. BIOWIN, also available from US EPA: http://www.epa.gov/opptintr/exposure/pubs/episuite.htm). These approaches introduce another uncertainty into the outcome of the calculation, especially if precise data is not available for the degradation rate constants in compartments that serve as a "sink" for a specific chemical.

4.5.4 Applications and limitations

The principal utility of multimedia models, as a first step in exposure assessment, is to determine to what extent intermedia partitioning may occur. If it appears that no significant partitioning into secondary compartments is expected, further exposure assessments may focus on the primary compartment(s) only. As intermedia transfer is usually relatively slow, its effect on the fate of chemicals is significant only over longer periods of time, i.e. if the spatial scale is large or the chemical does not degrade rapidly.

This brings us to one of the major applications of these models, which is the exposure assessment of chemicals on regional (usually 10^4 to 10^5 km^2) and larger spatial scales. These models are particularly useful for calculating the predicted environmental concentration ($PEC_{regional}$) especially of chemicals with a very diffuse release pattern. Results from Level III multimedia models are used in EU risk assessments for new and existing chemicals [2,18]. In addition to calculating the regional concentration of a chemical, the results of Level III models can also be used as input for local models. When using such models, the actual concentration is greatly underestimated if the concentration of the chemical in air or water from "outside" is set to zero, especially in relation to high production volume chemicals with a widely distributed use pattern. Regional concentrations estimated from the release rates for a larger region fed into a regional multimedia model can then be used as boundary concentrations in local model calculations.

One of the key processes in multimedia models is the partitioning between aqueous and solid phases. Most models follow in the footsteps of the original Mackay models and estimate solids-water partitioning from the octanol-water partition coefficient (K_{ow}). This

Table 4.3. Hierarchy of multimedia models with corresponding information requirements and model output (adapted from Mackay and Paterson [57]). With permission.

Level	Type	Information needed	Outcome
I	equilibrium, conservative[a]	- physicochemical properties - model environment parameters - amount of chemical in the system	- distribution of the chemical between the compartments
II	equilibrium, non-conservative	- Level I + - overall discharge rate - transformation and advection rates in different compartments	- distribution between compartments - environmental lifetime
III	steady-state, non-conservative	- Level II + - compartment specific discharge rates - realistic intermedia transfer rates	more accurate estimation of: - lifetime - chemical quantities - concentrations in different compartments
IV	non-steady-state, non-conservative	- as Level III	- time-dependent concentrations - time to steady-state - clearance time

[a] Conservative or non-conservative in the sense that degradation of the chemical is (or is not) possible.

Table 4.4. Typical data requirements for multimedia models.

Essential model input data	Supporting substance properties
Henry's law constant Sediment-water partition coefficient Soil-water partition coefficient	Molecular weight Water solubility Octanol-water partition coefficient Vapour pressure
Half-life air Half-life water Half-life sediment Half-life soil	(Estimated) constant OH-radical attack Readily biodegradable (yes/no)

means that the models are particularly useful for organic chemicals whose K_{ow} values can be accurately measured or estimated (Section 9.3.1). Applying these models to ionisable compounds, surface-active chemicals, polymers, or inorganic compounds (including metals) should be done with great care. However, the models can be used for these chemicals, provided certain adaptations to specific physicochemical properties are made. Mackay and Diamond, for instance, used an "equivalent" based model to describe the fate of lead in the environment [76], while in the example calculation for cadmium (Section 4.5.5) parameters such as soil-water and sediment-water partition coefficients or the fraction of the chemical associated with aerosols, must be specifically entered by the user in order to overrule the standard estimation routines.

Naturally, representing the environment in the form of a unit world or unit region with homogeneous boxes is a major simplification of reality. However, this extreme degree of simplification in this model concept is both a weakness and a strength at the same time. By disregarding spatial variation, the modelling effort can focus on intermedia distribution and understanding the ultimate fate of a chemical. The concentrations calculated with multimedia models should therefore be interpreted as "spatially-weighted averages" of the concentrations

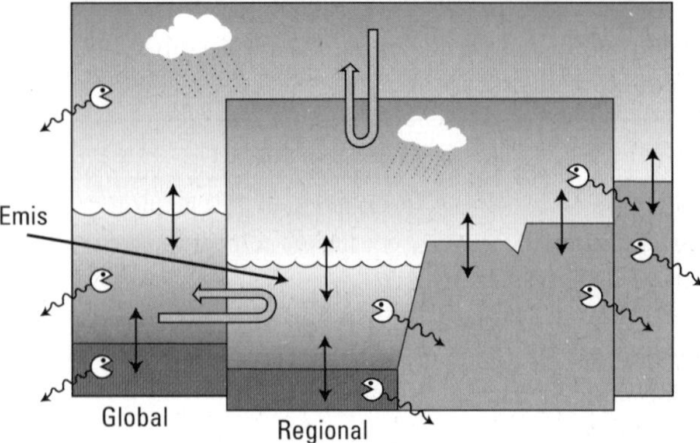

Figure 4.12. Concept of "nested" multimedia modelling. From [64,65].

that would be expected in real situations. However, the assumption of homogeneity brings with it a considerable risk that potentially more localized effects may be overlooked. The disadvantage of zero-dimensionality becomes evident with larger areas since, other than for air, it is difficult to identify any large-scale situations where the homogeneity of compartments would seem to be a realistic assumption. To overcome this problem the SimpleBox has introduced the concept of "nesting" [64,65]. In a nested model the input and output flows of a regional or smaller scale model are connected to a continental scale model which in turn, is connected to a global scale model. In this way, the specific environmental characteristics of the region can be taken into account when the overall fate of the chemical is assessed. Figure 4.12 illustrates this concept.

While spatial scale nesting was originally introduced as a tool for assessing the overall persistence of a chemical in the environment, the concept soon found wider application in regional exposure assessment in EUSES [64,65].

Testing the validity of multimedia models is difficult and, until recently, had not been seriously addressed [77]. If a common evaluation environment with agreed fixed environmental characteristics is used, validation of the outcome becomes almost paradoxical since this generic environment does not actually exist in reality. However, the regional generic characteristics can be modified at a later stage and region-specific information on environmental parameters, as well as information on specific discharge rates can be introduced in order to "validate" a specific model setting [58,78].

4.5.5 Application of multimedia models

Multimedia fate models of the Mackay type have been produced by different authors, most of them for their own scientific use. Many of these have been documented and made available for end users, e.g. HAZCHEM [24], SimpleBox [61,64,65], CemoS [62], CalTOX [63], ChemCAN [66], EQC [71], ChemRange [67], ELPOS [69], Globo-POP [70], CliMoChem [71], BETR North America [72], BETR World [73], IMPACT 2002 [74] and MSCE-POP [75]. The similarities between these models are more striking than the differences. When fed the same input, the models were shown to yield the same results [77]. The main differences lie in the number of compartments or sub-compartments included and how they are handled in terms of computer calculation. How the compartments are typically modelled is described in Box 4.5.

Calculation of exposure concentrations

Examples of how to perform Level I, II and III calculations for a range of different chemicals have been presented by Mackay and [5,6,57,58,59,66,67].

To illustrate the utility of Level III and IV type multimedia modelling, let us consider the use of three chemicals, 1,1,1-trichloroethane, dieldrin and cadmium, in a system resembling The Netherlands, as simulated with SimpleBox [64,65]. The system parameters are summarized in Table 4.5.

Let us assume that the background concentrations of these chemicals in air and water outside The Netherlands are equal to the quality standards or objectives set for environmental protection - this is equivalent to assuming successful environmental management practice in the rest of the world! After 10 years, with these background concentrations, domestic emissions of 1000 tonnes/y for each chemical start to occur: dieldrin to water, cadmium to air, and 1,1,1-trichloroethane to air, water and soil simultaneously (ratio 1:1:1). This situation continues for 40 years and then suddenly stops. What concentrations may be expected in the different environmental compartments, how are the chemicals distributed, and how long does it take to return to the original situation after the emissions stop?

In order to evaluate the change in concentrations of the three chemicals in the different environmental compartments some chemical-specific information is needed. This is summarized in Table 4.6. Intermedia partition coefficients for dieldrin and 1,1,1-trichloroethane can be estimated from their physicochemical properties; partition coefficients for cadmium, however, need to

Box 4.5. How to handle compartments in a multimedia model

Air

Air is a bulk compartment, consisting of a gas phase, an aerosol phase and a rainwater phase. The concentration of substances in air is influenced by air flow (wind), evaporation from water and soil, wet and dry deposition, and degradation.

Water

In the earlier Mackay models, the various physical states of chemicals in water (dissolved, sorbed to suspended matter and biota) were often modelled as distinct compartments. More recent models, such as SimpleBox/EUSES treat water as a bulk compartment, in which the phases (including colloidal material or the "third phase") are in true equilibrium. The presence of suspended matter and biota influences the fate of chemicals in a very similar way to aerosols and rainwater in the atmosphere. These phases bind the chemical, thus preventing it from taking part in mass-transfer and degradation processes in the water phase. Suspended matter acts as a physical carrier to the chemical in exchange across the sediment-water interface.

Suspended matter

Suspended matter refers to all abiotic colloidal or macromolecular materials (debris of organisms, humic material, dissolved organic matter, third phase, etc.) that are not truly dissolved. Treatment of suspended matter as a separate compartment has the advantage that the material balance for the suspended matter compartment, which may be important for the fate of chemicals that tend to partition into this phase, is explicitly considered. Factors influencing the amount of suspended matter are "import" and "export" to water. Suspended matter may also be produced in the system itself, by growth of small aquatic organisms (bacteria, algae). Sewage treatment plant effluent is another source of suspended matter. Finally, there is a continuous exchange of particles across the sediment-water interface through sedimentation and resuspension. The balance of these suspended matter mass flows determines the magnitude and direction of the particle exchange between sediment and water, and thus the mass flow of the chemical associated with the particles.

Biota

Biota refers to all living organisms in water, from bacteria to mammals. The biomass content of water is usually low in comparison to the mass of other forms of suspended matter. As a result, biota usually play an insignificant role with regard to the overall fate of chemicals.

Sediment

Sediment is usually treated as a bulk compartment, consisting of a water phase and a solid phase. Equilibrium between these two phases is assumed. If the sedimentation of particles from the water column is greater than the resuspension (net sedimentation), this top layer is continuously being refreshed.

Soil

Soil is the most stationary and, as a result, the most spatially inhomogeneous of all environmental compartments. There are many different soil types and uses. The fate of chemicals happens to be determined largely by soil properties that greatly vary (e.g. porosity, water content, organic matter content). Soil use is also a factor in determining whether it may be directly loaded with a chemical. One soil compartment is not sufficient to reflect the role of soil in the multimedia fate of chemicals. Therefore, different types of soil can be specified, e.g. natural soil, agricultural soil and soil used industrially. This differentiation of the soil compartment into subcompartments makes it possible to identify the effects of individual emissions to soil. Usually only the topsoil layer is considered. This layer is assumed to be homogeneous in the sense that the concentration of a chemical does not vary with depth. It is assumed that soil can be treated as bulk compartments, consisting of a gas phase, a water phase and a solid phase. The different soil phases can be assumed to be in equilibrium at all times.

be entered directly by the user. Similarly, the fraction associated with aerosols and the scavenging ratio of cadmium are entered manually since the "normal" estimation routines for these parameters do not apply to metals. Dieldrin has a very long reaction half-life in the environment; for cadmium, of course, no degradation is assumed.

The Level III mode of the SimpleBox program is

Table 4.5. Parameters used for steady-state calculations with SimpleBox.

Parameter	Value in SimpleBox	Parameter	Value in SimpleBox
Area of the system	$3.8x10^4$ km^2	organic carbon content in suspended matter	0.1
Area fraction of water	0.125	atmospheric mixing height	1000 m
Area fraction of natural soil	0.415	mixing depth of water[a]	3 m
Area fraction of agricultural soil	0.45	mixing depth of sediment[a]	0.03 m
Area fraction of industrial/urban soil	0.01	average annual precipitation	792 $mm.y^{-1}$
Mixing depth of natural soil[a]	0.05 m	wind speed	5 $m.s^{-1}$
Mixing depth of agricultural soil[a]	0.2 m	residence time air[b]	0.40 d
Mixing depth of industrial/urban soil[a]	0.05 m	residence time water[b]	54.5 d
Organic carbon content in soil	0.029	fraction of rain water infiltrating soil	0.4
Organic carbon content in sediment	0.029	fraction of rain water running off soil	0.5
Concentration suspended solids	15 $mg.L^{-1}$	Temperature	285 K (12°C)

[a] The mixing depth represents the thickness of the soil, water or sediment box.

[b] Residence time for air or water represents the time needed for air or water to flush through the air or water compartments, respectively.

then used to generate the concentrations and intermedia distribution at steady-state. The concentrations in and distribution over the environmental compartments at steady-state are summarized in Table 4.7.

The mass flows that support these steady-states are also shown in Figure 4.13. The model calculation emphasizes the high volatility of 1,1,1-trichloroethane. Approximately all emissions to soil and water go to air through diffusive transport. Of the total mass in the system, however, a high percentage still resides in the soil. Remarkably, the relatively high volatility of dieldrin causes more than half of the total load of the water compartment to be transported to air, from where it is exported out of the system. The high hydrophobicity and low biodegradation rates of the chemical produce relatively high concentrations in sediment and soil. Cadmium, of course, does not degrade at all. When emissions go to air the most important fate process is advection out of the air compartment. However, due to atmospheric deposition, some 10% of the total load of the atmosphere is transported to soil and water. Atmospheric deposition to soil leads to a build-up of cadmium in the soil, from where it is eventually leached to the ultimate sink: the deeper groundwater. It should be borne in mind that this build-up may be slow. If, as in the case of cadmium in soil, all mass flows are small, it may take an extremely long time before the steady-state is achieved. This can be demonstrated with Level IV calculations using the SimpleBox model. Figure 4.14 shows the change in concentrations in the different compartments according to the above emission scenario relative to the background concentrations which result when there are no domestic emissions. For cadmium, the compartments air, water and sediment are expected to respond relatively quickly, whereas a near linear increase in the concentration in soil is predicted over the 40-year exposure period. After reducing the emissions, the soil concentration of cadmium shows little response (Figure 4.13C). For dieldrin exposure for 40 years is almost long enough to reach a steady-state, even in the "slow" soil compartment; after reducing the emission to 10% of its original value, the concentrations decrease at the same rate (Figure 4.13B). For trichloroethane the situation is completely different. The steady-state situation is reached so quickly that plotting the concentrations against time on a 100-year scale would yield a block diagram. Therefore, the Level IV calculation was repeated over a time-scale of one year. The results as presented in Figure 4.13A show that concentrations in air, water and soil reach steady-state within one month. For sediment this takes a little longer, though probably not much longer than a year.

These results demonstrate the usefulness of Level III and Level IV multimedia box model calculations. Where steady-state calculations can give information on the concentrations and distribution in the environment at a constant emission scenario, the results of a Level IV

Table 4.6. Input parameters used in the multi-media model calculations for 1,1,1-trichloroethane, dieldrin and cadmium.

		1,1,1-Trichloroethane	Dieldrin	Cadmium
Background (air)	$g.m^{-3}$	10^{-8}	10^{-9}	10^{-9}
Background (water)	$g.L^{-1}$	10^{-8}	10^{-7}	10^{-7}
Emission (air)	$tonnes.y^{-1}$	333	-	1000
Emission (water)	$tonnes.y^{-1}$	333	1000	-
Emission (soil)	$tonnes.y^{-1}$	333	-	-
K_h (air-water)	-	1.1	$1.7x10^{-4}$	10^{-10a}
Frac (aerosol)	-	0.0	0.25	0.9
Scavenging ratio	-	0.96	$5.5x10^{4}$	10^{5}
K_p (susp.solids)	$L.kg^{-1}$	$3.1x10^{1}$	$6.3x10^{2}$	10^{4}
K_p (sediment)	$L.kg^{-1}$	$1.6x10^{1}$	$3.2x10^{2}$	10^{4}
K_p (soil)	$L.kg^{-1}$	$1.6x10^{1}$	$3.2x10^{2}$	10^{3}
Half-life (air)	d	200	200	-
Half-life (water)	d	1000	1000	-
Half-life (sediment)	d	1000	1000	-
Half-life (soil)	d	2000	100000	-

[a] Substitute for zero-value.

Table 4.7. Steady-state distribution of 1,1,1-trichloroethane, dieldrin and cadmium in The Netherlands, calculated with SimpleBox [64,65]. Numbers in parentheses represent mass, as a percentage of the total mass in the environment at steady-state.

	1,1,1-Trichloroethane		Dieldrin		Cadmium	
Air ($g.m^{-3}$)	$3.9x10^{-8}$	(19%)	$1.5x10^{-8}$	(0%)	$2.7x10^{-8}$	(0%)
Water ($g.L^{-1}$)	$4.5x10^{-8}$	(8%)	$5.1x10^{-6}$	(3%)	$2.1x10^{-7}$	(0%)
Suspended matter ($g.kg^{-1}$)	$1.2x10^{-6}$	(0%)	$2.8x10^{-3}$	(0%)	$2.1x10^{-3}$	(0%)
Sediment ($g.kg^{-1}$)	$7.5x10^{-7}$	(1%)	$2.1x10^{-3}$	(7%)	$2.1x10^{-3}$	(0.5%)
Soil ($g.kg^{-1}$)	$1.6x10^{-6}$	(73%)	$6.1x10^{-4}$	(90%)	$9.2x10^{-3}$	(99.5%)

calculation elucidate the time scale in which this situation may be reached. In addition, changes in the emission scenario as a result of evolving risk reduction strategies can be evaluated in this way.

Calculation of overall persistence in the environment and long-range transport potential

It is clear that the physical and chemical properties of substances greatly influence their concentrations and distributions in the environment. Not only does this have implications for the risks posed to humans and ecosystems, there are other ethical and scientific consequences to be considered [79]. Slow degradation and great mobility mean that substances disperse throughout the entire globe. This has been recognized internationally. Two international conventions: the UNEP Stockholm Convention [80] and the UN ECE POP protocol [81] now regulate substances on the basis of their persistence in the environment and long-range transport potential. Both of these are indirect or "derived" substance properties.

Persistence reflects the resistance of a substance to degradation. This is indicated by the dynamic response to changes in emissions, as shown in the previous paragraph. Alternatively, persistence can be quantified by the degradation half-life or reactive residence time during an emission episode [82,83]. As degradation half-lives in air, water and soil differ greatly, it needs to be decided which one to use, or how to combine the different single-medium half-lives. Calculation of "overall persistence in the environment" (P_{ov}) as the reciprocal of the overall degradation rate constant, k_{ov}, or the mass-weighted

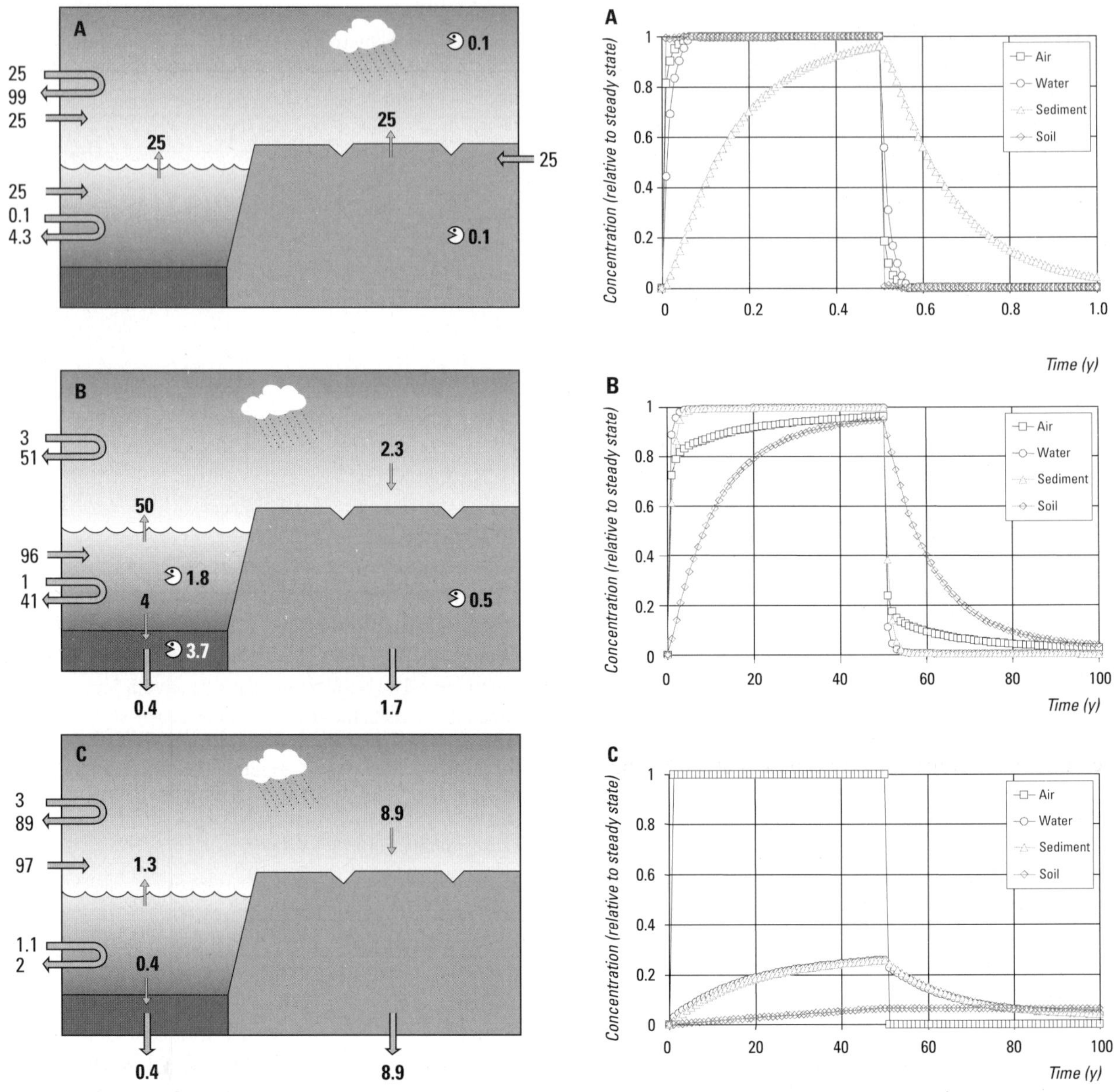

Figure 4.13. Steady-state mass flows of trichloroethane (A), dieldrin (B) and cadmium (C), as a percentage of the total throughput of the system.

Figure 4.14. Change in concentrations of trichloroethane (A), dieldrin (B) and cadmium (C) after a change in emission rates. Note the shorter time scale in graph A.

average reactive residence time in the environmental media, has been proposed for this purpose [82-84]:

$$P_{ov} = \frac{1}{k_{ov}} = \frac{\Sigma_i M_i}{\Sigma_i M_i \cdot k_i} \quad (4.33)$$

In this derivation of P_{ov}(d), k_i's are the first-order degradation rate constants in pure media (d^{-1}) and M_i's (kg) are the masses in the media at steady-state. According to this derivation, other substance properties than degradation half-lives (partition coefficients and mass-transfer velocities) play a role in determining the "derived property" P_{ov}. Applied to the calculation results of the previous paragraph, this would yield Pov values of 2.8 years, 20.8 years and ∞ for trichloroethane, dieldrin and cadmium, respectively.

The long-range transport potential (LRTP) reflects the tendency of a substance to be transported away from the location where it was emitted. There are different ways to capture this in a "derived property" [83,84]. One is to take the fraction of the total emission exported out of an open regional environment, as shown in the previous paragraph:

$$\mathrm{LRTP} = \frac{adv_{\mathrm{air}} + adv_{\mathrm{water}}}{\mathrm{E}} \qquad (4.34)$$

with adv_{air} and adv_{water} denoting the advective mass flows ($kg.d^{-1}$) by air and water, respectively and E ($kg.d^{-1}$) the sum of emissions. The LRTP values (dimensionless) for trichloroethane, dieldrin and cadmium would be 0.99, 0.92 and 0.91, respectively, based on example model used. Another method is to use the Lagrangian characteristic travel distance. The distance travelled (km) by a parcel in the period that the original mass is reduced exponentially to 37% (=1/e) of its original value is calculated as [83,84]

$$\mathrm{LRTP} = \frac{\mathrm{u}}{\mathrm{k}_{\mathrm{ov}}^{*}} \qquad (4.35)$$

in which u is the average velocity ($km.d^{-1}$) at which the parcel travels. Here, k_{ov}* considers non-reactive losses to ultimate sinks such as sediment burial, groundwater or deeper ocean layers as well as abiotic and biotic degradation processes.

What P_{ov} and LRTP have in common is that they cannot easily be determined by observation, but must be calculated from substance properties that can be measured (degradation rate constants, partition coefficients, mass-transfer velocities), using a multimedia environmental fate model. This has raised the concern that the choice of model could play a role in the calculation result, which would be undesirable if P_{ov} and LRTP are to be used as a property of the substance in a regulatory context. This issue has been thoroughly studied by an international group of modelling experts for the OECD [84]. The experts concluded that indeed the absolute values of P_{ov} and LRTP obtained from different models differ greatly, as a result of different modelling objectives and model parameterization. However, the rankings of substances obtained appeared to be relatively insensitive to the model choice: models tend to put chemicals in roughly the same order of P_{ov} and LRTP. If properly processed, output of any well-designed multimedia model can be used to derive P_{ov} and LRTP [85,86]. This was concluded from a comparison of the performance of existing models with respect to P_{ov} and LRTP calculation, which demonstrated that a simplified version of existing models could be constructed that differed as little from the existing models as the models differed among themselves. This consensus model is available from the OECD on their website [87].

4.6 FURTHER READING

For further reading on mathematical fate modelling, the reader is referred to specialized textbooks on the subject. The following titles are especially recommended: the early *Modelling the Fate of Chemicals in the Aquatic Environment*, edited by Dickson, Maki and Cairns [3], the two volumes of *Environmental Exposure from Chemicals*, edited by Neely and Blau [4], *Chemodynamics and Environmental Modelling* by Trapp and Matthies [5] and *Multimedia Environmental Models* by Mackay [6]. Overviews of available models for exposure assessment have been produced by the Organization for Economic Co-operation and Development [7] and the European Centre for Ecotoxicology of Chemicals [8].

REFERENCES

1. Lassiter RR. 1982. Testing models of the fate of chemicals in aquatic environments. In: Dickson KL, Maki AW, Cairns J Jr, eds., *Modelling the Fate of Chemicals in the Aquatic Environment*. Ann Arbor Science, Ann Arbor, MI, pp. 397-407.
2. Commission of the European Communities. 1994. Guidance document for the risk assessment of existing chemicals in the context of EC regulation 793/93. Commission of the European Communities, Directorate General of the Environment, Nuclear Safety and Civil Protection, Brussels, Belgium.
3. Dickson KL, Maki AW, Cairns J Jr. 1982. *Modelling the Fate of Chemicals in the Aquatic Environment*. Ann Arbor Science. Ann Arbor, MI.
4. Neely WB, Blau GE. 1985. *Environmental Exposure from Chemicals*, Vol. I and II. CRC Press, Boca Raton, FL.
5. Trapp S, Matthies M. 1998. *Chemodynamics and Environmental Modeling. An Introduction*. Springer, Heidelberg, Germany.
6. Mackay D. 2001. *Multimedia Environmental Models. Second Edition*. CRC Press LLC, Boca Raton, FL.
7. Organization for Economic Co-operation and Development. 1989. Compendium of environmental exposure assessment methods for chemicals. Environment Monographs 27. OECD, Paris, France.
8. European Centre for Ecotoxicology and Toxicology

of Chemicals. 1992. Estimating environmental concentrations of chemicals using fate and exposure models. Technical Report 50. ECETOC, Brussels, Belgium.

9. Gifford FA. 1961. Use of routine meteorological observations for estimating the atmospheric dispersion. *Nucl Safety* 2:47-57.
10. Turner DB. 1970. Workbook of atmospheric dispersions estimates. EPA Ref. AP-26 (NTIS PB 191-482). US Environmental Protection Research, Triangle Park, NC.
11. Green AE, Singhal RP, Venkateswar R. 1980. Analytical extensions of the Gaussian plume model. *J Air Pollut Control Assoc* 30:773-776.
12. Seinfeld H. 1986. *Atmospheric Chemistry and Physics of Air Pollution*. Wiley, New York, NY.
13. Werkgroep Verspreiding Luchtverontreiniging. 1984. Parameters in het lange-termijn model verspreiding luchtverontreiniging; Nieuwe aanbevelingen. Toegepast Natuurwetenschappelijk Onderzoek (TNO), Delft, The Netherlands [in Dutch].
14. Van Jaarsveld JA. 1990. An operational atmospheric transport model for priority substances; Specification and instructions for use. RIVM Report 222501002. National Institute for Public Health and the Environment (RIVM), Bilthoven, The Netherlands.
15. Van Den Hout KD, Van Dop H. 1985. Interregional modelling. In: Zwerver S, Van Ham J, eds., *Interregional Air Pollution Modelling*. Plenum Press, New York, NY.
16. Szepesi DJ. 1989. *Compendium of Regulatory Air Quality Simulation Models*. Akademiai Kiado, Budapest, Hungary.
17. Toet C, De Leeuw FAAM. 1992. Risk assessment system for new chemical substances: Implementation of atmospheric transport of organic compounds. RIVM Report 679102008. National Institute for Public Health and the Environment (RIVM), Bilthoven, The Netherlands.
18. European Commission. 2003. Technical guidance document in support of Commission Directive 93/67/EEC on risk assessment for new notified substances, Commission Regulation (EC) no. 1488/94 on Risk Assessment for existing substances and Directive 98/8/EC of the European Parliament and of the Council concerning the placing of biocidal products on the market. European Chemicals Bureau, Joint Research Centre, Ispra (VA), Italy.
19. US Environmental Protection Agency. 1987. PDM3 Documentation. Exposure Evaluation Division, Office of Toxic Substances, Washington, DC.
20. Rapaport RA. 1988. Prediction of consumer product chemical concentrations as a function of publicly owned treatment type and riverine dilution. *Environ Toxicol Chem* 7:107-115.
21. De Nijs T, De Greef J. 1992. Ecotoxicological risk evaluation of the cationic fabric softener DTDMAC II. Exposure modelling. *Chemosphere* 24:611-627.
22. Versteeg DJ, Feijtel TCJ, Cowan CE, Ward TE, Rapaport RA. 1992. An environmental risk assessment for DTDMAC in The Netherlands. *Chemosphere* 24:641-662.
23. European Commission. 2004. European Union System for the Evaluation of Substances 2.0 (EUSES 2.0). Prepared for the European Chemicals Bureau by the National Institute of Public Health and the Environment (RIVM), Bilthoven, The Netherlands (RIVM Report 601900005). Available via the European Chemicals Bureau, Ispra, Italy. http://ecb.jrc.it.
24. European Centre for Ecotoxicology and Toxicology of Chemicals. 1994. HAZCHEM, A mathematical model for use in risk assessment of substances. Special report No.8. ECETOC, Brussels, Belgium.
25. Rapaport RJ, Caprara RJ. 1988. PG ROUT: A national surface water quality model. Presented at the 61st Annual Conference of the Water Pollution Control Federation, 2-6 October 1988, Dallas, TX.
26. Caprara RJ, Rapaport RA. 1991. PG ROUT: A steady-state national water quality model. Proc. Nat. Conf. on Integrated Water Information Management. 4-9 August 1991, Atlantic City, NY, pp. 134-141.
27. Feijtel TCJ, Boeije G, Matthies M, Young A, Morris G, Gandolfi C, Hansen B, Fox K, Holt M, Koch V, Schröder R, Cassani G, Schowanek D, Rosenblom J, Niessen H. 1997. Development of a Geography-referenced Regional Exposure Assessment Tool for European Rivers – GREAT-ER. *Chemosphere* 34:2351-2374.
28. Matthies M, Berlekamp J, Koormann F, Wagner JO. 2001. Georeferenced regional simulation and aquatic exposure assessment. *Wat Sci Technol* 43(7):231-238.
29. European Centre for Ecotoxicology and Toxicology of Chemicals. 1994. Environmental exposure assessment Technical Report. 61. ECETOC, Brussels, Belgium.
30. Brock Neely W. 1982. The definition and use of mixing zone. *Environ Sci Technol* 16:519A-5121A.
31. Csanady GT 1973. *Turbulent Diffusion in the Environment. Geophysics and Astrophysics Monographs* Vol. 3. D. Reidel Publ. Co., Dordrecht, The Netherlands.
32. Fischer HB, Imberger J, List EJ, Koh RCY, Brooks RH 1979. *Mixing in inland and coastal waters*. Academic Press, New York, NY.
33. Noppeney RM. 1988. Gevoeligheidsonderzoek alarmmodel Rijn: De invloedslengte van samenvloeiingen bij dispersie. Mededeling 20. Delft Technical University

Faculty of Civil Engineering, Delft, The Netherlands [in Dutch].

34. Doneker RL, Jirka GH. 1988. CORMIX1. An expert system for mixing zone analysis of toxic and conventional single port aquatic discharges. DeFrees Hydraulics Laboratory, Dep. Env. Eng., Cornell University, Ithaca, New York, NY.

35. De Greef J, Van De Meent D. 1989. Beoordelingssysteem nieuwe stoffen: Transportroutines, een receptuur voor het schatten van de snelheid van het transport in oppervlaktewater. RIVM Report 958701001. National Institute for Public Health and the Environment (RIVM), Bilthoven, The Netherlands [in Dutch].

36. Jørgensen SE. 1984. *Modelling the Fate and Effect of Toxic Substances in the Environment. Developments in Environmental Modelling*, 6. Elsevier Sci. Publ., Amsterdam, The Netherlands.

37. Jørgensen SE, Gromiec MJ. 1989. *Mathematical Submodels in Water Quality Systems. Developments in Environmental Modelling, 14*. Elsevier Sci. Publ., Amsterdam, The Netherlands.

38. Burns LA, Cline DM, Lassiter RR. 1982. Exposure analysis modelling system (EXAMS): User manual and systems documentation. US Environmental Protection Agency, Athens, GA.

39. Burns LA, Cline DM. 1985. Exposure analysis modelling system, Reference manual for EXAMS II. EPA-600/3-85-038. US Environmental Protection Agency, Athens, GA.

40. Ambrose RB, Wool TA, Connolly JP, Schanz RW. 1988. WASP 4, A hydrodynamic and Water Quality Model. Model Theory, User's Manual and Programmer's Guide. EPA/600/3-87/039. U.S. Environmental Protection Agency, Athens, GA.

41. Delft Hydraulics Laboratory. 1990. DELWAQ Version 3.0 User Manual. Delft Hydraulics Laboratory, Delft, The Netherlands.

42. Beck A, Scheringer M, Hungerbühler K. 2000. Fate Modelling within LCA: The Case of Textile Chemicals, *Intern J LCA* 5:335-344.

43. Struijs J, Stoltenkamp J, Van De Meent D. 1991. A spreadsheet-based model to predict the fate of xenobiotics in a municipal wastewater treatment plant. *Wat Res* 25:891-900.

44. Struijs J, Van Den Berg T. 1992. Degradation rates in the environment: Extrapolation of standardized tests. RIVM Report 679102012. National Institute for Public Health and the Environment (RIVM), Bilthoven, The Netherlands.

45. International Institute for Applied System Analysis. 1991. Chemical time bombs: Definitions, concepts and examples; Basis document 1. In: Stigliani WM, ed., *ER-91-16*. IIASA, Laxenburg, Austria.

46. Stigliani WM, Doelman P, Salomons W, Schulin R, Schmidt GRB, Van Der Zee S. 1991. Chemical time bombs: predicting the unpredictable. *Environment* 33:4-9, 26-30.

47. Carsel RF, Smith CN, Mulkey LA, Dean JD, Jowise P. 1984. Users manual for the pesticide root zone model (PRZM). EPA-600/3-84-109. US Environmental Protection Agency, Athens, GA.

48. Carsel RF, Nixon WB, Ballantine LB. 1986. Comparison of pesticide root zone model predictions with observed concentrations for the tobacco pesticide Metalaxyl in unsaturated zone soils. *Environ Toxicol Chem* 5:345-353.

49. Smith CN, Parrish RS, Brown DS. 1990. Conducting field studies for testing pesticides leaching models. *Intern J Environ Anal Chem* 39:3-21.

50. Klein M. 1991. Application and validation of pesticide leaching models. *Pestic Sci* 31:389-398.

51. Bonazountas M, Wagner J. 1984. SESOIL: A seasonal soil compartment model. Office of Toxic Substances, US Environmental Protection Agency, Washington, DC.

52. Hettrick DM, Travis CC, Leonard SK, Kinerson RS. 1988. Qualitative validation of pollutant transport components of an unsaturated soil zone model. ORNL-TM-10672. Oak Ridge National Laboratory, Oak Ridge, TN.

53. Matthies M, Behrendt H. 1991. Pesticide transport modelling in soil for risk assessment of groundwater contamination. *Toxicol Environ Chem* 31-32:357-365.

54. Van Der Linden AMA, Boesten JJTI. 1989. Berekening van de mate van uitspoeling en accumulatie van bestrijdingsmiddelen als functie van hun sorptiecoëfficiënt en omzettingsnelheid in bouwvoormateriaal. RIVM Report 72800003. National Institute for Public Health and the Environment (RIVM), Bilthoven, The Netherlands [in Dutch].

55. Swartjes FA, Van Der Linden AMA, Van Den Berg R. 1993. Dutch risk assessment system for new chemicals: Soil-groundwater module. RIVM Report 679102015. National Institute for Public Health and the Environment (RIVM), Bilthoven, The Netherlands.

56. Mackay D. 1979. Finding fugacity feasible. *Environ Sci Technol* 13:1218-1223.

57. Mackay D, Paterson S. 1981. Calculating fugacity. *Environ Sci Technol* 15:1006-1014.

58. Mackay D, Paterson S, Cheung B, Brock Neely W. 1985. Evaluating the environmental behaviour of chemicals with a Level III fugacity model. *Chemosphere* 14:335-374.

59. Mackay D, Paterson S, Shiu WY. 1992. Generic

models for evaluating the regional fate of chemicals. *Chemosphere* 24:695-717.

60. Frische R, Klöpffer W, Rippen G, Günther K-L. 1984. The environmental segment approach for estimating potential environmental concentrations. I. The model. *Ecotoxicol Environ Saf* 8:352-362.
61. Van De Meent D. 1993. SIMPLEBOX, a generic multimedia fate evaluation Model. RIVM Report 672720001. National Institute for Public Health and the Environment, Bilthoven, The Netherlands.
62. Scheil S, Baumgarten G, Reiter B, Schwartz S, Wagner JO, Matthies M, Trapp S. 1994. CEMO-S: Eine objectorientierte Software zur Expositionsmodellierung. In: Totsche K, Matthies M, ed., *Eco-Informa* '94, Vol. 7. Wien, Austria, pp. 391-404 [in German].
63. McKone TE, Enoch KG. 2002. CalTOX™, A Multimedia Total Exposure Model. Spreadsheet User's Guide Version 4.0. Report LBNL-47399. Lawrence Berkeley National Laboratory, Berkeley, CA. http://eetd.lbl.gov/ied/era/.
64. Brandes LJ, den Hollander H, Van de Meent D. 1996. SimpleBox 2.0: a nested multimedia fate model for evaluating the environmental fate of chemicals. RIVM Report 719101029. National Institute for Public Health and the Environment (RIVM), Bilthoven, The Netherlands. http://www/rivm.nl.
65. Den Hollander H A, Van Eijkeren JCH, Van de Meent D. 2004. SimpleBox 3.0: Multimedia Mass Balance Model for Evaluating the Fate of Chemical in the Environment. RIVM Report 601200003/2004. National Institute for Public Health and the Environment (RIVM). Bilthoven, The Netherlands. http://www/rivm.nl, http://ecb.jrc.it.
66. Webster E, Mackay D, Di Guardo A, Kane D, Woodfine D. 2004. Regional differences in chemical fate model outcome. *Chemosphere* 55:1361-1376.
67. Mackay D, Di Guardo A, Paterson S, Cowan CE. 1996. Evaluating the environmental fate of a variety of types of chemicals using the EQC model. *Environ Toxicol Chem* 15:1627-1637.
68. Scheringer M. 1996. Persistence and spatial range as endpoints of an exposure-based assessment of organic chemicals. *Environ Sci Technol* 30:1652-1659
69. Beyer A, Matthies M. 2002. Criteria for Atmospheric Long-Range Transport Potential and Persistence of Pesticides and Industrial Chemicals. Umweltbundesamt, ed.. Erich-Schmidt-Verlag, Berlin, Germany.
70. Wania F, Mackay D. 2000. The Global Distribution Model. A Non-Steady-State Multi-Compartmental Mass Balance Model of the Fate of Persistent Organic Pollutants in the Global Environment. Technical Report and Computer Program on CD-ROM. 64.
71. Wegmann F, Möller M, Scheringer M, Hungerbühler K. 2004. Influence of Vegetation on the Environmental Partitioning of DDT in Two Global Multimedia Models. *Environ Sci Technol* 38:1505-1512.
72. MacLeod M, Woodfine DG, Mackay D, McKone TE, Bennett DH, Maddalena R. 2001. BETRNorth America: A regionally segmented multimedia contaminant fate model for North America. *Environ Sci Pollut Res* 8:156-163.
73. Toose L, Woodfine DG, MacLeod M, Mackay D, Gouin J. 2004. BETR-World: a geographically explicit model of chemical fate: application to transport of a-HCH to the Arctic. *Environ Pollut* 128:223–240.
74. Pennington DW, Margni M, Ammann C, Jolliet O. 2005. Multimedia fate and human intake modeling: spatial versus nonspatial insights for chemical emissions in Western Europe. *Environ Sci Technol* 39:1119 -1128.
75. Gusev A, Mantseva E, Shatalov V, Strukov B. 2005. Regional Multicompartment Model MSCE-POP. EMEP/MSC-E Technical Report 5/2005. Meteorological Synthesizing Centre - East, Moscow, Russia. www.msceast.org.
76. Mackay D, Diamond M. 1989. Application of the QWASI (quantitative water air sediment interaction) fugacity model to the dynamics of organic and inorganic chemicals in lakes. *Chemosphere* 18:1343-1365.
77. Cowan CE, Mackay D, Feijtel TCJ, Van De Meent D, Di Guardo A, Davies J, Mackay N. 1995. The multimedia fate model: A vital tool for predicting the fate of chemicals. SETAC Press, Pensacola, FL.
78. Berding V, Matthies M. 2002. European scenarios for EUSES regional distribution model. *Environ Sci Pollut* Res 9(3):193-198.
79. Scheringer M. 2002. *Persistence and spatial range of environmental chemicals.* Wiley-VCH Verlag, Weinheim, Germany.
80. UNEP. 2001. Stockholm Convention on Persistent Organic Pollutants; Geneva, Switzerland.
81. United Nations Economic Commission for Europe. 1979. Convention on Long-range Transboundary Air Pollution and its 1998 Protocols on Persistent Organic Pollutants and Heavy Metals. ECE/EB.AIR/66, ISBN 92-1-116724-8. UNECE, Geneva, Switzerland.
82. Webster E, Mackay D, Wania F. 1998. Evaluating environmental persistence. *Environ Toxicol Chem* 17:2148-2158.
83. Van de Meent D, McKone TE, Parkerton T, Matthies M, Scheringer M, Wania F, Purdy R, Bennett D. 2000. Persistence and transport potentials of chemicals in a multi-media environment. In: Klečka G et al, eds. *Persistence and long-range transport of chemicals in the environment,* pp. 169-204. SETAC Press, Pensacola, FL.

84. Organization for Economic Co-operation and Development. 2004. Guidance document on the use of multimedia models for estimating overall environmental persistence and long-range transport. OECD Environment, Health and Safety Publications, Series on Testing and Assessment 45. OECD, Paris, France.
85. Fenner K, Scheringer M, MacLeod MJ, Matthies M, McKone TE, Stroebe M, Beyer A, Bonnell M, Le Gall AC, Klasmeier J, Mackay D, Van de Meent D, Pennington D, Scharenberg B, Suzuki N, Wania F. 2005. Comparing estimates of persistence and long-range transport potential among multimedia models. *Environ Sci Technol* 39:1932-1942.
86. Klasmeier J, Matthies M, MacLeod M, Fenner K, Scheringer M, Stroebe M, LeGall AC, McKone T, Van de Meent D, Wania F. 2006. Application of multimedia models for screening assessment of long-range transport potential and overall persistence. *Environ Sci Technol* 40:53-60.
87. Organization for Economic Co-operation and Development. 2007. (Q)SAR Application Toolbox. OECD, Paris, France. www.oecd.org.

5. HUMAN EXPOSURE ASSESSEMENT

J.G.M. Van Engelen, P. J. Hakkinen, C. Money, M.G.J. Rikken and T.G. Vermeire

5.1 GENERAL INTRODUCTION

Humans may be exposed to a variety of substances from multiple exposure routes. In Chapter 5 we will distinguish between exposure through the environment (Section 5.2), exposure from use of consumer products (Section 5.3), and exposure at the workplace (*occupational exposure*; Section 5.4). In this chapter information is provided on how to perform an exposure assessment for each of these human populations. This information pertains to the general principles, the data needed and how to perform the actual quantitative assessment, based on either measured or modelled data.

Detailed information on exposure assessment is provided in the EU technical guidance document (TGD) and will be given in the technical guidance document for the preparation of chemical safety reports under REACH [1,2]. This guidance is published on the website of the European Chemicals Bureau (http://ecb.jrc.it/).

Exposure assessment is an essential element of risk assessment. The components of risk assessment for human health are: (1) hazard identification, (2) assessment of the external and internal exposure, (3) effects assessment or dose-response assessment, and (4) risk characterization, i.e. comparison of estimated exposure and appropriate no effect levels for man (Chapter 1). The subject of Chapter 4 was environmental exposure assessment. Some of the predicted environmental concentrations (PECs) discussed in Chapter 4 will be used as input for the assessment of human exposure. Chapters 6 and 7 will provide the reader with sufficient background information to perform the latter two steps of a risk assessment.

Exposure of man occurs, first of all, externally. External exposure can be defined as the concentration of a substance reaching a receptor, i.e. the epithelium of the gastrointestinal tract in the case of ingestion, the pulmonary epithelium in the case of inhalation, and the skin with dermal contact. Internal exposure or uptake can be defined as the quantity of a substance which has been absorbed, i.e. which has passed the receptors and entered the systemic circulation. Bioavailability, then, is defined as the fraction of the external dose which has been absorbed.

One important subject when talking about exposure is the term “exposure scenario”. It is important to highlight that there are currently two definitions of the term: exposure scenario. The first definition is provided by OECD and IPCS [3] and the second one is used in REACH [2]. The definitions are given in Box 5.1. These definitions are fundamentally different. According to OECD/IPCS [3] an exposure scenario is a combination of facts, assumptions, and inferences that define a discrete situation where potential exposures may occur. These may include the source, the exposed population, the time-frame of exposure, the micro-environment, and the activities. According to this definition exposure scenarios are often created to aid exposure assessors in estimating exposure. The definition of an exposure scenario (ES) under REACH (see also Chapters 2 and 12) is different from the definition of the IPCS as it encompasses an integral approach to control risks, i.e. risk reduction is explicitly included.

Under REACH, an exposure scenario (ES) describes a control strategy for substances, giving

Box 5.1. Definitions of exposure scenario according to the OECD/IPCS and REACH [2,3]

OECD/IPCS: An exposure scenario is a set of conditions or assumptions about sources, exposure pathways, amount or concentrations of agent(s) involved, and exposed organism, system or (sub)population (i.e. numbers, characteristics, habits) and used to aid in the evaluation and quantification of exposure(s) in a given situation.

REACH: An exposure scenario means the set of conditions, including operational conditions and risk management measures that describe how the substance is manufactured or used during its life cycle and how the manufacturer or importer controls, or recommends downstream users to control exposures of humans and the environment. These exposure scenarios may cover one specific process or use or several processes or uses as appropriate

C.J. van Leeuwen and T.G. Vermeire (eds.), Risk Assessment of Chemicals: An Introduction, 195–226.

realistic operational conditions for the manufacture of a substance or identified use(s) of a substance, a group of substances or a preparation. The REACH exposure scenario prescribes appropriate risk management measures (RMMs) that serve to effectively manage health, environmental and safety risks from the chemical during its entire life cycle. Further detailed information is provided in Chapter 2.

Based on the TGD, this chapter describes three different subpopulations: humans that may be exposed to substances via the environment, through use of consumer products (consumer exposure) or via substances in the workplace (occupational exposure).

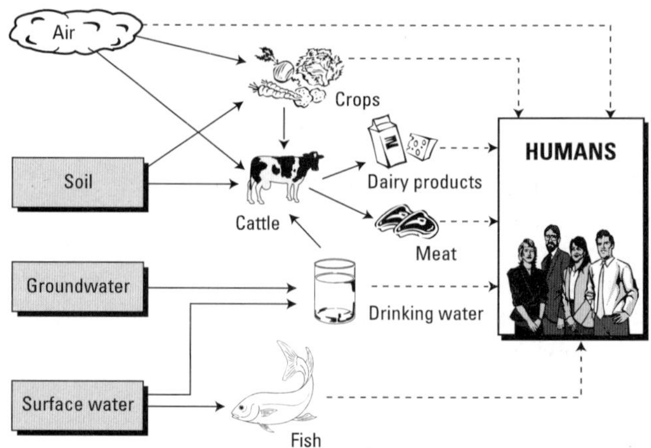

Figure 5.1. Diagram of the main exposure routes generally considered in human exposure assessment through the environment.

5.2 HUMAN EXPOSURE THROUGH THE ENVIRONMENT

5.2.1 Introduction

The exposure of human beings is an important part in the risk assessment of chemicals. Man can be exposed through the environment directly via inhalation, soil ingestion and dermal contact, and indirectly via food products and drinking water (Figure 5.1). Monitoring data of known quality that are representative for the exposed population are preferred over estimated exposure values calculated using models. Monitoring data can be applied to assess the indirect exposure of consumers via residual amounts of pesticides on treated foods (including meat, fish, dairy products, fruit and vegetables) or water. Monitoring data (air, water, soil) are often available to assess direct human exposure to metals. However, there is a need for sufficiently accurate models because there is little field data available on exposure levels and experimental data on bioconcentration. For *a priori hazard assessments* (i.e. new chemicals introduced on the market) a modelling approach is the only solution. Models can be used to estimate human exposure to environmental concentrations that are either measured or estimated with single or multimedia models. Some examples of available models are EUSES [4], CalTOX [5], ACC-HUMAN [7], E-FAST [8] and UMS [8]. It should be stressed here that while reliable field data are always preferable, the quality of such data and relevance to the population to be protected should be carefully considered. Assessment of human exposure through the environment can be divided into three steps:

- Assessing concentrations in intake media (air, soil, food, drinking water).
- Assessing total daily intake of these media.
- Combining concentrations in the media with total daily intake (and, if necessary, using a factor for bioavailability through the route of uptake concerned).

There are a large number of different models available to estimate concentrations in food products. Most often the concentration in food is estimated with simple partitioning models that are usually highly dependent on the octanol-water partition coefficient (K_{ow}). Although the theoretical basis for these models is sometimes limited, they provide practical tools for risk assessment, especially as they are often applicable to a wide range of substance properties. These models are used to estimate bioconcentration (BCF), biotransfer (BTF) and bioaccumulation (BAF) factors, defined as fixed concentration ratios. The use of fixed ratios implies that these models assume a steady-state. Hence, the period of exposure must be long enough, and the exposure level constant enough, to reach a steady-state.

In this section, methods for a general exposure assessment will be presented. Furthermore, some of the problems and limitations of models, and the importance of exposure scenarios will be discussed. The section ends with an example that illustrates how exposure methods can be integrated in the overall assessment.

5.2.2 Choice of exposure scenario

Since human behaviour shows an appreciable degree of variety, exposure will also vary greatly over the population. Every exposure assessment will inherently be extremely limited in its predictive ability for an entire population. As the choice of exposure scenario will have a major influence on the results of an exposure assessment, this choice should be steered by the assessment goal (i.e. the part of the population which is to be protected), by the emission and distribution patterns of the substance (locally around a plant, or diffusely over a larger entire region), and, of course, by policy objectives.

Indirect exposure assessments can be performed using model estimates or measured data. In a first step models can be used which often are based on generic scenarios and conservative assumptions. When available, a more accurate estimate of the indirect exposure can be developed in a second step using representative measured data of known quality. Reliable and relevant measured data are always preferable given the large uncertainties in the (Q)SARs (see Chapters 8-11). In this way uncertainty can be reduced for critical exposure routes.

The estimated concentration in each intake medium and the intake or consumption rates used depend on how conservative the models and assumptions used are. The result can vary from average to worst case. The target for indirect exposure of humans can be set at the exposure level of an average individual in a region. This implies that regional concentrations for air, water and soil can be used as input concentrations and average diets are assessed for the region under consideration. This regional approach accounts for the fact that people do not consume their total food basket from the immediate vicinity of a point source. In a more worst case approach, the subject receives his total consumption from the contaminated area for each food product and lives near the point source. This exposure scenario is less worst case than it might appear at first glance because, generally, only one or two of all possible routes dominate the total exposure estimation.

Besides an individual approach, groups or locations at risk can also be defined, for example, people living near a sewage treatment plant who consume a lot of fish products. However, these "groups at risk" may turn out to be different for substances with different distribution routes, which leads to the risk of ending up with a large number of such groups and uncertainty about the relevance and completeness of the groups selected. The ideal solution would be a method which can predict the percentage of the population exceeding a certain intake criterion, e.g. the total daily intake (TDI) or acceptable daily intake (ADI). This, in fact, means that an uncertainty analysis for the exposure assessment should be performed which implies that statistical information on consumption habits and concentrations in the diet are needed. Slob [9] describes a statistical exposure model and uses it to achieve a distribution of long-term intakes of chemicals in food in the population. The output of this model describes the long-term intake of and the variation between individuals as a function of age.

The type of emission or distribution model that provides the input for the indirect exposure assessment will, of course, determine the relevance of this assessment. Point source releases have a major impact on environmental concentrations on a local scale and also contribute to the concentrations on a larger regional scale. Local emission and distribution estimates will yield an exposure assessment for a specific worst case exposure location over a shorter period of time. Regional distribution models, like the multimedia models, will produce highly averaged concentrations over an entire region. In this case, an exposure assessment can be seen as an indication of the potential average exposure of the inhabitants of the region due to long-term continuous, diffuse emissions. Regional concentrations can provide a background concentration which can be incorporated in the concentration of the local assessment. This implies that first a regional and then a local calculation must be done, in sequence. It is not always appropriate to use regional data as background for the local situation. In the example that there is only one local source of a substance, this source is also responsible for the background concentration in the region, so this must not be counted twice at the local scale.

5.2.3 Exposure through food products

Food crops

Plants or plant products like vegetables, fruit and grains form the greater part of the food consumed by humans as well as herbivores that are part of the food chain for humans. The contamination of plants will, therefore, greatly affect the total daily intake of a substance. When trying to predict concentrations in plant tissues a number of important conceptual problems need to be considered:

- There are hundreds of different plant species in the group of food crops. Plants are extremely inhomogeneous with respect to physiology, rooting depth, leaf area, growth period or lipid (wax) content, for example. The considerable variation between plant species, and even within different varieties of the same species, can account for major differences in uptake.
- The uptake and distribution in plants is affected by environmental conditions, including temperature, water content, and organic and mineral matter in the soil.
- Many different plant tissues are consumed which

cannot be compared with each other; root crops (e.g. carrot), leaf crops (e.g. lettuce), tuberous crops (e.g. potato) and fruits (e.g. apple).
- A large proportion of the crops are produced outside the country concerned and imported. Many crops are produced in greenhouses in a controlled environment and with a different level of exposure to chemicals than crops from the field. A regional exposure assessment is more appropriate to account for the fact that not all the total food basket is consumed from the immediate vicinity of a point source (see also Section 5.2.2).

In view of these conceptual problems, it is clear that models can only give a very rough approximation of concentrations in food crops.

Plants can be exposed to chemicals via contaminated soil and groundwater, irrigation water, dry and wet deposition from the air, uptake from surrounding gas or vapour, direct resuspension contaminated soil particles on leaves caused by rain splash, erosion or direct application, as in the use of pesticides, for example. Uptake from soil is, generally, a passive process that is determined by the transpiration stream of the plant, for the purpose of accumulation in leaves, or by physical sorption in the case of roots. Briggs et al. [10] showed that the uptake of organic chemicals depends on the equilibrium between the concentration in the aqueous phase inside the root and in the surrounding solution, and that sorption takes place on hydrophobic root solids. The modelling approach of Trapp and Matthies [11] starts from the same idea and takes into account the uptake into the whole plant from soil, pore water and air and the elimination via growth dilution. It has been shown that the uptake of many compounds into plant roots from the soil solution is inversely proportional to water solubility and directly proportional to the hydrophobicity. Transfer to the shoots is more efficient for chemicals of intermediate solubility and intermediate hydrophobicity ($0 < \log K_{ow} < 3.5$), which results in a high transpiration-stream concentration factor (TSCF). The gaseous exchange between leaves and air can be described by a leaf-air partition coefficient, as described by Trapp and Matthies [11]. Chemicals that enter or exit plants through the stomata are most likely to be chemicals with a high vapour pressure. Chemicals with low vapour pressure and water solubility tend to sorb strongly to aerosols and soil particles (Chapter 3, Section 3.2.2). These contaminants may be deposited on above ground plant parts by soil particle resuspension (rain splash), providing a route of exposure to humans. The mass loading for the direct resuspension of soil particles on leaves can be 0.2 to > 20% of soil per dry weight plant [12,13]. For the indirect human exposure it is important to account for the part that might not be washed off. As a rough estimate, a fraction of 1% of soil per dry weight plant accounts for soil not washed off [14,15]. Exposure patterns, bioavailability and accumulation processes in terrestrial ecosystems are highly complicated. A few plant-related aspects are discussed in more detail in Chapter 3, Section 3.3.5, such as root and foliar uptake, factors influencing bioaccumulation, and plant bioaccumulation models.

Fish

Fish residing in contaminated surface water are able to take up appreciable amounts of substances through the gills or through the intake of food. The processes of bioconcentration and bioaccumulation were extensively discussed in Chapter 3. The predictive models for these processes, as described in Chapter 9, can be used to estimate concentrations in fish used for consumption on the basis of concentrations in the surface water. It should be noted, however, that these relationships are normally only valid within a certain range of the physicochemical properties and, moreover, do not apply to surfactants, ionizing substances, dissociating substances, inorganics, or in most cases to chemicals that are metabolized. It is generally agreed that there is a linear relationship of bioconcentration for organic chemicals with a log K_{ow} smaller than approximately 6 and which are not transformed [16-18]. The linear model of bioconcentration is inaccurate for chemicals with a log K_{ow} exceeding approximately 6 [17,19]. For these chemicals the bioconcentration data tend to decrease with increasing log K_{ow}. In the meantime several authors have developed more complex mechanistic models that integrate bioconcentration, biomagnification, growth and elimination [20-25]. Overall, the results of estimation methods should be carefully evaluated, especially as most relationships do not account for possible metabolism. Bioavailability and bioaccumulation processes are discussed in more detail in Chapter 3, Section 3.3.

Drinking water

Drinking water is usually sourced from surface water or groundwater. Groundwater can be contaminated through leaching from a polluted soil surface, surface water can be contaminated through direct emission or indirect emission, for instance, via a sewage treatment plant. Humans can be exposed to contaminants in drinking water via direct consumption, inhalation of vapours when showering or by dermal contact via bathing water or showering.

Hrubec and Toet [26] carried out a preliminary study to evaluate the predictability of the fate of organic chemicals during drinkingwater treatment. The contamination of drinking water prepared from surface water largely depends on the efficiency of the drinking water treatment process. The results of their evaluation were used in the TGD [1], where it is recommended that the removal of the dissolved fraction of organic compounds from the surface water be estimated by means of purification factors. Purification factors are based on measured removal percentages of about ten organic compounds, mainly pesticides, at surface water treatment sites in The Netherlands [26]. These data reflect a worst case situation, because the lowest available removal percentage was chosen for each compound and purification step. Additionally, the accuracy of the predicted removal efficiencies in the different purification steps according to physicochemical properties is rather low. This is mainly due to uncertainties in the removal predictability of the most effective treatment processes, such as activated carbon filtration. Therefore, purification factors are estimated quite conservatively [27]. The degree of pollution of groundwater sources largely depends on the removal of organic chemicals from the soil. The effect of the treatment processes used for the purification of groundwater, which are generally not intended for the removal of organic chemicals, can be ignored. After treatment at a drinking water plant, the drinking water can become contaminated by the permeation of contaminated water through synthetic drinking-water pipes. This exposure route has been investigated on a limited scale and seems only to be important for a limited number of compounds (e.g. monocyclic aromatic hydrocarbons, aliphatic chlorinated hydrocarbons) and also only a small proportion of the total water mains is made of polyethylene or PVC. The model concept for the permeation of drinking water pipes is extensively described by Vonk [28,29] and van der Heijden [30].

Meat and milk

Meat and milk are other important food products for humans. Lipophilic substances especially, are known to accumulate in meat, and are subsequently transferred to milk. Cattle can be exposed to substances in grass or other feed, via adhering soil, drinking water, and through the inhalation of air. In assessing concentrations in meat and milk there is the advantage that only a few animal species have to be considered (usually cows or pigs) with a limited diet (usually only grass with adhering soil is considered for cows). Biotransfer factors (BTF), which are defined as the steady-state concentration in beef divided by the daily intake of the chemical, are typically used to estimate these concentrations. Travis and Arms [31] carried out a linear regression analysis of the log BTF for meat and milk and log K_{ow}. The theoretical relevance of these relationships is limited, but they are of great practical relevance in risk assessment, since 28 (milk) and 36 (beef) organic chemicals with a wide range of log K_{ow} values (from 1.34 (beef) and 2.81 (milk) to 6.9) were used. Kenaga [32] found other regression equations, based on concentrations in fat of cattle and pigs. These models had to rely on simple empirical correlations, with no mechanistic basis which limited their range of application and predictive capability. Therefore, several authors have developed and more closely examined other complex mechanistic models.

Dowdy et al. [33] developed molecular connectivity indices (MCI) to predict the biotransformation factor of organic compounds in meat and milk. The MCI is a non-empirical parameter derived from the molecular structure. McLachlan [34] developed a simple pharmacokinetic fugacity model that describes the fate of trace organic pollutants in lactating cows. The model consists of three compartments: the digestive tract, blood and fat. Diffusive transport is possible between the digestive tract and blood and between blood and fat compartments. Transformation may occur in the digestive tract or blood. The model considers three advective flows: feed, faeces and milk. Storage is included in the fat compartment. The model can be used in a steady-state or non-steady-state situation. The steady-state model is promising and performs very well for very hydrophobic and non-metabolized compounds. The CK_{ow} models for meat and milk of Rosenbaum [35] are based on the model approach proposed by McLachlan [34], but substantial alterations were made to address the inability of this model to capture the behaviour of low-K_{ow} compounds. They compared the CK_{ow} model with the approach of McLachlan [34] and Travis and Arms [31] with measured data. They concluded that the CK_{ow} model provided a better scientific basis and significantly increased reliability in biotransfer modelling for meat and milk. Czub and McLachlan [24] adapted the steady-state model of McLachlan [34] and also included inhalation, exhalation and urination. A good agreement between predicted and experimental data can also be found with Physiologically-Based PharmacoKinetic (PBPK) models [36,37] although this has only been demonstrated for non-metabolized substances. The most important limitation of PBPK models is that the parameters found for the combination of animals and compounds cannot be used for other animals and compounds, preventing a more generic approach [38].

Ingestion of mother's milk
The ingestion of mother's milk by nursing infants is a potential source of exposure to toxic substances. Lipid soluble chemical compounds accumulate in body fat and may be transferred to nursing infants in the lipid portion of breast milk. Lactating women can transfer to breast milk their intake of chemicals from all routes (ingestion, inhalation, and dermal contact) and the total intake of infants may be via the ingestion of breast milk. Thus, the population of nursing infants may be at risk, especially when lipophilic compounds are assessed. All direct and indirect exposure routes to the mother must be considered in determining the chemical concentration in breast milk. A multimedia total exposure model for hazardous-waste sites, CalTOX [5] and a fugacity based mechanistic model ACC-HUMAN [7], address this exposure route.

Applicability of combined routes
Against the theoretical background of estimating exposure through food products, all underlying assumptions together drastically reduce the applicability of the models. Although the regression equations used sometimes show a wide range, the joint range can be quite small [39]. This is illustrated by Figure 5.2 which shows the range of applicability of the various regression equations of the indirect exposure routes of EUSES [4]. The joint range is quite small, with a log K_{ow} ranging from 3.0 to 4.5. For substances with a log K_{ow} outside this range, the regression equations can result in uncertain and possibly misleading estimations.

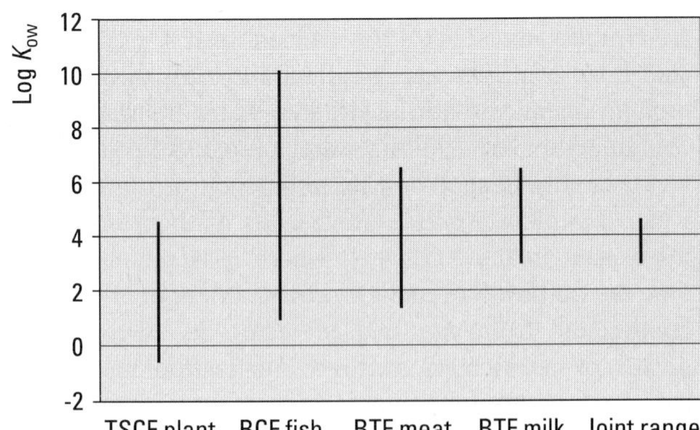

Figure 5.2. Regression ranges of the indirect exposure module of EUSES [39]. With permission.

5.2.4 Direct exposure through the environment

Direct exposure of humans through the environment can be caused by inhalation of air, dust or aerosols, ingestion of soil and dermal contact. Examples of direct exposure via the dermal route are, for instance, contact with soil during gardening, swimming in surface water or showering with chlorinated drinking water (chlorination byproducts). Modelling direct exposure is relatively simple because the concentration in intake or contact media can be derived directly from distribution models, as discussed in Chapter 4. Only the defined exposure scenario and quantification of the absorption and uptake from the external exposure are of importance (see Sections 5.3 and 5.4). Except for the inhalation of outdoor air, all other direct exposure routes, both indoors and outdoors, are more applicable to the risk assessment of a contaminated local environment than to the assessment of a regional environment. Direct exposure modelling can be done in a manner similar to the approaches described for consumer exposure in Section 5.3.

5.2.5 Derivation of the total daily uptake and sample calculations

The total daily uptake by man is calculated by combining concentrations in the different intake media, i.e. drinking water, air, fish, crops, meat and milk, with the daily intake values of the population to be protected. The following general formula is applied to calculate the doses from the different media:

$$DOSE_{\text{medium x}} = \frac{C_{\text{medium x}} - IH_{\text{medium x}}}{\text{BW}} \qquad (5.1)$$

$DOSE_{\text{medium x}}$ = daily dose via intake of a specific medium (mg/kgbw·d)
$C_{\text{medium x}}$ = concentration of the chemical in this medium (mg/kg or mg/m^3)
$IH_{\text{medium x}}$ = daily intake for this medium (kg/d or m^3/d)
BW = (average) human bodyweight (kg).

In the case of inhalation of contaminated air a correction factor for the bioavailability for inhalation (typically 0.75) has to be included in this formula. By adding up the different doses for the individual intake media, the total human dose can be calculated and compared with the no observed adverse effect level (NOAEL), the ADI, or TDI (see also Chapter 6). An example of how human exposure through the environment can be calculated is given below.

Table 5.1. Physicochemical properties of PCB and 2-propanol.

Substance	Molecular weight [g/mol]	Log K_{ow} [-]	Vapour Pressure [Pa]	Water solubility [mg/L]	Kp in soil[a] [L/kg]	Henry's Law constant[a] [Pa.m^3/mol]
PCB	290	6.5	0.25	0.05	4640	1450
2-propanal	60	0.1	4400	1.10^5	0.24	2.55

[a] Estimated from the given properties

Examples for an exposure assessment through the environment

Human exposure through the environment can best be demonstrated by means of two substances with different physicochemical properties. One substance, a PCB congener with 4 chlorine atoms, is highly hydrophobic and has a low vapour pressure and a low water solubility. The other substance is 2-propanol, which is hydrophilic, has a high vapour pressure and is highly watersoluble. The physicochemical properties required to perform the calculations are summarized in Table 5.1.

Calculations are performed with EUSES [4], based on the equations incorporated in the TGD [1]. The environmental concentrations, necessary as input for the calculation procedure, are assumed to be equal for both substances for a better comparison. For air, a total concentration of 10 mg/m^3; for surface water, a dissolved concentration of 0.5 mg/L; and for agricultural soil, a total concentration of 1.0 mg/kg$_{wwt}$ are assumed. It is obvious that PCB, being more hydrophobic, will be more strongly sorbed onto soil particles than 2-propanol. These environmental concentrations are subsequently used to calculate the intake of the chemicals by humans through different routes.

Air

In air, partitioning between aerosols and the gas-phase takes place. The vapour pressures of both substances, although relatively low for PCB, are high enough to expect that more than 99% of both substances to be in the free gas-phase (according to Junge's equation, see Section 3.2.2, Equation 3.9).

Drinking water

Surface water is regarded as the only source for drinking water. For PCB, a purification factor of 0.125 results in a drinking water concentration of 0.0625 mg/L. This purification factor is relatively high because during treatment a considerable amount is expected to be either adsorbed or volatilized. 2-propanol has a conservative purification factor of 1. Hence, its concentration in drinking water is expected to be similar to the concentration in surface water.

Fish

The hydrophobic PCB has an enormous potential for bioaccumulation in fish. Based on its octanol-water partition coefficient a BCF of 43,700 L/kg is estimated, resulting in a concentration in fish of 2.18.10^4 mg/kg. 2-propanol, however, has a low bioconcentration potency. The log K_{ow} of 2-propanol is outside the valid domain for use of this BCF (see Section 5.2.3). Therefore, the minimum log K_{ow} of the valid domain is used to calculate a BCF of 1.41 L/kg, which results in a concentration in fish of 0.7 mg/kg.

Crops and grass

The modelling approach proposed by Trapp and Matthies [11] is used to estimate levels in plants due to uptake from pore water and air (gas phase). This approach integrates uptake from pore water and air into a consistent, one-compartment model. The sink term in the model is formed by diffusive transfer from leaf (foliage) to air, elimination in the plant tissue, and dilution by growth. The source term is formed by the uptake and translocation from soil and the gaseous uptake from air. Aerosol deposition is not considered in the model, but as our example substances were not bound to aerosols in air, deposition of aerosols onto plants can be neglected.

The concentration in root tissue is determined mainly by physical sorption and is calculated by: $(K_{plant\text{-}water} \cdot C_{porewater})/RHO_{plant}$. Root uptake and translocation to higher parts of the plant is described by Briggs et al. [10,40] by defining the transpiration-stream concentration factor (TSCF). The TSCF is the ratio between the concentration in the transpiration stream and the concentration in pore water. Uptake from the gas phase is described by a foliage-air partitioning coefficient (see also Chapter 3, Section 3.3.5).

On the one hand, PCB is extremely hydrophobic

and is therefore more easily taken up by roots than 2-propanol. While, on the other hand, the PCB pore water concentration is much lower than 2-propanol. This results in comparable concentrations in root of 5.2 mg/kg_{wwt} and 2.9 mg/kg_{wwt} for PCB and 2-propanol, respectively. PCB is only slowly translocated in the plant because of its hydrophobicity. This is shown by a low TSCF of 0.038. The TSCF for PCB is calculated with a K_{ow} of 4.5 (and not 6.5), which is the maximum of the valid domain for use of the TSCF. 2-propanol is more easily translocated in the plant with a TSCF of 0.25. PCB is more easily taken up from air and pore water, because of its lipophilic properties. A $K_{leaf\text{-}air}$ of $2.5.10^4$ m^3/m^3 and $K_{plant\text{-}water}$ of $1.5.10^4$ m^3/m^3 leads to a PCB concentration in the leaves of crops and grass of 0.348 mg/kg_{wwt}. For 2-propanol, these partition coefficients are lower, with a value of 615 m^3/m^3 and 0.66 m^3/m^3, respectively. The 2-propanol concentration in the leaves of crops and grass is 0.01 mg/kg_{wwt}. The net result is that despite the higher pore water concentration for 2-propanol, lower concentrations are estimated in leaves and grass than for PCB.

Meat and milk

Concentrations in meat and milk are estimated with BTFs (d/kg) according to the regression equations described by Travis and Arms [31]. Cows are exposed through eating grass with adhering soil, and inhaling air. Equation 5.2 can be used to calculate the concentrations in meat or milk:

see below (5.2)

The hydrophobic PCB is estimated to accumulate appreciably more in the meat and milk of cattle than 2-propanol, even though the concentrations in crops are lower. BTFs of 0.08 and 0.025 (d/kg) result in relatively high concentrations in meat (2.28 mg/kg) and milk (0.72 mg/kg) for PCB. For 2-propanol these values are several orders of magnitude lower, resulting in concentrations in meat and milk in the ng/kg range. It must be noted that the log K_{ow} of 2-propanol is outside the valid domain for use of the BTF for meat and milk (see Section 5.2.3). Therefore, the minimum log K_{ow} of the valid domain is used to calculate a BTF for meat and milk for 2-propanol.

Total daily intake by humans

In this example humans are assumed to obtain their total consumption from contaminated media only. The human intake rates represent the highest country-average intake across all EU Member States for each food product [40]. Therefore, this exposure scenario can be seen as a worst case. The dose obtained from each medium can be calculated according to equation 5.1. For inhaled air a bioavailability of 75% is assumed. The results for the two compounds are summarized in Table 5.2.

From Table 5.2 it may be concluded that the human exposure through the environment is higher for PCB than for 2-propanol. Human exposure to PCB is caused mainly by the consumption of contaminated fish. Exposure to 2-propanol is caused by two major exposure routes: drinking water and root crops. Of course, a large degree of uncertainty is attached to the absolute figures for the TDI. However, it is relatively safe to assume that under similar environmental concentrations, exposure to PCB will be higher than exposure to 2-propanol.

5.3 CONSUMER EXPOSURE ASSESSMENT

5.3.1 Introduction

Individuals in and around residences come in contact with a variety of substances from various potential sources. The focus in this chapter is on consumer products, which includes consumer products such as household cleaning products, personal care products, clothing, furniture, toys, etc. Consumer product exposure assessment approaches and results have been published for many years, e.g., one of the first publications was an industry exposure study in 1970 exploring variations in exposure arising from consumer usage of a laundry detergent product, specifically assessing inhalation exposure while a laundry powder detergent is dispensed into the washing machine [41]. Additional noteworthy publications relating to consumer exposures followed in the 1970's and 1980's, [42-45]. Further to this, US-EPA [46] published detailed information on databases, tools, and a systematic approach to estimating exposures to a given chemical in consumer products, and included a listing of consumer product categories along with potential exposure pathways and mechanisms. The first edition of

$$C_{\text{meat or milk}} = BTF_{\text{meat or milk}} \cdot (C_{\text{leaf}} \cdot INTAKE_{\text{grass}} + C_{\text{soil}} \cdot INTAKE_{\text{soil}} + C_{\text{air}} \cdot INTAKE_{\text{air}}) \quad (5.2)$$

Table 5.2. Human intake rates and TDI from different routes for the PCB and 2-propanol example.

Medium	Intake rate	Dose [mg/kg.d]	Percentage of total intake	Dose [mg/kg.d]	Percentage of total intake
		PCB		2-propanol	
Drinking water	0.002 m^3	0.00179	0.00497%	0.0143	41.5%
Fish	0.115 kg_{wwt}/d	35.9	99.8%	0.00116	3.37%
Leaf crops (incl. fruit/cereals)	1.20 kg_{wwt}/d	0.00596	0.0166%	0.000177	0.515%
Root crops	0.384 kg_{wwt}/d	0.0287	0.0798%	0.0159	46.3%
Meat	0.301 kg_{wwt}/d	0.00978	0.0272%	$1.0.10^{-7}$	0.000297%
Milk	0.561 kg_{wwt}/d	0.00577	0.0161%	$1.9.10^{-6}$	0.00553%
Air	20 m^3/d	0.00286	0.00796%	0.00286	8.30%
Total		35.9	100%	0.0344	100%

US- EPA's Exposure Factors Handbook was published in 1989 (however, a section specific to consumer products and residential exposure factors was not added until the second edition in 1997) [47,48].

Consumer exposure assessment activities in the 1990's included reports and publications from OECD [49], the European Commission [50-52], WHO [53] and RIVM [54,55], and the Carnegie Mellon University [56]. More recent noteworthy efforts include work by ISEA and SRA [57], further research in 2000 and 2001 by Carnegie Mellon scientists [58,59] on the impact of consumer behaviour on exposures, work by RIVM and others in 2001 on the potential for meaningful intra- and inter-individual variations in the use of a product [60]. A first European Exposure Factors Sourcebook was published by ECETOC [61], RIVM published a series of "fact sheets" with scenarios and related information for various categories of products [62], and the European Commission published the Technical Guidance Document [1]. Further, the European Chemical Industry Council (CEFIC) funded a project, ExpoFacts, in which a database containing European data on exposure factors was created [63]. Reports of the USEPA-sponsored voluntary children's chemical evaluation programme (VCCEP) include a critical review of global information for the selected chemicals and exposures from the use of consumer products [64].

Two other industry initiatives providing useful information on exposure assessment is the European HERA project which focused on household cleaning products [65] and the US industry-led Alliance for Chemical Awareness website [66]. Finally, the European Commission-sponsored EIS-ChemRisks (the European Information System on "Risks from chemicals released from consumer products/articles") was developed to build knowledge and develop European and global infrastructure, methods and tools for understanding exposure to consumer products. The EIS-ChemRisks Exposure Assessment Toolbox allows users to access information and make queries from reference databases to find exposure scenarios, data, exposure factors, testing methods, and exposure models and algorithms [67].

5.3.2 Consumer exposure scenarios

In this section we will first address the development of consumer exposure scenarios according to the first definition of OECD/IPCS [3]. In Section 5.3.6 we will consider risk management measures for consumers which, in fact, provided the basis for the development of ES under REACH for chemicals with relevant exposure to consumers.

Building a consumer exposure scenario

An exposure scenario is often constructed on the basis of a logical, step-by-step analysis of:

- The factors and events known or postulated to affect how and when a substance of interest is released from a consumer product or other source into the environment or microenvironment.
- The transport, transformation, and fate of the substance of interest in various media (e.g., room air).
- The contact between the substance and consumers and other people.

- The concentrations of the substance in the relevant carrier media (e.g., room air), and
- The dose potentially entering the human body.

Each category of products and articles will need a set of scenarios to cover all key "real life situations," or "reasonably foreseeable exposures."

To assess the exposure to chemicals or substances present in consumer products, information is needed on two sets of parameters: contact parameters and concentration parameters. The contact parameters denote where, how long and how often contact with the consumer occurs. This will require estimations or knowledge about the extent, duration, and frequency of exposure associated with a particular type of usage of the product, and whether the exposure might be one event, a series of repeated events or a continuous exposure (e.g., concentrations in indoor air resulting from storage and use of a product). The data used might include behavioural observation studies conducted in homes, and activity diaries that consumers are asked to complete. In the absence of more substantive information, expert judgments and assumptions might be used. The concentration parameters are needed to estimate the concentration of a substance in a medium that might come into contact with the body. This is not necessarily equal to the concentration of the substance in the consumer product, e.g., a product might be diluted, mixed, undergo evaporation, etc., before the substance of interest actually reaches the human body. The routes of exposure can be dermal (e.g., cleaning agents, cosmetics, shampoos, or clothing), inhalation (e.g., hair spray) or by ingestion (e.g., swallowing of tooth paste). By combining the contact parameters with the concentration estimates, exposure or dose can be estimated.

The source of the chemical or substance could be a component of a synthesized material (e.g., a consumer product formulation), or a "product/assembly" (e.g., a component of a piece of clothing or furniture). The source provides the molecules of interest for the assessment, either as volatiles (as the original compound or released from a material/matrix), molecules in an aqueous or other liquid media, and/or as particles (as the original compound or released from a material/matrix). Molecules from a source that are not already in the local environment being assessed could be transported from the non-local environment to the local one (e.g., from outside of a residence to indoors, or from one room to another), either creating an exposure or perhaps adding to an existing source of exposure.

The local environment is where the dermal and/or inhalation and/or oral exposure(s) to the human subject(s) occurs. The source term is developed into a delivered dose in the scenario by the use of:

- Human-related exposure factors including: a) the behaviour and preferences of the user (e.g., activities, what product they choose to use, or what article they choose to wear) and possibly by the behaviour and preferences of others in the house and b) physiological characteristics (e.g., age, skin surface area, breathing rate, etc.).
- Residential exposure factors (e.g., the volume of the house or room, the types of appliances, etc.), and possibly by environmental factors (e.g., a strong wind could diminish the amount of the substance of interest reaching the local environment, and a warm or cold outside temperature will influence the residential air exchange rate depending on whether windows are open or not, and whether the house has the heating or air conditioning system in operation, etc.).

The combination of boundary conditions and human factors and product or article-specific factors, plus other factors noted below, determines qualitatively, quantitatively, and in terms of time, the evolution of the exposure process.

If a chemical or substance is used in more than one consumer product, or if the product is expected to be used in more than one way for a task, or for more than one type of task, it may be necessary to assess the exposure for each case. In addition, if the substance is used in different consumer products or has different modes of use, the exposure assessment could examine those uses for which the highest exposures are expected to occur on a regular basis. The aggregate exposure expected from the use of the same substance in different products may also be considered. When doing aggregate exposure assessments it is important to understand the correlation between the various uses, time-activity patterns, co-use patterns and non-users in the population. This is an important area of research on how to do aggregate exposure accurately rather than always taking the worst case of assuming all exposures happen at the same time.

Exposure assessment can be approached in steps, progressing from less refined, more conservative assessments, to more refined, data-rich assessments, as considered necessary for each risk assessment. An advantage of using scenarios is that initial exposure estimates can be developed with very little data; with the possible disadvantage of having a high level of uncertainty associated with the need to include assumptions and inferences given the limited data. These screening-level scenarios are often constructed to represent worst case exposures that would fall beyond the upper end of the expected exposure distribution. Parameters, such as emission rates, dispersion characteristics, concentrations in consumer products, human inhalation and consumption patterns, uptake rates, metabolism, and exertion, are either estimated from available data or are represented by "defaults" or other values.

Scenarios can have limitations, e.g., a lack of data for some exposure factors, and non-validated default assumptions. These limitations can contribute significantly to uncertainties in the exposure and risk assessment. To be able to quantify these uncertainties better, while gathering, analysing and utilizing exposure information it is important to consider:

- The potential for meaningful intra- and inter-individual variations in the use of consumer products.
- The potential contribution of non-consumer product sources (e.g., outdoor; smoking) to the exposure to a chemical in the residential environment.
- That an unexpected exposure factor (e.g., poor eyesight) in a scenario could have a key impact on the exposure from a consumer product.

Assessment of dermal exposures

There can be extensive dermal (i.e., contact with skin) exposure of consumers to substances in products (e.g., clothing, furniture, and toys). These could also include direct contact with laundry detergents and hard surface cleaning products during use, indirect contact with cleaning product residues (e.g., laundry detergent residues in washed clothing), contact with dislodgeable residues of a chemical after use (e.g., crawling infant contact with carpet cleaner residue on a carpet); and direct contact with materials that are intentionally applied to the skin (e.g., clothing, and various types of cosmetics and personal care products). Assessment of dermal exposure consists of two distinct steps:

- Estimation of the amount of chemical that comes into contact with the skin and can potentially be absorbed through the skin.
- Determination of the fraction of this external exposure that actually penetrates the skin and is taken up (is bioavailable).

The first step involves specification of the frequency and intensity of the contact with a product and the details of the release of the active ingredient from the matrix in which it is contained (e.g., diffusion through a watery solute to the skin, migration from a solid matrix onto the skin). The fraction of a substance that is available for absorption (to be determined in the second step) from this layer through the skin is generally difficult to estimate. It depends on the solubility of the substance in water and fat, its polarity and molecular size, environmental factors and skin dependent variables.

As a rule of thumb, bioavailability can be assumed to be 0 for substances with a log K_{ow} below -1 and over 5 or a relative molar mass over 700. The ratio of the LD50dermal/ LD50oral may also provide information on dermal absorption, a high ratio being indicative of poor absorption. In all other cases total absorption should be assumed.

Various additional factors must be taken into account in determining dermal exposures:

- Human exposure factors. Besides body weight, which varies between and within age and gender categories, it is necessary to develop an exposure scenario that specifies the amount of skin surface area exposed. Total surface area statistics can be used with a fraction taken to represent the exposed area, or exposed body parts can be specified (e.g., both hands) and body part surface area data used.
- Frequency and duration of exposure. The duration of exposure should represent the anticipated contact time with the skin prior to washing or removal.
- Concentration of the chemical on the skin. It is the estimation or measurement of vapour phase or aqueous-phase concentration of a given agent in contact with the skin. For example, aqueous-phase exposures are usually expressed as $\mu g/cm^3$ of aqueous solution.

Assessment of inhalation exposures

Developing data about relative indoor emissions from various types of consumer products has been an important area of research, especially in view of the variations in consumer products, consumer behaviour, and housing conditions encountered across the world. It is important to be able to estimate primary emissions of chemicals from consumer products using monitoring or testing methods, and/or via modelling. Emission rates of most chemicals in consumer products are greatest when product are new; emissions are likely to continue at low levels for longer periods with products such as carpets and pressed-wood products. Of particular interest are long-term exposures to VOC emissions from room air fresheners and bathroom deodorants which are intended to maintain an elevated concentration of deodorant in the room.

Exposure factors that are commonly considered when assessing inhalation exposures to chemicals in the home are given in Table 5.3. Many of these factors are associated with a wide range of variability across affected

Table 5.3. Exposure factors considered in inhalation exposure in the residence.

Source characteristics	Perhaps the most important factors determining the impact of chemical sources in the residence on inhalation exposures are the nature of the source (e.g., a consumer product, or a residential construction material, such as floor or wall surface), how it is released (e.g., as a respirable aerosol, non-respirable aerosols, or as a vapour release, and the source strength (roughly proportional to the concentration of the chemical in the source or product).
Human exposure factors	These include body weight, which varies between and within age and gender categories, and inhalation rates, which vary primarily by age, gender, and activity level.
Physicochemical properties	These include factors such as molecular weight and vapour pressure that determine the rate of evaporation into air of a chemical in an applied material (e.g., paint), or its release from aqueous solution (e.g., the role of Henry's law constant in determining the release of volatile organics from tap water used in the home).
Residential building factors	The basic characteristics of the room(s) and building in which residential exposures occur, as well as the ventilation configuration (i.e., number of windows and doors open, the rate of mechanical ventilation and air mixing, rate of infiltration of outside air), will determine the extent and rate of dilution of the chemicals of interest in a specific indoor air setting.

populations. A number of indoor air modelling tools are available for use in assessing inhalation exposures of a variety of contaminants from a variety of sources. Some are oriented more towards assessment of exposures to chemicals from consumer products when the specific emission term is not known.

Assessment of ingestion exposures
Ingestion of chemical residues can occur in the home from chemical residues in, e.g., packaging material (and subsequent leaching to food items). In addition, consumer products can lead to ingestion exposures via accidental exposure and incidental residues such as a cleaning agent residue on plates and silverware following product use. Another important pathway for incidental ingestion exposure is hand-to-mouth behaviour in infants and toddlers, and the mouthing of clothing, other textiles (e.g., blankets and furniture) and toys. For adults, hand-to-mouth behaviour is also possible for some scenarios, along with some product-to-mouth behaviour associated with certain types of products, e.g., pencils and pens.

Other routes of exposure
Besides dermal, inhalation, and oral exposures as the three major routes of exposure, other routes of exposure must be considered in special cases, e.g., the intradermal or intravenous routes. Intradermal exposure occurs when the integrity of the skin is disrupted by the use of consumer products (e.g., by earrings or tattoos). Intravenous exposure may occur during the use of medical devices (e.g., an infusion device from which migration of monomers or other substances takes place).

5.3.3 Primary and secondary exposures associated with consumer products

One way to characterize consumer exposure is by looking at the different populations and subpopulations that are actually exposed to the products. Primary exposure to substances occurs to the individual actively using the product or article containing the substance. Examples of primary exposure are wearing textiles, or the use of household cleaning products. Secondary exposure occurs to non-users or bystanders; these are individuals who do not actively use the products but are indirectly exposed to substances released during or after product use by another person (the user).

Examples of secondary exposure of non-users include exposure to paints, and cleaners, etc., during or after use by the user, and exposure to household articles and appliances (e.g., flame retardants in furniture, plasticizers in building materials) which have been treated with the substance. Secondary exposure scenarios also include contact with the substance following the use of professional products in the home, e.g., from paints after painting in the home by a professional painter. According to this definition, the user of a product may be subject to both primary and secondary exposure and, as a consequence, will often have the highest exposure, whereas the non-user or bystander has only secondary exposure. Such secondary exposures may be of less immediate concern than primary exposure unless this occurs to specific subgroups of the population that may experience higher exposures because of their specific behaviour (e.g., children crawling on the floor).

5.3.4 Accessing exposure factors and data

An exposure assessor needs to utilize various exposure factors to calculate exposures to a substance from the intended and reasonably foreseeable uses and misuses of a consumer product. For example, exposure factors could include the concentration of the substance in the product formulation, the amount of product applied, and the skin surface area of application.

The exposure factors can be based on actual data (e.g., measurements under actual or simulated consumer product usage, or measurements for chemicals judged to be similar in chemical properties to the chemical of interest), computer software estimations of the factor (e.g., how much might be released into residential air based on the volatility of the substance and the temperature of the consumer task), predictions, or expert judgments. A key challenge confronting all exposure assessors is the need to choose suitable values for important exposure factors. Basically, there are two general types of information sources to which the assessor can turn:

- Primary sources, which are studies or collections of studies reported in the scientific literature (e.g., peer-reviewed journals), and
- Secondary sources, which are compilations that summarize existing data from primary sources and recommend values for important human exposure factors.

Two weaknesses in the data used to assess exposure to chemicals in consumer products are: (1) product usage and (2) product contact. The diversity of consumer products does not allow for a single set of information sources, handbooks or databases to be consulted. Rather, it is necessary to explore which information sources apply to the substance of interest. There is only limited information available about chemicals in consumer products, e.g.:

- Product registers. These are available in some countries (e.g., Switzerland, the Nordic countries, Italy, and Germany) and may provide information on whether the substance under consideration is present in marketed consumer products.
- Poison information centres which have product information.

Figure 5.3. An example of an observation study.

- Safety data sheets and information brochures from industry.

Expert judgment and review of the original data could be needed when deciding whether to use an exposure factor value for a specific assessment, e.g., for the use of US EPA-published exposure factors outside the US. Other information sources on habits and customs of use that may be useful include:

- Specific information on use durations and contact frequencies. This information for consumer products is often lacking. An estimate of these parameters can be derived from time budget data (sometimes called human activity patterns) where available. Time budgets comprise information on the behaviour of a population during a day, week or year. Because time budgets may vary geographically, it is useful to check if the national statistical agencies have gathered such data on a regional basis.
- The directions provided by the manufacturer. These provide information on the recommended use, but usually not on the way products may be handled before or after actual use, nor on reasonably foreseeable misuse. Although information on the latter might be available from Poison Control Centres and case studies reported in the literature, such data might represent the more extreme misuses of the product and may not be very informative about the normal range of uses.
- Information accompanying computer programs for the exposure assessment. This may also be useful sources of data.
- Information from manufacturers. Some countries require manufacturers of certain products (e.g., cosmetics, toys, pharmaceuticals, food contact materials, pesticides) to provide data useful for estimating exposure.

Table 5.4. Concise overview of some consumer exposure models.

Model	Short description	Reference
ConsExpo	The program offers a number of generally applicable exposure models and a database with data on exposure factors for a broad set of consumer products. Evaluations for multi-route exposures. Deterministic and probabilistic assessments are supported.	www.consexpo.nl [62]
MCCEM	The program models time varying indoor air concentrations and inhalation exposures in different rooms of a residence. Includes various source and sink models. Combines time- dependent air concentrations with time-activity patterns.	www.epa.gov/oppt/exposure/pubs/mccem.htm [69]
WMPaint	Special purpose model to estimate exposure from solvents in latex and alkyd paints. Emission models are based on small chamber emission data.	http://www.epa.gov/oppt/exposure/pubs/wpem.htm [70]
Promise (probabilistic methodology for improving solvent exposure assessment)	Tool is designed for estimating single exposure events in occupational settings and in certain consumer-type applications. The program offers tools for probabilistic simulation, implementing a large number of statistical distribution functions.	www.americansolventscouncil.org/resources/promise.asp [71]
Lifeline aggregate and cumulative exposure/risk assessment software	Advanced tool to characterize population-based aggregate and cumulative exposures and risks from pesticide residues. The sources of exposure included in the program are diet, home environments, drinking and tap water, residential pesticide products. Contains large databases with US-specific data.	http://www.thelifelinegroup.org/lifeline [72]

- In-home observation data (see Figure 5.3), diary studies, recall studies, and/or objective measurement studies conducted by industry, trade associations, academic researchers, or government organizations. Some of this information has been published.
- Published literature.

5.3.5 Issues to be considered when performing an exposure assessment

Tiered approach

Exposure assessments are usually developed in a tiered (or phased) approach. The assessment of the exposure of consumers conducted following an iterative, tiered procedure starts with an initial "screening". This screening is needed to identify if the substance under investigation is actually used as or in consumer products or whether the expected (by estimations based on crude, worst case assumptions) consumer exposure is so low that it can be neglected further in the risk characterization phase (tier 0). If this is the case, no further assessment is needed and the conclusion can be mentioned in the chemical safety assessment. If use as or in consumer products has been identified and the exposure is not considered to be negligible as described above, then a rough quantitative exposure assessment will be desirable (tier 1). The results of this quantitative assessment are taken forward to the risk characterization where they are combined with the results of the effects assessment in order to decide whether or not there is any concern for the consumers exposed to the substance.

Several exposure scenarios are presented in the TGD, each with a different equation for the exposure calculation (also called first tier models (Table 5.4)), which calculate a realistic worst case exposure. Higher tier models are also available, in ConsExpo 4.0 [62,68], for example. For guidance on how to calculate exposure using first and higher tier models, the reader is referred to the TGD or the manual of the respective models.

5.3.6 Risk management measures for consumers

Risk management measures are generally used but not specifically indicated as such. REACH [2] will use exposure scenarios as a means of communication to instruct the user of a substance how to deal with it in

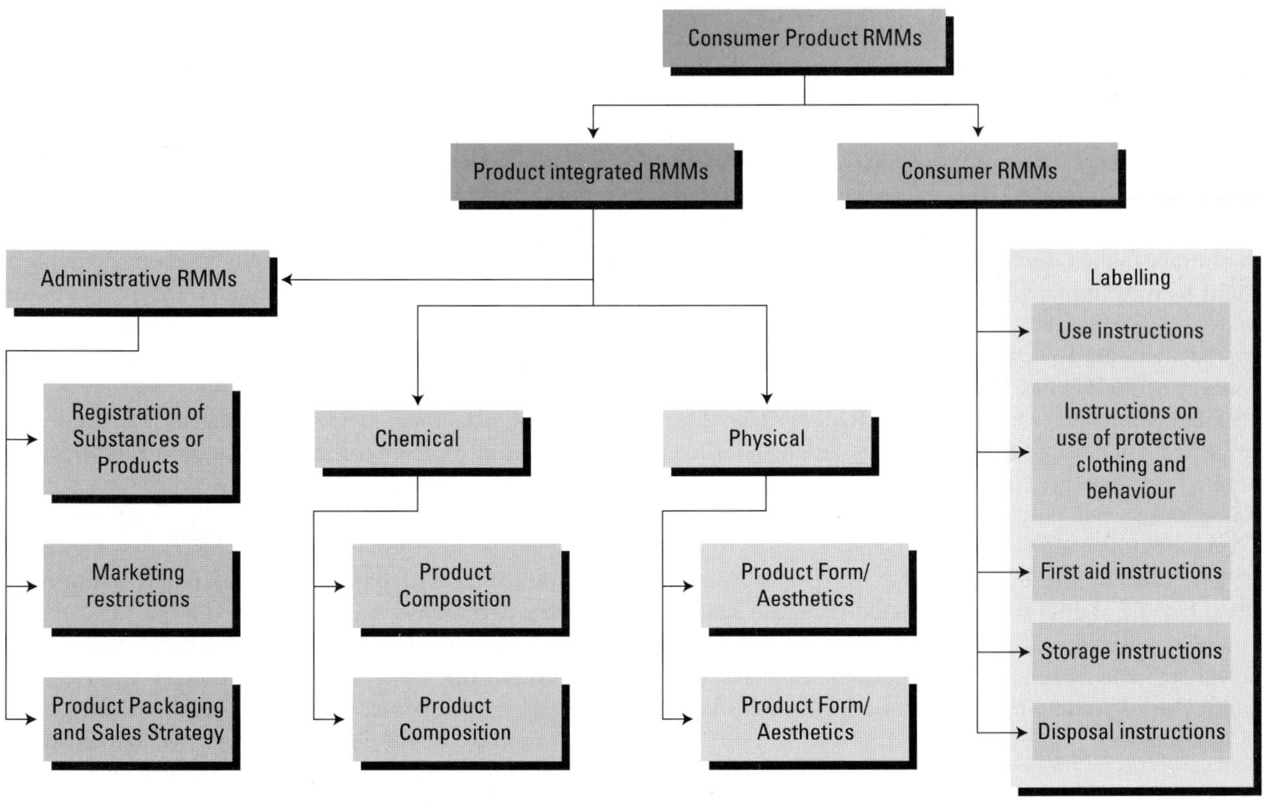

Figure 5.4. RMMs related to exposure to consumer products are divided into "product integrated" and "consumer" measures [73].

such a way that risks are adequately controlled, for all foreseeable environmental, occupational and consumer exposures. Producers and importers will have to report implemented RMMs as part of the ES. On the basis of published and additionally developed RMMs examples on package design, physico-chemical properties, product form, labelling for safe use/storage/disposal, modification of the product composition, etc., RMMs pertaining to consumers which aim to minimize exposure, maximize safety and avoid harm to consumers altogether have been described recently [73] and are presented in Figure 5.4.

One basic channel to provide direct information on RMMs for consumer products is labelling. Labelling is related to both "product intrinsic" RMMs and to "consumer use–related" measures: in particular, health and safety (product identifier, product handling and use, hazard statements, precautionary statements and pictograms, warning words and messages). The label needs to be sufficiently detailed and relevant to the use of the product. There are two main approaches to providing information to consumers through labelling. These are: 1) based on the likelihood of injury (i.e., risk communication), and 2) based on the "right to know" principle in providing information to consumers solely based on the product's hazards. Consumer product RMMs can be categorized into:

- Product-integrated RMMs. These RMMs mostly reflect technical measures to be applied during the pre-design phase of a product prior to its actual use by consumers. This category is grouped into chemical and physical RMMs. [41, 74-78]. These RMMs should largely integrate the foreseeable identified uses during the entire lifecycle of the product from manufacturing to disposal. The administrative RMMs are part of the product-integrated RMMs and mostly refer to organizational risk reduction and restriction strategies related to the products' foreseeable uses and misuses (see Chapters 1, 2 and 12 and [59]).
- Consumer RMMs. These RMMs include labelling and mostly refer to the product-related risk and safety instructions, communication and education directed to the product users The effectiveness or real application of this type of RMMs depends on the awareness and willingness of the actual user and such measures are therefore difficult to control by the product manufacturers/importers or producers (see

Chapters 1, 2 and 12 and [56,58,59]).
The degree of exposure depends on the use of a product and the most suitable RMMs consequently depends on each foreseen use of a substance; whether it is used as such, in a preparation, or in a product. Many uses can be identified for consumer products. These may be intended but could also be unintended uses. To be able to classify the foreseeable product usages, there is a greater need to share more information on the products' usages across the substance/product supply chain. As demonstrated by the literature review of Bruinen de Bruin [73], the concept of RMMs has been known for several decades, but the degree to which specific RMMs have been identified, applied and described varies among consumer product manufacturer/producers. Under REACH it will be a challenge not only to qualify but also to quantify the RMMs described above for consumers.

In order to illustrate the use of RMMs for the development of ESs we will provide two case studies, i.e., one on waterproofing spray products and one on the design of safe enzyme-containing laundry detergents, which can be seen as an example of effective RMMs use of industry.

Case study 1: Waterproofing spray products
Yamashita et al. [79, 76, 77], studied the role of mist particle diameter to the toxicity of waterproofing sprays in mice. When a waterproofing spray is used, the solvent vaporizes and the water-repelling agent remains on the surface of the sprayed area, thereby providing a waterproofing effect. Yamashita et al. [79] found that when the solvent was replaced by a less toxic, but more volatile, chemical in the waterproofing spray, the inhalation exposure increased. It was hypothesized that faster evaporation resulted in smaller aerosol particles over time. Better volatility of the solvent was believed to have affected the diameter of the particles and therefore changed the inhaled amount. To test this hypothesis, on the basis of reported intoxication in human beings after using waterproofing sprays, Yamashita et al. [76,77] grouped 12 sprays into either a toxic or a non-toxic group. The products in the toxic group generated mists with a smaller mean particle size than those in the non-toxic group, 42.3 ± 9.8 µm vs. 86.8 ± 28.6 µm, respectively. Yamashita et al. [76,77] mentioned that the particle diameter is most responsible for the amount and location of the deposition within the human respiratory tract. This study revealed that the percentages of particles ≤ 10 µm (PM10) was significantly higher and that the mean particle diameter was significantly smaller in the toxic group than in the non-toxic group. Yamashita et al. [77] showed that the diameter of the spray particles have a major influence on the ultimate toxicity of commercial waterproofing sprays. The studies described showed how the risk management measure of modifying the spray particle diameter produced – in this case increasing the particle diameter – can reduce the amount of inhaled particles and therefore reduce the health risk due to inhalation. However, by replacing a chemical with a less toxic but more volatile chemical, the particle size unforeseeably changed and in this case decreased. This led to increased inhalation exposure of the less toxic chemical. These studies underline the importance of a good understanding and careful planning of any risk management measure throughout the products' lifecycle and the importance of repeating the exposure assessment process after a risk management measure is applied.

Case study 2: The design of safe enzyme-containing laundry detergents
A report by the (US) Soap and Detergent Association [78] describes the association's risk assessment guidance for enzyme-containing laundry detergents. The purpose of the SDA guidelines is to provide a framework for manufacturers of detergent products to conduct appropriate risk assessments and to develop risk management programmes to help ensure the safety of new products containing enzymes. SDA [78] recommended that companies using enzymes take a responsible approach to how they manage enzymes and the safety of their use in order to avoid any unwarranted authority restriction on the use of enzyme technology in other consumer applications. In order to control the risks of exposure to enzyme-containing detergents, SDA [78] described RMMs such as "product modification", "product use restrictions via labelling", or a "decision not to market the enzyme-containing product". Among the options given for modifying the product are: 1) changing the matrix or delivery of the enzyme product, 2) reducing the enzyme concentration in the product, 3) substituting other ingredients that may affect the potency of the enzyme, or 4) a combination of these approaches. SDA [78] presented a practical example of how modification of the enzyme form reduced exposure during use. Nowadays, enzymes are encapsulated to limit consumer and worker exposure. In the mid 1960's to early 1970's, however, the exposure to enzymes present as an ingredient in unprotected detergent powder was estimated to be 212 ng/m^3 during use. By redesigning this detergent powder, exposures continually decreased over time from 1.01 ng/m^3 (1970), to 0.042 ng/m^3 (1984), to 0.0057 ng/m^3 (1993) for granulated, prilled, and double-

Box 5.2. Words of caution

"Determining significant exposure routes and pathways (for a substance in a consumer product) and selecting values required to estimate exposure for each pathway can involve intensive effort... Most consumer products are used in a variety of circumstances. The main criterion in selecting a standard scenario (for the US EPA document) to represent use of a consumer product was that exposure resulting from the activity is estimated conservatively. For products in which inhalation is a significant route of exposure, a single event involving use of the largest mass of product results in the most conservative estimate of exposure to the product for a single event. However, this event might not occur as frequently on an annual basis or over the lifetime of an individual as another event in which a smaller mass of product is used. Consequently, the single event involving use of the largest mass of product might not result in the most conservative estimate of exposure over the lifetime of an individual or on an annual basis. Therefore, the standard scenario selected to represent use of each product not only involved use of a large quantity of product during a single event, but also represented a circumstance that was judged to be likely to be repeated relatively frequently over the lifetime of the individual." (from a US EPA report [43])

coated prill, respectively. SDA [78] reported the long-prevalent RMM awareness within the detergent industry about reducing exposure by changing or modifying the product design.

5.3.7 Presenting and reviewing the results in relation to uncertainty

In presenting the assessment results, a balanced and impartial treatment of the information should be the goal, with the key assumptions highlighted. The data sources for these key assumptions need to be cited and any adjustments in the data should be discussed. The range of possible values for each exposure factor should be discussed along with a recommended default value when specific data relevant to the assessment of interest are not available.

The strategy for selecting default values could be to express them as a range, from a central value to a high-end value of their distribution. Where statistical distributions are known, the central value corresponds to the mean and the high-end value corresponds to the 90th or 95th percentile. Where statistical data are not available, judgement can be used to select central and high-end values. The range of values is intended to represent variations that occur across a population.

Characterization of the uncertainty will generally include a qualitative discussion of the rationale used in selecting specific scenarios. The discussion should allow the reader to make an independent judgement about the validity of the conclusions reached by the assessor, by describing the uncertainty associated with any inferences, extrapolations, and analogies used and the weight of evidence that led the assessor to particular conclusions.

Some questions a reviewer or presenter of a consumer exposure assessment should ask to avoid errors that could either under or overestimate exposures are:

- Have unrealistically conservative exposure parameters been used in the scenarios? The exposure assessor must conduct a reality check to ensure that the exposure cases used in the scenario(s) (except bounding estimates) could actually occur. Is the scenario chosen also the worst case scenario? (see Box 5.2).
- Have exposures derived from "not detected" levels been presented as actual exposures? For some exposure assessments it may be appropriate to assume that a chemical reported as not detected is present at either the detection limit or one-half the detection limit. The exposure estimates derived from these non detected levels, however, should be clearly labelled as hypothetical since they are based on the conservative assumption that chemicals are present at or below the detection limit, when, in fact, they may not be present at all. Exposures, doses, or risks estimated from data using substituting values of detection limits for "not detected" samples must be reported as "less than" the resulting exposure, dose, or risk estimate.
- Are the results presented with an appropriate number of significant figures? The number of significant figures should reflect the uncertainty of the numeric estimate. If the likely range of the results spans several orders of magnitude, then using more than one significant figure implies more confidence in the results than is warranted.
- Have the calculations been checked for computational errors?
- Are the factors for intake rates, etc., used appropriately? Exposure factors should be checked to ensure that they correspond to the site or situation being evaluated.

- Have the data gaps been noted, and have the uncertainties been adequately addressed? Exposure assessment is an inexact science, and the confidence in the results may vary tremendously. It is useful to highlight key data gaps, and to include an uncertainty assessment that places any uncertainties in perspective.
- Have all important populations and subpopulations been assessed, e.g., if children will use a product or wear the clothing being assessed, are they included in the exposure assessment?
- Would it be useful for risk assessment purposes to consider performing an aggregate exposure assessment that considers exposures to the substance of interest from the different types of products that a consumer might use, and possibly exposures to the substance via outside air, residential water, food, etc.
- If Monte Carlo simulations were used, were correlations among input distributions known and properly accounted for? Is the maximum value simulated by this method in fact a bounding estimate? Was Monte Carlo simulation necessary? (A Monte Carlo simulation randomly selects the values from the input parameters to simulate an individual. If data, e.g., from monitoring, already exist to show the expected exposures for a range of individuals covering the population or subpopulation being assessed, it makes little sense to use Monte Carlo simulation).
- The rationale for selection of any conceptual or mathematical models that are used should be discussed. This discussion should address the verification and validation status of the models, whether they have been shown to be appropriate for use with the product and/or substance of interest, how well they represent the situation being assessed (e.g., average versus high-end estimates), and any plausible alternatives in models that might be available.
- In addition, although incomplete analysis is essentially unquantifiable as a source of uncertainty, it should not be ignored. As a minimum, the rationale for excluding particular exposure scenarios should be noted, along with noting whether these decisions were made with a high, medium, or low level of confidence.

5.4 OCCUPATIONAL EXPOSURE

5.4.1 Introduction

History

The hazards and risks presented by work have now been studied and written about in Europe for almost 2000 years. In the 15th century, Agricola wrote about the disease experienced by German mine workers and surmized how this related to the working conditions in the industry at the time [80]. Later, Ramazzini described the diseases encountered in a variety of rural and urban occupations in northern Italy [81]. But even before then, references to diseases known to be commonly encountered in certain trades and occupations, particularly mining, can be found in the writings of the ancient Greeks and Egyptians. However, it was not until shortly after the start of the industrial revolution, at the start of the 19th century, that the writings moved from being collections of ad hoc personal observations to studies based upon a more systematic assessment of workplace risks. In the UK, concern over the impact that working conditions in the fast developing textile and coal mining industries were having on the general health (and hence employability) of the population led to the establishment of public commissions of enquiry on textile mills (1831) and mines (1842). These not only provided the first substantive evidence of the relationship between work and disease, but also showed how the intensity and frequency of exposure was inextricably linked to the incidence and severity of disease. Concurrently, early pioneers in the field of what is now known as occupational medicine, used modern methods of scientific enquiry to begin to describe the broader relationship between work and health [82]. Indeed, one of Thackrah's aphorisms represents an early basis for the conduct of the discipline: "In many of our occupations, the injurious agents might be immediately removed or diminished. Evils are suffered to exist, even when the means of correction are known and early applied. Thoughtlessness or apathy is the only obstacle to success. But even where no adequate remedy immediately presents itself, observation and discussion will rarely fail to find one".

The substantive and detrimental impact that industrialization could have on the well-being of its citizens inevitably led to regulatory intervention by government. As early as 1802, the UK had legislated on the working hours and conditions of "apprentices" (children as young as 6 years old who were sent to work in mills and factories). As a growing awareness of the risks presented by work was highlighted through the increasing application of methods of scientific enquiry, so the system of workplace health and safety regulation developed and expanded, such that by the early 20th century many European countries had a substantial body of "factories" legislation [83]. This general awareness was also reinforced and spread through the subject matter of popular writers of the period [84-86].

By the start of the 20th century, a substantial body of experience had begun to accumulate on the hazards presented by different substances and the risks experienced in different occupations. The experiences ranged from the effectiveness of a variety of approaches to reducing exposure to hazardous agents, to different forms of regulatory intervention. For example, the 19th century saw practical implementation of the principal of substitution (white phosphorous in the match making industry, white lead in paint production), as well as the first writings on precautionary approaches to the management of workplace risks [87].

The concern for workplace health and safety was also a major consideration when the European Coal and Steel Community (ECSC) was founded in 1951 under the Treaty of Paris. Among its other aims, the ECSC sought to "promote the improvement of the living and working conditions of the labour force … so as to make possible the equalization of such conditions in an upward direction". These sentiments were similarly echoed in the Treaty of Rome at the start of the process of creating what is now the European Union. Indeed the European Economic Community, introduced health and safety legislation as early as 1962, although the basis of the broad legislative framework (including the expectations for the control of chemical risks) today originates in the Health and Safety Framework Directive of 1980 [88].

The framework for workplace risk assessment and management

As the industrial revolution developed, so too did the understanding of the relationship between exposure and disease. Within the mining industries of the UK and Germany, in particular, studies were undertaken that begun to quantitatively describe the relationship between the magnitude of exposure and disease. Although such studies focused on gross disease outcomes, they enabled exposure levels to be identified that could then be used to help describe "safe" working conditions. This work, in turn, catalyzed the need to develop suitable atmospheric sampling and analytical methods to monitor the levels of key hazardous substances [89].

The process of identifying and managing workplace risks thus gradually shifted from one entirely based on observation and personal experience to one which also incorporated science and regulatory standards. By the 1930's, the process by which workplace risks could be identified, evaluated and controlled was firmly established [90]. However, despite these developments, it was not until the 1940's that any systematic attempt was made to develop a comprehensive series of "safe" workplace exposure limits for commonly encountered hazardous agents [91]. Since then, a number of processes for setting occupational exposure limits (OEL) have been established, including that of the EU [92, 93, 94] and those found at the Member State level. Today, OELs are now available for most important commercial chemicals, as well as other physical and biological hazards.

As a consequence, the process of risk assessment of workplace health risks is undertaken within a developed framework of guidance, including strategies for collecting and evaluating information and data on exposures to chemicals. There is an expectation of employers [95] that hazards will be identified, evaluated and controlled in a systematic and proactive manner that documents findings and shares these with workers and regulatory authorities. Because of the diversity of work, no single process is identified or recommended that will fulfil the expectation. Rather, it is envisaged that the level and detail will be a function of the complexity of the work process and activities, together with the magnitude of the associated risks.

Today a range of information is available to help employers meet their obligations, including a variety of approaches to evaluating and managing risks. Although these approaches can vary at Member State level, they can generally be characterized as follows:

- All exposures, whatever the hazard, need to be adequately controlled to manage risks.
- All health risks (chemical, physical, biological, etc.) need to be evaluated in combination.
- The process of evaluation needs to be recorded.
- The level of detail required is a function of the risk and the complexity of the work activity.
- The results need to be communicated to workers and made available to regulatory agencies.

General guidance on the considerations that need to be taken into account is invariably available at the Member State level. Specific guidance on the general standards expected to manage the risks is often available for particular industrial sectors and particular work activities, e.g., welding, painting, etc., or particular chemicals (or classes of them). This guidance is often supplemented by guidance from chemical suppliers and trade groups.

Whilst the totality of this advice is comprehensive, it is not readily accessible by smaller companies. More recently, therefore, generic tools for evaluating risks have been developed that aim to enable users of chemicals to implement approaches that efficiently tier their resource to target exposures of concern, including, in some instances, the provision of tailored exposure control advice [96-99]. Using a minimum of information, these conservative tools combine generalized evaluations of the hazard with exposure modelling to predict risk (as well as delivering advice on risk management measures commensurate with the risk level).

As a result, because of the effort and skill involved in exposure sampling, the quantitative measurement of workplace exposures is generally only undertaken when risks might be considered elevated (for example, where the risks might reasonably be expected to be one quarter of the OEL) and/or where there is a need to demonstrate the effectiveness of the implemented exposure controls. The exception to this is in those Member States that make exposure monitoring mandatory when workers are exposed to defined chemicals (normally those exhibiting carcinogenic or mutagenic properties). However, when exposure monitoring is undertaken, there is extensive guidance available on the necessary sampling strategies, sampling techniques, analytical methods, the interpretation of results and record keeping. Thus the validity of workplace risk assessment can be confirmed or rejected by reference to data deriving from actual or near analogous situations.

Role of chemicals supply information

A fundamental requirement for any successful workplace risk assessment is the ability to understand the hazards of any chemical being handled. Although information on chemical hazards and risks has been available for decades, much of it through national factory and labour inspectorates (as well as voluntarily provided through trade organizations and individual companies), chemical suppliers have not been formally required to provide this until quite recently. From 1970, chemical suppliers in Europe were required to label chemicals (but not chemical preparations) using a standard system of classification, including risk and safety phrases. However, it was not until 1988 that this system was extended to cover chemical preparations (which form the bulk of marketed chemical products) and it has only been since 1991 that a safety data sheet has also had to be provided to purchasers of chemicals.

At the same time as requiring the hazards of chemicals to be communicated, regulation has also intervened to prohibit or limit the supply of certain substances that present particularly high risks. In the UK, legislation prohibiting the use of specific chemicals in certain sectors of industry can be traced back to the late 19th century [87] but it was not until the 1960's that total prohibitions were introduced on certain carcinogenic substances (and even later for asbestos). No concerted approach to regulate the supply of chemicals at the European level was made until the introduction of the Existing Substances Regulation in 1993. This regulation [100], in many respects a forerunner to REACH, requires that the risks arising from the supply and use of chemicals are evaluated in order to identify whether there is a need to restrict their supply (via either legal measures or voluntary action). Technical guidance that describes how workplace risks are evaluated under the regulation has been developed [1]. However, the amount of information considered necessary to evaluate risk in the supply chain is less than for workplace legislation: the former focuses on the 'macro' risks affecting populations whilst the latter requires examination of specific risks affecting individuals at the enterprise level. One is not a substitute for the other. The two, in tandem, serve to complement one another [101], although somewhat counter-intuitively, supply chain based assessment processes can often be the more conservative of the two, because of the need to ensure that the overall process delivers the minimum false negatives for the overall population.

5.4.2 The workplace exposure scenario

In risk assessments carried out under health and safety legislation, the term "exposure scenario" (ES) has commonly been used to describe the particular situation that is the focus of the risk assessment. This may vary from a general assessment of the use of a chemical in an industry sector or activity, to one that is specific to a workplace. According to the OECD/IPCS definition [3], the scenario would tend to include consideration of all key variables that affect the risk (including non-chemical hazards and measures in-place to control such risks). This contrasts with the definition under REACH which describes a control strategy for substances, giving realistic operational conditions for manufacture of a substance or identified use(s) of a substance, a group of substances or a preparation (Box 5.1). The REACH exposure scenario prescribes appropriate measures that serve to effectively manage health, environmental and safety risks from the chemical during its manufacture or use for a given set of operational conditions. The appropriateness of these measures for a specific workplace can vary, however. For local reasons, the measures identified may be invalid and equivalent or better approaches available. Thus there is a need to use the information contained within the REACH exposure scenario as one information source when determining (and documenting) the adequacy of worker health protection strategies in the workplace.

It should be further noted that in workplace exposure assessments, special consideration is not generally given to vulnerable groups. The assumption is made that children and elderly people do not form part of the workforce. A further assumption made is that the requirements for pre-employment medical examinations and routine health surveillance serve to manage the "additional concerns" represented by sensitive working groups, such as asthmatics on medication. As such, acceptable workplace exposure levels are not generally determined by the vulnerability of the workers. The exception to this rule, however, is in the case of reprotoxic and teratogenic risks when particular consideration needs to be given to pre, in utero and post-natal exposures but where specific European legislation to manage such risks has only existed since 1992 [98].

Exposure emissions, sources and models

Exposures at work are invariably and simultaneously to several hazardous agents that extend beyond just chemical substances. These agents can be grouped into sources that principally derive from the task, such as chemicals, physical stresses (radiation, noise), biological agents (bacteria, proteins) and those that have a wider origin in terms of how the work itself is organized (ergonomic and psychological stresses). As such, these exposures with the most health significance could be derived from chemical substances that are not covered by REACH (such as those produced in small amounts, used as agrochemicals or pharmaceuticals, or handled as intermediates) but they are often of a non-chemical origin. Moreover, exposure to chemicals is seldom to the single substance. More usually, exposure is to several substances, arising either from the chemicals being used or from chemicals that are formed from the production processes, e.g., rubber fumes, welding fumes. Workplace exposure assessments under the EU Health and Safety Framework Directive therefore need to take into account all the chemicals which a worker is likely to be exposed to, as well as all other co-stresses. Exposure assessments under REACH only examine the scenarios presented by the use of the registerable substance.

There are a large number of variables that contribute to the nature and magnitude of workplace exposures. However not all determinants have a similar weight in their influence on exposure. For example, although the temperature in a room may be relevant for the evaporation of volatile materials, it is generally not considered to be a major determinant. On the other hand, whether or not an installation using volatile substances is enclosed is a major determinant that cannot be disregarded. A further consideration is that not all determinants are easily assessed and therefore may only be theoretically useful in the process of exposure assessment, e.g., the roughness of surfaces may affect dermal exposure, but its influence can hardly be evaluated in a risk assessment process. More recently, work has been undertaken that allows exposure determinants to be clustered, generally for specific types of work (e.g., spray painting) or specific forms of exposure (e.g., dermal exposure). These allow the number of exposure determinants for any scenario to be reduced to the principal components. They also enable exposure scenarios to be grouped by similar determinants. These developments now form the basis of what are now termed generic exposure/risk assessment and management approaches [98] which form the basis for a number of such tools for advising on workplace exposure assessment and control.

Exposure is caused by and results from a sequence of steps. Briefly it can be described as: emission, transmission, imission and exposure. Although authors have attempted to describe these and the empirical nature of their relationships in order to provide a better understanding for modelling workplace exposures [103, 104], because the number of determinants is so large, and the fact that many of the relationships are not constant, no single model has been developed that can reliably predict workplace exposures. Rather, workplace exposure modelling has developed in a somewhat piecemeal manner. Single models are available which do appear to be reasonably accurate, but they are only applicable within a narrow domain of use, usually just a specific process. At the same time, other less accurate models are available that offer (in most instances) conservative predictions of actual exposures (and, again, within a defined domain). The most well-known example of the general model is EASE [105].

It is therefore possible to determine workplace exposures by sampling, through the use of data from analogous activities, or the use of suitable models. The preferred basis for any evaluation is actual data for the specific scenario, although this is seldom undertaken on a routine basis in practice. More frequently, use is made of relevant data obtained from the use of the substance (or other substances with similar physicochemical properties) in comparable circumstances to the one

Table 5.5. Strengths and weakness of exposure prediction approaches.

Data Type	Advantage	Disadvantages
Measured data	• True picture of actual exposures • The "gold standard' for regulatory comparisons • Provides most accurate personal exposure estimates • Demonstrates the effectiveness of exposure controls (especially concerning substances of high concern)	• Can be expensive to undertake properly • Representative sampling substantially increases costs • Requires access to skill resources • Data remains valid only as long as working conditions remain unchanged • If sampling strategy unrepresentative, may deliver false negatives
Analogous data	• Accurate picture of exposure provided read across based on representative data • Ability to fill gaps quickly (and reliably) and target where sampling may be critical • Ability to quickly evaluate the impact of key control strategies, e.g. substitution	• Requires expert judgement if reliable read across to be executed • Read across invariably an approximation of true exposures
Modelled data	• The least costly option • Can be applied to a range of situations (actual and potential) • Useful as a screening/targeting tool • Ability to quickly evaluate impact of some exposure control strategies, e.g. extract ventilation	• Available models often not extensively validated • Limited number of models available • Models not always easy to understand or operate • Models invariably conservative in their predictions (high false positive rates) • Only provides grouped and not personal exposure estimates

under study. Moreover, as improved exposure models become available, increasing use is also being made of these. Each approach to exposure prediction has its own strengths and weaknesses (which are summarized in Table 5.5) and none of them is perfect. Because of this, it is worth combining all available data to develop weight-of-evidence-like exposure assessments. However in doing this, the differences in data quality and the consequences that this has in terms of the certainty of the prediction, also need to be taken into account. The available data needs to be evaluated within a comparative data framework if consistent assessments are to be created across different substances and types of exposure [106, 107].

Routes and patterns of exposure

Substances in the workplace may enter the body via inhalation, by passing through the skin (dermal exposure), via ingestion or, in certain cases, by direct inoculation. For chemicals, the two most important routes are inhalation and dermal exposure. Within the workplace, exposure is usually described in relation to "external exposure". This is most often defined as the amount inhaled (as represented by the airborne concentration of the substance in the breathing zone of a worker) and/or the amount in contact with the skin. The process for workplace risk assessment does not usually refer to "exposure" as the concentrations of a chemical within the body (apart from the specific case of data obtained from biological monitoring).

Exposure can be considered as the result of a single event; a series of repeated events; or from a continuous exposure source. Therefore, as well as estimating of the level of exposure, the assessor needs to address other parameters such as the duration and frequency of exposure and the numbers of the exposed workforce. It is also appropriate to consider the effect that exposures determined by the tasks that comprise any job have on the overall exposure of the worker. This is particularly so for acute effects.

Inhalation exposure

Exposure by inhalation is expressed as the concentration of the substance in the breathing zone atmosphere and is usually presented as an average concentration over a reference period, e.g., 8 hours for a full shift. If the substance of concern has acute health effects or if exposure is of intermittent and short duration, then there may also be interest in evaluating exposure over shorter periods. One convention in these circumstances is to assess exposure as a time-weighted average over 15 minutes. The assessment can also be based on exposure during specific tasks which have to be carried out. Information on peak exposures can be important for assessing acute effects, however, measurement of these types of exposures is often difficult to undertake in practice.

Dermal exposure

Although the main route of exposure for most substances is by inhalation, some substances may have the ability to penetrate intact skin and be absorbed into the body. Two terms can be used to describe dermal exposure:

- Potential dermal exposure is an estimate of the amount of contaminant landing on the outside of work clothing and on the exposed surface of the skin and is usually described by the sum of the exposure estimates for the affected body parts.
- Actual dermal exposure is an estimate of the amount of contamination actually reaching the skin. It is mediated by the efficiency and effectiveness of clothing garments and programmes to minimize the transfer of contamination from work wear to the skin.

Although actual dermal exposure is the most accurate determinant of likely dermal risks, potential dermal exposure is the most frequently used indicator within the risk assessment process, as it is the easiest to measure. Absorption through the skin can result from localized contamination, e.g. from a splash on the skin or clothing, or in some cases from exposure to high air concentrations of vapour. Dermal exposure can be influenced by the amount and concentration of the substance, the area of skin exposure and the duration and frequency of exposure.

Although there is agreement that dermal exposure should be expressed in terms of the mass of contaminant per unit surface area of the skin exposed, at present, there is no consensus on how dermal exposure is either best measured or assessed. Although the determinants of dermal exposure are similar to those for inhalation exposures [99], they have not been characterized to the same degree as inhalation determinants, and particularly so for the use of chemicals in general industry. Therefore, while models to predict dermal exposure have been developed [108], they are only generally applicable within a narrow domain of use. Much research is now being directed at improving the basis by which occupational dermal exposures and risks can be reliably evaluated. The recently completed EU RiskOfDerm project [109] developed a tool that aims to help chemical users in this respect [110], although the basis for its validation is still limited.

Ingestion exposure

The consideration of ingestion as a route of exposure is usually confined to those substances that accumulate within the body or which have serious, acute effects. There are no accepted methods for quantifying oral exposure. Rather, any potential risk is controlled by the adoption of good hygiene practices such as segregating working and eating facilities and adequate washing prior to eating. These matters are normally dealt with as general welfare provisions in national health and safety legislation and ingestion exposure is therefore not normally considered further in the assessment of workplace exposure at the chemicals supply level. However, the potential for exposure via ingestion should be borne in mind when considering uncertainties in the exposure assessment as a whole.

5.4.3 The process for exposure assessment

Workplaces are dynamic environments and are constantly changing. New substances or products may be used from time to time, new processes introduced and workers will be engaged, leave or tackle new jobs. Exposures therefore alter over the course of time and, while an exposure assessment may be sufficient to account for some amount of variation, its conclusions need to be re-affirmed on a regular basis. It is generally recommended that workplace risk assessments are reviewed every 3-5 years unless other substantive developments occur in the meantime, when the assessment should be reviewed straightaway.

Because of the scope of uses and risks, a strategic and systematic approach to evaluating risks is required to ensure that resources are targeted at those scenarios most likely to present the most concern. The processes for evaluation are generally of two types.

1. The first collects all relevant information on exposure determinants and then uses this to predict exposure for the scenario. The approach includes the collection of measured exposure data. The predicted exposure is then compared to some reference value (such as an occupational exposure limit) in order to determine the magnitude of the risk and the need for any exposure controls. This approach is empirically-driven and is the one that has been traditionally taken in established industry. As such, there is extensive

guidance available to describe it [111-113] and assist in its execution.

2. The second approach examines the nature of the control measures that are in place for a particular workplace situation and then compares these to measures known to provide satisfactorily control risks for the scenario in question. The "acceptable standard" is a combination of exposure controls known to offer an acceptable level of risk management (so-called "generic controls"). The "standard" may be specific to a substance but is more often "banded" for similar categories of substances. The approach is not driven by exposure measurements, but by the collection of relevant exposure information. For this reason, it is generally considered to be most suitable for adoption by smaller organizations that do not have ready access to the skilled technical resources required to implement data-driven approaches [96,114]. A similar approach is to compare the controls that are in place with those that are generally considered to constitute a good (or best) practice for that sector/activity. The Stoffenmanager tool [115] is based on such an approach.

Despite the above differences, it is increasingly accepted that the most efficient practice is to combine the benefits of both processes in a single strategy that uses the generic approach as a ready, conservative screening method to identify potential scenarios of concern that can then be evaluated in a more targeted manner at a more fundamental level (Figure 5.5). A preliminary screening method (based on general surrogates of industrial exposure and hazard, and termed Tier 0 in Figure 5.5) has been advocated for the prioritization of supply chain risks. While this may be useful to help inform regulatory bodies, it is too crude an approach to be reliably applied at the workplace level. Such a strategy will be more accessible for smaller enterprises (and worker representatives), based on information that is generally already available through chemical suppliers. It has the additional benefit of delivering practical advice which will help to improve working conditions [116]. Thus, rather than immediately collecting the detailed information demanded by empirical approaches, a tiered strategy can be adopted which prioritizes specific information needs and allows for iteration within the process prior to the steps of increasing sensitivity; improving efficiency; and ensuring resources are targeted to the most relevant determinants. Such an integrated approach to workplace exposure and risk evaluation is also envisaged under REACH (and is summarized in Table 5.6).

5.4.4 Exposure and risk management measures

One of the major determinants of exposure is the effectiveness of the measures in place at a workplace that are intended to control an individual's exposure to hazardous substances. These measures fall into two types: hardware controls and software controls. "Hardware" describes measures that are engineered into the work process and which are specifically there to reduce or control exposure and whose effectiveness is not directly dependent upon their use by a worker. For inhalation exposures, different forms of extraction ventilation are the most commonly used form of hardware. But such controls would also extend to measures intended to separate workers from hazardous areas or automated or interlocked controls. "Software" describes the management practices and other procedural controls intended to control exposure and manage risks. Software measures are by far the most numerous forms of control. These embrace the various management systems and procedures implemented to ensure health and safety and cover aspects such as worker training, hazard communication, maintenance of hardware, worker health surveillance, audits, etc. Because they depend on human intervention, they only remain effective for as long as any procedures are followed. Accordingly, the use of personal protective equipment (PPE) falls within the category of "software".

Because hardware measures are both the most reliable form of exposure control and, in many instances, the most effective, much guidance is available describing the types available, their design and how they can most appropriately be applied [117, 118]. The guidance reflects the general hierarchy of preferences that have historically been applied in the selection of control measures and which is now embodied in EU regulation [119]. More recently, the focus on health and safety standards in smaller enterprises has highlighted the need for alternative approaches that enable readily implementable, cost effective solutions to be made available to the sector, without the need for access to skilled resources [120,121]. This has lead to the development of tools that identify a package of control solutions (incorporating both hardware and software) depending on the circumstances in which a chemical is used [122, 123]. Compared to historical guidance that has mainly focused on empirical approaches to risk assessment, the major advantage of these tools is the fact that they provide an output that defines the package of controls necessary to adequately manage the risk. The most notable example is the COSHH Essentials approach [97].

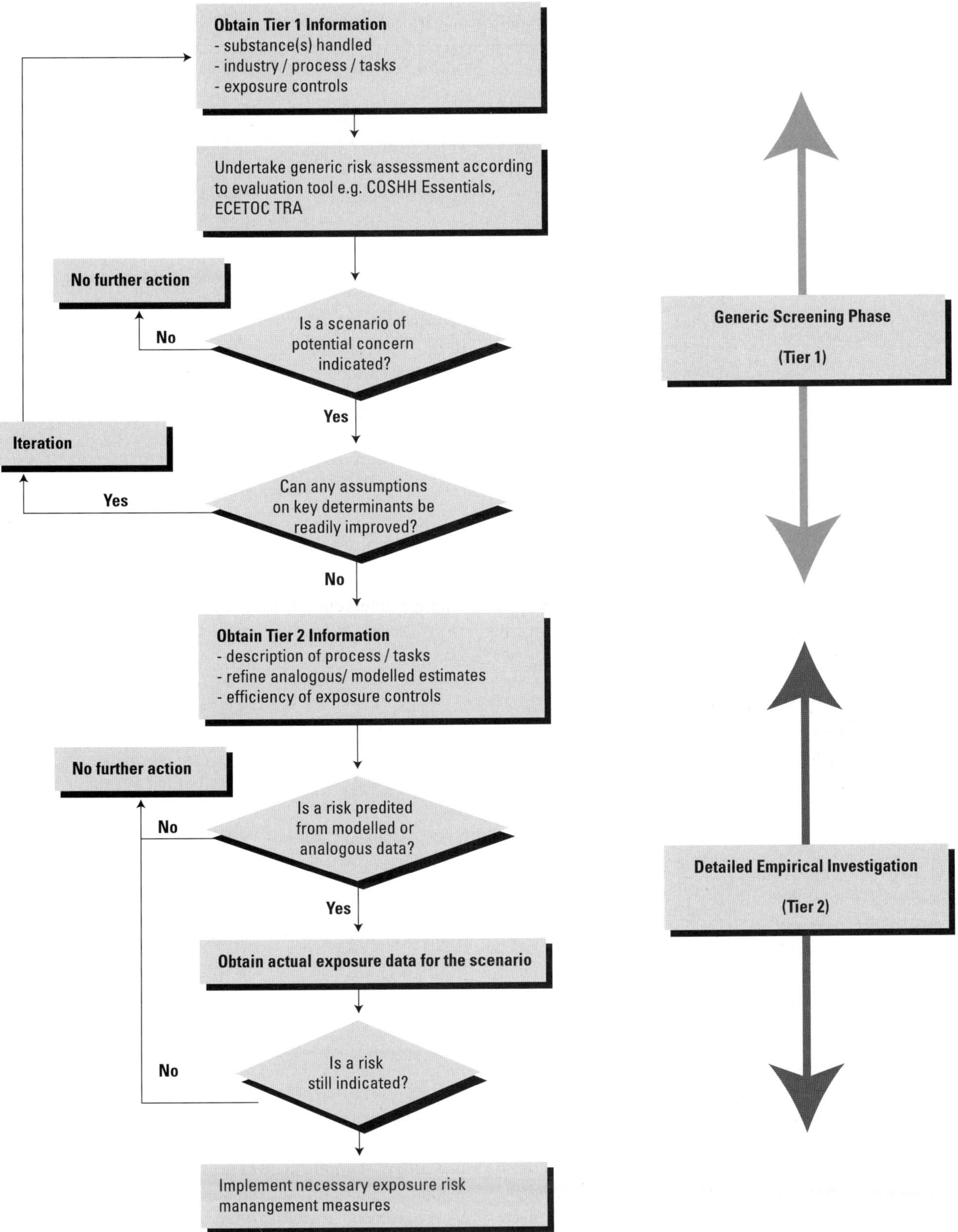

Figure 5.5. An integral strategy for the determination of workplace risks.

Table 5.6. Exposure information demands for workers.

Stage	Necessary Exposure Determinants (Workers)	Available Workplace Tools	Role of Measured Data
Tier 0 (Only relevant for supply chain prioritization. No role for reliable workplace exposure assessment.)	• Main use category • Basic use description for industrial and professional uses • Significant exposure routes • Pattern of exposure	• ECETOC TRA Tier 0 • VCI Exposure Categories • UIC DT 63 and 80 • SOMS Quick scan	Not required - estimates of exposure are descriptive rather than quantified. Annex IV.6 now defines basic information needs
Tier 1	• Physical state of the substance • Physical state of the product handled (specifically for dermal exposure) • Vapour pressure (for liquids) • "Dustiness" (for solids) • Presence or absence of local exhaust ventilation (LEV) • Duration of activity • Description of the "exposure process", covering factors such as: o energy exerted on the substance or product o surface area of source in contact with air o very limited amounts handled	• ECETOC TRA Tier1 • EASE • COSHH Essentials • UEC concept of the German agencies	Limited - RA based on models. Data grouped at a sector/ES level could also be used to refine generic model estimates
Tier 2	• Percentage of a substance in a preparation • Amount of substance used or use rate • Type and size of packaging • General "exposure control level" in the relevant industry • Viscosity of product used (for dermal exposure)	• CEMAS • Stoffenmanager • RISKOFDERM toolkit • EPA ChemSteer software suite • ILO Chemical Control Toolkit • CIA Safe Handling of Colorants	Desirable - real data will reduce uncertainty especially for ESs with low PE/DNEL ratio, i.e., would be advisable for ESs where restriction/ improved RRM are an option
Tier 3 (Authorization and restriction)	Likely expectation that detail of ES relates to: • Several other determinants may be relevant for more specific assessments and scenarios	• Workplace specific CAD assessments	Critical - real data likely to be necessary to demonstrate satisfactory performance of RRM (including possible role of health surveillance and biomonitoring)

The effectiveness of exposure control measures varies substantially. It is affected by the measures themselves, where (and how) they have been installed, and loss of efficiency over the course of time. This demands that suitable measures are applied in the workplace which both demonstrate their effectiveness at the time of introduction, and ensure satisfactory performance over the course of time. In this latter respect, personal protective equipment where the actual performance delivered is often substantially less than that cited by equipment manufacturers and suppliers, requires particular attention [124,125].

Higher tier exposure assessment

Although most uses of chemical substances should not present a significant risk to health, some activities can and do. In such circumstances, additional levels of risk control are appropriate. In Europe, additional legislative measures apply when workplace exposures to carcinogenic and mutagenic substances [126] occur. Enhanced levels of exposure control and risk management are also applied to some reprotoxic substances [102]. In such circumstances, EU practice asks two questions: firstly, is the use of the substance necessary or could it be replaced with another (safer) material (the principle of "substitution")? Secondly, if the substance cannot be substituted, then exposure to it should be reduced to as low a level as is practicable through the use of hardware controls. Apart from efforts to reduce exposure, additional attention is also expected to be given to the management systems ("software") necessary to ensure exposures remain acceptably low. Such measures include specific worker education and training; routine monitoring of personal exposures to chemicals (where appropriate, supported by biological monitoring); increased frequency of maintenance; and regular health surveillance of those at risk.

Under REACH, similar considerations apply. For those substances which are sufficiently hazardous to require authorization, the effectiveness of the measures considered (by the manufacturer/importer) to be sufficient to manage the risk will need to be demonstrated. In most cases this implies that suitable detailed information demonstrating the effectiveness of the risk management measures for key exposure scenarios will be available. This will generally need to be supported by quantitative data. Similar considerations may be expected to apply to the use of non-authorizable substances where the risks are elevated to such a degree that restriction of their use appears to be the most advisable course of action.

5.4.5 Discussion

Workplace exposure assessment under REACH is a less sensitive and less robust process than that established under EU workplace health and safety legislation. Most notably, because it only examines the risks presented from exposure to a specific supplied chemical, whereas the true nature of workplace risk is much broader and more complex than this. The REACH process is less sensitive for a number of reasons. Most importantly, the REACH CSA is intended to identify "macro level" issues that are best managed through the regulation of the supply chain, i.e., via restrictions, classification and labelling, and improved communication, via the safety data sheet.

Risk assessments undertaken by chemical suppliers (although essential in helping to inform the extent to which regulatory intervention may be advisable at the supply level, or where voluntary measures may be appropriate at the supplier level) are therefore unlikely to provide sufficiently accurate and detailed information to serve as a substitute for employer health and safety requirements. However, the process by which exposure scenarios (including their associated recommended risk management measures) are communicated will improve downstream users' access to relevant information on the safe use of the chemicals they purchase and, as a consequence, help to improve their ability to assess and manage risks at the downstream user level.

5.5 FURTHER READING

1. Whitmyre GK, Driver JH, Hakkinen PJ. 1997. Assessment of residential exposures to chemicals. In: Molak V, ed. *Fundamentals of risk analysis and risk management.* CRC Lewis Publishers, Boca Raton, FL.
2. Whitmyre G, Dang W, Driver J, Eberhart M, Fell L, Hakkinen PJ, Jayjock M, Kennedy P, Osimitz T. 2001. Consumer products and related sources. In: Baker S, Driver J, McCallum D, eds. *Residential Exposure Assessment. A Sourcebook.* Kluwer Academic/Plenum Publishers, New York, NY, pages 201-244.
3. European Centre for Ecotoxicology and Toxicology of Chemicals. 2004. *Targeted Risk Assessment.* Technical Report No. 93, ECETOC, Brussels, Belgium.
4. Money C. 2003. European approaches in the development of approaches for the successful control of workplace health risks. *Ann Occup Hyg* 47:533-540.
5. European Chemicals Bureau. 2007. Technical guidance document on preparing the chemical safety report under REACH. REACH Implementation Project 3.2. Report prepared by CEFIC, RIVM, the Federal Institute for Risk Assessment (BfR), Federal Institute for Occupational Safety and Health (BAuA), Ökopol, DHI Water & Environment and TNO Chemistry. European Commission, Joint Research Centre, Ispra, Italy (http://ecb.jrc.it/REACH/). (In preparation).

REFERENCES

1. Commission of the European Communities. 2003. Technical Guidance Document in support of Commission Directive 93/67/EEC on risk assessment for new notified substances. Commission Regulation (EC) No. 1488/94

on risk assessment for existing substances and Directive 98/8/EC of the European Parliament and of the Council concerning the placing of biocidal products on the market. Joint Research Centre, European Chemicals Bureau, Brussels, Belgium.

2. Commission of the European Communities. 2006. Regulation (EC) No. 1907/2006 of the European Parliament and of the Council of 18 December 2006 concerning the Registration, Evaluation, Authorisation and Restriction of Chemicals (REACH), establishing a European Chemicals Agency, amending Directive 1999/45/EC and repealing Council Regulation (EEC) No. 793/93 and Commission Regulation (EC) No. 1488/94 as well as Council Directive 76/769/EEC and Commission Directives 91/155/EEC, 93/67/EEC, 93/105/EC and 2000/21/EC. *Off J Eur Union* L 396/1 of 30.12.2006.
3. Organization for Economic Co-operation and Development. 2003. Description of selected key generic terms used in chemical hazard/risk assessment. Joint project with the international programme on chemical safety (IPCS) on the harmonization of hazard/risk assessment terminology. OECD Environment, Health and Safety Publications. *Series on Testing and Assessment* 44. OECD, Paris, France.
4. Commission of the European Communities. 2004. European Union System for the Evaluation of Substances 2.0 (EUSES 2.0). Prepared for the European Chemicals Bureau by the National Institute for Public Health and the Environment (RIVM), Bilthoven, The Netherlands. Available from European Chemicals Bureau, http://ecb.jrc.it.
5. McKone TE.1993. CalTOX, A multimedia total exposure model for hazardous-waste sites UCRL-CR-111456PtI-IV. US Department of Energy, Lawrence Livermore National Laboratory, Government Printing Office, Washington, DC.
6. Czub G, McLachlan MS. 2004. A food chain model to predict the levels of lipophilic organic contaminants in humans. *Environ Toxicol Chem* 23 (10):2356-2366.
7. US Environmental Protection Agency. 2006. Exposure and Fate Assessment, Screening Tool (E-FAST) Version 2.0 Documentation Manual. US EPA, Office of Pollution Prevention and Toxics, Exposure Assessment Branch, Washington, DC.
8. UMS, 1993. Umweltmedizinische Beurteilung der Exposition des Menschen durch altlastbedingte Schadstoffe (UMS) Anslussbericht "Weiterentwicklung und Erbrobung des Bewertungsmodells zur Gefahrenbeurteilung bei Altlasten" von der Arbeitsgemeinschaft Fresenius Consult GmbH und focon-Ingenieurgesellschaft mbH F&E-Vorhaben 10340107 [in German].
9. Slob W, Bakker MI. 2004. Probabilistic calculation of intake of substances via incidentally consumed food products. Supplement to the handbook for modelling of intake of substances via food. RIVM report 320103003. National Institute for Public Health and the Environment (RIVM), Bilthoven, The Netherlands.
10. Briggs GG, Bromilow RH, Evans AA. 1982. Relationships between lipophilicity and root uptake and translocation of non-ionised chemicals by barley. *Pestic. Sci.* 13:495-504.
11. Trapp S., Matthies M. 1995. Generic one-compartment model for uptake of organic chemicals by foliar vegetation. *Environl Sci Technol* 29: 2333-2338. Erratum vol. 30:360.
12. Smith KEC, Jones KC. 2000. Particles and vegetation: implications for the transfer of particle-bound organic contaminants to vegetation. *Sci Total Environ* 246:207-236.
13. Sheppard SC, Evenden WG. 1992. Contaminant enrichment of sparingly soluble contaminants (U, Th and Pb) by erosion and by soil adhesion to plants and skin. *Environ Geochem Health* 14:121-131.
14. Trapp S, Matthies M. 1998. *Chemodynamics and environmental modeling. An introduction.* Springer-Verlag, Berlin Heidelberg New York 1998. ISBN 3-540-63096-1.
15. Rikken MGJ, Lijzen JPA, Cornelese AA. 2001. Evaluation of model concepts on human exposure. Proposals for updating the most relevant exposure routes of CSOIL. RIVM report 711701022. National Institute for Public Health and the Environment (RIVM), Bilthoven, The Netherlands.
16. Veith GD, DeFoe DL, Bergstedt BV. 1979. Measuring and estimating the bioconcentration factor of chemicals in fish. *J. Fish Res. Board Can.* 36:1040-1048.
17. Bintein S, Devillers J, Karcher W. 1993. Nonlinear dependence of fish bioconcentration on n-octanol/water partition coefficient. *SAR QSAR Environ Res* 1(1):29-39.
18. Devillers J, Domine D, Bintein S, Karcher W. 1998. Fish bioconcentration modeling with log P. *Toxicol Methods* 8(1):1-10.
19. Connell DW, Hawker DW. 1988. Use of polynomial expressions to describe the bioconcentration of hydrophobic chemicals by fish. *Ecotoxicol Environ Safety* 16(3):242-257.
20. Thomann RV. 1989. Bioaccumulation model of organic chemical distribution in aquatic food chains. *Environ Sci Technol* 23(6):699-707.
21. Thomann RV, Connolly JP, Parkerton TF. 1992. An equilibrium model of organic chemical accumulation in

aquatic food webs with sediment interaction. *Environ Toxicol Chem* 11:615-629.

22. Gobas FAPC. 1993. Gastrointestinal magnification: the mechanism of biomagnification and food chain accumulation of organic chemicals. *Environ Sci Technol* 27(13):2855-63.
23. Campfens J, Mackay D. 1997. Fugacity-based model of PCB bioaccumulation in complex aquatic food webs. *Environ Sci Technol* 31 (2):577-583.
24. Czub G, McLachlan MS. 2003. A food chain model to predict the levels of lipophilic organic contaminants in humans. Environ Toxicol Chem 23(10):2356-2366.
25. Kelly BC, Gobas APC, McLachlan MS. 2004. Intestinal absorption and biomagnification of organic contaminants in fish, wildlife and humans. *Environ Toxicol Chem* 23(10):2324-2336.
26. Hrubec J, Toet C. 1992. Predictability of the removal of organic compounds by drinking-water treatment. RIVM report 714301007. National Institute for Public Health and the Environment (RIVM), Bilthoven, The Netherlands.
27. Rikken MGJ, Lijzen JPA. 2004. Update of risk assessment models for the indirect human exposure. RIVM report 601516011/2004. National Institute for Public Health and the Environment (RIVM), Bilthoven, The Netherlands.
28. Vonk MW. 1985. Permeatie van organische verbindingen door leidingmaterialen. mededeling nr 85, KIWA, Nieuwegein [in Dutch].
29. Vonk MW. 1985. Permeatie van organische verbindingen door leidingmaterialen. *H_2O* 18:529-538 [in Dutch].
30. Heijden BG van der. 1985. Enkele ervaringen met de permeatie van organische stoffen door kunststof drinkwaterleidingen. *H_2O* 18, nr. 5:88-95 [in Dutch].
31. Travis CC, Arms AD. 1988. Bioconcentration of organics in beef, milk and vegetation. *Environ Sci Technol* 22:271-274.
32. Kenaga EE. 1990. Correlation of bioconcentration factors of chemicals in aquatic and terrestrial organisms with their physical and chemical properties. *J Am Soc* 14:553-556.
33. Dowdy DL, McKone TE, Hsieh PH. 1996. Prediction of Chemical Biotransfer of Organic Chemicals from Cattle Diet into Beef and Milk Using the Molecular Connectivity Index. *Environ Sci Technol* 30(3): 984-989.
34. McLachlan MS. 1994. Model of the Fate of Hydrophobic Contaminants in Cows. *Environ Sci Technol* 28(13):2407-2414.
35. Rosenbaum R. 2006 Multimedia and food chain modelling of toxics for comparative risk and life cycle impact assessment. Ecole Polytechnique Fédérale, Lausanne, Switzerland.
36. Derks HJGM, Berende PLM, Olling M, Everts H, Liem AKD 1994. Pharmacokinetic modeling of polychlorinated dibenzo-p-dioxins (PCDDs) and furans (PCDFs) in cows. *Chemosphere* 28 (4):711-715.
37. Freijer JI, van Eijkeren JHC, Sips AJAM. 1999. Model for Estimating Initial Burden and Daily Absorption of Lipophylic Contaminants in Cattle. RIVM report 643810005. National Institute for Public Health and the Environment (RIVM), Bilthoven, The Netherlands.
38. Eijkeren JCH van, Jager DT, Sips AJAM. 1998. Generic PBPK-modelling of lipophilic contaminants in the cow. RIVM report 679102042. National Institute for Public Health and the Environment (RIVM), Bilthoven, The Netherlands.
39. Schwartz S, Berding V, Trapp S, Matthies M. 1998. Quality Criteria for environmental Risk Assessment Software - Using the Example of EUSES. *Environ Sci Pollut Res* 5:217-222.
40. European Centre for Ecotoxicology and Toxicology of Chemicals. 1994. Assessment of non-occupational Exposure to chemicals. Technical Report No. 58. ECETOC, Brussels, Belgium.
41. Hendricks MH. 1970. Measurement of enzyme laundry product dust levels and characteristics in consumer use. *J Am Oil Chem Soc* 47(6):207-211.
42. Becker D. 1979. Methodology for Estimating Direct Exposure to New Chemical Substances. Office of Toxic Substances, US Environmental Protection Agency, Washington, DC.
43. Versar Inc. 1986. Standard scenarios for estimating exposure to chemical substances during use of consumer products. Volumes I and II. Prepared for US Environmental Protection Agency. (http://www.epa.gov/opptintr/exposure/docs/Versar_1986_Standard_Scenarios Volume_I.pdf and http://www.epa.gov/opptintr/exposure/docs/Versar_1986_Standard_Scenarios_Volume_II.pdf)
44. US Environmental Protection Agency. 1987. National usage survey of household cleaning products. Westat, Inc., Rockville, Maryland. Prepared for USEPA, Exposure Evaluation Division, Office of Toxic Substances, Office of Pesticides and Toxic Substances, Washington, DC, USA Contract Number 68-02-4243. (http://www.epa.gov/opptintr/exposure/docs/Westat_1987a_Household_Cleaning_Products.pdf).
45. US Environmental Protection Agency. 1987. Household solvent products. A national usage survey. Westat, Inc., Rockville, Maryland. Prepared for USEPA, Exposure Evaluation Division, Office of Toxic Substances, Office of Pesticides and Toxic Substances, Washington, DC,

USA. Contract Number 68-02-4243. (http://www.epa.gov/opptintr/exposure/docs/Westat_1987b_Household_Solvent_Products.pdf)

46. US Environmental Protection Agency. 1987. Methods for Assessing Exposure to Chemical Substances, Volume 7. Methods for Assessing Consumer Exposure to Chemical Substances. USEPA, Exposure Evaluation Division, Office of Toxic Substances, Office of Pesticides and Toxic Substances, Washington DC, USA. EPA Contract 68-02-3968. (http://www.epa.gov/opptintr/exposure/docs/USEPA_1987c_Methods_for_Assessing_Exposure_Volume_7.pdf)
47. US Environmental Protection Agency. 1990. Exposure Factors Handbook. US-EPA, Washington, DC. (EPA/600/8-89/043)(http://risk.lsd.ornl.gov/homepage/EFH_1989_EPA600889043.pdf).
48. US Environmental Protection Agency. 1997. Exposure Factors Handbook. EPA No. 600C99001. US EPA, National Service Center for Environmental Publications, Cincinnati. (http://cfpub.epa.gov/ncea/cfm/recordisplay.cfm?deid=12464&CFID=404517&CF).
49. Organization for Economic Co-operation and Development. 1993. Occupational and consumer exposure assessments. *OECD Environmental Monographs* No. 70. OECD, Paris, France.
50. Commission of the European Communities. 1992. Council Directive 92/32/EEC of 30 April 1992 amending for the seventh time Directive 67/548/EEC on the approximation of the laws, regulations and administrative provisions relating to the classification, packaging and labelling of dangerous substances *OJEC* L 154, 5.6.1992, p. 1–29
51. Commission of the European Communities. 1994. Commission Regulation (EC) No. 1488/94 of 28 June 1994 laying down the principles for assessment of risks to man and to the environment of existing substances in accordance with Council Regulation (EEC) No. 793/93. *OJEC* L 161.
52. Commission of the European Communities. 1996. Technical Guidance document in support of commission directive 93/67/EEC on risk assessment for new notified substances and commission regulation (EC) No. 1488/94 on risk assessment for existing substances. Part 1. Brussels, Belgium.
53. International Programme on Chemical Safety. Principles for the Assessment of Risks to Human Health from Exposure to Chemicals. IPCS, World Health Organization, Geneva, Switzerland. *Environmental Health Criteria* 210. http://www.inchem.org/documents/ehc/ehc/ehc210.htm#SubSectionNumber:5.5.3.
54. Vermeire TG, van der Poel P, van de Laar RTH, Roelfzema H. 1993. Estimation of consumer exposure to chemicals: application of simple models. *Sci Total Environ* 136:155-176.
55. Van Veen MP. 1996. Een datamodel voor een blootstellingsanalyse van consumentenproducten [A Data Model for an Exposure Assessment Database for Consumer Products]. RIVM. Report 612810004. National Institute for Public Health and the Environment (RIVM), Bilthoven, The Netherlands.
56. Kovacs DC, Small MJ, Davidson CI, Fischoff B. 1997. Behavioral factors affecting exposure potential for household cleaning products. *J Exp Anal Environ Epidemiol* 7:505-520.
57. Baker S, Driver J, McCallum D, eds. 2001. *Residential exposure assessment. A sourcebook.* Kluwer Academic/ Plenum Publishers, New York, NY. ISBN 0-306-46517-5.
58. Riley DM, Small MJ, Fischhoff B. 2000. Modeling methylene chloride exposure-reduction options for home paint stripper users. *J Exp Anal Environ Epidemiol* 10:240-250.
59. Riley DM, Fischhoff B, Small M, Fischbeck P. 2001. Evaluating the effectiveness of risk-reduction strategies for consumer chemical products. *Risk Anal* 21:357-369.
60. Weegels MF, van Veen MP. 2001. Variation of consumer contact with household products: a preliminary investigation. *Risk Anal* 21:499-511.
61. Zaleski R, Gephart L. 2000. Exposure factors sourcebook for European populations, with focus on UK data. ECETOC Technical Report No. 79. European Centre for Ecotoxicology and Toxicology of Chemicals, Brussels, Belgium.
62. www.rivm.nl/consexpo
63. www.ktl.fi/expofacts
64. www.tera.org/peer/VCCEP/VCCEPIntroduction.html
65. www.heraproject.com
66. www.chemicalawareness.org/index.html
67. www.jrc.cec.eu.int/eis-chemrisks
68. Delmaar JE, Park MVDZ, van Engelen JGM. 2005. ConsExpo - Consumer exposure and uptake models - Program manual. RIVM report 320104004. National Institute of Public Health and the Environment (RIVM), Bilthoven, The Netherlands.
69. www.epa.gov/oppt/exposure/pubs/mccem.htm
70. http://www.epa.gov/oppt/exposure/pubs/wpem.htm
71. www.americansolventscouncil.org/resources/promise.asp
72. http://www.thelifelinegroup.org/lifeline
73. Bruinen de Bruin Y, Hakkinen P, del Pozo C, Reina V,

Papameletiou D. 2006. Risk Management Measures for Chemicals in Consumer Products. EUR number 22278 EN, ISBN 92-79-01972-4. European Commission, Joint Research Centre, Institute for Health and Consumer Protection, Physical and Chemical Exposure Unit, Ispra, Italy, in preparation.

74. Petersen DW. 1989. Profile of accidental ingestion calls received via a toll-free line on detergent product labels. *Vet Hum Toxicol* 31:125-127.
75. Petersen DW. 1989. Lemon aesthetics in hand dishwashing detergents do not influence reported accidental ingestion frequency and volume. *Vet Hum Toxicol* 31:257-258.
76. Yamashita M, Tanaka J, Yamashita M, Hirai H, Suzuki M, Kajigaya H. 1997. Mist particle diameters are related to the toxicity of waterproofing sprays: comparison between toxic and non-toxic products. *Vet Hum Toxicol* 39 (2):71-74.
77. Yamashita M, Yamashita M, Tanaka J, Hirai H, Suzuki M, Kajigaya H. 1997. Toxicity of waterproofing spray is influenced by the mist particle size. *Vet Hum Toxicol* 39 (6):332-334.
78. Soap and Detergent Association. 2005. Risk assessment guidance for enzyme-containing products. www.cleaning101.com/files/SDA_Enzyme_Risk_Guidance_October_2005.pdf.
79. Yamashita M and Tanaka J. 1995. Pulmonary collapse and pneumonia due to inhalation of a waterproofing. aerosol in female CD-1 Mice *J Toxicol Clin Toxicol* 33 (6):631-637.
80. Agricola G. 1556. *De Re Metallica.* Translated by Hoover HC and Hoover LH. 1912. Mining Magazine, London, UK.
81. Rammazzini B. 1713. *De Morbis Artificum.* Translated by Wright WC. 1964. Printed in the New York Academy of Medicine, History of Medicine Series No. 23, Hafner.
82. Thackrah CTH. 1832. *The effects of arts, trades and professions and of civic states on health and longevity with suggestions for the removal of many of the agents which produce disease and shorten the duration of life.* Longman, Leeds, UK.
83. Hutchins BL, Harrison A. 1903. *A history of factory legislation.* Frank Cass, London, UK.
84. Gaskill E. 1848. *Mary Barton: a tale of Manchester life.* Thomas Nelson, UK.
85. Dickens C. 1860 *The uncommercial traveller.* Chapman & Hall, London, UK.
86. Zola E. 1885. *Germinal.* Penguin Classics.
87. Bartrip P. 2002. The Home Office and the dangerous trades: regulating occupational disease in Victorian and Edwardian Britain. *Clio Medica* 68. Wellcome Series in the History of Medicine, Amsterdam, The Netherlands.
88. Commission of the European Communities. 1980. Council Directive 80/1107/EEC on the protection of workers from the risks related to exposure to chemical, physical and biological agents at work, *OJ* L 327, 3rd December 1980.
89. Haldane JS. 1912. *Methods of air analysis.* Charles Griffin, London, UK.
90. Drinker P, Hatch T. 1936. *Industrial dust: hygiene significance, measurement and control.* McGraw-Hill, New York, USA.
91. Piney M. 2001. OELs and the effective control of substances hazardous to health in the UK. Health and Safety Executive (HSE), London, UK.
92. Meldrum M. 2001. Setting occupational exposure limits for sensory irritants: the approach in the European Union. *Am Industr Hyg Assoc J* 62:730-732.
93. Ziegler-Skylakakis K. 2004. Approaches for the development of occupational exposure limits for man-made mineral fibres (MMMFs). *Mutat Res* 553:37-41.
94. Bolt HM, Thier R. 2006. Biological monitoring and Biological Limit Values (BLV): the strategy of the European Union. *Toxicol Lett* 162:119-124.
95. Commission of the European Communities. 1998. Council Directive 98/24/EC on the protection of the health and safety of workers from the risks related to chemical agents at work. *OJ*, L131, 5th May 1998.
96. Russell RM, Maidment SC, Brooke I, Topping MD. 1998. An introduction to a UK scheme to help small firms control health risk from chemicals. *Ann Occup Hyg* 42:367-376.
97. Health and Safety Executive. 1999. COSHH Essentials. HSE, London, UK.
98. Money C, de Rooij C, Floch F, Jacobi S, Koundakjian P, Lanz S, Penman M, Rodriguez C, Veenstra G. 2003. A structured approach to the evaluation of workplace health risks. *Policy and Practice in Health and Safety* 2:44-65.
99. European Centre for Ecotoxicology and Toxicology of Chemicals. 2004. Targeted Risk Assessment. Technical Report No. 93. ECETOC, Brussels, Belgium.
100. Commission of the European Communities. 1993. Council Regulation 793/93 on the evaluation and control of the risks of existing substances. *OJ* L084, 5th April 1993.
101. Northage C, Marquart H. 2001. Occupational exposure information needs for regulatory risk assessment of existing chemicals. *Appl Occup Environ Hyg* 16:315-318.
102. Commission of the European Communities. 1992. Council Directive 92/85/EEC on the introduction of measures to encourage improvements in the safety and

health at work of pregnant workers and workers who have recently given birth or are breastfeeding. *OJ* L 348, 28th November 1992.

103. Schneider T, Vermeulen R, Brouwer DH, Cherrie JW, Kromhout H, Fogh CL. 1999. Conceptual model for assessment of dermal exposure. *Occup Environ Med* 56:765-773.
104. Cherrie JW, Schneider T. 1999. Validation of a new method for structured subjective assessment of past concentrations. *Ann Occup Hyg* 43:235-245.
105. Tickner J, Friar J, Creely K S, Cherrie JW, Pryde DE, Kingston J. 2005. The development of the EASE model. *Ann Occup Hyg* 49:103-110.
106. Money C, Margary SA. 2002. Improved use of workplace exposure data in the regulatory risk assessment of chemicals within Europe. *Ann Occup Hyg* 46:279-285.
107. Tielemans E, Marquart H, De Cock J, Groenewold M, van Hemmen J. 2002. A proposal for evaluation of exposure data. *Ann Occup Hyg* 46:287-297.
108. Van-Wendel-de-Joode B, Brouwer DH, Vermeulen R, Van Hemmen JJ, Heederik D, Kromhout H. 2003. DREAM: a method for semi-quantitative dermal exposure assessment. *Ann Occup Hyg* 47:71-87.
109. Van Hemmen JJ, Auffarth J, Evans PG, Rajan-Sithamparanadarajah B, Marquart H, Oppl R. 2003. RISKOFDERM: risk assessment of occupational dermal exposure to chemicals. *Ann Occup Hyg* 47:595-598.
110. Oppl R, Kalberlah F, Evans PG, van Hemmen JJ. 2003. A toolkit for dermal risk assessment and management: an overview. *Ann Occup Hyg* 47:629-640.
111. Leidel NA, Busch KA, Lynch JR. 1977. Occupational exposure sampling strategy manual. DHEW (NIOSH) Publication 77-173. National Institute of Occupational Safety and Health, Cincinnati, USA.
112. Guest IG, Cherrie JW, Gardner RJ, Money CD. 1993. Sampling strategies for airborne contaminants in the workplace. British Occupational Hygiene Society Technical Guide No. 11, H and H Scientific Consultants, Leeds, UK.
113. Mulhausen JR, Damiano J. 1998. A strategy for assessing and managing occupational exposures (2nd Edition). American Industrial Hygiene Association Press, Fairfax, USA.
114. Wiseman J, Gilbert F. 2002. COSHH Essentials: survey of firms purchasing this guidance. Health & Safety Executive Contract Research Report 434. HMSO, Norwich, UK.
115. Groenewold M. 2004. Reducing the risk of chemical exposure for workers in industry. TNO Leads in Life Sciences 25 p.9 (see http://www.stoffenmanager.nl/).
116. Hudspith B, Hay AWM. 1998. Information needs of workers. *Ann Occup Hyg* 42:401-406.
117. American Conference of Governmental Industrial Hygienists. 2004. Industrial ventilation: a manual for control (25th edition). ACGIH , Cincinnati, USA.
118. Lipton S, Lynch J. 1994. Handbook of health hazard control in the chemical process industry. Wiley Interscience, New York, NY.
119. Commission of the European Communities. 1998. Council Directive 98/24/EC on the protection of the health and safety of workers from the risks related to chemical agents at work. *OJ* L131, 5th May 1998.
120. Briggs D, Crumbie N. 2000. Characteristics of people working with chemical products in small firms. Health & Safety Executive Contract Research Report 278. HMSO, Norwich, UK.
121. Walters D, Grodzki K. 2006. *Beyond limits*. Elsevier, The Hague, The Netherlands.
122. Money C. 1992. A structured approach to occupational hygiene in the design and operation of fine chemical plant. *Ann Occup Hyg* 36:601-607.
123. Chemical Industries Association. 1993. Safe handling of colourants 2. CIA, London, UK.
124. Shackleton S, Piney MD. 1984. A comparison of two methods of measuring personal noise exposure. *Ann Occup Hyg* 28:373-390.
125. Brouwer DH, Marquart H, van Hemmen JJ. 2001. Proposal for an approach with default values for the protection offered by PPE, under European new or existing substance regulations. *Ann Occup Hyg* 45:543-553.
126. Commission of the European Communities. 2004. Directive 2004/37/EC of the European Parliament and of the Council on the protection of workers from the risks related to exposure to carcinogens or mutagens at work. *OJ* L 158 , 30th April 2004.

6. TOXICITY TESTING FOR HUMAN HEALTH RISK ASSESSMENT

T.G. VERMEIRE, A.J. BAARS, J.G.M. BESSEMS, B.J. BLAAUBOER, W. SLOB, AND J.J.A. MULLER

6.1 INTRODUCTION

Research into the toxic effects of substances on humans can be traced back to the ancient centres of civilization in Egypt, Greece and China, where toxic chemical substances were used as poisons and sometimes as medicines. “Toxicology is the scientific discipline involving the study of actual or potential danger presented by the harmful effects of substances in living organisms and ecosystems, of the relationship of such harmful effects to exposure and of the mechanism of action, diagnosis, prevention and treatment of intoxications” [1]. Paracelsus’ saying: “Dosis sola facit venemum” (it is the dose which makes the poison) is well-known and depicts a property inherent to almost every chemical: at a certain dose, effects are inevitable. In risk assessment, the determination of the harmful or adverse effects and the relationship with exposure is one of the key steps towards the characterization of the risk. A large number of steps is involved between the administration of the external dose and the final toxic effect (Figure 6.1).

The science of human toxicology includes both the production and gathering of toxicity data in biological systems, and the subsequent evaluation and interpretation of these data, with the aim of predicting possible risk, or lack of risk, to humans. Toxicity testing is mandatory and the scope depends on the anticipated use. The toxicity testing of environmental chemicals initially focused on determining safe levels of human exposure to toxic chemicals. This testing has now expanded from

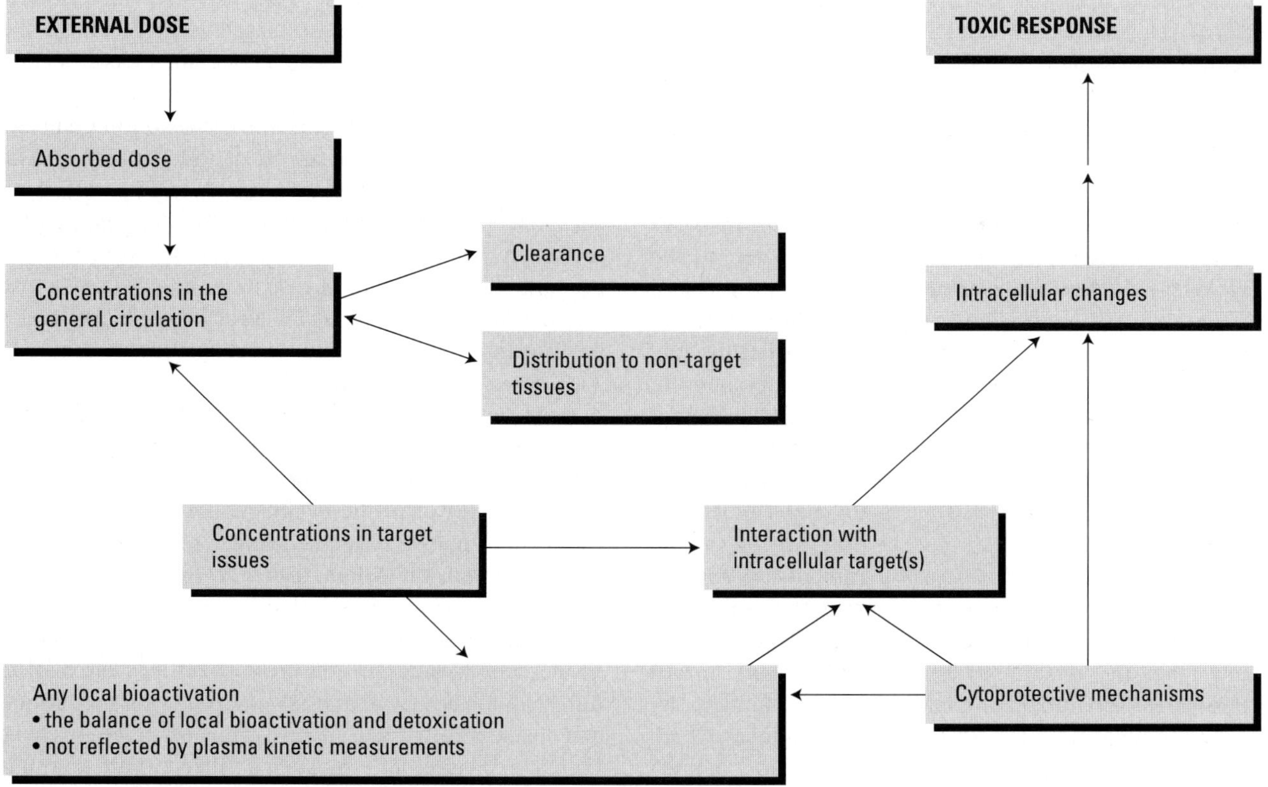

Figure 6.1. Processes leading to the generation of a toxic response [2].
Note:”Concentrations” refers to the relevant active form delivered by the general circulation and may be the parent compound or an active metabolite produced in another tissue and delivered to the target tissue or organ

C.J. van Leeuwen and T.G. Vermeire (eds.), Risk Assessment of Chemicals: An Introduction, 227–280.

simple acute and subacute tests to careful consideration of data on acute, subacute and chronic toxicity, specific toxicity such as carcinogenicity, mutagenicity, reproductive toxicity and, more recently, immunotoxicity, neurotoxicity, dermal toxicity and other organ tests. In addition to these toxicity studies, data on the mechanisms of action at the tissue, cellular, subcellular and receptor levels, as well as toxicokinetic data, greatly facilitate the interpretation of toxicity data and the assessment of the potential hazard to humans. "Protocol toxicology" and "receptor toxicology" are essential to provide the optimum context for risk prediction [3,4] and this requires international harmonization.

Toxicology is becoming increasingly complex. It takes a considerable amount of effort to determine the toxicity of just one agent, let alone the enormous variety of agents currently available. The large number of chemicals involved requires rules to be able to select priority chemicals and testing strategies. This is because of the time and cost that testing requires as well as for animal welfare reasons. As discussed in Chapter 11, such testing strategies increasingly include basic steps which rely on alternative estimation methods, such as quantitative structure-activity relationships (QSARs), structure-activity relationships (SARs) and *in vitro* tests, rather than on immediate testing on experimental animals. The starting point of such strategies should be the regulatory information requirements.

This chapter will discuss the toxicological methods used in risk assessment against the background of the Technical Guidance Documents on Risk Assessment of the European Commission [5]. After exploring general aspects of toxicology, Section 6.3 will consider the fate of chemicals in humans and shed light on methods to determine how chemicals are absorbed, distributed, metabolized and excreted, aspects commonly referred to as the toxicokinetics of chemicals. This will provide a foundation for the discussion of toxicity studies and their evaluation in Section 6.4, to establish the action of chemicals at the target tissue, commonly referred to as the toxicodynamic properties. Section 6.5 will show how toxicological information is used for classification and labelling, dose-response assessment and hazard assessment of mixtures. The chapter concludes by discussing the characterization of risks to humans which is determined by combining the knowledge obtained from the hazard assessment and the exposure assessment. The evaluation of all data with regard to their adequacy and completeness is very important and Chapter 8 will address this in general terms.

6.2 GENERAL ASPECTS OF TOXICITY

Toxicity

Toxicity is the capacity of a chemical to cause injury to a living organism. In theory, small doses can be tolerated due to the presence of systems for physiological homeostasis, i.e., the ability to maintain physiological and psychological stability, or compensation, i.e., physiological or psychological adaptation. Examples of this are metabolic detoxification, cellular adaptation and repair. Repair is a reaction to injury, causing irreversible tissue alteration. Above a given chemical-specific threshold the ability of organisms to compensate for toxic stress becomes saturated, leading to loss of homeostasis and adverse effects, which may be reversible or irreversible, and ultimately fatal.

Toxicity and hazard evaluation

The assessment of adverse effects starts with an evaluation of non-human and human data. Non-human data include animal data (Section 6.4), *in vitro* data (Section 6.4.10) and non-testing data such as (Q)SARs (Chapter 10). The data evaluation will result in the identification of hazards, which includes classification and labelling (Section 6.5.2), and establishing the relationship between the dose or concentration and the incidence and severity of an effect (Section 6.5.3). The latter process will preferably result in a no-effect or acceptable effect level for humans, by applying one or more extrapolation steps (Section 6.5.4). Where it is not possible to determine the quantitative dose-response relationship for effects, this should be explained and a semi-quantitative or qualitative analysis carried out. In such cases it is generally sufficient to evaluate whether the substance has an inherent capacity to cause such an effect. The route, duration and frequency of human exposure to a substance during normal use should be a principle factor in the evaluation of hazards: hazards which may not be expressed under one exposure may become apparent under another.

Adversity of effects

In the determination of a critical effect it is essential to differentiate between non-adverse and adverse effects and decide whether any adverse effect observed is related to the exposure, i.e. substance-related. For example, in repeated-dose toxicity testing the average values of selected parameters are compared with the average values of these parameters in concurrent untreated control animals. Adverse effects can then be defined in purely statistical terms as statistically significant changes

($P < 0.05$) relative to control values. This approach is too narrow: other factors also need to be considered such as the presence or absence of a dose and time-effect relationship or a dose and time-response relationship, the biological relevance of an effect, the reversibility of an effect, and the normal biological variation in effects as shown by historical control values. Guidance in selecting adverse effects from a particular subchronic or chronic animal test can be obtained from publications of the Health Council of The Netherlands [6], IPCS [7-10], USEPA [11] and OECD [12] (Box 6.1).

Further to the discussion on differentiating between adverse and non-adverse effects, as with classification (see Section 6.5.2), the question which should be asked is: at what dose or concentration does the substance cause "serious damage to health"? According to the guidance provided by the EC serious damage to health is considered to include death, clear functional disturbance or morphological changes which are toxicologically significant. Irreversibility of lesions is a key factor in this assessment. The response of cells and tissues to chemical injury at the intracellular level, i.e., biochemical, functional, and structural changes, or extracellular level, i.e., metabolic and regulatory changes, can be categorized as either degeneration, inflammation or proliferation. The outcome of these pathological changes depends on the combinations in which they occur, their potency, and their duration. Depending on these factors, initial injury such as mild cell degeneration or proliferation can, for example, regenerate to become normal or eventually result in irreversible injury such as neoplasia. Therefore, even assuming that it is always possible to detect chemical injury at the intracellular level in a 28-d test – which is by no means a valid assumption – and taking into account the guidance referred to above, direct advice by experienced pathologists and toxicologists is essential for correct evaluation of the degree of damage to health.

6.3 INTERNAL HUMAN EXPOSURE

6.3.1 Experimental biokinetics

Introduction

The intended use and production volume of a substance define the toxicity testing that is required. Although regulations may differ to some extent between countries, and even within countries depending on the regulatory bodies, they all require that chemical substances introduced into the human environment directly or indirectly must not constitute any significant risk to humans. When considering the hazard of a substance

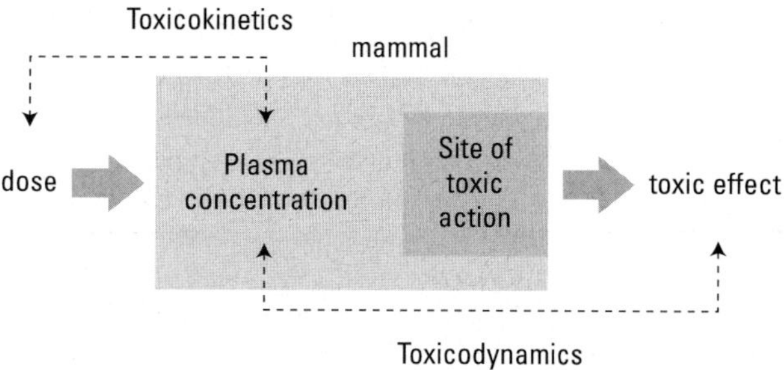

Figure 6.2. Toxicokinetics

it is assumed that all effects are produced through an interaction with the substance at a target site. A response is only produced if sufficient amounts of a chemical or its active metabolites reach a receptor, thus underpinning the importance of information on the absorption, distribution, biotransformation and excretion (ADME) that determine the fate of the chemical in the body, resulting in the *internal dose* at the target site (Figure 6.2). Toxic effects are a manifestation of the internal dose or concentration at the target site, as well as the duration of exposure at that site. Sufficient understanding of the interplay of these processes is essential for human hazard assessment. Up until recently, information on the fate of industrial chemicals in mammals, if available at all, was mostly limited to absorption, measured or predicted from physicochemical properties such as molecular weight and log K_{ow} and thereby only giving an indication of *internal exposure*. In some cases, little extra information was given, e.g., on distribution, major metabolites found or the extent of excretion of the substance. Generally, this information was interpreted too much in isolation, e.g., "the oral absorption of this substance is 50% and two major urinary metabolites are found".

The science of biokinetics is an integrative and interdisciplinary science that covers both the various processes of ADME as well as exposure and effects. By taking a holistic approach to ADME processes parameters can be established and then applied in various extrapolations for human risk assessment. As applies to dynamic processes in general, biokinetic studies should answer these questions: what and how much; where; and when (or rate)? Answers can be obtained by knowing the amount, concentration, time-scale and direction. The importance of the latter is illustrated here. The information that a compound has a half-life of 6 hours is not complete as the question of "where" remains open. If *"in plasma"* is added, the reader knows that it takes 6 hours before the plasma level will decrease by 50%. This brings us to a definition often used (although there are many other similar ones):

Box 6.1. Weight-of-evidence in hazard assessment

In view of the number of factors to be taken into account, expert judgement is an essential part of the assessment process. Certain decision-supporting rules can be applied:

- Effects can be ranked in order of severity. An attempt was made by the USEPA in 1986 [11]. The result, adapted to the hazard evaluation of subchronic tests using OECD or EC protocols and slightly expanded, is shown in Table 6.1. The borderline between adverse and non-adverse effects can be drawn somewhere in the upper part of the table. It should be emphasized here that the degree of severity of an effect very much depends on duration and frequency of exposure and the site and characteristics of the particular change observed. Therefore, Table 6.1 should be used with caution.

Table 6.1 Ranking of physiological and pathological effects in order of severity

Effect	Severity
Biochemical/haematological change with no pathological change and no change in organ weight; or a change in organ weight with no pathological and biochemical/haematological change	least severe ↑
Biochemical/haematological change with no pathological change and with a change in organ weight	
Enzyme induction and subcellular proliferation or other changes in organelles but no other apparent effects	
Biochemical/haematological change with slight pathological changes	
Hyperplasia, hypertrophy or atrophy with change in organ weight	
Reversible cellular changes: cloudy swelling, hydropic change or fatty changes	
Necrosis, or metaplasia with no apparent reduction in organ functions; any neuropathy without apparent behavioral, sensory, or physiological changes	
Necrosis, atrophy, hypertrophy, or metaplasia with a detectable reduction in organ functions; any neuropathy with a measurable change in behavioral, sensory, or physiological activity; reduced body weight gain; clinical symptoms	
Necrosis, atrophy, hypertrophy, or metaplasia with definitive organ dysfunction; any neuropathy with gross changes in behavioral, sensory, or motor performance	
Pronounced pathological changes with severe organ dysfunction; any neuropathy with loss of behavioral or motor control or loss of sensory ability	↓
Death or pronounced life-shortening	most severe

- More weight is attached to changes in parameters which increase in severity or response with increasing dose.
- More weight is attached to changes in parameters which are correlated to other changes observed. Examples are an increase in blood urea accompanied by an increase in kidney weight, an increase in liver weight accompanied by slight pathological changes such as fatty changes, or an increase in creatine phosphokinase combined with increases in lactate dehydrogenase and/or a-hydroxybutyrate dehydrogenase (indicative of myocardial damage). The lowest effect doses for

these effects need not agree. A survey of changes in biochemical parameters associated with actions in particular target organs is presented by Gad and Weil [13], Woodman (liver) [14], and Stonard (kidneys) [15].

- More weight is attached to changes in the functional status of physiological or neurological processes, e.g., abnormal behaviour, if correlated to histopathological (peripheral nerve lesions), or biochemical changes (changes in blood acetylcholinesterase activity).
- More weight is attached to changes in, or changes related to, organs and tissues known to be a target of the substance. For example, a change in urinary volume certainly gains in biological significance if the kidney is known to be the target organ.
- More weight is attached to a parameter which shows a statistically significant change compared with control values than to a parameter which only shows a tendency towards a change. However, a tendency cannot be ignored when a dose-effect or dose-response relationship is apparent or when other changes are found which could be related.
- More weight is attached to effects which appear to be irreversible during or following exposure.
- Changes that occur with a low incidence and that are perhaps not even dose-related but occur only in treated animals cannot be immediately dismissed as biologically irrelevant. Expert opinion is indispensable here.
- A change in a single haematological or biochemical parameter unsupported by other correlated haematological, biochemical or pathological changes may be biologically important, e.g., in the case of acetylcholinesterase measurements. More weight is attached to such a change if it is statistically significant and dose-related. The study protocols usually only prescribe blood sampling at the end of a test. Therefore, time trends, which may help in the interpretation of certain effects, cannot be observed.
- Generally a statistically significant decrease in body-weight gain cannot be considered an adverse effect if it is coupled with reduced food consumption.
- Organ weight changes should always be examined on an absolute organto-body-weight basis. Organ-to-body-weight ratios (relative organ weights) can be misleading if a change in body weight occurs. Increased relative organ weights may be the result of adaptation to chemical stress: e.g., increased liver weight may be due to stimulated protein synthesis which enables the liver to metabolize the foreign substance faster.
- The incidence of spontaneous changes is often highly variable among control groups of the same species and strain in different studies. For reference data on biochemical and haematological values see Clampitt [16], Wolford et al. [17], Loeb and Quimby [18] and Derelanko and Hollinger [19]. "Historical control values", i.e., data on the normal variation of a change in the test species, can be used in the interpretation of the biological significance of the changes observed, but should be used with great care. The historical control data ideally should be from the same species, strain, age, sex, supplier, and laboratory, and should come from contemporary control animals not older than a few years. If the authors of a report rely on historical control data in their interpretation of effects these should be provided together with the information necessary to assess their quality, including information on the time frame.

Biokinetics – *time-course of a chemical in a living organism, i.e., increase or decrease of substance at site of measurement due to transport or due to formation or breakdown. The term "toxicokinetics" is also often used synonymously.*

Concentration-time curves of the substance in plasma/blood are important outcomes of kinetic studies. These time-courses are like surrogate endpoints used to describe actual concentration at the target tissue. *Internal dose* or *internal exposure* can be assessed by calculating the area under the plasma concentration-time curve (*AUC*). *AUCs* are now often used as a central concept for various extrapolations, such as determining the linearity of internal exposure in dose escalation studies, interspecies extrapolation, etc. *AUCs* are the result of the four processes in ADME. The primary parameters used to characterize the four interdependent processes in the kinetics of a compound are shown in Table 6.2. These are useful for various comparisons (inter and intraspecies, dose, dose regimens, routes, exposure duration, concentration, surface area dose, etc.)

Secondary parameters, derived from the primary ones, that are equally useful in human risk assessment, are elimination half-life ($t_{1/2}$), area under the curve (*AUC* which is a useful overall indicator for exposure), maximum plasma concentration (C_{max}; Figure 6.3) and average plateau concentration ($C_{ss,av}$; Figure 6.4). Note that plasma elimination half-life (or elimination half-life) is the time it takes for a plasma concentration to be reduced by 50%. It does not automatically define the cause of the decrease as this may be due to excretion

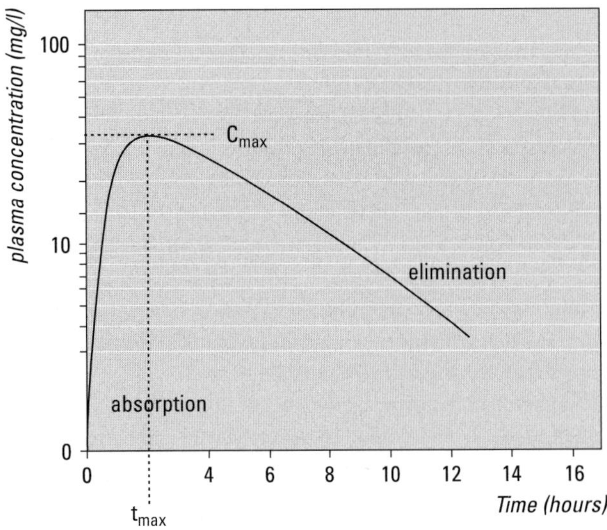

Figure 6.3. Absorption and elimination (http://coo.lumc.nl/TRC/).

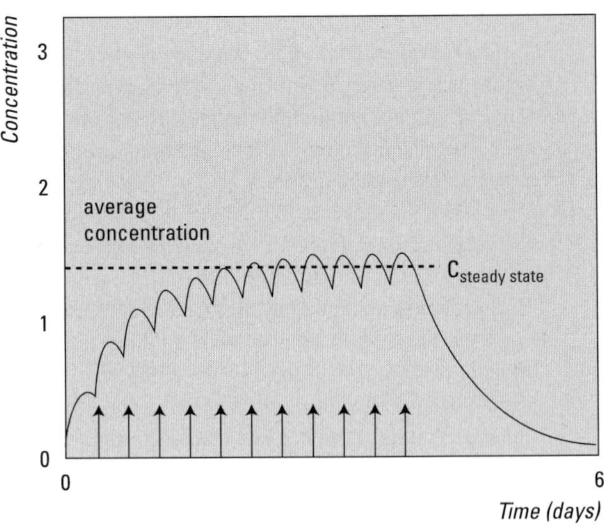

Figure 6.4. Steady-state (http://coo.lumc.nl/TRC/).

or metabolic clearance and is therefore not necessarily a measure for excretion of all compound-related material. For further information on this issue, the reader is referred to the Rowland and Tozer's textbook *Clinical pharmacokinetics* [20]. The following play a pivotal role in toxicity testing and risk assessment: linearity, bioavailability versus absorption, distribution and accumulation, metabolism and route-to-route extrapolation (Figure 6.5).

Linearity

In toxicity studies, doses administered normally range from a low dose where no adverse effects are to be expected up to a middle or high dose causing adverse effects. Importantly, dosing should not reach levels where one or more kinetic processes reach saturation. Saturation means that processes are no longer linear or, more precisely, not linearly dependent on the dose. This can be illustrated by plasma levels that increase linearly with increasing dose levels under non-saturated conditions and tend to increase less than linearly when absorption is saturated or more than linearly when metabolism is saturated [21].

Extrapolations *– Saturation of any of the four ADME processes influences various extrapolations as they can no longer be made on a linear basis.*

Bioavailability versus absorption

There is a broad range of definitions for bioavailability *(F)*. Traditionally bioavailability is defined as the fraction of a dose that reaches the systemic circulation. Absorption is seen as only the passage of a membrane (GIT lining, lung epithelium or epidermis) which can be followed

Table 6.2. Primary parameters of ADME.

Process	Primary parameter
Absorption	absorption rate constant (k_a) and bioavailability (F)[1]
Distribution	apparent volume of distribution $(V_D = A / C)$[2] as an indicator of the tissues involved
Metabolism	intrinsic clearance, described by V_{max} and K_M
Excretion	sum of biliary excretory and renal clearance (CL), irreversible loss of compound from the body

[1] F = Fraction of dose reaching the systemic circulation. It should be noted that bioavailability has a different meaning in environmental toxicity issues, where the bioavailable fraction is the fraction of the total amount of a chemical present in a specific environmental compartment that, within a given time span, is either available or can be made available for uptake by organisms, micro-organisms or plants. Substances that are irreversibly bound to, e.g. soil or sediment, are not bioavailable.

[2] A = amount in body at equilibrium, C = concentration in blood.

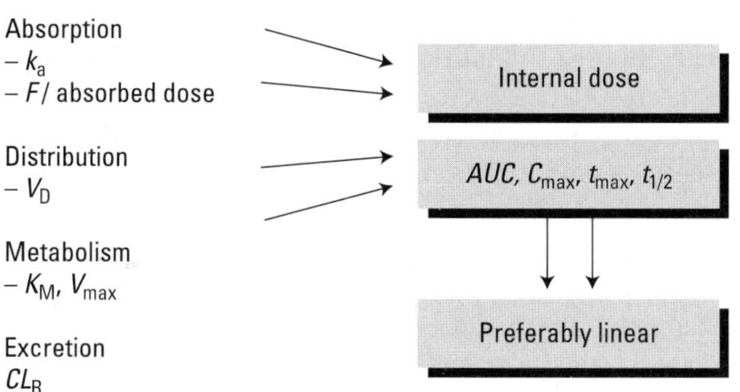

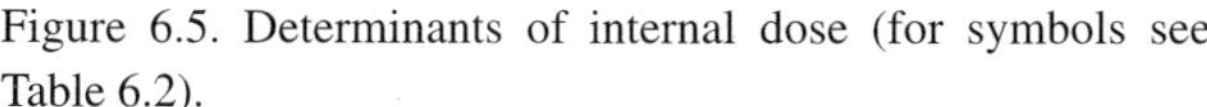

Figure 6.5. Determinants of internal dose (for symbols see Table 6.2).

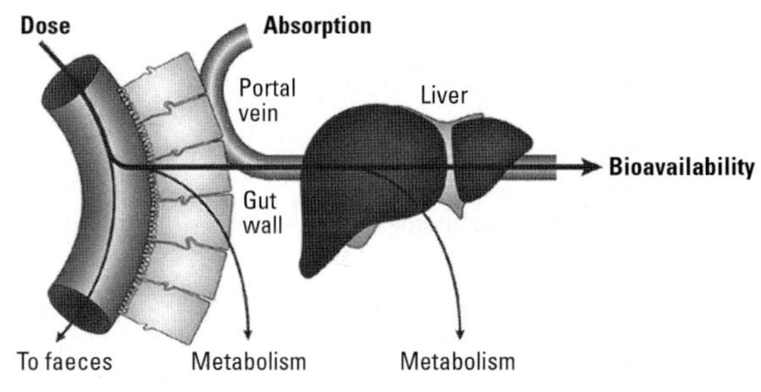

Figure 6.6. Bioavailability [22]. With permission.

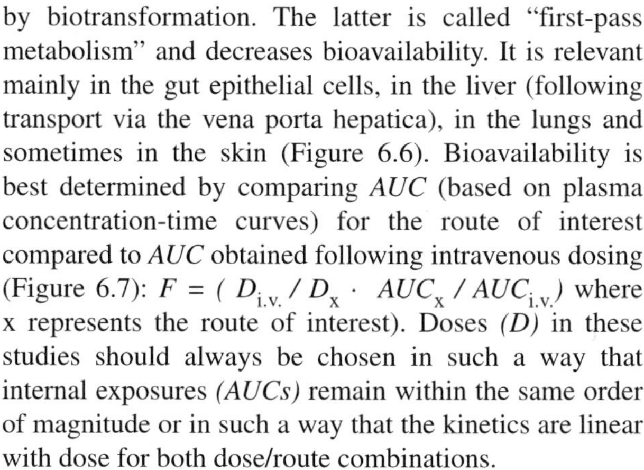

by biotransformation. The latter is called "first-pass metabolism" and decreases bioavailability. It is relevant mainly in the gut epithelial cells, in the liver (following transport via the vena porta hepatica), in the lungs and sometimes in the skin (Figure 6.6). Bioavailability is best determined by comparing *AUC* (based on plasma concentration-time curves) for the route of interest compared to *AUC* obtained following intravenous dosing (Figure 6.7): $F = (D_{i.v.} / D_x \cdot AUC_x / AUC_{i.v.})$ where x represents the route of interest). Doses *(D)* in these studies should always be chosen in such a way that internal exposures *(AUCs)* remain within the same order of magnitude or in such a way that the kinetics are linear with dose for both dose/route combinations.

Extrapolations – *Extrapolations for systemic effects should preferably be based on the use of systemic bioavailability and not only absorption via a membrane.*

Route-to-route extrapolations – *Should preferably be based on bioavailability of parent substance or a presumed toxic metabolite rather than on absorption of radiolabel.*

Distribution and accumulation

Distribution refers to the reversible transfer of a substance from one location to another within the body and as such is dependent on perfusion and the partitioning in the tissues that are perfused by the blood. Partitioning is defined by the tissue: blood partitioning constants are dependent on the physicochemical properties of the substance and the physicochemical properties of the tissue, such as lipid and protein content. As with all kinetic parameters, there is a rate and an extent of distribution. Rate is mostly defined by the perfusion rate. Extent is described by the term volume of distribution (V_D). Small hydrophilic substances tend to partition over the aqueous phase of the body and as such have a volume of distribution equal to the volume of water in the body (V_D = approx. 40 L). Large hydrophilic substances cannot pass cell membranes and remain in the extracellular water (V_D = 16 L). More hydrophobic substances distribute to the fatty tissues where they can reach concentrations that are much higher than the blood or plasma concentration, indicated by a large "apparent" volume of distribution (e.g., V_D = 500 L for digoxin). "Apparent" volume here means the volume of plasma that would be needed to accommodate the total amount of substance in a concentration equal to the actual plasma concentration.

Two closely interrelated terms that are linked to distribution but for which definitions are not so clear, at least within human risk assessment, are persistence and (bio)accumulation. The term persistence is actually not a common term in human risk assessment and both terms receive more attention in environmental risk assessment. Nevertheless, for the sake of clarity, some attention will be devoted to it here. In environmental hazard and risk assessment, criteria have been set for deciding when a substance is persistent and/or bioaccumulates. Consequently, persistence and bioaccumulation are undesirable properties. In human toxicology, both terms are neutral descriptions of dynamic phenomena, without interpretation. Persistence describes the length of time that the substance remains in the body, quantified, e.g., by terminal half-life $(t_{1/2,\ t-\infty})$. Accumulation is the increase in the plasma (and automatically tissue) concentration due to repeated administration or exposure but depends on the dosing regime (frequency and extent), bioavailability, distribution, and clearance. Persistence in environmental hazard assessment is defined as an intrinsic property of the substance and, as such, a *hazard*,

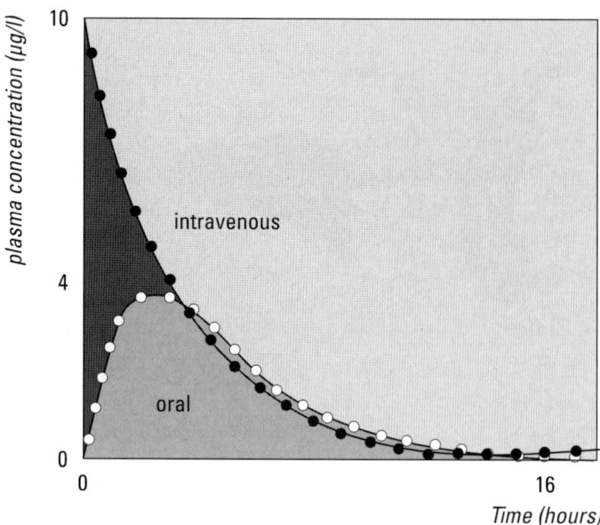

Figure 6.7. AUC oral (○ ○ ○) versus AUC intravenous (● ● ●). Source: http://coo.lumc.nl/TRC/.

dependent on physicochemical properties (mainly log K_{ow} that determines partitioning into fat tissue), which can trigger further toxicological testing. Persistence *per se* is a "single exposure parameter". Accumulation is more a property that depends on the intrinsic properties in combination with the exposure and, as such, is more a *risk term*, meaning that successive additions result in increasing concentrations. As a rule of thumb, compounds exhibiting a half-life of more than 10% of the expected lifetime of an organism tend to have persistency properties. Accumulation can only occur with multiple exposures. Without knowing the intrinsic toxic potency and exposure, persistence and accumulation are rather neutral terms in human risk assessment.

Interspecies extrapolation *– Differences in the lipid and protein content of tissues may result in species dependent distribution. Further, species specific excretion capacity may result in increased* V_D *and increased tissue concentrations (e.g., saturation of renal excretion in dog for phenoxyacetic acid compound).*

Intraspecies extrapolation *– Females often have increased fat content compared to males, possibly resulting in a different* V_D.

Metabolism and route-to-route extrapolation

Metabolism is defined as the conversion of a substance into a chemically distinct form. It is often regarded as synonymous with biotransformation, although biotransformation generally includes a qualitative description of the metabolites formed (metabolic pathways). For purely quantitative biokinetic parameters, such as bioavailability, only metabolism in the sense of metabolic clearance is important, i.e., "first pass" metabolic disappearance of a substance, regardless of what metabolite is formed. Biotransformation generally converts substances to metabolites that can be excreted more easily than the parent substance by making them more hydrophilic and thus more water soluble (urinary excretion). Two main groups of reactions are discerned: Phase-I reactions, such as oxidation, reduction, and hydrolysis; and Phase-II reactions consisting mainly of conjugation reactions (glucuronide, sulphate, acetyl and glutathione conjugation). A more detailed overview is presented in Section 3.6.

Interspecies extrapolation *– Activating or deactivating metabolism may be very species-specific.*

Intraspecies extrapolation *– Biotransformation in neonates and very young animals and children is qualitatively and quantitatively different from adults and young adults; variation due to age, pregnancy, lifestyle.*

Route-to-route extrapolation *(different first-pass effects) – Activation or deactivation by first-pass metabolism may be route-specific, thereby making route-specific metabolism the Achilles' heel of route-to-route extrapolation.*

The risk assessment of workers exposed via skin or inhalation is often based on oral toxicity studies via route-to-route extrapolation, even though this extrapolation has not been scientifically evaluated in terms of the level of uncertainty. Only route-specific differences in absorption (often based on radiolabel studies) are taken into account when an internal $NAEL_{oral}$ is derived from an external $NOAEL_{oral}$ by taking oral absorption into account. Subsequently, an *internal exposure* is determined for worker exposure and compared to the internal $NAEL_{oral}$ to provide a Margin of Safety (MOS). Possible extensive route-specific biotransformation, e.g., an oral first-pass effect is not taken into account.

6.3.2 (Bio)kinetic modelling

Introduction

In order to extract the maximum amount of information from the raw data of a kinetic experiment, mathematical techniques have been developed that enable the biological behaviour of the system to be described in terms of a model. Two types of models can be distinguished, i.e., compartment models and physiologically-based kinetic models (PBK models).

Compartment models
Compartment models are often used to assess kinetic parameters that cannot easily be obtained from measured data such *AUC* or C_{max}. An important concept in modelling the concentration-time profiles of compounds and their metabolites are compartments. A very simple model is the one-compartment model which assumes the mammalian organism to be a well-mixed compartment (see also the environmental compartment models in Chapter 3). This works only for a very limited number of substances without lipophilic and protein-binding properties which are liable to passive uptake and elimination. Two compartments or more may be needed to describe experimental kinetics properly. Moreover, it is important to be aware that the extent of the data determines the number of compartments that can be reliably distinguished. This group of models cannot be used for interspecies extrapolation as a complete new model with new data has to be developed for each species [23-25].

Physiologically-based kinetic (PBK) modelling
Since the 1980's, advances in computer technology have enhanced PBK modelling which is based on three groups of parameters: physiological parameters, such as blood flow, partitioning parameters and metabolism parameters. Compartments are also used in this type of modelling, but in contrast to compartment modelling, the number of compartments is physiologically-based and not dictated by the data available. The model is coded by a series of differential equations. These equations are solved to provide the blood concentration-time profile and a series of other parameters if needed. PBK modelling is used for various extrapolations such as inter and intraspecies extrapolations, dose extrapolations and route-to-route extrapolations [26-28]. Although many PBK models are quite complex, various open source models are available that can be used for training in biokinetics, e.g., what will be the effect of changing one of the parameters?

6.4 TOXICITY STUDIES

6.4.1 General aspects

Conducting a toxicity study, though apparently simple, has a number of caveats to it. Most of them are described in the "yellow bible" of the World Health Organization (WHO) entitled *Principles and Methods for Evaluating the Toxicity of Chemicals*, Part I [8] and related volumes (Environmental Health Criteria 70, 104, 141). Reference guidelines for toxicological tests can be found in the OECD *Guidelines for testing of chemicals* [29].

Important issues include chemical properties, route of exposure, dose selection, selection and care of animals, environmental variables such as caging, diet, temperature, humidity, parameters, data acquisition, presentation and interpretation of results [29]. Other important issues are: good laboratory practice (GLP), personnel requirements and animal welfare.

Toxicity studies should be properly planned, designed, conducted, presented and interpreted. International harmonization of test guidelines is of prime importance to ensure the generation of high quality, mutually accepted toxicity data. The OECD has a major role here, as discussed in Chapter 16. This chapter will first focus on important general issues in toxicity testing and subsequently discuss the experimental tests considered most relevant for risk assessment.

Test substance
Before the start of the study, all available information concerning the test substance should be gathered. Essential information includes chemical structure, composition, method of analysis, information on purity, nature of impurities and their quantity, stability of the substance; some physicochemical properties, such as lipophilicity, acid dissociation constant, ionization, particle size, molecular shape and, if applicable, density, vapour pressure and reactivity. In addition, the administered dose or concentration in every prepared batch of feed or drinking water should be measured as accurately as possible, in order to know the exact exposure and to be able to detect mistakes in the preparation of diets as early as possible. A detailed plan of collection of the samples to be analyzed should be made and followed strictly in order to avoid uncertainties concerning actual exposure.

An often difficult decision to make is what grade of purity of a substance should be studied. For practical purposes and adequate extrapolation to humans, it is usually best to use a technical grade product standardized to specifications used (or to be used) in commerce. As stated above, the nature and quantity of the impurities should be known. When the test substance is administered in the diet, this may be done as a fraction of the total diet, or as a sufficient quantity to achieve predetermined dose levels in mg/kg_{bw} per day. In this case it will be necessary to adjust the dietary concentration on a weekly or biweekly basis. When the dietary concentration is kept constant, it should be remembered that the actual dose received in early growth is over twice as much (as expressed in mg/kg_{bw}) compared to the total dosage [30].

Dose selection

The selection of the dose level depends on the type of study. In acute LD50 or LC50 studies the dose levels should be spaced in such a way as to produce a suitable dose-response curve. In a limit test only one dose will be administered that should not cause mortality [29]. The alternative Fixed Dose procedure aims at a dose producing overt toxicity, but no mortality, whereas the Acute Toxic Class Method uses pre-defined dose levels [29]. Repeated dose studies and continuous exposure studies require careful selection and spacing of the doses in order to obtain the maximum amount of information possible. Since the determination of dose-responses for any observed effects is one of the objectives of such studies, the number of dose levels is usually at least three (low, middle high) in addition to control groups. Increments between doses vary between factors of 2 and 10. Too large dose intervals will result in imprecise No Observed Adverse Effect Levels (NOAEL, Section 6.5.3). The high dose level should produce evidence of toxicity, but little mortality (below 10%). The largest administered dose should not compromise biological interpretability of the observed responses [12]. The mid-dose should produce slight toxicity and the low dose no toxicity. The dose at which no adverse effects are observed will be required to derive the NOAEL. Tests already performed, such as acute and other short-term toxicity studies, can help in this selection. Biokinetic studies describing the behaviour of a compound over a range of doses can significantly improve dose selection [31].

Animal species

Interspecies and intraspecies variation is a fact of life even when exposure route and pattern are the same. The selection of an appropriate animal species and strain for toxicity studies is influenced by a considerable number of factors which are well documented [8,31,32]. Knowledge of and experience with the laboratory animal to be used is of prime importance, since it provides the investigator with the inherent strengths and weaknesses of the model. The guiding principle in the choice of species is that it should resemble humans as closely as possible in terms of absorption, distribution, metabolic pattern, excretion and effect(s) at the site. Other important aspects are sensitivity, convenience and uniformity in response. How to deal with inter and intraspecies differences in dose-response assessment is further discussed in Section 6.5.4.

However, the reality is that for economic and logistic reasons, usually small laboratory rodents (mostly rats) of both sexes are used, although for specialized toxicity testing guinea pigs, rabbits, dogs and non-human primates may be used as well, and are sometimes even required by regulatory agencies. Small rodents provide the possibility of obtaining data on a sufficient number of animals for valid statistical analysis.

Randomly bred as well as highly inbred strains are used, the latter being preferred in specialized toxicity studies. The species and strain used should be well defined, available, economically effective (low cost) and disease free. For specialized toxicity studies, it may be preferred to use animals under model disease conditions to be able to identify the interactions between exposure and the relevant disease. The necessity of using both sexes in toxicity testing seems obvious: retrospective analysis of toxicity studies with chemicals indicates that in more than half of such tests, sex-related differences occur, which are decisive in establishing a NOAEL (Section 6.5.3) [3]. Finally, once the test species has been selected, transport, logistics, quarantine, disease surveillance and (random) allocation of animals to experimental groups need appropriate attention.

Test duration

The response of an organism to exposure to a potentially toxic substance will depend on the magnitude and duration of exposure. Acute or single-dose toxicity refers to the adverse effects occurring within a short time (usually within 14 days) after the administration of a single dose (or exposure to a given concentration) of a test substance, or multiple doses given within 24 hours. In contrast, repeated dose toxicity comprises the adverse general exposure to a substance for a part of the expected lifespan, or for the major part of the lifespan in the case of chronic exposure. For example, standard tests with rats are the 28-days subacute test, the 90-day semi-chronic (subchronic) test and the lifetime chronic test. The strategy for testing substances for repeated-dose toxicity is based on step-by-step tests, starting with shorter test durations. The need for longer-term testing will depend on the information obtained from all acute and repeated-dose tests carried out, the biokinetic profile of the substance and the expected duration of human exposure.

Diet

Research over the last 10 to 15 years has contributed considerably to our knowledge of the impact that diet may have on toxicity test results. Acute toxicity is generally more severe in animals that fasted and may differ as much as 2 to 3-fold by the oral route and up to 10 to 20-fold by inhalation [8]. Composition of the diet influences physiology and as a consequence, the response to a chemical substance. In addition, different levels of macro and micronutrients influence the biotransformation of substances and/or enzyme activity. In fact, a certain diet may also indirectly cause toxicity effects due to acid/base balance disturbances induced by the diet. De Groot [33] provoked bladder changes with a cereal-based stock diet containing 6% monosodium glutamate (MSG), while such changes did not occur at all with 6% MSG in a purified casein diet.

Overnutrition (easily attained by the present day practice of feeding animals *ad libitum*) is associated with increased chronic progressive nephropathy, corticomedullary nephrocalcinosis and increased incidence of multiple endocrine disturbances [4]. Food restriction reduces such confounding effects and, at the same time, considerably reduces spontaneous tumour incidence in long-term experiments [34,35]. Since toxicity testing, as described above, is carried out at dosages where effects are expected, such relatively high doses may also influence the palatability of diets, especially when a substance is administered in the diet. Growth retardation, due to reduced food intake, may also either reduce or increase toxicity. Finally, feed and drinking water should be appropriately tested for the presence of naturally occurring toxins and contaminants.

Other environmental variables

Housing conditions, such as caging, grouping and bedding, temperature, humidity, circadian rhythm, lighting and noise, may all influence animals' response to toxic substances. International bodies like the OECD [29] and WHO [8,9] have made valid suggestions in the relevant guidelines for maintaining good standards of housing and care. The variables referred to should be kept constant and controlled. However, little is known about the actual influence of these variables on the outcome of tests.

Parameters studied

Methods of investigation have changed dramatically in the past few decades. A better understanding of physiology, biochemistry and pathology has led to more and more parameters being studied in order to obtain information about functional and morphological states. While a few clinical chemical measurements were sufficient 20 years ago, today numerous measurements are assessed. In the same way, limited histopathology after a thorough general examination has been replaced by extensive general and very extensive microscopic examination. However, this increased number of parameters does not guarantee better information. On the contrary, it gives the toxicologist a feeling of false confidence and does not sufficiently take into consideration the importance of other parameters not tested at all in routine practice, such as endocrine parameters or atherogenic indicators. Reevaluation of the relevance of the parameters studied, preferably in a retrospective analysis of toxicity studies, is urgently required if toxicology is to remain a credible science. In general, biochemical organ function, physiological measurements, metabolic and haematological information and usually extensive general and histopathological examination must be assessed in routine toxicity testing, the extent of which will depend on the type of study.

Electronic data processing

Today's information technology and automation provides the investigator with better than ever integrated computerized data storage and retrieval systems. For standard general toxicity studies and specific toxicity studies, such as reproduction studies, data acquisition systems are essential. Moreover, histopathology requires semi-quantitative and quantitative measurements, which is greatly facilitated by image analysis. Electronic data processing systems have become indispensable in toxicity testing and provide the best way of achieving the accuracy required by the internationally accepted GLP regulations. Computerized data processing, however, should be properly validated and quality controlled. The use of information systems in the planning and conduct of toxicity studies, together with facilities for complex statistical analysis or metaanalysis and graphic representation, is very beneficial.

Presentation of results

Toxicity studies must be reported in a great detail in order to comply with GLP regulations [36]. The presentation of individual data is necessary to enable regulating agencies to evaluate data in depth. In addition, summary tables for the evaluation of results are required. Every change or incident during testing should be reported: animals killed or which died during the test, influencing circumstances, climatic changes, different light/dark requirements, or clinical observations. In the presentation of pathology data, it is imperative to include details such as number of animal necropsies, number of animals, and number of tissues or organs examined microscopically for each group and sex. The latter is essential, since whatever measures are taken, in practice the loss of small organs or tissues may occur. Any lesions found should always be expressed in terms of the actual quantity of organs or tissues examined. Finally, in the presentation of the results, proper attention should be devoted to the available information for the study and the reasons for performing the experiments, and a well-balanced evaluation of the results, ideally discussed in the light of existing knowledge, should always be included.

Interpretation and evaluation of the results

Although it will be impossible to describe in detail all the aspects involved in the process of interpreting and evaluating the results, some general aspects will be discussed here in order to provide insight into their relevance. A clear and objective interpretation of the results of toxicity studies is important in terms of the clear definition of the experimental objectives, the design and proper conduct of the study and a careful and detailed presentation of the results. While acute single-dose toxicity testing provides some insight into acute effects and possible target organs, repeated-dose subchronic and chronic experiments are capable of determining the nature of toxicity in much more detail and establishing a dose-response relationship and a no effect level.

Relevant data for evaluation include: group weight gain, body weight plotted against time, absolute and relative organ weights, food intake, water intake, biochemical and haematological effects, clinical signs and – a very important cornerstone in toxicity testing – the histopathological examination. In the interpretation of the data, we should be very well aware of confounding factors. Statistical evaluation in toxicity testing, although extremely important, is still very conventional. Too few are aware of the fact that a statistical significance of $P < 0.05$ need not to be of toxicological significance in circumstances where lots of data are evaluated. In fact, this level of significance may be expected in 1 out of 20 parameter sets evaluated [37].

New concepts in statistics are needed, such as techniques for non-continuous variables, non-linear regression with and without normally distributed errors, power analysis and meta-analysis. Interpretation and evaluation is the key to the science of toxicology. It requires common sense, a critical approach, vast experience and a cautious attitude. Obviously, differences in interpretation will occur. As long as interpretations are well described and argued, such differences are acceptable.

Good laboratory practice

Non-clinical toxicological or safety assessment studies that are to be part of a safety submission for the marketing of regulated products, are required to be carried out according to the principles of GLP [36, 38-41, Chapter 16]. These regulations concern food additives, animalfeed additives, medical and electronic devices for human use, human and animal drugs, and biological products as well as environmental chemicals, such as pesticides, fungicides, industrial chemicals, etc. These regulations were imposed because toxicity data was misused in the past which could lead to false safety assessments involving risk to humans and the environment. The regulations were imposed at a later stage for environmental testing.

The quality control measures laid down in the regulations have improved the quality of data considerably. Quality encompasses two elements: quality control and quality assurance. Quality-control procedures are meant to minimize mistakes or errors, to make it possible to correct them and to maximize the accuracy and validity of the collected data. Quality assurance covers the steps taken (i.e., inspections and audits) to verify that planning, procedures and quality control have been carried out according to the regulations. Quality-control procedures include all the processes of toxicity testing, including post-experimental pathological procedures. Sometimes, the procedures used even surpass the requirements of GLP, such as histopathological examination, blind scoring, non-blind scoring, classifications, nomenclature and quantitative measurement of degrees of lesions.

Besides acknowledged beneficial effects, GLP regulations have increased the cost of toxicity studies. This is reflected in the production costs and subsequently the cost to the consumer. One undesirable but, nevertheless, realistic side effect is that technological innovation and scientific progress is discouraged when GLP is considered an aim in itself. The rigid application of GLP in toxicity testing or the abolition of some of the less relevant procedures might be to the advantage of the scientific community [3]. However, compliance with GLP facilitates international acceptance of studies, thus reducing the need for duplication and encouraging better documentation procedures.

Personnel requirements

GLP regulations require the use of qualified personnel at every level. On-the-job training to gain experience is usually required, since educational establishments are obviously geared towards teaching, but insufficiently developed to provide practical experience. Teaching on the subject of toxicity has improved tremendously over the last two decades and accreditation procedures have been implemented in many industrialized countries.

Animal welfare

The use of animals for experimentation in general, and for toxicity testing in particular, has been criticized for many years. The increasing numbers of animals used in experiments to fulfil statistical requirements and the degree of discomfort to animals in some toxicity studies, causes great concern among animal welfare organizations and has started debate everywhere.

Toxicologists should participate in this debate, since they are particularly well placed to be able to judge the need for experimentation on the one hand, and to consider the animal's discomfort, on the other. Certain procedures and practices, such as the LD50 test, have been rightfully under fire (see acute toxicity testing) and are being replaced by alternative procedures which are also much more relevant to toxicity testing. Essentially, it is every toxicologist's responsibility to reduce the number of animals used in toxicity testing, to reduce stress, pain and discomfort as much as possible, and to seek alternatives, especially for the use of *in vitro* techniques.

The toxicologist is also responsible for the reduction, refinement and replacement (the 3Rs of Russel and Burch [42]) of animal use in experimentation, as described in the Council Directive of the European Communities [43] and the EU Community Action Plan on the Protection and Welfare of Animals [44]. Fundamental research to support the considered use of alternative methods should be encouraged. Since full replacement of *in vivo* testing in the near future is doubtful, "Intelligent Testing Strategies", intelligent combinations of *in vitro*, (Q)SARs, and *in vivo* methods applying risk-decision theory and weight-of-evidence approaches, could offer a way forward [45]. This is explained further in Chapter 11.

Human data

Human data will be available only for substances already on the market. When adequate human data from epidemiology studies, controlled experiments with volunteers or case reports are available, this can be highly useful in the hazard identification process. However, NOAELs derived from human studies are rare. If both animal data and human data are available, as a general rule, well reported relevant human data for any given endpoint is to be preferred for the risk assessment [5]. Exemptions from this general rule are studies conducted with human volunteers. The use of human volunteer studies has been subject to controversy. They can provide valuable data, but they have to be performed within strict ethical guidelines. Results from such studies should be used only in justified cases (e.g., tests which were conducted for the authorization of a medical product or when effects in already available human volunteer studies have been observed to be more severe than deduced from prior animal testing). However, the potential differences in the sensitivity of human studies and studies in animals should be taken into account in risk assessment on a case-by-case basis. In relation to hazard identification, the relative lack of sensitivity of human data may cause particular difficulty: negative data from studies in humans will not usually be used to override the classification of substances which have been classified on the basis of data from studies in animals, unless the classification is based on an effect which clearly would not be expected to occur in humans.

Uncertainty and variability

Uncertainty analysis refers to true uncertainty about a parameter with a fixed value (e.g., a distribution coefficient), characterized by a random variable with an identified probability distribution which can be reduced by further research, or inter-individual variability of parameters distributed empirically within a defined population (e.g., body weight) which cannot be reduced by further research [46]. As already illustrated above, there are many sources of uncertainty of both types in the toxicity testing of substances. For example, an effect may not be noticed because the number of animals is too small, the period of observation too short, the dose level too low or too high – because, as in the latter case, the metabolic pathway may differ – or the experimental design is too limited in scope, or simply due to inaccuracy. On the other hand, false positives may be the result of low standard deviations. An excellent review of possible sources of uncertainty in animal tests was carried out by IPCS [8]. Other major sources of uncertainty are more related to variability in the extrapolation from experimental studies to human populations. Important factors here are differences between species, e.g., between rats and humans, and within species, e.g., variability between humans [47]. Section 6.5.4 will further explain the difference between uncertainty and variability and will discuss methods for including these aspects in effects assessment and risk characterization.

Table 6.3. Some criteria for the classification of chemicals on the basis of LD50 values from acute oral toxicity data expressed as mg/kg_{bw}.

United Nations	toxic 1	toxic 2	toxic 3		
Solids	< 5	< 50	< 500		
Liquids	< 5	< 50	< 2000		
World Health Organization	extremely hazardous	highly hazardous	moderately toxic	slightly toxic	
Solids	< 5	< 50	< 500	< 5000	
Liquids	< 20	< 200	< 2000	< 2000	
European Communities	very toxic	toxic	harmful		
	< 25	< 200	< 2000		
USA	supertoxic	highly toxic	very toxic	moderately toxic	slightly toxic
	< 5	< 50	< 500	< 5000	< 15000

6.4.2 Acute effects: acute toxicity

Acute toxicity refers to the adverse effects occurring within a short time (usually within 14 days) after the administration of a single dose (or exposure to a given concentration) of a test substance, or multiple doses given within 24 hours. In the assessment of substances for toxic characteristics acute toxicity is usually a first step in providing information on relative toxicity. Information that can be obtained is the nature of the effects, the dose-response and sex differences. Among other things, this information will be used for classification of the substance (Section 6.5.2).

Under REACH no information is required below 1 tonne per annum. Testing for acute oral toxicity is required from 1 tonne per annum unless an inhalation test is available. From 10 tonnes per annum at least one other route needs to be tested. Classification and evaluation of acute toxicity will rely on the information available, *in vitro* data, (Q)SAR and read-across.

Testing of acute toxicity used to be predominantly concerned with lethality as the effect of interest. This involves the determination of the median lethal dose (LD50) or concentration (LC50), according to OECD Guidelines 401 (oral), 402 (dermal) and 403 (inhalation). The LD50 is defined as the statistically derived expression of a single dose that can be expected to be lethal to 50% of the test animals. Regulations still classify substances according to their acute toxicity in terms of identified dose, thus providing a need for such studies (Table 6.3). The use of these types of studies, however, has been seriously questioned. Efforts have been made to develop better, more sensible guidelines for acute toxicity [48,49]. This has led to the Fixed Dose Procedure (OECD Guideline 420) as a suitable alternative for the oral LD50 test [29,49,50]. Other alternatives to the standard LD50 test are the stepwise acutetoxic-class method (OECD Guideline 423) [51] and the sequential oral Up-and-Down method (OECD Guideline 425) [52], which, however, are still based on mortality as an endpoint, although acute toxic phenomena other than death can be included. A study comparing the results obtained with the conventional LD50 test, the Fixed Dose Procedure and the Up-and-Down Procedure was published in 1995 [53]. The standard acute oral test (OECD Guideline 401) was abolished in 2002.

An acute LD50 test is performed by administrating graduated doses to groups of experimental animals and the subsequent observation of signs of toxicity and death. All animals dying during the experiment and all surviving animals are autopsied. Gross examination and, where indicated, histopathological examination are performed. When acute toxicity is established by the inhalatory route (OECD Guideline 403), we speak of the lethal concentration (LC50), being the concentration of a substance in air that causes death following a certain period of exposure. In practice, most toxicologists maintain the time constant (i.e., 1 or 4 hours) and vary

Table 6.4. Evaluation and interpretation of results of acute toxicity tests (fixed dose procedure).

Dose	Results	Interpretation
5 mg/kg$_{bw}$	less than 100% survival	compounds which are *very toxic*
	100% survival; but evident toxicity	compounds which are *toxic*
	100% survival; no evident toxicity	see results at 50 mg/kg
50 mg/kg$_{bw}$	less than 100% survival	compounds which may be *toxic* or *very toxic*; see results at 5 mg/kg
	100% survival; but evident toxicity	compounds which are *harmful*
	100% survival; no evident toxicity	see results at 500 mg/kg
500 mg/kg$_{bw}$	less than 100% survival	compounds which may be *toxic* or *harmful*; see results at 50 mg/kg
	100% survival; but evident toxicity	compounds considered as having no significant
	100% survial; no evident toxicity	acute toxicity see results at 2000 mg/kg
2000 mg/kg$_{bw}$	less than 100% survival	see results at 500 mg/kg
	100% survival; with or without evident toxicity	compounds which do not have significant acute toxicity

the concentration of the test substance. As is the case with oral or dermal LD50 tests, the post-exposure observation period is 14 days. Inhalatory tests may be considered for substances which are volatile (e.g., vapour pressures above 0.01 Pa [5]) or with low particle size (e.g., mean mass aerodynamic diameter less than 100 μm [5]) or occur as aerosols.

In the Fixed Dose Procedure [29], acute toxicity is tested in a stepwise approach, in which far fewer animals are used, discomfort to animals is limited and much more relevant information on the acute toxicity (target organs and nature of toxicity) is obtained than in the classical LD50 test. In a preliminary study, various doses are administered to single animals of just one sex in a sequential manner. The information this "sighting study" provides on the dose-toxicity relationship and the estimated lethal dose usually will not require more than five animals. The main study is then carried out with groups of five animals of each sex at one of the preset dose levels (5, 50, 200 or 2000 mg/kg$_{bw}$). The dose used is derived from the sighting study and is that dose which is likely to produce evident toxicity, but not death. When the main study establishes evident toxicity, but no mortality is required, no further study is needed. If the initial dose level does not produce evident toxicity, the next higher pre-selected dose level should be used. However, when animals die or have to be destroyed due to severe toxicity at the initial dose level, the next lower pre-set dose level is used for the study. Evaluation and interpretation for classification is done on the basis of Table 6.4.

In the evaluation of acute toxicity studies it is not usual to derive acute No Observed Adverse Effect Levels (NOAELs). Information on toxic signs and the dose levels at which these occur can be useful for risk characterization. In particular, the slope of the dose-response curve may indicate the extent to which reduction of exposure will reduce the response: the steeper the slope, the greater the reduction in response for a particular finite reduction in exposure [5].

6.4.3 Acute local effects: irritation and corrosivity

Changes at the site of first contact (skin, eyes, mucous membranes of the gastro-intestinal tract or the respiratory tract) are called local effects and include irritation and corrosiveness. Irritant substances are non-corrosive substances which cause inflammation as evidenced by erythema and oedema of the skin and corneal opacity, iridal effects and conjunctival redness or swelling for the eye. Corrosive substances may destroy living tissues. According to EU criteria, skin corrosion produces full thickness destruction of skin tissue, and persistent ocular changes and colouration are regarded as severe ocular lesions [54].

Under REACH no information is required below 1 tonne per annum and no *in vivo* testing up to 10 tonnes per annum. Classification and evaluation for acute toxicity will then rely on the information available, such as physicochemical properties and on *in vitro* data, (Q)SARs and read-across. Further information on testing strategies for irritation can be found in Chapter 11.

The aim of the tests is to establish the likelihood of an acute irritant or corrosive response occurring in humans in relation to the route, pattern and extent of exposure. This information will be used for classification of the substance (Section 6.5.2). Testing methods are usually applied to rabbits according to OECD Guidelines 404 for dermal irritation/corrosion and 405 for eye irritation/ corrosion [29]. These OECD guidelines recommend a weight-of-evidence approach and sequential testing strategy based on available human and experimental data, structure-activity relationships, acidity and buffer capacity and validated *in vitro* or *ex vivo* tests for corrosion/irritation (Box 6.2 [29]). There is no international guideline for respiratory irritation.

In the *in vivo* irritation/corrosion tests the substance is applied in a single dose to the skin or in the eye of an experimental animal. Untreated skin areas or the untreated eye serve as the control. The degree of irritation/corrosion is read and scored at specified intervals and is further described in order to provide a complete evaluation of the effects. The duration of the study should be sufficient to evaluate the reversibility or irreversibility of the effects observed.

Animal testing for irritation and corrosion does not allow a dose-response assessment and will not provide a No or Lowest Observed Effect Level. These, however, may be derived from already available experimental studies in which a range of concentrations was used or from human data.

6.4.4 Sensitization

Skin sensitization, which results in allergic contact dermatitis, is a very common form of allergy. Following skin exposure and penetration, it develops in two phases: induction (sensitization) and elicitation. During induction, a primary immune response is triggered following a reaction between the chemical allergen and skin protein. This results in sensitization, and if the sensitized individual comes in contact with the same chemical allergen again in a later stage, a more pronounced secondary response is induced at the contact site [55]. Respiratory hypersensitivity is a term used to describe asthma and other related respiratory conditions to which both immunological and other mechanisms may apply [5].

Under REACH no information is required below 1 tonne per annum, whereas all substances produced or imported at >1 tonne per year, have to be tested for skin sensitization. Below 1 tonne per year, classification and evaluation for skin sensitization will then rely on the information available, (Q)SAR and read-across.

Tests minimally aim to establish the potential for sensitization and possibly a dose-response. According to OECD Guidelines for testing of chemicals [29], standardized predictive sensitization testing is performed exclusively *in vivo* in two species, i.e., guinea pig (OECD 406, Guinea Pig Maximization Test, GMPT, and Buehler Test) and mouse (OECD 429, Local Lymph Node Assay, LLNA). The guinea pig has been the animal of choice for several decades but the tests employing this species are increasingly being replaced by the mouse LLNA test. There are currently no internationally recognized test methods for predicting the ability of chemicals to cause respiratory hypersensitivity. According to the OECD Guidelines, (Q)SARs and *in vitro* models do not have to be considered for a sensitization test strategy because they are not sufficiently developed. Information on alternative testing strategies for skin sensitization can be found in Chapter 11.

In the guinea pig tests, the animals are initially exposed to the test substance by intradermal injection and/or epidermal application (induction exposure). The GMPT test uses an adjuvant which potentiates sensitization, the Buehler test does not. Following a rest period of 10 to 14 days, during which an immune response may develop, the animals are exposed to a challenge dose. The extent and degree of skin reaction to the challenge exposure in the test animals is compared with that demonstrated by control animals which undergo sham treatment during induction and receive the challenge exposure. The basic principle underlying the mouse LLNA test is that sensitizers induce a primary proliferation of lymphocytes. This proliferation is proportional to the dose applied (and to the potency of the allergen) and provides a simple means of obtaining an objective, quantitative measurement of sensitization. The mouse LLNA test assesses proliferation of lymphocytes in the lymph node draining the chemical application site as a dose-response. The proliferation in test groups is compared to that in vehicle treated controls. The proliferation ratio in treated groups compared to that in the vehicle controls, termed the Stimulation Index, is determined. This ratio must be at least three before a test substance can be further evaluated as a potential skin sensitizer. Critical points in all tests are the concentrations used at induction and challenge, the nature of the vehicles and the ability of chemicals to penetrate the skin.

Box 6.2. OECD Testing and evaluation strategy for dermal irritation/corrosion [29]

TESTING AND EVALUATION STRATEGY FOR DERMAL IRRITATION/CORROSION

	Activity	Finding	Conclusion
1	Existing human and/or animal data showing effetcs on skin or mucous membranes	Corrosive	Apical endpoint; considered corrosive. No testing is needed.
		Irritating	Apical endpoint; considered to be an irritant. No testing is needed.
		Not corrosive/not irritating	Apical endpoint; considered not corrosive or irritating. No testing is needed.
	↓ *No information available, or available information is not conclusive* ↓		
2	Perform SAR evaluation for skin corrosion/irritation	Predict severe damage to skin	Considered corrosive. No testing is needed.
		Predict irritation to skin	Considered an irritant. No testing is needed.
	↓ *No prediction can be made, or predictions are not conclusive or negative* ↓		
3	Measure pH (consider buffering capacity, if relevant)	pH ≤ 2.0 or ≥ 11.5 (with high buffering capacity, if relevant)	Assume corrosivity. No testing is needed.
	↓ *2.0 < pH < 11.5 or pH ≤ 2.0 or ≥ 11.5 with low/no buffering capacity, if relevant* ↓		
4	Evaluate systemic toxicity data via dermal route (Can be considered before Steps 2 and 3)	Highly toxic	No further testing is needed.
		Not corrosive or irritating when tested to limit dose of 2000 mg/kg body weight or higher, using rabbits	Assume not corrosive or irritating. No further testing is needed.
	↓ *Such information is not available or is non-conclusive* ↓		

continue on next page

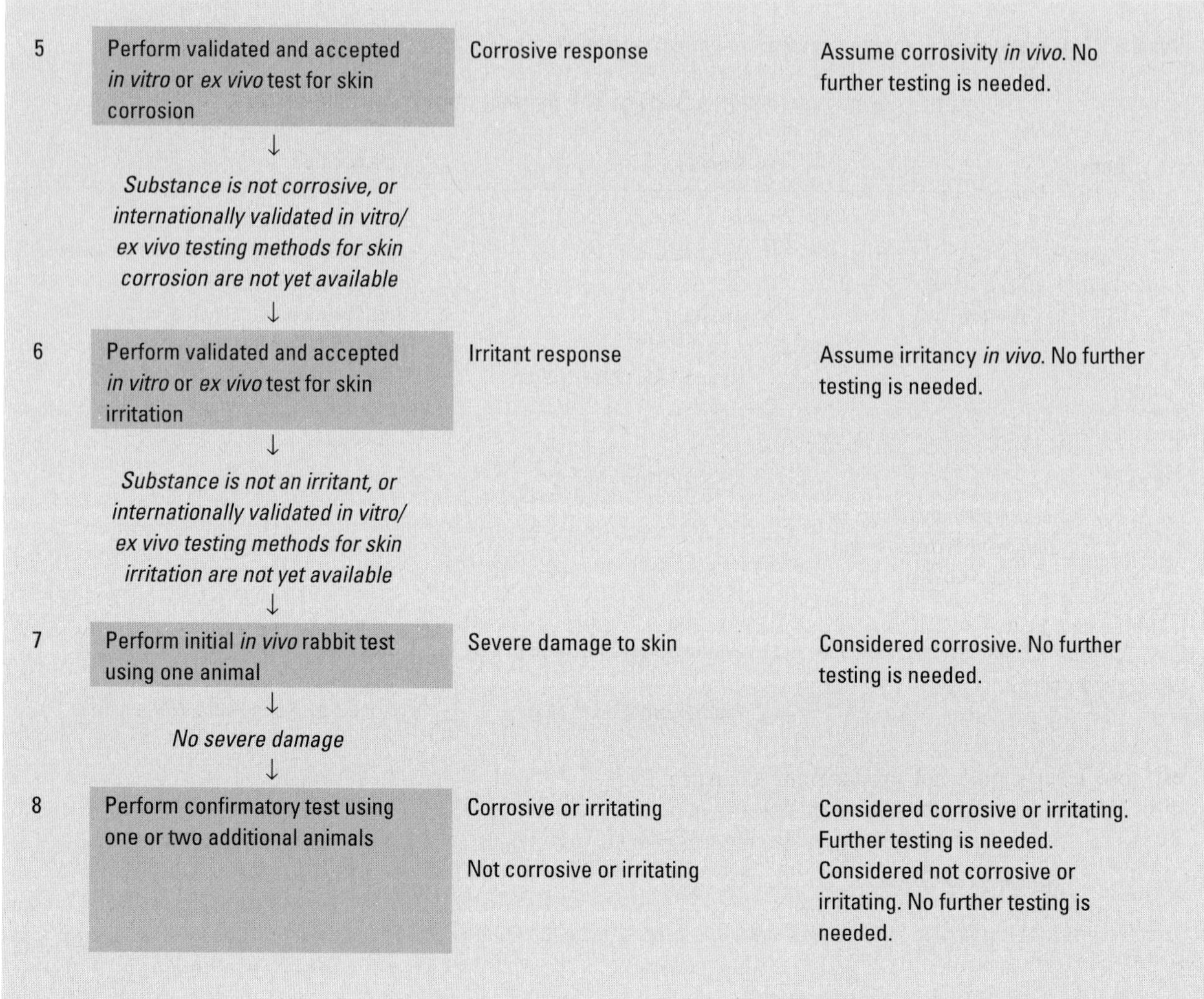

5	Perform validated and accepted *in vitro* or *ex vivo* test for skin corrosion	Corrosive response	Assume corrosivity *in vivo*. No further testing is needed.
	↓ *Substance is not corrosive, or internationally validated in vitro/ ex vivo testing methods for skin corrosion are not yet available* ↓		
6	Perform validated and accepted *in vitro* or *ex vivo* test for skin irritation	Irritant response	Assume irritancy *in vivo*. No further testing is needed.
	↓ *Substance is not an irritant, or internationally validated in vitro/ ex vivo testing methods for skin irritation are not yet available* ↓		
7	Perform initial *in vivo* rabbit test using one animal	Severe damage to skin	Considered corrosive. No further testing is needed.
	↓ *No severe damage* ↓		
8	Perform confirmatory test using one or two additional animals	Corrosive or irritating	Considered corrosive or irritating. Further testing is needed.
		Not corrosive or irritating	Considered not corrosive or irritating. No further testing is needed.

There is evidence that there are dose-response relationships for both skin sensitization and respiratory hypersensitivity. The dose-response generated by the LLNA makes this test more informative than guinea pig tests, which often employ one single concentration of the test material at both induction and challenge. All *in vivo* tests will give some information on potency, though it is necessary to be careful when extrapolating this to humans [5].

6.4.5 Repeated-dose toxicity

Acute toxicity studies only deal with the adverse effects of a single dose and only provide information concerning the possible hazard to humans in the event of an acute incidental high exposure. Much more common though, is human exposure at lower levels and in a repeated fashion. Repeated exposure for shorter or longer periods of time may not produce immediate effects, but delayed effects may very well be induced due to accumulation of the chemical in the body or due to other mechanisms.

Repeated-dose toxicity is the adverse general toxicological effects occurring as a result of repeated daily exposure via different routes for various fractions of the expected lifespan, up to a complete lifespan. Typical examples of repeated-dose tests with experimental animals are the 28-day sub-acute test, the 90-day sub-chronic test and the lifetime chronic test. Effects can be both local (i.e., at the site of first contact) and systemic (i.e., normally distant from the site of first contact).

Regulatory requirements usually follow a tiered approach with the complexity and duration of studies required increasing with production volume or exposure potential, for example. The minimum requirement usually is the 14 or 28-day test with rats. If such a study is properly planned and designed, and if relevant critical parameters are studied, the results of such a study will provide a fair basis for an initial toxicological evaluation over a limited time scale. Under REACH, a repeated-dose toxicity study, at least a 28-day test, is required from a yearly production or import level of 10 tonnes.

The aim of these tests is to provide information on the adverse effects (Section 6.2) likely to arise from repeated exposure of target organs, which could lead to classification. Furthermore, these tests should provide information on dose-response relationships, leading to the identification of No Adverse Effect Levels or Adverse Effect Levels, such as the NOAEL or the Benchmark Dose/Concentration (see Section 6.5.3). The dose at which no adverse effects are observed will be required to derive the NOAEL. Relevant OECD guidelines include TG 407-413, 417, 424 and tests 4.5.1-4.5.3 listed in Chapter 16 (Table 16.1 [29]).

The design of repeated-dose studies may vary, but usually consists of the repeated administration of a series of 3 to 4 doses or concentrations, with an increment of 2-10 between doses/concentrations, for a specified period of time. From the dose range chosen, the highest dose should have a clear adverse effect level with limited mortality, although preferably without mortality. The lowest dose level should not produce any evidence of toxicity, whereas ideally the medium dose(s) should produce minimal (and intermediate) observable adverse effects. Oral tests include administration by gavage, via the feed or via the drinking water. Usually, one dose per day is administered in oral and dermal tests 5 or 7 times per week. Exposure periods in inhalation tests may vary from several hours per day up to continuous exposure and from several days to 7 days per week. The most commonly used species are rats for rodents and dogs for non-rodents. The tests in most cases are carried out with young animals in their growth spurt to reach maturity. This period of life is considered to be a period sensitive to exogenous agents [8]. The tests should be carried out with both sexes and in groups of at least 5 to 10 per sex for rodents and 4 for non-rodents. If interim sacrifices are included for specific analyses, the number per group should be raised accordingly. In addition, a satellite group may be treated with the high concentration level and observed for reversibility, persistence, or delayed occurrence of toxic effects for a post-treatment period of appropriate length. A negative, concurrent control group should always be included and should be handled in an identical manner as the test groups. In gavage tests and in some inhalation tests, a vehicle control group is required, too. The animals in the tests are inspected daily for clinical signs, and body weight and food consumption are monitored (usually weekly). Clinical examination usually consists of haematological parameters, clinical biochemical data and urine analysis and is performed at the end of the study. At termination, extensive gross necropsy should be performed, the weights of the major organs determined and organs and tissues should be preserved for histopathological examination (Table 6.5).

Human data which is adequate to serve as the sole basis for dose-response assessment are rare in view of uncertain exposure, mixed exposure, low incidence of effects, small number of exposed individuals, heterogeneous populations and long latency periods between exposure and disease. In the evaluation of animal tests, preference should be given to tests using species with similar toxicokinetics and toxicodynamics as in humans. If this species cannot be identified, the most sensitive animal should be selected. Preference should also be given to tests with the most appropriate route and duration and frequency of exposure in relation to these characteristics in humans. It should be possible to identify a NOAEL or a Benchmark Dose/Concentration. Evaluation of the adversity of the effects, as discussed in Section 6.2, is crucial in the determination of the NOAEL or Benchmark Dose (Section 6.5.3).

6.4.6 Genotoxicity

Genotoxicity refers to potentially harmful effects on genetic material. It includes mutagenicity which can be defined as the induction of permanent transmissible changes in the amount or structure of the genetic material. Genotoxicity tests also provide indications of other DNA damage through unscheduled DNA synthesis, sister chromatid exchange, strandbreaks, adduct formation, mitotic recombination and numerical chromosome aberrations (aneuploidy). Genotoxicity testing is very useful in pre-screening for potential genotoxic carcinogenicity. In addition, it serves the purpose of establishing whether or not substances have the potential to induce heritable germ cell mutations at the gene or chromosome level. Such tests are described by Mason et al. [56] and Weisburger and Williams [57] and others.

Under REACH no information is required below 1 tonne per annum. The basic requirement is a bacterial Ames test for mutations for up to a yearly amount of 10 tonnes. Requirements increase if positive results are obtained and with increasing tonnage levels. Information on testing strategies for genotoxicity can be found in Chapter 11.

There are 15 OECD guidelines available for both *in*

Table 6.5. Repeated dose studies (28d, 90d, chronic); OECD Guidelines 407-413 and 452).

Conditions	chemical identification of substance, its purity and chemical characteristics
Route	oral (gavage, diet, drinking water or capsules), dermal, inhalatory
Experimental animals	rat (mouse, dog, rabbit, guinea pig)
Number of animals	28 and 90d: 5 to 10 of each sex per group1 chronic: 20 rats (4 to 5 dogs) per sex per group
Dose levels	control and at least 3 dose levels with an increment of 2 to 10 satellite groups may be added e.g., for interim kills, observation of reversibility, persistence or delayed occurrence of toxic effects a limit test may be performed using a control and one high dose level
Examinations:	
• physical measurements	- temperature, humidity, homogeneity and stability of test substance, food and water consumption and, for inhalation studies, air flow, concentrations, particle size
• clinical observations	- body weight - changes in: skin, fur, eyes, mucous membranes, occurrence of secretions and excretions, behaviour, respiratory, circulatory, autonomic and central nervous systems, somatomotor activity - sensory reactivity to stimuli, assessment of grip strength and motor activity - ophthalmologic examinations (90d/chronic)
• haematology	- haematocrit, haemoglobin concentration, erythrocyte count, total and differential leukocyte count, platelet count, measure of blood clotting time
• clinical biochemistry	- investigation of organ function, carbohydrate metabolism, electrolyte balance - serum salts (Ca, P, Na, K, Cl), serum enzymes (such as alanine aminotransferase, aspartate aminotransferase, alkaline phosphatase, gamma glutamyl transpeptidase, sorbitol dehydrogenase, ornithine decarboxylase), cholesterol, glucose, urea, creatinine, total protein, albumin, total bilirubin (may be extended to lipids, hormones, acid/base balance, methaemoglobin, cholinesterase activity) - urinalysis (not routinely in 28 d tests): appearance, volume, osmolality or specific gravity, pH, protein, glucose, blood cells
• pathology	- gross necropsy including external surfaces, orifices, cranial, thoracic and abdomical cavities and contents, organ weights - histopathological changes of all preserved organs and tissues at highest dose level and in controls; if indicated also at intermediate dose levels
Results	information concerning effects of repeated dose exposure on parameters studied, target organ(s); if possible, mechanism of toxicity and NOAEL

[1] For a range finding test 5 animals per group may be sufficient.

vitro and *in vivo* methods. The most commonly used are OECD TG 471-476, 483, 486-487 (Table 16.10 [29]).

In vitro genotoxicity testing usually involves at least two, but preferably three, different endpoints at several levels of biological complexity: one being an assay in a prokaryote to detect gene mutations, then an assay in a mammalian system to detect chromosomal damage and finally, an assay in a eukaryote or, preferably an assay to detect DNA damage, or an assay to detect adduct formation (Table 6.6). An assay in a prokaryote usually involves bacteria such as *Salmonella thyphimurium* (Ames assay) or *Escherichia coli* where reverse mutations are used as an indication of genotoxic potential (OECD Guideline 471). The principle behind this test is to detect reverse mutations of a strain of bacteria which are growth dependent and where reverse mutation leads to independent growth which can be detected on a feeding layer devoid of growth factor. Today these tests are standardized and well validated. Over 200 discrete *in vitro* genotoxicity assays have been described but most of them are insufficiently developed and validated to be used, and most are redundant to the *Salmonella* assay.

Table 6.6. Genotoxicity tests.

Gene mutation assays	
• Tests with prokaryotes	- *Salmonella typhimurium* reverse mutation assay (OECD Guideline 471) - *Escherichia coli* reverse mutation assay (OECD Guideline 471)
• Test with eukaryotes	- *Saccharomyces cerevisiae* gene mutation assay (OECD Guideline 480) - *in vitro* mammalian cell gene mutation assay (OECD Guideline 476) - *in vivo* sex linked recessive lethal assay in *Drosophila melanogaster* (OECD Guideline 477)
Chromosomal damage assays	
• *In vitro* tests	- mammalian cytogenetic test (OECD Guideline 473) - chromatid exchange assay in mammalian cells (OECD Guideline 479) - micronucleus test (draft OECD Guideline 487)
• *In vivo* tests	- mammalian bone marrow cytogenetic test for chromosomal analysis (OECD Guideline 475) - micronucleus test (OECD Guideline 474)
DNA damage/repair/adduct formation assays	
• *In vitro* tests	- DNA adduct formation 32-post coupling [58] - DNA repair synthesis in mammalian cells in vitro (OECD Guideline 482) - DNA repair test in primary liver cells [60]
• *In vivo* tests	- Unscheduled DNA synthesis (UDS) test with mammalian cells *in vivo* (OECD Guideline 486 - Alkaline single-cell gel electrophoresis assay for DNA strand breaks (Comet assay) [60]

To detect chromosome aberrations the *in vitro* mammalian cytogenetic test (OECD Guideline 473) or the *in vitro* sister chromatic exchange assay in mammalian cells (OECD Guideline 479) are available. There is an ongoing discussion regarding whether the use of an *in vitro* micronucleus assay is preferable, because it detects chromosomal aberrations as well as aneuploidy and is less affected by cytotoxic effects of the test substance. A OECD draft guideline is available at this point in time. In eukaryotic systems, yeast cells (OECD Guideline 480) or preferably, somatic cells are used (OECD Guideline 476). A system using mammalian cells *in vitro* where unscheduled DNA synthesis is measured as an indication of genotoxicity (OECD Guideline 482) can also be used. The principle of detection in eukaryotic systems is, in fact, the same as in prokaryotic systems. However, the principle of detection of unscheduled DNA repair is based on the ability of mammalian cells to repair damaged DNA to a certain extent and to detect such unscheduled DNA repair by autoradiographic methods, further to incorporation of tritiated thymidine [58]. Unscheduled synthesis can be differentiated from "scheduled" DNA synthesis as follows: normal cell duplication leads to heavily labelled cells which are easily discernible from cells showing unscheduled repair, which are only lightly labelled with autoradiographically detectable silver grains. Finally, chromosome aberration tests detect structural losses or changes in chromosomes, which can be studied by arresting the cells in mitosis and by counting the number of abnormal chromosomes or the exchange of chromatids in a statistically sufficient number of mitoses or cells.

When two out of three tests are positive, the genotoxicity of a substance is established. When all three tests are negative, there is good evidence that the substance has no genotoxic properties. However, it is necessary to determine whether the doses or concentrations used were high enough and maintained at a sufficient level (e.g., in the case of volatile chemicals). It is also important to take into account the reactivity of the chemicals and their metabolic pattern in the test system. When genotoxic properties are detected, prior to undertaking an extensive and costly bioassay, it is advisable to perform *in vivo* genotoxicity tests, such as the mouse bone marrow micronucleus test (OECD Guideline 474), the in vivo cytogenetic assay (OECD Guideline 475), the rodent liver genotoxicity test (OECD 486), or the much less sensitive rodent dominant lethal test (OECD Guideline 478). When these tests are negative and it is clear that the substance did reach the target organs, the likelihood of the substance being non-genotoxic in rodents is high, while a positive response may make it very likely that the substance in question will be a genotoxic rodent carcinogen. In the testing strategies for genotoxicity, there are few *in vivo* gene mutation tests. A number of novel methods based on endogenous reporter genes or transgenic reporter gene are in various stages of development, but these have not been sufficiently validated [59].

In the mouse bone marrow micronucleus test, micronuclei, derived from substance-treated mice, are counted in a statistically sufficient number of bone marrow cells and compared with those from control mice. A suitable rodent liver genotoxicity test is the test in which rats are treated with the substance concerned and liver cells in primary culture are exposed to tritiated thymidine in order to detect increased unscheduled DNA repair (OECD 486). Protocols for the *in vivo* alkaline single-cell gel electrophoresis assay for DNA strand breaks are available [60] and a validation study is ongoing. In the dominant lethal test (OECD 478), a serial mating technique is used in which substance-treated males are mated with single virgin females for one oestrus cycle. By replacing the virgin female with another the breeding study is continued for 70 days which is long enough to cover all stages of spermatogenesis. The detection of early embryonic deaths in the females is an indication for dominant lethality. This type of test also provides information about fertility.

The strengths and weaknesses of short-term genotoxicity tests have been discussed by many authors and are elegantly described by Ashby [61]. As indicated above, genotoxicity tests cannot detect all carcinogens because (human and animal) carcinogens can be divided into those which interact with DNA and those which have a different mechanism of action not involving interaction with DNA [62] (Section 6.4.7). Thus, genotoxicity testing only provides information about possible genotoxic potential, and substances with this potential may be suspected of being carcinogenic, but the final proof can only be obtained from animal experimentation.

One of the major failures of the past is that scientists did not clearly understand that there can only be a correlation between animal carcinogenicity and genotoxicity for those animal carcinogens which act in interaction with DNA. Therefore, correlation studies involving all animal carcinogens and *in vitro* genotoxicity tests are, by definition, false. For this reason "detection" rates for carcinogens in *in vitro* systems vary from 45-75% depending on the number of non-genotoxic carcinogens included in the study. Nevertheless, when used for pre-screening, genotoxicity tests can provide a very relevant and cost-effective tool for identifying mutagens, and thus potentially genotoxic carcinogens. Since genotoxic substances are not generally permitted in the human environment, the detection of such properties usually prohibits further industrial development of the substance and further animal testing. Only in cases where the substance is considered very important and beneficial will further testing be undertaken to find out whether the substance is indeed a carcinogen and, if so, to what extent a certain human exposure poses a risk.

After *in vitro* testing and before long-term animal testing, *in vivo* genotoxicity testing is sometimes advocated, because if the result of these tests are negative, the chances of the genotoxic substance being an animal carcinogen, and thus probably a human carcinogen, are smaller and further testing with long-term bioassay may well result in non-carcinogenicity. In the same way, a positive outcome of an *in vivo* genotoxicity test may in certain specific cases prevent further testing.

6.4.7 Carcinogenicity

Substances are defined as carcinogenic if they induce or increase the incidence of tumours (benign or malignant), cause malignancy or shorten the time of tumour occurrence when inhaled, ingested dermally applied, or injected. This effect may be route-specific. Carcinogens may be identified either from epidemiological studies, from animal experiments and/or other relevant data or studies [54].

Under REACH, a carcinogenicity test may be required at tonnage levels from 1000 tonnes per annum, if there is widespread dispersive use of the substance or evidence of frequent or long-term human exposure, and the substance is classified as category 3 mutagenic or there is evidence from repeated-dose studies that the substance is able to induce hyperplasia or preneoplastic lesions.

It is now generally accepted that the induction of cancer in animals and man involves several consecutive but independent events. Cancer growth results from heritable alterations in a cell which obtains a selected growth advantage, and growth as a clonal expansion. The steps involved in cancer causation and development are depicted in Figure 6.8 [63]. The first step is the alteration of the cellular DNA by a reactive form of the carcinogen (initiation). This reaction leads to translocation and amplification of specific genes, protooncogenes, which translate into a distinct expression of the properties of the altered cell. The altered or initiated cell, usually

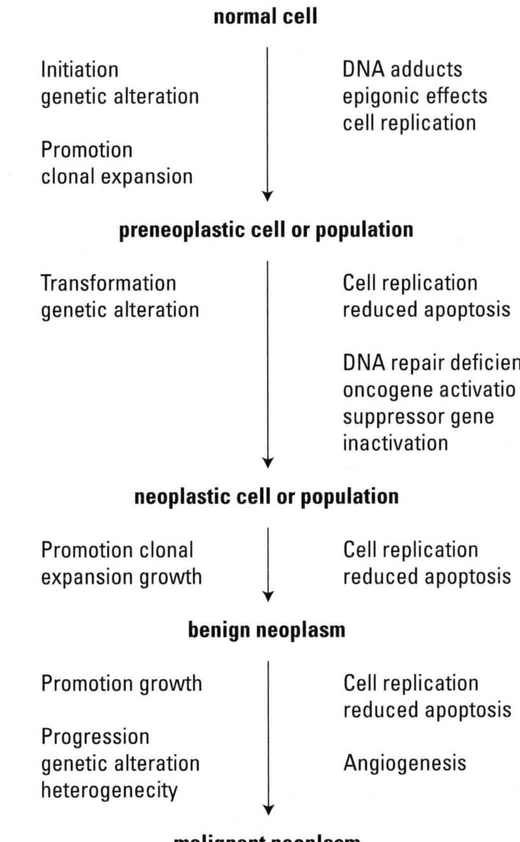

Figure 6.8. Sequence of carcinogenesis [63]. With permission.

called a latent tumour cell or neoplastic cell, may stay dormant or, under specific circumstances (e.g., under the influence of growth promoting agents), may proliferate into preneoplastic clonal expansions and ultimately progress to become cancer.

Since the alteration of DNA is a prerequisite first step, the detection of such properties provides an efficient, rapid and economical way of detecting carcinogenic potential. These tests are usually in vitro tests, where the induction of mutations is detected in prokaryotic or eukaryotic cell systems or by unscheduled DNA repair in *in vitro* bioassays. Carcinogens acting via genetic alteration are called genotoxic carcinogens; this contrasts with non-genotoxic carcinogens, which do not damage DNA but become active in the development of cancer after the first step of cancer causation. They exert their influence in the promotion or progression phase, where they require genetically altered cells. The exact mechanism of action of such non-genotoxic carcinogens is as yet only partially elucidated, but the end result is usually an increased proliferation in specific tissues. This can be caused by excessive secretion of hormones, or by injury, or can be receptor-mediated (e.g., peroxisome proliferation). Non-genotoxic carcinogens usually affect only one organ and, because of the nature of their indirect mechanism of action, there is a threshold for their action. This contrasts with genotoxic carcinogens for which, theoretically, a threshold cannot be expected since, in principle, every molecule which reacts with cellular DNA may reach a target cell and transform it into a quiescent, latent, neo-plastic cell, which may ultimately develop into cancer. However, it is also recognized that for certain genotoxic carcinogens a threshold may exist for the underlying genotoxic effect.

In a cancer bioassay, genotoxic as well as non-genotoxic carcinogens can be detected since the endpoint of this assay is the development of cancer. As it is important for the purposes of risk assessment to know about the genotoxicity or non-genotoxic properties of a substance, genotoxicity testing with *in vitro* assays will be relevant. These assays are described in Section 6.4.6. Depending on the results, anticipated use and duration of exposure, further testing may or may not be necessary. Guidelines used for carcinogenicity testing are OECD 451 or 453 (combined chronic/carcinogenicity test) listed in Table 16.1 [29].

A chemical which is found to be genotoxic in a short-term series of tests with various endpoints, is unlikely to be acceptable for human exposure, thus making further testing generally unnecessary, unless either the use or exposure is unavoidable. In these cases a carcinogenicity study is warranted to obtain information on the carcinogenic potential and the dose-response relationship in order to carry out a proper quantitative risk assessment (Section 6.5). For non-genotoxic substances carcinogenicity testing is usually required, although the relevance of such tests may be questionable when human exposure is far below the NOAEL or BMD [64].

When non-genotoxic substances are investigated it is also important to study the tumour-enhancing (or promoting) properties. As promotion is an organ-specific phenomenon, such a study should focus on lesions (e.g., hyperplasia or increased cell turnover) found in target organs in toxicity studies. Limited *in vivo* bioassays are extremely useful to indicate possible tumour-enhancing properties. The available organs for such studies include skin, lung, mammary gland, liver, stomach, colon and bladder [65].

Carcinogenicity studies

Although a carcinogenicity test specifically designed to detect carcinogenicity can be carried out, the combined

Table 6.7. Carcinogenicity studies.

Conditions	chemical identification of substance, its purity and chemical characteristics
Route	oral (gavage, diet, drinking water or capsules), inhalatory, dermal
Experimental animals	rat, mouse (dog, monkey)
Number of animals	50 rodents per sex per group; for non-rodents usually not more than 7 to 20 per group
Dose levels	control and at least 3 dose levels, more dose levels for proper quantitative risk assessment, satellite groups may be added
Duration of exposure	majority of expected lifespan inhalation: intermittent (e.g. 6 h/day, 5 d/wk) or continuous
Examinations	see table 6.5. recommended for microscopic examinations: (a) all grossly visible tumours or lesions suspected of being tumours in all groups; (b) all preserved organs and tissues of: (a) all animals that die or are killed during the study, and (b) animals of the highest dose group and controls. (c) if a significant difference is observed in hyperplastic, pre-neoplastic or neoplastic lesions between the highest dose and control groups, microscopic examination should be made on that particular organ or tissue of all animals in the study; (d) in case the results of the experiment give evidence for substantial alteration of the animals' normal longevity or the induction of effects that might affect a neoplastic response, the next lower dose level should be examined as described above; and (e) the incidence of tumours and other suspect lesions normally occurring in the strain of animals used (under the same laboratory conditions – i.e. historical control) is desirable for assessing the significance of changes observed in exposed animals.
Results	information on carcinogenic properties, tumour incidences in relation to dose, latency period, tumour multiplicity, potential for metastasis

chronic toxicity/carcinogenicity bioassay is more commonly used, in which the effects of a substance, whether of a neoplastic or non-neoplastic nature, can be determined.

The assay (Table 6.7) is almost exclusively carried out in rats and mice, where both sexes should be used [32]. The test begins with weanling or post-weanling animals and covers the animals' life span of at least two years (rats) or 18 months (mice). Since information concerning dose-response is crucial, a sufficient number of dose groups is required. At least three dose levels and a control group should be used with 50 animals per sex per group. The lowest dose should not interfere with growth and development and must not cause effects, whereas the group receiving the highest dose should show signs of toxicity. The highest dose should not exceed a concentration of 5% of the diet unless macro-nutrients are being examined. The intermediate dose should be in the mid range between the high and low doses. It is not uncommon to add a satellite high-dose group (20 animals per sex) to induce frank toxicity and a satellite control group (10 animals per sex) to evaluate the pathology of effects other than neoplasia (usually after 12 months' experimentation). As described above, under chronic toxicity testing, caging, care, feed and water supply (diet) must be optimum and well-controlled.

The rate of exposure to the substance should be comparable to the anticipated human exposure. The frequency of exposure usually depends on the route of exposure. In oral studies the substance is given daily, unless it is administered by gavage, in which event exposure is usually restricted to 5 times a week, as usually occurs in inhalation studies, where exposure will generally be limited to 6 hours per day. Careful daily clinical examination is required and appropriate action should be taken to minimize loss of animals during the study due to autolysis or cannibalism. Body weights are measured weekly during the first 13 weeks and once every 4 weeks thereafter. Food and drinking water intake are determined weekly during the first 13 weeks and thereafter 3 monthly. Blood tests (Section 6.4.5) are performed after 3, 6, 18 and 24 months on 20 animals per sex per group and a differential blood count is performed on samples of animals from the highest dose group and the controls, and at lower dose levels when indicated. Urine analysis of 10 animals per sex per group should be performed at the same intervals. At 6 month intervals clinical chemical analysis to the same extent as described for chronic toxicity testing should be carried out. At the end of the experiment a 50% survival rate is expected for mice at 18 months and rats at 24 months. Complete gross examination is performed and histopathological examination is carried out on all tissues and organs from the highest dose group and the control group. Where indicated, the tissues and organs of lower dose groups should be examined and all tumours or lesions suspected of being tumours should be examined histopathologically.

Positive carcinogenic findings in animals require careful evaluation to determine their relevance to humans. Of key importance is the mechanism of tumour induction. The International Programme on Chemical Safety has developed a conceptual framework to provide a structured and transparent approach for the assessment of the overall weight-of-evidence for a postulated mode of induction for each tumour type observed [65, 66]. There is a scientific consensus that some tumours seen in rodents arising from specific non-genotoxic mechanisms, are not relevant to humans. The International Agency for Research on Cancer (IARC) has provided detailed characterization for some of these mechanisms and has identified the key biochemical and histopathological events which should be observed in order to conclude that the tumours arose because of one of these mechanisms and can therefore be dismissed as not relevant to humans [67,68]. Human data may provide direct information on the potential carcinogenicity of the substance. When human data of sufficient quality are available, this is preferable to animal data as no interspecies extrapolation is necessary and exposure scenarios are likely to be more realistic.

Limited *in vivo* studies

Limited, medium-term *in vivo* studies have been developed to study the tumour enhancing properties of chemicals. Such studies are able to bridge the gap between *in vitro* and *in vivo* screening methods for genotoxicity and long-term carcinogenicity studies. They employ a known initiator or a genotoxic carcinogen in a subcarcinogenic dose, followed by administration of the substance to be examined. Several organ systems have been proposed and used [69] such as skin, lung, stomach, mammary gland, kidney, thyroid, pancreas, intestines and urinary bladder. These studies are used to obtain information on carcinogenic action as well as to determine dose-response relationships [70].

Various short and medium-term carcinogenicity assays with neonatal or transgenic animals have been developed which serve as a tool for studying *in vivo* induction of cancer and provide essential information about the predisposing factors to specific genetic alterations in carcinogenesis. Examples are the rat liver foci model, the $XPA^{-/-}$ and the p53+/- knockout mouse models, the Tg.AC and Tg.rasH2 transgenic mouse models and the neonatal mouse model. An evaluation of studies with transgenic mice has recently been published by the International Life Sciences Institute. These models were generally accepted to be useful as screens for hazard identification but not as a complete replacement for the two-year bioassay [71].

To study tumour enhancing properties various *in vitro* tests have been proposed [72], but these tests have been insufficiently validated. They are based on the determination of clinical properties common to a group of promoting agents, such as loss of cell-to-cell communication and outgrowth of partially transformed cells.

6.4.8 Reproductive and developmental toxicity

The term "reproductive toxicity" is used to describe the adverse effects induced on any aspect of mammalian reproduction. It covers all phases of the reproductive cycle, including impairment of male or female reproductive function or capacity and the induction of non-heritable adverse effects in the progeny, from conception to sexual maturity, such as death, growth retardation, structural and functional effects [5]. Reproductive toxicity is not the same as teratology, which refers to the study of (structural) malformations, although teratogenicity (embryo/foetotoxicity) is part of reproductive toxicity. Although all stages in the reproductive cycle may be vulnerable to directly or indirectly induced effects, the more rapid developmental stages may be more vulnerable than others. The following developmental stages, i.e., gametogenesis, conception, the embryonic period from conception to the end of major organogenesis, the foetal period, being the end of embryogenesis to the birth of the progeny, the neonatal period and the developmental period until adulthood, may all be involved in chemical toxicity. Not infrequently, there is a delay between the moment of exposure and the manifestation of the effect and this is especially the case in gamete formation and maturation. Gametogenesis occurs very early in embryogenesis, whereas germ cell formation in females occurs only before birth. Certain adverse effects in such cells can be induced before birth, but are not expressed before the germ cell is fertilized which undergoes the developmental period until sexual maturity, thus passing one generation.

Under REACH, testing for reproductive toxicity is not required at tonnage levels below 10 tonnes per annum. At 10-100 tonnes per annum, *in vivo* screening for reproductive/developmental toxicity (OECD 421 or 422, see Chapter 16) is requested if there is no evidence from available information on structurally related substances, from (Q)SAR estimates or from *in vitro* methods, that the substance may be a developmental toxicant. Where there is serious concern about the potential for adverse effects on fertility or development, a prenatal developmental toxicity test (OECD 414) or a two-generation reproductive toxicity study may be proposed, instead of the screening study. Otherwise these studies will be requested at 100 tonnes per annum. Reproductive toxicity testing need not be conducted for known genotoxic carcinogens or germ-cell mutagens since the results of reproductive toxicity testing are unlikely to influence the outcome of the risk assessment. The general objectives of reproductive toxicity testing are to establish:

- Whether administration of the substance to males and/or females prior to conception and during pregnancy and lactation causes adverse effects on reproductive function or capacity.
- Whether administration of the substance during the period of pre or post-natal development induces non-heritable adverse effects in the progeny.
- Whether the pregnant female is potentially more susceptible to general toxicity.
- The dose-response relationship for any adverse effects on reproduction.

Tests for adverse reproductive effects

The available OECD tests and important characteristics are shown in Table 6.8 [73]. Reproductive toxicity can essentially be detected during each stage of development. Examples of effects are given in Table 6.8. Detection of reproductive toxicity in animal experiments is usually done in four segments:

a. fertility and general reproductive performance;
b. embryotoxicity and teratogenicity;
c. pre and postnatal development;
d. multigeneration studies.

a. Fertility and general reproductive performance

This involves the treatment of males and females before mating for a period sufficiently long to cover the different stages of spermatogenesis or follicular development. Pregnancy, location and development to sexual maturity is followed and recorded in comparison with controls.

b. Embryotoxicity and teratogenicity

Embryotoxicity and teratogenicity are investigated by treating pregnant mammals (usually rats and or rabbits) during embryogenesis. Foetuses are recovered just before delivery and examined for morphological and structural malformations. If embryotoxic and teratogenic effects occur in such tests only at the level of maternal toxicity, the relevance of the found effect as a real effect is debatable.

c. Pre and postnatal development

In this case treatment of pregnant mammals is restricted to the latter third of pregnancy and during parturition and lactation in order to examine adverse effects during that particular period.

d. Multigeneration studies

The simple two-generation reproduction toxicity test (OECD Guideline 416) provides an excellent cost-effective way of testing that all reproductive functions are normal. The test cannot identify the origin of the adverse effects, but it is good for detection, although usually poor in the characterization of effects.

Short-term *in vivo* studies (e.g., Chernoff/Kavlock tests [74]), studies in non-mammalian species or *in vitro* studies will not, in the absence of more definitive data, provide a basis for a firm decision about the reproductive toxicity of a substance. In 2002, three *in vitro* tests for embryotoxicity were considered to be validated by the ECVAM Scientific Advisory Committee and ready for consideration for regulatory acceptance and application: the Whole Embryo Culture (WEC), the Micromass (MM) and the Embryonic Stem Cell Test (EST). However, these tests were also not considered to be replacements for the current *in vivo* testing as a whole, but to be used as part of a tiered testing strategy [73].

Data from repeated-dose toxicity studies in which there are marked adverse effects on the reproductive organs (usually the testes) can also be used to identify a substance as being toxic to reproduction. Data from such studies cannot be used to identify a substance as being of no concern in relation to reproduction. It is essential to distinguish between a specific effect on reproduction as a consequence of an intrinsic property of the substance and an adverse reproductive effect which is a non-specific consequence to general toxicity (e.g., reduced food

Table 6.8 Reproductive toxicity studies.

OECD Guideline 414 Prenatal development toxicity study
OECD Guideline 415 One-generation reproduction toxicity study
OECD Guideline 416 Two-generation reproduction toxicity study
OECD Guideline 421 Reproduction/developmental toxicity screening test
OECD Guideline 422 Combined repeated dose toxicity study with the reproduction/ developmental toxicity screening test

Time and targets at which a substance initiates its toxicity	Examples of adverse effects on
Adult toxicity	- libido - behaviour - endocrine function - mating - gamete production - reproductive life span
Maternal toxicity (changing physiology and metabolism during pregnancy and lactation)	- susceptibility - ability to nurse - milk quality and quantity
Developmental toxicity	
• Pre-implantation and implantation	- fertilization - movement of fertilized ova - implantation - survival of ova
• Embryonic development	- growth and differentiation - organ development - survival
• Placental development	- growth - organ function
• Foetal development	- growth and differentiation - organ function - survival
• Postnatal development (neonatal, pre-weaning, post-weaning, puberty)	- birth weight - organ function - hormone function - immune function - CNS and peripheral NS function - sexual function - other cellular functions (e.g. transplacental carcinogenesis) - survival

or water intake, maternal stress). Hence, reproductive toxicity should be assessed alongside parental toxicity in the same study. However, developmental toxicity occurring in the presence of maternal effects does not itself imply a causal relationship between the two and therefore it is not appropriate to discount developmental toxicity that occurs only in the presence of maternal toxicity. If a causal relationship can be established, it may be concluded that developmental toxicity does not occur at lower doses than the threshold for maternal toxicity, although the substance can still be considered as a developmental toxicant. In the absence of proven causality, the nature and severity of the developmental versus the maternal effects may well warrant the conclusion that a substance should be considered as a specific developmental toxicant when the effects are only observed in the presence of maternal toxicity.

If it is possible to identify a NOAEL from well-reported and reliable human studies, this value may be used preferentially in the risk characterization. However, it is expected that this will rarely be the case.

6.4.9 Specific studies and toxicogenomics

Although in a well-executed toxicity study all types of effects can usually be detected, the focus of the study will not, *per se*, be directed towards effects on, for example, the immune system, the central nervous system or particular related behaviour, or endocrine effects.

Immunotoxicity

Immunotoxicology in particular has received growing attention over the last decade, since it is well recognized today that chemicals may influence the immune system in a variety of ways and interact with immune responsiveness and thus health maintenance. Immunotoxic responses may occur when the immune system is the target of the chemical insult; this in turn can result in either immunosuppression and a subsequent decreased resistance to infection and certain forms of neoplasia, or immune dysregulation which exacerbates allergy or autoimmunity. Alternatively, toxicity may arise when the immune system responds to an antigenic specificity of the chemical as part of a specific immune response (i.e., allergy or autoimmunity) [75]. Numerous papers have been published on the subject and were well reviewed by Van Loveren and Vos [76]. A tiered approach is advocated, in which it is suggested that the first tier should be set up according to OECD Guidelines 407 and 408 Repeated-dose toxicity tests. A very detailed overview is given in the Environmental Health Criteria Document of IPCS [75]. The objective of the first tier is to identify potential immunotoxicity by including specific parameters such as complete blood cell count and a differential white blood cell count, organ weights of thymus, lymph nodes, spleen and histopathological examination of thymus, spleen, lymph nodes, Peyer's patches and bronchus associated lymphoid tissue (BALT). The measurement of serum Immunoglobulin M (IgM), IgG and IgA concentrations is also suggested. If indications of immunotoxicity are found, further specific test systems should be applied to identify immunotoxic properties and to detect the lowest level at which any effect will occur (i.e., cell-mediated immunity, humoral immunity, macrophage function, natural killer cell function or host resistance). If there are no indications of immunotoxicity in the 28-day (or 90-day) toxicity test, and none from SAR either, no further specific investigation for immunotoxicity will normally be required. Currently there are few methods for specific investigation of immuntoxic effects which are regarded as sufficiently validated for routine use [75,77]. The plaque forming assay or the equivalent using the ELISA method (Enzyme-linked Immunosorbent Assay) are recommended to identify altered T-cell dependent humoral responses [78,79]. Of particular value for risk assessment are so-called "host resistance models", in which the clinical relevance of immunotoxicity can be evaluated [75,80].

In contrast to the potential suppressive effects on the immune system, the hazards of immune sensitization, eventually resulting in chemically-induced allergy and autoimmunity, can not be fully assessed in the current tiered approach. Apart from the animal models for skin sensitization and respiratory hypersensitivity, no validated models are available for the testing of oral sensitization by chemicals or "novel" proteins. In drug development immune-mediated hypersensitivity reactions have been reported to be the most frequent cause of failure of drugs during clinical development [81]. Recently Nierkens et al. [82] published an oral exposure model in mice using reporter antigens to predict chemical-induced hypersensitivity reactions.

Neurotoxicity

Neurotoxicity is the induction by a chemical of adverse effects in the central or peripheral nervous system, or in sense organs. Anger [83] claimed that neurotoxic effects are not unusual: in 24% of chemicals for which threshold limit values in the working environment have been set, neurotoxic effects were the sole or partial reason for regulation. Neurotoxicity may be indicated by the following signs: morphological (structural) changes in the central or peripheral nervous system or in special sense organs; neurophysiological changes (e.g., electroencephalographic changes); behavioural (functional) changes; neurochemical changes (e.g., neurotransmitter levels). The subject was reviewed by OECD (OECD, 2004). The first indications of adverse neurotoxicological effects can usually be detected by the classical acute and repeated-dose toxicity studies (OECD 402, 403, 420, 423, 407 and 408, Table 16.1). These tests examine a number of simple nervous system endpoints (e.g., clinical observations of motor and autonomous nervous system activity, and histopathology of nerve tissue), which should be regarded as the starting point for the evaluation of a substance suspected to cause neurotoxicity. SAR considerations may prompt the introduction of additional parameters to be tested in standard toxicity studies. When available information provides indications of possible neurotoxic effects, additional endpoints may be included in the initial standard tests(s) in order to obtain in-depth information about a specific type of neurotoxic effect. Alternatively, existing information may indicate a need to conduct a neurotoxicity study (OECD 424) with specific tests to assess a suspected neurotoxic effect. Specific tests may include neurobehavioural, neuropathological, neurophysiological and neurochemical methods. Organophosphorous compounds (OPs) are often potent inhibitors of various types of esterases such as actylcholinesterase and neuropathy target esterase (NTE) and are also capable of inducing pathological lesions in the nervous system known as "delayed neurotoxicity", characterized by the delayed onset of flaccid paralysis and distinct neuropathological lesions of the peripheral nerves, spinal cord and brain. [84]. OECD Guidelines 418 and 319 have been developed to detect substances causing delayed neurotoxicity. For the evaluation of cholinesterase inhibition of OPs and other substances, the reader is referred to the WHO/FAO Joint Meeting of Experts on Pesticide Residues (JMPR) recommendations on the Interpretation of Cholinesterase Inhibition [85].

Endocrine disruption

An endocrine disrupter is an exogenous substance that causes adverse health effects in an intact organism or its progeny through alterations in the function of the endocrine system. Thus, endocrine disruption is a mechanism rather than an adverse health effect. Concern about endocrine disruption has resulted in the development of newly proposed test guidelines which specifically address effects on hormone homeostasis and on male and female reproductive organs. With respect to endocrine disruption, the two-generation study (OECD 416) is currently the most complete study available. In both this study and the developmental toxicity study (OECD 414), additional endocrine-sensitive parameters can be studied on a case-by-case basis, when endocrine disruption is an issue of concern. A number of possible improvements have been identified, many of which are most relevant to mammalian test designs [86]. These include:

- Extension of organ weight and histopathology requirements for gonads and accessory sex organs.
- Pathological examination of offspring, where appropriate.
- Measurement of sex hormone blood levels.
- Detailed assessment of spermatogenesis and/or semen quality.
- Monitoring of oestrus cyclicity.
- Enhancement of current monitoring of physical and behavioural development.
- Learning and memory functions in offspring.
- Possible investigation of accessory sex organ secretary products.

New and revised test guidelines to detect endocrine disruptors are being discussed within OECD. In 2006, the following projects were in progress:

- Peer-review of the rodent uterotrophic assay to detect oestrogenic effects.
- Validation of the rodent Hershberger assay to detect androgenic effects.
- Consideration of enhancements to the existing OECD TG 407 (Repeated-Dose Oral Toxicity).
- Further enhancement of TG 416 (two-generation reproduction test).

Toxicogenomics

Toxicogenomics is defined as a study of the response of a genome to hazardous substances, using "omics" technologies such as genomic-scale mRNA expression (transcriptomics), cell and tissue-wide protein expression (proteomics), and metabolite profiling (metabolomics), in combination with bioinformatic methods and conventional toxicology. In relation to chemical hazard/risk assessment, this emerging science could provide tools for:

- Improving the understanding of mechanisms of toxicity.
- Identifying biomarkers of toxicity and exposure.
- Offering ways to reduce, refine and replace costly animal intensive methods in chemical screening and testing.
- Reducing uncertainty in the grouping of chemicals for assessments, (Q)SARs, inter-species extrapolation, effects on susceptible populations, etc..
- Assessing the effects of chemical mixtures and combinations of stressors.

Currently, toxicogenomic approaches are recognized as not yet sufficiently developed for risk assessment decisions or to replace existing approaches [87-89]. However, they can be used to provide supportive evidence on a case by case basis [87]. More research is required if toxicogenomics is to become a tool routinely used in toxicology. This research should define further correlations between genomic and the more traditional hazard assessment data. Micro-array techniques need to be standardized and data available for analysis held in open-access databases [90].

6.4.10 *In vitro* tests for toxicity

The area of toxicological risk evaluation presently relies on a range of animal experiments. Many of these tests are standard procedures in the form of guidelines formulated by the OECD [29]. The use of these procedures has resulted in the relatively safe use of chemicals in industry, or as agrochemicals, drugs, household chemicals or cosmetics. However, large numbers of laboratory animals have been used and distress has been caused to many of these animals [91]. This has resulted in discussions on the ethical, scientific and financial feasibility of the process of toxicity testing. In their book, The Principles of Humane Experimental Technique, Russell and Burch introduced the terms "Replacement, Reduction and Refinement", or the three Rs [42]. This was much later followed by legal regulations concerning the humane use of animals in experimentation within the European Community [43]. Thus, animal studies should be justified and the available alternatives carefully considered.

Apart from these ethical objections to the use of animals, the reliance on animal data in toxicology also encounters scientific criticism. Animal models are, in many cases, motivated by the assumed fundamental biological comparability of the integrated system of intact mammalian organisms. However, the use of animal data to predict a compound's toxicity in humans is always prone to some degree of uncertainty. This is the result of qualitative or quantitative differences between physiological and biochemical processes in animals and humans, as well as in compound-specific parameters regarding uptake, distribution, biotransformation and excretion, which may also differ in a qualitative way [92]. These deviations may result in wide differences in the concentration of a compound at the target tissue in different species for the same external dose. Moreover, appreciable species differences in the mode of toxic action of compounds may occur. Further difficulties may result from the extrapolations that have to be made from a rather small, homogeneous group of laboratory animals to the very heterogeneous general human population. In risk assessments attempts are made to overcome these uncertainties by introducing assessment factors, e.g., no observed (toxic) effect levels (NOELs) determined in animal experiments are divided by these factors to account for interspecies and/or interindividual differences when establishing safety standards for human exposure [4].

Some procedures in toxicity testing are very time-consuming and expensive, e.g., a "classical" carcinogenicity study. Therefore, the economic aspect of toxicity testing using intact animals is also a factor of concern. In the development of drugs or pesticides the use of combination chemistry, together with high-throughput systems, to select possible compounds for further development also requires the more direct use of toxicity studies. Here, it would be very useful if the mechanisms of toxic action could also be taken into account. The use of studies in *in vitro* systems may well be an advantage here.

Non-animal test methods

Over the last decades an increasing number of test systems have been developed that do not rely on the use of intact animals, but make use of biological systems at a lower level of organisation than the organism: isolated organs, cell cultures, subcellular systems. These *in vitro* systems have been very useful in studying the molecular basis of a chemical's biological activity, including its mechanism(s) of toxic action [93]. There are now *in vitro* models for many different organ systems, including systems for studying effects in cells or tissues derived from the liver, kidney, neuronal system, lungs, muscle, etc. The European Centre for the Validation of Alternative Methods (ECVAM) has produced a range of reports summarizing the state-of-the-art for these systems. An overview of these reports can be found on the ECVAM website (http://ecvam.jrc.cec.eu.int/index.htm) and in a summary report by Worth and Balls [94].

Another important development is the prediction of biological reactivity on the basis of a compound's physicochemical properties, such as structure, molecular size, reactive groups, etc. [95]. One application of this knowledge is in the construction of SARs (Chapter 10).

Knowledge of a compound's mechanism of toxic action, either derived from *in vitro* systems or based on its structure, can provide a basis for hazard identification. In many areas of industrial development of new chemicals or products these approaches are widely applied, mainly for screening purposes.

Presently, a wide variety of *in vitro* systems is available or is being developed to study different forms of cytotoxicity [94]. Cytotoxicity can be defined as the adverse effects resulting from interference with structures and/or processes essential for cell survival, proliferation, and/or function [96]. These effects may involve the integrity of membranes and the cytoskeleton, cellular metabolism, the synthesis and degradation or release of cellular constituents or products, ion regulation, and cell division. This offers the concept of "basal cell functions" that virtually all cells possess (mitochondria, plasma membrane integrity, etc.). In this concept a wide range of toxic reactions are the consequence of non-specific alterations in these cellular functions, which may then lead to effects on organ-specific functions and/or death of the organism [96]. Based on this concept, *in vitro* tests can be used for screening and as potential replacements for *in vivo* toxicity testing, especially for acute (lethal) toxicity.

A further category of cellular toxicity tests will need to describe more specific functional disturbances of specialized differentiated cell systems. While the first type of test will focus on cytotoxicity as can be measured by parameters such as cell death or adverse effects on the household functions that are general to all cell types, this second type of test will be designed to quantify parameters that will reflect tissue or organ-specific toxicity [97]. It could be argued that these more specific forms of toxicity testing are more important for the non-lethal toxic effects and for chronic toxicity.

Depending on the aim of the study either basal cytotoxicity or organ or tissue-specific toxicity can be measured. As indicated above, a good overview of the currently available test systems can be found in the reports by ECVAM. The methods described in these reports also comprise, among many other systems, the use of hepatocyte cultures in biotransformation and hepatotoxicity studies, the use of cytotoxicity parameters in phototoxicity studies, neurotoxicity in neuronal cells, skin preparations for irritancy, corrosivity or absorption studies. A number of these methods have now been (or will soon be) adopted by the OECD. One set of *in vitro* tests has been used in regulatory procedures for over three decades: bacterial mutagenicity tests to determine genotoxicity.

Furthermore, *in vitro* systems can be used to study early cellular responses that may form the basis for predicting toxic responses in the *in vivo* situation. These biomarkers of toxic effects can be applied in a hazard characterization of the compounds under investigation. Examples of such early cellular responses are: oxidative stress and glutathione homeostasis, cellular stress responses, changes in enzyme activity, cytokine responses, etc. [98]. The increasing possibilities to use cell and tissue cultures to measure these biomarkers of effect are now becoming complemented by the potential use of information derived from genomics, transcriptomics, proteomics and metabonomics [99,100].

Extrapolation of *in vitro* toxicity data to the *in vivo* situation

As our understanding of toxic mechanisms steadily increases, the role of *in vitro* methods in this is obvious. However, a hazard assessment cannot easily be made without further knowledge of the compound's behaviour in the integrated system of an intact organism. Therefore, results obtained from *in vitro* studies in general, are often not directly applicable to the *in vivo* situation. One difference between the *in vitro* and the *in vivo* situation is the absence of the processes of absorption, distribution, metabolism and excretion (i.e., biokinetics) that govern the exposure of the target tissue of the organism *in vivo* [101]. The concentrations to which *in vitro* systems are exposed may not correspond to the actual situation at the target tissue after *in vivo* exposure. In addition, metabolic activation and/or saturation of specific metabolic pathways may also become relevant in terms of the toxicity of a compound *in vivo* thus leading to misinterpretation of *in vitro* data if such information is not taken into account. Therefore, predictive studies on the biological activity of compounds require the *integration* of data on the mechanisms of action with data on biokinetic behaviour.

Use of kinetic models in combination with *in vitro* systems for the prediction of *in vivo* dose.

Biokinetic modelling describes the absorption, distribution, metabolism and elimination of xenobiotics as a function of dose and time within an organism. Such models can be divided into two main classes: data-based compartmental ("classical") models and physiologically-based compartmental models [102]. Over the last 15 years, the feasibility of this modelling approach has been greatly increased due to the availability of computer techniques that allow for the simultaneous, numerical solution of differential equations [103].

The physiologically-based biokinetic (PB-BK) models are well suited to be combined with *in vitro* techniques for measuring kinetic parameters. These models describe the compartments with respect to the known anatomy and physiology of the organism. Compartments correspond to relevant anatomical structures such as liver or kidney, or tissue types such as fat or muscle. The distribution of a compound throughout the body is described by tissue-blood partition coefficients (PCs) and, if applicable, by any active transport processes. Metabolism and elimination of a compound can be described by either a linear clearance rate or a saturable Michaelis–Menten term in the organs capable of biotransformation (e.g., the liver, lungs, intestine or kidneys). The pulmonary ventilation rate and blood–air PC play a role in the uptake and exhalation of volatile compounds.

While many species-specific anatomical and physiological data are now available from the literature [104,105], compound-specific parameters for PB-BK models, like tissue-blood PCs and the Michaelis–Menten constants *V*max and *K*m, can be obtained either by fitting these parameters to experimental data obtained *in vivo* or, in some cases, based on results from *in vitro* data and physicochemical parameters for the chemicals under investigation.

Thus, the physiological as well as the chemical-specific parameters will be used in a set of differential equations that describe the biokinetic behaviour of a compound in the PB-BK model. Once the compound is taken up in the systemic circulation, the kinetic processes of distribution, metabolism and excretion of a compound are independent of the exposure route. Thus, it is possible to extrapolate from one exposure route to another. Besides route-to-route extrapolation, PB-BK models also facilitate extrapolation of dose and animal species beyond the conditions of laboratory studies [102].

The use of this technique of integrating *in vitro* data with PB-BK models to estimate toxic doses *in vivo* has shown promising results in a number of studies, e.g., for the neurotoxic effects of acrylamides [106], for some industrial chemicals [107] and for reproductive toxicity effects [108].

6.5 HUMAN HEALTH HAZARD ASSESSMENT

6.5.1 Introduction

In hazard assessment, the data available will first of all be evaluated with regard to quality and completeness. Both human and non-human studies need to be considered as well as *in vitro* and (Q)SAR data. Relevant aspects of data availability and data evaluation (i.e., validity, reliability, and relevance), are discussed in Section 8.5. Further to this evaluation, the usefulness of the data for hazard and risk assessment needs to be addressed, using a weight-of-evidence approach (Section 8.5.1). The data selected thus will be used further to determine the possible adverse effects to which humans could be exposed. One of the first end results of the hazard assessment will be the classification and labelling of the substance. The next goal of hazard assessment is the identification of exposure levels above which humans should not be exposed through dose (concentration) – response (effect) assessment. These aspects will be covered in this section.

6.5.2 Classification and labelling

The object of classification and labelling is "to identify all the physicochemical, toxicological and ecotoxicological properties of substances and preparations which may constitute a risk during normal handling or use. Having identified any hazardous properties, the substance or preparation must then be labelled to indicate the hazard(s) in order to protect the user, the general public and the environment." [54]. Classification should be based on a set of well-defined criteria. It is stressed here that classification and labelling pertains to intrinsic properties revealed in the hazard identification process, but not to hazard or risk assessment. Exposure considerations fall outside the scope of this exercise. "Any classification based on biological data can never be treated as final. Experts may differ in opinion and most borderline cases can be reclassified in an adjacent class. Variability or inconsistency in toxicity data due to differences in the susceptibility of test animals, or to the experimental techniques and materials used, can also result in differing assessments. The classification criteria are guidelines intended to supplement but never to substitute for specialist knowledge, sound clinical judgement or experience with a compound. Reappraisal might be necessary from time to time" [109].

Classification and labelling can be considered to be the first risk management tool for chemicals and is based on the results of hazard identification and/or effect or dose-response assessment but not on exposure assessment. Classification and labelling is not based on the results of the risk characterization because this is based on actual or predicted exposure levels and not on potential exposure levels during normal handling or use. For example, exposure to a carcinogenic substance is normally reduced to levels at which no risk is expected. However, the users of this substance still have to be warned about the carcinogenic property of the substance and the safety measures required because the hazardous properties of the substance have not changed.

Examples of international classification systems are:

1. *The general classification and labelling requirements for dangerous substances and preparations of the European Communities [54].*
   This system classifies substances and preparations on the basis of physicochemical properties, toxicological properties (acute toxicity, irritation, sensitization, repeated-dose toxicity), specific effects on human health, including carcinogenicity, mutagenicity and reproductive toxicity, and environmental effects

(acute toxicity, persistence, bioaccumulation, atmospheric effects). The classification and labelling of substances and preparations is based on the available data. There is no requirement for additional testing under these directives. The classification results in labelling using none or one or more of seven symbols (Chapter 1, Figure 1.8), 59 R (risk) phrases and 62 S (safety) phrases. This labelling is the first, and often the only, information on the hazard of the substance or preparation and on the required safety measures that reaches the user. Further, classification has consequences related to several regulatory fields, such as worker health and safety, transport, major industrial accidents, consumer products, waste and pollution.

Classification and labelling of substances and preparations under the EU system is done by the person placing the substance or preparation on the market (self-classification). However, this legislation also envisages harmonized classification and labelling of substances based on proposals by the member states. This has resulted in a list of classified substances (Annex I to 67/548/EEC). The classification process is mandatory and must also be used for self-classification of preparations containing one or more of these substances. This list, covering approximately 8000 substances is included in the European legislation but is also available as a searchable database on the website of the European Chemical Bureau (http://ecb.jrc.it/). As the classification in this list is based on the available data, the absence of a chemical in the list could mean that either the substance has no hazardous properties or that the substance was never evaluated for inclusion in Annex I. The absence of a certain hazard for a chemical in Annex I could mean that either the substance does not have this hazardous property or that there are no data available to determine whether classification for this hazardous property is necessary.

The classification and labelling of preparations can be based either on tests with the preparation or on the composition of the preparation. The directive provides simple rules to determine the classification of the preparation based on the weight/weight percentage and classification of each of the components.

2. *The WHO recommended classification of pesticides by hazard and guidelines to classification 1992-1993 [109].*

   This classification is based primarily on acute oral and dermal toxicity, as expressed by the LD50 test. No specific labelling is prescribed, except for general recommendations, e.g., to use the symbols which are usually applied to substances with a high degree of hazard. Information on the classification of individual pesticides is available on the WHO website (http://www.who.int/ipcs/publications/pesticides_hazard/en/index.html).

3. *Global Harmonised System for classification and labelling (GHS) [110].*

   This United Nations system is meant to harmonize existing classification and labelling systems. The system was completed in 2002 and will be revised regularly. It covers physicochemical properties, toxicological properties (acute toxicity, irritation, sensitization, specific target organ systemic toxicity, carcinogenicity, mutagenicity and reproductive toxicity), and aquatic toxicity for substances and mixtures. The system is currently being implemented in national and EU legislation. The GHS has many similarities with the EU classification and labelling system but there are also some differences. For example, different symbols are used and additional categories are introduced for some endpoints. Information on the GHS can be found at: http://www.oecd.org/department/0,2688,en_2649_34371_1_1_1_1_1,00.html and http://www.unece.org/trans/danger/publi/ghs/ghs_welcome_e.html.

   In the EU, GHS will be implemented together with the new REACH chemical legislation. Classification is important for REACH because an exposure assessment and a risk characterization is only required for substances which are classified as dangerous or meet a number of other criteria.

4. *Classification of carcinogens of the International Agency for Research on Cancer (IARC)[111].*

   The IARC evaluates the carcinogenic risk of chemicals, agents, mixtures or conditions of human exposure. These evaluations are available as monographs and result in classification into one of the following groups:

   Group 1 – The agent (mixture) is carcinogenic to humans.

   Group 2A – The agent (mixture) is probably carcinogenic to humans.

   Group 2B – The agent (mixture) is possibly carcinogenic to humans.

Group 3 – The agent (mixture or exposure circumstance) is not classifiable as to its carcinogenicity to humans.

Group 4 – The agent (mixture) is probably not carcinogenic to humans.

Classification by IARC does not result in any legal obligations. Information on the IARC classification system, including a list of classified substances, is available on the IARC website: (http://monographs.iarc.fr/index.php).

Box 6.3 shows how dieldrin is classified under these classification systems. The differences between the different systems could be due to differences in criteria, differences in the available data at the time of evaluation, or differences in the interpretation of the data between groups of experts. Table 6.3 shows the classification criteria of the four systems for chemicals on the basis of acute LD50 values.

6.5.3 Dose-response assessment

Dose-response evaluation for threshold effects: the NOAEL approach

The dose-response data resulting from toxicity studies need to be evaluated and, in general, the aim of such an evaluation is to derive a "safe" dose, i.e., a dose that does not result in biologically significant effects. This dose is called a "Reference Point" (RP) or a "Point of Departure" (PoD). In current approaches, a distinction is made between toxic effects that show a dose threshold below which adverse effects are assumed not to occur and effects lacking such a threshold. This Section discusses the NOAEL approach, which has been the standard approach for evaluating dose-response data for threshold effects. The next Section discusses the evaluation of endpoints for which no dose-threshold is assumed, in particular, tumours that are caused by a genotoxic mechanism.

NOAEL stands for "No Observed Adverse Effect Level". Briefly, this is the highest dose at which no (adverse) effects were observed in the available toxicity studies. Figure 6.9 illustrates the NOAEL principle for a single endpoint. The procedure to assess it is as follows:

- For those endpoints that show a (dose-related) change, determine the lowest dose that differs (statistically significantly) from the controls. This is the LOAEL (Lowest Observed Adverse Effect Level) for that endpoint (see Figure 6.9).
- For each of these endpoints, assess the dose below the LOAEL, this is the NOAEL for that endpoint (see Figure 6.9).

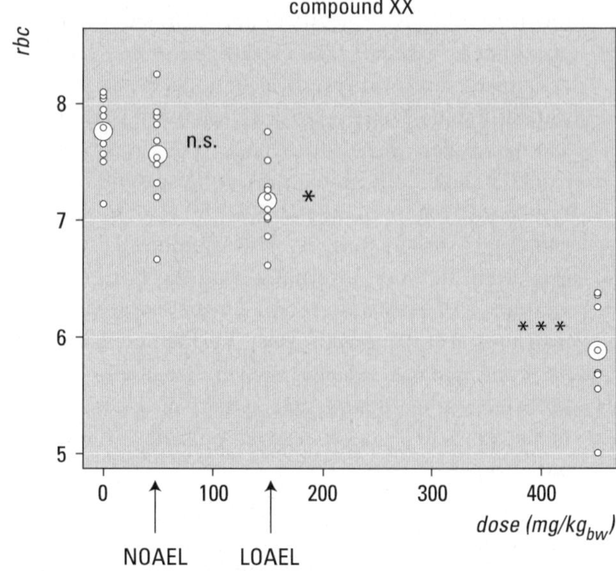

Figure 6.9. Illustration of the LOAEL and NOAEL for a decrease in red blood cell counts observed in an OECD toxicity study. The small marks indicate the observations in individual animals, the larger marks indicate the group means.
n.s.: not significantly different from the controls.
*: significantly different from the controls.
***: highly significantly different from the controls.

- Determine the lowest NOAEL over all endpoints in the study, this is the overall NOAEL for that study.
- Determine the lowest of the NOAELs of the available studies, this is the (overall) NOAEL for that chemical. The study and endpoint associated with the NOAEL for the chemical are called "critical study" and "critical endpoint", respectively.

This procedure implies that a NOAEL can only be derived from a study (and endpoint) that showed effects at higher doses. Further, it should be noted that the procedure does not rule out toxicological judgement. For instance, the critical effect found may not be relevant, or less so, for humans (e.g., kidney effects in rats, effects in the forestomach). Or, particular effects may be considered adaptive (and reversible) without being adverse. Therefore, the word "adverse" in the term NOAEL is essential.

Similarly, the word "observed" is essential. By definition, the effect at the NOAEL does not differ statistically significantly from the controls. This only means that the effect was not large enough to be detected by the statistical test. Or, conversely, the statistical test apparently was not sensitive enough to detect the effect

Box 6.3. Example: classification and labelling of dieldrin (CAS: 60-57-1)

EC [54]: symbols T+ and N, risk phrases R25-27-40-48/25-50/53, meaning:

- T+, R27 = very toxic (skull and cross bones symbol) in contact with skin (LD50 dermal, rat or rabbit $\leq$ 50 mg/kg$_{bw}$).

T+

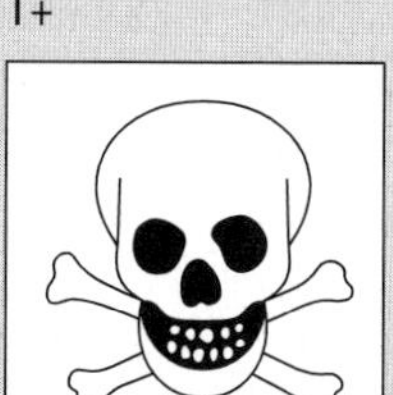

- T, R25 = toxic if swallowed (25 < LD50 oral, rat $\leq$ 200 mg/kg$_{bw}$).
- T, R48/25 = toxic with danger of serious damage to health by prolonged oral exposure (serious damage to health to be caused at levels below 5 mg/kg$_{bw}$·d).
- R40, category 3 carcinogen = possible risk of irreversible effects (concern to man owing to possible carcinogenic effects but for which the information available is not adequate for making a satisfactory assessment. There is some evidence from appropriate animal studies, but it is insufficient for a higher category).
- N, R50, R53 = dangerous to the environment (dead tree and fish symbol), very toxic to aquatic organisms (L(E)C50 fish or *Daphnia* or algae $\leq$ 1 mg/L), may cause long-term adverse effects in the aquatic environment (substance not be readily biodegradable or the log $K_{ow} \geq 3.0$, unless the experimentally determined bioconcentration factor $\leq$ 100).

N

WHO [109]: Class-1b, highly hazardous, oral LD50 for the rat is between 5 and 50 mg/kg$_{bw}$.

GHS [110[: An official EU classification is not available for any substance at the moment because GHS has not yet been introduced in EU legislation. However, based on proposed legislation and translation of the current EU classification, the following GHS classification may be expected:

- R25 becomes Acute toxicity, Category 2 or 3 (oral) with skull and crossbones pictogram, signal word "Danger" and hazard statement "Fatal if swallowed (Cat. 2)" or "Toxic if swallowed (Cat. 3)".
- R27 becomes Acute toxicity, Category 1 (dermal) with skull and crossbones pictogram, signal word "Danger" and hazard statement "Fatal if swallowed".
- R40 becomes Carcinogenicity, Category 2 with health hazard pictogram, Signal word "Warning" and hazard statement "Suspected of causing cancer".
- R48/25 becomes STOST[1] (repeated exposure), Category 1 with health hazard pictogram, signal word "Danger" and hazard statement "Causes damage to organs through prolonged or repeated exposure".
- R50/53 becomes Hazardous to the aquatic environment, chronic 1 with environmental hazard pictogram, signal word "Warning" and hazard statement "Very toxic to aquatic life with long lasting effects".

IARC [102]: Group 3: The agent (mixture or exposure circumstance) is not classifiable in terms of its carcinogenicity to humans.

[1] STOST = Specific Target Organ Systemic Toxicity

(in statistical terms: the power of the test was too low). Therefore, it can only be concluded that the effect at the NOAEL is smaller than the detectable effect size of the statistical test (and hence, of the particular study). Put another way, the size of the effect at the NOAEL could be anywhere between zero and the detectable effect size. In practice, this point is often overlooked, and the NOAEL is simply considered as a dose where the effect has been shown to be zero. This is unfortunate, since in some cases the detectable effect size is not negligible, and biologically significant effects cannot be excluded.

Apart from the fundamental problem that a NOAEL is often unjustly regarded as a no effect level, various other disadvantages of the NOAEL have been identified in the literature [47,103], the most important of which are briefly summarized here. Since the detectable effect size of a study depends on the number of animals used, the value of the NOAEL does as well. In fact, its value tends to be higher when fewer animals are used, while the opposite would be more appropriate (greater uncertainty should be paired with more conservatism). Further, the NOAEL can only be one of the applied doses. Both these points imply that the NOAEL strongly depends on the study design (choice of dose levels and number of animals per dose). As a consequence, replicating a particular toxicity study using another study design, but which is otherwise identical, is likely to result in another value for the NOAEL. This uncertainty in the value of a NOAEL is probably large, but how large cannot be quantified. This is another disadvantage of the NOAEL approach, as in risk assessment quantifying the uncertainties involved is crucial for deriving protective human exposure limits. Finally, the NOAEL approach does not make full use of the dose-response information as a whole.

Dose-response evaluation for non-threshold effects

For carcinogens that act by a genotoxic mechanism it could theoretically be argued that each single molecule has a very small probability of giving rise to a DNA adduct, and that this adduct has a very small probability of causing a mutation. This single mutation could possibly occur in a gene that is potentially related to the carcinogenic process, thereby increasing the probability of generating of a malignant cell. In reality, the process of carcinogenesis is much more complex, but the basic idea expressed here is that the onset of tumours appears to be stochastic in nature: it cannot be predicted, even if we understood precisely all the biological processes involved (just as we cannot predict the outcome of tossing a single coin). From this perspective decreasing the dose will lead to an ever decreasing tumour probability, and hence an ever decreasing tumour incidence in a population of animals (or humans). In other words, a dose-threshold below which tumours cannot be evoked at all by the chemical appears to be implausible.

Due to the lack of a dose-threshold the NOAEL approach has been considered unsuitable for genotoxic carcinogens. Since it is assumed that there is a non-zero risk at any (low) dose, an evaluation of tumour incidence data (evoked by a genotoxic carcinogen) can only try to determine a dose where the risk is acceptably small, e.g. 10^{-6}, i.e., one in a million (over a lifetime). The latter low risk level has also been denoted as a *de minimis* risk. The problem is that most tumour dose-response data originate from animal studies, which normally use dose groups consisting of 50 or maybe 100 animals per dose. Therefore, a *de minimis* risk is far below the range of observation: in animal studies an observable risk would be in the order of 10^{-1} (one in 10), five orders of magnitude higher than a *de minimis* risk of 10^{-6}. To assess a dose associated with a risk five orders of magnitude lower that the range of observation is a clear, and extreme, case of extrapolation. The term commonly used for this problem is low-dose extrapolation although, strictly speaking, the term "low-risk extrapolation" would better cover the essence of the problem.

The low-dose extrapolation problem is handled differently by different countries. At the one extreme, some countries tend to regard the extrapolation of risk levels observable in animal studies to risk levels acceptable for humans as impossible, and they tend to omit any quantitative evaluation of the tumour incidence data. Instead, the ALARA principle is adopted in the case of genotoxic carcinogens. At the other extreme, in the US, the default approach has been to fit a dose-response model (in particular, the linearized multi-stage model, or LMS model) to the tumour incidence data, and to use the fitted curve to estimate the dose associated with a specified low risk level, usually 10^{-6}. This estimated dose (or rather its lower confidence bound) is then called the VSD ("virtually safe dose").

Both these extreme approaches are now vanishing. The former approach (ALARA) is recognized as unnecessarily weak, for instance, because it treats all genotoxic carcinogens as exactly the same (i.e., under the ALARA principle), even when there are data indicating that one compound gives much more reason for concern than another. The latter approach (extrapolation using a fitted model) is now increasingly recognized as an unwarranted extrapolation method. Currently, and internationally, there is a tendency towards the BMD

(Benchmark Dose) approach, including for genotoxic carcinogens. In this approach a dose-response model is fitted to the tumour incidence data, and the fitted model is used to estimate a dose associated with a risk level that is within the observation range, typically a 10% risk. The estimated dose at 10% risk is called the BMD10, and its lower confidence bound, the BMDL10. See the next Section for a further discussion of the BMD approach in a broader context.

Obviously, a 10% cancer risk level would be unacceptable for the human population, and a BMDL10 is considered as a RP or PoD for further evaluation. In current practice there are two ways to proceed:

- Linear extrapolation. When the genotoxic process of carcinogenesis is simply a cascade of (independent) stochastic events (such as: formation of DNA adduct, no repair at cell division, mutation in relevant gene), where each event has a constant (small) probability, the tumour probability would be proportional to the dose (number of molecules) in the low dose region. In reality, there are various biological phenomena that appear to have the effect of making the dose-response more sigmoidal [63]. For instance, it is known that more than one mutation is required to turn a normal cell into a malignant cell. Or, enzyme saturation may lead to a more than proportional increase in internal dose levels when the dose is increased. Therefore, it may be assumed that a tumour incidence dose-response would be sublinear (sigmoidal), rather than linear. This is only a qualitative statement, however, and of not much use for risk assessment, except that it can be said that linear extrapolation would lead to a conservative estimate of a low-risk dose level (for an illustration of this, see Figure 6.10). Therefore, linear extrapolation from the BDML10 (i.e., lower confidence bound of the estimated dose at 10% risk) to a given low risk level may be assumed to result in a conservative estimate of the associated dose. The danger of this method is that low-risk estimates obtained by linear extrapolation are sometimes presented as realistic values (e.g., by reporting the number or yearly deaths due to exposure to the chemical). It should always be clearly indicated that the derived risk values should be considered as upper bound estimates, based on a conservative assumption (of linear dose-response), while it is not possible to state *how* conservative the estimates may be.
- The MOE (Margin of Exposure) approach (see also ILSI [63]). In this approach the estimated human exposure is divided by the RP, usually the BMDL10, and the resulting ratio reflects the interval between the human exposure and the dose with a "known" risk level. The MOE can be used to compare various genotoxic carcinogens, for example, to help risk managers in prioritizing chemicals that require attention. Further, it has been suggested that particular values of the MOE may be formulated that could be associated with low, intermediate, or high levels of concern. For instance, EFSA [115] suggested that an MOE higher than 10,000 might be regarded as a low level of concern. As yet, there is no consensus on this value, however. The advantage of the MOE approach is that calculations and explicit quantitative statements on risk levels far below the range of observation are avoided. It should be noted, however, that the low dose extrapolation problem is now hidden in the value of the MOE considered as a "low level of concern".

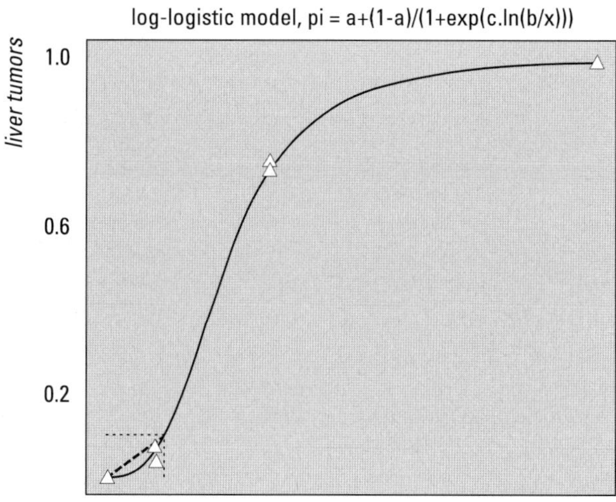

Figure 6.10. Sigmoidal dose-response relationship. Linear extrapolation from the BMD10 overestimates the risk.

Variations on these two approaches are possible by using another RP than the BMDL10. For example, linear extrapolation may be performed from the first statistically significant dose (simple Dutch method). In the MOE approach various summary statistics may be used to quantify exposure in the human population, e.g., the average (median) exposure, or a particular percentile, such as the 95th percentile, representing the exposure that is exceeded by 5% of the population, or an upper confidence bound for the 95th percentile, to take into account the impact of random errors in the data.

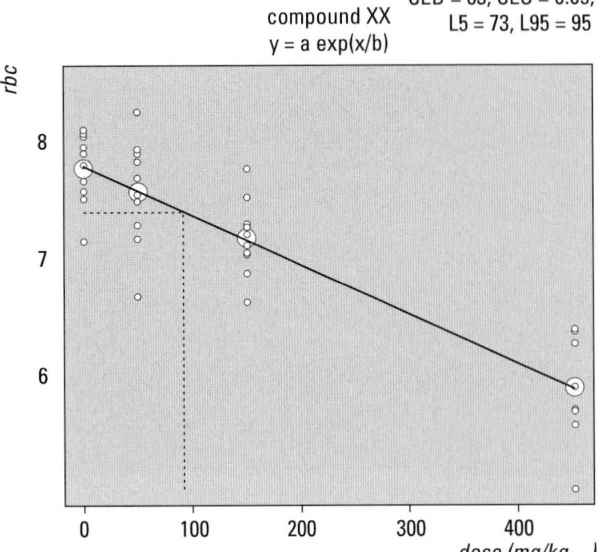

Figure 6.11. Illustration of the Benchmark Dose approach applied to the same data as in Figure 6.9. A curve, in this case an exponential function, is fitted to the data, and this curve is used to assess the CED (vertical dashed line) at a CES of 5% (horizontal dashed line). Next the confidence interval for the CED is calculated (see L-5 and L-95, denoting the lower and upper bound of the 90% confidence interval). The lower bound of this confidence interval (CEDL, or BMDL) is normally used as a RP (PoD) in risk assessment.

Dose-response evaluation: the BMD approach

Given the disadvantages of the NOAEL approach, an alternative method for deriving a RP (PoD) from toxicity data has been proposed by Crump [113]: the Benchmark Dose approach.

The BMD is defined as a dose level that is associated with a pre-specified (small) change in response (denoted as BMR, or Benchmark response) compared with the controls, given some endpoint showing a dose-related response. The value of the BMD is estimated from dose-response data by fitting a dose-response model to the observations. To take the experimental errors in the data into account, the lower confidence bound of the BMD estimate (denoted as BMDL) is normally used as the RP (PoD). Figure 6.11 provides an illustration of the BMD approach.

While the BMD approach was originally intended as an alternative to the NOAEL, i.e., to be used for threshold effects, it can equally well be used for non-threshold effects (see previous Section). In fact, the threshold assumption is not a very useful assumption in evaluating toxicity data: whether it exists or not, it can never be measured, simply because of the fact that zero effects cannot be measured. The BMD approach recognizes this, and makes use of a non-zero effect size (the BMR) as a surrogate for a zero response. Thus, one of the most important questions faced by risk assessors is: what value of the BMR to choose for the various endpoints measured in toxicity studies? This will be discussed for quantal and continuous data consecutively.

BMR for quantal endpoints

Quantal dose-response data reflect a dose-related change in incidence, and the BMR is defined as a particular change in incidence, for instance a 5% increase in incidence compared to the controls. An increase in incidence compared to the background can be done in various ways, and the most common of these are:

- Subtract the responding fraction at a given dose from that in the controls, the result is called additional risk.
- Divide the responding fraction at a given dose by that in the controls, the result is called relative risk.
- Take the additional risk, and divide that by the fraction of non-responding animals in the controls, the result is called extra risk.

Relative risk is often used by epidemiologists, while additional and extra risk are typically used by toxicologists and risk assessors (using animal data). The question of what value to take for the BMR (additional or extra risk) to derive a BMD is difficult to answer from a toxicological point of view. Therefore, it has been suggested that a BMR level should be used which, on average, will result in BMD values that are similar to the NOAELs derived from the same data. Various studies have been performed to investigate this, and the results indicated that a BMR of 5% or 10% would result in BMDs that are, on average, similar to the NOAELs. However, this only applies on average; in individual cases the values may be quite different.

BMR for continuous endpoints

For continuous endpoints, two approaches are currently used for defining a BMR. In one approach, the BMR is defined in the same way as for quantal data, i.e., in terms of extra risk. The variation in the controls is considered to provide information on animal responding "abnormally", for instance, by assessing the response level that is exceeded by 5% of the animals. Then, the dose where this same response level is exceeded by 15% of the animals is a BMD at a BMR of 10%.

In the other approach, the BMR is defined as a particular change in the level of the endpoint, for

Table 6.9. Some terms used for quantal and continuous response data in the context of the BMD approach.

	Quantal response data	Continuous response data	
Type of pre-specified effect	additional/extra* risk	extra* risk	percent change in average level
Terms for pre-specified effect	BMR	BMR**	CES, or BMR
Terms for associated dose	BMD(L)	BMD(L)	CED(L) , or BMD(L)

* Extra risk is additional risk divided by the non-responding fraction of the population.

** The BMR in terms of extra risk is used for continuous data in the "hybrid approach" [116,117].

example, a 5% decrease in red blood cell (RBC) counts, or a 20% decrease in AChE activity. This definition of a BMR is also called a Critical Effect Size (CES), to distinguish it from the BMR in terms of extra risk. The associated BMD in this case is usually denoted as CED (Critical Effect Dose). While extra risk reflects a change in the population, CES reflects the change in a biological parameter in an individual. Table 6.9 summarizes some of the terms used in the BMD approach.

Ideally, for continuous endpoints, the choice of an appropriate value for the BMR (CES) would be based on toxicological information indicating what particular effect size (in a given endpoint) can be considered as starting to be adverse to the organism. Such information is currently not available for most endpoints [114]. Therefore, as long as this is the case, a more pragmatic approach must be adopted, by selecting a CES which is as low as possible, for example, but which is still within the range of observation. Based on experience with dose-response modelling of toxicity data, it has been suggested to use a CES of 5% as a default value. This value appears to be within the range of observation for most endpoints encountered in regular toxicity studies. Further, an effect size of 5% is smaller than the detectable effect size under the NOAEL approach for most toxicity data. This does not imply, however, that the CED05 (i.e., CED at a CES of 5%) is generally smaller than the NOAEL derived from the same data. The reason for this is that the NOAEL approach is less efficient, and therefore unnecessarily conservative. Bokkers and Slob [118] show that for a sample of around 250 datasets, the NOAELs and CEDs at a CES of 5% are similar.

Selection of the model

The value of the BMD (and BMDL) resulting from a dose-response analysis depends on the model used. Therefore, the question of which model to use for describing the dose-response data is important. In general, the dose-response models used for describing dose-response data are relatively simple mathematical functions which are chosen for practical or historical reasons. They do not reflect the underlying mechanisms involved in the interaction between the organism and chemical determining the dose-response relationship. In incidental cases (e.g. for compounds) efforts are made to develop "biologically-based" dose-response models, but so far no models have been found which capable of predicting the dose-response relationship prior to the dose-response data. Therefore, dose-response modelling is, in fact, a statistical analysis of the available data which aims to describe the information provided by the data in such a way that errors in the data are smoothed out. Thus, all models that appear to adequately describe the dose-response data are appropriate models, and could be used in the BMD approach. In practice this could lead to a situation where various models describing the data might well result in different curves (and, possibly, different BMDs). There is no way to decide which of the models is the right one (if any), and this type of uncertainty is often referred to as "model uncertainty". It should be kept in mind, however, that this uncertainty is in fact caused by limitations in the data. For instance, the dose-response data may not contain a sufficient number of doses (or, more precisely, a sufficient number of observed response levels). With good data sets different models (that adequately describe the data) should result in similar curves.

The important property which a dose-response model must have is that it is flexible enough to follow the data, but at the same time it should not be too flexible. The flexibility of the model is reflected by the number of parameters in the model. Therefore, finding an appropriate model can be partly viewed as the task of determining the proper number of parameters needed to be in the model. For this reason, it is convenient to use nested families of models, as illustrated in Figure 6.12. The simplest model in this family (M1) has only one parameter *(a)*: the average response. This model reflects

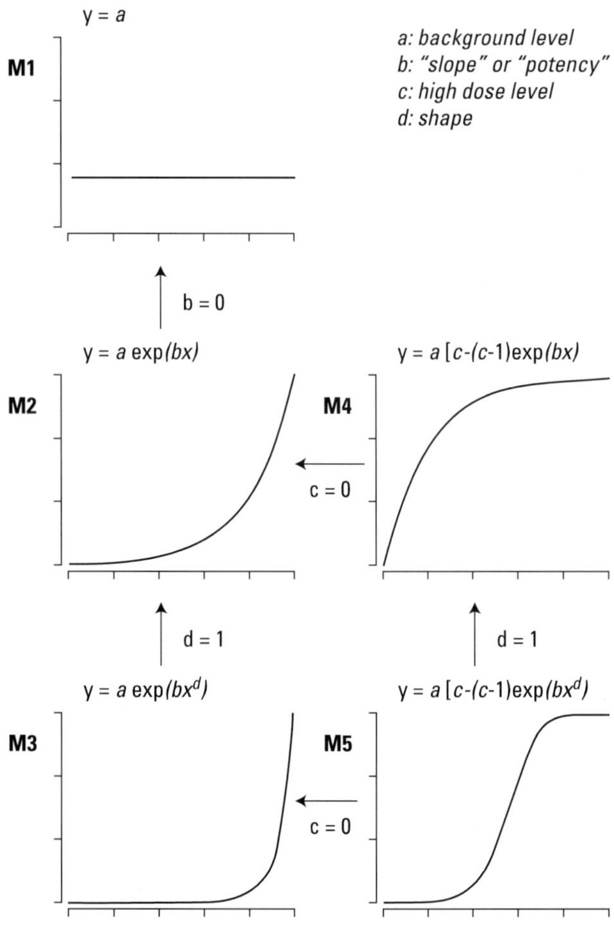

Figure 6.12. Illustration of a nested family of models [119]. With permission of Oxford University Press.

the situation that the response does not change with dose. The next model (M2) is an exponential function, which has one parameter more *(b)*, reflecting the steepness of the exponential curve. This model can be extended with an additional parameter *d*, which makes the curvature of the exponential function more flexible, resulting in M3. M2 can also be extended by a parameter *c*, to make the curve level off at higher doses (M4). And finally, M5 contains all four parameters.

Nested families of models, like the one in Figure 6.12, can be used for model selection using the following principle. When any of the models is extended by including an additional parameter, this should lead to a statistically significant improvement of the fit, otherwise it should be left out. Thus, by comparing the fits of the various models it can be decided which parameters are useful for inclusion in the model. Models containing too many parameters (over-parameterization) could result in curves that are not supported by the data. For a further description of the model selection process in nested families of models, see Slob [119].

The so-called "saturated model" plays a special role. This model does not assume any dose-response relationship: it simply consists of the average response levels. These form the parameters of the model, and they are estimated using the same statistical assumptions (e.g., on the distribution) as used in fitting the models. When the saturated model results in a significantly better fit than a particular fitted dose-response model, the model is significantly rejected. The "goodness of fit" may be tested in this way. However, it is important to be aware that this test assumes that the study is perfectly randomized regarding all experimental treatments (including time of feeding, dosing, and Section). In practice, this is normally not the case.

6.5.4 Default assessment factors

Introduction

The derivation of a NOAEL or BMD of a particular substance is only the starting point in the process of deriving a human health-based limit value (MPR: maximum permissible risk level (Box 6.4), under REACH defined as DNEL: derived no effect level) for this substance. To achieve this it is necessary to deal with the differences between the experimental effect data, generally obtained in animals, and the human situation, taking into account variability and uncertainty. Generally this is done by applying "assessment factors" (AFs). These are individual factors which depend on the available data set of the substance. Each AF quantifies one step in the process of extrapolation from experimental data to the human situation. Ideally, each of these factors is based on substance-specific information [2]. However, in practice this is rarely possible, mostly due to limitations of the dataset and lack of human data. Hence, quite often default AFs need to be used.

The most important aspects of the extrapolation process are:

- Interspecies differences.
- Intraspecies differences.
- Differences in exposure duration.
- Issues related to dose-response.
- Quality of the database.

Interspecies differences

Because animal studies are almost inevitably the starting point in human hazard assessment, and because it is quite unlikely that humans have exactly the same

Box 6.4. Maximum Permissible Risk level

Examples of oral MPRs for non-carcinogenic substances are the ADI (acceptable daily intake, for substances deliberately added to food items) and TDI (tolerable daily intake, for substances unintentionally present in food items), both are expressed in mg/kg bw/day and defined as the daily intake of a chemical which, during the entire lifetime, appears to be without appreciable risk on the basis of all known facts at the time. The RfD (Reference Dose) is similar to the ADI/TDI, but is more strictly defined. Inhalation MPRs are defined in a similar way and expressed as concentrations in air. An example of another health-based limit value is the AOEL (acceptable operator exposure level): the level that has no harmful effects on the health of operators (people working with the substance).
MPRs for carcinogenic substances are usually defined as the daily dose, taken during the entire lifetime that will cause $1:10^4$, $1:10^5$ or $1:10^6$ additional cancer cases during the entire lifetime.

sensitivity to any particular substance as experimental animals, the potential difference in sensitivity needs to be addressed. Unless specific human data are known (in which case in general no extrapolation is needed) [120], the default assumption is that humans are more sensitive than experimental animals. Assuming that the pivotal toxic effect of the particular substance is the same in all mammals (in most toxicity studies rodents are used), inter-species differences have to be attributed to (1) toxicokinetic and/or (2) toxicodynamic differences [47,121,122].

The most important quantitative factor in the expression of toxicologically relevant effects is the toxicokinetic behaviour of the test substance in the test animal: its absorption, distribution, metabolic conversion, and excretion. In general, this can be extrapolated from test animal to humans by allometric scaling of the critical dose. It has been demonstrated that generally equitoxic doses, expressed in mg per kg body weight (bw) per day, scale with body weight to the power of 0.75. This results in default allometric scaling factors for different animal species when compared with humans. They are derived with the expression presented in Equation 6.1. The default allometric AFs for common experimental animals are listed in Table 6.10 [47,121, 123,124,125].

Toxicodynamic differences between the test animal and humans are the potential differences in intrinsic susceptibility of the animal compared with humans. An AF of 2.5 is applied as the default. So, in case of a rat study, the overall default AF for interpecies differences is 10 (4 for toxicokinetic differences times 2.5 for toxicodynamic differences) [125].

It has to be borne in mind, however, that substance-specific information might demand the use of other AFs to cover interspecies differences [2]. This might be the case if detailed metabolic data of a particular substance in different animal species (including man) are available, for example. In vitro metabolic data may be helpful in this.

Intraspecies differences (inter-individual differences)
In contrast to experimental animals, which are genetically highly homogeneous, humans differ in sensitivity due to biological factors such as age, gender, health and nutritional status, metabolic polymorphisms, etc. [126]. It is generally assumed that a default assessment factor of 10 covers the vast majority of the human population including, e.g., children, the elderly, and the sick. It is thus assumed that the most susceptible individual is at most 10 times more susceptible to the toxic effects of a particular substance compared with the least susceptible individual in the human population. Based on the evaluation of a large volume of data it has been suggested that this default AF should be divided into two default AFs, each with a value of 3.16 [121, 127]. One of these to cover toxicokinetics, the other to cover toxicodynamics. The purpose of this is to allow for specific

$$\frac{bw_{\text{human}} / bw_{\text{animal}}}{(bw_{\text{human}} / bw_{\text{animal}})^{0.75}} = (bw_{\text{human}} / bw_{\text{animal}})^{0.25} \tag{6.1}$$

Table 6.10. Default assessment factors to cover toxicokinetic interspecies differences.

Species	Body weight (kg)	Allometric assessment factor
Mouse	0.03	7
Rat	0.25	4
Guinea pig	0.8	3
Rabbit	2	2.4
Monkey	4	2
Dog	18	1.4
Human	70	1

AFs if sufficient data are available. Indeed, as with interspecies differences, with intraspecies differences too, substance-specific information may demand that other AFs be used to cover these differences [2, 126].

Default assessment factor to cover intraspecies differences: 10.

Exposure duration

Normally the aim of the risk assessment process is to protect human individuals against the potentially toxic effects of a chemical following lifelong exposure. Since adequate human data are almost always lacking, any risk assessment is ideally based on animal experiments of chronic (i.e., lifelong) duration (note, however, that besides this, specific information is needed with respect to, e.g., neurotoxicity, reproductive toxicity, etc.). If such chronic experiments are not available, AFs have to be used to extrapolate from experiments of shorter duration [47,123]. In toxicity testing the following exposure periods are distinguished:

Acute:	a single exposure (oral), or up to 24 h exposure (inhalation)
Sub-acute:	28 days of daily exposure
Semi-/sub-chronic:	90 days of daily exposure
Chronic:	1.5-2 years of daily exposure (for rodents)

The default AFs to extrapolate from short to long test periods are listed in Table 6.11. [125].

Sometimes risk assessments are performed for shorter periods than lifelong exposure. In these cases too, the choice of AF depends on the quality of the database, the characteristics of the key study and the intrinsic properties of the substance under consideration. It is practically impossible to set default values for AFs in such cases: they have to be selected on a case-by-case basis.

Dose-response relationship

Since even the best toxicity study will never cover more than the parameters observed, there is always an intrinsic uncertainty with respect to the question: "does the NOAEL of a particular substance as observed in toxicity testing in experimental animals really represent the true no adverse effect level of this substance in humans too?" This question can of course never be fully answered but, never-the-less, it should not be forgotten!

Occasionally a pivotal toxicity study did not succeed in deriving a NOAEL, simply because even at the lowest dose significant toxic effects were observed. Consequently, this dose is the LOAEL (lowest observed adverse effect level) of this particular test. However, remember that if there are no additional toxicity data, it is impossible to be sure whether this is the true "lowest effect level"! Even so, in such a case it is necessary to extrapolate to the NOAEL from this LOAEL. In general, AFs between 3 and 10 are used for this extrapolation, depending on the data [48,116,118]. However, if possible the BMD approach is preferred over the LOAEL to NOAEL extrapolation (see Section 6.5.3 of this chapter).

A BMD which has been calculated as the lower confidence limit of the dose that produced a 5% response (BMD_{05}) is, on average, assumed to be comparable to a NOAEL (see Section 6.5.3). If other BMD indicators are used (e.g., a BMD_{10}) it has to be considered on a case-by-case basis whether an additional dose-response AF is needed.

Alternative data

In the framework of REACH the use of alternative data is considered acceptable if sufficiently justified. Examples

Table 6.11. Default assessment factors to cover exposure duration.

Extrapolation	Assessment factor
Semi/subchronic to chronic	2
Subacute to chronic	6
Subacute to semi/subchronic	3
Acute to subacute/subchronic/chronic	not possible

of such alternatives are *in vitro* data, (Q)SAR data and read-across of chemical categories. However, this does create additional uncertainty, which may be addressed by the application of an extra AF. Generally an AF between 2 and 10 is applied, but larger AFs are certainly not excluded. The risk assessor has to discuss and decide this on a case-by-case basis [125].

Route-to-route extrapolation

In the human-toxicological evaluation aimed at deriving a NOAEL or BMD, toxicity data for all routes of interest for a particular compound (i.e., oral, inhalation, and, if applicable, also dermal) are considered. This full dataset is needed to obtain a complete picture of the toxicological properties of the compound. In practice, however, the available datasets are often limited. Consequently, when oral data are insufficient to derive an oral NOAEL/BMD, route-to-route extrapolation is done based on inhalation data. Similarly, if inhalation data are lacking, route-to-route extrapolation can be applied using oral data. Such extrapolations are based on conversion of the oral dose in mg/kg bw/day to the inhalation dose expressed as the concentration of the substance in air together with the breathing volume. The latter is by default set at 20 m^3 per day (24 h) for a healthy adult for "light exercise". A conservative assumption of a retention factor of 100% by inhalation is also often applied. More precise data can be used if desired, e.g., for heavy physical labour a value of 3.9 m^3 per h as the mean for adult males and females. In addition, potential differences in absorption have to be taken into account [47,123,125]. It must be emphasized, however, that route-to-route extrapolation is a rather unreliable method to derive any limit value.

Data evaluation: quality of the database

Depending on the size and quality of the database from which a DNEL is to be derived, the resulting limit value has a certain reliability. Basically the reliability score is the result of expert judgement of the database from which the limit value is derived. This judgement involves:

- The size of the database. Any specific toxicity of a particular substance is better defined if observed in different studies, by different investigators, in different animals, with different study designs. Thus, if only studies in one experimental animal species are available, or if only a very small number of studies is available, the resulting DNEL will at best be of medium reliability. In this context it should be noted that more recent studies may be expected to have involved modern research methods and good laboratory practice, but that older studies are not by definition less reliable.
- The design of a particular study. It should allow the significance of a particular toxic effect, and its dose-effect relationship to be established. If possible a toxic effect should be supported by histopathological data, macroscopic observations, and research *(in vivo* or *in vitro)* on the molecular mechanism of the effect, etc. Thus, poorly designed studies will result in a DNEL with low reliability (if the database does not contain other, better designed and more extensive studies).
- The severity of the pivotal toxic effect. Obviously, a change in some biochemical parameter, which is often reversible, is of much less severity than, e.g., an irreversible change in kidney function. As noted above for the study design, a pivotal toxic effect also gains strength if it is supported by other data (histopathological data, macroscopic observations, etc.).
- In general, a DNEL can be qualified as highly reliable if it results from an evaluation by an internationally renowned committee of experts, particularly because these committees only derive a DNEL if a fairly complete database is available.
- The extent of international consensus regarding the nature and the severity of a specific toxic effect of a particular compound also indicates the faith (or lack of faith) which the international expert community has in the toxicological characterization of this substance.

The result of the above considerations is that in certain situations, the use of an additional AF may be deemed necessary [47,121,123,125]. It is not possible to define a default value for such an additional AF. Its size has to be discussed and decided on by the risk assessor.

Overall assessment factor

The overall AF for deriving a MPR/DNEL for a particular substance is obtained by multiplication of the individual

AFs as discussed in the previous sections. This is given in Equation 6.2 (see below).

6.5.5 Exposure to mixtures of chemicals

Introduction

The assessment of human-toxicological risks resulting from exposure to chemical substances is generally done by a substance-specific approach on the basis of chronic (i.e., long-term or lifelong) exposure. This approach was chosen in the past because the responsible authorities considered safety to be the most important aspect, i.e. the primary goal is to prevent health risks.

Exposure to mixtures of substances with a threshold for toxicity

Although exposures to only one substance do occur, exposures to mixtures of chemicals are quite common. In such mixtures the chemicals can exhibit combined effects, e.g., joint similar or joint dissimilar action. However, these are not interactions *per se*, because one substance does not alter the activity of the other.

The toxicity of a particular substance is an intrinsic characteristic of that substance. Hence it is difficult to evaluate a mixture of substances as such: ideally each of the components has to be assessed individually. Consequently the human-toxicological evaluation of a particular exposure scenario basically breaks down into individual risk assessments for each of the chemicals present. The final evaluation then has to be done on the basis of the compound that produces the greatest risk. However, often a detailed evaluation is not always feasible or even necessary. Joint action or interaction of combinations of chemicals have been defined in three basic concepts: (1) simple similar action, (2) simple dissimilar action, and (3) interaction.

Simple similar action

Simple similar action is also known as simple joint action, or dose-addition. Each of the chemicals in the mixture acts in the same way, by the same mechanism(s), differing only in potencies. Thus the additive effect can be described by summation of the doses of each individual component in the mixture after correction for the differences in potencies. The method by which this can be done is known as the toxic equivalency factor (TEF) approach, a method used for mixtures of compounds with related structures, sharing a similar toxic mechanism. This approach has been established for the dioxins (i.e., the polychlorinated dibenzo-p-dioxines, the polychlorinated dibenzofurans, and the coplanar polychlorinated biphenyls – PCDDs, PCDFs and dioxin-like PCBs, respectively). Each congener has been allocated a TEF expressing its toxic potency as a fraction of the potency of the most toxic congener, 2,3,7,8-tetrachloro-dibenzo-p-dioxin (TCDD) [128]. For each dioxin mixture the toxic potency can now be calculated by multiplying the concentration of each congener in the mixture with its TEF and adding up the resulting figures, resulting in a total toxic potency expressed in toxic equivalents (TEQs) of TCDD. The formula by which this is done is as follows:

$$D_{sum} = \sum_{i=1}^{n} D_i \times TEF_i \quad (6.3)$$

in which D_{sum} is the sum dose, and D_i and TEF_i are the dose and toxic equivalence factor of the ith component of the mixture, respectively. Basically this is the general formula for the proper application of dose-addition for compounds with similar action without interaction [129,130].

Another, more general approach is the method using hazard indexes as originally proposed in the US EPA mixture guidelines. Here the hazard quotients (the quotient of actual exposure and health-based exposure limit) are calculated for each individual component of the mixture. One possibility is then to add up all the quotients, resulting in an overall hazard quotient for the mixture. If the resulting quotient is > 1, there is an actual risk. But since this approach assumes a similar mode of action of all the components in the mixture, it is basically identical to the TEF approach. Another possibility is to take the largest hazard quotient as an indicator for the overall toxicity of the mixture. But this implicitly presumes simple dissimilar action with full positive correlation of susceptibility, as outlined above.

In its Manual for the Assessment of Joint Toxic Action of Chemical Mixtures 2004 the ATSDR [130]

$$\text{MPR or DNEL} = \frac{\text{overall NOAEL or BMD}}{AF_1 \times AF_2 \times \ldots \times AF_n} \quad (6.2)$$

outlined some refinements of the hazard index approach. In this method the hazard index of each chemical in the mixture is based on the target-organ toxicity dose (TTD) of each of these chemicals. Separate hazard index sums are thus estimated for all toxic endpoints of concern. This approach accommodates the assessment of mixtures whose components do not all have the same critical effect. For a full application of this method TTDs for each endpoint of concern are obviously needed – or have to be developed – for the chemicals that affect an endpoint at a dose higher than that for the critical effect of the same chemical.

Simple dissimilar action

In simple dissimilar action (simple independent action, independent joint action, response or effect-addition) the nature, mechanism and/or site of action of the chemicals in the mixture are different. Thus each chemical exerts its own individual toxic effect, and does not alter the effects of other chemicals in the mixture. This does not mean that two compounds can not each cause, e.g., kidney damage, it only means that the mechanisms of such seemingly similar effects are different and do not interact. The same applies, of course, for a mixture of two – not interacting – compounds each having a different effect. Thus, one way or another, such effects are added together. Note that following exposure to a mixture of chemicals the resulting sum of the effect(s) might differ from one human to another, due to inter-individual differences in susceptibility to each of the substances in the mixture.

When a population (or a group of animals) most sensitive to a particular chemical in the mixture is also most sensitive to all other chemicals in this mixture, the susceptibilities to the chemicals in the mixture are said to be fully and positively correlated. Thus, the hazard posed by a mixture assuming simple dissimilar action with full, positive correlation of susceptibility, is simply the hazard posed by the most dangerous component of the mixture. Consequently, there is no addition of responses at all. In contrast, simple dissimilar action with full negative correlation of susceptibility, leads to full response addition. Of course, intermediate forms of simple dissimilar action with incomplete or partial correlations of susceptibility will be the rule rather than the exception, and will lead to partial response addition.

Interaction

Interaction describes the combined effects of two or more substances in a mixture resulting in an effect stronger than the simple sum of effects of the individual substances (synergism, potentiation, supra-additivity), or weaker than expected (inhibition, antagonism, sub-additivity). The term "interaction" is used here in an empirical way, just to emphasize the difference with "additivity". Interaction might be of a physical-chemical or biological nature, and might occur in the toxicokinetic and/or in the toxicodynamic phase.

The major concern with interactions is supra-additivity, because interactions resulting in sub-additivity would have fewer possible health implications compared with an exposure to each of the substances separately. But with supra-additive interactions it is even conceivable that the exposure to each substance separately has no effect, because the levels of exposure are below their adverse effect levels, while exposure to the mixture does result in an adverse effect, just because of the supra-additivity of the substances in the mixture.

Exposure to mixtures of genotoxic carcinogenic substances

Mixtures of genotoxic carcinogens (which are assumed not to have a threshold) are commonly approached by assuming response addition (see above), because in expressing carcinogenic risks no distinction is made with respect to types of cancer [129,130]. Consequently, the carcinogenic risk resulting from a mixture of several different carcinogenic substances can be treated by the response-addition approach. In other words, the estimated cancer risks of the individual components of the mixture are added up, resulting in the carcinogenic risk of the mixture according to the formula [130]:

$$CR_{total} = \sum_{i=1}^{n} CR_i \qquad (6.4)$$

in which CR_{total} is the carcinogenic risk of the mixture, and CR_i is the carcinogenic risk of the ith component of the mixture

Consequently it is quite conceivable that in a mixture of, e.g., three carcinogens the concentration of each individual carcinogen is below the MPR level, while the cancer risk of the three together is above this level.

State-of-the-art and outlook

Studies with well-defined mixtures of chemicals, only a few of which have been done, have shown that in most cases exposure to the mixture at low doses (i.e., doses below the toxic level of the individual substances) appears to be of no health concern. Moreover, the probability of increased health hazards due to additivity or potentiating interaction seems to be small. However, mixtures of chemicals with a similar working mechanism do inevitably show dose-addition. Hence in the risk assessment process attention must be focused on substances that share a common mode of action. Understanding of the biokinetics and the toxic mechanisms involved is needed for reliable hazard characterization.

6.5.6 Concluding remarks

Assessing dose-response relationships with the BMD approach, combined with present-day knowledge on toxicodynamics and toxicokinetics, provides new avenues for safety and risk assessment. If internal doses at the target site can be determined and if the differences in sensitivity of cells of different species at the cellular level can be addressed at the cellular or even genome level, such information will have a profound influence on safety and risk assessment. Increasingly, toxicologists will have to quantify the risk of exposure, rather than just establish safe levels. This requires quantitative risk assessment. Quantitative risk assessment will also enable us to compare risks with other common and uncommon, voluntary and involuntary risks. It will provide us with tools to predict risks when humans are inadvertently exposed to chemicals. The sophisticated application of the BMD and assessment factors (based on toxicokinetic and toxicodynamic considerations) is already an important step in making risk assessments more uniform and more transparent. Ongoing research, of which "omics" is a promising example, will continue to improve risk assessment procedures. In the future, quantitative risk assessment and risk prediction may provide us with new and better ways of comparing toxicological risks with other risks that human beings face, thus enabling our society to conclude whether such risks are acceptable or not.

6.6 RISK CHARACTERIZATION FOR HUMAN HEALTH

6.6.1 General aspects

Risk characterization is the step in the risk assessment process where the results of the exposure assessment (daily intake) and the effects assessment (NOAEL, BMD) are compared. If possible, an uncertainty analysis is carried out which, if it results in a quantifiable overall uncertainty, produces an estimation of the risk. Several questions should be answered before any such comparison is made:

- What is the target population to be protected?
- What is the time scale of exposure?
- What is the spatial scale of exposure?
- Which route(s) of exposure is or are relevant?
- Are sufficient toxicity data available to derive a meaningful toxicological parameter corresponding to the time scale and the route(s) of exposure as established in the exposure assessment?
- What degree of uncertainty is acceptable?

Exposure of the general population through the environment is an example of long-term exposure on a local or regional scale. Man is exposed through the environment directly via inhalation, soil ingestion and dermal contact, and indirectly via food products and drinking water (Figure 5.1).

Due to human behaviour, exposure and intake will vary greatly within the population. Often human intake is estimated by multiplying the average concentration in each intake medium with the average intake or consumption rate [131]. However, taking the "average individual" as a default leaves potentially half of the population less protected or even unprotected [132]. The only way to characterize the human risk properly is to apply a method which predicts the percentage of the population exceeding a certain intake criterion, e.g., the TDI or ADI. Consequently, an uncertainty analysis for the intake assessment should be performed, which implies that statistical information on consumption habits and concentrations in the diet are needed [131-134].

6.6.2 Humans exposed via the environment

Direct exposure of humans through the environment can be caused by inhalation of air, dust, or aerosols, ingestion of soil (a significant route for children [135]), and dermal uptake, for instance, due to contact with soil or during bathing/swimming. These exposure assessments are outlined in Chapter 5. Based on the exposure assessment, and a number of basic human data [136,137,Table 6.12], the actual intake of the substance of interest can be estimated. The general aspects of exposure of humans to food (plant and plant products, animal and dairy products, and fish) and drinking water are also discussed in Chapter 5. The resulting total body burden can be expressed as a total oral intake.

The total oral, inhalation and dermal intakes are then compared with the appropriate toxicological parameter derived from preferably long-term studies or at least subchronic studies. The most frequently used parameter for non-genotoxic substances is still the NOAEL, but

Table 6.12. Default parameters for consumers and workers.

Humans via the environment, consumers		Workers	
Lifespan	75 years	Working life	40 years
Body weight (male & female)	70 kg	Working day length	8 h
Food intake	1.4 kg/day	Working days per week	5
Water intake	2.0 L/day	Working weeks per year	48
Breathing volume (light activity)	20 m^3 per 24 h	Body weight (male & female)	70 kg
		Breathing volume, light work	10 m^3 per 8 h

the use of the BMD approach is increasing. The studies selected are usually tests with experimental animals for which a NOAEL/BMD (mammal) is derived. If reliable human data can be used to derive a NOAEL/BMD, this value is to be preferred. Risk characterization for genotoxic substances takes place by comparing the acceptable risk level with the estimated total daily intake.

6.6.3 Workers

Occupational exposure is the result of a complex combination of both dependent and independent variables. These include the different physical states of the substance under consideration resulting in different exposure pathways, the immense variety of labour activities in which one comes into contact with chemicals, the processes and activities carried out, and the wide variety of individual, organizational and cultural attitudes with respect to what is acceptable practice and what is not. The potential for occupational exposure is discussed extensively in Chapter 5.

In general, exposure via inhalation and dermal contact are the primary routes of occupational exposure. Exposure through ingestion is also possible but this route is considered to be much more dependent on personal factors, and on the provision of hygiene facilities and more effective supervision than the other two routes. Exposure by inhalation is defined as the concentration of a substance in the breathing zone and is normally expressed as an average concentration over some reference period (usually 8 h for long-term and 15 minutes for short-term exposure). In addition, dermal exposure and uptake has to be taken into consideration. Together these occupational exposures result in a total intake (in general by inhalation because in most cases breathing is the most important route of exposure), which again is to be compared with the appropriate toxicological parameter as outlined above. With respect to the "8 h time-weight average" (the mean exposure level during an 8 h working shift that is considered acceptable), and assuming 48 working weeks per year over a working period of 40 years, this results in a default correction factor of 2.8 (7/5 x 52/48 x 75/40) in the calculation of the occupational exposure level corresponding to a certain risk level on the basis of (experimental animal) lifetime exposure data [125].

It has to be noted that actual exposure levels may well be lower or substantially lower than the reasonable worst-case estimated levels, while it also is unlikely that a worker will be exposed to worst-case estimated levels during his or her entire working life. Some default parameters for workers are provided in Table 6.12 [125].

6.6.4 Consumers

The previous Sections described the indirect exposure of humans to chemicals somehow emitted into the environment and contaminating air, food or drinking water, or more specifically to chemicals in the working environment. But people can also be directly exposed to chemicals through consumer products. Consumer exposure to hazardous substances is of particular concern because the exposed population may include people of all ages, both sexes and in all states of health, , on the one hand, while there are very few ways of controlling or monitoring the extent of exposure compared to the occupational situation, on the other. Consumers can be exposed to individual substances, preparations (mixtures or solutions composed of two or more substances, such as cosmetics, paints, and household detergents), and to substances embedded in a solid or semi-solid matrix. They can also be exposed to substances migrating from package material into a food matrix. They may receive chemical doses via the oral, dermal or inhalation routes. Various exposure estimations are discussed in Chapter 5.

These exposures result in a certain intake which can be compared with the appropriate toxicological parameter as outlined above. Some default parameters for consumers are given in Table 6.12 [125].

6.6.5 Physicochemical properties

Under various regulatory frameworks, human risks arising from physicochemical properties, such as flammability, explosivity, oxidizing potential and particle size, need to be assessed. This assessment addresses the likelihood that an adverse effect will be caused under reasonably foreseeable conditions of use in the workplace or by consumers. This subject goes beyond the scope of this book. For further information the reader is referred to the TGD [5] and the OECD test guidelines [29]. Secondary sources for physicochemical properties have been compiled by the OECD [138] (see also Table 8.4).

6.7 FURTHER READING

1. Hayes AW, ed. 2001. *Principles and methods of toxicology, 4th edition.* Taylor & Francis, New York. ISBN 1-56032-814-2.
2. Klaassen CD, ed. 2001. *Casarett and Doull's Toxicology: the basic science of poisons, 6th edition.* McGraw-Hill. ISBN 0-07-112453-5.
3. Niesink RJM, De Vries TDJ, Hollinger MA, eds. 1996. *Toxicology: principles and applications.* CRC Press, Boca Raton, FL. ISBN 0-8493-9232-2.
4. Gad SC. 2000. *In vitro toxicology, 2nd edition.* Taylor & Francis, New York.
5. Paustenbach DJ, ed. 2002. Human and ecological risk assessment: theory and practice. John Wiley & Sons, New York. ISBN 0-471-14747-8.
6. Lipscomb JC, Ohanian EV, Eds. 2007. *Toxicokinetics and Risk Assessment.* Taylor and Francis Publishers, Informa Healthcare Publishers, New York, NY.

REFERENCES

1. Duffus JH. 1993. Glossary for chemists of terms used in toxicology, IUPAC. *Pure Appl Chem* 65:2003-2122.
2. International Programme on Chemical Safety. 2005. Chemical-specific adjustment factors for interspecies differences and human variability: guidance document for use of data in dose/concentration-response assessment. *Harmonization Project Document* No. 2. IPCS, Geneva, Switzerland. ISBN 9241546786.
3. Kroes R, Feron VJ. 1984. General toxicity testing: sense and non-sense, science and policy. *Fundam Appl Toxicol* 4:298-308.
4. Kroes R, Feron VJ. 1990. Toxicity testing: strategies and conduct. In: Clayson DB, Munro IC, Shubik P, Swenberg JA, eds, *Progress in Predictive Toxicology.* Elsevier Scientific Publishers, Amsterdam, The Netherlands, pp. 15-39.
5. Commission of the European Communities. 2003. Technical Guidance Document on Risk Assessment in support of Commission Directive 93/67/EEC on risk assessment for new notified substances, Commission Regulation (EC) No. 1488/94 on risk assessment for existing substances and Directive 98/8/EC of the European Parliament and of the Council concerning the placing of biocidal products on the market. European Commission, Brussels, Belgium. (http://ecb.jrc.it/existing-chemicals/).
6. Health Council of The Netherlands. 1989. Toxicological assessment of chemicals. Report A89-6. Health Council of The Netherlands, The Hague, The Netherlands [in Dutch].
7. International Program on Chemical Safety. 1999. Principles for the assessment of risks to human health from exposure to chemicals. *Environmental Health Criteria* 210. IPCS, Geneva, Switzerland.
8. International Program on Chemical Safety. 1978. Principles and methods for evaluating the toxicity of chemicals. Part I. World Health Organization, *Environmental Health Criteria* 6. IPCS, Geneva, Switzerland.
9. International Program on Chemical Safety. 1987. Principles for the safety assessment of food additives and contaminants in food. World Health Organization, *Environmental Health Criteria* 70. IPCS, Geneva, Switzerland.
10. International Program on Chemical Safety. 1990. Principles for the toxicological assessment of pesticide residues in food. World Health Organization, *Environmental Health Criteria* 104. IPCS, Geneva, Switzerland.
11. US Environmental Protection Agency. 1986. Superfund public health evaluation manual. EPA/540/1-86/060. Office of Emergency and Remedial Response, US EPA, Washington DC.
12. Organization for Economic Co-operation and Development. 2002. Guidance notes for analysis and evaluation of chronic toxicity and carcinogenicity studies. OECD Environment, Health and Safety Publications, *Series on Testing and Assessment* No. 35 and *Series on Pesticides* No. 14. Environment Directorate, OECD, Paris, France.

13. Gad SC, Weil CS. 1994. Statistics for toxicologists. In: A.W. Hayes, ed, *Principles and Methods of Toxicology.* Third Edition, Raven Press, New York, NY. ISBN 0-7817-0131-7.
14. Woodman BB. 1988. Assessment of hepatic function and damage in animal species. A review of the current approach of the academic, governmental and industrial institutions represented by the Animal Clinical Chemistry Association. *J Appl Toxicol* 8:249-254.
15. Stonard MD. 1990. Assessment of renal function and damage in animal species. A review of the current approach of the academic, governmental and industrial institutions represented by the Animal Clinical Chemistry Association. *J Appl Toxicol* 10:267-274.
16. Clampitt RB. 1978. An investigation into the value of some clinical biochemical tests in the detection of minimal changes in liver morphology and function in the rat. *Arch Toxicol Suppl* 1:1-13.
17. Wolford ST, Schroer RA, Gohs FX, Gallo PP, Brodeck M, Falk HB, Ruhren R. 1986. Reference range database for serum chemistry and haematology values in laboratory animals. *J Toxicol Environ Health* 18:161-188.
18. Loeb WF, Quimby FW, eds. 1989. *The Clinical Chemistry of Laboratory Animals.* Pergamon Press, New York, NY.
19. Derelanko MJ, Hollinger MA, eds. 2001. *Handbook of toxicology, 2nd edition.* Taylor & Francis, ISBN 0849303702.
20. Rowland M, Tozer TN, eds. 1995. *Clinical Pharmacokinetics. Concepts and Applications, 3rd ed.* Williams and Wilkins, Media, Pennsylvania, ISBN 0-683-07404-0.
21. Barton HA, Pastoor TP, Baetcke K, Chambers JE, Diliberto J, Doerrer NG, Driver JH, Hastings CE, Iyengar S, Krieger R, Stahl B and Timchalk C. 2006. The Acquisition and Application of Absorption, Distribution, Metabolism, and Excretion (ADME) Data in Agricultural Chemical Safety Assessments, *Crit Rev Toxicol* 36:9–35.
22. Van de Waterbeemd H, Gifford E. 2003. ADMET in silico modelling: towards prediction paradise? *Nature Rev Drug Discovery* 2:192-204.
23. International Programme on Chemical Safety. 1986. Kinetic models, In: Principles of toxicokinetic studies, *Environmental Health Criteria* 57. IPCS, Geneva, Switzerland. http://www.inchem.org.
24. Renwick AG. 1994. Toxicokinetics Pharmacokinetics in toxicology. In: Hayes AW, ed, *Principles and Methods of Toxicology*, 3rd ed., Raven Press, New York, NY. ISBN 0-7817-0131-7.
25. Medinsky MA, Klaassen CD, 1996. Toxicokinetics, In: Klaassen CD, ed, *Casarett and Doull's Toxicology: The basic science of poisons*, 5th ed., McGraw-Hill, New York, NY. ISBN 0-07-105476-6.
26. Krishnan K, Andersen ME, 1994. Physiologically based pharmacokinetic modeling in toxicology, In: Hayes AW, ed, *Principles and Methods of Toxicology*, 3rd ed., Raven Press, New York, NY. ISBN 0-7817-0131-7.
27. Reddy MB, Yang RSH, Clewell HJ, Andersen ME. 2005. *Physiologically Based Pharmacokinetic Modeling*. Wiley, Hoboken, NJ. ISBN 0-471-47814-8.
28. US Environmental Protection Agency (EPA). 2006. Approaches for the Application of Physiologically Based Pharmacokinetic (PBPK) Models and Supporting Data in Risk Assessment. National Center for Environmental Assessment, Washington, DC; EPA/600/R-05/043F. Available from: National Technical Information Service, Springfield, VA, and online at http://epa.gov/ncea.
29. Organization for Economic Co-operation and Development. 1993. Guidelines for testing of chemicals. OECD, Paris, France. www.oecd.org.
30. Lucas Luyckx N. 1994. Intake of test compounds in long-term rat studies - considerations for risk assessment. *Regul Toxicol Pharmacol* 20:96-104.
31. Barton HA, Pastoor TP, Baetcke K, Chambers JE, Diliberto J, Doerrer NG, Driver JH, Hastings CE, Iyengar S, Krieger R, Stahl B, Timchalk C. 2006. The acquisition and application of absorption, distribution, metabolism and excretion (ADME) data in agricultural chemical safety assessments. *Crit Rev Toxicol* 36:9-35.
32. Montesano R, Bartsch H, Vainio H, Wilbourn J, Jamasaki H, eds. 1986. Long-term and Short-term Assays for Carcinogens: A Critical Appraisal. *IARC Scientific Publications* 83. International Agency for Research on Cancer, Lyon, France.
33. De Groot AP, Feron VJ, Immel HR. 1988. Induction of hyperplasia in the bladder and epithelium of rats by a dietary excess of acid or base: implications for toxicity/ carcinogenicity testing. *Fd Chem Toxicol* 26:425-434.
34. Roe FJC. 1981. Are nutritionists worried about the epidemic of tumours in laboratory animals? *Proc Nutr Soc* 40:57-65.
35. Roe FJC. 1988. Pathology of polyols. *Human Toxicol* 7:62-63.
36. Food and Drug Administration. 1978. Non-clinical laboratory studies. Good laboratory practice regulations. Department of Health Education and Welfare. *Fed Reg* 43, 59986. FDA, Washington, DC.
37. Hasemann JK, Winbush JS, McDonnell HW. 1986. Use of dual control groups to estimate false positive rats in laboratory animal carcinogenicity studies. *Fundam Appl Toxicol* 7:573-584.
38. Organization for Economic Co-operation and

Development. 1982. Principles of good laboratory practice. Final report of the group of experts on good laboratory practice. OECD, Paris, France.

39. US Environmental Protection Agency. 1983. Toxic substances control. Good laboratory practice standards. Final Rule. *Fed Reg* 48:53923.

40. US Environmental Protection Agency. 1983. Pesticides programs: Good laboratory practice standards. Final Rule. *Fed Reg* 48:53963.

41. Commission of the European Communities. 1987. Guideline of the Council for changing guideline 75/318/EC concerning good laboratory practice. *Off J Eur Communities* L15/31 and L15/34.

42. Russel WMS, Burch RL. 1959. *The principles of human experimental technique*. London, Methuen.

43. Commission of the European Communities. 1986. Council Directive 86/609/EEC of 24 November 1986 on the approximation of laws, regulations and administrative provision of the member states regarding the protection of animals used for experimental and other scientific purposes. *Off J Eur Communities* L 358/1-6.

44. Commission of the European Communities. 2006. Communication from the Commission to the European Parliament and the Council on a Community Action Plan on the Protection and Welfare of Animals 2006-2010. Brussels, 23 January 2006. COM(2006)13final.

45. Institute of Health and Consumer Protection. 2005. REACH and the need for Intelligent Testing Strategies. IHCP, Joint Research Centre, Ispra, Italy. EUR 21554EN.

46. Hoffman FO, Hammonds JS. 1994. Propagation of uncertainty in risk assessments: the need to distinguish between uncertainty due to lack of knowledge and uncertainty due to variability. *Risk Analysis* 14:707-712.

47. Vermeire T, Stevenson H, Pieters M, Rennen M. 1999. Assessment factors for human health risk assessment: a discussion paper. *Crit Rev Toxicol* 29:439-490.

48. Kroes R. 1986. Proposed revision of certain OECD guidelines for toxicity testing. In: Proceedings of an international seminar on chemical testing and animal welfare. The National Chemical Inspectorate, Solna, Sweden, pp. 143-151.

49. Van Den Heuvel MJ, Dayan AD, Shillaker RO. 1987. Evaluation of the BIS approach to the testing of substances and preparations for their acute toxicity. *Hum Toxicol* 6:279-291.

50. Van den Heuvel MJ, Clark DG, Fielder RJ, Koundakjian PP, Oliver GJ, Pelling D, Tomlinson NJ, Walker AP. 1990. The international validation of a fixed dose procedure as an alternative to the classical LD50 test. *Fd Chem Toxicol* 28:469-482.

51. Schlede E, Mischki U, Doll R, Kayser D. 1992. A national validation study of the acute toxic-class method - an alternative to the LD50 test. *Arch Toxicol* 66:455-470.

52. Bruce R.D. 1985. An Up-and-Down Procedure for Acute Toxicity Testing. *Fundam Appl Toxicol* 5:151-157.

53. Lipnick RL, Cotruvo JA, Hill RN, Bruce RD, Stitzel KA, Walker AP, Chu I, Goddard M, Segal L, Springer JA, Myers RC. 1995. Comparison of the Up-and-Down, Conventional LD50 and Fixed Dose Acute Toxicity Procedures. *Fd Chem Toxicol* 33:223-231.

54. Commission of the European Communities. 2001. Commission Directive 2001/59/EC of 21 November 2001 adapting to technical progress for the 28th time Council Directive 67/548/EEC of the approximation of the laws, regulations and administrative provisions relating to the classification, packaging and labelling of dangerous substances. *Off J Eur Communities* L225/1.

55. Kimber I, Basketter DA, Gerberick GF, Dearman RJ. 2002. Allergic contact dermatitis. *International immunopharmacology* 2:201-211.

56. Mason JM, Langenbaan R, Shelby MD, Zeiger E, Tennant RW. 1990. Ability of short-term tests to predict carcinogenesis in rodents. *Ann Rev Pharmacol Toxicol* 30:149-160.

57. Weisburger JH, Williams GM. 1991. Critical effective methods to detect genotoxic carcinogens and neoplasm promoting agents. *Env Health Perspectives* 90:121-126.

58. Randerath K, Randerath E, Danna TF, Van Golen KL, Putman KL. 1989. A new sensitive ^{32}P-postlabeling assay based on the specific enzymatic conversion of bulky DNA lesions to radio-labeled dinucleotides and nucleoside 5'-monophosphates. *Carcinogenesis* 10:1231-1239.

59. Gorelick NJ. 1995. Overview of mutation assays in transgenic mice for routine testing. *Environ Mol Environ* 25:218-230.

60. Tice RR, Agurell E, Anderson D, Burlinson B, Hartmann A, Kobayashi H, Miyamae Y, Rojas E, Ryu J-C, Sasaki F. 2000. Single cell gel/Comet asay: guidelines for *in vitro* and *in vivo* genetic toxicology testing. *Environ Mol Mutagen* 35:206-221.

61. Ashby J. 1990. Genotoxicity testing: to what extent can it recognize mutagens and carcinogens. In: Clayson DB, Munro IC, Shubik P, Swenberg JA, eds, *Progress in Predictive Toxicology*. Elsevier Sci Publ, Amsterdam, The Netherlands, pp. 185-205.

62. Purchase IFH. 1994. Current knowledge of mechanisms of carcinogenicity, genotoxins versus non-genotoxins. *Human Exp Toxicol* 13:17-28.

63. O'Brien J, Renwick AG, Constable A, Dybing E, Müller

DJG, Schlatter J, Slob W, Tueting W, van Benthem J, Williams GM, Wolfreys A. 2006. Approaches to the risk assessment of genotoxic carcinogens in food: A critical appraisal. *Fd Chem Toxicol* 44:1613-1635.

64. Kroes, R. 1987. Contribution of toxicology towards risk assessment of carcinogens. *Arch Toxicol* 60:224-228.
65. Sonich-Mullin C, Fielder R, Wiltse J, Baetcke K, Dempsey J, Fenner-Crisp P, Grant D, Hartley M, Knaap A, Kroese D, Mangelsdorf I, Meek E, Rice JM, Younes M. 2001. IPCS conceptual framework for evaluating a mode of action for chemical carcinogenesis. *Regul Toxicol Pharmacol* 34(2):146-152.
66. Boobis AR, Cohen SM, Dellarco V, McGregor D, Meek ME, Vickers C, Willcocks D, Farland W. 2006. IPCS Framework for Analyzing the Relevance of a Cancer Mode of Action for Humans. *Crit Rev Toxicol* 36:781-835.
67. International Agency for Research on Cancer, IARC. 1999. Results of short- and medium-term tests for carcinogens and data on genetic and related effects in carcinogenic hazard evaluations. *IARC Scientific Publications* No. 146 , Lyon, France.
68. International Agency for Research on Cancer, IARC. 1999. Species Differences in Thyroid, Kidney and Urinary Bladder Carcinogenesis. *IARC Scientific Publications* No. 147, Lyon, France.
69. Kroes R. 1987. Contribution of toxicology towards risk assessment of carcinogens. *Arch Toxicol* 60:224-228.
70. Feron VJ, Schwartz M, Krewski D, Hemminki K. 1999. Long and medium-term carcinogenicity studies in animals and short-term genotoxicity tests. In: Moolgavkar S, Krewski D, Zeise L, Cardis E, Moller H, eds, Quantitative estimation and prediction of human carcinogenic risks, Chapter 5, pp. 103-12. *IARC Scientific Publications*, International Agency for Research on Cancer, Lyon, France.
71. ILSI Health and Environmental Sciences Institute. 2001. Alternatives to Carcinogenicity Testing Project. ILSI/ HESI, *Toxicol Pathol* 29, Supplement.
72. Yamasaki H. 1990. Gap junctional intercellular communication and carcinogenesis. *Carcinogenesis* 11: 1051-1058.
73. Commission of the European Communities. 2005. TAPIR 3.3 A project for the information requirements of REACH. Final TAPIR Report RIP 3.3-1. Submitted to European Chemicals Bureau, Joint Research Centre by Cefic, Environment Agency of England and Wales, ECETOC, INERIS, KemI, TNO.
74. Chernoff N, Kavlock RJ. 1982. An *in vivo* teratology screen utilising pregnant mice. *J Toxicol Environ Health* 10:541-550.
75. International Programme on Chemical Safety. 1996. Principles and Methods for Assessing Direct Immunotoxicity Associated with Exposure to Chemicals. IPCS, World Health Organization (WHO), International Programme on Chemical Safety (IPCS), *Environmental Health Criteria* 180, Geneva, Switzerland.
76. Van Loveren H, Vos JG. 1989. Immunotoxicological considerations: a practical approach to immunotoxicity testing in the rat. In: Dayan AD, Paine AJ, eds, *Advances in Applied Toxicology*. Taylor and Francis, London, UK, pp. 143-163.
77. Richter-Reichhelm HB, Stahlmann R, Smith E, Van Loveren H, Althoff J, Bass R, Corsini E, Dayan A, Dean JH, Descotes J, Emmendörffer A, Eppler R, Hall AJ, Herrman JL, Lovik M, Luster MI, Miller FW, Riecke K, Schöning G, Schulte A, Smialowicz RJ, Ulrich P Vohr HW, Vos JG, White Jr KL. 2001. Approaches to risk assessment of immunotoxic effects of chemicals. Meeting Report. *Toxicology* 161:213-228.
78. Van Loveren H, Verlaan APJ, Vos JG. 1991. An enzyme linked immunosorbent assay of anti-sheep red blood cell antibodies of the classes IgM, G, and A in the rat. *Int J Immunopharmacol* 13:689-695.
79. Temple L, Kawabata TT, Munson AE, White Jr KL. 1993. Comparison of ELISA and plaque-forming cell assays for measuring the humoral immune response to SRBC in rats and mice treated with benzoapyrene or cyclosphosphamide. *Fundam Appl Toxicol* 21:412-419.
80. Van Loveren H. 1995. Host resistance models. *Hum Expl Toxicol* 14:137-140.
81. Dean JH. 2000. The nature and importance of drug allergy. *Toxicology* 148:11-12.
82. Nierkens S, Aalbers M, Bol M, van Wijk F, Hassing I, Pieters R. 2005. Development of an oral exposure mouse model to predict drug-induced hypersensitivity reactions by using reporters antigens. *Toxicol. Sci.* 83:273-281.
83. Anger WK. 1984. Neurobehavourial testing of chemicals: impact on recommended standards. *Neurobehav Toxicol Teratol* 6:147-153.
84. Abou-Donia MB, Lapadula DM. 1990. Mechanisms of organophosphorus ester-induced delayed neurotoxicity: type I and type II. *Ann Rev Pharmacol Toxicol* 30:405-440.
85. Joint Meeting on Pesticide Residues. 1998. Report of a consultation on interpretation of inhibition of acetylcholinesterase activity. JMPR, Geneva, Switzerland, 8-9 January 1998.
86. Organization for Economic Co-operation and Development. 2001. Appraisal of test methods for sex-hormone disrupting chemicals. Detailed Review Paper. *Series on testing and assessment* No. 21. OECD, Paris,

France, ENV/JM/MONO(2002)8.
87. International Programme on Chemical Safety. 2003. Workshop report toxicogenomics and the risk assessment of chemicals for the protection of human health. Federal Institute for Risk Assessment, Berlin 17-19 November 2003. IPCS, Geneva, Switzerland, IPCS/Toxicogenomics/03/1.
88. Organization for Economic Co-operation and Development. 2005. Report of the OECD/IPCS Workshop on toxicogenomics, Kyoto 13-15 October 2004. OECD, Paris, France, ENV/JM/MONO(2005)10.
89. US Environmental Protection Agency. 2004. Potential Implications of Genomics for Regulatory and Risk Assessment Applications at EPA. Prepared for the US EPA by members of the Genomics Task Force Workgroup, EPA's Science Policy Council, Washington DC..
90. Sarrif A, Joost HM, van Delft JHM, Gant TW, Kleinjans JCS, Vliet E. 2005. Toxicogenomics in genetic toxicology and hazard determination – concluding remarks. *Mutat Res, Fundamental and Molecular Mechanisms of Mutagenesis* 575:116-117.
91. Anonymous. 1986. European Convention for the Protection of Vertebrate Animals used for Experimental and Other Scientific Purposes. Council of Europe, Strasbourg, France. p. 51.
92. Relius, HW. 1987. Extrapolation from animals to man: prediction, pitfalls and perspectives. *Xenobiotica* 17:255–265.
93. Blaauboer BJ, Balls M, Barratt M, Casati S, Coecke S, Mohamed MK, Moore J, Rall D, Smith KR, Tennant R, Schwetz BA, Stokes WS, Younes M. 1998. 13th Meeting of the Scientific Group on Methodologies for the Safety Evaluation of Chemicals (SGOMSEC): alternative testing methodologies and conceptual issues. *Environ Health Perspect* 106:413-415.
94. Worth AP, Balls M. 2002. Alternative (non-animal) methods for chemicals testing: current status and future perspectives. *Altern. Lab. Anim.* 30 suppl 1:1-125.
95. Barratt MD. 1995. The role of structure–activity relationships and expert systems in alternative strategies for the determination of skin sensitization, skin corrosivity and eye irritation. *Altern Lab Anim* 23:111-122.
96. Ekwall, B. 1983. Screening of toxic compounds in mammalian cell cultures. *Ann. New York Acad. Sci.* 407:64-77.
97. Forsby A, Pilli F, Bianchi V, Walum E. 1995. Determination of critical cellular neurotoxic concentrations in human neuroblastoma (SH-SY5Y) cell cultures. *Altern Lab Anim* 11:800-811.
98. Eisenbrand G, Pool-Zobel B, Baker V, Balls M, Blaauboer BJ, Boobis A, Carere A, Kevekordes S, Lhuguenot J-C, Pieters R, Kleiner J. 2002. Methods of *In vitro* Toxicology. *Food Chem Toxicol* 40:193-236.
99. Nuwaysir EF, Bittner M, Trent J, Barrett JC, Afshari CA, 1999. Microarrays and toxicology: the advent of toxicogenomics. *Molecular Carcinogenesis* 24:153–159.
100. Harries HM, Fletcher ST, Duggan CM, Baker VA. 2001. The use of genomics technology to investigate gene expression changes in cultured human liver cells. *Toxicol Vitro* 15:399–405.
101. Blaauboer BJ. 2002. The necessity of biokinetic information in the interpretation of in vitro toxicity data. *ATLA* 30, suppl 2:85-91.
102. Andersen ME. 1991. Physiological modelling of organic compounds. *Ann Occup Hyg* 35:309-321.
103. Clewell HJ, Andersen ME. 1986. A multiple dose-route physiological model for volatile chemicals using ACSL/PC. In: Cellier, F.D. ed, *Languages for Continuous Simulation*. Society for Computer Simulation, San Diego, CA. p. 95.
104. Leung HW. 1991. Development and utilization of physiologically based models for toxicological applications. *J Toxicol Environ Health* 32:247-267.
105. Brown R, Foran J, Olin S, Robinson, D, 1994. Physiological Parameter Values for PBPK Models. International Life Sciences Institute, and Risk Science Institute, Washington, DC.
106. DeJongh J, Nordin-Andersson M, Ploeger B, Forsby A. 1999. Estimation of systemic toxicity of acrylamide by integration of in *vitro* toxicity data with kinetic simulations. *Toxicol Appl Pharmacol* 158:261-268.
107. Gubbels-van Hal WM, Blaauboer BJ, Barentsen HM, Hoitink MA, Meerts IA, van der Hoeven JC. 2005. An alternative approach for the safety evaluation of new and existing chemicals, an exercise in integrated testing. *Regul Toxicol Pharmacol* 42:284-295.
108. Verwei M, van Burgsteden JA, Krul CAM, van de Sandt JJM, Freidig AP. 2006. Prediction of in vivo embryotoxic effect levels with a combination of in vitro studies and PBPK modelling. *Toxicol Letters* 165:79-87.
109. International Program on Chemical Safety. 2005. The WHO recommended classification of pesticides by hazard and guidelines to classification 2004. World Health Organization, IPCS, Geneva, Switzerland. ISBN 9241546638.
110. United Nations. 2003. Globally Harmonized System of classification and labelling of chemicals (GHS). United Nations, Geneva, ST/SG/AC.10/30, ISBN 92-1-116840-6.
111. International Agency for Research on Cancer. 1987.

Overall Evaluations of Carcinogenicity: An Updating of IARC Monographs Volumes 1 to 42. *IARC Monographs on the evaluation of carcinogenic risks to humans.* Supplement 7. Lyon, IARC Press.

112. International Agency for Research on Cancer. 2004. IARC Some drinking-water disinfectants and contaminants, including arsenic. *Monographs on the evaluation of carcinogenic risks to humans.* Volume 84 Lyon, IARC Press.
113. Crump KS. 1984. A new method for determining allowable daily intakes. *Fund Appl Toxicol* 4:845-871.
114. Dekkers S, Telman J, Rennen MAJ, Appel MJ, de Heer C.2006. Within-animal variation as an indication of the minimal magnitude of the Critical Effect Size for continuous toxicological parameters applicable in the Benchmark Dose approach. *Risk Analysis* 26:867-880.
115. Barlow S, Renwick AG, Kleiner J, Bridges JW, Busk L, Dybing E, Edler L, Eisenbrand G, Fink-Gremmels J, Knaap A, Kroes R, Liem D, Müller DJG, Page S, Rolland V, Schlatter J, Tritscher A, Tueting W, Würtzen G. 2006. Risk assessment of substances that are both genotoxic and carcinogenic: Report of an International Conference organized by EFSA and WHO with support of ILSI Europe. *Fd Chem Toxicol* 44:1636-1650.
116. Gaylor DW and Slikker W. 1990. Risk assessment for neurotoxic effects. *NeuroToxicology* 11:211-218.
117. Crump KS. 1995. Calculation of Benchmark Doses from Continuous Data. *Risk Anal* 15:79-89.
118. Bokkers BGH, Slob W. 2007. Deriving a Data-Based Interspecies Assessment Factor Using the NOAEL and the Benchmark Dose Approach. *Crit Rev Toxicol*, in press.
119. Slob W. 2002. Dose-response modeling of continuous endpoints. *Toxicol Sc.* 66:298-312.
120. Van den Brandt P, Voorrips L, Hertz-Picciotto I, Shuker D, Boeing H, Speijers G, Guittard C, Kleiner J, Knowles M, Wolk A, Goldbohm A. 2002. The contribution of epidemiology. *Fd Chem Toxicol* 40:387-424.
121. Dybing E, Doe J, Groten J, Kleiner J, O'Brien J, Renwick AG, Schlatter J, Steinberg P, Tritscher A, Walker R, Younes M. 2002. Hazard characterisation of chemicals in food and diet: dose response, mechanisms and extrapolation issues. *Fd Chem Toxicol* 40:237-282.
122. Dorne JLCM, Walton K, Renwick AG. 2005. Human variability in xenobiotic metabolism and pathway-related uncertainty factors for chemical risk assessment: a review. *Fd Chem Toxicol* 43:203-216.
123. Vermeire T, Pieters M, Rennen M, Bos P. 2001. Probabilistic assessment factors for human health risk assessment - a practical guide. Report no. RIVM 60151005/TNO V3489, National Institute for Public Health and the Environment, Bilthoven, The Netherlands, and Netherlands Organisation for Applied Scientific Research, Zeist, The Netherlands.
124. Renwick AG, Barlow SM, Hertz-Picciotto I, Boobis AR, Dybing E, Edler L, Eisenbrand G, Greig JB, Kleiner J, Lambe J, Müller DJG, Smith MR, Tritscher A, Tuijtelaars S, Van den Brandt PA, Walker R, Kroes R. 2003. Risk characterisation of chemicals in food and diet. *Fd Chem Toxicol* 41:1211-1271.
125. European Chemicals Bureau. 2005. Technical guidance document on risk assessment, human risk characterisation; revised chapter, final draft 16 Nov. 2005. European Chemicals Bureau, Ispra, Italy.
126. Nebert DW. 2005. Inter-individual susceptibility to environmental toxicants - a current assessment. *Toxicol Appl Pharmacol* 207:S34-S42.
127. Dorne JLCM, Renwick AG. 2005. The refinement of uncertainty/safety factors in risk assessment by the incorporation of data on toxicokinetic variability in humans. *Toxicol Sci* 86:20-26.
128. Van den Berg M, Birnbaum LS, Denison M, De Vito M, Farland W, Feeley M, Fiedler H, Hakansson H, Hanberg A, Haws L, Rose M, Safe S, Schrenk D, Tohyama C, Tritscher A, Tuomisto J, Tysklind M, Walker N, Peterson RE. 2006. The 2005 World Health Organization re-evaluation of human and mammalian toxic equivalency factors for dioxins and dioxin-like compounds. *Toxicol Sci*, published on line July, 2006.
129. Health Council. 2002. Exposure to combinations of substances: a system for assessing health risks. Report no. 2002/05, Health Council of the Netherlands, The Hague, The Netherlands.
130. Agency for Toxic Substances and Disease Registry. 2004. Guidance Manual for the Assessment of Joint Toxic Action of Chemical Mixtures. ATSDR Atlanta, GA.
131. Kroes R, Müller D, Lambe J, Löwik MRH, Van Klaveren J, Kleiner J, Massey R, Mayer S, Urieta I, Verger P, Visconti A. 2002. Assessment of intake from the diet. *Fd Agricul Toxicol* 40:327-385.
132. Nusser SM, Carriquiry AL, Dodd KW, Fuller WA. 1996. A semi-parametric transformation approach to estimating usual daily intake distributions. *J Am Stat Assoc* 91:1440-1449.
133. Slob W, Bakker MI. 2004. Probabilistic calculation of intake of substances via incidentally consumed food products. Supplement to the handbook for modelling of intake of substances via food. RIVM report 320103003. National Institute for Public Health and Environmental Protection (RIVM), Bilthoven, The Netherlands.
134. Slob W. 1993. Modelling long-term exposure of the whole population to chemicals in food. *Risk Anal* 13:525-

530.

135. Calabrese EJ, Stanek EJ, James RC, Roberts SM. 1999. Soil ingestion. A concern for acute toxicity in children. *J Environ Health* 61:18-23.

136. U.S. Environmental Protection Agency. 1989. Exposure factors handbook. US EPA, Office of Health and Environmental Assessment, Washington DC. EPA/600/8-89/043.

137. European Centre for Ecotoxicology and Toxicology of Chemicals. 2001. Exposure factors sourcebook for European populations (with focus on UK data). Technical report no. 79. ECETOC, Brussels, Belgium.

138. Organization for Economic Co-operation and Development. 2004. Use of secondary sources for physico-chemical properties. OECD, Paris, France. Environment Directorate, Joint Meeting of the Chemical Committee of the Working Party on Chemicals, Pesticides and Biotechnology, ENV/JM/EXCH(2004)7.

7. ECOTOXICOLOGICAL EFFECTS

T.P. TRAAS AND C.J. VAN LEEUWEN

7.1 INTRODUCTION

Ecotoxicology is the study of toxic effects of substances on species in ecosystems and involves knowledge of three main disciplines: toxicology, ecology and chemistry (Figure 7.1). Truhaut [2] coined the term ecotoxicology and included effects on humans in his definition, man being part of ecosystems. The current tendency is to include the effects of chemicals on all species in the biosphere in the definition of ecotoxicology [3]. However, in this section, we will not consider effects on man. Environmental risk assessment (ERA) shares many methodological aspects with human health risk assessment (HRA). However, there are a number of fundamental differences between ERA and HRA related to the scope of ERA which covers ecosystems and the biosphere. Fundamental aspects of ERA are discussed in the next section.

Ecotoxicological effects are changes in the state or dynamics at the organism level, or at other levels of biological organization, resulting from exposure to a chemical. These levels may include the sub-cellular level, the cellular level, tissues, individuals, populations, communities and ecosystems, landscapes and finally, the biosphere. The number and variety of interactions increases dramatically with increasing levels of biological complexity.

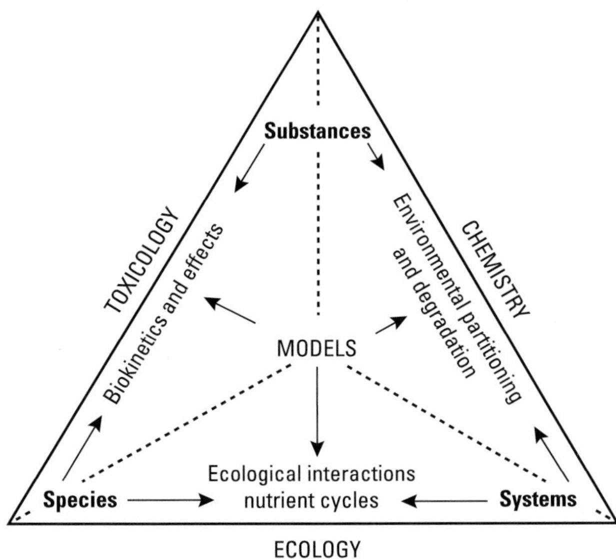

Figure 7.1. Ecotoxicology is a multi-disciplinary study into the toxic effects of substances on species in complex systems [1]. With permission.

Chemists are primarily interested in molecules and fate processes, toxicologists in biokinetics, modes of toxic action and effects in one or a number of standard test species, whereas ecologists are interested in the structure and function of ecosystems, effects, interactions and recovery at the population and ecosystem level, as well as in population genetics, biogeography, physiology and evolution. Due to the complexity of ecosystems, models are needed to describe the interactions between substances and species (toxicology), between substances and systems (chemistry) and between species in systems (ecology), as well as to account for the overall integration of these interactions (Figure 7.1). These models require input from mathematics, statistics and informatics.

Although the scientific backgrounds, interests and goals of the scientific disciplines differ, a synthesis of these disciplines is observed in the context of risk assessment. Normally, a sequence of research problems can be identified in the process of environmental risk assessment: the preliminary, the refined and the comprehensive stages [4]. Given the wide variety of research questions and topics (Table 7.1), this synthesis does not take place automatically. This chapter aims to illustrate how these disciplines can be integrated in ecotoxicology and are key to our methods for the risk assessment of chemicals.

This chapter will concentrate on ecotoxicological approaches used for the risk assessment of industrial chemicals. In Section 7.2 we will address some fundamental aspects of ERA. In Sections 7.3-7.5 we will introduce the core aspects of aquatic toxicity, sediment toxicity and terrestrial toxicity. For the aquatic environment the focus will be on freshwater species rather than on marine species. Readers interested in site-specific risk assessment, in effects beyond the population level, or in marine ecotoxicology, are referred to Suter [5], Suter et al. [6] and Hoffman et al. [7]. Two other subjects, i.e. factors modifying toxicity and mixture toxicity are presented in Sections 7.6 and 7.7. Sections 7.8 and 7.9 focus on ecotoxicogenomics and endocrine disruption. How PNECs are derived is presented in Section 7.10, whereas the assessment of PBT and

C.J. van Leeuwen and T.G. Vermeire (eds.), Risk Assessment of Chemicals: An Introduction, 281–356.

Table 7.1. "Disciplines" of ecotoxicology and some of their research topics.

Chemistry	Toxicology	Ecology	Mathematics
Exposure assessment	effects assessment	community structure	environmental fate models
Transport	modes of toxic action	community functions	pharmacokinetic models
Partitioning	bioaccumulation	population dynamics	LC50 and NOEC statistics
Transformation	biotransformation	nutrient/energy cycling	species-species extrapolation
SARs/QSARs	extrapolation	various interactions	population and ecosystem models

vPvB substances is given in Section 7.11. Section 7.12 provides some concluding remarks. Selected references are provided in Section 7.13.

7.2 FUNDAMENTAL ASPECTS OF ERA

7.2.1 Taxonomic diversity

ERA deals with millions of species rather than just one, as in the case of HRA. Estimates of the total number of species on earth vary from 10 to 100 million [8], and approximately 1.5 million species have been taxonomically classified. Some of the large taxonomic groups are given in Table 7.2. The majority of phyla is found in the marine environment. The largest taxonomic groups are the insects, spermatophytes, molluscs and fungi. The mainly marine phyla of porifera and echinodermata (5000 species each) belong to smaller taxonomic groups. Among the vertebrates (a total of 45,000 species), fish species account for 23,000, amphibians 2500, reptiles 5000, birds 8500, and mammals 4500 species respectively per taxon.

In ERA, effects on species from a few taxonomic groups are studied using a limited set of tests. This raises the following question: how do we select species for testing from among the 1.5 million taxonomically classified species? The current minimum requirement for ecotoxicological testing in risk assessment with fish, daphnids and algae is nothing but a gross simplification of an ecosystem. In practice it would be impossible to test a representative sample (e.g., 1%) of such a variety of species. In fact, the current trend in ERA is to generate more information from less testing. The practice in ERA is to be pragmatic: species are selected on the basis of their ecological function (trophic level), their morphological structure, and their route of exposure [10].

Table 7.2. Numbers of classified species of some large taxonomic groups of the plant and animal kingdom [9].

Regnum vegetabile		Regnum animalia	
Algae	20,000	Protozoa	46,000
Lichens	20,000	Porifera	5,000
Fungi	100,000	Coelenterata	10,000
Bryophyta	23,000	Plathyhelminthes	12,000
Pterydophyta	11,000	Nematoda	10,000
Spermatophyta	250,000	Mollusca	120,000
		Annelida	8,000
		Arachnida	30,000
		Crustacea	35,000
		Insecta	750,000
		Diplopoda	7,200
		Echinodermata	5,000
		Chordata	45,000

Table 7.3. Selection criteria for an ecotoxicity test.

Chemistry
The species should be representative in terms of:
– ecological function (trophic level)
– route of exposure
– morphology
The species should:
– be easy to keep under laboratory conditions
– be easy to feed and to breed
– have a large reference database
The test should be:
– applicable to a wide range of chemicals
– short, predictive, sensitive and cheap
– statistically sound, i.e. produce a quantifiable concentration-effect relationship within the test period
– useful for risk assessment
– internationally validated by various laboratories
– standardized, i.e. give reproducible results when carried out according to good laboratory practice (GLP)
– accepted by the regulatory and scientific communities

Practical aspects (Table 7.3) are important, as are social, economic and recreational factors.

7.2.2 Toxicological endpoints

In ERA, the goal is to protect populations and ecosystems, rather than individuals of certain species. It may be assumed that by protecting most of the species, the functioning of ecosystems is also protected [11]. Suter [5] postulated that ecological endpoints should satisfy five criteria (Table 7.4). This, however, leaves open the question of how to achieve an acceptable level of ecosystem protection. In routine toxicity testing, only a very limited number of species are tested and protection of all other species is assumed by extrapolating the results from toxicity testing on important endpoints: survival, growth and reproduction. This extrapolation should also protect ecological interactions, habitat factors, keystone species and functional groups. The terms "unacceptable" and "important" are value judgements and often lead to much debate. Stephan [12] has given seven major unacceptable effects that pollutants can directly or indirectly have on important species (Table 7.5).

Due to the large taxonomic diversity, life cycles vary greatly. Reproduction and growth depend on the species itself, time (e.g., food availability) and space (e.g., climatic conditions, soil-type, etc.). A further complication is the fact that species such as amphibians and insects undergo metamorphosis during transition from the larval to the adult stage. This affects their intrinsic sensitivity to pollutants, but may also affect the routes and magnitude of exposure.

Table 7.4. Criteria for selecting ecological endpoints [5]. With permission.

1. Biological relevance
2. Public relevance
3. Unambiguous operational definition
4. Accessibility to prediction and measurement
5. Susceptibility to the hazardous agent

Table 7.5. Unacceptable effects according to Stephan [12]. With permission.

1. Unacceptable reduction in survival
2. Unacceptable reduction in growth
3. Unacceptable reduction in reproduction
4. Unacceptable level of avoidance
5. Unacceptable percentage of gross deformities or visible tumours in organisms
6. Unacceptable concentrations of toxic residues in consumed tissues
7. Unacceptable flavour in consumed tissues

To what extent then should ecosystems be protected? To protect all species in an ecosystem is problematic for two reasons. It is impossible to guarantee that the most sensitive species is tested [13] and the associated cost would be tremendous. Furthermore if testing results in very conservative Predicted No Effect Concentrations (PNECs) this would imply a ban on most human activities which would not be acceptable to society as a whole.

In practice, the protection of species and ecosystem function is assumed by establishing either the most sensitive species of the relevant toxicity data and applying safety factors, or a relevant statistic of the toxicity data set, such as a certain cut-off percentage p when the toxicity data are described by a theoretical distribution function; known as species sensitivity distributions (see Section 7.10.2). In both cases, additional assessment factors can be applied to extrapolate from single-species laboratory data to a multi-species ecosystem [14]. Some science-policy papers [15,16] have explained the use of a

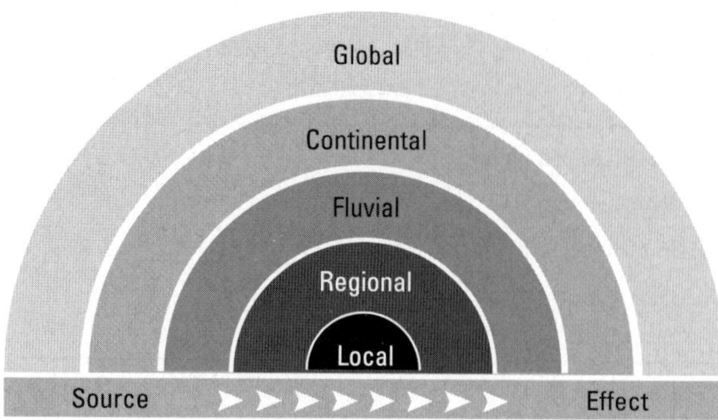

Figure 7.2. Five levels of scale at which environmental problems occur [15].

cut-off percentage as follows: the protection of all species at all times and places is not deemed necessary because ecosystems can tolerate some stress and occasional adverse effects. A reasonable level of protection can be provided by using a small cut-off percentage *p* of the species, pragmatically defined as 5% based on chronic toxicity data [17]. There are several problems with the cut-off percentage, e.g., the possibility that economically or ecologically important species fall within this 5% category.

Very pragmatic choices have been made in ERA to protect species, ecosystems, ecosystem functions or processes that have been successful in improving ecosystem quality, as demonstrated by cases such as the river Rhine [18].

7.2.3 Spatial scales

An understanding of the scale of environmental problems is key to effective risk assessment and remediation (Figure 7.2). Scale is linked to the area that a species needs to be able to maintain a stable population and the likelihood of exposure to a chemical in that area. Viable home ranges can range from very small for a microbe, to entire oceans for a blue whale.

Pollutant emissions may occur on a local scale, but due to redistribution and transport, the effects may become apparent on a global scale. The large-scale distribution of chemicals, however, should not be confused with the occurrence of effects, i.e., adverse effects may be restricted to certain sensitive populations or ecosystems which may occupy relatively small areas. Several examples can illustrate this scale dependency.

Pollution caused by heavy metals, many pesticides, and industrial chemicals exhibit their effects at the *fluvial scale* and/or *regional scale*. Indoor pollution caused by consumer products and air pollution in cities are examples of pollution on the *local scale*.

Acidification is the process whereby harmful effects occur as a result of pollution from the atmosphere with acid-forming substances and ozone. Acidification leads to damage to forests, heath land, aquatic ecosystems, agriculture, buildings and materials. Acidification arises from acid-forming substances such as sulfur dioxide, nitrogen oxides and ammonia. Oxygen radicals are formed from volatile organic compounds and nitrogen oxides. These react with the oxygen present in the air to form ozone. The harmful effects of ozone in the populated environment appear to be very similar to those of acid-forming substances and exert their influence on a *continental scale*. Other spatial examples include the impact of long-range transport and the effects of persistent organic pollutants (POPs) such as DDT and PCBs [19].

Finally, some pollutants can exert effects on the entire biosphere. Although the ozone layer in the stratosphere only contains minute quantities of ozone it has an important function. It absorbs ultraviolet (UV) radiation from the sun, which is harmful to man and ecosystems. As has become clear in recent years, the ozone layer is being depleted by a number of substances, such as chlorofluorocarbons (CFCs), which exert their effects on a *global scale*.

7.2.4 Temporal scales

Over the last two decades, our awareness of the importance of the spatio-temporal aspects of environmental pollution has increased. ERA deals with the sustainability of ecosystems, with large-scale effects, long-term processes and long recovery times (Figures 7.3 and 7.4). In HRA we are mainly concerned with individuals with a maximum exposure period of approximately 70 years. The generation time of man (approximately 25 years) is long compared to many other species (Table 7.6). This certainly goes for politicians, whose "generation time" is even shorter (approximately 5 years), while they make decisions that sometimes affect many generations to come [1]!

In ERA we are concerned with effects on a variety of temporal scales. Time scales are relatively long in relation to higher levels of biological organization [5,20], biological processes or evolutionary processes (Figure 7.3). This wide variety poses specific problems in ecotoxicity testing. In ERA, the hazard of a chemical is initially deduced from short time (acute) toxicity tests (e.g., see Section 7.3.3). However, depending on the

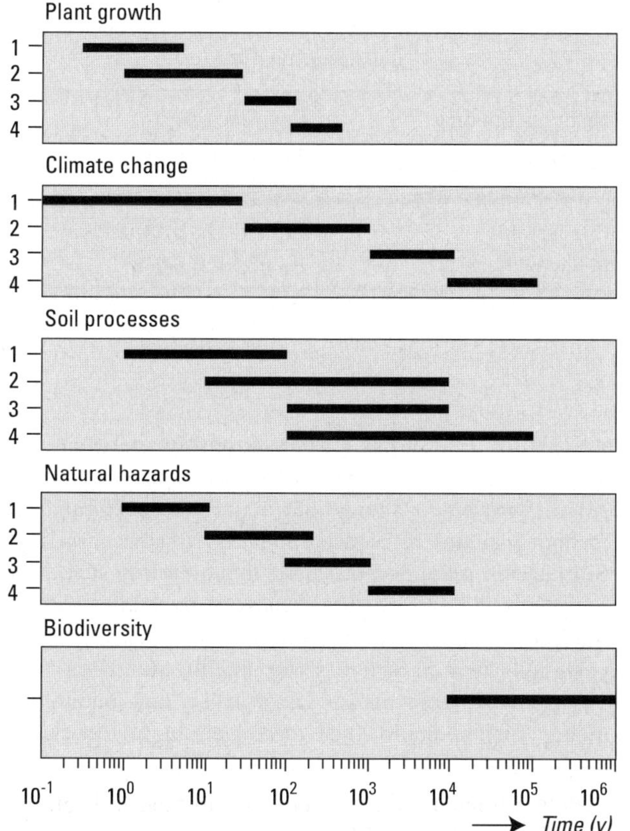

Figure 7.3. Timescales of processes affecting sustainability of ecosystems. *Plant growth:* (1) length of one growth cycle of annual crops, including rotation up to 5 years, (2) length of one growth cycle of perennial crops, (3) length of growth cycle of production forest and (4) average biomass turnover rates of tropical rainforest. *Climate change:* (1) time scales of meteorological fluctuations: decades (smallest time unit used in simulation models of plant growth), seasonal and annual changes variations up to 30 years, the minimum record length for reliable assessment of climatic parameters, (2) historical climate changes (cf. Little Ice Age 1500-1850 AD) (3) Holocene (cf. climatic optimum 6000 years BC) and (4) Pleistocene, stadial/interstadial and glacial/interglacial oscillations. *Soil processes:* (1) time needed for complete erosion of topsoil, (2) time needed for severe nutrient depletion by leaching in humid tropics, (3) the same for the temperate zone and (4) time needed for formation of fully developed topsoil. *Natural hazards:* (1) frequency intervals between moderate floods in alluvial areas, (2) the same for major disastrous floods, (3) frequency intervals for andesitic volcanic ash falls and (4) the same for destructive volcanic eruptions. *Biodiversity:* time needed for restoration of macrofauna and macroflora biodiversity by evolution after major disturbance. From Fresco and Kroonenberg [20]. With permission. Copyright Elsevier.

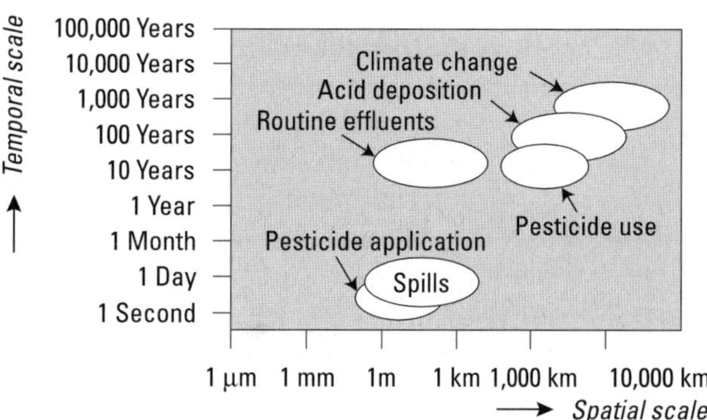

Figure 7.4. Chemical anthropogenic hazards on spatial and temporal scales. From Suter[5]. With permission.

typical generation time of a species and the mode of action (MOA) of the substance concerned, the distinction between short-term and long-term (chronic) toxicity is sometimes arbitrary. When specific hazards are identified, e.g., when a substance has effects on growth, reproduction and development, longer toxicity tests can be performed covering a partial or a full life cycle. The effects of some chemicals may be observed in the sensitive early life stages (ELS), such as embryonic development or neonates, for which ELS tests are developed. The impact of these effects on the viability of the population can be analyzed (see Sections 7.3.4 and 7.3.5).

7.2.5 Complexity of exposure

In ERA, exposure assessment is often restricted to external exposure: concentrations in media such as water, soil, sediment and air. These external concentrations, predicted environmental concentrations (PECs), are related to external effect levels (PNECs). In addition, ERA deals with a wide variety of species and factors influencing actual exposure which complicates exposure assessment (Table 7.7). When comparing the effects of chemicals under different exposure conditions or on different species, the bioavailability of the chemical needs to be taken into account (see Chapter 3).

In some cases it is more useful to determine the internal exposure concentration in species that have been exposed. Internal effect concentrations can then be compared with "critical body residues" (CBRs) associated with the onset of mortality for specific classes of chemicals, such as narcotics or polar narcotics. Chemicals with the same mode of action will have a relatively narrow range of critical body concentrations [21,22]. The internal dose can be estimated from external exposure with toxicokinetic models [23] or

Table 7.6 Generation times for some species.

Species	Generation time
Bacteria	≈ 0.1 d
Green algae (*Chlorella* sp.)	≈ 1 d
Waterflea (*Daphnia* sp.)	≈ 10 d
Snails (*Lymnaea* sp.)	≈ 100 d
Rats	≈ 1 y
Politicians	≈ 5 y
Man	≈ 25 y

Table 7.7 Summary of factors contributing to the complexity of exposures in ERA.

Niche-partitioning	Exposure time
Abiotic factors	Non-linearity
Surface/volume area	Consumption patterns
Life history	Feeding and growth rate
Behaviour	Biotransformation

with relatively simple partitioning models [24,25]. This may make it possible to move from a purely descriptive external exposure to internal exposure with toxicological relevance. Once this step is taken, extrapolation to chemicals with similar modes of action (MOA) immediately becomes possible [26-28]. It should be noted that the use of the CBR concept for general risk assessment needs improvement. Most notably the CBR distribution for narcotics is still quite wide, both between different chemicals and between species or phyla. Uncertainty is reduced by using lipid normalization, but is also related to the quality and interpretation of some of the original studies [29]. It is also not always easy to define the MOA of a substance using the currently available tools [30,31]. Hopefully, the development of structural alerts, read across and other methods (Chapters 9-11) will help to improve the use of this concept in risk assessment.

Niche partitioning

Once a chemical enters the environment, partitioning and degradation processes take place (Chapter 3). Species in specific ecological niches may be exposed intensively, depending on the chemical's fate and behaviour. Benthic species, for instance, that burrow in the sediment, such as the lugworm, are in intense contact with pollutants that partition to the sediment particles that they ingest. Exposure assessment in sediments and soils is complicated and, in many cases, predictions can only be made for certain groups of chemicals under a variety of assumptions such as equilibrium-partitioning between pore water and soil or sediment [32,33]. For the sake of simplicity, environmental exposure models often assume a homogeneous distribution of chemicals in a limited number of narrowly defined compartments. Nature, however, is not homogeneous but heterogeneous with many niches occupied by a great variety of species adapted to these niches.

Figure 7.5 shows the relationship between the atmospheric fallout of pollutants and concentrations in eel in Sweden. It shows that aerial transport can lead to high residues of bioaccumulating substances in fish. Similar observations were made in monitoring studies on pesticides in rain [35]. These monitoring studies showed the presence of high concentrations of some volatile pesticides in rain where water quality standards were exceeded by more than a factor 100. Thus, habitats or niches, such as shallow lakes which are highly dependent on rain, may be intensively exposed. This also applies to lichens, bryophytes and fungi living on trees or in the top horizons of soils with a wide variety of bacteria, plant and animal species present in these niches. Persistent organic pollutants (POPs) can be transported and deposited in vulnerable ecological niches. In the Canadian arctic, POPs have been shown to accumulate in the food chain from lichen to caribou to wolf [36].

Exposure time

In routine toxicity testing in ERA, the classical dose or concentration-response model is used where exposure time is kept constant. The exposure time in such tests depends on the species and its generation time (Table 7.6). Exposure time is an important variable, often crucial to toxicity (Sections 7.3.1 and 7.3.2). If the temporal dynamics of the endpoint that is studied are included, the statistical power of the test increases and effects can be expressed as functions of both exposure concentration and exposure time [37]. This then allows additional toxicokinetic parameters of the tested species to be estimated such as the elimination rate of a chemical [38], but can also provide input in models for population dynamics (Section 7.3.5). Despite the obvious advantage of gaining more insight with the same toxicity tests, little progress has been made with applications in regulatory toxicity testing.

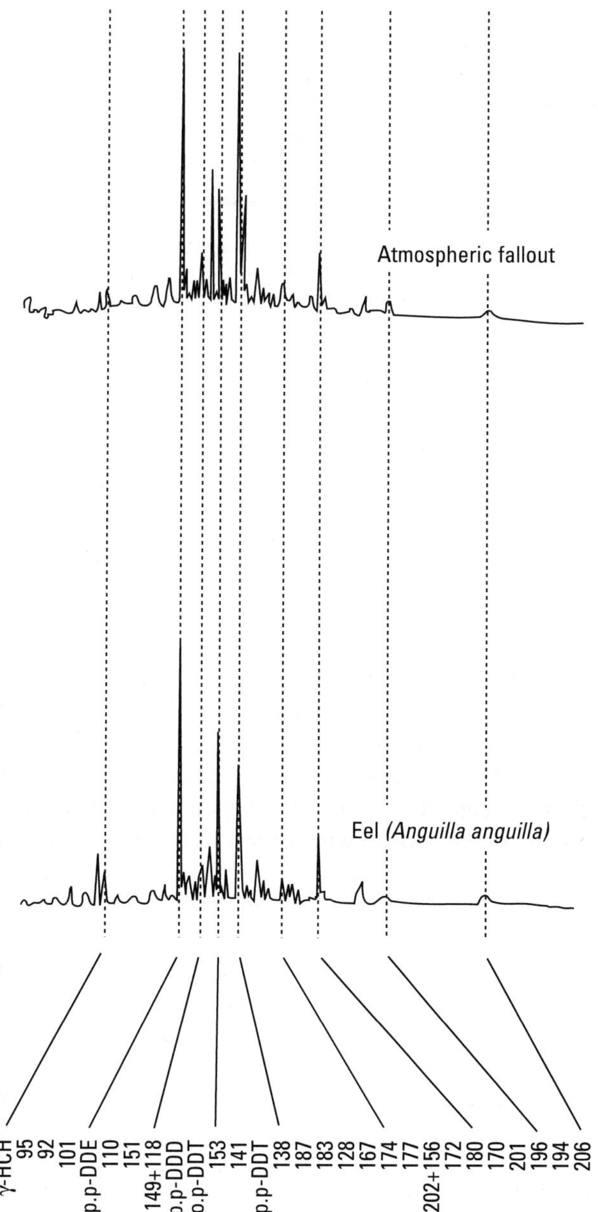

Figure 7.5. Composition and distribution of persistent pollutants in atmospheric fall-out and eel in Swedish lakes. Numbers represent 25 PCB congeners. From Larsson, Hamrin and Okla [34]. With permission. Copyright Elsevier.

Abiotic factors

The magnitude of external exposure is subject to large spatio-temporal fluctuations. These fluctuations may be caused by varying emissions of the chemicals but a number of abiotic factors may also be involved, such as soil type and climate, e.g., wind speed, temperature, humidity, and rainfall (see Section 7.6). The geographic and temporal variations in river flows may also differ and affect the actual exposure situation. It is clear that photochemically degradable chemicals or readily biodegradable chemicals may do much more harm in cold, northern climatic regions than in tropical areas with much more sunshine.

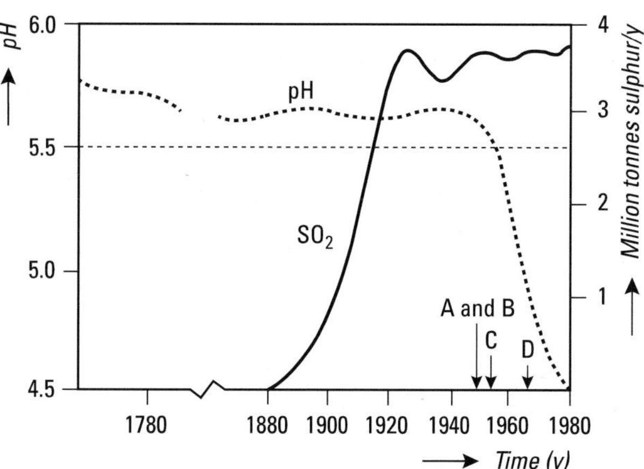

Figure 7.6. Non-linear changes in pH due to emissions of SO2 and extinction of smallmouth bass (A), lake whitefish (B), longnose sucker (C) and lake trout (D) in the period from 1760-1980. Modified from Stigliani [39]. With kind permission of Springer Science and Business Media.

Non-linearity

Long-term observations of emissions and exposure to chemicals show that unpredictable changes may occur (Figure 7.6). These changes are related to changes in a number of capacity controlling properties (CCPs) e.g., the cation or anion exchange capacity, pH, redox potential, organic matter content, soil texture, salinity and microbial activity [39]. Acidification, climate change, pollution-induced reductions in microbial activity, and lowering of the groundwater table are among the factors that may alter CCPs. These changes in CCPs may alter the bioavailability of pollutants by several orders of magnitude (Section 7.6 and Chapter 4) and may thus lead to unexpectedly strong ecological effects.

Surface area/volume ratio

So far we have dealt with factors modifying external exposure. Let us now turn to internal exposure, for which chemicals have to pass external barriers (Chapter 3). Chemicals can pass through biological barriers, e.g., the cell membrane, lungs, gills, skin, cuticle, etc. by diffusion. For soil-dwelling species with firm cuticles or exoskeletons, such as many arthropods, direct dermal

Table 7.8. The relationship between surface area and volume of species. For the sake of simplicity, the shape of species is taken to be cubic.

Edge (mm)	Surface area (mm^2)	Volume (mm^3)	Surface/ Volume ratio	Examples
0.001	$6x10^{-6}$	10^{-9}	6000	cells/bacteria
0.01	$6x10^{-4}$	10^{-6}	600	algae (*Chlorella* sp.) and fungi (*Penicillium* sp.)
0.1	$6x10^{-2}$	10^{-3}	60	protozoans (*Paramecium* sp.)
1	6	1	6	nematodes and crustaceans (e.g. *Ceriodaphnia dubia*)
10	$6x10^{2}$	10^{3}	0.6	earthworms/small fish (e.g. guppy)
100	$6x10^{4}$	10^{6}	0.06	rainbow trout/pigeon
1000	$6x10^{6}$	10^{9}	0.006	sharks/cows

uptake of pollutants from the soil does not seem to be an important exposure route. However, research carried out with spiders suggested that dermal exposure to contaminated soil may be important even for these species. For soft-bodied organisms living in close contact with the soil, such as earthworms, uptake via the skin is important [40]. This means that earthworms and probably other soft-bodied soil organisms such as protozoans, tardigrades, nematodes and enchytraeids, take up chemicals mainly via the body wall [41]. For these species, the toxicity of chemicals in the soil is mainly determined by the pore water concentration (Section 7.5). This pore water concentration can be derived from the total concentration using sorption data [32,33].

Dermal uptake via diffusion depends on the permeability of this barrier to the chemical, the concentration gradient and the surface area over which diffusion takes place. The larger the surface area and concentration gradient, the greater the transport rate. The transport rate is inversely related to the length of the diffusion path (Fick's law). This equation can be written as follows:

$$M = DA\,(C_1 - C_2)/L \tag{7.1}$$

where

M = the rate of diffusion (mol/s)
D = the diffusion coefficient (m^2/s)
A = the surface area over which diffusion takes place (m^2)
$(C_1\text{-}C_2)/L$ = the concentration gradient, i.e., the difference in concentrations (mol/m^3) divided by the length (L) of the diffusion path (m)

Diffusion is more efficient in cells or tissues with short diffusion paths, i.e., in tissues where the surface area/volume ratio is high (unicellular species or specialized tissues such as the lungs or gills). In nature, the permeability of external biological barriers varies widely. The same applies to surface area/volume ratios which depend on the size of the species (Table 7.8). Thus, diffusion, i.e., exposure, is much faster in small species than in large species. Large species often have special adaptations to accelerate diffusion processes for gas exchange, e.g., internal or external gills or lungs. These adaptations also affect the rate of chemical uptake. Furthermore, the toxic effects of chemicals which affect cell membranes, e.g., surface-active chemicals, will be greater in small species or tissues with high surface area/volume ratios. This is why many surface-active chemicals have bactericidal and algicidal properties and are relatively toxic to fish.

Consumption patterns

The consumption pattern of species (including man) differs widely. There are omnivorous, carnivorous and herbivorous species and many food specialists, such as caterpillars, mites, ticks and some bird species. Their average daily consumption patterns, i.e., the food chains, are largely unknown, both qualitatively and quantitatively. In order to illustrate the importance of consumption patterns to exposure assessment, fish consumption will be used as an example. In Table 7.9 a comparison is made between man and a fish-eating bird, the cormorant (Figure 7.7). The average daily consumption of fish (wet weight; wwt) in The Netherlands and Japan is 10 g and 96 g, respectively [42], while the cormorant's daily intake is 400 g to 750 g.

Table 7.9. Fish consumption patterns and daily intakes of hexachlorobenzene (HCB) in The Netherlands (NL), Japan and in the cormorant *(Phalacrocorax carbo).*

	NL	Japan	Cormorant	
			male	female
Body weight (kg)	70	70	2	3
Fish consumption (kg_{wwt}/d)	0.01	0.1	0.5	0.5
Fish consumption (70 kg_{bw})[a]	0.01	0.1	17.5	11.6
Intake of HCB[b] (mg/kg_{bw}·d)	0.03	0.3	50	33.3

[a] Fish consumption expressed in terms of the body weight of man (70 kg).
[b] The Swedish product standard for HCB (200 μg/kg fish) was used for the calculations [32].

When the fish consumption of cormorants is expressed in terms of human body weight, it can be concluded that their daily consumption is enormous (11.6 to 17.5 kg fish per day). It is more than 100 times the average daily fish consumption in Japan and more than 1000 times the average in The Netherlands. The second conclusion is that the exposure of food specialists to pollutants can be extremely high, which should be taken into account in risk assessment for secondary poisoning (Section 7.10.3).

Life histories

There is an overwhelming variety of species (Table 7.2). Many plants and especially some parasitic fungi, such as rusts (Uredinales) and smuts (Ustilaginales), have very complicated life histories. The same applies to parasitic nematodes, mites and insects. Many insects, such as butterflies (Lepidoptera), stoneflies (Plecoptera), mayflies (Ephemeroptera), dragonflies (Odonata) and midges (Diptera), undergo a metamorphosis with concomitant changes in the niches they occupy. Many amphibian and oviparous fish species go through a number of different embryonic and post-embryonic stages each with their own exposure patterns (Figure 7.8). Particularly the early life stages appear to be very sensitive to pollutants. Frogs, toads and many insect species undergo transformations which take them from an aquatic to a terrestrial life-cycle stage. This has consequences for both their direct exposure routes (exposure via air, water and soil) and their indirect exposure routes, i.e., their food consumption patterns. In other words, life-history patterns are extremely important in ecotoxicological testing. The diversity in life histories is huge. Unfortunately, qualitative and quantitative information is often not used or lacking.

Figure 7.7. A food specialist: the cormorant *(Phalacrocorax carbo).* Courtesy of P. Van Der Poel, Huizen, the Netherlands.

Feeding and growth rates

Many abiotic factors can modify the feeding and growth rates of species and may also determine the type of diet and hence exposure. Feeding rates determine the uptake rate of chemicals, whereas individual growth rates or rates of cell division may be seen as "internal dilution processes" for body burdens of chemicals. For many species data on feeding and growth rates are lacking but reasonable approximations are available in the literature [43,44].

Behaviour

The behavioural responses of organisms to toxicants may modify subsequent exposure. The most commonly reported example is avoidance of contaminated food, soil or water. However, toxicants may also go unnoticed or attract organisms. Migration, hibernation, isolation, breeding and the formation of resistant structures such as plant seeds or the winter eggs (ephippia) of daphnids all affect the actual exposure of organisms. There is little behavioural data for many species, but avoidance behaviour is now recognized in several guidelines.

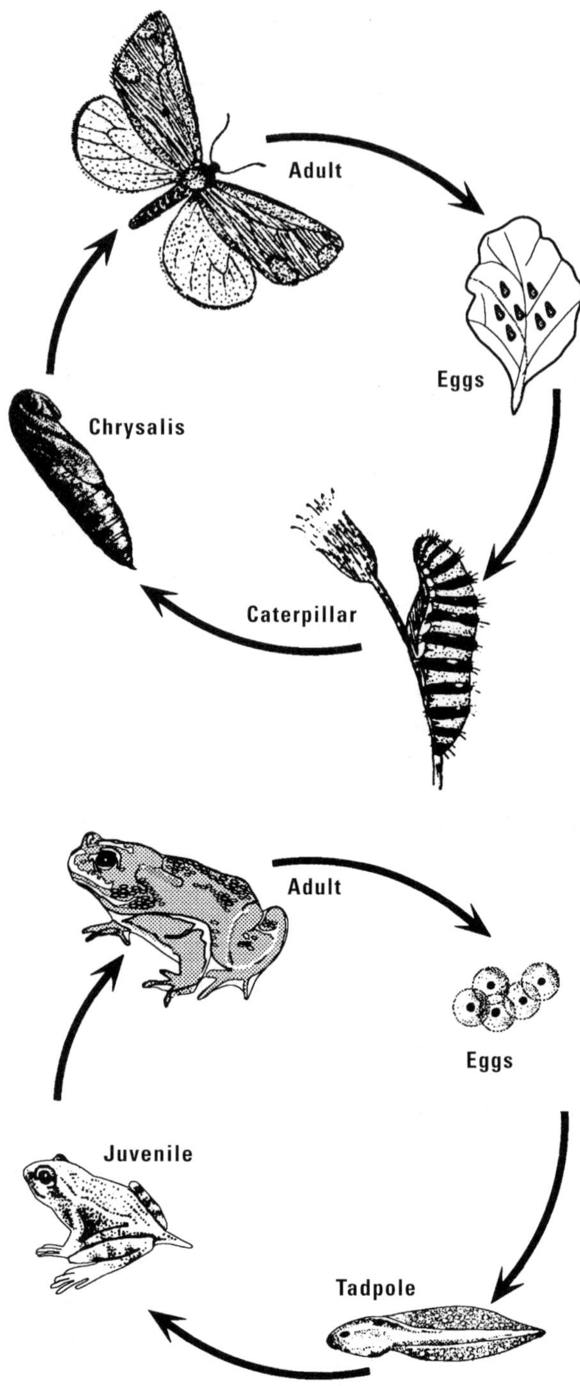

Figure 7.8. Life cycles of an insect and amphibian species with concomitant changes in exposure patterns.

Biotransformation

The biotransformation of toxicants is of essential importance. It may result in detoxification and elimination of the metabolite, or it may enhance toxicity through the formation of toxicologically active metabolites (Chapter 3). Biotransformation patterns vary between organisms and may be modified by a number of abiotic factors, such as temperature. However, the absence of empirical biotransformation rates and information on formed metabolites is mostly lacking which hampers risk assessment. This is generally recognized as an important field of study. Information on rates and routes of biotransformation has been compiled to provide a basis for models that predict which metabolites are formed [45].

Conclusions

In conclusion, there are many factors in ERA which are crucial for the calculation of external and internal exposure concentrations. In ERA there is no single PEC but a variety of PECs. These PECs are species dependent and are influenced by a large number of biotic and abiotic factors. Most of them are unknown and where they are known, they are often not quantified. This lack of information prevents the calculation of actual received dose or internal exposure concentrations. It is mainly for this reason that in ERA exposure predictions are restricted to predictions of external concentrations (PECs) in media such as soil, sediment, air and water. Exposure predictions may go beyond this level for only a few species.

7.3 AQUATIC TOXICITY

7.3.1 Exposure systems

In aquatic toxicology, exposure is of particular importance. Contrary to mammalian or avian toxicology, where the toxicant is often administered directly to the organism via food or injection which leads to a known internal dose, exposure in the aquatic environment is much more complicated. In most aquatic toxicity tests the toxicant is dissolved in the test medium. The test organisms build up an internal concentration through the skin and particularly through the gills, by partitioning between water and the organism (Chapter 3). Because the internal concentration of the toxicant is usually not known, toxicity is expressed as external concentration in the exposure medium, rather than as internal concentration.

Because the actual concentration of the chemical together with the duration of exposure is of prime importance in determining whether an adverse effect will occur or not, concentration and exposure time must be considered carefully. Maintaining stable exposure concentrations is a problem in aquatic toxicity testing which is why particular attention is devoted to this subject. Exposure to volatile chemicals, degradable chemicals, adsorptive, highly bioaccumulative chemicals and chemicals with low water solubility poses great problems in practice. Therefore, various methods have been developed for exposing aquatic organisms to such substances in order to test for ecotoxicological effects, with varying degrees of success. Four general types of toxicant delivery systems are used in toxicity testing: (1) static, (2) renewal, (3) flow-through, and (4) food.

Static exposure systems

Static exposure systems are much simpler in design and operation than flow-through systems. They generally consist of exposure vessels in which the test organisms are subjected to the same test solution for the duration of the test. The test substance is administered once only and the solution is not changed or renewed. Such systems are only generally used for acute tests with a few exceptions, and then generally for technical reasons (e.g., the alga growth inhibition test) [46]. The advantages of this type of exposure system are its simplicity, reduced handling stress to the organisms compared with renewal techniques, and low cost. Static systems are generally used where:

- The test substance is known to be highly soluble and stable in aqueous solution.
- The test substance is not expected to be toxic at the limit test concentration.
- A multi-component test substance is tested using a water-accommodated fraction.
- A very small quantity of the test compound is available.
- Disposal of the test solutions is critical.

Nevertheless, there are a number of problems that commonly arise in static systems:

- Decrease in the concentration of the test material through loss due to evaporation, transformation, sorption, biodegradation or bioaccumulation in the test species. If the exposure concentrations deviate by more than 80-120% of nominal, they should be expressed relative to the geometric mean of the measured concentrations at the start and end of the test.
- Low dissolved oxygen concentrations occur if the test material has a high biochemical oxygen demand (BOD) or as a result of the accumulation and microbial degradation of faecal material. This can be circumvented by the use of aeration and oxygen measurements, unless the substance is expected to be volatile.
- Starvation, where feeding is not possible because it could interfere with the bioavailability of the toxicant.

Owing to these limitations, static exposure systems are generally used in short-term tests (< 96 h), with non-volatile or slowly degradable chemicals with a low bioaccumulation potential and a low loading (biomass/volume of water) of test organisms. Box 7.1 shows the consequences of high loading of test vessels. The results of simulation studies can be seen in Figure 7.9.

Renewal exposure systems

Renewal or semi-static exposure systems are a compromise between flow-through and static exposure systems. The apparatus used is essentially the same as in a static system; however, instead of exposing the test organisms to the same solution throughout the test, the test organisms are periodically transferred to fresh solutions or a proportion of the solution is removed and renewed with fresh test solution. Renewal exposure systems allow feeding and the test can be prolonged indefinitely. Renewal exposure systems are mainly used with small organisms (e.g., *Daphnia* spp.) that could be flushed out of flow-through systems or are very sensitive to currents (e.g., copepods). They are also useful when only a limited amount of test material is available but a prolonged test is required. Although static-renewal systems circumvent some of the disadvantages of static systems, some disadvantages remain:

- Frequent handling of the test organisms increases stress and the possibility of injury.
- The concentration of the test material may not be constant throughout the test.
- It is more labour-intensive than static tests.

Flow-through

Flow-through or continuous-flow exposure systems are designed to expose the test organisms to a relatively constant concentration of the toxic material and control water flowing into and out of the exposure chambers. The flow may be continuous or intermittent. Flow-through systems are able to maintain a constant concentration of the test material, a constant water temperature, and maintain the dissolved oxygen concentration in water at between 60 and 100% saturation. In the case of fish, the flow should preferably be 6 litres of test water per gram of fish per day. In addition, a flow rate of at

Box 7.1. Consequences of high loading of test vessels for the exposure concentration under static exposure conditions

Basic information

n-octanol-water partition coefficient (K_{ow})	= 100,000
Mass of fish (M)	= 0.001 kg
Fat content of fish	= 5%
Bioconcentration factor (BCF) ≈ 0.05 x K_{ow}	= 5000
Volume of test vessel (V)	= 1L
Test concentration (C_w) in water (at t = o)	= 1 mg/L

Mass balance equation at t = o

No bioaccumulation, total mass of toxicant in water: C_w x V = 10 mg

Mass balance after prolonged exposure

We assume t ≈ ∞ and no losses due to (a) volatization, (b) biotic or abiotic degradation and (c) adsorption to the wall of the test vessels. Therefore the chemical only partitions between fish and water. The mass balance equation then becomes:

(mass in fish)	+ (mass in water)	= total mass
(BCF x M x C_w)	+ (C_w x V)	= 1 mg
(5000 x 0.001 x C_w)	+ (C_w x 1)	= 1 mg

Result

In this example C_w becomes 0.16 mg/L. With a higher loading of fish (0.01 kg/L) or when testing superlipophilic chemicals with a K_{ow} of e.g., 1,000,000, the concentration in water would drop to 0.02 mg/L.

least five times the test chamber volume per 24 hours is needed. Many types of toxicant delivery systems have been designed for use in flow-through exposure systems. Peristaltic and syringe pumps are widely used for the delivery of concentrations of a toxic chemical to aquatic organisms [47]. Another common system is the proportional diluter, a gravity-fed system, which delivers a series of more or less constant concentrations of the test material. First developed by Mount and Brungs [48], the proportional diluter has been modified and improved for a wide variety of applications.

Flow-through systems are the preferred method for aquatic toxicity studies on fish, particularly if the test substance is not stable or is poorly soluble. For volatile substances they should be used in conjunction with a closed system. In some cases a headspace inside such a system is acceptable depending on the Henry constant of the substance. The major disadvantage of flow-through systems and proportional diluters is their complexity; they require considerable attention and maintenance if they are to function properly, such as frequent verification of the actual concentrations of the test compound. However, once functional the fluctuation of test substance concentration is generally much lower than in static or semi-static tests, thereby increasing confidence in the results of the study.

Food

Highly bioaccumulative substances are usually poorly water soluble which is problematic in the standard test systems for both toxicity tests and bioaccumulation tests designed to determine the bioconcentration factor. For substances with a log K_{ow} > 4.5 (decimal logarithm of the *n*-octanol-water partition coefficient), it is difficult to achieve a constant exposure level that is high enough to easily measure toxicity or bioaccumulation. Test concentrations that exceed the solubility level or are supplemented with a large amount of solvent (OECD Guidelines recommend a maximum of 100 mg/L) may result in an underestimation of true toxicity levels due to physical effects. In risk assessment, tests with effect levels above the solubility level are considered to be invalid.

In dietary tests, fish are fed chemical-spiked food at a fixed concentration over a specific period of time, depending on the expected half-life of the chemical. At the end of the food exposure period, the remaining animals are provided with uncontaminated diet and

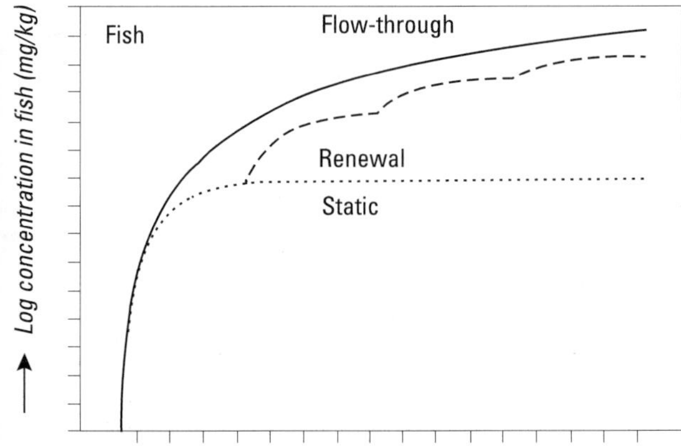

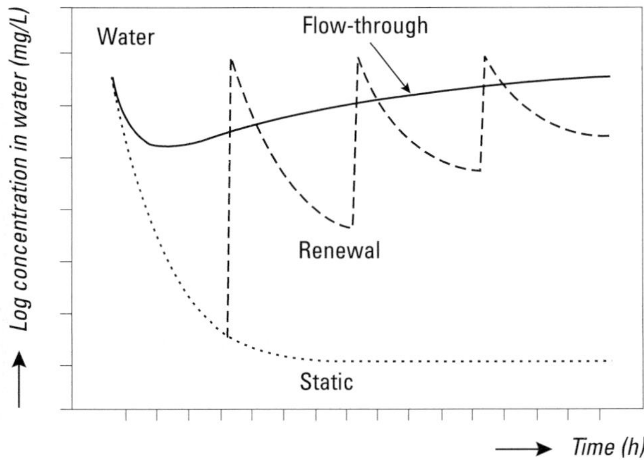

Figure 7.9. Simulations of concentrations of a non-volatile persistent chemical in fish and water in a static, a renewal and a flow-through system. It is assumed that no losses occur due to volatilization, adsorption and degradation of the chemical. Courtesy of D. De Zwart, National Institute for Public Health and the Environment, Bilthoven, The Netherlands.

analyzed to establish a depuration curve. From these data, the half-life, dietary assimilation and bioaccumulation factor can be easily derived [49]. It is also possible to determine whether the (final) toxicity at a given exposure level was observed under steady-state conditions or not. It should be noted that there is currently no standard procedure for performing a dietary-based toxicity test for fish, although exposure to contaminants in food may be commonplace in *Daphnia* reproduction tests due to the time of exposure of the food source (algae) to the test substance.

7.3.2 Analysis of toxicity tests

Lethal or sublethal effects of chemicals are typically analyzed in a setup with a series of containers or tanks with increasing concentration of a chemical and must include a control. The response of the organism to the increasing concentration of the chemical is used to determine the endpoint of interest. For acute toxicity tests, mortality is expressed as the median lethal concentration (LC50), which is the estimated concentration of the test material that will kill or immobilize 50% of the test organisms in a predetermined period of time. Similarly, median effect concentrations (EC50) can be calculated for any specified effect. For EC values, the endpoint has to be specifically defined. If an asymptote has been reached in the toxicity-time curve (Figure 7.10), the final value is called the incipient or ultimate LC50, or threshold lethal concentration. Because this value eliminates the influence of time of exposure, the result can only be compared to similar L/EC50 values and its use cannot easily be extended to determining ecological significance in terms of population effects.

A variety of methods can be used to calculate LC50 and EC50 values and their confidence limits, of which the non-parametric and the parametric methods are most commonly applied. The most common parametric methods are based on transforming the concentration levels so that the transformed concentration-mortality relationship has a known concentration-effect relationship [50,51]. The nonparametric methods, such as the Spearman Karber method, use the monotonicity of the concentration-mortality curve to generate an empirical curve from which LC50 and EC50 estimates can be obtained. Reviews are provided by Hoekstra [52] and Newman and Unger [3].

Another summary statistic that is commonly used in toxicity tests for regulatory testing is the no observed effect concentration (NOEC). This assumes that there is a concentration (threshold) of a toxicant below which no adverse effect is expected (Figure 7.11). The threshold concentration-response curve climbs at the threshold concentration; the response is zero up to that point and increases beyond that point. The NOEC is determined by hypothesis testing, e.g., by the Williams test or a post-analysis of variance (ANOVA), such as Dunnet's multiple comparison test [50]. In a statistical analysis of variance, the NOEC is determined by comparing the responses of the exposure concentrations with the control (unexposed) responses to test the zero hypothesis that they are the same as the control responses. Such an analysis will produce the lowest observed effect concentration (LOEC), i.e., the lowest concentration whose mean response differs significantly from the control. The NOEC is defined as the test concentration directly beneath the LOEC. In some cases no effects are observed

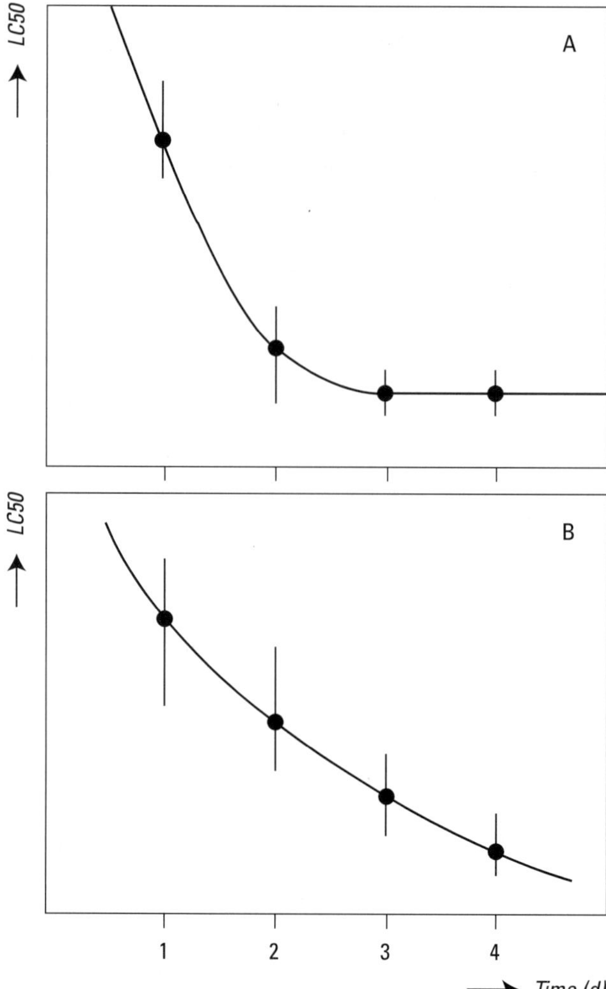

Figure 7.10. LC50 values and their 95% confidence limits vary with time. Two LC50-time curves are shown. The incipient LC50 for chemical A, i.e. the concentration of a chemical which is lethal to 50% of the test organisms as a result of exposure for periods sufficiently long for acute toxicity essentially to have ceased, is reached within 3 days. For chemical B the asymptotic part of the LC50-time curve is not reached and prolonged testing is necessary to estimate its incipient LC50.

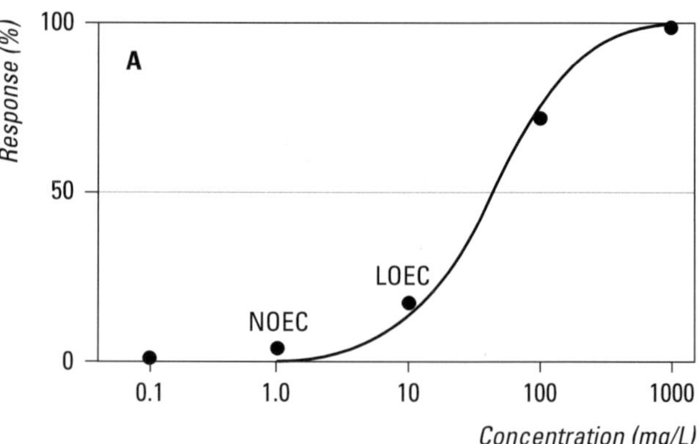

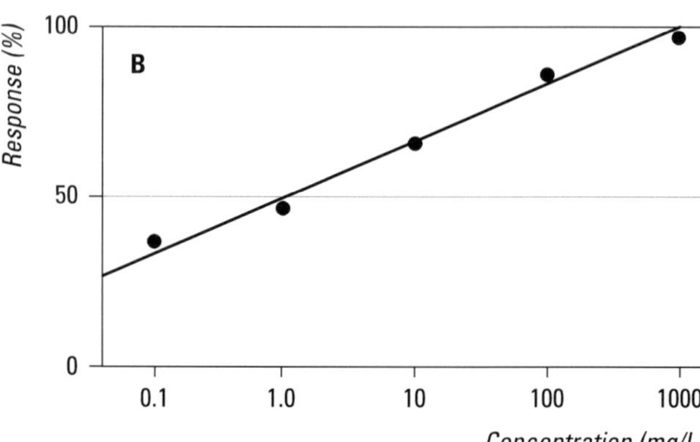

Figure 7.11. Typical examples of concentration-response functions for chemicals with (A) and without (B) a threshold. For chemicals with a threshold a No Observed Effect Concentration (NOEC) can be determined. The nonthreshold concentration-effect curve shows that there is no threshold concentration below which exposure is relatively harmless. Most carcinogens and mutagens are non-threshold chemicals.

at the highest test concentration. It is statistically incorrect to designate this as a NOEC and it should be reported as no effect at the highest concentration tested. When this value coincides with the solubility limit of the test substance this should be specified. Sometimes results are reported as the maximum allowable toxic concentration (MATC), which is the geometric mean of the LOEC and NOEC.

When no threshold is observed experimentally, it implies one of the following:

- There is no theoretical basis for the existence of a threshold, as in the case of genotoxic carcinogens and mutagens. Zero response occurs only at zero dose or concentration (see also Chapter 6, Section 6.5.3).
- Although there might be a threshold, experimental limitations have kept it from being identified.

The nonthreshold concentration-effect curve shows that there is no threshold concentration below which exposure is relatively harmless. As the concentration increases so does the probability of an adverse effect. The relationship between concentration and response is a straight line (Figure 7.11).

The NOEC based on hypothesis testing suffers from a number of disadvantages [37,53]. First, the ANOVA design is more concerned with avoiding having

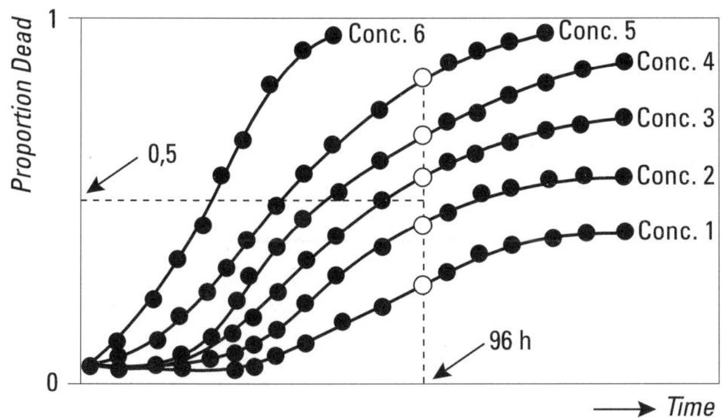

Figure 7.12. Time-concentration-response diagram to illustrate the increased power of analyzing toxicity data when using time to death data, instead of only 96-h data to calculate the LC50. From Newman and Unger [3]. With permission.

to state that a concentration is toxic when it is not (with an arbitrary Type I error not based on biological significance) than with avoiding having to state that a concentration is not toxic when it is (Type II error). However, in risk assessment, we are concerned with the latter [5]. In addition, the NOEC itself does not give any information on the concentration-effect curve. The NOEC can only be one of the tested concentrations for which no confidence limits can be calculated.

These disadvantages can be partly overcome in several ways: first, by regression analysis of the concentration-effect relationship. The great advantage of this is that after deriving an equation for the concentration-effect relationship, a concentration can be estimated which corresponds to a specified degree of an adverse effect. This key aspect is utilized in the benchmark dose approach (BMD), described in Section 6.5.3 of this book. In this approach, adapted here for ecotoxicology, the NOEC is replaced by a critical effect concentration derived from the concentration-effect curve, corresponding to a prescribed small effect considered non-adverse, such as an EC5 or EC10. The BMD approach, however, is not always an improvement if the variation in response between animals in relatively small dose groups is large [54].

The second approach is to utilize the temporal dynamics of effects in a time-response approach (reviewed in [3]). Instead of only reporting the survival of test animals at a single point in time (e.g., 96 hours), survival can be monitored during the entire experiment (Figure 7.12). The advantages of time-response approaches are that, due to increased statistical power, additional biological factors can be taken into account such as sex, temperature or acclimation history. To fit survival-time data, an analysis is needed that differs from estimating the LC50. Several nonparametric and parametric methods can be used to fit the survival curves [3,50]. In the parametric models, the shape of the survival curve is described by a hazard model. Hazard models are used to analyze a variety of phenomena, ranging from mechanical component failure to cancer incidence. A special application of the hazard model is used to analyze the mortality probability, related to accumulation of the chemical in an organism [38,55]. Essentially, it combines the CBR concept [21] with time-response modelling. Although internal concentration can be treated as a hidden variable in this approach, if available, it improves the modelling of the accumulation-related increase in mortality over time (Figure 7.13).

Regardless of the outcome of a statistical test, it is still necessary to draw a separate conclusion about the biological importance of the observed effect [5]; hence, a statistically significant effect is not the same as a relevant biological effect.

7.3.3 Short-term toxicity

Introduction

Laboratory toxicity tests with fish, invertebrates or algae are usually single-species tests in which the toxicity of a chemical is measured through mortality, decreased growth rate and lowered reproductive capacity. These tests have been highly standardized and applied to a select group of organisms. A distinction should be made between acute and chronic tests. Acute toxicity can be defined as the severe effects suffered by organisms from short-term exposure to toxic chemicals. The objective of acute toxicity testing is to determine the concentration of a particular chemical that will elicit a specific response or measurable end-point from a test species in a relatively short period of time, usually 2 to 7 days. In chronic toxicity tests, effects are studied over prolonged periods of exposure, often over entire life cycles and usually the endpoints are primarily sublethal (such as growth) or measurements of reproductive output. Subchronic studies are of longer duration than acute exposure but generally do not exceed a period equivalent to one-third of the time taken for a species to reach sexual maturity. Short test duration is not synonymous with acute toxicity. This can best be explained by using the algae growth inhibition test as an example. Both acute and chronic endpoints can be obtained from toxicity tests with algae because algae have relatively short life cycles (Table 7.6) so the EC50 is used as an acute endpoint and the NOEC/EC10 as a

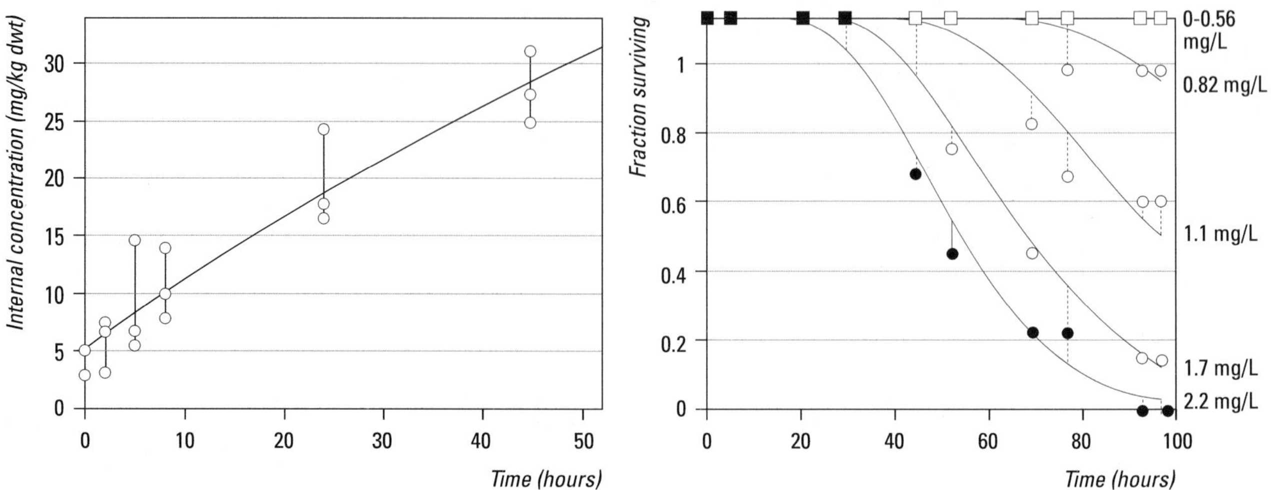

Figure 7.13. Combining body residue analysis (left panel, at 0.1 mg/L) with time-to-death mortality data for cadmium (right panel, 0-2.2 mg/L) in a dynamic energy budget model for *Daphnia magna*. From Jager et al. [55]. With kind permission of Springer Science and Business Media.

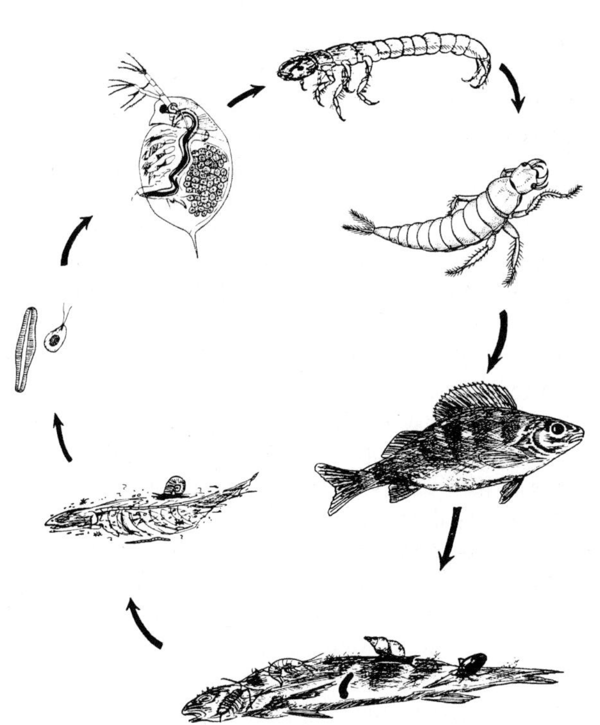

Figure 7.14. Simplified aquatic food chain consisting of primary producers (various species of algae), herbivores (daphnids), primary carnivores (caddisfly larvae), secondary carnivores (beetle larvae), tertiary carnivores (fish) and decomposers, i.e., detritus-feeding animals such as snails and amphipods and bacteria and fungi (species not drawn to the same scale).

chronic endpoint. Furthermore, short-term or episodic exposure may lead to chronic effects, for example the occurrence of neurotoxic effects in organisms after shortterm exposure to certain organophosphate insecticides.

Acute toxicity tests have two general applications in environmental risk analysis. One application is in determining acute toxicity. The objective here is to provide a basic set of data for three trophic levels (algae, daphnids, fish) which can be used in conjunction with a large assessment factor to estimate PNECs of a specific chemical. For risk assessment, the concentration-response curves can also be used to determine the biological response of the species at a given environmental concentration.

The second type of application is toxicological screening. The purpose of screening is to determine whether the chemical or solution being tested is biologically active with respect to the endpoint being measured. Essentially, screening tests provide "yes or no" answers, i.e., a chemical is toxic or non-toxic, mutagenic or non-mutagenic, and so on, at the concentration tested, usually a regulatory threshold.

Tests with animals

Freshwater invertebrate and fish species commonly used for acute toxicity studies [47,56,57] are chosen to represent different functional groups such as herbivores, carnivores or decomposers (Figure 7.14). The endpoints measured in these studies can include any response that an organism or population may exhibit as a

Table 7.10. Characteristics of the acute immobilization test with daphnids [59].

Test species	*Daphnia magna, Daphnia pulex* or any other suitable *Daphnia* species less than 24 h old
Test duration	usually 48 h
Test system	static, semi-static or renewal test in tubes or beakers, with at least 2 ml of test solution per animal
Feeding	no food
Light/temperature	light-dark cycle of 16:8 h at constant temperature between 18-22°C
Endpoints	immobility
Parameter	EC50

Table 7.11. Characteristics of the fish acute toxicity test [60].

Test species	juveniles of various fish species, e.g. guppy *(Poecilia reticulata)*, zebrafish *(Brachydanio rerio)*, fathead minnow *(Pimephales promelas)* or rainbow trout *(Oncorhynchus mykiss)*
Test duration	usually 96 h
Test system	static and renewal test (maximum loading 1.0 g fish/L) and flow-through (higher loading can be acceptable)
Feeding	no food
Light/temperature	light-dark cycle of 16:8 h at constant temperature between 20-25°C (warm water fish species) and 13-17°C (cold water fish species)
Endpoints	survival
Parameter	LC50

result of chemical exposure. However, the end-point most commonly used in acute toxicity studies using invertebrates, such as daphnids and fish, is death (LC50) or immobilization (EC50). These end-points are easily determined, have obvious biological and ecological significance and are amenable to concentration-effect analysis (see also Section 6.5.3). The characteristics of routine acute toxicity tests with daphnids and fish are presented in Tables 7.10 and 7.11 respectively.

The general set up of short-term toxicity tests usually consists of five test concentrations, a control, a solvent control if needed, and 10 to 20 organisms for each concentration or control. Although short-term toxicity testing is generally seen as a simple routine matter, it is relatively complicated (Table 7.12). Therefore, highly standardized test protocols have been developed by international organizations. Recommended procedures are provided by e.g., the Organization for Economic Co-operation and Development (OECD; see Chapter 16), the International Organization for Standardization (ISO), the US Environmental Protection Agency (USEPA), the American Society for Testing and Materials (ASTM), and Environment Canada. Through the harmonization of test guidelines [58-60], the OECD plays an important role in the international arena of chemicals control. Most industrialized countries adopt OECD test guidelines once they are officially approved.

Tests with plants

The development of testing procedures to study the toxic effects of chemicals on aquatic plants has centred on unicellular algae and duckweed [46, 61]. A short summary is given in Table 7.13.

Because of their short generation times, phytotoxic effects can be measured over several generations in a relatively short period. The parameters generally measured in phytotoxicity studies are photosynthesis and population growth. Effects on photosynthesis can be measured by a number of well-established methods including O_2 production, $^{14}CO_2$-uptake, photosynthetic pigment concentration, ATP production, and cell counts [61-63].

Effects assessment using growing populations requires repeated counting of cells or fronds (leaf-like part of a plant), or determination of biomass over a period of time, several times the generation period of the organisms. In the data analysis the main emphasis is on the inhibitory effects on the population growth rate [63, 64] (see Section 7.3.5 for an explanation of basic population dynamics). The growth rate in the exponential

Table 7.12. Important aspects of a test protocol with fish.

Biological aspects
Ecology of test species
Acclimation
Treatment of unhealthy fish
Age at testing
Feeding
– type of food
– amount of food
– frequency of feeding
Loading (density)
Sample size
Randomization
Duration of test
Control mortality
Physical aspects
Temperature
Light/dark regime
Holding facilities
Materials
Shape/volume of test vessel
Chemical aspects
Source of water
– dissolved oxygen (DO)
– pH
– hardness
– particulate matter
– complexants
– impurities
Carrier solvent
– type
– concentration
Test compound
– solubility
– stability
– volatility
– BOD
– bioaccumulation potential
– chemical detection method
Exposure conditions
– static/renewal/flow-through
– replacement time
– stability of DO and pH
– test concentration: nominal or measured
– test concentration: stability over time

phase of growth (Figure 7.15) for each toxicant concentration is calculated with linear regression:

$$r = [ln\,(N_2 / N_1)]\ /[t_2 - t_1] \tag{7.2}$$

where

r = the exponential growth rate (1/d) from time t_1 to t_2
N_2 = cell number in the exponential growth phase at time t_2
N_1 = cell number in the exponential growth phase at time t_1
t_2, t_1 = time (d)

The effects of a chemical on inhibition of the average specific growth rate can be calculated as follows:

$$\%\ \text{inhibition} = \frac{r_0 - r_t}{r_0} \cdot 100 \tag{7.3}$$

where r_0 is the growth rate in the control and r_c is the growth rate in the presence of the toxicant at concentration c. For the biomass increase (yield) in the same period, a similar equation is used. As an alternative, growth inhibition can also be modelled with a threshold model [65,66]. This model assumes no effects at concentrations below a certain threshold, and the hazard is modelled as proportional to the concentration above the threshold. An example of this method is given for *Daphnia magna* in Figure 7.21 in Section 7.3.5.

Microbial tests

Bacterial processes are extremely important with regard to nutrient cycling, secondary productivity, biodegradation and metabolism, as are the ecological consequences of toxicity-induced stimulation or inhibition of these processes (Figure 7.16).

Some tests are performed with isolated bacterial and fungal species to study the bactericidal or fungicidal effects (concentrations causing mortality) and the bacteriostatic or fungistatic properties (concentrations preventing growth and proliferation of cells without killing them). Species often used are *Vibrio fischeri*, a saltwater species used in the Microtox test [56], *Pseudomonas putida*, *P. fluorescens* and the ciliate *Tetrahymena sp*.

In microbial ecotoxicology, attention is primarily focused on functional approaches, i.e., the effects on microbial processes. Functional tests include the study of toxic effects on the carbon cycle, especially the effects of heterotrophic bacteria on mineralization (the biological process of transforming organic matter by

Table 7.13. Characteristics of the microalgae or cyanobacteria growth inhibition test [46].

Test species	unicellular green algae, *(Pseudokirchneriella subcapitata* or *Desmodesmus subspicatus)*, diatoms *(Navicula pelliculosa)* or cyanobacteria *(Anabaena flos-aque* or *Synechococcus leopoliensis)*.
Test duration	72 h (short-term chronic test)
Test system	static test in Erlenmeyer flasks with 100 ml test solution on a rotary or oscillatory shaker
Medium	synthetic nutrient-enriched medium
Light/temperature	constant light at constant temperature between 21-24°C
Endpoints	inhibition of population growth/biomass and yield
Parameter	EC50 (50% inhibition of growth or yield), EC10/EC20, NOEC

complete oxidation into carbon dioxide, water and other inorganic compounds). Other tests focus on the nitrogen cycle: nitrogen fixation (the process of fixing molecular nitrogen into organic matter), ammonification (the release of ammonium from organic matter), nitrification (the conversion of ammonia to nitrite and nitrate), and denitrification (the anaerobic conversion of nitrate to atmospheric nitrogen and N_2O). Specific enzyme activities can be measured as well, but in practice they are of little value for monitoring adverse effects.

In addition to their function in ecosystems, microbial activity in sewage treatment plants (STP) is essential. To protect the microbial activity of STPs, microbial toxicity tests can be used to derive a no effect level for microorganisms. The current presence of pharmaceuticals for veterinary and human use, including antibiotics, raises concerns about microbial inhibition in STPs [67,68]. Routine tests with bacterial strains or inocula are generally carried out to study the inhibition of respiration, nitrification, growth or changes in bioluminescence [69].

An example of a routine respirometric test is given in Table 7.14. The EC50 values in such tests can be derived by non-linear regression [70]. In some cases, specific biodegradability tests (OECD guidelines 301-302) can also be used to derive NOECs for microbial toxicity. A more detailed discussion of biodegradation and how it can be predicted is given in Chapters 3 and 9 (Sections 3.5 and 9.4.3).

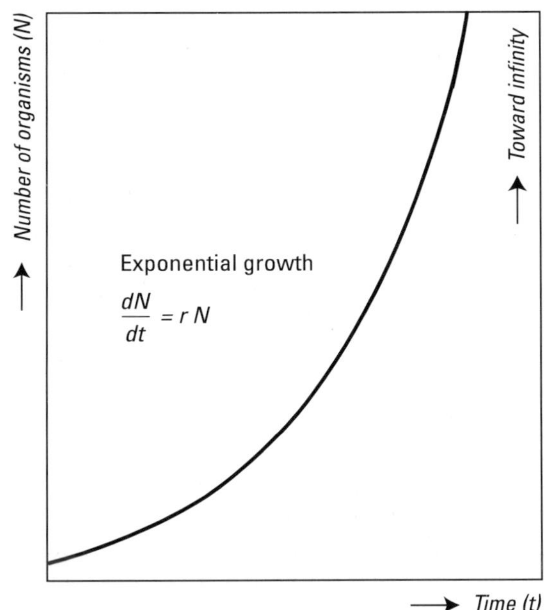

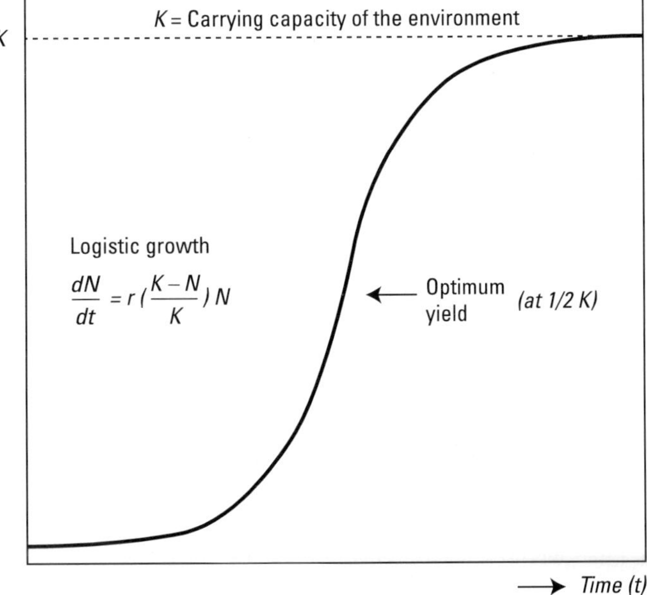

Figure 7.15. Basic forms of population growth: exponential growth (left) and logistic growth (right).

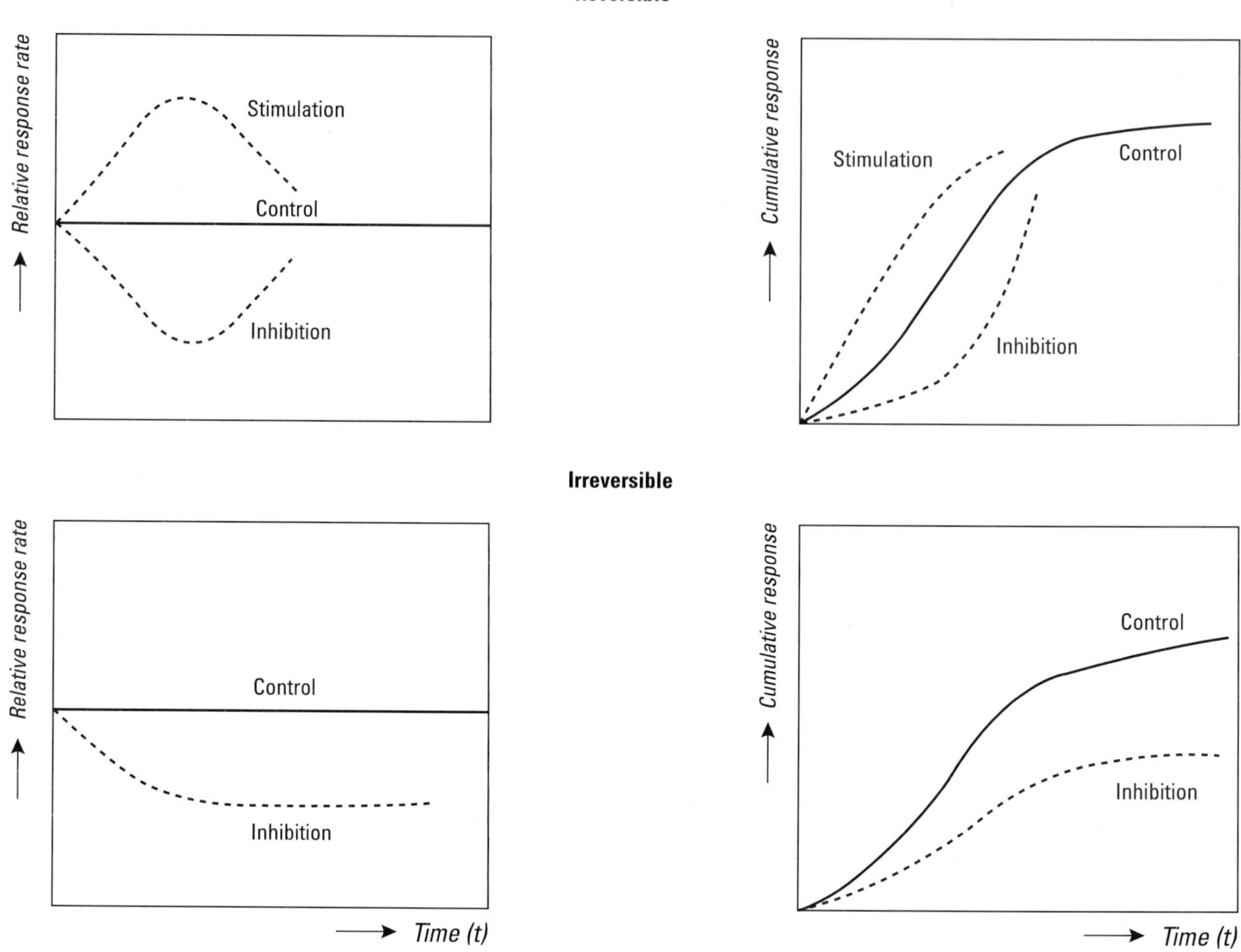

Figure 7.16. Types of microbial responses.

7.3.4 Long-term toxicity

Introduction

The aim of chronic toxicity testing is to determine whether prolonged exposure to chemicals will have significant adverse effects on ecosystems. For the aquatic environment, this is accomplished by estimating chronic toxicity threshold concentrations for a number of selected species inhabiting the ecosystem. From these data, the chronic threshold concentration for the aquatic ecosystem or PNEC can be predicted with fixed or calculated extrapolation or uncertainty factors (Section 7.10).

Apart from survival, chronic toxicity studies are based on end-points like individual growth (body length and body weight), abnormal development (teratogenicity), hatching time, hatchability, reproduction (total number of young, brood frequency, etc.) and behavioural aspects, etc. These data are then subjected to concentration-effect modelling or hypothesis testing to derive the NOEC.

Three categories of tests are commonly used to predict the chronic effects of toxic chemicals on aquatic organisms (Table 7.15). Data from these categories of tests can be used to estimate the PNECs (Section 7.10).

Partial and full life-cycle tests

In life-cycle tests, groups of test organisms are exposed to a series of concentrations of the test chemical over one or more generations. In fact, most algal toxicity studies are life-cycle tests, but this term is generally used in the context of fish and invertebrate studies. Life-cycle tests begin with the eggs, larvae or juveniles and continue until the test organisms have (or should have) reproduced. The tests can continue through several generations, if desired. Chemical concentrations range from those

Table 7.14. Characteristics of the activated sludge respiration inhibition test [69].

Test species	inoculum from aerobic sewage sludge
Test duration	3 h
Test system	static test in a BOD bottle
Medium	synthetic medium with inorganic nutrients and peptone, meat extract and urea
Light/temperature	dark at 20°C
Endpoints	inhibition of the respiration rate (oxygen consumption of micro-organisms expressed as mg O_2/L·h)
Parameter	EC50 (50% inhibition of the respiration rate)

Table 7.15. Types of chronic toxicity studies.

Life-cycle toxicity tests measure the effects of chronic exposure to a chemical on reproduction, growth, survival, and other parameters over one or more generations of a population of test organisms
Sensitive life stage tests measure the effects of chronic exposure on survival and growth of the toxicologically most sensitive life stages of a species, for example, eggs and larvae of fish
Sublethal chronic toxicity tests measure the effects of chemicals on various biochemical or physiological functions or on histology of individual organisms

with significant adverse effects on survival, growth and reproduction to at least one which has no significant effect on these parameters, compared with the controls. The species that can be used in life-cycle toxicity tests are limited to those which can complete their life cycles under laboratory conditions. Rand [47] has listed those animal species most commonly used in life-cycle toxicity tests.

Due to the cost and length of time required for full life-cycle tests for some species, certain routine reproduction tests do not cover the entire life cycle but are partial life-cycle tests. Only the most important partial life-cycle tests with invertebrates will be discussed here, together with some basic principles of population dynamics. For a more extensive review of invertebrate studies see Persoone and Janssen in Calow [56].

Tests with daphnids

The best known partial life-cycle test is the chronic reproduction test with daphnids (Figure 7.17). According to Persoone and Janssen in Calow [56] there are five reasons for selecting this species:

1. They are broadly distributed in freshwater bodies and are found in a wide range of habitats.
2. They form an important link in many aquatic food chains (they graze on primary producers and are food for many fish species).
3. They have a relatively short life cycle and are relatively easy to culture in the laboratory.
4. They are sensitive to a broad range of aquatic contaminants.
5. Their small size means that only small volumes of test water and little bench space are required.

The entire lifespan of *D. magna* or *D. pulex* takes approximately 60 days [72]. The life-cycle test with *D. magna* or *D. pulex* takes 21 days (Table 7.16). After approximately 10 days the first brood will appear and subsequent broods will normally be produced at intervals of 2 to 3 days (Figure 7.18). The young are separated from the parents and the total number of offspring is treated as a reproduction parameter.

Another widely accepted chronic reproduction test, especially in the US, is the test with *Ceriodaphnia dubia*. This 7-day bioassay was developed by Mount and Norberg [73]. It is a cost-effective bioassay, and is frequently used as an invertebrate bioassay in the USA [74]. In the *Ceriodaphnia* reproduction test, three broods are normally produced on days 3, 5 and 7. The experimental design and the statistical analysis of the data are comparable with the *Daphnia* test, but the test has several advantages, although there is no OECD guideline for *Ceriodaphnia*. *Ceriodaphnia* are distributed widely throughout Europe, Asia and North America, are easy to culture and the exposure period is much short-

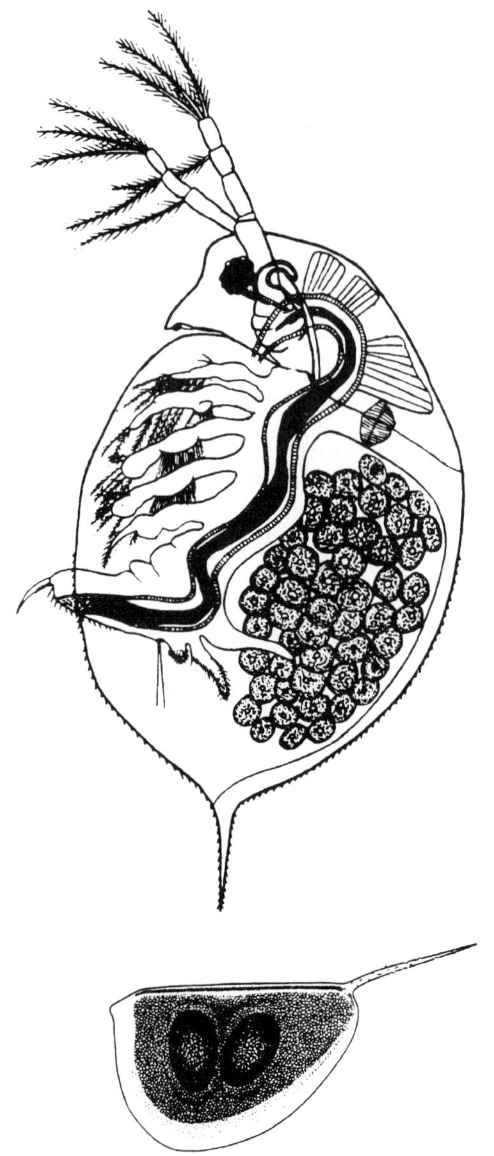

Figure 7.17. *Daphnia* with normal eggs in the brood pouch and ephippia (winter eggs).

er (1 week instead of 2 to 3 weeks). Acute and chronic sensitivity to a broad array of substances was found to be comparable to that of *Daphnia* sp. [75].

Tests with sensitive life stages

Considerable time and expense is involved in conducting ELS toxicity tests, especially for fish. Methods have been developed for utilizing tests with the most sensitive life stages to predict chronic toxicity threshold concentrations. Figure 7.19 gives a representation of the various life stages of an oviparous fish: salmon. After gametogenesis (the production of sperm and egg cells) fertilization takes place. This is accompanied by swelling of the egg through water uptake. The egg membrane becomes relatively impermeable. Within the egg membrane the fertilized egg cell divides and differentiates through a number of different embryonic stages (embryogenesis) until the eggs hatch. Once the eggs are hatched an alevin with a yolk sac or a yolksac larva (also known as eleuthroembryo) appears. The alevin feeds itself using its internal food source: the yolk deposited in the yolk sac. During further development this yolk is resorbed and the so-called swim-up fry start to catch and ingest food, progressing to the juvenile and finally adult stages. At this reproductive stage maturation of the ovary and testes occurs, producing mature egg and sperm cells (gametogenesis). Many other fish species develop in a similar manner, although some show marked differences, e.g., viviparous fish, such as guppies.

According to McKim [76] and Van Leeuwen et al. [77] it is generally the early life stages of fish which are most sensitive to chemical toxicants. This susceptibility results from a potential for exposure and responsiveness: the intrinsic susceptibility or sensitivity, in its strictest sense. McKim [76] showed that estimates of chronic toxicity threshold concentrations calculated from ELS tests were not significantly different from those calculated from entire life-cycle toxicity tests. An OECD test guideline for fish ELS studies is available [78].

A standard fish ELS test (FELS, Table 7.17) starts with freshly fertilized eggs, which implies that FELS tests exclude any potential effect of a chemical on the process of gametogenesis, or on the process of fertilization (Figure 7.19). FELS tests are terminated after swim-up fry have been fed for a given period of time. The length of the feeding period is also species-dependent.

The great advantage of FELS tests is that they save time and money compared with full or partial life-cycle studies with fish. Thus, estimates of chronic toxicity thresholds can be made for more chemicals and for a wider variety of species from different habitats and trophic levels than are possible with life-cycle toxicity tests. However, as several life stages are covered in the test it is still a sensitive assay, and may be preferable to the juvenile fish growth test (OECD guideline 215) which determines effects on juvenile fish growth during a 28-day exposure period. Although compared with full life-cycle studies, embryolarval tests reduce the time required to produce information on the toxicity of chemicals, they remain laborious. To further reduce the exposure time, short-cut methods are needed. This is why several procedural variations are used [58,79]. The short-

Table 7.16. Characteristics of the *Daphnia* reproduction test [71].

Test species	neonates of *Daphnia magna*
Test duration	21 d
Test system	semi-static (renewal at least three times a week) or flow-through, with 50-100 ml test solution per animal
Feeding	green unicellular algae obtained from a laboratory culture, ration between 0.1 and 0.2 mg C/(daphnid.day)
Light/temperature	light/dark cycle of 16:8 h at constant temperature of 18-22°C
Endpoints	reproduction (total number of offspring), parent survival and time to production of first brood
Parameter	LC50, EC50 and NOEC

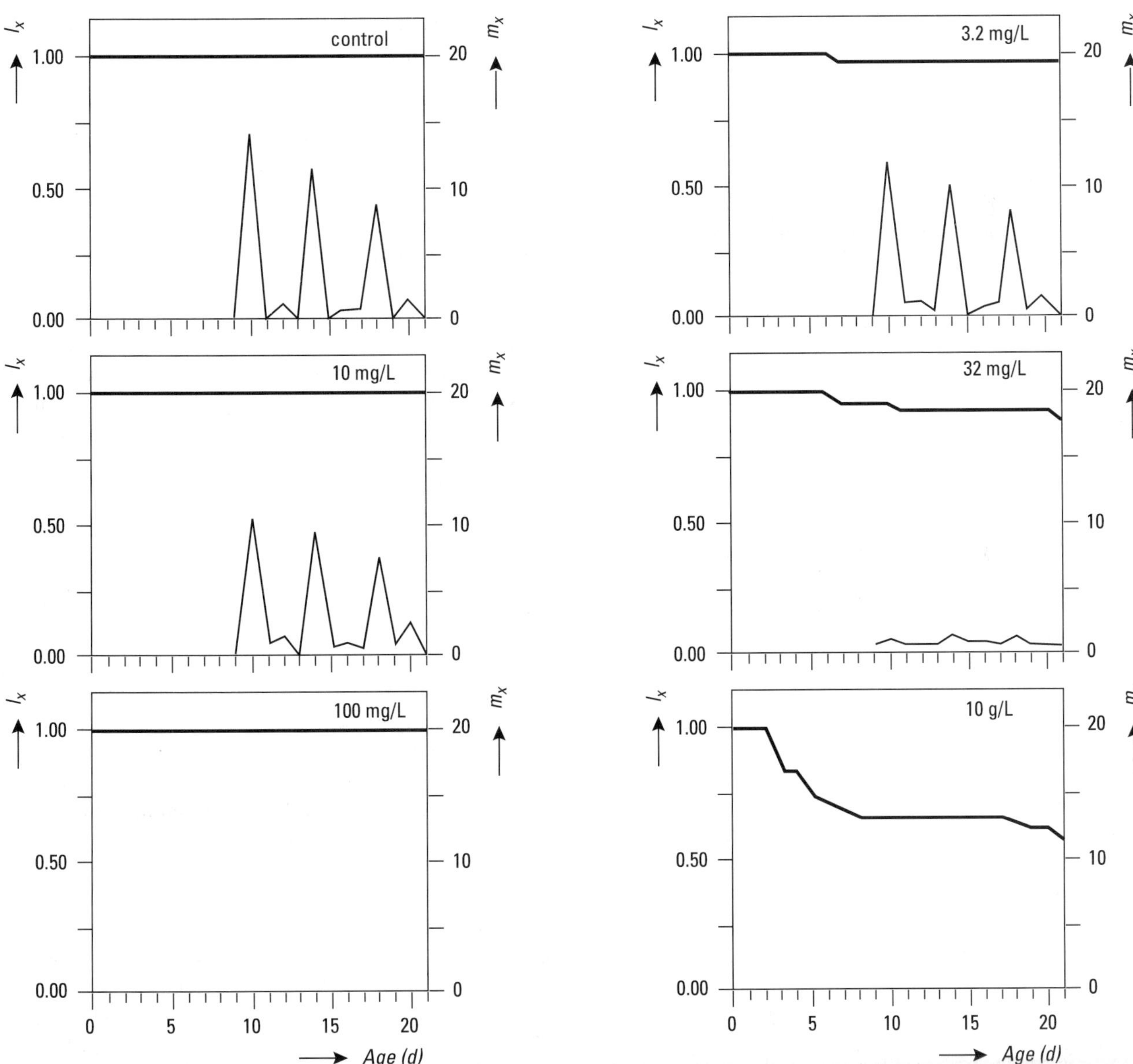

Figure 7.18. Life table of *Daphnia magna* at various concentrations of bromide: survivorship curve l_x (—) and fertility curve m_x, (–). From Van Leeuwen, Rijkeboer and Niebeek [66]. With kind permission of Springer Science and Business Media.

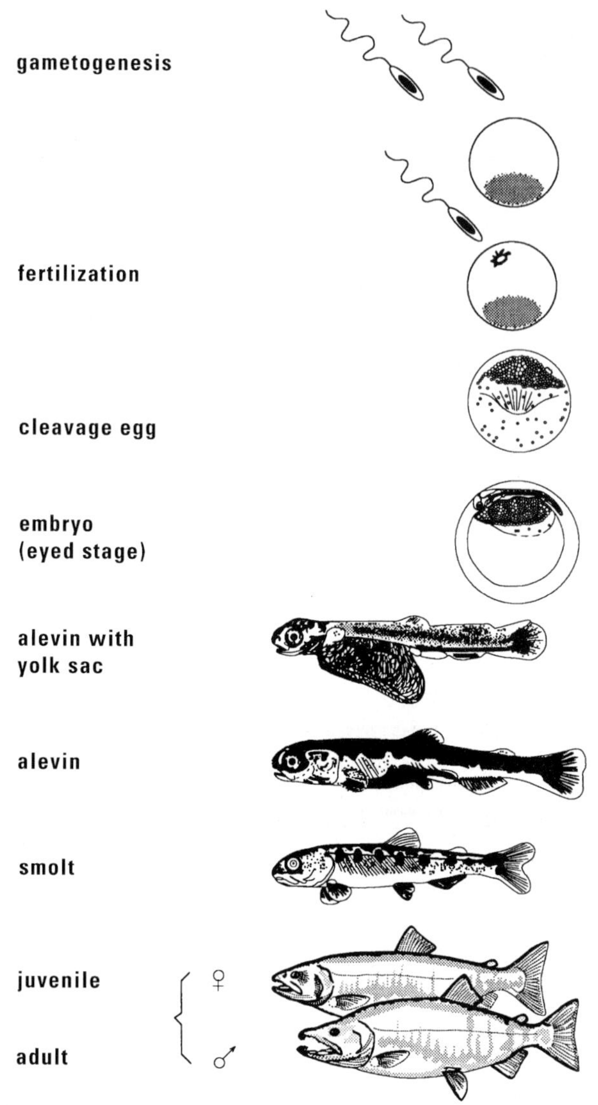

Figure 7.19. Life cycle of a salmonid fish.

term toxicity test on embryo and sacfry stages (OECD guideline 212) can be used as an alternative to a full FELS test. It can be considered a chronic test because it covers sensitive life stages from egg to sacfry. However, it is expected to be less sensitive than the full FELS test.

7.3.5 Population dynamics

Population dynamics are relevant to the study of toxic effects in ecosystems and their recovery after exposure. Although the focus of ERA is on the protection of populations, relatively little attention has been devoted to population dynamics. The effects of chemicals on population dynamics are increasingly integrated in strategies for higher-tier toxicity testing [80,81]. The same principles of population dynamics apply to both sediment-dwelling organisms and terrestrial populations (Sections 7.4 and 7.5) and are discussed in this section. Some examples will be given to illustrate the current lack of ecological realism in single-species toxicity tests.

The most commonly performed population toxicity experiments are tests with algae and *D. magna*, although soil invertebrates such as *Orchesella cincta* or *Folsomia candida* or the nematode *Plectus acuminatus* may also be used [82-86]. These experiments focus on effects on either the exponential or logistic growth of populations. When a population is subject to a constant schedule of birth and death rates it will gradually approach a fixed or *stable-age distribution*, whatever the initial age distribution may have been, and will then maintain this stable age distribution indefinitely (Lotka theory). When the population has reached this stable age distribution, it will increase in numbers according to the simplest model for population growth: *exponential growth*. The exponential growth model has a constant per capita growth rate *(r)*, which is independent of the population density (Figure 7.15), resulting in unbounded exponential growth. For some periods of time, exponential growth can be observed for fast growing micro-organisms, algae and daphnids (Figure 7.20). Unbounded growth is not found in nature for prolonged periods of time. A simple model which captures the essential features of an environment with finite resources is *logistic growth* (Figure 7.15). Population growth decreases as population density increases. Here the effective per capita growth rate has the density dependent form $r(1\text{-}N/K)$: this is positive if $N<K$, negative if $N>K$, and thus leads to a generally stable equilibrium value at $N=K$. K may be thought of as the *carrying capacity* of the environment, as determined by food, space, predators, or other things; r is the *intrinsic growth rate*, free from environmental constraints.

What is the relationship between age and population growth? We know that both birth and death rates vary with age. In fact, there are four basic concepts. First, every population has an age distribution that indicates the proportion of the population in various age classes. Second, every population has a growth rate. Third, in every population there is a regime of *age-specific mortality*, often depicted by the survivorship curve, which describes the probability of surviving from age 0 to age x or beyond. The fourth concept is that of *age-specific fertility*, often represented by the fertility, fecundity or maternity function. The study of population

Table 7.17. Characteristics of the fish early-life-stage test [78].

Test species	zebrafish *(Brachydanio rerio)*, fathead minnow *(Pimephales promelas)*, rainbow trout *(Oncorhynchus mykiss)* and a variety of other species
Test duration	all tests begin with freshly fertilized eggs and may end at the early fry stage, but the test duration depends on the species and the temperature of the water (normally 28 d for fathead minnow and zebrafish and 60-90 d for rainbow trout)
Test system	renewal or flow-through test
Feeding	with commercial fish food starting at the transition of the yolk sac larval stage and the swim-up fry stage
Light/temperature	species-dependent
Endpoints	survival, growth (length and weight) and developmental (teratogenic) effects, time till hatching and end of hatching, yolk resorption, histopathology and behavioural effects may be included as well
Parameter	LC50, EC50, LOEC and NOEC

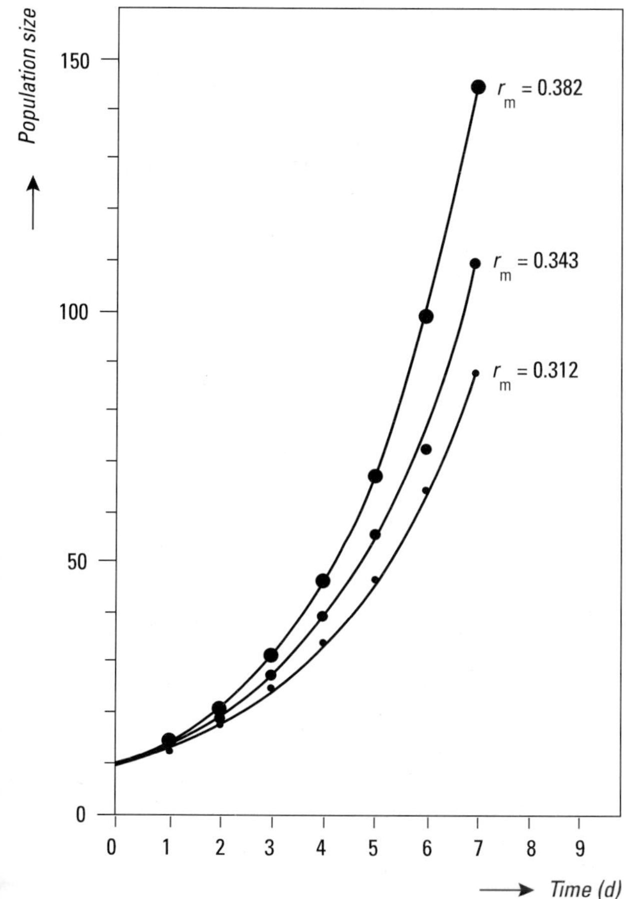

Figure 7.20. Exponential growth of *Daphnia magna* populations with different intrinsic rates of increase, derived from Box 7.2.

growth in relation to age structure is called *demography*. Figure 7.18 shows an example of a demographic study with survivorship and fertility curves for *D. magna*.

Population dynamics concepts can also be incorporated in tests with cohorts (isolated generations), i.e., tests in which a fixed number of individuals are exposed from the juvenile through the adult period. In cohort studies separate measures of age-specific survival and fecundity are combined in a life-table and used to estimate the intrinsic rate of natural increase. The intrinsic rate of increase *(r)*, the growth rate of an exponentially increasing population, can be calculated with the Euler-Lotka equation:

$$\sum_{x=0}^{\infty} l_x m_x e^{-rx} = 1 \qquad (7.4)$$

where l_x is the probability of surviving to age x, m_x is the age-specific fecundity (number of female offspring per surviving adult at age x) and x is time expressed in days.

In Box 7.2, three hypothetical life-table experiments (A, B and C) are shown. The experiments which started with newborn (< 24 h) daphnids show no mortality (l_x remains 1 in all three experiments). The total number of young produced in four broods in each experiment over a period of 21 days was 70. In standard *Daphnia* toxicity protocols, effects on reproduction would not have been detected, because the total number of newborn daphnids is the only measure of toxic effects on reproduction. But if we use basic population dynamics, dramatic adverse ecological effects can be demonstrated. If the intrinsic rate of increase is calculated by successive approximation using Equation 7.4, three different values for r are obtained (Box 7.2.). If these values are substituted in the equation for exponential growth (Figure 7.15), it appears

Box 7.2. Demonstration of the consequences of a delay in reproduction on the intrinsic rate of increase *(r)* in three hypothetical 21-d life-table studies (A, B and C) with *Daphnia magna*.
Note that there is no parental mortality (l_x remains 1) for all three experiments.

Time (d)	l_x	m_x (A)	m_x (B)	m_x (C)
1	1	0	0	0
2	1	0	0	0
3	1	0	0	0
4	1	0	0	0
5	1	0	0	0
6	1	0	0	0
7	1	10	0	0
8	1	0	10	0
9	1	0	0	10
10	1	0	0	0
11	1	15	0	0
12	1	0	15	0
13	1	0	0	15
14	1	0	0	0
15	1	20	0	0
16	1	0	20	0
17	1	0	0	20
18	1	0	0	0
19	1	25	0	0
20	1	0	25	0
21	1	0	0	25
T[a]		70	70	70
r[b]		0.382	0.343	0.312

[a] T is total number of young after 21d.

[b] r is calculated by successive approximation from Equation 7.4. For the examples A, B, and C the following set of equations is obtained:

Life-table study A: $10 \times e^{-7r} + 15 \times e^{-11r} + 20 \times e^{-15r} + 25 \times e^{-19r} = 1$ (r = 0.382)

Life-table study B: $10 \times e^{-8r} + 15 \times e^{-12r} + 20 \times e^{-16r} + 25 \times e^{-20r} = 1$ (r = 0.343)

Life-table study C: $10 \times e^{-9r} + 15 \times e^{-13r} + 20 \times e^{-17r} + 25 \times e^{-21r} = 1$ (r = 0.312)

that population growth is greatly affected (Figure 7.20). The special importance of toxicity-induced effects on the age of first reproduction is often overlooked [87]. Apart from the age of first reproduction, there are several other aspects which greatly influence population growth (Table 7.18). Some of these parameters can be measured in the laboratory, others cannot.

Experiments with populations may provide a good alternative to life-cycle studies, especially when the population-level response depends on the sensitivity of different life-cycle stages or variables. The integration of effects on different life-history traits can only be studied in this way. Population toxicity studies begin with small, exponentially growing populations. The underlying assumption for projecting future growth, with either the exponential or the logistic growth model, is that the population has a stable age structure. At low population densities, growth will proceed exponentially and the stable-age structure can be calculated with:

$$c_x = (l_x e^{-rx}) \Big/ \left(\sum_{x=0}^{\infty} l_x e^{-rx} \right) \tag{7.5}$$

Table 7.18. Factors affecting population growth.

Age at first reproduction
Brood size
Brood frequency
Length of reproductive period
Condition of neonates
Emigration and immigration
Predation and competition e.g. for food and space

where c_x is the proportion of the total population in the xth age class. If r is 0, the stable age distribution will have exactly the same shape as the survivorship curve. Furthermore, the equation shows that as r increases, the younger age classes become an increasing proportion of the population. An example of a population toxicity experiment is shown in Figure 7.21. The test began with exponentially growing populations of 20 daphnids, composed of different ages. The stable-age distribution was calculated with Equation 7.5 (where $r = 0.3$ and l_{21}

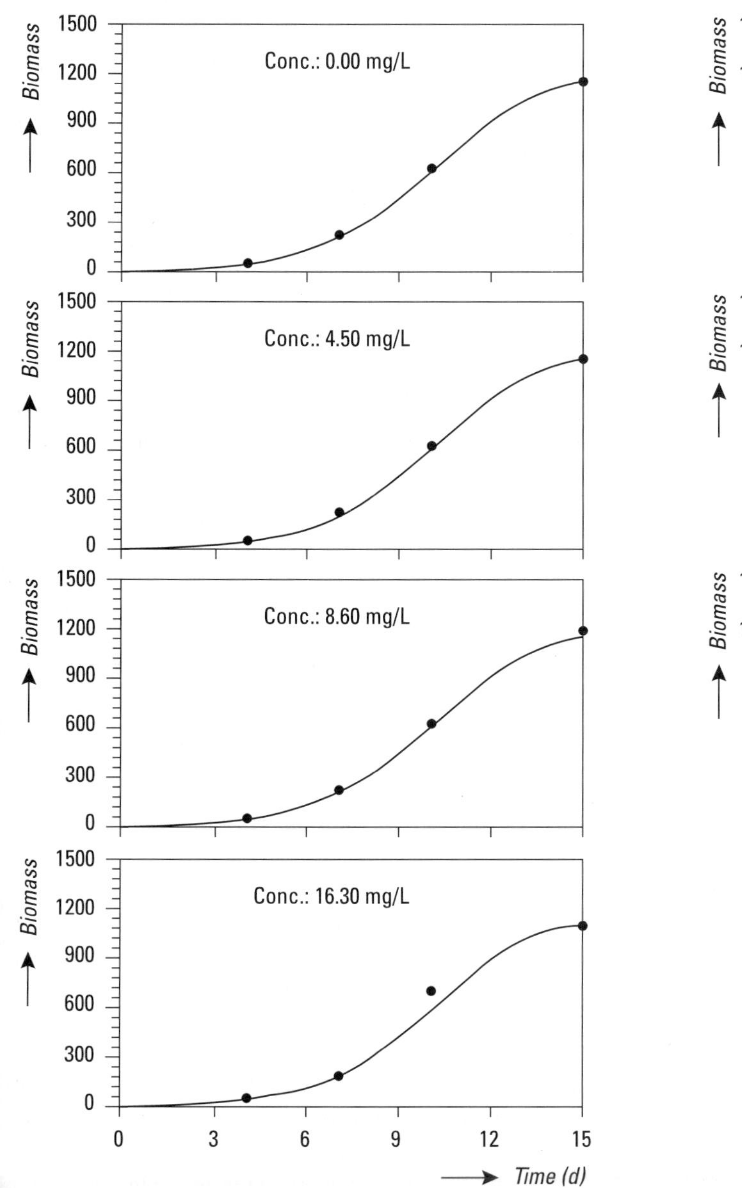

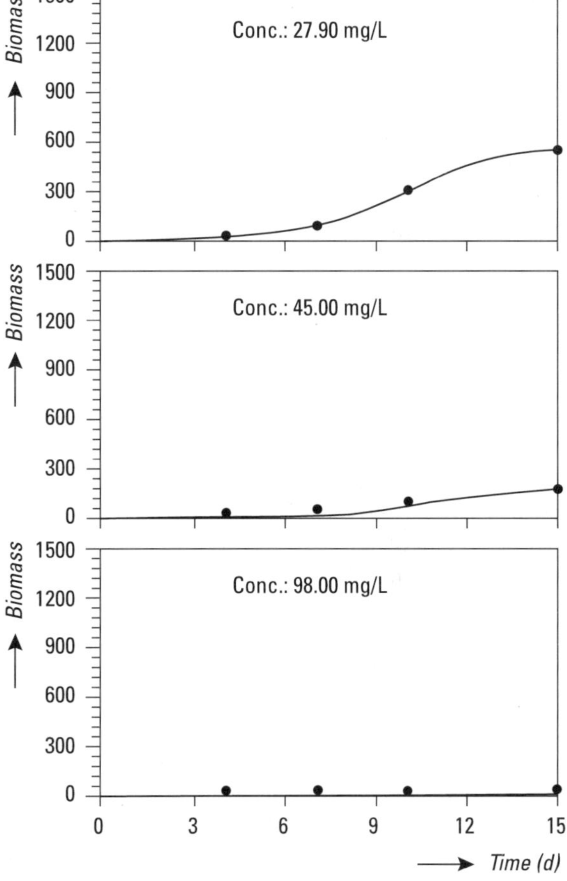

Figure 7.21. The effects of bromide on the logistic growth of *Daphnia magna* populations [66]. Biomass represents the number of daphnids per test container, circles represent the observed and lines the expected values based on model calculations after Kooijman et al. [88]. With permission. Copyright. Elsevier.

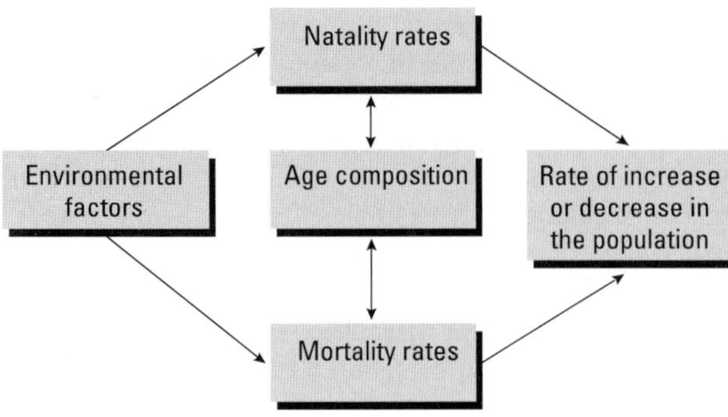

Figure 7.22. The relationship between environmental factors, age distribution and population growth.

= 1, Box 7.2, study C), giving the following stable-age distribution: 5, 4, 3, 2, 2, 1, 1, 1, and 1 daphnid(s) aged 0 to 1, 1 to 2, 2 to 3, 3 to 4, 4 to 5, 5 to 6, 6 to 7, 7 to 8, and 8 to 9 day(s), respectively.

Population growth is often far from logistic when tests are prolonged after populations have attained their numerical maxima. This may be ascribed to the absence of an instantaneous reaction to changes in population density, i.e., time lags are likely to occur which causes populations repeatedly to "overshoot" and then "undershoot" their equilibrium densities. In field situations environmental conditions are never constant and logistic growth can only be observed over short periods of time. Furthermore, in natural populations the age structure is almost constantly changing, because populations do not increase for long in an unlimited fashion. These relationships are shown in Figure 7.22.

Toxic effects in ecosystems often include a change in species composition [88-90]. The problem of life-history strategy may be viewed as that of the optimum allocation of an organism's energy to growth, maintenance and reproduction [38]. In evolutionary terms an organism will devote energy to growth and maintenance only if this will increase its reproductive contribution to future generations. Growth is important in many species because fecundity increases with size, and competition for territories may favour larger individuals. When explaining ecosystem changes, many authors refer to life-history strategies such as the *r* and *K*-species theory. If *K*-species are more sensitive than *r*-species then in contaminated ecosystems *r*-species will replace *K*-species. The *r* and *K*-theory originates from MacArthur and has been formulated most clearly by Roughgarden [91]:

1. In a population that is repeatedly reduced to a low density by some exogenous factor, a genotype with a high intrinsic rate of population growth *(r)* will prevail.
2. In a population constantly occurring in a state of high density with strong competition, a genotype with a high equilibrium density *(K)* will prevail, even if it has a low intrinsic growth rate.

There are examples showing that the theory is correct, but as is often the case in ecotoxicology, there are also examples showing that the theory cannot be generalized. Interactions with other species and differences in the sensitivity of species play an important role in the dynamics of species in a community experiencing toxic stress [89,92]. This shows that detailed case studies are needed to disentangle the various biological and toxicological factors that influence the response of interacting populations to chemical stress. To improve the predictability of the effect of substances on ecosystems, these studies should go hand-in-hand with theory and model development [92].

7.3.6 Multi-species studies

Ideally, PNECs should be determined through studies of exposed ecosystems that are representative of the ecosystems to be protected. However, full scale field tests are expensive and complex. Intermediate methods between laboratory and field tests may contribute to more effective and cost-effective higher-tier risk assessment. Uncertainty about the ecological effects of a chemical can be addressed with indoor microcosm experiments, outdoor micro/mesocosm tests, or a combination of these. Multi-species tests (MS tests) aim to determine fate processes and how these affect bioavailability to different species, and species interactions. They are ecologically more relevant than single-species tests (SS tests), but may be harder to interpret due to system-specific conditions that may not be easy to extrapolate to different conditions. MS tests require that the natural conditions of ecosystems are well documented and that many basic ecological, physicochemical and toxicological data are available (Chapter 1). Unfortunately, in most cases, these data are not available (see Figure 12.1 in Chapter 12). This lack of information hampers the interpretation of observations in MS tests. In many cases PNECs are based on single-species toxicity data. SS tests have some major advantages:

- They are rapid, easy to conduct and not too expensive.
- They can be standardized.

Table 7.19. Some shortcomings of single-species tests.

1. They utilize genetically homogeneous laboratory stock test populations
2. They examine only the responses of individuals, which are averaged to give a mean response for the test species instead of population responses
3. They use species of unknown relative sensitivities and species that may not be indigenous to the receiving ecosystem
4. They are mostly conducted under experimental conditions that are not similar to natural habitats
5. Distribution and degradation processes are often ignored
6. Indirect toxic effects resulting from various ecological interactions are not taken into account
7. Toxic effects on basic ecological processes are often not studied
8. They do not consider recovery rates of populations or ecosystems
9. Cumulative effects of multiple stresses coupled with varying chemical/physical properties are often not studied

- They are relatively easy to replicate.
- Their interreplicate and intertest variability is usually lower compared to micro/mesocosms.
- They are valuable screening tools.
- They are an appropriate way to determine toxicological effects on survival, growth, reproductive success, behaviour and a variety of other individual characteristics.

SS tests also have some serious limitations (Table 7.19) which impair the proper scientific assessment of chemical impacts on ecosystems [93,94].

Comprehensive, system level tests, or MS tests, are complex by their nature. There are a wide variety of potential measurements that can be made, which are generally subdivided into fate, functional parameters and structural parameters. Knowledge of the physicochemical and toxicological properties and fate influencing the exposure conditions is essential. In order to understand the potential routes of exposure of various species, it is useful to obtain a mass balance of the chemical under investigation. Thus, measurements of the chemical in soil, water, sediment and biota should be made. Measures of critical by-products, degradation products or metabolites of the chemical should be included if they are expected to be toxic, or if their measurement is required to complete the mass balance.

The relevant endpoints for effects assessment are structure and function, and may include effects on genetic variability, or the probability of extinction of certain species. Ecosystem functions, such as respiration or primary production (Figure 7.23) are often regarded as less sensitive than structural parameters (species composition), because the species responsible for an ecosystem function may be replaced by less sensitive species capable of maintaining the same functional processes (functional redundancy). However, some herbicides can have a pronounced direct effect on oxygen production in an ecosystem, which is noticeable at lower concentrations than the ensuing changes in species composition [96]. This indicates that both functional and structural aspects need to be considered in MS tests.

Measurements at the species level are often focused on representatives of various trophic levels, functional groups, or otherwise important species. At community level, measurements often include species composition and abundance, presence of important taxa, biodiversity indices, and other functions.

In a survey of MS tests, Emans et al. [97] and De Jong et al. [98] listed a number of criteria to evaluate MS tests (Table 7.20). The frequency of sampling should be sufficient to allow development of time-concentration relationships in the critical phases. The spatial and temporal distribution of samples will depend on the test system, the objective of the study and the chemical. Many MS tests do not meet the quality criteria listed above [97,99], but experience in this field is rapidly growing. MS tests are used in higher-tier risk assessment of chemicals, mainly of pesticides [80,81,100-104], but they also have some drawbacks (Table 7.21).

Multivariate techniques are recommended to analyze the effects of chemicals on model ecosystems. Different approaches (e.g., Principal Component Analysis, Similarity Analysis) are available in general statistical packages or as specific programs that have their own pros and cons. The Principal Response Curves is a multivariate technique especially designed for the analysis of microcosm and mesocosm data at a community level [105,106]. Its advantage over other techniques is that its output is easy to communicate. The differences in community composition between

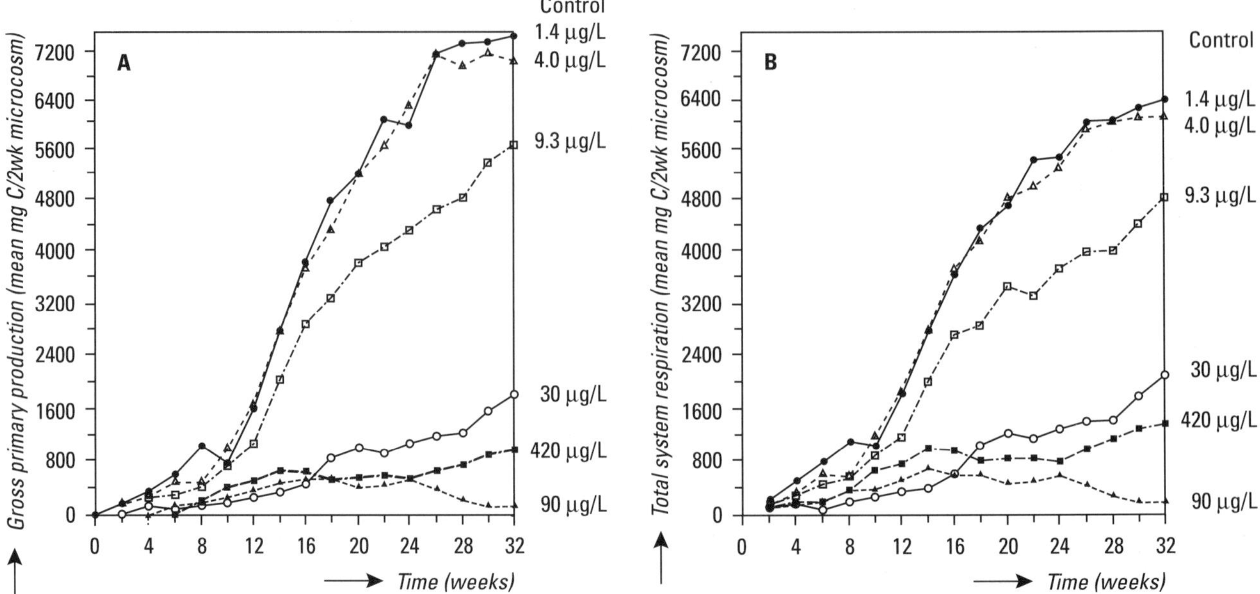

Figure 7.23. Bi-weekly gross primary production (A) and total system respiration (B) as measured in a microcosm test with copper. From Hedtke [95]. With permission. Copyright Elsevier.

the control and the treatments are displayed as a graph. This graph makes it possible to identify which species are affected most, how much the effect differs from the control, and how this evolves in time over the course of the experiment (Figure 7.24).

Modelling studies have shown that selection pressure, differences in interspecies sensitivity and competition between species for resources play an important role in understanding the effects of chemicals in micro or mesocosms. Long-term toxicant effects on sensitive species will influence the competition between species for resources and will lead to the replacement of sensitive species by more competitive species that are more tolerant to the chemical, with a decrease in species diversity [89], as observed in metal-stressed nematode communities [107]. Single pollution events disturb the competition and predator-prey relationship between species [92], leading to changes in species composition that are sometimes are predictable based on laboratory toxicity data [108].

Table 7.20. Criteria for the validity of multi-species tests [97,98]. With permission.

1. The system should represent a realistic community
2. The experimental setup and conditions should be well described, including physicochemical parameters such as pH, temperature and hardness
3. Several taxonomic groups should be exposed to well-described test concentrations for a longer period
4. In each experiment several concentrations should be tested, consisting of one control and at least two test concentrations
5. Each test concentration should have at least one replicate
6. The concentration of the test compound should be measured several times during the experiment
7. Apart from effect parameters like population density and biomass, effect parameters on higher integration levels such as species diversity and species richness should be determined. The endpoints should be in accordance with the mode of toxic action
8. A distinct concentration-effect relationship should be obtained
9. A reliable multi-species NOEC should be derived

Table 7.21. Some difficulties of using multi-species tests [104]. With permission.

1. Costs.
   These are relatively high.
2. Standardization.
   Standardization (harmonization) of MS tests is difficult because the type of study to be performed depends on the question to be answered and may differ from chemical to chemical, from application to application, and from site to site. This will hinder the *mutual acceptance of data*, and therefore increase the costs to industry. The need for standardization is doubtful when we consider the lack of ecological and environmental realism it would imply. However, no standardization at all may lead to hidden subjectivity related to, amongst other things, the taxa included in the experiments (e.g. macrophytes, fish and amphibian species).
3. Lumping of variables.
   In view of the laboriousness of monitoring all species in a community, in MS tests often only lumped variables (e.g. functions, total algal biomass and oxygen production) are observed and many other effects, i.e. reduction in species diversity, may thereby escape notice. As a result extinction of species, i.e. "genetic erosion" may occur. Multivariate statistical methods can be of help to study the diversity and density of all species present in a MS test [105,106]
4. Rapid divergence.
   Some experimental communities tend to diverge rapidly in their development so that only coarser kinds of acute effects stand a reasonable chance of being detected [111].
5. Stability of exposure concentration.
   Stress in MS tests usually decreases rapidly after inoculation, because the toxic chemical is (biologically) degraded or becomes less available in other ways. In single-species tests, the level of stress is usually kept fixed by continuous or intermittent replacement of the test media. Supplying a continuous dose in an MS test resolves the problem only partially [110], but the administration of the chemical can be made to mimic the actual field exposure situation.
6. Adaptation.
   There may be processes which modifies the susceptibility of species (e.g. adaptation or selection of resistant individuals). The quantitative importance of such processes is hard to assess and interpret. Individuals that survive because of their resistance to one chemical may be more vulnerable to another [111]. This process does allow to study the impact of a chemical on community composition and related secondary effects.
7. Replication.
   If any effects are found in the variables observed, there is the problem of disentangling them from the scatter or of avoiding errors of the second kind in the statistical analysis of the results [100,102,111]. Experimental standardization and improved statistical techniques have greatly improved the replicability of MS systems [105,106].
8. Extrapolation.
   As one MS study cannot be representative of all ecosystems, caution is needed when extrapolating the results to other communities or ecosystems. In this respect MS tests do not differ from standardized SS tests.

Types of multi-species tests

MS tests encompass a broad range of bioassays, ranging from small laboratory microcosms made up of artificial assemblages of a few species [107], to more natural microcosms up to 10 m^3 or mesocosms up to 10^4 m^3, or even larger natural systems such as an entire lake or a section of a watercourse which is deliberately contaminated with a chemical to determine concentrationeffect relationships [56,102]. Essentially, there are two basic types of MS tests (Table 7.22). Microcosm and mesocosm tests provide ways of studying potential pollutants in systems which simulate parts of the natural environment (i.e., the macrocosm) but which are also open to experimental manipulation.

Microcosms are often used to study contaminant effects on community structure and function. Microcosms can be used indoors or outdoors. Due to their size (from a few litres to several hundreds of litres), some aspects of natural systems may not be mimicked in full, such as presence of all trophic levels. Nevertheless, essential characteristics such as diversity, competition and primary productivity can easily be studied. The standardized

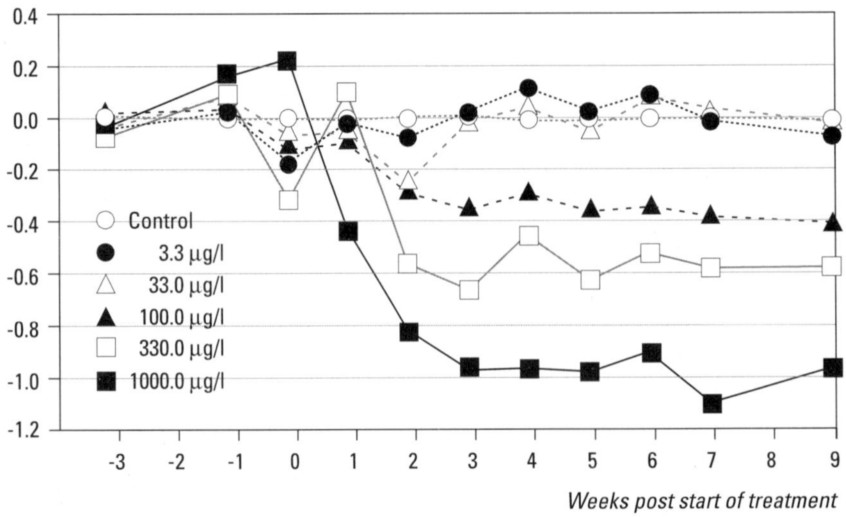

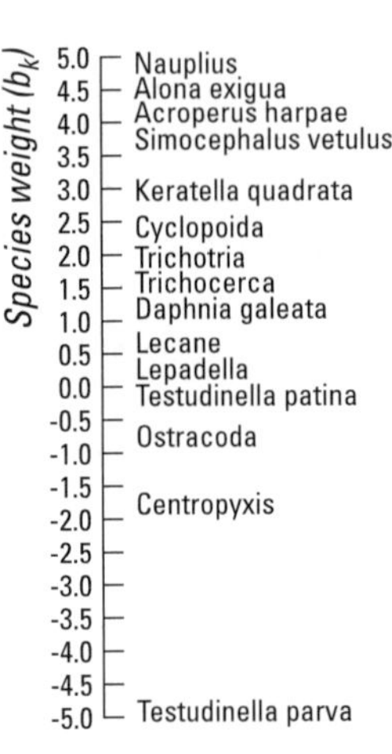

Figure 7.24. Multi-variate statistical analysis (principal response curves) of a freshwater microcosm zooplankton community, after treatment with the fungicide Carbendazim. The curves represent the time course of the effect of treatment on the zooplankton community. The species weight (b_k) explains the affinity of the taxon with the curve. Taxa indicated with a positive species weight are expected to decrease in abundance, relative to the controls, and taxa with negative weights are expected to increase. From van den Brink et al. [90]. With permission. Copyright Elsevier.

aquatic microcosm (SAM) originally developed by Taub [110], has been extensively evaluated. It is a multitrophic level community test with more than 15 biotic components (microalgae, pelagic and benthic invertebrates). The test system is static and consists of a series of 3-L glass jars, containing an artificial medium and a sand substrate (Figure 7.25). The test is carried out in triplicate with a total of 6 treatments over 63 days. The abundance of algae, macro-invertebrates and micro-invertebrates as well as nutrient dynamics and chemical fate, are among the recommended endpoints. The design and analysis of microcosm tests has greatly improved, such that former concerns about cause-effect relationships, replication and divergence of test results [104] have been sufficiently addressed [80, 105,106].

The difference between mesocosms and microcosms is mainly their size. Many mesocosm studies have been performed in artificial ponds, enclosures in lakes or oceans, or artificial streams. Artificial ponds (10m x 5m x 1m) contain sediment and are colonized by algae, macrophytes and macro-invertebrates, and can be stocked with invertebrates and fish [80,102,103]. Studies normally take approximately five months. Endpoints include chemical fate processes, dissolved oxygen, algal biomass, composition and the abundance of phytoplankton and zooplankton, as well as macro-invertebrates, snail reproductive success, and fish survival and growth. Experimental stream ecosystems are used to study lotic ecosystems, with an emphasis on the specific benthic macrofauna. For an overview, see Kennedy et al. in [7].

Concluding remarks

SS tests and MS tests both have their place in ERA. The use of MS tests in regulatory testing is of greatest value if used in combination with tests that can provide data

Table 7.22. Types of multi-species tests [80,103].

Microcosms: experimental tanks/ponds or bench-top systems with a water volume of between 10^{-3} and 10 m^3, or experimental streams less than 15 m in length

Mesocosms: outdoor experimental tanks/ponds with a water volume between 1 to 10^4 m^3 or experimental streams greater than 15 m in length

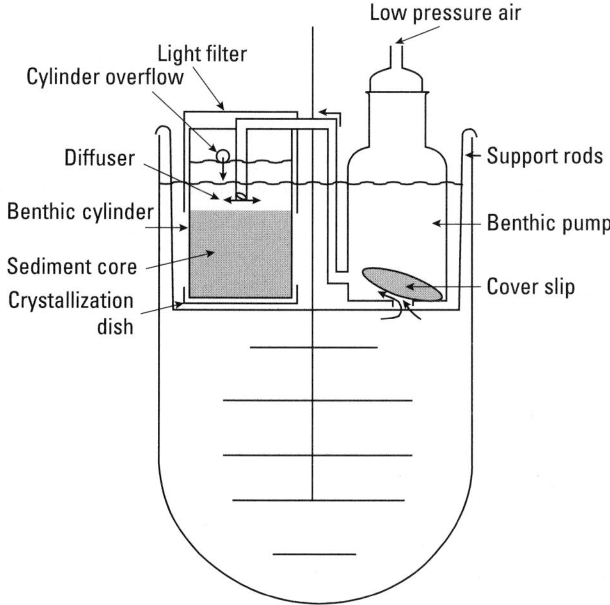

Figure 7.25. Diagram of a single test vessel in an experimental microcosm test system.

on fate, population interactions and ecosystem processes. Adequate fate and population models can be used to fill the gap between SS and MS testing, i.e., between environmental chemistry, toxicology and ecology. This will also serve the regulatory needs.

Multi-species indoor or outdoor tests can play an important part in elucidating the role of environmental factors that may modify the exposure and susceptibility of species. The decision whether or not a field study is required should be based on data obtained from preliminary and refined effects assessment and on data used for exposure assessment, e.g., degradation rates and partition coefficients between air, water and soil, or sediment. Therefore, these studies should not be seen in isolation but should be incorporated in a tiered scheme of testing, consisting of various stages: preliminary, refined and comprehensive assessment (Section 7.10.4).

It should be noted that the bioavailability of the test compound and the distribution of sensitivities of a number of important species should be known before a field test can be carried out. If, based on this information, there is some degree of risk or uncertainty, a field study may be necessary. Such a study should include sensitive and representative organisms. Several guidance documents are available for tiered testing strategies, with MS tests or field tests incorporated in higher tier testing [80,81,103].

7.4 SEDIMENT TOXICITY

7.4.1 Introduction

Most of the experimental work in aquatic toxicology has focused on the potential effects of dissolved pollutants on pelagic organisms. It has been well established that pollutants entering the aquatic environment partition to suspended particles and sediment [112,113], depending on their partition coefficients. Sediment constitutes an important compartment of aquatic ecosystems. Where a substance is likely to be found in sediments at harmful levels, due to its known chemical and toxicological properties and use pattern, a risk assessment addressing its fate and effects on benthic organisms should be performed. This should, in particular, consider the chemical's association and degradation in sediment and its toxicity to benthic organisms. Due to chemical loading of the sediment, tumours or liver neoplasm's have been observed in bottom feeding fish like carp [114], probably caused by polyaromatic hydrocarbons (PAH). Other sublethal effects, such as deformities in the mandibles or antennae of relatively tolerant sediment-dwelling taxa like chironomids [115] and setal abnormalities in oligochaetes [116] have also been observed. These effects are indicative of chemically polluted sediment areas.

Sediment and soil have a large capacity to retain environmental contaminants, especially persistent hydrophobic organic chemicals, or positively charged divalent or trivalent ions. Consequently, soil and sediment may act as a sink for, and a source of, toxic chemicals through the sorption of contaminants to particulate matter. Sediment can serve as a historical record of change due to both manmade pollution and natural environmental causes [117]. Surface water contamination disperses over time and space, and the chemical that is sorbed to the sediment can become a hazard to aquatic communities (both pelagic and benthic) which may be undetected from observations of contaminant concentrations in the water column. Even if the quality of the overlying water is improved, e.g., due to emission reduction, the polluted sediment may still act as a long-term threat to the organisms exposed to it [118].

In the following sections the principles of assessing the toxicity of chemicals in sediment are discussed with reference to the source of exposure, i.e., from sediment or interstitial water. Several methods for measuring this toxicity have been proposed. Internationally agreed water-sediment test guidelines are described only for chironomids, [119,120], but ASTM guidelines are available for other species [121]. Sediment toxicity tests

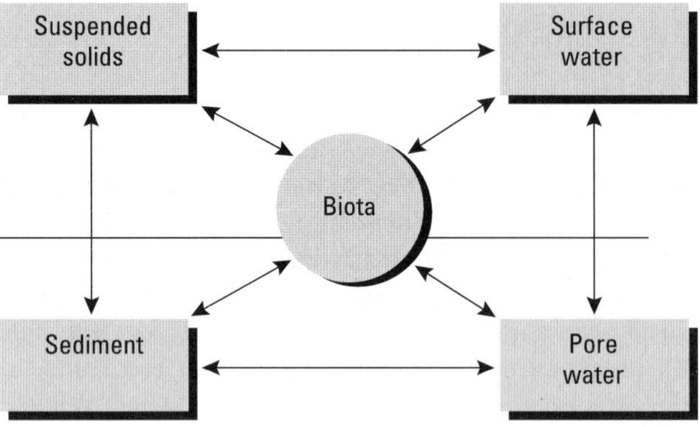

Figure 7.26. Compartments and their interrelationships in the sediment-water system [122]. With permission.

often differ with regard to the source and preparation of the sediment. The biological procedures and toxicological responses, however, are similar to those for water column (nektonic) organisms such as fish and daphnids. The emphasis is on the predictive goal, i.e., how can we derive sediment quality objectives (SQOs) or LC50, EC50, or NOEC values for chemicals or groups of chemicals?

7.4.2 Exposure systems

The design of test systems to determine the toxicity of chemicals to benthic organisms needs to take into account how the test chemical is introduced and distributed. It must also address sediment characteristics and the habitat, physiology and feeding modes of the test organism. Figure 7.26 illustrates how the different compartments of a sediment-water system might contribute to the contaminant uptake by sediment inhabiting benthic, epibenthic and pelagic organisms. Experimental information on the relative significance of the various uptake pathways for aquatic organisms is somewhat conflicting. Benthic and epibenthic organisms like polychaetes and many arthropod species can accumulate strongly adsorbing inorganic compounds like cadmium, and very lipophilic organic compounds like PCBs and PAHs. The main routes of exposure of organic contaminants for soil inhabiting invertebrate species can be pore water or ingestion of soil particles or for sediment dwellers, pore water, overlying water and ingestion of suspended solid or sediment particles [123-126].

Using equilibrium partitioning models, and taking into account environmental factors and sediment parameters, such as particle size and organic carbon content, NOECs for sediments can be estimated from (pore) water NOECs for individual compounds [26,32,33,124,127]. These models are presented in more detail in Chapter 3. Quantitatively the mode and kinetics of contaminant uptake from sediment can vary considerably between species, depending on factors such as feeding, habitat, activity and metabolism of the organism, developmental stage, season and history of exposure. Furthermore, the biological community itself strongly influences the physicochemical environment in the sediment and thus the bioavailability of contaminants by various processes:

- Primary productivity influences pH conditions which in turn influence metal chemistry.
- Sulfate reduction to sulfide by bacteria facilitates metal sulfide formation [128].
- Biological activity influences redox conditions and metal redox conversions.
- Production or degradation of organic matter may influence complexation of the contaminants.
- Bioturbation influences sediment-water exchange processes and redox conditions.
- Oxygen consumption leads to anaerobic conditions which can favour dehalogenation reactions [129] but inhibit mineralization processes.

Thus the bioavailability of pollutants depends on several abiotic as well as biotic factors. This is one of the main obstacles in experimental ecotoxicological research with sediments. Experimental manipulation of sediments may drastically influence the bioavailability of the test compound and therefore its toxicity. A basic understanding of a chemical's fate is, therefore, a prerequisite. Various exposure systems are available for testing sediment toxicity. Exposure systems can be entirely aquatic, but when experiments are carried out with pore water such systems should be used with care, depending on the species used and the duration of the study. Chronic studies using sediment organisms in sedimentfree conditions can cause severe stress (e.g., *Lumbriculus variegatus*, annelid worms, will tie themselves in knots when tested without substrate).

7.4.3 Effects assessment

This section provides a short review of methods which can be used to derive NOECs for sediments. It includes test methods for assessing the toxicity of sediment and setting quality objectives and methods for determining the toxicity of chemicals to sediment-dwelling organisms. Eight methods were evaluated at an OECD workshop [112] as being potentially useful for deriving sediment quality objectives (Table 7.23). The first three methods

Table 7.23. Potentially useful methods for deriving sediment quality objectives according to the OECD [112].

1. Equilibrium partitioning
2. Interstitial water quality
3. Spiked sediment toxicity
4. Reference concentrations
5. Apparent effects threshold
6. Screening-level concentrations
7. Sediment quality triad
8. Tissue residues

listed in Table 7.23 were recommended by the OECD [112] for the development of numerical sediment quality criteria on the basis of seven evaluation criteria (Table 7.24) and are discussed further. The remaining five were not recommended for this purpose, which does not mean that they are inadequate. The section ends with specific considerations for site-specific effects assessment.

Sediment quality objectives

The equilibrium partitioning (EP) approach derives NOECs or sediment quality objectives (SQOs) from aquatic NOECs or water quality objectives (WQOs) by predicting interstitial water concentrations and appropriately normalized sediment concentrations [32,33,124]. This method can be used as a screening method to assess the risk to sediment organisms. If exposure levels exceed the SQO, tests with benthic organisms should be part of a refined risk assessment. The formula for deriving the SQO for a particular chemical is:

$$\mathrm{SQO} = K_{\mathrm{p}} \cdot \mathrm{WQO} \tag{7.6}$$

where SQO is the sediment quality objective (mg/kg dry wt), K_{p} is the solids-water partition coefficient (L/kg dry wt), and WQO is the effects-based water quality objective (mg/L). For comparative reasons, the solids-water partition coefficient is often adjusted with respect to organic carbon (OC) content (f_{oc} = %OC/100) and an organic carbon partition coefficient is thus defined:

$$K_{\mathrm{p}} = K_{\mathrm{oc}} \cdot f_{\mathrm{oc}} = C_{\mathrm{s}} / C_{\mathrm{w}} \tag{7.7}$$

where K_{oc} is the OC-normalized K_{p}, and C_{s} and C_{w} are the chemical concentrations in solids and water, respectively. Typical values for OC content in sediment are in the range of 4 to 6%. The standard value for f_{oc} in sediment is set at 0.05. Where only the organic matter (OM) content is known, f_{oc} can be derived as follows: $f_{\mathrm{oc}} = 0.6 \times f_{\mathrm{om}}$. For neutral organic chemicals K_{oc} is often estimated from the K_{ow} using Log K_{oc} = Log K_{ow} – 0.21 [130]. The EP approach is based on the observation that interstitial water concentrations are more closely correlated than bulk sediment concentrations with toxicity to or bioaccumulation of environmental contaminants in benthic organisms (Figure 7.27). The EP method [32,124] assumes that:

1. The concentrations in sediment and interstitial water are in equilibrium.
2. The concentrations in any of these phases can be predicted using appropriate partition coefficients and concentrations in one phase.
3. The effect concentrations in sediment can be predicted using adequate partition coefficients and effect concentrations.
4. The WQO provides an appropriate effect concentration for deriving sediment quality objectives.

Table 7.24. Evaluation criteria for methods for deriving sediment quality objectives according to the OECD [112].

1. *Chemical specificity:* can the method be used to derive a concentration for a specific chemical?
2. *Causality:* are the observed effects caused by a specific chemical?
3. *Chronic effects:* does the method consider chronic toxicity endpoints?
4. *Bioaccumulation:* does the method consider food chain accumulation and ingestion of contaminated sediment for benthos and fish?
5. *State of development:* is the method validated, used and ready for use?
6. *Bioavailability:* how generally applicable is the method across sediment types? Are sediment quality objectives a function of the bioavailable phase?
7. *Applicability:* is the method applicable to bedded sediment or suspension?

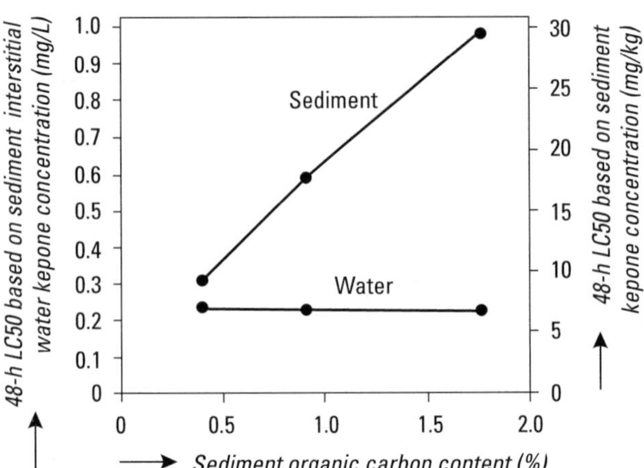

Figure 7.27. Plot of 48-h LC50 values based on interstitial water and sediment kepone concentrations versus sediment organic carbon content for the midge *Chironomus tentans*. From Ziegenfuss, Renaudette and Adams [131]. With permission.

The advantage of the EP method is that its theoretical basis is well established. It has been tested for non-ionic hydrophobic chemicals and metals [25,26,32,124] and has been applied for derivation of SQOs in the absence of test data (see Chapter 11, Section 11.3.3). The procedure for normalizing sediment concentrations requires a model for chemical partitioning, K_p, which can relate solid phase and liquid phase concentrations. At present there are models available for non-ionic organic chemicals, certain metals and a few ionic organic chemicals. The EP method can be applied to all chemicals (including metals) for which a series of aquatic NOECs or WQOs (also known as water quality criteria or standards) are present, and for which reliable K_p values are available. K_p values for most heavy metals vary with environmental conditions (Section 7.6), which complicates extrapolation to other conditions and the general application of the method. It can be used for marine and freshwater sediments and between sites. Examples of the calculation of SQOs are given in Chapters 4 and 11.

The *interstitial water quality* method is similar to the EP method except that interstitial water concentrations are measured instead of being predicted. This is difficult where substances are present at low levels or have low solubility. Non-depleting solid phase extraction methods are used to measure freely dissolved concentrations of lipophilic substances in interstitial water [132-134]. The *tissue residue* approach seeks to relate chronically acceptable chemical concentrations in benthic organisms to chemical concentrations in sediment using the EP approach. It has the same relative strengths and weaknesses as the interstitial water toxicity method but is less developed in terms of sediment quality criteria. One of its strengths is that it allows a predictive approach using sediment exposure-driven bioaccumulation [26,27,33].

Table 7.25. Test phase systems studied in predictive and empirical sediment toxicity studies.

1. Elutriate (water-extractable)
2. Extractable (solute other than water)
3. Interstitial or pore water phase
4. Whole sediments
5. *In situ*

In the *spiked sediment toxicity* approach, LC50s or NOECs in sediment are derived from experimental dose-response data generated in the laboratory. The SQO is then derived based on the experimental data (see Section 7.10). The test organisms are exposed to sediments spiked with a range of concentrations. The toxicological endpoints are those normally studied in aquatic toxicity tests. With this method it is assumed that spiked sediment in the laboratory behaves similarly and shows similar effects to natural *in situ* sediments. The major limitation of this method is that this assumption may not be true (Section 7.4.2). The advantages of the method are that it is chemical specific, demonstrates a clear causality, and reflects the bioavailability of the test compound. The use of artificial sediment is preferred over the use of natural sediment because of the reproducibility of results in spiking studies, unless site-specific effects are studied. Spiking the sediment can occur via the sediment itself (preferred) or via the water phase. In both cases, care should be taken with regard to equilibration time between the different phases, especially for poorly soluble compounds [120,121]. The sediment should be characterized in terms of particle size, organic matter content, and cation or anion exchange capacity. For natural sediment, additional parameters can be reported such as pH, ammonium and nitrogen content.

Site-specific approaches

The practical aspects of sediment sampling, storage, the collection of interstitial water, elutriates, spiking, sediment dilution and other conditions of exposure are discussed by the OECD [112] and Burton et al. [121,135]. The *sediment quality triad* is designed to evaluate the overall quality of the sediment of specific sites [136,137]. It compares: (a) chemical concentrations

Table 7.26. Representative freshwater and sediment toxicity tests [121,122,135].

Biological level	Assay/organism/community	Endpoint
Amphibians	*Xenopus laevis*	embryolarval survival, terata
Fish	*Salmo gairdneri*	embryolarval survival, growth, terata
	Pimephales promelas	embryolarval survival, growth, terata
	Brachydanio rerio	embryolarval survival, growth, terata
Zooplankton	*Colpidium campylum*	growth
	Brachionus sp.	survival
	Protozoan colonization	structure indices, respiration
	Daphnia magna	survival, reproduction
	Ceriodaphnia dubia	survival, reproduction
Benthic invertebrates	*Panagrellus redivivus*	survival, growth, moulting
	Caenorhabditis elegans	survival
	Tubifex tubifex	survival
	Stylodrilus heringianus	survival, avoidance, reworking rate, growth
	Hyalella azteca	survival, growth, reproduction
	Pontoporeia hoyi (*Diporeia* sp.)	survival, avoidance
	Corbicula fluminea	survival, growth
	Anodonta imbecilis	survival
	Chironomus tentans	survival, growth, emergence
	C. riparius	survival, growth
	Hexagenia limbata	survival, moulting frequency
	macrobenthic community	community/population indices
Microbes	*Vibrio fischeri*	luminescence
	alk. phosphatase	enzyme activity
	dehydrogenase	enzyme activity
	β-Galactosidase	enzyme activity
	β-Glucosidase	enzyme activity
Phytoplankton	Pseudokirchneriella subcapitata	population growth, ^{14}C-uptake
	natural phytoplankton	fluorescence, structure-species abundance
Macrophytes	*Lemna minor*	growth (fond number), chlorophyll-*a*, biomass
	Hydrilla verticillata	shoot length, root length, dehydrogenase-activity, chlorophyll-*a*, peroxidase

in sediment, (b) toxicological responses in the laboratory, and (c) benthic community health. It has some major advantages as it combines chemical and biological observations under laboratory and field conditions. It is widely used, but it is not chemical specific, i.e., unknown mixtures of chemicals and other stressors are often considered and therefore it cannot show causality. The test phases normally used in sediment toxicity studies (Table 7.25) each have their own strengths and weaknesses. These problems can be partly overcome by performing a toxicity identification and evaluation (TIE), linked to a triad analysis. In a TIE, toxicity-based fractionation procedures are used to identify specific contaminants as causative toxicants [138]. Sediment pore water is isolated from sediment and toxicity experiments are conducted with it. This can involve concentrating the isolate (e.g., using solid phase extraction methods) and subsequent dilutions. Solid phase micro-extraction methods offer a relatively quick and cheap way of characterizing the pollution profile of organic substances that are bioavailable and also give an indication of the bioaccumulation potential [139]. Depending on the scope of the TIE research, one or several test species can be used to test the toxicity of pore water [140].

In their excellent review on freshwater sediment toxicity, Burton et al. [121] describe a number of

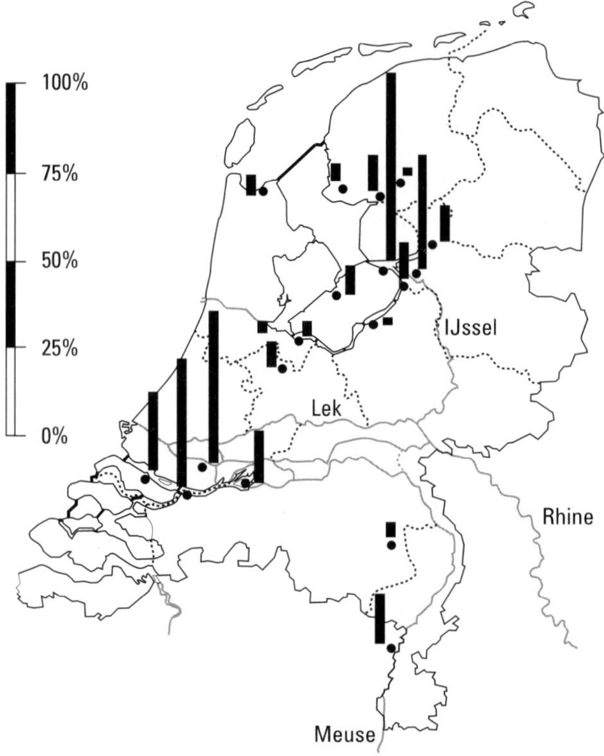

Figure 7.28. High frequencies of mandible malformations in midge larvae as observed in sediments in The Netherlands. Data from Van Urk and Kerkum [115].

approaches, the practical difficulties and various species used to assess freshwater sediment toxicity (Table 7.26). From Table 7.26, it may be concluded that standard test species like bacteria, algae, daphnids and fish are commonly used to assess the toxicity of aqueous fractions of contaminated sediments. Tests with pore water extracts or elutriates are often carried out and semistatic sediment-water systems are also used. In these systems direct contact with contaminated sediment particles is taken into account.

Some problems are associated with TIE with regard to the influence of extraction techniques and their relevance to the bioavailability in the in-situ situation. The same applies when only pore water exposure is considered, ignoring exposure to sediment by contact or ingestion, which could be especially relevant for lipophilic substances with low water solubility. Various tiered decision-making frameworks for sediment contamination have been developed to address these problems [141,142]

7.4.4 Sediment toxicity testing

Good quality water and sediment are typified by species of macro-invertebrate taxa like mayflies (Ephemeroptera), caddisflies (Trichoptera) and stoneflies (Plecoptera). By contrast, dominance of the tubificid oligochaete species *Limnodrilus hoffmeisteri* and *Tubifex tubifex* is recognized as an indication of polluted sediment. Based on the results of field studies [115], midge larvae (Chironomidae) are used as test organisms for spiked sediments (Figure 7.28). There are also internationally harmonized guidelines available [119,120]. A test guideline with the benthic oligochaete *Lumbriculus* using spiked sediment is available in draft form [143].

Other benthic organisms with different morphological structures and different environmental behavioural and trophic properties are also used, such as oligochaetes, polychaetes, nematodes, bivalves, burrowing mayflies and crustaceans, such as amphipods and isopods, as well as several plant species [58, 121]. ASTM has published a number of standard techniques for 10 and 28-day tests with marine and freshwater amphipods (e.g., *Hyalella azteca*) polychaetes, oligochaetes *(Tubifex tubifex)* and mayflies, that are reviewed in Burton et al. [121].

Table 7.27. Characteristics of a chironomid toxicity test [119,120].

Test species	Chironomus riparius, *C. tentans* or other species, e.g. *C. yoshimatsui.*
Test duration	20-65 d , depending on species and growth rate
Test system	Static (in some cases semi-static or flow-through) with elutriates or sediment and water at a layer depth ratio of 1:4 in 600 ml glass beakers
Feeding	at least three times a week with a commercial fish food
Endpoints	larval emergence, growth and survival
Temperature	20-25°C, depending on species.
Parameter	EC50, LC50 and NOEC/LOEC

Table 7.28. Susceptibility of Chironomus riparius life stages to dieldrin (nominal concentrations in mg/L) [122].

Parameter	Life stage	Concentration
96-h LC50	egg	>100
96-h LC50	2nd larval stage	5.2
96-h LC50	3rd larval instar	12.6
96-h LC50	4th larval instar	17.9
23-d NOEC	egg-4th larval instar	0.1

Short and long-term toxicity tests compare the survival, growth, reproduction or other toxicological endpoints among a range of benthic, epibenthic and pelagic organisms. These organisms are exposed to experimentally contaminated (spiked) solid or liquid phases in a test system in order to determine the effects of chemicals or groups of chemicals on sediment-dwelling organisms. As an example of a sediment toxicity test, the subchronic test with *Chironomus riparius*, is given in Table 7.27. As with fish, chironomids exhibit differences in susceptibility at various life stages (Table 7.28).

In conclusion, sediment toxicity studies have matured. Much work has been done to standardize the assessment of sediment toxicity, both for routine toxicity testing with relevant benthic organisms and for the assessment of the toxic effects of a polluted sediment. Some pitfalls remain. Sediment is a very heterogeneous environment and exposure and bioavailability in the field may differ from that in the laboratory. Manipulation of sediment may drastically influence bioavailability and thus toxicity, indicating that risk assessment requires careful consideration of the physicochemical processes at work in sediments.

7.5 TERRESTRIAL TOXICITY

7.5.1 Introduction

Soil contamination is widespread and thousands of polluted sites have been identified in industrialized countries. The importance of soil as a key component of ecosystems is now widely recognized. Several countries have already established soil quality objectives and programs for site-specific risk assessment [144].

Due to their public appeal, adverse effects of contamination of the terrestrial environment are often discussed in terms of the decline and recovery of populations of rare plant species, such as orchids, mammals such as otters *(Lutra lutra)*, bats (e.g., *Myotis dasycneme*), and various species of birds, such as terns (*Sterna* sp.), eider ducks *(Somateria mollissima)*, cormorants *(Phalacrocorax carbo)*, partridges *(Perdix perdix)*, peregrine falcons *(Falco peregrinus)*, goshawks *(Accipiter gentilis)* or little owls *(Athene noctua)*. Environmental protection is often directed towards protecting soil organisms, but protecting soil functions is at least as important in view of the sustainability of land use.

From an ecological point of view, the main functions of soil are those associated with the decomposition of organic matter, mineralization of nutrients, and synthesis of humic substances. Essential parts of the carbon, nitrogen, phosphorus and sulfur cycles take place in the soil. The root zone (rhizosphere) in particular, is closely involved in soil processes. Soil organisms mainly contribute to litter breakdown. This is done by soil invertebrates and soil microbes in concert. The vertical distribution of species varies greatly. The highest density of species is found in the topmost layer of the soil profile. Apart from the role of soil in nutrient cycles, the soil formation process (Figure 7.3) is essential for supporting plant life and in stabilizing mineral particles.

Although the terrestrial environment is crucial for the human population, the soil has only recently become an important topic for ecotoxicologists. First we will turn our attention to exposure assessment in experimental systems.

7.5.2 Exposure systems

Soil contains solid, liquid and gas compartments, each of different and varied composition. The solid compartment is composed of mineral particles and organic material, the liquid one is made up of water with dissolved nutrients and dissolved organic carbon, while the gas compartment consists of different gases and volatile organic substances. These constituents are arranged in a certain order and according to particle size in a certain texture and structure. Particle size influences the total surface area. Soil is an extremely heterogeneous environment, both horizontally and vertically. As a consequence, physical, chemical, and biological characteristics vary, thus creating a wide variety of habitats for soil-dwelling species. This complexity and heterogeneity greatly affect actual exposure situations.

In toxicity tests with terrestrial organisms, different exposure systems are used depending on the way in which organisms are exposed. The three major uptake routes are: (1) ingestion and oral uptake of food or soil

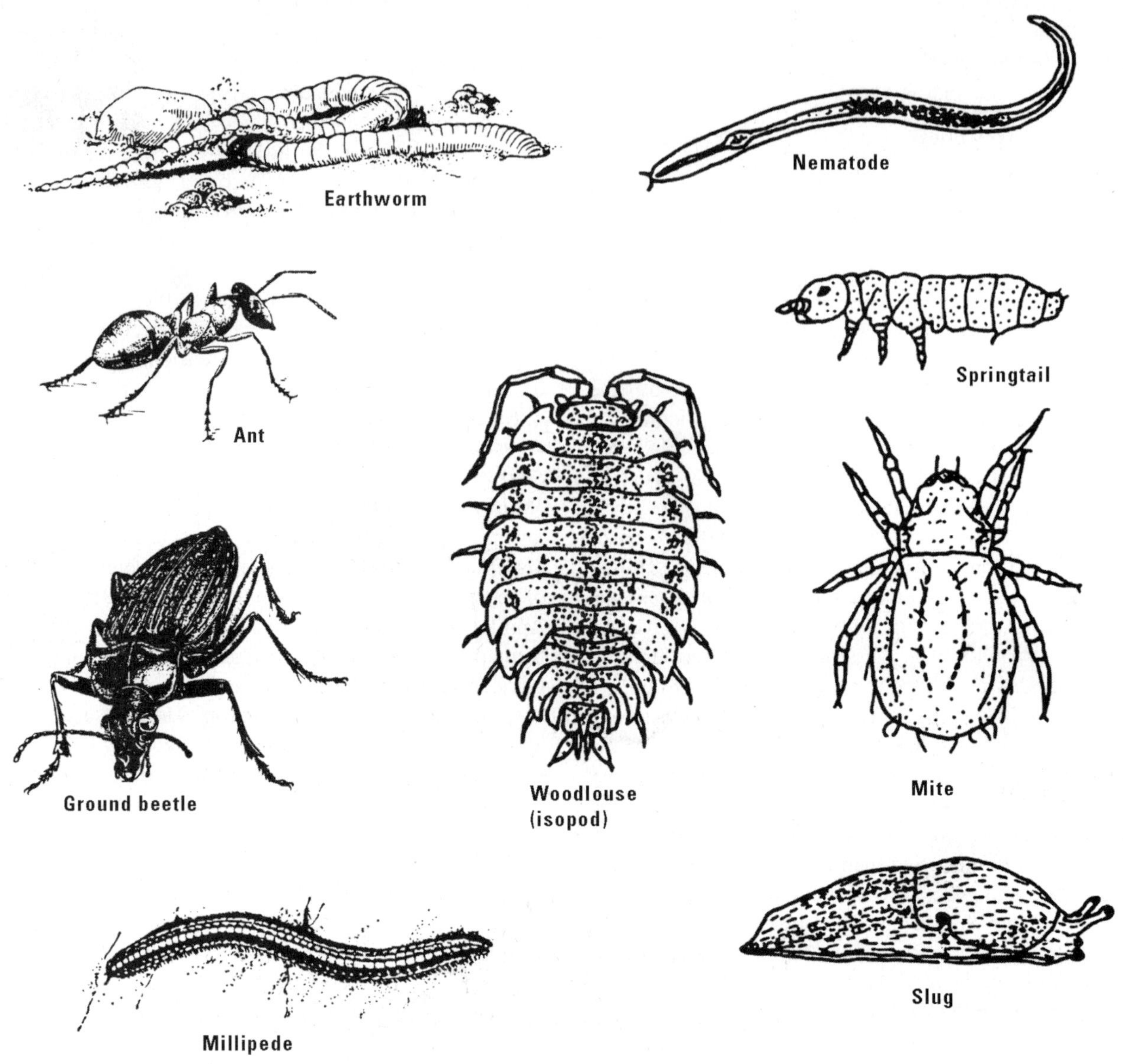

Figure 7.29. Some representatives of the soil invertebrate fauna (not drawn to the same scale). Modified from Van Straalen and Van Gestel in Calow [56]. With permission.

particles, (2) dermal uptake of pollutants from the soil or plant surfaces, and (3) respiration (via stomata, tracheae and lungs).

Effects on micro-organisms are mostly studied by exposing the indigenous microflora of a clean soil by introducing the test chemical into a soil sample [145] or by isolating micro-organisms or microbial communities and testing them in artificial substrates. In invertebrate toxicity tests a species-specific exposure method is often used in relation to the expected uptake route in the field. Tests with birds or mammals can be used to study oral, inhalation or dermal toxicity. The bioavailability of the chemical tested will differ for each exposure pathway, which is reviewed in the next section.

Chemicals mixed with the soil

For soil-dwelling species such as bacteria, fungi, soil invertebrates (e.g., protozoans, earthworms, enchytraeids, mites, and nematodes; see Figure 7.29), and most vascular plant species, tests in soil seem to best simulate

Table 7.29. Composition of OECD artificial soil [146,151].

Industrial quartz sand	70%
Kaoline clay	20%
Sphagnum peat	10%
Water content (% of the water holding capacity)	40-60%
pH (by addition of $CaCO_3$)	7.0±0.5

natural exposure routes. Although arthropods normally do not ingest mineral soil, many live in close contact with it and take up chemicals from the soil/air interface. This is thought to be mediated via a water film. The way the substrate is prepared allows the dose to be expressed as a concentration per mass unit of dry soil, i.e., in mg/kg. The substrates used in soil tests vary from natural materials taken from the field, to soils artificially created out of commercially available materials [146]. The type of soil will have a major influence on bioavailability, i.e., the distribution of chemicals over the solid, gas and pore water phases, and will greatly affect toxicity.

As with aquatic sediments (Figure 7.27) the soil toxicity of many organic chemicals is often directly related to the organic matter content of the soil. This has been demonstrated for earthworms and a variety of other species [147,148]. With heavy metals, bioavailability may depend on soil pH, organic matter content, cation exchange capacity (CEC) and clay content [149]. Quantification of the contribution made by each factor to the toxicity and bioavailability of metals to earthworms is difficult [147], but modern statistical and mathematical models have greatly helped to improve our understanding [41,150]. For purposes of standardization, the use of artificial soil is recommended, for earthworms, enchytraeids and soil arthropods (Table 7.29).

Direct and indirect application

Topical dosing is generally applied to mammals and arthropods [152]. The toxic solution is applied directly to a predetermined area of the body surface after immobilization of the animal. This method of topical application allows the dose to be expressed as an absolute amount per animal. From a toxicological point of view, this is a preferred exposure method, since any effect can be directly related to the dose; disrupting factors such as consumption, movement and other activity can be eliminated as sources of variation. However, in a field situation the actual dose received is usually unknown.

Arthropods and plant seeds may also be dosed by contact with a chemical through immersion in a solution. The time of immersion is standardized. The dipping technique is easy to carry out but it has the disadvantage that the dose received is unknown. Effects are expressed in terms of the concentration of the chemical in the dipping solution. Plants and invertebrates, such as earthworms, nematodes and protozoans, may also be tested in aqueous solutions of the test chemical. In these tests, the species are treated as aquatic organisms and the aqueous phase is considered to be the most important route of exposure and the interaction with the soil solid phase or air is neglected. Both topical and whole-body exposure techniques are mainly restricted to laboratory research methods. According to Van Straalen and Van Gestel in Calow [56] other routes of exposure may also be important in the field situation [147].

An important exposure route for species on agricultural land treated with sprayed chemicals is residual uptake. Surfaces coated with films of pesticides will act as a source of uptake by organisms as they move over the surface, especially high surface-activity species, such as predatory mites, spiders, beetles, and springtails. In experiments organisms may be present during application of the toxicant, in which event the effect is caused both by direct and residual exposure. More frequently, however, the treated surface is allowed to dry and organisms (often arthropods) are placed on the treated surface for the test. Surfaces used in such tests include plant leaves, sand, natural soils, or artificial substrates such as filter paper or glass [153]. The bioavailability of the residue depends very much on the nature of the substrate, its tendency to adsorb the chemical, and its moisture content. For the purposes of standardization the use of an inert material that will neither adsorb nor react with the chemical, i.e., glass or sand, is recommended, but effective doses established in this way are very difficult to apply to field situations, as inert surfaces do not resemble the natural situation [154].

Chemicals added to food

Dietary uptake of chemicals via food is a well-known exposure route of mammals and birds (Chapter 5). Dietary uptake is a direct route for chemicals sprayed on leaf surfaces, acting as stomach poisons in phytophagous invertebrates, as well as for chemicals associated with dead organic matter which have an effect on saprotrophs. Dietary exposure may also occur via the food chain, e.g., predatory birds or mammals feeding on fish (Table 7.9) or mammals, herbivores feeding on various species of plants, or microbivorous arthropods feeding on fungi that concentrate chemicals from the soil. In these toxicity tests, chemicals are homogeneously mixed with the food, and the effective dose is expressed per dry mass of food. Other dietary routes are via drinking water or sucrose solutions. If the test animals take in the amount fed completely, or if consumption can be determined by weighing the food left, the dose can be expressed in mg/kg body weight. This allows comparison with doses taken up via other routes, e.g., topical application. The food used in feeding experiments largely depends on the species. The uptake efficiency of chemicals added to the diet is highly variable. Effective concentrations are difficult to compare between species because they will be influenced by the type of food used and the physiological condition of the animal. Avoidance of contaminated food is a common response in some arthropods but is also found for birds [155].

Exposure via the air

When organisms are tested for their susceptibility to gaseous air pollutants, exposure units must be airtight before they are flushed with a known concentration of the chemical in air. Plant exposure to gases is controlled in open top chambers, which is the preferred method [156]. Several pesticides, especially those applied as fumigants, exert their effects through aerial exposure and are tested in this way on various plant and animal species such as flies, fleas, and ticks, etc. [148,153]. When some pesticides are sprayed on a surface, actual exposure may actually be via the air, as the chemical evaporates from the surface film to reach toxic concentrations above the surface.

7.5.3 Effects assessment

Soil toxicity tests have been developed as a means to provide hazard information for the terrestrial environment. This hazard information can be used to derive soil quality objectives, similar to the way in which sediment quality objectives are derived. Sites that are polluted may need remedial treatment to prevent risks to man or the environment. Due to the high costs associated with soil remediation, it is essential to have efficient laboratory test methods to indicate potential hazards and use in-situ bioassays to determine the risks at specific sites. Several large-scale programmes have been devised to provide such methods [157-159]. The triad framework for site-specific risk assessment for sediments can be adapted for soils. By combining information on measured concentrations, toxicological responses in the field or laboratory and soil community composition, conclusions can be drawn on the risks to the soil ecosystem [159].

Situations where sediments are removed from water bodies and deposited on land for reasons of water management or remediation, require special attention. Once these sediments are put on land, the physicochemical conditions change dramatically due to water loss and the predominantly aerobic conditions, thereby influencing biodegradation and the mobility of the contaminants.

7.5.4 Soil toxicity testing

Microbial tests

Bacteria are by far the most numerous organisms in soil, varying from 10^6 to 10^9 cells/g [56]. Although bacteria are dominant in soil, fungi, as a group, also play an essential role in the decomposition of organic matter. This is because of their ability to develop in the soil by means of hyphae and through the use of enzymes capable of degrading a variety of persistent substances such as lignin. In the same way as for sediments, toxicity tests can be carried out with micro-organisms from a clean soil. Soil functional microbial tests are carried out with freshly sampled soil containing an active microflora consisting of numerous species. These functional tests are thought to be more representative of the soil ecosystem. The microbial processes studied are essentially the same as those described in Section 7.3.3.

For site-specific risk assessment to deal with chronic soil pollution, the ability of micro-organisms to develop pollution-induced community tolerance (PICT) is studied relative to control sites [160-162]. Microbial communities can develop tolerance to specific chemicals, due to the loss of sensitive species and genetic or physiological adaptation. The ability of bacteria to use a variety of specific substrates in micro-well plates is compared between control sites and polluted sites. When the bacteria from a polluted site show a higher metabolic activity on specific substrates, after pre-exposure to specific chemicals, than bacteria from a control site, community tolerance is increased and a strong causal link between the pollutants and microbial functions of a specific site is established.

Vascular plants

The available toxicity data for terrestrial plants are highly diverse. Plant tests have been reviewed by Kapustka and Reporter in Calow [56] and Klaine et al. [62]. Two plant groups have been used extensively in developing rapid partial life-cycle tests, *Arabidopsis* and *Brassica*.

Substances can be taken up by the plant via the soil, via soil splash on the leaves, and through direct

Table 7.30. Characteristics of the terrestrial plant growth test [163].

Test species	a minimum of three species should be selected for testing, at least one from each of the following categories: rye grass *(Lolium perenne)*, rice *(Oryza sativa)*, oat *(Avena sativa)*, wheat *(Triticum aestivum)*, sorghum *(Sorghum bicolor)* (category 1), mustard *(Brassica alba)*, rape *(Brassica napus)*, radish *(Raphanus sativus)*, turnip *(Brassica rapa)*, Chinese cabbage *(Brassica campestris)* (category 2) and vetch *(Vicia sativa)*, mung bean *(Phaseolus aureus)*, red clover *(Trifolium pratense)*, fenugreek *(Trifolium ornithopodioides)*, lettuce *(Lactua sativa)* and cress *(Lepidium sativum)*
Test duration	plants are harvested usually 14-21 d after 50% emergence in the controls
Test system	static system, the test substance is dissolved in a solvent and mixed with natural soil or applied to soil surface
Light/temperature	suitable for growth
Endpoints	emergence and growth (wet weight)
Parameter	LC50 (emergence) and EC50 (growth)

deposition on leaves and other above-ground parts of the plant. Standard tests differentiate between the two main exposure pathways [163-164]. This recognizes the need to evaluate the effects of plant protection products that are sprayed on non-target plant species.

The most common type of phytotoxicity test is the seedling emergence and growth test [163]. Four to five plant species are commonly used (Table 7.30). The seed germination tests, often promoted as representing a sensitive and critical stage in the life cycle, is rather insensitive to many toxicants. This is caused by two factors: first, many chemicals are not taken up by the seed; and second, the embryonic plant derives its nutritional requirements internally from the seed storage materials, essentially making it isolated from the environment. The early growth test yields relevant information on exposure via the soil. The endpoints that are reported after a 14 to 21-day growth period are biomass of the plant [165], as well as shoot height and visible detrimental effects.

The vegetative vigour test [164] evaluates the effect of a spray application of a substance on shoot weight or shoot height, after a 21 to 28-day growth period from treatment. In addition, visual differences with the control with regard to chlorosis, necrosis, wilting or deformations can be reported.

Soil invertebrates

Harmonized soil toxicity test using invertebrates are available for earthworms, enchytraeids, Collembola, snails and insect larvae. Earthworms are commonly used because of their great ecological importance. International guidelines include the OECD acute earthworm toxicity test [146] and the OECD reproduction test [151]. For these two tests, *Eisenia andrei* or *E. fetida* are recommended. These are not actual soil-dwelling species, but are commonly found in compost and dung heaps, and can be cultured easily in the laboratory on a substrate of horse manure or cow dung. According to the guidelines, other soil-dwelling species may also be used. Instead of *E. fetida*, it is suggested to use the soil inhabiting *Apporectodea caliginosa* to improve the ecological relevance of the reproduction test. However, due to its slow reproduction cycle and the need to collect test individuals from the field, this species is not recommended for routine toxicity testing [166]. Therefore the two *Eisenia* species are recommended for practical reasons.

The acute toxicity screening test consists of the 2-day filter paper contact test, and the 14-day artificial soil test, scored on the survival endpoint. The filter paper contact test is a toxicity-screening test, but it has no predictive value for the effect of chemicals in the soil. The results obtained from the artificial soil test (Table 7.31) can easily be applied to natural soils using sorption data (Section 7.4.3). This test is reasonably capable of predicting effects in the field.

The earthworm reproduction test lasts for 8 weeks [151]. The parent animals are exposed for four weeks and then removed from the system and mortality and growth determined (Table 7.32). After another 4 weeks, the total number of off-spring produced is recorded. Another long-term invertebrate test is the 6-week enchytraeid reproduction test, with a similar test design as for the earthworm reproduction test [167].

Several other soil invertebrate toxicity tests have been developed for a number of major soil invertebrate taxa: oribatid mites, nematodes, isopods, staphilinid beetles,

Table 7.31. Characteristics of the acute artificial soil test with earthworms [146].

Test species	*Eisenia fetida* and *E. andrei*
Test duration	14 d
Test system	static test in test jars with 750 g (wet weight) of OECD artificial soil
Light/temperature	low light intensity (400-800 lux) at 20°C
Endpoints	survival
Parameter	LC50

Table 7.32. Characteristics of the reproduction test with earthworms [151].

Test species	*Eisenia fetida* and *E. andrei*
Test duration	pre-incubation (at least one day), exposure of adults to treated soil (4 weeks) followed by incubation of cocoons in untreated soil (4 weeks)
Test system	static test in test jars with OECD artificial soil
Light/temperature	low light intensity (400-800 lux) at 20°C
Food	oatmeal, cow or horse manure (dried and ground)
Endpoints	survival and growth of adults (after 4 weeks of exposure) and reproduction i.e. the total number of offspring per adult worm (after a further four weeks)
Parameter	NOEC and/or ECx (EC10, EC50) for reproduction, LC50, % of initial weight.

centipedes, millipedes, Collembola, and interactions between nematodes and between predatory mites and nematodes [166]. Some of these species are listed in Table 7.33. Additional information on test procedures can be found in Van Gestel and Van Straalen [148] and in Løkke and Van Gestel [166].

Beneficial arthropods

A special group of invertebrates are the "beneficial" arthropods that may improve the productivity of agricultural soils. There is commercial interest in designing and applying plant protection products in such a way that "beneficials" are least affected. Among the beneficials are pollinators such as the honey bee *(Apis mellifera)* and predatory and parasitic species that attack pest species. There is an internationally harmonized test guideline available for the effects of substances on honey bees [152] which is based on the guideline of the European and Mediterranean Plant Protection Organization (EPPO).

The Hymenoptera contain a large number of parasitic species. The female insect deposits an egg in or on a host (usually an insect egg or larva), which is then gradually eaten as the offspring develop. Within the order of the Coleoptera, the families Carabidae (ground beetles), Staphylinidae (rove beetles) and Coccinellidae (ladybirds) contain representatives that are commonly found on agricultural land and are recognized for their predation of pests (Figure 7.30). Among the various arthropod groups, other predators such as spiders and predatory mites are also important. The families Erigonidae and Linyphiidae (money spiders) are important groups with a great species diversity. Guidelines for evaluating the side effects of pesticides to non-target arthropods have been published [153] which may be useful for other categories of substances as well.

The array of methods used in testing terrestrial invertebrates is wide because different tests have been developed with different aims. Many methods still differ in relation to the medium to which the chemical is applied (different types of soil, contact surfaces), and the influence on bioavailability (see Section 7.6). This is why the OECD will continue to review and further harmonize terrestrial ecotoxicology guidelines.

Tests with birds and mammals

There has long been public concern about the effects of pollutants on mammals and birds. Bird and mammalian toxicity tests therefore have a much longer tradition than tests with soil invertebrates, for example. Mammalian toxicity data are required mainly to determine the potential risk to humans. Toxicity is determined using

Table 7.33. Overview of selected laboratory tests using terrestrial invertebrates, evaluated according to three criteria[a] according to Van Gestel and Van Straalen [148]. With permission.

Tests species		A	B	C
Protozoans	*Colpoda cuculus*	+	+	–
Nematodes	*Plectus acuminatus*	+	+	–
Isopods	*Porcellio scaber*	+	–	±
	Trichoniscus pusillus	+	–	±
Mites	*Platynothrus peltifer*	+	–	±
Collembola	*Folsomia candida*	+	+	+
	Orchesella cincta	+	±	±
Enchytraeidae	*Enchytraeus albidus*	±	+	+
Lumbricidae	*Eisenia fetida*	–	+	+
Molluscs	*Helix aspersa*	+	±	±
Hymenopteran parasites	*Encarsia formosa*	±	+	–
	Trichogramma cacoeciae	+	±	–
Beetles	*Bembidion lampros*	+	±	±
	Aleochara bilineata	+	±	±
Predatory mites	*Phytoseiulus persimilis*	±	+	–
Spiders	*Oedothorax apicatus*	±	–	±
Honey bees	*Apis mellifera*	+	+	±

[a] A: ecological relevance, B: potential for standardization and culture by different laboratories, C: potential to derive environmental quality criteria from the test results.

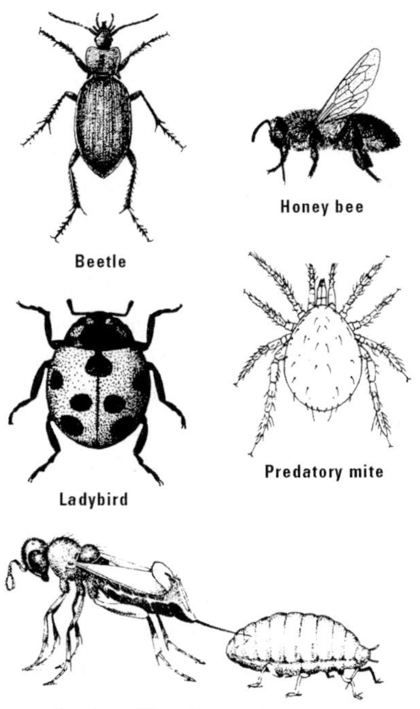

Figure 7.30. Some representative "beneficial" arthropods (not drawn to the same scale).

terrestrial mammals such as the laboratory rat and at least one other species (mouse, rabbit, guinea pig or dog) in order to test for skin and eye irritation and skin sensitization, and to determine acute, repeated dose and reproductive toxicity. As the principles of laboratory toxicity testing of wild mammals and mammals used for HRA do not differ, see Chapter 6 for more information.

In ERA it can be important to obtain toxicity data for birds, since rats and mice do not necessarily provide reliable surrogate data. Many carbamate and some organophosphate insecticides are distinctly more toxic to birds than to mammals. This reflects biochemical and physiological differences between these two taxonomic groups. These additional mammalian laboratory toxicity studies may reduce the uncertainty in ERA, but from an ecological, ethical and cost point of view the need seems questionable, particularly in view of the enormous uncertainty surrounding ERA for other taxonomic groups (Table 7.2).

Bird toxicity studies may be required for the notification of new chemicals or for environmental risk assessment of existing chemicals. In most countries these tests are obligatory for the registration of pesticides. In the US, the USEPA requires a series of tests on pesticides

Table 7.34. Observations in the USEPA avian oral dose LD50 test [168].

- Survival, body weight and food consumption
- Gross necropsies (optional). When performed, all dead birds should be examined, as well as a sufficient number of survivors in order to provide a characterization of gross lesions. Inspections of the gastro-intestinal tract, liver, kidneys, heart, and spleen should be made
- Other signs of intoxication should be described as to what was observed when and for how long

Table 7.35. Characteristics of the short-term OECD avian dietary test [171].

Test species	the Mallard duck *(Anas platyrhynchos)*, the northern bobwhite quail *(Colinus virginianus)*, the Japanese quail *(Coturnix coturnix japonica)*, pigeons *(Columba livia)*, ring-necked pheasant *(Phasianus colchicus)* and red-legged partridges *(Alectoris rufa)*
Test duration	usually 11 d, acclimatization (3 d), exposure to a diet containing the test substance (5 d) and exposure to the basal diet free of the test substance (for a minimum of 3 additional days)
Test levels	5 dietary levels and 2 control groups
Food	commercial food type
Observations	mortality, body weights, food consumption, signs of toxicity, and tissues from poisoned birds or from birds killed at the end of the test may be subjected to pathological, biochemical and residue examination
Parameter	LC50 and, if appropriate, an estimated NOEL

arranged in a tiered system that may progress from basic laboratory studies to applied field studies. Typically, the initial requirement is for two avian laboratory tests, an acute oral LD50 study and a dietary LC50 study. Additional avian reproduction toxicity data may be required with high PEC/NEC ratios or in the event of frequent application or persistence, which may result in long-term exposure. Given the urgent need for reduction of the use of animals, it has been proposed to reduce the number of tests to a modified acute toxicity test and a reproductive toxicity test, and use more efficient test protocols [168].

Determination of the avian single-dose oral LD50 follows the same principles described for mammals (Chapter 6). The species normally tested are the Mallard duck *(Anas platyrhynchos)*, the northern bobwhite quail *(Colinus virginianus)* or the Japanese quail *(Coturnix coturnix japonica)*. Normally five treatment levels are tested plus an additional control group. If necessary, a vehicle control group is included as well. The number of birds per treatment level is 10. In the single-dose oral LD50 test (Table 7.34) food is withheld from all birds for at least 15 hours prior to oral dosing. After administration of the test material the birds should have free access to a standard ration of food and water. Food consumption is monitored. The observation period is 14 days. This period must be extended if toxic signs persist or birds continue to die on the last day of the observation period. The results are expressed as LD50 (mg/kg$_{bw}$). If possible, NOELs should be reported as well. To reduce animal use in this test, a more efficient test protocol has been proposed [169], based on an "up and down" procedure [170]. This procedure is based on a stepwise reduction of the dose to two birds with a fixed factor. Each time the two birds survive, the dose is increased with a factor x. If one bird dies, the dose is reduced by a factor $\sqrt{x}$, or increased by a factor $\sqrt{x}$ if the previous dose was lower. The procedure stops when two deaths occur. The LD50 is then calculated as the geometric mean of the relevant doses. Although the results are less precise than when classical methods are used, far fewer birds are needed.

The avian dietary toxicity test is part of the OECD test guidelines. The aim of this test guideline is to determine the acute dietary LC50 (Table 7.35). A practical problem with the dietary study is the incorporation of the test chemical in food. This can raise certain difficulties related to uniform mixing of the substance in the diet, its volatility and, for pesticides, its formulation. The dietary study might be omitted by using information from a reproduction study for dietary exposure [172].

Long-term toxicity testing is occasionally carried out with birds where a long-term effect is suspected, or with

Table 7.36. Characteristics of the OECD avian reproduction test [172].

Test species	recommended species: the Mallard duck *(Anas platyrhynchos)*, the northern bobwhite quail *(Colinus virginianus)* and the Japanese quail *(Coturnix coturnix japonica)*
Test duration	approximately 34 weeks, exposure to a diet containing the test substance (for a minimum of 20 weeks), collection of eggs (over a 10-week period), followed by incubation and hatching of the eggs, the young are maintained for 2 weeks
Test levels	a minimum of 3 dietary concentrations and 1 control group
Food	commercial food type
Observations	mortality and signs of toxicity, body weights of adults and of the young at 14 days of age, food consumption of adults and young, gross pathological examination of adult birds, egg production, cracked eggs, egg shell thickness, viability, hatchability and effects on young birds, the residue analysis of selected tissues is optional
Parameter	NOEC (mg/kg diet)

chemicals that produce a delayed effect (e.g., certain organophosphorous pesticides). An avian reproduction test guideline is also provided by the OECD (Table 7.36). The aim of the reproduction test is to determine the NOEL (mg/kg diet) for the parameters studied. If a carrier is used for test diets, the same vehicle should be added to the diets of birds in the control group. After an exposure period of 20 weeks, the birds are induced by photoperiod manipulation to lay eggs. The eggs are collected, artificially incubated and hatched. Currently, improvements to this test are being discussed with a 10 week exposure period and fewer birds per treatment.

In test guidelines for birds, particular emphasis is placed on lethal effects. However, animal welfare concerns are pushing towards a reduction in test animals and improved statistical design. The trade-off is that with reduced testing set ups, the ability to detect the effects of chemicals may be less than with classical methods. Clearly, an optimum balance needs to be found between statistical power, animal welfare and sufficient safety for the environment. Examples of how bird and mammalian toxicity studies can be used to derive PNECs for soil and water are given in Section 7.10.3.

Multi-species tests

Model ecosystems, microcosms or micro-ecosystems are designed to simulate certain aspects of real ecosystems and therefore go beyond standardized tests for standard setting or testing of polluted soil. These systems can be used to study effects on individual species, predator-prey relationships, competition for resources, soil functions and biodiversity [173,174].

Multi-species tests with bacteria, plants and invertebrates have not reached the stage of international harmonization. Several terrestrial model ecosystems (TMEs) have been described [174-176] and an attempt has been made to standardize them [177]. The TMEs contain a soil column, soil micro-organisms, invertebrates, sometimes plants or even a small tree. The system may be either closed or open to the ambient air, and may contain intact core samples from a natural habitat or reasonably standardized soil. The effects of pre-treatments, such as drying, sterilizing, inoculation, litter type, age of litter, etc., can have a significant impact on the behaviour of the system and need to be thoroughly investigated.

Different types of TMEs (or integrated soil microcosms) exist. They may be composed of intact soil columns with intact soil cores, indigenous invertebrates and mixed plant flora, or they may be assembled systems consisting of sieved soil, selected introduced and indigenous invertebrates, and perhaps a single plant species [173,174]. The natural situation is approached more closely in the first type, but the second type offers more possibilities for replication under controlled laboratory conditions. Rainfall may be simulated and leachate can be collected and analyzed. Activities of saprotrophic invertebrates can, for example, be easily assessed in terms of system functions [175] such as leaf litter fragmentation and nutrient conversion (Figure 7.31). Uncertainties attached to laboratory-field extrapolation can be partly avoided by carrying out experiments under semi-field or field trial conditions [56], or by using more complex TMEs as a bridge between laboratory and field testing (Figure 7.32).

TMEs and associated modelling [178] offer good potential for improving our ability to predict effects on soil. More scientific research is needed to understand

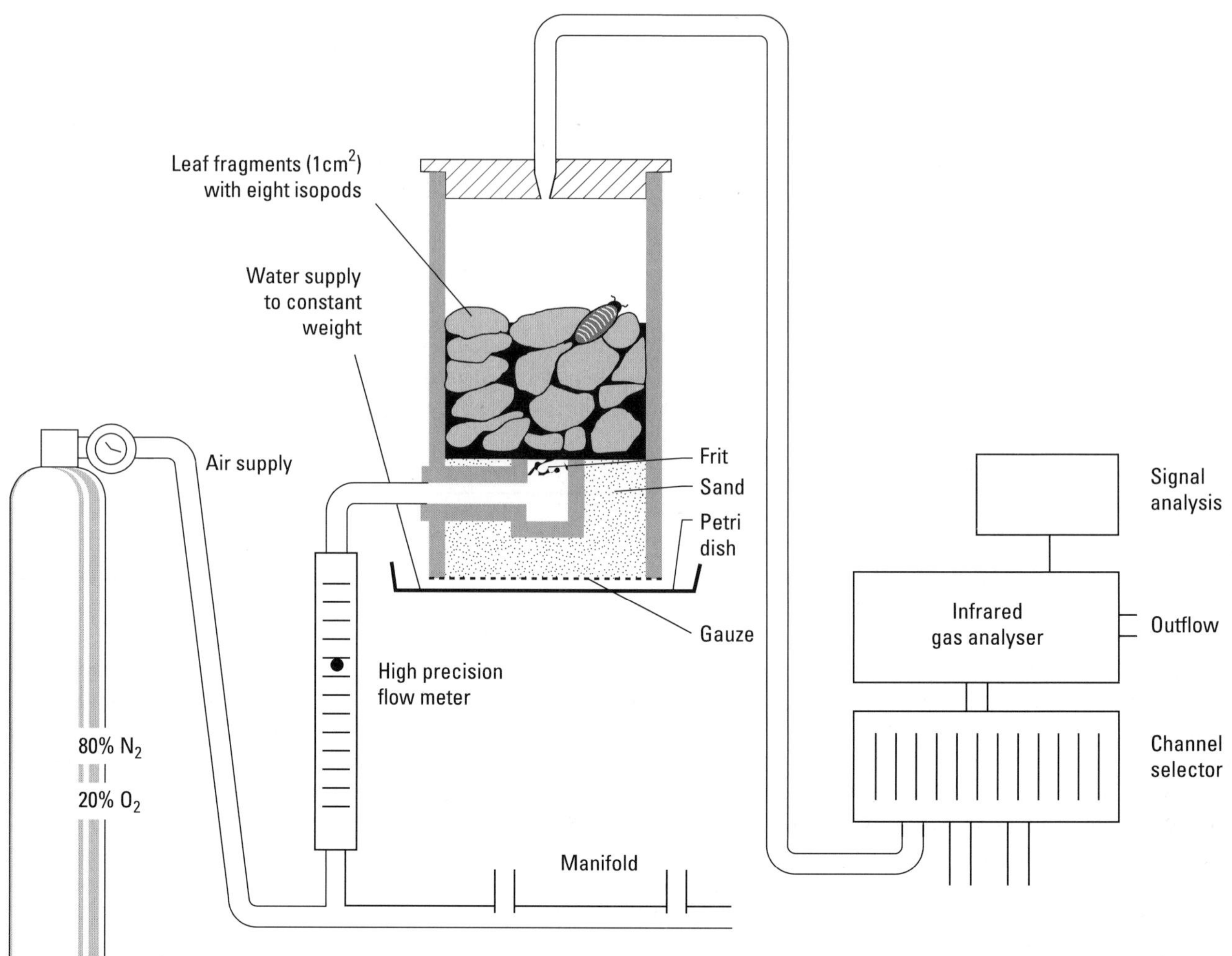

Figure 7.31. Diagram of equipment designed to measure CO_2 production in a flow-through system. The microcosm contains an amount of litter supplemented with isopods, placed on the basal layer of sand. From Van Wensem [175]. With kind permission of Springer Science and Business Media.

the complexity of terrestrial ecosystems and improve the assessment tools for regulatory soil ecotoxicology.

7.6 FACTORS MODIFYING TOXICITY

7.6.1 Introduction

Modifying factors can be defined as any characteristic of the organism or its environment that affects the toxicity of a particular chemical. The initial topics discussed in the previous sections on aquatic, sediment and terrestrial toxicity included exposure systems. Exposure and exposure systems are extremely important. Exposure systems affect the direct or indirect bioavailability of the test chemicals. Exposure systems affect the behaviour of the exposed test species and the behaviour of the test species may affect the bioavailability of the chemical. Even in very simple artificial laboratory test systems, modifying factors can dominate the results of the toxicity test. Numerous modifying factors have also been summarized in Table 7.12. An extensive review of the literature on modifying factors has not been attempted; only major factors are described in more detail in the following sections. Bioavailability issues are also addressed in Chapter 3.

7.6.2 Abiotic factors

Oxygen concentration

Lloyd [179] published a guide for estimating the lethal level of ammonia, based on research with rainbow trout

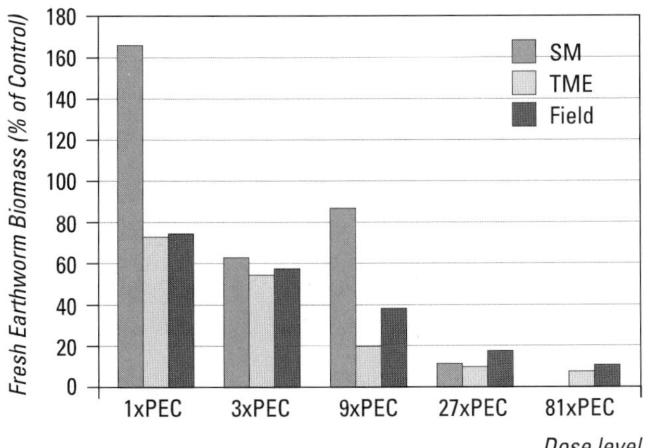

Figure 7.32. Effects of a fungicide on earthworm biomass in a comparison between a soil microcosm (SM), a terrestrial model ecosystem (TME) and a field experiment. From Edwards [174]. With permission. Copyright Elsevier.

Table 7.37. Effects of oxygen on the toxicity of ammonia to rainbow trout. From Lloyd [179].

Oxygen saturation (%)	LC50 (mg/L N)
81	42
62	34
41	25
30	21

Table 7.38. Effects of pH on the toxicity (LC50 in µmol/L) of chlorophenols to fish. From Hermens (unpublished results).

Chemical	pH = 6	pH = 8
4-Chlorophenol	60	71
Pentachlorophenol	0.44	3.4

and the chemical behaviour of ammonia in water. It was shown that low oxygen saturation levels increased the aquatic toxicity of ammonia (Table 7.37). Depletion of oxygen also favours the activity of anaerobic bacteria, reducing conditions (speciation of heavy metals), and affects the breakdown of organic chemicals.

Redox potential (Eh)

Decreasing redox potential (more reducing conditions) mobilizes oxide-sorbed toxic chemicals as it dissolves iron and manganese oxides. Increasing redox potential (more oxidizing conditions) mobilizes heavy metals by dissolving metal sulfides [179]. The influence of acid volatile sulfide (AVS) has been studied extensively [180]. Sulfides of cadmium, copper, nickel, lead, and zinc have lower sulfide solubility product constants than sulfides of iron and manganese, which are naturally formed as a product of the bacterial oxidation of organic matter in sediments. Manganese and iron will be displaced when metals are in a sediment with manganese and iron monosulfides. Because these sulfides have low solubility, sediments with an excess of AVS will have very little metal activity in the interstitial water and the expected toxicity will be low. The metal/AVS ratio is indicative of toxicity. The vast majority of sediments found in the environment have metal/AVS ratios <1.0 and toxicity is predicted to be low. For sediments with metal/AVS ratios >1.0 toxicity is less certain.

Temperature

Temperature affects the solubility of chemicals in water, it influences the form of some chemicals (e.g., ammonia), and governs the amount of oxygen dissolved in water. It also affects biochemical processes such as mineralization. Temperature also affects the activity of cold-blooded animals up to a certain maximum, which is species dependent. There is no single pattern for the effects of temperature on the toxicity of pollutants. The toxicity of metals (e.g., zinc) generally increases with increasing temperature, whereas the aquatic toxicity of pesticides can be positively, negatively [47] or not correlated with temperature [181].

Hydrogen ion concentration

The behaviour of weak acids and bases depends on the extent to which they exist in the neutral or charged state. This is determined by the pK_a value of the chemical and the pH. The pH affects the toxicity of ionized chemicals. Generally, chemicals are more toxic in their neutral unionized state. Pentachlorophenol (pK_a = 4.69) and, to a lesser extent, 4-chlorophenol (pK_a = 9.37) are more toxic at low pH values (Table 7.38). These chlorophenols are weak acids. In normal pH ranges they are dissociated ($HA \Leftrightarrow H^+$ and A^-) in water. The presence of the ionized toxic form increases with pH as $\log [A^-] / [HA] = pH - pK_a$, resulting in a higher LC50. Similarly, the toxicity of ammonia (pK_a = 9.35) increases with pH as the proportion of ammonia in the toxic unionized state (NH_3) increases. Lowering pH also increases heavy metal solubility which enhances bioavailability and thus ecotoxicity (Figure 7.33). Lowering pH also reduces the cation exchange capacity of soil, and alters the soil

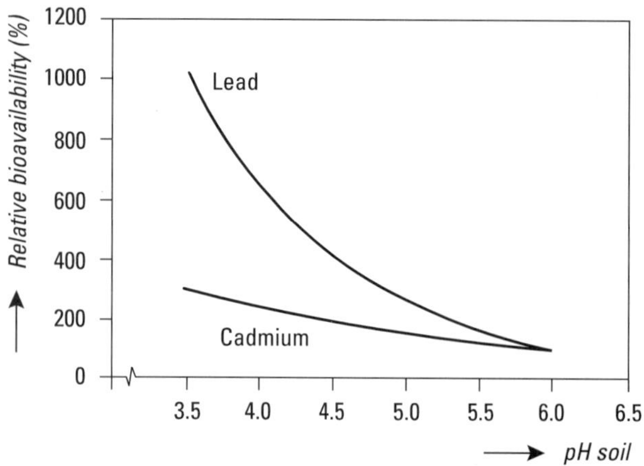

Figure 7.33. Estimated relationship between the relative bioavailability of cadmium and lead for the earthworm *Lumbricus rubellus* and soil pH, with pH 6.0 taken as 100%. From Bergema and Van Straalen [182]. With permission.

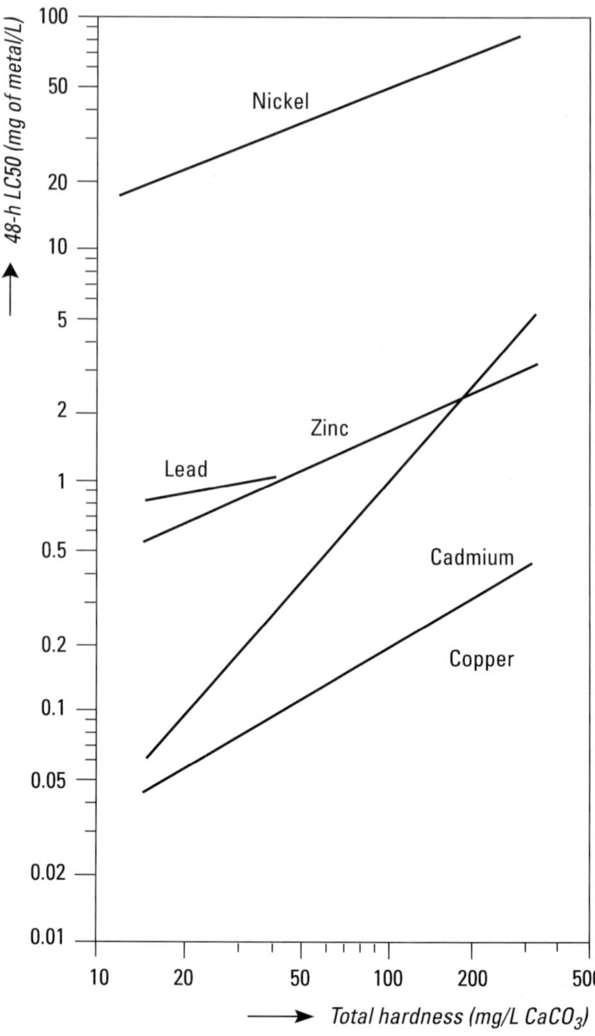

Figure 7.34. Relationship between total hardness of water and 48-h LC50 of some heavy metals in rainbow trout. From Brown [183]. With permission. Copyright Elsevier.

microbial population. Changes in microbial activity affect the biodegradability of chemicals and may also affect their bioavailability.

Water hardness

Calcium and, to a lesser extent, magnesium are the predominant dissolved cations in fresh water and are chiefly responsible for water hardness. Water hardness affects the speciation of heavy metals in a complicated manner. The aquatic toxicity of heavy metals such as cadmium, copper, lead, zinc and nickel decreases with increasing hardness (Figure 7.34).

Cation or anion exchange capacity (CEC or AEC)

Soil with a low CEC or AEC has a poor capacity to retain cations (e.g., metals) or anions (e.g., organic anions) by sorption. CEC and AEC are important soil properties which depend on inorganic clay mineral content and type, organic matter content, and soil pH [184].

Clay and organic matter

High clay and organic matter (OM) content reduce the bioavailability of many organic chemicals and heavy metals, and thereby toxicity (Figure 7.27). Decreasing OM content reduces CEC, soil buffering capacity, the sorption of toxic organics, and soil water-holding capacity, it also alters physical structure (e.g., increases soil erodibility) and decreases microbial activity. Clay and OM content are among the most important soil and sediment capacity-controlling properties. In fact, they determine the cation exchange capacity. Regressions or "reference lines", as they are known, were originally developed to correct for background concentrations in different soil types, but are now applied as a bioavailability correction [185]. In The Netherlands, these relationships are used as correction factors to compare measured concentrations of heavy metals and organic chemicals (C_{obs}) in different types of soils, each with their own respective quality standards:

$$QS \geq C_{obs} \frac{a + 25b + 10c}{a + bL + cH} \tag{7.8}$$

where QS is the quality standard for "standard soil", i.e., soil or sediment with a 25% clay content (w/w) and a 10% organic matter content (w/w), C_{obs} is the observed concentration of the contaminant in soil, L is the measured percentage of clay (fraction < 2μ) in the soil or sediment, H is the measured percentage of organic matter (humus) in the soil under investigation, and a, b and c are constants whose values depend on the specific contaminant under consideration. For example, for cadmium these values are 0.4, 0.007 and 0.021, respectively. The constants for the various metals are derived from measurements in undisturbed soil taken from nature reserves, and therefore are not indicative of bioavailability. It should be noted that the pH, an important factor which determines the bioavailability of metals, is not included in the equation. For organic chemicals the clay content is not considered important and standard soil is simply defined as soil with 10% organic matter.

Salinity

Increasing salinity can make toxic chemicals more soluble by altering the ion exchange equilibrium, increasing soluble complexation, and decreasing chemical thermodynamic activities in solution. It can also reduce microbial activity. For metals, toxicity increases with decreasing salinity. For organophosphorous insecticides, the opposite was found [186].

7.6.3 Biotic factors

Biotic characteristics also constitute important modifying factors. Food availability influences the energy budget of species. The allocation of energy to maintenance, growth and reproduction can be affected by both toxicants and food availability [38]. Lack of food generally makes species more sensitive to the effects of chemicals [181,186].

Sometimes it is difficult to distinguish between biotic and abiotic factors because in practice complicated interactions take place. Jagers op Akkerhuis [187] showed a strong positive correlation between spider activity and deltamethrin-induced toxicity. He demonstrated that pesticide toxicity was determined by walking activity through increased residual uptake via the cuticle. Walking activity itself was greatly affected by temperature and humidity.

The most important biotic factors are the test species themselves. There are clear differences in sensitivity. These can be explained on the basis of taxonomy, (i.e., morphological and physiological differences), trophic level, (i.e., the niche they occupy), and the exposure routes of the chemicals. Life stage (Section 7.3.4) and size, (i.e., surface/volume ratio (Section 7.2.5)), intrinsic rates of increase (*r* or *K*-species, Section 7.3.5) all affect the susceptibility of species. The same is true in relation to a number of factors, such as nutrition, health, population density, parasitism and acclimation, all of which can be controlled during toxicity testing (Table 7.12).

7.6.4 Biotic ligand models to predict toxicity of metals

All of the abiotic factors mentioned above have an influence on the form in which metals are present, and this affects the bioavailability of metals and their effect on organisms. Predicting the toxicity of metals has evolved to much more than adjusting for the influence of Mg^{2+} and Ca^{2+} content. Chemical equilibrium models can be used to predict in which forms metals are present in the water column, also called metal speciation. Different water characteristics lead to differences in metal speciation that in turn affects the acute toxicity of the metals. The method is an extension of the gill surface interaction model and the free ion activity model (FIAM), where the free ion is responsible for the toxicity. The free ion concentration is calculated with chemical equilibrium models, but other reactive metal species can also bind to the critical sites and thus need to be incorporated [188,189]. Biotic ligand models (BLMs) have been developed to predict the effect of these complex abiotic-biotic interactions on metal accumulation and toxicity (Figure 7.35). The development of BLMs is reviewed by Niyogi and Wood [189] and Paquin et al. [190]. The term “biotic ligand” refers to a discrete receptor or site of action in an organism where accumulation of metal leads to toxic effects. Acute toxicity of the metal is related to the critical metal accumulation at the biotic ligand. However this critical concentration or critical burden may be receptor-specific instead of a whole-tissue concentration or burden [189].

The BLM can provide an estimate of the amount of metal accumulation at the biotic ligand site for a variety of chemical conditions and metal concentrations. BLMs for various metals have been developed for algae, daphnids and fish [191,192]. The current focus of BLM research is on predicting chronic toxicity [193,194], since for risk assessment purposes the application of a pragmatic acute-chronic ratio to a mechanism-based BLMs is not appropriate.

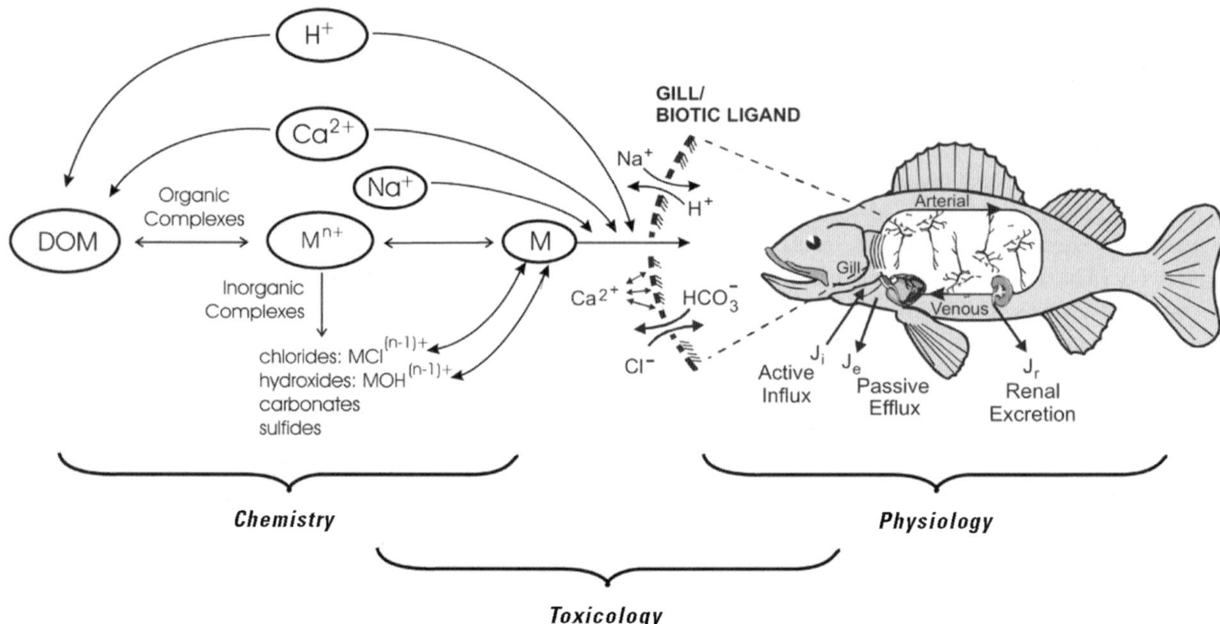

Figure 7.35. Diagram of the biotic ligand model (BLM) and the relation between chemistry, physiology and toxicology in the BLM approach. From Paquin et al. [190]. With permission. Copyright Elsevier.

7.7 MIXTURE TOXICITY

7.7.1 Mixture toxicity scales

Much of the information available on the ecotoxicity of substances relates to chemicals tested simply under laboratory conditions, or considered separately in field studies. Yet, it is uncommon to find an aquatic or terrestrial ecosystem which is polluted by a single toxicant. Usually several harmful substances are present together in significant quantities in polluted soil, sediment or surface water. This possibility of organisms being exposed to several chemicals simultaneously requires consideration of the possible interactions between the chemicals themselves and their effects on the organisms.

Table 7.39 gives four types of joint action with respect to quantal responses. A joint action is defined as similar or dissimilar depending on whether the sites of primary action of the two chemicals are the same or different, and as interactive or non-interactive depending on whether one chemical does or does not influence the biological action of the other. Other terminology is given in Figure 7.36. In most practical applications of mixture toxicity, the concepts of concentration-addition and response addition are most frequently used.

Concentration-addition is used for chemicals with a similar mode of action. The joint effect of such a mixture is calculated with concentration-addition rules [188,189]. Almost every hydrophobic chemical can exert at least a narcotic, non-specific toxicity often called baseline toxicity [30,33]. The toxicity of mixtures of narcotic chemicals can be calculated using concentration-addition rules. For chemicals with different modes of toxic action that do not interact at the target site or receptor, mixture toxicity can be described by response addition [195]. For mixtures where interaction between the tested chemicals does occur, the theory is well developed [196] but toxicological confirmation is not strong.

Table 7.39. Four types of joint action of chemicals according to Plackett and Hewlett [195]. With permission.

	Similar joint action	Dissimilar joint action
Interaction absent	simple similar action or *concentration addition*	independent action or *response addition*
Interaction present	complex similar action	dependent action

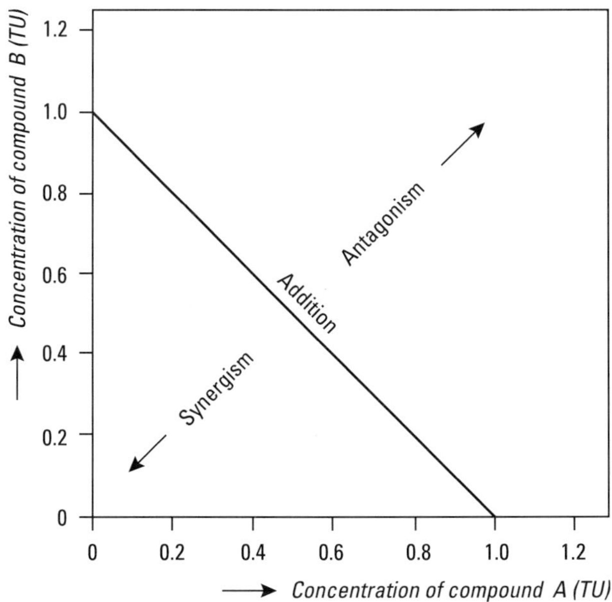

Figure 7.36. Possible toxicological interactions in a mixture of two chemicals.

Quantitative structure-activity relationships (QSARs) can be used in the classification of many pollutants into a small number of groups of compounds with a similar mode of toxic action (see also Chapters 10 and 11). When this classification is applied to mixtures of more than two chemicals, problems can arise because the different pairings can fall into different classes of joint action, and other joint actions may be possible between different pairs. Therefore, a mathematical description of the joint toxicity of a mixture of *n* compounds ($n \geq 2$) is possible only in a few cases.

Effects in LC50 experiments can be predicted in mixture toxicity studies by using compounds with a similar mode of action (concentration-addition). In a mixture of two compounds 50% mortality will be produced if both compounds are present at a concentration of 0.5 of their respective LC50 values. The ratio of a chemical's concentration and its LC50, i.e., c/LC50, is termed the toxic unit (TU) [197]. In a mixture of 10 chemicals the same effect will be observed when each chemical is present at a concentration of 0.1 TU, i.e., an equitoxic mixture with for each chemical equal fractions of their LC50s. Thus, concentration-addition means that the LC50 of a mixture *M* is described by the sum of the concentrations of *n* individual compounds (expressed as fractions of their LC50), equalling unity, and the effect would be equal to the effect of 1 TU. In mathematical terms this can be expressed as:

Table 7.40. Classification of mixture toxicity [198].

M	Classification of mixture potency
> n	antagonism
n	no-addition
1 to n	partial addition
1	addition
<1	supra-addition

$$M = \sum_{x=1}^{n} \frac{c_i}{LC50_i} = 1 \tag{7.9}$$

It should be noted that the effective concentration of the mixture M is assumed to be unity in equitoxic mixture studies. In field situations, equitoxic mixtures never occur, but the principle holds for studying the effect of mixtures.

When compounds in a mixture have different modes of action the situation becomes much more complicated. The effect can be predicted in only one case, i.e., when the compounds in the mixture have dissimilar modes of toxic action, and the tolerances are fully positively correlated (independent action). When an LC50 experiment is performed 50% mortality will be observed if one of the compounds is present at a concentration which equals its LC50. In equitoxic mixtures this means that every single compound will be present at its LC50 concentration, which means that the sum of the concentrations equals the number of compounds *(n)* present in the mixture *(M=n)*. In fact, there is no combined effect at all; the toxicity of the mixture does not exceed that of the compound present at the highest toxic concentration. This situation is therefore called "no addition". When the compounds interfere toxicologically, i.e., interaction occurs, the toxicity of the mixture may vary from partially additive, if *M* lies between 1 and *n*, to antagonistic or supra-additive, when *M* is either greater than *n* or smaller than 1.0 (Table 7.40). Antagonism occurs when one compound diminishes the toxic effects of another. Supra-addition or potentiation is the opposite effect: one compound increases the toxicity of another.

Most studies on the combined effects of compounds have been performed with mixtures of only a few compounds. The exception to this are aquatic toxicological studies in which complete additivity has been proven for mixtures of many compounds (n = 50) with a similar mode of action [199].

Table 7.41. Toxicity of equitoxic mixtures of chemicals having similar modes of toxic action[a]

Mixture	Species	Criterion	n	M	Reference
1	*P. reticulata*	14-d LC50	50	0.9	[199]
2	*D. magna*	48-h EC50	50	1.2	[200]
3	*D. magna*	16-d LC50	25	1.5	[200]
4	*D. magna*	16-d EC50[b]	25	1.5	[200]
5	*D. magna*	16-d EC50[c]	25	0.6	[201]
6	*P. reticulata*	14-d LC50	11	1.0	[199]
7	*P. reticulata*	14-d LC50	17	1.1	[202]
8	*P. reticulata*	14-d LC50	9	1.0	[202]

[a] Mixtures 1-5 comprise chemicals with a limited chemical reactivity (narcotic chemicals), mixtures 6-8 show results of experiments with chlorophenols, anilines and reactive organic chemicals, respectively.

[b] Reproduction.

[c] Growth.

7.7.2 Mixture toxicity studies

Many mixture toxicity studies have been carried out with only two chemicals. This information was published in a report by the European Inland Fisheries Advisory Commission [198]. Mixture toxicity experiments have also been performed with 50 narcotic compounds with fish *(Poecilia reticulata)* and daphnids *(D. magna)*. The results are summarized in Table 7.41 for equitoxic mixtures. The experiments affirm the assumption of concentration-additivity (M = 1). Mixtures 3, 4 and 5, however, show an M value which deviates quite considerably from 1, but the value suggests additivity rather than non-additivity (Table 7.40). Deneer et al. [203] showed that in mixtures consisting of narcotic chemicals, compounds present at concentrations as low as 0.0025 TU will still contribute to joint toxicity. The concentration-addition model was still valid at these very low concentrations. Clear examples of concentration-addition were also provided in fish toxicity studies with equitoxic mixtures of chlorophenols, anilines and reactive organic chemicals (Table 7.41). The toxicity of a mixture of 18 triazine herbicides to algae was very well predicted by concentration-addition (Figure 7.37).

Table 7.42 shows the results of experiments with mixtures of compounds with different modes of action. In cases where chemicals have strictly dissimilar modes of action, Faust et al. [207] showed that response-addition rules could predict algal toxicity due to a mixture of 16 chemicals (Figure 7.38). The joint effect of chemicals where response-addition rules are expected to apply are often underestimated by these rules. Complex, larger mixtures often show a tendency to behave as mixtures with a similar mode of action, best described by concentration-addition rules [208,209]. This may be caused by the absence of true response-addition of the compounds involved. It may also be caused by the combined effects of non-specific chemical activity, e.g., several neurotoxic pesticides also show minimal toxicity due to their anaesthetic action. This combined effect may result in partial response additivity which is hard to distinguish from concentration-addition in large mixtures. In studies with *D. magna* carried out by Enserink et al. [210], the combined effect of equitoxic mixtures of

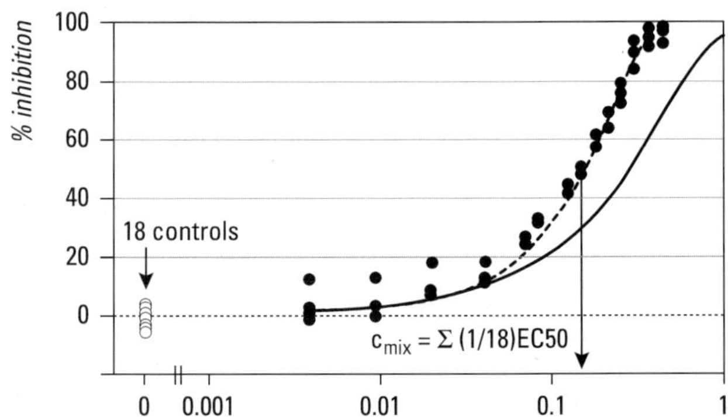

Figure 7.37. Observed and predicted toxicity to algae of a mixture of 18 triazine herbicides, mixed in the ratio of their individual EC50 values. Dashed line: prediction according to concentration-addition; solid line: prediction according to response-addition (independent action). Open symbols: controls, filled symbols: observed toxicity. From Faust et al. [204]. With permission. Copyright Elsevier.

Table 7.42. Toxicity of equitoxic mixtures of chemicals having different modes of toxic action.

Mixture	Species	Criterion	n	M	Reference
9	*P. reticulata*	14-d LC50	8	1.1-1.7[a]	[205]
10	*P. reticulata*	14-d LC50	24	2.3	[205]
11	*D. magna*	16-d LC50	14	1.1	[206]
12	*D. magna*	16-d NOEC[b]	14	1.9	[206]

[a] Results of 5 experiments with different mixtures.
[b] Reproduction.

eight heavy metals was near complete concentration-addition. These results were in accordance with effects of equitoxic mixtures of six metals on survival, body weight and reproduction of *Ceriodaphnia dubia* [211].

In conclusion, concentration-addition strictly applies to chemicals with similar modes of toxic action and response-addition (independent action) predicts well for strictly dissimilarly acting chemicals [205]. Chemicals with different modes of toxic action can often almost behave according to concentration-addition. The most important conclusion from experimental studies on the combined effects of heavy metals and organic chemicals is that mixture toxicity is a reality. Chemicals exert their detrimental effects in equitoxic mixtures at very low concentrations, at fractions of 0.0025 of their LC50, i.e., at or below their NOEC [203]. This raises major questions about the quality criteria set for single compounds. In fact, it has been shown that mixtures

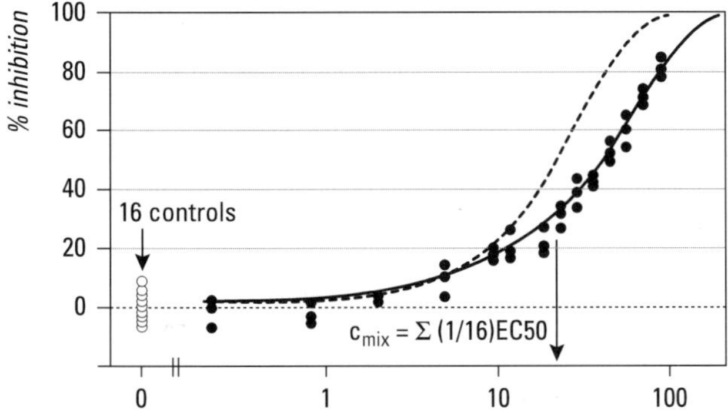

Figure 7.38. Observed and predicted toxicity to algae of a mixture of 16 dissimilarly acting substances, mixed in the ratio of their individual EC50 values. Dashed line: prediction according to concentration-addition; solid line: prediction according to response-addition (independent action). Open symbols: controls, filled symbols: observed toxicity. From Faust et al. [207]. With permission. Copyright Elsevier.

of heavy metals at levels of their water quality criteria induce adverse effects on crustaceans and fish [210,211]. The consequences for the risk management of chemicals should then be taken into account.

A very pragmatic way of dealing with mixtures in the context of quality objectives is to calculate the ratio of the ambient concentration and the quality objective for each compound. The sum of these fractions is the scaled risk quotient *RQ* for the mixture:

$$\mathrm{RQ} = \frac{C_a}{\mathrm{EQO}_a} + \frac{C_b}{\mathrm{EQO}_b} + \frac{C_c}{\mathrm{EQO}_c} + \cdots \qquad (7.10)$$

with C_x ambient concentrations for substance x, and EQO_x, the quality objective for substance x. If the RQ is > 1, the environmental quality objective for the mixture is exceeded. If RQ <1, the environmental quality objective is not exceeded. This procedure is often used, but can lead to misleading results for the following reasons. It is assumed that the quality objectives (if present) are derived in a comparable way for each substance. In reality, the EQOs are often not based on comparable data sets and similar effects. It is inappropriate for substances that have dissimilar modes of action and therefore do not follow simple concentration-addition rules. This limits the practical use of this rule for structurally similar substances that have the same mode of toxic action. The risk to ecosystems further to exposure to mixtures can be assessed by linking mixture toxicity rules to the SSD concept [212]. The SSD concept is explained in Section 7.10.2.

7.8 ECOTOXICOGENOMICS

Experimental data can be supplemented or even replaced by the use of *in vitro* tests, (Q)SARS, and read-across methods, as explained in Chapters 6, 10 and 11. The rapidly developing application of molecular biology in ecology and ecotoxicology *("genomics")* holds promise for developing alternatives to *in vivo* testing as well [213]. In addition, genomics could be used for prioritization of chemicals, guiding experimental design and providing insight into the molecular and mechanistic background to toxicological effects [214,215].

The field of genomics aims to elucidate how the genome of a species translates into biological functions. Genomics consists of many disciplines and methods including sequencing, identifying the function of specific genes, gene expression by studying mRNA transcription *("transcriptomics")*, protein expression *("proteomics")*, and metabolite characterization *("metabolomics")*. The term ecotoxicogenomics was coined to cover the application of these methods to ecotoxicology [216]. At present, the available methods do not yet allow the use of genomics in regulatory testing. However, this may change [215,217]. The challenge faced in ecotoxicogenomics is to get a grip on the relationship between the toxicological stimulus, gene transcription and expression and the ensuing metabolic changes, and the relationship between dose/concentration and effect.

Gene expression is usually altered as a result of toxicity. Single gene biomarkers exist for classes of chemicals, such as induction of hepatic vitellogenin mRA by oestrogen-like compounds, or up regulation of cytochrome P450 1A by binding of planar aromatic compounds to the aryl hydrocarbon receptor [215]. The significance of changes in gene expression in terms of concentration response for risk assessment purposes may be difficult to interpret, since the mRNA is not always transcribed and many proteins are modified after translation. This is essentially the same "significance" debate on the use of biomarkers in risk assessment [218].

Transcriptomics deals with studying changes in genome-wide expression through quantification of mRNA, possibly extending to many thousands of genes at the same time [216]. With this method, transcripts that are up and down regulated as a result of experimental conditions can be identified. There is a need to analyze the transcriptome in reaction to non-toxicological and toxicological stimuli in order to interpret the toxicological "fingerprint" compared to the control organisms. A more advanced method is where the transcriptome can be unambiguously related to specific genes (profiling), but this requires that the genome of the species is sequenced.

Proteomics refers to the total evaluation of protein profiles in a cell or specific tissue. This can provide the linkage between gene regulation and the phenotypical changes in response to a chemical or class of chemicals.

Metabolomics describes the overall characterization of the dynamic metabolic reaction to a toxic or physiological stimulus. Both proteomics and metabolomics refer to functional entities within the tissue of the cell and offer a more integrated assessment than that based on genes or gene products [215,216]. The challenge is to use these methods to provide insight into the mode of toxic action, act as evidence for the absence of effects (decision for no further testing), or to replace or complement further testing [214]. The application of ecotoxicogenomics in regulatory testing has been elaborated by Tyler et al. [215] and Ankley et al. [217].

7.9 ENDOCRINE DISRUPTION

Endocrine disrupting compounds cause functional changes of the endocrine system through a variety of mechanisms. It is one of the aspects of reproductive toxicology. Endocrine disruption may result in adverse effects in an organism or its progeny. Effects on reproduction and development are especially of interest due to their possible effects at the population level (see previous section). Following the seminal book on endocrine disruption by Colborn et al. [219], many new tests have been developed (for an overview, see [220]) with an emphasis on the steroid sexual hormone system of vertebrates, but none has yet been approved by the OECD at the time of writing this chapter. The goal of these studies is to determine if modulation of endocrine activity leads to serious long-term adverse effects that cannot be detected with other toxicity tests. Some effects on early life stages could lead to delayed population effects that can only be detected in life-cycle tests with one or more consecutive generations.

Endocrine disruption can be studied in *in vitro* studies, mammalian screening assays or human health studies for repeated-dose, carcinogenicity and reproductive toxicity (see Chapter 6). The difficulty with endocrine disruption in ecotoxicology is that species from different phyla have different endocrine systems that may not react to a chemical in the same way that mammals do. *In vitro* screening assays have been developed that are mostly based on cell lines or receptors from mammalian tissue, such as estrogenic androgenic, progestagenic and thyroidal receptor binding assays. An example of a fish-specific estrogen activity assay is the induction of the egg yolk protein vitellogenin in cultured hepatocytes from fish liver. Batteries of *in vitro* screening assays have been used to identify endocrine modulating effects in both man and wildlife for a specific group of chemicals, such as brominated flame retardants [221].

Table 7.43. Stages in risk assessment and required effects information [4].

Tiers	Stages	Effects data
Tier-1	preliminary or initial	short-term toxicity
Tier-2	refined	chronic toxicity
Tier-3	comprehensive	(semi) field data

Table 7.44. Assessment factors applied in aquatic effects assessment [14].

Available information	Assessment factor[a]
At least one short-term L(E)C50 for each trophic level (base set: algae, *Daphnia* and fish)	1000
One long term NOEC(fish or *Daphnia*) in addition to base set	100
Two long term NOECs from two trophic levels (fish, *Daphnia* or algae) in addition to base set	50
Three long term NOECs from three trophic levels (fish, *Daphnia* or algae) in addition to base set	10
Species sensitivity distribution method	5-1, case by case
Field data or model ecosystems	case by case

[a] Many additional rules are available to cover different situations, leading to adaptations of the appropriate assessment factor [14].

In vivo screening assays are based on changes in vitellogenin levels to signal estrogenic or anti-estrogenic effects, or androgenic effects (21-d fish screening assay, draft OECD guideline). Other proposed fish tests are tests on sexual development, reproduction and a two-generation full life-cycle test, allowing effects on the F2 generation to be studied. Amphibian metamorphosis of *Xenopus laevis* is under the influence of thyroid hormones, and has been proposed as a 21-d study. Adverse effects on thyroid activity can be developmental disturbance, histopathological effects on the thyroid or thyroid hormone levels [222].

Confirmatory tests are all based on reproduction studies and have been proposed for *D. magna* (enhanced OECD 211 guideline), copepods and mysids. These tests can detect effects on invertebrate hormone systems such as ecdysteroids.

Apart from improvements in test design, endocrine modulating effects need to be identified efficiently and with sufficient coverage of phyla in the animal kingdom.

7.10 DERIVATION OF PNECs

According to the OECD [4] effects assessment can be divided into three stages, depending on the type of information available (Table 7.43). *Preliminary effects assessment* is the stage at which only reliable QSAR estimates or a few LC50 or EC50 values from short-term studies are available. *Refined or intermediate effects assessment* can take place if a few NOECs from chronic tests are available and, finally, *comprehensive effects assessment* is the stage at which field studies, multi-species toxicity studies (or many chronic test results) are available. At each stage different methods may be applied to arrive at a PNEC for the environment. PNECs are derived based on a number of important assumptions. These assumptions are critical to this analysis although their validity has not been thoroughly substantiated:

1. The species selected for testing are representative of the sensitivities of species found in ecosystems.
2. The chronic toxicity threshold determined for the most sensitive species is the chronic toxicity threshold for ecosystems.
3. Species and species-level properties of ecosystems are the most sensitive to toxic chemicals.

Effects assessment does not go beyond the preliminary stage for most chemicals because of the lack of basic toxicity data. This means that, in practice, precise predictions about effects at the ecosystem level can hardly ever be made. Yet, PNECs can always be predicted even at the preliminary stage. This means that chemicals can always be compared on the basis of little data, provided that the assessments are carried out consistently (Chapter 1). Effects assessment involves many uncertainties and

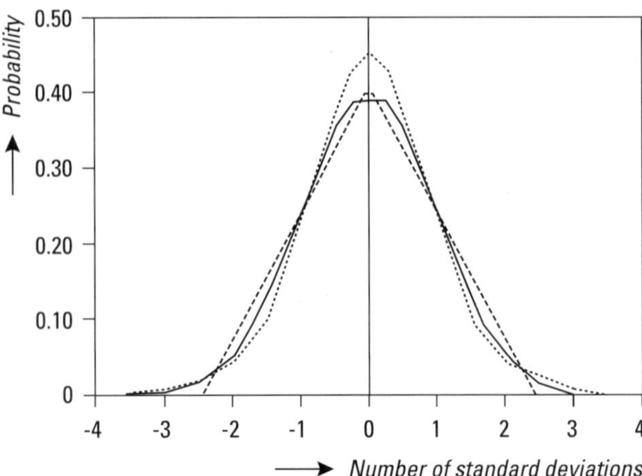

Figure 7.39. Probability density functions for standard log-logistic (·······), log-normal (——) and triangular distributions (- - -) (m = o, s = l) of species' sensitivities. Modified from Health Council of The Netherlands [10].

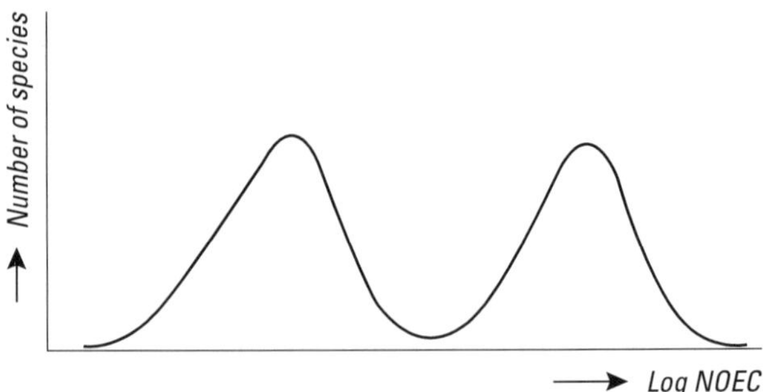

Figure 7.40. For chemicals with a specific mode of toxic action, e.g. certain herbicides and insecticides, a bimodal distribution of species sensitivities may be found instead of a log-logistic or log-normal distribution. In such cases HC5 calculations can be made for both the target and non-target species.

many extrapolations are made from a few species to many species, from acute to chronic effects, from the laboratory to the field, etc. [1,5,223].

7.10.1 Preliminary effects assessment using assessment factors

For the estimation of PNECs, assessment factors or uncertainty factors can be applied to the available toxicity data to account for the different sensitivities of other, untested species in ecosystems. When only a limited set of toxicity data is available, a *constant assessment factor* is used to adjust the effects concentration (laboratory LC50, EC50, NOEC, etc.) to PNECs for ecosystems (Table 7.44). Assessment factors may be used to extrapolate from concentrations with acute effects to NOECs, from a few NOECs to a representative sample, and from the lowest chronic NOEC to the field situation. For each extrapolation step a factor of 10 is suggested. If a data set contains LC50 values for algae, daphnids and fish, the PNEC is estimated from the lowest LC50/(10 x 10 x 10).

The assessment factors presented in Table 7.44 are largely based on a report dating from 1984 [224] which have subsequently been updated in different regulatory frameworks, e.g., in the EU [14]. Although the assessment factors may differ between these frameworks, there is agreement about the magnitude of these factors.

Assessment factors are not based on any theoretical model but are based on experience with chemical effects assessment. They are useful but provide only an approximate means of deriving PNECs. Assessment factors should be used with care with *acute data* since specific modes of toxic action may not be detected in acute toxicity tests (e.g., pesticides, neurotoxicants, cell division inhibitors), or for chemicals with high log K_{ow} values that significantly bioaccumulate. The test results must be evaluated to confirm, for example, that the test concentration does not exceed solubility limits and that the duration of the test is sufficiently long in relation to the log K_{ow} value or LC50-time curve (Figure 7.10). The assessment factor approach is suggested for extrapolation of a limited set of laboratory toxicity data not only for aquatic species, but also for terrestrial and sediment invertebrate species and for birds and mammals.

7.10.2 Refined effect assessment using species sensitivity distributions

PNECs can be calculated using assessment factors (Table 7.44), but they can also be calculated with species sensitivity distributions (SSDs). The variation in sensitivity of species to a contaminant, described by a statistical or empirical distribution function of responses is called a species sensitivity distribution (SSD). The input for these calculated extrapolation models are LC50s or NOECs from a number of representative species. For the derivation of environmental quality objectives, it is common to use NOECs. Especially for data-rich substances, the SSD method can be used to analyze patterns in species sensitivity and derive quality objectives based on statistical theory instead of fixed assessment factors. The use of SSD in ecotoxicology is reviewed by Posthuma et al. [225]. The SSD approach is based on five critical assumptions:

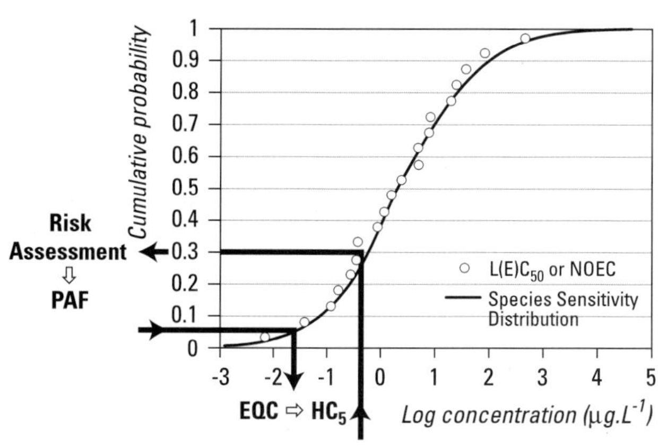

Figure 7.41. Cumulative species sensitivity distribution (SSD), with the toxicity data for different species (dots) and the fitted SSD (line). SSDs can be used in two ways: for calculating risk at a specific concentration (expressed as potentially affected fraction, PAF), or by calculating an environmental quality criterion (EQC) for a certain cut-off value, e.g. the 5th percentile (HC5). From Posthuma et al. [225]. With permission.

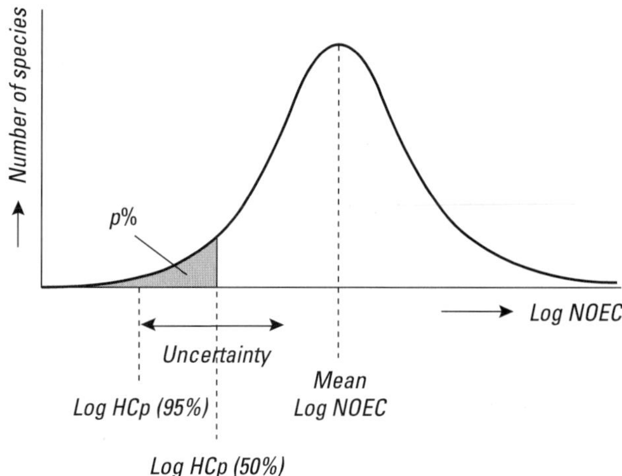

Figure 7.42. The normal density function and estimation of the concentration at which the NOEC of no more than 5% of the species within an ecosystem is exceeded (HC5). The HC5 can be calculated at two levels of confidence: 50% and 95%. The 50% confidence estimate of the HC5 is the "most probable" estimate, whereas the left 95% confidence limit of the HC5 is the "safer" value [226]. With permission. Copyright Elsevier.

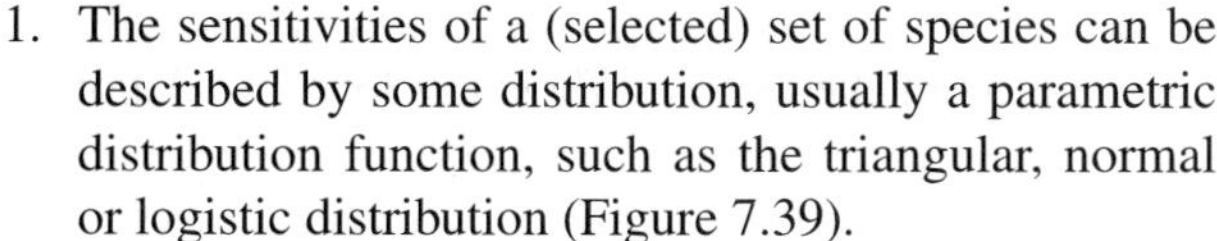

1. The sensitivities of a (selected) set of species can be described by some distribution, usually a parametric distribution function, such as the triangular, normal or logistic distribution (Figure 7.39).
2. Since the true distribution of toxicity endpoints is not known, the SSD is estimated from a sample of toxicity data.
3. The distribution should adequately describe the observed sensitivity of species. In the case of chemicals with a specific action, a bi-model pattern of target species versus non-target species is often observed (Figure 7.40). In which case, it could be more appropriate to use the target species distribution for the SSD calculation.
4. The SSD can be used for setting or deriving environmental quality objectives, and for risk assessment using measured or predicted environmental concentrations (Figure 7.41). The 5th percentile of a chronic toxicity distribution has often been chosen as a concentration which is protective for most species in a community, but the cut-off value of 5 is a policy decision. This concentration is called the HC5. The complementary value of p has become known as the 95 % (100-p) protection criterion. This is considered to be an acceptable approach for protecting the structure and function of aquatic ecosystems [10,14,16,225].

The available SSD methods use different assumptions regarding the shape of the species sensitivity distribution (SSD). Aldenberg and Jaworska [226] assume a log-normal distribution, but a log-logistic distribution [227] or log-triangular distribution is also possible (Figure 7.39). The methods of Aldenberg and Jaworska were based on earlier models by Aldenberg and Slob [227], Kooijman [228], Van Straalen and Denneman [185], and Wagner and Løkke [229].

The log HC5 is estimated with:

$$\log \mathrm{HC5} = \bar{x} - K_s \cdot s \tag{7.11}$$

where

HC5	=	the hazardous concentration for 5% of species
$\bar{x}$	=	the sample mean of log NOEC data for m species
K_s	=	the one-sided extrapolation constant for a logistic or normal distribution, dependent on m
s	=	the sample standard deviation of log NOEC values for m species.

The uncertainty in the estimated HC5 can be calculated at a lower (95%), a median (50%) and a higher (5%)

Box 7.3. The calculation of HC5 and FCV values with different extrapolation techniques, using experimental chronic and subchronic NOECs (mg/L) of sodium bromide (NaBr), dimethoate and pentachlorophenol (PCP) for 11 different test species [232]. With permission. Copyright Elsevier.

Test species	NaBr	Dimethoate	PCP
Pseudomonas fluorescens (bacteria)	3200	320	1.0
Microcystis aeruginosa (bacteria)	3200	32	1.0
Scenedesmus pannonicus (algae)	3200	100	0.1
Lemna minor (higher plants)	3200	32	1.0
Daphnia magna (crustaceans)	10	0.032	0.1
Culex pipiens (insects)	100	0.32	3.2
Hydra oligactis (hydrozoans)	1000	100	0.032
Lymnea stagnalis (mollusks)	10	10	0.0032
Xenopus laevis (amphibians)	32	1.0	0.032
Poecillia reticulata (fish)	32	0.1	0.1
Oryzias latipes (fish)	320	0.32	0.032

The HC5 values are calculated according to Aldenberg and Jaworska [226] using the software ETX 2.0 [230], and the FCV values according to Erickson and Stephan [233] using the software ETX 1.3a [231]. All species were used in the calculations.

Results [mg/L]	NaBr	Dimethoate	PCP
HC5 (5-95% confidence limits)	4.3 (0.29-21)	0.019 (0.00057-0.16)	0.004 (0.00041-0.016)
FCV	5.45	0.019	0.0023

confidence level (Figure 7.42). The corresponding K_s values for each confidence level depend on the toxicity data sample size and are implemented in software for calculating the HC5 [230,231]. The K_s values for a log-logistic distribution [227] do not differ very much from those for the log-normal distribution at the same level of confidence [226]. Consequently, the calculated HC5 values are in the same range. Sample calculations are shown in Box 7.3. It is generally recognised that a diversity of taxonomical groups needs to be considered for deriving HC5 values (Table 7.45).

Erickson and Stephan [233] presented a method based on the triangular distribution to estimate a final chronic value (FCV) which applies the 5% cut-off to taxonomic genera, instead of species. Therefore, the FCV is an estimate of the 5th percentile concentration of chronic toxicity values for genera. The FCV is preferably calculated from chronic NOEC values for at least eight different animal families. Chronic values for species are combined to estimate mean chronic values for each genus. From the cumulative distribution of these

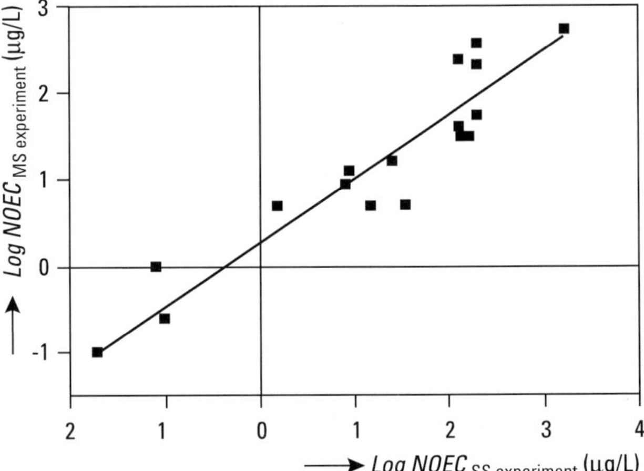

Figure 7.43. Model-II regression of $NOEC_{ms\ experiment}$ on $NOEC_{ss\ experiment}$ for similar or related species, corresponding effects parameters and similar exposure concentrations, based on 17 data pairs: log $NOEC_{ms\ experiment}$ = 0.750 x log $NOEC_{ss\ experiment}$ + 0.263; r = 0.935. From Van Leeuwen, Van De Plassche and Canton [104]. With permission.

Table 7.45. Information requirements for using statistical extrapolation in the EU based on species sensitivity distributions [14], based on similar requirements in the USA [233] except for algae and higher plants.

1. Class Osteichthyes, frequently tested species including salmonids, minnows, bluegill sunfish, channel catfish, etc)
2. A second family in the phylum Chordata (may be in the class Osteichthyes or an amphibian, etc.)
3. A crustacean (e.g. cladoceran, copepod, ostracod, isopod, amphipod, crayfish etc.)
4. An insect (e.g. mayfly, dragonfly, damselfly, stonefly, caddisfly, mosquito, midge, etc.)
5. A family in a phylum other than Arthropoda or Chordata (e.g. Rotifera, Annelida, Mollusca, etc.)
6. A family in any order of insect or any phylum not already represented
7. Algae
8. Higher plants

genus means, the HC5 is estimated from the lowest four genus means by a non-parametric or graph method. As a variation on the original method, where only data for specified animal families were used (Table 7.45), single-species data (of plants and animals) may be used as input in the equation for comparison with the other methods, in which event the calculated FCV is considered to be equivalent to the HC5.

The method has some advantages over the other SSD methods. Deviations from the assumed distribution restricted to the upper part of the distribution will have little impact on the calculation if only the lowest data in a sample are used. Another advantage of using only the lowest data is that it allows the inclusion of test results with "greater than" values, which are excluded in other approaches [233]. A comparison of HC5 values with the FCV is made in Box 7.3. It shows that the differences are relatively small at the same level of confidence.

Criticisms have addressed statistical issues, how representative SSDs are of species in different ecosystems, the inability of SSDs to deal with species interactions and issues related to environmental quality [234,235].

Verification of SSD methods with species sensitivities in microcosms, mesocosms or semi-field studies showed that HC5 values predicted from single-species tests generally do not significantly differ from NOECs derived from field studies [97,236]. Species tested in multi-species experiments appeared to be equally sensitive as similar or related species in single-species experiments when tested for corresponding parameters (Figure 7.43). These results were partly confirmed by the comparison of laboratory SSDs for chlorpyrifos with field-derived SSDs (Figure 7.44). This shows that the use of SSDs for standard setting can produce field-relevant results and makes better use of the available data instead of focusing on the lowest available test result for a specific endpoint.

7.10.3 Effects of secondary poisoning

Only methods for direct toxic effects have so far been described. A point of major concern is the effect of biomagnification (accumulation in the food chain) which may lead to indirect toxicity, i.e., to secondary poisoning. Most birds and mammals in aquatic and terrestrial ecosystems are predating organisms at the end of a food chain and thus may be exposed to high concentrations in their diet (Table 7.9). A simple approach has been developed to estimate NOECs for predating animals such as fish-eating and worm-eating birds and mammals [237,238]. Where no data are available on toxicity for

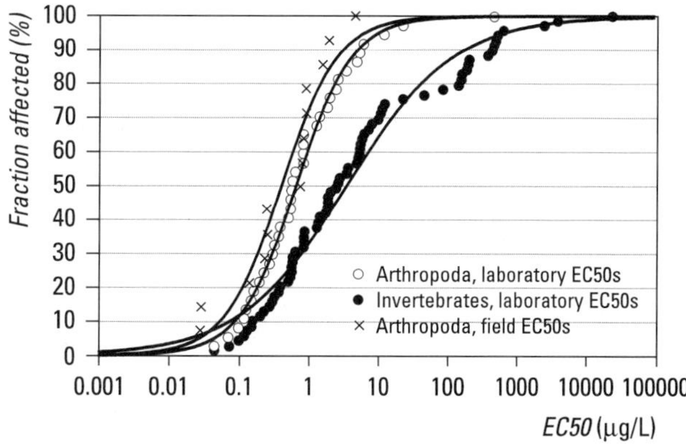

Figure 7.44. Comparison of SSD curves for chlorpyrifos between the laboratory and a semi-field test, based on acute LC/EC50 values. From Van den Brink et al. [108]. With permission.

Table 7.46. Relationship between mg/kg diet (dry laboratory chow diets) and mg/kgbw according to Lehman [14,239].

Animal	Body weight (kg)	Food consumption (g/d)	Conversion factor mg/kg diet to $mg/kg_{bw}·d$
Mouse	0.02	3	0.15
Rat (young)	0.10	10	0.10
Rat (old)	0.40	20	0.05
Guinea pig	0.75	30	0.04
Rabbit	2.0	60	0.03
Dog	10.0	25	0.025

wild mammal or bird species, the subchronic toxicity for laboratory mammals (mg/kg_{bw}) is used and converted to concentrations in the diet (mg/kg diet) using a conversion factor based on the consumption rate of the species. This method is only advisable when no other toxicity data for birds or mammals are available:

$$NOAEL_{diet} = NOAEL / F \qquad (7.12)$$

where $NOAEL_{diet}$ is the estimated dose expressed as the concentration in the diet (mg/kg diet), NOAEL is the chronic or subchronic value for laboratory mammals expressed in $mg/kg_{bw}·d$ and F is the consumption rate ($kg_{diet}/kg_{bw}·d$). According to Lehman [239] the conversion factor is approximately 0.10, but this factor varies considerably, depending on the animal studied (Table 7.46). If only acute toxicity data are available an application factor of 90 to 3000 (Table 7.47) may be used to extrapolate from acute to chronic toxicity, but it should be stressed that this may lead to large errors. Different approaches are available to derive a $NOAEL_{diet}$ for taxonomic groups such as birds and mammals, based on either an assessment factor approach [240] or an SSD-based approach [241].

To avoid secondary poisoning, the concentration of chemicals in the food should be below the NOAEL in dietary toxicity tests with animals that are representative of fish-eating or worm-eating birds or mammals. The diet is assumed to consist completely of fish or earthworms. The NOAEL is considered the maximum concentration in food which will not lead to adverse effects.

The maximum concentration in the food of fish-eating predators can be converted to a maximum concentration in water that will protect predators, based on the bioconcentration and biomagnification in the food chain:

$$NOEC_{pred} = \frac{NOAEL}{BCF_{fish} \cdot BMF_1} \qquad (7.13)$$

where the $NOEC_{pred}$ is the external no observed effect concentration for fish or worm-eating birds or mammals expressed as mg/L (water) or mg/kg (soil). The BCF is expressed as L/kg wwt for fish. The BMF [-] in this equation is a correction for the fact that fish can accumulate substances from food as well (Table 7.48), thereby exceeding the level if fish were to be exposed in the water phase only (see Chapter 3 on bioaccumulation). To account for food chains in the marine environment, this route can be extended by one extra biomagnification step. This step represents the biomagnification from fish to fish-eating birds and mammals that serve as prey for top-predators. Thus the route of exposure is uptake by aquatic organisms (e.g., small fish), biomagnification by fish, biomagnification by fish-eating predators and finally, consumption by the top-predator. Equation 7.13 is extended with an additional BMF and then becomes:

$$NOEC_{pred} = \frac{NOAEL}{BCF_{fish} \cdot BMF_1 \cdot BMF_2} \qquad (7.14)$$

BCFs are derived from experimental data, or where data are lacking, from estimates. The BCF in fish can be estimated using QSARs from Chapter 9 [14]; e.g.

$$\log BCF_{fish} = 0.85 \cdot \log K_{ow} - 0.70 \qquad (7.15)$$

For substances with a log K_{ow} higher than 6, a parabolic equation can be used.

see next page (7.16)

Table 7.47. Assessment factors used to derive PNECs for birds and mammals to assess the effects of e.g. secondary poisoning [14].

Available information	Duration of test	Assessment factor (oral) applied to the lowest value
LC50 bird	5 d	3000
NOEC bird	chronic	30
NOEC mammal, food	28d	300
	90 d	90
	chronic	30

Table 7.48. Default BMF-values [-] for organic substances, used in assessment of secondary poisoning [14].

log K_{ow} of substance	BCF (fish)	BMF_1	BMF_2
< 4.5	< 2000	1	1
4.5 - < 5	2000-5000	2	2
5 – 8	> 5000	10	10
> 8 – 9	2000 – 5000	3	3
> 9	< 2000	1	1

The biomagnification factor (BMF) is preferably measured, but defaults can be used for organic substances estimated from the relationship between K_{ow}, BCF and the BMF of the substance (Table 7.48). Similar equations can be derived for protecting worm-eating birds or mammals [14] based on measured or estimated BCFs for earthworms (see Equation 3.58, Chapter 3).

Comparison of the $NOEC_{pred}$ with PNECs for surface water or soil can reveal whether secondary poisoning could constitute a critical pathway. This occurs when these values for fish-eating birds and mammals are lower than the PNECs for direct toxic effects in water or soil. Examining secondary poisoning by using worm and fish-eating birds and mammals is a clear simplification of food webs occurring in nature (Figure 7.45). Large errors may occur for superlipophilic chemicals that are not well predicted by existing relationships (Equation 7.16, Chapters 3 and 9). Furthermore, the use of these simple models does not mean that other birds or mammals feeding on other species are not at risk, even though the value for $NOEC_{pred}$ should be protective. Therefore, the $NOEC_{pred}$ values for fish-eating birds and mammals should be considered as indicative of secondary poisoning. Alternative approaches to bioaccumulation in risk assessment are discussed in Chapter 3.

7.10.4 Comprehensive assessment

In ecotoxicology discussions on the advantages and disadvantages of single-species testing are still relevant [13,96-106,111,234]. Acute toxicity tests are the first step towards understanding the toxic effects of chemicals in ecosystems. Chronic tests are the second step and provide a reference point closer to the actual NEC at ecosystem level (Table 7.43). Much aquatic ecotoxicological research has been devoted to finding the most susceptible species [13], but the responses have been shown to be chemical specific, i.e., dependent on the nature of the chemical. It is not surprising that research carried out to select the most suitable combination of aquatic species [232] has led to the conclusion that the toxic potential of a chemical can be reasonably predicted from a test set with an alga, a crustacean and an egglaying fish species. Nevertheless, NOEC values obtained from SS tests are often in the same order of magnitude as those derived from more labour-intensive and expensive MS tests [97,236].

$$\log BCF_{fish} = -0.20 \cdot \log {K_{ow}}^2 + 2.74 \log K_{ow} - 4.72 \qquad (7.16)$$

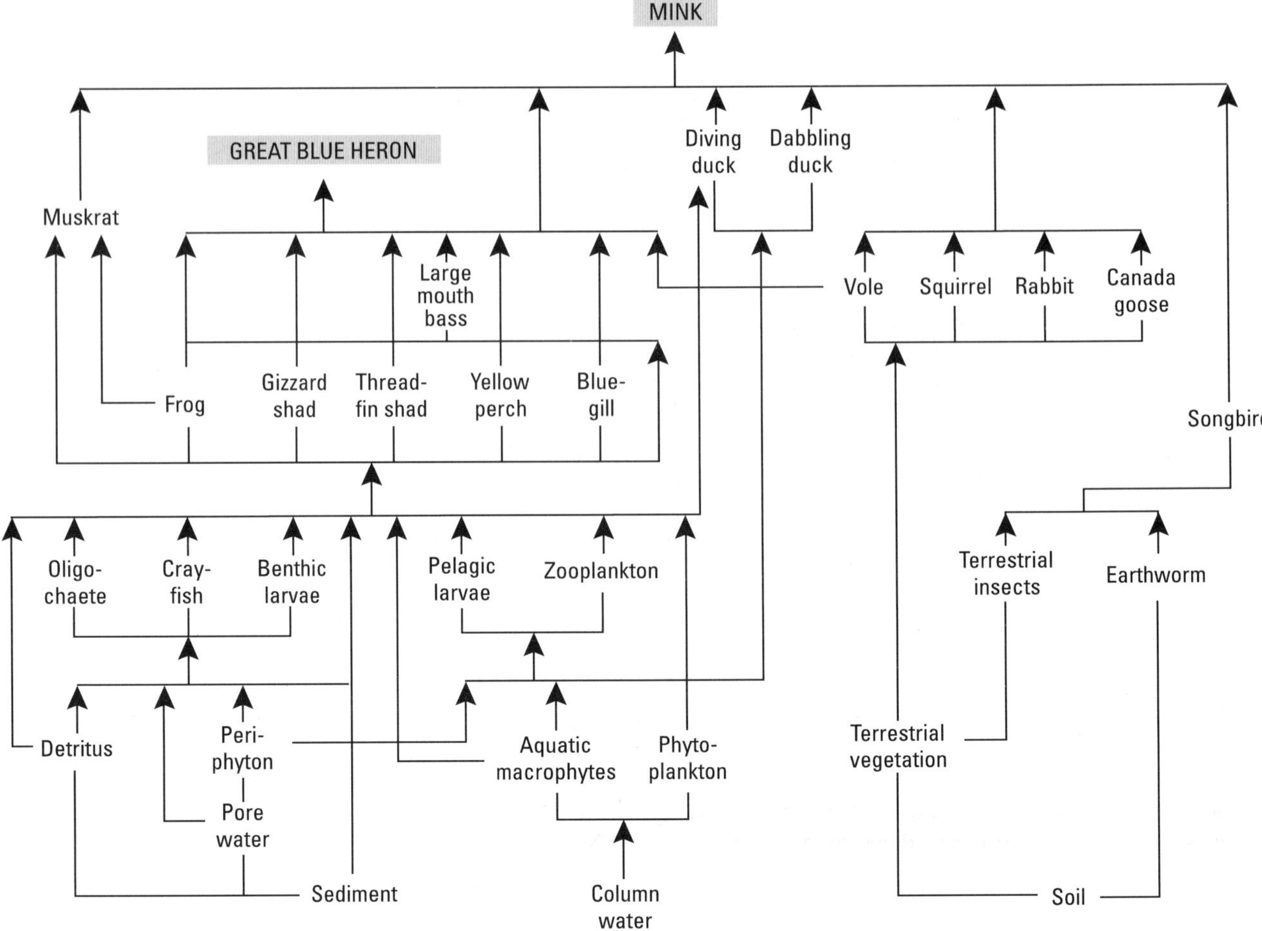

Figure 7.45. Diagrammatic representation of the routes of exposure of mink and great blue heron to contaminated terrestrial and aquatic environments. From Macintosh et al. [242]. With permission.

The studies normally carried out at the comprehensive stage are system level tests (MS tests). The best MS tests are field studies, but experimental microcosm and mesocosm studies provide a more cost-effective and efficient alternative. When the margin of safety is small, i.e., where the PNEC is close to the PEC, the effects of chemicals may need to be tested in more complex studies, as in higher tier testing for pesticides. Comprehensive tests may be appropriate when the economic consequences of a preliminary risk management decision are too great. In either case, additional information from MS tests should assist in environmental decision making because [80,81,94]:

- The overall impact of a chemical on populations within a community may be different from what was predicted from laboratory single-species tests due to poorly understood *interactions* between populations and their environment.
- Ecosystem studies may provide an opportunity to evaluate the ability of populations and communities within the affected ecosystem to recover from stress. The rate at which *recovery* occurs is a measure of the permanence of the effect.
- Ecosystem studies often provide *more realistic exposure conditions* with regard to the bioavailability and fate of the chemical, e.g., volatilization, adsorption and degradation. In this way better information can be obtained about the predicted environmental exposure concentration. MS tests thus provide more realistic evaluations of fate and effects.

It has been argued in many papers that (semi-)field studies may provide the ultimate answer in effects assessment (Section 7.3.6). Their role in the risk management is still relatively limited. Effects in the field are difficult to interpret and much depends on the questions that need answering. Field tests can only

Table 7.49: PBT and vPvB criteria according to Annex XIII of the REACH Regulation [246]

Property	PBT criteria	vPvB criteria
Persistence[1]	$T_{1/2}$ > 60 days in marine water, or $T_{1/2}$ > 40 days in fresh or estuarine water, or $T_{1/2}$ > 180 days in marine sediment, or $T_{1/2}$ > 120 days in fresh or estuarine sediment, or $T_{1/2}$ > 120 days in soil.	$T_{1/2}$ > 60 days in marine, fresh or estuarine water, or $T_{1/2}$ > 180 days in marine, fresh or estuarine sediment, or $T_{1/2}$ > 180 days in soil.
Bioaccumulation[2]	BCF > 2000 L/kg	BCF > 5000 L/kg
Toxicity	NOEC < 0.01 mg/L for marine or freshwater organisms, or substance is classified as carcinogenic (category 1 or 2), mutagenic (category 1 or 2), or toxic for reproduction (category 1, 2 or 3), or there is other evidence of chronic toxicity, as identified by the classifications: T, R48, or Xn, R48 according to Directive 67/548/EEC.	-

[1] The assessment of persistence in the environment is based on available half-life data collected under adequate conditions, which must be described by the registrant.

[2] The assessment of bioaccumulation must be based on measured data on bioconcentration in aquatic species, which may be freshwater or marine species.

provide clear answers to specific questions, but very often the questions cannot be formulated clearly because of the relatively limited amount of standard physicochemical and ecotoxicological data available, although much progress has been made [58].

Because so little is known about the variations in ecosystem susceptibility to chemicals (community to community extrapolation), it is not correct to propose a single extrapolation factor between a MS test and other ecosystems. The use of microcosms provides a reasonable alternative, which has many advantages over true field studies [96-99]. Effects assessment for most chemicals will, in most cases, still be based on extrapolation methods used for preliminary and refined effects assessment, i.e., on acute and chronic single-species toxicity data. From the validation of several of these extrapolation methods, by comparing MS NOECs with extrapolated data [97,108,236], it appears that the use of extrapolation methods leads to equal or lower rather than higher values than the MS NOECs derived from field studies and microcosm studies.

The general view is that, if little data are available, which is the case for more than 99% of chemicals, only a preliminary effects assessment is possible, in which event assessment factors can be used (Table 7.45). For more data-rich substances, it is now accepted to use the extrapolation methods of Aldenberg and Jaworska [226] and the USEPA [16] as a good basis for determining PNECs.

Modern statistical methods could make even more use of the existing data, by mining the information hidden in the sensitivity of species for chemicals that we *do know* a lot about. These sensitivity patterns can be used to estimate the uncertainty for those chemicals where we lack this information [241,243].

Both laboratory and field work are needed to provide more insight into the complexity of ecosystems and to improve the way in which PNECs are derived for environmental risk assessment.

7.11 ASSESSMENT OF PBTs AND vPvBs

PBT substances are chemicals that pose specific risks to ecosystems and human health, due to their persistence in the environment, their bioaccumulative properties in food webs and their toxicity. A special class of chemicals are those that are very persistent and very bioaccumulative (vPvB). For these substances, it is recognized that accumulation in the environment and food webs is highly likely, but unpredictable levels could occur in man or the environment over long time periods. PBTs such as the insecticide endosulfan, the aromatic hydrocarbon anthracene, and the flame retardant octobromodiphenyl-ether, have been associated with negative health and ecological effects, due to chronic exposure to these substances. This experience with PBT/vPvB substances has shown that they can give rise to specific concerns that may arise due to their potential to accumulate in parts of the environment:

- The effects of such accumulation are unpredictable in the long-term.
- Such accumulation is practically difficult to reverse as cessation of emission will not necessarily result in a reduction in chemical concentration.

PBTs and vPvBs distribute between air, water and soil or sediment. These properties also mean that these substances can reach remote areas and contaminate food webs in pristine areas. Many persistent chemicals have been found in the arctic due to this process of *long range transport*, for which screening models are available [244].

The combination of persistence and bioaccumulation, which can give rise to toxic effects after a longer time and over a greater spatial scale than chemicals without these properties, makes PBTs a group of special concern. Methods and tools, such as the PBT profiler, have been developed to screen chemical inventories for PBT properties. These screening tools can be used in the absence of chemical-specific data [245].

The properties of the PBT/vPvB substances lead to a increased uncertainty in the estimation of risk to human health and the environment by applying quantitative risk assessment methodologies. For PBT and vPvB substances a "safe" concentration in the environment is difficult to establish with sufficient reliability. Therefore, a separate PBT/vPvB assessment is required under REACH [246] in order to take these specific concerns into account. Registrants are required to perform this specific PBT/vPvB assessment in the context of their chemical safety assessment. A general introduction on the assessment of PBTs and vPvBs is given in Chapter 12, Section 12.3.5.

The PBT and vPvB assessment in the REACH regulation [246] consists of a screening assessment and a definitive assessment [220]. The screening assessment for biodegradation uses a limited set of biodegradation tests or model predictions. Bioaccumulation is screened based on the *n*-octanol-water partition coefficient, physicochemical indicators such as molecular weight, and maximum diameter and octanol solubility. Toxicity is screened based on the available aquatic and bird or mammalian toxicity data or estimated toxicity.

If the substance fulfils the criteria for a potential PBT/vPvB (Table 7.49), a definitive assessment should be conducted. The definitive assessment should be based on measured data for biodegradation, bioaccumulation and long-term toxicity tests for aquatic organisms and by evaluating the classification of the substance for human health hazards. Detailed guidance for the assessment of PBT and vPvB substances will become available in the Technical Guidance Document [220].

7.12 CONCLUDING REMARKS

Chemical substances and their by-products are being release into the environment, on a worldwide basis, at increasing levels. It is estimated that up to 90% of these chemicals have not been adequately evaluated for their aquatic toxicity towards algae, daphnids and fish (Chapter 1 and 12). Terrestrial and sediment toxicity data are also very scarce. Few ecotoxicological studies have been reported addressing adverse effects at higher levels of biological organization, i.e. the population or ecosystem level.

Applying the current risk assessment paradigm and meeting the associated data-generation requirements, combined with the increased need to evaluate the potential effects of thousands of industrial chemicals [246], are big challenges for the chemical industry, national and international regulatory agencies and associated stakeholders [214]. The long-term solution to these challenges will not be to generate more hazard data more quickly but rather to determine which specific effects data, groups of chemicals, and exposures are essential for assessment and appropriate management of the risks. Testing to cover all data gaps according to a generalized checklist approach ("box ticking") should be prevented [214]. In fact, a complete review of all available scientific evidence data will not provide clear, definitive answers to the risk management questions that regulators must address. Uncertainty inevitably remains. Steensberg [247] has discussed this before: "We, correctly, believe that we have not understood anything at a fundamental level unless we have understood the mechanism of causation. And we often think, incorrectly, that such understanding is a prerequisite of wise action. It is often necessary to make a decision on the basis of knowledge sufficient for action but insufficient to satisfy the intellect."

Regulators and ecotoxicologists must avoid the cult of a search for complete answers to the pressing health and ecological problems before it takes action against them. To do otherwise is to erect a barrier to timely and intelligent action [247]. That is why the next part (Part IV) of this book (Chapters 8-11) is entirely devoted to data, data estimation methodologies and testing strategies.

7.13 FURTHER READING

1. European Chemicals Bureau 2007. Technical guidance documents on preparing the Chemical Safety Assessment (in prep). European Commission, Joint Research Centre, Ispra, Italy.
2. Hoffman DJ, Rattner BA, Burton GA Jr, Cairns J Jr. 2003. *Handbook of Ecotoxicology*. 2nd edition, Lewis publishers, Boca Raton, FL.
3. Newman MC. 1995. *Quantitative Methods in Aquatic Ecotoxicology.* Lewis publishers, Boca Raton FL.
4. Newman MC, Unger MA. 2003. *Fundamentals of Ecotoxicology.* 2nd edition. Lewis publishers, Boca Raton, FL.
5. Rand GM. 1995. *Fundamentals of Aquatic Toxicology.* 2nd edition. CRC Press, Washington, DC.

REFERENCES

1. Van Leeuwen CJ. 1993. About and beyond ecotoxicological limits. *H_2O* 11:282-292 [in Dutch with a summary in English].
2. Truhaut R. 1977. Ecotoxicology: objectives, principles and perspectives. *Ecotoxicol Environ Saf* 1:151-173.
3. Newman MC, Unger MA. 2003. *Fundamentals of Ecotoxicology,* 2nd edition. Lewis publishers, Boca Raton, FL.
4. Organization for Economic Co-operation and Development. 1989. Report of the OECD workshop on ecological effects assessment. OECD Environment Monographs 26. OECD, Paris, France.
5. Suter GW. 1993. *Ecological Risk Assessment.* Lewis Publ, Chelsea, MI.
6. Suter GW II, Efroymson RA, Sample BE, Jones DS. 2000. *Ecological Risk Assessment for Contaminated Sites.* Lewis Publishers, Boca Raton, FL.
7. Hoffman DJ, Rattner BA, Burton GA Jr, Cairns J Jr. 2003. *Handbook of Ecotoxicology.* 2nd edition, Lewis publishers, Boca Raton, FL.
8. Ehrlich PR, Wilson EO. 1991. Biodiversity studies: Science and policy. *Science* 253:758-762.
9. Central Bureau for Statistics. 1989. Number of plant and animal species. *Kwartaalbericht Milieu* 4:15-21, The Hague, The Netherlands [in Dutch].
10. Health Council of the Netherlands. 1989. Assessing the risk of toxic chemicals for ecosystems. Report No. 1988/28E, The Hague, The Netherlands.
11. Van Leeuwen CJ. 1990. Ecotoxicological effects assessment in the Netherlands, recent developments. *Environ Management* 14:779-792.
12. Stephan CE. 1986. Proposed goal of applied aquatic toxicology. In: Poston TM, Purdy R, eds, *Aquatic Toxicology and Environmental Fate* (Ninth Volume). STP 921. American Society for Testing and Materials, Philadelphia, PA, pp 3-10.
13. Cairns J Jr. 1986. The myth of the most sensitive species. *Bioscience* 36:670-672.
14. Commission of the European Communities. 2003. Technical Guidance Document in support of Commission Directive 93/67/EEC on risk assessment for new notified substances, Commission Regulation (EC) No 1488/94 on risk assessment for existing substances and Directive 98/8/EC of the European Parliament and of the Council concerning the placing of biocidal products on the market. Joint Research Centre, European Chemicals Bureau, Brussels, Belgium.
15. National Environmental Policy Plan. 1989. To choose or to lose 1990-1994. Second Chamber of the States General, session 1988-1989, 21137, Nos 1-2. The Hague, The Netherlands.
16. Stephan CE, Mount DI, Hansen DJ, Gentile JH, Chapman GA, Brungs WA. 1985. Guidelines for deriving numerical national water quality criteria for the protection of aquatic organisms and their uses. Report PB85-227049, US Environmental Protection Agency, Springfield, VA.
17. Van Straalen NM, Van Leeuwen CJ. 2002. European history of species sensitivity distributions. In: Posthuma L. Suter GW, Traas TP, eds, *Species Sensitivity Distributions in Ecotoxicology.* Lewis, Boca Raton FL, pp 19-34.
18. Nienhuis PH, 2, Buijse AD, Leuven RSEW, Smits AJM, de Nooij RJW Samborska EM. 2002. Ecological Rehabilitation of the lowland basin of the river Rhine (NW Europe). *Hydrobiologia* 478:53-72.
19. Porta M, Zumeta E, 2002. Implementing the Stockholm Treaty on Persistent Organic Pollutants. *Occ Environ Med* 59:651-652.
20. Fresco LO, Kroonenberg SB. 1992. Time and spatial scales in ecological sustainability. *Land Use Policy,* July:155-168.
21. McCarty L and Mackay D. 1993. Enhancing ecotoxicological modeling and assessment: body residues and modes of toxic action. *Environ Sci Technol* 27:1719-1728.
22. Sijm DTHM, Hermens JLM. 1999. Internal effect concentrations: link between bioaccumulation and ecotoxicity for organic chemicals. In: Beek B, ed, *The Handbook of Environmental Chemistry*, Volume 2- J. Bioaccumulation. New aspects and developments. Springer-Verlag, Berlin, GDR, pp. 167-199.
23. Escher BI, Hermens JLM. 2002. Modes of action in

ecotoxicology: their role in body burdens, species sensitivity, QSARs, and mixture effects. *Environ Sci Technol* 36:4201-4217.

24. Vaes WHJ, Urrestarazu Ramos E, Verhaar HJM, Hermens JLM. 1998. Acute toxicity of nonpolar versus polar narcosis: is there a difference? *Environ Toxicol Chem* 17:1380-1384.
25. DiToro DM, McGrath J, Hansen DJ. 2000. Technical basis for narcotic chemicals and polycyclic aromatic hydrocarbon criteria. I Water and tissue. *Environ Toxicol Chem* 19:1951-1970.
26. DiToro DM, McGrath J. 2000. Technical basis for narcotic chemicals and polycyclic aromatic hydrocarbon criteria. II Mixtures and sediments. *Environ Toxicol Chem* 19:1971–1982.
27. Traas TP, Van Wezel AP, Hermens JLM, Zorn M, Van Hattum AGM, Van Leeuwen CJ. 2004. Environmental quality criteria for organic chemicals predicted from internal effect concentrations and a food web model. *Environ Toxicol Chem* 23:2518-2527.
28. Verbruggen EMJ. 2004. Environmental Risk Limits for mineral oil (total petroleum hydrocarbons) Report 601501021, National Institute for Public Health and the Environment (RIVM), Bilthoven, The Netherlands [available via www.rivm.nl].
29. Barron MG, Hansen JA, Lipton J. 2002. Association between contaminant tissue residues and effects in aquatic organisms. *Rev Environ Contam Toxicol* 173:1-37.
30. Verhaar HJM, Van Leeuwen CJ, Hermens JLM. 1992. Classifying environmental pollutants. 1: Structure-activity relationships for prediction of aquatic toxicity. *Chemosphere* 25:471-491.
31. Russom CL, Bradbury SP, Broderius SJ, Hammermeister DE, Drummond RA.1997. Predicting modes of toxic action from chemical structure: acute toxicity in the fathead minnow *(Pimephales promelas). Environ Toxicol Chem* 16:948–967.
32. Van Der Kooy LA, Van De Meent D, Van Leeuwen CJ, Bruggeman WA. 1991. Deriving quality criteria for water and sediment from the results of aquatic toxicity tests and product standards: Application of the equilibrium partitioning theory. *Water Res* 25:697-705.
33. Van Leeuwen CJ, Van Der Zandt PTJ, Aldenberg T, Verhaar HJM, Hermens JLM. 1992. Application of QSARs, extrapolation and equilibrium partitioning in aquatic effects assessment. I Narcotic industrial pollutants. *Environ Toxicol Chem* 11:267-282.
34. Larsson P, Hamrin S, Okla L. 1991. Factors determining the uptake of persistent pollutants in an eel population *(Anguilla anguilla L). Environ Pollut* 69:39-50.
35. Hamers T, van den Brink OJ, Mos L, van der Linden SC, Legler J, Koeman JH, Murk AJ. 2003. Estrogenic and esterase-inhibiting potency in rainwater in relation to pesticide concentrations, sampling season and location. *Environ Pollut* 123:47-65.
36. Kelly BC, Gobas FAPC. 2003. An arctic terrestrial food-chain bioaccumulation model for persistent organic pollutants. *Environ Sci Technol* 37:2966-2974.
37. Crane M, Newman MC, Chapman PF, Fenlon J. 2002. *Risk Assessment with Time-to-Event Models.* Lewis publishers, Boca Raton, FL.
38. Kooijman SALM. 2000. *Dynamic Energy and Mass Budgets in Biological Systems.* Cambridge University Press, Cambridge, UK.
39. Stigliani WM. 1989. Changes in valued "capacities" of soils and sediments as indicators of non-linear and time-delayed environmental effects. *Environ Monit Assessm* 10:245-307.
40. Vijver MG, Vink JPM, Miermans CJH, Van Gestel CAM. 2003. Oral sealing using glue: a new method to distinguish between intestinal and dermal uptake of metals in earthworms. *Soil Biol Biochem* 35:125-132.
41. Jager T, Fleuren RHLJ, Hoogendoorn AM, De Korte G. 2003. Elucidating the routes of exposure for organic chemicals in the earthworm, *Eisenia andrei* (Oligochaeta). *Environ Sci Technol* 2003:3399-3404.
42. Organization for Economic Co-operation and Development. 1993. Report of the OECD workshop on application of simple models for exposure assessment. OECD Environment Monographs 69. OECD, Paris, France.
43. Peters RH. 1983. *The Ecological Implications of Body Size.* Cambridge University Press, Cambridge, UK.
44. Hendriks AJ. 1999. Allometric scaling of rate, age and density parameters in ecological models. *Oikos* 86:293-310.
45. Jaworska J, Dimitrov S, Nikolova N, Mekenyan O, 2002. Probabilistic assessment of biodegradability based on metabolic pathways: CATABOL system. *SAR QSAR Environ Res* 3:307-323.
46. Organization for Economic Co-operation and Development. 2006. Freshwater Alga and Cyanobacteria, growth inhibition test. Guideline for testing of chemicals, No. 201. OECD Paris, France.
47. Rand GM. 1995. *Fundamentals of Aquatic Toxicology*, 2nd edition. CRC Press, Washington, DC.
48. Mount DI, Brungs WA. 1967. A simplified dosing apparatus for fish toxicology studies. *Water Res* 1:21-29.
49. Anonymous, 2004. Fish dietary bioaccumulation study protocol, based on a version adapted by the TC NES subgroup on PBTs of the original protocol, January 20,

2004. ExxonMobil Biomedical Science, Inc (EMBSI).

50. Newman MC. 1995. *Quantitative Methods in Aquatic Ecotoxicology.* Lewis publishers, Boca Raton FL.
51. Slob W. 2002. Dose-response modelling of continuous endpoints. *Toxicol Sci* 66:298-312.
52. Hoekstra J. 1991. Estimation of the LC50, a review. *Environmetrics* 2:139-152.
53. De Bruijn J, Hof M. 1997. How to measure no effect. Part IV: How acceptable is the ECx from an environmental policy point of view? *Environmetrics* 8:263-267.
54. Slob W, Moerbeek M, Rauniomaa E, Piersma AH. 2005. A statistical evaluation of toxicity study design for the estimation of the benchmark dose in continuous endpoints. *Toxicol Sci* 84:167-185.
55. Jager T, Heugens EWH, Kooijman SALM. 2006. Making sense of ecotoxicological test results: towards application of process-based models. *Ecotoxicology* 15:305-314.
56. Calow P. ed 1993. *Handbook of Ecotoxicology.* Blackwell Sci Publ, London, UK.
57. Adams WJ and Rowlands CD. 2003. Aquatic toxicology test methods. In Hoffman DJ, Rattner BA, Burton GA Jr, Cairns J Jr, eds, *Handbook of Ecotoxicology*, 2nd edition. Lewis publishers, Boca Raton FL, pp 19-43.
58. Organization for Economic Co-operation and Development. 2006. 16th Addendum to the OECD Guidelines for the Testing of Chemicals. OECD, Paris, France.
59. Organization for Economic Co-operation and Development. 2004. *Daphnia* sp., acute immobilization test. Guideline for testing of chemicals, No. 202. OECD, Paris, France.
60. Organization for Economic Co-operation and Development. 1992. Fish, acute toxicity test. Guideline for testing of chemicals, No. 203. OECD, Paris, France.
61. Organization for Economic Co-operation and Development. 2006. *Lemna* sp. Growth inhibition test. Guideline for testing of chemicals, No. 221. OECD, Paris, France.
62. Klaine SJ, Lewis MA, Knuteson SL. 2003. Phytotoxicity. In: Hoffman DJ, Rattner BA, Burton GA Jr, Cairns J Jr, eds, *Handbook of Ecotoxicology*, 2nd edition. Lewis Publishers, Boca Raton FL, pp. 191-218.
63. Weyers A, Vollmer G. 2000. Algal growth inhibition: effect of the choice of growth rate or biomass as endpoint on the classification and labelling of new substances notified in the EU. *Chemosphere* 41:1007-1010.
64. Eberius M, Guido Mennicken G, Ilka Reuter I, Vandernhirtz J. 2002. Sensitivity of different growth inhibition tests-just a question of mathematical calculation? theory and practice for algae and duckweed. *Ecotoxicology* 11: 293-297.
65. Kooijman SALM, Bedaux JJM. 1996. *The Analysis of Aquatic Toxicity Data.* VU University press, Amsterdam, The Netherlands.
66. Van Leeuwen CJ, Rijkeboer M, Niebeek G. 1986. Population dynamics of *Daphnia magna* as modified by chronic bromide stress. *Hydrobiologia* 133:277-285.
67. Hirsch R, Ternes T, Haberer K, Kratz K. 1999. Occurrence of antibiotics in the aquatic environment. *Sci Total Environ* 225:109-118.
68. Kolpin DW, Furlong ET, Meyer MT, Thurman EM, Zaug SD, Barber LB, Buxton HT. 2002. Pharmaceuticals, Hormones, and other organic wastewater contaminants in U.S. streams, 1999-2000: A national reconnaissance. *Environ Sci Technol* 36:1202-1211.
69. Organization for economic co-operation and development. 1984. Activated sludge respiration inhibition. Guideline for testing of chemicals, No. 209. OECD, Paris, France.
70. Gendig C, Domogala G, Agnoli F, Udo Pagga U, Strotmann UJ. 2003. Evaluation and further development of the activated sludge respiration inhibition test. *Chemosphere* 52:143-149.
71. Organization for Economic Co-operation and Development. 1998. *Daphnia magna* reproduction test. Guideline for testing of chemicals, No. 211. OECD, Paris, France.
72. Van Leeuwen CJ, Niebeek G, Rijkeboer M. 1987. Effects of chemical stress on the population dynamics of *Daphnia magna*: A comparison of two test procedures. *Ecotoxicol Environ Saf* 14:1-11.
73. Mount DI, Norberg TJ. 1984. A seven-day life cycle cladoceran toxicity test. *Environ Toxicol Chem* 3:425-434.
74. American Society for Testing and Materials. 1989. Standard guide for conducting three-brood renewal toxicity tests with *Ceriodaphnia dubia*. In: Annual Book of ASTM Standards, Vol 11.01, E 1295. American Society for Testing and Materials, Philadelphia, PA, pp. 879-897.
75. Versteeg DJ, Stalmans M, Dyer SD, Janssen C. 1997. *Ceriodaphnia* and *Daphnia*: a comparison of their sensitivity to xenobiotics and utility as a test species. *Chemosphere* 34:869-892.
76. McKim JM. 1985. Evaluation of tests with early life stages of fish for predicting long-term toxicity. *J Fish Res Can* 34:1148-1154.
77. Van Leeuwen CJ, Griffioen PS, Vergouw WHA, Maas-Diepeveen H. 1985. Differences in susceptibility of early life stages of rainbow trout *(Salmo gairdneri)* to environmental pollutants. *Aquat Toxicol* 7:59-78.
78. Organization for Economic Co-operation and Development. 1992. Fish early-life stage toxicity test.

Guideline for testing of chemicals 210. OECD, Paris, France.

79. Environment Canada. 1992. Biological test method: Toxicity tests using early life stages of salmonid fish (rainbow trout, coho salmon, or atlantic salmon). Environmental Protection Series, Report EPS 1/RM/28, Ottawa, Ontario, Canada.
80. Campbell PJ, Arnold DJS, Brock TCM, Grandy NJ, Heger W, Heimbach F, Maund SJ and Streloke M. 1999. Guidance document on higher-tier aquatic risk assessment for pesticides (HARAP), SETAC-Europe/ OECD/EC Workshop. Lacanau Océan, France, SETAC-Europe, Brussels, Belgium.
81. Giddings J, Heger W, Brock T, Heimbach F, Maund S, Norman S, Ratte H, Schäfers C and Streloke M (Eds). 2002. Community-level aquatic system studies – Interpretation criteria (CLASSIC). Fraunhofer Institute, Schmallenberg, Germany; SETAC, Pensacola, FL.
82. Allen JD, Daniels RE. 1982. Life table evaluation of chronic exposure of Eurytemora affines (Copepoda) to kepone. *Mar Biol* (Berlin) 66:179-184.
83. Van Straalen NM, Schobben JHM, De Goede RGM. 1989. Population consequences of cadmium toxicity in soil microarthropods. *Ecotoxicol Environ Saf* 17:190-204.
84. Crommentuijn T. 1997. Life-table study with the springail Folsomia candida (Willem) exposed to cadmium, chlorpyrifos and triphenyltin hydroxide. In: Van Straalen NM, Løkke H, eds, *Ecological Risk Assessment of Contaminants in Soil*. Chapman & Hall, London, UK, pp. 275-291.
85. Smit CE, Stam EM, Baas N, Hollander R, Van Gestel CAM. 2004. Effects of dietary zinc exposure on the life history of the parthenogenetic springtail *Folsomia candida* (Collembola: Isotomidae). *Environ Toxicol Chem* 23:1719-1724.
86. Kammenga JE, Van Koert PHG, Koeman JH, Bakker J. 1997. Fitness consequences of toxic stress evaluated within the context of phenotypic plasticity. *Ecol Appl* 7:726-734.
87. Forbes VE, Calow P. 1999. Is the per capita rate of increase a good measure of population-level effects in ecotoxicology? *Environ Toxicol Chem* 18:1544-1556.
88. Kooijman SALM, Hanstveit AO, Oldersma H. 1983. Parametric analyses of population growth in bioassays. *Water Res* 17:527-538.
89. Klepper O, Traas TP, Schouten AJ, Korthals GW, De Zwart D. 1999. Estimating the effect on soil organisms of exceeding no-observed effect concentrations (NOECs) of persistent toxicants. *Ecotoxicology* 8:9-21.
90. Van den Brink PJ, Hattink J, Bransen F, Van Donk E, Brock TCM. 2000. Impact of the fungicide carbendazim in freshwater microcosms II. Zooplankton, primary producers and final conclusions. *Aquatic Toxicol* 48:251-264.
91. Roughgarden J. 1971. Density-dependent natural selection. *Ecology* 52:453-468.
92. Traas TP, Janse JH, Van den Brink PJ, Brock TCM, Aldenberg T. 2004. A freshwater food web model for the combined effects of nutrients and insicticide stress and subsequent recovery. *Environ Toxicol Chem* 23:521-529.
93. National Research Council. 1981. Testing for effects of chemicals on ecosystems. National Academy Press, Washington, DC.
94. Organization for Economic Co-operation and Development. 1992. Report of the OECD workshop on the extrapolation of laboratory aquatic toxicity data to the real environment. OECD Environment Monographs 59. OECD, Paris, France.
95. Hedtke SF. 1984. Structure and function of copper-stressed aquatic microcosms. *Aquat Toxicol* 5:227-244.
96. Cuppen JGM, Van den Brink PJ, Van der Woude H, Zwaardemaker N, Brock TCM. 1997. Sensitivity of macrophyte-dominated freshwater microcosms to chronic levels of the herbicide Linuron II. Community metabolism and invertebrates. *Ecotox Environ Saf* 38:25-35.
97. Emans HJB, Okkerman PC, Van De Plassche EJ, Sparenburg PM, Canton JH. 1993. Validation of some extrapolation methods used for effect assessment. *Environ Toxicol Chem* 12:2139-2154.
98. De Jong FMW, Brock TCM, Foekema EM, Leeuwangh P. 2007. Guidance for summarizing of aquatic micro- and mesocosm studies. Dutch platform for the Assessment of Higher Tier Studies. RIVM Bilthoven, The Netherlands, in prep.
99. Brock TCM, Van Wijngaarden RPA, Van Geest, PJ. 2000. Ecological risks of pesticides in freshwater ecosystems. Part 2: Insecticides. Report 089, Alterra, Wageningen, The Netherlands.
100. La Point TW, Perry JA. 1989. Use of experimental ecosystems in regulatory decisionmaking. *Environ Management* 13:539-544.
101. Crossland NO. 1990. The role of mesocosm studies in pesticide registration. Brighton Crop Protection Conference. *Pests and Diseases* 6B-1:499-508.
102. Crossland NO, Wolff CJM. 1988. Outdoor ponds: Their construction, management, and use in experimental ecotoxicology. In: Hutzinger O, ed, *The Handbook of Environmental Chemistry*, Vol. 2/D, Springer-Verlag, Berlin, Germany, pp 51-69.
103. Organization for Economic Co-operation and

Development. 2006. Guidance document on simulated freshwater lentic field tests (outdoor microcosms and mesocosms). OECD series on testing and assessment number 53. OECD, Paris, France.

104. Van Leeuwen CJ, Van De Plassche EJ, Canton JH. 1994. The role of field tests in hazard assessment. In: Hill IR, Heimbach F, Leeuwangh P, Matthiessen P, eds, *Freshwater Field Tests for Hazard Assessment of Chemicals*. Lewis Publ, Chelsea, MI, pp 339-453.
105. Van den Brink PJ, Ter Braak CJF. 1998. Multivariate analysis of stress in experimental ecosystems by principal response curves and similarity analysis. *Aquatic Ecol* 32:163-178.
106. Van den Brink PJ, Ter Braak CJF. 1999. Principal Response Curves: analysis of time-dependent multivariate responses of a biological community to stress. *Environ Toxicol Chem* 18:138-148.
107. Korthals GW, Alexiev AD, Lexmond TM, Kammenga JE, Bongers T. 1996. Long-term effects of copper and pH on the nematode community in agroecosystems. *Environ Toxicol Chem* 15:979-985.
108. Van den Brink PJ, Brock TCM, Posthuma L. 2002. The value of the species sensitivity distribution concept for predicting field effects: (non-) confirmation of the concept using semi-field experiments. In: Posthuma L. Suter GW, Traas TP, eds, *Species Sensitivity Distributions in Ecotoxicology*. Lewis, Boca Raton, FL, pp 155-193.
109. Hamers T, Krogh PH. 1997. Predator-Prey Relationships in a two-species toxicity test system. *Ecotox Environ Saf* 37:203-212.
110. Taub FB. 1989. Standardized aquatic microcosm: Development and testing. In: Boudou A, Ribeyre F, eds, *Aquatic Ecotoxicology*. CRC Press Inc. Boca Raton, FL, pp 47-94.
111. Kooijman SALM. 1985. Toxicity at population level. In: Cairns J Jr, ed, *Multispecies Toxicity Testing*. Pergamon Press, New York, NY, pp 143-164.
112. Organization for Economic Co-operation and Development. 1992. Report of the OECD workshop on effects assessment of chemicals in sediment. Environment Monographs 60. OECD, Paris, France.
113. Giesy JP, Hoke RA. 1989. Freshwater sediment toxicity bioassessment: Rationale for species selection and test design. *J Great Lakes Res* 15:539-569.
114. Malins DC, McCain BB, Brown DW, Varanasi U, Krahn MM, Myers MS, Chan S. 1987. Sediment-associated contaminants and liver diseases in bottom-dwelling fish. *Hydrobiologia* 149:67-74.
115. Van Urk G, Kerkum FCM. 1987. Chironomid mortality after the Sandoz accident and deformities in chironomid larvae due to sediment pollution in the Rhine. *Aqua* 4:191-196.
116. Milbrink G. 1980. Oligochaete communities in pollution biology. In: Brinkhurst RO, Cook DG, eds, *Aquatic Oligochaete Biology*. Plenum Press, New York, NY, pp 433-456.
117. Beurskens JEM, Barreveld GAJ, Mol HL, Van Munster B, Winkels HJ. 1993. Geochronology of priority pollutants in a sedimentation area of the river Rhine. *Environ Toxicol Chem* 12:1549-1566.
118. Traas TP, Stäb JA, Kramer PRG, Cofino WP, Aldenberg T. 1996. Modeling and risk assessment of tributyltin accumulation in the food web of a shallow freshwater lake. *Environ Sci Technol* 30:1227-1237.
119. Organization for Economic Co-operation and Development. 2004. Sediment-water Chironomid toxicity test using spiked sediment. Guideline for testing of chemicals 218. OECD, Paris, France.
120. Organization for Economic Co-operation and Development. 2004. Sediment-Water Chironomid Toxicity Test Using Spiked Water. Guideline for testing of chemicals 219. OECD, Paris, France.
121. Burton GA Jr, Denton DL, Ho K, Ireland DS. 2003. Sediment toxicity testing: issues and methods. In Hoffman DJ, Rattner BA, Burton GA Jr, Cairns J Jr, eds, *Handbook of Ecotoxicology*, 2nd edition.. Lewis publishers, Boca Raton, FL, pp. 111-150.
122. Van De Guchte C, Maas-Diepeveen H. 1988. Screening sediments for toxicity: A water concentration related problem. Proceedings of the 14th Annual Aquatic Toxicity Workshop, 2-4 November 1987. Can Tech Rep Fish Aquat Sci 1607, Toronto, Canada, pp 81-91.
123. Connell DW, Bowman M, Hawker D. 1988. Bioconcentration of chlorinated hydrocarbons from sediment by oligochaetes. *Ecotoxicol Environ Saf* 16:293-302.
124. Di Toro DM, Zarba CS, Hansen DJ, Berry WJ, Swartz RC, Cowan CE, Pavlou SP, Allen HE, Thomas NA, Paquin PR. 1991. Technical basis for establishing sediment quality criteria for non-ionic organic chemicals by using equilibrium partitioning. *Environ Toxicol Chem* 10:1541-1583.
125. Thomann RV, Connolly JP, Parkerton TF. 1992. An equilibrium model of organic chemical accumulation in aquatic food webs with sediment interaction. *Environ Toxicol Chem* 11:615-629.
126. Kaag NHBM, Foekema EM, Scholten MCT, Van Straalen NM. 1997. Comparison of contaminant accumulation in three species of marine invertebrates with different feeding habits. *Environ Toxicol Chem* 16:837–842.
127. US Environmental Protection Agency. 1989. Briefing report to the EPA Science Advisory Board on the

equilibrium partitioning approach to generating sediment quality criteria. Office of Water Regulations and Standards, Criteria and Standard Division, Washington, DC.

128. Wang F, Chapman PM. 1999. Biological implications of sulfide in sediment - a review focusing on sediment toxicity. *Environ Toxicol Chem* 8:2526–2532.
129. Dolfing J, Beurskens EM. 1995. The microbial logic and environmental significance of reductive dehalogenation. Adv Microbial Ecology 14:143-206.
130. Karickhoff SW, Brown DS, Scott TA. 1979. Sorption of hydrophobic pollutants on natural sediments. *Water Res* 13:241-248.
131. Ziegenfuss PS, Renaudette WJ, Adams WJ. 1986. Methodology for assessing the acute toxicity of chemicals sorbed to sediments: Testing the equilibrium partitioning theory. In: Poston TM, Purdy R, eds, *Aquatic Toxicology and Environmental Fate.* Vol 9 STP 921. American Society for Testing and Materials, Philadelphia, PA, pp 479-493.
132. Mayer P, Tolls J, Hermens JLM, Mackay D. 2003. Equilibrium sampling devices. *Environ Sci Technol* 37:184A-191A.
133. Kraaij, P Mayer, FJM Busser, M Van Het Bolscher M, Seinen W, Tolls J, Belfroid AC. 2003. Measured pore-water concentrations make equilibrium partitioning work - A data analysis. *Environ Sci Technol* 37:268-274.
134. Ter Laak TL, Agbo SO, Barendregt A, Hermens JLM. 2006. Freely dissolved concentrations of PAHs in soil pore water: Measurements via solid-phase extraction and consequences for soil tests. *Environ Sci Technol* 40:1307-1313.
135. Burton GA Jr. 1992. Assessing the toxicity of freshwater sediments. *Environ Toxicol Chem* 10:1585-1627.
136. Chapman PM. 1986. Sediment quality criteria from the sediment quality triad: An example. *Environ Toxicol Chem* 5:957-964.
137. Chapman PM. 1996. Presentation and interpretation of Sediment Quality Triad data. *Ecotoxicology* 5:327-339.
138. Ankley GT, Schubauer-Berigan MK. 1995. Background and overview of current sediment toxicity identification evaluation procedures. *J Aquat Ecosys Health* 4:133-149.
139. Leslie HA, Ter Laak TL, Busser FJM, MHS Kraak MHS, Hermens JLM 2002. Bioconcentration of organic chemicals: is a solid-phase microextraction fiber a good surrogate for biota? *Environ Sci Technol* 36:5399-5404.
140. Kosian, PA, Makynen EA, Monson PD, Mount DR, Spacie A, Mekenyan OG, Ankley GT. 1998. Application of toxicity-based fractionation techniques and structure-activity relationship models for the identification of phototoxic polycyclic aromatic hydrocarbons in sediment pore water. *Environ Toxicol Chem* 17:1021-1033.
141. Chapman PM, Anderson J. 2005. A decision-making framework for sediment contamination. *Integr Environ Assess Manag* 1:163-173.
142. Simpson SL, Batley GE, Chariton AA, Stauber JL, King CK, Chapman JC, Hyne RV, Gale SA, Roach AC, Maher WA. 2005. *Handbook for Sediment Quality Assessment.* CSIRO, Bangor NSW, Australia.
143. Organization for Economic Co-operation and Development. 2006. Draft test guideline on sediment-water *Lumbriculus* toxicity test using spiked sediment. OECD, Paris, France.
144. Nortcliff S. 2002. Standardisation of soil quality attributes. *Agric Ecosys Environ* 88:161-168.
145. Rutgers M, Van't Verlaat IM, Wind B, Posthuma L, Breure AM. 1998. Rapid method for assessing pollution-induced community tolerance in contaminated soil. *Environ Toxicol Chem* 17:2210-2213.
146. Organization for Economic Co-operation and Development. 1984. Earthworm acute toxicity tests. Guideline for testing of chemicals, no 207. OECD, Paris, France.
147. Van Gestel CAM. 1992. The influence of soil characteristics on the toxicity of chemicals for earthworms: A review. In: Greig-Smith PW, Becker H, Edwards PJ, Heimbach F, eds, *Ecotoxicology of Earthworms*. Intercept Ltd, Andover, UK., pp 44-54.
148. Van Gestel CAM, Van Straalen NM. 1994. Ecotoxicological test systems for terrestrial invertebrates. In: Donker MH, Eijsackers H, Heimbach F, eds, *Ecotoxicology of Soil Organisms.* Lewis Publ, London, UK, pp 205-229.
149. Van Gestel CAM, Rademaker MCJ, Van Straalen NM. 1995. Capacity controlling parameters and their impact on metal toxicity in soil invertebrates. In: Salomons W, Stigliani WM, eds, *Biogeodynamics of Pollutants in Soils and Sediments.* Springer Verlag, Berlin, Germany, pp. 171-192.
150. Jager T. 2004. Modeling ingestion as an exposure route for organic chemicals in earthworms (Oligochaeta). *Ecotoxicol Environ Saf* 57:30-38.
151. Organization for Economic Co-operation and Development. 2004. Earthworm reproduction test. Guideline for testing of chemicals, No. 222. OECD, Paris, France.
152. Organization for Economic Co-operation and Development. 1998. Honeybees, acute contact toxicity test. Guideline for testing of chemicals 214. OECD, Paris, France.
153. Candolfi MP, Blümel S, Forster R, Bakker FM, Grimm

C, Hassan SA, Heimbach U, Mead-Briggs MA, Reber B, Schmuck R, Vogt H. 2000. Guideline to evaluate side-effects of plant-protection products to non-target arthropods. IOBC/WPRS, Gent, Belgium.

154. Pekar S, Haddad CR. 2005. Can agrobiont spiders (Araneae) avoid a surface with pesticide residues? *Pest Manage Sci* 61:1179-1185.

155. Luttik R. 1998. Assessing repellency in a modified avian LC50 procedure removes the need for additional tests. *Ecotox Environ Saf* 40:201-205.

156. Binnie J, Cape JN, Mackie N, Leith ID. 2002. Exchange of organic solvents between the atmosphere and grass - the use of open top chambers. *Sci Total Environ* 285:53-67.

157. EPA. 1997. Ecological risk assessment guidance for Superfund. Interim Final Report 540-R-97-006, EPA Washington, DC.

158. Swartjes FA. 1999. Risk-based assessment of soil and groundwater quality in the Netherlands: standards and remediation urgency. *Risk Anal* 19:235-1249.

159. Jensen J, Mesman M, eds. 2006. Ecological risk assessment of contaminated land - Decision support for site specific investigations. Report 711701047, National Institute for Public Health and the Environment (RIVM), Bilthoven, The Netherlands [available via www.rivm.nl].

160. Blanck H. 2002. A critical review of procedures and approaches used for assessing pollution-induced community tolerance (PICT) in biotic communities. *Human Ecol Risk Asess* 8:1003-1034.

161. Boivin MEY, Breure AM, Posthuma L, Rutgers M. 2002. Determination of field effects of contaminants-significance of pollution-induced community tolerance. *Human Ecol Risk Asess* 8:1035-1055.

162. Schmitt H, Van Beelen P, Tolls J, Van Leeuwen CJ. 2004. Pollution-induced community tolerance of soil microbial communities caused by the antibiotic sulfachloropyridazine. *Environ Sci Technol* 38:1148-1153.

163. Organization for Economic Co-operation and Development. 2006. Terrestrial plant test: seedling emergence and seedling growth test. Guideline for testing of chemicals 208. OECD, Paris, France.

164. Organization for Economic Co-operation and Development. 2006. Terrestrial plant test: vegetative vigour test. Guideline for testing of chemicals 227. OECD, Paris, France.

165. Boutin C, Elmegaard N, Kjær C. 2004. Toxicity testing of fifteen non-crop plant species with six herbicides in a greenhouse experiment: Implications for risk assessment. *Ecotoxicology* 13:349-369.

166. Løkke H, Van Gestel CAM. 1998. *Handbook of soil invertebrate toxicity tests*. John Wiley & Sons, Chichester, UK.

167. Organization for Economic Co-operation and Development 2004. Enchytraied reproduction test. Guideline for testing of chemicals 220. OECD, Paris, France.

168. US Environmental Protection Agency. 1985. Avian single-dose oral LD50. USEPA Hazard Evaluation Division, standard evaluation procedure 540/9-85-007, Washington, DC.

169. Hart A, Balluff D, Barfknecht R, Chapman PF, Hawkes T, Joermann G, Leopold A, Luttik R. 2001. *Avian effect assessment: a framework for contaminants studies*. Society of Environmental Toxicology and Chemistry (SETAC), Pensacola FL.

170. Organization for Economic Co-operation and Development. 2001. Guidance document on acute oral toxicity testing. OECD Environment, Health and Safety Publications Series on Testing and Assessment, No. 24. OECD, Paris, France.

171. Organization for Economic Co-operation and Development. 1984. Avian Dietary toxicity test. Guideline for testing of chemicals, No. 205. OECD, Paris, France.

172. Organization for Economic Co-operation and Development. 1984. Avian reproduction test. Guideline for testing of chemicals, No. 206. OECD, Paris, France.

173. Römbke J, Heimbach F, Hoy S, Kula C, Scott-Fordsmand J, Sousa J,. Stephenson G & Weeks J, eds. 2003. Effects of plant protection products on functional endpoints in soil (EPFES), Lisbon 24-26 April 2002. Society of Environmental Toxicology and Chemistry (SETAC), Pensacola FL.

174. Edwards CA. 2002. Assessing the effects of environmental pollutants on soil organisms, communities, processes and ecosystems. *Eur J Soil Biol* 38:225-231.

175. Van Wensem J. 1989. A terrestrial micro-ecosystem for measuring effects of pollutants on isopod-mediated litter decomposition. *Hydrobiologia* 188/189:507-516.

176. Salminen J, Anh BT, Van Gestel CAM. 2001. Indirect effects of zinc on soil microbes via a keystone enchytraeid species. *Environ Toxicol Chem* 20:1167–1174.

177. Knacker T, van Gestel CAM, Jones SE, Soares AMVM, Schallnaß HJ, Bernhard Förster B, Edwards CA. 2004. Ring-testing and field-validation of a terrestrial model ecosystem (TME) – an instrument for testing potentially harmful substances: conceptual approach and study design. *Ecotoxicology* 13:9-27.

178. Dekker SC, Scheu S, Schröter D, Setälä H, Szanzer M, Traas TP. 2005. Towards a new generation of dynamic

soil decomposer food web models. In: De Ruiter P, Wolters V, Moore JC, eds, *Dynamic food webs: multispecies assemblages, ecosystem development, and environmental change*. Academic Press, Burlington, USA, pp. 258-266.

179. Lloyd R. 1961. The toxicity of ammonia to rainbow trout (*Salmo gairdneri* Richardson). *Water and Waste Treatm J* 8:278-279.

180. Berry WJ, Hansen DJ, Mahony JD, Robson DL, Di Toro DM, Shipley BP, Rogers B, Corbin JM, Boothman WS.1996. Predicting the toxicity of metal-spiked laboratory sediments using acid-volatile sulfide and interstitial water normalizations. *Environ Toxicol Chem* 15:2067–2079.

181. Heugens EHW, Hendriks AJ, Dekker T, Van Straalen NM, Admiraal W. 2001. A review of the effects of multiple stressors on aquatic organisms and analysis of uncertainty factors for use in risk assessment. *Crit Rev Toxicol* 31:47-284.

182. Bergema WF, Van Straalen NM. 1991. Ecological risks of increased bioavailability of cadmium and lead as a consequence of soil acidification. Report TCB91/04-R. Technical Committee on Soil Protection, The Hague, The Netherlands [in Dutch].

183. Brown VM. 1968. The calculation of the acute toxicity of mixtures of poisons to rainbow trout. *Water Res* 2:723-733.

184. Hesterberg D, Stigliani WM, Imeson AC. 1992. Chemical time bombs: linkages to scenarios of socioeconomic development. Report 20. International Institute for Applied System Analysis, Laxenburg, Austria.

185. Van Straalen NM, Denneman CAJ. 1989. Ecotoxicological evaluation of soil quality criteria. *Ecotoxicol Environ Saf* 18:241-251.

186. Heugens EHW, Tokkie LTB, Kraak MHS, Hendriks AJ, Van Straalen NM. 2006. Population growth of *Dapnia magna* influenced by multiple stressors – joint effects of temperature, food and cadmium. *Environ Toxicol Chem* 25:1399–1407.

187. Jagers op Akkerhuis G. 1994. Effects of walking activity and physical factors on the short term toxicity of deltamethrin spraying in adult epigeal money spiders (Linyphiidae). In: Donker MH, Eijsackers H, Heimbach F, eds, *Ecotoxicology of Soil Organisms* Lewis Publ, London, UK, pp 323-338.

188. Di Toro DM, Allen HE, Bergman HL, Meyer JS, Paquin PR, Santore RC. 2001. Biotic ligand model of the acute toxicity of metals. 1. Technical basis. *Environ Toxicol Chem* 20:2383–2396.

189. Niyogi S, Wood CM. 2004. Biotic ligand model, a flexible tool for developing site-specific water quality guidelines for metals. *Environ Sci Technol* 38:6177-6192.

190. Paquin PR, Gorsuch JW, Apte S, Batley GE, Bowles KC, Campbell PGC, Delos CG, Di Toro DM, Dwyer RL, Galvez F, Gensemer RW, Goss GG, Hogstrand C, Janssen CR, McGeer JC, Naddy RB, Playle RC, Santote RC, Schneider W, Stubblefield WA, Wood CM, Wu KB 2002. The biotic ligand model: a historical overview. *Comp Biochem Physiol Pt C* 133: 3-35.

191. Campbell PGC, Errecalde O, Fortin C, Hiriart-Baer VP, Vigneault B. 2002. Metal bioavailability to phytoplankton-applicability of the biotic ligand model. *Comp Biochem Physiol Pt C* 133:189-206.

192. Santore RC, Mathew R, Paquin PR, DiToro DM. 2002. Application of the biotic ligand model to predicting zinc toxicity to rainbow trout, fathead minnow, and Daphnia magna. *Comp Biochem Physiol Pt C* 133:271-285

193. Borgmann U, Norwood WP, Dixon DG. 2004. Re-evaluation of metal bioaccumulation and chronic toxicity in Hyalella azteca using saturation curves and the biotic ligand model. *Environ Pollut* 131:469-484

194. De Schamphelaere KAC, Janssen CA. 2004. Development and field validation of a biotic ligand model predicting chronic copper toxicity to *Daphnia magna. Environ Toxicol Chem* 23:1365-1375.

195. Plackett RL, Hewlett PS. 1952. Quantal responses to mixtures of poisons. *J Roy Stat Soc B* 14:141-163

196. Greco WR, Bravo G, Parsons JC. 1995. The search for synergy: a critical review from a response surface perspective. Pharmacol Rev 47:331-385.

197. Sprague JB. 1973. The ABCs of pollutant bioassay using fish. In: Cairns J Jr, Dickson KL, eds, *Biological Methods for the Assessment of Water Quality*. STP 528, American Society for Testing and Materials, Philadelphia PA, pp 6-30.

198. European Inland Fisheries Advisory Commission. 1987. Revised report on combined effects on freshwater fish and other aquatic life. EIFAC Technical Paper 37 Rev. 1. FAO, Rome, Italy.

199. Könemann H. 1981. Fish toxicity tests with mixtures of more than two chemicals: A proposal for a quantitative approach and experimental results. *Toxicology* 19:229-238.

200. Hermens J, Canton H, Janssen P, De Jong R. 1984. Quantitative structure-activity relationships and mixture toxicity studies of chemicals with anaesthetic potency: Acute lethal and sublethal toxicity to *Daphnia magna. Aquat Toxicol* 5:143-154

201. Hermens J, Broekhuyzen E, Canton H, Wegman R. 1985. Quantitative structure-activity relationships and mixture toxicity studies of alcohols and chlorohydrocarbons:

Effects on growth of *Daphnia magna. Aquat Toxicol* 6:209-217.

202. Hermens J, Leeuwangh P, Musch A. 1984. Quantitative structure-activity relationships and mixture toxicity studies of chloro- and alkylanilines at an acute toxicity level to the guppy *(Poecilia reticulata). Ecotoxicol Environ Saf* 8:388-394.
203. Deneer JW, Sinnige TL, Seinen W, Hermens JLM. 1988. The joint acute toxicity to *Daphnia magna* of industrial organic chemicals at low concentrations. *Aquat Toxicol* 12:33-38.
204. Faust M, Altenburger R, Backhaus T, Blanck H, Boedeker W, Gramatica P, Hamer V, Scholze M, Vighi M, Grimme LH. 2001. Predicting the joint algal toxicity of multi-component s-triazine mixtures at low-effect concentrations of individual toxicants. *Aquatic Toxicol* 56:13-32.
205. Hermens J, Leeuwangh P. 1982. Joint toxicity of mixtures of 8 and 24 chemicals to the guppy *(Poecilia reticulata). Ecotoxicol Environ Saf* 6:302-310.
206. Hermens J, Canton H, Steyger N, Wegman R. 1984. Joint effects of a mixture of 14 chemicals on reproduction of *Daphnia magna. Aquat Toxicol* 5:315-322.
207. Faust M, Altenburger R, Backhaus T, Blanck H, Boedeker W, Gramatica P, Hamer V, Scholze M, Vighi M, Grimme LH. 2003. Joint algal toxicity of 16 dissimilarly acting chemicals is predictable by the concept of independent action. *Aquatic Toxicol* 63:43-63.
208. Pedersen F, Petersen GI. 1996. Variability of species sensitivity to complex mixtures. *Water Sci Technol* 33:109-119.
209. Faust M, Altenburger R, Backhous T, Boedeker W, Scholze M, Grimme LH. 2000. Predictive assessment of the aquatic toxicity of multiple chemical mixtures. *J Environ Qual* 29:1063-1068.
210. Enserink EL, Maas-Diepeveen JL, Van Leeuwen CJ. 1991. Combined effects of metals: An ecotoxicological evaluation. *Water Res* 25:679-687.
211. Spehar RL, Fiandt JL. 1986. Acute and chronic effects of water quality criteria-bases metal mixtures on three aquatic species. *Environ Toxicol Chem* 5:917-931.
212. De Zwart D, Posthuma L. 2005. Complex mixture toxicity for single and multiple species: proposed methodologies. *Environ Toxicol Chem* 24:2665-2676.
213. Van Straalen NM, Roelofs D. 2006. *An Introduction to Ecological Genomics*, Oxford University Press, Oxford UK.
214. Bradbury S, Feijtel T, Van Leeuwen K. 2004. Meeting the scientific needs of ecological risk assessment in a regulatory context. *Environ Sci Technol* 38:463a-470a.
215. Tyler CR, Filby A, Iguchi, T, Kramer, V, Larsson, J, van Aggelen G, van Leeuwen, C, Viant, M and Tillitt, D. 2007. Molecular biology and risk assessment: evaluation of the potential roles of genomics in regulatory ecotoxicology. In: *Application of Genomics to Tiered Testing*. SETAC, Pellston, MI (submitted).
216. Snape JR, Maund SJ, Pickford DB, Hutchinson TH. 2004. Ecotoxicogenomics: the challenge of integrating genomics into aquatic and terrestrial ecotoxicology. *Aquatic Toxicol* 67:143–154.
217. Ankley GT, Daston GP, Degitz SJ, Denslow ND, Hoke RA, Kennedy SW, Miracle AL, Perkins EJ, Snape J, Tillitt DE, Tyler CR, Versteeg D. 2006. Toxicogenomics in regulatory ecotoxicology. *Environ Sci Technol* 40:4055-4065.
218. Hutchinson TH, Ankley GT, Segner H, Tyler CR. 2006. Screening and testing for endocrine disruption in fish-biomarkers as "signposts," not "traffic lights," in risk assessment. *Env Health Persp* 114:106-114.
219. Colborn T, Dumanoski D, Meyers JP. 1996. *Our Stolen Future. How we are threatening our fertility, intelligence and survival-A scientific detective story*. Penguin Books, Dutton, NY.
220. European Commission. 2007. Technical guidance documents on preparing the Chemical Safety Assessment, in prep. European Chemicals Bureau, Joint Research Centre, Ispra, Italy.
221. Hamers T, Kamstra JK, Sonneveld E, Murk AJ, Kester MHA, Andersson PL, Legler J, Brouwer A. 2006. In vitro profiling of the endocrine-disrupting potency of brominated flame retardants. *Toxicol Sci* 92:157-173.
222. Sørmo EG, Jüssi I, Jüssi M, Braathen M, Skaare JU, Jenssen BM. 2005. Thyroid hormone status in gray seal *(Halichoerus grypus)* pups from the baltic sea and the atlantic ocean in relation to organochlorine pollutants. *Environ Toxicol Chem* 24:610–616.
223. Risk Assessment Forum. 1992. Framework for ecological risk assessment. Report 630/R-92/001. US Environmental Protection Agency, Washington, DC.
224. US Environmental Protection Agency. 1984. Estimating "concern levels" for concentrations of chemical substances in the environment. Environmental Effect Branch, Health and Environmental Review Division, Washington, DC.
225. Posthuma L, Suter GW II, Traas TP. 2002. *Species Sensitivity Distributions in Ecotoxicology*. Lewis publishers Boca Raton, FL.
226. Aldenberg T, Jaworska JS. 2000. Uncertainty of the hazardous concentration and fraction affected for normal species sensitivity distributions. *Ecotoxicol Environ Saf* 46:1-18.
227. Aldenberg T, Slob W. 1993. Confidence limits for

hazardous concentrations based on logistically distributed NOEC toxicity data. *Ecotoxicol Environ Saf* 25:48-63.

228. Kooijman SALM. 1987. A safety factor for LC50 values allowing for differences in sensitivity among species. *Water Res* 21:269-276.
229. Wagner C, Løkke H. 1990. Estimation of ecotoxicological protection levels from NOEC toxicity data. *Water Res* 25:1237-1242.
230. Van Vlaardingen PLA, Traas TP, Wintersen AM, Aldenberg T. 2004. ETX 2.0 - A Program to calculate Hazardous concentrations and fraction affected, based on normally distributed toxicity data. Report 601501028, National Institute for Public Health and the Environment (RIVM), Bilthoven, the Netherlands [available from www.rivm.nl].
231. Aldenberg T. 1993. ETX 1.3a. A program to calculate confidence limits for hazardous concentrations based on small samples of toxicity data. Report 719102015. National Institute for Public Health and the Environment (RIVM), Bilthoven, the Netherlands.
232. Slooff W, Canton JH. 1983. Comparison of the susceptibility of 11 freshwater species to 8 chemical compounds. II (Semi-)chronic toxicity tests. *Aquat Toxicol* 4:271-282.
233. Erickson RJ, Stephan CE. 1984. Calculating the final acute value for water quality criteria for aquatic organisms. Report 600/X-84-040. Environmental Research Laboratory-Duluth, Office of Research and Development, USEPA, Duluth, MN.
234. Chapman PM, Fairbrother A, D. Brown D. 1998. A critical evaluation of safety (uncertainty) factors for ecological risk assessment. *Environ Toxicol Chem* 17:99-108.
235. Power M, L.S. McCarthy, LS. 1997. Fallacies in ecological risk assessment practices. *Environ Sci Technol* 31:370A-374A.
236. Versteeg DJ, Belanger SE, Carr GJ. 1999. Understanding single species and model ecosystem sensitivity, A data based comparison. *Environ Toxicol Chem* 18:1329-1346.
237. Romijn CAF, Luttik R, Van De Meent D, Slooff W, Canton JH. 1993. Presentation and analysis of a general algorithm for risk assessment on secondary poisoning. *Ecotoxicol Environ Saf* 26:61-85.
238. Romijn CAF, Luttik R, Van De Meent D, Slooff W, Canton JH. 1991. Presentation and analysis of a general algorithm for risk assessment on secondary poisoning. Part II. Terrestrial food chains. Report 679102007, National Institute of Public Health and Environmental Protection, Bilthoven, The Netherlands.
239. Lehman AJ. Untitled. 1954. Assoc. *Food Drug Off Quart Bull* 18:66.
240. Mineau P, Baril A, Collins BT, Duffe J, Joerman G, Luttik R. 2001. Pesticide acute toxicity reference values for birds. *Rev Environ Contam Toxicol* 170:13-74.
241. Aldenberg T, Luttik R. 2002. Extrapolation factors for tiny toxicity data sets from Species Sensitivity Distributions with known standard deviation. In: Posthuma L, Suter GW II, Traas TP, eds, *Species Sensitivity Distributions in ecotoxicology*. Lewis, Boca Raton FL, pp 103-118.
242. Macintosh DL, Suter II GW, Hoffman FO. 1992. Model of PCB and mercury exposure to mink and great blue heron inhabiting the off-site environment downstream from the US Department of Energy Oak Ridge Reservation. ORNL/ER-90. Oak Ridge National Library, Oak Ridge, TN.
243. Luttik R, Aldenberg T. 1997. Extrapolation factors for small samples of pesticide toxicity data: special focus on LD50 values for birds and mammals. *Environ Toxicol Chem* 16:1785-1788.
244. Klasmeier J, Matthies M, MacLeod M, Fenner K., Scheringer M, Stroebe M, Le Gall AC, McKone T, Van de Meent D, Wania F. 2006. Application of multimedia models for screening assessment of long-range transport potential and overall persistence. *Environ Sci Technol* 40:53-60.
245. EPA. 2006. PBT profiler, Ver 1.203 . September 21, 2006, http://www.pbtprofiler.net/. Developed by the Environmental Science Center under contract to U.S. Environmental Protection Agency.
246. Commission of the European Communities. 2006. Regulation (EC) No 1907/2006 of the European Parliament and of the Council of 18 December 2006 concerning the Registration, Evaluation, Authorisation and Restriction of Chemicals (REACH), establishing a European Chemicals Agency, amending Directive 1999/45/EC and repealing Council Regulation (EEC) No 793/93 and Commission Regulation (EC) No 1488/94 as well as Council Directive 76/769/EEC and Commission Directives 91/155/EEC, 93/67/EEC, 93/105/EC and 2000/21/EC. Off J Eur Union L 396 of 30.12.2006.
247. Steensberg J. 1989. Environmental health decision making. The politics of disease prevention. Thesis. Supplementum 42 to the *Scandinavian Journal of Social Medicine*, Almqvist & Wiksell International, Copenhagen, Denmark.

8. DATA: NEEDS, AVAILABILITY, SOURCES AND EVALUATION

P.G.P.C. ZWEERS AND T.G. VERMEIRE

8.1 INTRODUCTION

The recognition that a core set of data is needed for prioritization and risk assessment of chemicals goes back to the 1970's. Among other things, this resulted in the OECD Council Act on the Minimum Premarketing Set of Data (MPD) for new chemicals [1] and the equivalent Screening Information Data Set (SIDS) for high production volume chemicals (see also Chapter 16). It is equally important to remember that it is not feasible to conduct complete and comprehensive testing on every chemical and therefore tiered approaches are needed to identify those chemicals which require further testing to guarantee safe use (see Chapter 11).

The data sets required always try to strike a balance between the interests of the public at large, regulators, industry and the scientific community. This balancing of interests takes place based on the need to reduce the number of experimental animals used, scientific views, acceptable degree of uncertainty and complexity, time to generate data and costs. This chapter will present various views on data needs from different perspectives and compare them with data availability. Although the focus of this chapter will again be on industrial chemicals, other regulatory frameworks in the EU will also be discussed for comparison.

When considering testing needs to meet testing requirements, data generation depends on what data are already available and on the quality of these data. Hence, the first step in an intelligent approach to testing is to gather the available information. Key sources of information on chemicals are presented in this chapter as well as methods for retrieving that information. All risk assessments have inherent uncertainties which are mainly determined by the quality of the input data. Additional uncertainty may arise when input data have to be estimated, e.g., by using models or (quantitative) structure-activity relationships ((Q)SARs). While specific elements of the evaluation of test results are discussed in previous chapters, this chapter, however, discusses the general aspects of evaluating data quality. Scoring methods to rank the quality of experimental and model input data are addressed.

Overall, this chapter looks at the subject of data needs and data availability against the background of EU regulatory frameworks. It is believed that these approaches offer a model for chemicals management worldwide. This chapter focuses on information on physicochemical properties and hazards, and only touches on use and exposure. For details of the latter, the reader is referred to Part II of this book (Chapters 2-5).

8.2 DATA REQUIREMENTS IN EU REGULATORY FRAMEWORKS

8.2.1 Introduction

The data required for a risk assessment or chemical safety assessment (CSA) of a substance can be categorized as data on the identity of the substance, its physicochemical, toxicological and ecotoxicological properties, and its uses, emissions and exposures (Figure 8.1). The minimum set of data required for a risk assessment depends on the use category, the regulation involved and the goal of the risk assessment. There is international agreement on the need to test for acute toxicity, including local effects, repeated dose toxicity, developmental and reproductive toxicity, mutagenicity, ecotoxicity and environmental fate. The categories which can be classified by their use are: new and existing industrial chemicals, biocides and pesticides, veterinary medicines, pharmaceuticals, cosmetics, food additives and food contact materials. The minimum data requirements have been specified in detail for some of these categories.

There are both similarities and differences in the filing requirements of different regulatory frameworks. Data on the physicochemical properties of the substance are required under all regulatory frameworks, although the level of detail can vary considerably. Instead of test data, waiving arguments for not performing specific studies can also be submitted by industry. These arguments should be scientifically sound and thoroughly evaluated. In the following sections the differences in the specified minimum data requirements for general industrial chemicals are discussed in detail for the various regulatory frameworks.

8.2.2 Industrial chemicals

The general data requirements for all manufactured, imported and marketed industrial chemicals in the EU are specified in Annex VI of REACH (see Chapter 12).

C.J. van Leeuwen and T.G. Vermeire (eds.), Risk Assessment of Chemicals: An Introduction, 357–374.

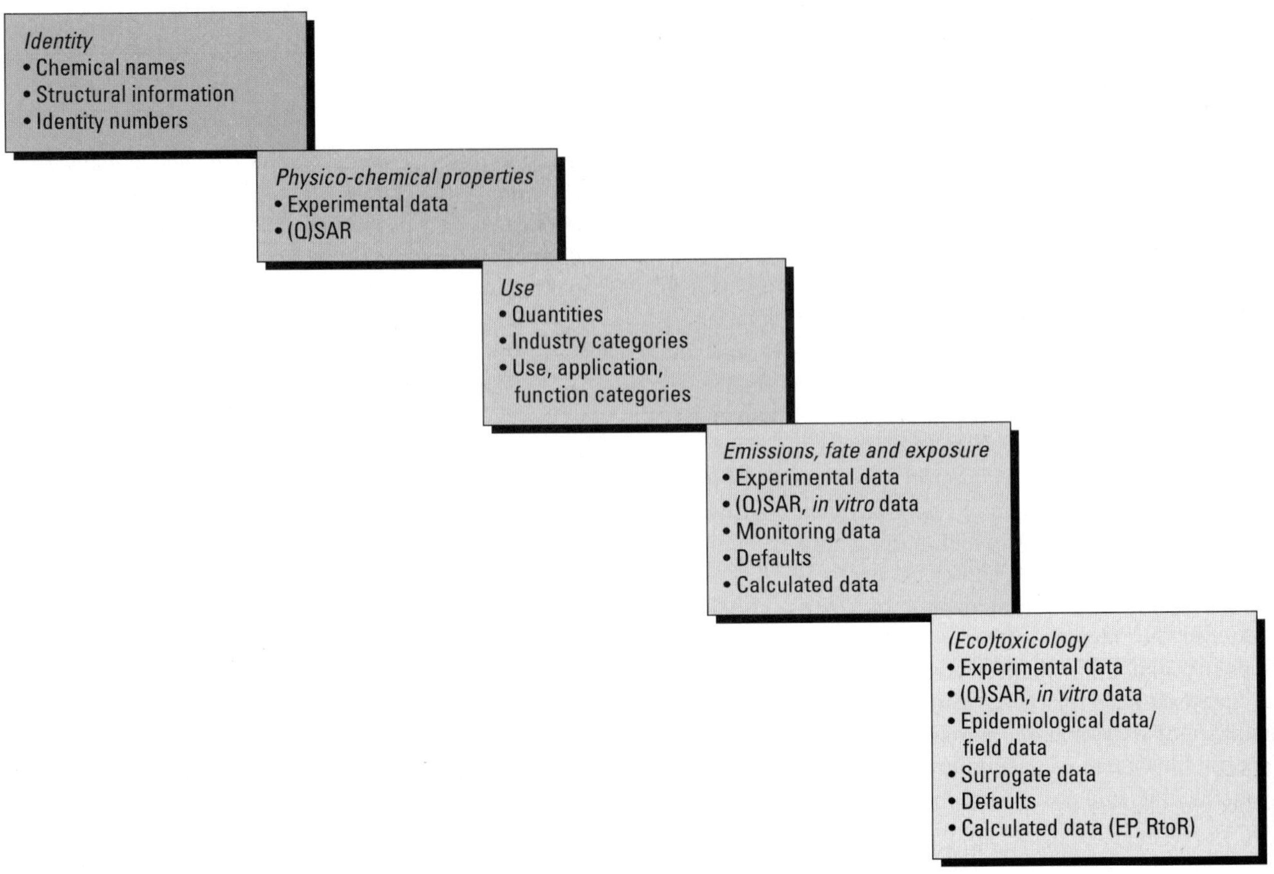

Figure 8.1. Data types in risk assessment of chemicals.
(Q)SAR = Quantitative Structure-Activity Relationship (Chapters 9-11).
EP = Equilibrium Partitioning (Chapters 3 and 7).
RtoR = Route-to-Route extrapolation (Chapter 6).

Additional data requirements for tonnage levels from >1 tonne, >10 tonnes, >100 tonnes, and >1000 tonnes, are summarized in Annexes VII, VIII, IX and X of REACH (Table 8.1), respectively.

Two types of industrial chemicals were identified prior to REACH, i.e. new and existing chemicals. For new chemicals the data requirements are based on the OECD MPD, and can increase with increased tonnage levels [1]. Information should be provided by the industry from a production or import level of 10 kg/year onwards. Data requirements gradually increase when tonnage thresholds of 100 kg/year, 1 tonne/year, 10 tonnes/year, 100 tonnes/year, up to a level of >1000 tonnes per year are reached or when cumulative volumes of 5 times these thresholds are reached. For existing chemicals the SIDS is generally used for high production volume chemicals (HPVCs). HPVCs are chemicals produced or imported in quantities in excess of 1000 tonnes per year. In the EU four priority lists were adopted for HPVCs for which a risk assessment report was developed based on existing information and risk-based data.

In the EU, the MPD evolved into the "base-set" for the risk assessment of new and existing chemicals within the scope of EC Directive 67/548/EEC on new substances [2] and for priority existing substances within the scope of EC Council Regulation 793/93 [3] (Table 8.2). The EC base-set does not include exposure data. For many substances, specifically new chemicals, there is little or no information on actual emissions to the environment, doses or concentrations in consumer products, when relevant, and at the work place. Moreover, measured concentrations very often vary significantly with regard to time and spatial scales, which limits their usefulness for risk assessment. Therefore, exposure

Table 8.1. Data requirements as defined in REACH.

ANNEX VII (≥1 TONNE)	
Physical and chemical information	
7.1	State of the substance (at 20 °C/101.3 kPa)
7.2	Melting/freezing point
7.3	Boiling point
7.4	Relative density
7.5	Vapour pressure
7.6	Surface tension
7.7	Water solubility
7.8	Partition coefficient *n*-octanol/water
7.9	Flash-point
7.10	Flammability
7.11	Explosive properties
7.12	Self-ignition temperature
7.13	Oxidising properties
7.14	Granulometry
Toxicological information	
8.1	Skin irritation or skin corrosion
8.2	Eye irritation
8.3	Skin sensitization
8.4.1	Mutagenicity (gene mutation in bacteria)
8.5.1	Acute toxicity (oral route)
Ecotoxicological information	
9.1.1	Short-term toxicity invertebrates *(Daphnia)*
9.1.2	Growth-inhibition plants (algae)
9.2.1.1	Ready biodegradability

ANNEX VIII (≥10 TONNES)	
Toxicological information	
8.1.1	Skin irritation *(in vivo)*
8.2.1	Eye irritation *(in vivo)*
8.4.2	Cytogenicity in mammalian cells *(in vitro)*
8.4.3	Gene mutation in mammalian cells *(in vitro)*
8.5.2	Acute toxicity (inhalation)
8.5.3	Acute toxicity (dermal)
8.6.1	Repeated dose toxicity (28 days)
8.7.1	Reproductive/developmental toxicity screening test; OECD 421 or 422)
8.8.1	Toxicokinetics
Ecotoxicological information	
9.1.3	Short-term toxicity fish
9.1.4.	Activated sludge respiration inhibition test
9.2.2.1	Hydrolysis as a function of pH
9.3.1	Adsorption/desorption screening test

ANNEX IX (≥100 TONNES)	
Physical and chemical information	
7.15	Stability in organic solvents and identity of relevant degradation products
7.16	Dissociation constant
7.17	Viscosity
Toxicological information	
8.6.1	Repeated dose toxicity (28 days)
8.6.2	Sub-chronic toxicity (90 days)
8.7.2	Pre-natal developmental toxicity; OECD 414
8.7.3	Two-generation reproductive toxicity study
Ecotoxicological information	
9.1.5	Long-term toxicity invertebrates *(Daphnia)*
9.1.6	Long-term toxicity to fish
9.1.6.1	Fish early-life stage test
9.1.6.2	Fish short term toxicity embryo and sac fry
9.1.6.3	Fish juvenile growth test
9.2.1.2	Ultimate degradation in surface water
9.2.1.3	Soil simulation testing
9.2.1.4	Sediment simulation testing
9.2.3	Identification of degradation products
9.3.2	Bioaccumulation in aquatic species (fish)
9.3.3	Further information on adsorption/desorption
9.4.1	Short-term terrestrial toxicity (invertebrates)
9.4.2	Effects on soil micro-organisms
9.4.3	Short-term toxicity to terrestrial plants

ANNEX X (≥1000 TONNES)	
Toxicological information	
8.6.3	Long-term repeated toxicity (≥12 months)
8.7.2	Developmental toxicity; OECD 414
8.7.3	Two-generation reproductive toxicity
8.9.1	Carcinogenicity study
Ecotoxicological information	
9.3.4	Further fate and behaviour in the environment of the substance and/or degradation products
9.4.4	Long-term toxicity on invertebrates
9.4.6	Long-term toxicity on plants
9.5.1	Long-term toxicity to sediment organisms
9.6.1	Long-term or reproductive toxicity to birds

Table 8.2. Base-set requirements.

1. Identity: trade name, chemical name, formulae composition, spectra, methods of analysis
2. Quantity, functions, applications
3. Precautionary measures, emergency measures
4. Physical properties:
   a. melting point, boiling point
   b. relative density
   c. vapour pressure
   d. surface tension
   e. solubility in water
   f. n-octanol/water partitioning coefficient
   g. particle size
5. Chemical properties:
   a. flash point
   b. (auto)flammability
   c. explosive properties
   d. oxidizing properties
6. Toxicological properties:
   a. acute toxicity (2 routes)
   b. skin/eye irritation
   c. sensitization
   d. subacute toxicity
   e. genotoxicity (2 tests)
7. Ecotoxicological properties
   a. acute toxicity (algae, fish, Daphnia)
   b. inhibition bacteria
   c. ready biodegradability
   d. hydrolysis
8. Methods rendering the substance harmless

doses and concentrations must often be predicted on the basis of information on default emission rates and physicochemical properties.

8.2.3 Pesticides and biocides

The objective of the evaluation of pesticide residues in crops and animal products is to establish acceptable dietary exposures based on Good Agricultural Practice (GAP), which leads to the derivation of Maximum Residue Levels (MRLs) for man indirectly exposed via the environment. The risks to the environment in general and non-target species (flora and fauna) should also be assessed.

The minimum data requirements for this assessment are specified under Directive 91/414/EC [4]. What triggers the data requirements, as specified below, is the use of the pesticide formulation on food crops or on crops for animal feed. The toxicological test data requirements in addition to the data requirements in Annexes VII and VIII of Table 8.1 are:

- Subchronic and chronic/carcinogenicity repeated dose toxicity studies (via all relevant routes of exposure).
- Reproduction study (two-generation study).
- Two developmental toxicity studies (in rat and rabbit).
- Toxicokinetic studies (single and repeated dose, oral route in rat) and data (on dermal adsorption).
- Specific tests where relevant (e.g., *in vivo* mutagenicity tests if a positive effect is observed for one of the *in vitro* tests, delayed neurotoxicity for organophosphates or choline esterase inhibition).

An overview of ecotoxicological test data requirements in addition to the data requirements listed in Annexes VII and VIII of Table 8.1 is given below. What has to be borne in mind that only those tests which are *relevant* (i.e., when the pesticide is expected to occur in the environmental compartment or prey of interest) have to be performed:

- Additional degradation tests in surface water, sediment and soil, with identification of degradation products.
- Long-term aquatic toxicity tests at three taxonomic levels for freshwater species.
- Bioaccumulation study in fish.
- Soil adsorption/desorption.
- Short and long-term toxicity studies with plants and earthworms.
- Soil respiration and nitrification tests.
- Long-term toxicity tests with sediment organisms.
- Long-term or reproductive toxicity to birds.

The exposure data which have to be determined are residue levels in raw agricultural crops and animal products (i.e., meat, milk and eggs), calculated from residue trials in crops based on GAP and from animal feed studies. Data on metabolism in plants and livestock is also needed to derive the residue level of concern.

In the EU, the data requirements for the active ingredient in biocides (non-agricultural pesticides) are laid down in annexes to Directive 98/8/EC [5]. Since the focus of this section is to identify the data requirements for different uses of substances rather than give a complete and detailed overview of data requirements for all types of substances, the data requirements for biocides are not listed. However, it must be understood that the data requirements for pesticides and biocides are far more extensive than for industrial chemicals due to their intended use. Obviously, these types of chemicals are used intentionally to give cause an adverse effect on target organisms and therefore may give rise to adverse effects in non-target organisms as well. Therefore both the human toxicological and the ecotoxicological profile have to be explored in much more detail.

8.2.4 Veterinary medicines and feed additives

For the active ingredients of veterinary medicines and additives in animal nutrition both the ADI (Acceptable Daily Intake) and species-specific Maximum Residues Limits (MRLs) are assessed under the EU regulations. Relevant guidance is given in EU Directive 2001/82 [6] and EU Regulation 2377/90 (for MRL assessment) [7] for veterinary medicines and in EC regulation 1831/2003 [8] for feed additives. The test requirements for veterinary medicines and additives in animal nutrition are similar. The required set of physicochemical data is limited (e.g., water solubility, dissociation constants, UV-visible absorption spectrum, melting temperature, vapour pressure and the *n*-octanol/water partition coefficient). The human toxicological data requirements are as described in Table 8.1 for REACH Annex VII with the following additional tests:

- Subchronic and chronic/carcinogenicity repeated dose toxicity studies (oral exposure for both a rodent and non-rodent species with a 28-day and 90-day duration, and a 2-year duration for the most sensitive species. A carcinogenicity study is obligatory, but derogation or a combined chronic/carcinogenicity study is possible.
- Reproductive toxicology study (two-generation study).
- Two developmental toxicity studies (in rat, if negative also in rabbit).
- Toxicokinetic studies (for both experimental and target animals).
- Specific tests if relevant (e.g., *in vivo* mutagenicity tests if a positive effect is observed for one of the *in vitro* tests and on possible effects on micro-organisms involved in milk processing).
- Additional information (e.g., effect on intestinal flora for antibiotics, on the possibility of developing drug resistance in consumers, and on pharmacological effects).

Exposure scenarios have to be developed for users and man exposed indirectly via the environment (see Chapters 2 and 5). Ecotoxicological test requirements depend on the exposure level following a risk-based tiered approach. When relevant, the following tests are required:

Tier A:

- Soil adsorption/desorption.
- Soil biodegradation or degradation in aquatic systems.
- Optional: photolysis and hydrolysis.
- Acute aquatic toxicity tests at three taxonomic levels for freshwater or marine species.
- Terrestrial studies (on nitrogen transformation, plants and earthworms).
- Studies with dung fly and dung beetle larvae.

Tier B:

- A bioconcentration study in fish.
- Long-term aquatic toxicity tests of three taxonomic levels and a sediment invertebrate toxicity test for freshwater or marine species.
- An extended nitrogen transformation test and two toxicity studies with other plant species.

8.2.5 Cosmetics

The relevant EU legislation for cosmetics risk assessment is laid down in Directive 76/678/EEC [9]. Technical guidance for the risk assessment procedure is given in the Notes of Guidance for testing (NOG) of Cosmetic Ingredients for their Safety Evaluation [10], updated basic requirements for toxicological dossiers to be evaluated by the SCCNFP (EU Scientific Committee on Cosmetic and Non-Food Products intended for consumers) [11] and in a memorandum concerning the actual status of alternative methods to the use of animals in the safety testing of cosmetic ingredients [12]. Besides default calculations of Margins of Safety (MOS) under use conditions, it is also concluded whether the use is safe or unsafe with respect to genotoxicity and carcinogenicity.

The data requirements for substances used in cosmetic products largely depend on the ingredients used, the formulation of the finished product and the degree and the route of consumer application and should therefore be adapted case-by-case on a risk basis.

Besides the physical and chemical information, toxicological data are also required on:

- Acute toxicity (preferably oral exposure).
- Irritation and sensitization (human data preferred).
- Subacute and subchronic repeated dose toxicity (oral or dermal exposure).
- Chronic/carcinogenicity, reproduction and developmental studies (needs to be established case-by-case).
- Mutagenicity (*in vitro* tests obligatory - *in vivo* tests may be required depending on results).
- Specific tests for specified types of consumer

products, e.g., phototoxicity and photo-irritation for sunscreens; photomutagenicity and photogenotoxicity data for sunscreens.

Systemic human exposure is calculated using dermal absorption rate and standard use assumptions. Risk evaluation for cosmetics is based on consumers as the protection target. As a consequence, no ecotoxicological data are required.

8.2.6 Food additives

For food additives an Acceptable Daily Intake (ADI) should be derived, based on toxicological data, for the additive itself. Guidance is given in Environmental Health Criteria 70 [13] and 170 [14]. In general a standard set equivalent to the base-set (Table 8.2) is required. Derogation is possible, e.g., where application is limited, but should be scientifically sound and evaluated on a case-by-case basis. An ADI may be omitted in such cases. If the data set is incomplete, an additional assessment factor may be applied, and the authorization is provisional pending the requested data. The test requirements (besides the physicochemical data) are:

- Acute, subacute, subchronic and chronic/carcinogenicity repeated dose toxicity studies (all with oral exposure).
- Biokinetics.
- Mutagenicity tests (3 *in vitro* studies and an *in vivo* test).
- Reproduction and developmental study.

Industry should indicate the amount of substance in food, based on proper use for exposure characterization. Genotoxic carcinogens are not permitted. For most additives only a chronic risk assessment is required, taking oral exposure via all possible sources into account.

The risk evaluation for food additives is from a human toxicological perspective. As a consequence, no ecotoxicological data are required.

8.2.7 Food contact materials

Following food additives, the active ingredient is also taken into account for food contact materials, for which a Total Daily Intake (TDI) should be derived. Exposure via food and beverages is considered at three levels, depending on the level of migration (level A: < 0.05 mg/kg food; level B: 0.05-5 mg/kg food; level C: 5-60 mg/kg food). Guidance is given in the form of a Practical Guide and a Note for Guidance SANCO D3/LR D (04.2003) [15] which can be found on the website: http://cpf.jrc.it/webpack/. The test requirements depend on the migration level. In addition, only chronic repeated dose toxicity data are needed. Polymers with a molecular weight in excess of 1000 D are unlikely to be absorbed by the gastrointestinal tract; therefore a reduced data set may be required, depending on the molecular weight distribution and the amount of residual monomer. The test requirements (besides physicochemical data) are:

- Subchronic (at level B), chronic/carcinogenicity studies (at level C) (all with oral exposure).
- Toxicokinetics (at level B).
- Mutagenicity tests (3 *in vitro* studies, all at level A; *in vivo* tests may be required depending on results).
- Reproduction and developmental screening study (at level C).
- Specific tests (for hydrolysable substances, data are required for both the parent compound and its hydrolysis products, for biocides additional information on microbiological properties is required and additional studies may be required if specific biological effects are expected based on structural alerts or expert judgement).

For the exposure assessment, data are needed on the migration of the substance into food and/or beverages. Genotoxic substances and genotoxic carcinogens are not permitted. A threshold approach is applied for non-genotoxic carcinogens.

The risk evaluation for food contact materials is from a human toxicological perspective. As a consequence, no ecotoxicological data are required.

8.3 DATA AVAILABILITY

Two essential types of data can be distinguished for risk assessment purposes. Firstly, exposure data are needed which can be either measured (release data or monitoring in the environment) or estimated making use of both mathematical models and (Q)SARs. Besides exposure data, effects data on all defined human toxicological and ecotoxicological endpoints are also needed. Effects data can be experimentally derived from animal studies or human studies, making use of volunteers. However, epidemiological data can provide a good alternative for both human as well as environmental toxicological effects data. In this section general guidance is given on how to deal with data gaps in effects data.

Data availability on substances is very often determined by its regulatory context. For industrial chemicals, the REACH Annexes VII to X (Table 8.1) list specific rules under which the tonnage-triggered standard information may be omitted, replaced by other information, provided at a different stage or adapted in another way. If the specified conditions are met these Annexes allow adaptations. However, the fact and the reasons for each adaptation should be clearly indicated in the registration. In addition to these specific rules, the required standard information set may be adapted in accordance with the general rules in Annex XI. In this case too, the fact and the reasons for each adaptation should be clearly indicated in the registration. In some cases, the rules set out in Annexes VII to X may require certain tests to be undertaken earlier than or in addition to the tonnage-triggered requirements.

For the risk assessment of new industrial chemicals, agricultural pesticides and biocides in the EU, complete data sets will generally be available, consisting of at least the EC base-set of data (see Table 8.2). Data obtained by industry at production and industrial use sites relating to environmental and occupational exposure are available, especially for industrial chemicals. Besides these "regulatory" risk assessments, many assessments are performed by scientists in government, industry and the private sector to answer immediate questions on the potential risks to a variety of human and/or environmental targets. The availability of data will greatly differ and also depend on the available human and financial resources. In 1998, USEPA reported that the hazard data availability for nearly 2,900 high production volume chemicals (HPVCs) with a production or import at or above 1 million pounds (436 tonnes) per year in the US was generally rather limited [16]. Analysis revealed that no basic toxicity information, i.e., neither human health nor environmental toxicity, is publicly available for 43% of the HPVCs manufactured in the US and that a full set of basic toxicity data is available for only 7% of these chemicals. Similarly, in the EU only 14% of HPVCs were found to have data at the level of the base-set, 65% have less than the base-set level, and 21% have no data available at all [17]. From these and similar studies it may be concluded that there are serious data gaps. Estimates by the European Commission in 2003 indicated that the impact of the proposed REACH system would result in the most number of tests being required for the endpoints skin sensitization (for approximately 35% of all substances), eye irritation (approximately 24% of which 5% are *in vivo* tests) and the *in vivo* mutagenicity study (approximately 22%) [18]. For all other endpoints, new testing would be required for less than 20% of the substances.

Nowadays, most new data generated within the scope of existing regulations are produced in accordance with the principles of good laboratory practice (GLP) [19] and internationally recognized test guidelines, such as those of the EC [20], OECD [21] and USEPA [22]. In addition, when a compound has been on the market for a longer period of time and is more widely studied, the proportion of reports based on guidelines will diminish and other, more specialized and unusual studies will take their place. Consequently, the selection of tests critical to a certain risk assessment will become more complicated and more dependent on expert judgement. As the available data may thus be from "guideline tests", "non-guideline tests", and from "non-testing methods" (e.g. (Q)SAR, read across etc.), the evaluation of quality, adequacy (see Section 8.5), and completeness of data will have to rely on weight-of-evidence (WoE) approaches, that combine and weigh the data.

The hazard assessment should be based on all available and relevant information and, as a minimum, on the basis of the information required, while additional information may be needed as a result of the outcome of the exposure estimation and risk characterization. If the risk for a given use is not sufficiently controlled, additional data may need to be collected or generated to refine the assessment. Available relevant information on a substance is collected or generated for physicochemical properties, environmental fate and toxicity data for human health and the environment. Table 8.1 lists the minimum hazard data required under REACH at the different tonnage triggers. In addition to this, hazard data should be supplemented with relevant available information, e.g., from the published literature or from previous testing experience. A systematic approach is required to check the available relevant information and to avoid overlooking relevant data or duplication of effort (see Section 8.4).

Alternative information on hazards that can be used instead of *in vivo* test data may be available or generated based on the outcome of the risk assessment. Such information includes results of *in vitro* tests and information obtained through the use of non-testing methods. Information obtained through the use of non-testing approaches ((Q)SAR, read-across from analogue substances, categorization, etc.) is considered valid, if the approach and the results meet validity criteria including transparency, applicability domains, mechanistic understanding, and formal validation.

More information can be found in additional guidance on intelligent testing strategies (ITS, Chapter 11), while Chapters 9 and 10 provide an overview of relevant (Q)SARs currently available. ITS describes how available information can be combined in a WoE procedure to draw conclusions for each endpoint. In some cases, minimal or negligible exposure and/or risk can be expected for certain target groups in the assessment. When such low-risk exposure situations are found, it may be possible to waive hazard data. Additional guidance on this issue is also available in the section on ITS.

Wherever possible, data gaps in the risk assessment (also with respect to exposure) should be filled with estimated data, using generally agreed procedures, or by default values (see also Section 2.4). Default values are intended to substitute unknown or uncertain values and are determined by a combination of evaluation of available databases and expert judgement. The use of defaults is considered a reasonable way to deal with uncertainty, as long as the principles for choosing defaults and the rationale for the values chosen are transparent. In addition, guidance on when and how to depart from these principles should be available. The procedure should take into account the protection of public health and the environment, ensuring scientific validity, minimizing serious errors in estimating risks and maximizing incentives for research in such a way that it results in orderly and predictable process [23]. For example, when a default assumption is used to address a data gap in a risk assessment it should be indicated where the default was used and explained why it is considered a reasonable approach. Data gaps also include secondary data such as partition coefficients, which are derived from the available data. Estimation of parameter values using sufficiently validated quantitative structure-activity relationships ((Q)SARs) on physicochemical properties, fate and (eco)toxicity endpoint (Chapters 9 and 10) is preferred over the use of defaults, even though expert judgement has to be used to estimate these parameters. The major advantage of estimation methods is that more specific use is made of existing knowledge to support the decision-making process.

The OECD has produced a number of reports on QSARs, i.e., on physicochemical parameters [24], biodegradation [25], aquatic toxicity [26] and exposure assessment [27], as well as a comparison of toxicity estimates and actual data, carried out jointly by the US Environmental Protection Agency (USEPA) and the Commission of the European Communities [28]. Recently, the OECD published an overview of regulatory uses and application [29] and the principles of validation [30].

8.4 DATA SOURCES

Toxicological and ecotoxicological information can be obtained from specialized libraries and documentation centres worldwide. However, the use of IT technologies is more common practice nowadays. Most bibliographical searches for primary toxicological information will now start by interrogating on-line bibliographical databases, such as Current Contents, TOXLINE and more specialized databases, such as CANCERLIT (for cancer-related references) and AGRICOLA (for pesticides). Secondary sources containing factual toxicity and ecotoxicity data can be obtained from a number of databases; a selection of publicly available sources is listed in Table 8.3. Useful data sources on physicochemical properties are shown in Table 8.4. (Q)SAR expert systems, handbooks and databases are described in Chapters 9 and 10. Many of these sources are available on-line.

The formal training and experience of information specialists in selecting and extracting this type of information from databases is the only guarantee that the searches will be sufficiently exhaustive and cost-effective. Intelligent search profiles can increase the relative number of relevant hits as well as decrease the overall number of hits. For instance: searching for recent publications on the toxicity of cadmium in water in Current Contents will be cumbersome if only the key words “cadmium” and “toxicity” are used (> 3440 hits), but less so if “water” is also entered (approx. 1100 hits) and even less so if also “>2000” is used (approx. 700 hits). Sanderson [36,37] rightly points out that it is important not to restrict oneself to on-line computer-based sources. The reliability of the retrieved data varies according to the protocols under which they were collected. Depending on the resources available, it is recommended that the primary sources of retrieved data are consulted to select the input data for risk assessments. A good alternative is the use of secondary sources peer-reviewed by national or international organizations.

An exhaustive standard volume on information sources in toxicology is available [38]. Bradbury et al. [39] provided a good overview of information sources in ecotoxicology.

8.5 DATA EVALUATION: QUALITY AND SELECTION

8.5.1 Introduction and definitions

The aim of this section is to provide general guidance for the evaluation of both testing and non-testing data used in the risk assessment of chemicals (Figure 8.1). Important aspects of this evaluation are determining the reliability, relevance and adequacy of the available information

Table 8.3. Sources of secondary information on the toxicity and ecotoxicity of chemicals.

Source (alphabethically)	Description	Location
Peer reviewed secondary sources[1]		
ATSDR Toxicological profiles	Comprehensive review of all human toxicity endpoints, data on human exposure via all routes	www.atsdr.cdc.gov
California - Office of Environmental Health Hazard Assessment reports (OEHHA)	Comprehensive reviews and fact sheets on human toxicity endpoints, database with ecotoxicological data	http://www.oehha.ca.gov
CCRIS	Database with carcinogenicity and mutagenicity test results	toxnet.nlm.nih.gov
EFSA opinions	Review of all human toxicity endpoints, data on human exposure via food	www.efsa.europa.eu/en/science/
EU Risk Assessment Reports	In depth review all (eco)toxicity endpoints, all exposure scenarios quantified	ecb.jrc.it/esis/
European Chemical Industry Ecology & Toxicology Centre reports	Comprehensive reviews of all (eco)toxicity endpoints	www.ecetoc.org
FAO/WHO Joint-Expert Committee on Food Additives (JECFA) reports	Comprehensive review of oral human toxicity endpoints, comprehensive review of exposure via food	inchem.org/
FAO-WHO Joint Meeting on Pesticide Residues (JMPR) reports	Comprehensive review of oral human toxicity endpoints, comprehensive review of residue levels in food	inchem.org/
FAO/WHO Pesticide Data Sheets	Summary evaluations of human toxicity data and recommendations for safe use and disposal of pesticides	inchem.org/
GENE-TOX	Database with peer reviewed genetic toxicology test data	toxnet.nlm.nih.gov
Hazardous Substances Data Bank (HSDB)	Database with comprehensive, peer reviewed (eco)toxicology data and physicochemical data	toxnet.nlm.nih.gov
IARC Monographs	Summary evaluations of carcinogenicity taken from IARC monographs	inchem.org/
International Toxicity Estimates for Risk (ITER) of Toxicology Excellence in Risk Assessment (TERA)	Human limit values with links to background review documents of ATSDR, Health Canada, WHO IARC, US NSF, NL RIVM and US EPA	www.tera.org toxnet.nlm.nih.gov
IPCS Environmental Health Criteria	Comprehensive review of all (eco)toxicity endpoints, data on exposure via all routes	inchem.org/
IPCS Concise International Assessment Documents (CICADs)	Summary review of all (eco)toxicity endpoints	inchem.org/
IPCS Health and Safety Guides	Summary information in non-technical language on all (eco)toxicity endpoints and on exposure with practical advice on medical and administrative issues	inchem.org/
N-CLASS	Database with EU classification of chemicals with links to background information on environmental effects	http://apps.kemi.se/nclass/default.asp

OECD eChemPortal	Access to data submitted to government chemical review prgrammes at national, regional and international levels[2]	http://webnet3.oecd.org/echemportal
OECD Screening Information Data Sets (SIDS)	Comprehensive review of all (eco)toxicity endpoints	www.chem.unep.ch/irptc/sids/oecdsids/sidspub.html inchem.org
RIVM Review Soil contaminants	Brief review of all (eco)toxicity endpoints, estimate of total background exposure (non-soil-related)	www.rivm.nl/bibliotheek/index-en.html
US EPA Integrated Risk Information System (IRIS)	Comprehensive review of all toxicity endpoints	toxnet.nlm.nih.gov
Secondary sources		
Household Products Database	Database on potential health effects of chemicals in 5000 household products and composition data	toxnet.nlm.nih.gov
IUCLID	Database with (eco)toxicity data on industrial chemicals	ecb.jrc.it/esis/
Sax's Dangerous Properties of Industrial Materials	Hazard information on 20,000 chemicals	[31]
US EPA ecotoxicological database	Database providing single chemical toxicity information for aquatic and terrestrial life	http://cfpub.epa.gov/ecotox/
US National Toxicology Program	Database with abstracts of carcinogenicity and reproductive toxicity studies	http://ntp.niehs.nih.gov:8080/

[1] Peer reviewed secondary sources have been critically reviewed by independent experts.
[2] Participating databases are CHRIP (Japan), ESIS (EU), HPVIS (USA), INCHEM (WHO), OECD HPV, SIDS IUCLID, SIDS UNEP

Table 8.4. Sources of data on physicochemical properties.

Source (alphabetically)	Description	Location/reference
ChemFinder	Database with comprehensive chemical information and many useful links	http://chemfinder.cambridgesoft.com/
CRC Handbook of Chemistry and Physics	Standard source for physicochemical data	[32]
Hazardous Substances Data Bank (HSDB)	Database with comprehensive, peer reviewed (eco)toxicology data and physicochemical data	toxnet.nlm.nih.gov
EpiWin database		[33]
Handbook of physical-chemical properties and environmental fate		[34]
Property estimation methods		[35]

(OECD manual for investigation of HPV chemicals [40]). Definitions for these terms were published by Klimisch et al. [41] and will be given below. Specific guidance on the evaluation of tests for specific endpoints, (Q)SARs and exposure data and can be found in other chapters.In the evaluation of data there are four steps:

1. *Evaluation of the validity* of the method used for the generation of data for a specific endpoint. This especially applies in the absence of any formal, generally accepted guideline for a specific test (e.g., a toxicity test, a biodegradation test, an *in vitro* endocrine disruption test) or a non-testing method

(e.g., a (Q)SAR). The method can be evaluated against an official guideline [21] or against general validation principles, such as the OECD validation principles for (Q)SARs [30].

2. *Evaluation of the reliability* of an individual result of applying the method. Klimisch et al. [41] defined reliability as: "evaluating the inherent quality of a test report or publication relating to preferably standardised methodology (e.g., OECD guidelines shown in Table 16.1 and EC guidelines as described in Annex V to Directive 67/548/EEC) and the way the experimental procedure and results are described to give evidence of the clarity and plausibility of the findings". The OECD Manual for Investigation of HPV chemicals [40] deals primarily with determining the reliability of data. This essentially relates to how the study was carried out. This information is needed to enable robust study summaries to be prepared before relevancy and adequacy (see points 3 and 4) can be considered. Careful consideration must be made of the quality of the study, e.g., the method used, the reporting of the results, the conclusions drawn. There are several reasons why existing study data may be of variable quality. Klimisch et al. [41] have suggested the following (see also Box 8.1):
   - The use of different test guidelines (compared with today's standards).
   - The improper (or absence of) characterization of the test substance (in terms of purity, physical properties, etc.).
   - The use of crude techniques/procedures which have since become refined.
   - The fact that certain information may not have been recorded (or possibly even measured) for a given endpoint which has since been recognized as important.
   - Studies should preferably also comply with the principles of Good Laboratory Practice (GLP), (see Section 16.2.2 for the OECD principles of GLP). Section 8.5.2 gives two approaches for making a selection of test data on the basis of reliability.

Non-testing data also need to be evaluated for their reliability. For example, for the reliability of a QSAR prediction it is essential to make sure that the chemical of interest is within the applicability domain of the model and is similar to one or more chemicals in the training set (Chapter 10). (Q)SARs have been used in regulatory assessment of chemical safety in some OECD member countries for many years, but universal principles for their regulatory applicability are lacking. In 2004, the OECD member countries agreed on the principles for validating (Q)SAR models for their use in regulatory assessment of chemical safety. The agreed principles provide member countries with a basis for evaluating the regulatory applicability of (Q)SAR models and will contribute to their enhanced use for more efficient assessment of chemical safety [30]. The OECD also summarized the experience of OECD member countries in the regulatory use of (Q)SAR models in chemicals assessment [29]. A report by the European Chemicals Bureau provides preliminary guidance on how to characterize (Q)SARs based on the OECD validation principles [42].

Box 8.1. Data quality and selection in practice

Questions to ask while determining:

Reliability of result:
- Is a complete test report available?
- Is the test reported in a peer reviewed journal?
- Is the test conducted in accordance with a well-established guideline?
- Are deviations from the guideline sufficiently supported?
- Is the test conducted in compliance with GLP principles?
- Are the identity, purity and source of the substance fully known?
- Is the test species properly identified?
- Is the exposure fully characterized?
- Are the methods of analysis performed correctly?
- Is a statistical analysis performed and is it done correctly?
- Are other tests on structural analogues available confirming the results obtained?

Relevance:
- Is the species relevant with regard to the target population?
- Is the route of exposure relevant to the population and exposure scenario under consideration?
- Is the exposure regime relevant for the population under consideration?
- Is the effect of concern relevant with regard to the target population?

3. *Evaluation of the relevance* of data for risk assessment. Relevance covers the extent to which data and test results are appropriate for a particular

Box 8.2. Expert judgement

Expert judgement is required throughout the risk assessment process [43], including for:

- Evaluation of the reliability and relevance of individual studies
- Identification of critical gaps in the data set
- Decision on which effects are adverse and which not
- Selection of exposure scenarios which reflect the actual (likely) use and abuse of the risk source
- Choice of extrapolation methods (between species, routes of administration, etc.)
- Determination of the degree of uncertainty in the estimate of risk
- Consideration of animal welfare and other ethical issues associated with experimentation
- Examination of the consistency of the risk assessment with that for other comparable risk

hazard identification or risk characterization [41]. Box 8.1 shows the relevant questions for this stage of data evaluation and Section 8.5.3 provides further explanation.

4. *Evaluation of the adequacy* of data. Adequacy can be defined as "the usefulness of data for hazard/risk assessment purposes"[41]. When there is more than one result for a particular endpoint, a weight-of-evidence procedure should determine which result(s) will be used in the risk characterization. In general, the greatest weight will normally be attached to the study that is the most reliable and relevant. Robust study summaries are prepared for the highest quality or "key" studies. Box 8.1 provides relevant questions for this stage of data evaluation and Section 8.5.3 provides further details. Expert judgement is fundamental to the evaluation and selection of data for risk assessment [43]. Box 8.2 shows some of the important elements in expert judgement.

The WoE approach in expert judgement is often mentioned in the risk assessment literature without adequate documentation. The definition of WoE is used in different ways and both qualitative and quantitative weighting methods are in use. An important issue in WoE is the influence of values on expert judgement that needs to be recognized and made explicit as far as possible [44]. In many cases, it is not clear which methods were used, how they were applied to the scientific evidence, what the results were and how these were used to make decisions in a specific risk assessment. The scientific evidence gathered in the course of a risk assessment is often of a variable nature: strong studies versus weak studies, *in vitro* studies versus *in vivo* studies, animal studies versus human studies, etc. The WoE approach taken should make it clear which interpretative methods were used and how these were applied to scientific evidence and expert judgement.

Quality assurance of the evaluation process and the subsequent steps in risk assessment is usually performed through peer review. Peer review can be defined as an "in-depth critique of assumptions, calculations, extrapolations, alternate interpretations, methodology and acceptance of criteria employed, and conclusions drawn in the original work"[45]. In Europe, important scientific peer review to inform regulatory decision-making is taking place in various scientific committees (Table 8.5). However, peer review of risk assessment documents should take place at all levels of organization. Careful and transparent peer review can enhance the credibility of risk assessments for all stakeholders. Crucial aspects are the level of support from the management layer, the selection of members of such a scientific committee with regard to level and type of expertise, the transparency of the selection process, a clear description of the charge of the committee, the independent nature of the members of the committee and the way comments and recommendations are followed up (see, for example [46]).

8.5.2 Scoring system for reliability

The reliability of the data is a key consideration which can be used to relatively quickly filter out unreliable studies and focus further resources on those considered most reliable. Without knowledge of how the study has been conducted all other considerations (e.g., relevance, adequacy) may be irrelevant. Two major approaches have been proposed to assist in the initial screening of study reports to screen and set aside unreliable study data. Both are compatible and may be used either alone or in combination by a person compiling a dossier (for registration or another purpose) and considering data quality.

One approach is that developed by Klimisch et al. [41]. This approach was developed as a scoring system for reliability, particularly for ecotoxicology and health studies; however, it can be extended to physicochemical and environmental fate and pathway studies. The second approach was developed in 1998 as part of the US EPA HPV Challenge Programme [47].

Klimisch's scoring system [41] can be used to categorize the reliability of a study as follows:

Table 8.5. Scientific Committees (SC) in the EU involved in risk assessment.

Organization	Scientific Committee (SC)
European Chemicals Agency	No SC prescribed in the Regulation (see Box 12.8)
European Environment Agency (EEA)	1 SC
European Food Safety Authority (EFSA)	9 scientific panels: Panel on genetically modified organisms (GMO), Panel on contaminants in the food chain (CONTAM), Panel on animal health and welfare (AHAW), Panel on biological hazards (BIOHAZ), Panel on additives and products of substances used in animal feed (FEEDAP), Panel on Plant Protection and their Residues (PPR), Panel on food additives, flavourings, processing aids and materials in contact with food (AFC), Panel on plant health (PLH), Panel on dietetic products, nutrition and allergies (NDA) and an overall EFSA SC dealing with common issues for the 8 panels, new technologies and cross-cutting issues.
European Medicines Agency (EMEA)	4 SCs: SC for Medicinal Products for Human Use (CHMP), SC for Medicinal Products for Veterinary Use (CVMP), SC for Orphan Medicinal Products, and SC for Herbal Medicinal Products (HMPC).
European Centre for Diseases Prevention and Control (ECDC)	In the process of establishing a SC.
Directorate General for Health and Consumer Protection (DG SANCO)	3 SCs: SC on Consumer Products (SCCP), SC on Health and Environmental Risks (SCHER), SC on Emerging and Newly-identified Health Risks (SCENIHR).
Directorate General Employment	3 SCs: SC on Occupational Exposure Limits (SCOEL), Senior Labour Inspectors Committee (SLIC), Advisory Committee on Safety and Health at Work (ACSH).

1 = reliable without restrictions: "studies or data... generated according to generally valid and/or internationally accepted testing guidelines (preferably performed according to GLP) or in which the test parameters documented are based on a specific (national) testing guideline...or in which all parameters described are closely related/comparable to a guideline method."

2 = reliable with restrictions: "studies or data...(mostly not performed according to GLP) in which the test parameters documented do not totally comply with the specific testing guideline, but are sufficient to accept the data or in which investigations are described which cannot be subsumed under a testing guideline, but which are nevertheless well documented and scientifically acceptable."

3 = not reliable: "studies or data...in which there were interferences between the measuring system and the test substance or in which organisms/test systems were used which are not relevant in relation to the exposure (e.g., not physiological pathways of application) or which were carried out or generated according to a method which is not acceptable, the documentation of which is not sufficient for assessment and which is not convincing for an expert judgment."

4 = not assignable: "studies or data...which do not give sufficient experimental details and which are only listed in short abstracts or secondary literature (books, reviews, etc.)."

The use of Klimisch's codes provides a useful tool for organizing studies for further review. For example, it would enable someone reviewing studies to focus on the most reliable studies first, to allow time to be devoted later to considering the relevance and adequacy of only reliable studies. Studies which fail to meet essential criteria for reliability can thus be set aside at the beginning.

The second approach developed by the US EPA [47] provides more information than the Klimisch system by describing the key reliability criteria for each type of data making up the dossier. For all physicochemical, environmental fate and (eco)toxicological studies the test substance identification, test temperature and a full reference/citation should be included. Besides this, information should be included on environmental fate and (eco)toxicological studies, as well as the controls, dose/concentration levels and duration of exposure. Statistics, species, strain, number, gender, age of organism and route/type of exposure should also be detailed for the latter. These criteria address the overall scientific integrity and validity of the information in a study, i.e., reliability. This approach is consistent with the Klimisch approach - any study, which does not meet the stated criteria would also not be assignable under the Klimisch system. Such studies may, however, be considered later as supplementary information to the overall assessment of a particular endpoint, particularly where there is no single key study.

8.5.3 Determination of relevance and adequacy

The use of expert judgement is the most important principle in considering relevance and adequacy. The assessment of relevance and adequacy requires more detailed consideration and is very much related to preparing the dossier (including robust study summaries, as appropriate). It can therefore be regarded as a second tier in the data evaluation.

To evaluate the relevance of the available data, it is necessary to judge, *inter alia*, if an appropriate species has been studied, if the route of exposure is relevant for the population and exposure scenario under consideration, and if the substance tested is representative for the substance as supplied. To be able to assess this the substance must be properly identified and any significant impurities described. Relevant human data of adequate quality may, of course, sometimes be the best available data but, more often, the available human, animal and other data are considered together in order to reach a conclusion about the relevance to humans of effects observed in animal studies.

Evaluation of the relevance of data from animal studies for humans is aided by use of data on the toxicokinetics, including metabolism, of a substance in both humans and the animal species used in the toxicity tests, if available, even when relatively limited. Clear, well-documented evidence on mode of action for a species-specific effect/response (e.g., light hydrocarbon-induced nephropathy in the kidney of male rats) can be used as justification for the conclusion that a particular effect is not expected to occur in humans exposed to the substance. In the absence of such information (on the substance itself or, if it can be scientifically justified, on a close structural analogue), "threshold" adverse effects observed in animal studies will normally be assumed to occur in humans also exposed to the substance above a certain level of exposure. In any event, the dose-response relationship(s) in the animal studies (or the severity of the effect, when only a single dose was tested) are also assessed as a part of the risk assessment process. These assessments are taken into account at the risk characterization stage when a judgement is made about the likelihood of an adverse effect occurring in humans at a particular level of exposure. As an illustration, within the scope of a project on the harmonization of approaches to the assessment of risk from exposure to chemicals, the International Program on Chemical Safety has provided a framework for the analysis of the relevance of a cancer mode of action for humans [48].

Interpretation of the relevance and adequacy of data derived from tests conducted *in vitro* should take into account whether the results observed may be expected to occur *in vivo* (e.g., based on a knowledge of the toxicokinetics of the substance). According to the validation procedures established by the European Centre for the Validation of Alternative Methods (ECVAM), the relevance of an alternative (non-animal) test, such as an *in vitro* test, is assessed according to the scientific basis of the test system (scientific relevance) and the predictive capacity (predictive relevance) of the prediction model, which is an algorithm for extrapolating from *in vitro* data to an *in vivo* endpoint [49].

In general, the results of *in vitro* tests provide supplementary information which can, for instance, be used to facilitate the interpretation of the relevance to humans of data from animal studies, or to gain a better understanding of a substance's mechanism of action. Toxicogenomic approaches are recognized as not yet sufficiently developed for direct replacement of existing approaches, but which could provide supportive evidence on a case-by-case basis. As such, these advanced technologies hold great promise for future application based on solid experience, proven demonstration and sufficient harmonization [50].

Further to the classification of studies and data based on the criteria for reliability and relevance, the data with highest adequacy (i.e., usefulness) can be selected for hazard/risk assessment purposes.

8.5.4 Quality of modelled data

Risk assessment more and more relies on models for the assessment of fate, exposure and effects. Models represent or describe a real-world system in a schematic or simplified way. As such, models replace actual data.

It is essential to know the uncertainty and variability in the answers that models give us. The quality of models is determined by the way the processes are described in mathematical formulae and the way in which the model is analyzed. Questions to be answered here are;

1. Has the model been adequately described?
   The description should include the aim of the model, the model structure, the domain (boundary conditions), the mathematical formulae and their limitations, the assumptions made and default parameter values used and their basis, and the results of model analyses. If the model has been computerized, a user manual should be available as well.
2. Has the model been adequately analyzed?
   Model analysis includes investigation of the correctness of theories, concepts and assumptions, the correctness of mathematical or numerical calculations, the representativeness of scenarios used, the relevance of and uncertainty in measured or estimated input parameters, the sensitivity of the model to changes in parameter values, and the validation status. Validation is the ultimate determination of the reliability, uncertainty and usefulness of a model within a specified domain, by comparing the model output to measured data (see, for example [51]).

8.6 CONCLUDING REMARKS

In Chapter 1 a distinction was made between (1) data, (2) information, and (3) knowledge. Data was defined as basic observations or measurements. Information was defined as products of analysis and interpretation, such as provided in a risk assessment report. Knowledge was seen as information in a context that allows it to be transformed into action. We summarized this in a "knowledge pyramid" (Figure 1.15.) In Chapter 8 we focused on data. The emphasis was, in fact, on physicochemical properties and hazard data, and not on exposure-related information. From Section II of this book (Chapters 2-5) it is clear that exposure-related information on uses, emissions and fate of chemicals is crucial to understanding exposure to chemicals [52, 53].

As shown in this chapter, risk assessment strongly hinges on the availability, evaluation and selection of data. The availability of data is largely governed by regulatory frameworks. Data gaps can be filled with non-testing approaches such as QSARs, read-across and default values. In the implementation phase of REACH and other legislative frameworks, we need to be pragmatic in dealing with gaps in hazard and exposure data. The following chapters of this book (Chapters 9-11) therefore describe some pragmatic approaches to generating hazard data for chemicals. We want to emphasize that dealing with data gaps in this way is not the ultimate answer to speed up the risk assessment and risk management process. It is only a first step.

The evaluation and selection of data really is an arduous task which needs to be done by experts with a background in the subjects relevant in risk assessment. They should be trained in dealing with large amounts of data, summarizing, analyzing and interpreting these. Very often these experts are generalists and, depending on the resources available, will need help from scientists specialized in, e.g., analytical and environmental chemistry, ecotoxicology, human toxicology, carcinogenicity, reproductive toxicology, QSARs, etc. Together, they can perform the necessary critical evaluation for the risk assessment in terms of the validity of the methods used, the reliability of the results and the relevance and adequacy of the data for the risk assessment. Finally, quality assurance by peer review is another essential step in this process. As pointed out in this chapter, the WoE approach in expert judgement is often mentioned in the risk assessment literature without adequate documentation. The WoE approach taken should make clear which interpretative methods were used and how these were applied to scientific evidence and expert judgement.

As stated previously (in Section 1.8 of Chapter 1), our current ability to generate more data often exceeds our ability to evaluate them. Further simplifications of information and information flows are necessary to efficiently manage chemicals in the near future. Managing risk implies that we are able to manage the dilemma of simplicity and complexity in a pragmatic, cost-effective and timely manner and understand the context in which risk assessments are performed.

8.7 FURTHER READING

1. Wexler P, Hakkinen PJ, Kennedy GL, Stoss FW, eds. 2000. *Information resources in toxicology, third edition*. Academic Press. ISBN 0-12-744770-9.
2. Wexler P *et al.* ed. 2001/2002/2003. Special issues on digital information and tools. *Toxicology* 157(1-2), pp 1-164; *Toxicology* 173(1-2), pp. 1-189; *Toxicology* 190(1-2), pp 1-139.
3. Patterson J, PJ Hakkinen, AE Wullenweber. 2002. Human Health Risk Assessment: Selected Internet and World Wide Web Resources. *Toxicol* 173:123-143.
4. Dourson ML, Lu FC. 1995. Safety/Risk assessment of

chemicals compared for different expert groups. *Biomed Environ Sci* 8:1-13.

5. Risk Analysis special issue on peer review. 2006. *Risk Anal* 26:5-43.

REFERENCES

1. Organization for Economic Co-operation and Development. 1982. Decision of the Council concerning the Minimum Pre-marketing Set of Data in the Assessment of Chemicals. OECD, Paris, France, (82) 196/Final.
2. Commission of the European Communities. 1967. Council Directive 67/548/EEC of 16 August 1967 on the approximation of the laws, regulations and administrative provisions relating to the classification, packaging and labelling of dangerous substances. *Off J Eur Communities* L196.
3. Commission of the European Communities. 1993. Council Regulation (EC) 793/93 of 23 March 1993 on the evaluation and control of the risks of existing substances. *Off J Eur Communities* L84.
4. Commission of the European Communities. 1991. Council Directive of 15 July 1991 concerning the placing of plant protection products on the market (91/414/EEC). *Off J Eur Communities* L230.
5. Commission of the European Communities. 1998. Council Directive 98/8/EC of 16 February 1998 concerning the placing of biocidal products on the market. *Off J Eur Communities* L123.
6. Commission of the European Communities. 2001. Council Directive 2004/28/EC of 31 March 2004 amending Directive 2001/82/EC on the Community code relating to veterinary medicinal products. *Off J Eur Communities* L136.
7. Commission of the European Communities. 1990. Council Regulation (EEC) No. 2377/90 of 26 June 1990 laying down a Community procedure for the establishment of maximum residue limits of veterinary medicinal products in foodstuffs of animal origin. *Off J Eur Communities* L224.
8. Commission of the European Communities. 2003. Council Regulation (EC) 1831/2003 of 22 September 2003 on additives for use in animal nutrition. *Off J Eur Communities* L268.
9. Commission of the European Communities. 1990. Council Directive 76/678/EEC of 27 July 1976 on the approximation of the laws of the Member States relating to cosmetic products. *Off J Eur Communities* L262.
10. Scientific Committee on Cosmetic Products and Non-food products intended for Consumers. 2000. Notes of guidance for testing of cosmetic ingredients for their safety evaluation, 4th revision, adopted by the SCCNFP during the plenary meeting of 24 October 2000. SCCNFP/0321/00, Final.
11. Scientific Committee on Cosmetic Products and Non-food products intended for Consumer. 2000. Updated basic requirements for toxicological dossiers to be evaluated by the SCCNFP, revision, adopted by the SCCNFP during the plenary meeting of 17 December 2000. SCCNFP/0633/02.
12. Scientific Committee on Cosmetic Products and Non-food products intended for Consumers. 2002. The actual status of alternative methods to the use of animals in the safety testing of cosmetic ingredients, adopted by the SCCNFP during the plenary meeting of 4 June 2002. SCCNFP/0546/02. Final.
13. International Program on Chemical Safety. 1987. Principles for the safety assessment of food additives and contaminants in food. World Health Organization, *Environmental Health Criteria* 70. IPCS, Geneva, Switzerland
14. International Program on Chemical Safety. 1994. Assessing human health risks of chemicals: derivation of guidance values for health-based exposure limits. World Health Organization, *Environmental Health Criteria* 170. IPCS, Geneva, Switzerland.
15. European Commission, Health & Consumer Protection Directorate-General (SANCO D3/LR D). 2003. Note for Guidance for Food Contact Materials.
16. US Environmental Protection Agency. 1998. Chemical Hazard Data Availability Study; What do we really know about the safety of high production volume chemicals? US EPA, Office of Pollution Prevention and Toxics, Washington DC, USA.
17. Allanou R, Hansen B, Van der Bildt Y. 1999. Public availability of data on EU High Production Volume Chemicals. European Chemicals Bureau, Joint Research Centre, Ispra, Italy, EUR18996EN.
18. Pedersen F, de Bruijn J, Munn S, van Leeuwen K. 2003. Assessment of additional testing needs under REACH; Effects of (Q)SARs, risk based testing and voluntary industry initiatives. Institute of Health and Consumer Protection, JRC Report EUR 20863.
19. Organization for Economic Co-operation and Development. 1982. Final report of the Expert Group on Good laboratory Practice. Geneva, Switzerland.
20. Commission of the European Communities. 1992. Council Directive of 30 April 1992 amending for the seventh time Directive 67/548/EEC on the approximation of the laws, regulations and administrative provisions relating to the classification, packaging and labelling

of dangerous substances (92/32/EEC). *Off J Eur Communities* L154.

21. Organization for Economic Co-operation and Development. 1981-1994. Guidelines for testing of chemicals. OECD, Paris, France.
22. US Environmental Protection Agency. 1982-1986. Standard evaluation procedures. Office of Pesticide Programs, Washington, DC.
23. US Environmental Protection Agency. 2004. An examination of EPA risk assessment principles and practices. Risk Assessment Task Force, Washington DC, USA.
24. Degner P, Jackel H, Muller M, Nendza M, Von Oepen, B. 1991. Application of structure-activity relationships (SARs) to the estimation of properties important in exposure assessment. *OECD Environment Monographs* 67. OECD, Paris, France.
25. Degner P, Muller M, Nendza M, Klein W. 1991. Estimating biodegradability of chemicals by computer assisted reactivity simulation. Report prepared for the OECD Hazard Assessment Advisory Body. Fraunhofer Institut für Umweltchemie und Ökotoxikologie, Schmallenberg, Germany.
26. Organization for Economic Co-operation and Development. 1992. Report of the OECD Workshop on quantitative structure-activity relationships (QSARs) in aquatic effects assessment. *OECD Environment Monographs* 58. OECD, Paris, France.
27. Organization for Economic Co-operation and Development. 1993. Application of structure-activity relationships to the estimation of properties important in exposure assessment. *OECD Environment Monographs* 67. OECD, Paris, France.
28. US Environmental Protection Agency, Commission of the European Communities. 1993. Joint project on the evaluation of (quantitative) structure-activity relationships. Final report. Brussels, Belgium.
29. Organization for Economic Co-operation and Development. 2006. Report on the regulatory uses and applications in OECD member countries of (quantitative) structure-activity relationship [(Q)SAR] models in the assessment of new and existing chemicals. OECD Environment Directorate, Joint Meeting of the Chemical Committee of the Working Party on Chemicals, Pesticides and Biotechnology, ENV/JM/MONO(2006)25, *OECD Series on Testing and Assessment* No. 58. OECD, Paris, France
30. Organization for Economic Co-operation and Development. 2004. The report from the expert group on (Quantitative) Structure Activity Relationship (Q)SAR on the principles of the validation of (Q)SARs. *OECD Series on Testing and Assessment* No. 49. ENV/JM/MONO(2004)24. OECD, Paris, France
31. Lewis RJ. 1996. *Sax's dangerous properties of industrial materials, 9th edition.* Van Nostrand-Reinhold, New York, USA (1998 CD-ROM version available).
32. Lide DR, ed. 1999. *CRC Handbook of Chemistry and Physics.* CRC Press, Boca Raton, FL (CD-ROM version available).
33. SRC. 1997. EPIWIN, Estimation Program Interface for Microsoft Windows 3.1. Syracuse Research Corporation, North Syracuse, NJ.
34. Mackay D, Shiu W-Y, Ma K-C. 1999. *Illustrated Handbook of Physical-chemical Properties and Environmental Fate for Organic Chemicals,* Volumes I-V. Lewis Publishers, Boca Raton, FL.
35. Boethling RS, Mackay D. 2000. *Handbook of property estimation methods for chemicals.* Environmental and Health Series. Lewis Publishers, Boca Raton, FL. ISBN 1-56670-456-1.
36. Sanderson DM. 1986. Methods of data retrieval - computer. In: Richardson ML, ed. *Toxic hazard assessment of chemicals.* Royal Society of Chemistry, London, UK pp. 15-23.
37. Sanderson DM. 1986. Methods of data retrieval - manual. In: Richardson ML, ed. *Toxic Hazard Assessment of Chemicals.* Royal Society of Chemistry, London, UK pp. 24-28.
38. Wexler P, Hakkinen PJ, Kennedy GL and Stoss FW, eds. 2000. *Information Resources In Toxicology, Third Edition.* Academic Press, ISBN 0-12-744770-9.
39. Bradbury SP, Russom CL, Ankley GT, Schultz TW, Walker JD. 2003. Overview of data and conceptual approaches for derivation of quantitative structure-activity relationships for ecotoxicological effects of organic chemicals. *Environ. Toxicol. Chem.* 22 (8):1789-1798.
40. Organization for Economic Co-operation and Development. *Manual for the investigation of HPV chemicals.* OECD, Paris, France. Available from www.oecd.org.
41. Klimisch HJ, Andreae E, Tillmann U. 1997. A systematic approach for evaluating the quality of experimental and ecotoxicological data. *Regul Toxicol Pharmacol* 25:1-5.
42. Worth AP, Bassan A, Gallegos A, Netzeva TI, Patlewicz G, Pavan M, Tsakovska I, Vracko M. 2005. The characterization of (Quantitative) Structure-Activity Relationships: preliminary guidance. Ispra, Italy, Joint Research Centre, European Chemicals Bureau, EUR 21866 EN.
43. European Commission 2000. First report on the harmonisation of risk assessment procedures. The

Report of the Scientific Steering Committee's Working Group on Harmonisation of Risk Assessment Procedures in the Scientific Committees advising the European Commission in the area of human and environmental health. Published on the Internet 20.12.2000.

44. Weed DL. 2005. Weight of evidence: a review of concepts and methods. *Risk Anal* 25:1545-1557.
45. US National Research Council. 1998. Peer review in environmental technology development programs. US NRC, Washington, DC.
46. Patton DE, Olin SS. 2006. Scientific peer review to inform regulatory decision making: leadership responsibilities and cautions. Risk Analysis 26:5-16.
47. US Environmental Protection Agency. 1999. Determining the Adequacy of Existing Data, Guidance for the HPV Challenge Program, Draft dated 2/10/99.
48. Boobis AR, Cohen SM, Dellarco V, McGregor D, Meek ME, Vickers C, Willcocks D, Farland W. 2006. IPCS Framework for analyzing the relevance of a cancer mode of action for humans. *Crit Rev Toxicol* 36:781-792.
49. Worth AP, Balls M. 2001. The importance of the prediction model in the development and validation of alternative tests. *ATLA* 29:135-143.
50. Organization for Economic Co-operation and Development/International Programme on Chemical Safety. 2005. Report of the OECD/IPCS Workshop on Toxicogenomics, Kyoto, 13-15 Oct. 2004. OECD Environment Directorate, Joint Meeting of the Chemical Committee of the Working Party on Chemicals, Pesticides and Biotechnology, ENV/JM/MONO(2005)10, *OECD Series on Testing and Assessment* No. 50. OECD, Paris, France.
51. Cullen AC, Frey AC. 1998. *Probabilistic techniques in exposure assessment. A handbook for dealing with variability and uncertainty in models and inputs.* Plenum Press, New York, London. ISBN 0-306-45957-4.
52. Haigh N, Baillie A. 1992. Final report on chemicals control in the European Community in the 1990s. Institute for European Environmental Policy, London, UK.
53. Van Leeuwen CJ, Bro-Rasmussen F, Feijtel TCJ, Arndt R, Bussian BM, Calamari D, Glynn P, Grandy NJ, Hansen B, Van Hemmen JJ, Hurst P, King N, Koch R, Müller M, Solbé JF, Speijers GAB, Vermeire T. 1996. Risk assessment and management of new and existing chemicals. *Environ Toxicol Pharmacol* 2:243-299.

9. PREDICTING FATE-RELATED PHYSICOCHEMICAL PROPERTIES

G. Schüürmann, R.-U. Ebert, M. Nendza, J.C. Dearden, A. Paschke and R. Kühne

9.1 INTRODUCTION

The environmental fate of chemical substances is determined by partitioning between environmental compartments, and by transport and degradation processes. In this context, long range transport potential and persistence are important characteristics of the compound behaviour [1-7]. Besides specialized models to address the compound fate in individual environmental compartments, multimedia fate models have become popular in exposure and fate assessment on global and regional scales. A main application of such models is the screening-level prediction of the fate of environmental chemicals under standardized emission scenarios, and more recently the focus has shifted to more detailed process descriptions including time-dependent concentration levels of compounds and consideration of spatial resolution. For a more detailed account of multimedia fate modelling, the reader is referred to Chapter 4.

In general, environmental compartments are modelled in terms of thermodynamic phases. The compound-specific disposition for a certain environmental behaviour is thus determined by physicochemical properties governing partition processes, by molecular reactivity that drives abiotic degradation, and by the accessibility of the compound for microbial degradation. It follows that compound properties are essential input values for models to simulate their environmental behaviour and fate. As an example, multimedia fate models typically exploit the fugacity of chemicals [8] in the respective compartments. Fugacity is a central thermodynamic characteristic that in turn can be estimated from, or related to, some of the compound properties introduced in the following sections of this chapter.

The key role of compound properties in environmental fate modelling implies corresponding requirements for the quality of input data. Substance-specific data have to be provided with appropriate accuracy in order to obtain meaningful results [9-12]. Recently, efforts have been made to improve and harmonize the application of compound properties in environmental fate modelling [13,14]. Following an increased awareness of the fact that xenobiotics in the environment also include compounds with more polar and complex chemical structures, developments are on the way [15] to introduce into fate models a more detailed description of compound-matrix interactions through the LSER approach in terms of the Abraham equation [16] (see also below).

Despite the huge number of experimental property data already available, there is still a need for methods to predict fate-relevant properties from molecular structure. This holds true for both the hazard and risk evaluation of existing compounds, and for predictive analyses of the expected environmental profile associated with chemical structures of compound candidates not yet synthesized. For a meaningful application of property estimation methods, knowledge on the conditions of usage and the prediction capability is required to avoid misinterpretations. This becomes apparent when using poorly predicted parameter values as input for sophisticated model equations, as was demonstrated recently in the context of LSER models to predict partition properties [17]. A further problem is the appropriate characterization of the application domain, the latter of which needs careful consideration and can be defined at different levels of sophistication [18].

In the following, we will discuss the mechanistic background and performance of existing models to predict fate-relevant compound properties from molecular structure or otherwise available information. For each of the nine properties under consideration, an introduction into the underlying thermodynamics is given, before methods selected according to different criteria such as mechanistic basis, simplicity and availability are evaluated with respect to their prediction capability. To this end, we also used data sets established in our ChemProp software [19] over the last decade to perform comparative analyses of the performance of prediction methods for the partition properties: *n*-octanol-water partition coefficient (K_{ow}), Henry's law constant (H or its dimensionless form as air-water partition coefficient K_{aw}), sorption constant normalized to organic carbon content (K_{oc}), vapour pressure (P_v), bioconcentration factor (BCF), water solubility (S_w), and indirect photolysis as a dominant loss process in the troposphere. For hydrolysis and biodegradation as two further degradation pathways, the analysis is confined to a critical evaluation of respective literature findings.

In view of the increased importance of *in silico* methods to screen and evaluate the hazard and risk potential of chemical substances, guidelines have been

C.J. van Leeuwen and T.G. Vermeire (eds.), Risk Assessment of Chemicals: An Introduction, 375–426.

introduced for the validation of QSAR models to be used in the regulatory context. These OECD principles for QSARs [20] are important as the next steps towards including *in silico* models in integrated testing strategies for the hazard and risk evaluation of chemical substances with respect to human health and the environment. The present study, however, does not attempt to apply these criteria in a comprehensive way, but focuses on the meaning and environmental applications of physicochemical compound properties, as well as on the mechanistic basis and performance of existing prediction methods.

9.2 TYPES OF MOLECULAR PROPERTIES USED IN ESTIMATION METHODS

The simplest form of estimation methods employs correlations between related bulk compound properties. Thermodynamically, bulk properties are determined by the behaviour of a huge ensemble of molecules in a pure solute compound, in a solution of this compound in a solvent, or in a system of several phases containing different solvents or phases. Prominent examples are the prediction of water solubility (S_w) of liquids from their octanol-water partition coefficient (K_{ow}), and of the octanol-air partition coefficient (K_{oa}), from K_{ow} and the air-water partition coefficient (K_{aw}). The nature of the relationship may be derived from theoretical considerations, or may be partly or fully empirical such as multilinear regression equations.

1D molecular descriptors

In contrast to bulk properties, one-dimensional (1D) molecular descriptors exist for isolated molecules and do not depend on the ensemble, but in turn may (and often do) reflect some aspect of the bulk behaviour. Well-known examples are molecular weight and the alkyl chain length in terms of the number of C atoms. 1D molecular descriptors can be obtained from atom and bond counts, and thus are easily accessible from the molecular structure of the compounds.

2D molecular descriptors

Two-dimensional (2D) molecular descriptors are based on the connectivity (topology) of the molecule. Accordingly, knowledge of the chemical structure is necessary to calculate them, but no atom coordinates are required. 2D molecular descriptors may relate either to the total molecule (e.g., topological indices) or to substructures (fragments). 2D molecular descriptors are fairly easily calculated, and a huge number of descriptor definitions is available. A comprehensive overview including respective software packages is given in [21]. Most of the models discussed in the following sections rely on 2D molecular descriptors.

Fragment methods

An important group of 2D descriptor models are fragment methods, which are also called increment methods or group contribution methods. While in general group contribution models may be either linear or non-linear, typical fragment models follow the multilinear approach:

$$\log \text{Property} = \sum_i n_i F_i + \sum_j n_j C_j + \sum_k n_k I_k + d \qquad (9.1)$$

In Equation 9.1, F_i denotes the (calibrated) increment value of fragment (molecular substructure) *i* that occurs ni times ($n_i = 0, 1, 2, ..$) in the compound of interest, C_j represents the (fitted) correction term associated with fragment *j* again multiplied by the associated number of occurrences (n_j), I_k is the (fitted) increment value of substructural feature *k* used as an indicator variable where the associated factor n_k is only 0 (absence of fragment *k*) or 1 (presence of fragment *k*, regardless of how often *k* occurs in the compound of interest), and *d* is a regression constant.

For the application of a given fragment method, the molecule is entirely separated into substructures defined by the respective model. The allocation of atoms to substructures is one-to-one such that for each atom there is one unambiguous allocation, and the intrinsic model application domain rejects compounds with atoms or atom groups not allocated to any of the predefined model-specific substructures.

Only a few group contribution models are confined to employing fragments of this kind. Typically, additional terms are required: correction factors C_j are substructures similar to fragments and usually account for interactions between different functional groups. In contrast to fragments, correction factors are applied in addition and independently from each other. Certain atoms can be allocated to none, one or more than one correction factor. As with fragments, n_j is 0 if the substructure associated with the correction factor does not occur in the compound of interest.

Indicator variables I_k can be interpreted as a special type of correction factor. Here, only the presence or absence of the relevant structural feature *k* is important, without making a distinction between one and multiple occurrences in a given molecule. Accordingly, I_k is applied if the associated substructure is present at least once in the molecule.

Finally, models according to Equation 9.1 may, but need not, have a non-zero intercept *d*. In the case of logarithmic models, *d* directly relates to the unit of the property. Unit conversion here does not change any other model parameter.

Topological indices

Another popular type of 2D descriptors are topological indices. These are graph invariants calculated by a respective algorithm from the atom and bond connection and the topological distance matrix. The most familiar type of topological indices are connectivity indices. The first connectivity index was introduced by Randić [22]. Then, the concept was developed further by Kier and Hall [23], and for some time now there have also been electrotopological indices (E-states) available that encode 2D electronic structure information in a more sophisticated way [24,25].

Generally, topological indices account for molecular branching, size, shape and also (through consideration of atom and bond types) for some aspects of the molecular electronic structure. However, a direct physical or chemical interpretation is usually not straightforward, and thus models built from topological indices are often considered as mainly empirical. For a detailed explanation of such indices as well as an overview of less common 2D descriptors, the reader is referred to [21] and associated literature references.

3D molecular descriptors

In general, three-dimensional (3D) parameters refer to the 3D geometry of a molecule. In principle, simple 3D descriptors can be obtained by arithmetic expressions from atomic coordinates of the molecules. However, most 3D descriptors currently used rely on quantum chemistry, and thus require suitable software packages for their calculation. A more detailed outline of quantum chemical descriptors and their mechanistic background is given in [26], and some related information can also be found in [21].

For any 3D parameter, proper molecular geometries are required to obtain proper descriptor values. 3D coordinate calculations range from simple library-based algorithms (with pre-defined values for bond lengths, bond angles and dihedral angles, e.g., CORINA [27]) over more sophisticated force fields to quantum chemical geometry optimization schemes. As discussed in [26], the final descriptor values may depend significantly on the level of theory (semi-empirical vs. *ab initio* schemes).

Conformational flexibility

3D descriptors usually refer to a particular conformation of the molecule, which is typically selected as the minimum energy conformation. At a given temperature, however, conformationally-flexible compounds may occur in a variety of geometric arrangements (Boltzmann distribution). Moreover, intermolecular interactions may be energetically more favourable through a conformation different from the minimum energy conformation, resulting in an overall gain in free energy. The group of Mekenyan [28,29] developed a computerized tool to include conformational flexibility in QSAR models by generating a range of energetically reasonable conformers and selecting the relevant descriptors according to certain schemes (e.g., energy-weighted average parameter value, maximum or minimum parameter value sampled across all considered conformers, parameter values referring to specifically selected conformers, etc.).

LSER approach

Linear solvation energy relationships (LSERs) were introduced by Kamlet and Taft [30] employing "solvatochromic parameters". Solvatochromism represents the dependence of the UV/VIS absorption band of a solute on the polarity of the solvent (bathochromic or hypsochromic shifts of the wavelength), and originally LSER parameters were developed to describe such solvatochromic effects through respective Hammett relationships. Abraham [16] further developed this approach, and the most recent form of the LSER equation used to model solvation-dependent properties [31] (where solvation may also refer to the solvation of a compound in its own bulk phase),

$$\log \text{Property} = e\,E + s\,S + a\,A + b\,B + v\,V + c \qquad (9.2)$$

is now also called the Abraham equation (for an explanation of the symbols see next paragraph). LSER equations represent a special type of linear free energy relationships (LFERs), and have become widely-used to predict partition equilibria because of their mechanistically oriented approach to decomposing the intermolecular interactions in fundamental types, such as dispersion interactions (London forces), dipole-dipole interactions (Keesom forces) and interactions between dipoles and induced dipoles (Debye interactions), and H-bonding interactions.

Besides the intercept *c* (regression constant), Equation 9.2 contains five terms related to particular types of interactions. Each of these terms is given as product of a *compound descriptor* depending on the solute but independent of the solvent (or partitioning system), and a *phase parameter* depending on the solvent but independent of the solute. In the first term, compound descriptor E denotes the excess molar refraction of the solute, and e the respective solvent-specific parameter. By definition, E relates the molar refraction of the solute to the corresponding molar refraction of a hydrocarbon with equal molecular size, and the associated solute-solvent interaction is understood to be mediated through the n and π electrons of the compounds.

Parameter S represents the solute polarity (dipolarity) and polarizability, and s the respective interaction property of the solvent. A and B denote the H-bond donor and acceptor strength (H-bond acidity and basicity) of the solute, and the counterpart parameters of the solvent are its H-bond basicity a and H-bond acidity b. Depending on the property of interest, the organic phase may be pure (dry) or wet (saturated with the strong H-bond donor water, such as octanol in the two-phase octanol-water system). Accordingly, there are two slightly different types of the solute H-bond acceptor strength, B^{H} and B^{O}, available that are supposed to apply for dry (B^{H}) and wet (B^{O}) organic phases, respectively.

The cavity term contains the characteristic McGowan volume V [32] of the solute and its solvent counterpart v, and is theoretically restricted to liquid phases or liquid-liquid systems, while for gas-liquid and gas-solid systems it should be replaced by the solute parameter L (logarithmic hexadecane-air partition coefficient) and an associated phase parameter l. In practice, however, cross applications may result in better calibration statistics, an example of which is shown below in the context of modelling log K_{aw} (see Section 9.3.2).

For a given target property with experimental values, introduction of the solute descriptors E, S, A, B and V or L yields a regression fit of the phase parameters e, a, b, v or l, the latter of which characterize the phase system in terms of the associated types of interaction forces. Once a respective Abraham equation has been established (e.g., for log K_{ow}), the target property (in this example log K_{ow}) can be predicted for untested solutes provided their solute specific Abraham parameters (E, S, ..) are available. Note further that until recently, the following different notation was used for the solute descriptors: R_2 (now E), $\Sigma\pi_2^H$ (now S), $\Sigma\alpha_2^H$ (now A), $\Sigma\beta_2^H$ and $\Sigma\beta_2^O$ (now B^H and B^O), V_x (now V), and log L^{16} (now L).

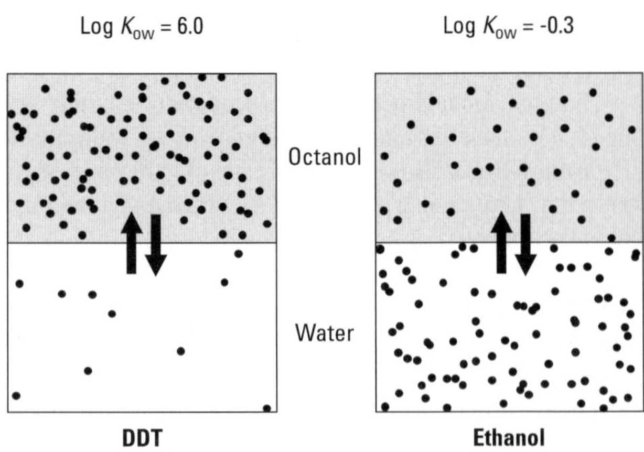

Figure 9.1. Illustrations of the octanol-water partition coefficient, together with examples of a hydrophobic (DDT = *p,p'*-dichlorodiphenyl trichloroethane) and a hydrophilic (ethanol) compound.

9.3 THERMODYNAMIC PARTITIONING BETWEEN ENVIRONMENTAL PHASES

9.3.1 Octanol-water partition coefficient

For many years, the equilibrium partitioning of an organic substance between octanol and water has been used as measure of the hydrophobicity of the compound in aqueous solution [33-39]. In terms of concentration values such as mol/L, the *n*-octanol-water partition coefficient, K_{ow}, is defined by:

$$K_{ow} = \frac{C_o}{C_w} \tag{9.3}$$

where C_o and C_w denote the compound concentration in *n*-octanol and water, respectively. As such, K_{ow} is dimensionless. A theoretical prerequisite of K_{ow} as a thermodynamic property is its independence from the absolute values of C_o and C_w in accord with Nernst's law. In practice, however, K_{ow} values are measured at low concentration values in order to meet this condition (at least approximately). Note further that K_{ow} is usually only weakly dependent on temperature, and that experimental and calculated values typically refer to 25°C and 1 atm as standard conditions.

For organic compounds, K_{ow} values typically range from 10^{-4} to 10^{8} and thus cover at least 12 orders of magnitude. Accordingly, hydrophobicity is often expressed as log K_{ow}, which denotes the decadic logarithm of the octanol-water partition coefficient. In

Figure 9.1, typical measurement situations are shown for compounds with $K_{ow} > 1$ ($\log K_{ow} > 0$) and $K_{ow} < 1$ ($\log K_{ow} < 0$), respectively.

Experimental methods

Log K_{ow} values up to ca. 4-5 can be experimentally determined by the shake-flask method [40]. An indirect method for the same log K_{ow} range is based on RP-HPLC and can be used within certain compound classes, provided reliable K_{ow} values are available for reference compounds [41]. Here, the stationary phase (usually C_{18}) is nonpolar and the mobile phase is polar (water with methanol or acetonitrile), such that increasing hydrophobicity results in prolonged retention times. Note that the partitioning process occurring in RP-HPLC may not appropriately model the compound distribution between octanol and water. For superhydrophobic compounds ($\log K_{ow} > 6$), the direct slow-stirring method [42,43] is recommended, because through this approach the formation of micro-emulsions in the aqueous phase and thus an overestimation of aqueous-phase concentrations of test chemicals can be avoided. Literature values with $\log K_{ow} > 6$ may contain substantial errors and thus should be checked carefully. A more detailed discussion on the experimental methods available is given in [44].

Dissociation and protonation

By definition, K_{ow} implies that the solute under consideration has identical speciation in both phases. Accordingly, processes such as complex formation, dissociation and protonation in at least one of the two phases (often in water) lead to more complex distribution phenomena. For acids AH and bases B that can donate or accept one proton, the unionized and ionized fractions f_u and f_i depend on pK_a and pH according to the Henderson-Hasselbalch relationship as follows:

$$f_u^{acid} = f_i^{base} = \frac{1}{1 + 10^{\mathrm{pH}-\mathrm{p}K_a}} \tag{9.4}$$

$$f_i^{acid} = f_u^{base} = \frac{1}{1 + 10^{\mathrm{p}K_a-\mathrm{pH}}} \tag{9.5}$$

Equation 9.4 quantifies the undissociated compound fraction AH of the acid as well as the protonated (ionized) form BH^+ of base B. In the latter case, pK_a refers to BH^+ as the conjugate acid of B. Correspondingly, Equation 9.5 yields the fraction of the dissociated (ionized) form A^- of AH as well as the unprotonated (neutral) form of B.

Taking the weak acid phenol and the base NH_3 as examples, the pK_a values of 9.9 (phenol) and 9.25 (NH_4^+) imply that in dilute aqueous solution at pH 7, the fractions of phenol and phenolate anion are 99.9 % and 0.1 %, respectively, while there is an equilibrium between 0.6 % NH_3 and 99.4% NH_4^+ as conjugate acid. The overall distribution of acids and bases is given by

$$D_{ow} = f_u \cdot K_{ow} + f_i \cdot \left(K_i + \sum_k K_{ip}(k) \right) \tag{9.6}$$

In Equation 9.6, K_i and $K_{ip}(k)$ denote individual partition coefficients of the ionized compound (A^- or BH^+) and of their possible ion pairs k formed with metals (e.g., from metal salts) or anionic ligands if present in the aqueous solution [45].

Because the neutral compound fraction of acids and bases is lowered by dissociation and protonation, respectively, their effective hydrophobicity in terms of the neutral species partitioning,

$$D_{ow}^{u} = f_u \cdot K_{ow} \tag{9.7}$$

is increasingly lowered with decreasing pK_a (increasing acidity) of the acidic form of the solute, where $f_u = f_u^{acid}$ for acids, and $f_u = f_u^{base}$ for bases. Accordingly, Equation 9.7 can be used to estimate the effective hydrophobicity of ionogenic compounds in terms of values that account for both K_{ow} and f_u.

Experimental K_{ow} values for acids and bases usually refer – as far as possible – to the unionized compound fraction, implying respective selections of the measurement pH. Correspondingly, prediction methods as discussed below, yield values of the unionized compound fraction of acids and bases, which should be kept in mind when performing predictive fate assessment of environmental chemicals based on properties that in turn are estimated from K_{ow} values. A special OECD guideline for the K_{ow} determination of ionizable substances is awaiting approval [46] which uses the pH-metric technique [47] based on the comparison of two linked potentiometric titrations, one in aqueous solution only and the other in the presence of the secondary octanol phase.

Environmental applications

Molecular hydrophobicity in terms of K_{ow} is often used to estimate the sorption of neutral organic compounds into soil organic matter [48,49]. Figure 9.2 shows a linear relationship between log K_{ow} and log K_{oc} respectively, (where K_{oc} = sorption constant normalized to organic carbon content, see Section 9.3.3 for more details),

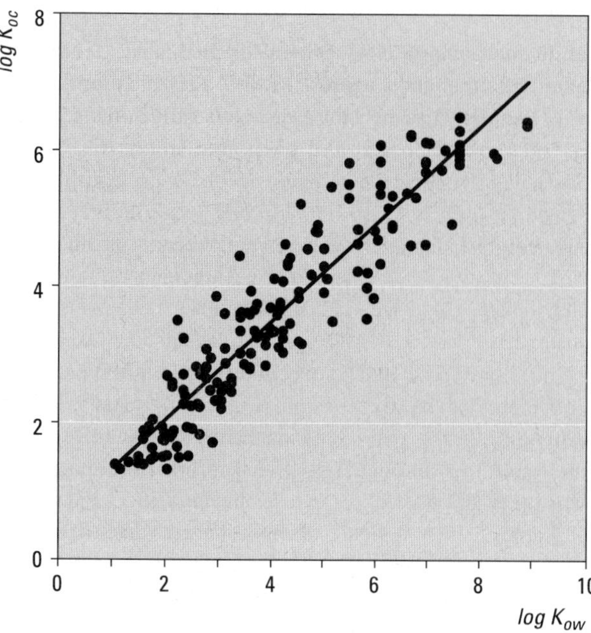

Figure 9.2. Log K_{oc} vs. log K_{ow} for 167 nonpolar and weakly polar compounds. The regression line is according to Equation 9.8 ($r^2 = 0.87$, $rms = 0.52$).

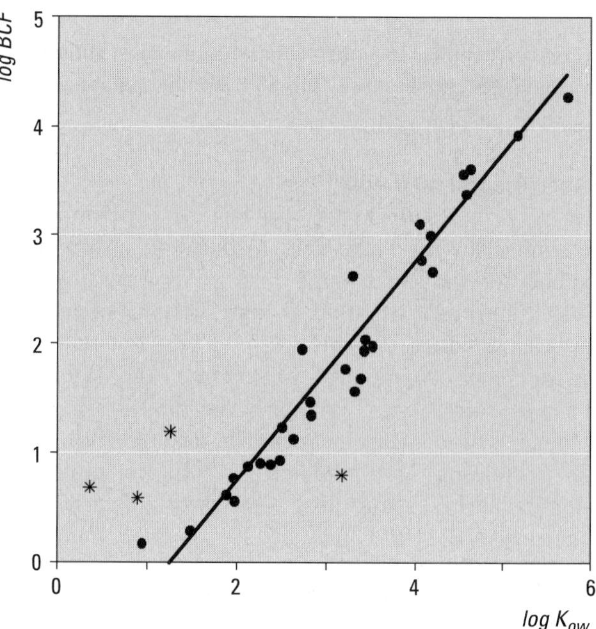

Figure 9.3. Log BCF vs. log K_{ow} for 35 nonpolar narcotic compounds. The regression line is according to Equation 9.9 ($r^2 = 0.94$, $rms = 0.29$), excluding the four compounds plotted with the star symbol.

$$\log K_{oc} = 0.71 \log K_{ow} + 0.62 \quad (9.8)$$

$n = 167$, $r^2 = 0.87$, $q_{cv}^2 = 0.87$, $rms = 0.52$

(n = number of compounds, r^2 = squared correlation coefficient, q_{cv}^2 = predictive squared correlation coefficient from leave-one-out cross-validation, rms = root-mean-square error of calibration) for 167 nonpolar and weakly polar compounds, where K_{ow} works reasonably well as a surrogate for K_{oc}. As discussed below in Section 9.3.3, increasing complexity of the molecular structure through increasing the presence of functional groups and, in particular, of hydrogen bonding sites, generally decreases the suitability of K_{ow} to predict K_{oc}.

Another important application is the estimation of bioconcentration factors BCF of waterborne organic substances for fish or other aquatic species [50-53] (see Chapter 3 and Section 9.3.5 for a discussion of bioaccumulation processes). An illustration of such relationships between log K_{ow} and log BCF is given in Figure 9.3, confined to compounds classified as nonpolar narcotics [54] (with 4 outliers excluded for the regression line and statistics):

$$\log \text{BCF} = 1.00 \log K_{ow} - 1.25 \quad (9.9)$$

$n = 31$, $r^2 = 0.94$, $q_{cv}^2 = 0.93$, $rms = 0.29$

Equation 9.9 is almost identical to the log BCF – log K_{ow} relationship of Mackay [50] published 25 years ago (where slope and intercept were 1.0 and –1.32, respectively). As outlined below in Section 9.3.5, the intercept of such regression equations can be related to the lipid content of the organism. The respective value resulting from Equation 9.9 is 5.6 %, which agrees well with the range of experimental values for the lipid content of fish species (see below).

A further and very important application of log K_{ow} is its use to predict the minimum or baseline toxicity of organic compounds to aquatic species when considering acute exposure regimes [55] (see also Chapters 10 and 11). In Figure 9.4, LC_{50} values (LC_{50} = lethal concentration 50%) for fathead minnow and a 96 h exposure time are plotted against log K_{ow} for 526 compounds taken from the Duluth dataset [56]. The subset of 70 nonpolar narcotics (selected according to the narcotics list of [54]) yields the following regression statistics:

$$\log LC_{50}\ [\text{mol/L}] = -0.91 \log K_{ow} - 1.17 \quad (9.10)$$

$n = 70$, $r^2 = 0.91$, $q_{cv}^2 = 0.91$, $rms = 0.38$

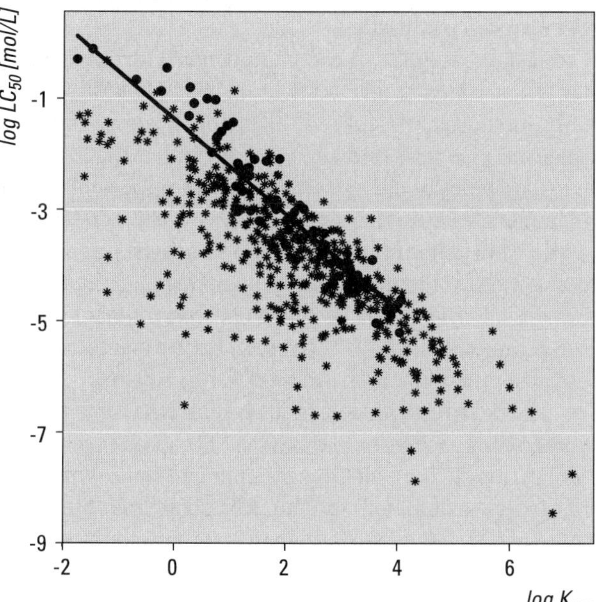

Figure 9.4. LC_{50} vs. log K_{ow} for 526 compounds. The regression line for 70 nonpolar narcotics (filled circles) is according to Equation 9.10 ($r^2 = 0.91$, $rms = 0.38$).

The respective plot (together with LC_{50} data not fitting to this equation) is shown in Figure 9.4. Interestingly, respective narcotic-type relationships of the general form:

$$\log LC_{50} = a \cdot \log K_{ow} + b \tag{9.11}$$

can also be used as a starting point to estimate the lethal body burden or dose of the contaminant [57]. Under narcosis theory, the LD_{50} (lethal dose 50%) as the dose counterpart of the externally measured LC_{50} can be estimated through:

$$LD_{50} = f_L \cdot 10^b \tag{9.12}$$

for a given lipid content f_L (that typically varies between 0.03 and 0.20, corresponding to 80% – 97% water content). Assuming $f_L = 0.05$ as an often-used estimate and a fish density of 1 g/cm^3 (= 1 kg/L), applying Equation 9.12 to the intercept (–1.17) of Equation 9.10 yields LD_{50} = 3.4 mmol/kg as the narcotic dose for fathead minnow *(Pimephales promelas)*, which is indeed within the experimental range of 2-8 mmol/kg determined for both this species as well as for the guppy *(Poecilia reticulata)* [58].

Prediction methods

Methods to predict log K_{ow} from molecular structure or compound-related information include fragment methods (Equation 9.1) such as the widely used KOWWIN [59], Abraham-type linear free energy relationships [31], and non-linear models such as ALOGPS [60].

ALOGPS [60] is a neural network model based on 75 E-state indices [25] and five hidden neurons, making up a total of ca. 380 parameters. The model was developed with log K_{ow} data for 12908 compounds from the Physprop database [61], and is publicly available for online applications.

ADME/logP [62] is a fragment scheme according to Equation 9.1, accounting for both intermolecular and intramolecular interactions through respective fragment values and correction factors. KOWWIN [59] contains 144 fragments and 290 correction factors and represents a reductionist-type fragment method. According to the manual, the training set contained 2474 compounds yielding $r^2 = 0.981$ and $rms = 0.22$, and subsequent validation using 10331 compounds resulted in $r^2 = 0.94$ and $rms = 0.47$. The fragment scheme by Marrero and Gani [63] was developed using a training set of 9560 compounds with log K_{ow} values.

Finally, there are three variants of the Abraham LSER approach according to Equation 9.2. The original regression equation to predict log K_{ow} from excess molar refraction E, dipolarity and polarizability S, hydrogen bond acidity A, hydrogen bond basicity B and McGowan volume V was published in 1994 [64]:

see below (9.13)

$n = 613$, $r^2 = 0.994$, $rms = 0.116$, $F = 2316$

For the derivation of Equation 9.13, only experimental values have been employed for the input parameters E, S, A and B^O, while V was calculated from molecular structure employing a simple increment scheme [32].

According to the Abraham approach, the two major

$$\log K_{ow} = 0.562\,E - 1.054\,S + 0.032\,A - 3.460\,B^O + 3.814\,V + 0.088 \tag{9.13}$$

factors governing log K_{ow} are molecular size *(V)* and H-bond basicity (B^O referring to wet octanol) of the solute. Increasing size increases log K_{ow}, because the energy penalty for cavity formation is much greater in water than in octanol. By contrast, increasing H-bond acceptor strength decreases log K_{ow}, because water as the solvent is a much stronger H-bond donor than octanol as the solvent. Smaller contributions to log K_{ow} come from *S* and *E*, while the solvents octanol and water are not discriminative for the solute's H-bond donor acidity *(A)*. Increasing dipolarity/polarizability *(S)* and thus increasing susceptibility of the solute to interact with the dipole and induction forces decreases log K_{ow}, while *n* and π electrons *(E)* support partitioning into octanol.

Besides using experimental values as input parameters, the LSER model of Equation 9.13 has also been used in combination with predicted values for *E, S, A* and B^O according to the Platts scheme [65] as implemented in ChemProp [17]. The third LSER variant employs Abraham parameters predicted by the 2006 version of the ABSOLV software (ADME-Boxes) [62]. In all method combinations, the McGowan volume is calculated using the above increment scheme of Abraham and McGowan [32].

CLOGP [66] is one of the most prominent commercial log K_{ow} calculation programs. The first CLOGP prototype was published in 1979 [67], and its constructionist increment methodology has been discussed thoroughly in the literature [34,35,37,68-71], although we are not aware of a publication specifying the actual number of model parameters. In a footnote to Leo's chapter in the handbook of Boethling and Mackay [72], the regression equation:

$$\log K_{ow} = 0.956 \text{ CLOGP} + 0.084 \qquad (9.14)$$

$$n > 10000,\ r^2 = 0.970,\ rms = 0.278$$

was presented for version 4.0 of CLOGP available at that time, and the KOWWIN manual [73] reports the following calibration statistics in terms of r^2 and *rms* (that actually corresponds to Equation 9.14, but with different slopes and intercepts):

KOWWIN: $n = 12805,\ r^2 = 0.95,\ rms = 0.435$
CLOGP: $n = 11735,\ r^2 = 0.91,\ rms = 0.59$

The difference in the number of compounds was due to the fact that CLOGP (now BioByte's Windows version 1.0) was unable to handle 1070 of the 12,805 compounds of the KOWWIN database due to missing fragments.

Performance statistics

Our test set includes 14,899 compounds with experimental log K_{ow} values from –5.08 to 11.29. KOWWIN is the only method applicable to all of these compounds, and achieves $r^2 = 0.88$ and $rms = 0.66$. With ALOGPS, only two compounds are missing when simply applying all increment values where possible ($r^2 = 0.92$, $rms = 0.51$), while consideration of the method-specific application domain (AD) feature eliminates another 243 compounds ($r^2 = 0.94$, $rms = 0.45$). The Marrero and Gani scheme can be applied to 12,409 compounds, and is inferior to the other fragment methods ($r^2 = 0.84$, $rms = 0.76$).

The LSER approach is a special case. For the 769 compounds with experimental Abraham parameters (except that the McGowan characteristic volume *V* is always calculated by the Abraham and McGowan scheme [32]), the LSER statistics are excellent ($n = 769$, $r^2 = 0.99$, $rms = 0.21$), although application of the Platts scheme to predict excess molar refraction *E*, dipolarity and polarizability *S*, hydrogen bond acidity *A* and hydrogen bond basicity *B*, results in pretty poor statistics ($n = 14707$, $r^2 = 0.58$, $rms = 1.75$). Interestingly, the ABSOLV prediction of the Abraham parameters leads to significantly improved statistics ($n = 14802$, $r^2 = 0.76$, $rms = 1.00$). Unfortunately, the underlying calculation method has – so far – not been published and thus is not accessible for independent analysis.

At present, the LSER approach is not on a par with reference models such as KOWWIN, ALOGPS and ADME/LogP. While experimental Abraham parameters are available for only 5 % of the compounds, employing predicted Abraham parameters yields only moderate statistics, reflecting their currently limited quality, as discussed elsewhere [17,74].

Overall, the results indicate that the freely available KOWWIN scheme is a very good choice for estimating log K_{ow} from molecular structure. A further advantage of this method is its transparency with respect to a given calculation result. Due to its fragmental methodology, every calculation can be broken down into contributions associated with substructural features, also providing guidance on how the structure could be modified in order to increase or decrease the log K_{ow} most efficiently.

9.3.2 Henry's law constant

The volatilization of waterborne compounds into air is governed by external factors such as wind and temperature, and by the degree of intrinsic affinity of the compound to water as solvent. Under thermodynamic conditions at a given temperature and pressure, the equilibrium partitioning of a compound between the gas phase and the water phase is determined by Henry's law.

For dilute solutions, the partial pressure above solution due to the solute, p, is directly proportional to its concentration in solution, and the proportionality constant H is Henry's law constant:

$$H = \frac{p}{C_w} \tag{9.15}$$

As can be seen from Equation 9.15, the dimension of H is energy/mol (resulting from [force/area] · [volume/mol]), and typical units include (atm · L)/mol and its SI equivalent (Pa · m^3)/mol.

In the aqueous solution, C_w can be increased until saturation, where C_w becomes equal to the water solubility S_w of the compound. For convenience, we assume that S_w is still sufficiently low to allow application of Henry's law, which is typically the case for hydrophobic compounds. Note that high S_w values may cause deviations from Henry's law.

An increase in the substance amount in water above S_w would result in the formation of a separate phase, consisting of the pure substance (provided the solubility of water in this substance phase is sufficiently low and thus can be neglected). At $C_w = S_w$, the initial separate phase is in equilibrium with both the dissolved phase and the gas phase. Accordingly, the partial pressure p associated with S_w equals the vapour pressure of the (approximately) pure substance, P_v at the given temperature and pressure (because the pure substance phase is *thermodynamically* also in equilibrium with the gas phase). It follows that H can also be written as:

$$H = \frac{P_v}{S_w} \tag{9.16}$$

Note that in Equation 9.15, C_w (dissolved compound concentration) and p (partial pressure above solution) may vary within the limits of Henry's law (that applies to dilute solutions with attendant implications for C_w), while in Equation 9.16, P_v (vapour pressure) and S_w (water solubility) are substance-specific constants with fixed values for a given temperature and external pressure, bearing in mind that S_w should be sufficiently small to comply with the conditions for Henry's law.

Experimental methods

The experimental measurement of Henry's law constants includes both direct and indirect methods. The latter rely on the determination of the solute's limiting activity coefficient in water that can easily be converted into Henry's law constant, provided reliable vapour pressure data are available. For a more detailed overview, the reader is referred to the literature [75-78].

With the inert gas stripping method that is suited for fairly volatile compounds [79,80], K_{aw} is determined through analysis of the volatilization loss of the aqueous solution in the "bubble column" or by quantification of the solute amount stripped-off over a certain time period. Nowadays, static headspace analysis [81], with its elegant variants called "equilibrium partitioning in closed systems" [82,83] and "phase ratio variation" [84,85] can also be used for a wide range of compounds, including less volatile substances. This is due to the availability of solid-phase microextraction tools [86] in combination with a drastic increase of the GC-MS sensitivity. However, with low volatile compounds such as pesticides and when great accuracy is needed, analysis of both the air and water phase is recommended. There are several publications with larger volumes of experimental data for Henry's law constant [75,87-89].

Air-water partition coefficient

The air-water partition coefficient K_{aw} is defined as the ratio of the compound concentration in the gas phase, C_a, and in the water phase, C_w under thermodynamic equilibrium conditions:

$$K_{aw} = \frac{C_a}{C_w} \tag{9.17}$$

K_{aw} is also termed the dimensionless form of Henry's law constant, and the relationship with H is given by:

$$H = RT \cdot K_{aw} \tag{9.18}$$

Equation 9.18 reveals that the temperature dependence of H can be broken down into an intrinsic contribution affecting K_{aw} and an additional contribution from the temperature as an external factor.

Note further that log K_{aw} can be estimated from the logarithmic octanol-water and octanol-air partition coefficients:

$$\log K_{aw} = \log K_{ow} - \log K_{oa} \tag{9.19}$$

K_{oa} and K_{ow} refer to dry and wet octanol, respectively, which is the reason for the approximate nature of Equation 9.19. In practice, however, there are substantially more K_{aw} values than K_{oa} values, so that the scope of Equation 9.19 is more for situations where direct prediction methods for log K_{aw} have known difficulties or are not applicable.

Temperature dependence

As mentioned above, Henry's law constant depends on temperature both directly and through K_{aw} (see Equation 9.18). The latter is governed by the standard enthalpy and entropy of desolvation. For the environmentally relevant temperature range, these can be taken as essentially constant, and in this case a convenient van't Hoff form of the temperature dependence of log K_{aw} is:

see below (9.20)

where T_{ref} denotes a reference temperature, such as 298 K (25°C) to which most laboratory data refer.

Recently, an increment model based on Equation 9.20 was introduced to predict the temperature dependence of Henry's law constant based on molecular structure [90]. For 456 organic compounds covering the atom types C, H, N, O, F, Cl, Br, I and S, 46 fragments yielded an r^2 value of 0.81 with a standard error of 7.1 kJ/mol for predicting the standard enthalpy of desolvation. This model was then applied together with experimental H values at 25°C for 462 compounds with 2119 experimental Henry's law constants at temperatures below 20°C. The resultant q^2 (predictive squared correlation coefficient) and *rms* were 0.99 and 0.21 log units, respectively.

As a general rule, Henry's law constant increases with increasing temperature, mainly because of the corresponding increase in the vapour pressure, as can be seen from Equation 9.16. Interestingly, temperature effects on the multimedia fate of substances are most pronounced for polar compounds and when the primary discharge compartment is air [10]. Note, however, that temperature also affects degradation rates and thus absolute environmental concentrations of xenobiotics, even in cases when the distribution between the compartments remains essentially unchanged [10].

Prediction methods

According to Equation 9.16, Henry's law constant can be predicted as the ratio of vapour pressure (P_v) over water solubility (S_w). As noted above, this approach is theoretically confined to relatively small water solubilities, because Henry's law applies to dilute solutions. In practice, Equation 9.16 is often used, keeping in mind that the input properties P_v and S_w can in turn also be estimated from molecular structure where experimental values are missing.

Besides this thermodynamic approach, models to predict Henry's law constant from molecular structure include increment methods, structure-activity relationships employing connectivity indices or other molecular parameters, Abraham-type LSERs, and quantum chemical continuum-solvation models. For an extended discussion of more than 40 methods published in the last 32 years, the reader is referred to a review published a few years ago [91].

The development of increment methods for predicting log H or log K_{aw} goes back to 1975 when Hine and Mookerjee introduced both a bond contribution method (34 fragments calibrated with 263 compounds) and a group contribution method (49 primarily calibrated fragments augmented by 20 additional fragments to correct for systematic errors and by two correction factors, employing 212 compounds) [92]. Meylan and Howard extended the bond contribution scheme to 59 fragments and included 15 correction factors [93].

In the recent version of their HENRYWIN software [94], the bond contribution method HENRYWIN-Bond consists of 97 fragments, 36 correction factors and 3 indicator variables, and the group contribution scheme HENRYWIN-Group extending the corresponding Hine and Mookerjee method, contains the 69 original fragments and an additional 31 fragments with estimated increment values.

The model by Nirmalakhandan and Speece [95] in its later update [96] contains a modified first-order valence-corrected connectivity index ${}^1\chi^v$ (the associated δ values are set as equal to the number of bonds of the given heavy atom). It has a "polarity term" Φ that in turn consists of 17 atomic and structural fragments calibrated to yield the best fit for log K_{aw} in combination with the other model parameters, and an indicator variable to account for hydrogen bonding, referring to electronegative heteroatoms (O, N, halogen) attached to CH as well as to aromatic and acetylenic moieties:

$$\log K_{aw}(T) = \log K_{aw}(T_{ref}) - \frac{\Delta H^0_{des}}{2.3R}\left(\frac{1}{T} - \frac{1}{T_{ref}}\right) \quad (9.20)$$

see below (9.21)

original model [95]: $n = 180$, $r^2 = 0.99$, $rms = 0.262$
refined model [96]: $n = 462$, $r^2 > 0.95$
(no other statistics given)

Note that the difference between the original and refined model lies in the actual definition of the polarity term Φ, which in turn represents an increment model calibrated to yield the best fit for log K_{aw} when inserted in Equation 9.21. The Abraham model to predict log K_{aw} reads [97]:

see below (9.22)

$n = 408$, $r^2 = 0.995$, $rms = 0.151$, $F = 16810$

and includes the McGowan characteristic volume V (ignoring that for this gas-liquid system, L would be theoretically preferred over V, see Section 9.2 above).

Performance statistics

Our test set consists of 2070 organic compounds with experimental data for log K_{aw} at 25°C from –16.50 to 3.14. Except for dimethyl diselenide, HENRYWIN-Bond can be applied to all compounds, and yields the best overall statistics ($r^2 = 0.87$, $rms = 1.18$). Interestingly, the associated *rms* is twice as large as achieved for the best log K_{ow} prediction methods (see previous subsection). As demonstrated earlier, standard errors of 0.5 log units may already result in variations by a factor of 2.5 with respect to the relative multimedia distribution of compounds [9], apart from the impact of temperature variation as mentioned above [10]. HENRYWIN-Group can be applied to only 56% of the compounds ($r^2 = 0.88$, $rms = 0.91$), which also applies to the Nirmalakhandan and Speece model which has rather poor statistics ($r^2 = 0.66$, $rms = 2.07$).

The LSER approach with experimental Abraham parameters can be applied to only 36% of the compounds with typically more simple structures, where it performs very well ($r^2 = 0.96$, $rms = 0.47$). LSER-Platts (Abraham parameters predicted according to the Platts methods [65] as implemented in ChemProp [17]) can formally handle 2040 of the 2070 compounds, but only with poor statistics ($r^2 = 0.73$, $rms = 2.63$). LSER-ABSOLV (Abraham parameters predicted by unpublished algorithms as implemented in ADME-Boxes/ABSOLV [62]) can be applied to all except six compounds, but is still not competitive to HENRYWIN-Bond ($r^2 = 0.87$, $rms = 1.68$).

At present, HENRYWIN-Bond is the method of choice regarding both the application range and the overall performance. In the near future, a new method for predicting log K_{aw} will become available which is based on 2D molecular parameters and increments (Schüürmann et al., to be published). For the present set of 2070 compounds, the r^2 value is around 0.95, and the *rms* is below 0.6 log units.

9.3.3 Sorption constant

Waterborne compounds may accumulate in solid phases depending on the balance of their affinities for water and solid matter. The association of compounds (sorbates) with solid material (the sorbent) is called sorption, which covers adsorption both onto or near the surface, and absorption via penetration into the volume of the material. Adsorption is triggered by the availability of adsorption sites at the surface, which in turn is proportional to the surface area. By contrast, the capacity for absorption in solid phases is usually sufficiently large to impose no limit under environmental conditions. While sorption may occur from both the gas phase and solution, the remainder of this section is confined to the phase transfer of compounds from aqueous solution.

At the molecular level, there are different types of interactions that contribute to the association of compounds with solid matter. Sorption forces include the three van der Waals components (dispersion forces, dipole-dipole interactions, and interactions between dipoles and induced dipoles), hydrogen bonding, and in case of ionic species also Coulomb forces (attractive interaction between charges of opposite sign).

For the quantification of sorption, the equilibrium distribution of the compound between the solid phase and water is usually related to the mass of the sorbent:

$$K_d = \frac{X_s}{C_w} \tag{9.23}$$

where X_s = number of sorbate molecules per kg sorbent,

$$\log K_{aw} = -0.468\ {}^1\chi^v + 1.005\ \Phi - 1.258\ I + 1.29 \tag{9.21}$$

$$\log K_{aw} = -0.577\ E - 2.549\ S - 3.813\ A - 4.841\ B^H + 0.869\ V + 0.994 \tag{9.22}$$

and C_w = compound concentration in aqueous solution. Accordingly, a typical unit of K_d is L/kg.

For neutral organic compounds, the organic matter (OM) fraction of solid phases is the major compartment for sorption. Moreover, OM typically contains around 58% organic carbon (OC), which means that for the sorption of organic non-electrolytes, 1 g of OC corresponds to 1.724 g of OM. Analyses of 32 soils and 36 sediments showed a range of fractional OC content, f_{oc}, from 0.0016 (0.16 %) to 0.061 (6.1 %) [98]. The sorption constant normalized to organic carbon, K_{oc}, is related to K_d according to:

$$K_{oc} = \frac{1}{f_{oc}} K_d \tag{9.24}$$

Thus, K_{ow} has the same dimension as K_d (volume/mass). In the event of missing experimental values for the OC content of the solid matter, f_{oc} values of 0.01 to 0.03 are often taken as an estimate except for specifically organic-rich material such as peat, where f_{oc} is around 0.5 [99]. Sometimes, sorption constants are normalized to OM in terms of corresponding K_{om} values. Here, the standard conversion to K_{oc} proceeds via:

$$K_{oc} = 1.724\ K_{om} \tag{9.25}$$

(see above OC-OM relationship) if no suitable experimental data are available.

Sorption reduces the mobility of environmental chemicals due to the fixation of part of the compounds to solid matrices. This affects both the transport in streams and the volatilization from water into air. Note that Henry's law constant refers to the dissolved compound concentration (see Section 9.3.2 above). It follows that with increasing particulate or dissolved organic matter content (the latter of which consists mainly of humic and fulvic acids), the effective or apparent air-water partition coefficient decreases due to the increasingly reduced fraction of compound in the dissolved state. For comprehensive accounts of sorption processes of xenobiotics in the environment, there are textbooks [100,101], reviews [102-104] and discussions of the underlying thermodynamics [105,106] available.

Experimental methods

Direct methods for measuring sorption coefficients include static and dynamic procedures. The former, usually called batch equilibration tests [107], are performed on soil-sediment suspensions with various concentrations of the test chemical at a constant temperature. Careful selection of the appropriate experimental conditions (solid-to-water ratio, concentration levels, efficient agitation, duration of experiment, phase separation, etc.) is necessary to obtain reliable data at distribution equilibrium. Application of SPME fibres [86] or other micro-extraction tools that enable non-depletive measurements in static soil-sediment-water systems [108] offer an elegant means of generating sorption coefficients at lower expense in terms of both sorbent/chemical consumption and the duration of the experiment [109]. In dynamic leaching tests, the chemical is pumped through a packed soil column, and the concentration of the chemical in the leachate is monitored over time and evaluated against the leaching behaviour of a conservative (non-sorbed) tracer [49]. Although column tests are closer to natural sorption processes in soils, equipment costs are higher, and such experiments are time-consuming and subject to a variety of experimental artefacts [110]. An indirect estimation method based on the calibration of HPLC retention times (similar to that used for log K_{ow} estimation) is standard [111]. An extensive review of experimental method for the determination of sorption coefficients can be found in [112].

Prediction methods

The prediction of the intrinsic compound affinity for soil sorption has focused on K_{oc} or its decadic logarithm, log K_{oc}. The classical approach is to use octanol as a surrogate for the organic carbon in soils, sediments and suspended matter, and to relate K_{oc} to the *n*-octanol-water partition coefficient, K_{ow} according to [48]:

$$K_{oc} = 0.41 \cdot K_{ow} \tag{9.26}$$

and its logarithmic form:

$$\log K_{oc} = \log K_{ow} - 0.39 \tag{9.27}$$

While Equation 9.26 is still used as the default to predict K_{oc} in software packages for simulating the multimedia fate of organic compounds [113], its actual application range is confined to neutral hydrophobic compounds where sorption is governed by unspecific dispersion interactions. Note further that the use of K_{ow} to predict K_{oc} implies that absorption into organic matter is the dominant process, and that adsorption phenomena such as the association of ionized chemicals at polar mineral surfaces are not covered by K_{ow}-based models.

As noted above, log K_{ow} has a limited scope as a global predictor for log K_{oc}. However, local calibrations of logarithmic K_{oc}-K_{ow} regression relationships for individual compound classes may yield better predictions, provided an unequivocal allocation of the compound of interest to the relevant submodel can be made. A related decision tree model has been developed by Sabljić and co-workers [114,115] that includes different regression equations based on log K_{ow}, and one regression equation employing the first-order connectivity index ${}^1\chi$.

There are two more models that combine increment values associated with certain structural features, and molecular parameters that apply to all compounds. One is PCKOCWIN [116] which updates a previously published model [117] with polar fragment corrections:

$$\log K_{oc} = 0.53\ {}^1\chi + \sum_k n_k I_k + 0.62 \tag{9.28}$$

$n = 189$, $r^2 = 0.955$, $rms = 0.230$, $me = 0.182$

Both the original model [117] and the PCKOCWIN version [116] include 18 correction factors as mentioned in its manual.

The other model was developed recently and employs molecular weight covering dispersion interactions and the cavity formation energy (that are both related to molecular size), bond connectivity ε as a measure of the geometric accessibility of compounds for intermolecular interactions [118,119], the molecular E-state [25] to correct ε for differences in polarity and polarizability, 21 fragment correction terms, and 4 indicator variables [74]:

see below (9.29)

$n = 571$, $r^2 = 0.852$, $rms = 0.469$, $me = 0.361$

Huuskonen has developed a model to predict log K_{oc} from log K_{ow} in combination with molecular weight (MW), the number of aromatic 5- or 6-atom rings (NAR), the number of rotational bonds (ROT, implemented as the number of non-ring single bonds except H-R), and I_{acid} as an indicator for carboxylic groups [120]:

see below (9.30)

$n = 403$, $r^2 = 0.86$, $rms = 0.43$, $F = 491$

This Huuskonen model suggests that sorption to soil organic matter increases with increasing numbers of aromatic moieties, and decreases with increasing conformational flexibility and the presence of acidic groups. For the same 403 compounds, further regression models have been derived employing log S_w without ($r^2 = 0.80$, $rms = 0.51$) and with additional molecular parameters ($r^2 = 0.85$, $rms = 0.44$) [120].

The Abraham LSER calibrated by Poole and Poole [121]:

see below (9.31)

$n = 131$, $r^2 = 0.955$, $rms = 0.248$, $F = 655$

employs H-bond basicity (B^O) calibrated for wet octanol. According to Equation 9.31, increasing molecular size increases sorption to soil organic matter, H-bond basicity supports partition to water (because water is a strong H-bond donor as discussed above (see Section 9.3.1), and n and π electrons *(E)* favour phase interactions with soil organic matter.

Performance statistics

For the comparative analysis, experimental log K_{oc} data have been collected from different literature sources [114,122-127] as previously described [74]. Of the initial set of 733 compounds, three aliphatic amines (dimethyl amine, trimethyl amine and *n*-butyl amine) have reported log K_{oc} values significantly above all model predictions, suggesting a systematic error in these data. Omission of these data led to the current total set of 730 compounds with log K_{oc} ranging from –0.31 to 6.50. Note that 571 of these compounds had served as a training set for the Schüürmann model (Equation 9.29), and that the training sets of most other presently discussed models probably contained substantial portions of the current data set.

$$\log K_{oc} = 0.00321\ \mathrm{MW} + 0.255\ \varepsilon - 0.0139\ \text{E - state} + \sum_j n_j C_j + \sum_k n_k I_k \tag{9.29}$$

$$\log K_{oc} = 0.48 \log K_{ow} + 0.26\ \mathrm{NAR} - 0.07\ \mathrm{ROT} + 0.002\ \mathrm{MW} - 0.77\ I_{acid} + 0.56 \tag{9.30}$$

$$\log K_{oc} = 0.74\ E - 0.31\ A - 2.27\ B^O + 2.09\ V + 0.21 \tag{9.31}$$

Simple calibration with log K_{ow} (calculated from KOWWIN [59]) results in r^2 = 0.75 and rms = 0.64, which serves as a reference for more elaborated models, bearing in mind that K_{ow} has only a limited ability to mimic the partitioning between water and organic matter when more specific solute-phase interaction forces come into play.

Inclusion of additional 2D structure information (Huuskonen [120]) improves the performance somewhat (r^2 = 0.79, rms = 0.63), and slightly better statistics are achieved with locally calibrated log K_{ow} regression equations according to the Sabljic decision tree model [114,115] (n = 723, r^2 = 0.79, rms = 0.58).

In the group of models that combine fragmental increments with 2D molecular descriptors, the Schüürmann model [74] yields the best overall statistics (r^2 = 0.83, rms = 0.53) and in fact outperforms all other methods included in the present analysis. Interestingly, the PCKOCWIN [116] statistics are inferior to using log K_{ow} as the only predictor.

The LSER model using experimental Abraham parameters (LSER-exp) can be applied to only 190 compounds (26%), where it provides very good statistics (r^2 = 0.91, rms = 0.50). The introduction of calculated descriptors according to Platts [65] and ABSOLV [62] increases the (formal) application range to the total compound set, and decreases the performance significantly (r^2 = 0.71, rms = 0.79; ABSOLV: r^2 = 0.77, rms = 0.67).

Overall, the analysis demonstrates that for priority setting, as well as for predictive analyses of the multimedia fate of compounds, models are available to provide reasonable estimates of log K_{oc}. In this context, it should be borne in mind that the Karickhoff relationship (Equations 9.26, 9.27 and 9.8) which has great merits for structurally quite simple hydrophobic compounds, is not the best choice when addressing more complex chemical structures.

9.3.4 Vapour pressure

The vapour pressure P_v of a compound characterizes the extent to which it evaporates from its pure phase into air. More precisely, the vapour pressure is the pressure of the pure chemical vapour that is in thermodynamic equilibrium with the pure chemical in its solid or liquid state. For a given compound, vapour pressure depends on temperature, and increases with increasing temperature. A common illustration of the latter is the heating of a substance until boiling. At the boiling point (T_b), P_v equals the ambient pressure; accordingly, the normal boiling point is the temperature where $P_v(T_b)$ = 1 atm (101.325 kPa).

As discussed in Section 9.3.2 above, vapour pressure is related to Henry's law constant (which governs evaporation from aqueous solution) through the compound's water solubility (Equation 9.16). Moreover, P_v affects the rate of desorption from organic matter and thus the gas-particle partition coefficient of airborne organic compounds. As a general rule, P_v decreases with increasing molecular size because of the correspondingly increasing attractive strength of van der Waals interactions.

Experimental methods

Classical thermodynamic methods for determining the vapour pressure of a pure component are described in [128] together with recommendations on their respective P_v ranges of applicability. The reader is also referred to [112,129] for a general overview of the possible techniques. However, low vapour pressures (< 10^{-3} Pa) are difficult and time-consuming to determine using classical methods, i.e., effusion techniques and the gas saturation method. More rational indirect methods for this purpose are based on either measuring evaporation rates or gas chromatographic retention times [130,131] and require the use of several reference compounds, whose vapour pressures are accurately known over the whole temperature range investigated. Serious error sources associated with these indirect methods (for low volatile compounds) are the extrapolation of results from high-temperature measurements to ambient conditions, and the selection of appropriate reference compounds with (sufficiently accurate) vapour pressure data [132].

Temperature dependence

Liquid-gas phase transitions are governed by the vapour pressure of the compound in its liquid state, P_v^L. Its temperature dependence is given by the Clausius-Clapeyron equation:

$$\frac{d(\ln P_v^L)}{dT} = \frac{\Delta H_{vap}}{RT^2} \tag{9.32}$$

where ΔH_{vap} is the enthalpy of vaporization. With respect to the solid-gas phase transition, a corresponding equation applies for the sublimation vapour pressure, P_v^s, employing the enthalpy of sublimation, ΔH_{sub}. Deviations of the compound vapour from the ideal behaviour can be taken into account by introducing the difference in the compressibility of the two phases:

$$\Delta Z = \frac{P_v \Delta V}{RT} \tag{9.33}$$

leading to ($\Delta Z\, RT^2$) instead of (RT^2) in the denominator of Equation 9.32. Integration of the Clausius-Clapeyron equation from some reference temperature (e.g., boiling point T_b) to the ambient temperature of interest (e.g., T = 25°C = 298 K) forms the basis of an important class of prediction methods for P_v, as outlined in the subsequent section.

Prediction methods

A review of a variety of related models covering increment methods and quantitative structure-property relationships based on topological or other descriptors is available in the recent literature [133]. Our present analysis will be confined to methods for predicting log P_v at 25°C based on the integration of the Clausius-Clapeyron equation in various approximations. Most of these methods address the liquid-gas phase transition through calculation of the enthalpy of vaporization, ΔH_{vap}, in Equation 9.32 to predict P_v^s.

For solids, P_v^L refers to the (hypothetical) subcooled liquid state, omitting the energy associated with melting. This can be accounted for by considering the fugacity ratio of the compound in its pure solid and subcooled liquid state, f^s / f^L. For low vapour pressures (which is a reasonable assumption for many organic compounds), these can be taken as reasonable estimates of the fugacities, and further approximations yield the solid-liquid fugacity ratio as the difference between the respective logarithmic vapour pressures [134]:

$$\log P_v^s = \log P_v^L - \frac{\Delta S_m}{2.3R}\left(\frac{T_m}{T} - 1\right) \tag{9.34}$$

In Equation 9.34, T and the melting point T_m must be given in Kelvin. Walden's rule [135] to estimate the average of the entropy of melting as ΔS_m = 56.5 J/(mol K) converts the term $\Delta S_m/(2.3R)$ to the number 2.95. Because $T_m > T$, the equation shows that at a given temperature, the vapour pressure of a solid is smaller by a term proportional to the entropy of melting than the vapour pressure of its (hypothetical) subcooled liquid state, and that this difference increases with increasing melting point.

The simplest approach for integrating Equation 9.32 (from T_b to T, keeping in mind that $P_v(T_b)$ = 1 atm) is to assume that ΔH_{vap} is constant over the temperature range of interest. The resultant expression contains the two constants $A = \Delta H/RT_b$ and $B = \Delta H/R$, and the introduction of a further parameter C yields the Antoine equation [136]:

$$\log P_v^L = A - \frac{B}{T - C} \tag{9.35}$$

that forms the basis of the Grain-Antoine method [137] for predicting log P_v^L. In this method, the Antoine equation is modified to:

see below (9.36)

where $\Delta H_b = \Delta H_{vap}(T_b)$ denotes the enthalpy of boiling, and T and the boiling point T_b have to be specified in Kelvin. For the practical application of Equation 9.36, ΔH_b and parameter C are estimated from T_b and a structure-specific parameter, the Fishtine constant K_F [138]. The Grain-Antoine method thus requires the boiling point as input, which in turn can also be estimated from molecular structure. In our performance analysis as summarized below, T_b was estimated using the methods of Constantinou [139], Stein and Brown [140], Meissner (taken from [69]), Joback and Reid [141], and Miller based on Lydersen (also taken from [69]) in that order, applying selection criteria based on the expected reliability according to our experience (as implemented in ChemProp [19]).

For solids, Equation 9.36 is introduced in Equation 9.34 to convert P_v^L (that in this case is the P_v value for the subcooled liquid) into P_v^s. For the method variant to predict vapour pressure from structure without any experimental data input, the melting point is estimated in a similar manner by the models of Marrero and Gani [142], Constantinou [139], and Joback and Reid [141]. The latter are similar to the respective fragment models for the boiling point, and the method of Marrero and Gani is also a fragment method.

The Grain-Watson method [137] employs a

$$\log P_v^L\,[\text{atm}] = \frac{\Delta H_b\,(T_b - C)^2}{0.97\,RT_b^{\,2}}\left(\frac{1}{T_b - C} - \frac{1}{T_b - C}\right) \tag{9.36}$$

simplified form of the Watson correlation to evaluate the temperature dependence of the enthalpy of vaporization:

$$\Delta H_{vap}(T) = \Delta H_b (3 - 2T_{rb})^m \tag{9.37}$$

where $T_{rb} = T/T_b$ is called reduced temperature. The exponent m is 0.19 for liquids, and for solids the following setting is applied [137]: $m = 0.36$ if $T/T_b > 0.6$, $m = 0.80$ if $0.6 \quad T/T_b \quad 0.5$, and $m = 1.19$ if $T/T_b < 0.5$. Introduction of Equation 9.37 into the Clausius-Clapeyron equation and integration twice by parts [137] yields:

see below (9.38)

as the final result of the Grain-Watson method to predict $\log P_v$.

Sage and Sage [143] suggested modified versions of the Grain-Antoine and Grain-Watson models. In the Sage-Grain-Antoine model, ΔH_b is calculated by:

$$\Delta H_b = K_F R T_b (2.3) \log (82.06\, T_b) \tag{9.39}$$

However, according to our current test set, this approach appears to be inferior compared to the original Grain-Antoine model. In the Sage-Grain-Watson model, m depends linearly on the ratio of T and T_b:

$$m = 0.4133 - 0.2575 \frac{T}{T_b} \tag{9.40}$$

The second difference concerns the temperature dependence of the conversion from the subcooled liquid state to the solid state in the case of solids. Instead of applying the fugacity ratio approach according to Equation 9.34, the mathematical form used for quantifying the impact of $T_b/T \equiv T_{rb}$ on $\log P_v^L$ (Equation 9.38) is also used for evaluating the compound-specific entropy of melting, resulting in:

see below (9.41)

with m' obtained from Equation 9.40 by replacing T_b with T_m, and $T_{rm} = T/T_m$. This correction does not apply for $T \quad T_m$. While both Sage and Sage modifications of the models Grain-Antoine and Grain-Watson applied $\Delta Z = 1$ for the compressibility term (see Equation 9.33 above), we recommend the original setting of Grain-Watson ($\Delta Z = 0.97$) which turned out to be superior for our present data set.

A third method was developed by Mackay and co-workers [144] and is again based on several approximations for the temperature dependence of ΔH_{vap} when integrating the Clausius-Clapeyron equation. The final result is:

see below (9.42)

employing Trouton's rule to estimate the entropy of boiling (ΔS_b = 88 J/[mol K], resulting in $\Delta S_b/(2.3R) = 4.6$). The melting point term applies only for solids (where $T_m > T$ for the temperature of interest). Among the methods included in the present comparative analysis, the Mackay model is the only one that makes no use of structural information beyond T_b and T_m. Note that this model had been calibrated for hydrocarbons and halogenated compounds, which should be kept in mind when applying the method to more polar compounds that were not part of the training set [144].

$$\log P_v\,[\text{atm}] = \frac{\Delta H_b}{2.3 \cdot 0.97\, R T_b} \left(1 - \frac{(3 - 2T_{rb})^m}{T_{rb}} - 2m\,[3 - 2T_{rb}]^{m-1} \cdot (2.3) \log T_{rb}\right) \tag{9.38}$$

$$\log P_v\,[\text{atm}] = \log P_v^L\,[\text{atm}] + 0.6 \log (82.06\, T_m) \cdot \left(1 - \frac{(3 - 2T_{rm})^{m'}}{T_{rm}} - 2m'\,[3 - 2T_{rm}]^{m'-1} \cdot (2.3) \log T_{rm}\right) \tag{9.41}$$

$$\log P_v\,[\text{atm}] = -4.6 \left(\frac{T_b}{T} - 1\right) - 2.95 \left(\frac{T_m}{T} - 1\right) \tag{9.42}$$

$$\log P_v\,[\text{atm}] = -\frac{(88 + 0.4\,\tau + 1421\,\text{HBN}) + (91 - 1.2\,\tau)}{2.3\,R} \left(\frac{T_b}{T} - 1\right) + \frac{91 - 1.2\,\tau}{R} \log \left(\frac{T_b}{T}\right) - \frac{50 - 19.1 \log \sigma + 7.4\,\tau}{2.3\,R} \left(\frac{T_m}{T} - 1\right) \tag{9.43}$$

Yalkowsky and co-workers [145-147] elaborated the integration of the Clausius-Clapeyron equation further by introducing methods to estimate the phase-transfer entropies and heat capacities in a structure-specific manner. The resultant model contains a melting point term for solids and reads:

see previous page (9.43)

$n = 805$, $rms = 0.18$

where τ is the effective number of torsional bonds accounting for conformational flexibility, HBN encodes the number of hydrogen bonds associated with alcohol groups (OH), carboxylic groups (COOH) and primary amine groups (NH_2), and σ is the rotational symmetry number.

The EPISUITE model for predicting vapour pressure, MPBPVP [148], uses the Grain-Watson methods for solids, and the average of the Grain-Watson and Grain-Antoine predictions for liquids and gases. When no experimental boiling point is provided (which was the approach we selected), T_b is predicted using the Stein and Brown method [140] with some (unpublished) extensions in terms of additional fragments and correction factors. For solids, MBPVP offers one result employing an estimated melting point, and one with user-supplied experimental T_m. The MPBPVP manual [149] reports the following statistics:

MPBPVP (Grain-Watson + Grain-Antoin, T_b from Stein and Brown + extensions)

$n = 805$, $r^2 = 0.717$, $rms = 0.941$

ACD version 5.0 [150] contains a module to predict log P_v from molecular structure. The underlying methodology is not published and thus can not be described. There is no option to provide melting points for solids, and consequently this method could be applied only in the variant without experimental T_m.

While all the methods discussed contain a temperature-dependent term for P_v, their application for environmental chemicals has usually been confined to data at 25°C, as is the case in our present study. So far, little experience has been gained with the predictive quality of these methods for xenobiotics when considering temperature variation. However, for the group of hydrocarbons and halogenated hydrocarbons, a fragment scheme (23 substructural features) augmented by the standard melting point term (fugacity ratio, Equation 9.34, applying Walden's rule for ΔS_m) for solids has been combined with an artificial neural network to account for the non-linear dependence of P_v on temperature [151]. After collecting 8148 experimental P_v data for 1838 solids and liquids from the literature, calibration with 1200 compounds led to $r^2 = 0.995$ and $rms = 0.08$ log units, and external prediction to log $P_v(T)$ of the remaining 638 compounds yielded $q^2 = 0.990$ and $rms = 0.13$ log units.

Performance statistics

Our test set contains experimental P_v data at 25°C for 1672 compounds (780 solids with experimental melting points, 883 liquids, 9 gases) with a log P_v range from - 13.9 to 6.1 (P_v in Pa). Grain-Antoine ($n = 1622$, $r^2 = 0.87$, $rms = 1.37$) is superior to Grain-Watson ($n = 1601$, $r^2 = 0.85$, $rms = 1.76$) with respect to both statistics and the application range, and (except for a greater application range) is similar in performance to Sage-Grain-Watson ($n = 1601$, $r^2 = 0.89$, $rms = 1.34$). The Mackay model can be applied to still more compounds, but yields a significantly greater scatter ($n = 1639$, $r^2 = 0.90$, $rms = 2.33$). Sepassi-Yalkowsky ($n = 1618$, $r^2 = 0.90$, $rms = 1.25$) comes closest to MPBPVP ($n = 1649$, $r^2 = 0.92$, $rms = 1.13$) as the best-performing model overall for predicting log P_v at 25°C. MPBPVP with calculated T_m is the only method applicable to all compounds ($n = 1672$, $r^2 = 0.91$, $rms = 1.14$). ACD (the only method without use of T_m) cannot handle 54 compounds and provides statistics slightly inferior to Sepassi-Yalkowsky ($n = 1616$, $r^2 = 0.88$, $rms = 1.31$).

Note that all methods provide *rms* values above one log unit, which is much greater than most of the *rms* values reported for the individual model calibrations. At present, MPBPVP is recommended, also offering a variant where T_m values are missing.

9.3.5 Bioconcentration

Chemical substances may accumulate in organisms far beyond their average environmental concentration levels. The associated process is termed bioaccumulation, which in turn may include different mechanisms. Biomagnification relates to the compound uptake via food, and is quantified as the ratio of the compound concentration in the organism to the one in the diet. Bioconcentration takes place through direct uptake from the ambient environmental medium. In this section, we will confine ourselves to bioconcentration in aquatic systems.

For fish and other aquatic organisms, the

bioconcentration process can be described as phase transfer of the compound between aqueous solution and the organism. The bioconcentration factor, BCF, is defined as the ratio of the compound concentration in the organism, C_B (subscript B stands for biota), to the concentration in water, C_w at thermodynamic equilibrium:

$$BCF = \frac{C_B}{C_w} \tag{9.44}$$

Assuming first-order kinetics for the phase transfer of the compound between water and fish, the equilibrium constant BCF equals the ratio of the uptake rate constant, k_1 (uptake from water into fish), and the elimination rate constant, k_2 (elimination from fish to water):

$$BCF = \frac{k_1}{k_2} \tag{9.45}$$

(where k_1 and k_2 have the dimension time^{-1}). Equation 9.45 provides a kinetic definition of BCF, while Equation 9.44 represents its thermodynamic definition.

So far we have not specified the mechanisms of uptake and elimination that determine k_1 and k_2. In the simplest case, both uptake and elimination are treated as physical processes, confined to the respiratory pathway via gills, and to diffusion through the skin. Beyond this physical description of bioconcentration, the following additional processes are included in more elaborate treatments (see Chapter 3 for more details): uptake through the diet (k_D), elimination through a metabolic route (k_M), elimination through faecal egestion (k_E), and pseudo-elimination through growth dilution (k_G).

Generally, metabolism provides the most important factor in addition to respiratory (and diffusive) elimination, and typically reduces the BCF of metabolically active compounds. The effect can easily be seen upon introduction of k_M (elimination rate constant of the metabolic route) in the kinetic BCF definition:

$$BCF = \frac{k_1}{k_2 + k_M} \tag{9.46}$$

Because now both respiratory and metabolic elimination operate, the total elimination rate constant, $k_2 + k_M$, is larger than for the purely physical process, resulting in a corresponding lowering of the BCF [152,153]. The reduction in BCF can be traced back to the increased water solubility of the metabolites, which in turn are excreted more quickly than the more hydrophobic parent compound. As outlined below, the consideration of metabolism as a relevant additional elimination process has now entered into the QSAR modelling of BCF [154].

Experimental methods

OECD Guideline 305 [155] describes a flow-through fish test for the determination of BCF that can in principle also be applied to other aquatic test organisms (e.g., the oyster). The test consists of two phases: the exposure (uptake) and the post-exposure (depuration) phase. During the first part, separate groups of organisms of the same species are exposed to at least two aqueous concentrations of test substance until a steady-state is achieved. Then organisms are transferred for a depuration phase to water that contains no test chemical. The resulting uptake/elimination profiles obtained through successive sampling and analysis of organisms are used to estimate the corresponding rate constants and thus the BCF. To obtain meaningful results, care must be taken regarding the test organisms (species, weight, lipid content, tissue analyzed), exposure conditions (temperature control, pH, organic carbon content, etc.), analytical quality assurance and data evaluation (non-linear regression analysis). Recently, confidence criteria were published [156] for the evaluation of experimental BCF data.

Hydrophobicity model

The relationship between the bioconcentration potential of a compound and its hydrophobicity was introduced more than 30 years ago [157]. For a comprehensive account of this relationship and its limitations as well as of various extensions, the reader is referred to the literature [50-53,158-164].

The hydrophobicity model considers bioconcentration as compound partitioning from water into the lipid compartment of the organism. Further assumptions are that the phase transfer obeys first-order kinetics without physiological barriers, that the primary driving force of the diffusive uptake is the gradient between the compound concentrations in water and in the organism, that the only physiological factor affecting the bioconcentration factor is the lipid content of the organism, that metabolism and all other additional processes (uptake via food, faecal egestion, growth dilution) referred to above can be neglected, and that the partition equilibrium is determined only by the difference in the compound's affinity for water and octanol as a surrogate medium for the lipid phase, respectively. The relevant model ignores the compound concentration in the aqueous phase of the organism and therefore reads:

$$\log \mathrm{BCF} = a \cdot \log K_{ow} + b \quad (9.47)$$

Here, slope a is a measure of the suitability of octanol as a surrogate for the lipid compartment, and intercept b would equal the logarithm of the fractional lipid content and thus should be a negative number. Indeed, most (but not all) empirically calibrated log-log relationships between BCF and K_{ow} have slopes a in the value range 0.7 – 1, and intercepts b below 0 [51-53,165-167]. Where metabolism plays no important role, a linear increase in log BCF with increasing log K_{ow} is typically observed in the log K_{ow} range of 1-7 of neutral organic compounds [168].

Taking our regression equation confined to nonpolar narcotics (see Section 9.3.1, Equation 9.9) as an example: the slope of 1 suggests that for these types of compounds, octanol can be considered as a perfect surrogate of lipid with regard to membrane-water partitioning. The negative intercept (–1.25) corresponds to a lipid content of 5.6%, which is close to the estimate of 5% often used for fish. Thus, Equation 9.9 illustrates that nonpolar narcotics bioconcentrate in lipid membranes, and that the extent of their bioconcentration is governed by their hydrophobicity in terms of K_{ow}.

Mitigating factors

For more hydrophilic substances, the aqueous phase becomes more important as a storage compartment of the fish, resulting in a larger BCF than would be expected when considering the lipid compartment only. For compounds with log K_{ow} values above 7 that are also known as superhydrophobic compounds, log BCF no longer increases. Instead, there is evidence that in the high log K_{ow} range, log BCF decreases with increasing hydrophobicity, which was already noted many years ago [169-173]. Possible explanations for this deviation from the linear hydrophobicity model include:

- Decreased uptake rate because of steric hindrance due to large molecular size [173-175].
- Insufficient exposure time to reach equilibrium [152,176,177].
- Increased elimination rate through metabolic pathways [153,178,179].
- Increased elimination rate through faecal excretion [180,181].
- Limited fat solubility of highly hydrophobic compounds [182,183].
- Limited bioavailability due to sorption to dissolved or particulate organic matter [181,184-188].

In addition, growth dilution may play a role when considering the bioconcentration of strongly hydrophobic compounds in the field. Here, the BCF is affected by both the increased time needed to achieve partition equilibrium and the additional pseudo-elimination through the increase in body size over longer periods of time.

With respect to compound speciation in aqueous solution, bioconcentration refers to the freely dissolved compound fraction, which is understood to be represented by C_w (mol solute / L water). Here, three confounding factors require attention. First, for acids and bases their degree of ionization is governed by the difference between their pK_a and the ambient pH (Henderson-Hasselbalch equation). Because of the significant reduction in hydrophobicity upon dissociation or protonation, K_{ow} values would require an appropriate correction if still used to estimate the BCF of compounds with substantial degrees of ionization. A straightforward first-order approach would be to relate BCF to the unionized compound fraction (see Equations 9.4 and 9.5 in Section 9.3.1) as a bioavailable form. As pointed out earlier [45], however, literature BCF data on pentachlorophenol (log K_{ow} = 5.09, pK_a = 4.76) are significantly above values estimated from K_{ow} corrected for ionization, $D^u_{ow} = f_u K_{ow}$ (Equation 9.7 in Section 9.3.1, where f_u denotes the unionized compound fraction), suggesting that a more elaborate treatment is required to account for ionization when predicting BCF from hydrophobicity.

Second, for highly hydrophobic compounds sorption to dissolved or particulate organic matter (see Section 9.3.3 above) may compete with bioconcentration to a significant degree, which in turn depends on the amount of sorbent matrices prevalent in aqueous solution. In this case, the volume fraction of freely dissolved compound, f_w, is approximately given by Equation 9.48 (see below) where $[S]$ denotes the sorbent concentration in aqueous solution (kg sorbent / L water). On the right-hand side of Equation 9.48, the Karickhoff relationship (Equation 9.26) has been introduced to estimate K_{oc} through K_{ow}.

$$f_w = \frac{1}{1 + K_d[S]} = \frac{1}{1 + f_{oc}K_{oc}[S]} \approx \frac{1}{1 + 0.41\, f_{oc}K_{ow}[S]} \quad (9.48)$$

Third, experimental measurements above the water solubility result in too low BCF values if these are based on nominal concentrations. Because of this problem, BCF data for superhydrophobic compounds are frequently questionable, and here kinetic measurements should be used employing compound concentrations below their water solubility.

From the chemical perspective, areas to extend the hydrophobicity model are the consideration of bioavailability in terms of sorption and ionization, lipid solubility, molecular size as well as of biotransformation, which obviously also depends on the metabolic capacity of the organism of interest. As outlined below, ways to account for molecular size, ionization and metabolism have now been introduced into the QSAR prediction of BCF.

Prediction methods

The most prominent linear relationship between log BCF and log K_{ow} is the one-parameter equation derived by Mackay [50]:

$$\log BCF = \log K_{ow} - 1.32 \qquad (9.49)$$

$n = 44$, $r^2 = 0.95$, $rms = 0.25$

which assumes that octanol is a perfect surrogate for fish lipid (slope $a = 1$). The intercept of –1.32 corresponds to a lipid content of 4.8%. This Mackay equation is often used to estimate the physical bioconcentration potential of organic xenobiotics (keeping in mind that the hydrophobicity model systematically overestimates the BCF of metabolizing compounds). It is included as a default setting in various multimedia fate models such as CalTOX [189]. While Equation 9.49 is very suitable for predicting BCF values of baseline narcotics (see above), it was not designed to account for mitigating factors such as steric hindrance of membrane penetration or metabolism. Here, the hydrophobicity model provides an estimate of the maximum bioconcentration that would be expected from physical chemistry, while still disregarding the constraints of molecular size, lipid solubility and bioavailability.

The earlier Veith relationship [190] to estimate log BCF:

$$\log BCF = 0.85 \log K_{ow} - 0.70 \qquad (9.50)$$

$n = 55$, $r^2 = 0.90$

is used as default in EUSES (European Union System for the Evaluation of Substances) [191] for substances with log K_{ow} values up to 6. For more hydrophobic compounds, EUSES employs the parabolic equation:

see below (9.51)

$n = 43$, $r^2 = 0.78$

to mimic the observed reduction in BCF for superhydrophobic compounds.

Several other models have addressed the BCF decline at high log K_{ow} through non-linear regression equations with log K_{ow} as the only descriptor. The Bintein model [192] applied the bilinear approach [193-195] to derive the following equation:

see below (9.52)

$n = 154$, $r^2 = 0.90$, $rms = 0.347$, $F = 464$

The experimental data used for the calibration of Equation 9.52 refer to BCF values of five different families of freshwater fish. Dimitrov et al. [196] developed a non-linear model from BCF data of non-polar and polar narcotics through calibration of a Gaussian function:

see below (9.53)

$n = 443$, $r^2 = 0.73$, $rms = 0.65$, $me = 0.46$

The reasoning behind the compound selection was that in contrast to reactive toxicants, narcotic-type compounds reach steady-state concentrations in fish, which agrees

$$\log BCF = -0.20 (\log K_{ow})^2 + 2.74 \log K_{ow} - 4.72 \qquad (9.51)$$

$$\log BCF = 0.91 \log K_{ow} - 1.975 \log (6.8 \cdot 10^{-7} K_{ow} + 1) - 0.786 \qquad (9.52)$$

$$\log BCF = 3.321 \exp\left(-[\log K_{ow} - 6.348]^2 / 10.151\right) + 0.420 \qquad (9.53)$$

with partitioning as the driving force of bioconcentration. Among the 443 compounds were hydrocarbons, halogenated hydrocarbons, esters and amines, as well as polar narcotic phenols and anilines. In the following subsection discussing prediction performances, this model is called Dimitrov-Mekenyan model 1.

A somewhat different approach was taken by the group of Howard [168]. Subdivision of the compounds into subsets of non-ionic and ionic substances, and introduction of correction factors in terms of indicator variables I_k associated with substructural features, resulted in the following set of equations which represent the BCFWIN model [197]:

Non-ionic compounds:

see below (9.54)

$n = 610$, $r^2 = 0.73$, $rms = 0.67$, $me = 0.48$

where $n_k = 0$ or 1 reflect the absence or presence of the k-th substructural feature, respectively.

Ionic compounds (carboxylic acids, sulfonic acids and salts, quaternary N compounds):

$$
\begin{array}{ll}
\log K_{ow} < 5 & : \log \mathrm{BCF} = 0.50 \\
\log K_{ow} = 5..< 6 & : \log \mathrm{BCF} = 0.75 \\
\log K_{ow} = 6..< 7 & : \log \mathrm{BCF} = 1.75 \\
\log K_{ow} = 7..< 9 & : \log \mathrm{BCF} = 1.00 \\
\log K_{ow} > 9 & : \log \mathrm{BCF} = 0.50 \\
\text{alkyl chain} \geq C_{11} & : \log \mathrm{BCF}_{min} = 1.85
\end{array} \quad (9.55)
$$

$n = 84$, $r^2 = 0.62$, $rms = 0.41$, $me = 0.31$

Total statistics for non-ionic and ionic compounds:

$n = 694$, $r^2 = 0.74$, $rms = 0.65$, $me = 0.47$

Of the 12 correction terms used in Equation 9.54, nine decrease the BCF by –0.32 to –1.65 log units, and three increase the BCF by 0.48 to 1.40 log units. While these corrections were derived empirically through inspection of deviations from an initial linear equation fitted to a subset of non-ionic substances with log K_{ow} values in the range from 1 to 7:

$$\log \mathrm{BCF} = 0.86 \log K_{ow} - 0.39 \quad (9.56)$$

(no statistics given)

some of them reflect the reduction in bioconcentration due to known biotransformation reactions associated with certain functional groups. Note further that during model development, pK_a turned out to be statistically not significant as further model parameter.

An alternative to predicting the actual BCF is to predict its potential maximum value. As indicated above, the hydrophobicity model can also be regarded as predictor of maximum BCF (BCF_{max}) that equals the actual BCF only for compounds without mitigating factors (such as non-polar narcotics) in the log K_{ow} range from about 1-7. More elaborate models to quantify the maximum BCF over the whole log K_{ow} range are those of Nendza [166] and the Mekenyan group [198]. The Nendza model represents a bilinear equation:

see below (9.57)

constructed (without regression and thus without calibration statistics) such that the associated curve is located in the upper margin of the data distribution of log BCF vs. log K_{ow} for 132 compounds. The Dimitrov-

$$
\begin{array}{ll}
\log K_{ow} < 1 & : \log \mathrm{BCF} = 0.50 \\
\log K_{ow} = 1 \cdot\cdot 7 & : \log \mathrm{BCF} = 0.77 \log K_{ow} - 0.70 + \sum n_k I_k \\
\log K_{ow} > 7 & : \log \mathrm{BCF} = -1.37 \log K_{ow} + 14.4 + \sum n_k I_k \\
\log K_{ow} > 10.5 & : \log \mathrm{BCF} = 0.50 \\
\text{aromatic azo compound} & : \log \mathrm{BCF} = 1.0 \\
\text{organic tin and mercury} & : \log \mathrm{BCF}_{min} = 2.0
\end{array} \quad (9.54)
$$

$$\log \mathrm{BCF}_{max} = 0.99 \log K_{ow} - 1.47 \log (4.97 \cdot 10^{-8} K_{ow} + 1) + 0.0135 \quad (9.57)$$

Mekenyan BCF_{max} model was derived through re-calibration of their non-linear BCF model above to observed BCF_{max} values:

see below (9.58)

$n = 81$, $r^2 = 0.96$, $rms = 0.20$

Here, a set of compounds with upper borderline BCF values were selected such that for each K_{ow} interval of 0.5 log units, compounds with BCF values within 0.5 log units from the maximum BCF observed for the respective log K_{ow} interval were included in the training set.

In Figure 9.5, the curves of the linear Mackay model (Equation 9.49), the partly non-linear EUSES model (log K_{ow} 6: Equation 9.50 log K_{ow} > 6: Equation 9.51 [191]), and the non-linear BCF_{max} models of Nendza (Equation 9.57) and Dimitrov-Mekenyan (Equation 9.58) are plotted against log K_{ow} together with experimental log BCF data for 691 compounds taken from the EPISUITE database [149]. By deduction, all non-linear models predict a BCF decrease in the high log K_{ow} range. However, the bilinear model tends to overestimate BCF_{max} above a log K_{ow} of 5, while the Gaussian model accepts some underestimates, arguing that at least some of the excess BCF values are artefacts due to measurements based on radio-labelled compounds [154]. Note further that only the Gaussian model accounts for the decrease in slope of the data pattern in the range of low log K_{ow} values. Moreover, the only non-linear model in Figure 9.5 designed to predict BCF (and not BCF_{max}) is the EUSES model.

Combining the model for multicompartment diffusion with mitigating factors accounting for molecular size (including consideration of conformational flexibility) and ionization, as well as for biotransformation, resulted in a novel BCF prediction model [154]:

see below (9.59)

$n = 511$, $r^2 = 0.84$, $rms = 0.542$

In this Dimitrov-Mekenyan model 2, F_M is a factor reducing the BCF due to metabolism (which in turn is estimated from a library of biotransformation reactions with associated probabilities), $F_w = 6.22$ (related to the water content of the organism), F_i represents the mitigating factors referred to above of ionization (F_{acid}) and molecular size (F_{size}), and $a = 4.24 \cdot 10^{-7}$ and $n = 0.774$ are further model parameters. An applicability domain was also developed for this model according to a scheme introduced separately [18].

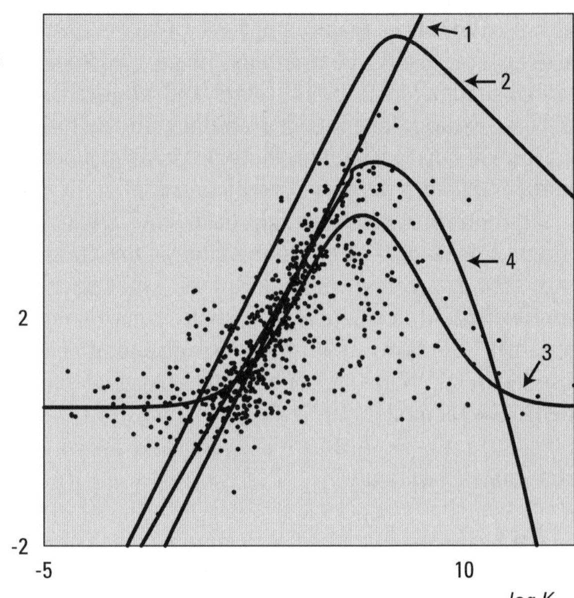

Figure 9.5. Log BCF vs. log K_{ow} for 694 compounds.
1 = Mackay relationship (Equation 9.49) [50]
2 = Bilinear Nendza model for log BCF_{max} (Equation 9.57) [166]
3 = Gaussian Dimitrov-Mekenyan model for log BCF_{max} (Equation 9.58) [196]
4 = EUSES model (Equations. 9.50, 9.51) [191]

Performance statistics

For the present comparative analysis, 691 compounds with experimental log BCF data from –1.43 to 5.59 were selected from the EPISUITE database [149]. Note that this data set is probably almost identical to the one used as the training set for the BCFWIN model [197], while

$$\log BCF_{max} = 3.93 \exp\left(-[\log K_{ow} - 6.61]^2 / 11.9\right) + 0.931 \quad (9.58)$$

$$\log BCF = \log\left(F_M \frac{K_{ow}^n}{(a\,K_{ow} + 1)^{2n}} - F_w\right) + \sum_i \log F_i \quad (9.59)$$

other BCF models discussed below were calibrated with somewhat different data sets.

The partly linear, partly parabolic EUSES model (n = 689, $r^2 = 0.45$, $rms = 1.41$), and bilinear Bintein model (n = 689, $r^2 = 0.44$, $rms = 1.37$) yield poor statistics, while the Gaussian-type Dimitrov-Mekenyan model 1 as a third method, employing only log K_{ow}, provides a significantly reduced scatter but still only moderate statistics ($n = 689$, $r^2 = 0.55$, $rms = 0.87$).

The best overall statistics are achieved with BCFWIN ($n = 690$, $r^2 = 0.73$, $rms = 0.67$) and the Dimitrov-Mekenyan model 2 ($n = 691$, $r^2 = 0.72$, $q^2 = 0.72$, rms = 0.66). Restriction of the latter model to its applicability domain yields significantly improved results (n = 448, $r^2 = 0.78$, $rms = 0.56$), but reduces the number of compounds by 35%.

Overall, the results demonstrate that for various compounds, consideration of mitigating factors improve the prediction performance significantly. The three non-linear K_{ow}-based models are significantly inferior to BCFWIN and the Dimitrov-Mekenyan 2 model, and EUSES yields the overall largest standard errors. The results further demonstrate that for predictive BCF assessment, mitigating factors such as biotransformation should be taken into account.

9.3.6 Water solubility

For organic compounds, the maximum amount that can be dissolved in the aqueous phase differs by several orders of magnitude. The water solubility (S_w) of a compound is the highest equilibrium concentration it can achieve as a dissolved species in aqueous solution. If the compound concentration in the water phase, C_w, has reached its maximum value, S_w, the solution is saturated with respect to this substance. In the saturated state, the solute is in equilibrium with its pure compound phase. A further increase of the compound amount beyond S_w will lead to the formation of this pure compound phase as a separate phase.

A more precise description of S_w would take into account the fact that both the organic and the water phase will dissolve in each other until the chemical potentials of the mutually saturated phases are equal in both phases. Accordingly, the separate organic phase formed through solute amounts above S_w may contain some amount of water. A well-known example is the octanol-water system used to quantify molecular hydrophobicity in terms of K_{ow} (see Section 9.3.1 above). At 25°C, octanol in equilibrium contact with water contains 2.3 mol/L water, and the water phase is saturated with $4.5 \cdot 10^{-3}$ mol/L octanol [199].

Relationship with environmental processes

The water solubility of organic compounds is an important determinant of their mobility in the environment. High S_w facilitates transport through the hydrological cycle, and generally biodegradation as well as abiotic degradation processes such as hydrolysis, oxidation and aquatic photolysis for waterborne compounds (see also Section 9.4 below). Because increasing S_w often (but not always) implies decreasing hydrophobicity in terms of K_{ow}, as discussed below, high S_w typically indicates a low tendency for sorption into suspended or particular organic matter, and a low tendency for bioconcentration in aquatic organisms. High water solubility is also correlated with low baseline toxicity towards aquatic species in a way that corresponds to the K_{ow}-based narcosis relationship (Equation 9.11, Section 9.3.1), keeping in mind that reactive and specific mechanisms may result in high aquatic toxicity of hydrophilic compounds (see Chapter 10). Moreover, water solubility is inversely related to Henry's law constant ($H = P_v/S_w$, Equation 9.16 in Section 9.3.2), which in turn governs the volatilization of waterborne solutes into air.

Experimental methods

For the determination of water solubility, preparation of the saturated solution is important. Three procedures are recommended depending on the S_w range in which the chemical under investigation is supposed to belong [200]: a) the direct dissolution method (with shaking/stirring) for substances with a $S_w > 10$ mg/L, b) the glass-beads method for 10 µg/L < S_w < 10 mg/L, where the test substance is coated to glass beads before adding (in excess) to agitated water, and c) the column elution method for $S_w < 10$ µg/L. OECD guideline 105 [201] describes regular use of methods a) and c). The generator column technique is a synonym for method c). Here, a suitable sorbent is coated with a layer of the test substance to enlarge the interfacial area for the dissolution process. For more information, particularly on the design and accurate use of a generator column apparatus for measuring extremely low aqueous solubilities of highly hydrophobic compounds, the reader is referred to [202,203].

Organic liquids with negligible water solubility

Consider an aqueous solution saturated with an organic liquid. At thermodynamic equilibrium, the chemical potential of the solute in the dissolved state equals its chemical potential in its organic liquid phase. A corresponding relationship holds for the solute activities in the two phases, a_w and a_L, resulting in:

$$\gamma_w x_w = \gamma_L x_L \tag{9.60}$$

for the activity coefficients in water and the liquid phase, γ_w and γ_L, and the associated mole fractions x_w and x_L, respectively. Where the solubility of water in the organic phase is negligible, the latter can be treated as pure compound phase with $x_L = 1$ and $\gamma_L = 1$ (applying Raoult's law mole-fraction based standard states), and thus also $a_L = 1$. Therefore:

$$x_w = \frac{1}{\gamma_w} \tag{9.61}$$

which can be expressed in logarithmic form as:

$$\log x_w = -\log \gamma_w \tag{9.62}$$

It follows that for organic liquids which exclude water from their own phase, the mole fraction solubility will be inversely related to the activity coefficient at saturated solution. Because water usually enters the organic phase in sometimes considerable concentrations (e.g., 2.3 mol/L water in octanol as mentioned above), the more hydrophobic the solute is the better Equations 9.61 and 9.62 hold true. Note that a more precise notation would refer explicitly to the state of the saturated solution, which we have omitted for the sake of simplicity.

Prediction of the water solubilities of organic liquids thus corresponds to predicting the inverse of their activity coefficients (at saturation) in aqueous solution. For liquids with relatively low water solubilities, x_w can be converted to C_w [mol/L] through the molarity of water (M_w = 55.56 mol/L = 1000 g/L divided by 18 g/mol):

$$C_w\,[\text{mol/L}] \approx x_w\, 55.56 \tag{9.63}$$

which leads to:

$$\log S_w\,[\text{mol/L}] = -\log \gamma_w^{sat} + 1.74 \tag{9.64}$$

where γ_w^{sat} denotes the activity coefficient of the solute at saturation concentration (which is now indicated by the superscript *sat*).

For sparingly soluble compounds, the limiting activity coefficient, γ_w^{∞}, (that refers to $x_w \rightarrow 0$) can be taken as an approximation of γ_w^{sat}. Thus, Equation 9.64 provides the thermodynamic basis for predicting water solubility through γ_w^{∞} [204], which can then be estimated with UNIFAC [205-207].

Organic liquids with no solubility limit

Ethanol and aceton are examples of liquids that are completely miscible with water. In this case, the solute mole fraction in water may reach the value 1 (where $x_w = x_L$, because there is no more water present), and consequently there is no well-defined S_w value. With such liquids, their hydrophobicity is better characterized through their limiting activity coefficients γ_w^{∞} that in this case are typically below 20 [71].

In practice, methods to predict the water solubility of such liquids may still be applied and would result in unusually high S_w values that may well exceed the molarity (inverse of molar volume) of the compounds (and thus their maximum amount in 1 L volume). Thus, a safeguard would be to cross-check prediction results with the solute's molarity, that in turn could be obtained through estimation of the molar volume [208] or via molecular weight and density which could also be estimated from calculation methods [208] or (if not applicable) through taking a rough average value for organic liquids.

Organic solids

As mentioned above, the dissolution of a solid can be broken down conceptually into two steps. First, melting of the solid at fixed temperature generates the subcooled liquid that may or may not be attainable physically. Second, mixing of the subcooled liquid with water (or any other solvent) yields the final solution. Equation 9.61 could be applied to the latter, now employing a superscript L to indicate this particular (hypothetical) subcooled liquid state of the solid:

$$x_w^L = \frac{1}{\gamma_w^L} \tag{9.65}$$

Becoming solid again, its solubility is reduced compared to its subcooled liquid state through the solid-liquid transfer also required. This reduction in solubility can be quantified by the ratio of the fugacities of the (pure) solid, f^s, and subcooled liquid, f^L, giving the overall mole fraction water-solubility of the solid, x_w, as:

$$x_w = \frac{1}{\gamma_w^L}\,\frac{f^s}{f^L} \tag{9.66}$$

with:

$$\log \frac{f^s}{f^L} = -\frac{\Delta S_m}{2.3R}\left(\frac{T_m}{T} - 1\right) \tag{9.67}$$

Equation 9.67 is essentially equivalent to Equation 9.34 (Section 9.3.4) that was used to address the solid-liquid transfer when evaluating vapour pressure (in Equation 9.34, f^s and f^L were approximated by the corresponding vapour pressures P_v^s and P_v^L of the solid and subcooled liquid state of the compound). The fugacity ratio in terms of Equation 9.67 further shows that in the two-step description of the dissolution of solid compounds, the generation of the subcooled liquid is not affected by solute-solvent interactions.

Introduction of Equation 9.67 into the logarithmic form of Equation 9.66 gives a relationship for the mole fraction solubility of the solute, x_w^{sat}, in water:

$$\log x_w^{sat} = -\log \gamma_w^{sat} - \frac{\Delta S_m}{2.3RT}(T_m - T) \qquad (9.68)$$

(here, γ_w^{sat} refers to the subcooled liquid state of the solid, although we have omitted the additional superscript L in order to avoid making the notation overly complex), and the conversion from mole fraction to mol/L (Equation 9.63) yields the corresponding expression for log S_w:

see below (9.69)

Equations 9.68 and 9.69 show the two components affecting the water solubility of solids. The phase transfer to the subcooled liquid state is governed by the entropy of melting and the difference between the melting point and the temperature of interest, and the liquid phase activity coefficient in aqueous solution determines the extent of subsequent mixing.

For the practical application of Equation 9.69, a common approximation is to introduce $\Delta S_m = 56.5$ J/(mol K) as an average value estimated according to Walden's rule [135], which was mentioned already in the context of predicting the vapour pressure of solids (Section 9.3.4). In this context, the term $\Delta S_m/(2.3\ RT)$ becomes 0.0099 0.01 at 25°C. Within structurally related compounds, water solubility decreases with increasing melting point, which increases with increasing molecular weight. Because high symmetry is usually associated with relatively high melting points, the water solubility of high-symmetry isomers is lower than that of their low-symmetry counterparts.

The log S_w prediction for solids via γ_w^∞ (of the subcooled liquid) and T_m is based on Equation 9.69, and typically applies Walden's rule for the melting point term [204-207]. As noted above, these methods rely on UNIFAC to estimate the limiting activity coefficient γ_w^∞.

Relationship with log K_{ow}

A prominent class of methods for predicting water solubility is based on log-log regression equations with the *n*-octanol-water partition coefficient [209]. The observed negative correlation between log S_w and log K_{ow} has a simple theoretical basis [210].

The usual definition of K_{ow} in terms of the solute concentrations in octanol (C_o) and water (C_w) ignores the fact that apart from the solute, the total molarity of wet octanol, $M_{o(w)}$, is greater than that of pure octanol, M_o. The latter is 6.36 mol/L (density 827 g/L divided by molecular weight 130 g/mol), and addition of 2.3 mol/L water results in $M_{o(w)} = 8.36$ mol/L (ignoring the impact of water on the density of the octanol phase). At low solute concentrations, C_o $x_o \cdot M_{o(w)}$ and C_w $x_w \cdot M_w$ (see Equation 9.63), and further elaboration of the K_{ow} ratio for liquid solutes yields:

see below (9.70)

In Equation 9.70, it is further assumed that $x_w = 1/\gamma_w$ (Equation 9.61) and a corresponding relationship for the octanol phase apply, and that the octanol phase forms an ideal solution for the solute ($\gamma_o = 1$). Approximation of γ_w through its value at saturation, γ_w^{sat}, and inclusion of a melting point term for solids finally yields:

see below (9.71)

where MP denotes the melting point in °C (as opposed to T_m that refers to Kelvin). Note that in the literature, Equation 9.71 often has the slightly different intercept 0.8 [211]. The reason for this is that in this case, the dry octanol molarity (6.36 mol/L) is used for the octanol

$$\log S_w\ [\text{mol/L}] = -\log \gamma_w^{sat} - \frac{\Delta S_m}{2.3RT}(T_m - T) + 1.74 \qquad (9.69)$$

$$K_{ow} = \frac{C_o}{C_w} \approx \frac{x_o M_{o\,(w)}}{x_w M_w} \approx \frac{\gamma_w \cdot (6.36 + 2.3)}{\gamma_o \cdot 55.56} \approx \gamma_w \cdot 0.156 \qquad (9.70)$$

$$\log S_w\ [\text{mol/L}] = -\log K_{ow} - 0.01\,(\text{MP} - 25) + 0.92 \qquad (9.71)$$

phase, which results in $K_{ow}\ \gamma_w \cdot 0.115$ and log (6.36) = 0.8 as the intercept.

The evaluation of K_{ow} according to Equation 9.70 forms the theoretical basis for most regression equations to predict log S_w from log K_{ow}. A slightly different approach was taken in the more recent work of Yalkowsky and co-workers [212]. The approximation of K_{ow} through the corresponding solubility ratio and assuming that the solute has a mole fraction of 0.5 in the octanol phase as well as a molecular weight and density identical to octanol leads to:

$$S_o\ [\text{mol/L}] = \frac{6.36}{2} = 3.18 \tag{9.72}$$

and:

see below (9.73)

The only difference with Equation 9.71 is the lower intercept (0.5 = log 3.18 vs. 0.92). Equation 9.73 is now called the general solubility equation [212-214].

Prediction methods

In the literature, a broad variety of water solubility prediction methods is discussed in several reviews [71,215-217].

In the following, we focus on five method types. The first approach [204,205,207] makes use of the inverse relationship between the compound mole fraction and its activity coefficient at saturation (Equation 9.61). It replaces the activity coefficient at saturation by the activity coefficient at infinite dilution, γ_w^∞, applies Walden's rule to provide an estimate of the average entropy of melting in case of solids, and uses UNIFAC to predict γ_w^∞ from the molecular structure. Among the different methods available in the literature [204,205,207], the Banerjee regression model yielded the best overall performance when combined with the Tiegs UNIFAC Revision 4 parameters [218] to predict γ_w^∞ [205]. The relevant equation for log S_w at 25°C reads:

see below (9.74)

$n = 50$, $r^2 = 0.96$

where (as usual) the melting point term applies only to solids, and UNIFAC is used to predict γ_w^∞ from molecular structure.

The second method type employs log K_{ow} to predict log S_w, which was originally introduced by Hansch and co-workers for organic liquids [209]. Here, the most recent model is given by the general solubility equation (GSE, Equation 9.73) of Jain & Yalkowsky [212].

EPISUITE contains several variants of the log K_{ow} approach to predict log S_w, considering models with and without MP as well as with and without correction factors C_i associated with certain structural features [219]. For a set of 1450 compounds, the final regression equations with MP, with molecular weight (MW) instead of MP, and with both MP and MW were:

see below (9.75)

$n = 1450$, $r^2 = 0.960$, $rms = 0.452$, $me = 0.348$

see below (9.76)

$n = 1450$, $r^2 = 0.934$, $rms = 0.585$, $me = 0.442$

see below (9.77)

$n = 1450$, $r^2 = 0.970$, $rms = 0.409$, $me = 0.313$

In each of the three equations, differently calibrated correction factors C_i are applied (12 terms for Equations 9.75 and 9.77 and 14 terms for Equation 9.76).

$$\log S_w\ [\text{mol/L}] = -\log K_{ow} - 0.01\ (\text{MP} - 25) + 0.5 \tag{9.73}$$

$$\log S_w\ [\text{mol/L}] = -0.782 \log \gamma_w^\infty - 0.01\ (\text{MP} - 25) + 1.2 \tag{9.74}$$

$$\log S_w\ [\text{mol/L}] = -1.0374 \log K_{ow} - 0.0108\ (\text{MP} - 25) + \sum C_i + 0.342 \tag{9.75}$$

$$\log S_w\ [\text{mol/L}] = -0.854 \log K_{ow} - 0.00728\ \text{MW} + \sum C_i + 0.796 \tag{9.76}$$

$$\log S_w\ [\text{mol/L}] = -0.96 \log K_{ow} - 0.0092\ (\text{MP} - 25) - 0.00314\ \text{MW} + \sum C_i + 0.693 \tag{9.77}$$

Recently, ESOL was introduced as further model combining log K_{ow} with additional molecular structure information [220]. The final equation employs only three parameters in addition to (the calculated) log K_{ow}: molecular weight, number of rotatable bonds (RB), and aromatic proportion (AP = proportion of heavy atoms in aromatic rings):

see below (9.78)

$n = 2874, r^2 = 0.72, rms = 0.97, F = 1865$

The compound set covered conventional environmental chemicals as well as larger numbers of pesticides and agrochemicals or agrochemical candidates (Syngenta in-house compounds). Initial training with 1305 compounds yielded r^2, *rms* and *me* values of 0.69 as well as 1.01 and 0.75 log units for ESOL, and 0.67 as well as 1.05 and 0.81 log units for the general solubility equation (GSE) that was run in comparison [220].

The commercial software package ACD [150] contains a module to predict log S_w. According to the ACD manual, the method consists of a suite of equations derived for individual classes of compounds. Besides log K_{ow}, the following additional parameters are used: boiling point, molecular weight, volume, number of H-bond donor and acceptor sites, and refractive index. Moreover, an additional consideration of internal H bonds is mentioned.

The third model type are fragment methods predicting log S_w directly from molecular structure. The model equation of AQUAFAC [221-225] contains 69 fragments (F_i) and 2 correction terms (C_j) and can be written as:

see below (9.79)

$n = 1296$ (no statistics given)

where the melting point term applies only to solids, n_i and n_j denote the number of occurrences of F_i and C_j, respectively, and ΔS_m (entropy of fusion at MP) is estimated through the rotational symmetry σ if possible, otherwise by applying Walden's rule.

Besides the previously mentioned WSKOWWIN models, EPISUITE [149] contains WATERNT [226] as an additional log S_w prediction method that employs fragments and correction factors. According to the manual, the methodology used for the derivation of WATERNT is the same as that used for their KOWWIN method [59], with the following statistics for a training set of 1000 compounds and a validation set of 3923 compounds:

see below (9.80)

training: $n = 1000, r^2 = 0.975, rms = 0.336, me = 0.18$
validation: $n = 3923, r^2 = 0.86, rms = 0.869, me = 0.70$

Model calibration was performed with 1000 compounds, and a later validation set contained 3923 compounds, but the model parameters are unpublished.

The final method included in the group of fragment schemes is ADME/logS [62], another commercial software package. The underlying model is based on molecular fragments with chain lengths of 3-5. It was derived using an automated algorithm with a training set of 7000 compounds.

Tetko [227] used 33 electrotopological indices (E-states) for a non-linear model employing an artificial neural network (NN) (33-4-1 architecture). While this NN model is not described in full detail, it can be used online via internet through the ALOGPS software [60], and the reported statistics [227] are:

33 E-states, NN architecture 33-4-1
training: $n = 879, r^2 = 0.95, q^2_{cv} = 0.91, rms = 0.47, rms_{cv} = 0.62$
validation: $n = 412, r^2 = 0.92, rms = 0.60$

The final method type considered is the LSER approach.

$$\log S_w\ [\text{mol/L}] = -0.63 \log K_{ow} - 0.0062\ \text{MW} + 0.066\ \text{RB} - 0.74\ \text{AP} + 0.16 \quad (9.78)$$

$$\log S_w\ [\text{mol/L}] = -\sum_i n_i F_i - \sum_j n_j C_j - \frac{\Delta S_m}{5.71}(\text{MP} - 25) \quad (9.79)$$

$$\log S_w\ [\text{mol/L}] = \sum_i n_i F_i + \sum_j n_j C_j + 0.24922 \quad (9.80)$$

The LSER analysis of water solubility goes back to early work [228,229] that has also been criticized because of the varying definition and statistical significance of some LSER parameters used at the time [230]. In 1999, Abraham and Le [231] introduced the following model to predict log S_w:

see below (9.81)

$n = 659$, $r^2 = 0.920$, $rms = 0.557$, $F = 1256$

According to Equation 9.81, water solubility increases with increasing H-bond donor and H-bond acceptor capability of the solute (A and B), and with increasing dipolarity/polarizability (B). At the same time, increasing solute volume (V) and increasing excess molar refraction (E) decrease the solubility in water. For bipolar solutes that act both as H-bond donors and acceptors, respective donor-acceptor interactions in the solute phase (liquid or solid) reduce water solubility (interaction term $A \cdot B$). Note that in the original publication [231], the former notation of the LSER parameters was used (see Section 9.2 above). Interestingly, the melting point term gave no substantial improvement of the prediction capability, and so was not included in the final model equation.

Besides relying on a particular method, an alternative strategy is to select for each compound of a given list the presumed best-performing method. A simple but powerful set of selection criteria can be applied for this, as follows [232]. First, for every compound, a set of k (in our case: $k = 5$) structurally most similar compounds is selected from a database with experimental values. Second, the performances of the methods under consideration are tested for the local subsets associated with each of the compounds through comparison of calculated and experimental values. Third, the method showing the best performance on the structurally most similar compounds of a given compound of interest is used for the prediction of its water solubility, and this procedure is applied to all compounds for which S_w values are needed. An appropriate algorithm has been computerized, and the overall improvement in prediction capability was demonstrated [232] with seven literature methods for a data set with 1876 compounds (which was also used for the present study).

Performance statistics

Our test set contained 1876 compounds, and was taken mainly from earlier compilations [231,233,234]. UNIFAC can be applied to only 606 compounds (32%) and thus is not a contender because of its severely restricted application range. The reason is that for many compounds, the relevant UNIFAC interaction terms to predict the activity coefficient at infinite dilution, γ_w^∞, are missing.

Among the methods employing log K_{ow}, the EPISUITE WSKOWWIN models provide the best performance (without MP: $n = 1876$, $r^2 = 0.87$, $rms = 0.76$; with MP: $n = 1766$, $r^2 = 0.91$, $rms = 0.66$), and is only slightly inferior to ACD that uses an unpublished algorithm ($n = 1875$, $r^2 = 0.89$, $rms = 0.69$). The general solubility equation (GSE) of Yalkowsky and Jain ($n = 1766$, $r^2 = 0.84$, $rms = 0.86$) is inferior to WSKOWWIN, except for the application range superior to ESOL ($n = 1876$, $r^2 = 0.79$, $rms = 0.98$) that uses no melting point term.

ADME/logS (that employs an unpublished algorithm) shows the best overall performance of the fragment methods ($n = 1876$, $r^2 = 0.90$, $rms = 0.67$) and can be applied to all 1876 compounds. The next best statistics are achieved by WATERNT ($n = 1876$, $r^2 = 0.84$, $rms = 0.88$) that is also not published in detail. AQUAFAC can handle only 76% of the compounds, and in this respect is inferior to the other fragment methods ($n = 1417$, $r^2 = 0.87$, $rms = 0.83$).

The Tetko neural network model ALOGS [227] employing 33 E-states yields the best overall statistics ($n = 1875$, $r^2 = 0.89$, $rms = 0.66$) and a still slightly better result when restricting the predictions to the model-specific application domain ($n = 1828$, $r^2 = 0.90$, $rms = 0.66$).

The last method type considered is the LSER approach. The use of experimental Abraham parameters is possible for only 36% of the compounds ($n = 671$, $r^2 = 0.95$, $rms = 0.56$). With LSER-Platts (employing published increment methods [65] to predict the Abraham parameters) and LSER-ABSOLV (that employs respective but unpublished prediction methods [62]), only moderate to poor results are achieved (Platts: $n = 1874$, $r^2 = 0.54$, $rms = 1.73$; ABSOLV: $n = 1876$, $r^2 = 0.58$, $rms = 1.60$).

$$\log S_w\ [\text{mol/L}] = -1.004\,E + 0.771\,S + 2.168\,A + 4.238\,B - 3.362\,(A \cdot B) - 3.987\,V + 0.518 \qquad (9.81)$$

According to the present analysis, the generally best results can be expected with the log K_{ow}-based models WSKOWWIN and ACD, the fragment models ADME and WATERNT, and the E-state neural network ALOGPS. Among these models, WSKOWWIN is the only method with a published algorithm. Because WSKOWWIN-MP is superior to WSKOWWIN-MW as noted earlier [219], experimental melting point information should be used if available, and in this case the general solution equation (GSE) is preferred over all other fully published methods except WSKOWWIN.

9.4 DEGRADATION

9.4.1 Indirect photolysis in the air

At the earth's surface, the sunlight spectrum ranges from 290 nm to about 1500 nm. In the UV (low-wavelength end) and visible region, the radiation energy corresponds to $n \rightarrow \pi^*$, $\pi \rightarrow \pi^*$ and $n \rightarrow \sigma^*$ transitions between respective electronic energy levels of molecules. For most organic compounds, however, the absorption intensity for solar radiation is low, and so direct photolysis plays only a minor role as an abiotic degradation pathway of organic contaminants. By contrast, the tropospheric lifetime of organic compounds is largely governed by reactions with oxidants, which are generated through photochemical pathways. Only a few compound classes are inert under tropospheric conditions, especially the perhalogenated alkanes.

Tropospheric oxidants

Oxidants are electrophilic compounds that gain electron density upon reaction with a substrate. In the troposphere, generally the most important oxidative agent attacking organic compounds is the hydroxyl radical, $OH\cdot$ [101,235,236]. For substances with unsaturated carbon-carbon bonds such as olefinic and acetylenic compounds, reaction with ozone, O_3, may also provide substantial or sometimes even dominant contributions to their atmospheric degradation [101,237,238]. Moreover, reactions with the nitrate radical, $NO_3\cdot$, are potentially important loss processes for alkenes (>C=C<) and alkynes (–C≡C–) as well as for organosulfur compounds, phenols and certain nitrogen compounds [101,235,238,239].

The measurement of degradation rate constants in the gas phase is difficult, time-consuming and costly. Experiments are generally limited to compounds with at least some water solubility ($S_w > 1$ µmol/L) and volatility ($P_v > 1$ Pa).

Rate law of indirect photolysis

For a given compound in the troposphere, its decomposition through indirect photolysis is governed by its concentration, c, and the concentration of the relevant oxidant, [Ox]:

$$\frac{dC}{dt} = k^{(2)} \cdot [Ox] \cdot C \qquad (9.82)$$

where $k^{(2)}$ denotes the second-order rate constant. When [Ox] can be taken as approximately constant, which is usually assumed for the oxidants $OH\cdot$, $NO_3\cdot$ and O_3, $k^{(2)} \cdot [Ox] \equiv k$ represents a pseudo-first-order rate constant, resulting in:

$$\frac{dC}{dt} \approx k \cdot C \qquad (9.83)$$

Typical values of 24 h average tropospheric oxidant concentrations are $[OH\cdot] = 1.5 \cdot 10^5$ molecules/cm^3, $[NO_3\cdot] = 1.25 \cdot 10^8$ molecules/cm^3, and $[O_3] = 7.5 \cdot 10^{11}$ molecules/cm^3. It is further assumed that the degradation rate of organic compounds is determined by their primary reaction with the oxidant. Then, the lifetime τ associated with the loss process due to any single oxidant is given by:

$$\tau = \frac{1}{k} \qquad (9.84)$$

and the overall lifetime due to reactions with all relevant oxidants can be calculated as:

$$\tau_{tot} = \frac{1}{k_{tot}} = \frac{1}{\sum_i k_i} \qquad (9.85)$$

Note that despite the lower average $OH\cdot$ concentration, its dominance as an oxidative agent is due to the fact that in the majority of cases, the second-order rate constant $k^{(2)}$ (Equation 9.82) of compound–$OH\cdot$ reactions is higher by orders of magnitude than the ones associated with $NO_3\cdot$ and O_3.

Prediction methods

In the literature, several reviews about models to predict rates of indirect photolysis in the troposphere are available [235,240-243]. The Atkinson method [244-248] contains an extensive list of fragment constants to estimate degradation rate constants due to reactions with OH radicals (k_{OH}), and is the most widely applied model in the environmental context. It includes H-atom abstraction from aliphatic C–H and O–H, addition at alkenes, alkynes and aromatic rings, and reactions with N-, S- and P-containing functional groups. The underlying principle is to account for molecular fragments that are likely to be attacked by reactive species.

For a given compound, its total hydroxyl reaction rate constant (k_{OH}) is calculated based on a consideration of all sites suitable for H-atom abstraction ($k_{H\text{-abs}}$), all unsaturated bonds where OH addition may take place (k_{add}), and all N, S and P functional groups reactive for interaction with OH ($k_{N,S,P}$):

$$k_{OH} = k_{H\text{-abs}} + k_{add} + k_{N,S,P} \quad (9.86)$$

Note that in Equation 9.86, each of the rate constants on the right-hand side may represent a sum of individual site-specific rate constants. For alkenes and alkynes, the Atkinson method also contains fragments for the rates of reactions with O_3 [237], and an extended version based on a larger body of experimental data has been computerized in the AOPWIN module of EPISUITE [249].

According to recommendations and results from validation exercises [241,250-253], the Atkinson method is the current method of choice for estimating atmospheric oxidation rate constants and associated half-lives. The latest computerized version of the AOPWIN program of the US EPA's EPISUITE [149] can be downloaded free of charge. The program is very easy to use and when run separately (without EPIWIN), AOPWIN can show the calculation equations. The accuracy of the estimates must be evaluated by the user on a case-by-case basis. Some indications on the predictive power of the program for certain chemical classes can be obtained from the tables of measured and calculated rate constants within the manual documentation of AOPWIN.

In 1993, Klamt introduced an alternative approach to predict atmospheric degradation rate constants employing semi-empirical quantum chemistry [254,255]. The original name was the MOOH model, referring to molecular orbital OH model. Here, local charge and energy parameters are used to quantify the site-specific molecular susceptibility for oxidative attack through $OH^{\bullet}$, and the final regression equations were obtained through non-linear optimization.

The initial version was confined to H-atom abstraction from aliphatic C–H ($k_{H\text{-abs}}$), and to OH· addition to double, triple and aromatic bonds, and achieved r^2 = 0.98 and *rms* = 0.15 log units with a training set of 159 compounds, and application to a validation set of 38 compounds resulted in *rms* = 0.20 log units [254]. Later, the method was extended to OH· reactions with oxygen-containing compounds, providing *rms* = 0.20 log units for 93 training compounds [255].

Both the Atkinson and the Klamt method exclusively consider substructural units that are known to be liable to transformations. If none of these predefined structural features is present in the molecule, zero degradation is assumed. As such, both models provide – with quite different methodologies and currently also different application ranges – a worst-case estimate of the atmospheric lifetime of organic compounds.

Performance statistics

For a set of 720 compounds with experimental log k_{OH} values, the Atkinson method as implemented in AOPWIN [249] achieved r^2 = 0.95, *rms* = 0.246 log units, and a mean error *(me)* of 0.138 log units [241]. With respect to the non-logarithmic k_{OH}, 90% of the predicted values were within a factor 2 of experimental data, and 95% within a factor 3.

Our present data was taken from the literature and is part of an ongoing study (Anna Böhnhardt, UFZ). It contains 886 compounds with experimental log k_{OH} values from -15.7 to -9.1 (with k_{OH} in cm^3 $molecule^{-1}$ s^{-1}). Of these, 140 compounds contain N, P or S, and here only the Atkinson method can be applied (see Equation 9.86), while the Klamt method is confined to H-abstraction through OH· from aliphatic C–H and O–H ($k_{H\text{-abs}}$) as well as to OH· addition to double, triple and aromatic bonds (k_{add}), as outlined above.

For all 886 compounds, the Atkinson method yields r^2 = 0.92 and *rms* = 0.35. Restriction to the subset of 140 N, S and P compounds decreases both r^2 (0.87) and *rms* (0.31). For the remaining 746 compounds where both models are applicable, Atkinson (r^2 = 0.92, *rms* = 0.33) is superior to Klamt (r^2 = 0.89, *rms* = 0.43). While the former appears to perform better for oxygen-containing and aromatic compounds, the latter is superior for alkenes, and there is only a slight difference for aliphatic hydrocarbons.

The results confirm that, at present, the Atkinson method provides good k_{OH} estimates for most compounds, and is the current method of choice. Note, however, that the Klamt method, though computationally more demanding, contains far fewer parameters, and was derived using a much smaller compound set. Accordingly, the Klamt method offers room for improvement with extension to OH· reactions with N, P and S functionalities, as well as to loss processes through reaction with other oxidants.

9.4.2 Hydrolysis in the water phase

Water is ubiquitous in the environment. The hydrosphere

($361 \cdot 10^6$ km^2) covers ca. 71% of the earth's surface ($510 \cdot 10^6$ km^2), of which 97% is the oceans, 2.4% the polar caps, 0.57% groundwater and 0.03% lakes and streams (surface freshwater). Soil contains pore water, and air humidity. Moreover, water is the major constituent of biological cells. Accordingly, environmental chemicals are very likely to encounter water, and the question is under what conditions and at what rate a chemical reaction with water will take place.

The degradation of a compound through reaction with water is called hydrolysis, and can be written as a pseudo-1-step process in the form:

$$RX + H_2O \rightarrow ROH + HX \quad (9.87)$$

where X denotes some leaving group. Note that the actual course of the reaction may be quite complex. It may include intermediates and will generally depend on the pH of the water as outlined below. Usually, hydrolysis is considered relevant under ambient conditions for halogenated aliphatics, carboxylic acid esters, carboxylic acid anhydrides, organophosphorus esters, epoxides, nitriles, carbamates and sulfonylureas [101,236,256,257].

Reaction kinetics

For a given compound RX at concentration [RX] in water, the overall rate law which applies to the hydrolysis reaction of Equation 9.87 is:

see below (9.88)

where k_{hyd} is the pseudo-first-order hydrolysis rate constant, k_n the pseudo-first-order rate constant for the reaction with neutral water (but not necessarily at neutral pH), and $k_a^{(2)}$ and $k_b^{(2)}$ the second-order rate constants for the reactions with H^+ (acid-catalyzed route) and OH^- (base-catalyzed route), respectively. Note that $k_n = k_n^{(2)}$ $[H_2O]$ with the respective second-order rate constant $k_n^{(2)}$, assuming that the H_2O concentration is constant.

The pseudo-first-order rate constant k_{hyd} refers to a specific pH, that in turn determines $[H^+]$ and $[OH^-]$ and thus will allow the second-order rate constants for the acid and base-catalyzed reaction contribution to be converted to pseudo-first-order counterparts.

Prediction methods

Models to predict hydrolysis rate constants (Table 9.1) are available only for a few compound classes and are based mostly on limited sets of experimental data [243,258,259] except for SPARC (see below) [260-262]. For more detailed discussions, several reviews are available in the literature [256,259,263].

Hydrolysis rates are subject to substituent effects, and consequently Hammett/Taft LFERs were applied to their estimations. Alkaline hydrolysis rates depend on the electronic and steric features of the leaving group (alcohol moiety), and generally decrease with the size and branching of the alcohol and increase with electron-withdrawing groups on the alcohol. The respective LFERs are strictly limited to homologous compounds because of the class-specific sensitivity of the chemical reaction centre, as indicated by the greatly varying slopes and intercepts of the regression functions.

The TGD [250] recommends five class-specific hydrolysis models for brominated alkanes, esters, carbamates and benzonitriles (Equations given in bold in Table 9.1). The HYDROWIN program of the US EPA's EPI Suite [264] calculates second-order acid or base-catalyzed hydrolysis rate constants at 25° C for esters, carbamates, halomethanes, alkyl halides and epoxides. Acid and base-catalyzed half-lives are calculated for pH 7 and/or pH 8. The prediction methodology was developed by the EPA's Office of Pollution Prevention and Toxics based on data by Mill et al. [265]. The class-specific equations are given in the HYDROWIN help file, but no statistics are provided. The available hydrolysis QSARs are limited in their application range, but are generally assumed to be acceptable when operated in accordance with their restrictions. It should be noted, however, that none of the models has been cross-validated or externally validated.

SPARC [260-262] employs a different methodology based on a parameterization of intra- and intermolecular interactions, that in turn are linked by the appropriate thermodynamic relationships to provide estimates of reactivity parameters under the desired conditions. Without specifying the model equations in full detail and without providing all data sets used for their derivation, the following statistics have been reported:

$$-\frac{d[RX]}{dt} = k_{hyd}\,[RX] = \left(k_n + k_a^{(2)}\,[H^+] + k_b^{(2)}\,[OH^-]\right)[RX] \quad (9.88)$$

Table 9.1. Examples of QSAR models for the estimation of hydrolysis rates for specific chemical classes ($k_{hyd.}$ in $mol^{-1}s^{-1}$).

Chemical class		r^2	n	Ref.
Benzonitriles (p-subst.)	**$\log k = 1.64\ (\pm 0.42)\ \sigma(para) - 1.37\ (\pm 0.17)$** TGD-model: Masanuga et al., 1993 [321] Endpoint: first-order rate constant for neutral and pseudo first-order for base-promoted hydrolysis, extrapolated from high temperatures (85 °C) Units: k in day^{-1} Descriptor: Hammett sigma (para) constants (for refs and tabulations: see footnotes) Chemical domain: p-substituted benzonitriles	**0.858**	**14**	**[250]**
Benzoic esters	$\log k_{OH} = 1.17\ \sigma + 2.26$ Endpoint: 2nd-order alkaline hydrolysis rate constant Units: k_{OH} in $M^{-1}\ s^{-1}$ Descriptor: Hammett sigma constants (for refs and tabulations: see footnotes) Chemical domain: benzoic esters	0.992	18	[258]
Phosphoric acid esters	$\log k_{OH} = 1.4\ \Sigma\ \sigma - 0.47$ Endpoint: 2nd-order alkaline hydrolysis rate constant Units: k_{OH} in $M^{-1}\ s^{-1}$ Descriptor: Hammett sigma constants (for refs and tabulations: see footnotes) Chemical domain: phosphoric acid esters	0.990	4	[258]
Phosphoric acid esters	$\log k_{OH} = -9.65\ q(P) + 2.85\ E_{S(alcohol)} + 4.89$ Endpoint: 2nd-order hydrolysis rate constant Units: k_{OH} in $M^{-1}\ s^{-1}$ Descriptors: q(P): net charge on P calculated by CNDO/2 and Taft steric substituent constants (for refs and tabulations: see footnotes) Chemical domain: phosphoric acid esters	0.896	19 (?)	[322]
Phthalate esters	$\log k_{OH} = 4.59\ \sigma^* + 1.52\ E_S - 1.02$ Endpoint: 2nd-order alkaline hydrolysis rate constant at 30 °C in water Units: k_{OH} in $M^{-1}\ s^{-1}$ Descriptors: Hammett sigma constants and Taft steric substituent constants (for refs and tabulations: see footnotes) Chemical domain: phthalate esters	0.975	5	[323]
N-Methyl-*N*-phenyl-carbamates	$\log k_{hyd.} = -0.26\ (\pm 0.001)\ pK_{a(alcohol)} - 1.3\ (\pm 0.1)$ Endpoint: 2nd-order alkaline hydrolysis rate constant at 25 °C in water Units: $k_{hyd.}$ in $M^{-1}\ s^{-1}$ Descriptor: pK_a for the resulting alcohol Chemical domain: *N*-Methyl-*N*-phenylcarbamates	1.0	3 (!)	[324]
N-Phenyl-carbamates	$\log k_{hyd.} = -1.15\ (\pm 0.02)\ pK_{a(alcohol)} + 13.6\ (\pm 0.2)$ Endpoint: 2nd-order alkaline hydrolysis rate constant at 25 °C in water Units: $k_{hyd.}$ in $M^{-1}\ s^{-1}$ Descriptor: pK_a for the resulting alcohol Chemical domain: *N*-phenylcarbamates	0.99	20	[324]
N-Methyl-carbamates	$\log k_{hyd.} = -0.91\ (\pm 0.03)\ pK_{a(alcohol)} + 9.3\ (\pm 0.4)$ Endpoint: 2nd-order alkaline hydrolysis rate constant at 25 °C in water Units: $k_{hyd.}$ in $M^{-1}\ s^{-1}$ Descriptor: pK_a for the resulting alcohol Chemical domain: *N*-Methylcarbamates	0.99	6	[324]

Chemical class		r^2	n	Ref.
N,N-Dimethyl-carbamates	$\log k_{hyd.} = -0.17\ (\pm 0.04)\ pK_{a(alcohol)} - 2.6\ (\pm 0.4)$ Endpoint: 2nd-order alkaline hydrolysis rate constant at 25 °C in water Units: $k_{hyd.}$ in $M^{-1}\ s^{-1}$ Descriptor: pK_a for the resulting alcohol Chemical domain: *N,N*-Dimethylcarbamates	0.80	7	[324]
Carbamates (R2 = H)	$\mathbf{\log k_{OH} = 2.39\ \sigma^*(R1, R2) + 0.96\ \sigma(X1) + 7.97\ \sigma^*\ (R3) + 2.81\ \sigma(X2) + 0.275}$ TGD-model: Drossman et al., 1988 [325] Endpoint: 2nd-order alkaline hydrolysis rate constant at 25 °C in water Units: k_{OH} in $M^{-1}\ s^{-1}$ Descriptors: Hammett sigma constants (for refs and tabulations: see footnotes) Chemical domain: carbamates, X1R1N(R2)C(O)OR3X2, where R2=hydrogen, R1=alkyl or phenyl and R3=alkyl or phenyl with X1 and X2 their respective substituents	**0.973**	**62**	**[250]**
Alkyl/Phenyl-carbamates	$\mathbf{\log k_{OH} = 7.99\ \sigma^*(R3) + 0.31\ \sigma(X2) + 3.14\ E_S(R1, R2) + 0.442}$ TGD-model: Drossman et al., 1988 [325] Endpoint: 2nd-order alkaline hydrolysis rate constant at 25 °C in water Units: k_{OH} in $M^{-1}\ s^{-1}$ Descriptors: Hammett sigma constants and Taft steric substituent constants (for refs and tabulations: see footnotes) Chemical domain: carbamates, X1R1N(R2)C(O)OR3X2, where R2=alkyl or phenyl NOT hydrogen, R1=alkyl or phenyl and R3=alkyl or phenyl with X1 and X2 their respective substituents	**0.903**	**18**	**[250]**
Esters	$\mathbf{\log k_{OH} = 0.98\ E_S(R) + 0.25\ E_S(R') + 2.24\ \sigma^*(R) + 2.24\ \sigma^*(R') + 2.09\ \sigma\ (X) + 1.21\ \sigma(X') + 2.69}$ TGD-model: Drossman et al., 1988 [325] Endpoint: 2nd-order alkaline hydrolysis rate constant at 25 °C in water Units: k_{OH} in $M^{-1}\ s^{-1}$ Descriptors: Hammett sigma constants and Taft steric substituent constants (for refs and tabulations: see footnotes) Chemical domain: any chemical that contains an ester bond –C(=O)-O-; developed with alkyl/aryl - alkyl/aryl esters X-R-C(=O)-O-R'-X', where R, R' can be alkyl or aryl substituents and X, X' any other substituents	**0.974**	**103**	**[250]**
Aromatic nitriles (m, p-subst.)	$\log k_{corr.} = 0.54 \log K_{ow} + 0.57\ \sigma - 5.28$ Endpoint: pseudo first-order reaction rate constant of reductive hydrolysis, corrected for the fraction of compound sorbed to sediment: $k_{corr.} = k_{exp.}\ (1 + \rho\ K_d)$ with $k_{exp.}$: pseudo first-order reaction rate constant in anaerobic sediment/water systems at 22 °C; r: sediment concentration in g g^{-1}; K_d: distribution coefficient ($K_d = K_{oc}$ * OC with OC: measured sediment organic carbon fraction; K_{OC} calculated from $\log K_{oc} = 0.544 \log K_{ow} + 1.377$ (= Equation 4-8 in Lyman 1990 [326]). NOTE: $k_{corr.}$ represents a sediment-catalyzed transformation and rates of hydrolysis in the overlying water phase will be much lower than predicted using these QSARs. Units: $k_{corr.}$ and $k_{exp.}$ in min^{-1} Descriptors: $\log K_{OW}$ and Hammett sigma constants (for refs and tabulations: see footnotes) Chemical domain: meta- and para-substituted benzonitriles	0.925	17	[327]

Chemical class		r^2	n	Ref.
Aromatic nitriles (o-subst.)	$\log k_{corr.} = -0.46 \log K_{ow} + 1.26\,\sigma - 4.56$ Endpoint: pseudo first-order reaction rate constant of reductive hydrolysis, corrected for the fraction of compound sorbed to sediment: $k_{corr.} = k_{exp.}(1 + \rho K_d)$ with kexp.: pseudo first-order reaction rate constant in anaerobic sediment/water systems at 22 °C; r: sediment concentration in g g^{-1}; K_d: distribution coefficient ($K_d = K_{oc}$ * OC with OC: measured sediment organic carbon fraction; K_{oc} calculated from $\log K_{oc} = 0.544 \log K_{ow} + 1.377$ (Equation 4-8 in Lyman 1990 [326]). NOTE: $k_{corr.}$ represents a sediment-catalyzed transformation and rates of hydrolysis in the overlying water phase will be much lower than predicted using these QSARs. Units: $k_{corr.}$ and $k_{exp.}$ in min^{-1} Descriptors: log K_{ow} and Hammett sigma constants (for refs and tabulations: see footnotes) Chemical domain: ortho-substituted benzonitriles	0.981	7	[327]
Brominated alkanes	$\mathbf{\log k(i)/k(o) = -11.9\ (\pm 3.5)\ \sigma(I)}$ TGD-model: Vogel and Reinhard, 1986 [328] Endpoint: The parameter k(i) is the pseudo first-order alkaline hydrolysis rate constant, and k(o) is the corresponding constant for CH3-Br hydrolysis. Units: k(i) and k(o) in s^{-1} Descriptor: Taft's polar sigma(I) constants (for refs and tabulations: see footnotes) Chemical domain: saturated linear and branched bromoalkanes with phenyl, chloro and bromo substituents.	**0.770**	**16**	**[250]**

Equations given in bold: class-specific hydrolysis models recommended by the TGD [250]
r^2: squared correlation coefficient
n: number of compounds analyzed
σ, σ^*: Hammett substituent constants, to retrieve σ- and σ^*-values see, e.g., [258] (some values), [329] and [68]
q(P): net charge on P calculated by CNDO/2
E_s: Taft steric substituent constant, to retrieve E_s-values see, e.g., [258] (some values), [68]
σ(I): Taft polar sigma (I) constant, to retrieve σ(I)-values see, e.g., [68]

Carboxylic acid esters with 2nd-order hydrolysis rate constants [261]:

base-catalyzed:
$\log k_{hyd} \equiv \log k_b^{(2)}$ $n = 654$, $r^2 = 0.96$, $rms = 0.37$

acid-catalyzed:
$\log k_{hyd} \equiv \log k_a^{(2)}$ $n = 667$, $r^2 = 0.97$, $rms = 0.37$

general base-catalyzed:
$\log k_{hyd} \equiv \log k_b^{(2)}$ $n = 150$, $r^2 = 0.97$, $rms = 0.39$

Phosphate esters with temperature-dependent 2nd-order hydrolysis rate constants [262]:

base-catalyzed:
$\log k_{hyd} \equiv \log k_b^{(2)}$ $n = 83$, # data = 225, $r^2 = 0.93$, $rms = 0.401$

SPARC is available for online use [266], and the hydrolysis options include acid and base-catalyzed transformation rates and also neutral catalysis for variable conditions (temperature, solvent or mixture, catalyst). The full output delivers the mechanistic information from the calculations. SPARC has been considered the most accurate predictive model for hydrolysis (Tratnyek PG 2002, personal communication).

For compounds outside the range of applicability of existing QSARs, a qualitative approach based on structural analogy has been proposed [267], as outlined in [256].

9.4.3 Biodegradation

Micro-organisms can be seen as environmental catalysts. They represent the principal biotic force to degrade organic compounds. Bacteria form the biochemically most active family of micro-organisms, and further microbial species include fungi, protozoans, yeasts, and algae [268]. One m^2 of grassland soil contains up to 10^{15} bacterial cells, each of which weighs approximately 1 pg = 10^{-12} g.

Biodegradation can be an effective mechanism in transforming organic compounds in water, soil and sediment. It is a complex multi-step process involving uptake, intracellular transport and enzymatic reactions in micro-organisms. Microbial transformation reactions are usually the only processes by which a xenobiotic organic compound may be mineralized in the environment, while abiotic reactions commonly yield other organic degradation products.

Depending on the ambient conditions, different modes and rates of biodegradation may predominate. These are influenced by factors related to the chemical substrate, the micro-organisms and the environment [268-270]. In different environmental conditions, a given chemical may be biodegraded by different pathways, resulting in different degrees of persistence. Bacteria have a variety of enzyme systems, but specific enzymes for transforming xenobiotics are generally absent. The enzymes involved are either always present at certain concentrations and activities (constitutive enzymes) or have to be induced, expressed or transferred by plasmids during an adaptive lag phase (inducible enzymes). Accordingly, the xenobiotic degradation is typically incidental to normal metabolism, and called co-metabolism.

Primary metabolic reactions are mediated mostly by non-specific enzyme systems, which catalyze oxidation, reduction and hydrolysis, with the different transformations occurring either consecutively or simultaneously (competitive) at different sites of the substrate. For large compounds with a molecular weight > 500 Dalton, which cannot react with the intracellular bacterial enzymes because of hindered membrane transfer, biodegradation is generally negligible, and in condensed environmental phases abiotic degradation (hydrolysis, redox transformations) may be the only degradation pathway.

Experimental data

Most biodegradation data are from screening tests with yes vs. no results for passing a predefined threshold. In screening tests, media with defined mineral salts are employed, and the loss rate of the compound is measured directly (e.g., decrease in UV absorption intensity) or indirectly through monitoring of consumption parameters such as BOD (biological oxygen demand), COD (chemical oxygen demand) or other processes, such as CO_2 production or DOC (dissolved organic carbon) disappearance [271]. Although kinetic data from laboratory and field biodegradation tests are becoming increasingly available, they are still too scarce and heterogeneous for modelling purposes. Experimental data can be found in books [269,272], and large compilations are available through the database BIODEG [273] and from the MITI-I protocol [274] each of which covers ca. 900 compounds.

The various testing protocols for the experimental determination of biodegradability differ considerably in the duration of the adaptation and incubation periods, kind of inoculum used (e.g., pure cultures, surface water, sewage water, soil), inoculum size and density, treatment and concentration of the test compounds, parameters measured (BOD, COD, CO_2 production, DOC etc.) and the different pass-levels evaluated (pass-levels from 15 % to 90 % are used to classify degradable and non-degradable substances). The different test conditions may result in different classifications of biodegradability. In other words, a compound evaluated as readily biodegradable in one test may be considered non-degradable in another.

The MITI-I test represents a screening test for ready biodegradability in an aerobic aqueous medium. To this end, 100 mg/L of the test substance is inoculated and incubated with 30 mg/L sludge, and BOD is monitored during a 28-d exposure period. The compound is classified as readily biodegradable if BOD 60% of ThOD (theoretical oxygen demand) holds true. Otherwise, the chemical is classified as not readily biodegradable. The MITI-I test was developed in Japan, and is described as OECD guideline 301C [275]. More detailed information about biodegradation test protocols is available in the literature [271,276,277].

The categorical experimental classification limits the adequate ranking particularly of compounds of intermediate biodegradability. Because of the inherent diversity of the end-points and parameters concerned, biodegradability is not a uniform principal property of chemical contaminants. Analogous to the strictly empirical nature of experimental biodegradability assessments, it is necessary to realize that biodegradability is not a well-defined parameter. The significance of biodegradation estimates cannot be expected to exceed that of the underlying data, and such estimates should be regarded rather as indicators of probabilities towards greater or lesser biodegradability.

Qualitative biodegradation rules

Most models to predict biodegradation (see reviews [240,271,278]) indicate the biodegradability associated with certain substructural features. These include chain length, degree of branching, saturation state of the carbon chain and oxidation state of the terminal groups for acyclic compounds. The type, number and position of substituents, and also the number of rings, are relevant for microbial transformations of aromatic compounds. Physicochemical properties affect the degradability of the chemicals relating to processes such as transport into the microbial cell. Electronic parameters can be used to explain different transformation mechanisms caused by differing polarity of chemicals. The electron density on the aromatic ring, which depends on ring substituents, governs the ease of ring cleavage of an aromatic system.

Because of the multitude of processes involved in biodegradation, no single-descriptor model can accurately predict the biodegradability of a broad range of chemicals. If the reaction mechanism (in terms of explicit reaction equation and stoichiometry) of the degradation process is known, thermodynamic relationships may be used; these have the advantage of allowing for variable environmental conditions such as the presence of water, concentrations of reactants and products, temperature and redox conditions [279].

As a general trend, functional groups such as carboxyl-, hydroxyl- and methyl groups tend to increase biodegradability, while nitro-, amino-, cyano- and halogen groups tend to decrease biodegradability. Note, however, that biodegradability greatly depends on the experimental conditions, and that structure-activity relationships refer to the conditions under which the data used for the model derivation have been generated.

Prediction methods

Despite considerable uncertainties regarding the suitability of available biodegradation data sets for modelling, numerous QSARs on biodegradation have been reported. For more detailed information, the reader is referred to reviews [240,271,278] and recent applications of models to larger data sets [274,280-282].

Most currently available models are based on substructural features as indicators for a certain biodegradability category. The associated modelling techniques, such as cluster and discriminant analyses, usually employ indicator variables for the absence or presence of the predefined features of the structures. In this way, the models may indicate substructures or properties contributing to the biodegradability of compounds, but they do not (are not intended to) relate to underlying, rate-limiting processes. Non-linear substructure models are supposed to account for the contribution of group interactions to the degradability of the substance. A principal limitation is that substructure models cannot be applied to compounds with structural elements missing from the original data set.

A general observation is that most models reveal a marked difference in recognition of degradable and non-degradable compounds. This imbalance in recognition of degradable and non-degradable compounds indicates a major problem with the application of many biodegradation models: their predictions that compounds are readily degradable are often not reliable. Only if a compound is predicted as non-degradable by such models is there a good chance that it really is non-degradable, and the predicted result may be used with some confidence.

So far, current biodegradation models focus on the estimation of ready/not-ready biodegradability in screening tests. This is because most experimental data are from such tests (e.g., MITI-I). There are far fewer data that are both quantitative and environmentally relevant, such as measured half-lives or rate constants. However, individual transformations and pathways are well documented in the literature. This allows for development of explicitly mechanistic models, making use of established group-contribution approaches, hierarchic rule-based expert systems and probabilistic evaluation of possible transformation pathways.

The following description of several algorithms and software is partly based on a previous comparative evaluation of model performances [283] and on the most recent review of broadly applicable methods for predicting biodegradation [278].

BIOWIN [284]. The BIOWIN group-contribution method of the US EPA's EPI Suite features six linear and non-linear regression models based on weight-of-evidence evaluations of screening test results. The original BIODEG version contained 35 structural fragments and was calibrated with data for 264 chemicals in the BIODEG database [285,286]. A later revision contained five new or redefined substructures and molecular weight as an additional descriptor, now using a data set of 295 compounds. Here, both linear and non-linear regression models were derived to predict primary biodegradation (now BIOWIN1 = BIODEG linear model, and BIOWIN2 = BIODEG non-linear model), and a corresponding pair of models employing the same set of descriptors (fragments and molecular weight) was developed to predict ultimate biodegradation based on semi-quantitative estimates (hours, days, weeks, months, longer than months) for 200 compounds made by a panel of 17 experts (BIOWIN3 = survey ultimate biodegradation linear model, and BIOWIN4 = survey ultimate biodegradation non-linear model) [287]. Finally, re-calibration of BIOWIN1 and BIOWIN2 to MITI-I data for 884 organic compounds resulted in a further linear and non-linear model, now including 42 fragments and molecular weight [274] (BIOWIN5 = MITI linear model, BIOWIN6 = MITI non-linear model). The three linear models take the general form:

$$P = \sum a_i F_i + b \cdot \mathrm{MW} + c \tag{9.89}$$

where P denotes the probability of a given compound being readily biodegradable, a_i and b are the regression coefficients of fragments F_i and molecular weight, respectively, and c is a constant. The three associated non-linear models read:

$$P = \frac{\exp\left(\sum a_i F_i + b \cdot \mathrm{MW} + c\right)}{1 + \exp\left(\sum a_i F_i + b \cdot \mathrm{MW} + c\right)} \tag{9.90}$$

External validation with MITI-I data [283,288] and PMN (premanufacture notice) data [280] yielded acceptable predictions of non-readily biodegradable compounds (91–95% correct predictions), but major deficiencies in reliably predicting readily biodegradable compounds (36–70% correct predictions).

Application of the BIOWIN model suite to 374 PMN substances revealed that combining BIOWIN3 (survey ultimate biodegradation linear) with BIOWIN5 (MITI linear) significantly reduced false positives for ready biodegradability and the overall misclassification rate [281]. For 63 pharmaceuticals, similar improvements were obtained with the combination of BIOWIN3 and BIOWIN6 (MITI non-linear). Moreover, a recent external validation with 110 newly notified substances confirmed a high accuracy for ready biodegradability in contrast to an only moderate performance for not ready biodegradability [282]. Overall accuracy of predictions > 90 % can be achieved, but only when the two MITI (linear + non-linear) and the two Survey (ultimate linear + ultimate non-linear) models agree.

PLS model [289]: Based on 894 substances with biodegradation assessed according to the MITI-I test protocol (388 ready, 506 not-ready), Loonen and co-workers [289] developed a multivariate PLS model for the prediction of ready biodegradability. The chemicals were characterized by a set of 127 predefined structural fragments [290]. The model was evaluated by means of internal cross-validation and repeated external validation [283] (% correct: ready: 84%, not-ready: 86%). However, this PLS group contribution method is not generally available.

MultiCASE / META [291,292]: The commercial MultiCASE/META system combines a group-contribution model and an expert system to simulate aerobic biodegradation pathways [293]. The META expert system features 70 transformations that match 13 biophores. Substructures that inhibit biodegradation (biophobes) are neglected. MLR combined with the high number of structural descriptors gives a fair chance of overfitting the data, leading to reduced performance in external validation [283] (% correct: ready: 73%, not-ready: 80%).

CATABOL [294-296]: The commercial knowledge-based expert system assesses entire biotransformation pathways, not only parent structures. The probabilistic predictions are parameterized for the MITI-I test and based on a hierarchy of > 550 principle transformations in sequential and branched pathways. Considering spontaneous biotic and abiotic as well as catabolic transformations, CATABOL finally generates one as the most probable pathway. Through the analysis of the pathway and its critical steps based on individual transformation probabilities, CATABOL can identify potentially persistent catabolic intermediates and estimate their molar amounts. Cross-validation with MITI-I data [278] by four times leaving out 25% of the data resulted in q^2_{cv} = 0.86 and 82% and 91% readily/non-readily correct classifications, respectively. Development of CATABOL is ongoing and new transformations are being added.

PredictBT [297]: This freely accessible system predicts biotransformation pathways using substructure searching, a rule-base and atom-to-atom mapping. PredictBT assesses multiple pathways (metabolic maps) based on 262 biotransformation descriptors for 251 biotransformation rules. The rules are designed by expert knowledge and are generalized based on metabolic logic. The system is fully transparent and comprehensible, a list of all rules is published on the website. The generalized transformations also enable new pathways to be found besides those already stored in the system. Internal validation [298] revealed correct predictions for 98% of the primary degradation steps and 72% of the pathway branches. The pathway prediction system is further expanded to predict what biodegradation pathways are more or less likely under certain environmental conditions.

Overall, the models still perform better in predicting non-readily biodegradable compounds. This can partly be explained by the fact that the presence of a biodegradation retarding fragment will prevent mineralization, while

Table 9.2. Decadic logarithms of categorical half-lives according to [299] in h.

Half-life Category	Log $t_{1/2}$ of lower boundary	Log $t_{1/2}$ of average	Log $t_{1/2}$ of upper boundary
1	–	0.70[a]	< 1.00
2	1.00	1.23	< 1.48
3	1.48	1.74	< 2.00
4	2.00	2.23	< 2.48
5	2.48	2.74	< 3.00
6	3.00	3.23	< 3.48
7	3.48	3.74	< 4.00
8	4.00	4.23	< 4.48
9	4.48	4.74[a]	+

[a] No average, but representative category value according to [299].

a biodegradation enhancing fragment can indicate a possible metabolic step, but does not necessarily lead to complete mineralization. The reason for non-ready biodegradability may be a structural fragment that is not present in the parent compound, but in (one of the) metabolites from the multitude of transformation processes that lead to ultimate degradation. As a consequence, it remains that only if a compound is predicted as non-degradable by the models, is there a good probability that it really is non-degradable and the predicted result may be used with some confidence.

The degradation rates of chemicals, either determined in the laboratory under ideal conditions or estimated from structural information, can only indicate which of several possible transformation pathways is most likely to occur. Because the degradability assessed in this manner may not correspond to the degradation occurring in the environment, extrapolation from laboratory data to the field is extremely difficult. Under given local conditions, different degradation reactions, abiotic and also biotic, at varying rates may pre-dominate, yielding different metabolites with differing fates. The assessment of the transformation of chemicals can be valid only for a stringently defined environmental scenario. An extrapolation to more general conditions is feasible only in terms of comparing different compounds on a relative scale, which means identifying the substances with the least likelihood of persistence. These inherent uncertainties have to be accounted for when evaluating the degradability of chemicals, no matter what the provenance of the data.

9.4.4 Compartmental half-lives

In environmental fate and exposure assessments, level II, III and IV multimedia models require compartmental half-lives ($t_{1/2}$) for compounds in air, water, soil and sediment. For each of the four compartments, $t_{1/2}$ will reflect the overall degradation rate due to all processes that are relevant, such as biodegradation, hydrolysis and photolysis:

$$t_{1/2} = \frac{\ln 2}{k_{\text{tot}}} = \frac{\ln 2}{\sum_i k_i} \tag{9.91}$$

Several methods are available to translate experimental findings and modelling results into transformation rate constants and half-lives. The Technical Guidance Document of the European Commission (TGD) [250] provides conversion tables and equations for degradation in air, surface water, sediment and soil, and elimination in sewage treatment plants (Box 9.1). Mackay et al. [299] devised a semi-quantitative scale of compartmental half-lives, associated with nine half-life categories (Table 9.2). The half-life categories are defined through lower and upper boundaries (e.g., category 4 covers log $t_{1/2}$ values from 2.00 to < 2.48, corresponding to half-lives from 100 h to < 302 h). EPI Suite [149] predictions for process-specific half-lives provide either transformation rate constants (AOPWIN, HYDROWIN) or semi-quantitative ratings of the time required for primary and ultimate biodegradation (BIOWIN): 5 - hours; 4 - days; 3 - weeks; 2 - months; 1 - longer. Table 9.3 summarizes the conversion of BIOWIN ratings (as predicted by the ultimate-primary models) to the time required to achieve ultimate or primary biodegradation.

Box 9.1. Overview of rate constants *(k)* and compartmental half-lives ($t_{1/2}$) according to the TGD [250]

AIR (k_{air} [d^{-1}], $t_{1/2}$ (air) [d])

$k_{air} = k_{OH}^{(2)} \times [OH^{\bullet}] \times 24 \times 3600$

with: $k_{OH}^{(2)}$ = 2nd-order rate constant for the reaction with OH-radicals [cm^3 $molec^{-1}$ s^{-1}]

$[OH^{\bullet}]$ = concentration of OH radicals in the lower atmosphere: 5×10^5 [molec cm^{-3}]

$t_{1/2.}$(air) = ln 2 / k_{air} = compound half-life in the air

SURFACE WATER (k_{water} [d^{-1}], $t_{1/2}$ (water) [d])

$k_{water} = k_{hydr} + k_{photo} + k_{bio}$

with: k_{hydr} = 1st-order rate constant for hydrolysis in surface water [d^{-1}]

k_{photo} = 1st-order rate constant for (indirect) photolysis in surface water [d^{-1}]

k_{bio} = 1st-order rate constant for biodegradation in surface water [d^{-1}]

$t_{1/2}$ (water) = ln 2 / k_{water} = compound half-life in surface water

k_{bio} [d^{-1}] and $t_{1/2}$ (bio) [d] extrapolated from results from standardized biodegradation tests:

Test result	k_{bio} [d^{-1}]	$t_{1/2}$ (bio) [d]
Readily biodegradable	4.7×10^{-2}	15
Readily biodegradable, but failing 10-d window	1.4×10^{-2}	50
Inherently biodegradable	4.7×10^{-3}	150
Not biodegradable	0	

SOIL (k_{soil} [d^{-1}], $t_{1/2}$ (soil) [d])

k_{soil} = ln 2 / $t_{1/2}$ (soil)

$t_{1/2}$ (soil) [d] extrapolated from soil sorption constant (k_{soil}) and results from standardized biodegradation tests:

k_{soil} [L kg^{-1}]	Readily biodegradable	Readily biodegradable, but failing 10-d window	Inherently biodegradable
100	30	90	300
100 < k_{soil} 1000	300	900	3000
1000 < k_{soil} 10000	3000	9000	30000
etc.	etc.	etc.	etc.

SEDIMENT (k_{sed} [d^{-1}], $t_{1/2}$ (sed) [d])

$k_{sed} = \ln 2 / t_{1/2.soil} \times f_{aerob}$

with: f_{aerob} = fraction of sediment compartment that is aerobic: 0.10 [m^3 m^{-3}]

$t_{1/2}$ (sed) = ln 2 / k_{sed} = compound half-life in sediment

SEWAGE TREATMENT PLANTS (k_{STP} [h^{-1}])

k_{STP} [h^{-1}] extrapolated from results from standardized biodegradation tests:

Test result	k_{STP} [h^{-1}]
Readily biodegradable	1
Readily biodegradable, but failing 10-d window	0.3
Inherently biodegradable, meeting specific criteria	0.1
Inherently biodegradable, not meeting specific criteria	0
Not biodegradable	0

Table 9.3. Conversion of BIOWIN ratings to the time required to achieve ultimate or primary biodegradation according to [284].

Predicted rating	Time required for biodegradation
5.0	hours
4.5	hours – days
4.0	days
3.5	days – weeks
3.0	weeks
2.5	weeks – months
2.0	months
1.0	longer

Note that the two semi-quantitative scales are inversely related with different numbers of categories. Interconversions between the different scales are ambiguous and subject to considerable variability (Table 9.4), particularly because the half-life boundaries between TGD [250], the nine semi-quantitative categories [299] and the BIOWIN ratings [284] do not correspond. A generic relationship between process-specific and compartmental degradation rates, which would enable process-specific half-life information to be exploited in a systematic and comprehensive manner in the context of multimedia fate modelling, is still lacking [300].

Despite the need for compartmental degradation half-lives as input for models of the multimedia fate and life-time of organic compounds, QSAR methodology has focused so far on predicting individual processes, as outlined in Sections 9.4.1–9.4.3. Only recently, have structural similarities of substances with existing data on compartmental half-lives been used for extrapolation to relevant data for untested compounds [300]. The k-nearest neighbours (KNN) model on atom-centred fragments (ACFs) estimates medium-specific half-lives for substances of interest as average values of compounds with sufficiently similar chemical structures. In the corresponding model equation:

$$\log t_{1/2}(A, j) = \sum_{B} w_B \log t_{1/2}(B, j) \tag{9.92}$$

where A denotes the compound of interest, j the relevant compartment (air, soil, water, sediment), B any of the k (in our case: $k = 3$) most similar reference compounds with known compartmental half-lives, and w_B is a weighting factor (either $1/n$ as a simple arithmetic weighting, or a weighting taking into account ACF similarity) [300]. Equation 9.92 thus estimates $\log t_{1/2}$ for a given compound as (a suitably weighted) average of the $\log t_{1/2}$ values of k most similar compounds.

The reasoning behind this approach is that molecular susceptibility to biotic and abiotic degradation is associated with certain structural features, such that a properly designed similarity measure provides a reasonable guide by which similarity in structure can be related to similarity in degradability [300].

9.5 HANDBOOKS, SOFTWARE AND WEB RESOURCES

Comprehensive compilations of environmentally important properties are available in several data books by Mackay et al. [88], Tomlin [301], Verschueren [302], and Lide [303]. There are various computerized resources for compound data. Besides the large commercial offline compilations of MedChem [304] and AQUASOL [305], there are also online databases on the WWW provided with free access (NIST Chemistry WebBook [306], Chemfinder WebServer [307], ChemID plus [308], Physprop – demo version [61], ARS Pesticide Properties Database [309]) and on a commercial basis (e.g., Beilstein [310], and FIZ-STN [311]).

The classic handbook on property estimation methods by Lyman et al. [69] was originally published in 1982,

Table 9.4. Comparison of biodegradation half-lives in surface waters as specified in the TGD (Overview in Box 9.1 [250]) with half-life categories (Table 2 in [299]) and BIOWIN ratings (Table 4 in [284]).

Test result	$t_{1/2}$ (bio) [d] [250]	half-life category [299]	BIOWIN rating [284]
Readily biodegradable	15	1 – 4	3.5 – 5.0
Readily biodegradable, but failing 10-d window	50	5	2.5 – 3.5
Inherently biodegradable	150	6	2.0 – 2.5
Not biodegradable		7 – 9	< 2.0

and reprinted in 1990. An update focusing on the most relevant properties was published by Boethling and Mackay in 2000 [104], and further textbooks covering a similar range of properties have appeared from Baum in 1998 [312] and from Reinhard & Drefahl in 1999 [313]. Two other textbooks deal specifically with methods to predict water solubility [211,314], and there are also textbooks providing more fundamental information on the physicochemical properties of liquids and gases and their estimation [208,315]. In addition, reviews such as those from Mackay et al. [4] and Boethling et al. [316] address the current role of organic compound properties and their estimation in the context of environmental fate assessment. With regard to inorganic compounds, there is also a handbook available that discusses estimation methods for properties of environmental interest [317].

Some computerized versions of property estimation methods are publicly available and can be downloaded free of charge, with the EPISUITE [149] being a prominent example. Commercial products covering a number of compound properties are ADME-Boxes [62] and ACD [150]. The SPARC [266] system and the web version of EPISUITE [318] can be accessed free via the WWW.

To calculate compound properties, chemical structures are required as input, except if the software includes a database that enables access to structural representations through the CAS number or compound name. On the web, chemical structure information can be obtained together with other information from various freely accessible databases such as Chemistry WebBook [306], ChemID plus [308], NCI [319] and PubChem [320].

9.6 FURTHER READING

1. Baum EJ. 1998. *Chemical Property Estimation.* CRC Press, Boca Raton, FL.
2. Boethling, Robert S., Mackay, Donald, eds. 2000. *Handbook of Property Estimation Methods for Chemicals: Environmental and Health Sciences.* CRC Press, Boca Raton, FL.
3. Lyman WJ. 1990. Octanol/water partition coefficient. In: Lyman WJ, Reehl WF, Rosenblatt DH, eds, *Handbook of Chemical Property Estimation Methods*, 3rd ed. American Chemical Society, Washington, DC, pp 1-1-1-54.
4. Poling BE, Prausnitz JM, O'Connell JP. 2000. *The Properties of Gases and Liquids.* McGraw-Hill Book Company, New York, NY.
5. Reinhard M, Drefahl A. 1999. *Handbook for Estimating Physicochemical Properties of Organic Compounds.* John Wiley, New York, NY.

Acknowledgments

We would like to thank Ovanes G. Mekenyan and his group for their cooperation by running their BCF model [154] with our data set, Igor V. Tetko for providing us with the ALOGPS software [60], and Anna Böhnhardt (UFZ) for the log k_{OH} calculations. Technical support for the present study provided by Barbara Wagner, Daniel Stosch, Daniel Exner, and Armin Vollmer (all UFZ) is greatly appreciated.

Financial support provided by the European Union (European Commission, FP6 Contract No. 003956, Integrated Project NoMiracle, and FP6 Contract No. 022674, Project CAESAR) is gratefully acknowledged.

REFERENCES

1. Mackay D, Di Guardo A, Paterson S, Cowan CE. 1996. Evaluating the environmental fate of a variety of types of chemicals using the EQC model. *Environ Toxicol Chem* 15:1627-1637.
2. Webster E, Mackay D, Wania F. 1998. Evaluating environmental persistence. *Environ Toxicol Chem* 17:2148-2158.
3. Hertwich EG, McKone TE. 2001. Pollutant-specific scale of multimedia models and its implications for the potential dose. *Environ Sci Technol* 35:142-148.
4. Mackay D, Hubbarde J, Webster E. 2003. The role of QSARs and fate models in chemical hazard and risk assessment. *QSAR Comb Sci* 22:106-112.
5. Cahill TM, Mackay D. 2003. Complexity in multimedia mass balance models: when are simple models adequate and when are more complex models necessary? *Environ Toxicol Chem* 22:1404-1412.
6. Wania F, Dugani CB. 2003. Assessing the long-range transport potential of polybrominated diphenyl ethers: a comparison of four multimedia models. *Environ Toxicol Chem* 22:1252-1261.
7. Stroebe M, Scheringer M, Hungerbühler K. 2004. Measures of overall persistence and the temporal remote state. *Environ Sci Technol* 21:5665-5637.
8. Mackay D. 1991. *Multimedia Environmental Models.* Lewis Publishers, Chelsea, MI.
9. Kühne R, Breitkopf C, Schüürmann G. 1997. Error propagation in fugacity level-III models in the case of uncertain physicochemical compound properties. *Environ Toxicol Chem* 16:2067-2069.
10. Breitkopf C, Kühne R, Schüürmann G. 2000. Dependence of multimedia level-III partitioning and residence times of compounds on physicochemical properties and system parameters of water-rich and water-poor environments. *Environ Toxicol Chem* 19:1430-1440.

11. MacLeod M, Fraser AJ, Mackay D. 2002. Evaluating and expressing the propagation of uncertainty in chemical fate and bioaccumulation models. *Environ Toxicol Chem* 21:700-709.
12. Fenner K, Scheringer M, Hungerbühler K. 2004. Prediction of overall persistence and long-range transport potential with multimedia fate models: robustness and sensitivity of results. *Environ Pollut* 128:189-204.
13. Cole JG, Mackay D. 2000. Correlating environmental partitioning properties of organic compounds: the three solubility approach. *Environ Toxicol Chem* 19:265-270.
14. Schenker U, MacLeod M, Scheringer M, Hungerbühler K. 2005. Improving data quality for environmental fate models: a least-squares adjustment procedure for harmonizing physicochemical properties of organic compounds. *Environ Sci Technol* 39:8434-8441.
15. Breivik K, Wania F. 2003. Expanding the applicability of multimedia fate models to polar organic chemicals. *Environ Sci Technol* 37:4934-4943.
16. Abraham MH. 1993. Scales of solute hydrogen-bonding: their construction and application to physicochemical and biochemical processes. *Chem Soc Rev* 22:73-83.
17. Schüürmann G, Ebert R-U, Kühne R. 2006. Prediction of physicochemical properties of organic compounds from 2D molecular structure - Fragment methods vs. LFER models. *Chimia* 60:691-698.
18. Dimitrov S, Dimitrova G, Pavlov T, Dimitrova N, Patlewicz G, Niemela J, Mekenyan O. 2005. A stepwise approach for defining the applicability domain of SAR and QSAR models. *J Chem Inf Model* 45:839-849.
19. Schüürmann G, Kühne R, Kleint F, Ebert R-U, Rothenbacher C, Herth P. 1997. A software system for automatic chemical property estimation from molecular structure. In: Chen F, Schüürmann G, eds, *Quantitative Structure-Activity Relationships in Environmental Sciences - VII*. SETAC Press, Pensacola, FL, pp. 93-114.
20. Organization for Economic Co-operation and Development. The report from the expert group on (quantitative) structure-activity relationships [(Q)SARs] on the principles for the validation of (Q)SARs. OECD, Paris, France.
21. Todeschini R, Consonni V. 2000. *Handbook of Molecular Descriptors. Methods and Principles in Medicinal Chemistry*. Wiley-VCH, Weinheim, Germany.
22. Randić M. 1975. Characterization of molecular branching. *J Am Chem Soc* 97:6609-6615.
23. Kier LB, Hall LH. 1976. *Molecular Connectivity in Chemistry and Drug Research*. Academic Press, New York, NY.
24. Kier LB, Hall LH. 1990. An electrotopological-state index for atoms in molecules. *Pharm Res* 7:801-807.
25. Kier LB, Hall LH. 1999. *Molecular Structure Description*. Academic Press, San Diego, CA.
26. Schüürmann G. 2004. Quantum chemical descriptors in structure-activity relationships - calculation, interpretation, and comparison of methods. In: Cronin MTD, Livingstone DJ, eds, *Predicting Chemical Toxicity and Fate*. CRC Press, Boca Raton, FL, pp. 85-149.
27. Molecular Networks GmbH. 2006. CORINA 3.4. Molecular Networks GmbH - Computerchemie, Erlangen, Germany.
28. Mekenyan O, Nikolova N, Schmieder P, Veith G. 2004. COREPA-M: A multi-dimensional formulation of COREPA. *QSAR Comb Sci* 23:5-18.
29. Mekenyan O, Pavlov T, Grancharov V, Todorov M, Schmieder P, Veith G. 2005. 2D-3D migration of large chemical inventories with conformational multiplication. Application of the genetic algorithm. *J Chem Inf Model* 45:283-292.
30. Kamlet MJ, Abboud JL, Taft RW. 1981. An examination of linear solvation energy relationships. *Prog Phys Org Chem* 13:485-630.
31. Abraham MH, Whiting GS, Doherty RM, Shuely WJ. 1990. Hydrogen bonding. Part 13. A new method for the characterisation of GLC stationary phases — the laffort data set. *J Chem Soc Perkin Trans* 2:1451-1460.
32. Abraham MH, McGowan JC. 1987. The use of characteristic volumes to measure cavity terms in reversed phase liquid chromatography. *Chromatographia* 23:243-246.
33. Hansch C, Fujita T. 1964. r-s-p Analysis. Method for correlation of biological activity and chemical structure. *J Am Chem Soc* 86:1616-1626.
34. Leo A, Hansch C, Elkins D. 1971. Partition coefficients and their uses. *Chem Rev* 71:525-616.
35. Leo AJ. 1993. Calculating log P_{oct} from structures. *Chem Rev* 93:1281-1306.
36. Lyman WJ. 1990. Octanol/water partition coefficient. In: Lyman WJ, Reehl WF, Rosenblatt DH, eds, *Handbook of Chemical Property Estimation Methods*, 3rd ed. American Chemical Society, Washington, DC, pp. 1-1-1-54.
37. Leo A. 1990. *Medicinal Chemistry and Biological Chemistry, Part IV*. Wiley-VCH, Weinheim, Germany, pp. 295-319.
38. Taylor KR. 1990. *Medicinal Chemistry and Biological Chemistry, Part IV*. Wiley-VCH, Weinheim, Germany, pp. 241-294.
39. Leo AJ. 1990. Methods of calculating partition coefficients. In: Hansch C, et al, eds, *Comprehensive Medicinal Chemistry*, 1st ed. Pergamon Press, Oxford, UK, pp. 295-319.
40. Organization for Economic Co-operation and

Development. 1996. OECD Guideline for the Testing of Chemicals 107: Partition coefficient (*n*-octanol/water): Shake flask method. OECD Environment Directorate, Paris, France.

41. Organization for Economic Co-operation and Development. 1989. OECD Guideline for the Testing of Chemicals 117: Partition coefficient (*n*-octanol/water): High performance liquid chromatography (HPLC) method. OECD Environment Directorate, Paris, France.
42. Tolls J, Bodo K, de Felip E, Dujardin R, Kim YH, Moeller-Jensen L, Mullee D, Nakajima A, Paschke A, Pawliczek JB, Schneider J, Tadeo JL, Tognucci AC, Webb J, Zwijzen AC. 2003. Slow-stirring method for determining the *n*-octanol/water partition coefficient (P_{ow}) for highly hydrophobic chemicals: Performance evaluation in a ring test. *Environ Toxicol Chem* 22:1051-1057.
43. Organization for Economic Co-operation and Development. 2006. OECD Guideline for the Testing of Chemicals 123: Partition coefficient (*n*-octanol/water): Slow-stirring method. OECD Environment Directorate, Paris, France.
44. Sangster J. 1997. *Octanol-Water Partition Coefficients: Fundamentals and Physical Chemistry.* Wiley, Chichester, UK.
45. Schüürmann G. 1998. Ecotoxic modes of action of chemical substances. In: Schüürmann G, Markert B, eds, *Ecotoxicology*. John Wiley and Spektrum Akademischer Verlag, New York, NY, pp. 665-749.
46. Organization for Economic Co-operation and Development. 2000. OECD Guideline for the Testing of Chemicals 122: Partition coefficient (*n*-octanol/water): pH-Metric method for ionisable substances (proposal). OECD Environment Directorate, Paris, France.
47. Slater B, McCormack A, Avdeef A, Comer JEA. 1994. pH-metric log P: 4. Comparison of partition-coefficients determined by HPLC and potentiometric methods to literature values. *J Pharm Sci* 83:1280-1283.
48. Karickhoff SW. 1981. Semi-empirical estimation of sorption of hydrophobic pollutants on natural sediments and soils. *Chemosphere* 10:833-846.
49. Doucette WJ. 2003. Quantitative structure-activity relationships for predicting soil-sediment sorption coefficients for organic chemicals. *Environ Toxicol Chem* 22:1771-1788.
50. Mackay D. 1982. Correlation of bioconcentration factors. *Environ Sci Technol* 16:274-278.
51. Connell DW. 1988. Bioaccumulation behavior of persistent organic-chemicals with aquatic organisms. *Rev Environ Contam Toxicol* 102:117-154.
52. Connell DW. 1990. *Bioaccumulation of Xenobiotic Compounds.* CRC Press, Boca Raton, FL.
53. Barber MC. 2003. A review and comparison of models for predicting dynamic chemical bioconcentration in fish. *Environ Toxicol Chem* 22:1963-1992.
54. Van Leeuwen CJ, van der Zandt PTJ, Aldenberg T, Verhaar HJM, Hermens JLM. 1992. Application of QSARs, extrapolation and equilibrium partitioning in aquatic effects assessment. 1. Narcotic industrial pollutants. *Environ Toxicol Chem* 11:267-282.
55. Moore DRJ, Breton RL, MacDonald DB. 2003. A comparison of model performance for six quantitative structure-activity relationship packages that predict acute toxicity to fish. *Environ Toxicol Chem* 22:1799-1809.
56. Russom CL, Bradbury SP, Broderius SJ, Hammermeister DE, Drummond RA. 1997. Predicting modes of toxicity action from chemical structure: acute toxicity in the fathead minnow *(Pimephales promelas). Environ Toxicol Chem* 16:948-967.
57. Schüürmann G, Somashekar RK, Kristen U. 1996. Structure-activity relationships for chloro- and nitrophenol toxicity in the pollen tube growth test. *Environ Toxicol Chem* 15:1702-1708.
58. Sijm DTHM, Schipper M, Opperhuizen A. 1993. Toxicokinetics of halogenated benzenes in fish - lethal body burden as a toxicological end-point. *Environ Toxicol Chem* 12:1117-1127.
59. Meylan WM. 2004. KOWWIN 1.67. Syracuse Research Corporation, Syracuse, NY.
60. Tetko IV, Tanchuk VYu, Kasheva TN, Villa AEP. 2001. Internet software for the calculation of the lipophilicity and aqueous solubility of chemical compounds. *J Chem Inform Comput Sci* 41:246-252.
61. Beauman JA, Howard PH. 1996. Physprop database. Syracuse Research Corporation, Syracuse, NY. http://www.syrres.com/esc/databases.htm.
62. Advanced Pharma Algorithms Inc. 2006. ADME Boxes 3.5 Build 12. Toronto, ON, Canada.
63. Marrero J, Gani R. 2002. Group-contribution-based estimation of octanol/water partition coefficient and aqueous solubility. *Ind Eng Chem Res* 41:6623-6633.
64. Abraham MH, Chadha HS, Whiting GS, Mitchell RC. 1994. Hydrogen Bonding. 32. An analysis of water-octanol and water-alkane partitioning and the Δlog *P* parameter of Seiler. *J Pharm Sci* 83:1085-1100.
65. Platts JA, Butina D, Abraham MH, Hersey A. 1999. Estimation of molecular linear free energy relation descriptors using a group contribution approach. *J Chem Inform Comput Sci* 39:835-845.
66. Daylight Chemical Information Systems, Inc. 1998. CLOGP 4.61. Irvine, CA.
67. Chou JT, Jurs PC. 1979. Computer-assisted computation

of partition-coefficients from molecular-structures using fragment constants. *J Chem Inform Comput Sci* 19:172-178.

68. Hansch C, Leo AJ. 1979. *Substituent Constants for Correlation Analysis in Chemistry and Biology.* Wiley, New York, NY.
69. Lyman WJ, Reehl WF, Rosenblatt DH, eds. 1990. *Handbook of Chemical Property Estimation Methods.* American Chemical Society, Washington DC.
70. Pliska V, Testa B. 1996. *Lipophilicity in Drug Action and Toxicology.* Wiley-VCH, Weinheim, Germany.
71. Mackay D. 2000. Solubility in water. In: Boethling RS, Mackay D, eds, *Handbook of Property Estimation Methods for Chemicals: Environmental and Health Sciences*, 1st ed. CRC Press, Boca Raton, FL, pp. 125-138.
72. Leo A. 2000. Octanol/water partition coefficient. In: Boethling RS, Mackay D, eds. CRC Press, Boca Raton, FL, pp. 89-114.
73. Meylan WM, Howard, PH. 1994. SRC-TR-94-024. Validation of water solubility estimation methods using Log K_{ow} for application in PCGEMS & EPI. Syracuse Research Corporation, NY, USA.
74. Schüürmann G, Ebert R-U, Kühne R. 2006. Prediction of the sorption of organic compounds into soil from molecular structure. *Environ Sci Technol* 40:7005-7011.
75. Staudinger J, Roberts PV. 1996. A critical review of Henry's law constant for environmental applications. *Crit Rev Environ Sci Technol* 26:205-297.
76. Sangster J. 2003. The experimental measurement of Henry´s law constant. In: Fogg P, Sangster J, eds, *Chemicals in the Atmosphere – Solubility, Sources and Reactivity.* Wiley, Chichester, UK, pp. 53-67.
77. Raal JD, Ramjugernath D. 2005. Measurements of limiting activity coefficients: non-analytical tools. In: Weir RD, de Loos TW, eds, *Measurement of the Thermodynamic Properties of Multiple Phases.* IUPAC & Elsevier, Amsterdam, The Netherlands, pp. 339-356.
78. Dohnal V. 2005. Measurements of limiting activity coefficients using analytical tools. In: Weir RD, de Loos TW, eds, *Measurement of the Thermodynamic Properties of Multiple Phases.* IUPAC & Elsevier, Amsterdam, The Netherlands, pp. 359-377.
79. Mackay D, Shiu WY, Sutherland RP. 1979. Determination of air-water Henry's law constants for hydrophobic compounds. *Environ Sci Technol* 13:333-337.
80. Hovorka S, Dohnal V. 1997. Determination of air-water partitioning of volatile halogenated hydrocarbons by the inert gas stripping method. *J Chem Eng Data* 42:924-933.
81. Kolb B, Ettre LS. 2006. *Static Headspace Analysis – Theory and Practice.* Wiley, Hoboken, NJ.
82. Gosset JM. 1987. Measurement of Henry's law constants for C_1 and C_2 chlorinated hydrocarbons. *Environ Sci Technol* 21:202-208.
83. Kolb B, Welter C, Bichler C. 1992. Determination of partition coefficients by automatic equilibrium headspace gas chromatography by vapor phase calibration. *Chromatographia* 34:235-240.
84. Ettre LS, Welter C, Kolb B. 1993. Determination of gas-liquid partition coefficients by automatic equilibrium headspace-gas chromatography utilizing the phase ratio variation method. *Chromatographia* 35:73-84.
85. Robbins G, Wang S, Stuart JD. 1993. Using the static headspace method to determine Henry's law constants. *Anal Chem* 65:3113-3118.
86. Pawliszyn J. 1997. *Solid Phase Microextraction – Theory and Practice.* Wiley-VCH, New York, NY.
87. Suntio LR, Shiu WY, Mackay D, Seiber JN, Glotfelty D. 1988. Critical review of Henry's law constants for pesticides. *Rev Environ Contam Toxicol* 103:1-59.
88. Mackay D, Shiu WY, Ma KC, Lee SC. 2006. *Handbook of Physical-Chemical Properties and Environmental Fate for Organic Chemicals.* CRC Press, Boca Raton, FL.
89. Eastcott L, Shiu WY, Mackay D. 1988. Environmentally relevant physical-chemical properties of hydrocarbons: A review of data and development of simple correlations. *Oil Chem Pollut* 4:191-216.
90. Kühne R, Ebert R-U, Schüürmann G. 2005. Prediction of the temperature dependency of Henry's Law constant from chemical structure. *Environ Sci Technol* 39:6705-6711.
91. Dearden JC, Schüürmann G. 2003. Quantitative structure-property relationships for predicting Henry's law constant from molecular structure. *Environ Toxicol Chem* 22:1755-1770.
92. Hine J, Mookerjee PK. 1975. The intrinsic hydrophilic character of organic compounds. Correlations in terms of structural contributions. *J Org Chem* 40:292-298.
93. Meylan WM, Howard PH. 1991. Bond contribution method for estimating Henry's law constants. *Environ Toxicol Chem* 10:1283-1293.
94. Meylan WM. 2000. HENRYWIN 3.1. Syracuse Research Corporation, Syracuse, NY.
95. Nirmalakhandan NN, Speece RE. 1988. QSAR model for predicting Henry's constant. *Environ Sci Technol* 22:1349-1357.
96. Nirmalakhandan N, Brennan RA, Speece RE. 1997. Predicting Henry's Law constant and the effect of temperature on Henry's Law constant. *Wat Res* 31:1471-1481.
97. Abraham MH, Andonian-Haftvan J, Whiting GS, Leo A, Taft RS. 1994. Hydrogen bonding. Part 34. The factors that influence the solubility of gases and vapors in water

at 298 K, and a new method for its determination. *J Chem Soc Perkin Trans* 2:1777-1791.

98. Kile DE, Chiou CT, Zhou HD, Li H, Xu OY. 1995. Partition of nonpolar organic pollutants from water to soil and sediment organic matters. *Environ Sci Technol* 29:1401-1406.
99. Chiou CT, Kile DE. 1998. Deviations from sorption linearity on soils of polar and nonpolar organic compounds at low relative concentrations. *Environ Sci Technol* 32:338-343.
100. Chiou CT. 2002. *Partition and Adsorption of Organic Contaminants in Environmental Systems.* Wiley, Hoboken, NJ, USA.
101. Tinsley IJ. 2004. *Chemical Concepts in Pollutant Behaviour.* John Wiley, New York, NY.
102. Cornelissen G, Gustafsson Ö, Bucheli TD, Jonker MTO, Koelmans AA, van Noort PCM. 2005. Extensive sorption of organic compounds to black carbon, coal, and kerogen in sediments and soils: mechanisms and consequences for distribution, bioaccumulation, and biodegradation. *Environ Sci Technol* 18:6881-6895.
103. Burkhard LP. 2000. Estimating dissolved organic carbon partition coefficients for nonionic organic chemicals. *Environ Sci Technol* 34:4663-4668.
104. Boethling RS, Mackay D, eds. 2000. *Handbook of Property Estimation Methods for Chemicals: Environmental and Health Sciences.* CRC Press, Boca Raton, FL.
105. Spurlock FC, Biggar JW. 1994. Thermodynamics of organic chemical partition in soils. 1. Development of a general partition model and application to linear isotherms. *Environ Sci Technol* 28:989-995.
106. Chiou CT. 1995. Thermodynamics of organic-chemical partition in soils - comment. *Environ Sci Technol* 29:1421-1422.
107. Organization for Economic Co-operation and Development. 2000. OECD Guideline for the Testing of Chemicals 106: Adsorption-desorption using a batch equilibration method. OECD Environment Directorate, Paris, France.
108. Heringa MB, Hermens JLM. 2003. Measurement of free concentrations using negligible depletion-solid phase microextraction (nd-SPME). *Trac-Trends in Analytical Chemistry* 22:575-587.
109. Hawthorne SB, Grabanski CB, Miller DJ, Kreitinger JP. 2005. Solid-phase microextraction measurement of parent and alkyl polycyclic aromatic hydrocarbons in milliliter sediment pore water samples and determination of K-DOC values. *Environ Sci Technol* 39:2795-2803.
110. MacIntyre WG, Stauffer TB, Antworth CP. 1991. A comparison of sorption coefficients determined by batch, column, and box methods on a low organic-carbon aquifer material. *Ground Water* 29:908-913.
111. Organization for Economic Co-operation and Development. 2001. OECD Guideline for the Testing of Chemicals 121: Estimation of the adsorption coefficient (K_{oc}) on soil and on sewage sludge using HPLC. OECD Environment Directorate, Paris, France.
112. Delle Site A. 2001. Factors affecting sorption of organic compounds in natural sorbent/water systems and sorption coefficients for selected pollutants. A review. *J Phys Chem Ref Data* 30:187-439.
113. Vermeire T, Rikken M, Attias L, Boccardi P, Boeije G, De Bruijn J, Brooke D, Comber M, Dolan B, Fischer S, Heinemeyer G, Koch V, Lijzen J, Müller B, Murray-Smith R, Tadeo J. 2005. European Union system for the evaluation of substances: the second version. *Chemosphere* 59:473-485.
114. Sabljić A, Güsten H, Verhaar H, Hermens J. 1995. QSAR modelling of soil sorption. Improvements and systematics of log K_{oc} vs. log K_{ow} correlations. *Chemosphere* 31:4489-4514.
115. Sabljić A, Güsten H, Verhaar H, Hermens J. 1996. QSAR modelling of soil sorption. Improvements and systematics of K_{oc} vs. log K_{ow} correlations (Vol 31, pp. 4489, 1995). *Chemosphere* 33:2577.
116. Meylan WM. 2000. PCKOCWIN 1.66. Syracuse Research Corporation, Syracuse, NY.
117. Meylan WM, Howard PH, Boethling RS. 1992. Molecular topology/fragment contribution method for predicting soil sorption coefficients. *Environ Sci Technol* 26:1560-1567.
118. Estrada E. 1995. Edge adjacency relationships and a novel topological index related to molecular volume. *J Chem Inform Comput Sci* 35:31-33.
119. Kier LB, Hall LH. 2000. Intermolecular accessibility: the meaning of molecular connectivity. *J Chem Inform Comput Sci* 40:792-795.
120. Huuskonen J. 2003. Prediction of soil sorption coefficient of a diverse set of organic chemicals from molecular structure. *J Chem Inform Comput Sci* 43:1457-1462.
121. Poole SK, Poole CF. 1999. Chromatographic models for the sorption of neutral organic compounds by soil from water and air. *J Chromatogr A* 845:381-400.
122. Sabljić A. 1987. On the prediction of soil sorption coefficients of organic pollutants from molecular structure: application of molecular topology model. *Environ Sci Technol* 21:358-366.
123. Nguyen TH, Goss KU, Ball WP. 2005. Polyparameter linear free energy relationships for estimating the equilibrium partition of organic compounds between water and the natural organic matter in soils and sediments. *Environ Sci Technol* 39:913-924.
124. Lohninger H. 1994. Estimation of soil partition

coefficients of pesticides from their chemical structure. *Chemosphere* 29:1611-1626.

125. Howard PH. 1995. Chemfate database 1.3. Syracuse Research Corporation, Syracuse, NY.
126. Baker JR, Mihelcic JR, Sabljic A. 2001. Reliable QSAR for estimating Koc for persistent organic pollutants: correlation with molecular connectivity indices. *Chemosphere* 45:213-221.
127. Tao S, Piao H, Dawson R, Lu X, Hu H. 1999. Estimation of organic carbon normalized sorption coefficient (K_{oc}) for soils using the fragment constant method. *Environ Sci Technol* 33:2719-2725.
128. Organization for Economic Co-operation and Development. 1995. OECD Guideline for Testing of Chemicals No. 104: Vapour pressure. OECD Environment Directorate, Paris, France.
129. Verevkin SP. 2005. Phase changes in pure component systems: liquids and gases. In: Weir RD, de Loos TW, eds, *Measurement of the Thermodynamic Properties of Multiple Phases.* IUPAC & Elsevier, Amsterdam, The Netherlands, pp. 5-26.
130. Koutek B, Cvacka J, Streinz L, Vrkocova P, Doubsky J, Simonova H, Feltl L, Svoboda V. 2001. Comparison of methods employing gas chromatography retention data to determine vapour pressures at 298 K. *J Chromatogr A* 923:137-152.
131. Letcher TM, Naicker PK. 2004. Determination of vapor pressures using gas chromatography. *J Chromatogr A* 1037:107-114.
132. Paschke A, Schröter U, Schüürmann G. 2005. Indirect determination of low vapour pressures using solid-phase microextraction - application to tetrachlorobenzenes and tetrachlorobenzyltoluenes. *J Chromatogr A* 1072:93-97.
133. Dearden JC. 2003. Quantitative structure-property relationships for prediction of boiling point, vapor pressure, and melting point. *Environ Toxicol Chem* 22:1696-1709.
134. Prausnitz JM, Lichtenthaler RN, de Azevedo EG. 1986. *Molecular Thermodynamics of Fluid-Phase Equilibria.* Prentice Hall PTR, Englewood Cliffs, NJ.
135. Walden P. 1908. Über Schmelzwärme, spezifische Kohäsion und Molekulargröße bei der Schmelztemperatur. *Zeitschrift für Elektrochemie* 14:713-728. [in German]
136. Antoine C. 1888. Tensions de vapours: nouvelle relation entre les tensions et les temperatures. *Compt Rend* 107:681-684. [in French]
137. Grain CF. 1990. Vapor pressure. In: Lyman WJ, Reehl WF, Rosenblatt DH, eds, *Handbook of Chemical Property Estimation Methods,* 3rd ed. American Chemical Society, Washington DC, USA, pp. 14-1-14-20.
138. Fishtine SH. 1963. Reliable latent heats of vaporization. *Ind Eng Chem* 55:47-60.
139. Constantinou L, Gani R. 1994. A new group contribution method for the estimation of properties of pure compounds. *AIChe Journal* 40:1697-1710.
140. Stein SE, Brown RL. 1994. Estimation of normal boiling points from group contribution. *J Chem Inform Comput Sci* 34:581-587.
141. Joback KG, Reid RC. 1987. Estimation of pure-component properties from group-contributions. *Chem Eng Commun* 57:233-243.
142. Marrero J, Gani R. 2001. Group-contribution based estimation of pure component properties. *Fluid Phase Equilibria* 183-184:183-208.
143. Sage ML, Sage GW. 2000. Vapor pressure. In: Boethling RS, Mackay D, eds, *Handbook of Property Estimation Methods for Chemicals: Environmental and Health Sciences,* 1st ed. CRC Press, Boca Raton, FL pp. 53-65.
144. Mackay D, Bobra A, Chan DW, Shiu WY. 1982. Vapor pressure correlations for low-volatility environmental chemicals. *Environ Sci Technol* 16:645-649.
145. Mishra DS, Yalkowsky SH. 1991. Estimation of vapor pressure of some organic compounds. *Ind Eng Chem Res* 30:1609-1612.
146. Myrdal PB, Yalkowsky SH. 1997. Estimating pure component vapor pressures of complex organic molecules. *Ind Eng Chem Res* 36:2494-2499.
147. Sepassi K, Myrdal PB, Yalkowsky SH. 2006. Estimating pure-component vapor pressures of complex organic molecules: Part II. *Ind Eng Chem Res* 45:8744-8747.
148. Meylan, WM. 2000. MPBPWIN 1.41. Syracuse Research Corporation, Syracuse, NY.
149. Meylan, WM. 2000. EPIWIN 3.12. Syracuse Research Corporation, Syracuse, NY.
150. Advanced Chemistry Development Inc. 2001. ACD/PhysChem Batch 5.13. Advanced Chemistry Development Inc., Toronto, ON, Canada.
151. Kühne R, Ebert R-U, Schüürmann G. 1997. Estimation of vapour pressures for hydrocarbons and halogenated hydrocarbons from chemical structure by a neural network. *Chemosphere* 34:671-686.
152. Hawker DW, Connell DW. 1985. Relationship between partition coefficient, uptake rate constant, clearance rate constant and time to equilibrium for bioaccumulation. *Chemosphere* 14:1205-1219.
153. De Wolf W, De Bruijn JHM, Seinen W, Hermens JLM. 1992. Influence of biotransformation on the relationship between bioconcentration factors and octanol water partition-coefficients. *Environ Sci Technol* 26:1197-1201.
154. Dimitrov S, Dimitrova N, Parkerton T, Comber M, Bonnel M, Mekenyan O. 2005. Base-line model for identifying the bioaccumulation potential of chemicals. *SAR QSAR Environ Res* 16:531-554.

155. Organization for Economic Co-operation and Development. 1996. OECD Guideline for Testing of Chemicals No. 305: Bioconcentration: Flow-through fish test. OECD Environment Directorate, Paris, France.
156. Arnot JA, Gobas FAPC. 2006. A review of bioconcentration factor (BCF) and bioaccumulation factor (BAF) assessments for organic chemicals in aquatic organisms. *Environ Rev* 14:257-297.
157. Neely WB, Branson DR, Blau GE. 1974. Partition coefficient to measure bioconcentration potential of organic chemicals to fish. *Environ Sci Technol* 8:1113-1115.
158. Spacie A, Hamelink JL. 1982. Alternative models for describing the bioconcentration of organics in fish. *Environ Toxicol Chem* 1:309-320.
159. Gobas FAPC, Mackay D. 1987. Dynamics of hydrophobic organic chemical bioconcentration in fish. *Environ Toxicol Chem* 6:495-504.
160. Hawker DW, Connell DW. 1988. Octanol-water partition coefficient of PCB congeners. *Environ Sci Technol* 22:382-387.
161. Clark KE, Gobas FAPC, Mackay D. 1990. Model of organic-chemical uptake and clearance by fish from food and water. *Environ Sci Technol* 24:1203-1213.
162. Mackay D, Fraser A. 2000. Bioaccumulation of persistent organic chemicals: mechanisms and models. *Environ Pollut* 110:375-391.
163. Baron MG. 1990. Will water-borne organic chemicals accumulate in aquatic animals? *Environ Sci Technol* 1612-1618.
164. Gobas FAPC, Morrison HA. 2000. Bioconcentration and biomagnification in the aquatic environment. In: Boethling RS, Mackay D, eds, *Handbook of Property Estimation Methods for Chemicals: Environmental and Health Sciences,* 1st ed. CRC Press, Boca Raton, FL, pp. 189-231.
165. Schüürmann G, Klein W. 1988. Advances in bioconcentration prediction. *Chemosphere* 17:1551-1574.
166. Nendza M. 1991. QSARs of bioconcentration: validity assessment of log Pow/log BCF correlations. In: Nagel R, Loskill R, eds, *Bioaccumulation in Aquatic Systems.* Wiley-VCH, Weinheim, Germany, pp. 43-66.
167. Connell DW. 1991. Extrapolating Test Results on Bioaccumulation between Organism Groups. In: Nagel R, Loskill R, eds, *Bioaccumulation in Aquatic Systems.* Wiley-VCH, Weinheim, Germany, pp. 133-149.
168. Meylan WM, Howard PH, Boethling RS, Aronson D, Printup H, Gouche S. 1999. Improved method for estimating bioconcentration/bioaccumulation factor from octanol/water partition coefficient. *Environ Toxicol Chem* 18:664-672.
169. Zitko V, Carson WG. 1977. Uptake and excretion of chlorinated diphenyl ethers and brominated toluenes by fish. *Chemosphere* 6:293-301.
170. Sugiura K, Ito N, Matsumoto N, Mihara Y, Murata K, Tsukakoshi Y, Goto M. 1978. Accumulation of polychlorinated biphenyls and polybrominated biphenyls in fish - limitation of correlation between partition-coefficients and accumulation factors. *Chemosphere* 7:731-736.
171. Tulp MTM, Hutzinger O. 1978. Thoughts on aqueous solubilities and partition-coefficients of PCB, and mathematical correlation between bioaccumulation and physicochemical properties. *Chemosphere* 7:849-860.
172. Könemann H, Van Leeuwen K. 1980. Toxicokinetics in fish: accumulation and elimination of six chlorobenzenes by guppies. *Chemosphere* 9:3-19.
173. Bruggeman WA, Opperhuizen A, Wijbenga A, Hutzinger O. 1984. Bioaccumulation of super-lipophilic chemicals in fish. *Toxicol Environ Chem* 7:173-189.
174. Zitko V, Hutzinger O. 1976. Uptake of chlorobiphenyls and bromobiphenyls, hexachloromobenzene and hexabromobenzene by fish. *Bull Environ Contam Toxicol* 16:665-673.
175. Opperhuizen A, Van der Velde EW, Gobas FAPC, Liem DAK, van der Steen JMD. 1985. Relationship between bioconcentration in fish and steric factors of hydrophobic chemicals. *Chemosphere* 14:1871-1896.
176. Hawker DW, Connell DW. 1986. Bioconcentration of lipophilic compounds by some aquatic organisms. *Ecotox Environ Saf* 11:184-197.
177. Connell DW, Hawker DW. 1988. Use of polynomials to describe the bioconcentration of hydrophobic chemicals by fish. *Ecotox Environ Saf* 16:242-257.
178. Sijm DTHM, Opperhuizen A. 1988. Biotransformation, bioaccumulation and lethality of 2,8-dichlorodibenzo-p-dioxin: a proposal to explain the biotic fate and toxicity of PCCD's and PCDF's. *Chemosphere* 17:83-99.
179. Opperhuizen A, Sijm DTHM. 1990. Bioaccumulation and biotransformation of polychlorinated dibenzo-para-dioxins and dibenzofurans in fish. *Environ Toxicol Chem* 9:175-186.
180. Gobas FAPC, Muir DCG, Mackay D. 1988. Dynamics of dietary bioaccumulation and faecal elimination of hydrophobic organic chemicals in fish. *Chemosphere* 17:943-962.
181. Gobas FAPC, Clark KE, Shiu WY, Mackay D. 1989. Bioconcentration of polybrominated benzenes and biphenyls and related superhydrophobic chemicals in fish - Role of bioavailability and elimination into the feces. *Environ Toxicol Chem* 8:231-245.

182. Banerjee S, Baughman GL. 1991. Bioconcentration factors and lipid solubility. *Environ Sci Technol* 25:536-539.
183. Chessells M, Hawker DW, Connell DW. 1992. Influence of solubility in lipid on bioconcentration of hydrophobic compounds. *Ecotox Environ Saf* 23:260-273.
184. McCarthy JF. 1983. Role of particulate organic-matter in decreasing accumulation of polynuclear aromatic-hydrocarbons by daphnia-magna. *Arch Environ Contam Toxicol* 12:559-568.
185. McCarthy JF, Jimenez BD. 1985. Reduction in bioavailability to bluegills of polycyclic aromatic-hydrocarbons bound to dissolved humic material. *Environ Toxicol Chem* 4:511-521.
186. Black MC, McCarthy JF. 1984. Dissolved organic macromolecules reduce the uptake of hydrophobic organic contaminants by the gills of rainbow trout (Salmo gairdneri). *Environ Toxicol Chem* 7:593-600.
187. Servos MR, Muir DCG, Webster GRB. 1989. The effect of dissolved organic-matter on the bioavailability of polychlorinated dibenzo-para-dioxins. *Aquat Toxicol* 14:169-184.
188. Schrap SM, Opperhuizen A. 1990. Relationship between bioavailability and hydrophobicity - Reduction of the uptake of organic-chemicals by fish due to the sorption of particles. *Environ Toxicol Chem* 9:715-724.
189. McKone TE, Hall D, Kastenberg WE. 1997. CalTOX 2.3. University of California, Berkeley, CA.
190. Veith GD, de Foe DL, Bergstaedt DV. 1979. Measuring and estimating the bioconcentration factor of chemicals in fish. *J Fish Res Board Can* 36:1040-1048.
191. European Commission. 1996. Technical Guidance Document in support of the Commission Regulation (EC) 1488/94 on risk assessment for existing chemicals, Part III. Luxemburg, Belgium. Office for Official Publications of the European Communities. Luxembourg.
192. Bintein S, Devillers J, Karcher W. 1993. Nonlinear dependence of fish bioconcentration on *n*-octanol/water partition coefficient. *SAR QSAR Environ Res* 1:29-39.
193. Kubinyi H. 1976. Quantitative structure-activity-relationships .4. Nonlinear dependence of biological-activity on hydrophobic character - new model. *Arzneimittel-Forschung/Drug Research* 26:1991-1997.
194. Kubinyi H. 1979. Non-linear dependence of biological-activity on hydrophobic character - bilinear model. *Farmaco-Edizione Scientifica* 34:248-276.
195. Kubinyi H. 1979. Lipophilicity and drug activity. Prog Drug Res 23:97-198.
196. Dimitrov S, Breton R, MacDonald D, Walker JD, Mekenyan O. 2002. Quantitative prediction of biodegradability, metabolite distribution and toxicity of stable metabolites. *SAR QSAR Environ Res* 13:445-455.
197. Meylan, WM. 2000. BCFWIN 2.15. Syracuse Research Corporation, Syracuse, NY.
198. Dimitrov SD, Dimitrova NC, Walker JD, Veith GD, Mekenyan OG. 2003. Bioconcentration potential predictions based on molecular attributes - an early warning approach for chemicals found in humans, birds, fish and wildlife. *QSAR Comb Sci* 22:58-68.
199. Leo A, Hansch C. 1971. Linear free-energy relationships between partitioning solvent systems. *J Org Chem* 36:1539-1544.
200. Kishi H, Hashimoto Y. 1989. Evaluation of the procedures for the measurement of water solubility and normal-octanol water partition-coefficient of chemicals results of a ring test in Japan. *Chemosphere* 18:1749-1759.
201. Organization for Economic Co-operation and Development. 1995. OECD Guideline for Testing of Chemicals No. 105: Water solubility. OECD Environment Directorate, Paris, France.
202. Roberts GF, Oliver BG. 1997. The preparation and validation of generator column for highly hydrophobic materials. In: Clement RE, Keith LH, Siu KWM, eds, *Reference Materials for Environmental Analysis.* CRC-Lewis, Boca Raton, FL, pp. 113-127.
203. Dohanyosova P, Dohnal V, Fenclova D. 2003. Temperature dependence of aqueous solubility of anthracenes: accurate determination by a new generator column apparatus. *Fluid Phase Equilibria* 214:151-167.
204. Arbuckle WB. 1983. Estimating activity coefficients for use in calculating environmental parameters. *Environ Sci Technol* 17:537-542.
205. Banerjee S. 1985. Calculation of water solubility of organic compounds with UNIFAC-derived activity coefficients. *Environ Sci Technol* 19:369-370.
206. Arbuckle WB. 1986. Using UNIFAC to calculate aqueous solubility. *Environ Sci Technol* 20:1060-1064.
207. Chen F, Holten-Andersen J, Tyle H. 1993. New development of the UNIFAC model for environmental applications. *Chemosphere* 26:1325-1354.
208. Poling BE, Prausnitz JM, O'Connell JP. 2000. *The Properties of Gases and Liquids.* McGraw-Hill Book Company, New York, NY.
209. Hansch C, Quinlan JE, Lawrence GL. 1968. The linear free-energy relationship between partition coefficients and the aqueous solubility of organic liquids. *J Org Chem* 33:347-350.
210. Yalkowsky SH, Valvani SC. 1980. Solubility and partitioning I: Solubility of nonelectrolytes in water. *J Pharm Sci* 69:912-922.
211. Yalkowsky SH, Banerjee S. 1992. *Aqueous Solubility,*

Methods of Estimation for Organic Compounds. Marcel Dekker, New York, NY.

212. Jain N, Yalkowsky SH. 2001. Estimation of the aqueous solubility I: Application to organic nonelectrolytes. *J Pharm Sci* 90:234-252.
213. Ran Y, Yalkowsky SH. 2001. Prediction of drug solubility by the general solubility equation (GSE). *J Chem Inform Comput Sci* 41:354-357.
214. Ran Y, Jain N, Yalkowsky SH. 2001. Prediction of aqueous solubility of organic compounds by the general solubility equation (GSE). *J Chem Inform Comput Sci* 41:1208-1217.
215. Taskinen J. 2000. Prediction of aqueous solubility in drug design. Curr Opin Drug Discov Develop 3:102-107.
216. Dearden JC. 2006. *In silico* prediction of aqueous solubility. *Expert Opin Drug Discov* 1:31-52.
217. Delaney JS. 2005. Predicting aqueous solubility from structure. *Drug Discovery Today* 10:289-295.
218. Tiegs D, Gmehling J, Rasmussen P, Fredenslund A. 1987. Vapor-liquid equilibria by UNIFAC group contribution. 4. Revision and extension. *Ind Eng Chem Process Des Dev* 26:159-161.
219. Meylan WM, Howard PH, Boethling RS. 1996. Improved method for estimating water solubility from octanol/water partition coefficient. *Environ Toxicol Chem* 15:100-106.
220. Delaney JS. 2004. ESOL: Estimating aqueous solubility directly from molecular structure. *J Chem Inform Comput Sci* 44:1000-1005.
221. Myrdal PB, Ward GH, Dannenfelser RM, Mishra DS, Yalkowsky SH. 1992. AQUAFAC 1: Aqueous function of group activity coefficients; application to hydrocarbons. *Chemosphere* 24:1047-1061.
222. Myrdal PB, Ward GH, Simamora P, Yalkowsky SH. 1993. AQUAFAC: Aqueous functional group activity coefficients. *SAR QSAR Environ Res* 1:53-61.
223. Myrdal PB, Manka AM, Yalkowsky SH. 1995. AQUAFAC 3: Aqueous functional group activity coefficients; application to the estimation of aqueous solubility. *Chemosphere* 30:1619-1637.
224. Lee YC, Myrdal PB, Yalkowsky SH. 1996. Aqueous functional group activity coefficients (AQUAFAC) 4: Applications to complex organic compounds. *Chemosphere* 33:2129-2144.
225. Pinsuwan S, Myrdal PB, Yalkowsky SH. 1997. AQUAFAC 5: aqueous functional group activity coefficients; application to alcohols and acids. *Chemosphere* 35:2503-2513.
226. Meylan WM. 2002. WATERNT 1.01. Syracuse Research Corporation, Syracuse, NY.
227. Tetko IV, Tanchuk VYu, Kasheva TN, Villa AEP. 2001. Estimation of aqueous solubility of chemical compounds using E-state indices. *J Chem Inform Comput Sci* 41:1488-1493.
228. Taft RW, Abraham MH, Dougherty RM, Kamlet MJ. 1985. The molecular properties governing solubilities on nonelectrolytes in water. *Nature* 313:384-386.
229. Kamlet MJ, Doherty RM, Abraham MH, Carr PW, Doherty RF, Taft RW. 1987. Linear solvation energy relationships. 41. Important differences between aqueous solubility relationships for aliphatic and aromatic solutes. *J Phys Chem* 91:1996-2004.
230. Yalkowsky SH, Pinal R, Banerjee S. 1988. Water solubility: A critique of the solvatochromic approach. *J Pharm Sci* 77:74-77.
231. Abraham MH, Le J. 1999. The correlation and prediction of the solubility of compounds in water using an amended solvation energy relationship. *J Pharm Sci* 88:868-880.
232. Kühne R, Ebert R-U, Schüürmann G. 2006. Model selection based on structural similarity – method description and application to water solubility prediction. *J Chem Inf Model* 46:636-641.
233. Kühne R, Ebert R-U, Kleint F, Schmidt G, Schüürmann G. 1995. Group contribution methods to estimate water solubility of organic chemicals. *Chemosphere* 30:2061-2077.
234. Huuskonen J. 2000. Estimation of aqueous solubility for a diverse set of organic compounds based on molecular topology. *J Chem Inform Comput Sci* 40:773-777.
235. Atkinson R. 2000. Atmospheric oxidation. In: Boethling RS, Mackay D, eds, *Handbook of Property Estimation Methods for Chemicals: Environmental and Health Sciences.* CRC Press, Boca Raton, FL, pp. 335-354.
236. Larson RA, Weber EJ. 1994. *Reaction Mechanism in Environmental Organic Chemistry.* CRC Press, Boca Raton, FL.
237. Atkinson R, Carter WPL. 1984. Kinetics and mechanisms of the gas-phase reactions of ozone with organic-compounds under atmospheric conditions. *Chem Rev* 84:437-470.
238. Atkinson R. 1994. Gas-phase tropospheric chemistry of organic-compounds. *J Phys Chem Ref Data Monogr* 2:1-216.
239. Atkinson R. 1991. Kinetics and mechanisms of the gas-phase reactions of the NO3 radical with organic-compounds. *J Phys Chem Ref Data* 20:459-507.
240. Sabljic A, Peijnenburg W. 2001. Modeling lifetime and degradability of organic compounds in air, soil, and water systems - (IUPAC Technical Report). *Pure & Appl Chem* 73:1331-1348.
241. Meylan WM, Howard PH. 2003. A review of quantitative structure-activity relationship methods for the prediction

of atmospheric oxidation of organic chemicals. *Environ Toxicol Chem* 22:1724-1732.
242. Nendza M. 2004. Prediction of persistence. In: Cronin MTD, Livingstone DJ, eds, *Predicting chemical toxicity and fate*. CRC Press, Boca Raton, FL, pp. 315-331.
243. Nendza M. 1998. *Structure-Activity Relationships in Environmental Sciences*. Chapman & Hall, London, UK.
244. Atkinson R. 1986. Kinetics and mechanisms of the gas-phase reactions of the hydroxyl radical with organic-compounds under atmospheric conditions. *Chem Rev* 86:69-201.
245. Atkinson R. 1987. A structure-activity relationship for the estimation of rate constants for the gas-phase reactions of OH radicals with organic-compounds. *International Journal of Chemical Kinetics* 19:799-828.
246. Atkinson R. 1988. Estimation of gas-phase hydroxyl radical rate constants for organic-chemicals. *Environ Toxicol Chem* 7:435-442.
247. Kwok ESC, Atkinson R. 1995. Estimation of hydroxyl radical reaction rate constants for gas-phase organic compounds using a structure-reactivity relationship: an update. *Atmos Environ* 29:1685-1695.
248. Kwok ESC, Aschmann SM, Atkinson R. 1996. Rate constants for the gas-phase reactions of the OH radical with selected carbamates and lactates. *Environ Sci Technol* 30:329-334.
249. Meylan WM. 2000. AOPWIN 1.91. Syracuse Research Corporation, Syracuse, NY.
250. European Commission. 2003. Technical Guidance Document on risk assessment in support of Commission Directive 93/67/EEC on risk assessment for new notified substances, Commission Regulation (EC) No. 1488/94 on risk assessment for existing substances, Directive 98/8/EC of the European Parliament and of the Council concerning the placing of biocidal products on the market. European Commission, Joint Research Centre, Ispra, Italy.
251. Güsten H, Klasinc L, Dubravko M. 1984. Prediction of the abiotic degradability of organic-compounds in the troposphere. *J Atmos Chem* 2:83-93.
252. Müller M, Klein W. 1991. Estimating atmospheric degradation processes by SARs. *Sci Total Environ* 109:261-273.
253. Organization for Economic Co-operation and Development. 1993. Application of structure-activity relationships to the estimation of properties important in exposure assessment. OECD Environment Monograph No. 67. OECD, Paris, France.
254. Klamt A. 1993. Estimation of gas-phase hydroxyl radical rate constants of organic compounds from molecular orbital calculations. *Chemosphere* 26:1273-1289.
255. Klamt A. 1996. Estimation of gas-phase hydroxyl radical rate constants of oxygenated compounds based on molecular orbital calculations. *Chemosphere* 32:717-726.
256. Wolfe NL, Jeffers PM. 2000. Hydrolysis. In: Boethling RS, Mackay D, eds, *Handbook of Property Estimation Methods for Chemicals: Environmental and Health Sciences*. CRC Press, Boca Raton, FL, pp. 311-334.
257. Mabey W, Mill T. 1978. Critical review of hydrolysis of organic compounds in water under environmental conditions. *J Phys Chem Ref Data* 7:383-415.
258. Harris JC. 1990. Rate of hydrolysis. In: Lyman WJ, Reehl WF, Rosenblatt DH, eds, *Handbook of Chemical Property Estimation Methods,* 3rd ed. American Chemical Society, Washington, DC, pp. 7-1-7-48.
259. Peijnenburg WJGM, Dobbs AJ, Malz FR, Waldman M, Wolfe NL, Solomon KR, Tetlow JA, Dolejs P, Pitter P, Ewald M, Dobrevsky ID, Trobisch KH, Dobolyi E, Khan MMT, Stevenson CD, Boybay M. 1991. The use of quantitative structure-activity-relationships for predicting rates of environmental hydrolysis processes. *Pure & Appl Chem* 63:1667-1676.
260. Karickhoff SW, Mcdaniel VK, Melton C, Vellino AN, Nute DE, Carreira LA. 1991. Predicting chemical-reactivity by computer. *Environ Toxicol Chem* 10:1405-1416.
261. Hilal SH, Karickhoff SW, Carreira LA, Shrestha BP. 2003. Estimation of carboxylic acid ester hydrolysis rate constants. *QSAR Comb Sci* 22:917-925.
262. Whiteside TS, Hilal SH, Carreira LA. 2006. Estimation of phosphate ester hydrolysis rate constants. I. Alkaline hydrolysis. *QSAR Comb Sci* 25:123-133.
263. Mill T. 1993. Environmental chemistry. In: Suter GW, II, ed, *Ecological Risk Assessment*. Lewis Publishers, Chelsea, MI, pp. 91-127.
264. Meylan WM. 2004. HYDROWIN 1.67. Syracuse Research Corporation, Syracuse, NY.
265. Mill T, Haag W, Penwell P, Pettit T, Johnson H. Environmental fate and exposure studies. Development of a PC-SAR for hydrolysis: esters, alkyl halides and epoxides. SRI Project 3247. Summary report: Tasks 32 and 50. SRI International, Menlo Park, CA.
266. Hilal S, Karickhoff S, Carreira L. 1995. A rigorous test for SPARC's chemical reactivity models: estimation of more than 4300 ionization pK_as. *Quant Struct -Act Relat* 14:348-355.
267. Kollig HP, Ellington JJ, Karickhoff SW, Kitchens BE, Kollig HP, Long JM, Weber EJ, Wolfe NL. Environmental fate constants for organic chemicals under consideration for EPA's hazardous waste identification projects. EPA/600/R-93/132. US Environmental Protection

Agency, Washington, DC.

268. Crosby DG. 1998. *Environmental Toxicology and Chemistry.* Oxford University Press, New York, NY.

269. Manahan SE. 1990. *Environmental Chemistry*. Lewis Publishers, Chelsea, MI, USA.

270. Alexander M. 1998. *Biodegradation and Bioremediation*. Academic Press, San Diego, CA.

271. Howard P.H. 2000. Biodegradation. In: Boethling RS, Mackay D, eds, *Handbook of Property Estimation Methods for Chemicals: Environmental and Health Sciences.* CRC Press, Boca Raton, FL, pp. 281-310.

272. Pitter P, Chudoba J. 1990. *Biodegradability of Organic Substances in the Aquatic Environment.* CRC Press, Boca Raton, FL.

273. Syracuse Research Corporation. BIODEG. Syracuse, NY. http://www.syrres.com/esc/biodeg.htm.

274. Tunkel J, Howard PH, Boethling RS, Stiteler W, Loonen H. 2000. Predicting ready biodegradability in the Japanese Ministry of International Trade and Industry test. *Environ Toxicol Chem* 19:2478-2485.

275. Organization for Economic Co-operation and Development. 1994. OECD Guideline for Testing of Chemicals No. 301: Ready biodegradability. OECD Environment Directorate, Paris, France.

276. Boethling RS. 1993. Biodegradation of xenobiotic compounds. In: Corn M, ed, *Handbook of Hazardous Materials.* Academic Press, New York, NY, pp. 55-67.

277. Cowan CE, Federle TW, Larson RJ, Feijtel T. 1996. Impact of biodegradation test methods on the development and applicability of biodegradation QSARs. *SAR QSAR Environ Res* 5:37-49.

278. Jaworska JS, Boethling RS, Howard PH. 2003. Recent developments in broadly applicable structure-biodegradability relationships. *Environ Toxicol Chem* 22:1710-1723.

279. Govers HAJ, Parsons JR, Krop HB, Cheung CL. Thermodynamic descriptors for (bio-)degradation. Proceedings of the workshop "Quantitative structure activity relationships for biodegradation". Report No. 719101021. National Institute of Public Health and Environmental Protection, Belgirate, Italy.

280. Boethling RS, Lynch DG, Thom GC. 2003. Predicting ready biodegradability of premanufacture notice chemicals. *Environ Toxicol Chem* 22:837-844.

281. Boethling RS, Lynch DG, Jaworska JS, Tunkel JL, Thom GC, Webb S. 2004. Using BIOWIN™, bayes, and batteries to predict ready biodegradability. *Environ Toxicol Chem* 23:911-920.

282. Posthumus R, Traas TP, Peijnenburg WJGM, Hulzebos EM. 2005. External validation of EPIWIN biodegradation models. *SAR QSAR Environ Res* 16:135-148.

283. Rorije E, Loonen H, Müller M, Klopman G, Peijnenburg WJGM. 1999. Evaluation and application of models for the prediction of ready biodegradability in the MITI-I test. *Chemosphere* 38:1409-1417.

284. Meylan WM. 2000. BIOWIN 4.01. Syracuse Research Corporation, Syracuse, NY.

285. Howard PH, Hueber AE, Boethling RS. 1987. Biodegradation data evaluation for structure/biodegradability relations. *Environ Sci Technol* 6:1-10.

286. Howard PH, Boethling RS, Stiteler WM, Meylan WM, Hueber AE, Beauman JA, Larosche ME. 1992. Predictive model for aerobic biodegradability developed from a file of evaluated biodegradation data. *Environ Toxicol Chem* 11:593-603.

287. Boethling RS, Howard PH, Meylan WM, Stiteler WM, Beauman JA, Tirado NF. 1994. Group contribution method for predicting probability and rate of aerobic biodegradation. *Environ Sci Technol* 94:459-465.

288. Langenberg JH, Peijnenburg W, Rorije E. 1996. On the usefulness and reliability of existing QSBRs for risk assessment and priority setting. *SAR QSAR Environ Res* 5:1-16.

289. Loonen H, Lindgren F, Hansen B, Karcher W, Niemelä J. 1999. Prediction of biodegradability from chemical structure modeling of ready biodegradation test data. *Environ Toxicol Chem* 18:1763-1768.

290. Eakin DR, Hyde E, Palmer G. 1974. Use of computers with chemical structural information - ICI CROSSBOW system. *Pestic Sci* 5:319-326.

291. MultiCASE Inc. 2007. MultiCASE. Beachwood, OH. http://www.multicase.com.

292. MultiCASE Inc. 2007. METACPC 1.3. Beachwood, OH.

293. Klopman G, Tu M. 1997. Structure-biodegradability study and computer-automated prediction of aerobic biodegradation of chemicals. *Environ Toxicol Chem* 16:1829-1835.

294. Jaworska J, Dimitrov S, Nikolova N, Mekenyan O. 2002. Probabilistic assessment of biodegradability based on metabolic pathways: CATABOL System. *SAR QSAR Environ Res* 13:307-323.

295. Laboratory of Mathematical Chemistry, University Bourgas. 2007. OASIS Software. Bourgas, Bulgaria. www.oasis-lmc.org.

296. Laboratory of Mathematical Chemistry, University Bourgas. 2007. CATABOL 5.10.0. Bourgas, Bulgaria.

297. UM-BBD. 2005. PredictBT. University of Minnesota, Predictive Biogradation Project, Minneapolis, MN.

298. Hou BK, Ellis LBM, Wackett LP. 2004. Encoding microbial metabolic logic: predicting biodegradation. *J Industr Microbiol Biotech* 31:261-272.

299. Mackay D, Shiu WY, Ma KC. 1992. *Illustrated Handbook*

of Physical-Chemical Properties and Environmental Fate for Organic Chemicals. Lewis Publishers, Chelsea, MI.
300. Kühne R, Ebert R-U, Schüürmann G. 2007. Estimation of compartmental half-lives of organic compounds - structural similarity vs. EPI-suite. *QSAR Comb Sci.* 26:542-549.
301. Tomlin CDS, ed. 2006. *The Pesticide Manual. A World Compendium.* British Crop Protection Council, Farnham, UK.
302. Verschueren K. 2001. *Handbook of Environmental Data on Organic Chemicals.* John Wiley & Sons, Inc., New York, NY.
303. Lide DR, ed. 2005. *Handbook of Chemistry and Physics.* CRC Press, Boca Raton, FL.
304. Leo DA. MedChem database. Daylight Chemical Information Systems, Inc., Irvine, CA.
305. Yalkowsky SH, Dannenfelser RM. 1990. AQUASOL dATAbASE of aqueous solubility 5. University of Arizona, Tucson, AZ.
306. US Secretary of Commerce. NIST Chemistry WebBook. http://webbook.nist.gov.
307. CambridgeSoft Corporation. Chemfinder WebServer. CambridgeSoft, Cambridge, MA. http://www.chemfinder.com.
308. United States National Library of Medicine. ChemID plus. http://chem.sis.nlm.nih.gov/chemidplus.
309. US Department of Agriculture - Agricultural Research Service. The ARS Pesticide Properties Database. http://www.arsusda.gov/acsl/services/ppdb.
310. Beilstein Informationssysteme GmbH. Beilstein online database. http://www.beilstein.com.
311. Fachinformationszentrum Chemie (FIZ) Berlin (D). Infotherm - thermophysical properties database. http://www.chemistry.de/infotherm/servlet/infothermSearch.
312. Baum EJ. 1998. *Chemical Property Estimation.* CRC Press, Boca Raton, FL.
313. Reinhard M, Drefahl A. 1999. *Handbook for Estimating Physicochemical Properties of Organic Compounds.* John Wiley, New York, NY.
314. Grant DJW, Higuchi T. 1990. *Solubility Behavior of Organic Compounds.* John Wiley, New York, NY.
315. Van Krevelen DW. 1997. *Properties of Polymers.* Elsevier, Amsterdam, The Netherlands.
316. Boethling RS, Howard PH, Meylan WM. 2004. Finding and estimating chemical property data for environmental assessment. *Environ Toxicol Chem* 23:2290-2308.
317. Bodek I, Lyman WJ, Reehl WF, Rosenblatt DH, eds. 1988. *Environmental Inorganic Chemistry. Properties, Processes and Estimation Methods.* Pergamon Press, Elmsford, NY.
318. Meylan WM. EPISUITE. Syracuse Research Corporation, Syracuse, NY. http://www.syrres.com/esc/est_soft.htm.
319. Frederick/Bethesda Data and Online Services. 2002. NCI Database. http://129.43.27.140/ncidb2.
320. United States National Center for Biotechnology Information. 2005. PubChem. Bethesda, MD. http://pubchem.ncbi.nlm.nih.gov/search.
321. Masunaga S, Wolfe NL, Carriera L. 1993. Transformation of parasubstituted benzonitriles in sediment and in sediment extract. *Water Sci Technol* 28:123-132.
322. Johnson H, Kenley RA, Rynard C, Golub MA. 1985. Qsar for cholinesterase inhibition by organophosphorus esters and CNDO/2 calculations for organophosphorus ester hydrolysis. *Quant Struct -Act Relat* 4:172-180.
323. Wolfe NL, Steen WC, Burns LA. 1980. Phthalate-ester hydrolysis - linear free-energy relationships. *Chemosphere* 9:403-408.
324. Wolfe NL, Zepp RG, Paris DF. 1978. Use of structure-reactivity relationships to estimate hydrolytic persistence of carbamate pesticides. *Wat Res* 12:561-563.
325. Drossman H, Johnson H, Mill T. 1988. Structure activity relationships for environmental processes 1: Hydrolysis of esters and carbamates. *Chemosphere* 17:1509-1530.
326. Lyman WJ. 1990. Adsorption coefficient for soils and sediment. In: Lyman WJ, Reehl WF, Rosenblatt DH, eds, *Handbook of Chemical Property Estimation Methods,* 3rd ed. American Chemical Society, Washington, DC, pp. 4-1-4-33.
327. Peijnenburg WJGM, Debeer KGM, Denhollander HA, Stegeman MHL, Verboom H. 1993. Kinetics, products, mechanisms and QSARs for the hydrolytic transformation of aromatic nitriles in anaerobic sediment slurries. *Environ Toxicol Chem* 12:1149-1161.
328. Vogel TM, Reinhard M. 1986. Reaction products and rates of disappearance of simple bromoalkanes, 1,2-dibromopropane, and 1,2-dibromoethane in water. *Environ Sci Technol* 20:992-997.
329. Perrin DD, Dempsey B, Serjeant EP. 1981. *pK_a Prediction for Organic Acids and Bases.* Chapman and Hall, Cambridge, UK.

10. PREDICTING TOXICOLOGICAL AND ECOTOXICOLOGICAL ENDPOINTS

A.P. WORTH, T.I. NETZEVA AND G. PATLEWICZ

10.1 INTRODUCTION

In the last few decades, society has become increasingly concerned about the possible impacts of chemicals to which humans and environmental organisms are exposed. In many industrialised countries, this has led to the implementation of stringent chemicals legislation and to the initiation of ambitious risk assessment and management programmes (see Chapter 1). However, it has become increasingly apparent that the magnitude of the task exceeds the availability of resources (experts, time, money) if traditional test methods are employed. This realization, coupled with increasing attention to animal welfare concerns, has prompted the development and application of various (computer-based) estimation methods in the regulatory assessment of chemicals.

Estimation methods include "structure-activity relationships" (SARs) and "quantitative structure-activity relationships" (QSARs), which are collectively (and confusingly) referred to as (Q)SARs. These are theoretical models that can be used to predict the physicochemical, biological (e.g. toxicological) and environmental fate properties of molecules from the knowledge of chemical structure. In addition to the (Q)SARs that have been reported in the scientific literature (more than 20,000 models), a number of "expert systems" have been developed, generally as commercial products. The term "expert system" refers to a heterogeneous collection of computer-based estimation methods, which are based on the integrated use of databases (containing experimental data) and/or rule bases (containing SARs, QSARs and other decision rules).

In the context of chemical risk assessment, the information on chemicals provided by (Q)SARs and related estimation methods, collectively referred to as "non-test methods", can be used in combination with information from test methods by applying stepwise and/or weight-of-evidence approaches in the context of integrated (or intelligent) testing strategies (Chapter 11).

This chapter provides an overview of currently available (Q)SARs and expert systems for predicting human health and ecotoxicological endpoints, and supplements the review on estimation methods for physicochemical properties and fate parameters (Chapter 9). The review is preceded by an explanation of how (Q)SARs are developed and validated, and is followed by an explanation of how the models can be applied for regulatory purposes.

10.2 DEVELOPMENT AND VALIDATION OF SARS AND QSARS

10.2.1 Development of (Q)SAR models

The development of (Q)SARs is part science and part art: it requires expertise in molecular modelling and statistics, coupled with expert judgement based on an understanding of chemistry, biology and toxicology. The provision of guidance on how to develop (Q)SARs is beyond the scope of this Chapter, so the reader is referred to some useful articles [1-4].

As explained in Chapter 9, a (Q)SAR model consists of three main elements: a) the feature(s) of the chemical on which the model is based (descriptors or structural fragments), b) the property or effect for which predictions are made (endpoint, response), and c) the algorithm that converts the descriptors (or substructures) into the endpoint (response) of interest.

A multitude of descriptors have been reported in the literature, and a variety of software programmes have been developed to automatically generate descriptors directly from chemical structure. Extensive reviews on descriptors are given by Kier and Hall [5], Dearden [6], Netzeva [7], and Todeschini and Consonni [8]. Commonly-used software packages include EPI Suite (Syracuse Research Corporation, NY, USA), DRAGON (Talete srl, Milan, Italy), Adriana.Code (Molecular Networks GmBH, Erlangen, Germany), Molconn–Z (Edusoft, CA, USA), TSAR (Accelrys Inc., CA, USA), MDL QSAR (Elsevier MDL, CA, USA), MOPAC (CAChe Group, OR, USA) and QSAR Builder (Pharma Algorithms, Ontario, Canada).

The relationship in a (Q)SAR model takes one of three forms: a) a theoretical QSAR model, in which the equation is based on fundamental physical principles, b) statistical (empirical) SAR or QSAR model, developed by applying a statistical method to the training set of chemicals and c) a decision rule based on experience or expert judgement.

Theoretical QSAR models are limited to models for physicochemical or kinetic properties, given the current status of "QSAR science". An example would be the Arrhenius rate equation (see Chapter 9).

C.J. van Leeuwen and T.G. Vermeire (eds.), Risk Assessment of Chemicals: An Introduction, 427–465.

Many statistical (empirical) models also have a theoretical basis, even if they are not strictly derived from first principles. Such models include QSARs developed by the so-called Hansch approach, named after Corwin Hansch, who is widely regarded as the founder of modern QSAR. In several classic articles [9-11], Hansch demonstrated that the biological activity of a group of congeneric chemicals could generally be described by a simple model:

$$\log 1/C_m = a\pi + bE + cS + \text{constant} \qquad (10.1)$$

In this equation, C_m is the molar concentration of the compound that produces a defined biological response; π is a hydrophobicity term (originally the hydrophobic contribution of the substituent but now more typically the logarithm of the *n*-octanol-water partition coefficient (log K_{ow}), E is an electronic term (originally, the Hammett electronic descriptor (σ)), and S is a steric term (originally, Taft's steric parameter (E_S)) and a, b and c are the appropriate coefficients.

A rationale for Equation 10.1 was later given by McFarland [12]. He hypothesised that the relative activity of a biologically active chemical depends on: a) the probability (Pr1) that the chemical reaching its site of action (the toxicokinetic phase); b) the probability (Pr2) that the chemical interacts with the appropriate target molecule (e.g. receptor) at this site (the toxicodynamic phase); and c) the external concentration (C) or dose to which the organism is exposed.

For a given level of effect, the number of molecular interactions, or the concentration of the target molecules (C_t), will be constant. So, C_t can be written as:

$$C_t = c \cdot \text{Pr1} \cdot \text{Pr2} \cdot C = \text{constant} \qquad (10.2)$$

where c is a constant. Logarithmic transformation of Equation 10.2 yields:

$$\log 1/C_m = a(\log \text{Pr1}) + b(\log \text{Pr2}) + c' \qquad (10.3)$$

According to this explanation, chemical toxicity can be regarded as the consequence of uptake, distribution and elimination (toxicokinetics) and the interaction of the toxicant with the molecular target at the site of action (toxicodynamics). Pr1 is dependent on hydrophobicity, whereas Pr2 is dependent on steric and electronic effects. Thus, Equation 10.3 is equivalent to Equation 10.1.

10.2.2 OECD PRINCIPLES FOR (Q)SAR VALIDATION

In Chapter 16 a short description is given about the chemicals programme of the Organization for Economic Cooperation and Development (OECD) and their project on (Q)SARs (Section 16.3.3). Within OECD and other international regulatory programmes, there is general agreement that models should be "scientifically valid" or "validated" if they are to be used for the regulatory assessment of chemicals. In the EU, the concept of "scientifically valid model" is incorporated into the legal text of the REACH regulation [13]. The term "validated model" is also found in the guidance on the application of EU and GHS (Globally Harmonized System of Classification and Labelling of Chemicals) classification schemes. Since the concept of validation is incorporated into legal texts and regulatory guidelines, it is important to clearly define what it means, and to describe what the validation process entails.

The first step towards a harmonised definition of (Q)SAR validation, in the context of chemical risk assessment, was made during an international workshop on the "Regulatory Acceptance of QSARs for Human Health and Environment Endpoints", organised by the International Council of Chemical Associations (ICCA) and the European Chemical Industry Council (CEFIC), held in Setubal, Portugal, on 4-6 March, 2002 [14-18]. During this workshop, a set of six principles were proposed for assessing the validity of (Q)SARs.

Subsequently, an Expert Group established by the OECD carried out an extensive assessment of the six principles (referred to as the "Setubal principles") by applying them to a range of different (Q)SARs, including literature-based models and models in expert systems [19]. On the basis of this assessment, the OECD Expert Group on (Q)SARs reworded the six principles and combined two of the principles into a single principle, to produce a set of five principles. In November 2004, this set of five principles was adopted at the policy level by the OECD Member Countries and the European Commission, and it was decided that an OECD guidance document on (Q)SAR validation should be written, to explain how the principles should be applied with practical approaches. As an input to the OECD guidance document, preliminary guidance on the characterization of (Q)SARs has been developed by the ECB (European Chemicals Bureau) [20].

The OECD principles for (Q)SAR validation state that in order "to facilitate the consideration of a (Q)SAR model for regulatory purposes, it should be associated with the following information:

1. A defined endpoint.
2. An unambiguous algorithm.
3. A defined domain of applicability.
4. Appropriate measures of goodness-of-fit, robustness and predictivity.
5. A mechanistic interpretation, if possible.

The principles should be understood in the context of some explanatory notes that explain the intent of the principles (see Box 16.2 in Chapter 16).

Intent of the OECD principles

The principles for (Q)SAR validation identify the types of information that are considered useful for the regulatory review of (Q)SARs. The principles constitute the basis of a conceptual framework, but they do not in themselves provide criteria for the regulatory acceptance of (Q)SARs. The definition of acceptance criteria, where considered necessary, is the responsibility of individual authorities within the Member Countries.

According to Principle 1, a (Q)SAR should be associated with a "defined endpoint", where endpoint refers to any physicochemical, biological or environmental effect that can be measured and therefore modelled. The intent of this principle is to ensure transparency in the endpoint being predicted by a given model, since a given endpoint could be determined by different experimental protocols and under different experimental conditions. Ideally, (Q)SARs should be developed from homogeneous datasets in which the experimental data have been generated by a single protocol. However, this is rarely feasible in practice, and data produced by different protocols are often combined.

According to Principle 2, a (Q)SAR should be expressed in the form of an unambiguous algorithm. The intent of this principle is to ensure transparency in the description of the model algorithm. In the case of commercially-developed models, this information is not always made publicly available.

According to Principle 3, a (Q)SAR should be associated with a "defined domain of applicability". The need to define an applicability domain expresses the fact that (Q)SARs are reductionist models which are inevitably associated with limitations in terms of the types of chemical structures, physicochemical properties and mechanisms of action for which the models can generate reliable predictions. This principle does not imply that a given model should only be associated with a single applicability domain. In fact, the boundaries of the domain can vary according to the method used to define it and the desired trade-off between the breadth of model applicability and the overall reliability of predictions. Methods for defining the applicability domains of (Q)SARs are described elsewhere [21-24].

According to Principle 4, a (Q)SAR should be associated with "appropriate measures of goodness-of–fit, robustness and predictivity." This principle expresses the need to provide two types of information: a) the internal performance of a model (as represented by goodness-of-fit and robustness), determined by using a training set; and b) the predictivity of a model, determined by using an appropriate test set. There is no absolute measure of predictivity that is suitable for all purposes, since predictivity can vary according to the statistical methods and parameters used in the assessment. In addition, no thresholds for these measures are likely to be set, since each model and prediction made will be evaluated on a case-by-case basis to determine the prediction is "fit for purpose". Statistical methods for assessing goodness-of-fit, robustness and predictivity are described extensively in several books (e.g. [4, 25]) and articles (e.g. [26]).

According to Principle 5, a (Q)SAR should be associated with a "mechanistic interpretation", wherever such an interpretation can be made. Clearly, it is not always possible to provide a mechanistic interpretation of a given (Q)SAR. The intent of this principle is therefore to ensure that there is an assessment of the mechanistic associations between the descriptors used in a model and the endpoint being predicted, and that any association is documented. Where a mechanistic interpretation is possible, it can add strength to the confidence in the model already established on the basis of Principles 1-4. (Q)SARs that are based on mechanistically interpretable and plausible descriptors are sometimes called "mechanistically-based (Q)SARs". As demonstrated by Cronin et al. [27], such models often include descriptors for partitioning behaviour (hydrophobicity) and reactivity (electrophilicity or nucleophilicity).

Definition of (Q)SAR validation

At the time of writing, there is no internationally accepted definition of (Q)SAR validation. The following definition has been proposed by the ECB [20]:

"The validation of a (Q)SAR is the process by which the performance and mechanistic interpretation of the model are assessed for a particular purpose."

In this definition, the "performance" of a model refers to its goodness-of-fit, robustness and predictive ability, whereas "purpose" refers to the scientific purpose of the (Q)SAR, as expressed by the defined endpoint and applicability domain. The first part of the definition (performance) refers to "statistical validation", whereas the second part (mechanistic interpretation) refers to the assignment of physicochemical meaning to the descriptors (where possible) and to the establishment of a hypothesis linking the descriptors with the endpoint.

The scientific purpose of a (Q)SAR may or may not have an association with possible regulatory applications. Thus, the purpose of a (Q)SAR could be for predicting a particular endpoint (along a continuous or categorical scale) for a particular class of chemicals, irrespective of whether the endpoint is required by any particular legislation or whether the class of chemicals is contained with a given regulatory inventory.

10.3 PREDICTION OF HUMAN HEALTH ENDPOINTS

Since extensive reviews of (Q)SARs for human health endpoints can be found elsewhere [18,28], the emphasis of this section is to provide a few examples, to give an impression of the different types of models and approaches used to develop them.

10.3.1 Acute toxicity

This section provides an overview of (Q)SARs for skin and eye irritation and corrosion, as well as (Q)SARs for acute systemic toxicity. For more-detailed information, the reader is referred to reviews on models for acute dermal and ocular toxicity [29,30], and for acute toxicity [31,32].

Skin irritation and corrosion

The traditional regulatory approach for assessing skin irritation and corrosion is the Draize rabbit skin test [33]. For skin corrosion testing, this has been replaced in EU legislation and OECD Test Guidelines by several *in vitro* tests, and it is also likely to be replaced for the purpose skin irritation.

Dermal irritation is often graded according the Draize grading scale in which subjective scores from 0-4 are assigned, depending on the extent of erythema/eschar formation and of edema observed. Regulatory classification schemes are based on cut-off values along these parameters, although different schemes vary according to the way in which the parameters are used. The erythema/eschar and edema scores are sometimes combined and averaged into a Primary Irritation Index (PII). Skin corrosion is fundamentally a categorical response: it either occurs or it does not, but sometimes a distinction is made between severe corrosives and mild/moderate corrosives, depending on the exposure duration needed to elicit the response (destruction of the epidermis through to the dermis).

Compared with other human health endpoints, relatively few (Q)SARs have been reported for dermal lesions. Some (Q)SARs attempt to model the basic parameters or the PII, whereas other attempt to model the regulatory classification of irritation or corrosion.

The simplest model for corrosion is based on the acidity/basicity (pH value) of the chemical, and takes the form of a simple decision rule:

IF $pH < 2$ or $pH > 11.5$ THEN predict to be corrosive

This model is included in OECD testing strategy for skin irritation and corrosion [34] and can be used as a means of derogating from further assessment. The use of pH as a predictor of corrosion [35], and its possible contribution to tiered testing strategies [36], have been analysed by the first author.

Within the OECD testing strategy the occurrence of structural analogues that exhibit corrosion (or irritation) potential can also be used to predict the effect in the substance of interest and derogate from further assessment. However, negative data from structural analogues cannot be used to make predictions.

For defined classes of chemicals, QSARs and related models have been reported for discriminating between corrosives and non-corrosives [37,38], and between skin irritants and non-irritants [39,40]. A few models for predicting the PII have also been published [41,42].

Among the expert systems, TOPKAT incorporates models to discriminate severe irritants from non-severe irritants, as well as mild/moderate irritants from non-irritants [43]. Another system, developed by the German Federal Institute for Risk Assessment (BfR), uses physicochemical exclusion rules to predict the absence of an effect in combination with structural inclusion rules to predict the presence of an effect [44,45]. The performance of the physicochemical rule base has been assessed by Rorije and Hulzebos [46].

Eye irritation and corrosion

The conventional test for the regulatory assessment of eye irritation and corrosion is the Draize rabbit eye test [33]. Eye irritation is graded according to a variety of responses in the conjunctivae, cornea and iris. These scores are sometimes converted into weighted averages, such as the mean average score (MAS), or used in classification schemes. A considerable research effort has been directed at the development of alternative (non-animal) approaches to replace the Draize eye test, including a wide variety of (Q)SAR methods.

Among the expert systems, models based on structural fragments have been developed by Enslein [47], and incorporated in the TOPKAT software. For non-ring compounds, TOPKAT discriminates severe irritants from other chemicals, and non-irritants from other chemicals. For ring-compounds, it also distinguishes between severe irritants from other chemicals, but also distinguishes non-irritants and mild irritants (collectively) from moderate and severe irritants. Another fragment-based model is based on the Multi-CASE approach [48].

The BfR decision support system also makes predictions of eye irritation/corrosion, by applying a physicochemical rule base to predict the absence of effects along with a structural rule base to predict the presence of effects [44,45]. The performance of the physicochemical rule base has been assessed by Tsakovska et al. [49].

In the open scientific literature, (Q)SARs have been based on continuous (e.g. molar eye scores) or categorical (e.g. EU classifications) measures of eye irritation (e.g. [50-52]). The use of solvatochromic parameters in regression models has been explored by Abraham et al [53]. Other workers have used neural network approaches (e.g. [54]). Based on the findings of earlier QSAR analyses, a new classification approach, called embedded cluster modelling (ECM), was proposed by Worth and Cronin [55] as a means of generating elliptic models in two or more dimensions, so that irritants can be identified as those chemicals located within the boundaries of the ellipse. The statistical significance of these "embedded clusters" can be verified by cluster significance analysis [56]. Another approach, called Membrane-Interaction QSAR analysis, has been developed by Kulkarni and Hopfinger [57] as a means of incorporating molecular dynamic simulations to generate membrane–solute interaction properties.

Acute systemic toxicity

Acute toxicity studies involve either a single administration of a test chemical or several administrations within a 24 hour period. Most acute toxicity studies determine the median lethal dose (LD50) of the chemical, the statistically derived single dose expected to kill 50% of animals in the experimental group. When the route of exposure is inhalation, the endpoint is generally the median lethal concentration (LC50). The information generated by acute toxicity testing is generally considered to be useful for assessing chemical effects on humans as well as on mammalian wildlife. The main regulatory purpose of acute toxicity testing is to classify chemicals according to their intrinsic toxicity.

(Q)SARs for acute mammalian toxicity have not been used for the regulatory assessment of chemicals. Nevertheless, there are reasonable prospects for using such models in the near future, since there has been considerable development of these methods over the past 10 years. In principle, one could foresee the use of these models for providing mechanistic and supplementary information, for predicting the start dose for *in vivo* testing, thereby reducing the overall number of animals used, or even for replacing testing (perhaps in combination with *in vitro* data). From the scientific perspective, the development of QSARs for inhalational toxicity is likely to be more successful than for other routes of exposure, because a steady-state situation is more readily obtained and suitable descriptors exist for the partitioning of gases.

Some (Q)SAR studies have focused on the modelling of *in vivo* LD50 or LC50 values, whereas other studies have focused on the modelling of cytotoxicity in *in vitro* systems, or of effects observed on subcelluar fractions. The idea for using cytotoxicity assays to predict *in vivo* toxicity arises from the concept of "basal cell cytotoxicity" proposed by Ekwall [58]. It was suggested that for most chemicals, toxicity is a consequence of non-specific alterations in cellular functions, implying that the determination of the cytotoxic potential of a chemical can often be used to extrapolate to its toxic potential *in vivo*. A large number of studies have confirmed that reasonable correlations exist between cytotoxicity and *in vivo* toxicity [59-61].

There are three main types of models of *in vivo* endpoints: a) traditional (Q)SARs, which tend to be "local" models (i.e. applicable to defined chemical classes), b) expert systems, which tend to be of a "global" nature (i.e. widely applicable to diverse chemistries), and c) artificial neural network (ANN) models (which can be both local and global models). Expert systems tend to be proprietary, whereas traditional local (Q)SARs tend to be published in the open literature. Some ANN models are proprietary, whereas others are published in the scientific literature.

An example of a literature based model for acute toxicity is provided by Cronin et al. [27], who obtained the following QSAR for the toxicity of a set of pyrines to female mice (it should be noted that the reciprocal of this equation was cited in the source publication).

Box 10.1 Common statistical parameters associated with QSARs

QSAR models are generally reported with a set of standard statistics. Commonly used statistical parameters are:

a. n is the number of chemicals in the training set
b. r^2 is the coefficient of (multiple) determination
c. r^2_{cv} (or q^2) is a cross-validated r^2
d. r^2_{adj} is the r adjusted for the number of degrees of freedom, i.e. number of variables
e. s is the standard error of estimate
f. F is the Fischer statistic

The coefficient of determination (r^2) estimates the proportion of the variation in y (in the case of Equation 10.4, y=logLD50) that is explained by the regression. If there is a perfect linear relationship between the dependent and independent variables, then r^2 is equal to 1. The standard error of estimate (s) measures the dispersion of the observed values about the regression line: the smaller the value of s the higher the reliability of the prediction. The cross-validated coefficient of determination provides a measure of how stable (robust) the regression model is. Ideally, the r^2_{cv} value should not be substantially lower than the r^2. The adjusted r^2 value (r^2_{adj}) is a variant of the r^2 that takes into account the number of variables in the QSAR equation. This statistic can be generated in various ways, depending on the type of cross-validation used. The Fischer statistic (F) provides a measure of the statistical significance of the regression model. The F-value is defined as the ratio between explained and unexplained variance for a given number of degrees of freedom. The higher the F-value, the greater the probability is that the equation is significant

see below (10.4)

$$n = 20, r^2 = 0.85, r^2_{cv} = 0.82, s = 0.19, F = 54.1$$

where LUMO is the lowest unoccupied molecular orbital and K_{ow} is the *n*-octanol-water partition coefficient.

The descriptors in this model emphasise the importance of electrophilicity (reactivity) and hydrophobicity (partitioning) in the mechanism of toxic action. The statistics reported alongside the model are explained in Box 10.1

An example of an expert system is TOPKAT, which contains a module for the prediction of rat oral LD50 values. The module contains 19 statistically significant and cross-validated QSAR regression models, based on a variety of structural, topological and electropological descriptors. The TOPKAT LD50 models are based on experimental values of approximately 4000 chemicals taken from the RTECS® (Registry of Toxic Effects of Chemical Substances) database (see 10.5.2 for further details). Since RTECS® lists the most toxic value when multiple values exist, the TOPKAT model tends to overestimate the acute toxicity.

10.3.2 Skin sensitization

There are a number of methods available for the prospective identification of skin sensitizing chemicals. Guinea pig tests, and in particular the guinea pig maximization test (GPMT) and the occluded patch test of Buehler, have been used extensively for the identification of skin sensitization hazard. These are described in OECD Guideline 406 [62] and in EU Annex V (B.6). The GPMT is an adjuvant type test in which the allergic state (sensitization) is potentiated by the use of Freund's Complete Adjuvant (FCA). The Buehler test is a non-adjuvant method involving topical application of the test chemical for the induction phase rather than the intradermal injection used in the GPMT. Both the GPMT and the Buehler test have demonstrated the ability to detect chemicals with moderate to strong sensitization potential as well as those with relatively

$$\log LD50 = 0.660\ LUMO - 0.380 \log K_{ow} - 1.81 \qquad (10.4)$$

weak sensitization potential. These guinea pig methods provide information on skin responses which are evaluated for each animal after several applications of the substance, and on the percentage of animals sensitized. For classification purposes, a response of at least 30% of the animals is considered positive for the GPMT and at least 15% for the Buehler test.

In recent years, the murine local lymph node assay (LLNA) described in OECD Guideline 429 [63] and EU Annex V (B.42) has emerged as the preferred method for skin sensitization. In addition to identifying the hazard, the LLNA can also provide a reliable measure of relative skin sensitizing potency. The LLNA identifies potential skin sensitizing chemicals as a function of events associated with the induction phase of skin sensitization (the clonal expansion of the T-lymphocytes), the vigour of the response correlates closely with the extent to which sensitization will develop. Potency is measured by derivation of an estimated concentration of substance required to induce a three-fold stimulation index (SI) value (EC3) as compared to concurrent vehicle controls. A substance is classified as a sensitizer if it induces a three-fold SI or greater at one or more test concentrations [64,65].

The skin sensitization potential of a chemical is related to its ability to react covalently with skin proteins. Reactions can either occur directly or indirectly after the chemicals are activated by metabolism or chemically. Consideration of the chemical properties of a wide variety of known sensitizers and comparison with non-sensitizers has led to the conclusion that binding to a protein takes place by the protein acting as a nucleophile and the sensitizer acting as an electrophile [66]. The sorts of reactions typical of known sensitizers include saturated aldehydes leading to the formation of a Schiff base, α,β-unsaturated carbonyl groups reacting via Michael addition or dinitrohalobenzenes which are able to react via an SNAr mechanism. Another consideration is penetration into the viable epidermis of the skin [67].

For direct-acting chemicals, sensitizing ability may be modelled using the relative alkylation index (RAI), a mathematical model derived by Roberts and Williams [68]. The underlying hypothesis driving this model is that the extent of sensitization produced at induction and challenge is dependent on the degree of covalent binding. The RAI was derived from differential equations modelling the competition between the carrier haptenation reaction in a hydrophobic environment and removal of the sensitizer through partitioning into polar lymphatic fluid. The general form of the RAI is:

$$\text{RAI} = \log \text{D} + \text{a} \log k + \text{b} \log K_{ow} \qquad (10.5)$$

Thus the degree of haptenation increases with the increasing dose (D) of sensitizer, with increasing reactivity (as quantified by the rate constant or relative rate constant k for the reaction of the sensitizer with a model nucleophile) and with increasing hydrophobicity (as quantified by log K_{ow}). This RAI model has been used to evaluate a wide range of different datasets of skin sensitizing chemicals, including sulphonate esters [69], sultones [68], primary alkyl bromides [70], and acrylates [71]. For example, the following equation was derived for primary alkyl bromides [70]:

$$\text{pEC3} = 1.61 \log K_{ow} - 0.09 (\log K_{ow})^2 - 7.4 \qquad (10.6)$$

$$n = 9, r = 0.97, s = 0.11, F = 50.0$$

In this equation, pEC3 is log(1/EC3*) where EC3* is (EC3/MW).

The RAI approach continues to be used to develop mechanistically based QSARs. Examples of more recent models include those developed for Schiff base and Michael acceptor aldehydes [72-74], aldehydes and 1,2-diketones [75-77]. In [74], QSARs were developed for Schiff base aldehydes (Equation 10.7) and Michael acceptors (Equation 10.8).

see below (10.7)

$$n = 13, r^2 = 0.73, r^2_{adj} = 0.64, s = 0.270, F = 8.16$$

$$\text{pEC3} = 0.17 + 0.30 \log K_{ow} + 0.93\,\sigma^* \qquad (10.8)$$

$$n = 14, r^2 = 0.87, r^2_{adj} = 0.85, s = 0.165, F = 37.7$$

In Equations 10.8 and 10.9, the Taft constant (σ^*) provides a means of quantifying the inductive effects of the alkyl groups that are attached to the carbonyl groups in the Schiff base aldehydes and Michael acceptors; these effects are used to account for the electrophilic reactivity of the carbonyl group. Specifically, the σ^* value is the Taft substituent constant for the alkyl group

$$\text{pEC3} = 0.55 + 0.14 \log K_{ow} + 0.51\text{R}\sigma^* \text{(beta)} + 1.07\, \text{R}'\sigma^* \qquad (10.7)$$

attached to the carbonyl group; R σ*(beta) is the Taft substituent constant for the beta alkyl group R; and R'σ* is the Taft substituent constant for the alpha alkyl group.

Statistical models involve the development of empirical QSARs by application of statistical methods to sets of biological data and structural descriptors. Several examples using LLNA data have been reported in the recent literature; these have been extensively characterised and evaluated with respect to the OECD Principles in [78].

In Miller et al. [79], a set of 87 LLNA data were considered and after removal of 20 outliers, 67 chemicals were analysed. The so-called Codessa (Comprehensive Descriptors for Structural and Statistical Analysis) descriptors were calculated by using the Codessa software (Semichem, Inc., Shawnee Misson, KS, USA) and several correlations were derived from these descriptors. The best model was the following.

see below (10.9)

$n = 50, r^2 = 0.773, r^2_{adj} = 0.763, r^2_{cv} = 0.738, F = 79.9$

In Equation 10.9, $FPSA2_{ESP}$ is the fractional positively charged surface area descriptor based on electrostatic potential charge, and $E_{HOMO\text{-}LUMO}$ is the energy gap between Highest Occupied Molecular Orbital (HOMO) and Lowest Unoccupied Molecular Orbital (LUMO).

Federowicz et al. [80] used stepwise logistic regression on a set of 54 LLNA-tested chemicals and 1204 molecular descriptors. Combinations of no more than four descriptors were assessed for the ability to correctly discriminate between sensitizers and non-sensitizers, and the best combination was selected.

Estrada et al. [81] used linear discriminant analysis to relate topological (specifically TOPS-MODE) descriptors to skin sensitization data as measured in the LLNA. A set of 93 diverse chemicals and their associated LLNA EC3 values were collated. The EC3 values were categorised into bands of potency. Two QSAR models were developed. The first discriminated strong/moderate sensitizers (EC3 < 10%) from all other chemicals and the second discriminated weak sensitizers (10% < EC3 < 30%) from extremely weak and non-sensitizing chemicals (EC3 > 30%).

Enslein et al. [82] developed QSAR models for dermal sensitization by using guinea pig data for 315 chemicals. Two suites of models were proposed: one for aromatics (excluding chemicals with 1 benzene ring) and the other for aliphatics and chemicals with 1 benzene ring. Instead of adopting a hypothesis-based approach, a variety of descriptors were computed for the chemicals selected, and stepwise two-group discriminant analysis was used to identify relevant descriptors and build the models. The first set of models discriminated between non-sensitizers and sensitizers, according to whether the calculated probability for the submitted structure was less than 0.30 (non-sensitizer) or was greater than 0.7 (sensitizer). The second set of models resolved the potency: weak/moderate vs. strong where a probability of 0.7 or more indicated a strong sensitizer and a probability below 0.30 indicated a weak or moderate sensitizer. Probability values between 0.30 and 0.70 were associated with an "indeterminate region" in which reliable predictions could not be made. An optimum prediction space (OPS) algorithm was incorporated into the model to ensure predictions were only made for chemicals within the model domain. This model was incorporated into the Toxicity Prediction by Komputer Assisted Technology (TOPKAT) expert system. The current version of TOPKAT (Version 6.2) has been supplemented with data from a further 20 studies.

The skin sensitization knowledge base in Derek was initially developed in collaboration with Unilever in 1993, using its historical database of guinea pig maximization test (GPMT) data for 294 chemicals, and it contained approximately 40 alerts [83]. The knowledge base has since undergone extensive improvements as more data have become available and scientific knowledge has increased [84]. The current version of Derek (version 9.0.0) contains 70 alerts for skin sensitization and photoallergenicity [85].

10.3.3 Chronic toxicity

Chronic toxicity occurs as a result of exposure to repeated, non-lethal doses, causing damage over a long period of time. Observations are related to growth, reproduction and survival, and are typically expressed as the lowest observed adverse effect level (LOAEL). The LOAEL is the lowest exposure level at which biologically

$$EC3 = 9.16\ FPSA2_{ESP} + 4.29\ E_{HOMO\text{-}LUMO} - 45.89 \tag{10.9}$$

Table 10.1. Accuracy of the TOPKAT submodels for the prediction of LOAEL. After Venkatapathy et al. [87]. With permission. Copyright 2004 American Chemical Society.

Class	No of chemicals	adj R^2	% of chemicals predicted within a factor of				95% of chemicals predicted within a factor of
			2	3	4	5	
acyclics	73	0.87	73	92	97	100	4
alicyclics	39	0.98	94	100			3
heteroaromatics	68	0.85	78	92	98	100	4
multiple benzenes	83	0.78	70	92	96	97	4
single benzenes	130	0.79	66	88	94	98	5

significant increases in the severity of adverse effects are observed. There have been relatively few attempts to model chronic toxicity by QSAR methods, probably because it is not really a single endpoint, but an umbrella term for many different effects, which can occur in different organs and tissues over different time scales.

The TOPKAT software includes regression models for predicting the rat oral chronic lowest observed adverse effect levels (LOAEL), even though it is questionable whether the LOAEL is a defined endpoint in terms of the OECD validation principles. The initial oral rat chronic LOAEL model in TOPKAT, developed by Mumtaz et al. [86], was a 44-descriptor model based on 234 chemicals of diverse structures and developed by stepwise regression analysis. The goodness-of-fit of the model was tested by predicting the LOAELs for each compound in the training set and comparing it with its experimental LOAEL. A comparison of the calculated and experimental chronic LOAELs showed that about 55% of the compounds were predicted within a factor of two and more than 93% of the compounds were predicted within a factor of five.

The TOPKAT model was subsequently refined by including additional data in the training set. The expanded training set of 393 chemicals was used to develop models for five chemical subclasses: acyclics, alicyclics, heteroaromatics, single benzenes, and multiple benzenes. The predictive performance of the five submodels was subsequently assessed Venkatapathy et al. [87], by using 343 chemicals from the USEPA's Office of Pesticide Program (OPP) database and 313 chemicals from several other USEPA databases. The results of this assessment are summarised in Table 10.1.

10.3.4 Mutagenicity and carcinogenicity

Carcinogenicity and mutagenicity are among the toxicological effects that cause particular concern for human health. Whereas the mutagenic potential of chemicals can be assessed with relatively simple test methods, carcinogenicity testing in rodents is long (usually two years for rats and 18 months for mice), expensive, and requires a large number of animals (rats and mice of both sexes, 3 dose levels and a control group, and at least 50 animals per sex per group). Genotoxicity testing is widely used as a screen for potential carcinogenicity and teratogenic potential. However, genotoxicity testing provides only information regarding genotoxic carcinogens and does not detect hazard associated with non-genotoxic mechanisms of carcinogenicity.

Among the most commonly used *in vitro* tests for mutagenicity (genotoxicity) are the bacterial reverse mutation tests with *Salmonella typhimurium* (Ames test) and *Escherichia coli*. Since bacterial cells differ from mammalian cells in factors such as uptake and metabolism, the tests conducted *in vitro* generally require the use of an exogenous source of metabolic activation.

In the Ames test, bacterial cultures are exposed to the test substance in the presence and in the absence of an exogenous metabolic activation system. After 2 or 3 days of incubation, revertant colonies are counted and compared to the number of spontaneous revertant colonies on control plates. The principle of the bacterial reverse mutation test is that it detects chemicals that induce mutations which revert mutations present in the tester strains and restore the functional capability of the bacteria to synthesise an essential amino acid. The revertant bacteria are detected by their ability to grow in the absence of the amino acid required by the parent tester strain. The endpoint, which is most often used in the (Q)SAR analysis of mutagenicity, is the logarithm of the number of revertant colonies on replicated plates.

(Q)SAR models for mutagenicity and carcinogenicity include both structure-activity relationships (SARs, structural alerts) and quantitative structure-activity

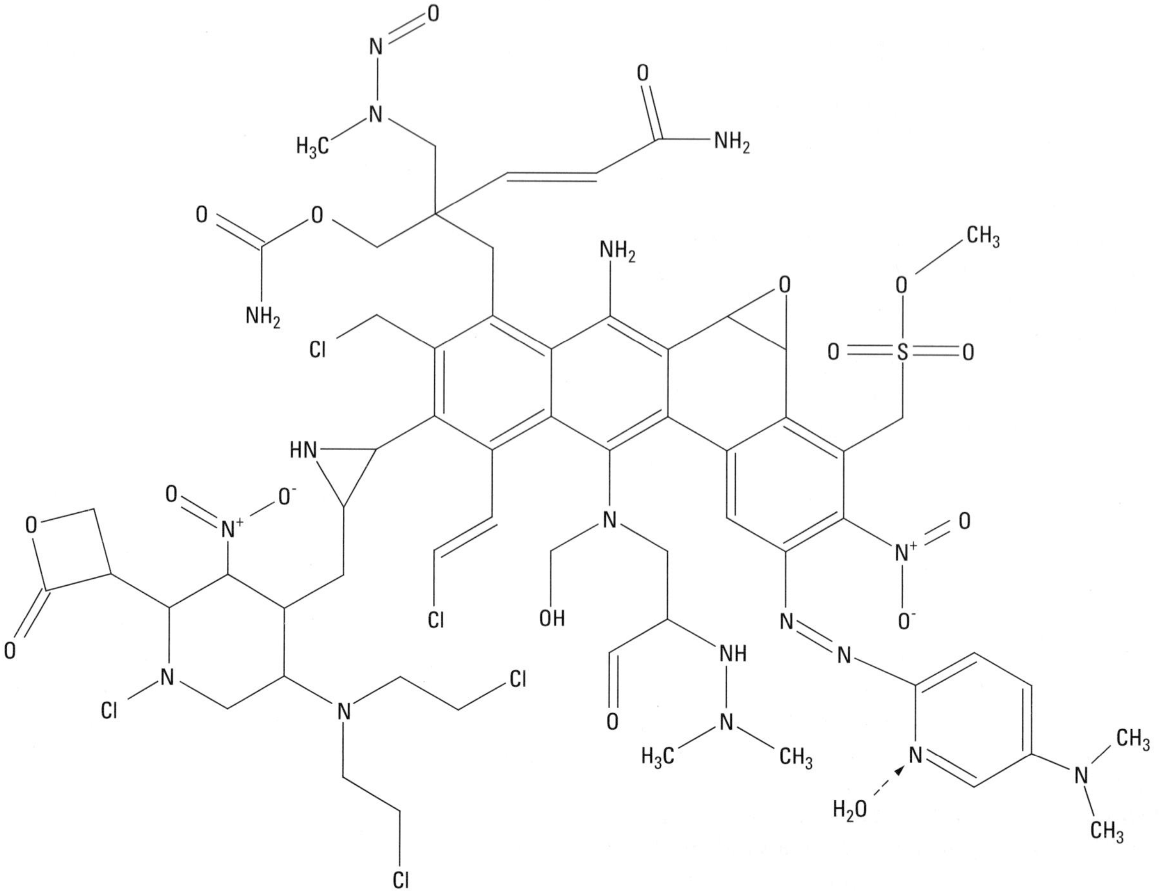

Figure 10.1. Ashby's poly-carcinogen model. Modified after [88]. The model includes also an alert for halogenated methanes C(X)4, where X = H, F, Cl, Br, I in any combination. With permission. Copyright Elsevier.

relationships (QSARs). Well recognised structural alerts for genotoxicity and genotoxic carcinogenicity are carbonium ions (alkyl-, aryl-, benzylic-), nitrenium ions, epoxides and oxonium ions, aldehydes, polarised α,β-unsaturated fragments, peroxides, free radicals, and acylating intermediates [88,89]. Examples of structural alerts for mutagenicity have been combined in Ashby's poly-carcinogen model [88]. The supermutagen model was one of the first attempts to relate molecular structure to toxicity for a number of fragments, and is illustrated in Figure 10.1.

It is useful for identifying potential carcinogens, but it does not embody an exhaustive list of all possible structural alerts for genotoxic carcinogenicity. A comparative exercise on the prediction of rodent carcinogenicity, including Tennant and Ashby method [90], emphasised the importance of the expert opinion when classifying chemicals according to their carcinogenic potential [91]. It appeared that the approach of using structural alerts in combination with expert interpretation guaranteed the highest true positive rate in classifying 44 compounds from the US National Toxicology Program (NTP) and a medium rate of false positive results; as illustrated by the position of the model in a Receiver Operating Characteristic (ROC) graph (see Box 10.2 and Figure 10.2).

A common feature of genotoxic substances is that they can bind covalently to DNA and cause direct DNA damage. Since they are usually electrophilic chemicals, or chemicals that can be metabolised to electrophilic products, it has been relatively easy to identify structural alerts. In contrast, non-genotoxic carcinogens lack a common mechanism of action and this has made it more difficult to identify structural alerts.

Rule-based methods incorporate current knowledge, viewpoints and mechanistic assumptions [92]. A drawback of a rule base consisting only of structural alerts is that the modulatory effects of other functional groups are difficult to account for. As a result, false positive chemicals can be identified. This "excessive" sensitivity is due to the fact that various alerts act as "class-identifiers". They point out the presence of an

Box 10. 2 Receiver Operating Characteristic (ROC) graph

In the Receiver Operating Characteristic (ROC) graph, the true positive rate (sensitivity) is plotted against the false positive rate (1-specificity), as illustrated in Figure 10.2. The graph is a convenient means of representing and comparing the performances of different classification models. The diagonal line of this graph represents random prediction, where the top left corner represents the ideal performance of a classification model. Thus, if a model is in the upper left triangle, this is an indication that it predicts better than chance, and the closer a system to the upper left corner, the better its predictive performance.

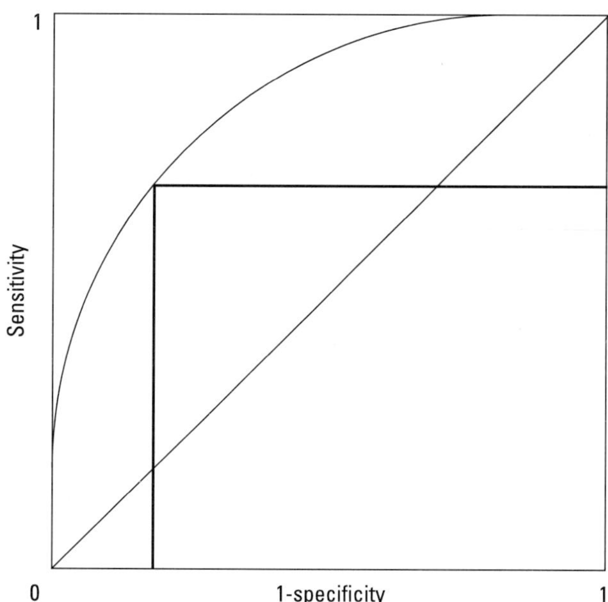

Figure 10.2. Receiver Operating Characteristic (ROC) graph. The coordinates are indicative of the performance of the models: (0,1) represents perfect classification whereas the diagonal indicates models with no discriminatory power. See Box 10.2.

alerting chemical functionality, but they are not able to make gradations within each potentially harmful class [93].

Quantitative models for mutagenicity and carcinogenicity can be developed to predict presence or absence of hazard (classification models), or to predict a numerical value associated with that hazard (continuous models). Examples of classification models can be found in Livingstone et al. [94] and Contrera et al. [95]. An excellent overview of different models and approaches for prediction of mutagenicity and carcinogenicity can be found in Benigni [96]. The general models differ significantly from classical QSARs, ranging from expert panel judgements, to computerised expert systems, to quantitative approaches derived specifically for sets of non-congeneric series.

Models based on non-congeneric series are associated with a number of problems, such as modelling multiple or overlapping mechanisms of action with a single model, defining the applicability domain of the model, assigning confidence levels to the predictions, and determining the mechanistic significance of the model descriptors [92]. In contrast, models that take into account chemical reactivity and known or postulated mechanisms of action provoke more confidence that models built on a purely statistical background.

An example of class-based linear regression model for mutagenicity was developed by Debnath et al. [97] and given in Equation 10.10.

see below (10.10)

$n = 67, r = 0.877, s = 0.708$

where log TA100 is the mutagenic potency (revertants/nmol), HOMO is the energy of the highest occupied molecular orbital, and LUMO is the energy of the lowest unoccupied molecular orbital. The model has a clear mechanistic basis, with the determiniants of genotoxicity being expressed by hydrophobicity (represented by log K_{ow}), and reactivity (represented by HOMO and LUMO). The model also has a defined applicability domain, being developed for a set of aromatic amines.

Nowadays, a considerable number of QSARs for specific chemical classes are available. These include aromatic amines, nitroaromatic compounds, N-nitrosocompounds, polycyclic aromatic amines, halogenated aliphatics, as reviewed by Benigni [98]. Compared with many general models based on heterogeneous datasets, class-based models offer a few advantages, such as providing a stronger mechanistic

$$\log \text{TA100} = 0.92\, K_{ow} + 1.17\, \text{HOMO} - 1.18\, \text{LUMO} + 7.35 \qquad (10.10)$$

basis. However, they are also relatively narrow in terms of their applicable across the chemical space of interest (e.g. the chemical space of REACH chemicals).

10.3.5 Reproductive toxicity

Reproductive toxicity is another endpoint of particular concern. Reproductive toxicity includes adverse effects on fertility in males and females and on developmental toxicity (toxicity to the embryo or foetus), and covers any effect interfering with normal development before or after birth, from conception to sexual maturity. The detection of reproductive toxicity in animal experiments is based on the assessment of fertility and general reproductive performance, embryotoxicity and teratogenicity, prenatal and postnatal development, and multigeneration effects. Reproductive toxicity is therefore an umbrella term for a set of diverse and complex endpoints, each of which is likely to involve multiple mechanisms of action. Because of the financial costs and animal welfare implications of reproductive toxicity, there has been considerable impetus to find alternative (non-animal) methods, including QSARs, capable of reducing the need for animal testing. However, because of the diversity and complexity of the effects to be modelled, many of which are poorly understood, it is also one of the most challenging areas of predictive toxicology [99].

Several expert systems incorporate structural alerts for developmental toxicity (teratogenicity), including Derek and HazardExpert. Quantitative models for developmental toxicity include those in TOPKAT and MultiCASE. Currently, TOPKAT contains three models for teratogenicity, each applicable to different chemical group. The training set contains more than 170 chemicals with observations in rat oral studies. MultiCASE offers a larger variety of endpoints including teratogenicity in the mouse, rat, hamster, rabbit and even humans, with training sets varying in size between 40 and 1400 chemicals. Other QSAR studies of developmental toxicity have focused on defined and restricted sets of compounds.

In a study by Pearl et al. [100], a comparison was made between the predictivities of MultiCASE, TOPKAT and Derek for a set of 105 compounds tested in *in vivo* rodent teratogenicity studies (34 teratogens and 71 non-teratogens). TOPKAT demonstrated a concordance of 50%, with a larger percentage of false positive than false negative predictions. MultiCASE demonstrated a greater concordance (66%), with a higher percentage of false negative than false positive predictions. Derek identified a few positive chemicals for which there were structural alerts in its rule base. The authors concluded that by combining the use of different software programs, it is possible to increase the overall level of predictivity.

A specific mechanism of reproductive toxicity is associated with the disruption of the endocrine function. One of the most studied and best understood pathways of endocrine disruption is the direct ligand-receptor interaction with the nuclear hormone receptor superfamily, including the estrogen and androgen receptors, although other pathways have also been investigated. The interactions of chemicals with nuclear hormone receptors form the basis of numerous (Q)SAR models, which have been developed by using a wide spectrum of methods, including automatic docking [101], comparative molecular field analysis (CoMFA) [102,103], classical QSAR and hologram QSAR [104], the common reactivity pattern approach [105,106], molecular quantum similarity analysis [107], decision forests [108], and various combinations of tools ranging from simple rejection filters to more sophisticated tools used for chemical identification and lead optimization [109].

As an example of a transparent and easily interpretable model, Netzeva et al. [110] developed a decision tree classification model for relative estrogenic gene activation, based on *in vitro* data from a recombinant yeast assay. The model, which has been further developed and assessed for external predictivity [111], is shown in Figure 10.3.

10.3.6 Biokinetic parameters

Information on the absorption, distribution, metabolism and excretion (ADME) of chemicals describe the fate of a chemical in a biological organism. ADME information is important in evaluating systemic exposure and, in combination with toxicity hazard data, can be used to determine therapeutic and safe levels of chemicals (e.g. drugs) for humans.

A preliminary assessment of ADME properties is possible by using (Q)SAR approaches [112]. In this section, we illustrate this by focusing on three properties: aqueous solubility, ionization and hydrophobicity. Furthermore, since the dermal route of exposure is one of most relevant for industrial chemicals, we also describe some of the most common models for predicting skin penetration.

Aqueous solubility

Water-soluble compounds are often rapidly absorbed and eliminated (via the kidneys in urine). Compounds of low solubility may be deposited at the dosage site and be

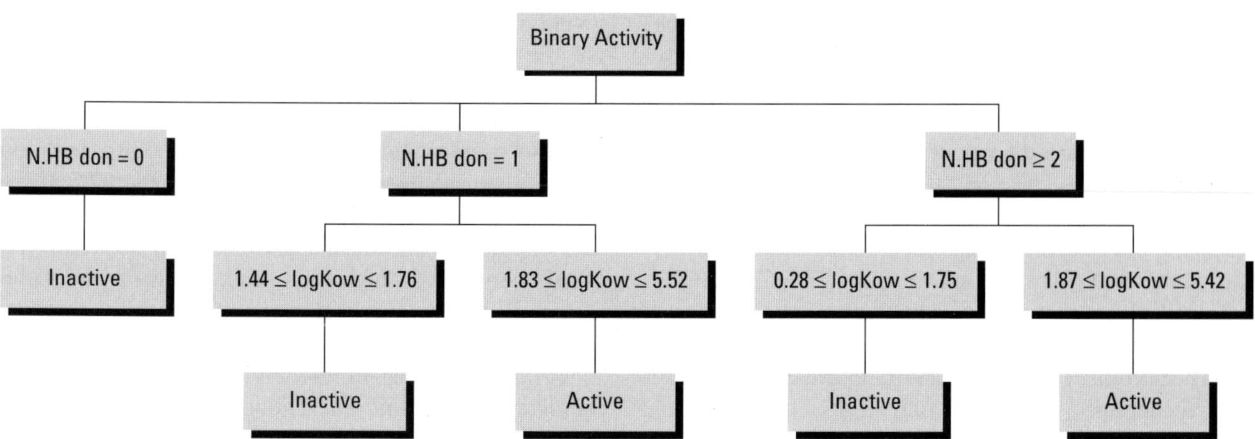

Figure 10.3. Classification model for estrogenicity. N.Hdon is the number of hydrogen bond donors in the molecule and log K_{ow} is the decimal logarithm of the *n*-octanol-water partition coefficient [111]. With permission from Taylor & Francis Ltd.

poorly absorbed. Aqueous solubility has been extensively studied and a large number of computational methods have been reported [113]. Predictive models for aqueous solubility have been based on diverse descriptors, such as experimentally based descriptors, molecular properties, and structural features, which are correlated to activity by means of various statistical techniques. Specific approaches include mobile order thermodynamics [114], linear solvation energy relationships [115], electrotopological state indices [116]. Many examples of QSARs for aqueous solubility exist in the literature [117-122], although this remains a difficult parameter to predict (see also Chapter 9).

Ionization

Passive diffusion across membranes occurs only for the non-ionized form of a chemical. Therefore pH is a factor affecting absorption (and elimination). For example, gastrointestinal absorption will depend on the ionization of a chemical at given pH (pH of 2 in stomach vs. pH of 6 in intestine). Several commercial software packages exist to predict acidity or basicity (pK_a/pK_b). Examples include Advanced Chemistry Development (Toronto, Ontario, Canada) (http://www.acdlabs.com/products/phys_chem_lab/pka), Pallas (Compudrug, Budapest, Hungary) and ChemSilico (http://www.chemsilico.com/CS_products/products.html).

Hydrophobicity

The passage across membranes is favored for lipophilic compounds. Accumulation in fat (if biotransformation is slow) is also possible. Passage through the skin is highly dependent on the hydrophobicity of a compound. Log K_{ow} is a measure of the hydrophobicity of a chemical (Chapter 9). Numerous QSARs have been developed for its estimation. Commonly used methods are based on fragment constants. Available packages include ClogP and KoWwin. The ClogP program (Biobyte Corp, CA, USA) is based on the Hansch and Leo calculation procedure [123] which comprises two parts. A summation of fragmental values for each of the composing atoms or groups is performed and followed by the application of correction values associated with factors such as chain length, ring size, branching and unsaturation (http://www.biobyte.com). The Syracuse Research Corporation (SRC) KoWwin method (http://www.syrres.com/esc/kowwin.htm) is also based on a group contribution method. The method uses structural fragments and correction factors [124].

Other methods for estimating log K_{ow} are based on QSARs developed by algorithms such as regression analysis or neural networks. These are based on large heterogeneous datasets employing a wide range of different descriptors, including topological indices. Examples include models developed by Devillers [125, 126] and Moriguchi [127]. AutoLogP [125] was derived from a heterogeneous set of 800 substances collected from the literature which were then described by means of the autocorrelation method [128] using the fragmental constants of Rekker and Manhold [129] and resulting in 66 atomic and group contributions. A review of these prediction methods is available in [112].

Skin penetration

The absorption of a chemical through the skin can be regarded as a passive diffusion process governed by Fick's first law of diffusion at steady-state [130] where the absorption rate (flux) is proportional to the concentration gradient across the membrane. Immediately following skin exposure, there is a period of time, the lag time (ranging from minutes to days). Once the system reaches steady-state, the rate of absorption is proportional to the applied concentration i.e. a first- order rate process, which can represented by the following equation:

$$Js \propto C \qquad (10.11)$$

In Equation 10.11, Js is the rate of absorption or the steady-state flux (μg/cm²/h) and C is the concentration of the penetrant.

If the concentration gradient of the chemical across the skin is considered and a proportionality constant is added then the steady-state flux per unit area and concentration leading to Kp can be derived:

$$Js = Kp \cdot \Delta C \qquad (10.12)$$

In Equation 10.12, Kp is the permeability coefficient (cm/h) and ΔC is the concentration gradient across the membrane. The parameter Kp is of particular importance since it is independent of the applied dose and can also be calculated using the following equation:

$$Kp = (A/SA) \cdot C \cdot t \qquad (10.13)$$

The parameter A is the amount of penetrant absorbed through the skin and SA is the surface area to which the dose is applied and C is the concentration of the compound and t is the elapsed time [131].

The ability of chemicals to penetrate the skin depends on a number of factors, including lipophilicity, size and solubility [132]. Many workers have attempted to use these simple parameters in an effort to build simple QSARs for percutaneous absorption (see also Chapter 5).

The first large dataset of skin permeability values measured in a single species was published by Flynn [133]. A total of 97 *in vitro* permeability coefficients for 94 compounds were measured by using human skin. The dataset was a compilation from at least 15 different literature sources and hence is subject to a high degree of experimental error including inter-laboratory variability and variability arising from the use of skin from different sources and locations on the body. Flynn [133] proposed a number of algorithms to predict skin permeability based on the premise that skin permeability is mainly a function of partitioning between aqueous and non-aqueous layers and thus can be effectively described by hydrophobicity (i.e. log K_{ow}) and molecular size (expressed by molecular weight [MW] or molecular volume [MV]). Following this approach, Potts and Guy used log K_{ow} in combination with either MW or MV to predict the skin permeabilities [134]. The Flynn dataset has also been analysed by a number of other workers, as described in several recent reviews [135-137]. There are a number of problems with the original Flynn dataset [133], in particular some of the Kp values for the steroid compounds have been subsequently found to be incorrect. The review by Geinoz et al. [137] describes this in further detail.

A large number of models have been developed for skin penetration. The models reasonably explain the skin permeabilities of compounds falling in the lower molecular size range and in the middle range of lipophilicity, but diverge for very hydrophilic (in particular, charged) and very lipophilic compounds (i.e. those substances least well-represented in the training sets). More experimental work is needed to generate data for classes that are poorly represented. In addition, the models currently available only make predictions for Kp in one vehicle system (water). More research is therefore needed to account for the penetration of chemicals in other vehicles, solvents and formulations [138].

10.3.7 Chemical metabolism and biotransformation

Metabolic fate depends on a range number of variables related both to the compound itself and to the biological system. In this section, we describe some of the computer systems available for predicting xenobiotic metabolism as well as some of the reaction databases of enzymatic biotransformations.

COMPACT

The computer-optimized molecular parametric analysis of chemical toxicity (COMPACT) system was developed at the University of Surrey (UK) by Lewis and co-workers [139]. COMPACT has modules that assess the ability of xenobiotics to form enzyme substrates complexes and undergo metabolic activation by the CYP1A and CYP2E subfamilies of cytochrome P450s.

META

The META system has been developed by Klopman and co-workers [140-142] at Case Western Reserve University (OH, USA). It is an expert system capable of predicting the sites of potential enzymatic attack and the nature of the chemicals formed by such metabolic transformations. The program uses dictionaries of biotransformation operators which are created by experts in the field of xenobiotic metabolism to represent known metabolic pathways. A query structure is entered and the program applies biotransformation operators according to the functional groups detected. After each biotransformation a stability check is performed on the reaction product by using quantum mechanical calculations to detect unstable atom arrangements. The program then evaluates the stable metabolites formed and attempts to transform them further until water soluble metabolites that are deemed to be excretable are formed.

MetabolExpert

MetabolExpert is composed of a database, a knowledge base and several prediction tools [143]. The basic biotransformation database contains 179 biotransformations, 112 of which are derived from Testa and Jenner [144], and the others are based on frequently occurring metabolic pathways [145]. The transformation knowledge-base is composed of "if-then" rules derived from the literature by experts.

METEOR

Meteor is a computer system which uses a knowledge-base of structure-metabolism rules to predict the metabolic fate of a query chemical structure. The system is developed and marketed by LHASA Ltd (Leeds, UK) and evolved from the Derek system for toxicity prediction [146]. Meteor's biotransformation rules are generic reaction descriptors rather than simple entries in a reaction database. To limit over prediction, Meteor has an integrated reasoning engine based on a system of non-numerical argumentation, which uses a repository of higher level reasoning rules. The reasoning model allows the system to evaluate the likelihood of biotransformation taking place and to make comparisons between potentially competing biotransformations. The user can choose to analyse queries at a number of available search levels. At the "high likelihood" level, only the more likely biotransformations are requested for display. The system is also supplied with a knowledge base editor so that users can add their own (proprietary) rules. The metabolic tree can be searched and metabolites of specific molecular mass and or molecular formula highlighted. The generated tree is also structure-searchable. Individual biotransformations can be viewed with generalised graphical descriptions of their scope. It is possible to generate sequences automatically and to generate metabolites from an individually chosen biotransformation. It is possible to search for either phase-I or phase-II biotransformations only. Additionally, Meteor is provided with a link to ClogP (Biobyte Corp, CA, USA) to identify biotransformations that are not likely to occur, due to very low lipophilicity.

TIMES

The Tissue MEtabolism simulator (TIMES) involves the use of a heuristic algorithm to produce plausible biotransformation pathways from a query molecule by using rules developed from a comprehensive library of biotransformations [147]. The generation of metabolites by TIMES can be limited to the most likely ones or can be extended to include less likely ones. The developers have also integrated reactivity models for various macromolecular interactions, for example for mutagenicity and sensitization, to simulate the generation of reactive metabolites by specific metabolising systems, such as S9.

MDL Metabolite

MDL Metabolite (http://www.mdli.com) comprises a database, a registration system and a browsing interface. The database is the only source that uses information from multiple studies to assemble structural metabolic database entries for particular parent compounds. The prime focus is on xenobiotic compounds and biotransformations of medicinal drugs. Experimental data is abstracted from *in vitro* and *in vivo* studies. In addition to structural information, the database contains enzyme information, species information, physiological activity, parent compound toxicity, bioavailability, analytical methodology, route of administration, excretion routes, quantitative and qualitative yield, CAS number of parent compound and references to the original literature.

The Accelrys Biotransformation database

This database, available as a CD ROM from Accelrys (http://www.accelrys.com), comprises biotransformations of chemical entities, including pharmaceuticals, agrochemicals, food additives and environmental and industrial chemicals. The database is indexed with original citations, test systems and a variety of keywords for generic searching and is fully cross referenced to all seven volumes of the Biotransformations book series edited by David Hawkins [148].

University of Minnesota Biocatalysis/Biodegradation Database

The University of Minnesota Biocatalysis/Biodegradation Database (UM-BBD, http://umbbd.ahc.umn.edu) contains compound, enzyme, reaction and pathway information for microbial catabolism of primarily anthropogenic materials. It has been available on the web for over 10 years, and has grown from 4 to almost 150 pathways. It currently contains information on over 900 compounds, over 600 enzymes, over 1000 reactions and about 350 microorganism entries. Along with pathway data, Biochemical Periodic Tables (http://umbbd.ahc.umn.edu/periodic) and a Biodegradation Pathway Prediction System (PPS) (http://umbbd.ahc.umn.edu/predict) are also available.

10.4 PREDICTION OF ECOTOXICITY ENDPOINTS

In this section, we provide a few examples of (Q)SARs for ecotoxicological effects, to give an impression of the different types of models and approaches used to develop them. Extensive reviews of (Q)SARs for ecotoxicological endpoints can be found elsewhere [17,149].

The endpoints used in the hazard and risk assessment of chemicals in the aquatic environment are usually based on effect concentrations in a few species, typically algae, crustaceans (e.g. *D. magna*) and fish. Usually, simple *in vivo* effects, such as survival, immobilization, or inhibition of growth and reproduction are measured and expressed as the concentration of test chemicals causing 50% of the predefined effect (EC50). In some cases, point estimates are used, whereas in other cases, the response is categorical (e.g. toxic/non toxic or low/medium/high toxicity). Estimation methods such as (Q)SARs have been developed to predict the endpoints on continuous or categorical scales. In general, where continuous numeric data exist, it is recommended to develop and use continuous QSARs, but the use of categorical QSARs (classification models) might be appropriate, depending on the purpose and availability of data.

The quality of the QSAR estimate depends on number of factors, including the quality of the underlying toxicity data and the quality of the model, bearing in mind also the scope (applicability domain) of the model. Schultz and Cronin [150] proposed some essential and desirable features associated with ecotoxicity QSARs. The quality (variability) of the biological data used in development of the QSAR inevitably places a limitation on the quality of the ultimate model. High quality toxicity data usually come from standardized as says with a well-defined endpoint and generated in a consistent manner. For example, the fathead minnow *(Pimephales promelas)* mortality database of USEPA-Duluth [151] is widely regarded as one of the highest quality ecotoxicity databases available. A lower level of quality is associated with compilations of toxicity data of differing (and often unchecked) sources [152]. Even within high quality databases, it is well established [153] that there is generally less error associated with chemicals acting by non-reactive mechanisms of action (e.g. narcosis) than for more specifically acting chemicals (e.g. those that are reactive or become reaction following metabolic activation). Given the importance of data quality for QSAR modelling, Schultz et al. [154] attempted to associate a confidence factor with some of the more frequently used databases.

10.4.1 Classification of chemicals by modes of toxic action

Compared with some toxicological endpoints, the modes of toxic action in ecotoxicology are well understood, and consequently (Q)SARs for aquatic toxicity endpoints tend to be mechanistically based. These modes of action have been reviewed extensively [155-158].

Several modes of action in acute fish toxicity tests were identified by McKim et al. [159], and subsequently by Bradbury et al. [160]. The different modes of action are distinguished experimentally in terms of the fish acute toxicity syndromes (FATS), which are defined combinations of respiratory, cardiovascular, and physiological responses observed in rainbow trout after acute exposure to a chemical. McKim et al. [159] distinguished six different modes of toxic action, whereas Verhaar et al. [161] recognized four modes of action associated with different structural classes, and Russom et al. [162] suggested seven modes of toxic action (see Table 10.2). Nendza and Müller [163] considered nine mechanisms (including also inhibition of photosynthesis, associated with algae only, and estrogenic activity as a specific mechanism in fish).

The Verhaar classification system was challenged to predict the class of new chemicals with measured toxicity data. It was observed that the system generally provides adequate predictions but additional research is needed to refine the rules for classification of certain chemicals [164]. Recently, the Verhaar classification scheme has been used in a study to address the environmental threshold of no toxicological concern for freshwater systems [165]. While offering a convenient and simplistic picture of the way the chemicals exert toxicity in fish, these structure-based classifications might not be sufficient to allow development of mechanism-based QSAR models for all four identified mechanisms.

In recent studies by Pavan et al. [166,167], in which the abilities of different models to predict acute fish toxicity were assessed with reference to an OECD Screening Information Data Set (SIDS) of 177

Table 10.2 Modes of toxic action according to McKim et al. [159], Verhaar et al. [164], and Russom et al. [162].

Verhaar et al. [161]	McKim et al. [159]	Russom et al. [162]
Inert chemicals	Non-polar narcosis	Non-polar narcosis
Less inert chemicals	Polar narcosis	Polar narcosis
	Uncoupling of oxidative phosphorilation	Uncoupling of oxidative phosphorylation
Reactive chemicals	Respiratory membrane irritation	Respiratory inhibition
		Electrophile/nucleophile reactivity mechanisms
Specifically acting chemicals	Acetyl cholinesterase inhibition	Acetyl cholinesterase inhibition
	CNS seizure mechanisms	CNS seizure mechanisms

compounds, more than 15 mechanisms of toxic action in fish were distinguished. Some of the mechanisms were represented by small numbers of chemicals (e.g. isocyanate based reactivity), whereas other mechanisms (in addition to those proposed by McKim [159] and Verhaar [161]) were well represented.

The required level of differentiation into separate mechanisms of toxic action is a matter of debate, and ultimately depends on the needs of the model user. Having said this, different experts often use different terms and criteria and for different mechanisms, which is impeding the practical application of the mechanistic concept.

10.4.2 QSARs for narcotic chemicals

The narcosis mode of action is associated with altered structure and function of the cell membranes. The overall response of fish to narcotic chemicals include dramatic slowing of all respiratory-cardiovascular functions and classic anaesthesia effects such as loss of reaction to external stimuli, loss of equilibrium, decline in respiratory rate, and medullary collapse [159]. In a consensus classification of 177 SIDS chemicals more than 50% of these industrial chemicals were classified as narcotics [166, 167]. Each organic compound can, in principle, act as narcotic. Therefore, this mode of action is considered a baseline or minimal effect, and QSAR equations for this type of chemicals can be used to predict minimum toxicity [168]. It is conventional to distinguish between non-polar and polar narcosis, the latter being slightly more toxic than the former. Both mechanisms can be modelled solely by the octanol-water partition coefficient; however, the linear regression equations have slightly different slopes and intercepts. The most commonly used models for non-polar narcosis (Equation 10.14) and polar narcosis (Equation 10.15) are recommended in the EU technical guidance document on risk assessment [169]:

$$\text{Log LC50} = -\,0.85 \log K_{ow} - 1.39 \qquad (10.14)$$

$$n = 58,\ r^2 = 0.94,\ q^2 = 0.93,\ s = 0.3$$

$$\text{Log LC50} = -\,0.73 \log K_{ow} - 2.16 \qquad (10.15)$$

$$n = 86,\ r^2 = 0.90,\ q^2 = 0.90,\ s = 0.33$$

In these models, LC50 is the concentration (in moles per litre) causing 50% lethality in fathead minnow following an exposure for 96 hours; q^2 is the coefficient of determination between the observed and predicted observations in the leave-one-out procedure. Graphically, the two equations are presented in Figure 10.4.

Quantitative relationships between the hydrophobicity and toxicity of non-polar narcotics have been reported by many authors [e.g. 170]. In a study by Lessigiarska et al. [171], a large dataset containing fish, algae and *Daphnia* toxicity data was taken from the new chemical database of the ECB (http://ecb.jrc.it), and used to derive interspecies correlations, QSARs for non-polar narcosis, and QSARs for toxicity to algae *(Scenedesmus subspicatus)* and fish (rainbow trout). This study highlighted some of the difficulties in obtaining high-quality for modelling, even when the data were carefully selected from a database containing records based on standardised protocols.

It can be argued that chemicals with log K_{ow} values higher than approximately 3.0 can be modelled equally well by either Equation 10.14 or Equation 10.15, since the regression lines start to converge and the log LC50 (in moles per litre) value results obtained from the two equations are not significantly different. In fact, it is reasonable to develop a general narcosis model (Equation 10.16) from the dataset [172]:

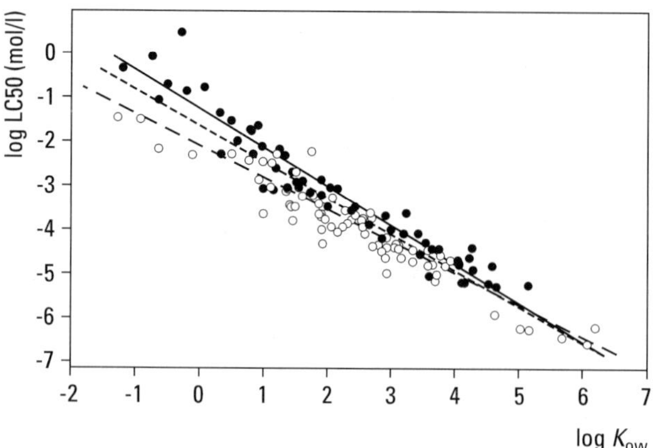

Figure 10.4. QSAR regression lines for non-polar narcosis (solid line, solid circles) and polar narcosis (dashed line, empty circles) to fathead minnow. The dotted line represents the general narcosis model developed from the combination of the two subsets [149].

$$\text{Log LC50} = -0.81 \log K_{ow} - 1.74 \quad (10.16)$$

$$n = 144, r^2 = 0.88, q^2 = 0.87, s = 0.45$$

This model should be used with caution for chemicals with relatively low log K_{ow} values. Similar models have been developed by Roberts and Costello [173] for non-polar and polar narcosis to guppy:

$$\text{Log LC50} = -0.84 \log K_{ow} - 1.12 \quad (10.17)$$

$$n = 8, r^2 = 0.97, F = 199, s = 0.24$$

$$\text{Log LC50} = -0.76 \log K_{ow} - 1.98 \quad (10.18)$$

$$n = 10, r^2 = 0.89, F = 0.90, s = 0.29$$

In the original paper [173], the authors use the logarithm of the inverse LC50 (in moles per litre) values, which changes the sign of the coefficients in the equations. The pair wise similarities between 10.14 and 10.17, and between 10.15 and 10.18 can be explained by the fact that the mechanism of narcosis is not specific and depends only on the ability of the chemical to be absorbed, the mechanism of absorption (generally passive diffusion through biological membranes) being comparable for different species. Similar equations (in terms of coefficients) have been published by Hansch for carp and goldfish [156]. Thus, the log K_{ow} models for non-polar narcosis to four species suggest that there is little interspecies variability for this mode of action. Other studies, however, show that different species might demonstrate different susceptibilities to aromatic narcotic chemicals, which is more evident at log K_{ow} values less than 4 [174].

The distinction between non-polar and polar narcosis has long been debated. In 1977, Kaufmann [175] published an extensive review on the biophysical mechanisms of unaesthetic action in which he defined narcosis as a reversible state of arrested activity of protoplasmic structures resulting from exposure to the appropriate xenobiotic. In 1989, Bradbury et al. [176] showed that the polar narcotic effect can be distinguished from non-polar narcosis by various electrophysiological and biochemical variables in fish. Vaes et al. [177] demonstrated that the two classes of chemicals can be modeled together by a high-quality QSAR ($r^2 = 0.98$, $q^2 = 0.97$) using L-adimyristoyl phosphatidyl-choline (DMPC)-water partition coefficients (log KDMPC) instead of log K_{ow}. However, the relatively small number of chemicals from both mechanisms (8 non-polar and 11 polar narcotics) in this study might preclude from drawing any strong conclusions. Escher and Hermens [155] supported the same point of view, arguing that octanol is not an optimal surrogate for biological membranes, even though, for practical applications, log K_{ow}-based QSARs for baseline toxicity are useful in establishing water and sediment quality criteria. Roberts and Costello [173], however, argued that the two mechanisms should be considered separately, referring to a difference in the physicochemical behavior of the narcotics: the non-polar narcotics act by three-dimensional partitioning (i.e. able to move in all directions in the hydrocarbon-like interior of the membrane), whereas the polar narcotics act by two-dimensional partitioning (i.e. there is a binding between a functional group on the narcotic and the polar phosphatidyl choline head groups at the membrane surface). It could be argued that this discussion is rather academic. From a practical point-of-view, the selection of the incorrect model for narcosis (or selection of a combined model) could result in a worst-case error of one log unit at log $K_{ow} = -1$, whereas a much bigger error would be made if a reactive chemical is treated as a narcotic.

Other mechanisms of narcosis have also been described in the literature. For amines, other than aniline derivatives, an enhanced toxicity was observed in fathead minnow and a separate "amine narcosis" model was developed [178]:

$$\text{Log LC50} = -0.67 \log K_{ow} + 0.81 \quad (10.19)$$

$$n = 61, r^2 = 0.86, s = 0.53$$

Table 10.3. Examples of chemicals classes exhibiting enhanced reactivity.

Chemical class	Chemical structure	Example chemicals	Descriptors in a model
α,β-Unsaturated aldehydes	RCH=CHCHO	2-Butylacrolein	$\log K_{ow}$, the difference between the partial charges over the C and O atoms in the carbonyl group
α,β-Unsaturated ketones	$R_1COCH=CHR_2$	3-Penten-2-one	E_{LUMO}, the sum of the partial charges over the carbon atoms at both ends of the double/triple bond
α-Halogenated esters and nitriles	$R_1C(X)C(=O)R_2$ $R_1C(X)C(\#N)R_2$	4-Bromobutylacetate 4-Bromobutyronitrile	E_{LUMO}, maximum acceptor superdelocalisability in a molecule, ellipsoidal volume
Acrylates	$R_1OC(=O)CH=CH_2$	Ethyl acrylate	Toxicity can be accepted as a constant
Isothiocyanates	RN=C=S	Propylisothiocyanate	Toxicity can be accepted as a constant

In the original paper [178], the authors use the logarithm of the inverse LC50 values. In this model, the LC50 value is in mmol/L. In addition, the acute toxicity of aliphatic and aromatic mono and diesters to *P. promelas* has been modelled by a separate "ester narcosis" model, due to enhanced toxicity [179]. The LC50 is again in mmol/L.

$$\text{Log LC50} = -0.64 \log K_{ow} + 0.64 \quad (10.20)$$

$n = 14, r^2 = 0.95, s = 0.22, F = 207$

In the original paper [179], the inverse logarithm is used. The explanation for the deviation from the baseline model was that the *in vivo* hydrolysis of esters is significant and leads to greater toxicity than observed for non-polar narcotics.

10.4.3 QSARs for other modes of action

When chemical reactivity is involved in the mechanism of acute toxicity, hydrophobicity becomes an insufficient (or even redundant) factor in the QSAR model. A general extension of the log K_{ow}-based QSAR models for more reactive chemicals is provided by the general equation:

$$\log C = a\,(\text{penetration}) + b\,(\text{interaction}) + c \quad (10.21)$$

Most often the toxic potency of organic molecules for aquatic toxicity endpoints can be modelled by two factors: hydrophobicity (represented by log Kow) and reactivity (presented by various quantum-mechanical indices, such as orbital energies, partial charges, and/or superdelocalisability indices). One example of this approach is the model for 96h acute toxicity to fathead minnow (Equation 10.22), developed for aromatic narcotics as well as for non-specific (soft) electrophiles [167] rederived from [180]:

see below (10.22)

$n = 114, r^2 = 0.78, q^2 = 0.76, s = 0.48$

Where LC50 is in mol/L and E_{LUMO} is the energy of the lowest unoccupied molecular orbital. Models of this type are referred to as "response-surface" models [181].

Based on earlier studies of Deneer et al. [182,183], Escher and Hermens [150] observed that hydrophobic electrophiles tend to deviate less from the baseline models than hydrophilic electrophiles. This might be due to the tendency of hydrophobic chemicals to occupy preferably the membranes and decreases the concentration of the chemical in the cytosol, where the reactivity takes place. Recently, it was shown that the more electrophilic the

$$\log(\text{LC50}) = -0.57 \log K_{ow} + 0.45\, E_{LUMO} - 2.44 \quad (10.22)$$

chemical, the larger the residual from the response-surface model [184]. This might be due to a shift in the mechanism from soft, non-specific electrophilicity, to more specific mechanisms where particular electrophilic centres are responsible for strong irreversible interaction between the exogenic chemical and the biological molecules. Tools have been developed to address the conformational variability of the electronic descriptors as descriptors of chemical reactivity [185]. Another source of variability when calculating electronic descriptors is the choice of quantum-mechanical method. For the modelling of whole body phenomena such as acute toxicity to fish, the more time-consuming and computer-intensive *ab initio* methods do not offer significant advantages compared to the more traditionally used and less computer-intensive semi-empirical methods [186]. It has been argued that above certain value of E_{LUMO}, equal to about 1.7 eV, chemicals can be regarded as non-electrophilic [187]. This finding, however, does not exclude the possibility for biotic or abiotic activation, or enhanced toxicity as a result of nucleophilicity (expressed by the energy of the highest occupied molecular orbital [E_{HOMO}], instead of the energy of the lowest unoccupied molecular orbital E_{LUMO}).

Although the response-surface model is probably the best generalization for modelling the toxicity of narcotics and non-specific soft electrophiles, it is not applicable to all reactive mechanisms of toxic action. The relationship between chemical structure and possible mechanisms accounting for excess toxicity has been discussed by Lipnick [188]. Numerous studies have focused on the development of QSARs for reactive chemicals [189-193].

Some examples of chemical classes that exhibit enhanced reactivity and therefore might be considered separately from the response-surface models are given in Table 10.3. The aldehyde moiety is associated mainly with Schiff-base formation [188]. This mechanism can also be associated α,β-unsaturated aldehydes, even though these can also act as Michael-type acceptors, a mechanism typically associated with the α,β-unsaturated ketones [193]. Another example of a reactive mechanism that require the use of separate QSAR models is the nucleophilic substitution (SN2) mechanism, expected for a-halogenated esters and nitriles [187]. In the case of acrylates and isothiocyanates, if the reactive group is not sterically hindered, the variability in toxicity is so little that meaningful QSAR models cannot be developed, so the toxicities of these homologous series can be regarded as constant.

Several trends can be drawn from the literature on QSARs for aquatic toxicity. One is that the higher the reactivity, the higher the variability in toxicity. Another is that the higher the reactivity, the lower the significance of hydrophobicity as a predictor of toxicity. Thus, QSARs progress from models based on only log K_{ow} (for the narcotics), through response-surface models (for soft electrophiles), to log K_{ow}-independent models. Furthermore, models are lacking for the most reactive chemicals. For the modelling of lower levels of reactivity, it can be speculated that whole molecule electronic descriptors are sufficient, whereas for higher levels of reactivity, atom-specific and bond-specific electronic descriptors might be more applicable. The group of reactive chemicals in the Verhaar scheme (Table 10.2) is a broad category, and even the examples given in Table 10.3 represent only a few of the many possible mechanisms of action.

There have been numerous attempts to model aquatic toxicity without direct consideration of the mechanism of action. Thus, QSARs have been developed for different chemical classes, which may or may not represent single mechanisms. A number of log K_{ow}-based models for different chemical classes are incorporated in the ECOSAR program, developed by the US Environmental Protection Agency. In addition, chemical class-specific QSARs including descriptors other than, or in addition to log K_{ow}, can be found in numerous papers [194-199]. Other QSARs have been developed by multivariate techniques applied to a large range of theoretical descriptors [186, 200-202].

Some specific mechanisms of action such as acetylcholinesterase (AChE) inhibition and central nervous system stimulation were recognized together with other mechanisms such as narcosis, oxidative phosporylation uncoupling, respiratory membrane irritation and respiratory blocking [159]. The most-striking demonstration of AChE inhibition in fish is the respiratory cardio-vascular syndrome, including immediate oxygen uptake decrease and reduction in the heart rate. Visible signs of CNS activity include tremors and convulsions, accompanied by cessation of ventilatory and cardiac activity. The CNS activity is rather mode than mechanism of action, the QSAR modelling parameters depending on the processes that result in CNS affection. The AChE inhibition is a strongly specific mechanism, including ligand-receptor type interaction, which can be modelled by descriptors of molecular size, charge distribution/polarity and hydrophobicity [197]. The specificity of chemical binding to AChE active site is also amenable to more sophisticated methods for modelling such as 3D QSAR analysis [198]

There is growing interest in a special group of models,

which are not strictly QSARs but so-called quantitative activity-activity relationships (QAARs) [202-204] and quantitative structure-activity-activity relationships (QSAARs) [61, 205-207]. These types of models are based on one or both of the following premises: the mechanisms of action underlying a given endpoint are generally similar in different organisms, and that similar mechanisms sometimes underlie different endpoints (e.g. mutagenicity and sensitization). An advantage of these models is that they can be derived without a precise knowledge of the underlying mechanisms of action. A limitation, however, is that experimental data are required to extrapolate between the two organisms (and/or endpoints).

In conclusion, there has been significant progress in the modelling of narcotic mechanisms of aquatic toxicity, while the QSAR modelling of reactive chemicals remains a challenge for several reasons. Firstly, the scarcity of high quality toxicity data is impeding meaningful QSAR investigations of reactive mechanisms. Secondly, the classification of chemicals according to mechanism of action is often subjective and not always unambiguous. Thirdly, the description of reactivity is not trivial either from a theoretical or experimental point-of-view. There are several tools developed to assist in prediction of aquatic toxicity, some of which are described below (e.g. ECOSAR, TOPKAT, CASE, OASIS/TIMES). Moore et al. [208] published a comparative analysis on model performance of six software packages that predict acute toxicity to fish. The overall conclusion was that all software packages predict well the toxicity of narcotics but more research is needed to provide recommendations for chemicals with other mechanisms of toxic action. As a whole, TOPKAT had excellent model performance for substances within its optimum prediction space (OPS) but only small percent of the tested substances fell within the TOPKAT OPS, thus limiting the utility of the program. A computational neural network was recommended if the chemical was outside the TOPKAT OPS.

10.5 COMPUTER PROGRAMS AND DATABASES

The trend towards the development and use of intelligent testing strategies (see Chapter 11) relies upon access to *in silico* tools and a means of efficiently retrieving existing information on chemicals and their analogues. In this section, we provide some examples from the many existing tools and databases.

10.5.1 Computer programs

Computer-based expert systems enable predictions of the toxicity of chemicals to be obtained directly from the chemical structure. All are built upon some experimental toxicity data with rules derived from the data [209]. The rules are based on mathematical induction (e.g. QSARs) and/or expert judgment (e.g. SARs describing reactive chemistry). Examples of QSAR rule-based systems include TOPKAT and MCASE. Knowledge-based systems include DEREK, OncoLogic® and HazardExpert, whereas other systems, such as TIMES and ECOSAR, are hybrids.

Derek

Derek is a knowledge-based expert system created with knowledge of structure-toxicity relationships and an emphasis on the need to understand mechanisms of action and metabolism [210]. It is marketed and developed by Lhasa Ltd, a not-for-profit company and educational charity (http://www.lhasalimited.org/index.php).

Within Derek, there are over 504 alerts covering a wide range of toxicological endpoints. An alert consists of a toxicophore (a substructure known or thought to be responsible for the toxicity) and is associated with literature references, comments and examples. The Derek knowledge base covers a broad range of toxicological endpoints, but its main strengths lie in the areas of mutagenicity, carcinogenicity and skin sensitization. All the rules in Derek are based either on hypotheses relating to mechanisms of action of a chemical class or on observed empirical relationships. Information used in the development of rules includes published data and suggestions from toxicological experts in industry, regulatory bodies and academia. The toxicity predictions are the result of two processes. The program first checks whether any alerts in the knowledge base match toxicophores in the query structure. The reasoning engine then assesses the likelihood of a structure being toxic. There are nine levels of confidence: certain, probable, plausible, equivocal, doubted, improbably, impossible, open, contradicted. The reasoning model considers the following information: a) the toxicological endpoint; b) the alerts that match toxicophores in the query structure; c) the physicochemical property values calculated for the query structure; and d) the presence of an exact match between the query structure and a supporting example within the knowledge base.

A further application of Derek is its integration with the Meteor system to enable predictions of toxicity for both parent and metabolites [211].

ONCOLOGIC

The Cancer Expert System or OncoLogic® is an expert system that assesses the potential of chemicals to cause cancer. OncoLogic® was developed under a cooperative agreement between the USEPA's Office of Pollution Prevention and Toxics (OPPT) and LogiChem, Inc. It predicts the potential carcinogenicity of chemicals by applying the rules of SAR analysis and incorporating what is known about the mechanisms of action and human epidemiological studies. OncoLogic® has the ability to reveal its line of reasoning just as human experts can. After supplying the appropriate information about the structure of the compound, an assessment of the potential carcinogenicity and the scientific line of reasoning used to arrive at the assessment outcome are produced. This information provides a detailed justification of a chemical cancer causing potential. The Cancer Expert System is comprised of four subsystems that evaluate fibres, metals, polymers, and organic chemicals of diverse chemical structures. The OncoLogic® Cancer Expert System was previously distributed exclusively by LogiChem, Inc. The USEPA has recently purchased the right to the system and is currently updating the system for free distribution to the public (available by contacting Dr Yin-tak Woo; email: woo.yintak@epa.gov).

TOPKAT

TOPKAT is a statistical system consisting of a suite of QSAR models for a range of different endpoints. There are currently 16 modules for the following endpoints: aerobic biodegradability, Ames mutagenicity, *D. magna* EC50, developmental toxicity, fathead minnow LC50, FDA rodent carcinogenicity, NTP rodent carcinogenicity ocular irritancy, log K_{ow}, rabbit skin irritancy, rat chronic LOAEL, rat inhalation toxicity LC50, rat Maximum Tolerated Dose (MTD), rat oral LD50, skin sensitization, and weight-of-evidence rodent carcinogenicity.

TOPKAT models are typically based on the analysis of large datasets of toxicological information derived from the literature. The molecular descriptors used include structural, topological and electro topological indices. The QSARs are developed by regression analysis for continuous endpoints and by discriminant analysis for categorical data [18, 212].

The CASE family of methods

The CASE methodology and all its variants have been developed by Klopman and Rosenkranz [18]. There are a multitude of models for a variety of endpoints and hardware platforms. There are many forms of the CASE models, and the software is variously called CASE, MULTICASE, MCASE, CASETOX and TOXALERT, depending on the endpoint and the hardware platform.

The CASE approach uses a probability assessment to determine whether a structural fragment is associated with toxicity. To achieve this, molecules are split into structural fragments up to a certain path length. Probability assessments determine whether fragments significantly promote or inhibit toxicity. Models are created by including structural fragments in regression analyses.

HazardExpert

HazardExpert is a rule-based system that uses known toxic fragments from the results of *in vivo* experiments. It predicts a number of endpoints, including mutagenicity and carcinogenicity [18]. The knowledge base was developed from a list of toxic fragments reported by more than twenty experts. In addition to toxicity, HazardExpert also estimates toxicokinetic effects on the basis of predicted physicochemical values. A further application is its integration with the Metabolexpert system to generate predictions of toxicity for both parent chemicals and their metabolites.

TIMES

The Tissue MEtabolism Simulator (TIMES) is a heuristic algorithm intended to generate plausible metabolic maps from a comprehensive library of biotransformations and abiotic reactions. The TIMES platform has been used to predict skin sensitization, mutagenicity, and ER/AR binding affinities of chemicals, while accounting for metabolic activation [147]. Recently, it has incorporated models to predict the toxicity to aquatic species (OASIS/TIMES). OASIS/TIMES uses a response-surface approach for modelling acute toxicity for two types of toxicochemical domains: reversible (non-covalent) acting chemicals and irreversible covalent bioreactive chemicals.

ECOSAR

ECOSAR uses a number of class-specific log K_{ow}-based QSARs in order to predict the toxicity of chemicals to aquatic organisms (fish, daphnids, green algae). The QSARs are developed for chemical classes based on measured test data that have been submitted by industry to the US Environmental Protection Agency (USEPA). ECOSAR produces warnings in several occasions (e.g. when the water solubility is very low, or when the prediction is outside the range of log K_{ow} in the training set). The software is freely available from the USEPA (downloadable from http://www.epa.gov/oppt/exposure /docs/episuitedl.htm)

EPI Suite

The EPI (estimation program interface) Suite program integrates a number of estimation models for the prediction of environmental and physical/ chemical properties in one convenient interface. EPI Suite is freely available from the USEPA website (http://www.epa.gov/oppt/exposure/docs/episuitedl.htm). These models include KowWin (for estimating log K_{ow}), AopWin (for predicting gas-phase reaction rates), HenryWin (for Henry's Law constant), MPBPVP (for predicting melting point, boiling point, and vapour pressure), WsKow (for estimating water solubility and log K_{ow}), Hydro (for estimating hydrolysis rate constants for specific organic classes), DermWin (for estimating the dermal permeability coefficient (K_p)), ECOSAR (described above) and BCFWin (for estimating the bioconcentration factor). EPI Suite also estimates a chemical's rate of volatilization from a model river and lake to the atmosphere as well as its expected fate in a sewage treatment plant and level-III fugacity model.

Toxtree

Toxtree, developed by Ideaconsult Ltd, is able to estimate different types of toxic hazard by applying structural rules. Currently, Toxtree includes options for applying the Cramer decision tree and the Verhaar scheme. It is freely available from the ECB website (http://ecb.jrc.it/QSAR).

The Cramer classification scheme (tree) is probably the best known approach for structuring chemicals in order to make estimations of the so-called threshold of toxicological concern (TTC; [213]). The tree relies primarily on chemical structures and estimates of total human intake to establish priorities for testing. The procedure uses recognised pathways for metabolic deactivation and activation, toxicity data and the presence of a substance as a component of traditional foods or as an endogenous metabolite. Substances are classified into one of three classes:

a. Class-1 contains substances of simple chemical structure with known metabolic pathways and innocuous end products which suggest a low order of oral toxicity.
b. Class-2 contains substances that are intermediate. They possess structures that are less innocuous than those in Class-1 but they do not contain structural features that are suggestive of toxicity like those in Class-3.
c. Class-3 contains substances with structures that permit no strong initial impression of safety and may even suggest a significant toxicity.

The Verhaar scheme is a widely used scheme for determining the mode of action of chemicals that display aquatic toxicity. It divides chemicals into four groups: non-polar narcotics, polar narcotics, reactive chemicals and specifically-acting chemicals [161].

ASTER

ASTER (ASsessment Tools for the Evaluation of Risk) was developed by the USEPA Mid-Continent Ecology Division (Duluth, MN, USA) to assist regulators in performing ecological risk assessments. ASTER is an integration of the AQUIRE (AQUatic toxicity Information REtrieval) toxic effects database and a QSAR-based expert system. When empirical data are not available mechanistically-based predictive models are used to estimate ecotoxicology endpoints, chemical properties, biodegradation, and environmental partitioning. ASTER is designed to provide high quality data for discrete chemicals, when available in the associated databases and QSAR-based estimates when data are lacking. The QSAR system includes a database of measured physicochemical properties such as melting point, boiling point, vapour pressure, and water solubility as well as more than 56,000 molecular structures stored as SMILES (Simplified Molecular Input Line Entry System) strings for specific chemicals (http://www.epa.gov/med/Prods_Pubs/smiles.htm). ASTER is currently not publicly available but technical support is available upon request (see http://www.epa.gov/med/Prods_Pubs/aster.htm). AQUIRE is now a component of ECOTOX (described below).

TerraQSAR™

TerraQSAR™ (http://www.terrabase-inc.com) is a collection of computation programs for the prediction of biological effects and physicochemical properties of organic compounds.. The available models developed using a probabilistic neural network (PNN) methodology include: DM 24-hr EC50 for *D. magna*, E2-RBA estrogen receptor binding affinity (RBA), FHM 96-h LC50 for *P. promelas*, LOGP octanol-water partition coefficient, OMAR mouse and rat oral LD50, RMIV rat and mouse *intravenous* LD50 as well as SKIN a skin irritation potential model.

10.5.2 Databases

Numerous databases (including many freely accessible ones) are available on-line and facilitate the searching and retrieval of information on chemicals and their structural analogues. Some of these databases are summarised in this section.

Danish (Q)SAR database
The Danish Environmental Protection Agency (EPA) constructed a database of (Q)SAR predictions made by some 70 models for about 166,000 organic chemicals for a wide range of different endpoints. A collaborative project was set up between the Danish EPA and the ECB to develop an internet-accessible version of this database. This is capable of performing different types of searching, including structure (substructure/exact match) searching, ID (CAS number, name) searching and parameter (endpoint) searching. The (Q)SAR models encompass endpoints for physicochemical properties, fate, ecotoxicity, absorption, metabolism and toxicity. The Danish (Q)SAR database can be accessed from the ECB website (http://ecb.jrc.it/QSAR).

ChemFinder
Chemfinder is a free on-line chemical searching tool (http://www.chemfinder.com). The index provides chemical structures, physical properties, and hyperlinks to other data sources such as RTECS and TOXNET. A subscription product is also available for purchase.

ChemIDPlus
ChemIDPlus provides access to structure and nomenclature information for the identification of chemical substances cited in the National Library of Medicine (NLM) databases. The database contains over 368,000 chemical records, of which 200,000 include chemical structures. ChemIDPlus is accessible from TOXNET (http://toxnet.nlm.nih.gov).

Distributed Structure-Searchable Toxicity (DSSTox)
The Distributed Structure-Searchable Toxicity (DSSTox) Database Network is a project of USEPA's Computational Toxicology Program, to facilitate the building of a public data foundation for improved structure-activity and predictive toxicology capabilities. The DSSTox website (http://www.epa.gov/nheerl/dsstox) provides a public forum for publishing downloadable, standardized toxicity data files that include chemical structures.

The Registry of Toxic Effects of Chemical Substances (RTECS®)
The Registry of Toxic Effects of Chemical Substances (RTECS®) is a commercially available database which is compiled, maintained, and updated by MDL Information Systems, Inc., under the authority of the US government. Toxicity information and other data used in the preparation of safety directives and hazard evaluations are available for over 162,100 chemicals and substances.

ECOTOX database
ECOTOX is a comprehensive database, which provides information on adverse effects of single chemical stressors to ecologically relevant aquatic and terrestrial species. ECOTOX includes more than 400,000 test records covering 5,900 aquatic and terrestrial species and 8,400 chemicals. The primary source of ECOTOX data is the peer-reviewed literature, with test results identified through comprehensive searches of the open literature. All pertinent information on the species, chemical, test methods, and results presented by the authors are abstracted into the ECOTOX database. ECOTOX also includes third-party data collections from the EPA, US Geological Survey, Russia, and OECD Member Countries summarising research that is either published in non-English journals or not available in the open literature. ECOTOX is available on the EPA's public web page (http://www.epa.gov/ecotox).

ECETOC Aquatic Toxicity (EAT) database
The ECETOC Aquatic Toxicity (EAT) database includes information on the toxicity of substances to aquatic species in fresh and saline waters. The latest version of the database (EAT 3) contains more than 5460 entries on almost 600 chemicals and provides the most comprehensive compilation of highly reliable ecotoxicity data published in the scientific press in the period 1970-2000. For each entry, there are over 50 fields of information on the substance, test species, test conditions, test description, endpoint, results and source references. The database is available with ECETOC Technical Report No. 91 [215].

KEGG
The Kyoto Encyclopedia of Genes and Genomes (KEGG) is a freely available bioinformatics resource being developed by Kyoto University and the University of Tokyo (http://www.genome.jp/kegg). The KEGG project was initiated in May 1995, with a view to providing a tool that helps to understand the basic principles and practical applications of the relationships between genomic information and higher order functional information.

KEGG consists of: a) the PATHWAY database providing information on molecular interaction networks such as pathways and complexes, b) the GENES database providing information about genes and proteins generated by genome sequencing projects, c) the LIGAND database providing information about chemical compounds and metabolic pathway information, d) limited amounts of experimental gene expression data in the EXPRESSION and BRITE databases, and e) the

SSDB database, containing information about amino acid sequence similarities among all protein-coding genes in the complete genomes.

Ambit

Ambit is freely available software for data management and QSAR applications, including databases and tools for searching and applicability domain assessment. It was developed by Ideaconsult Ltd (Sofia, Bulgaria) with funding from the CEFIC LRI project, and is available from http://ambit.acad.bg. Search options include searching by name, CAS number, SMILES, substructures and structure-based similarity, and by descriptor ranges. It can also apply grouping approaches based on mechanistic understanding, such as the Verhaar classification scheme. The suite of software tools includes a module for QSAR applicability domain assessment, Ambit Discovery.

SciFinder

SciFinder is a commercially available research tool providing access the world's largest collection of biochemical, chemical, chemical engineering, medical, and other related information (http://www.cas.org/SCIFINDER). It provides a means of using a single source to obtain scientific information in journals and patent literature from around the world. It is possible to explore the database by chemical name, structure, substructure, biological sequence and reaction, as well as by research topic, author, and company.

ISSCAN

ISSCAN (http://progetti.iss.it/ampp/hhhh/hhhh.php?id=233#top) from the Istituto Superiore di Sanita, Rome, Italy (developed by Romualdo Benigni and Cecilia Bossa) is a database specifically designed as an expert decision support tool and includes carcinogenicity classification "calls" to facilitate the application of structure activity relationships (SAR). The database originated from in-house experience and expertise in developing SAR. The database contains information on chemicals that were experimentally tested in long-term rodent studies (rats and mice), no epidemiological data is included.

Carcinogenic Potency Database

The carcinogenic potency database (CPDB) is a unique and widely used international resource of results from 6,153 chronic, long-term animal cancer tests on 1,485 chemicals. CPDB provides a standardized and easily accessible database with qualitative and quantitative analyses of both positive and negative experiments that have been published in the general literature through 1997 and by the National Cancer Institute/National Toxicology Program through 1998. For each experiment, information is included on species, strain, and sex of test animal; features of experimental protocol such as route of administration, duration of dosing, dose level(s) in mg/kg body weight/day, and duration of experiment; target organ, tumour type, and tumour incidence; carcinogenic potency (TD50) and its statistical significance; shape of the dose-response, author's opinion as to carcinogenicity, and literature citation. It was developed by Lois Swirsky Gold and co-workers and is publicly available (http://potency.berkeley.edu/cpdb.html).

NTP

The National Toxicology Program (NTP) makes available data from more than 500 two-year, two species, toxicology and carcinogenesis studies collected by the NTP and its predecessor, the National Cancer Institute's Carcinogenesis Testing Program. The NTP database also contains the results collected on approximately 300 toxicity studies from shorter duration tests and from genetic toxicity studies, which include both *in vitro* and in vivo tests. In addition, test data from the immunotoxicity, developmental toxicity and reproductive toxicity studies are continually being added to this database. Data from these various studies are indexed by Chemical Name and CASRN and organised into five Study Areas. These data are publicly viewable and accessible from the NTP Website (http://ntp.niehs.nih.gov/ntpweb), hosted at the National Institute of Environmental Health Sciences (NIEHS).

Leadscope

Leadscope is a software tool developed and commercialised by Leadscope Inc. (http://www.leadscope.com). It possesses a unique chemical hierarchy containing over 27,000 chemical fingerprints which represent functional groups, chemical groupings and pharmacophores. The software can be purchased with a toxicity database and/or known drugs database. The toxicity database contains integrated information on over 160,000 chemical structures from multiple sources including the FDA PAFA Database, NTP, RTECS®, and the DSSTox Carcinogenicity Potency Database (CPDB). The database covers a range of endpoints including acute and multiple dose studies, such as subchronic liver, carcinogenicity, genetic toxicity, reproductive and irritation. The database can be searched by structure (such as substructure or similarity), type of study, toxic effect, species, sex, dosage, duration and route of exposure. Results can be viewed and exported in convenient formats, such as Excel files.

10.6 REGULATORY APPLICATIONS

10.6.1 Introduction

It is useful to distinguish between the scientific purposes for which (Q)SARs have been developed and the regulatory applications for which they could be applied. This is because (Q)SARs are not always developed with regulatory applications in mind, and not all of the 20,000+ models reported in the scientific literature make predictions of regulatory endpoints of interest.

As illustrated above and in Chapter 9, (Q)SARs can be developed for a wide range of scientific purposes (i.e. endpoints and associated applicability domains). For convenience, (Q)SARs can be grouped into five categories according to whether the models make predictions of: a) physicochemical properties; b) (eco)toxic potential or potency; c) environmental distribution and fate parameters; and d) biokinetic parameters, such as those involved in the absorption, distribution, metabolism and excretion of chemicals.

In principle, (Q)SARs could be used for the following regulatory purposes in the implementation of chemicals legislation:

a. To provide information for use in priority setting procedures, which are used to expedite the risk assessment process for chemicals of concern.
b. To guide the experimental design of an experimental test or testing strategy.
c. To improve the evaluation of existing test data, e.g. by helping to choose valid and representative data or by providing mechanistic information.
d. To support the grouping of chemicals into categories, so that not all members of the category need to be tested for every endpoint.
e. To fill a data gap needed for classification and labelling.
f. To fill a data gap needed for risk assessment.

In the first three applications (a-c), there is no replacement of test data by the (Q)SAR, which is generally used to provide *supplementary* information to experimental data. In the second three applications (d-f), the (Q)SAR is being used to replace test data for an endpoint. This does not mean that testing is not carried out, since the (Q)SAR information could still be used in addition to test data, especially when the regulatory decision is based on multiple endpoints (often from multiple tests).

In practise, the regulatory use of (Q)SARs varies considerably between different countries and between different authorities in the same country. There do not appear to be any systematic reviews of the ways in which (Q)SARs and estimation methods in any particular regulatory framework, but various publications have identified examples. For example, an extensive collection of case studies on the regulatory use of (Q)SARs in different OECD Member Countries has been published by the OECD [215]. A summary of the use of (Q)SAR by US governmental organizations is given by Walker [216], and some perspectives based on the USEPA experience have been published by Tunkel et al. [217]. In this section, some examples are taken from EU legislation on new and existing substances.

10.6.2 Priority setting

In the scientific literature, "ranking" and "priority setting" are sometimes used synonymously. In the regulatory assessment of chemicals, it is useful to make a distinction between ranking methods, which are mathematically based and which can be automated in the form of computer-based algorithms, and priority setting procedures, which include additional considerations, such as expert judgement and concerns by regulatory authorities. In practise, priority setting for the regulatory assessment of chemicals often involves both automated and non-automated steps.

In the context of EU legislation on existing substances (Council Regulation EEC 793/93), four steps have been carried out: data collection, priority setting, risk assessment and risk reduction (if necessary). The priority setting procedure has focussed on ranking the approximately 2500 High Production Volume Substance on the EU list of 101,195 Existing Substances (EINECS). For this purpose, the EU Risk Ranking Method (EURAM) method was developed [218-220]. EURAM was designed to select suitable data from the International Uniform Chemical Information Database (IUCLID). It provides a ranking of substances based on their potential risk to humans and the environment. It calculates an Environment Score (ES), based on environmental exposure and effects scores, and a Human Health Score (HS), based on human exposure and effects scores. The EURAM rankings were not directly used to set priorities for testing, but they were used as the basis for technical discussions leading to the preparation of priority lists, which also took into account national priorities of the EU Member States.

For the purposes of the REACH regulation, priority setting procedures are foreseen for the use in the evaluation and authorization procedures (see Chapter 12). Therefore, technical work is being carried out in the context of the REACH-implementation projects (RIPs) to develop proposals for ranking methods, taking into account the challenges and data requirements of the new legislation.

From the scientific perspective, two main types of ranking methods are distinguished: total order and partial order methods. Total order ranking methods are scalar techniques that can be used to rank chemicals on the basis of more than one criterion. The different criteria values are combined into a global ranking index, and chemicals are ordered sequentially according to the numerical value of the ranking index. Partial order ranking methods are vectorial approaches that recognise that different criteria are not always in agreement, but can be conflicting. Methods for total and partial order ranking have been described by Pavan [221].

10.6.3 Classification and labelling

The "EU Labelling Guide" (Annex VI of Directive 67/548/EEC) contains criteria that are based largely on the interpretation of experimental test results. Nevertheless, Section 1.6.1 of the Annex recognises that "validated" QSARs can be used for the classification and labelling of substances with the following wording:

"For substances the data required for classification and labelling may be obtained: ...The results of validated structure-activity relationships and expert judgement may also be taken into account where appropriate."

The use of a QSAR in Annex VI can be illustrated by the use of predicted log K_{ow} values in the classification of long-term aquatic hazard (bioaccumulation). When valid test data on the preferred predictor of bioaccumulation (fish BCF) are not available, the BCF value can be calculated by using a QSAR or by using a decision rule based on the (experimental or calculated) log Kow value, provided that the QSAR is considered valid for the chemical in question. Classifications based on log Kow values are more conservative than those based on experimental BCF data (i.e. application of log Kow-based trigger results in the classification of more chemicals).

The use of SARs in Annex VI is illustrated can be illustrated by the assumption that an isocyanate is likely to be a respiratory sensitizer, unless there is evidence to the contrary. Similarly, organic peroxides are assumed to be skin irritants, unless evidence suggests otherwise. In addition, read-across from structural analogues that are known sensitizers or carcinogens can be used as supporting evidence for classifications based on sensitization or carcinogenicity.

The EU List of Dangerous Substances, Annex I of Directive 67/548/EEC, also contains a significant number of group entries, in which a classification is assigned to the entire group. There are more than 90 group entries covering more than 1900 chemicals. In addition, by the time of 21st Adaptation to Technical Progress (ATP) of the Directive, 149 coal-derived complex substances had been assigned to 41 groups, and 543 oil-derived complex substances had been assigned to 22 groups in Annex I.

Official EU classifications in Annex I are produced according a consensus process in which the EU Member State authorities agree on the classification. However, the classification criteria in Annex VI are also implemented by the manufacturer and/or importer to provisionally classify and label chemicals, and a number of industry sectors have published guidance for the "self-classification" of chemicals within their responsibility.

To support the self-classification process, the Danish EPA published an "advisory list for self-classification of dangerous substances". The list of suggested hazard classifications was derived by using predictions from (Q)SAR models obtained or developed by the Danish EPA for the following endpoints: acute oral toxicity, skin sensitization, mutagenicity, carcinogenicity and danger to the aquatic environment. The QSAR models were used to make predictions for the approximately 47,000 discrete organic substances in the EINECS. This Danish Advisory List contains 20,624 chemical substances with suggested classifications for one or more of the dangerous properties, and is searchable via the internet [222]. The Danish (Q)SAR database (described above) is also accessible via the ECB website (http://ecb.jrc.it/QSAR).

10.6.4 Risk assessment

For the purposes of risk assessment in the EU, (Q)SARs have generally not been used as stand-alone methods, but in conjunction with available test data, although the reliance on (Q)SAR estimates has depended on the nature of the endpoint.

For physicochemical properties, predictions have seldom been made, because experimental data (which are generally preferred) have been available. In a few cases, physicochemical properties have been predicted. For example, a QSAR was used to estimate the vapour pressure of V6 (2,2-Bis(chloromethyl) trimethylene bis(bis(2-chloroethyl)phosphate)) due to practical difficulties in performing the test. The validity of the QSAR estimate was established by using measured data on the structural analogues TCPP (Tris(2-chloro-1-methylethyl) phosphate) and TDCP (Tris[2-chloro-1-(chloromethyl)ethyl] phosphate).

QSAR estimates have been used routinely for predicting key environmental fate parameters of organic substances, partly because the experimental determination of these parameters can be difficult and/or expensive, and partly because the information is not normally required in the regulatory submissions. For example, the AOPWIN program (Syracuse Research Corporation (SRC), NY, USA) has been used to derive atmospheric degradation rate constants, and log K_{ow} has been used as a predictor of the solid-water partitioning coefficient. (NB AOPWIN is available as a part of the EPI Suite software freely downloadable from http://www.epa.gov/oppt/exposure/docs/episuitedl.htm). For a few chemicals (e.g. trichloroethylene, nonylphenol), QSAR-generated BCF values have been used instead of a range of measured values.

There are a few cases in which aquatic toxicity has been predicted. For example, no aquatic toxicity data were available for 1,3-butadiene due to the physical nature of the substance (volatile, carcinogenic and flammable), so the toxicity was estimated by using a QSAR from the EU Technical Guidance Document (TGD) on risk assessment [169]. The validity of this estimate was established by comparing predictions made by the QSAR for two structurally similar substances, isoprene (2-methyl-1,3-butadiene) and 1,3-pentadiene, for which experimental data were available. In another case, QSARs were used to estimate the acute and chronic aquatic toxicity for octabromodiphenyl ether and decabromodiphenyl ether. The estimates suggested that toxicity would not be expressed below the limit of water solubility, which provided argumentation against the need to perform chronic toxicity tests on these two substances.

Grouping approaches have been used in the context of the Existing Substances Regulation for both registration and risk assessment. Examples include metals and their compounds (e.g. chromium (VI), nickel, cadmium and zinc) as well as petroleum substances. The hydrocarbon block method is a grouping method for evaluating the environmental fate and effects of complex hydrocarbon mixtures. Individual hydrocarbons with similar properties are grouped into "Hydrocarbon Blocks" and a surrogate chemical from the block is selected to represent the properties of the whole block [169]. The properties of individual blocks are then used to predict the environmental fate and effects of the complex hydrocarbon substance. The method has been used to predict both environmental fate properties and effects on environmental species, for example in risk assessments for gasoline and naphtha.

10.6.5. PBT and vPvB assessment

The assessment of PBT (Persistence Bioaccumulation and Toxicity) and vPvB (very Persistent and very Bioaccumulative) potential (referred to hereafter as PBT assessment) is treated separately, because in the EU, the identification of such potential is not part of the classification and labelling process. QSARs for biodegradation (persistence), bioconcentration and bioaccumulation are covered in Chapter 9.

PBT assessment in the EU has been carried out in accordance with the strategy and criteria proposed in the TGD on risk assessment [169], and in the framework of the European Commission's "interim strategy for the management of PBT and vPvB substances" [223]. The work has been carried out by the PBT working group, which is a subgroup of the technical committee on new and existing substances (TCNES). In general, QSARs have been used in combination with experimental data, but have also been used on their own for the selection of PBT candidates where experimental data did not exist or was considered unreliable, and alongside experimental data to confirm PBT status. An initial screening exercise, based on the use of both experimental and QSAR data for persistency, bioaccumulation and toxicity (aquatic and mammalian), led to the selection of 125 candidate PBTs with tonnages in the range 10-1000 metric tonnes. The TGD criteria for identifying PBT candidates on the basis of QSAR estimates alone are similar, but not identical, to the use of screening test data [169], and this inconsistency has led to some discussion in the PBT Working Group.

The subsequent assessment of the candidate PBTs, using both existing experimental data and QSAR predictions in a weight-of-evidence approach, has led to some chemicals being deselected from the list, whereas others have been confirmed as PBTs, or targeted for further assessment. For persistence, the EPIWIN models available within the EPI Suite (SRC, NY, USA) have been used, in addition to a MultiCASE model developed by the Danish EPA. For bioaccumulation, the BCFWIN model has been used, in addition to the TGD BCF model and METABOL. For toxicity, QSARs for short-term aquatic toxicity to algae, fish and *Daphnia* have been used, generally when test data were available for one or more of the three organisms, but lacking for the remaining ones. QSARs for chronic mammalian toxicity, reproductive toxicity and mutagenicity have been proposed, but have not been decisive for T assignment. Read-across has been used on a case-by-case basis (e.g. 1,2,4-trichlorobenzene and 1,2,3 trichlorobenzene) and grouping approaches have also been used (e.g. diarylide pigments, in which different functional groups attached to a common substructure are thought to account for differences in bioconcentration). In addition to single substances, QSARs (and experimental data) have been used to evaluate whether constituents of multi-component mixtures fulfil the PBT criteria.

10.7 CONCLUDING REMARKS

In the past decade, there has been considerable progress in the development of methods for generating (Q)SAR models, as well as in the development of the models themselves. In this chapter, we have tried to capture the current status of (Q)SARs for toxicological, ecotoxicological endpoints, and biokinetic parameters by focusing on selected examples, and referring the reader to more extensive reviews on the subject.

Many models have not been developed specifically with the needs of chemical risk assessment in mind (e.g. many models have been developed from pharmaceutical or agrochemical datasets), and many models, especially literature-based models, tend to be quite specific in their scope of applicability. This does not mean that they are not useful models, but simply that the domain of applicability of any model needs to be understood and considered when using it assess a specific chemical. The need to assess the applicability domain, along with other key characteristics of (Q)SAR models, is captured by the OECD Principles for QSAR Validation, as described in this chapter. These principles form the basis of a conceptual framework according to which (Q)SARs should be characterised to facilitate their acceptance for regulatory purposes. In this chapter, we have also reviewed some of the current experience under EU legislation of applying (Q)SARs and related estimation approaches for the regulatory purposes of classification and labelling, risk assessment and PBT/vPvB assessment. This experience is expected to increase significantly under REACH.

Under the REACH legislation, the registrants of chemicals will be able to use (Q)SARs to replace or reduce testing, and they will need to justify the use of the (Q)SAR-generated information. For this purpose, (Q)SAR Reporting Formats are currently being developed with reference to the OECD validation principles. It should be noted, however, that the relationship between scientific validity and regulatory acceptability is complex, and is very much context-dependent. Thus, the compliance of a (Q)SAR model with the OECD principles does not necessarily imply that it will be accepted. Conversely, it is foreseeable that a model may also be accepted for a particular use even if not all of the principles have been fulfilled. The acceptability of (Q)SAR estimates will ultimately need to be decided on a case-by-case basis, taking into account the endpoint being predicted, the availability of other information contributing to a weight-of-evidence assessment, and the possible consequences of making decisions based on inaccurate estimates (compared with the possible consequences of not making a decision at all, due to lack of information).

Increasingly, we expect that (Q)SARs will be used as part of weight-of-evidence approaches and in the context of Integrated Testing Strategies. This means that the strengths and limitations of different models (including (Q)SARs and *in vitro* models) can be combined in such a way that the limitations of one model are compensated for by another. Thus, there will be an increasing need to develop (Q)SAR models that are fit for a specific purpose in a testing strategy, rather than to simply make the best of existing models. This implies the need to adapt currently available models and to develop new ones. It also implies a greater need for transparency in the documentation of models, so that the models are in principle reproducible and therefore verifiable and modifiable. This is not currently the case with many commercial models, but there are signs that some commercial developments are beginning to embrace this philosophy.

10.8 FURTHER READING

1. Benigni R., ed. 2003. *Quantitative Structure-Activity Relationship (QSAR) Models for Mutagens and Carcinogens.* CRC Press, Boca Raton, FL, USA.
2. Cronin MTD, Livingstone DJ, eds. 2004. *Predicting Chemical Toxicity and Fate.* CRC Press, Boca Raton, FL, USA.
3. Netzeva TI, Worth AP, Aldenberg T, Benigni R, Cronin MTD, Gramatica P, Jaworska JS, Kahn S, Klopman G, Marchant CA, Myatt G, Nikolova-Jeliazkova N, Patlewicz GY, Perkins R, Roberts DW, Schultz TW, Stanton DT, van de Sandt JJM, Tong W, Veith G., Yang C. 2005. Current Status of Methods for Defining the Applicability Domain of (Quantitative) Structure-Activity Relationships. The Report and Recommendations of

ECVAM Workshop 52. *Altern Lab Anim* 33:155-173.

4. Organization for Economic Cooperation and Development. 2006. Report on the Regulatory Uses and Applications in OECD Member Countries of (Quantitative) Structure-Activity Relationship [(Q)SAR] Models in the Assessment of New and Existing Chemicals. OECD, Paris, France.
5. Worth AP, Bassan A, Gallegos A, Netzeva TI, Patlewicz G, Pavan M, Tsakovska I, Vračko M. *2005. The Characterisation of (Quantitative) Structure-Activity Relationships: Preliminary Guidance. JRC report EUR 21866 EN. European Chemicals Bureau, Ispra, Italy. Accesible from: http://ecb.jrc.it.*

REFERENCES

1. Livingstone DJ. 2004. Building QSAR models: a practical guide. In *Predicting Chemical Toxicity and Fate.* Cronin MTD, Livingstone DJ, eds. CRC Press, Boca Raton, FL, USA. pp. 151-170.
2. Cronin MTD, Schultz TW. 2003. Pitfalls in QSAR. *Journal of Molecular Structure (Theochem)* 622:39-51.
3. Kubinyi, H. 1993. QSAR: *Hansch analysis and related approaches.* VCH, Weinheim, Germany.
4. Livingstone DJ. 1995. *Data Analysis for Chemists: applications to QSAR and chemical product design.* Oxford University Press, Oxford, England.
5. Kier LB, Hall LH. 1986. *Molecular connectiviy in structure-activity analysis.* Res. Studio Press Ltd, Letchworth, UK.
6. Dearden JC. 1990. Physicochemical properties. In *Practical Applications of Quantitative structure-activity relationships (QSAR) in Environmental Chemistry and Toxicology.* Karcher W, Devillers J, eds. Kluwer, Dordrecht, The Netherlands, pp. 25-59.
7. Netzeva TI. 2004. Whole molecule and atom based topological descriptors. In: Cronin MTD, Livingstone DJ, eds., *Predicting Chemical Toxicity and Fate.* Taylor and Francis, London, pp. 61-83.
8. Todeschini R, Consonni V. 2000. *Handbook of Molecular Descriptors.* Wiley-VCH, Weinheim, Germany.
9. Hansch C, Maloney PP, Fujitya T, Muir RM. 1962. Correlation of biological activity of phenoxyacetic acids with Hammett substituent constants and partition coefficients. *Nature* 194:178-180.
10. Hansch C, Fukita T. 1964. Rho-sigma-pi analysis. A method for the correlation of biological activity with chemical structure. *J American Chemical Society* 86:1616-1626.
11. Leo A, Hansch C, Church C. 1969. Comparison of parameters currently used in the study of structure-activity relationships. *J Med Chem* 12:766-771.
12. McFarland JW. 1970. On the parabolic relationship between drug potency and hydrophobicity. *J Med Chem* 13:1192-1196.
13. Commission of the European Communities. 2003. Proposal for a Regulation of the European Parliament and of the Council concerning the Registration, Evaluation, Authorisation and Restriction of Chemicals (REACH), establishing a European Chemicals Agency and amending Directive 1999/45/EC and Regulation (EC) {on Persistent Organic Pollutants} / Proposal for a Directive of the European Parliament and of the Council amending Council Directive 67/548/EEC in order to adapt it to Regulation (EC) of the European Parliament and of the Council concerning the registration, evaluation, authorisation and restriction of chemicals. Brussels, 29 October 2003, Brussels, Belgium
14. Commission of the European Communities. 2006. Regulation (EC) No 1907/2006 of the European Parliament and of the Council of 18 December 2006 concerning the Registration, Evaluation, Authorisation and Restriction of Chemicals (REACH), establishing a European Chemicals Agency, amending Directive 1999/45/EC and repealing Council Regulation (EEC) No 793/93 and Commission Regulation (EC) No 1488/94 as well as Council Directive 76/769/EEC and Commission Directives 91/155/EEC, 93/67/EEC, 93/105/EC and 2000/21/EC. *Off J Eur Union*, L 396/1 of 30.12.2006.
15. Jaworska JS, Comber M, Auer C, van Leeuwen CJ. 2003. Summary of a workshop on regulatory acceptance of (Q)SARs for human health and environmental endpoints. *Environ Health Perspect* 111:1358-1360.
16. Eriksson L, Jaworska JS, Worth AP, Cronin MTD, McDowell RM, Gramatica P. 2003. Methods for reliability, uncertainty assessment, and applicability evaluations of classification and regression based QSARs. *Environ Health Perspect* 111:1361-1375.
17. Cronin MTD, Walker JD, Jaworska JS, Comber MHI, Watts CD, Worth AP. 2003. Use of quantitative structure-activity relationships in international decision-making frameworks to predict ecologic effects and environmental fate of chemical substances. *Environ Health Perspect* 111:1376-1390.
18. Cronin MTD, Jaworska JS, Walker JD, Comber MHI, Watts CD, Worth AP. 2003. Use of quantitative structure-activity relationships in international decision-making frameworks to predict health effects of chemical substances. *Environ Health Perspect* 111:1391-1401.
19. Organization for Economic Cooperation and Development. 2004. The Report from the Expert Group on (Quantitative) Structure-Activity Relationships

[(Q)SARs] on the Principles for the Validation of (Q)SARs. ENV/JM/TG(2004)27/REV. OECD, Paris, France, 17pp.

20. Worth AP, Bassan A, Gallegos A, Netzeva TI, Patlewicz G, Pavan M, Tsakovska I, Vracko M. 2005. The Characterisation of (Quantitative) Structure-Activity Relationships: Preliminary Guidance. JRC report EUR 21866 EN. European Chemicals Bureau, Ispra, Italy. Accesible from: http://ecb.jrc.it.
21. Netzeva TI, Worth AP, Aldenberg T, Benigni R, Cronin MTD, Gramatica P, Jaworska JS, Kahn S, Klopman G, Marchant CA, Myatt G, Nikolova-Jeliazkova N, Patlewicz GY, Perkins R, Roberts DW, Schultz TW, Stanton DT, van de Sandt JJM, Tong W, Veith G., Yang C. 2005. Current Status of Methods for Defining the Applicability Domain of (Quantitative) Structure-Activity Relationships. The Report and Recommendations of ECVAM Workshop 52. *Altern Lab Anim* 33:155-173.
22. Jaworska J, Nikolova-Jeliazkova N, Aldenberg T. 2005. QSAR applicability domain estimation by projection of the training set in descriptor space: a review. *Altern Lab Anim* 33:445-459.
23. Nikolova-Jeliazkova N, Jaworska J. 2005. An approach to determining applicability domains for QSAR group contribution models: an analysis of SRC KOWWIN. *Altern Lab Anim* 33:461-70.
24. Dimitrov S, Dimitrova G, Pavlov T, Dimitrova N, Patlewicz G, Niemela J, Mekenyan O. 2005. A stepwise approach for defining the applicability domain of SAR and QSAR models. *J Chem Inf Model* 45:839-849.
25. Draper NR, Smith H. 1991. *Applied Regression Analysis.* John Wiley and Sons, New York, USA.
26. Worth AP, Cronin MTD. 2003. The use of discriminant analysis, logistic regression and classification tree analysis in the development of classification models for human health effects. *J Mol Struct (Theochem)* 622:97-111.
27. Cronin MTD, Dearden JC, Duffy JC, Edwards R, Manga N, Worth AP, Worgan ADP. 2002. The importance of hydrophobicity and electrophilicity descriptors in mechanistically-based QSARs for toxicological endpoints. *SAR QSAR Environ Res* 13:167-176.
28. Cronin MTD, Livingstone DJ. 2004. *Predicting Chemical Toxicity and Fate.* CRC Press, Boca Raton, FL, USA. 472 pp.
29. Patlewicz G, Rodford R, Walker JD. 2003. Quantitative Structure-Activity Relationships for Predicting Skin and Eye Irritation. *Environ Toxicol Chem* 22:1862-1869.
30. Gallegos Saliner A., Patlewicz G, Worth AP. 2006. Review of literature-based models for skin and eye irritation and corrosion. JRC report EUR 22320 EN. European Chemicals Bureau, Ispra, Italy. Accesible from: http://ecb.jrc.it.
31. Lessigiarska I, Worth AP, Netzeva TI. 2005. Comparative Review of QSARs for Acute Toxicity. JRC report EUR 21559 EN. European Chemicals Bureau, Ispra, Italy. Accesible from: http://ecb.jrc.it.
32. Tsakovska I, Lessigiarska I, Netzeva T, Worth AP. 2006. Review of QSARs for mammalian toxicity. JRC report EUR 22846 EN. European Chemicals Bureau, Ispra, Italy. Accesible from: http://ecb.jrc.it.
33. Draize JH, Woodard G, Calvery HO. 1944. Methods for the study of irritation and toxicity of substances applied topically to the skin and mucous membranes. *J Pharmacol Exp Therapeutics* 82:377-390.
34. Organization for Economic Cooperation and Development. 2001. OECD Attachment to the Test Guideline 404: A Sequential Testing Strategy for Skin Irritation and Corrosion. OECD, Paris, France.
35. Worth AP, Cronin MTD. 2001. The use of pH measurements to predict the potential of chemicals to cause acute dermal and ocular toxicity. *Toxicology* 169: 119-131.
36. Worth AP. 2004. The tiered approach to toxicity assessment based on the integrated use of alternative (non-animal) tests. In: Cronin MTD, Livingstone DJ, eds., *Predicting Chemical Toxicity and Fate*. CRC Press, Boca Raton, FL, USA, pp. 389-410.
37. Barratt MD. 1996. Quantitative structure-activity relationships for skin corrosivity of organic acids, bases and phenols: principal components and neural network analysis of extended sets. *Toxicol in Vitro* 10:85-94.
38. Barratt MD. 1996. Quantitative structure-activity relationships for skin irritation and corrosivity of neutral and electrophilic organic chemicals. *Toxicol in Vitro* 10:247-256.
39. Smith JS, Macina OT, Sussman NB, Luster MI, Karol MH. 2000. A robust structure-activity relationship (SAR) model for esters that cause skin irritation in humans. *Toxicol Sci* 55:215-222.
40. Smith JS, Macina OT, Sussman NB, Karol MH, Maibach HI. 2000. Experimental validation of a structure-activity relationship model of skin irritation by esters. *QSAR* 19:467–474.
41. Hayashi M, Nakamura Y, Higashi K, Kato H, Kishida F, Kaneko H. 1999. A quantitative structure-activity relationship study of the skin irritation potential of phenols. *Toxicol in Vitro* 13:915-922.
42. Kodithala K, Hopfinger AJ, Thompson ED, Robinson MK. 2002. Prediction of skin irritation from organic chemicals using membrane-interaction QSAR analysis. *Toxicol Sci* 66: 336-346.

43. Enslein K, Borgstedt HH, Blake BW, Hart JB. 1987. Prediction of rabbit skin irritation severity by structure activity relationships. *In Vitro Toxicology* 1:129-147.
44. Gerner I, Zinke S, Graetschel G, Schlede E. 2000. Development of a decision support system for the introduction of alternative methods into local irritation/ corrosivity testing strategies. Creation of fundamental rules for a decision support system. *Altern Lab Anim* 28:665-698.
45. Gerner I, Graetschel G, Kahl J, Schlede E. 2000. Development of a decision support system for the introduction of alternative methods into local irritation/ corrosivity testing strategies. Development of a relational database. *Altern Lab Anim* 28:11-28.
46. Rorije E, Hulzebos E. 2005. Evaluation of (Q)SARs for the prediction of skin irritation/corrosion potential. Physicochemical exclusion rules. Final report for ECB contract IHCP.B430206. European Chemicals Bureau, Joint Research Centre, Ispra, Italy. Accesible from: http:// ecb.jrc.it.
47. Enslein K, Blake BW, Tuzzeo TM, Borgstedt HH, Hart JB, Salem H. 1988. Estimation of rabbit eye irritation scores by structure activity equations. *In vitro Toxicology* 2:1-14.
48. Rosenkranz HS, Zhang YP, Klopman G. 1998. The development and characterisation of a structure-activity relationship model of the Draize eye irritation test. *Altern Lab Anim* 26:779-809.
49. Tsakovska I, Netzeva T, Worth AP. 2005. Evaluation of (Q)SARs for the prediction of eye irritation/corrosion potential - physicochemical exclusion rules. JRC Report EUR 21897 EN, 42pp. European Chemicals Bureau, Ispra, Italy. Accesible from: http://ecb.jrc.it.
50. Barratt MD. 1997. QSARs for the eye irritation potential of neutral organic chemicals. *Toxicol in Vitro* 11:1-8.
51. Cronin MTD, Basketter DA, York M. 1994. A quantitative structure-activity relationship (QSAR) investigation of a Draize eye irritation database. *Toxicol in Vitro* 8: 21-28.
52. Sugai S, Murata K, Kitagaki T, Tomita I. 1991. Studies on eye irritation caused by chemicals in rabbits—II. Structure-activity relationships and *in vitro* approach to primary eye irritation of salicylates in rabbits. *J Toxicol Sci* 16:111-130.
53. Abraham MH, Kumarsingh R, Cometto-Muñiz JE, Cain WS. 1998. A Quantitative structure-activity relationship (QSAR) for a Draize eye irritation database. *Toxicol in Vitro* 12:201-207.
54. Patlewicz GY, Rodford RA, Ellis G, Barratt MD. 2000. A QSAR model for the eye irritation of cationic surfactants. *Toxicol in Vitro* 14:79-84.
55. Worth AP, Cronin MTD. 2000. Embedded cluster modelling: a novel quantitative structure-activity relationhip for generating elliptic models of biological activity. In: Balls M, van Zeller A-M, Halder ME, eds., *Progress in the Reduction, Refinement and Replacement of Animal Experimentation.* Elsevier Science, Amsterdam, The Netherlands, pp. 479-491.
56. Cronin MTD. 1996. The use of cluster significance analysis to identify asymmetric QSAR datasets in toxicology. An example with eye irritation data. *SAR QSAR Env Res* 5:167-175.
57. Kulkarni A, Hopfinger AJ, Osborne R, Bruner LH, Thompson ED. 2001. Prediction of eye irritation from organic chemicals using membrane-interaction QSAR analysis. *Toxicol Sci* 59:335-345.
58. Ekwall B. 1983. Screening of toxic compounds in mammalian cell cultures. *Ann N Y Acad Sci* 407:64-77.
59. Ekwall B, Clemedson C, Crafoord B, Ekwall B, Hallander S, Walum E, Bondesson I. 1998. MEIC evaluation of acute systemic toxicity. Part V. Rodent and human toxicity data for the 50 reference chemicals. *Altern Lab Anim* 26 Suppl. 2:571-616.
60. Ekwall B. et al. 1998. MEIC evaluation of acute systemic toxicity. Part VI. The prediction of human toxicity by rodent LD50 values and results from 61 *in vitro* methods. *Altern Lab Anim* 26 Suppl. 2:617-658.
61. Lessigiarska I, Worth AP, Netzeva TI, Dearden JC, Cronin MTD. 2006. Quantitative structure-activity-activity and quantitative structure-activity investigations of human and rodent toxicity. *Chemosphere* 65:1878-1887.
62. Organization for Economic Cooperation and Development. 1992. Guidelines for Testing of Chemicals No. 406. Skin sensitisation. OECD, Paris, France.
63. Organization for Economic Cooperation and Development. 2002. Guidelines for Testing of Chemicals No. 429. Skin sensitisation: the local lymph node assay. OECD, Paris, France.
64. Basketter DA, Pease Smith CK, Patlewicz GY. 2003. Contact allergy: the local lymph node assay for the prediction of hazard and risk. *Clin Exp Dermatol* 18:218-221.
65. Gerberick GF, Ryan CA, Kimber I, Dearman RJ, Lea LJ, Basketter DA. 2000. Local lymph node assay: validation assessment for regulatory purposes. *American J Contact Dermatitis* 11:3-18.
66. Dupuis G, Benezra C. 1982. *Allergic contact dermatitis to simple chemicals: a molecular approach.* Marcel Dekker Inc, New York & Basel.
67. Aptula AO, Patlewicz G, Roberts DW. 2005. Skin sensitisation: reaction mechanistic applicability domains for structure-activity relationships. *Chem Res Toxicol*

18:1420-1426.

68. Roberts, DW, Williams DL. 1982. The derivation of quantitative correlations between skin sensitisation and physicochemical parameters for alkylating agents and their application to experimental data for sultones. *J Theor Biol* 99:807-825
69. Roberts, DW, Basketter DA. 2000. Quantitative structure-activity relationships: sulfonate esters in the local lymph node assay. *Contact Dermatitis* 42:154-61.
70. Basketter DA, Roberts DW, Cronin M Scholes EW. 1992. The value of the local lymph node assay in quantitative structure-activity investigations. *Contact Dermatitis* 27:137-142
71. Roberts DW. 1987. Structure-activity relationships for skin sensitisation potential of diacrylates and dimethacrylates. *Contact Dermatitis* 17:281-289.
72. Patlewicz G, Basketter DA, Smith CK, Hotchkiss SA, Roberts DW. 2001. Skin-sensitisation structure-activity relationships for aldehydes. *Contact Dermatitis* 44:331-336.
73. Patlewicz G, Wright ZM, Basketter DA, Pease CK, Lepoittevin J-P, Arnau EG. 2002. Structure-activity relationships for selected fragrance allergens. *Contact Dermatitis* 47:219-226.
74. Patlewicz G, Basketter DA, Pease CK, Wilson K, Wright ZM, Roberts DW, Bernard G, Gimenez Arnau E, Lepoittevin J-P. 2004. Further evaluation of quantitative structure-activity relationship models for the prediction of the skin sensitisation potency of selected fragrance allergens. *Contact Dermatitis* 50:91-97.
75. Roberts DW, York M, Basketter DA. 1999. Structure-activity relationships in the murine local lymph node assay for skin sensitisation: alpha, beta-diketones. *Contact Dermatitis* 41:14-7.
76. Roberts DW, Patlewicz G. 2002. Mechanism based structure-activity relationships for skin sensitisation – the carbonyl group domain. *SAR QSAR Env Res* 13:145-152.
77. Patlewicz G, Roberts DW, Walker JD. 2003. QSARs for the skin sensitisation potential of aldehydes and related compounds. *QSAR Comb Sci* 22:196-203.
78. Roberts DW, Aptula AO, Cronin MT, Hulzebos E, Patlewicz G. 2007. Global (Q)SARs for Skin Sensitisation – assessment against OECD principles. *SAR QSAR Env Res* 18:343-365.
79. Miller MD, Yourtee DM, Glaros AG, Chappelow CC, Eick JD, Holder AJ. 2005. Quantum mechanical structure-activity relationship analyses for skin sensitisation. *J Chem Inf Model* 45:924-929.
80. Fedorowicz A, Zheng L, Singh H, Demchuk E. 2004. Structure-activity models for contact sensitisation. *Int J Mol Sci* 5:56-66.
81. Estrada E, Patlewicz G, Chamberlain M, Basketter D, Larbey S. 2003. Computer-aided knowledge generation for understanding skin sensitisation mechanisms: The TOPS-MODE approach. *Chem Res Toxicol* 16:1226-1235.
82. Enslein K, Gombar VK, Blake BW, Maibach HI, Hostynek JJ, Sigman CC, Bagheri D. 1997. A quantitative structure-toxicity relationships model for the dermal sensitisation guinea pig maximization assay. *Food Chem Toxicol* 35:1091-1098.
83. Barratt MD, Basketter DA, Chamberlain M, Admans GD, Langowski JJ. 1994. An expert system rulebase for identifying contact allergens. *Toxicol in Vitro* 8:1053-1060.
84. Barratt MD, Langowski JJ. 1999. Validation and subsequent development of the Derek skin sensitisation rulebase by analysis of the BgVV list of contact allergens. *J Chem Inf Comput Sci.* 39:294-298.
85. Barratt MD, Castell JV, Miranda MA, Langowski JJ. 2000. Development of an expert system rulebase for the prospective identification of photoallergens. *J Photochem Photobiol B.* 58:54-61.
86. Mumtaz MM, Knauf LA, Reisman DJ, Peirano WB, DeRosa CT, Gombar VK, Enslein K, Carter JR, Blake BW, Huque KI, Ramanujam VMS. 1995. Assessment of effect levels of chemicals from quantitative structure-activity relationship (QSAR) models. I. Chronic lowest-observed-adverse-effect level (LOAEL). *Toxicol Lett* 79:131-143.
87. Venkatapathy R, Moudgal CJ, Bruce RM. 2004. Assessment of the oral rat chronic lowest observed adverse effect level model in TOPKAT, a QSAR software package for toxicity prediction. *J Chem Inf Comput Sci* 44:1623-1629.
88. Tennant RW, Ashby J. (1991). Classification according to chemical structure, mutagenicity to Salmonella and level of carcinogenicity of a further 39 chemicals tested for carcinogenicity by the U.S. National Toxicology program. *Mutation Research*, 257, 209-227.
89. Woo YT, Lai D, McLain JL, Manibusan MK, Dellarco V. 2002. Use of mechanism-based structure-activity relationships analysis in carcinogenic potential ranking for drinking water disinfection by-products. *Environ Health Perspect* 110 Suppl 1:75-87.
90. Tennant RW, Spalding J, Stasiewicz S, Ashby J. 1990. Prediction of the outcome of rodent carcinogenicity bioassays currently being conducted on 44 chemicals by the National Toxicology Program. *Mutagenesis* 5:3-14.
91. Benigni R. 1997. The first U.S. National Toxicology Program exercise on the prediction of rodent carcinogenicity: definitive results. *Mutat Res* 387:35-45.

92. Richard AM. 1994. International commission for protection against environmental mutagens and carcinogens. Application of SAR methods to non-congeneric data bases associated with carcinogenicity and mutagenicity: issues and approaches. *Mutat Res* 305:73-97.
93. Benigni R. 2004. Chemical structure of mutagens and carcinogens and the relationship with biological activity. *J Exp Clin Cancer Res* 23:421-424.
94. Livingstone DJ, Greenwood R, Rees R, Smith MD. 2002. Modelling mutagenicity using properties calculated by computational chemistry. *SAR QSAR Env Res* 13:21-33.
95. Contrera JF, Matthews EJ, Kruhlak NL, Benz RD. 2005. In silico screening of chemicals for bacterial mutagenicity using electrotopological E-state indices and MDL QSAR software. *Regul Toxicol Pharmacol* 43:313-23.
96. Benigni R, ed. 2003. *Quantitative Structure-Activity Relationship (QSAR) Models for Mutagens and Carcinogens.* CRC Press, Boca Raton, FL, USA.
97. Debnath AK, Debnath G, Shusterman AJ, Hansch C. 1992. A QSAR investigation of the role of hydrophobicity in regulating mutagenicity in the Ames test: 1. Mutagenicity of aromatic and heteroaromatic amines in *Salmonella typhimurium* TA98 and TA100. *Environ Mol Mutagen* 19:37-52.
98. Benigni R. 2005. Structure-activity relationship studies of chemical mutagens and carcinogens: mechanistic investigations and prediction approaches. *Chem Rev* 105:1767-1800.
99. Polloth C, Mangelsdorf I. 1997. Commentary on the application of (Q)SAR to the toxicological evaluation of existing chemicals. *Chemosphere* 35:2525-2542.
100. Pearl GM, Livingston-Carr S, Durham SK. 2001. Integration of computational analysis as a sentinel tool in toxicological assessments. *Curr Top Med Chem* 1:247-255.
101. Mizutani MY, Tomioka N, Itai A. 1994. Rational automatic search method for stable docking models of protein and ligand. *J Mol Biol* 243:310-326.
102. Tong W, Perkins R, Xing LI, Welsh WJ, Sheehan DM. 1997. QSAR models for binding of estrogenic compounds to estrogen receptor α and β subtypes. *Endocrinology* 138:4022-4025.
103. *Yu SJ, Keenan SM, Tong W, Welsh WJ.* 2002. Influence of the structural diversity of data sets on the statistical quality of three-dimensional quantitative structure-activity relationships (3D-QSAR) models: predicting the estrogenic activity of xenestrogens. *Chem Res Toxicol* 15:1229-1234.
104. Tong W, Lowis DR, Perkins R, Chen Y, Welsh WJ, Goddette DW, Heritage TW, Sheehan DM. 1998. Evaluation of quantitative structure-activity relationship methods for large scale prediction of chemicals binding to the estrogen receptor. *J Chem Inf Comput Sci* 38:669-677.
105. Schmieder PK, Aptula AO, Routledge EJ, Sumpter JP, Mekenyan OG. 2000. Estrogenicity of alkylphenolic compounds: A 3-D structure-activity evaluation of gene activation. *Environ Toxicol Chem* 19:1727-1740.
106. Serafimova R, Walker J, Mekenyan O. 2002. Androgen receptor binding affinity of pesticide "active" formulation ingredients. QSAR evaluation by COREPA method. *SAR QSAR Env Res* 13:127-134.
107. Gallegos Saliner A, Amat L, Carbó-Dorca R, Schultz TW, Cronin MTD. 2003. Molecular quantum similarity analysis of estrogenic activity. *J Chem Inf Comput Sci* 43:1166-1176.
108. Tong W, Xie Q, Hong H, Shi L, Fang H, Perkins R. 2004. Assessment of prediction confidence and domain extrapolation of two structure-activity relationship models for predicting estrogen receptor binding affinity. *Environ Health Perspect* 112:1249-1254.
109. Tong W, Fang H, Hong H, Xie Q, Perkins R, Sheehan DM. 2004. In: Cronin, MTD and Livingstone, DJ, eds, *Predicting Chemical Toxicity and Fate*. CRC Press, Boca Raton, FL, USA, pp. 285-314.
110. Netzeva TI, Gallegos Saliner A, Worth AP. 2006. Comparison of the applicability domain of a Quantitative Structure-Activity Relationship for estrogenicity with a large chemical inventory. *Environ Toxicol Chem* 25:1223-1230.
111. Saliner AG, Netzeva TI, Worth AP. 2006. Prediction of estrogenicity: validation of a classification model. *SAR QSAR Env Res* 17:195-223.
112. Duffy JC. 2004. Prediction of pharmacokinetic parameters in drug design and toxicology. In: Cronin MTD, Livingstone DJ, eds., *Predicting Chemical Toxicity and Fate.* CRC Press, Boca Raton, FL, USA. pp. 229-261.
113. Taskinen J. 2000. Prediction of aqueous solubility in drug design. *Current Opinion Drug Discovery 3*:102-107.
114. Ruelle P, Kesselring UW. 1998. The hydrophobic effect. 2. Relative importance of the hydrophobic effect on the solubility of hydrophobes and pharmaceuticals in H bonded solvents. *J Pharm Sci.* 87:998-1014.
115. Abraham MH, Le J. 1999. The correlation and prediction of the solubility of compounds in water using an amended solvation energy relationship. *J Pharm Sci.* 88:868-880.
116. Huuskonen J, Rantanen J, Livingstone D. 2000. Prediction of aqueous solubility for a diverse set of organic compounds based on atom type electrotopological

state indices. *Eur J Med Chem.* 35:1081-1088.

117. Huuskonen J. 2000. Estimation of aqueous solubility for a diverse set of organic compounds based on molecular topology. *J Chem Inf Comput Sci.* 40:773-777.

118. Kuhne R, Ebert R-U, Kleint F, Schmidt G, Schuurmann G. 1996. Group contribution methods to estimate water solubility of organic chemicals. *Chemosphere* 33:2129-2144.

119. Jain N, Yalkowsky SHJ. 2001. Estimation of the aqueous solubility I: application to organic nonelectrolytes. *Pharm. Sci.* 90:234-252.

120. Liu R, So S-S. 2001. Development of quantitative-property relationship models for early ADME evaluation in drug discovery. 1. Aqueous solubility. *J Chem Inf Comput Sci* 41:1633-1639.

121. Klopman G, Wang S, Balthasar DM. 1992. Estimation of aqueous solubility of organic molecules by the group contribution approach. Application to the study of biodegradation. *J Chem Inf Comput Sci* 32:474-482.

122. Klopman G, Zhu HJ. 2001. Estimation of the aqueous solubility of organic molecules by the group contribution approach. *J Chem Inf Comput Sci* 41:439-445.

123. Hansch C, Leo AJ. 1979. *Substituent constants for Correlation Analysis in Chemistry and Biology,* Wiley, New York, USA.

124. Meylan WM, Howard PH. 1995. Atom/fragment contribution method for estimating octanol-water partition coefficients. *J Pharm Sci.* 84:83-92.

125. Devillers J, Domine D, Karcher W. 1996. Prediction of *n*-octanol-water partition coefficients with AUTOLOGP. Abstracts of Papers of the American Chemical Society 212:83.

126. Devillers J, Domine D, Guillon C, Karcher W. 2000. Simulating lipophilicity of organic molecules with a back-propagation neural network. *J Pharm Sci.* 87:1086–1090.

127. Moriguchi I, Hirono S, Liu Q, Nakagome I, Matsushita Y. 1992. Simple method of calculating octanol/water partition coefficient. *Chem Pharm Bull.* 40:127-130.

128. Broto P, Moreau G, Vandycke C. 1984. Molecular-structures - Perception, auto-correlation descriptor and SAR studies - System of atomic contributions for the calculation of the normal-octanol water partition-coefficients. *Eur J Med Chem.* 19:71-78.

129. Rekker RF, Terlaak AM, Mannhold R. 1993. On the reliability of calculated log P-values - Rekker, Hansch/Leo and Suzuki approach. *QSAR* 12:152-157.

130. Barry BW. 1983. *Dermatological Formulations: Percutaneous Absorption. Drugs and the Pharmaceutical Sciences.* Vol 18. Marcel Dekker, New York, USA.

131. Smith CK, Hotchkiss SAM. 2001. *Allergic Contact Dermatitis.* Taylor and Francis, London, UK.

132. Howes D, Guy R, Hadgraft J, Heylings J, Hoeck U, Kemper F, Maibach H, Marty J-P, Merk H, Parra J, Rekkas D, Rondelli I, Schaefer H, Teuber U, Verbiese N. 1996. Methods for assessing percutaneous absorption. *Altern Lab Anim* 24:81-106.

133. Flynn GL. 1990. Physicochemical determinants of skin absorption. In: Gerrity TR, Henry CJ, eds., *Principles of route-to-route extrapolation for risk assessment.* Elsevier, New York, USA. pp 93-127.

134. Potts RO, Guy R. 1992. Predicting skin permeability. *Pharm Res* 9:663-669.

135. Moss GP, Dearden JC, Patel H, Cronin MTD. 2002. Quantitative structure-permeability relationships (QSPRs) for percutaneous absorption. *Toxicol in Vitro* 16:299-317.

136. Walker JD, Rodford R, Patlewicz G. 2003: Quantitative structure-activity relationships for predicting percutaneous absorption rates. *Environ Toxicol Chem* 22:1870-1884.

137. Geinoz S, Guy RH, Testa B, Carrupt PA. 2004. Quantitative structure-permeation relationships (QSPeRs) to predict skin permeation: a critical evaluation. *Pharm Res* 21:83-92.

138. Jones AD, Dick, IP Cherrie, JW Cronin, MTD,Van De Sandt JJM, Esdaile DJ, Iyengar S, ten Berge W, Wilkinson SC, Roper CS, Semple S, de Heer C, Williams FM. 2004. CEFIC Workshop on methods to determine dermal permeation for human risk assessment. Institute of Occupational Medicine, Research Report TM/04/07, December 2004.

139. Lewis DVF. 2001. COMPACT: a structural approach to the modelling of cytochromes P450 and their interactions with xenobiotics. *J Chem Technol Biotechnol.* 76:237-244.

140. Klopman G, Dimayuga M, Talafous J. 1994. META 1. A program for the evaluation of metabolic transformation of chemicals. *J Chem Inf Comput Sci* 34:1320-1325.

141. Talafous J, Sayre LM, Mieyal JJ, Klopman G. 1994. META 2. A dictionary model of mammalian xenobiotic metabolism. *J Chem Inf Comput Sci* 34:1326-1333.

142. Klopman G, Tu MH. 1999. META - a program for the prediction of products of mammalian metabolism of xenobiotics. In: Erhardt PW, ed., *Drug Metabolism, Databases in High Throughput Testing during Drug Design and Development.* Blackwell, Oxford, UK. pp. 271-276.

143. Darvas F. 1987. MetabolExpert, an expert system for predicting metabolism of substances. In: Kaiser K, ed., *QSAR in Environmental Toxicology.* Riedel, Dordrecht, The Netherlands. pp 71-81.

144. Testa B, Jenner P. 1976. *Drug Metabolism: Chemical and Biochemical Aspects.* Marcel Dekker, New York, USA.
145. Pfeifer S, Borchert H. 1963. *Biotransformation von Arzneimittelm.* VEB, Berlin, Germany.
146. Greene N, Judson P, Langowski J, Marchant CA. 1999. Knowledge based expert systems for toxicity and metabolism prediction: DEREK, StAR and METEOR. *SAR QSAR Env Res* 10:299-313.
147. Mekenyan OG, Dimitrov S, Schmieder P, Veith G. 2004. A systematic approach to simulating metabolism in computational toxicology. I. The TIMES heuristic modelling framework. *Curr Pharm Des* 10:1273-1293.
148. Hawkins DR. 1996. *Biotransformations.* Volumes 1-7, Royal Society of Chemistry, Cambridge, UK.
149. Verhaar HJM, Mulder W, Hermens JLM. 1995. QSARs for ecotoxicity. In: Hermens, J.L.M. ed., Overview of structure-activity relationships for environmental endpoints, Part 1: General outline and procedure. Report prepared within the framework of the project "QSAR for Prediction of Fate and Effects of Chemicals in the Environment", an international project of the Environmental; Technologies RTD Programme (DGXII/D-1) of the European Commission under contract number EV5V-CT92-0211.
150. Schultz TW, Cronin MTD. 2003. Essential and desirable characteristics of ecotoxicity QSARs. *Environ Toxicol Chem* 22:599–607.
151. Russom CL, Bradbury SP, Broderius SJ, Hammermeister DE, Drummond RA. 1997. Predicting modes of toxic action from chemical structure: Acute toxicity in the fathead minnow *(Pimephales promelas). Environ Toxicol Chem* 16:948–967.
152. Cronin MTD. 2004. Toxicological information for use in predictive modelling: quality, sources, and databases In: Helma C, ed., *Predictive Toxicology.* Marcel Dekker Inc., New York, USA. pp. 93-133.
153. Seward JR, Sinks GD, Schultz TW. 2001. Reproducibility of toxicity across mode of toxic action the *Tetrahymena* population growth impairment assay. *Aquat Toxicol* 53:33-47.
154. Schultz TW, Netzeva TI, Cronin MTD. 2004. Evaluation of QSARs for ecotoxicity: a method for assigning quality and confidence. *SAR QSAR Env Res* 15:385-397.
155. Escher BI, Hermens JLM. 2002. Modes of toxic action in ecotoxicology: their role in body burdens, species sensitivity, QSARs, and mixture effects. *Environ Sci Technol* 36:4201-4217.
156. Bradbury SP, Russom CL, Ankley GT, Schultz TW, Walker JD. 2003. Overview of data and conceptual approaches for derivation of quantitative structure-activity relationships for ecotoxicological effects of organic chemicals. *Environ Toxicol Chem* 22:1789-1798.
157. Schultz TW, Cronin MTD, Walker JD, Aptula AO. 2003a. Quantitative structure–activity relationships (QSARs) in toxicology: a historical perspective. *J Mol Struct (Theochem)* 622:1-22.
158. Schultz TW, Cronin MTD, Netzeva TI. 2003b. The present status of QSAR in toxicology. *J Mol Struct (Theochem)* 622:23-38.
159. McKim JM, Bradbury SP, Niemi GJ. 1987. Fish acute toxicity syndromes and their use in the QSAR approach to hazard assessment. *Environ Health Perspect.* 71:171-186.
160. Bradbury SP, Henry TR, Carlson RW. 1990. Fish acute toxicity syndromes in the development of mechanism-specific QSARs. In: Karcher W and Devillers J, eds., *Practical applications of Quantitative Structure-Activity Relationships (QSAR) in Environmental Chemistry and Toxicology.* Kluwer, Dordrecht, The Netherlands, pp. 295-315.
161. Verhaar HJM, Van Leeuwen CJ, Hermens JLM. 1992. Classifying environmental pollutants. 1: Structure-activity relationships for prediction of aquatic toxicity. *Chemosphere* 25:471-491.
162. Russom CL, Bradbury SP, Broderius SJ, Hammermeister DE, Drummond RA. 1997. Predicting modes of action from chemical structure: Acute toxicity in the fathead minnow *(Pimephales promelas). Environ Toxicol Chem* 16:948-967.
163. Nendza M, Müller M. 2000. Discriminating toxicant classes by mode of action: 2. Physicochemical descriptors. *Quant. Struct.-Act. Relat.* 19:581-598.
164. Verhaar HJM, Solbe J, Speksnijder J, Van Leeuwen CJ, Hermens JLM. 2000. Classifying Environmental pollutants. 3: External validation of the classification system. *Chemosphere* 40:875-883.
165. De Wolf W, Siebel-Sauer A, Lecloux A, Koch V, Holt M, Feijtel T, Comber M, Boeije G. 2005. Mode of action and aquatic exposure thresholds of no concern. *Environ Toxicol Chem* 24:479-485.
166. Pavan M, Worth AP, Netzeva TI. 2005. Preliminary analysis of an aquatic toxicity dataset and assessment of QSAR models for narcosis. JRC Report No. 21749 EN. European Chemicals Bureau, Ispra, Italy. Accesible from: http://ecb.jrc.it.
167. Pavan M, Worth AP, Netzeva TI. 2005. Comparative Assessment of QSAR Models for Aquatic Toxicity. JRC Report No EUR 21750 EN. European Chemicals Bureau, Ispra, Italy. Accesible from: http://ecb.jrc.it.
168. Veith GD, Call DJ, Brooke LT. 1983. Structure-toxicity relationships for the fathead minnow, Pimephales promelas: narcotic industrial chemicals. *Can J Fish*

Aquat Sci 40:743-748.

169. European Commission. 2003. Technical guidance documents on risk assessment in support of Commission Directive 93/67/EEC on risk assessment for new notified substances, Commission Regulation (EC) No. 1488/94 on risk assessment for existing substances, and Directive 98/8/EC of the European Parliament and of the Council concerning the placing of biocidal products on the market. JRC report EUR 20418 EN. European Chemicals Bureau, Ispra, Italy. Accesible from: http://ecb.jrc.it.
170. Hansch C, Kim D, Leo AJ, Novellino E, Silipo C, Vittoria A. 1989. Toward a quantitative comparative toxicology of organic compounds. *Crit Rev Toxicol* 19:85-226.
171. Lessigiarska I, Worth AP, Sokull-Klüttgen B, Dearden JC, Netzeva TI, Cronin MTD. 2004. QSAR investigation of a large data set for fish, algae and *Daphnia* toxicity. *SAR QSAR Env Res* 15:413-431.
172. Pavan M, Netzeva TI, Worth AP. 2006. Validation and applicability domain of a QSAR model for acute toxicity. *SAR QSAR Env Res* 17:147-171.
173. Roberts DW, Costello JF. 2003. Mechanisms of action for general and polar narcosis: a difference in dimensions. *QSAR Comb Sci* 22:226-233.
174. Di Marzo W, Galassi S, Todeschini R, Consolaro F. 2001. Traditional versus WHIM descriptors in QSAR approaches applied to fish toxicity studies. *Chemosphere* 44:401-406.
175. Kaufman RD. 1977. Biophysical mechanisms of anesthetic action: historical perspectives and review of current theories. *Anesthesiology* 46:49-62.
176. Bradbury SP, Henry TR, Niemi GJ, Carlson RW, Snarski VM. 1989. Use of respiratory-cardiovascular responses of rainbow trout *(Oncorhynchus mykiss)* in identifying acute toxicity syndromes in fish. Part 3: Polar narcotics. *Environ Toxicol Chem* 8:247-261.
177. Vaes WHJ, Urrestarazu Ramos E, Verhaar HJM, Hermens JLM. 1998. Acute toxicity of nonpolar versus polar narcosis: is there a difference? *Environ Toxicol Chem* 17:1380-1384.
178. Newsome LD, Johnson DE, Nabholz JV. 1993. Validation and upgrade of a QSAR study of the toxicity of amines to freshwater fish. ASTM Special Technical Publication, STP 1179 (Environmental Toxicology and Risk Assessment), 413-426.
179. Jaworska JS, Hunter RS, Schultz TW. 1998. Quantitative structure-toxicity relationships and volume fraction analyses for selected esters. *Arch Environ Contam Toxicol* 29:86-93.
180. Veith GD, Mekenyan OG. 1993. A QSAR approach for estimating the aquatic toxicity of soft electrophiles. *Quant. Struct.-Act. Relat.* 12:349-356.
181. Dimitrov SD, Mekenyan OG, Sinks GD, Schultz TW. 2003. Global modelling of narcotic chemicals: ciliate and fish toxicity. *Journal of Molecular Structure (Theochem)* 622:63-70.
182. Deneer JW, Seinen W, Hermens, JLM. 1988. The acute toxicity of aldehydes to the guppy. *Aquatic Toxicol* 12:185-192.
183. Deneer JW, Sinnige TL, Seinen W, Hermens JLM. 1988. A quantitative structure-activity relationship for the acute toxicity of some epoxy compounds to the guppy. *Aquat Toxicol* 13:195-204.
184. Schultz TW, Hewitt M, Netzeva TI, Cronin MTD. 2007. Assessing applicability domains of toxicological QSARs: definition, confidence in predicted values, and the role of mechanisms of action. *QSAR Comb. Chem* 26: in press.
185. Mekenyan O, Pavlov T, Grancharov V, Todorov M, Schmieder P, Veith G. 2005. 2D-3D migration of large chemical inventories with conformational multiplication. Application of the genetic algorithm. *J Chem Inf Model* 45:283-92.
186. Netzeva TI, Aptula AO, Benfenati E, Cronin MTD, Gini G, Lessigiarska I, Maran U, Vracko M, Schüürmann G. 2005. Description of the electronic structure of organic chemicals using semiempirical and ab initio methods for development of toxicological *QSARs. J Chem Inf Model* 45:106-114.
187. Schultz TW, Cronin MTD, Netzeva TI, Aptula AO. 2002. Structure-toxicity relationships for aliphatic chemicals evaluated with Tetrahymena pyriformis. *Chem Res Toxicol* 15:1602-1609.
188. Lipnick RL. 1991. Outliers, their origin and use in the classification of molecular mechanisms of toxicity. In: Hermens JLM and Opperbuizen A, eds., *QSAR in Environmental Toxicology* - IV. Elsevier, Amsterdam, The Netherlands, pp 131-153.
189. Schüürmann G. 1990. QSAR analysis of the acute fish toxicity of organic phosphorothionates using theoretically derived molecular descriptors. *Environ Toxicol Chem* 9:417-428.
190. Karabunarliev S, Mekenyan OG, Karcher W, Russom CL, Bradbury SP. 1996. Quantum chemical descriptors for estimating the acute toxicity of electrophiles to the fathead minnow. *Quant. Struct.-Act. Relat.* 15:302-310.
191. Karabunarliev S, Mekenyan OG, Karcher W, Russom CL, Bradbury SP. 1996. Quantum chemical descriptors for estimating of acute toxicity of substituted benzenes to the guppy *(Poecilia reticulata)* and fathead minnow *(Pimephales promelas). Quant. Struct.-Act. Relat.* 15:311-320.
192. Freidig AP, Hermens JLM. 2001. Narcosis and chemical reactivity QSARs for acute toxicity. *Quant. Struct.-Act.*

*Relat.*19:547-553.

193. Schultz TW, Netzeva TI, Roberts DW, Cronin MTD. 2005. Structure-toxicity relationships for the effects to *Tetrahymena* pyriformis of aliphatic, carbonyl-containing, α,β-unsaturated chemicals. *Chem Res Toxicol* 18:330-341.
194. Protic M, Sabljic A. 1989, Quantitative structure-activity relationships of acute toxicity of commercial chemicals on fathead minnows: effect of molecular size. *Aquat Toxicol* 14: 47-64.
195. Kulkarni SA, Raje DV, Chkrabarti T. 2001, Quantitative structure-activity relationships based on functional and structural characteristics of organic compounds. *SAR QSAR Env Res* 12:565-591.
196. Toropov AA, Benfenati E. 2004. QSAR modelling of aldehyde toxicity by means of optimisation of correlation weights of nearest neighbouring codes. *Journal of Molecular Structure (Theochem)* 676:165-169.
197. Nendza M, Müller M. 2000. Discriminant toxicant classes by mode of action: 2. Physicochemical descriptors. *Quant. Struct.-Act. Relat.* 19:581-598.
198. Sippl W, Contreras JM, Parrot I, Rival YM, Wermuth CG. 2001. Structure-based 3D QSAR and design of novel acetylcholinesterase inhibitors. *J. Comput. Aided Mol. Des.* 15:395-410.
199. Dimitrov S, Koleva Y, Schultz TW, Walker JD, Mekenyan O. 2004. Interspecies quantitative structure-activity relationship model for aldehydes: aquatic toxicity. *Environ Toxicol Chem* 23:463-70.
200. Huuskonen J. 2003. QSAR modelling with the electrotopological state indices: predicting the toxicity of organic chemicals. *Chemosphere* 50:949-953.
201. Papa E, Villa F, Gramatica P. 2005. Statistically validated QSARs, based on theoretical descriptors, for modelling aquatic toxicity of organic chemicals in *Pimephales promelas* (fathead minnow). *J Chem Inf Model.* 45:1256-66.
202. Cronin MTD, Dearden JC, Dobbs AJ. 1991. QSAR studies of comparative toxicity in aquatic organisms. *Sci Total Environ* 109/110:431-439.
203. Sinks GD, Schultz TW. 2001. Correlation of Tetrahymena and Pimephales toxicity: evaluation of 100 additional compounds. *Environ Toxicol Chem* 20:917-921.
204. Tremolada P, Finizio A, Villa S, Gaggi C, Vighi M. 2004. Quantitative inter-specific chemical activity relationships of pesticides in the aquatic environment. *Aquatic Toxicol* 67:87-103.
205. Zhao Y, Wang L, Goa H, Zhang Z. 1993. Quantitative structure-activity relationships. Relationship between toxicity of organic chemicals to fish and to *Photobacterium phosphoreum. Chemosphere* 26:1971-1979.
206. Dearden JC, Cronin MTD, Dobbs AJ. 1995. Quantitative structure-activity relationships as a tool to assess the comparative toxicity of organic chemicals. *Chemosphere* 31:2521-2528.
207. Cronin MTD , Netzeva TI, Dearden JC, Edwards R, Worgan ADP. 2004. Assessment and Modeling of the Toxicity of Organic Chemicals to *Chlorella vulgaris*: Development of a Novel Database. *Chem Res Toxicol* 17:545-554.
208. Moore DRJ, Breton RL, MacDonald DB. 2003. A comparison of model performance for six quantitative structure-activity relationship packages that predict acute toxicity to fish. *Environ Toxicol Chem* 22:1799-1809.
209. Dearden JC, Barratt MD, Benigni R, Bristol DW, Coombes RD, Cronin MTD, Judson PN, Payne MP, Richard AM, Tichy M, Worth AP, Yourick JJ. 1997. The development and validation of expert systems for predicting toxicity. *Altern Lab Anim* 25:223-252.
210. Sanderson DM, Earnshaw CG. 1991. Computer prediction of possible toxic action from chemical structure. The DEREK system. *Hum Exp Tox* 10:261-273.
211. Judson PN. 2002. Prediction of toxicity using DEREK for Windows. *J Tox Sci* 27:278.
212. Enslein K. 1988. An overview of SARs as an alternative to testing in animals for carcinogenicity, mutagenicity, dermal and eye irritation and acute oral toxicity. *Toxicol Ind Health* 4:479-498.
213. Cramer GM, Ford RA, Hall RL. 1978. Estimation of Toxic Hazard-A Decision Tree Approach. *J Cosmet Toxicol* 16:255-276.
214. European Centre for Ecotoxicology and Toxicology of Chemicals. 2003. Aquatic Hazard Assessment II Technical Report 91. ECETOC, Brussels, Belgium.
215. Organization of Economic Cooperation and Development. (2006). Report on the Regulatory Uses and Applications in OECD Member Countries of (Quantitative) Structure-Activity Relationship [(Q)SAR] Models in the Assessment of New and Existing Chemicals. OECD, Paris, France.
216. Walker JD. 2003. Applications of QSARs in toxicology: a US government perspective. *Journal of Molecular Structure (Theochem)* 622:167-184.
217. Tunkel J, Mayo K, Austin C, Hickerson A, Howard P. 2005. Practical considerations on the use of predictive models for regulatory purposes. *Environ Sci Technol* 39:2188-2199.
218. Van der Zandt PTJ, van Leeuwen CJ. 1992. A proposal for priority setting of existing chemical substances. VROM 92408/b/9-92, 1502/033. Ministry of Housing, Physical Planning and the Environment, The Hague, The

Netherlands.

219. Hansen BG, van Haelst AG, van Leeuwen K, van der Zandt P. 1999. Priority setting for existing chemicals: European Union risk ranking method. *Environ Toxicol Chem* 18:772-779.

220. van Haelst AG, Hansen BG. 2000. Priority setting for existing chemicals: automated data selection routine. *Environ Toxicol Chem* 19:2372-2377.

221. Pavan M. 2003. Total and Partial Ranking Methods in Chemical Sciences. PhD Thesis. University of Milano-Bicocca, Milan, Italy. Accessible from http://www.disat.unimib.it/chm.

222. Danish Environmental Protection Agency. 2006. Advisory list for self-classification of dangerous substances. DK-EPA, Copenhagen, Denmark. http://www.mst.dk/chemi/01050000.htm.

223. European Commission. 2001. An interim strategy for management of PBT and vPvB substances. Joint Meeting of the Competent Authorities for the Implementation of Council Directive 67/548/EEC (New Substances) and Council Regulation (EEC) 793/93 (Existing Substances). ENV/D/432028/01. NOTIO/36/2001. DG Environment, Brussels.

11. INTELLIGENT TESTING STRATEGIES

C.J. van Leeuwen, G.Y. Patlewicz and A.P. Worth

11.1 INTRODUCTION

In the context of regulatory programs for the safety evaluation of chemicals, there is a need for a *paradigm shift*. The challenge is to move in a scientifically credible and transparent manner from a paradigm that requires extensive hazard (animal) testing to one in which a hypothesis- and risk-driven approach can be used to identify the most relevant *in vivo* information [1]. So-called Intelligent or Integrated Testing Strategies (ITS) are a significant part of the solution to the challenge of carrying out hazard and risk assessments on large numbers of chemicals. ITS (Figure 11.1) are integrated approaches comprising multiple elements aimed at speeding up the risk assessment process while reducing costs and animal tests [1]. In this chapter a short overview will be presented about ITS, its rationale (Section 11.1) and its components (Section 11.2). Although some components of ITS, such as *in vitro* methods, (Q)SARs and read-across are already in use, the development and regulatory application of ITS has only just started. The development and implementation of ITS is a major challenge for at least the next decade. Thus, it is not possible at this stage to provide examples of fully developed ITS. Instead, a few examples of preliminary ITS for a number of fate, toxicological and ecotoxicological endpoints are presented to illustrate their potential role in the assessment of chemicals (Section 11.3). Summary and concluding remarks are described in Section 11.4. Selected references for further reading are provided in Section 11.5.

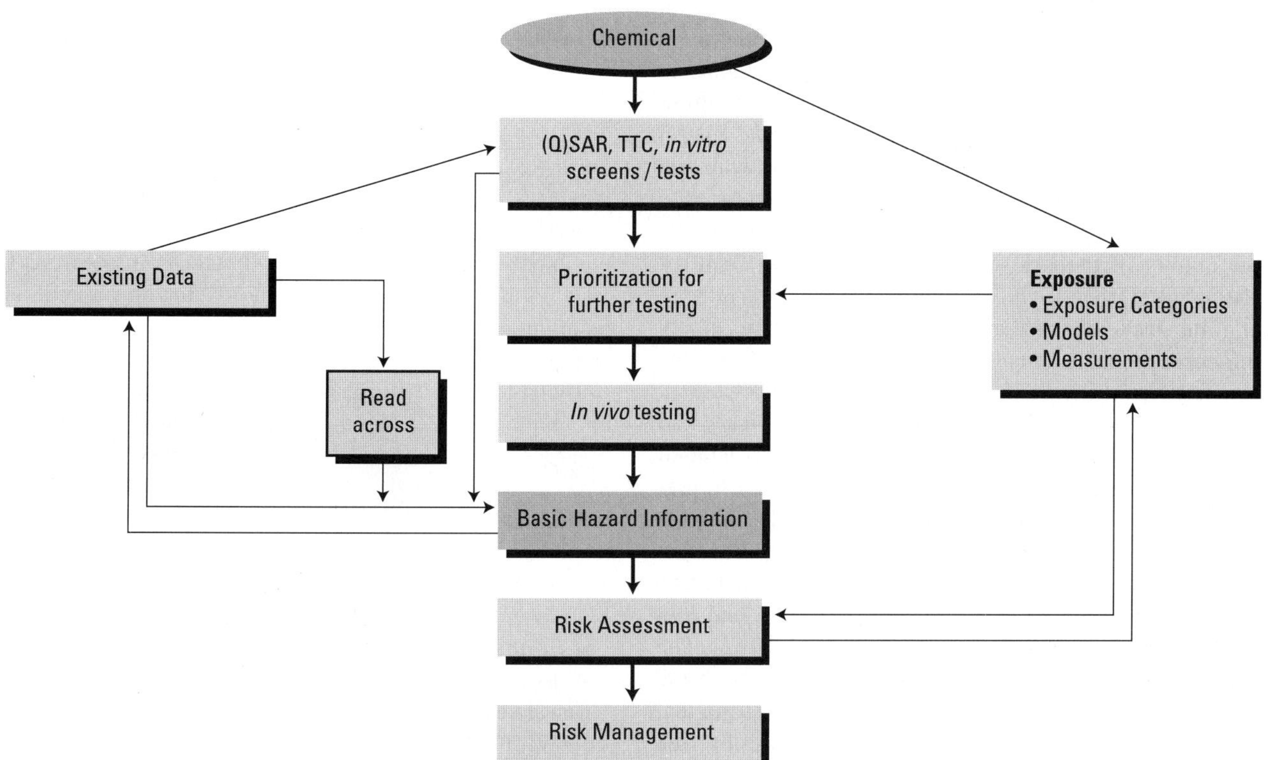

Figure 11.1. Combining use and exposure information and effects information obtained from QSARs, read-across methods, thresholds of toxicological concern (TTCs), and *in vitro* tests prior to *in vivo* testing is a more rapid, efficient, and cost-effective way to perform risk assessment of chemicals. From[1]. With permission.

C.J. van Leeuwen and T.G. Vermeire (eds.), Risk Assessment of Chemicals: An Introduction, 467–509.

11.1.1 Streamlining the hazard and risk assessment process

As currently undertaken for industrial chemicals, risk assessment is a tiered process distinguished by levels of increasing complexity, beginning with a preliminary categorization step, followed by a refined or screening assessment, and progressing to a full, comprehensive risk assessment [2-4]. For each tier, a minimum level of information is required (Chapter 1). For example, the OECD has established an international programme - called Screening Information Data Sets (SIDS) - for surveying high production volume chemicals (HPVCs) for potential effects. SIDS include the basic physicochemical, toxicological and ecotoxicological information needed to perform a preliminary assessment of a chemical's hazard and risk [5]. The development and harmonization of test guidelines [6] and hazard classification criteria is co-ordinated by the OECD.

Applying the current risk assessment paradigm and meeting the associated data-generation requirements, combined with the increased need to evaluate the potential effects posed by thousands of industrial chemicals, are major challenges for the chemical industry, national and international regulatory agencies, and associated stakeholders [7-9]. To address these challenges, governments have implemented several initiatives to overcome the lack of publicly available data on the hazardous properties of chemicals in order to accelerate the risk assessment process while enhancing the quality of the risk assessment and management of chemicals [1,10].

The lack of publicly available chemical safety information for industrial chemicals is not a new problem. In 1984, the U.S. National Research Council estimated that only 22% of U.S. HPV chemicals had "minimal" toxicity data available. In 1990, a detailed analysis of chemical control in the EU revealed a similar lack of information on use and toxicity [11]. Analyses by the European Chemicals Bureau (ECB; http://ecb.jrc.it/) found that only 14% of EU HPV chemicals had basic information at the level of the base set, 65% had less than the base set; and 21% had no data at all [12]. The base set is a minimum data package comprising basic physicochemical, toxicological and ecotoxicological properties. Similar observations were made by the USEPA [13]. Governmental agencies, the regulated community, and stakeholders face two major challenges:

1. Developing new approaches to improve animal welfare.
2. Streamlining the hazard and risk assessment process, i.e. increasing efficiency, cost-effectiveness, speed and focus for regulatory programs that require submission of defined studies.

11.1.2 Animal testing and testing costs

Two reasons for the development of ITS are animal welfare concerns and the need to reduce the cost of toxicological testing. The number of animals that are actually used in Europe for toxicological and other safety evaluations has been presented in the "Fourth Report" from the Commission [14]. In this report the estimated total number of animals used on an annual basis for experimental and other scientific purposes was 10.7 million. About 10% of these are used for toxicological and other safety evaluations (Figure 11.2). The remaining 90% of test animals are used for other purposes mainly for biological studies of a fundamental nature and research and development of human medicine, dentistry and veterinary medicine.

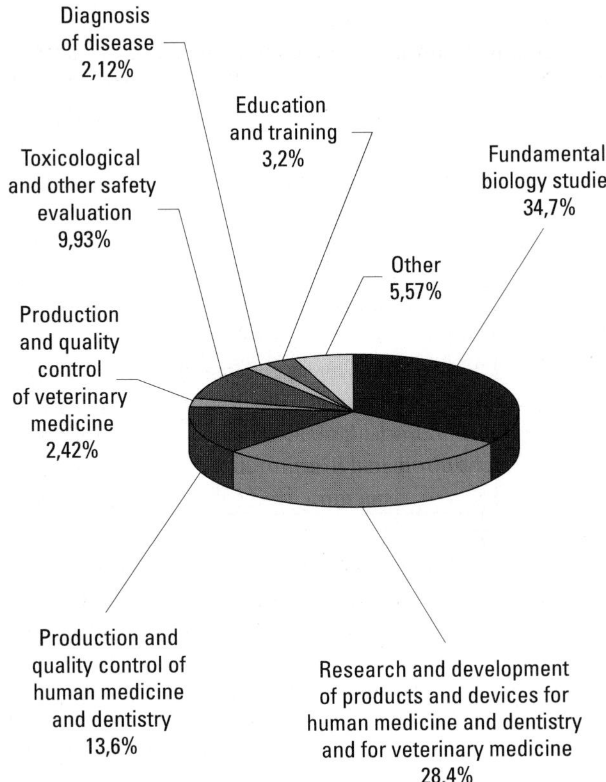

Figure 11.2. Purposes of experiments with animals. Taken from [14].

Table 11.1. Purposes of animal experiments for the safety evaluation of products [14].

Total number of animals	10,700,000	100%
Safety evaluations	1,066,000	10%
Agricultural chemicals	123,000	1%
Industrial chemicals	136,000	1%
Cosmetics	2,700	0.025%

The 10% of animals used for toxicological and other safety evaluations include safety evaluation of products and devices for human medicines and dentistry as well as veterinary medicines (Table 11.1). This includes the group of products and substances that fall under the scrutiny of authorities concerned with the safety of health and the environment, such as industrial chemicals, which equates to 1.27 %, corresponding to about 136,000 test animals for the year 2002. Animals used in the safety evaluation of cosmetics correspond to 0.025 % of the total use of animals.

In 2004, Van der Jagt et al. [15] published a report in which the direct testing costs and the number of vertebrate test animals were estimated, using a number of different scenarios. The predictions were based on the original Commission proposal for REACH of November 2003 [16]. In this report, the number of additional vertebrate test animals used as a consequence of REACH over a time period of 11 years (i.e. the time period for the full implementation of REACH) was estimated at 3.9 million (maximum scenario; assuming that alternative approaches to testing were not applied). This corresponded to approximately 350,000 animals per year, i.e. a 2% increase in the total number of vertebrate test animal used per year for experimental and other scientific purposes [15]. After this time period, the number should return to a base-line level comparable to today's situation. By then, however, the huge knowledge gap we currently face for widely used chemicals should be closed, enabling safer use of chemicals for generations to come. The overall direct testing costs (maximum scenario) were estimated at € 2.4 billion over a time period of 11 years (i.e. the time period for the full implementation of REACH). About 90% of these costs were attributed to human health related endpoints. It was estimated that by applying (Q)SARs and read-across methods that are currently available, the needs for animal tests could be reduced to 70% for individual endpoints resulting in significant savings in testing costs and animal use. The cost and animal saving potential of applying these techniques over a period of 11 years were estimated at:

- Cost savings: € 800-1130 million
- Animal savings: 1.3-1.9 million animals.

So, according to Van der Jagt et al. [15] the most likely scenario for REACH (based on the Commission proposal of 2003) is 2.6 million vertebrate animals (mammals, birds and fish) over a period of 11 years [15,16]. This corresponds to 240,000 animals per year, an increase of about 1% on the basis of total animal use as presented in Figure 11.2 and Table 11.1. The resulting costs (most likely scenario) for testing were estimated at € 1.5 billion. The report also showed that certain tests required for REACH stood out in terms of test animal use and costs. In terms of animals, it was estimated that about 72% would be required for carrying out three types of tests:

1. The two-generation reproductive toxicity study.
2. The developmental toxicity study.
3. The further mutagenicity (*in vivo*) study.

It was also concluded that these test requirements also account for the majority of the testing costs, estimated at 32%, 25% and 9%, respectively. In addition, these tests are not the easiest ones to replace, as noted by the SCCNFP, i.e. the Scientific Committee on Cosmetics Products and Non-Food Products intended for Consumers [18]. The need for different animal tests, in terms of percentages of the total animal use, is given in Figure 11.3.

We would like to emphasize that the data provided on costs and animal numbers are estimates based on available information at the time and as outlined in the original Commission proposal for REACH [16]. Within the REACH regulation, information requirements have been reduced. This will have consequences in terms of testing costs and animal numbers. The estimates of animal numbers and testing costs will change again once new ITS tools become available. So in the near future, depending on the progress made in the development and implementation of ITS, animal numbers and testing costs could decrease further.

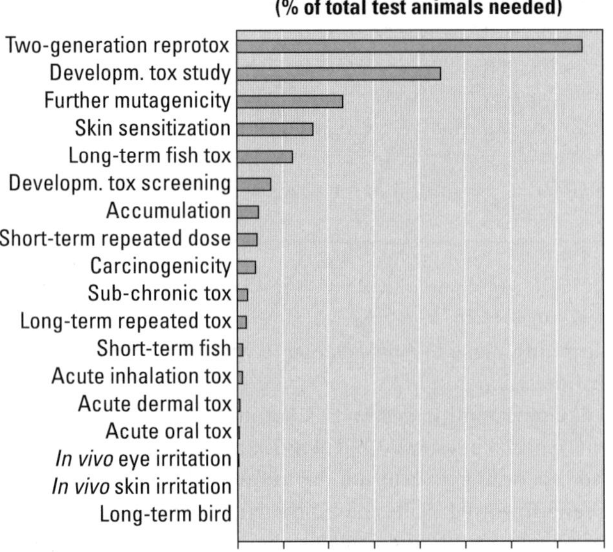

Figure 11.3. Relative needs for animal tests under REACH. From [15].

11.1.3 The need for integrated testing approaches

In the context of regulatory programs for the safety evaluation of chemicals, governmental agencies, the regulated community, and stakeholders face the challenges of generating and interpreting data for risk assessment in a cost-effective and efficient manner [1]. It should also be emphasized that although regulatory requirements often result in data for a wide array of endpoints, in many cases only a relatively small subset of all the *in vivo* data forms the basis of the final risk assessment. In other words: a lot of information is not used! For chemicals that lack toxicological and exposure data, the challenge is to create ways to efficiently and credibly predict toxic potency and exposure levels. These predictions should facilitate reasonable decisions to be made as to whether experimental studies are required to refine a risk assessment further. The underlying rationale is to:

- Minimise animal testing by using alternative methods and exposure information.
- Apply quicker and cheaper methods before slower and more expensive ones.
- Introduce risk-based approaches instead of hazard-driven, check-list approaches.
- Enable proper consideration of exposure as a key determinant of risk assessment.
- Maximise the use of up-to-date information from different sources in an integrated manner.
- Allow greater flexibility in introducing new tools and scientific knowledge.
- Allow more robust and focussed regulatory decisions using testing and non-testing approaches.

REACH [17] and other regulatory programs (Chapters 12-15) will require more information to address basic hazard information needs and risk assessment uncertainties across chemical classes. This information is needed to determine adverse effects and outcomes of concern, but the magnitude of the data gaps precludes the use of a traditional toxicity-testing approach. The challenge will be to increase the overall efficiency, cost-effectiveness and focus of the risk assessment process, while reducing the current reliance on animal tests. The long-term solution to these challenges will not be to generate more hazard data faster but rather to determine what specific effects data are essential for the assessment and appropriate management of risks for particular groups of chemicals and exposures. It should also be noted that the generation of ITS data is not without cost implications. Recently Combes et al. [19] articulated this idea of predictive modelling and intelligent testing, writing that "additional testing should only be required where essential information is missing, rather than testing to cover all data gaps according to a generalized, checklist approach". Then, researchers could also address questions concerning responsible use of animals in *in vivo* testing [19,20].

The scientific advances needed to meet these challenges will require further investment in computational chemistry, systems biology, toxicology, and exposure modelling [1,21]. The full potential of scientific advances in risk assessment will only be realized if developments from these disciplines are integrated in a concerted and systematic fashion.

11.2 COMPONENTS OF ITS

While the details of the different proposals for intelligent testing vary, a number of common components can be identified:

1. Chemical categories and read-across.
2. SARs and QSARs
3. Thresholds of toxicological concern (TTCs).
4. Exposure-based waiving.
5. *In vitro* methods.
6. Optimised *in vivo* tests.

The six ITS components can be subdivided into two main categories: *testing approaches* (components 5 and 6) and *non-testing approaches* (components 1-4). The toxicological information can be derived from: (1) chemical categories or read-across, (2) estimation methodologies such as SARs and QSARs, or (3) TTCs, threshold values for chemicals below which no significant risks are expected. If this basic hazard information of the chemicals is combined with adequate exposure information about the chemical, exposure-based waiving (4) can be applied. *In vitro* methods (5) and optimised *in vivo* testing (6) can also make significant contributions to the goals of ITS in obtaining reliable information on the (toxic) properties of chemicals with minimal use of animals.

Chemical categories, read across or analogue approaches, (Q)SARs and TTCs have the following characteristics:

- They are all based on the notion that similar compounds (usually structurally similar chemicals) have similar activities.
- They are used to predict properties of interest for (groups of) chemicals for which no or limited data exist.
- Fundamental to all (and sometimes quite limiting) are the size and the quality of the databases on which the methods were based. As with alternative test methods, non-testing methods rely heavily on the availability of *in vivo* information, i.e. the availability of high quality experimental toxicological data. The availability of high quality *in vivo* information can be a limiting factor both in the development of *in silico* and *in vitro* methodologies.

There are clear differences among the components of ITS in terms of financial investments and the time needed to develop and implement them for regulatory application. There are also multiple ways to obtain hazard and risk information by means of different combinations of these components. Some ways could be more efficient than others, depending on the underlying rationale of the strategy. An example of a (generic) testing strategy is depicted in Figure 11.1. Similar schemes have been published by other authors e.g. [22-24]. In the following sections, the different components of ITS are described briefly, starting with the non-testing approaches (Sections 11.2.1-11.2.4), followed by the testing approaches (Sections 11.2.5 and 11.2.6). Concluding remarks are presented in Section 11.2.7.

11.2.1 Chemical categories and read-across

Read-across and category approaches are probably the simplest tools to reduce animal testing under the REACH legislation but guidance for their regulatory application needs to be developed further. Firstly we will describe the current process of read-across, secondly the grouping of chemicals into groups or families of chemicals and thirdly the regulatory experiences gained so far. Finally we discuss the prospects of their application in the management of industrial chemicals.

Read-across

In read-across, one or more properties of a chemical of interest are inferred by comparison to a similar chemical or chemicals, for which the properties of interest are known (Figure 11.4). These properties may include physicochemical properties, environmental fate, toxicity and ecotoxicity. An assessment of similarity underpins the approach. The basic assumption is that similarity in structure implies similarity in activities or properties. The read-across can be qualitative or quantitative:

1. *Qualitative* read-across can be regarded as an application of SAR. The process involves: (a) the identification of a chemical substructure that is common to the two substances (which are therefore analogues) and (b) the assumption that the presence (or absence) of a property/activity for a substance can be inferred from the presence (or absence) of the same property/activity for an analogous substance. This assumption implies that analogues behave qualitatively similarly, and is usually the result of an expert judgement evaluation.
2. *Quantitative* read-across involves the identification of a chemical substructure that is common to the two substances (which are therefore analogues), and the assumption that the *known* value of a property for one substance can be used to estimate the *unknown* value of the same property for another substance. This assumption implies that the potency of an effect shared by different analogous substances is similar, and is also usually the result of an expert judgement evaluation [25,26].

The REACH legislation [17] considers read-across and (Q)SARs as important instruments for generating information on the intrinsic properties of chemicals. To a limited extent, read-across is already being used by some of the Member States, such as the UK. The Health and Safety Executive (HSE) follows a series of needs and principles that were laid out by Hanway and Evans in 2000 [27]. These comprise a number of steps which are applied on a case-by-case basis under the notification of new substances. Initially a chemical is evaluated with respect to its structural similarity e.g. whether specific functional groups are absent or present which

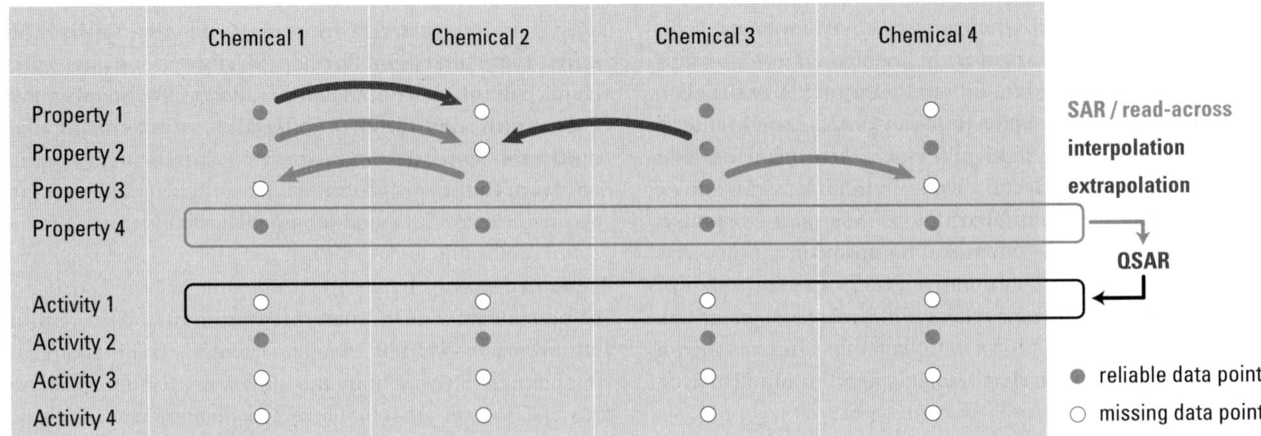

Figure 11.4. A chemical category can be represented graphically as a two-dimensional matrix in which different category members occupy different columns, and the different category endpoints occupy different rows. Data gaps may be filled by one or more of the following procedures: qualitative read-across, quantitative read-across, use of SARs and QSARs. Modified after [5].

may modify the likely activity/toxicity expressed. The purity and impurity profile are then considered to assess the impact of overall toxicity profile. Physicochemical properties such the log of the *n*-octanol/water partition coefficient (log K_{ow}), or aqueous solubility are compared since these types of properties may provide insights on the likely absorption and bioaccumulation characteristics of a chemical. The likely toxicokinetics are evaluated to consider the stability of the chemical and whether it will metabolise, decompose or hydrolyse in some manner. Derek for windows, the knowledge based expert system (http://www.lhasalimited.org) is used to identify potential structural alerts. Finally all available toxicological information is collated and assessed. All these steps are considered in turn and will form the basis of any read-across argument. In addition, some toxicity testing is usually requested to confirm the validity of the read-across. An acute oral toxicity study and an Ames test are typically conducted. If the results of these studies reflect toxicity differences, then further testing for other endpoints may be appropriate. Several principles have been drawn from this approach:

a. Acute oral toxicity testing and Ames testing have been effective in underpinning a read-across argument. However, in view of animal welfare concerns and the availability of standardised *in vitro* tests for cytotoxicity, it is questionable whether it is necessary to perform acute oral toxicity testing *in vivo* to support read-across.
b. For regulatory purposes, it is usually easer to read-across positive data, i.e. chemicals exerting adverse effects that can be linked to their mode of toxic action.
c. Within a series of structurally similar new substances, the two substances at either end of the series may be the only two that need to be fully tested in order to define the domain.

Whilst this approach has been demonstrated to be useful in reading across a number of chemicals on a case-by-case basis, no information is available as to the practical steps of how a read-across analogue is selected or whether any mechanistic considerations are accounted for.

The UK Environmental Agency has a similar step-by-step approach to read-across as the HSE. Structural similarity is assessed including an evaluation of whether there are any additional functional groups (or absence of groups) that might modify the toxicity. Then the purity and impurity profiles are assessed. An evaluation of the physicochemical properties is made including an assessment of how these properties may influence ecotoxicity. QSARs for determining the likely toxicity of analogues are used. Basic toxicity tests such as acute toxicity to *Daphnia* are conducted to confirm the validity of the read-across.

Chemical categories

A chemical category is a group or "family" of chemicals whose physicochemical, toxicological and ecotoxicological properties are likely to be similar or follow a regular pattern as a result of structural similarity. These structural similarities may create a predictable pattern in all or the following parameters: physicochemical properties, environmental fate, environmental effects, and/or human health effects. The similarities may be based on:

- A common functional group (e.g., aldehyde, epoxide, ester) related to specific activity.
- The likelihood of common precursors and/or breakdown products, via physical or toxicological processes, which result in structurally similar chemicals (e.g., the "family approach" of examining related chemicals such as acid/ester/salt).
- An incremental and constant change of key physicochemical properties across the category which determines other properties such as biological and (eco)toxicological effects (e.g., the methylene group difference between adjacent members of the alpha-olefins).

Within a category, different members may be selected for the endpoint desired. If the available test results show that the chemicals in a category behave in a similar or predictable manner, then interpolation and/or extrapolation may be used to assess the chemicals instead of conducting additional testing.

Chemical categories are "designed" on the basis of scientific considerations, including SAR, QSAR and read-across (where an endpoint value or classification for one chemical is used as the best estimate for a related chemical). Guidance on the formation and use for chemical categories for fulfilling data requirements has been published by the OECD as part of the OECD Manual for Investigation for HPV Chemicals [25,26]. This guidance is used, among others, for fulfilling the data requirements within the OECD HPV Chemicals Programme. The same guidance document is published by the US EPA for use within the US HPV Challenge Programme. The OECD guidance document was revised during 2004 and 2005 and addresses the following issues:

- Definitions and explanations of the chemical category concept.
- General approach for developing categories.
- Differences in grouping for different endpoints.
- Use of (Q)SARs for the development of a category.
- Guidance on different types of categories (i.e. chain-length, metabolic pathways, isomers and their mixtures, complex substances, metal and metal compounds).

The guidance document also provides a number of examples of categories that have been adopted within the OECD HPV Chemicals Programme or are currently under preparation. The OECD guidance is an excellent starting point in describing the principles and approaches of chemical categories but it lacks detail on the practical steps that would help a registrant to formulate a category, justify it and document it. In addition the category description examples provided are largely limited to chemical classes or structurally similar chemicals containing common functional groups.

Regulatory application of read-across and category approaches

Apart from the experiences in the UK, there is little regulatory experience in the application of category approaches and read-across for the assessment of chemicals in the EU. On the other hand, the EU countries have gained experience in these approaches in the context of the OECD existing chemicals programme where these approaches have been accepted for use when adequately justified scientifically.

The US EPA has a wealth of experience in the use of chemical categories. Their chemical categories are a good reflection of "knowledge within the EPA". The categories are typically biased towards environmental endpoints with limited available justification for how the category or its members were selected. Increased transparency could contribute considerably to increasing shared experience as a basis for more robust guidance. Prior to 1987, nearly 20% of Pre-Manufacture Notifications (PMNs) submitted underwent a detailed review ("standard review") by EPA, a highly resource-intensive effort that lasted most of the mandated 90-day PMN review period. In 1987, after several years of experience in the review of PMNs, EPA's Office of Toxic Substances (now the Office of Pollution Prevention and Toxics) had sufficient accumulated experience to group PMN chemicals with shared chemical and toxicological properties into categories, enabling both PMN submitters and EPA reviewers to benefit from the accumulated data and past decisional precedents allowing reviews to be facilitated. Candidate categories for the new chemicals review process are proposed by the EPA staff, based on experience reviewing PMNs on similar substances. At proposal, the database supporting the category is scrutinized for quality and for general applicability to other potential members of the category. Based on this analysis, a category statement is prepared describing the molecular structure. Boundary conditions such as molecular weight, equivalent weight, log K_{ow}, or water solubility, that would determine inclusion in (or exclusion from) a category, and standard hazard and fate tests to address concerns for the category are all considered. The categories may not be made up of the most hazardous chemicals, but rather include chemicals for which sufficient history has been accumulated so that hazard concerns and testing recommendations vary little from chemical to chemical within the category. The categories are not intended to be a comprehensive list of all substances. More than 50 such categories are documented by the US EPA [25,28].

The Canadian Environmental Protection Act (CEPA) from 1999 requires categorization of the approximately 23,000 substances on the Domestic Substances List (DSL) prior to a legally mandated deadline of September 14, 2006 (see also Chapter 15). The objective of categorization is to identify substances (on the basis of either exposure or hazard) that need further assessment [25,29,30]. The two phases of assessment are screening assessment and in-depth assessment. In order to efficiently identify and prioritize substances on the DSL that represent highest priorities from a human health perspective, a framework based on an iterative application of increasingly discriminating (i.e., simple and complex) tools for consideration of exposure and hazard was developed by Health Canada [29]. The "simple tools" are sufficiently robust to address all substances on the DSL based on limited information. The "complex tools" (the so-called ComHaz) are more discriminating. Stepwise application of these tools minimizes overemphasis on data-rich compounds, while making optimum and efficient use of available information. ComHaz involves a hierarchical consideration of various sources of information (including data, (quantitative) structure-activity analysis and comparison with analogues) for a range of endpoints of toxicity. Potentially relevant analogues are identified using a range of different tools as well as in-house expert opinion. The software tools include Accord and TOPKAT, internet solutions such as ChemID (http://chem.sis.nlm.nih.gov/chemidplus/), or Chemfinder (http://www.chemfinder.com), Leadscope's structural hierarchy (www.leadscope.com) and the US EPA's AIM or Analogue Identification Methodology (http://esc.syrres.com/analog/).

Environment Canada [30] uses the following general rules of thumb but recognises that there will always be exceptions. An analogue should preferably contain most, if not all, of the same structural features as the substance on the DSL:

- An analogue should have approximately the same molecular weight as the substance.
- An analogue should have water solubility similar to that of the substance of interest.
- For persistence, an analogue should have the same reactivity or stability as the DSL substance of interest.
- For an endpoint of interest, the relevant molecular descriptors of an analogue should be of comparable value to those of the substance.

Environment Canada uses the functionality of a number of tools such as, OASIS, TOPKAT, AI Expert to help in the identification of analogues and their assessment.

Challenges ahead

The best illustration of the potential advantages of using chemical categories in the implementation of EU legislation comes from experience obtained in the US High Production Volume Challenge Program, which aims to make publicly available basic health and environmental effect data for approximately 2800 high production volume chemicals. As of July 2004, data for 1266 chemicals had been submitted by industry to the US Environmental Protection Agency [31]. Of these chemicals, 81% were included in a category, and new testing was proposed for fewer than 10% of the endpoint data needed (five human health endpoints and three environmental endpoints). Details are given in Table 11.2. For the human health data, the table shows that 88% (44/50 x 100) of the missing data was estimated using QSAR and read-across. For the environmental data, 83% (35/42 x 100) of the missing data was estimated.

Although read-across is applied in the USA and Canada, the approach is in a different regulatory context to that under REACH. The challenge for REACH will be to provide explicit non-prescriptive guidance on how the non-testing approaches can be applied in practice, and to develop IT tools to make it easier to establish, justify, and document read-across, chemical category and (Q)SAR approaches. This guidance will need to explain and illustrate:

- The commonalities and differences between SARs, read-across and categories.
- How to justify and report qualitative and quantitative read-across (in terms of supporting information).
- How to build a category (practical details), including examples of qualitative and quantitative read-across.
- How to evaluate the robustness and applicability domain of a category or (Q)SAR model.
- How to justify and report a category proposal or a (Q)SAR model prediction, in terms of its underlying rationale, the scientific basis and validity, including an assessment of whether the predictions are within the applicability domain for the category/model or not.

Capacity building requires transparency. It will be absolutely necessary to gain more regulatory experience and acceptance for this approach in the EU member states for which transparent guidance is conditional [25].

Table 11.2. Experience from the US HPV Challenge Program. Information presented by Dr. C. Auer (US EPA) at the EU-US Transatlantic Conference on Chemicals, April 26-28, 2004, Charlottesville (VA) USA.

	Human health data	Environmental effects
Adequate studies	50%	58%
Estimation	44%	35%
Testing	6%	7%

Table 11.3. Regulatory applications of QSARs.

1.	Support priority setting of chemicals
2.	Guide experimental design (e.g. selection of tests / doses)
3.	Provide mechanistic information
4.	Support grouping of chemicals into chemical categories
5.	Fill in data gaps and complement test data for classification and labelling
6.	Fill in data gaps and complement test data for risk assessment

This comprises guidance on how categories are defined, how analogues are selected. Initial work on this has been carried out within the REACH Implementation Projects (http://ecb.jrc.it/), i.e. RIP 3.3-2 (Chapter 12). The guidance on chemical categories developed in RIP 3.3-2 expands on the current OECD guidance.

11.2.2 SARs AND QSARs

Structure-activity relationships (SARs) and quantitative structure-activity relationships (QSARs), collectively referred to as "(Q)SARs", are theoretical models that can be used to predict the physicochemical and biological properties of molecules. They are sometimes called "*in silico* models" because they can be applied by using a computer. A structure-activity relationship (SAR) is a (qualitative) association between a chemical substructure and the potential of a chemical containing the substructure to exhibit a certain biological effect (e.g. a toxicological endpoint). A quantitative structure-activity relationship (QSAR) is a mathematical model that relates a numerical measure of chemical structure (e.g. a physicochemical property) to a physical property or to a biological effect (e.g. a toxicological endpoint). This Section 11.2.2 is a relatively short section as detailed information on SARs and QSARs is presented in Chapters 9 and 10. Further information can also be found in the book edited by Cronin and Livingstone [32].

As described in Chapters 9 and 10, (Q)SARs can be used to provide: (1) physicochemical properties, (2) environmental distribution and fate, (3) biokinetic processes and (4) toxic potential and potency. For regulatory purposes (Q)SARs could potentially be used either to supplement experimental data (applications 1-6), or to replace testing (applications 4-6) as given in Table 11.3.

Over the past 25 years, many OECD member countries have established (Q)SAR tools to provide exposure and effects inputs in ranking and prioritization schemes for *in vivo* screening and testing programs [6,32-40]. In the United States and Canada, (Q)SARs (and read-across) are used extensively for regulatory purposes. In the USA, more than 40,000 new chemicals have been assessed by the US EPA using QSARs. Under the current EU legislation for new and existing chemicals, the use of (Q)SARs has been limited. In the case of new chemicals and biocidal products, the application of (Q)SARs (except for bioconcentration) has rarely been necessary, because the hazard information has been provided by means of compulsory testing. In the case of existing chemicals, however, (Q)SARs have been applied for priority setting, for advisory self-classification and labelling on environmental hazard and selected human health effects, and for the screening of persistent, bioaccumulating and toxic (PBT) candidates and other priority chemicals, and for the prediction of environmental effects of HPVCs. QSAR estimates have in some cases been accepted directly in the EU risk assessment programme on existing substances and indirectly by the EU, e.g. when participating in the OECD existing chemicals programme, especially where experimental data have been lacking or of questionable quality. This took place via the contribution of the USA to the OECD existing chemicals programme, where the USA applied read-across or (Q)SARs. (Q)SARs will be applied in REACH. An overview of the regulatory use of QSARs in OECD member countries has been published recently [40].

In practice, the availability of reliable (Q)SARs varies considerably between endpoints, so efforts are needed to develop models of better quality. This is especially true for human health endpoints. The need to develop better (Q)SAR models was also recognised in the Technical Guidance Document (TGD) for industrial chemicals and biocides [41] and more than 10 years ago, when the US EPA and the European Commission undertook a joint project. In this collaborative project, predictions for 144 substances were made by using the US EPA (Q)SARs for a wide variety of physicochemical properties, ecotoxicological effects and human health effects [38]. The study identified promising areas for the wider application of (Q)SARs, such as biodegradation and acute toxicity to fish and *Daphnia*, as well as areas where further work was needed, such as the development of (Q)SARs for human health effects. Since then, many scientific developments have taken place, especially in the pharmaceutical industry.

According to a monograph on commercially available QSAR models, the ECETOC Task Force on (Q)SARs [39] concluded that the predictions of human health related toxicological endpoints "are often poor because the endpoints are expressed through many different mechanisms, are receptor-mediated, involve multi-stage processes comprising ADME and are site-specific. At the present time, this complexity imposes severe limitations on the successful development of (Q)SARs suitable for non-congeneric sets of endpoints." The poor predictions are probably also related to the lack of availability of large homogenous high quality experimental data sets. These high quality toxicological data are crucial for the development of reliable QSARs (and *in vitro* methodologies; Section 11.2.5). Another reason may be the poor definition of some of the toxicological endpoints.

The observations from ECETOC [39], OECD [40], the European Commission [41] and its scientific committees [18,42] lead to the following conclusions:

1. There is a clear regulatory application of QSARs in countries such as the USA and Canada.
2. Regulatory acceptance of QSARs in the European Union is currently limited.
3. (Q)SARs will play an important role in REACH.
4. Further work is needed, especially in the area of human health related endpoints of toxicity, where new and better models need to be developed.

11.2.3 Thresholds of toxicological concern (TTCs)

Thresholds of toxicological concern (TTCs) are exposure threshold values for chemicals below which no significant risk to human health and/or the environment is expected to exist. The establishment of TTC is based on the analysis of toxicological and/or structural data of a broad range of different chemicals. TTCs might be used as substitute for substance-specific information in situations where there are limited or no information on toxicity of the compound and where human (or environmental) exposure is so low that undertaking toxicity studies is considered not warranted, due to animal welfare considerations and of the costs incurred in manpower and laboratory resources [43-45]. As such, TTCs are risk-based approaches where estimated (worst-case) effects of large groups of chemicals are compared with exposure information.

Under the current chemicals legislation, the TTC approach has rarely been used. An important issue to be addressed is that the eventual introduction of the TTC approach in the risk assessment process would have implications for the risk management. In other words, if the TTC approach is used to avoid testing, a decision would need to be taken on what safety advice (e.g. labelling) should be given, and for which exposure scenarios this advice would be applicable.

The TTC concept has been proposed as a tool in risk assessment both for defining when no testing is needed based on exposure information and as a substitute for effects values both for human health and the environment [48,49]. The potential use of the TTC concept for risk assessment is largely determined by the adequacy of the exposure information of the chemical in question (Section 11.2.4). A discussion on the (regulatory) acceptability in the context of the assessment of industrial chemicals still needs to take place. This was also concluded in a literature review of the Nordic Council [43]. They concluded that it is premature to use the concept due to limitations and uncertainties in the derivation of TTCs as well as to the fact that the TTC concept has not yet been evaluated for the diverse group of industrial chemicals.

Table 11.4. Structural classes of Cramer [52] for chemicals within the TTC concept. With permission. Copyright Elsevier.

Class I	Substances with simple chemical structures and for which efficient modes of metabolism exist, suggesting a low order of oral toxicity (e.g. glutamic acid, mannitol and propyleneglycol)
Class II	Substances which possess structures that are less innocuous than class I substances, but do not contain structural features suggestive of toxicity like those substances in class III (e.g. beta-carotene, maltol and allyl-compounds)
Class III	Substances with chemical structures that permit no strong initial presumption of safety or may even suggest significant toxicity or have reactive functional groups (e.g.nitrile, nitro-compounds, chlorobenzene and p-aminophenol)

The TTC concept has evolved from a lengthy history of attempts by scientists over the years, in regulatory authorities and elsewhere, to develop generic approaches to the safety assessment of large groups of chemicals or individual chemicals of unknown toxicity. The driving forces behind these efforts have been [47]:

- The continuing improvements in analytical capabilities which allow more and more chemicals to be identified in food at lower and lower concentrations.
- The widely accepted premise that exposure to very low amounts of chemicals is usually without harm.
- The view that time and attention devoted to a particular chemical should be in proportion to the risk to health.
- The limited toxicological resources worldwide, both in capacity for toxicity testing and for evaluation.
- The desire to minimise the use of animals.
- The ability to analyse large sets of existing toxicity data to make predictions about the behaviour of other structurally-related chemicals.

TTCs and food safety

The US Food and Drug Administration has adopted the concept for substances used in food contact articles [48]:

- If a substance or an impurity has not been shown to be a carcinogen in humans or animals and there is no reason, based on the chemical structure of the substance, to suspect that it is a carcinogen, a threshold of regulation is defined as a dietary concentration of 0.5 µg/kg diet or 1.5 µg/person/day assuming a consumption of 3 kg diet per day.
- If the substance contains an impurity that is a known carcinogen it is only allowed if the TD50 value (the dose that causes cancer in 50% of the animals corrected for tumours in the control animals) of the impurity based on chronic feeding studies is greater than 6.25 mg/kg/day.

This concept is based on statistical analysis of the Gold carcinogen database [49-51] of nearly 500 chemical carcinogens tested in animals using lifetime exposures. For the probabilistic analysis of the cancer studies, TD50 values were calculated and linearly extrapolated to a cancer risk of 1 in 1,000,000. These and further studies [47] came to the conclusion that there is a sound scientific basis for a general threshold of concern at 1.5 µg/person/day below which there is no significant risk to human health. This would imply that this threshold could be used for any chemical, including those of unknown toxicity. Further analysis may indicate that a higher TTC may be appropriate for chemicals which do not possess structural alerts for genotoxicity [20,47].

Other scientists, such as Munro, Kroes and Cramer [54-56], have further developed the TTC approach by analysing toxic, but non-carcinogenic, effects of more than 600 chemicals using oral toxicity tests in rats and rabbits with a wide range of structures and uses. The tests included chronic, subchronic, reproductive and developmental toxicity studies. From these, the most conservative NOEL for each chemical was selected according to their chemical structure using a decision tree approach leading to a categorizing of three different classes as shown in Table 11.4.

In order to derive human exposure thresholds, for each of these three different classes, the fifth percentile NOEL was selected. These human exposure thresholds or TTCs were calculated by multiplying the fifth percentile NOEL by 60, assuming an individual body weight of 60 kg and dividing them by a safety factor of 100 to ensure substantial margins of safety. Both the fifth percentile NOELs and the derived human exposure thresholds (TTCs) for the different classes of chemicals are given in Table 11.5.

The International Life Sciences Institute [47] brought together scientists around the world to further develop the TTC concept into a practical tool for assessing

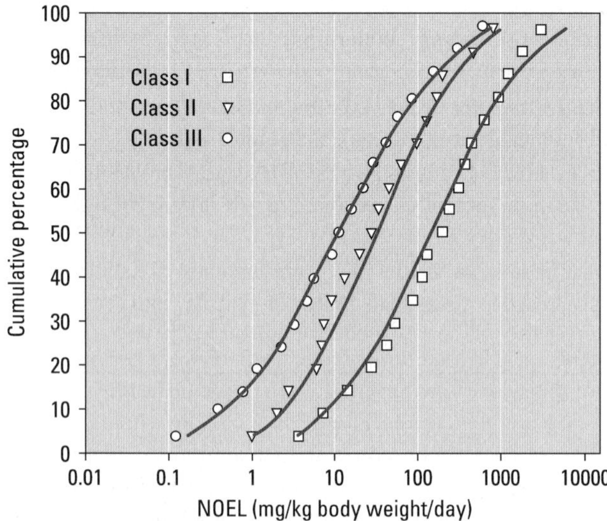

Figure 11.5. Structural chemical classes and NOELs. Symbols represent the NOELs and lines the fitted distributions. From Munro et al. [52]. With permission. Copyright Elsevier.

substances of unknown toxicity present at low levels in the diet. A decision tree has been proposed [44,47] which has a great potential for use in risk assessment (Figure 11.6).

Recently the ECB commissioned the development of a computer program (toxTree Version 1.00) to encode the Cramer classification scheme. The program is available as a free download from the ECB website (http://ecb.jrc.it/QSAR/).

TTCs for the environment

When comparing the developments of the TTC-concept in the area of food safety with those in the area of environmental safety, the similarities are striking. In the early nineties organic chemicals for which aquatic ecotoxicity were available have been classified into 4 distinct classes based on their chemical structures assuming different modes of action (MOA). Verhaar et al. [55] distinguished the following 4 classes: Class-I: inert chemicals (baseline toxicity), Class-II: relativity inert chemicals, Class-III: reactive chemicals, and Class-IV: specifically acting chemicals.

- Class-I: Inert chemicals are chemicals that are not reactive when considering overall acute effects, and that do not interact with specific receptors in an organism. The MOA of such compounds in acute toxicity is called (lethal) narcosis. Effect concentrations for a number of endpoints can be predicted using QSARs (Chapter 10 and Section 11.3.3).
- Class-II: Less inert chemicals are slightly more toxic than predicted by baseline toxicity estimations. These chemicals are often characterised as compounds acting by so-called polar narcosis, and can commonly be identified as possessing hydrogen bond donor acidity, e.g. phenols and anilines [56].
- Class-III: Reactive chemicals display an enhanced toxicity that is related to the phenomenon that these chemicals can react unselectively with certain chemical structures.
- Class-IV: Specifically acting chemicals exhibit toxicity due to (specific) interactions with certain receptor molecules (specific or receptor toxicity).

The Verhaar categorization scheme does not include metals, inorganics and ionisable organic chemicals. This Verhaar classification scheme was used by de Wolf et al. [46]. They analysed environmental toxicological databases (acute and chronic endpoints), substance hazard assessments and derived 95th percentile values. The TTC values were obtained by multiplying these 95th percentile values by appropriate application factors (10, 100, or 1000; Chapter 7). The derived values for MOA 1 to 3 were approximately 0.1 μg/L [20,46]. The derived MOA for specifically acting chemicals was about 3 orders of magnitude lower than the MOA for classes 1-3 [46].

Table 11.5. Fifth percentile NOEL and human exposure thresholds (TTCs) for three structural classes of chemicals [52]. With permission. Copyright Elsevier.

Structural class	Fifth percentile NOEL (mg/kgbw/day)	TTC (mg/person/day)
I	3.0	1.8
II	0.91	0.54
III	0.15	0.09

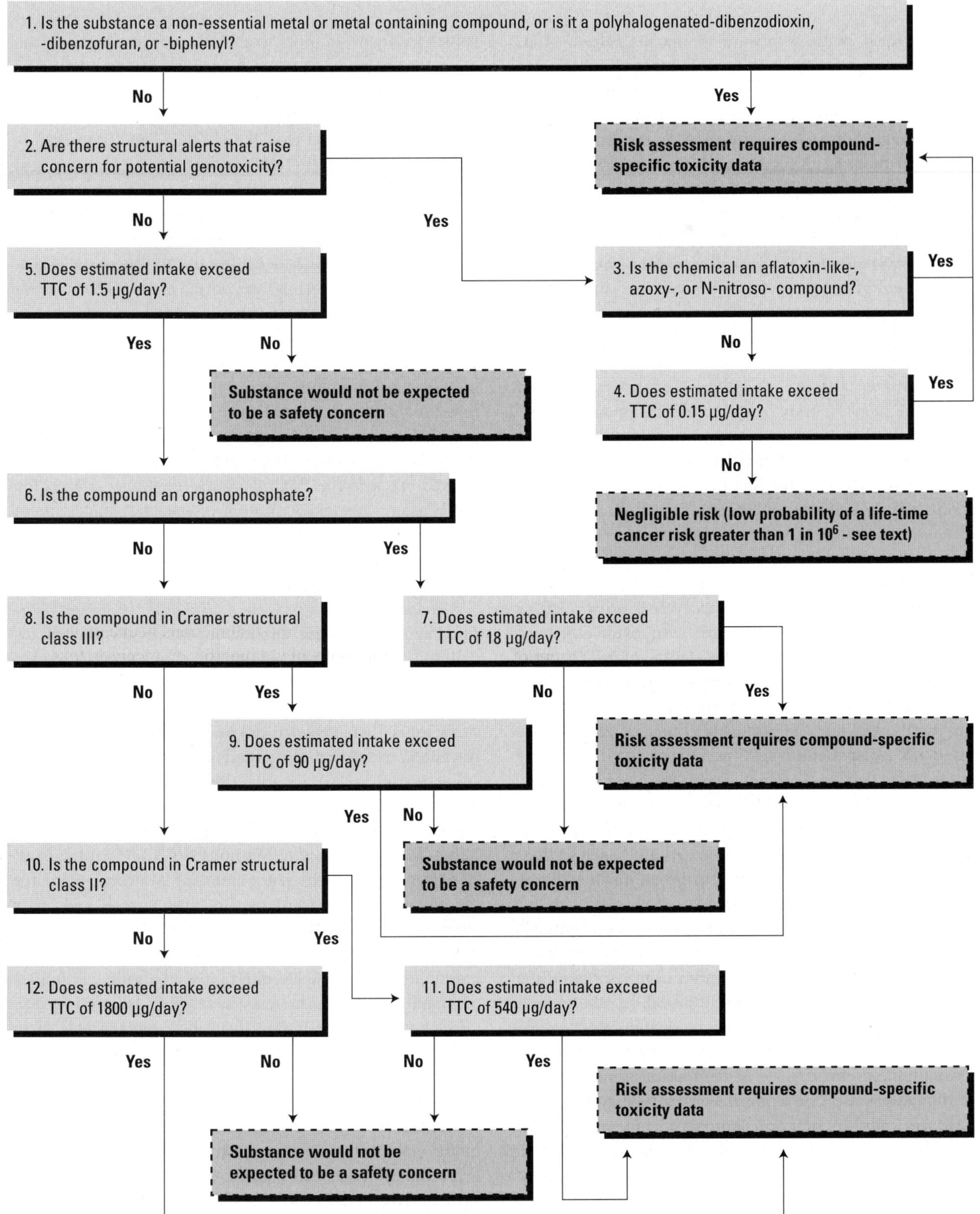

Figure 11.6. Decision tree proposed by ILSI Europe to decide whether substances can be assessed by the TTC approach. From Kroes et al. [44]. With permission. Copyright Elsevier.

Combined with exposure scenarios, these TTC values can form the basis of an ITS approach in a tiered risk-assessment scheme as depicted in Figure 11.1. If exposure information shows that TTCs will not be reached in the human body, in food or in the environment, this could be used as screening tool for priority setting, i.e. to classify chemicals as being of "low concern". If the measured or predicted exposure concentration comes close to the TTC, this could trigger the need to obtain further information on the toxicity of the chemical. The use of the TTC concept could be used to limit testing, provided that adequate information on the use of and exposure to chemicals is available. This is of particular importance when substances are used in new applications for which the exposure situations can be very different and hence the TTC may be exceeded. From a pragmatic perspective, the tools and approaches applied within TTC such as the Cramer structural classification scheme could be investigated for its applicability in the further development of chemical categories.

11.2.4 Exposure-based waiving

Increased realism in the exposure evaluation will allow stakeholders to give a low priority or even eliminate a higher number of substances that are of no (or low) concern [1,19,20]. In other words, if the exposure to a chemical (or a group of chemicals) can be predicted or measured adequately and the toxicological effects (e.g. based on experimental data or reliable estimates) are much lower than the (predicted) exposure concentration, further animal testing could be waived. In this concept, a decision to waive (exposure-based waiving) or trigger the generation of effects information (exposure-based testing) is a risk-based process for which the exposures for all applications and use scenarios need to be assessed.

Activities related to the categorization of chemicals in Canada [29] have shown that *early exposure profiling* has a considerable potential to more robustly define the nature of information on testing required. In other words tiered testing strategies, depending on the use profiles of (groups of) chemicals, can be developed and hold considerable promise to focus and reduce animal testing.

The application of exposure-based waiving requires extensive, highly detailed exposure information and should provide sufficient information to enable a thorough and reliable assessment of exposure throughout the life cycle [25]. This approach cannot be regarded as a "soft option". For it to be admissible, the exposure details need to be investigated and demonstrated rigorously, and the cost of doing this is not insignificant for many substances/uses. In order to achieve such an objective, the elements shown in Table 11.6 are considered necessary.

Further development of this approach is part of the current work on the REACH Implementation Projects (RIPs) co-ordinated by the ECB. The approach is promising, especially when combined with chemical categories, read-across, (Q)SAR and TTC approaches. It requires further investment in the development of exposure models and it requires precise information on the use pattern (life cycle) of the chemicals (e.g. downstream use information), which is one of the current bottlenecks. Product registers such as those available in the Nordic countries, e.g. Substances in Preparations in Nordic Countries (http://www.spin2000.net/spin.html), may be an important source for obtaining this exposure information. Once this information gap is closed (which is a challenge in its own right and a clear requirement of REACH), it is expected that many animal tests could be waived. As with the TTC approach mentioned above, the implications for risk communication and management would need to be further explored.

11.2.5 *In vitro* methods

In vitro tests include subcellular fractions, as well as cell and tissue cultures maintained for varying periods of time. These tests have been described in Chapters 6 and 7. The development of *in vitro* tests for different endpoints is being carried out by a wide range of research activities, and different methods are at different stages of standardization and documentation [57,58]. Information on the current status of *in vitro* tests can for example in the EU be obtained from the European Centre for the Validation of Alternative Methods (ECVAM) of the Joint Research Centre, which provides an EU focal point for the exchange of information on *in vitro* tests and other alternative tests. ECVAM's primary role is to co-ordinate independent validation studies of alternative tests. It also maintains a database on alternative methods. A number of stand-alone *in vitro* tests have been adopted by the OECD as official test guidelines. Full replacement of *in vivo* procedures is still, however, limited. The following successes have resulted in the last 15 years [59] and are listed in Table 11.7.

OECD *in vitro* test guidelines [6] have been used for many years to evaluate several genotoxicity endpoints and recently, guidelines for *in vitro* tests on phototoxicity, skin corrosion and skin absorption were adopted. In these cases, *in vitro* methods may be used to replace the animal testing currently required for hazard and risk assessment (Chapter 16). In this respect Council

Table 11.6. Relevant questions for generating exposure information to explore exposure-based waiving of testing [25].

Considerations	Comments
1. Where used	– Provide details of those industry sectors in which the substance is encountered together with supporting information on the broad areas of application/use. – Provide details of those consumer uses of the substance (including formulations) in a manner that readily enables the exposure patterns to be identified. – Provide details of the sites where the substance is manufactured or used.
2. How used	– Details are required *for each identified use* on how exposures are controlled. This information will include a description of potential release points, specific exposure management measures, general exposure controls etc.
3. How not used	– Details are required for any uses where exposure is managed by procedural means e.g. restrictions on supply, targeted risk management advice, bespoke exposure controls , etc.
4. 'Intensity' of the use	– Information should be provided that enables an assessment to be made of other factors that would affect how the nature of the exposure might be viewed e.g. numbers exposed, heterogeneity of the exposed population, including vulnerable groups, and likely frequency and duration of use, produced tonnage.
5. Predicted Exposures	– Details of the models that have been used, together with core assumptions/defaults, need to be provided for each use and exposure situation, such that the exposure prediction can be repeated by others.
6. Uncertainties	– Description of any major uncertainties in exposure estimates e.g. areas where a thorough understanding of exposure may be absent.
7. Supporting Data	– Where measured exposure data are included, then at a minimum these need to be described by - Number of samples (personal only in the case of worker exposures) - Frequency of sampling - Description of where/how samples were taken and if samples are representative of normal/ unusual operations - Limit of detection - Basic sample statistics e.g. mean, range, 90th% And supported (on request) by details of the sampling and analytical methodologies, together with the actual results.
8. Rationale	– Addresses both acute and chronic exposures. – Addresses exposures throughout the life cycle. – Provides clear basis for rationale e.g. margin of exposure when the predicted exposure/PEC is compared with PNEL and PNEC etc.

Table 11.7. Regulatory acceptance of tests that replace, reduce and/or refine existing animal methods over the last 15 years [59]. Reproduced with permission of ATLA.

Endpoint	Achievement
Skin corrosion	Development, validation and OECD acceptance of *in vitro* tests (human skin models, rat skin).
Phototoxicity	Development, validation and OECD acceptance of an *in vitro* test (3T3 NRU phototoxicty test).
Skin sensitization	Development, validation and OECD acceptance of a refinement and reduction alternative test (mouse local lymph node assay).
Skin penetration	Development and OECD acceptance of *in vitro* methodology (excised skin preparations).
Acute toxicity (systemic, local)	Development and OECD acceptance of animal testing approaches that refine and reduce animal use.

Directive 86/609/EEC is of importance, according to which "an experiment shall not be performed if another scientifically satisfactory method of obtaining the result sought, not entailing the use of an animal, is reasonably and practically available". Tests for skin corrosivity, phototoxicity and percutaneous absorption have been accepted by EU regulatory authorities and are incorporated into Annex V of Council Directive 67/548/EEC. For ocular and dermal irritation several well-studied *in vitro* tests are available but none of them have been formally validated or adopted as an OECD guideline. There has been little progress in the use of *in vitro* tests for predicting overall systemic toxicity (including reproductive toxicity).

A report was published in 2002 with the aim of providing a broad and objective picture of the current status and future prospects of alternative methods in the context of REACH [57]. A similar activity took place in order to estimate the time required to achieve full replacement of animal testing for cosmetics [58]. A draft timetable was developed by ECVAM and subsequently reviewed by the SCCNFP [18]. The SCCNFP raised serious doubts about the potential of *alternative* methods to *fully* replace *in vivo* experiments. This appraisal is also relevant for the discussion on REACH. The key conclusion of the SCCNFP was: "The SCCNFP wants to clearly give the message to the Commission that total abolishment of animal tests within 10 years is not feasible from an objective scientific point of view. Even the alternative strategies discussed in the document, which are estimated to take more than 10 years for further development, still include an animal test in the final tier." Similar comments about the full replacement potential of both *in vitro* and computer-based models (Section 11.2.2) have been expressed by the Scientific Committee on Toxicity, Ecotoxicity and the Environment (CSTEE) in 2004 [42]. The CSTEE concluded that "for the foreseeable future, the use of live animals in toxicity testing is essential to perform reliable risk assessments". A short summary of the SCCNFP advice is given below (Table 11.8).

The CSTEE [42] has commented on the reductionist nature of *in vitro* tests and (Q)SARs and the implications of this in predicting the effects of chemicals on animals. Since *in vitro* cell culture models cannot account for "unknown" mechanisms of action, which are detected in live animals (where all the relevant interactions occur), the predictive value of non-animal alternative tests is limited at present. The same is true for the computer-based models that work by incorporating existing knowledge into "expert systems", but which are unable to reliably predict unexpected or unanticipated mechanisms of action. Hence, it can be concluded that:

1. The European Commission, the Member States, industry, academia and animal welfare groups need to join forces to develop a programme aimed at the development and validation of alternative methods.
2. It would be at least 10 years before such a programme could bear fruit, and even then, the *full* replacement of animal testing would not be possible for most endpoints.
3. Although full replacement may not be possible for most endpoints, *in vitro* methods can contribute significantly to the reduction and refinement of animal tests when combined in an intelligent manner with other elements of ITS.

The optimistic predictions about the timetable for phasing out animal testing as provided by ECVAM in 2004 followed by critical reactions from the scientific committees of the European Commission [18,42] supporting more traditional toxicological approaches for risk assessment are a welcome illustration of the paradigm shift that is urgently needed. The maximization of up-to-date information from different sources in an integrated manner is the key challenge. It should embrace exposure-based considerations, new science and new tools in an integrative fit-for-purpose approach for risk assessment alongside more traditional tests. In the coming years, the advent of new "omics" technologies could dramatically increase the synergy between QSAR and *in vitro* assay methods [1]. Similar views have been expressed in the context of our understanding of diseases, complex biological systems and mechanisms of toxicity [60] and many new scientific challenges will be faced [59].

11.2.6 Optimized *in vivo* tests

Recently the CSTEE [42] summarized the need for animal testing in a clear manner: "Toxicological testing aims to predict possible adverse effects in humans when exposed to chemicals. Currently it is extensively based on animal testing to identify hazards and the dose-response relationships of chemicals. Ethical concerns have been raised by the use of laboratory animals. However, independent of ethical concerns, the primary objective of the risk assessment of chemical exposures is the protection of human health, wildlife and ecosystems".

There is hope that the number of animals used in *in vivo* tests can be reduced. This view has recently been published in a critical analysis of the OECD health effects test guidelines for *in vivo* testing [6,61]. The authors conclude that the opportunities for streamlining

Table 11.8. Proposed timeframe for full replacement of animal testing according to the SCCNFP [18].

Human health effects	Foreseeable time involved for full replacement
Acute toxicity	>2014
Skin irritation	>2014
Skin corrosion	<2004
Eye irritation	>2010
Skin sensitization	>2019
Skin absorption/penetration	>2006
Subacute/subchronic toxicity	>>2014 (not foreseeable)
Genotoxicity/mutagenicity	>2016
UV-induced effects	>>2019 (not foreseeable)
Toxicokinetics and biotransformation	>>2014 (not foreseeable)
Carcinogenicity	>>2014 (not foreseeable)
Reproductive toxicity	>>2014 (not foreseeable)

individual assays are very limited but they did put forward the view that *in vivo* testing can be made more efficient by: a) only performing tests that provide relevant data; b) eliminating redundant tests; c) using one sex; d) applying some tests simultaneously to the same animals and e) making greater use of screens and preliminary testing. Another contribution not mentioned in this review is to obtain more information from the dose-response curves in current bioassays. In this way, the numbers of animals in each dose group can be reduced considerably and improved statistical methodologies will allow us to derive more significant information for risk assessment purposes. A couple of initiatives have been taken to explore and implement the reduction in the use of animals in *in vivo* experiments. Areas where this can be achieved are e.g. systemic toxicity, sensitization, and fish tests [61-64]. A detailed discussion on the (regulatory) acceptability of these proposals still needs to take place.

11.2.7 Concluding remarks

Can we estimate the (relative) costs of developing and validating the different components of ITS? A full quantitative figure cannot be given but estimates can be provided. The costs for developing and validating the non-testing approaches such as chemical category approaches, TTCs and (Q)SARS vary according to the availability of high quality experimental toxicological data and according to the applicability of estimation methods to the chemicals being assessed. The costs for computational methods are negligible assuming that suitable training data sets and models are available. Probably about € 5-10 million is needed for an international QSAR toolbox and analogue identification tools. The reason for the relatively low costs is that methodologies have already been in use for a long time. In fact this is a capitalization of huge investments that have been made in the last decades in both the USA and Europe.

The US, Canadian and OECD experience on chemical categories, read-across and (Q)SARs shows that these methods can be applied now and at low cost for a significant percentage of chemicals (Table 11.2). Their application is principally related to priority setting for environmental and human health endpoints. The tools are sometimes applied for the derivation of hazard data in risk assessment [40]. In the REACH context, it is likely that this information will not always be suitable to meet the additional needs for risk assessment and classification and labelling. It is necessary to establish international collaboration, e.g. under the flag of OECD, to develop further guidance and a QSAR toolbox. Searching tools for chemical structures are crucial and are already available. The various methods have to be presented in guidance documents in a transparent and explicit manner and could then be implemented in computer-based systems at a later date. Participation of authorities and industry is essential. In fact, this work has recently started.

The costs for the development of *in vitro* tests are difficult to estimate and will also vary according to the availability of high quality experimental toxicological data. Estimations have been provided of about

€ 5 million per test (http://www.colipa.com/; information downloaded in January 2005). The number of tests will be at least a hundred and possibly several hundred. The costs for development of *in vitro* methods are estimated at about € 1-8 billion over the next 10-15 years. The costs for (pre)validation of a single *in vitro* method are estimated as € 300,000 per test by ECVAM giving a total estimate of € 150 million for the next 10 years for (pre)validation.

Exposure information is crucial information for any risk assessment. Use information, downstream use information, exposure factors, etc. are all important inputs for making reliable estimates of exposure both for human health and the environment (Section 11.2.4). Exposure-based waiving of animal tests [20] is a realistic opportunity, provided that a certain basic level of hazard information is available. For TTCs, it can be concluded that the science is more or less done. TTCs need to be put into the regulatory framework. This will require a thorough evaluation about the consequences for risk assessment, risk management and risk communication among stakeholders [43].

The timescale for developing and validating alternative methods will, according to the SCCNFP [18], take more than 10 years for most of the relevant toxicological endpoints, and even that will not lead to the complete replacement of animal tests. It is very likely that the most complex toxicological endpoints cannot be replaced at all [18,42]. *In vitro* tests may contribute significantly to the reduction and refinement of animal tests. It is very likely that the non-testing approaches may be developed relatively fast and at low cost. The testing approaches will probably take more time and are certainly more expensive. Exposure-based waiving requires adequate information to be provided on exposure of substances throughout their life-cycle. This is one of the big challenges of REACH. It is an area of considerable priority for the completion of exposure scenarios under REACH. In terms of time and cost needed to apply these ITS components, the following ranking can be given from fast and inexpensive to slow and costly:

(Categories/read-across = QSARs = TTC) > exposure-based waiving > *in vitro* tests > *in vivo* tests.

Category approaches, including read-across, (Q)SARs and TTCs can be developed and implemented relatively rapidly, provided that adequate guidance documentation is available. There is no need for a major investment in research. There is a discussion about the level of validation needed [65]. The real issue is that the methods must be valid [17] and must be accepted by the regulatory community. Already now more than a third of the substances assessed yearly in the OECD HPVC Programme [5] are assessed through the use of chemical categories and this fraction is estimated to increase significantly over the next few years as experience grows in Member Countries [25]. Transparent guidance needs to be developed on how to group chemicals, i.e. how to identify analogues. Read-across has been applied successfully in the US HPV Challenge Program over the last 5 years, and over the last 30 years under the US Toxic Substances Control Act (TSCA) [31]. (Q)SAR models can also be implemented rapidly for the limited numbers that are considered valid. If appropriate data is available, then in general (Q)SAR models have a short time for development and validation. The TTC concept can also be implemented relatively fast although the concept has not yet been evaluated for the diverse group of industrial chemicals [43]. Obtaining adequate exposure information may also be a bottleneck in the application of TTCs in risk assessment.

Transparent guidance, adequate use and release information, and models are also needed for the estimation of exposure. This information is essential for the development of exposure scenarios as required under REACH. It is a real challenge as the focus of REACH is on exposure and risk management (Chapters 2 and 12). It is not always easy to obtain adequate exposure information. This may render exposure-based waiving of further testing into a rather theoretical exercise and, at the same time, an area of primary focus considering the obligations under REACH in general, and ITS in particular! On the other hand it should be realized that "early exposure profiling" of large groups of chemicals has considerable potential to more robustly define the nature of the hazard information that needs to be generated. In other words, the development of test strategies for substances or groups of substances falling into various categories depending upon their potential for use and exposure is an interesting option for future exploration. It may focus the testing requirements and as such may make a considerable contribution to focus and reduce animal testing.

It should be noted that in certain cases, *in vivo* tests have to be undertaken in order to provide data for developing both *in vitro* and (Q)SAR methods. Mode of action based *in vitro* models can inform considerably the development of (Q)SAR models. In other cases, (Q)SAR models could be developed to (partially) model validated *in vitro* tests, thereby increasing the efficiency of *in vitro* testing.

In vivo methods have the longest development and validation time. For human-health related tests where the complexity is greatest, 15 years is a typical timescale, but most of the tests are currently available and not a lot of new test guidelines need to be developed. What is absolutely necessary is to refine (optimise) these methods with the aim to reduce the number of test animals per test.

11. 3 EXAMPLES OF ITS

In the previous sections, the advantages and limitations of each of the different components of ITS in terms of success rate, investment costs and time for development, validation and regulatory implementation has been shown. It was emphasized that the best way forward will be to develop "intelligent" testing strategies (ITS) for regulatory endpoints, comprising multiple elements including *in silico* approaches (including chemical categories, read-across, TTCs and QSARs), exposure-based waiving, *in vitro* tests and optimised *in vivo* tests. The strategies will need to be developed, assessed and refined iteratively. Furthermore, the future development and refinement of all component methods and models should be driven by the gaps in the strategies. It is therefore not possible to anticipate in detail where all future work will need to be focussed. However, it is clear that the benefits of integrating the use of all approaches will need to be continuously explored. In the forthcoming sections, a handful of examples of promising ITS are given.

11.3.1 Degradation

Degradation is an important endpoint. It is used for environmental hazard classification, for the assessment of PBT and vPvB chemicals [41] and for risk assessment, i.e. the exposure assessment of chemicals (Chapters 7 and 12). The major factors influencing the persistence of a chemical are: (1) its chemical structure, (2) the environmental conditions into which it is released and (3) the bioavailability of the compound. The ITS on degradation is aimed at providing a number of steps from more simple and non-expensive screening test methods followed by more complicated and expensive simulation tests methods. The following approach has been developed [25] and provides guidance for a cost-effective use of existing information and available resources.

1. Collection and review of available existing information including both non-test ((Q)SAR and read-across) and test information and release form. Where valid estimation methods are available, these may be considered as fulfilling standard data requirements when it can be assured that the prediction in question is within the applicability domain of the QSAR model. Non-standard test data may also be available that allows a prediction of the standard endpoints with sufficient confidence to allow decision-making.
2. Test for biodegradation normally start with ready biodegradability tests. Based on the physicochemical properties of the substance, enhanced/modified ready biodegradability test may be considered.
3. Test for abiotic degradation.
4. Assessment of environmental distribution and emission pattern for guiding the choice of the simulation test.
5. Simulation degradation test for the relevant compartment(s).
6. When primary degradation data are obtained it is essential to determine and assess the degradation products.

The overall scheme that has been proposed is given in Figure 11.7. The decision criteria for use of the ITS and the approaches based on chemical structure (SARs and QSARs) are presented in the original report [25]. Based on the analysis with only few chemicals, it is concluded that the result of the (Q)SAR predictions would most often (but not always) be in agreement with that based on screening test data. This is not surprising as generally applicable biodegradation models have been derived from and focus on the estimation of ready and nonready biodegradability in screening tests [25]. It was also concluded that some of the estimated degradation half-lives according to the simulation studies differ from those based on screening test data. In general the use of screening test data gave a more strict result than higher tier testing.

11.3.2 Bioconcentration

Depending on the sensitivity, specificity and reliability of ITS approaches and the regulatory decision to be taken, it may be relevant to use non-test information in a cautious way. The examples described in Chapter 9 and [41] are different estimation methodologies using the *n*-octanol-water partition coefficient (K_{ow}) as predictor for bioconcentration. For the estimation of this parameter in relation to hazard classification for the aquatic environment a simple QSAR is used only if valid test data on the more reliable predictor for bioconcentration is not available. Already for more than a decade there is the standard requirement of the K_{ow}, as a reliable predictor of

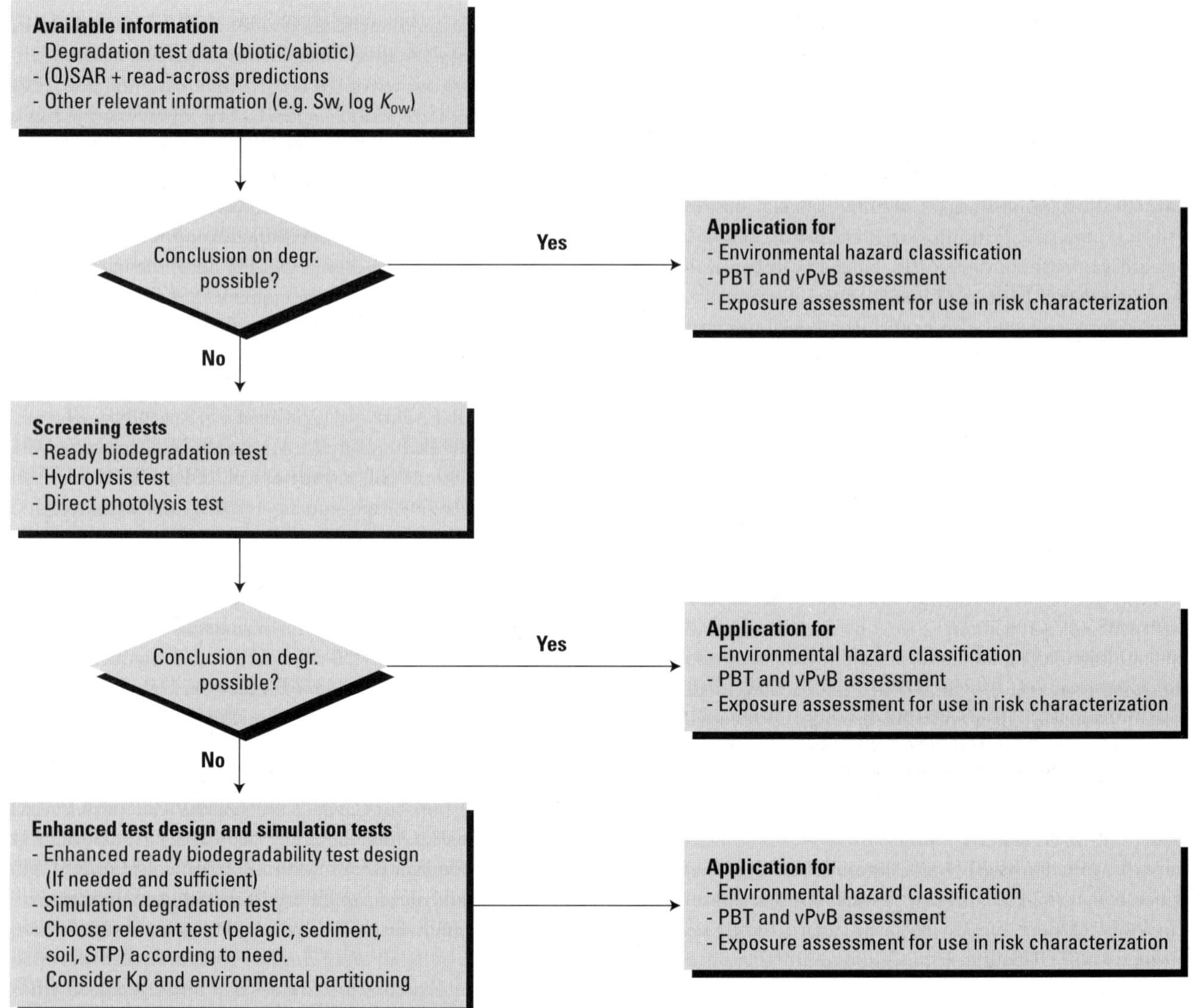

Figure 11.7. Overview of a decision scheme on degradation for the three regulatory needs: environmental hazard classification, PBT/vPvB assessment and exposure assessment for use in risk characterization. Modified after [25].

bioconcentration in fish for organic chemicals [32]. The K_{ow} can be calculated, using computer programmes, or be determined experimentally (Chapter 9). This estimate saves approximately 100 fish and € 50,000 per test [9,15]. ECETOC [64] has recently presented an integrated strategy for the assessment of bioconcentration (Figure 11.8). This ECETOC strategy reflects in a step-wise approach, the considerations given above. Furthermore, an *in vitro* verification is introduced to check the potential for metabolism (Tier 2). Before doing the OECD 305 *in vivo* flow-trough bioaccumulation test with fish (Tier 4), screening studies are proposed (Tier 3). Some of these screening bioconcentration studies have been published (e.g. static bioaccumulation tests with fish), whereas others, such as studies to determine kinetic data for bioaccumulation, are still under development [64].

11.3.3 Aquatic toxicity

For the endpoint of aquatic toxicity, the following regulatory needs have been identified: (1) environmental hazard classification, (2) effects assessment, including the derivation of a PNEC, and (3) PBT and vPvB assessment. These regulatory needs are often based on experimental data [6,12,17]. Testing for environmental effects often requires acute toxicity for freshwater fish and daphnids

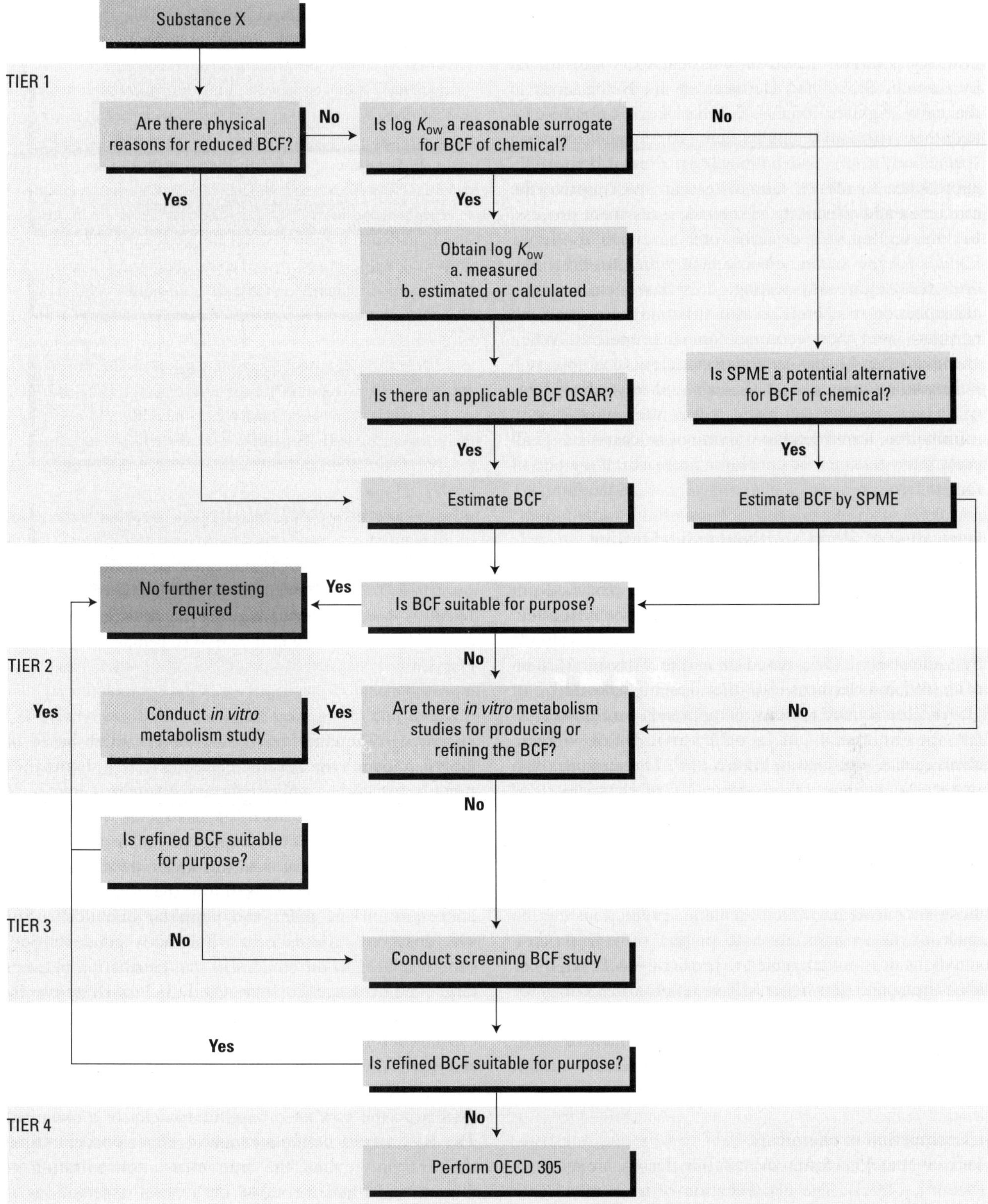

Figure 11.8. ITS for bioconcentration according to ECETOC [64]. With permission. SPME = Solid Phase Micro-Extraction.

as well as a growth inhibition test on freshwater algae (Chapters 7 and 8). This selection of species for the base set is largely based on classical work carried out by Canton, Slooff and Hermens in the Netherlands in the early eighties [66,67]. Further work on (Q)SARs and their regulatory application [40,63] and work on species sensitivity distributions [68], allow different ITS approaches to reduce animal testing and improve the efficiency and effectivity of the risk assessment process. In this section, we describe one approach to derive PNECs for the aquatic environment which, in principle, does not require any testing. It is based on a proper classification of chemicals into structural classes which correlate with their ecotoxicological properties. When chemicals can be classified appropriately, this approach will save millions of test animals and tens of millions of testing costs. Combined with relevant exposure information, it may lead to waiving of at least 30 % of all ecotoxicity testing (see concluding remarks at the end of this section).

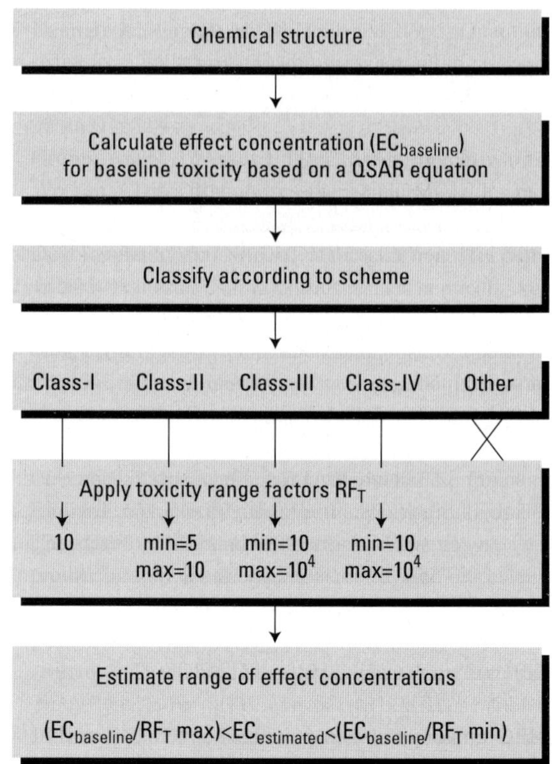

Figure 11.9. Approach for predicting aquatic toxicity of organic chemicals. From Verhaar et al. [55]. With permission. Copyright Elsevier.

Derivation of PNECs without animal testing

Although various kinds of estimation models have been developed for several classes of chemicals, they are not always well-defined and their applicability is limited. Therefore, a rather conservative approach was developed by Verhaar et al. [55], based on earlier work by McKim et al. [69] and Hermens [70]. Based on this knowledge of QSARs for aquatic toxicity, an approach was developed for the estimation of aquatic toxicity of organic chemicals, as outlined in Figure 11.9. The approach is a two-step procedure: (1) classification of chemicals [55] followed by (2): the prediction of aquatic toxicity of a variety of aquatic species and the PNEC using QSAR approaches [63,71].

The general philosophy of this approach is to select those chemicals for which "reliable" predictions can be made or, as an alternative, to predict only worst-case situations. It is not attempted to predict the ecotoxicity of each compound. It is better to have reliable predictions for a few chemicals and a clearly defined set of compounds for which no reliable predictions can be made, than to have a prediction for each chemical without information about its reliability.

Classification of chemicals

Verhaar et al. [55] distinguished four classes of chemicals (Section 11.2.3). The classification of a chemical into one of these four classes is performed on the basis of its molecular structure, according to rules formulated for each of the classes. For chemicals in class-I, reliable estimates of aquatic toxicity can be obtained, based on QSAR models for several endpoints [63]. Estimation models for class-I chemicals are given in Chapter 10 and [41]. Currently also for chemicals in classes-II relative reliable predictions can be made, as further work showed that there is no real difference between internal effect concentrations (based on the membrane lipid concentration) of polar and nonpolar chemicals [56]. The observed difference in whole-body concentrations can be related to differences in the distribution between target and nontarget components. For class-III no precise predictive models are available yet and only worst-case estimates for acute fish toxicity (LC50) can be made (Figure 11.10). The estimates are based on toxicity range factors (RF_T), which represent multiplication factors applied to the estimated baseline toxicity of a chemical. The RF_T values define a range of effect concentrations that extend beyond the true effect concentration of this chemical and are based on general observations of acute toxicities (LC50 for guppy) of members of these three classes of chemicals. As shown in Figure 11.9, the highest RF_T values are below a factor of 10^4.

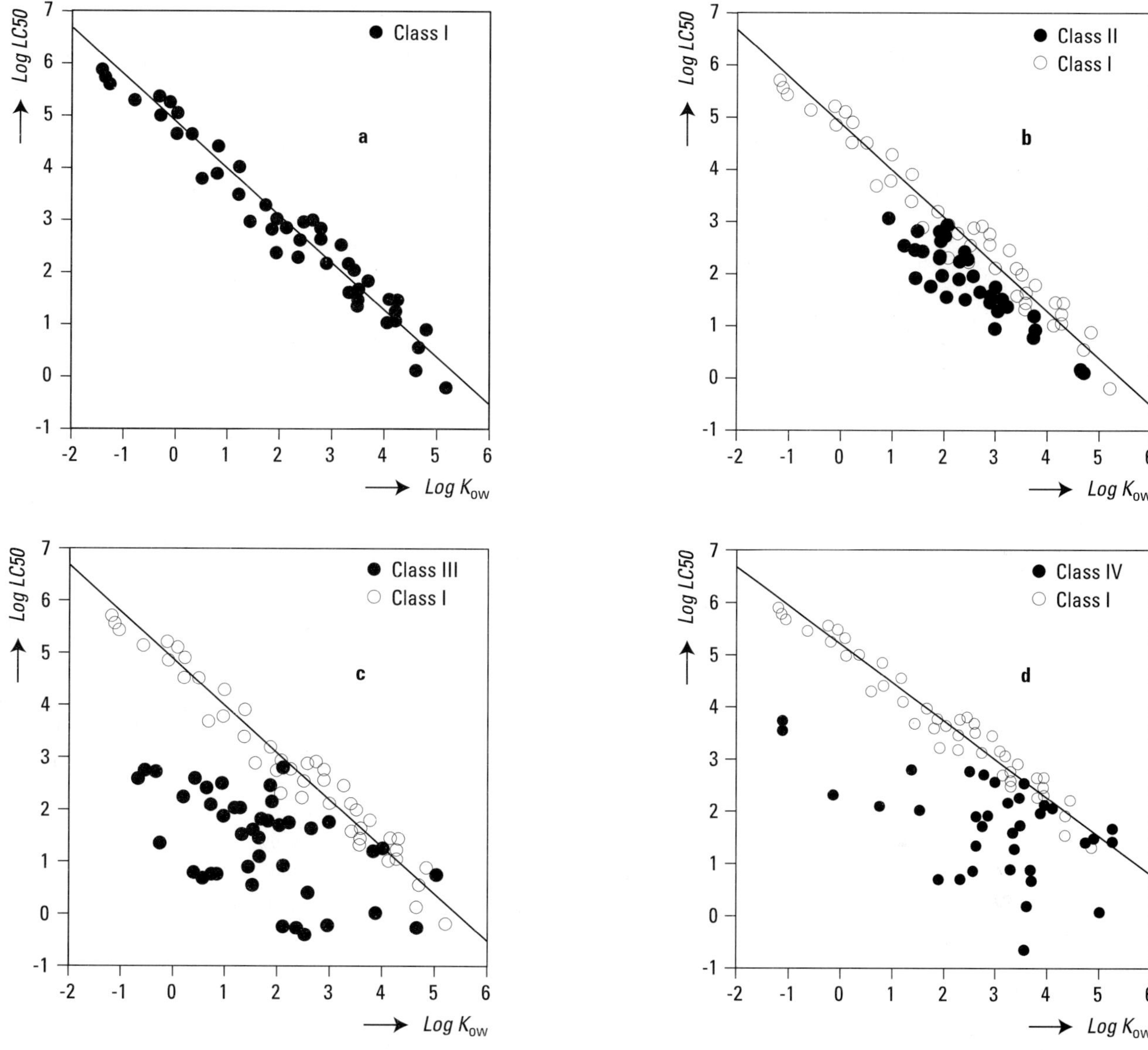

Figure 11.10. LC50 data for guppy *(Poecilia reticulata)* for class I, II, III and IV chemicals. Data compiled by Verhaar et al. [55]. With permission. Copyright Elsevier.

The classification has been applied to high production volume chemicals (HPVCs) [71,72]. The results are presented in Figure 11.11. Around 48% (977 chemicals) of the 2000 HPVCs could be classified according to this scheme. It appears that a little over 1000 HPVCs fall into non-classifiable categories like inorganics, polymers and ill-defined mixtures (e.g. petroleum products). The sub-set of class-I chemicals, which consists of 230 HPV chemicals, represents those chemicals that can be classified as "inert chemicals". About 70 chemicals are classified as class-II. The estimated parameters for the narcotic chemicals include acute and chronic toxicity to fish and daphnids, toxicity to algae, and HC5 values for surface water and sediment. The HC5 (hazardous concentrations for 5% of the species) is the 5th percentile of a chronic toxicity distribution. It is the outcome of a statistical extrapolation method based on species sensitivity distributions [68,73]. The HC5 concept is explained in more detail in Chapter 7. An example of an aquatic toxicity profile for a class-I chemical is given in Figure 11.12. For class-II chemicals a worst-case estimate of toxicity could be predicted using the log K_{ow} estimate of narcotic toxicity and applying toxicity range factor between 5 and 10. For all other classes of HPV

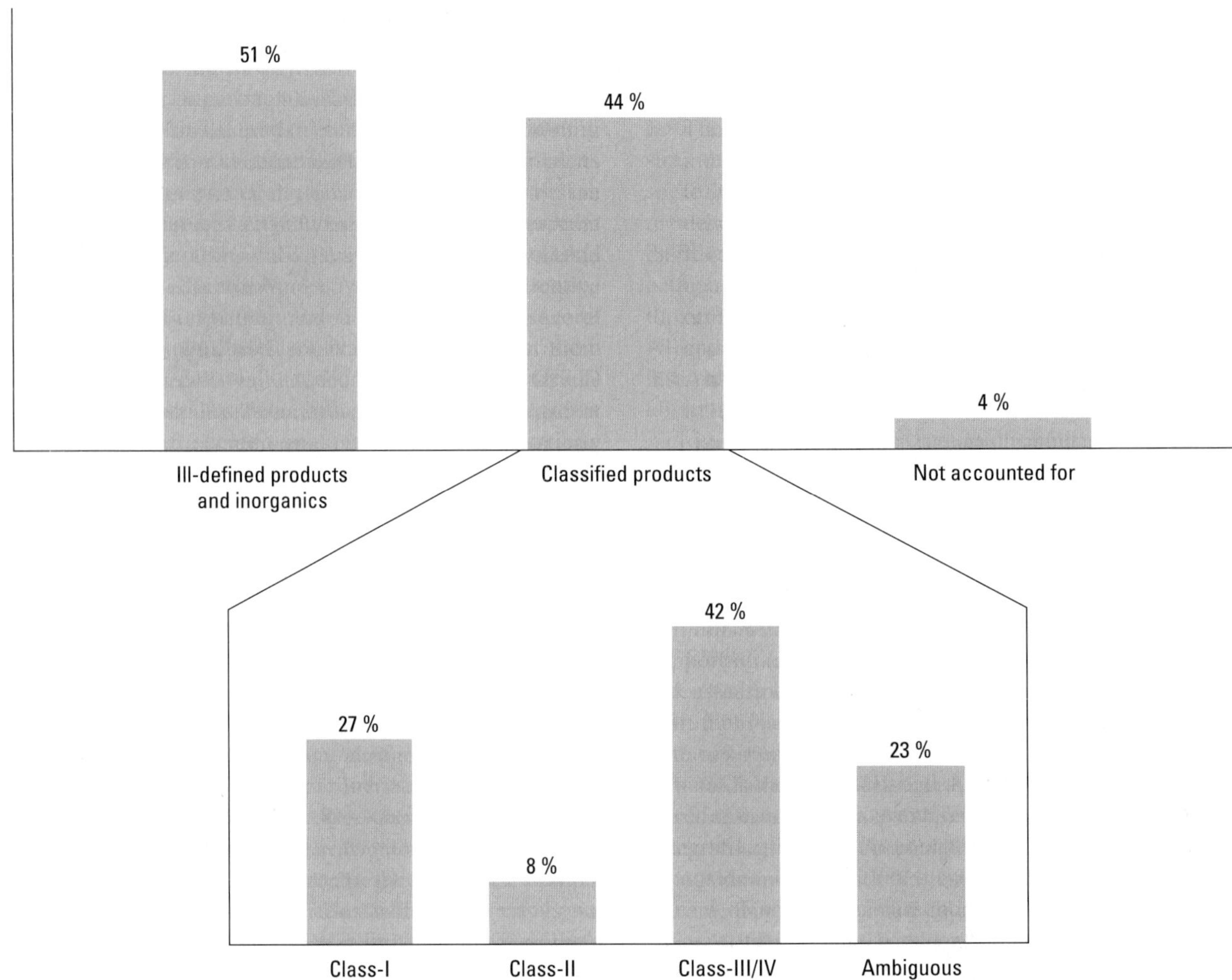

Figure 11.11. Distribution of high production volume chemicals (HPVCs) based on the classification by Verhaar et al. [71]. With permission from Taylor & Francis Ltd. (http://www.informaworld.com).

chemicals, only worst-case estimates for acute toxicity to fish have been presented [72]. A more in depth description of the approach is given below.

Estimation models for chemicals with a non-specific mode of toxic action

The class of relatively unreactive chemicals which, in acute toxicity tests act as narcotics, is the best-known class of compounds for which several QSARs have been established. Studies by Könemann [74] and Veith et al. [75] showed that external effect concentrations such as LC50s or NOECs for these chemicals depend on K_{ow}, as expressed by Equation 11.1.

$$\log \mathrm{C} = \mathrm{A} \log K_{ow} + \mathrm{B} \qquad (11.1)$$

Two examples, one for LC50 for guppy [74] and one for NOEC for *Daphnia magna* [76], are given in Equations 11.2 and 11.3, respectively.

$$\log \mathrm{LC50\ (mol/L)} = -\ 0.87 \log K_{ow} -\ 1.1 \qquad (11.2)$$

$$\log \mathrm{NOEC\ (mol/L)} = -\ 0.95 \log K_{ow} -\ 2.0 \qquad (11.3)$$

The lower value for the inter-cept in Equation 11.3 is due to the more sensitive endpoint (growth reduction versus survival). The experimental data on which Equation 11.2 is based are also given in Figure 11.10.

QSAR studies for the aquatic toxicity of this particular class of chemicals have been extensively discussed in several publications [37,70]. At first sight, it seems remarkable that QSAR equations for various kinds of species are so similar. However, the explanation is quite simple. As it is generally accepted that the

108-67-8

Name:	1, 3, 5-trimethylbenzene				
CAS:	108-67-8	EINECS:	203-604-4	EEG:	601-025-00-5
LogPstar:		CLOGP:	3.64	MW:	120.1950
		Class:	1		

LC50 fish	= 4.819e-2 mmol / L;	5.793e+0 mg / L.
NOEC fish	= 4.727e-3 mmol / L;	5.682e-1 mg / L.
LC50 daphnid	= 3.639e-2 mmol / L;	4.374e+0 mg / L.
NOEC daphnid	= 5.536e-3 mmol / L;	6.654e-1 mg / L.
EC50 algae	= 2.239e-2 mmol / L;	2.691e+0 mg / L.
HC5–50 % (water)	= 1.852e-3 mmol / L;	2.226e-1 mg / L.
HC5–50 % (sed.)	= 1.510e-1 mmol / L;	1.815e+1 mg / L.
HC5–95 % (water)	= 4.578e-4 mmol / L;	5.503e-2 mg / L.
HC5–95 % (sed.)	= 3.733e-2 mmol / L;	4.487e+0 mg / L.

Sediment HC5 values are based on a number of assumptions; before using these figures PLEASE refer to appropriate sections in the accompanying documents!

Figure 11.12. Example of a prediction set for a class-I chemical [72].

mechanism of narcosis is not a very specific process and that each compound has the same intrinsic activity, the external concentration of a compound *(C)* at a fixed effect (e.g. death) is only a function of the probability of a compound reaching its site of action. For many chemicals whose bioaccumulation is not influenced by biotransformation reactions, this probability correlates with the K_{ow}, which explains the influence of K_{ow} on the effect concentrations. McCarty [77,78] and Van Hoogen and Opperhuizen [79] showed that internal lethal concentrations of narcotic chemicals, i.e. chlorobenzenes, in fish are indeed quite constant (about 1-2.5 μmol/g fish). Because internal effect concentrations (lethal body burdens) are independent of K_{ow} and because BCFs increase with K_{ow}, it automatically follows that external effect concentrations (LC50s) will decrease with K_{ow}.

Since each organic compound can, in principle, act as an anaesthetic agent, narcosis is considered a baseline or minimal effect, and therefore, QSAR equations for these types of chemicals will predict minimum toxicity [74,75]. Van Leeuwen et al. [63] published QSAR equations for the NOEC values of 19 different species of bacteria, algae, fungi, protozoans, coelenterates, rotifers, molluscs, crustaceans, insects, fish and amphibians. These 19 different QSARs are presented in Figure 11.13. This study showed that differences in sensitivity are very small. The information from these 19 different QSAR equations was used to calculate HC5 values, using the extrapolation methods of Aldenberg and Slob [73] described in Chapter 7, leading to the following equation for the prediction of the HC5 (with 95% certainty) for narcotic chemicals in water [71]. This HC5 value (Equation 11.4) provides an adequate estimate of the PNEC for the aquatic environment. In the Technical Guidance Document [41] an additional small assessment factor is applied to derive the PNEC in case of data-rich chemicals.

see below (11.4)

Equilibrium partitioning theory [80] involving sediment, water and biota was applied to derive HC5 values for total water concentrations (water including suspended matter), aquatic sediment and aquatic organisms, on the basis of the concentration in the water phase, i.e. the $HC5_w$. The background to these equations is discussed in more detail in Chapter 3 and 7. Equation 11.5 was used to calculate the concentration of a chemical in surface water including suspended matter at a concentration of 3×10^{-5} kg/L with an organic carbon fraction of 0.1.

$$HC5(95\%) = -0.851 \log K_{ow} - 1.601 - 2.53 \times \sqrt{(1.81 \times 10^{-2} \log K_{ow}{}^2 + 0.294 + 4.76 \times 10^{-2} \log K_{ow})} \quad (11.4)$$

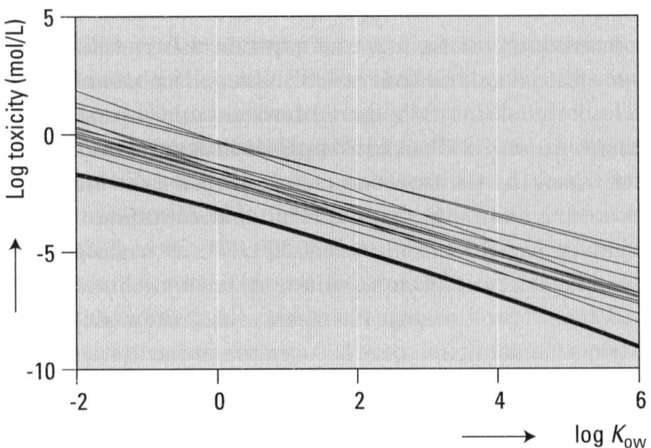

Figure 11.13. QSARs (thin line) for NOEC's of 19 different species and the HC5 (thick line). From van Leeuwen et al. [63]. Copyright ©1992. Reprinted by permission of Alliance Communications Group, Allen Press, Inc.

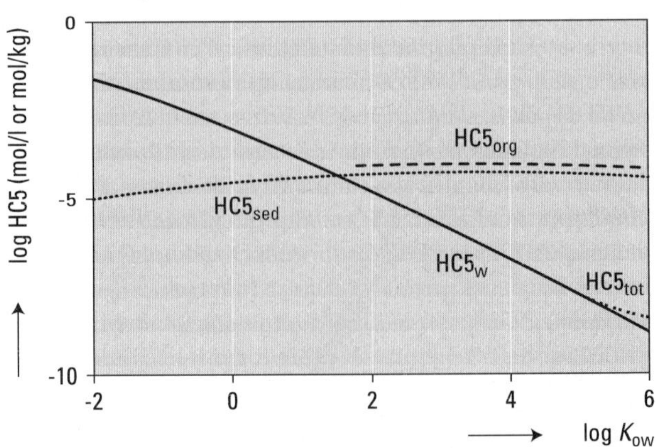

Figure 11.14. HC5 values (with 95% certainty) for the water phase ($HC5_w$), surface water including suspended matter at a concentration of 30 mg/L ($HC5_{tot}$), sediment with an organic carbon content of 5% ($HC5_{sed}$) and biota with a lipid content of 5% ($HC5_{org}$) as a function of the K_{ow}. HC5 values for sediment at log K_{ow}. <1.5 are speculative (see text). From van Leeuwen et al. [63]. Copyright ©1992. Reprinted by permission of Alliance Communications Group, Allen Press, Inc.

$$C_{tot} = C_w (1 + 1.85 \times 10^{-6} K_{ow}) \qquad (11.5)$$

where C_{tot} is the total concentration in water (mg/L), C_w is the concentration in the water phase (mg/L) and K_{ow} is the octanol-water partition coefficient. The concentration in sediment (C_{sed} in mg/kg) with an organic carbon content of 0.05 was calculated from:

$$C_{sed} = 0.031 \times C_w \cdot K_{ow} \qquad (11.6)$$

Similarly, the internal toxicant concentrations in biota can be calculated from the BCF. The BCF at equilibrium may be defined as the quotient of the concentration in the organism (C_{org} in mg/kg) and the concentration in water. When expressed on a lipid basis, BCF is equal to K_{ow} [80]. According to McCarty et al. [77], this equation does not hold for hydrophilic chemicals (log K_{ow} < 1.5) because with these compounds the amount in the hydrophilic (water) phase of the organism dominates the amount in the lipid phase. Therefore, they suggest the use of the following equation for the whole-body bioconcentration:

$$\text{BCF} = 1 + 0.05 K_{ow} \qquad (11.7)$$

by which the total body residue (C_{org}) can be calculated by simply multiplying this value by C_w:

$$C_{org} = C_w (1 + 0.05 K_{ow}) \qquad (11.8)$$

All the relationships expressed by Equations 11.4 -11.8, are drawn together in Figure 11.14. Effect concentrations for pollutants at the HC5 level for water including suspended matter ($HC5_{tot}$), sediment ($HC5_{sed}$) and biota ($HC5_{org}$) can now be obtained by substituting $HC5_w$ for C_w in Equations 11.5, 11.6, and 11.8, respectively. Although Equation 11.7 holds for many hydrophobic chemicals, deviations may be observed with chemicals that are easily metabolized or with very hydrophobic chemicals which require a long time to reach equilibrium [63].

Concluding remarks

The availability of reliable QSARs varies considerably between endpoints and for the same endpoint between substances depending on the domain of the model. Some QSAR models for aquatic toxicity recommended in the TGD are already accepted by regulators and have been used for environmental hazard classification and preliminary PBT assessment [41]. The QSARs for narcotic effects are reliable estimation methods. It should be noted that the approach may underestimate the possible delayed effects which are not of a narcotic mechanism of toxic action. On the other hand these effects may also be masked in current experimental tests for the assessment of aquatic toxicity. Combining the 19 QSARs into an equation to predict the PNEC for aquatic toxicity for narcotic chemicals (Equation 11.4) is a promising approach which needs further consideration for implementation in risk assessment.

The success of the approach strongly depends on a proper classification of the chemical. The approach of Verhaar et al. [55] is useful, but it should be remembered that the approach did not cover substances with multi-functional groups (and thus their potential for interactions) and was not comprehensive in its coverage of functional groups. Nevertheless, when appropriately classified, the approach can be applied successfully for about 25% of all existing single organic chemicals. Applying an extra toxicity range factor between 5 and 10 on the HC5, a preliminary estimate can be made for the PNEC for polar narcotic chemicals. This will increase the predictive capacity to 30% of all single organic chemicals. Thus, when chemicals can be classified appropriately, this approach will save millions of test animals and tens of millions of testing costs. Combined with relevant exposure information it may lead to waiving of at least 30 % of all ecotoxicity testing. There are three approaches to improve the classification of chemicals:

1. *Experimental verification with Daphnia and algae.* The classification of chemicals into Class-I (and Class-II) can be experimentally verified by using experimental data with algae *(Selenastrum capricornutum)* and *Daphnia* and comparing these with the predictions of acute toxicity using the QSAR equations for *D. magna* and *S. capricornutum* as given in the TGD [41]. The advantages are obvious: lower costs and no tests with vertebrates. Once (polar) narcosis is confirmed by this experimental validation, adequate estimates can be made of PNECs for the aquatic environment.
2. *Experimental verification with fish.* A second additional approach has been proposed recently [62,75,81]. This approach is relatively simple: fish tests would be performed only at one concentration, the lowest between the EC50 concentrations obtained in the tests with algae and daphnids. When fish would be more sensitive than algae and daphnids, testing with fish would be continued at lower concentrations using this step-down approach or limit test with a few fish. The work is based on the classical work in the early eighties of Canton, Slooff and Hermens [66,67]. They aimed at finding a useful combination of test methods to determine the aquatic toxicity of environmentally dangerous chemicals and showed that species had different sensitivities to different chemicals and that "the most sensitive species" did not exist. Based on a preliminary analysis, the results show that 54-71% reduction can be obtained in the number of fish used for industrial chemicals [81] and approximately 73% for pharmaceuticals [81]. This approach (Figure 11.15) for fish in acute toxicity testing can easily be extended to predict long-term aquatic toxicity and effects at the ecosystem level (PNECs for the aquatic environment) as was shown in the assessment of (polar) narcotic HPV chemicals [55,63,71,72]. It should be noted that the step-down limit test with fish is still under discussion.
3. *Refining the classification.* Another approach has been developed recently by von der Ohe et al. [82]. They have developed a more rigorous classification system to distinguish chemicals into two classes, i.e. narcotic chemicals and chemicals exerting excess toxicity using Daphnia. They have developed three simple discrimination schemes that enable the identification of excess toxicity from structural alerts based on the presence or absence of certain heteroatoms and their chemical functionality. It follows that such compounds would have little priority for experimental testing. Thus, the ability to identify – directly from chemical structure – compounds that are likely to be toxic only in the narcotic range would offer the possibility to reduce the need for experimental testing and thus provide an attractive component of an ITS.

11.3.4 Irritation and corrosivity

Testing for irritation and corrosion (Table 11.9) have been among the most criticised of all toxicity tests. This has resulted in a standardisation of protocols, a reduction in the number of animals used and the introduction of testing strategies that require testing in animals (rabbits) only when other means of determining irritation and corrosion have been addressed. According to the current OECD guidelines, substances should not be tested in animals for irritation/corrosion if they can be predicted to be corrosive from their physicochemical properties. In particular, substances exhibiting strong acidity (pH 2 or alkalinity (pH 11.5) should not be tested.

There are OECD test guidelines for in *vitro* skin corrosion under which substances may be classified as corrosive [6]. A negative result in these tests should be supported by a weight of evidence determination using other existing information, e.g. pH, (Q)SAR, human and/ or animal data. These tests do not provide information on skin irritation and, therefore, further information on non-corrosives as to their skin irritation is required. If the non-corrosive property of a substance in an *in vitro* test has not been possible to confirm by other data an *in vivo* test will have to be conducted. There are also no internationally adopted test guidelines for *in vitro* skin or eye irritation, nor are there fully validated methodologies.

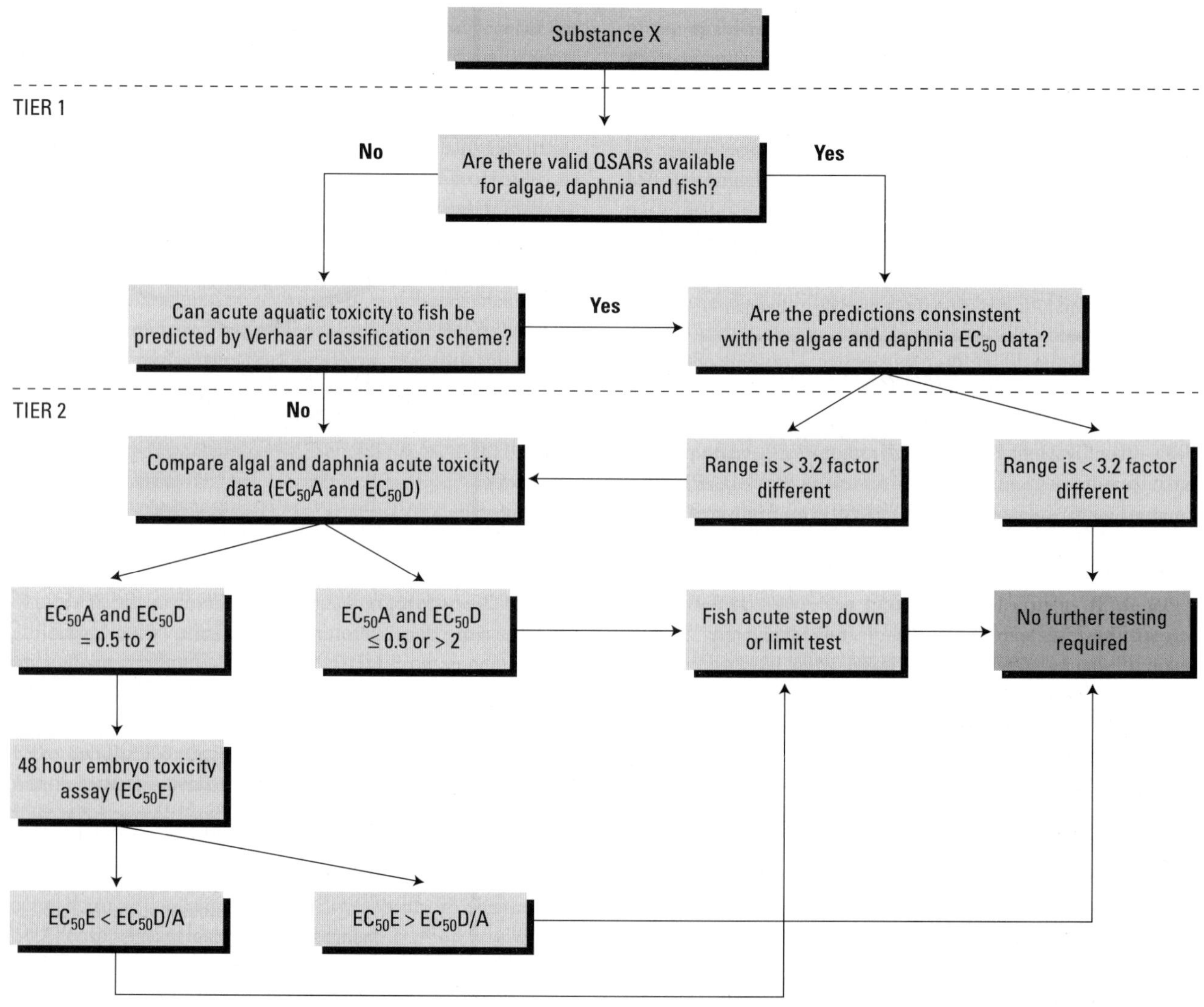

Figure 11.15. ITS for assessing acute toxicity according to ECETOC. A represents the algal test result, D represents the test result with *Daphnia* and E represents the fish embryo test. When the ratio of the two EC50 values is between 0.5 and 2.0 it should be assumed that the sensitivity of the two assays is similar and a 48 hour fish embryo toxicity assay is proposed as this would give a greater confidence in predicting an acute toxicity value for fish. Modified after [64]. With permission.

Therefore, choice of appropriate test methods and the interpretation of the results is presently difficult. An overview of currently available methodologies has been provided in Eskes and Zuang [58].

In the REACH legislation [17] data and test requirements for skin and eye irritation/corrosion have been listed. Before testing in accordance with these requirements, all relevant physicochemical and toxicological information e.g. acid or alkaline reactions, human and animal data, *in vitro* test data and (Q)SAR analysis, should be assessed. If these data are not available or they are inadequate for hazard and risk assessment, an *in vitro* skin corrosion study is normally required. Where the substance is corrosive in the *in vitro* study, it should be classified accordingly and no further testing for irritation conducted. However, if the substance is not corrosive in this study an *in vitro* test for skin irritation and normally an *in vitro* eye irritation study should be undertaken. If there are positive *in vitro* results from these studies the substance should be classified as being irritating to skin and eyes.

For substances with no or very few data, sequential

Table 11.9. Definitions of eye and dermal irritation and corrosion according to the OECD [6].

Endpoint	Definition according to the OECD
Dermal irritation	The production of reversible damage of the skin following the application of a test substance for up to 4 hours.
Dermal corrosion	The production of irreversible damage to skin; namely, visible necrosis through the epidermis and into the dermis, following the application of a test substance for up to four hours. Corrosive reactions are typified by ulcers, bleeding, bloody scabs, and, by the end of observation at 14 days, by discolouration due to blanching of the skin, complete areas of alopecia, and scars
Eye irritation	The production of changes in the eye following application of a test substance to the anterior surface of the eye, which are fully reversible within 21 days of application
Eye corrosion	The production of tissue damage in the eye, or serious physical decay of vision, following application of a test substance to the anterior surface of the eye, which is not fully reversible within 21 days of application

Table 11.10. General consecutive steps in the analysis of irritation and corrosion. Please note that more detailed examples of ITS for skin and eye toxicity are available [25].

Step	Information
1	Existing data on physicochemical properties
2	Existing human data
3	Existing animal data from irritation/corrosion studies
4	Existing data from general toxicity studies via the dermal route and from sensitization studies
5	Existing (Q)SAR data and read-across
6	Existing *in vitro* data
7	Weight-of-evidence analysis
8	New *in vitro/ex vivo* tests for corrosivity
9	New *in vitro/ex vivo* tests for irritation
10	New *in vivo* test for irritation

ITS are recommended for developing adequate and scientifically sound data for assessment/evaluation and classification of the corrosive and irritating properties of substances. Recently, a preliminary ITS has been proposed comprising of 10 steps (Table 11.10). In this strategy [25], it is recommended that the sequence is followed to step 6. In step 7 a weight-of-evidence analysis is done. It is important to note that before the weight of evidence analysis in step 7, no new *in vivo* tests should be conducted, but the assessment should be based on the existing data. Detailed information and guidance on the various steps of ITS both for skin irritation/corrosion and eye irritation has been developed [25].

A promising further development in this area is the refined use of QSARs, physicochemical property limits and chemical's structural alerts. Physicochemical property limits have been developed for the identification of chemical substances with no skin irritation or corrosion potential [83-85]. This approach is especially powerful when combined with structural alerts for the identification of chemicals with skin irritation or corrosion potential, as proposed in a two-step procedure called skin irritation corrosion rules estimation tool shown in Figure 11.16. In the first step, physicochemical property limits are used to identify chemicals with no skin corrosion or skin irritation potential. If the chemical's physicochemical properties

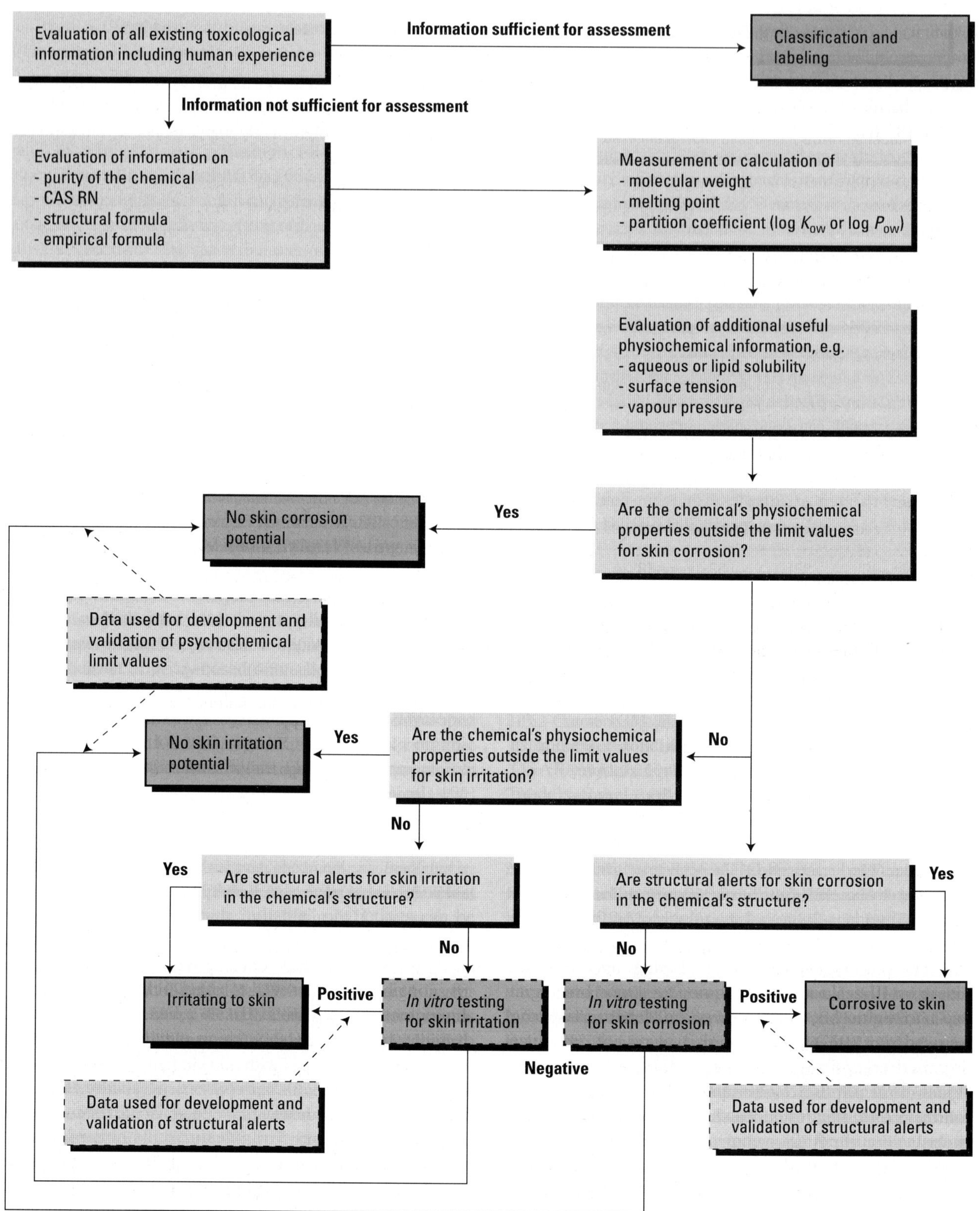

Figure 11.16. ITS for skin irritation and corrosion as proposed in the skin irritation corrosion rules estimation tool (SICRET). After Walker et al. [87]. With permission.

do not meet the prescribed limits to identify chemicals with no skin corrosion or skin irritation potential then the chemical's structural alerts are used in the second step to identify chemicals with skin corrosion or skin irritation potential [83-87].

Irritation and corrosion are frequently associated with pain. It is mainly for this reason that there has been such strong pressure to replace animal testing with alternative approaches. Much work is under progress and further investments, developments and discussions are needed to develop internationally agreed ITS methodology for these toxicological endpoints [25,58].

11.3.5 Skin sensitization

Skin sensitization results from a T-lymphocyte mediated immune response to a chemical allergen that comes into contact with the skin. A chemical penetrates the skin and binds to a carrier protein typically by a covalent bond to form an antigenic hapten-protein complex. This complex is processed by antigen presenting cells principally dendritic Langerhans cells (LC) of the epidermis. These cells then migrate to the draining lymph node where they present the chemical to T-lymphocyte to provide the stimulus for antigen-specific commitment and the production of memory and effector T-lymphocytes. Subsequent contact with sufficient dose of the chemical will then result in the expression of the clinical signs of allergic contact dermatitis (ACD) [88]. Thus chemicals need to overcome a number of hurdles in order to induce skin sensitization. These comprise:

- Penetration into the viable epidermis across the *stratum corneum*.
- Formation of a stable association with protein to create an immunogenic complex. This requires that a chemical is inherently protein-reactive, or can be transformed chemically or metabolically to a protein-reactive species. Typically the stable association is thought to be a covalent one
- Deliver dermal trauma sufficient to induce and up-regulate those epidermal cytokines that are necessary for the mobilization, migration and maturation of LC.
- Be inherently immunogenic such that a T- lymphocyte response of sufficient magnitude is stimulated.

If these hurdles are not successful then skin sensitization will either not occur or will be sub-optimal [89-92].

There are a number of methods available for the prospective identification of skin sensitizing chemicals. Guinea pig tests, in particular the guinea pig maximization test (GPMT) and the occluded patch test of Buehler, were used extensively for the identification of skin sensitization hazard. These are described in OECD Test Guideline 406. In recent years, the local lymph node assay (LLNA) has been adopted as the preferred test for skin sensitization (OECD Test Guideline 429). Here activity is measured as a function of lymph node cell proliferative response induced following topical exposure of mice to a test chemical. In this method, skin sensitizing chemicals are defined as those that provoke a 3-fold or greater increase in proliferation compared with vehicle-treated controls [6]. The EC3 value which represents the *effective concentration* of a skin sensitizing chemical required to induce a *3-fold increase* in the proliferative activity of draining lymph node cells compared with concurrent vehicle-treated controls has proven a valuable tool in the evaluation of skin sensitization risk assessment [93,94] as it provides a measure of relative potency. High EC3 values are associated with weak or extremely weak sensitizers, low values with extreme-strong sensitizers. More information on relative potency proposals can be found in [94].

In the REACH legislation [17] data and test requirements for skin sensitization have been listed in Annex V. Before testing in accordance with these requirements, all relevant physicochemical and toxicological information e.g. acid or alkaline reactions, human and animal data, *in vitro* test data and (Q)SAR analysis, should be assessed first. If these data are not available or they are inadequate for hazard and risk assessment, an *in vivo* skin sensitization study is normally required. This comprises conducting the LLNA for the substance in question. Alternative tests such as the GPMT may be used with justification if the LLNA is not adequate.

At the present time, there is no *in vitro* test to replace the LLNA. The current thinking is to develop an integrated framework where information from a number of alternative approaches is combined together to provide a comprehensive assessment of sensitizing potential. These include:

- Exploitation of the relationship between chemistry and skin sensitization (Reaction chemistry underpins the mechanistic attempts to predict skin sensitization from structural and physical properties (Q)SARs), some of which have been embedded into expert systems [85,91].
- Assessment of the ability of chemicals to form stable associations with proteins or peptides [95-98].
- The activation of dendritic cells (DC) or DC-like cells by exposure *in vitro* to chemicals [99-103].

In terms of developing a preliminary ITS for skin sensitization this might involve the following steps:

1. Assessment of *in silico* approaches to include identification of structural alerts, (Q)SARs and reaction chemistry. Chemicals that are sensitizers are typically electrophilic in nature – the key is to identify the electrophilic centre within a chemical).
2. A next step would be to consider the ability of a chemical to bind with protein. Work is on-going to develop *in vitro* assays to measure protein/peptide binding.
3. An assessment of epidermal bioavailability. Some chemicals may not penetrate the viable epidermis sufficiently to be of concern.
4. *In vitro* assays focusing on the activation of dendritic cells.
5. Modified LLNA. The LLNA is viewed as having made a significant contribution to animal welfare, achieved through a requirement for fewer experimental animals, coupled with a reduction in the trauma to which animals are potentially subject. A modified version of the LLNA using only a single high dose group, together with a concurrent vehicle control, may provide a potential approach for screening purposes when there is a need to evaluate the sensitizing activity of a large number of chemicals. The feasibility of this was evaluated based on a large compilation of existing LLNA data [104,105] and found to be promising.

Rather than a stepwise ITS, an alternative approach might be to combine different elements of information in a Bayesian Belief Network (BBN) weight-of-evidence approach. Here each piece of information (evidence) is evaluated and assigned a value depending on its type and quality. The BBN model provides a framework for assessing the relative contribution of different sources of information and for quantifying uncertainty which ultimately facilitates prediction for new substances with limited data. Such a BBN model has been reported in the literature which combines available test results from Buehler, Guinea Pig Maximization (GPMT), Human Repeat Insult Patch (HRIPT), Human Maximization (HMT) and Local Lymph Node (LLNA) tests together with Quantitative Structure-Activity Relationship (QSAR) predictions for sensitization and penetration [106]. A proposed ITS for skin sensitization is given in Figure 11.17.

11.3.6 Reproductive and developmental toxicity

Reproductive toxicity refers to the adverse effects of a substance on any aspects of the reproductive cycle, including the impairment of reproductive function, and the induction of adverse effects in the embryo, such as growth retardation, malformations, and death (Chapters 6 and 7). In Section 11.1.2 it was shown that reproductive toxicity is among the most expensive toxicological endpoints in terms of both animals and costs. It is also obvious that this is an endpoint with great toxicological relevance, i.e. it is a relevant endpoint for the protection of human health and the environment if substantial exposure would be measured or otherwise be demonstrated e.g. by the application of use and exposure categories or by occupational and/or consumer exposure models (Chapter 5). The main driver for the information requirements for reproduction for substances is the annual tonnage of manufacture in or import into the European Union [17]. The information requirements (and costs for testing) increase with tonnage. Therefore an ITS has been developed around two core objectives [53]:

1. To have sufficient data to classify (or to exclude from classification) a chemical as a reproductive and/or developmental toxicity hazard.
2. To have sufficient data to support risk assessment and risk management decisions.

This ITS has been developed in a relatively short period of time and is of a preliminary nature. It needs to be refined and further harmonized for regulatory application [25]. This ITS is based on a weight-of-evidence approach and designed to allow informed decisions on reproductive and developmental toxicity potential in a step-by-step manner. The ITS will enable the decisions to be made on the need for further testing or whether sufficient data already exist to meet the agreed objectives. The ITS provides a three-stage process for clear decision-making [25]:

Stage 1

Stage 1 is relevant for all tonnage levels. This is a series of questions relating to the existing classification for fertility and development as well as for carcinogenicity and mutagenicity. Exposure potential needs to be considered before deciding whether any further reproductive and/or developmental toxicity testing is required. Therefore, it is possible that chemicals that exhibit low toxicological activity, no systemic absorption and no relevant human exposure or those classified cat. 1 or 2 for fertility or development, or classified as a genotoxic carcinogen or a germ cell mutagen and with appropriate risk management measures in place may not progress beyond Stage 1.

Stage 2

This stage is relevant for lower tonnage substances. This comprises an in-depth evaluation of the existing

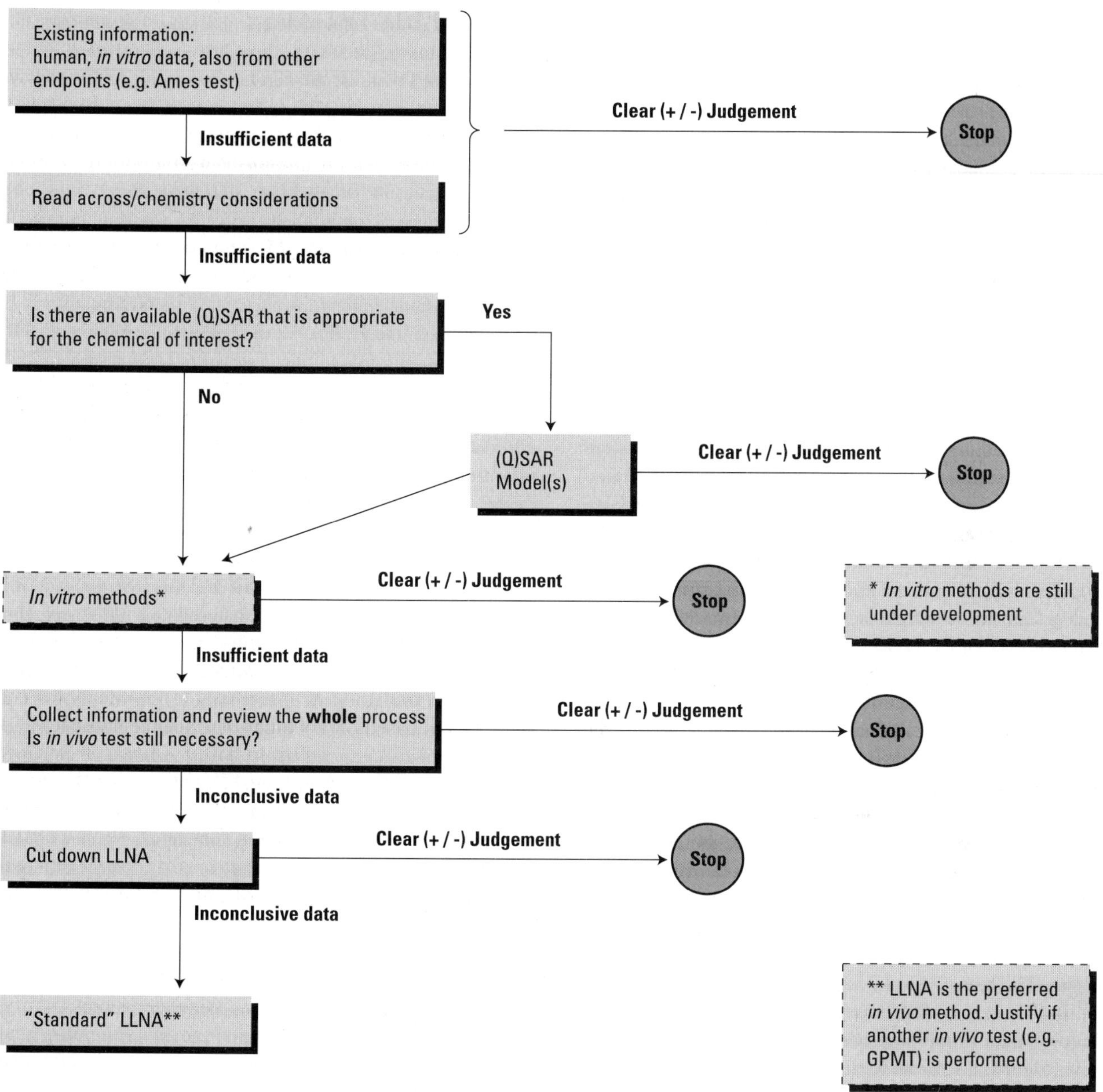

Figure 11.17. Proposed ITS for skin sensitization.

toxicology data base, consideration of reproductive and/or developmental toxicity alerts, (Q)SAR predictions and exposure potential. The aim of this stage is to determine the scope of reproductive and/or developmental toxicity testing in order to cover the identified alerts. It is possible that, following this review coupled to a weight-of-evidence analysis in Stage 1 or if sufficient data for risk assessment/risk management and classification purposes already are available, no further testing is necessary. However, for the majority of chemicals, it is likely that Stage 2 will be used to determine the scope of reproductive and/or developmental toxicity testing required in Stage 3.

Stage 3

Stage 3 describes the relevant reproductive and

developmental toxicity tests upon which classification, labelling and risk assessment decisions will be based for chemicals progressing beyond Stages 1 and 2.

A battery of four tests is proposed to enable classification, labelling and risk assessment/risk management decisions in Stage 3. It must be stressed that not all chemicals reaching Stage 3 will be assessed in all four tests. Instead, individual chemical testing requirements will be based on the nature of alerts identified in the Stages 1 and 2 and the tonnage level. The four tests protocols on which the ITS is based are the following [6]:

1. Reproduction and developmental toxicity screening test (OECD 421).
2. Combined repeat dose toxicity study with the reproduction/developmental toxicity screening test (OECD 422).
3. Prenatal developmental toxicity study (OECD 414).
4. Two generation reproduction study (OECD 416).

The core mantra of the proposed ITS for reproductive toxicity [25] is: (1) to ensure that adequate exposure and effects information is provided to permit informed decisions on classification, labelling and risk assessment, (2) to avoid unnecessary testing and usage of animals, and (3) to optimize the use of human and financial resources. For instance, it was noted that in case of alert(s) for reproductive or developmental toxicity in Stage 2, it may be more appropriate to bypass screening tests and go immediately to a two generation reproduction study (OECD 416) or to conduct a prenatal developmental toxicity study (OECD 414) in addition to the reproduction/developmental toxicity screening test (OECD 421). An example of the ITS for chemicals (between 100 and 1000 tonnes per year) is provided in Figure 11.18.

It is clear that further discussion is needed to develop and implement an ITS for reproductive and developmental toxicity [25]. Nevertheless, the outline of the stepwise preliminary approach provides a firm basis for future regulatory implementation of this ITS. In case of the ITS for reproductive and developmental toxicity it is clear that science provides the basis, but in the end the ITS of this and other toxicological endpoints is a risk management decision.

11.4 SUMMARY AND CONCLUDING REMARKS

Lack of public information of chemicals

The lack of public information on chemicals is a key issue. This problem and the slow progress in risk assessment and management [3] is not unique to Europe. Similar problems have been observed in non-EU countries, e.g. Canada and the USA. The rate at which the risks of HPVCs have been assessed in the EU is approximately 10 substances per year. With a total of about 2750 HPVCs this would lead to a period of more than 200 years before we can start with the assessment of 30,000 lower production volume chemicals. As a consequence, we know a lot about a few chemicals (< 5%), but we have very little public information on the properties and risks of most (> 95%) chemicals [12]. This lack of public knowledge about chemicals is one of the main drivers of the REACH legislation [17]. It is the basis for international collaboration on existing chemicals in the OECD [5], activities in Canada [29,30] and voluntary programmes in the USA [31]. Further information is given in Chapters 12-16.

Costs and animal saving potential of ITS

Estimates of the cost and animal saving potential of ITS have been given in Section 11.1.2. Estimates of the investments needed for the developments of the components of ITS have been given in Section 11.2.7. In terms of the time and cost needed to apply methods, the following ranking can be done from fast and inexpensive to slow and costly: (chemical categories/read-across = (Q)SARs = TTC) > exposure-based waiving > *in vitro* tests > *in vivo* tests. Applying ITS can reduce the needs for tests by up to 70% for individual endpoints resulting in significant savings in testing costs and use of animals [13,15]. The ITS approaches have a price as well and will take time from the organizations involved in their development and implementation.

Challenges ahead

Governmental agencies, the regulated community, and stakeholders face the challenge of generating and interpreting data for risk assessments in a cost-effective and efficient manner thereby minimising the use of laboratory animals [1]. Different elements of ITS have been described in Sections 11.2. It is clear that each different approach can contribute to this challenge, but that none of these individual components can meet this challenge individually. In other words intelligent testing strategies are needed (Figure 11.1). These integrated approaches can speed up the risk assessment process while reducing costs and animal tests [1].

While the details of the different proposals for intelligent testing vary, a number of common components can be identified. A further analysis of a few examples of ITS for different endpoints as described in Section 11.3

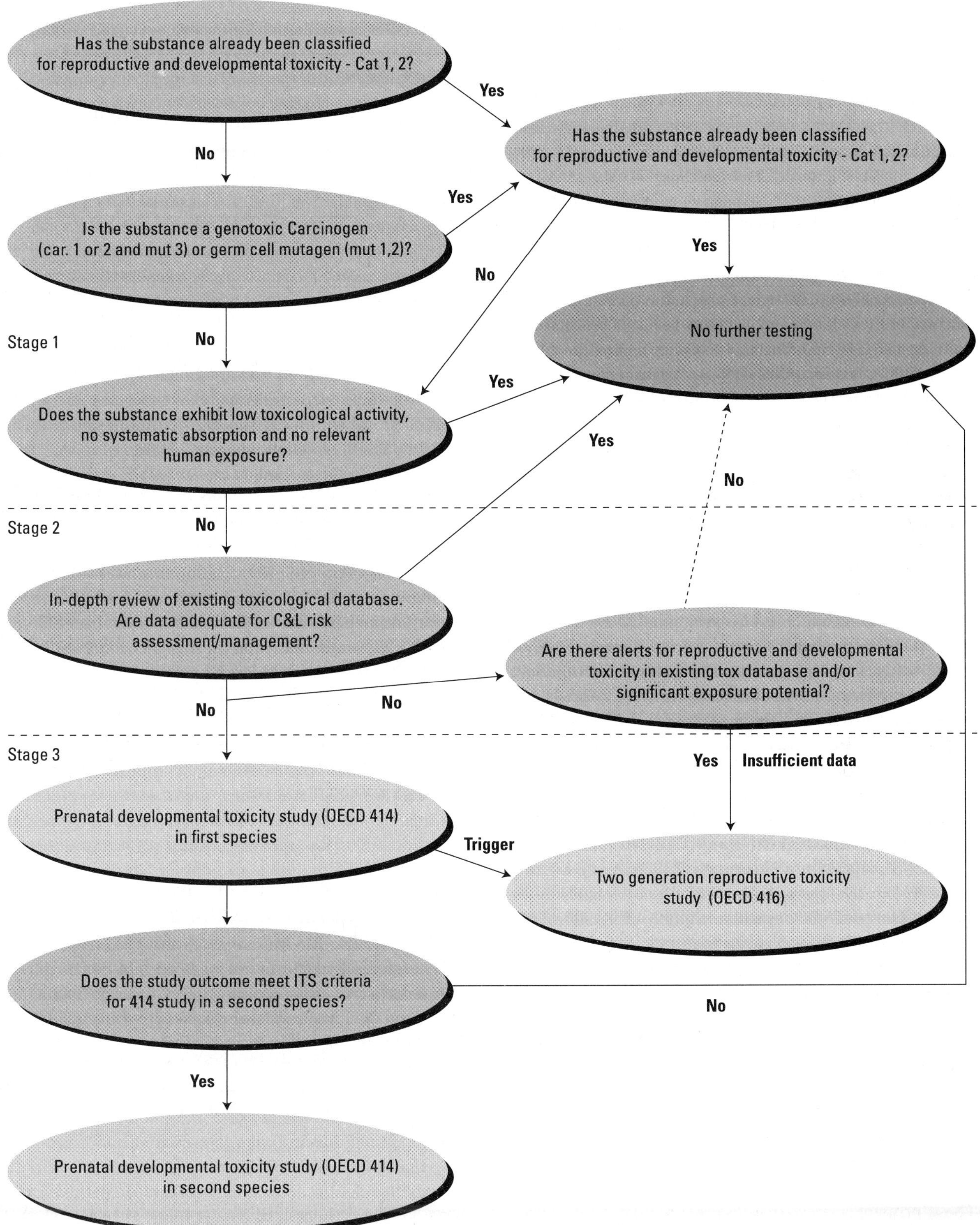

Figure 11.18. Proposed testing strategy for reproductive toxicity. Modified after [25].

has shown that no overarching scheme can cope with the diversity of all the scientific aspects for all endpoints and their dynamism, where the mix of scientific issues, techniques and approaches is so diverse [25]. There is no doubt that in the next decade many of the intelligent testing strategies will be further refined and, therefore, there is no doubt that the examples given in Section 11.3 will soon be replaced by better ones. We need to move forward on at least five fronts:

1. Paradigm shift

In the introduction of this chapter we stated that "the challenge is to move in a scientifically credible and transparent manner from a paradigm that requires extensive hazard (animal) testing (Figure 11.19) to one in which a hypothesis- and risk driven approach can be used to identify the most relevant *in vivo* information". This challenge is depicted in Figure 11.20.

2. Implement new techniques, where possible

New developments in molecular technologies [1,59,60,108] hold considerable promise to enhance exposure and risk assessments; methods include optimization of test methods and testing strategies. With today's computing power, thousands of chemicals can be processed readily in near real time to estimate properties associated with three-dimensional structures. As a result, it is now possible to rapidly predict the toxicological and ecotoxicological potential for endpoints associated with chemical reactivity (e.g., covalent binding to nucleophilic sites in DNA, RNA, or critical proteins), redox-cycling, and oxidative stress as well as for noncovalent interactions with membrane and protein receptors. The same holds fro the prediction of fate-related properties. With the establishment of increasingly diagnostic cellular and biochemical endpoints derived from well-characterized *in vitro* systems, defined and consistent toxicological responses can also now be generated. Development and validation of this capability is essential to formulating QSARs as well as standardized assay methods for empirically evaluating chemicals of concern. In the coming years, the advent of "omic" technologies could dramatically increase the synergy between QSAR and *in vitro* assay methods [1, 59,60].

3. Test methods

At the level of the test method development the "Solna principles" [109] need to be implemented efficiently. These principles stress that tests for regulatory purposes need to reflect the following:

- Biological relevance.

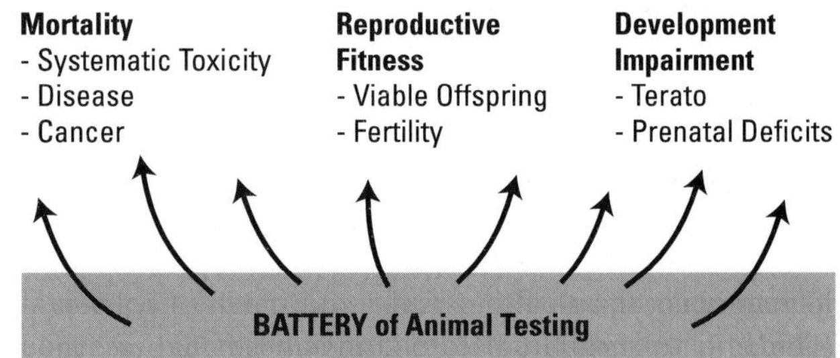

Figure 11.19. Risk managers focus on potential adverse outcomes of toxicological tests. The current testing strategies for chemicals are based on animal testing and generate many toxicity data. Only a fraction of these data will actually be used in risk assessment. Modified after Jones [107].

- Reliability/reproducibility.
- Regulatory acceptability.

Test methods and their inclusion in ITS are context-dependent. The relationship between the scientific level of validity and the regulatory acceptability of methods and ITS are not constant but vary with the purpose of the test or ITS. The regulatory implementation varies with the kind of regulatory decision to take. It is clear that science provides the basis for test methods and ITS, but in the end the adoption of test methods and testing strategies are multi-stakeholder risk management decisions where regulatory acceptance is key.

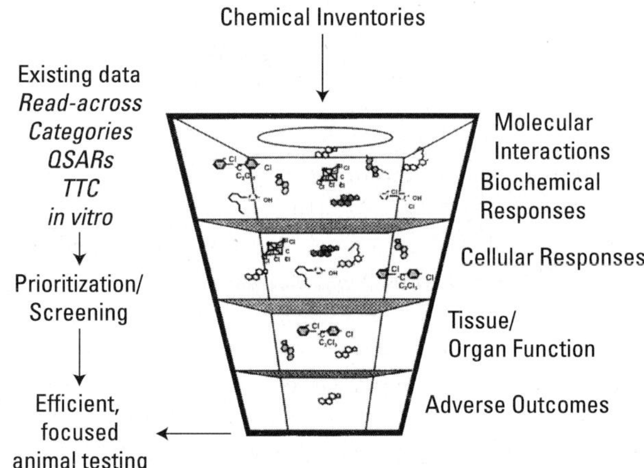

Figure 11.20. The new hypothesis-driven paradigm of risk assessment of chemicals will lead to a more efficient and focussed testing on animals. Modified after Jones [107].

Table 11.11. In order to deal with the challenges ahead a much broader approach, i.e. a 7-Rs strategy is needed.

Component	Comment
1. Risk	Apply a risk-based approach by focussing on both exposure and effects
2. Restriction	Apply waiving of testing when possible on the basis of adequate exposure information and minimal information on effects
3. Repetitive	Apply tiered assessment strategies by minimising animal testing and by applying quicker and cheaper methods before more refined and more expensive ones
4. Relatives	Focus on families or categories of chemicals by application of computational toxicology (analogue identification, read-across, (Q)SARs and TTCs)
5. Replacement	Full or partial replacement of information from animal tests
6. Refinement	Development/modification of animal tests based on less painful and distressing procedures
7. Reduction	Development of approaches that obtain the same amount of information with fewer animals

4. Broaden the scope

We also need to rethink and reframe the current 3-Rs strategy of replacement, reduction and refinement. It is too narrow. In fact, a much broader approach, a 7-Rs strategy, is proposed (Table 11.11 and Figure 11.21) in order to deal with the challenges ahead.

This 7-Rs strategy is a risk-based approach. As such the focus is on: (1) *risk*, by focusing both on exposure and effects. The guiding philosophy should be (2) *restriction* of testing (waiving of testing) on the basis of adequate exposure information and minimal information on effects. Risks should be assessed in a (3) *repetitive* manner (a tiered approach, going from simple, to refined or comprehensive risk assessment, if necessary). *In vivo* testing should only be carried as a last resort. Furthermore, there is a need to move away from the chemical-by-chemical approach as much as possible and to focus on (4) *relatives*, i.e. on families or categories of chemicals (a group-wise approach), by applying (Q)SARs, read-across, TTCs and exposure-categories. The overall strategy should also encompass the current 3-Rs strategy of (5) *replacement*, (6) *refinement* and (7) *reduction* of animal tests. In conclusion, the suggestion is to evolve the current 3-R strategy to a broader 7-Rs strategy by applying ITS. These strategies will need to be developed, assessed and refined in an iterative manner. Streamlining of these actions to focus upon the areas with greatest potential for success would serve to optimise the use of available resources. We need a pragmatic approach, integrating common sense, "fit for purpose" principles (proportionality) and precaution. The knowledge and the tools should be applied in a context-dependent manner to meet the challenges we currently face.

5. Capacity building

At the level of ITS multi-stakeholder collaboration needs to be established to further develop, refine and implement ITS in the next decade. The principle barrier to acceptance of alternative methods is the current lack of understanding both at the scientific, regulatory and political level. Therefore capacity building is essential. Capacity building cannot be implemented without collaboration. Collaboration cannot be established without communication and communication cannot take place without having some form of common language. Speaking the same "language" is essential. Communication requires transparency. As such, the

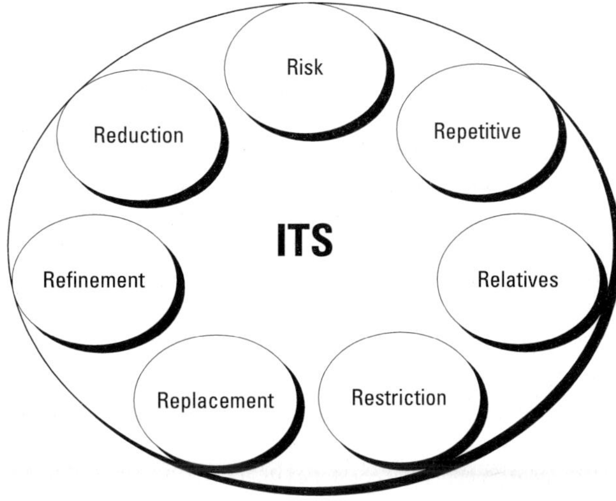

Figure 11.21. The 7Rs of intelligent testing strategies.

whole process can only start when stakeholders work together, communicate in a simple and transparent manner, understand the scientific and regulatory contexts of their collaboration and are willing to provide robust descriptions of the assumptions, the input data, the models and their guidance, as well as their output.

6. Realistic expectations
The current expectations to replace animal tests seem to be running ahead of scientific reality [18,42]. Experience from the past has shown that it requires a considerable amount of time to introduce and implement new concepts, new methods and new tools at the international level. But it is worth the effort! The development and implementation of intelligent testing strategies will not be a trivial exercise, either from a scientific or organizational perspective.

7. Long-term commitment
The development, assessment and implementation of ITS in the regulatory context are important areas for further work. It will need considerable and co-ordinated input from the industry, Commission, regulatory authorities, academia and other stakeholders. It will require a long-term commitment by all stakeholders involved in the process. Fortunately, the relevance of ITS is well understood. The Commission has released a considerable amount of money for an EU Framework Project on ITS and, recently, a European Partnership has been agreed with industry [110].

11.5 FURTHER READING

1. European Chemicals Bureau. 2007. Draft Technical Guidance Document to industry on the information requirements for REACH. Part 1: General issues (327 pp), Part 2. Endpoint specific issues (413 pp), Part 3. Endpoint specific issues (347 pp), and Part 4: Appendices (44 pp). European Commission, Joint Research Centre, Ispra, Italy (http://ecb.jrc.it/REACH/).
2 Worth AP, Balls M, eds. 2002. Alternative (non-animal) methods for chemical testing: current status and future prospects. A report prepared by ECVAM and the ECVAM Working Group on Chemicals. *ATLA* 30 (Suppl 1):1-125.
3 Eskes C, Zuang V, eds. 2005. Alternative (non-animal) methods for cosmetics testing: current status and future prospects. A report prepared in the context of the 7th amendment to the cosmetics directive for establishing the timetable for phasing out animal testing. *ATLA* 33 (Suppl 1):1-228.
4 European Centre for Ecotoxicology and Toxicology of Chemicals. 2004. Targeted risk assessment. Technical Report No. 93. ECETOC, Brussels, Belgium.
5 Cronin MTD and Livingstone DJ. 2004. *Predicting Chemical Toxicity and Fate*. CRC Press, Boca Raton, Florida, USA.

REFERENCES

1 Bradbury S, Feijtel T, Van Leeuwen K. 2004. Meeting the scientific needs of ecological risk assessment in a regulatory context. *Environ Sci Technol* 38/23: 463a-470a.
2 Organization for Economic Co-operation and Development. 1992. Report of the OECD workshop on the extrapolation of laboratory aquatic toxicity data to the real Environment. OECD Environment Monographs 59. OECD, Paris, France.
3 Van Leeuwen CJ, Bro-Rasmussen F, Feijtel TCJ, Arndt R, Bussian BM, Calamari D, Glynn P, Grandy NJ, Hansen B, Van Hemmen JJ, Hurst P, King N, Koch R, Müller M, Solbé JF, Speijers GAB, Vermeire T. 1996. Risk assessment and management of new and existing chemicals. *Environ Toxicol Pharmacol* 2:243-299.
4 Worth AP. 2004. The tiered approach to toxicity assessment based on the integrated use of alternative (non-animal) tests. In Cronin MTD, Livingstone D, eds, *Predicting Chemical Toxicity and Fate*. CRC Press, Boca Raton, FL, USA, pp 389-410.
5 Organization for Economic Co-operation and Development. OECD Existing Chemicals Programme. OECD, Paris, France (http://www.oecd.org/).
6 Organization for Economic Co-operation and Development. 2006. Guidelines for the testing of chemicals. OECD, Paris, France.
7 Hansen BG, van Haelst AG, van Leeuwen K, van der Zandt P. 1999. Priority setting for existing chemicals: the European Union risk assessment method. *Environ Toxicol Chem* 8:772-779.
8 Bodar CWM, Berthault F, De Bruijn JHM, Van Leeuwen CJ, Pronk MEJ, Vermeire TG. 2003. Evaluation of EU risk assessments existing chemicals (EC Regulation 793/93). *Chemosphere* 53:1039-1047.
9 Bodar WM, Pronk MEJ, Sijm DTHM. 2005. The European Union risk assessment on zinc and zinc compounds: the process and the facts. *Integrated Environ Assessm Managem* 1:301-319.
10 Pedersen F, De Bruijn J, Munn S, Van Leeuwen K. 2003. Assessment of additional testing needs under REACH. Effects of QSARs, risk based testing and voluntary industry initiatives. Report EUR 20863 EN, European

Commission, Joint Research Centre, Ispra, Italy.

11 Haigh N, Baillie A. 1992. Final Report on Chemical Control in the European Community in the 1990s. Institute for European Environmental Policy, London, UK.

12 Allanou R, Hansen BG, Van Der Bilt Y, 1999. Public availability of data on EU high production volume chemicals. Report EUR 18996 EN. European Commission, Joint Research Centre, Ispra, Italy.

13 US Environmental Protection Agency. 1998. Chemical Hazard Data Availability Study. Government Printing Office, Washington, DC.

14 Commission of the European Communities. 2005. Report from the Commission to the Council and the European Parliament: Fourth report on the statistics on the number of animals used for experimental and other scientific purposes in the Member States of the European Union. COM(2005) 7 final.

15 Van der Jagt K, Munn S, Tørsløv J, De Bruijn J. 2004. Alternative approaches can reduce the use of test animals under REACH. Addendum to the report "Assessment of additional testing needs under REACH. Effects of (Q)SARs, risk based testing and voluntary industry initiatives". Report EUR 21405. European Commission, Joint Research Centre, Ispra, Italy.

16 Commission of the European Communities. 2003. Proposal for a Regulation of the European Parliament and of the Council concerning the Registration, Evaluation, Authorisation and Restriction of Chemicals (REACH), establishing a European Chemicals Agency and amending Directive 1999/45/EC and Regulation (EC) {on Persistent Organic Pollutants} / Proposal for a Directive of the European Parliament and of the Council amending Council Directive 67/548/EEC in order to adapt it to Regulation (EC) of the European Parliament and of the Council concerning the registration, evaluation, authorisation and restriction of chemicals. Brussels, 29 October 2003, Brussels, Belgium.

17 Commission of the European Communities. 2006. Regulation (EC) No 1907/2006 of the European Parliament and of the Council of 18 December 2006 concerning the Registration, Evaluation, Authorisation and Restriction of Chemicals (REACH), establishing a European Chemicals Agency, amending Directive 1999/45/EC and repealing Council Regulation (EEC) No 793/93 and Commission Regulation (EC) No 1488/94 as well as Council Directive 76/769/EEC and Commission Directives 91/155/EEC, 93/67/EEC, 93/105/EC and 2000/21/EC. Off J Eur Union, L 396/1 of 30.12.2006.

18 Scientific Committee on Cosmetic Products and Non-Food products intended for Consumers. 2004. The report for establishing the timetable for phasing out animal testing for the purpose of the cosmetics directive issued by ECVAM (30/4/2004). SCCNFP/0834/04. Brussels, Belgium.

19 Combes R, Barratt M, Balls M. 2003. An overall strategy for the testing of chemicals for human hazard and risk assessment under the EU REACH System. *ATLA* 31:7-19.

20 European Centre for Ecotoxicology and Toxicology of Chemicals. 2004. Targeted risk assessment. Technical Report No. 93. ECETOC, Brussels, Belgium.

21 European Chemical Industry Council, European Association for Bioindustries and European Commission's DG Research. 2004. A European Technology Platform for Sustainable Chemistry. Brussels, Belgium.

22 Health Council of the Netherlands. 2001. Toxicity testing: a more efficient approach. Publication no. 2001/24E. The Hague, the Netherlands.

23 Hofer T, Gerner I, Gundert-Remy U, Liebsch M, Schulte A, Spielmann H, Vogel R, Wettig K. 2004. Animal testing and alternative approaches for the human health risk assessment under the proposed new European chemicals regulation. *Arch Toxicol* 78:549-564.

24 Worth AP, Fentem JH, Balls M, Botham PA, Curren RD, Earl LK, Esdaile DJ, Liebsch M. 1998. An evaluation of the proposed OECD testing strategy for skin corrosion. *ATLA* 26:709-720.

25 European Chemicals Bureau. 2005. Scoping study on the development of a technical guidance document on information requirements on intrinsic properties of substances (RIP 3.3-1). Report prepared by CEFIC, DK-EPA, Environmental Agency of Wales and England, ECETOC, INERIS, KemI and TNO. European Commission, Joint Research Centre, Ispra, Italy.

26 Organization for Economic Co-operation and Development. 2002. Manual for investigation of HPV Chemicals. Section 3.2. Guidance on the development and use of chemical categories. OECD, Paris, France.

27 Hanway RH, Evans PF. 2000. Read-across of toxicological data in the notification of new chemicals. *Toxicology Letters* 116, Suppl 1, 61.

28 US Environmental Protection Agency. 2002. TSCA New Chemicals Program (NCP). (http://www.epa.gov/oppt/newchems/pubs/chemcat.htm). USEPA, Office of Pollution Prevention and Toxics, Washington, DC.

29 Health Canada. 2005. A proposed integrated framework for the health-related components of categorization of the Domestic Substances List under CEPA 1999. Health Canada, Ottawa, Canada.

30 Environment Canada. 2003. Guidance manual for the

categorization of organic and inorganic substances on Canada's Domestic Substances List. Environment Canada, Ottawa, Canada.

31 US Environmental Protection Agency. 2004. Status and future directions of the High Production Volume Challenge programme. US Environmental Protection Agency. Office of Pollution Prevention and Toxics, Washington, DC.

32 Cronin MTD, Livingstone DJ. 2004. *Predicting Chemical Toxicity and Fate*. CRC Press, Boca Raton, Florida.

33 Bradbury SP, Russom CL, Ankley GT, Schultz TW, Walker JD. 2003. Overview of data and conceptual approaches for derivation of quantitative structure-activity relationships for ecological effects of organic chemicals. *Environ Toxicol Chem* 22:1789-1798.

34 Jaworska JS, Comber M, Auer C, van Leeuwen CJ. 2003. Summary of a workshop on regulatory acceptance of (Q)SARs for human health and environmental endpoints. *Environ Health Persp* 111/10:1358-1360.

35 US Environmental Protection Agency. 2004. A Framework for a Computational Toxicology Research Program in ORD. Report No. EPA/600/R 03/065. US Government Printing Office, Washington, DC.

36 Christensen FM, de Bruijn JHM, Hansen BG, Munn SJ, Sokull-Kluttgen B, Pedersen F. 2003. Assessment tools under the new European Union chemicals policy. *GMI* 41:5-19.

37 Organization for Economic Co-operation and Development. 1992. Report of the OECD Workshop on Quantitative Structure-Activity Relationships in Aquatic Effects Assessment. OECD Environment Monographs 58. OECD, Paris, France.

38 Organization for Economic Co-operation and Development. 1994. US EPA/EC Joint Project on the Evaluation of (Quantitative) Structure-Activity Relationships. OECD Environment Monograph No 88. OECD, Paris, France.

39 European Centre for Ecotoxicology and Toxicology of Chemicals. 2003. (Q)SARs: evaluation of the commercially available software for human health and environmental endpoints with respect to chemical management applications. ECETOC Technical Report No. 89. ECETOC, Brussels, Belgium.

40 Organization for Economic Co-operation and Development. 2006. Report on the regulatory uses and applications in OECD member countries of (quantitative) structure-activity relationship [(Q)SAR] models in the assessment of new and existing chemicals. Environmental Health and Safety Publications. Series on Testing and Assessment 58. OECD, Paris, France.

41 Commission of the European Communities. 2003. Technical Guidance Document in support of Commission Directive 93/67/EEC on risk assessment for new notified substances, Commission Regulation (EC) No 1488/94 on risk assessment for existing substances and Directive 98/8/EC of the European Parliament and of the Council concerning the placing of biocidal products on the market. Joint Research Centre, European Chemicals Bureau, Ispra, Italy.

42 Scientific Committee on Toxicity, Ecotoxicity and the Environment. 2004. Opinion of the Scientific Committee on Toxicity, Ecotoxicity and the Environment (CSTEE) on the BUAV-ECEAE report on "the way forward-Action to end animal toxicity testing". Adopted by the CSTEE during the 41st plenary meeting of 8 January 2004. Brussels, Belgium.

43 Nordic Council of Ministers. 2005. Threshold of Toxicological Concern (TTC). TemaNord 2005:559. Copenhagen, Denmark.

44 Kroes R, Renwick AG, Cheeseman M, Kleiner J, Mangelsdorf I, Piersma A, Schilter B, Schlatter J, van Schothorst F, Vos JG, Wurtzen G. 2004. Structure-based thresholds of toxicological concern (TTC): guidance for application to substances present at low levels in the diet. *Food Chem Toxicol* 42:65-83.

45 Kroes R, Kozianowski G. 2002. Threshold of toxicological concern (TTC) in food safety assessment. *Toxicol Letters* 127:43-46.

46 De Wolf W, Siebel-Sauer A, Lecloux A, Koch V, Holt M, Feijtel T, Comber M, Boeije G. 2004. *Environ Toxicol Chem* 24:479-485.

47 Barlow S. 2005. Thresholds of toxicological concern (TTC). A tool for assessing substances of unknown toxicity present at low levels in the diet. ILSI Europe Concise Monograph Series. International Life Sciences Institute, Brussels, Belgium.

48 US Food and Drug Administration. 1995. Food additives: thresholds of regulation for substances used in food-contact articles. Final Rule. *Fed Register* 60:36582-36596.

49 Gold LS, Sawyer CB, Magaw R, Backman GM, de Veciana M, Levinson R, Hooper NK, Havender WR, Bernstein L, Peto R, Pike M, Ames BN. 1984. A carcinogenesis potency database of the standardised results of animal bioassays. *Environ Health Perspect* 58:9-319.

50 Rulis AM, Hatan DG. 1985. FDA's priority-based assessment of food additives II. General toxicity parameters. *Regul Toxicol Pharmacol* 5:152-174.

51 Munro IC. 1990. Safety assessment procedures for direct food additives, an overview. *Regul Toxicol Pharmacol* 12:2-12.

52 Munro IC, Ford RA, Kennepohl E, Sprenger JG. 1996. Correlation of structural class with no-observed-effect levels: a proposal for establishing a threshold of concern. *Food Chem Toxicol* 34:829-867.

53 Kroes R, Galli C, Munro I, Schilter B, Tran L-A, Walker R, Wurtzen G. 2000. Thresholds of toxicological concern for chemical substances present in the diet: a practical tool for assessing the need for toxicity testing. *Food Chem Toxicol* 38:255-312

54 Cramer GM, Ford RA, Hall RL. 1978. Estimation of toxic hazard – a decision tree approach. *Food Cosmet Toxicol* 16:255-276.

55 Verhaar HJM, Van Leeuwen CJ, Hermens JLM. 1992. Classifying environmental pollutants. 1: Structure-activity relationships for prediction of aquatic toxicity. *Chemosphere* 25:471-491.

56 Escher BE, Hermens JLM. 2002. Modes of action in ecotoxicology: their role in body burdens, species sensitivity, QSARs, and mixture effects. *Environ Sci Technol* 36:4201-4217.

57 Worth AP, Balls M, eds 2002. Alternative (non-animal) methods for chemical testing: current status and future prospects. A report prepared by ECVAM and the ECVAM Working Group on Chemicals. *ATLA* 30 (Suppl 1):1-125.

58 Eskes C, Zuang V, eds 2005. Alternative (non-animal) methods for cosmetics testing: current status and future prospects. A report prepared in the context of the 7th amendment to the cosmetics directive for establishing the timetable for phasing out animal testing. *ATLA* 33 (Suppl 1):1-228.

59 Fentem JH. 2006. Working together to respond to the challenges of EU policy to replace animal testing. *ATLA* 34:11-18.

60 Fentem J, Chamberlain M, Sangster B. 2004. The feasibility of replacing animal testing for assessing consumer safety: a suggested future direction. *ATLA* 32: 617-623.

61 Combes RD, Gaunt I, Balls M. 2004. A scientific and animal welfare assessment of the OECD health effects test guidelines for the safety testing of chemicals under the European Union REACH system. *ATLA* 32:163-208.

62 Jaram S, Riego Sintes JM, Halder M, Baraibar Fentanes J, Sokull-Klüttgen B, Hutchinson TM. 2005. A strategy to reduce the use of fish in acute ecotoxicity testing of new chemical substances notified in the European Union. *Regul Toxicol Pharmacol* 42:218-224.

63 Van Leeuwen CJ, Van Der Zandt PTJ, Aldenberg T, Verhaar HJM, Hermens JLM. 1992. Application of QSARs, extrapolation and equilibrium partitioning in aquatic assessment: I. Narcotic industrial pollutants. *Environ Toxicol Chem* 11:267-282.

64 European Centre for Ecotoxicology and Toxicology of Chemicals. 2005. Alternative testing approaches in environmental safety assessment. ECETOC Technical Report No. 97. ECETOC, Brussels, Belgium.

65 Barratt MD. 2003. QSAR, read-across and REACH. *ATLA* 31:463-465.

66 Slooff W, Canton JH, Hermens JLM. 1983. Comparison of the susceptibility of 22 freshwater species to 15 chemical compounds. I. (Sub)acute toxicity tests. *Aquatic Toxicol* 4:113-128.

67 Slooff W, Canton JH.1983. Comparison of the susceptibility of 11 freshwater species to 8 chemical compounds. II. (Sub)acute toxicity tests. *Aquatic Toxicol* 4:113-128.

68 Posthuma L, Suter II GW, Traas TP, eds. 2002. *Species Sensitivity Distributions in Ecotoxicology.* Lewis Publ. Boca Raton, Florida, USA.

69 McKim JM, Bradbury SP, Niemi GJ. 1987. Fish acute toxicity syndromes and their use in the QSAR approach to hazard assessment. *Environ Health Persp* 71:171-186.

70 Hermens J. 1989. Quantitative structure-activity relationships of environmental pollutants. In O. Hutzinger, ed, *Handbook of Environmental Chemistry*, Vol. 2E. Springer, Berlin, Germany. pp. 111-162.

71 Verhaar HJM, Van Leeuwen CJ, Bol J, Hermens JLM. 1994. Application of QSARs in risk management of existing chemicals. *SAR QSAR Environ Res* 2:39-58.

72 Bol J, Verhaar, HJM, Van Leeuwen CJ, Hermens JLM. 1993. Predictions of the aquatic toxicity of high production volume chemicals. Ministry of Housing, Spatial Planning and Environment, The Hague, The Netherlands.

73 Aldenberg, T., W. Slob. 1993. Confidence limits for hazardous concentrations based on logistically distributed NOEC toxicity data. *Ecotoxicol Environ Saf* 25:48-63.

74 Könemann H. 1981. Quantitative structure-actvity relationships in fish toxicity studies. 1. Relationship for 50 industrial pollutants. *Toxicology* 19:209-221.

75 Veith GD, Call DJ, Brooke LT. 1983. Structure-toxicity relationships for the fathead minnow, *Pime-phales promelas*: Narcotic industrial chemicals. *Can J Fish Aquat Sci* 40:743-748.

76 Hermens J, Broekhuyzen E, Canton H, Wegman R. 1985. Quantitative structure-activity relationships and mixture toxicity studies of alcohols and chlorohydrocarbons: effects on growth of Daphnia magna. *Aquat Toxicol* 6:209-217.

77 McCarty LS. 1987. Relationship between toxicity and bioconcentration for some organic chemicals. I. Examination of the relationship. In K.L.E. Kaiser, ed,

QSAR in Environmental Toxicology-II. Reidel, Dordrecht, The Netherlands, pp. 207-220.

78 McCarty LS, Mackay D, Smith AD, Ozburn GW, Dixon DG. 1991. Interpreting aquatic toxicity QSARs: The significance of toxicant body residues at the pharmacological endpoint. *Sci Total Environ* 109/110:515-525.

79 Van Hoogen G., Opperhuizen A. 1988. Toxicokinetics of chlorobenzenes in fish. *Environ Toxicol Chem* 7:213-219.

80 Van Der Kooy LA, Van De Meent D, Van Leeuwen CJ, Bruggeman. WA 1991. Deriving quality criteria for water and sediment from the results of aquatic toxicity tests and product standards: application of the equilibrium partitioning theory. *Water Res* 25:697-705.

81 Hutchinson TM, Barrett S, Buzby M, Constable D, Hartmann A, Hayes E, Huggett D, Laenge R, Lillicrapp AD, Staub JO, Thompson RS. A strategy to reduce the use the numbers of fish in acute ecotoxicity testing of pharmaceuticals. *Environ Toxicol Chem* 22:3031-3036.

82 Von der Ohe PC, Kühne R, Ebert R, Altenburger R, Liess M, Schüürmann G. 2005. Structural alerts - a new classification model to discriminate excess toxicity from narcotic effect levels of organic compounds in the acute daphnid assay. *Chem Res Toxicol* 18:536-555.

83 Walker JD, Gerner I, Hulzebos E, Schlegel K. 2004. (Q)SARs for predicting skin irritation and corrosion: mechanisms, transparency and applicability predictions. *QSAR Comb Sci* 23:721-725.

84 Gerner I, Schlegel K, Walker JD, Hulzebos E. 2004. Use of physicochemical property limits to develop rules for identifying chemical substances with no skin irritation or corrosion potential. *QSAR Comb Sci* 23:726-733.

85 Patlewicz G, Rodford R, Walker JD. 2003. Quantitative structure-activity relationships for predicting skin and eye irritation. *Environ Toxicol Chem* 22:1862-1869.

86 Hulzebos E, Walker JD, Gerner I, Schlegel K. 2005. Use of structural alerts to develop rules for identifying chemical substances with skin irritation or corrosion potential. *QSAR Comb Sci* 24:332-342.

87 Walker JD, Gerner I, Hulzebos E, Schlegel K. 2005. The skin irritation corrosion rules estimation tool (SICRET). *QSAR Comb Sci* 24:378-384.

88 Smith CK, Hotchkiss SAM. 2001. *Allergic contact dermatitis: chemical and metabolic mechanisms*. Taylor & Francis Ltd, London, UK.

89 Dearman RJ, Kimber I. 2003. Factors influencing the induction phase of skin sensitization. *Am J Contact Dermatitis* 14:188-194.

90 Kimber I, Dearman RJ. 2003. What makes a chemical an allergen? *Ann Allergy Asthma Immunol* 90:28-31.

91 Smith Pease CK, Basketter DA, Patlewicz GY. 2003. Contact allergy: the role of skin chemistry and metabolism. *Clin Exp Derm* 28:177-83.

92 Pease Smith CK. 2003. From xenobiotic chemistry and metabolism to better prediction and risk assessment of skin allergy. *Toxicology* 192:1-22.

93 Lepoittevin J-P, Basketter DA, Goossens A, Karlberg A-T. (eds.) 1998. *Allergic Contact Dermatitis: The Molecular Basis*. Springer-Verlag, Berlin Germany.

94 Kimber I, Basketter DA, Butler M, Gamer A, Garrigue J-L, Gerberick GF, Newsome C, Steiling W, Vohr H-W. 2003. Classification of contact allergens according to potency: proposals. *Food Chem Toxicol* 41:1799-1809.

95 Aptula AO, Patlewicz G, Roberts DW. 2005. Skin sensitization: reaction mechanistic applicability domains for structure-activity relationships. *Chem Res Toxicol* 18:1420-1426.

96 Gerberick GF, Vassallo JD, Bailey RE, Chaney JG, Morrall SW, Lepoittevin JP. 2004. Development of a peptide reactivity assay for screening contact allergens. *Toxicol Sci* 81:332-343.

97 Divkovic M, Pease CK, Gerberick GF, Basketter DA. 2005. Hapten-protein binding: from theory to practical application in the in vitro prediction of skin sensitization. *Contact Dermatitis* 53:189-200.

98 Aptula AO, Patlewicz G, Roberts DW, Schultz TW. 2006. Non-enzymatic glutathione reactivity and in vitro toxicity: A non-animal approach to skin sensitization. *Toxicol In Vitro* 20:239-247.

99 Ryan CA, Gerberick GF, Gildea LA, Hulette BC, Betts CJ, Cumberbatch M, Dearman RJ, Kimber I. 2005. Interactions of contact allergens with dendritic cells: opportunities and challenges for the development of novel approaches to hazard assessment. *Toxicol Sci* 88:4-11.

100 Aeby P, Wyss C, Beck H, Grien P, Scheffler H, Goebel C. 2004. Characterization of the sensitizing potential of chemicals by in vitro analysis of dendritic cell activation and skin penetration. *J Invest Dermatol* 122:1154-1164.

101 Straube F, Grenet O, Bruegger P, Ulrich P. 2005. Contact allergens and irritants show discrete differences in the activation of human monocyte-derived dendritic cells: consequences for in vitro detection of contact allergens. *Arch Toxicol* 79:37-46.

102 Toebak MJ, Pohlmann PR, Sampat-Sardjoepersad SC, von Blomberg BME, Bruynzeel DP, Scheper RJ, Rustemeyer T, Gibbs S. 2006. CXCL8 secretion by dendritic cells predicts contact allergens from irritants. *Toxicol In Vitro* 20:117-124.

103 Vandebriel RJ, Van Och FMM, van Loveren H. 2005. In vitro assessment of sensitizing activity of low molecular

weight compounds. *Toxicol Appl Pharmacol* 207 (Suppl 2):142-148.

104 Kimber I, Dearman R, Betts CJ, Gerberick GF, Ryan CA, Kern PS, Patlewicz GY, Basketter DA. 2006. The local lymph node assay and skin sensitization: a cut-down screen to reduce animal requirements? *Contact Dermatitis* 54:181-185.

105 Gerberick GF, Ryan CA, Kern PS, Schlatter H, Dearman RJ, Kimber I, Patlewicz G, Basketter DA. 2005. Compilation of historical local lymph node assay data for the evaluation of skin sensitization alternatives. *Contact Dermatitis* 16:157-202.

106 Jefferies D, Aspinall L, Madrigal A-M, Safford B, Clapp C, Chamberlain M, Basketter DA. 2005. A Bayesian Network model to predict hazard potency for skin sensitization. Poster presented at Society of Toxicology. New Orleans, USA.

107 Jones J. 2006. National Pesticide Program. A new toxicology testing paradigm: meeting common needs. Presented to the National Research Council Committee on toxicity testing and assessment of environmental agents on January 19. Irvine, CA. Office of Pesticide Programs. US Environmental Protection Agency, Washington, DC.

108 Tyler CR, Filby A, Iguchi T, Kramer V, Larsson J, van Aggelen G, van Leeuwen K, Viant M and Tillitt D. 2006. Molecular biology and risk assessment: evaluation of the potential roles of genomics in regulatory ecotoxicology. Proceedings of a SETAC workshop. Pellston, MI, USA (in press).

109 Organization for Economic Co-operation and Development. 1996. Final report of the OECD workshop on harmonization of validation and acceptance criteria for alternative toxicological test methods. OECD, Paris, France.

110 European Commission. 2005. Reducing animal testing: Commission agrees partnership with industry. Europa rapid press releases IP/05/1375 of 07/11/2005.

12. THE MANAGEMENT OF INDUSTRIAL CHEMICALS IN THE EU

C.J. VAN LEEUWEN, B.G. HANSEN AND J.H.M. DE BRUIJN

12.1 INTRODUCTION

Chemicals are used to make virtually every man-made product and play an important role in the everyday life of people around the world. The chemical industry is the third largest industrial sector in the world. It is also a major economic force. Worldwide, it employs some ten million people and generates billions of euros in shareholder value and tax revenue for governments. Eighty percent of the production takes place in 16 countries, primarily in the OECD member countries. According to the OECD, the chemical industry accounts for 7% of global income and 9% of international trade [1]. The chemical industry is very important to the EU economy. According to the European Chemical Industry Council (http://www.cefic.be/factsandfigures/), the chemicals sector is the third largest manufacturing industry in the EU, encompassing 31,000 companies employing 1.9 million people. Internationally, the EU is the leading chemicals-producing area. In 2004, it accounted for 33% (€ 580 billion) of global sales (€ 1736 billion). The use and releases of chemicals increased enormously in the 20th century. It has become apparent that this increase was not without "cost" to health and the environment, particularly in the industrialized countries. This was clearly illustrated by Rachel Carson in the early 1960's, whose book Silent Spring [2] described the disastrous effects of the widescale use of pesticides on fish, birds and ecosystems.

Table 12.1. Regulation of some categories of chemicals in the EU.

New chemicals	Directive 67/548/EEC
Existing chemicals	Regulation (EEC) 793/93
Plant protection products	Directive 91/414/EEC
Biocides	Directive 98/8/EC
Veterinary drugs	Directive 2004/28/EC
Human drugs	Directive 2004/27/EC
Feed additives	Directive 70/524/EEC
Food additives	Directive 89/107/EEC
Cosmetics	Directive 2003/15/EC

The need to establish legally binding frameworks for the control of chemicals was soon recognized and started in the 1960's [3-5]. Various categories of substances can be identified and for each of them separate legislation has been developed in the European Union. Examples of some categories of substances are given in Table 12.1.

In 1967 the European Community adopted Council Directive 67/548/EEC on the classification, packaging and labelling of Dangerous Substances [6]. In the 60's the focus was on the hazards of chemicals, i.e., the inherent or intrinsic properties of chemicals having the potential to cause adverse effects. In subsequent years the Community's chemicals control legislation enlarged considerably, with sectoral legislation, mainly to combat water and air pollution and primarily aimed at the protection of human health (see also Chapter 1). Gradually the legislation governed different uses of chemicals (Table 12.1) and was aimed at preventing or reducing emissions of chemicals into the environment and protecting workers and consumers from exposure to chemicals. Faced with this exposure-driven management of chemicals an important paradigm shift took place: from hazard assessment to risk assessment. This shift from *intrinsic properties of chemicals* to *risks* of chemicals triggered the development of exposure assessment methodologies and risk assessment models.

It was also increasingly recognized that legislation to control chemicals should not only protect human health but also seek to protect the environment (both point source and diffuse pollution), taking into account the entire life cycle of a chemical. In the early 90's it was recognized that the allocation of resources to manage industrial chemicals was not "exposure-driven". It was not efficient as the main risks of industrial chemicals were caused not by new chemicals but by existing chemicals [3]. New legislation was developed as a result, focusing on the management of existing chemical substances.

Before the introduction of REACH (Registration, Evaluation, Authorization and restriction of Chemicals), there were four main instruments for the management of industrial chemicals in the EU. These were Directive 92/32/EEC (the 7th amendment of Directive 67/548/EEC), Directive 1999/45/EC on the classification, packaging and labelling of dangerous preparations, Regulation (EEC) 793/93 on the evaluation and control

C.J. van Leeuwen and T.G. Vermeire (eds.), Risk Assessment of Chemicals: An Introduction, 511–551.

Box 12.1. Core tools under REACH

- The Chemical Safety Assessment (CSA) is the tool used to **determine** which risk management measures and operational conditions are necessary to protect human health and the environment
- The Chemical Safety Report (CSR) is the tool used to **record/document** these measures and conditions
- The Safety Data Sheet (SDS) is the tool used to **communicate** these risk management measures. The measures need to be implemented by the downstream users. In this way risks are adequatly controlled

of existing chemical substances and Directive 76/769/EEC relating to restrictions on the marketing and use of certain chemical substances [7-10]. All these instruments were based on Article 95 of the new EU Treaty and were therefore established with the joint aims of preserving the internal market and assuring a high level of protection of humans and the environment. REACH will cover these and other instruments.

This chapter - Chapter 12 - is limited to the assessment and management of industrial chemicals in the EU (REACH). We will describe why REACH was developed (Section 12.1) and will introduce the general aspects of the REACH legislation including the data requirements (Section 12.2). In Sections 12.3 and 12.4 we will focus on the core tools for the implementation of REACH (Box 12.1), i.e., the chemical safety assessment (CSA), the chemical safety report (CSR) and the safety data sheets (SDSs).

In Section 12.5 a short overview will be presented of the REACH implementation projects (RIPs). REACH will have a substantial impact on the different stakeholders involved. A detailed overview of the roles and responsibilities of the different stakeholders involved in the REACH process is presented in Annex 12.1 at the end of this chapter. Selected references for further reading are provided in Section 12.6.

12.1.1 Background

At the United Nations Conference on Environment and Development (UNCED) in 1992 in Rio de Janeiro [11], agreement was reached on an action plan to reach sustainable development in a number of environmental policy areas. In this context, sustainable development was defined in the Brundtland report of 1987 [12] as "development, which meets the needs of the present without compromising the abilities of the future generations to meet their own needs". In Chapter 19 of UNCED's Agenda 21 on Chemicals, the need for closer cooperation in six key areas was identified to achieve sustainable use of chemicals worldwide (See Chapter 1, Table 1.2). On 31 May 2000 the Ministers of the Environment, gathered at the first Global Ministerial Environment Forum, stated that "we have at our disposal the human and material resources to achieve sustainable development, not as an abstract concept but as a concrete reality". Our efforts "must be linked to the development of cleaner and more resource efficient technologies for a Life Cycle Economy". In the 5th Environment Action Programme (1992-1999 "Towards sustainability") the EU aligned many of its chemicals management strategies to meet the challenges posed by UNCED [11].

The old EU legislative framework for industrial chemicals was an assembly of many different Directives and Regulations which developed historically. For example, there were different rules for "existing" and "new" chemicals. This distinction between so-called "existing" and "new" chemicals was based on a cut-off date of 1981. All chemicals that were put on the market before 1981 were called "existing" chemicals. In 1981, these numbered 100,106 different substances. Chemicals introduced onto the market after 1981 (about 3000 until 2007) were termed "new" chemicals [4]. While new chemicals had to be tested before they are placed on the market, there were no such provisions for "existing" chemicals. Thus, although some information exists on the properties and uses of existing substances, there was generally not sufficient information publicly available in order to assess and control these substances effectively. Let us have a look at the problems that were identified [13-19].

Lessons learned: the main problems

The main problems were related to the lack of progress in the area of existing chemicals and, in particular, the High Production Volume Chemicals (HPVCs), i.e., chemicals produced or imported in the EU in quantities of 1000 tonnes or more per year. For most (99%) of these chemicals, we do not have enough information about their effects and uses, and how they need to be handled to be safe. Another problem was the management of new chemicals. This work placed a relatively large burden on the Commission, Member States and Industry compared with existing chemicals. The main problems were [13-19]:

- Lack of progress. The European Inventory of Existing

Commercial Chemical Substances (EINECS) lists more than 100,000 chemicals. The number of existing substances marketed in volumes above 1 tonne is estimated at 30,000. The EINECS is a list of all chemicals either separately or as components in preparations supplied to a person in an EC Member State at any time between 1 January 1971 and 18 September 1981. Of these 30,000 existing chemicals marketed in volumes of 1 tonne or more, there are about 2750 HPVCs. Since the programme started in 1993, Member State rapporteurs completed the first draft risk assessment reports on 132 out of a total of 141 priority substances until 2007. For 79 of these 130 priority substances the scientific and technical discussions have been finalized and the conclusions agreed: 66 substances need risk reduction measures, 2 substances need further information and for 11 substances there is no need for further information and/or testing or for additional risk reduction measures (http://ecb.jrc.it/). The results of the whole evaluation have been published for 28 substances.

- The current knowledge on use, fate and (eco)toxicological properties remains insufficient for an adequate analysis of the risks of HPVCs other than priority substances and existing chemicals, in general (Figure 12.1). Concerns have been expressed that chemicals are marketed that may pose an immediate or future threat to human health and the environment.
- The allocation of responsibilities *(burden of proof)* is not appropriate: public authorities are responsible for evaluating safety and demonstrating the risk of substances rather than the businesses that manufacture, import or use the substances.
- Decisions on further testing of priority substances can only be taken via a lengthy committee procedure and can only be requested from industry after the authorities have proven that a substance may present a risk. Again, the burden of determining whether further information is needed rests on Member State authorities and not on the businesses.
- Information on the uses of substances is difficult to obtain and information about exposure arising from downstream uses is generally scarce [17]. The legislation required only the manufacturers and importers of chemicals to provide information, but did not impose any such obligations on downstream users (industrial users and formulators). As a result information on exposure and risks due to the various uses of chemicals is incomplete and, consequently, not all risks can be adequately assessed and managed.

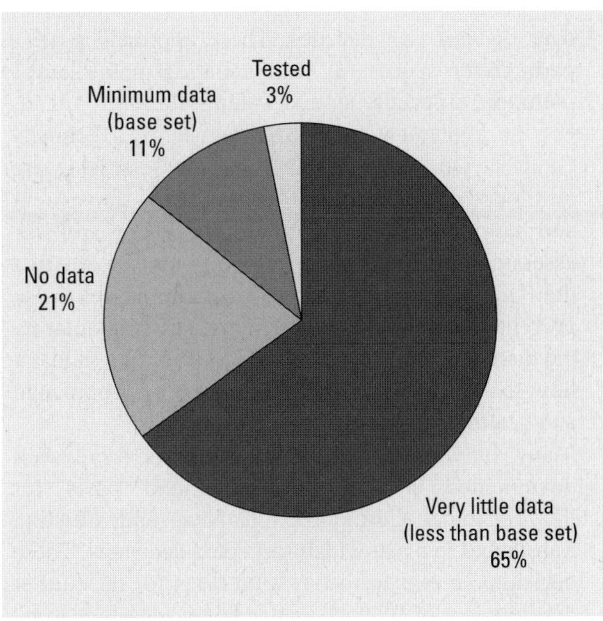

Figure 12.1. Current knowledge about HPVCs is very limited. Less than base-set information is available for 86% of these chemicals [19].

- New chemicals have to be notified and tested from volumes of as little as 10 kg per year. This has been a barrier to innovation within the EU chemicals industry by favouring the use of existing substances over new ones. The management of new notified substances was very resource intensive compared to existing chemicals. Currently, new chemicals represent only about 1% of the total volume of all industrial substances on the market.

Lessons learned: the main successes

Although the political discussion of the last decade has focused on the main problems with the existing legislation, it is important to recognize that positive lessons have also been learned during this implementation period. Some of the main successes of the approaches of the past were:

- The general approach to combating chemical pollution. Chemicals control in the EU has been implemented by: (1) exposure controls through technical measures or best available use practices, allowing for considerable national or local flexibility, and (2) through chemical-specific controls. These two concepts have, often in a coherent and integrated fashion, contributed significantly to improvements in the quality of human health and the environment over the last few decades. These approaches were particularly successful in combating point-source pollution, especially air and water pollution.

- For the implementation of the legislation, Technical Guidance Documents (TGDs) on risk assessment were developed [13]. These TGDs provide a transparent and harmonized method for further testing and risk assessment. Perhaps even more important is the fact that this transparent risk assessment process also prevented certain dangerous chemicals from entering the market. Furthermore, the TGDs informed industry how the authorities would assess the environmental and health risks of their chemicals.
- Many of the risk assessment reports were well documented and provided a sound basis for the Technical Committee on New and Existing Substances to make widely-accepted decisions. These decisions in turn formed a solid basis for the further development of EU risk management measures (e.g., restrictions of marketing and use) where these where considered indispensable.
- Despite the criticism of the slow functioning of the Existing Substances Regulation, the priority setting activities have been relatively successful. It was concluded that for the vast majority of priority substances for which comprehensive risk assessment reports have been finalized, further information and/or risk reduction measures need to be taken [14-16]. Furthermore, in relation to priority setting, it was possible to identify the HERO substances (substances with a High Expected Regulatory Outcome) allowing for the cost-effective selection of (potentially) problematic chemicals.
- The enforcement of compliance for new chemicals was relatively easy as the information needs were clearly specified in the legislation, with flexible waiving and derogation possibilities.
- Substantial and reliable data was generated to provide the basis for assessing and determining appropriate risk management measures (RMMs) for new substances.
- Two points should be mentioned concerning further reduction of animal testing. Firstly, risk management decisions can render requirements for further testing unnecessary. Secondly, data collection for priority substances under the regulation and other international programmes has proven to be successful in terms of encouraging industry to search for and submit previously unknown data, thereby preventing unnecessary tests being performed.

12.1.2 Towards solutions

In Section 12.1.1 the problems and successes of the current EU legislative framework have been identified. In 1998, the Environment Ministers met at an informal meeting in Chester, UK, to discuss the Community approach to the management of industrial chemicals. They recognized the need for a review of the industrial chemicals legislation as it was not living up to the expectations originally set and was not meeting the new challenges. These observations led to formal Council Conclusions in June 1999 and a Commission White Paper [17] outlining a future chemicals policy in February 2001. The development of a future chemicals policy was aligned with other environment policies in the 6th Environment Action Programme, i.e., *sustainable development*. The White Paper outlined a strategy for obtaining sustainable use of chemicals within one generation, i.e., within 20 years.

According to the White Paper the EU chemicals policy must ensure *a high level of protection of human health and the environment* as enshrined in the Treaty for the present generation and future generations, while also ensuring the efficient functioning of the internal market and the competitiveness of the chemical industry. Fundamental to achieving these objectives is the *precautionary principle* [20]. Whenever reliable scientific evidence is available that a substance may have an adverse impact on human health and the environment but there is still scientific uncertainty about the precise nature or the magnitude of the potential damage, decision-making must be based on precaution in order to prevent damage to human health and the environment. Another important objective is to encourage the substitution of dangerous by less dangerous substances, where suitable alternatives are available. It is also essential to ensure the efficient functioning of the internal market and the competitiveness of the chemical industry. EU policy on chemicals should provide incentives for technical innovation and the development of safer chemicals. Ecological, economic and social aspects of development have to be taken into account in an integrated and balanced manner in order to reach the goal of sustainability.

The Commission presented a new strategy in the White Paper. The proposed system was called REACH: Registration, Evaluation and Authorization of Chemicals. The White Paper identified seven political objectives that need to be balanced within the overall framework of sustainable development, creating one *single regulatory system* for all substances (Box 12.2).

Box 12.2. Political objectives of REACH as presented in the White Paper

1. Protection of human health and the environment
2. Maintenance and enhancement of the competitiveness of the EU chemical industry
3. Prevention of fragmentation of the internal market
4. Increasing transparency
5. Integration with international efforts
6. Promotion of non-animal testing
7. Conformity with EU international obligations under the WTO

Box 12.3. Main features of REACH

- **Responsibility** for all manufacturers and importers
- **Registration** of substances produced/imported above 1 tonne/year
- **Evaluation** by the Agency and Member States
- **Authorization** for substances of very high concern
- **Restrictions** - the safety net
- **Agency** to manage the system

The strategy for a future chemicals policy as presented in the White Paper [17] in 2001 was laid down in a draft proposal for a regulation in 2003. This proposal was submitted to all interested parties by means of an internet consultation in May 2003. Both the White Paper and the development of the draft proposal for REACH were discussed in great detail at both technical and political levels [21,22] with all stakeholders. In this respect there was adequate stakeholder participation and communication provided (see Chapter 1). More than 6000 separate comments were received on the draft proposal for REACH published in May 2003. These comments were taken into account by the Commission in preparing its final proposal. The final Commission proposal was submitted to the European Parliament and the Council in October 2003 [23].

12.1.3 The further development of REACH

The Commission's proposal [23] and many impact assessment studies [24-28] were discussed extensively. The European Parliament completed its first reading in November 2005 and adopted what has become known as the Sacconi-Nassauer compromise on registration. The Commission welcomed the registration compromise and concluded that a good balance was achieved between the health, environment and competitiveness goals of REACH. Further discussions, well-managed by the UK presidency, led to approval in the Council of the European Union in December 2005. After further discussions about the registration requirements and the authorization process, REACH was finally adopted on 18 December 2006 [29]. With the publication of the REACH legislation in the Official Journal, the process is still not complete. On the contrary the work only begins with the publication of the legislation! It is clear that the real challenge is the implementation of REACH. It will require advances in science and technology, efficient communication between all the stakeholders, pragmatism and, last but not least, effective monitoring of the progress made.

12.1.4 Aim of the REACH legislation

The purpose of REACH is to ensure a high level of protection of human health and the environment, as well as the free movement of substances, on their own, and in preparations and articles, while enhancing competitiveness and innovation. REACH replaces the current ineffective and inefficient system of about 40 existing Community Directives and Regulations on chemicals with different rules for new and existing substances, by a single regulation with one consistent approach to controlling risks. In this regard REACH implements one of the 19 recommendations that were made by an international group of stakeholders [18]. This group concluded that "in order to increase the effectiveness of chemical control there is a need to improve the integration of the myriad of directives and regulations on chemicals and waste".

To adequately control the risks arising from the manufacture, import, placing on the market and use of substances, REACH reverses the burden of proof, shifting it from the authorities to industry, when it comes to gathering information on chemical substances and using this information to assess the safety of chemicals and select appropriate RMMs. To reflect this new approach, the Regulation states in Article 1 (3) that it is based on the principle "that it is for manufacturers, importers and downstream users to ensure that they manufacture, place on the market or use such substances that do not adversely affect human health or the environment.

Its provisions are underpinned by the precautionary principle". The main features of the REACH legislation are summarized in Box 12.3. These aspects will be discussed in more detail in Section 12.2.

In conclusion, REACH should be regarded as a concrete step in the context of the implementation plan adopted at the Johannesburg World Summit on sustainable development. The European Union is aiming to achieve that, by 2020, chemicals are used and produced in ways that lead to the minimization of significant adverse effects on human health and the environment.

12.1.5 Costs and benefits

There has been a long-standing discussion over the potential costs and benefits of REACH and many attempts have been made to estimate them. In 2003, the European Chemicals Bureau of the Joint Research Centre of the European Commission (http://ecb.jrc.it/) estimated the direct testing costs of REACH [24]. According to the ECB the direct testing costs in addition to existing obligations and voluntary initiatives were estimated at € 1.6 billion for the most likely scenario during an 11-year period of implementation. This estimate ranged from € 1.2 to 2.4 billion depending on the assumptions in the uncertainty analysis. According to the ECB about 86% of the estimated costs of the most likely scenario will be needed to test for human health endpoints, while only about 14% will be needed for environmental endpoints. On the basis of this ECB report, the Commission staff published an extended impact assessment in 2003 in which the total costs for testing and registration, including Agency fees, were estimated at € 2.3 billion over a period of 11 years. In the extended impact assessment, a preliminary estimate of the total health benefits was provided of € 50 billion (over a period of 30 years).

Many impact assessments followed, with some reports estimating the costs of REACH at hundreds of billions of euros, while other studies performed for the Commission and the Nordic Council of Ministers [25] arrived at much lower estimates. In 2004 a report was presented summarizing 36 studies on the impact of REACH [26]. In this report the direct cost for the implementation of REACH during an 11-year period were estimated at about € 4 billion. The health benefits (estimated over a period of 30 years) ranged from € 18-54 billion. In 2006 a study was done at the request of the Commission's environment directorate in order to refine the environment and health benefits of REACH [27]. This study concluded that REACH would save a minimum of € 150-500 million by the year 2017, at the expected close of its 11-year roll-out period. By the year 2041, the savings would amount to € 9.0 billion, mostly in areas such as purification of drinking water, disposal of dredged sediment and incineration of sewage instead of disposal on farmlands. The estimated health savings (€ 50 billion) were in line with the earlier preliminary estimate provided in the extended impact assessment of the Commission. Overall, this and other available studies concluded that REACH would generate net benefits. Based on a case study approach [28] completed in 2005, a high level group of industry experts and the Commission undertook further work on the impact assessment of REACH and concluded that the costs of REACH are moderate and manageable. This study drew the following conclusions:

- There is limited evidence that higher volume substances are vulnerable to withdrawal following the REACH registration requirements. However, lower volume substances under 100 tonnes are most vulnerable to being made less or non profitable by the REACH requirements.
- There is limited evidence that downstream users will be faced with a withdrawal of substances of greatest technical importance to them.
- Small and medium enterprises (SMEs) could be particularly affected by REACH in relation to their more limited financial capacity and lower market power in terms of passing on costs.
- Companies have recognized some business benefits from REACH.

More detailed information on the impact assessments can be found on the website of the Commission (http://europa.eu.int/comm/enterprise/reach/eia_en.htm).

12.2 A BRIEF OVERVIEW OF REACH

12.2.1 Introduction

REACH creates one system for the evaluation of all industrial chemicals. The Regulation is divided into 141 articles divided over 15 titles and it has 17 annexes. It is impossible in the context of this book to discuss all the Articles, Titles and Annexes but as "structure follows function", the structure of REACH is presented in Tables 12.2 and 12.3 and Figure 12.2. The main features of REACH are:

1. *Registration.* Registration (Title II) requires industry to obtain relevant information on their substances and to use the acquired data to manage them safely. To reduce testing on vertebrate animals, *data sharing*

(Title III) is required for studies on such animals. Better *information* on exposure, hazards and risks and how to manage them will be passed down and up the supply chain (Title IV) and downstream users are explicitly brought into the system (Title V) by placing specific obligations on them.

2. *Evaluation.* The aim of evaluation (Title VI) is to prevent unnecessary testing with vertebrate animals, by having the Agency evaluate the *testing proposals* made by industry, and to check compliance with the registration requirements and if not, to ask industry for further information. *Substance evaluation* is coordinated by the Agency and involves the competent authorities who investigate chemicals with potential risks by asking industry for further information. This information may be used later to prepare proposals under restrictions or authorization.
3. *Authorization.* For substances of very high concern, authorization is required for their use and placing on the market (Title VII). Substances with properties of very high concern are substances classified as carcinogenic, mutagenic and toxic to reproduction (known as CMR substances). Also included are substances which are persistent, bioaccumulative and toxic (PBT substances) and substances which are very persistent and very bioaccumulative (vPvB substances). Substances of equivalent concern may be included as well. Applicants will have to demonstrate that risks associated with the use of these substances are adequately controlled. Risks associated with the use of these substances will be reviewed. An authorization will be granted if these substances are adequately controlled, or if the socio-economic benefits outweigh the risks to human health or the environment, and if there are no suitable alternative substitute substances or technologies that are economically and technically viable.
4. *Restrictions* (Title VIII). Restriction means any condition for or prohibition of the manufacture, use or placing on the market of certain dangerous substances. The restrictions procedure provides a safety net to manage risks that have not been adequately addressed by another part of the REACH system.
5. *European Chemicals Agency.* Fees and charges are the subject of Title IX. The Agency (Title X) manages the technical, scientific and administrative aspects of the REACH system, and ensures consistency in decision-making at Community level. It also plays a pivotal role in coordinating communication surrounding the Regulation and its implementation.

Furthermore, a *classification and labelling inventory* (Title XI) will help promote the harmonization of different classifications of a substance. For CMR substances (classified as carcinogenic, mutagenic and toxic to reproduction), as well as respiratory sensitizers, there may be a Community-wide agreement on classification by the authorities. Harmonized classification and labelling for other effects may also be proposed on a case-by-case basis if justification is provided which demonstrates the need for action at Community level. *Information* (Title XII) describes the reporting process of Member States, the Agency and Commission on the operation of the Regulation. It also describes access to information, i.e., a system of publicly available information over the internet, requests for access to information and specific rules on the protection of confidential business information. Cooperation with third countries and international organizations is included as well. The appointment of *Competent Authorities* in the Member States (MS-CA), their cooperation, and their role in communication to the public on the risks of substances is given in Title XIII. Member States will have to maintain a system of official controls and shall lay down the provisions or penalties applicable for infringement of the provisions of the Regulation and need to take all measures necessary to ensure that they are implemented. Title XIV describes this process of *enforcement*. Transitional and final provisions are given in Title XV.

It should be noted that the regulation does not apply to radioactive substances, substances subject to customs supervision, non-isolated intermediates, the transport of dangerous substances and waste. The Regulation also exempts the use (but *not* the production and formulation) of certain substances that are adequately regulated under other legislation, such as medicinal products for human or veterinary use, food additives, flavourings in foodstuffs, additives in animal feed and animal nutrition.

The REACH legislation is quite complex and has developed its own jargon. Some key definitions of REACH are given in Table 12.4.

12.2.2 Registration (Title II)

General aspects

There is a general obligation for manufacturers and importers of substances to submit a registration to the Agency for each substance manufactured or imported in quantities of 1 tonne or more per year. Failure to register means that the substance is not allowed to be manufactured or imported *(no data, no market).* The

Table 12.2. The structure of REACH [29].

Title I	General issues
Title II	Registration of substances
Title III	Data sharing and avoidance of unnecessary testing
Title IV	Information in the supply chain
Title V	Downstream users
Title VI	Evaluation
Title VII	Authorization
Title VIII	Restrictions on the manufacturing, placing on the market and use of certain dangerous substances and preparations
Title IX	Fees and charges
Title X	Agency
Title XI	Classification and labelling inventory
Title XII	Information
Title XIII	Competent authorities
Title XIV	Enforcement
Title XV	Transitional and final provisions

Table 12.3. The 17 annexes to REACH [29].

Annex I	General provisions for assessing substances and preparing chemical safety reports
Annex II	Guide to the compilation of safety data sheets
Annex III	Criteria for substances registered in quantities between 1 and 10 tonnes
Annex IV	Exemptions from obligations to register in accordance with Article 2(7)(a)
Annex V	Exemptions from obligations to register in accordance with Article 2(7)(b)
Annex VI	Information requirements referred to in Article 10
Annex VII	Standard information requirements for substances manufactured or imported in quantities of 1 tonne or more
Annex VIII	Additional standard information requirements for substances manufactured or imported in quantities of 10 tonnes or more
Annex IX	Additional standard information requirements for substances manufactured or imported in quantities of 100 tonnes or more
Annex X	Additional standard information requirements for substances manufactured or imported in quantities of 1000 tonnes or more
Annex XI	General rules for adaptation of the standard testing regime set out in Annexes VII to X
Annex XII	General provisions for downstream users to assess substances and prepare chemical safety reports
Annex XIII	Criteria for the identification of persistent, bioaccumulative and toxic substances, and very persistent and very bioaccumulative substances
Annex XIV	List of substances subject to authorization
Annex XV	Dossiers
Annex XVI	Socio-economic analysis
Annex XVII	Restrictions on the manufacture, placing on the market and use of certain dangerous substances, preparations and articles

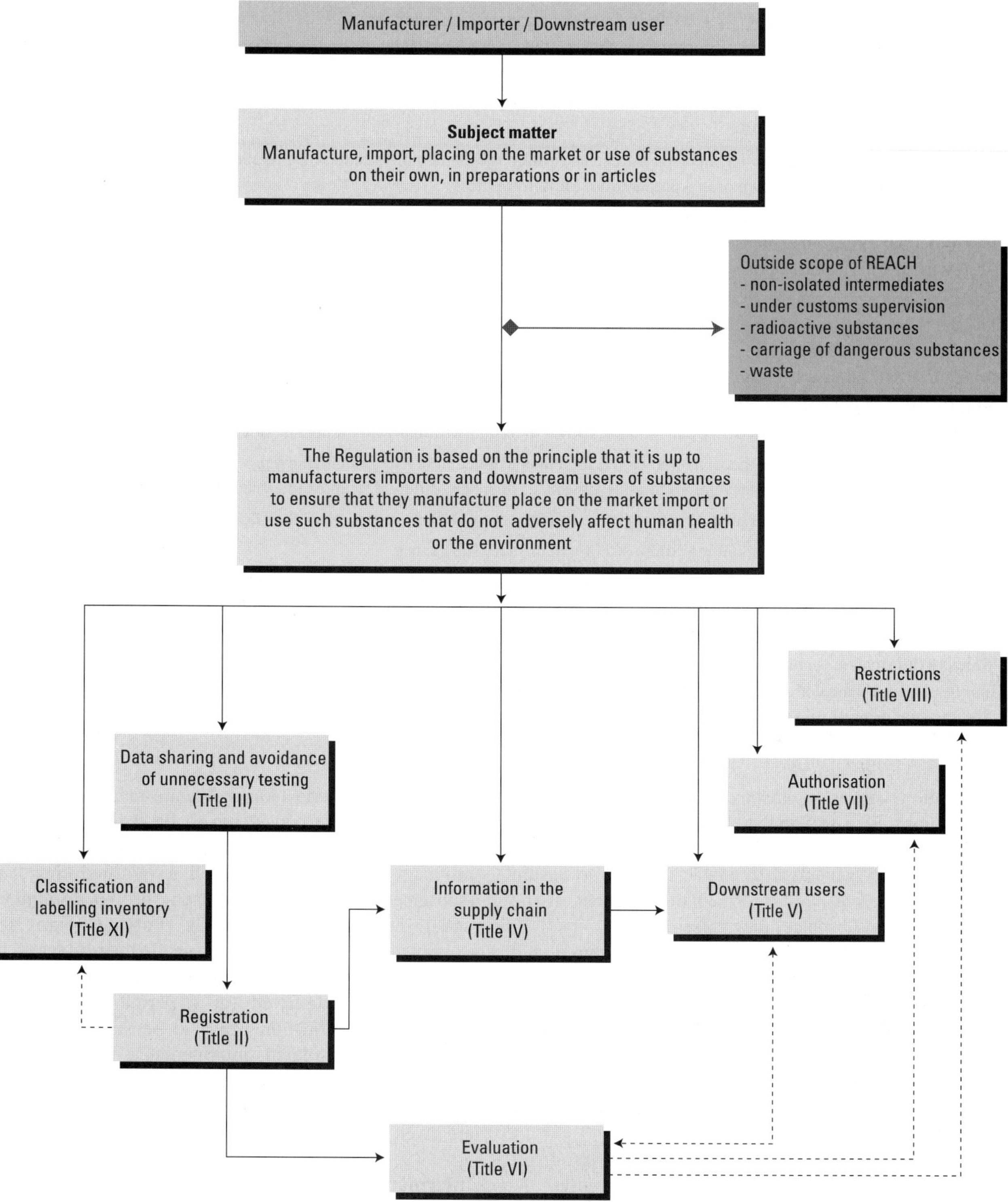

Figure 12.2. Structure of the REACH legislation.

Table 12.4. Definitions of some key terms used in the REACH legislation.

Actors in the supply chain means all manufacturers and/or importers and/or downstream users in a supply chain.

Competent authority means the authority or authorities or bodies established by the Member States to carry out the obligations arising from this Regulation.

Distributor means any natural or legal person established within the Community, including a retailer, who only stores and places on the market a substance, on its own or in a preparation, for third parties.

Downstream user means any natural or legal person established within the Community, other than the manufacturer or the importer, who uses a substance, either on its own or in a preparation, in the course of his industrial or professional activities. A distributor or a consumer is not a downstream user.

Exposure scenario means the set of conditions, including operational conditions and risk management measures, that describe how the substance is manufactured or used during its life-cycle and how the manufacturer or importer controls, or recommends downstream users to control, exposures of humans and the environment. These exposure scenarios may cover one specific process or use, or several processes or uses as appropriate.

Identified use means a use of a substance on its own or in a preparation, or a use of a preparation, that is intended by an actor in the supply chain, including his own use, or that is made known to him in writing by an immediate downstream user.

Intermediate means a substance that is manufactured for and consumed in or used for chemical processing in order to be transformed into another substance (hereinafter referred to as “synthesis”). There are different types of intermediates (see Glossary).

Phase-in substance is equivalent to an existing chemical. The precise definition is given in the Glossary.

Preparation means a mixture or solution composed of two or more substances.

Recipient of a substance or a preparation means a downstream user or a distributor being supplied with a substance or a preparation. *Recipient* of an article means an industrial or professional user being supplied with an article but does not include consumers.

Robust study summary means a detailed summary of the objectives, methods, results and conclusions of a full study report providing sufficient information to make an independent assessment of the study minimising the need to consult the full study report.

Supplier of a substance or a preparation means any manufacturer, importer, downstream user or distributor placing on the market a substance, on its own or in a preparation, or a preparation.

registration concerns substances on their own or in preparations. Registration thereby ensures:

- That sufficient relevant information is collected and if necessary generated to enable industry to ensure responsible and well-informed management of the risks which substances may present.
- A level playing field among EU manufacturers and importers, while minimizing the possibilities for free-riders.
- Transparency and accountability to stakeholders on the progress made by industry in meeting their responsibilities.
- That the Commission and Member States can monitor and enforce the requirements.
- That new animal tests are carried out as a last resort without compromising a high level of protection.

Certain categories of chemicals do not need to be registered, because they are exempted from REACH altogether (see Section 12.2.1). Others are exempted from the registration requirements, for example, as they are considered to cause minimum risk (Annex IV), fulfil certain criteria (Annex V), are regulated by other legislation, or are considered to be registered already. Examples include active substances in plant protection products and biocides, as well as notified chemicals (substances for which a notification has been submitted under Directive 67/548/EEC). Polymers are also exempted from registration. So-called “substances for product and process oriented research and development” (PPORD) are exempted for a period of five years (although a notification needs to be submitted to the Agency), but isolated intermediates, non-registered monomers and other substances in polymers need to be registered. The main features of the registration process are given in Box 12.4.

Information requirements

Manufacturers and importers of substances will need to obtain information on the substances they manufacture or import and use this information to assess the risks arising from the uses and to ensure that the risks which the substances may present are properly managed.

Box 12.4. Registration under REACH

- Aim: to ensure that industry adequately manages the risk arising from its substance (starting at 1 tonne/y)
- Method:
  - Manufacturer/importer obtains adequate data
  - Provides a registration dossier which includes a chemical safety report for substances above 10 tonnes/y
  - Submits to authorities (enforcement, transparancy)
  - Increased info requirements according to tonnage (testing proposal)
  - Reduced requirements for polymers and intermediates

Registration documents the performance of this duty and requires manufacturers and importers to submit:

- A technical dossier (for substances ≥ 1 tonne/y).
- A chemical safety report (for substances ≥ 10 tonnes/y).
- If the substance meets the criteria for classification as dangerous or is assessed to be PBT or vPvB, the CSA has to include an exposure assessment including exposure scenario(s), exposure estimation and risk characterization.

The technical dossier contains a general part (Annex VI to the Regulation) and standard information requirements on the properties of the chemical (Annex VII to the Regulation). Annex VI provides the format for information about the registrant, the identification of the substance, information on manufacture and uses(s), classification and labelling, as well as guidance on safe use and information on exposure for substances registered in quantities between 1 and 10 tonnes/y per manufacturer or importer. Annex VII provides the format for the physicochemical properties, the toxicological information and the ecotoxicological information for substances manufactured or imported in quantities of ≥ 1 tonne/y. Further information requirements on physicochemical properties and (eco)toxicological data vary according to the tonnage in which the substance is manufactured or imported and the needs of the chemical safety assessment, and are specified in Annexes VIII-X to the Regulation. A submission for registration needs to be accompanied by a fee.

The tonnage "trigger" is chosen "as it gives an indication of the potential for exposure". The Annexes apply cumulatively, i.e., the higher the quantities of the substances manufactured or imported per year, the more annexes apply. The physicochemical and (eco)toxicological information requirements are specified in a simplified manner in Table 12.5, and are as follows:

- 1 tonne or more per year according to Annex VII.
- 10 tonnes or more per year according to Annexes VII and VIII.
- 100 tonnes or more per year according to Annexes VII-IX.
- 1000 tonnes or more per year according to Annexes VII-X.

Any other relevant physicochemical, toxicological and ecotoxicological information that is available shall also be provided. Where there is an information gap that needs to be filled, new data have to be generated (Annexes VII and VIII), or a proposal and a time schedule for fulfilling the information requirements *(testing strategy)* has to be submitted (Annexes IX and X), depending on the tonnage. New toxicological tests on vertebrates shall only be conducted or proposed as a last resort when all other data sources have been exhausted.

For substances manufactured or imported between 1 and 10 tonnes/y, 6 steps are required. These are presented in Figure 12.3.

The CSA (see also Section 12.3) for substances manufactured or imported in quantities starting at 10 tonnes/y, documents the human, physicochemical and environmental hazard assessment including the assessment of whether the substance is a PBT or vPvB. If, as a result of this analysis, the manufacturer or importer concludes that the substance meets the criteria for classification as dangerous, or is assessed to be a PBT or vPvB, the CSA needs to include the following additional steps: (a) exposure assessment including the generation of exposure scenario(s) and exposure estimation and (b) risk characterization.

Exposure scenarios (ESs) are sets of conditions, including operational conditions and risk management measures, that describe how substances are manufactured or used during their life cycle and how the manufacturer or importer controls, or recommends how to control, exposures of humans and the environment. The ESs must include the appropriate RMMs which, when properly implemented, ensure that the risks from the uses of the substance are adequately controlled. ESs need to be developed to cover all "identified uses", which are the manufacturers' or importers' own uses, and uses which are made known to the manufacturer or importer by his downstream users and which the manufacturer or

Table 12.5. Data requirements as defined in REACH.

ANNEX VII (≥1 TONNE)	
Physical and chemical information	
7.1	State of the substance (at 20 °C/101.3 kPa)
7.2	Melting/freezing point
7.3	Boiling point
7.4	Relative density
7.5	Vapour pressure
7.6	Surface tension
7.7	Water solubility
7.8	Partition coefficient *n*-octanol/water
7.9	Flash-point
7.10	Flammability
7.11	Explosive properties
7.12	Self-ignition temperature
7.13	Oxidising properties
7.14	Granulometry
Toxicological information	
8.1	Skin irritation or skin corrosion
8.2	Eye irritation
8.3	Skin sensitization
8.4.1	Mutagenicity (gene mutation in bacteria)
8.5.1	Acute toxicity (oral route)
Ecotoxicological information	
9.1.1	Short-term toxicity invertebrates *(Daphnia)*
9.1.2	Growth-inhibition plants (algae)
9.2.1.1	Ready biodegradability

ANNEX VIII (≥10 TONNES)	
Toxicological information	
8.1.1	Skin irritation *(in vivo)*
8.2.1	Eye irritation *(in vivo)*
8.4.2	Cytogenicity in mammalian cells *(in vitro)*
8.4.3	Gene mutation in mammalian cells *(in vitro)*
8.5.2	Acute toxicity (inhalation)
8.5.3	Acute toxicity (dermal)
8.6.1	Repeated dose toxicity (28 days)
8.7.1	Reproductive/developmental toxicity screening test; OECD 421 or 422)
8.8.1	Toxicokinetics
Ecotoxicological information	
9.1.3	Short-term toxicity fish
9.1.4.	Activated sludge respiration inhibition test
9.2.2.1	Hydrolysis as a function of pH
9.3.1	Adsorption/desorption screening test

ANNEX IX (≥100 TONNES)	
Physical and chemical information	
7.15	Stability in organic solvents and identity of relevant degradation products
7.16	Dissociation constant
7.17	Viscosity
Toxicological information	
8.6.1	Repeated dose toxicity (28 days)
8.6.2	Sub-chronic toxicity (90 days)
8.7.2	Pre-natal developmental toxicity; OECD 414
8.7.3	Two-generation reproductive toxicity study
Ecotoxicological information	
9.1.5	Long-term toxicity invertebrates *(Daphnia)*
9.1.6	Long-term toxicity to fish
9.1.6.1	Fish early-life stage test
9.1.6.2	Fish short term toxicity embryo and sac fry
9.1.6.3	Fish juvenile growth test
9.2.1.2	Ultimate degradation in surface water
9.2.1.3	Soil simulation testing
9.2.1.4	Sediment simulation testing
9.2.3	Identification of degradation products
9.3.2	Bioaccumulation in aquatic species (fish)
9.3.3	Further information on adsorption/ desorption
9.4.1	Short-term terrestrial toxicity (invertebrates)
9.4.2	Effects on soil micro-organisms
9.4.3	Short-term toxicity to terrestrial plants

ANNEX X (≥1000 TONNES)	
Toxicological information	
8.6.3	Long-term repeated toxicity (≥12 months)
8.7.2	Developmental toxicity; OECD 414
8.7.3	Two-generation reproductive toxicity
8.9.1	Carcinogenicity study
Ecotoxicological information	
9.3.4	Further fate and behaviour in the environment of the substance and/or degradation products
9.4.4	Long-term toxicity on invertebrates
9.4.6	Long-term toxicity on plants
9.5.1	Long-term toxicity to sediment organisms
9.6.1	Long-term or reproductive toxicity to birds

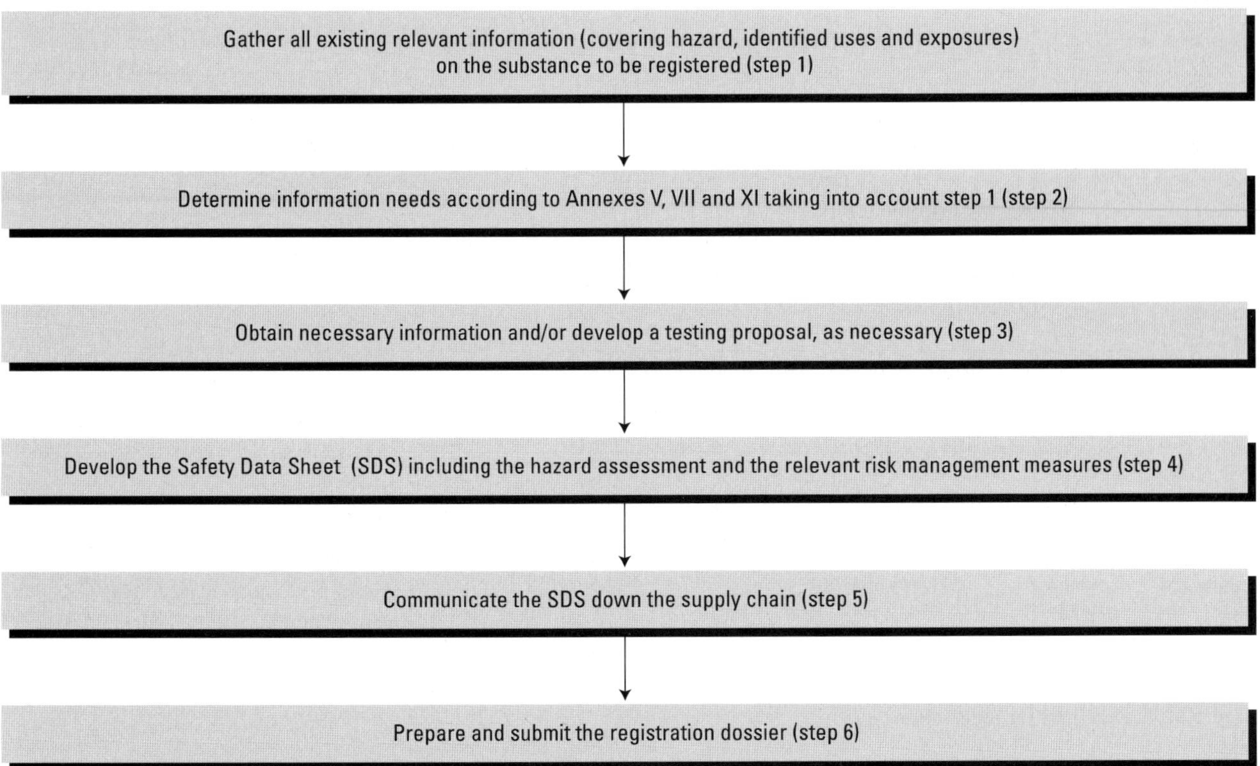

Figure 12.3. Steps required for the assessment of substances at or above 1 tonne per year.

importer includes in his assessment. Relevant ESs will need to be annexed to the SDS that will be supplied to downstream users and distributors.

Joint submission and sharing of information on substances should be provided for to increase the efficiency of the registration system, to reduce costs to industry and the authorities and to reduce testing on vertebrate animals. Joint submissions are the legislative implementation of what is known as the "one substance, one registration" or "OSOR" approach. Registrants need to assess together the available information on the properties of the substance and its classification, which is then submitted by a "lead registrant" on behalf of the others. The other registrants will separately provide their company, substance identity and use information and will refer to the joint submission for all other information. In certain specified cases and provided a justification is given (e.g., for reasons of confidentiality), a registrant may be able to submit information directly to the Agency ("opt out clause").

Registration deadlines

Registration is a phased approach. The first REACH obligation, *pre-registration for phase-in substances*, will take place in 2008 from 1 June to 1 December. It will be followed by *registration* in 3.5, 6 or 11 years, depending on the volume band or level of concern of the substance. This information and evidence demonstrating the safe use of the substance, need to be submitted in a registration dossier to the Agency.

Pre-registration is mandatory for each potential registrant of a phase-in substance, including intermediates, in quantities of 1 tonne or more per year. The basic information that needs to be submitted to the Agency includes the name of the substance including EINECS and CAS number, name and address of the contact person and the envisaged deadline for the registration and the tonnage band. The information requirements are specified in Article 28. "Phase-in" deadlines are presented in Figure 12.4.

Registration of substances in articles

Registration of substances in articles is obligatory for any substance contained in those articles if the following conditions are met: (a) the substance is present in quantities totalling over 1 tonne per producer or importer per year and (b) the substance is intended to be released under normal or reasonably foreseeable conditions of

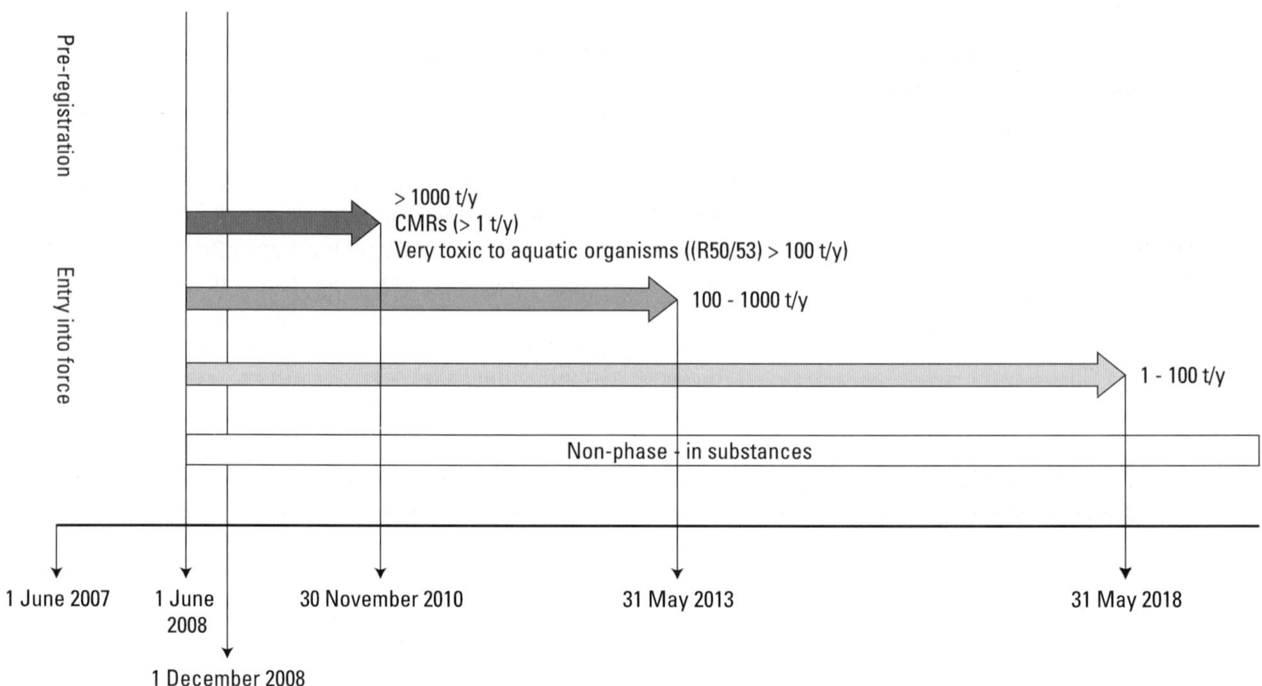

Figure 12.4. Timelines for registration.

use. Producers and importers of articles need to notify the Agency of the presence of substances of very high concern identified as a candidate for authorization in quantities of 1 tonne or more and above a concentration of 0.1% weight by weight (w/w). If the producer or importer of an article can exclude exposure to humans or the environment under reasonably foreseeable conditions of use, including disposal, such as by supplying appropriate instructions, then they do not have to notify. Further guidelines will be developed for this. The Agency may also take decisions requiring producers or importers to submit a registration for any substance in those articles, if all the following conditions are met: (a) the substance is present in quantities totalling over 1 tonne per producer or importer per year, (b) the Agency has grounds for suspecting that the substance is released from articles and the release presents a risk to human health or the environment.

12.2.3 Data sharing and avoidance of unnecessary testing (Title III)

Rules on data sharing are set out to reduce testing on vertebrate animals and to reduce costs to industry. Data obtained by animal testing are to be shared, in exchange for payment. Communication mechanisms are available to enable and encourage manufacturers and importers to reach agreements on the sharing of studies on vertebrate animals. A system has been established to help registrants to find other registrants with whom they can share data (pre-registration). Pre-registrants of the same phase-in substance are then required to share existing animal test data and agree on the generation of new animal test data in a *substance information exchange forum* (SIEF). The communication mechanisms may also be used for tests which do not involve vertebrate animals, since this will reduce costs. It is important to note that under REACH the generation of information by alternative means offering equivalence to prescribed tests and test methods is allowed, for example, when information comes from valid qualitative or quantitative structure-activity models or from structurally related substances. These SARs, QSARs and read-across methods, as they are known, are described in Chapters 9-11.

12.2.4 Information in the supply chain (Title IV)

The communication requirements of REACH ensure that not only manufacturers and importers but also their customers, i.e., downstream users and distributors, have the information they need to use chemicals safely. Information relating to health, safety and environmental

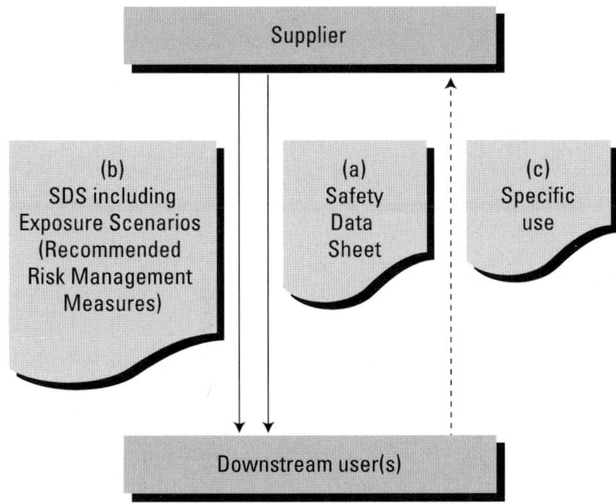

Figure 12.5. Managing the communication challenge within the supply chain. Safety data sheets (a) are the main instrument. When chemical safety assessments are required, exposure scenarios must be added to the SDS (b). Downstream users may provide information to assist in the preparation of a registration (c).

properties, risks and RMMs is required to be passed both down and up the supply chain, involving all actors in that supply chain. Commercially sensitive information is not required to be exchanged. The communication process within the supply chain is given in Figure 12.5.

The primary tool for information transfer is the well-established and familiar safety data sheet (see also Section 12.4) which has to be supplied for all dangerous substances. The provisions of the current Safety Data Sheets Directive (91/155/EEC) have been carried over into the REACH Regulation. As more information will be available as a result of registrations, the quality of safety data sheets will improve. Where chemical safety assessments are performed according to the registration requirements, relevant ESs have to be annexed to the SDS and will thus be passed down the supply chain. Downstream users may provide information to assist in the registration process. Their main responsibility is to manage the risks arising from their use(s) of substances. Therefore, they are responsible for assessing the risks arising from their uses of substances and of the uses of their customers, if these uses are not covered by the SDS received from their suppliers. New information on hazardous properties and information that challenges the quality of RMMs in the SDS has to be passed up the supply chain (see Section 12.2.5).

12.2.5 Downstream users (Title V)

Downstream users (see Table 12.4 for the definition) are required to consider the safety of their use(s) of substances, based primarily on information from their suppliers, and to apply appropriate RMMs. Downstream users will need to communicate effectively with their manufacturers or importers, to get the information they need in the SDS supplied to them (Figure 12.5). In particular they will have to check that their use(s) are "covered" by the safety data sheet, i.e., that they use a substance under the conditions described in the ESs and apply these conditions. To get the correct information, downstream users have the right to make their uses known to their manufacturers or importers so that the manufacturers or importers can include these uses in their CSA as "identified" uses. The downstream user need not to prepare a CSR if:

- The use of the substance as such or in a preparation is less than 1 tonne/y.
- A SDS is not required.
- A CSR is not required by his supplier (i.e., manufacture or import is less than 10 tonne/y).
- The downstream user implements or recommends an ES as communicated to him in the SDS.

A downstream user relying on the 1 tonne exemption still needs to consider the use(s) of the substance and identify, apply and recommend appropriate RMMs.

Downstream users can also choose to keep the uses confidential or decide to use a substance outside the conditions described in an ES communicated to them. In such cases they will have to perform a CSA, i.e., they will have to develop ESs for these intended uses, perform the exposure assessment and characterize the risks arising from these uses. If a downstream user is using a substance in quantities of more than 1 tonne/y not under the conditions described in the exposure scenario communicated to him in the SDS he will need to report his use to the Agency as a brief general description. The CSA itself does not have to be submitted with the report. These reports enable the authorities to evaluate substances if reported uses give rise for concern, and to take appropriate measures. In rare cases, the downstream user may propose additional testing if he considers this necessary to complete his CSA.

12.2.6 Evaluation (Title VI)

Evaluation under REACH is a structured process by which the Agency, with input from the Member State (MS) authorities, may examine registration dossiers.

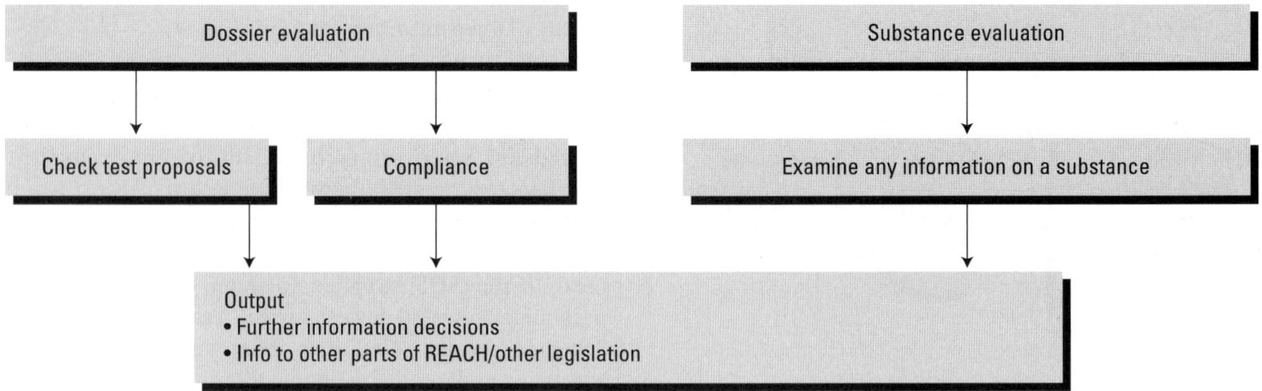

Figure 12.6. Evaluation under REACH.

The aim is to provide confidence that industry is meeting its obligations and it also serves as a tool to prevent unnecessary testing on vertebrate animals. The evaluation may result in a request for further information on substances. There are two types of evaluation with different aims (Figure 12.6).

1. *Dossier evaluation.* This deals with two aspects: (a) the testing proposals and (b) the compliance of the dossier with the registration requirements. The Agency is responsible for dossier evaluation. The aim of checking the testing proposals is to prevent unnecessary testing with vertebrate animals, i.e., the repetition of existing tests, and poor quality tests. Therefore, the Agency will check the testing proposals submitted as part of the registrations before such tests are performed. The Agency will also perform a minimum number of compliance checks (not less than 5% of the total number of dossiers received) on registration dossiers against the requirements laid down for registration in the Regulation.
2. *Substance evaluation.* The goal of substance evaluation is to clarify any suspicion of risk to human health or the environment, by requesting further information from industry. The Agency is responsible for coordinating the substance evaluation process but will rely on the MS competent authorities to perform the evaluations. A single EU-wide rolling plan for substance evaluation will be established by the Agency with input from the Member States. The Agency will develop criteria for prioritizing substances for substance evaluation and will select substances for the Community rolling action plan. The first draft rolling action plan to the Member States will be submitted by 1 December 2011 and will be updated annually. The MS competent authorities can use expert institutes to perform the substance evaluations. If a draft decision prepared by a competent authority of a Member State requesting further information on a substance is accepted by all other Member States' competent authorities, the Agency takes the decision accordingly. If an agreement cannot be reached, the decision-making will be referred to the Commission. The Agency is also responsible for ensuring the consistency of such decisions at the draft stage.

Evaluation may lead the authorities to the conclusion that action needs to be taken under the restriction or authorization procedures in REACH, or that information needs to be passed on to other authorities responsible for relevant legislation. The evaluation process will ensure that reliable and useful data is provided and made available to the relevant bodies by the Agency.

12.2.7 Authorization (Title VII)

The aim of authorization is to ensure the proper functioning of the internal market while assuring that the risks from substances of very high concern are properly controlled. Authorization in itself encourages *substitution:* substances are eventually replaced (substituted) by suitable alternative substances or technologies where these are economically and technically viable. In order to further encourage the development of safer substitutes, all applications for authorization by manufacturers, importers or downstream users will include an analysis of available alternatives considering their risks and the technical and economic feasibility of substitution. The essential aspects of authorization are given in Box 12.5.

Box 12.5. Authorization under REACH

- Deals with CMR (category 1 or 2), PBT, vPvB, and substances of "equivalent concern"
- Focus on PBT or vPvB properties, wide dispersive use and high volumes
- Prioritized by the Agency with input from the Member States
- Considerations:
  - the risks and adequate control of these risks
  - social and economic benefits/implications of a refusal to authorize
  - the analysis of alternatives submitted
  - available information on risks of any alternative substances or technologies
- Commission decision based on Agency opinion

For substances of very high concern, an authorization is required for their use and for them to be placed on the market. The substances required to be authorized are:

1. Carcinogenic, mutagenic or toxic for reproduction (CMRs; category 1 and 2).
2. Persistent, bioaccumulative and toxic (PBT) or very persistent and very bioaccumulative (vPvB).
3. Substances, such as those having endocrine disrupting properties or those having PBT or vPvB properties but which do not meet the criteria of PBT or vPvB substances, and for which there is scientific evidence of probable serious effects on humans or to the environment which give rise to an equivalent level of concern. These are identified on a case-by-case basis.

These substances have hazardous properties of such high concern that it is essential to regulate them centrally through a mechanism that ensures that the risks related to their actual uses are assessed, considered and then decided upon by the Community. This is justified because the effects of these substances on humans and the environment are very serious and normally irreversible. Substances that fall into these categories will be fed into the authorization system.

The authorization procedure consists of two steps:

- Step 1. Identification of substances. A decision is taken as to which substances will be included in the system, which uses of the included substances will be exempted from the authorization requirement (e.g., because sufficient controls established by other legislation are already in place), and what deadlines will have to be met, e.g., a sunset date after which other uses will no longer be permitted. This step is necessary to prioritize substances and to focus resources. The Agency will prepare such decisions and recommend substances for inclusion, and interested parties will have an opportunity to comment on such recommendations.
- Step 2. Granting of authorization. Once a substance is included on the list, the applicant needs to apply for an authorization for each use of the substance within the set deadlines. The Commission decides on granting the authorization based on the opinions of the Agency's committees for risk assessment (RA) and socio-economic analysis (SEA).

The burden of proof is placed on the applicant to demonstrate that the risk from the use of the substance is adequately controlled or that the socio-economic benefits outweigh the risks, taking into account available information on alternative substances or processes. The Agency's RA and SEA committees will give their draft opinions within ten months of the date of receipt of the application. The Risk Assessment Committee will provide a draft opinion about the assessment of the risk to health and/or the environment arising from the use(s) of the substance as described in the application and, if relevant, an assessment of the risks arising from possible alternatives. The Socio-economic Analysis Committee will provide a draft opinion for the assessment of the socio-economic factors and the availability, suitability and technical feasibility of alternatives associated with the use(s) of the substance as described in the application, when an application is made. The applicant has the opportunity to comment on these draft opinions. The Commission prepares a draft authorization within 3 months of receiving the opinions from the Agency. A final decision on granting an authorization is taken in accordance with the Committee procedure referred to in Article 133. When authorizations are granted for socio-economic reasons, these authorizations will normally be time-limited. Downstream users may use a substance for an authorized use provided they obtain the substance from a company which has been granted an authorization and that they observe the conditions of that authorization. Such downstream users will need to notify the Agency that they are using an authorized substance.

12.2.8 Restrictions (Title VIII)

The Restrictions procedure enables focused Community-wide regulation of the conditions for the manufacture, placing on the market or use of certain dangerous substances or the banning of any of these activities, if

Box 12.6. Restrictions under REACH

- Safety net
- Community wide concern
- Member States/Commission initiated
- Agency Committees examine:
  - the risks, and
  - the socio-economic aspects involved
- Commission - final decision

necessary. All activities involving a regulated substance which are not restricted, are allowed unless the substance is included in the authorization system. Any substance on its own, in a preparation or in an article, may be subject to Community-wide restrictions if it is demonstrated that risks are being not adequately controlled. Thus, the restrictions provisions act as a *safety net* (Box 12.6).

Proposals for restrictions will be prepared by Member States or by the Agency on behalf of the Commission in the form of a structured dossier (Table 12.6). This dossier is required to demonstrate that there is a risk to human health or the environment that needs to be addressed at Community level and to identify the most appropriate set of RMMs. Deadlines for the procedure to prepare a Commission decision are set out in the Regulation. Interested parties will have an opportunity to comment and the Agency will provide opinions on any proposed restriction. The existing restrictions set out in Directive 76/769/EEC (such as the ban on asbestos and restrictions on the uses of certain azo-dyes) are carried over into the REACH Regulation in a consolidated version in Annex XVII. Annex XVII will be amended by adopting new restrictions, or amending current restrictions. The decision will be taken by the Commission following the procedure specified in Article 133(4). Until 1 June 2013, a Member State may maintain any existing restriction in relation to Annex XVII provided that the restrictions have been notified. The Commission will compile and publish an inventory of these restrictions by 1 June 2009.

12.2.9 Fees and charges (Title IX)

The fees that are required will be specified in a separate Commission Regulation by 1 June 2008. The fees must

Table 12.6. Key information to be submitted in dossiers for proposed restrictions.

(1)	*Proposal*
(2)	*Information on hazards and risks*
(3)	*Information on alternatives* Available information on alternative substances and techniques shall be provided, including: • information on the risks to human health and the environment related to the manufacture or use of the alternatives. • availability, including the time scale. • technical and economical feasibility.
(4)	*Justification for restrictions at Community Level* • action is required on a Community-wide basis. • a restriction is the most appropriate Community-wide measure, which shall be assessed using the following criteria: a. *effectiveness:* the restriction must be targeted to the effects or exposures that cause the risks identified, capable of reducing these risks to an acceptable level within a reasonable period of time and proportional to the risk. b. *practicality:* the restriction must be implementable, enforceable and manageable. c. *monitorability:* the ability to monitor the result of the implementation of the proposed restriction.
(5)	*Socio-economic assessment* The socio-economic impacts of the proposed restriction may be assessed with reference to Annex XVI. To this end, the net benefits to human health and the environment of the proposed restriction may be compared to its net costs to manufacturers, importers, downstream users, distributors, consumers and society as a whole.
(6)	*Information on stakeholder consultation* Information on any consultation of stakeholders and how their views have been taken into account shall be included in the dossier.

Box 12.7. European Chemicals Agency

- Day to day management of REACH
  - Technical, scientific and administrative aspects
- Responsibilities:
  - Manage the registration process (Title II and III)
  - Dossier evaluation (compliance check and evaluation of testing proposals)
  - Co-ordination of substances evaluation (prioritization/set up a community-wide rolling action plan for evaluation)
  - Ensure a harmonized approach and take decisions
  - Through its expert committees it advises the Commission on:
    1. Priorities in setting up the authorization procedure
    2. Applications for authorizations for the use(s) of substances of very high concern
    3. Other risk reduction measures for dangerous substances (restrictions)
  - Provide technical and scientific guidance and tools to industry and MSs (e.g. TGDs and IT tools)
  - Secretariat for Forum and Committees
  - Deal with appeals (registration, R&D, evaluation, confidentiality)

be sufficient to cover the cost of the services delivered and will take into account whether the information from the registrants has been submitted jointly or separately. A reduced fee will be set for SMEs. Fees need not to be paid for the registration of substances in quantities between 1-10 tonnes/y where the registration dossier contains the full information in Annex VII. The Agency may make charges for other services it provides.

12.2.10 Agency (Title X)

The European Chemicals Agency in Helsinki manages the technical, scientific and administrative aspects of REACH. It ensures consistency in decision-making at Community level. The Agency manages the registration process and maintains databases. It has a clear responsibility for the evaluation of dossiers (compliance with the registration requirements and the evaluation of testing proposals). It plays a key role in ensuring that a harmonized approach is taken to evaluation by providing criteria to guide Member States' selection of substances for evaluation and by resolving disputes about requests for further information on substances arising from evaluation, and takes such decisions. It provides expert opinions and recommendations to the Commission in the authorization and restriction procedures and has duties with regard to communication and confidentiality. It also handles requests for exemptions from the registration requirement for product and process-oriented research and development, and facilitates the sharing of animal test data at the pre-registration stage by putting registrants of non-phase-in substances in touch with each other, and provides a database listing of what studies are available to members of each SIEF. Its tasks are summarized in Box 12.7.

In designing the structure of the European Chemicals Agency, the Commission followed the principles set out in its Communication on the operating framework for European Regulatory Agencies published in 2002. An overview of the structure of the Agency is given in Box 12.8.

Box 12.8. Structure of the Agency

- Management Board
- Executive Director, reporting to the Management Board
- Committee for Risk Assessment and a Committee for Socio-economic Analysis
- Member State Committee
- Forum for exchange of information on enforcement. This Forum coordinates a network of Member States authorities responsible for enforcement of the Regulation
- Secretariat that will provide technical, scientific and administrative support to the Committees and Forum and will undertake a number of other tasks
- Board of Appeal that will consider any appeals against the decisions of the Agency

12.2.11 Classification and labelling inventory (Title XI)

The classification and labelling of chemicals is based on the intrinsic properties of chemicals. It is an effective approach for managing the potential risks of chemicals (see Chapters 1, 6 and 7). Therefore, a requirement for industry to classify and label dangerous substances and preparations according to standard criteria has long been a key feature of the EU's chemicals legislation. REACH builds on the existing legislation. Responsibility for the classification of substances will normally rest

Box 12.9. The C&L Inventory

- Inventory:
  - contains C&L info for all marketed substances (no tonnage limit)
  - managed by Agency based on submissions from industry
- Industry needs to cooperate to resolve differences in C&L
- EU harmonization:
  - CMRs
  - respiratory sensitizers
  - other effects on a case-by-case basis

with industry, not on the authorities, except for hazards of the most concern. The classification and labelling inventory (Box 12.9) ensures that hazard classifications (and consequent labelling) of all dangerous substances manufactured in, or imported into, the EU are available to all with the aim of harmonizing the classifications. Industry will be required to submit all its classifications to the Agency, to be included in the inventory before 1 December 2010. Any divergences between classifications of the same substance should be removed over time either through cooperation between notifiers and registrants or by EU harmonized classifications for substances that are category 1, 2, and 3 CMRs, or respiratory sensitizers. Harmonized classification and labelling (C&L) for other effects may be included as well if there is justification for taking action at Community level. A globally harmonized system for the classification and labelling of chemicals (GHS; see Section 12.3.2) will be implemented in the near future.

12.2.12 Information (Title XII)

Access to information

Public access to European Parliament, Council and Commission documents will apply to documents held by the Agency. Non-confidential information on chemicals will be made available to enable those exposed to chemicals to make decisions on the acceptability of the related risks, for example. This is done in such a way that the interests of the public's "right to know" is balanced against the need to keep certain information confidential, such as details of the full composition of a preparation, precise use, function or application of a substance or preparation, etc. Some information will be published on the Agency's web page, some information will generally always be kept confidential, and some may be made available on request in accordance with the Commission's normal rules on access to information. The following information will be made publicly available by the Agency: IUPAC nomenclature, if applicable, the name of the substance as given in EINECS, the classification and labelling of the substance, physicochemical data, environmental fate and pathways, results of each toxicological and ecotoxicological study, any derived no effect level (DNEL) or predicted no effect level (PNEC), guidance on safe use, and analytical methods.

Reporting

Every five years, Member States need to submit to the Commission a report on the operation of the regulation in their respective territories, including sections on evaluation and enforcement (see Section 12.2.14). The first report must be submitted by 1 June 2010. Every five years the Agency has to submit to the Commission a report on the operation of this Regulation. The first report will be submitted by 1 June 2011. Furthermore, every three years the Agency will prepare a report for the Commission on the status of implementation and use of non-animal test methods. The first report will be submitted by 1 June 2011. Every five years the Commission has to publish a general report on the experience gained with the operation of this Regulation. The first report shall be published by 1 June 2012.

12.2.13 Competent authorities (Title XIII)

The Member States shall appoint the competent authority or competent authorities for the REACH-related tasks. These tasks, together with the tasks of the other stakeholders, are provided in Annex 12.1 to this chapter. The competent authorities will cooperate and share information necessary for the implementation of their tasks. They will have to communicate with the general public on risk arising from substances where this is considered necessary for the protection of human health and the environment. They will have to submit information on substances to the Agency and establish help desks to provide advice to manufacturers, importers, downstream users and other stakeholders.

12.2.14 Enforcement (Title XIV)

In the EU enforcement is a key task of the Member States. They are required to lay down provisions on penalties applicable for infringement of the provisions of REACH and to ensure that they are implemented.

These sanctions for non-compliance must be effective, proportionate and dissuasive. The Member States have to notify the Commission of those provisions no later than 1 December 2008. The Member States have to provide reports about the official inspections, the monitoring carried out, and the penalties provided for. This information will be made available to the Commission. The first report, on enforcement, has to be submitted by 1 June 2010 (see Section 12.2.12).

12.2.15 Transitional and final provisions (Title XV)

There are a number of transitional measures concerning already notified substances, existing substances, restrictions and the Agency. We will focus here on three other issues: (1) the free movement clause, (2) the safeguard clause and (3) the review.

Although Member States are not allowed to prohibit, restrict or impede the manufacturing, import, placing on the market or use of substances *(free movement clause)*, they are allowed to maintain or lay down national rules to protect workers, health and the environment in cases where the regulation does not match requirements on the manufacture, marketing or use of substances.

Furthermore, Member States maintain the right to take appropriate provisional measures in cases where urgent action is believed necessary for the protection of human health or the environment *(safeguard clause)*. On the basis of the information provided by the Member States, the Commission will take a decision within 60 days of receipt of this information.

Last but not least, the Commission will carry out a *review* of this Regulation 12 years after its entry into force, i.e., by 1 June 2019. On the basis of this review the Commission may present legislative proposals:

1. On whether or not to extend the scope of the obligation to perform a chemical safety assessment and to document it in a chemical safety report, to substances not covered by this obligation because they are not subject to registration or subject to registration but manufactured or imported in quantities of less than 10 tonnes per year.
2. On a practicable and cost-efficient way of selecting polymers for registration on the basis of sound technical and valid scientific criteria and, after publishing a report about (a) the risks posed by polymers in comparison with other substances and (b) the need, if any, to register certain types of polymers, taking into account competitiveness and innovation, on the one hand, and protection of human health and the environment, on the other.
3. To modify the information requirements for substances manufactured or imported in quantities of 1 tonne or more up to 10 tonnes per year per manufacturer or importer, taking into account the latest developments, for example in relation to alternative testing and (quantitative) structure-activity relationships ((Q)SARs).

Furthermore, the Annexes I, IV, V and XIII will be reviewed in 2008. Further reviews will be reported on the scope of the Regulation (2012), endocrine disrupters (2013), Article 33 on information on substances in articles (2019), and the testing requirements of Section 8.7 of Annex VIII (2019) with the aim of presenting legislative proposals or amendments, if necessary.

12.3 ELEMENTS OF THE CHEMICAL SAFETY REPORT

12.3.1 General introduction

The main purpose of the registration requirement and data sharing provisions of REACH is to establish a transparent, predictable and balanced framework within which industry ensures responsible and well-informed management of the risk which substances may present. This requires industry to collect sufficient information, if necessary by performing new tests, and to use this information to determine appropriate RMMs. These RMMs need to be implemented by manufacturers and importers. To achieve fair burden sharing with their customers, in their CSAs manufacturers and importers should address not only their own uses and the uses for which they place their substances on the market, but also all uses which their customers ask them to address.

In this section a short overview of the CSA will be given with a focus on risk characterization and the content of the CSR. More detailed information on risk assessment in general, exposure scenarios, fate, exposure assessment and hazard assessment methodologies is given in Chapters 1-7.

The CSA of a chemical substance aims to establish the safe conditions of manufacture and use of a substance for all life-cycle stages. Manufacturers, importers and downstream users of substances on their own or in preparations have to ensure that these are manufactured and can be used in such a way that human health and the environment are not adversely affected. The basic steps that a registrant will need to go through to prepare a registration dossier are given in Figure 12.7. For substances at or above 1 tonne/y a registration dossier has to be submitted (Figure 12.3). As explained in Section

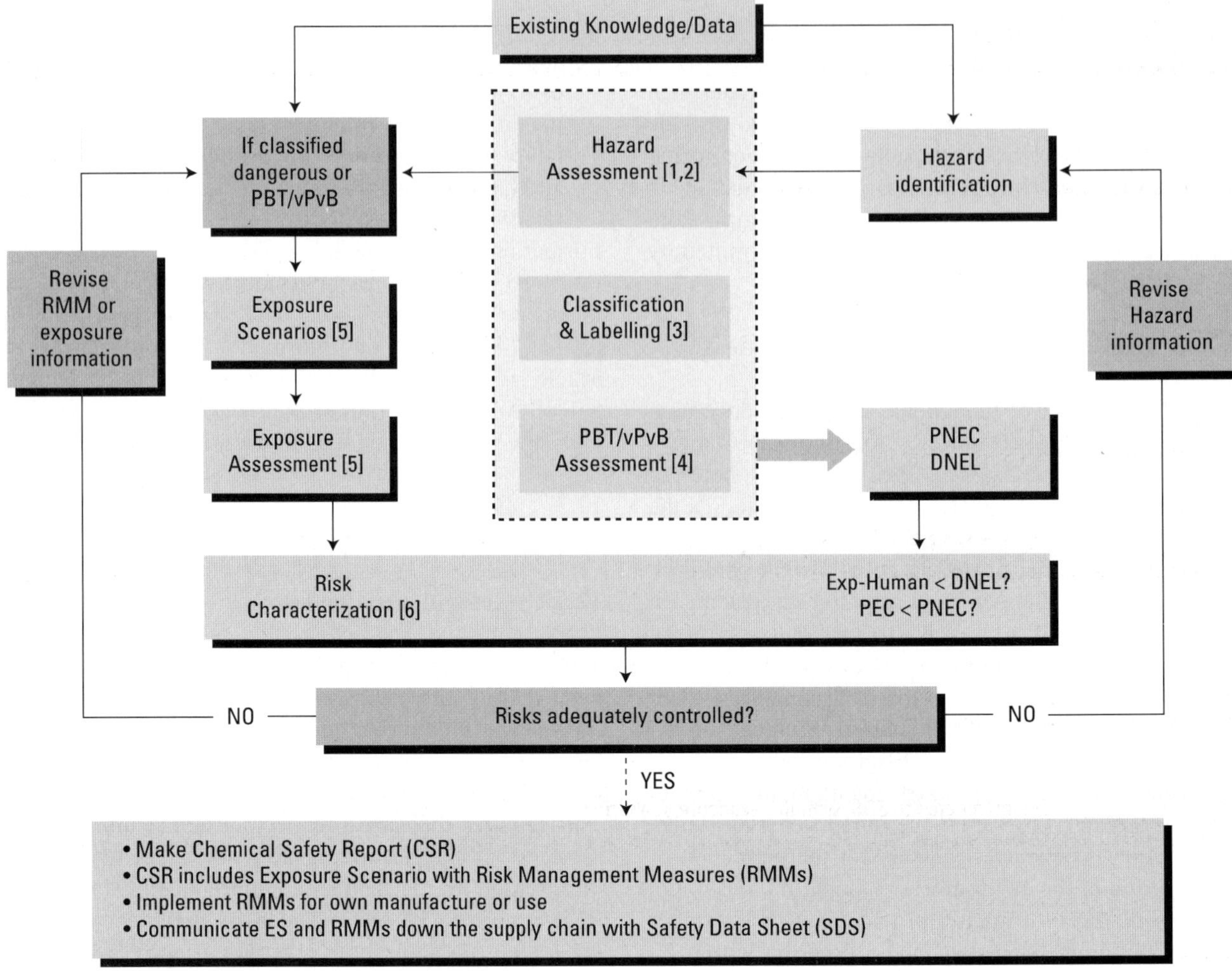

Figure 12.7. Overview of the steps in the CSA.

12.2.2, registrants of substances in volumes above 10 tonnes per year are obliged to conduct a CSA and document it in a CSR, which has to be submitted to the Agency as part of the registration dossier. The different steps required for the preparation of registration dossiers for substances ≥ 10 tonnes/y is given in Figure 12.8.

The CSA follows the different steps in risk assessment as discussed in Chapter 1. An overview of the steps for the CSA, CSR and SDS under REACH is presented in more detail in Figure 12.7. The assessment considers the use of the substance on its own (including any major impurities and additives), in a preparation and in an article, as defined by the identified uses. The assessment considers all stages in the life-cycle of the substance resulting from the manufacture and identified uses. The CSA includes the following steps (see also Figure 12.7):

1. Human health hazard assessment including C&L and the derivation of a DNEL, i.e., a level of exposure to the substance above which humans should not be exposed.
2. Physicochemical hazard assessment including C&L.
3. Environmental hazard assessment including C&L and derivation of a Predicted No Effect Concentration (PNEC).
4. The assessment of PBT and vPvB chemicals.

If, as a result of steps 1 to 4, the manufacturer or importer concludes that the substance or the preparation meets the criteria for classification as dangerous according to Directive 67/548/EEC or Directive 1999/45/EC or is assessed to be a PBT or vPvB, the CSA should also include:

5. Exposure assessment. This includes the generation of exposure scenario(s) or the generation of relevant

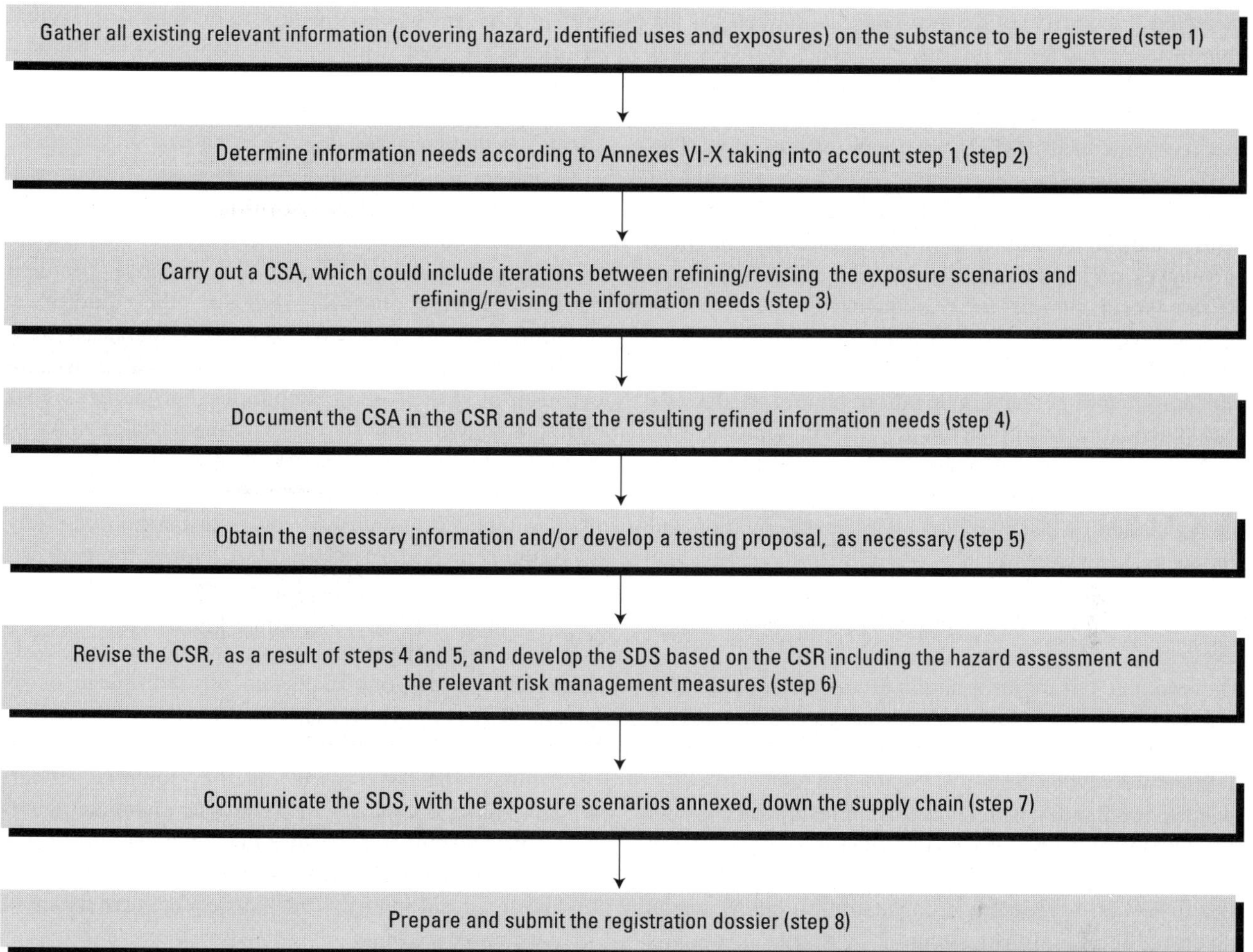

Figure 12.8. Steps required for the assessment of substances at or above 10 tonnes per year. Modified after [31].

use and exposure categories and an exposure estimation. This is an analysis of the reasonably foreseeable exposure of man and/or the environment to the substance taking into account implemented and recommended RMMs and the exposure assessment, i.e., the derivation of predicted environmental and/or occupational and/or consumer exposure concentrations.

6. Risk characterization. This shall be carried out for each exposure scenario.
7. The sequence is repeated in one or more subsequent iterations until safe use has been demonstrated.

It should be stressed that steps 5 and 6 are only necessary if a substance is classified as dangerous or is a PBT/vPvB substance. The iterations to be carried out might only be needed to achieve adequately controlled risks.

12.3.2. Environmental hazard assessment

Different types of information need to be collected or generated when compiling a CSA. The hazard assessment part of the CSA (steps 1 to 4 in Figure 12.7) has to be conducted based on all available information and, as a minimum, on the basis of the information required in accordance with Annexes VI-X of REACH (Table 12.5), while additional information may be needed as a result of the outcome of the exposure estimation and risk characterization (steps 5 and 6). If risk for a given use is not controlled, additional information may need to be collected or generated to see if the assessment can be refined.

Classification and labelling

The criteria for C&L are applied to the hazard data

collected previously. If the substance is dangerous, the appropriate category of hazard, the hazard symbols and the risk and safety (R and S) phrases need to be specified. The outcome of the C&L may have implications for the recommended RMMs or conditions of use. The GHS (see also Chapters 6 and 7), which provides a harmonized basis for globally uniform environmental, health and safety information on hazardous chemical substances and mixtures, was developed at UN level. At the World Summit for Sustainable Development in Johannesburg 2002, the European Commission, the EU Member States and stakeholders from industry and non-governmental organizations endorsed the UN recommendation to implement the GHS in domestic law. In 2006, the European Commission (http://ec.europa.eu/enterprise/reach/) took the necessary steps to implement the legislation by distributing a GHS draft regulation and impact assessment.

Derivation of no effect concentrations or no effect levels

The available hazard information on toxicity or risks due to certain physicochemical properties partly depends on the quantity of the substance manufactured or imported. If there are several tests addressing the same property, then the valid and most relevant test or tests which giving rise to the most concern in the situations encountered has to be used to determine various hazard endpoints. The PNEC is regarded as a concentration below which adverse effects will not occur. The PNEC is derived from toxicity test endpoints (LC50s or NOECs) using appropriate assessment factors or other extrapolation methodologies as described in the Technical Guidance Document [13] and Chapter 7.

12.3.3 Human health hazard assessment

The procedure for hazard assessment for human health is comparable to the approach applied for environmental hazard assessment. The DNEL for workers and the general population (including consumer exposure) is regarded as an exposure level (internal or external) below which an unacceptable effect will not occur and is based on a series of assessment steps related to the available toxicity data. The DNEL is derived from toxicity test endpoints (NOAELs or LOAELs) using appropriate assessment factors [13]. For the human endpoints a distinction needs to be made between threshold and non-threshold substances. For substances considered to exert their effect by a non-threshold mode of action it is generally assumed, as a default assumption, that even at very low levels of exposure residual risks cannot be excluded (see Chapter 6). The physicochemical properties of a chemical which may cause risks such as explosivity, flammability and oxidizing potential, are also taken into consideration.

12.3.4 Physicochemical hazard assessment

Substances which are dangerous because of their physicochemical hazard, trigger the additional requirements for CSA/CSR and SDS under REACH in the same way as substances which are dangerous because of their (eco)toxicological properties. Risk characterization with regard to human health must be carried out for substances which have been classified on the basis of certain physicochemical properties (explosivity, flammability or oxidizing potential), or if there are other reasonable grounds for concern. For every physicochemical property, the assessment needs to entail an evaluation of the likelihood (risk) that an adverse effect will be caused under reasonably foreseeable conditions of use in the workplace or by consumers. The safety assessment to be presented in the CSR has to document that the risks are adequately controlled. The assessment of the potential effects arising from the capacity of hazardous chemical agents to cause accidents, in particular fires, explosions or other hazardous chemical reactions covers:

- Hazards resulting from the physicochemical nature of the chemical agents.
- Risk factors identified in their storage, transport and use.
- The estimated consequences in the event of occurrence.

As the REACH legislation does not cover the major-accident hazards of certain industrial activities as defined by the so-called Seveso Directive (Council Directive 82/501/EEC) the accident scenarios to be considered are those which might occur in the workplace and those related to consumer use. As it is unreasonable to expect catastrophic consequences to result from such accidents, simplified assessments, based on questionnaires and/or checklists, can be used to evaluate whether the risks are adequately controlled [31].

12.3.5 PBT and vPvB assessment

The PBT and vPvB substances are required to undergo further evaluation as the potential for long-term effects arising from their persistence, bioaccumulation potential, and toxicity, are difficult to predict and because such accumulation would be practically difficult to reverse. The assessment of PBTs and vPvBs is a two-step procedure. Step 1 is the *comparison with the criteria* where a comparison is made between the available information submitted as part of the technical dossier, against the criteria given in Annex XIII of the REACH Regulation [29]. If the available information is not sufficient to be able to decide whether the substance meets the criteria in Annex XIII then other evidence, like monitoring data available to the registrant which gives rise to an equivalent level of concern, needs to be considered on a case-by-case basis. Step 2 is the *emission characterization*. If the substance fulfils the criteria, an emission characterization is conducted comprising the relevant parts of the exposure assessment (see Section 12.3.6). In particular it needs to contain an estimate of the amounts of the substance released to the different environmental compartments during all activities carried out by the manufacturer or importer and all identified uses, and an indication of the likely routes by which humans and the environment may be exposed to the substance [29]. The PBT and vPvB assessment is further detailed in Chapters 6 and 7.

12.3.6 Exposure assessment

The exposure assessment has to be carried out if a substance meets the criteria for classification as dangerous and or the PBT/vPvB criteria (or has properties of similar concern). In which case, the registrant must follow two consecutive steps: (1) generation of ESs and (2) an exposure estimation. In the subsequent risk characterization the registrant needs to show that risks can be adequately controlled for all identified uses throughout all stages of the chemical life cycle. If exposure estimation is not required, the registrant can finish the CSA by completing the CSR. An SDS is only required for dangerous substances. The SDS is not required if the substance does not meet the classification criteria.

Step 1. Development of exposure scenarios

Exposure scenarios are the core of the process of carrying out a chemical safety assessment. A chemical safety assessment may be iterative. The first assessment is based on the required minimum of all available hazard and exposure information. If the initial assumptions lead to a risk characterization indicating that risks are not adequately controlled, then it is necessary to carry out an iterative process with amendments to one or more of the factors determining the hazard(s) or exposure(s) with the aim of demonstrating adequate control. ESs consider the emissions during all relevant parts of the life-cycle of the substance under the assumption that the RMMs described in the ES have been implemented. The ESs are a new and crucial element in REACH. They play a dual role:

1. They enable a quantitative release and exposure estimation to be made by describing the determinants of exposure.
2. They are the communication tool to the user(s) on how to use the chemical in such a way that the risks are controlled (see Chapters 1 and 2).

The ES is a description of a control strategy for a substance, giving realistic operational conditions for its use for manufacture or identified use(s), or for a group of substances or a preparation, and prescribes the necessary RMMs that have to be in place during the manufacture or use of a substance, including service life and waste phase, under a given set of operational conditions.

The ES should be developed by the manufacturer or importer. When appropriate, downstream users or their organizations may be involved in or take over responsibility for the development of the ES. The ES is intended for risk management in the various life-cycle stages to ensure safe handling and adequate control of risk related to human health (workers and consumers) and the environment.

Step 2. Exposure estimation

Exposure is estimated for each ES. Exposure estimation follows a logical process of three consecutive steps:

a. Estimation of the emission during all relevant parts of the life-cycle of the substance resulting from manufacturing and each of the identified uses and cover, and where relevant, the waste stage.
b. Characterization of possible degradation, transformation, or reaction processes and estimation of environmental distribution and fate. These processes are discussed in detail in Chapter 3 of this book.
c. Estimation of exposure levels. This needs to be performed for all human populations (workers, consumers and people liable to exposure indirectly via the environment) and all environmental compartments for which exposure to the substance is known or reasonably foreseeable. Environmental concentrations and human daily intake doses or inhalation concentrations are calculated using models which take into account the transport and fate of the substance. Generic exposure models covering a wide range of applications are used in the exposure estimation. The exposure estimation aims

at "reasonable worst case" results by applying Tier 1 models: conservative, but not unrealistic, standard exposure models and, where possible, mean, median or typical parameter values. If monitoring data on exposure levels is available, interpretation of this data has to be given special consideration. Each relevant route of human exposure (inhalation, oral, dermal and combined through all relevant routes of exposure) has to be addressed. Such estimations need to take account of spatial and temporal variations in exposure patterns. Environmental exposure assessment methods are discussed in detail in Chapter 4, and approaches for the estimation of exposure of humans (consumer, occupational and indirect exposure via the environment) are the subject of Chapter 5.

12.3.7 Risk characterization for human health and the environment

Risk characterization in the context of the CSA is the (quantitative) estimation of the likelihood that adverse effect levels occur in man or the environment due to actual or predicted exposure to a chemical. In risk characterization, exposure levels are compared to suitable no effect levels to yield "risk characterization ratios" (RCRs) for each protection goal. RCRs are derived for all end-points and time scales, environmental and human. It should be noted that these RCRs have to be derived for all stages in the life-cycle of a compound. Safe use of substances is demonstrated when:

- RCRs are below one, both at local and regional level.
- The likelihood and severity of an event occurring due to the physicochemical properties of the substance as determined in the hazard assessment is negligible.

For those human health effects and environments for which it was not possible to determine a DNEL or a PNEC, a qualitative assessment of the likelihood that effects are prevented has to be carried out when implementing the ES. For PBT and vPvB substances, the manufacturer or importer has to implement, and recommend RMMs for downstream users that minimize exposure to humans and the environment. This assessment needs to be repeated iteratively until the outcome of the risk characterization which is recorded in the CSR shows that risks are adequately controlled (Figures 12.7 and 12.9).

A risk characterization needs to be carried out for each ES. It needs to consider the human populations (exposed as workers, consumers or indirectly via the environment and if relevant a combination thereof) and the environments for which exposure to the substance is known or reasonably foreseeable, under the assumption that the RMMs described in the ESs have been implemented. In addition, the overall environmental risk caused by the substance needs to be reviewed by integrating the results for all relevant environments and all relevant emission/release sources of the substance. The assessment consists of:

- A comparison of the exposure of each human population known to be or likely to be exposed with the appropriate DNELs.
- A comparison of the predicted environmental concentrations (PECs) in each environmental compartment with the appropriate PNECs.
- An assessment of the likelihood and severity of an event occurring due to the physicochemical properties of the substance.

The protection goals under REACH are human health and the environment [31], i.e., the aquatic environment (freshwater and marine ecosystems, including sediments as well as sewage treatment plants), micro-organisms, the terrestrial environment, predators (aquatic and terrestrial) and marine top-predators (Box 12.10).

For the human endpoints a distinction needs to be made between threshold and non-threshold substances. For threshold substances the RCR is the ratio of the estimated exposure (concentration or dose) and the DNEL. For substances considered to exert their effect by a non-threshold mode of action, especially mutagenicity and carcinogenicity, it is generally assumed, as a default assumption, that even at very low levels of exposure residual risks cannot be excluded. Because a DNEL cannot be established for these substances, this shall be clearly stated and fully justified. The derivation of DNELs and PNECs is described in Chapters 6 and 7. Further detailed information can be found in Chapters 6 and 7, the TGD [13] and the preliminary TGD for REACH [31].

Steps in the derivation of RCRs

The derivation of RCRs follows a normal sequence used in risk assessment (see Chapter 1) of hazard identification, exposure assessment, effects assessment and risk characterization. The RCRs are calculated as follows:

Step 1. For each ES it is necessary to collect the exposure values, measured or estimated, for the relevant time scales and spatial scales, environmental compartments, human populations and human routes of exposure (PEC), or dose (D).

Step 2. For each ES it is necessary to derive no effect levels (PNECs or DNELs) for the relevant

Box 12.10. Protection goals

Inland environmental proiection targets:

- aquatic ecosystems
- terrestrial ecosystems
- atmosphere
- predators (fish and worm-eating)
- micro-organisms in sewage treatment plants

Marine environmental protection targets:

- aquatic ecosystems
- predators and top predators

Human health:

- man exposed via the environment
- workers
- consumers

time scales, environmental ecosystems, human populations, endpoints of concern, and routes of exposure.

Step 3. The ratio of matching exposure and no effect levels is calculated for each combination (see Equation 12.1).

$$RCR = \frac{PEC}{PNEC} \text{ or } \frac{PEC(D)}{DNEL} \tag{12.1}$$

The risk assessment is a tiered approach, going from simple to more refined and comprehensive assessments (see Chapter 1). The risk assessment is normally based on the evidence that gives rise to the most concern. If on the basis of such a preliminary analysis the conclusion is drawn that the RCR is below 1, safe use of the substance has been demonstrated and the risk assessment can be considered to be complete. If RCRs are higher than 1, further steps in the risk assessment process are needed. Possible RCRs are listed in Table 12.7. Depending on the use, exposure and effects of the chemical, only a subset of these RCRs need to be calculated following a decision-tree approach. For chemicals for which risks are demonstrated, the risk characterization process may involve quite some work. This work can be facilitated and coordinated with computer tools and models such as EUSES [32].

12.3.8 Potential iterations of the CSA

The CSA may be refined in an iterative process until adequate control of risks has been demonstrated (Figures 12.7 and 12.9). A "tentative" ES describes how a process is conducted and which RMMs are or should be implemented under the assumption that risks are adequately controlled. The tentative ES forms the starting point for the exposure estimate and risk characterization, which has to be conducted as part of the CSA with the purpose of documenting adequate control of risks. If this cannot be demonstrated, further assessment is needed. This assessment may be iterative by revising the ES, the hazard information or the exposure estimate until it is shown that risks are adequately controlled during the process covered by the ES. A "tentative" ES can be developed from typical operational conditions for a process (e.g., industrial process or consumer use) and taking typically implemented RMMs as a starting point. It should be clear which RMMs are included as well as their mitigating effects [31].

The "final" ES is the outcome of the CSA process and is part of the CSR, which should document the adequate control of risks. The CSR has to be submitted to the Agency as part of the registration dossier. The ES includes information on operational conditions and RMMs for the process that may be subject to control and enforcement by the authorities. If adequate control of risks during manufacture and use can be demonstrated with the tentative ES, the tentative ES will become the final ES for the substance(s) and process(es) considered.

12.3.9 Risk management measures

Different approaches and strategies are available to reduce emissions and exposure during the life-cycle of a chemical. A short overview is presented in Table 12.8. These examples of RMMs which can be taken by the supplier or recommended down the supply chain have been proven to be adequate and practicable in the EU Member States. They are used in different industry/ downstream user sectors and are applied on the basis of voluntary initiatives or legal requirements.

Examples of substance-related measures are e.g., limiting the concentration of a substance in a preparation, design of the package (child resistant fastenings) or modification of the physical state (solids, powders, solutions). Examples of limitation of the marketing of a substance or product are uses in specific areas or countries or use by trained (certified) specialists.

Table 12.7. Possible RCRs for environmental and human health risk characterization [31].

Environment	
$RCRlocal_{water}$	RCR for local water compartment
RCRregwater	RCR for regional water compartment
RCRlocalwater,marine	RCR for local marine water compartment
RCRregwater,marine	RCR for regional marine water compartment
RCRlocalsoil	RCR for local soil compartment
RCRregsoil	RCR for regional soil compartment
RCRlocalsed	RCR for local sediment compartment
RCRlocalsed,marine	RCR for local marine sediment compartment
RCRregsed	RCR for regional sediment compartment
RCRregsed,marine	RCR for regional marine sediment compartment
RCRstp	RCR for sewage treatment plant
RCRoral,fish	RCR for fish-eating birds/mammals (freshwater environment)
RCRoral,fish,marine	RCR for fish-eating birds/mammals (marine environment)
$RCR_{oral,fish\ predator,marine}$	RCR for top-predators (marine environment)
RCRoral,worm	RCR for worm-eating birds and mammals
Humans exposed via the environment	
RCRman-envlocal,tot,i	MOS local, total exposure via all media, for endpoint of concern
RCRman-envlocal,inh,i	MOS local, exposure via air, for endpoint of concern
RCRman-envreg,tot,i	MOS regional, total exposure via all media, for endpoint of conc.
RCRman-envreg,inh,i	MOS regional, exposure via air, for endpoint of concern
RCRman-envlocal,tot,i	MOE local, total exposure via all media, non-thr.
RCRman-envlocal,inh,i	MOE local, exposure via air, non-thr.
RCRman-envreg,tot,i	MOE regional, total exposure via all media, non-thr.
RCRman-envreg,inh,i	MOE regional, exposure via air, non-thr.
Humans (occupational exposure)	
RCRworkerinh,acute	RCR acute, inhalatory exposure
RCRworkerinh,vapour,i	RCR for endpoint of concern, inhalatory exposure of vapour
RCRworkerinh,fibre,i	RCR for endpoint of concern, inhalatory exposure of fibers
RCRworkerinh,dust,i	RCR for endpoint of concern, inhalatory exposure of dust
RCRworkerder,acute	RCR acute, dermal exposure
RCRworkerder,i	RCR for endpoint of concern, dermal exposure
RCRworkertot,acute	RCR acute, total exposure
RCRworkertot-v/d,i	RCR for endpoint of concern, total exposure (vapour + dermal)
RCRworkerinh,vapour,nt	RCR inhalatory worker vapour exposure, non-thr.
RCRworkerinh,fibre,nt	RCR inhalatory worker fibre exposure, non-thr.
RCRworkerinh,dust,nt	RCR inhalatory worker dust exposure, non-thr.
RCRworkerder,nt	RCR dermal worker exposure, non-thr.
RCRworkertot-v/d,nt	RCR total worker exposure, non-thr.
Humans (consumers)	
RCRconsinh,acute	RCR acute, inhalatory exposure
RCRconsinh,i	RCR for endpoint of concern, inhalatory exposure
RCRconsder,acute	RCR acute, dermal exposure
RCRconsder,i	RCR for endpoint of concern, dermal exposure
RCRconsoral,acute	RCR acute, oral exposure
RCRconsoral,i	RCR for endpoint of concern, oral exposure
RCRconstot,acute	RCR acute, total exposure
RCRconstot,i	RCR for endpoint of concern, total exposure
RCRconsinh,nt	RCR inhalatory consumer exposure, non-thr.
RCRconsder,nt	RCR dermal consumer exposure, non-thr.
RCRconsoral,nt	RCR oral consumer exposure, non-thr.
RCRconstot,nt	RCR total consumer exposure, non-thr.

i = {repdose,carc,fert,mattox,devtox}

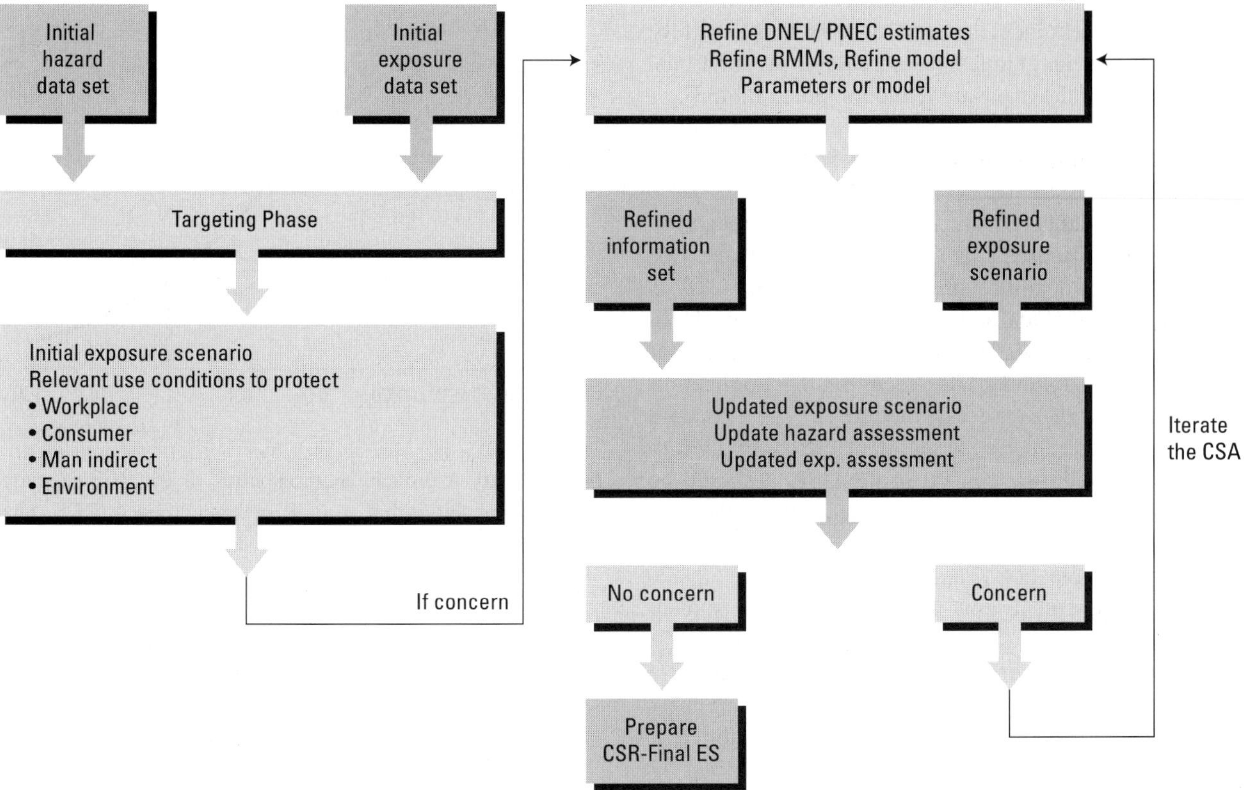

Figure 12.9. Iterating the CSA process [31].

Substances can be limited in their use by using them only in specified applications or restricting their uses with releases (e.g., in water). Technical measures can be taken to reduce exposure, such as ventilation, separation of people and sources, redesign of production/use processes. Organizational measures include, for instance, access restrictions to certain/specific workplaces, limiting time of operation/work activities, training, etc. Personal protection measures include protective clothing, gloves, gas/dust filter masks, etc. Further information is provided in the preliminary TGD [31] and Chapter 2. When implementing RMMs a distinction should be made between industrial use, professional use and use by consumers:

- In industrial use, workers generally have a high standard of competence, qualifications and skills and there is a high standard of supervision and control as well as a high standard for technical measures.
- In professional use, workers have various/different competences, qualifications and skills and there are lower or varying standards of supervision and control as well as lower standards for technical measures.
- In consumer use, there is no expertise/fewer skills for implementing measures and no technical measures and no individual protection (except for gloves or protective glasses in certain cases).

Typically available RMMs and their possible mitigating effect on releases or exposure have been presented in more detail in one of the REACH Implementation Projects [31]. In these reports information can be found about RMMs related to consumer, professional and industrial use, the performance of risk reduction

Table 12.8. Categories of risk management measures [31].

Product/substance related measures
Limitation of the marketing of a substance/product
Limitation of the use of a substance/product
Instructions/information/warnings
Technical measures
Organizational measures
Personal protection measures

measures, workplace protection factors, safety instructions for dangerous substances and preparations, and estimates of the exposure reduction potential [33].

12.3.10 Communicating uncertainty in the CSA

The issues of uncertainty, variability and precaution [34-37] were discussed in Chapter 1. There are two distinct ways of dealing with uncertainty in the CSA: a deterministic approach and a probabilistic approach. In the current deterministic approach, uncertainty is not explicitly but implicitly addressed by the application of reasonable worst-case assumptions in the exposure assessment and the use of assessment factors applied to the hazard data. In the context of the CSA it should be stressed that uncertainty affects all aspects of the risk assessment and subsequent steps, i.e., risk management and risk communication [31]:

- Hazard assessment: how uncertain is the measure of (no) effect?
- Exposure assessment: how uncertain is the exposure estimate?
- Risk characterization: how uncertain is the risk quotient or RCR?
- Risk assessment: how to take decisions in the light of uncertainty?
- Risk communication: how to communicate the uncertainty considerations?

Uncertainties (also the non-quantifiable uncertainties) should preferably be addressed as an integral part of the work during the assessment and not as an "add-on" in the CSR at the end of the assessment. To be able to make decisions on chemical safety, a decision to act or not act nevertheless needs to be based on a boundary or measure of (no) effect, even if uncertainty is incorporated in the outcome of the assessment. A transparent evaluation of uncertainty therefore should assist in communicating these uncertainties to improve decision-making in the light of the uncertainty associated with the outcome of the risk assessment [38-41]. Attempts have been made to come up with methods that are able both to quantify and to communicate the risks [42]. In the risk assessment methods in the envisaged TGD on preparing the CSR, risk is characterized by means of a deterministic RCR (Section 12.3.7; Equation 12.1). Although it is not a true measure of risk, interpreting uncertainty in REACH is best linked to the RCR and the likelihood that it is (or is not) exceeded. The RCR remains an important vehicle for demonstrating that risk is adequately controlled. Probably, the most optimal approach will be a step-wise or tiered approach [31]. Three different approaches have

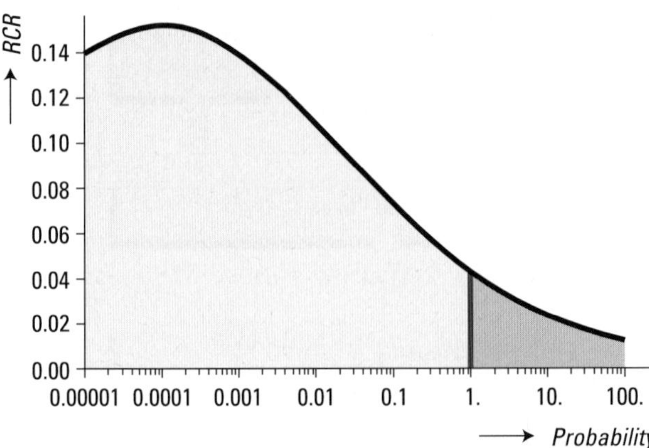

Figure 12.10. A probability distribution of the RCR (x-axis), with the probability that an RCR of 1 is exceeded (RCR ≥ 1). From [31].

been proposed for getting to grips with uncertainty in the CSA. They have different levels of complexity, resource intensity (time and or money) and data needs:

- Tier 1 is a qualitative uncertainty assessment using the deterministic approach linked to a scenario analysis.
- Tier 2 is a simple (semi-quantitative) analysis providing a probability distribution of the RCR (i.e., the probability that the RCR ≥ 1). See Figure 12.10.
- Tier 3 is a full quantitative probabilistic assessment.

Further discussions are needed to decide on the way forward with uncertainty analysis in the CSA.

In conclusion, it is essential to list the different sources of uncertainty and/or variability and assign them to classes of inputs to improve the identification of the main sources that can be addressed by refining the CSA, or that are intrinsic to the assessment [31]. As risk assessment inherently involves judgements, assumptions and uncertainty (see Chapter 1), it is essential to document this in the CSR in a way which is transparent. This will help to address the communication challenge of REACH in general and the CSR in particular.

12.3.11 Format of the CSR

The information and outcome of the risk assessment and risk management steps discussed in the previous sections of this chapter need to be reported in the specified format set out in detail in Annex 1 to the REACH legislation [23]. A short summary of this format is presented in Table 12.9.

Table 12.9. Simplified format of the chemical safety report [23].

Part A	
1.	Summary of risk management measures
2.	Declaration that risk management measures are implemented
3.	Declaration that risk management measures are communicated
Part B	
1.	Identity of the substance and physical and chemical properties
2.	Manufacture and uses
3.	Classification and labelling
4.	Environmental fate properties
5.	Human health hazard assessment
6.	Human health hazard assessment of physicochemical properties
7.	Environmental hazard assessment
8.	PBT and vPvB assessment
9.	Exposure assessment
10.	Risk characterization

12.3.12 Concluding remarks

The CSR has to be prepared and submitted in a standard format as set out in Table 12.9 and Annex I to the REACH Regulation. It can be seen that Part A reflects the focus of REACH on risk management, whereas Part B follows the current risk assessment approach documented in the TGDs [13]. The Agency will determine which word processing programs may be used to complete the CSR and make this known via its website. To facilitate the work, the development of an automatic CSR generation tool is being considered, which can generate the basic elements of the chemical safety report from data that are entered into the database IUCLID-5 as part of the REACH–IT system.

12.4 SAFETY DATA SHEETS

Under REACH there is a general obligation to pass on information up and down the supply chain regardless of the quantity of a substance that is manufactured, imported or used. The main purpose of the information through the supply chain provisions of REACH is to establish a comprehensive and transparent framework within which industry can transmit information on hazards and risks down the supply chain to ensure appropriate RMMs are implemented by downstream users.

The main tool used by industry is the Safety Data Sheet. The SDS is a key element in the hazard and risk management communication from chemical substance suppliers and formulators to downstream users (Figures 12.3 and 12.5).

REACH also requires that the downstream user plays an active role in providing new information back to the supplier on intrinsic properties and any other information that might call into question the appropriateness of RMMs included in the SDS concerning both chemical substances and preparations.

Suppliers of chemicals are required to inform their downstream users or distributors about hazards and measures to adequately control the risks of the substances they supply. With regard to obligations to pass on information down the supply chain, a distinction has to be drawn between dangerous and non-dangerous substances. Dangerous substances are substances classified as dangerous in accordance with Directive 67/548/EEC or 1999/45/EC. An SDS has to be provided for these chemicals (Section 12.4.1), as well as PBT or vPvB substances. The information requirements of the SDS will be discussed in Section 12.4.2. The information requirements for non-dangerous substances will be given in Section 12.4.3. The role of downstream users in providing information will be discussed in Section 12.4.4, and the information requirements for workers will be given in Section 12.4.5.

12.4.1 Safety data sheets for dangerous substances

In Article 31 and Annex II to the legislation REACH takes over the already existing duty of providing a SDS. Annex II is the guide to the compilation of the SDS. Safety data sheets will continue to contain information on the hazards of the substance or preparation, as well as information on the recommended RMMs required to adequately control any risks to health and the environment. In addition, for all those substances for which a chemical safety assessment is required, the information in the SDS must be consistent with the CSA and the relevant ESs for the recipient have to be annexed to the SDS. This obligation to annex exposure scenarios to the SDS has to be met by:

- Manufacturers or importers: for registered substances on their own, or in a preparation, manufactured or imported in quantities of 10 tonnes or more per year.
- Downstream users: for substances on their own, or in preparations, supplied to them that are manufactured or imported in quantities of 10 tonnes or more per year, for uses which the downstream user did not identify to his suppliers, as well as the relevant ESs from their suppliers.
- Distributors: if they have received such information and it is relevant for their customers.

All these actors in the supply chain need to make sure that they use the information derived in the CSA to compile the SDS and that the SDS is thus consistent with the CSA. In the case of preparations, Article 31 allows the option of developing a CSA for the preparation as a whole, instead of for all the individual substances in the preparation, to derive DNELs and PNECs and report the hazards and control measures for the preparation itself.

12.4.2 Information required in the SDS

There are a number of new mandatory items in the SDS under REACH. Firstly, consideration should be given to persistency, bioaccumulation and toxicity in order to check whether the substance belongs in the group of PBT or vPvB substances, as these substances also require SDSs even if they are not classified as dangerous under Directive 67/546/EEC. Secondly, the SDS should contain the PNECs for the environmental compartments and the DNELs for man. Finally, REACH requires that the relevant ESs are provided in an annex to the SDS as discussed above. The substance properties from the IUCLID dossier and the ESs provide the information for the CSA. Many of the results of the CSA, like classification and labelling as a hazardous substance, precautionary measures and emergency measures, packaging requirements and possibilities for rendering the substance harmless, are fed back into the appropriate section of the IUCLID-5 dossier and are included in the SDS and CSR. The format of the SDS is set out in detail in Annex II of REACH. A summary of the required information is presented in Table 12.10. A further explanation of the SDS requirements under REACH has been published recently [31].

Table 12.10. Information required in the SDS under REACH [23].

1.	Identification of the substance/preparation and of the company/undertaking
2.	Hazards identification
3.	Composition/information on ingredients
4.	First-aid measures
5.	Fire-fighting measures
6.	Accidental release measures
7.	Handling and storage
8.	Exposure controls/personal protection
9.	Physical and chemical properties
10.	Stability and reactivity
11.	Toxicological information
12.	Ecological information
13.	Disposal considerations
14.	Transport information
15	Regulatory information
16.	Other information
Annex: Exposure Scenarios	

These safety data sheets are the instrument used to deliver all information to the downstream user. It must be in a form which will allow him to check that his intended use and, more importantly, his associated control measures, comply with those described in the SDS provided to him by his upstream supplier. Furthermore, the downstream user has to check if RMMs included in the SDS are relevant to his conditions of use. The downstream user must then pass on the relevant information to his customers down the supply chain.

12.4.3 Information down the supply chain

Any supplier of a substance on its own or in a preparation who does not have to supply a SDS according to Article

31 will have to pass some basic information down the supply chain: the registration number(s), whether the substance is subject to authorization or details of any restriction, and other information which is necessary to enable appropriate RMMs to be taken.

12.4.4 Information up the supply chain

Any actor in the supply chain will need to inform the next actor or distributor of new information on the hazard of the substance, regardless of the uses concerned. For identified uses they will have to inform the next actor or distributor of information leading them to consider any communicated RMMs as not being appropriate. Information on the appropriateness of RMMs will need to be passed up the supply chain to the supplier who developed the recommended measures.

12.4.5 Information for workers and obligation to keep and make information available

Under REACH, workers and their representatives will have access to all information that is passed down the supply chain, for substances they use or may be exposed to during the course of their work. All information that will have been generated to fulfil any obligation under REACH will need to be kept for at least ten years after the last manufacture, import or use of the substance. The Agency and competent authorities of the Member State of any actor in the supply chain may request that this information be made available, in addition to any requests made under the registration or evaluation procedures.

12.5 CONCLUDING REMARKS

In the introduction to Chapter 12 we said that "it is clear that the real challenge is the implementation of REACH. It will require advances in science and technology, efficient communication between all the stakeholders, pragmatism and, last but not least, effective monitoring of the progress made". It is more than a communication challenge: it will require intense and continuous dialogue with, and cooperation between, all stakeholders (Annex 12.1). Successful implementation of REACH will involve more than 30,000 companies, more than 30,000 chemicals, all EU Member States, the European Chemicals Agency and NGOs. This was recognized by the Commission in early stages of this process. It was the reason why the European Chemicals Bureau was given the task in 2004 of coordinating a number of the REACH Implementation Projects (RIPs) in close collaboration with these stakeholders, long before the REACH legislation was finalized (Box 12.11).

Box 12.11. REACH implementation Projects

- RIP 1: REACH Process description
- RIP 2: REACH - IT
- RIP 3: Technical Guidance and Tools for Industry
- RIP 4: Technical Guidance and Tools for Authorities
- RIP 5/6: Setting up the Pre-Agency
- RIP 7: Commission preparations for REACH

At that stage, it was already clear that guidance was needed to facilitate this implementation challenge. As we write this chapter work on the RIPs is in progress. A summary of this planned guidance (especially RIP 3 and RIP 4) is presented in Figures 12.11 and 12.12. Further information on the development and progress of the RIPs can be obtained from the websites of the ECB (http://ecb.jrc.it/REACH/) and the Agency (http://ec.europa.eu/echa/).

It is envisaged that the results of the different RIP projects related to the development of guidance material will be integrated into one guidance package.

12.6 FURTHER READING

1. Commission of the European Communities. 2007. Corrigendum to Regulation (EC) No 1907/2006 of the European Parliament and of the Council of 18 December 2006 concerning the Registration, Evaluation, Authorisation and Restriction of Chemicals (REACH), establishing a European Chemicals Agency, amending Directive 1999/45/EC and repealing Council Regulation (EEC) No. 793/93 and Commission Regulation (EC) No. 1488/94 as well as Council Directive 76/769/EEC and Commission Directives 91/155/EEC, 93/67/EEC, 93/105/EC and 2000/21/EC (OJ L 396, 30.12.2006). OJ L136, volume 50, 29 May 2007.
2. European Chemicals Bureau. 2004. The REACH proposal process description. Reach Implementation Project 1. European Commission, Joint Research Centre, Ispra, Italy.
3. European Commission, 2007. Questions and answers on REACH. http://ec.europa.eu/enterprise/reach/docs/reach/TechnicalQA_Feb2007. pdf.
4. European Chemicals Agency. 2007. Guidance on the different processes under REACH, e.g. Guidance on IUCLID, Guidance on registration, pre-registration, data-sharing, dossier and substance evaluation, on how to comply with the provisions of the new Regulation on

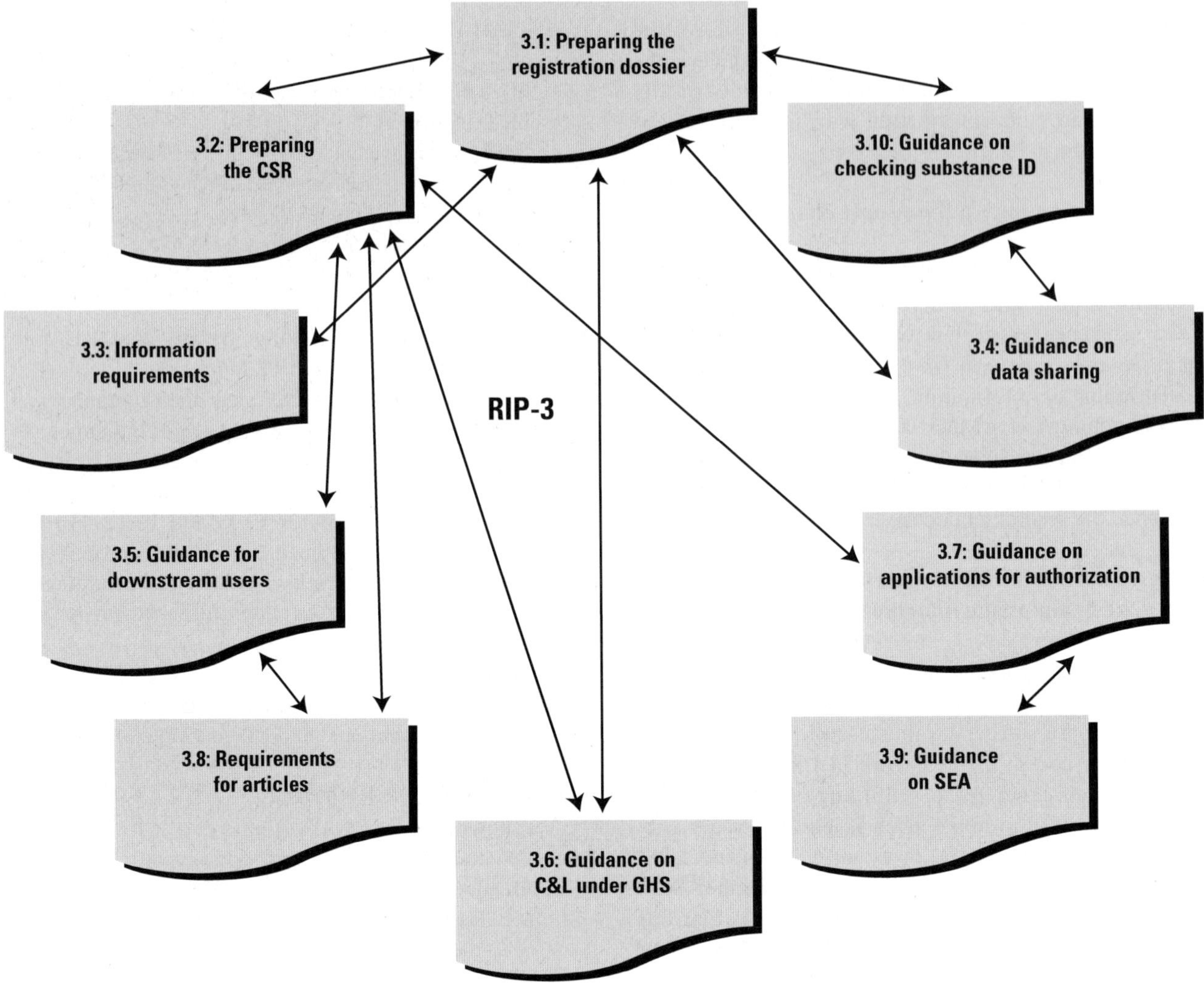

Figure 12.11. Guidance for industry. Diagram of RIP 3. The arrows indicate the main relationships between the different RIP projects.

Classification, Packaging and Labelling of substances and mixtures; Guidance for identification and naming of substances under REACH, for downstream users, for the preparation of an Annex XV dossier for restrictions, etc. (http://echa.europa.eu/reach_en.html).

5. Commission of the European Communities. 2003. Technical Guidance Document in support of Commission Directive 93/67/EEC on risk assessment for new notified substances, Commission Regulation (EC) No. 1488/94 on risk assessment for existing substances and Directive 98/8/EC of the European Parliament and of the Council concerning the placing of biocidal products on the market. Joint Research Centre, European Chemicals Bureau, Brussels, Belgium.

REFERENCES

1. Organization for Economic Co-operation and Development. 2001. Environmental outlook for the chemicals industry. OECD, Paris, France.
2. Carson R. 1962. *Silent Spring*. Houghton Mifflin, Boston, MA, US.
3. Haigh N Baillie A. 1992. Final report on chemicals control in the European Community in the 1990s. Institute for European Environmental Policy, London, UK.
4. Vermeire T, Van Der Zandt P. 1995. Procedures of hazard and risk assessment. In: Van Leeuwen CJ, Hermens JLM, eds. *Risk Assessment of Chemicals: An Introduction.*

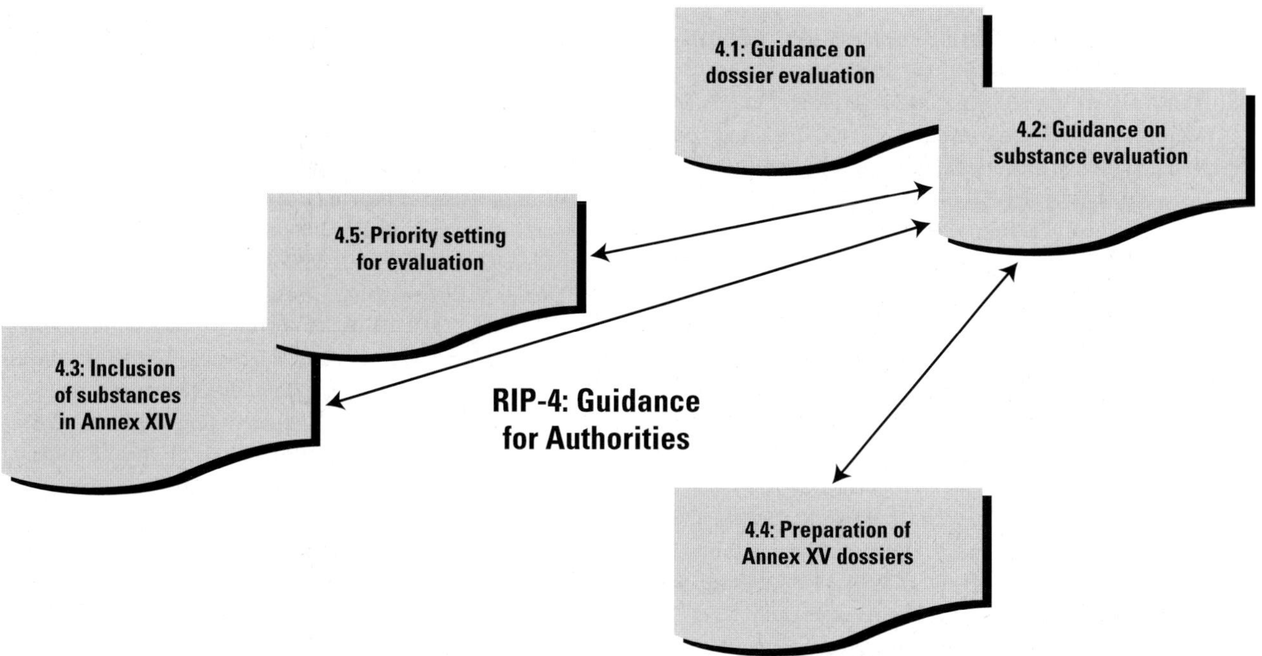

Figure 12.12. Diagram of the guidance for authorities. The arrows indicate the main relationships between the different RIP projects.

Kluwer, Dordrecht, The Netherlands, pp. 293-337.

5. European Environment Agency. 2001. Late lessons from early warning: the precautionary principle 1896-2000. Environmental issue report No 22. EEA, Copenhagen, Denmark.
6. Commission of the European Communities. 1967. Council Directive 67/548/EEC of 16 August 1967 on the approximation of the laws, regulations and administrative provisions relating to the classification, packaging and labelling of dangerous substances. *OJ* L96/1.
7. Commission of the European Communities. 1992. Council Directive 92/32/EEC of 30 April 1992 amending for the 7th time Directive 67/548/EEC on the approximation of the laws, regulations and administrative provisions relating to the classification, packaging and labelling of dangerous substances. *OJ* L154/1.
8. Commission of the European Communities. 1999. Council Directive 1999/45/EC of 31 May 1999 concerning the approximation of laws, regulations and administrative provisions of the member States relating to the classification, packaging and labelling of dangerous preparations. *OJ* L 200/1.
9. Commission of the European Communities.1993. Council Regulation 793/93/EEC of 23 March 1993 on the evaluation and control of the risks of existing substances. *OJ* L 084/1.
10. Commission of the European Communities. 1976. Council Directive 76/769/EEC of 27 July 1976 on the approximation of the laws, regulations and administrative provisions of the Member States relating to restrictions on the marketing and use of certain dangerous substances and preparations. *OJ*, L 262 (27.09.1976) p. 201.
11. United Nations. 1992. Rio Declaration, Agenda 21. United Nations Conference on Environment and Development (UNCED), Rio de Janeiro, Brazil.
12. United Nations. 1987. *Our Common Future.* World Commission on Environment and Development. Oxford University Press, Oxford, UK.
13. Commission of the European Communities. 2003. Technical Guidance Document in support of Commission Directive 93/67/EEC on risk assessment for new notified substances. Commission Regulation (EC) No. 1488/94 on risk assessment for existing substances and Directive 98/8/EC of the European Parliament and of the Council concerning the placing of biocidal products on the market. Joint Research Centre, European Chemicals Bureau, Brussels, Belgium.
14. Bodar C, De Bruijn J, Vermeire T, van der Zandt P. 2002. Trends in risk assessment of chemicals in the European Union. *Human Ecol Risk Assess* 8:1825-1843.
15. Bodar CWM, Berthault F, De Bruijn JHM, Van Leeuwen CJ, Pronk MEJ, Vermeire TG. 2003. Evaluation of EU

risk assessments of existing chemicals (EC Regulation 793/93). Chemosphere 53:1039-1047.

16. Bodar WM, Pronk MEJ, Sijm DTHM. 2005. The European Union risk assessment on zinc and zinc compounds: the process and the facts. *Integrated Environ Assessm Managem* 1:301-319.
17. Commission of the European Communities. 2001. White Paper. Strategy for a future chemicals policy. Com(2001) 88 final. Brussels, Belgium.
18. Van Leeuwen CJ, Bro-Rasmussen F, Feijtel TCJ, Arndt R, Bussian BM, Calamari D, Glynn P, Grandy NJ, Hansen B, Van Hemmen JJ, Hurst P, King N, Koch R, Müller M, Solbé JF, Speijers GAB, Vermeire T. 1996. Risk assessment and management of new and existing chemicals. *Environ Toxicol Pharmacol* 2:243-299.
19. Allanou R, Hansen BG, Van Der Bilt Y. 1999. Public availability of data on EU high production volume chemicals. Report EUR 18996 EN, European Commission, Joint Research Centre, Ispra, Italy.
20. Commission of the European Communities. 2002. Communication from the Commission on the precautionary principle. COM (2000) 1 final. Brussels, Belgium.
21. Schörling, I. 2004. The Only Planet Guide to the Secrets of REACH. REACH - What happened and why? European Parliament, Brussels, Belgium.
22. Wallstrom, M. 2004. Keynote speech by the European Commissioner for Environment. 2nd US-EU Chemicals conference, Charlottesville, VA.
23. Commission of the European Communities. 2003. Proposal for a Regulation of the European Parliament and of the Council concerning the Registration, Evaluation, Authorization and Restriction of Chemicals (REACH), establishing a European Chemicals Agency and amending Directive 1999/45/EC and Regulation (EC) {on Persistent Organic Pollutants} / Proposal for a Directive of the European Parliament and of the Council amending Council Directive 67/548/EEC in order to adapt it to Regulation (EC) of the European Parliament and of the Council concerning the registration, evaluation, authorization and restriction of chemicals. Brussels, 29 October 2003, Brussels, Belgium.
24. Pedersen F, De Bruijn J, Munn S, Van Leeuwen K. 2003. Assessment of additional testing needs under REACH. Effects of QSARs, risk based testing and voluntary industry initiatives. Report EUR 20863 EN, European Commission, Joint Research Centre, Ispra, Italy.
25. Ackerman F, Massey R. 2004. The true costs of REACH. A study performed for the Nordic Council of Ministers. TemaNord 557. Copenhagen, Denmark.
26. Witmond B, Groot S, Groen W, Donszelmann E. 2004. The impact of REACH. Overview of 36 studies on the impact of the new EU chemicals policy (REACH) on society and business. ECORYS and OpdenKamp Adviesgroep, Rotterdam/The Hague, The Netherlands.
27. Pedersen F, Samsøe-Petersen L, Gustavson K, Höglund L, Koundouri P, Pearce D. 2006. The Impact of REACH on the environment and human health. Report prepared for DG Environment. ENV.C.3/SER/2004/0042r. DHI Water & Environment. Horshølm, Denmark.
28. KPMG. 2005. REACH - further work on impact assessment. A case study approach. KPMG Business Advisory Services. Amsterdam, The Netherlands.
29. Commission of the European Communities. 2006. Regulation (EC) No. 1907/2006 of the European Parliament and of the Council of 18 December 2006 concerning the Registration, Evaluation, Authorization and Restriction of Chemicals (REACH), establishing a European Chemicals Agency, amending Directive 1999/45/EC and repealing Council Regulation (EEC) No. 793/93 and Commission Regulation (EC) No. 1488/94 as well as Council Directive 76/769/EEC and Commission Directives 91/155/EEC, 93/67/EEC, 93/105/EC and 2000/21/EC. *Off J Eur Union* L 396 of 30.12.2006.
30. European Chemicals Bureau. 2004. The REACH proposal process description. Reach Implementation Project 1. European Commission, Joint Research Centre, Ispra, Italy.
31. European Chemicals Bureau. 2005. Technical guidance document on preparing the chemical safety report under REACH. REACH Implementation Project 3.2. Report prepared by CEFIC, RIVM, the Federal Institute for Risk Assessment (BfR), Federal Institute for Occupational Safety and Health (BAuA), Ökopol, DHI Water & Environment and TNO Chemistry. European Commission, Joint Research Centre, Ispra, Italy (http://ecb.jrc.it/REACH/).
32. Commission of the European Communities. 2005. EUSES, the European Union System for the Evaluation of Substances. European Chemicals Bureau (ECB) and National Institute of Public Health and the Environment (RIVM). Available from the ECB. European Chemicals Bureau, Ispra, Italy. (http://ecb.jrc.it/REACH/).
33. Health and Safety Executive. 2006. COSHH: A brief guide to the Regulations. London, UK (http://www.coshh-essentials.org.uk/).
34. Scheringer M, Steinbach D, Escher B, Hungerbuhler K. 2002. Probabilistic approaches in the effect assessment of toxic chemicals. What are the benefits and limitations? *Environ Sci & Pollut Res* 9:307-314.
35. MacLeod M, Fraser AJ, MacKay D. 2002. Evaluating and expressing the propagation of uncertainty in chemical

fate and bioaccumulation models. *Environ Toxicol Chem* 4:700-709.

36. Verdonck AM, Van Sprang PA, Jaworska J, Vanrolleghem PA. 2005. Uncertainty and precaution in European environmental risk assessment of chemicals. *Water Sci Technol* 52:227-234.
37. Verdonck AM, Aldenberg T, Jaworska J, Vanrolleghem PA. 2003. Limitations of current risk characterization methods in probabilistic environmental risk assessment. *Environ Toxicol Chem* 22:2209-2213.
38. Slob W, Pieters. 1998. A probabilistic approach for deriving acceptable human intake limits and human health risks from toxicological studies: a general framework. *Risk Analysis* 18:787-798.
39. Bosgra S, Bos PMJ, Vermeire TG, Luit RJ, Slob W. 2005. Probabilistic risk characterisation: an example with di(2-ethylhexyl)phthalate. *Reg Toxicol Pharmacol* 43:104-113.
40. Jager T, Den Hollander HA, Van der Poel P, Rikken MGJ (2001b). Probabilistic environmental risk assessment for dibutylphtalate (DBP). *J Human Ecol Risk Ass* 7 (6):1681-1697.
41. Vermeire T, Jager T, Janssen G, Bos P, Pieters M. 2001. A probabilistic human health risk assessment for environmental exposure to dibutylphthalate. *J Hum Ecol Risk Ass* 7(6):1663-1679.
42. Jager T, Vermeire TG, Rikken MGJ, van der Poel P. 2001. Opportunities for a probabilistic risk assessment of chemicals in the European Union. *Chemosphere* 43:257-64.

ANNEX 12.1. THE MAIN STAKEHOLDERS IN REACH: THEIR ROLES AND DUTIES. MODIFIED AFTER [30].

I. Industry

A first glance at currently existing duties, which will continue to exist after REACH enters into force:

Manufacturers/importers/downstream users:

- Comply with any restrictions on the marketing and use of substances and preparations (restrictions as set out in Directive 76/769/EEC will be taken over by REACH in Annex XVII).
- Classify and label substances and preparations that are placed on the market according to Directive 67/548/EEC and Directive 1999/45/EC.
- Prepare Safety Data Sheets (SDSs) for substances and preparations (requirements in Directive 91/155/EEC will be taken over by REACH in Art. 31 and Annex II).
- Conduct risk assessments and reduce risks for any chemical agent occurring at the workplace (Directive 98/24/EC on chemical agents at work).

A glance at the duties after REACH takes effect:

(1) Manufacturers and importers of substances in quantities < 1 tonne/y:

- Classify and label substances and preparations that are placed on the market.
- Notify classification of dangerous substances with the Agency for the classification and labelling inventory for all substances placed on the market.
- Prepare and supply SDSs for substances and preparations as required by Art. 31 and Annex II to downstream users and distributors.
- Prepare and supply information on non-classified substances as required by Article 32 to downstream users and distributors.
- Conduct risk assessments and reduce risks for any chemical agent occurring at the workplace (Directive 98/24/EC on chemical agents at work).
- Comply with any restrictions on manufacture, placing on the market and use of substances and preparations as set out in Annex XVII.
- Apply for authorization for use(s) of substances listed in Annex XIV.
- In case of having relevant data act as data holder in Substance Information Exchange Fora (SIEF).

(2) Manufacturers of substances in quantities of ≥ 1 tonne/y:

- If your substance has a phase-in status, pre-register it to the Agency.
- In case your substance is a non phase-in substance send an inquiry to the Agency whether the registration has already been submitted for the same substance.
- Collect and share existing, and generate and propose to generate new, information on the properties and use conditions of substances.
- Prepare a technical dossier (note that special provisions apply for isolated intermediates).
- Prepare CSA and CSR (for each chemical ≥ 10 tonnes/y per manufacturer).
- Prepare CSA and CSR including exposure scenarios and risk characterization (for each chemical ≥ 10 tonnes/y per manufacturer, which are dangerous, PBT or vPvB).
- Implement appropriate RMMs for own manufacture and use.
- Submit registration for substances (≥ 1 tonne/y per manufacturer).
- Keep the information submitted in the registration up-to-date.
- Classify and label substances and preparations that are placed on the market.
- Notify/register classification of dangerous substances with the Agency for the C&L inventory for all substances placed on the market.
- Prepare and SDSs for substances and preparations as required by Art. 31 and Annex II to downstream users and distributors.
- Recommend appropriate RMMs in SDS.
- Communicate ESs developed in CSA as Annex to the SDS (≥ 10 tonnes/y per manufacturer).
- Prepare and supply information on non-classified substances as required by Article 32 to downstream users and distributors.
- Conduct risk assessments and reduce risks for any chemical agent occurring at the workplace (Directive 98/24/EC on chemical agents at work).
- Respond to any decision requiring further information as a result of the evaluation process.
- Comply with any restrictions on manufacture, placing on the market and use of substances and preparations as set out in Annex XVII.
- Apply for authorization for use(s) of substances listed in Annex XIV.

(3) Importers of substances and preparations in quantities of ≥ 1 tonne/y:

- If your substance has a phase-in status, pre-register it to the Agency.
- In case your substance is a non phase-in substance send an inquiry to the Agency whether the registration has already been submitted for the same substance.
- Collect and share existing, and generate and propose to generate new, information on properties and use conditions of substances.
- Prepare a technical dossier (note that special provisions apply for isolated intermediates).
- Prepare CSA and CSR developing when required exposure scenarios (≥ 10 tonnes/y per importer).
- Implement appropriate RMMs for own use.
- Submit registration for substances, on their own or in preparations (≥ 1 tonne/y per importer).
- Keep the information submitted in the registration up-to-date.
- Classify and label substances and preparations that are placed on the market.
- Notify/register classification of dangerous substances with the Agency for the C&L inventory for all substances placed on the market.
- Prepare and supply SDSs for substances and preparations as required by Art. 31 and Annex II to downstream users and distributors.
- Recommend appropriate RMMs in SDS.
- Communicate ESs developed in CSA as Annex to SDS (≥ 10 tonnes/y per importer).
- Prepare and supply information on non-classified substances as required by Article 32 to downstream users and distributors.
- Respond to any decision requiring further information as a result of the evaluation process.
- Comply with any restrictions on manufacture, placing on the market and use of substances and preparations as set out in Annex XVII.
- Apply for authorization for use(s) of substances listed in Annex XIV.

(4) Producers of articles:

- Under some circumstances register substances in articles (tonnage trigger ≥ 1 tonne/y in all articles per producer). Comply with C&L, pre-registration and inquiry obligations if relevant.
- Keep the information submitted in the registration up-to-date.
- Under some circumstances notify substances in articles (tonnage trigger ≥ 1 tonne/y per producer).
- When receiving SDS with ESs annexed for dangerous substances and preparations to be incorporated into the articles:
  1. If the use is covered by the ES, implement RMMs as set out in ES, or
  2. If the use is not covered by the SDS annex, inform supplier of the use (i.e., make use known with the aim of making it an identified use) and await new SDS with updated ES(s) or conduct own chemical safety assessment and (if downstream user tonnage ≥ 1 tonne/y) notify the Agency.
- Implement those RMMs as set out in SDS for dangerous substances and preparations which are applicable when incorporated into the articles.
- Conduct risk assessments and reduce risks for any chemical agent occurring at the workplace (Directive 98/24/EC on chemical agents at work).
- Respond to any decision requiring further information as a result of the evaluation process (only relevant for registered substances).
- Comply with any restrictions on manufacture, placing on the market and use of substances and preparations as set out in Annex XVII.
- Use substances authorized for incorporation into the articles as set out in the authorization or apply for authorization for use(s) of substances listed in Annex XIV.

(5) Importers of articles:

- Under some circumstances register substances in articles (tonnage trigger ≥ 1 tonne/y in all articles per importer). Comply with C&L, pre-registration and inquiry obligations if relevant.
- Keep the information submitted in the registration up-to-date.
- Under some circumstances notify substances in articles (tonnage trigger ≥ 1 tonne/y per importer).
- Respond to any decision requiring further information as a result of the evaluation process (only relevant for registered substances).
- Comply with any restrictions on manufacture, placing on the market and use of substances and preparations as set out in Annex XVII.
- Apply for authorization for use(s) of substances listed in Annex XIV.

(6) Downstream Users:

- Check if the substance is placed on the list of pre-registered substances published by the Agency. If not, and considered relevant, ask the Agency to add the substance to the list.

- In case of having relevant data act as data holder in Substance Information Exchange Fora (SIEF).
- Implement RMMs as set out in SDS.
- When receiving SDS with ESs annexed:
  1. If DU use is covered by the ES, implement RMMs as set out in ES annexes to SDS; or
  2. If DU use is not covered by the SDS annex, inform supplier of the use (i.e. make use known with the aim of making it an identified use) and await new SDS with updated ES(s) or conduct own chemical safety assessment and (if downstream user tonnage ≥ 1 tonne/y) notify the Agency.
- Prepare and supply SDS(s) and recommend appropriate RMMs in them and annex ES(s) for further downstream use.
- Prepare and supply information on non-classified substances as required by Article 32 to further downstream users and distributors.
- Pass on new information directly to their suppliers on the hazard of the substance and information that might call into question the RMM identified in the SDS for identified uses.
- Conduct risk assessments and reduce risks for any chemical agent occurring at the workplace (Directive 98/24/EC on chemical agents at work).
- Respond to any decision requiring further information as a result of the evaluation of testing proposals in downstream user reports.
- Comply with any restrictions on manufacture, placing on the market and use of substances and preparations as set out in Annex XVII.
- Use authorised substances as set out in the authorization (this info should be found in the suppliers' SDS) or apply for authorization for use(s) of substances listed in Annex XIV.
- Notify about using an authorised substance to the Agency.

II. Member States:

- Provide advice to manufacturers, importers, downstream users and other interested parties on their respective responsibilities and obligations under REACH (establish national helpdesks).
- Conduct substance evaluation of prioritised substances. Prepare draft decisions.
- Suggest harmonized C&L for CMRs and respiratory sensitizers.
- Identify substances of very high concern for authorization.
- Suggest restrictions.
- Nominate candidates for membership of Agency committees on risk assessment and socio-economic analysis.
- Appoint member for "Member State Committee" to resolve divergences of opinion on decisions following evaluation, consider proposals for harmonized classification and labelling, and identify substances for authorization.
- Provide adequate scientific and technical resources to the members of the committees they nominate.
- Appoint member to the "forum" and meet to discuss enforcement matters.
- Enforce REACH.

III. Agency:

Day to day management of technical, scientific and administrative aspects of REACH. Responsibilities:

- Provide technical and scientific guidance and tools for the operation of REACH in particular to assist the development of chemical safety reports by industry and especially by small and medium-sized enterprises (SMEs).
- Provide technical and scientific guidance on the operation of REACH for Member State competent authorities and provide support to the competent authorities' help desks.
- Receive and check requests for research and development (PPORD) exemptions.
- Pre-registration:
  1. Receive information and grant access to all manufacturers and importers who have submitted information on one substance. When foreseen decide about conflicting issues.
  2. Publish a list of pre-registered substance on the Agency website. Update the list on the request of downstream users.
- Operate the rules on data-sharing for non-phase-in substances.
- Registration: check completeness, require completion of registration and reject incomplete registrations.
- Evaluation:
  1. Ensure a harmonized approach. Set priorities and take decisions.
  2. Conduct dossier evaluation of registrations including testing proposals and other selected registrations.
  3. Substance evaluation: propose draft Community rolling action plans, coordinate the substance evaluation process.
- Substances in articles: take decisions on notifications.

- Authorization/restrictions: manage the process and provide opinions. Suggest priorities.
- Secretariat for forum and committees.
- Take decisions on access to submitted data.
- Publish certain specified data on a publicly accessible database.
- Deal with appeals - registration, R&D, evaluation, confidentiality, etc.

IV. Commission:

- Take decisions on further information needs under the evaluation process where there is no unanimous agreement by Member States.
- Include substances into the authorization system.
- Take decisions on granting or rejecting authorizations.
- Take decisions on restrictions.

V. All stakeholders including industry groups/ associations, NGOs, and the public:

Note: The following are possibilities for stakeholders:

- Access to non-confidential information via the Agency web-site.
- Request access to information.
- Evaluation: submit scientifically valid, relevant information and studies addressed by the testing proposal published on the Agency website.
- Authorization:
  1. Provide comments on substances which the Agency has proposed to be prioritised and on uses which are to be exempted from the authorization requirement.
  2. Provide information on possible alternatives.
- Restrictions:
  1. Provide comments on restriction proposals.
  2. Provide socio-economic analysis for suggested restrictions, or information to contribute to one.
  3. Provide comments on draft opinions from Agency's committee for risk assessment and committee for socio-economic analysis.

13. THE MANAGEMENT OF INDUSTRIAL CHEMICALS IN THE USA

C. AUER AND J. ALTER

13.1 INTRODUCTION

Within the United States Environmental Protection Agency (EPA), the Office of Pollution Prevention and Toxics (OPPT) is responsible for implementing the *Toxic Substances Control Act* (TSCA) of 1976 and the *Pollution Prevention Act* (PPA) of 1990 [1,2]. Before TSCA was passed, it was not known how many chemicals were in commerce in the U.S., where they were being produced and/or imported, and in what quantities they were being produced and/or imported.

TSCA established a National Program within EPA for oversight of chemicals and gave the Agency broad authority to protect against unreasonable risks of toxic substances. Chemicals otherwise regulated were excluded such as pesticides, some nuclear materials, firearms, ammunition, tobacco products, food, food additives, drugs, cosmetics, and medical devices. The statute directs EPA to collect information on new and existing chemicals and to impose testing requirements on those chemicals. It also provides ways for the Agency to impose regulations on chemical production, import, processing, distribution in commerce, use, and disposal. In summary, the law established the foundation for a systematic review process for evaluating new chemicals before they enter commerce and an array of tools for responding to potential risks from existing chemicals [3].

TSCA established for the first time as national policy that the development of adequate data with respect to the effect of chemical substances and mixtures on health and the environment is the responsibility of those who manufacture and process such chemical substances and mixtures, and that the EPA should ensure that adequate measures are taken to control the risks [4]. TSCA established authorities to:

- Compel testing of chemical substances/mixtures for which information is determined to be inadequate and provides a mechanism to share the costs of that testing (§4).
- Require submission of information on new chemicals (not already in commerce) to the Agency for its determination whether the new chemical may present unreasonable risk of harm (§5).
- Establish controls on use, or ban use, for existing chemicals (those already in commerce) if EPA determines that the chemical will present an unreasonable risk of harm (§6).
- Compel submission to EPA of data showing "substantial risk" from chemical substances (§8).
- To develop an inventory of chemical substances in commerce (§8).

EPA's approach to chemical safety has evolved into two basic functions, i.e. (1) as gatekeeper/guardian for chemical hazard and (2) as facilitator of environmental stewardship. The Agency acts as a *gatekeeper/guardian*, using its traditional regulatory authorities to control or keep risky new chemicals out of the market while assessing and managing the risks of existing chemicals. This approach is often referred to as "command and control". The organization's second function, which is newer and is expanding, is to *facilitate environmental stewardship* (Box 13.1).

Both functions contribute to accomplishing EPA's mission of ensuring that industrial chemicals do not pose unreasonable risks to human health or to the environment by promoting:

- Pollution prevention as the guiding principle for controlling industrial pollution.
- Safer new chemicals through a combination of regulatory and voluntary stewardship efforts.
- Risk reduction to control production, use or to otherwise minimize exposure to existing chemicals such as lead, asbestos, and polychlorinated biphenyls (PCBs).
- Increased understanding of risks by providing understandable, accessible, and useful information on chemical hazards, exposures, and risks to the broadest audience possible.

13.2 EPA RISK ASSESSMENT AND RISK MANAGEMENT UNDER TSCA

In all cases, the Agency is directed to conduct risk assessments and to consider the costs and the benefits of any risk management actions it may undertake. The Agency has a general process, but has also established substance-specific programs in response to individual chemicals for which new information becomes available. Below we will cover first the general process, then the chemical-specific programs.

C.J. van Leeuwen and T.G. Vermeire (eds.), Risk Assessment of Chemicals: An Introduction, 553–574.

Box 13.1 EPA acts as gatekeeper/guardian for chemical hazard and facilitator of environmental stewardship

EPA works to create and promote broader use of chemical assessment tools, partnerships, voluntary programs and educational and other initiatives. In all cases, the risk management decisions the Agency is tasked with are determined by its regulatory authority. In its environmental stewardship initiatives, the Agency seeks to elicit risk management and pollution prevention behavior from other actors consistent with its statutory mission.

13.2.1 TSCA risk assessment and risk thresholds for taking action

Control of new chemicals (TSCA §5)

New chemicals are chemicals which have not been made or used commercially before, so the statute makes an assumption that intending makers have invested less in the new materials than in materials already in production, and sets a "may present an unreasonable risk" threshold for action. This is a criterion which is relatively easy to reach. After review of applications to manufacture and use these materials the Agency can choose to take action to restrict or ban production or uses. Regulatory measures include issuance of a Consent Order or a Significant New Use Rule (SNUR), and use of various (e.g. low volume and low release/low exposure) exemptions [5].

Control of existing chemicals (TSCA §6)

The Agency can seek to restrict existing materials. The TSCA requires that restrictive measures be undertaken only after an assessment showing that the substance "will present unreasonable risk," on the assumption that interrupting an ongoing industrial process has higher social costs (there are jobs, plant, etc., invested in the material) than forbidding or restricting conditions under which an intending maker can initiate manufacture. Substance-specific §6 actions have been directed by Congress, e.g., PCBs [6].

Environmental stewardship

The Agency has undertaken several initiatives categorized as "environmental stewardship." Encouraging chemical makers to undertake these initiatives serve to complement the compulsory authorities in TSCA. These voluntary and semi-voluntary programs are described at Section 13.4; the threshold for action in these initiatives is agreement between the Agency and the private stakeholders. In some cases, voluntary action is the primary goal; in others the Agency makes clear that it intends to promulgate a regulatory "backstop" to address gaps or ensure complete coverage.

13.2.2 Risk assessment mechanisms

Risk assessment for chemical substances under TSCA consists of separately conducted hazard and exposure assessments, which are combined when complete to produce a risk assessment. In both the hazard and the exposure assessments, test data on the material is valued, if available. The statute does not have a specified Minimum Data Set (MDS) requirement for new chemicals; accordingly, EPA relies on structure-activity relationship (SAR) analysis to support decisions on new chemicals (see Chapters 9-11). Testing can be required when needed to confirm the SAR-based hazard assessment. The Agency uses a number of tools and models to estimate environmentally important properties for chemicals in the absence of measured data. Many of these models are publicly available [7]. Models available from EPA are given in Tables 13.1-13.3.

Risk assessment for new chemicals

In its review of TSCA § 5 submissions, EPA's new chemicals program (NCP) relies on the knowledge and experience of dedicated and committed engineers, scientists, information management specialists and regulators -- 65 full time staff in 2006 -- to identify and evaluate concerns regarding health and environmental effects, exposures and releases, and economic impacts. Submissions, known as *premanufacture notifications* (PMNs), proceed through an evaluation process to determine whether a more detailed review is required and to identify candidates for regulatory action. The Program focuses on the relatively few new chemicals of greatest concern, such as those which are structurally related to known toxic chemicals and those about which little is known.

Submissions are first passed through an administrative screen to ensure completeness and are then sent through a series of meetings for multi-disciplinary review. In these meetings, reviewers: confirm the chemical identity, predict or verify the chemical's properties, identify similar chemicals for which hazards are known, evaluate processing and use, identify impurities or by-products, and predict environmental and human health effects

Table 13.1. Models for estimating physicochemical properties and fate in the environment. All modules listed in this section are incorporated into USEPA EPISuite software which can be downloaded from the USEPA website [8].

MPBPVP™	Estimates melting point, boiling point, and vapour pressure of a chemical
KOWWIN™	Estimates K_{ow} (n-octanol-water partition coefficient), which describes partitioning between n-octanol (which represents lipids or body fat of biota) and water
WSKOWWIN™	Estimates K_{ow} and water solubility
HENRYWIN™	Calculates partitioning between air and water (Henry's Law Constant)
AOPWIN™	Estimates atmospheric oxidation
HYDROWIN™	Estimates chemical hydrolysis
BIOWIN™	Estimates aerobic biodegradation
PCKOCWIN™	Estimates adsorption of a chemical to the organic carbon portion of soil and sediment
BCFWIN™	Estimates fish bioconcentration factor
STPWIN™	Predicts removal of a chemical in a sewage treatment plant
LEV3EPI™	Level 3 fugacity model that predicts partitioning of chemicals between air, water, soil, and sediment

Table 13.2. Hazard and toxicity models.

OncoLogic™	Estimates the potential for a chemical to cause cancer in humans *OncoLogic is available from EPA upon request [9]*
Non-Cancer for **Health Effects Method**	A stepwise process, not computerized, to screen untested chemicals non-cancer health effects. *Available in the USEPA P2 Framework Manual June 2005 [10]*
ECOSAR	Predicts a chemical's toxicity to aquatic biota. *This model can be downloaded from the USEPA website [11]*
PBT Profiler	Screens chemicals for potential to persist, bioaccumulate, and be toxic (PBT). *The PBT Profiler can be accessed through the web [12]*

Table 13.3. Release, exposure, and risk models.

E-FAST	Estimates aquatic releases and chemical exposure concentrations for the general public. *This model can be downloaded from the USEPA website [13]*
ChemSTEER	Estimates environmental releases and worker exposures resulting from chemical manufacture, processing, and/or use in industrial and commercial workplaces. *This model can be downloaded from the U.S. EPA website [14]*

and exposure of the material, its major impurities and byproducts, and its potential environmental breakdown products. This process culminates in the focus meeting approximately two to three weeks after initial receipt of the PMN. Experts participating at the focus meeting review preliminary exposure, hazard and risk assessments and, applying the experience gained over the years, make an initial decision whether any further regulatory action should be undertaken. Nearly all of the submissions that are considered at the focus meeting reach a definite outcome (there is an array of potential actions). In recent years, the focus meeting has decided that approximately 20% of submissions require further analysis and review. The NCP's reviews have resulted in some action being taken to reduce the risks of approximately 5% of new chemicals that have been submitted for review. An additional approximately 5% are voluntarily withdrawn by the notifier, often in the face of EPA action.

Because TSCA does not require companies to submit testing on a new chemical, EPA relies on SAR analysis to identify potentially hazardous chemicals. These modeling and assessment techniques have worked well to identify problem chemicals, as indicated by the results of a 1993 study conducted jointly by the EPA and authorities in the European Union (EU) [15]. This study showed that EPA's SAR-based predictions agreed with the results of base set testing in about 60-90% of the cases, depending on testing endpoint. The Agency will require data development via testing when novel or moderate-high-concern substances are presented, or for new chemicals produced at high volumes (>100,000 kg), which have substantial exposure according to the criteria under the new chemicals exposure-based policy [16].

Risk assessment for existing chemicals

This process is less formalized than the new chemicals process. Absent a specific "submitter", as is the case with new chemicals, the events or information that trigger risk assessments for existing chemicals may include new monitoring or test data from within EPA, other public or private sector sources, or other countries.

Section 13.3.4 discusses certain OPPT actions that have been conducted under TSCA §6 authority, including those directed at PCBs and asbestos. EPA has regulated a number of substances under TSCA §6 via proposed and final rulemaking procedures, including metalworking fluids (40 CFR part 747) and hexavalent chromium chemicals (40 CFR part 749). In addition, polychlorinated biphenyls (PCBs) (40 CFR part 761), and asbestos (40 CFR part 763) risk management actions have also been promulgated under TSCA §6; however, in both cases statutory requirements were followed (TSCA §6(e) and TSCA §203 [part of Title II of TSCA], respectively).

Risk assessment for TSCA § 8 submissions

TSCA §8(e) requires that chemical manufacturers, processors, and distributors notify EPA immediately of new (e.g., not already reported), unpublished information on chemicals that reasonably supports a conclusion of substantial risk. TSCA §8(e) substantial risk information notices most often contain toxicity data but may also contain information on exposure, environmental persistence, or actions being taken to reduce human health and environmental risks. EPA considers TSCA §8(e) to be an important information-gathering tool that serves as an early warning mechanism. EPA screens all TSCA §8(e) submissions and identifies chemicals for further assessment, referral or follow-up with submitters for additional information on exposure and/or risk management. As of March 31, 2006, EPA has received 16,395 initial TSCA §8(e) submissions and approximately 8,200 supplemental or follow-up TSCA §8(e) submissions. EPA receives approximately 200 initial and 100 supplemental §8(e) submissions per year.

This "substantial-risk reporting" has heightened industry awareness of potential chemical risks, often resulting in manufacturers, importers, processors, and distributors taking action on their own to minimize exposure to hazardous substances. A "substantial risk" also pertains to emergency situations of environmental contamination that seriously threatens humans or the environment.

"For Your Information" (FYI) submissions are the voluntary adjunct to "substantial risk" notices submitted to EPA under TSCA §8(e). Similar to §8(e) submissions, FYI submissions may contain information on human exposure, epidemiology, toxicity test results, environmental monitoring, environmental fate, or other information pertinent to risk assessment. FYI submissions may contain negative or equivocal findings that submitters may wish to share with EPA and the public. In other cases, FYI submissions contain positive data but are submitted on an FYI basis because the submitter does not have a TSCA reporting obligation (is not a chemical manufacturer, processor or distributor) or does not believe the data are reportable under §8(e). EPA receives an average of about 30 FYI submissions per year [17].

Additional information submission requirements under TSCA §8

In addition to compiling the inventory, EPA has used its TSCA Section 8 authority to obtain health and safety test data and production, importation, use, release, and exposure data on a number of chemicals, including chemicals recommended by the *Interagency Testing Committee* (ITC) for test rule consideration.

13.2.3 Exposure assessment can modulate need for risk assessment

Several parts of the TSCA chemicals process enable feedback to adjust the level of risk assessment which the EPA must undertake: either feedback from predictable levels of exposure, or from membership in a class of chemicals known to be relatively benign. And intending makers can accept use conditions which lower expected exposure, thus enabling the Agency to predict that levels of exposure will be lower, posing less concern.

The Agency, once it receives a new chemical notice, can determine how much risk assessment work must be undertaken based on its structure-activity relationship to like materials, and the hazard posed by those similar materials. If the proposed material is like materials known to be of low concern, risk assessment work on the specific chemical substance can be minimized. The Agency has some tiered testing requirements - PMN applications for materials expected to be used at high volumes require more testing, and within the *low volume exemption category*, less scrutiny is applied to lower-volume materials. The low *release-low exposure exemption* for new chemicals specifies high levels of containment which leads to very low releases. The TSCA statute calls for the establishment of an *"Interagency Testing Committee"*, which sets priorities for testing of chemical substances already in commerce and identifies classes of chemicals deserving of scrutiny.

A last example of limited risk assessment is the *polymer exemption*, in which the Agency has set out characteristics it has identified in a large number of low concern polymers: an intending maker of a new polymer can make a determination that its polymer is within the ambit of the identified characteristics and initiate manufacture. After doing so, the maker must keep records covering manufacture of the material and its determination of inclusion in the characteristics, and must notify the Agency in January of the following year that it has done so.

13.3 COMMAND AND CONTROL MECHANISMS

13.3.1 TSCA inventory (TSCA §8)

One of the first requirements under TSCA (§8) was the creation of the *TSCA chemical substance inventory*. Chemicals included in the Inventory are considered "existing chemicals" and chemicals not included are considered "new chemicals." Since the late 1970s, the TSCA inventory has grown from about 60,000 chemicals in commerce to over 80,000 chemicals due to the addition of over 20,000 former new chemicals.

Under TSCA's broad information-gathering authorities, EPA established an Inventory Update Rule (IUR) in order to obtain updated basic U.S. production and importation information on nonpolymeric organic chemicals that are listed on the Inventory. The IUR was recently amended to require the reporting of additional exposure-related information for higher-volume chemicals and to include reporting on inorganic chemicals. This new information, together with the hazard information being developed by efforts in the U.S. such as the High Production Volume (HPV) Challenge Program and those by the Organization for Economic Cooperation and Development (OECD; see Chapter 16), such as the HPV Screening Information Data Set (SIDS) Program, will enable EPA, industry, environmental groups, and others to evaluate, understand, and take action to gather needed additional information or to reduce chemical risks.

13.3.2 Testing of chemicals (TSCA §4)

Section 4 of TSCA gives EPA authority to require manufacturers of existing chemicals to conduct testing for health and environmental effects, and establishes a mechanism for multiple manufacturers to share the costs of the testing. Testing requirements are imposed either after a public rulemaking process (test rule) or public negotiation of an Enforceable Consent Agreement. This section of the act also supports the voluntary high production volume chemicals initiative, to be discussed later under the "stewardship" heading.

13.3.3 Control of new chemicals (TSCA §5)

Section 5 of TSCA requires the submission of a premanufacture notice (PMN) 90 days prior to the manufacture or import of a new chemical substance. Therefore, prospective manufacturers or importers of a new chemical substance submit PMNs, or applications for §5(h) exemptions.

The goal of EPA's new chemicals program is to ensure that chemicals that may present an unreasonable risk are appropriately managed, including prevented from entering commerce. Additionally, if more data are needed to allow for an adequate assessment, the Agency has the authority to regulate a substance pending the development of the needed information if certain risk or exposure-based findings can be made. To date, EPA has reviewed more than 40,000 pre-manufacture notifications, and regulatory action has been taken on about 5% of the notified chemicals. Actions have included testing requirements and control measures on production, use, and disposal, as shown in Figure 13.1. As mentioned earlier, for another 5% of notified chemicals, companies have decided to withdraw their notices.

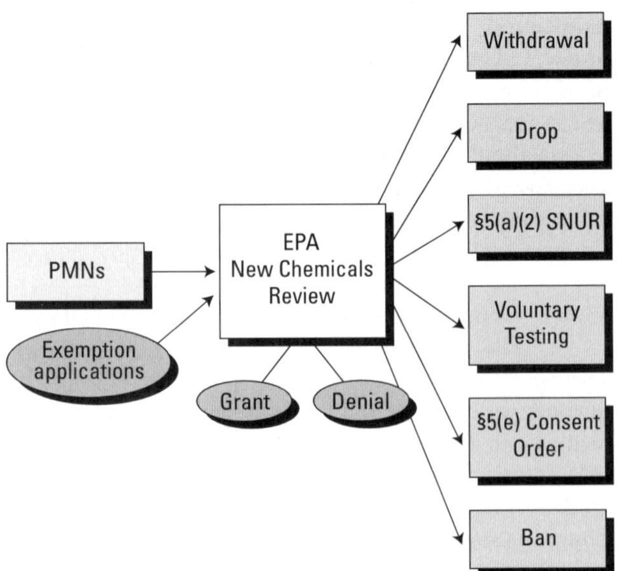

Figure 13.1. Input-output diagram of the EPA new chemicals review process.

13.3.4 Control of existing chemical substances (TSCA §6)

Under TSCA Section 6, EPA has the authority to prohibit or limit the manufacture, import, processing, distribution in commerce, use, or disposal of a chemical if there is a reasonable basis to conclude that the chemical presents or will present an unreasonable risk of injury to human health or the environment.

EPA must consider risks, costs, and benefits of a substance, as well as its alternatives to be regulated under Section 6 as posing an "unreasonable risk". Section 6 includes a menu of possible regulatory options, ranging from totally banning a chemical substance to requiring notices and warnings. TSCA requires that the EPA Administrator impose the "least burdensome" regulatory measure that still provides adequate protection. If EPA determines that a chemical is likely to present an unreasonable risk or serious or widespread injury to human health or the environment before normal rulemaking procedures can be completed, and it is in the public=s interest, the Agency may declare a proposed rule under Section 6 effective upon publication and until the effective date of the final action. Under TSCA Section 7, EPA may ask a court to require whatever action may be necessary to protect against chemicals that present an imminent and unreasonable risk of serious or widespread injury to health or the environment

In addition to using various control options, EPA also uses chemical advisories to warn the public about potential chemical hazards. These advisories discuss toxic effects and routes of exposure related to chemicals of concern and provide information that can help individuals or organizations voluntarily reduce risks. The Agency's advisories can be found at the USEPA website [18].

The Agency has been directed by Congress to control several specific chemical substances, as well.

13.3.5 PCBs, asbestos, and lead

When drafting TSCA, Congress singled out PCBs for immediate regulation. In response, EPA regulates the use, processing, distribution in commerce, and disposal of PCBs while also banning their manufacture [19]. The Agency has also been directed to take actions to restrict exposures to asbestos through legislation, the Asbestos Hazard Emergency Response Act (AHERA), to protect school children and staff from asbestos risks in schools [20]. As discussed later in this chapter, the Congress also enacted Title X - The Residential Lead-Based Paint Hazard Reduction Act of 1992 establishing requirements relating to lead paint [21].

13.4 EXAMPLES OF EPA'S GATEKEEPER/ GUARDIAN ROLE

EPA's chemical regulatory program has evolved over time and has moved toward a mix of regulatory and voluntary approaches to chemicals assessment. EPA's experience with its HPV Challenge Program (Section 13.4.1) provides a great deal of insight and direction for the Agency in shaping its future assessment and risk management efforts. This Program embodies the strengths of collaborative involvement of stakeholders, the use of data management technologies, and a commitment to making information on chemicals available publicly in an accessible, usable form. Other examples of risk assessment and management include perfluorinated acids (Section 13.4.2), polybrominated diphenyl ethers (Section 13.4.3), and lead (Section 13.4.4).

13.4.1 High production volume challenge program

The Environmental Protection Agency in partnership with environmental and industry groups created the High Production Volume (HPV) Challenge Program in an effort to make chemical hazard assessment data available to the public. The HPV Challenge Program was launched in 1998, as part of the *Chemical Right-to-Know* (ChemRTK) initiative, to ensure that a baseline set of hazard and environmental fate data on approximately 2,800 HPV chemicals could be easily accessed by the public. The American Chemistry Council, Environmental Defense, and the American Petroleum Institute participated with EPA in the launch of the Program. The program considered "HPV" to represent chemicals manufactured or imported in amounts equal to or greater than one million pounds (approximately 500,000 kg) per year in the United States [22].

A basic premise of the Program is that the public has a right to know about the hazards associated with chemicals in their environment. Everyone – including industry, environmental groups, animal welfare organizations, government groups, and the general public – can use the chemical data provided through the HPV Challenge Program to make informed decisions related to the human and environmental hazards of chemicals that they encounter in their daily lives.

The HPV Challenge Program was established in response to several studies in the late 1990's that showed that there were relatively few U.S. HPV chemicals for which an internationally agreed upon set of data for hazard screening was available to the public. The data set sought by the HPV Challenge Program is known as the Screening Information Data Set (SIDS) that was developed by the OECD (see Section 13.3.1 and Chapter 16). The SIDS provides an internationally agreed-upon set of test data for screening HPV chemicals for human and environmental hazards and environmental fate. This data set encompasses six main types of test data: acute toxicity, repeated-dose toxicity, genetic toxicity, reproduction and developmental toxicity, ecotoxicity, and environmental fate.

Commitments to sponsor chemicals and provide data to the HPV Challenge Program have come from companies and consortia both inside and outside the United States. As part of their commitment to the HPV Challenge Program, sponsors submit data summaries of existing information along with a test plan that proposes a strategy to fill data gaps. These documents are then posted to the HPV Challenge Program website. The majority of sponsors' materials were submitted to the Agency between 2001 and 2003.

Sponsors submit test plans for either individual chemicals or for a category of chemicals. A chemical category comprises a group of substances, usually similar in chemical structure, with a regular pattern of properties and effects. Data for chemicals in the category can be used to estimate the chemical properties and effects of other category members. A 120-day comment period begins when test plans and data summaries are posted to the Program website [23]. Industry, environmental groups, animal welfare groups, private citizens, etc. – can comment on the data summary and test plan submissions; EPA comments on all of the submissions as well. Environmental Defense has submitted comments on 87% of all posted test plans. Two animal welfare groups – People for the Ethical Treatment of Animals (PETA) and Physicians Committee for Responsible Medicine (PCRM) – submitted comments on 62% of all test plans, and private individuals and other groups submitted comments on fewer than 3% of all test plans. Sponsors can also indirectly submit data through the International Council of Chemical Associations (ICCA) HPV Initiative, which is a complementary program aimed at HPV chemicals from around the world.

A key EPA goal in managing the HPV Challenge Program has been to provide clear guidance to assist stakeholders in participating in the Program. EPA's "Guidance Documents" can be found at the USEPA website [24]. Guidance is provided for subjects such as category formation, developing robust data summaries, and assessing adequacy of existing data, to name a few. A number of EPA's guidance documents have achieved international acceptance through their incorporation into OECD guidance documents.

Extensive voluntary participation

As of April 2006, a total of 2,244 chemicals had been sponsored, with 1,383 chemicals sponsored directly in the Program by 370 companies and 103 consortia and an additional 861 chemicals sponsored indirectly in the ICCA HPV Initiative. For chemicals sponsored directly in the HPV Challenge Program, 394 test plans had

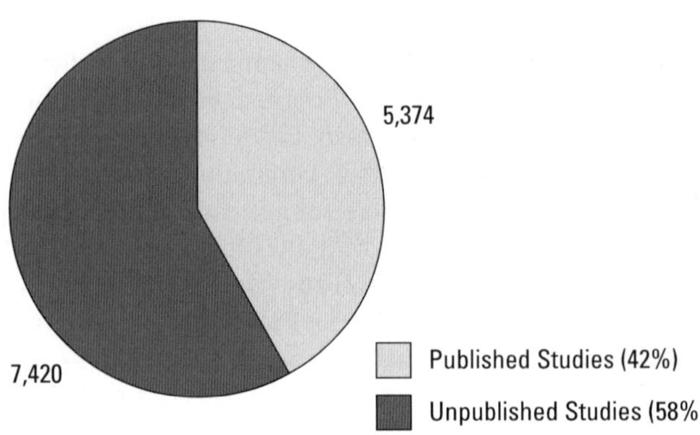

Figure 13.2. Sources of existing data submitted under the HPV Challenge Program.

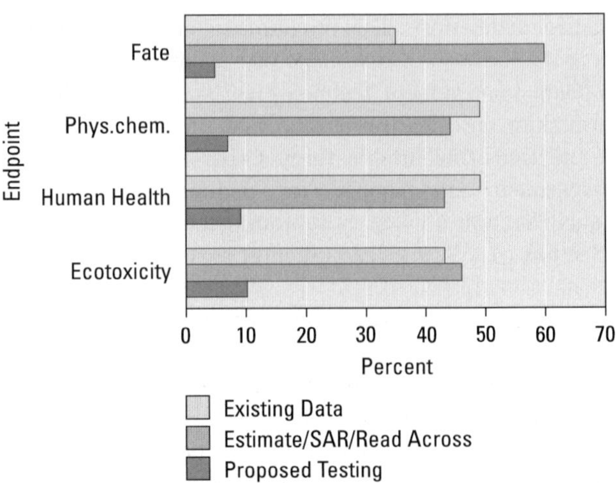

Figure 13.3. Endpoint data sources under the HPV Challenge Program.

been submitted for 1,335 (97%) of the 1,383 sponsored chemicals. Forty-eight (3%) of the sponsored chemicals were considered "overdue" because data had not been received for them [25].

EPA analyzed submitted test plans to determine how the health and environmental effects endpoints were addressed. Sponsors proposed to meet data needs through the use of existing scientifically adequate data, estimation techniques such as SAR, or proposed new testing. An examination of submitted data revealed that a significant amount of unpublished data have now been made public by the sponsors. This indicates that many sponsors made significant efforts in evaluating the hazards of their chemicals prior to the launch of the HPV Challenge Program, but often did not make the data available to the public. This has resulted in a limited amount of new testing. In fact, 58% of the existing data reported under the HPV Challenge Program, which represents over 7,400 studies, was not publicly available before the launch of the Program (Figure 13.2).

The category approach

One of the most significant results of the HPV Challenge Program has been the use of the category approach to address the SIDS endpoints (see also Chapter 11). In fact, 80% of all chemicals addressed in test plans have been included in a category. Categories require a supporting hypothesis of how the category chemicals relate to each other, as well as a description of how data for one chemical can be used to predict the toxicological responses of similar chemicals in the category. EPA and other stakeholders then comment on the reasonableness of the hypothesis, the adequacy of supporting data, and any proposed testing. Once the sponsor submits its final category analysis, EPA will either agree that the category "held," or will notify the sponsor that the sponsor may need to consider additional testing or restructure the category.

The following analysis is directly related to the above discussion and shows how sponsors have proposed to address the SIDS endpoints. EPA analyzed test plans to determine how the health and environmental effects endpoints were addressed. Three methods were available to meet the minimum data needs for each SIDS endpoint:

- Using existing scientifically adequate data.
- Using an estimation technique such as SAR or "read-across" categories.
- Proposing new testing.

Figure 13.3 illustrates that sponsors have made maximum use of EPA's guidance concerning the use of SAR and category proposals. Additionally, and in combination with the significant amount of unpublished data made available through the robust summaries, only a minimal amount of new testing has been proposed. Overall, for physicochemical and fate endpoints, as well as health and environmental effects, 10% or fewer of the endpoints are proposed to be addressed with new testing [26].

Follow-up HPV challenge program activities

As the data collection part of the HPV Challenge Program draws to a close, activities can focus more on public access and assessing the HPV data. A major component of public access is the HPV Information System (HPVIS), which is a searchable web application that provides the public with comprehensive and easy access to critical information on HPV chemicals [27].

Users can easily search for technical chemical property data that was submitted to the HPV Challenge Program, and the application also allows users to create reports based upon their queries.

With the conclusion of the HPV Challenge Program, EPA is developing approaches for identifying and prioritizing chemicals of concern for future management actions. A screening process will serve to help the Agency order its review of data in HPV Challenge Program submissions, and to provide structure to the review process for determining the hazard potential for substances examined under the program. The screening process consists of two tiers. In Tier-I of the process, key endpoint data are screened against predetermined criteria to establish a logical order in which EPA should review individual chemicals and categories. Criteria used in the Tier-I screen for health effects, ecological effects, and environmental fate endpoints are based on the hazard classification criteria used in the Globally Harmonized System (GHS) for the Classification and Labeling of Chemical Substances [28]. The Tier-I process ultimately assigns chemicals to first, second, or third priority groups for further review by OPPT.

In Tier-II, EPA conducts in-depth reviews of data submitted to the HPV Challenge Program for quality and completeness, develops screening-level hazard assessments for the chemicals, and informs industry and the public of its findings. Tier-II incorporates a full scientific review of all endpoint data in each HPV Challenge submission, and the key output of Tier-II is the screening-level characterization of the hazards of each chemical examined in the program. Following the screening process, screening-level hazard assessments developed in Tier- II may assist in any subsequent management activities that may be deemed appropriate for the chemicals reviewed.

The HPV Challenge Program, as noted earlier in this chapter, consists of approximately 2,800 chemicals that were considered HPV according to reporting under the Inventory Update Rule (IUR) of TSCA during the 1990 reporting year. However, since that time, other chemicals have become HPV. In 2005, the American Chemistry Council – in partnership with the Synthetic Organic Chemical Manufacturers Association and the Soap and Detergent Association, and with input from EPA and Environmental Defense – launched the Extended HPV (EHPV) Program. The EHPV Program is designed to secure data on 573 chemicals that were HPV on the 2002 IUR, but that were not HPV according to the 1990 IUR.

HPV chemicals that were not sponsored in the Program will be the subject of regulatory consideration (test rules) under TSCA Section 4 in order to ensure that a SIDS-level of data are available. They may also be made subject to TSCA Section 8 exposure information reporting rules and TSCA Section 8 health and safety data reporting rules as part of the test rule development process. The 1st Final HPV Test Rule was published in March 2006 (71 FR 13708, March 16, 2006) and contained 17 chemicals, and development of other rules is underway.

OPPT established the National Pollution Prevention and Toxics Advisory Committee (NPPTAC) under the Federal Advisory Committee Act (FACA) specifically to provide advice, information, and recommendations on the overall policy and implementation of programs managed by the Office. The Committee addresses OPPT's implementation of TSCA and PPA and one if its work groups was formed to address HPV issues. Specifically, the NPPTAC has provided advice and recommendations as to how OPPT should proceed with using the HPV data to identify and prioritize chemicals of concern, the design and implementation of the HPV Information System (HPVIS), and how to proceed with obtaining additional data for unsponsored (orphan) chemicals. Additional NPPTAC information can be found at the USEPA website [29].

13.4.2 Perfluorinated acids

Perfluorinated acids are manmade chemicals with uniquely valuable properties and functionality. Concerns began with perfluorooctyl sulfonate (PFOS) in the late 1990's when reporting under TSCA Section 8(e) indicated that PFOS was persistent, widespread in the environment, caused reproductive/developmental toxicity in animal studies, and was present in people at low levels. The EPA investigation expanded to include perfluorooctanoic acid (PFOA) and fluorinated telomers in 2000, based on their similarity to PFOS.

PFOS was widely used in surface treatment products on paper, carpet, and textiles, and as a high-tech industrial surfactant in multiple industries (aviation, semiconductor manufacture, imaging, fire fighting foam). The company 3M was the major global producer of PFOS and voluntarily phased out production worldwide between 2000 and 2002. Small quantities continue to be produced overseas by other companies. Significant New Use Rules under TSCA Section 5(a) were published to restrict the return of these chemicals to the US market. Currently, very limited high-tech surfactant uses continue for which there are no currently viable substitutes (67 FR 72854, December 9, 2002).

As with PFOS, data on PFOA also demonstrated persistence, bioaccumulation and toxicity in animal studies, and widespread presence in humans and the environment. PFOA is used as an essential polymerization aid in the manufacture of fluoropolymers, which are used in many industries and consumer products such as non-stick cookware coatings, architectural coatings, chemical and fire-resistant cabling, and waterproof, breathable fabrics. PFOA is not expected to be contained in final fluoropolymer products, except in trace amounts, and tests are currently being undertaken to test for the presence of PFOA in such products. PFOA may also be produced by the breakdown of other chemicals called fluorinated telomers. Telomers are widely used in industrial and consumer products, with uses similar to PFOS, including stain and water repellent surface treatments for carpets, textiles, and paper, and as surfactants in performance products and cleaners [30].

EPA released a draft preliminary risk assessment on PFOA in April 2003, and simultaneously initiated an enforceable consent agreement (ECA) process to pursue the development of additional information on the sources of PFOA in the environment and the pathways leading to human and environmental exposures. After receiving more information, EPA revised the draft risk assessment and released it for peer review by EPA's Science Advisory Board in 2005. The SAB published its report in May 2006, generally approving the EPA's approach and making recommendations for additional risk assessment work. The ECA process, risk assessment, and other data development activities (including telomer biodegradation studies) on PFOA are currently underway.

In January 2006, EPA Administrator Stephen Johnson invited the eight major companies in this industry to join in a global initiative called the 2010/15 PFOA Stewardship Program to:

- Commit to achieve, no later than 2010, a 95% reduction in *both* facility emissions to all media *and* product content of PFOA, PFOA precursor chemicals, and related higher homologue chemicals (measured from a year 2000 baseline).
- Commit to working toward elimination of PFOA, PFOA precursors, and related higher homologue chemicals from emissions and products by 2015.

All eight of the invited companies committed to the Program by March 2006, and are expected to:

- Submit baseline year 2000 data on emissions and product content by October 31, 2006.
- Report annual progress toward goals each succeeding October.
- Report progress in terms of both U.S. and global operations.

Under the Program, companies also commit to work cooperatively with EPA and others to develop scientifically credible analytical standards and laboratory methods to ensure comparability of data reporting [31].

EPA is creating PFOA guidance documents that will assist companies in reporting: (1) baseline year 2000 emissions and product content levels; and (2) annual achievements in a useable, trackable fashion.

13.4.3 Polybrominated diphenyl ethers

Polybrominated diphenyl ethers (PBDEs) is a group of brominated flame retardant chemicals that have been of increasing interest to scientists, government agencies, and the public. There are commercial PBDE products with different average amounts of bromination: penta-, octa-, and decaBDE. These chemicals are major components of commercial products often used as fire retardants in furniture foam (pentaBDE), plastics for personal computers and small appliances (octaBDE), and plastics for TV cabinets, consumer electronics, wire insulation, and backcoatings for draperies and upholstery (decabromodiphenylether, or decaBDE). In the event of a fire involving these products, PBDEs slow ignition and rate of fire growth, allowing more time for people to extinguish or escape the fire. However, findings that PBDEs are widely distributed in the environment and are present at increasing levels in people have raised concerns about the potential risks of PBDE exposure to human health and the environment.

Current information suggests strongly that PBDEs as a class are persistent and may bioaccumulate. The mechanisms or pathways by which the PBDEs move into and through the environment and humans are not known, but are likely to include releases from manufacturing of the chemicals, manufacturing of products like plastics or textiles, aging and wear of products like sofas and electronics, and releases at the end of product life (disposal, recycling). Studies have also been conducted in laboratory animals to gain a better understanding of the potential health risks of PBDEs. However, there remains much to learn about both exposure to PBDEs and the potential health effects; and there are different concerns for the different PBDEs.

Great Lakes Chemical Corporation, the sole U.S. producer of the commercial pentaBDE and octaBDE mixtures, discontinued production of these two products in the U.S. at the end of 2004. EPA issued a final TSCA Section 5 Significant New Use Rule (71FR 34015, June 13, 2006) for these PBDEs (including PBDE congeners that comprised these two products). This rule requires manufacturers and importers to notify EPA at least 90 days before commencing the manufacture or import of any one or more of these chemical substances [32].

As discussed later in this chapter, EPA is leading a Furniture Flame Retardancy Partnership to evaluate alternatives to PBDEs in furniture applications. The partnership includes furniture and fabric manufacturers, chemical manufacturers and environmental organizations. Its focus has been assessment of alternatives to pentaBDE to inform decisions on adoption of substitutes.

EPA also posted its Polybrominated Diphenyl Ethers (PBDEs) project plan in March 2006 [33]. The project plan identifies key areas in which EPA addresses concerns for PBDEs and other flame retardants. Many of these activities involve partnership and coordination with other federal agencies, industry, and NGOs (nongovernmental organizations).

13.4.4 Lead

Lead is a soft metallic element mined from rock and has been known since antiquity for its adaptability in making various useful items. Lead is also an extremely potent toxic substance and, according to both the Environmental Protection Agency and the U.S. Centers for Disease Control and Prevention (CDC), there is no known safe blood lead level [34]. Health effects associated with exposure to lead and lead compounds include, but are not limited to, neurotoxicity, developmental delays, hypertension, impaired hearing acuity, impaired hemoglobin synthesis, and male reproductive impairment [35]. Young children are particularly susceptible to the effects of lead.

In modern times, lead has been used in the manufacture of many different products, including paint, batteries, pipes, solder, pottery, and gasoline. In the last three decades of the 20th century, various agencies of the Federal government took independent actions to address lead exposure. In 1973, EPA issued regulations designed to gradually reduce the amount of lead in gasoline, largely for the purpose of reducing lead damage to auto exhaust catalytic converters. In 1978, the U.S. CPSC banned the use of paint containing more than 0.06% lead by weight. EPA lowered the maximum levels of lead permitted in public water systems in 1991.

The CDC set and lowered blood lead "levels of concern" several times, as new studies showed the impact of lead levels on children's health. The level of concern is the level where medical and environmental case management activities should be implemented. The U.S. Department of Housing and Urban Development (HUD) began to abate lead hazards in public housing that was being renovated or in structures occupied by a child with elevated blood lead levels. Because of these and other efforts, average blood lead levels in the U.S. declined more than 80 percent between the late 1970s and the early 1990s.

In 1991, the Secretary of the U.S. Department of Health and Human Services (HHS) characterized lead poisoning as the "number one environmental threat to the health of children in the United States" and CDC identified lead-based paint as the major source of high-dose lead poisoning in the United States.

In response to this persistent health threat, in 1992 Congress enacted Title X - The Residential Lead-Based Paint Hazard Reduction Act of 1992. Title X established a national goal of eliminating lead-based paint hazards in housing as expeditiously as possible and directed the Federal government to take a leadership role in building the infrastructure necessary to achieve this goal. The Act also established the roles and responsibilities for each of the relevant Federal agencies responsible for achieving this goal. Title X amended TSCA by adding Title IV - Lead Exposure Reduction. Title IV outlined four key program strategies for the Agency in establishing its lead program:

- Establish hazard standards for residential lead-based paint and for lead in residential dust and soil.
- Establish a program to give the public information about lead hazards generally and the steps they can take to protect themselves and their families from lead-based paint hazards specifically.
- Ensure that information about known lead-based paint or lead-based paint hazards is disclosed to individuals buying or renting pre-1978 housing, and that owners and occupants of pre-1978 housing are provided information on lead-based paint hazards before renovation activities take place
- Promulgate work practice standards for safely and effectively identifying and eliminating lead-based paint hazards, and require lead-based paint professionals to be trained and certified.

The Act directed the Agency to initiate a series of regulatory and non-regulatory actions to address each of these strategies. To date, significant progress has been made in implementing this program. The Agency promulgated a regulation that establishes hazard standards for lead in residential paint, dust, and soil. These health-based standards were designed to be protective of children under the age of six and are used as benchmarks for the Federal lead-poisoning prevention effort. The Agency has established, along with State and Federal partners, a comprehensive program aimed at educating the public about the hazards of lead. The Agency supports the National Lead Information Center that operates a bilingual toll-free hotline that handles 60,000 calls per year. The center also distributes 1.6 million documents annually. In addition to these National efforts, the program is focused on providing outreach materials targeted at vulnerable populations and at-risk communities [36].

To ensure that information about the potential hazards of lead-based paint is provided to homeowners and renters, the Agency has issued two significant regulations. The first requires renovation contractors to give a pamphlet on lead hazards to tenants and homeowners no more than 60 days prior to beginning work. The second, issued by EPA and the HUD, requires sellers or lessors to disclose information on the presence of lead to prospective buyers or tenants. To ensure that lead-based paint hazards are safely identified and remediated, the Agency has issued a regulation that requires individuals conducting lead-based paint activities in most pre-1978 housing to be trained and certified. The regulation requires that all lead-based paint activities must be conducted according to work practice standards and that all training must be conducted by accredited training providers.

The federal government has established a goal of eliminating childhood lead poisoning by the 2010 and EPA's lead-based paint program is designed to support the achievement of that goal. A recent journal article by CDC, EPA and HUD estimates that there were 310,000 children with elevated blood lead levels in 1999-2002 which reflects a 68% reduction in the number of children with elevated blood lead levels from the 1991-1994 time period [37].

13.5 EPA'S ROLE AS FACILITATOR OF STEWARDSHIP

13.5.1 Pollution prevention and voluntary partnerships

EPA complements its traditional role as a gatekeeper and guardian with innovative programs stressing pollution prevention and environmental stewardship. The Pollution Prevention Act of 1990 established that it is the national policy of the United States that pollution should be prevented, or reduced at the source whenever possible. The Act outlined an environmental management hierarchy that emphasizes an ethic of preventing pollution before it happens. If this is not possible, the Act declared that recycling is better than the treatment of waste, and environmentally safe disposal and release should be a last resort [38].

As a facilitator of environmental stewardship, EPA seeks to empower companies, states, tribes and the public by providing information, tools and incentives to develop, produce, supply, buy and use safer, greener chemicals as illustrated later. EPA recognizes that the strength of the information is not in its collection, but in its use. The Agency is working together with key stakeholders to make information both understandable and useful [39].

Voluntary partnerships are an important and effective tool in promoting and achieving environmental stewardship and sustainability. Partnerships can be driven by industry's desire to control or avoid risk management actions – voluntary partnerships can make regulations unnecessary or better focus regulations to achieve desired results at the lowest cost. Partnerships can also be driven by an industry desire to improve public perception. Moving to alternative (safer) chemicals or processes can be a high profile demonstration of an industry's environmental stewardship, especially when the chemical being replaced may present significant potential risks to human health or the environment. Partnerships can reduce regulatory and liability costs through reduced workplace exposures. In addition, industry can gain market advantage by producing greener products. In the United States, these partnerships have also been driven by international regulatory actions, and industry's desire to understand the hazard associated with alternatives that have been shown to be cost-effective substitutes. The following subchapters go into greater detail on several of EPA's specific programs and initiatives that carry out the facilitator role discussed above [40].

13.5.2 Examples of EPA as facilitator of voluntary partnerships

As explained in the previous sections, chemical risk management at EPA is accomplished through varying means, under a range of statutory authorities. Sometimes, however, the most cost and time-effective risk management is accomplished through voluntary partnerships with broad-based stakeholder groups that do not depend on the Agency's legal authorities. EPA acts as a convener of stature for these partnerships, bringing together groups that might not otherwise communicate and cooperate. EPA ensures a level playing field across industries and brings unique technical tools and expertise that enhance understanding of potential human health and environmental outcomes.

Voluntary chemical risk management at EPA is facilitated by the toxicological tools, models, and expertise that have been developed over the last three decades for new and existing chemical review under TSCA. OPPT uses these tools and models to provide estimates and predictions of risk, when empirical data are unavailable or insufficient. The tools and models focus on analyzing existing hazard information or predicting hazard when data are not available, estimating the potential for human exposure, and assessing risk by examining both hazard and exposure. These tools and models and EPA's risk assessment expertise can present great value to industry and other stakeholders and provide an incentive for participation in voluntary programs.

Integrating the Agency's toxicological tools, models, and expertise into voluntary partnership programs can inform decision-making. Availability of appropriate information can allow an industry to shift toward safer alternatives, or choose safer alternatives as they redesign or reformulate processes and products. However, safer alternatives are not always available and alternatives are not necessarily hazard-free for each of the human health and environmental endpoints considered. The information provided through tools such as SAR analysis can help chemical manufacturers and users design safer or greener chemicals or help them use hazardous chemicals more safely. The results can also help shape or enhance industry environmental stewardship programs, by facilitating long-term planning for moving to next-generation chemicals, and by better targeting animal testing to the toxicological endpoints with the strongest indication of hazard using SAR.

13.5.3 Design for the environment

The *Design for the Environment* (DfE) Program and *Sustainable Futures* (SF) Initiative are two voluntary programs in OPPT. These programs use the tools and models described above as an incentive for stakeholder participation. The focus of both programs is to provide industry and others with information on safer alternative chemicals or processes. These programs encourage risk reduction and risk management of industrial chemicals and promote cooperation between EPA and a wide variety of industry sectors [41].

DfE facilitates multi-stakeholder, collaborative partnerships to explore chemical risk management issues and help identify means to reduce risk. The success of DfE partnerships relies on multi-stakeholder participation and transparency, in which all viewpoints are considered in defining goals and methodologies. The diverse group of stakeholders DfE might convene includes industry representatives, NGOs, environmental groups, academic or research institutions, and other government agencies.

DfE partnerships encourage the use of safer chemicals and processes by conducting alternatives assessments and through the recognition of safer formulations. To conduct alternatives assessments, DfE partnerships evaluate the human health and environmental risks, performance, and cost of traditional and alternative chemicals and processes. Alternatives assessments serve industries that seek to make informed choices by considering characteristics and environmental risks and impacts of traditional and alternative chemicals and processes. Examples of successful partnerships are DfE's flame retardant alternatives assessment, lead-free solder life-cycle assessment, and the DfE program to recognize safer product formulations [42,43].

DfE's Furniture Flame Retardancy Partnership (FFRP) was initiated in response to stakeholder concerns with the occurrence of pentabromodiphenyl ether (pentaBDE) in the environment and human tissues. PentaBDE was the primary flame retardant in the manufacture of low-density, flexible polyurethane foam for furniture, with production levels of approximately 19 million pounds per year. This partnership was a broad multi-stakeholder effort, convened to develop and disseminate information on alternative technologies for achieving furniture fire safety standards. PentaBDE was voluntarily phased out at the end of 2004, making it critical to investigate and identify available flame retardant alternatives that could protect lives and property. One outcome of the partnership was a comprehensive report designed to help industry factor environmental and human health impacts into their decision-making as they choose alternatives for flame-retarding furniture foam [44].

A diverse group of stakeholders, including members of the furniture industry, chemical manufacturers, environmental groups, fire safety advocates, and government representatives, contributed to the FFRP and the development of the report. They provided critical information on chemicals and participated in the technical workgroups. These partners agreed that fire safety is critical and must be achieved in a way that minimizes risk to human health and the environment. Through this partnership, EPA reviewed 14 commercially available flame retardant formulations identified by leading chemical flame retardant manufacturers.

The FFRP developed an alternatives assessment methodology for evaluating these flame retardants, based on EPA's new chemicals program. This methodology included a screening level assessment of the chemicals,

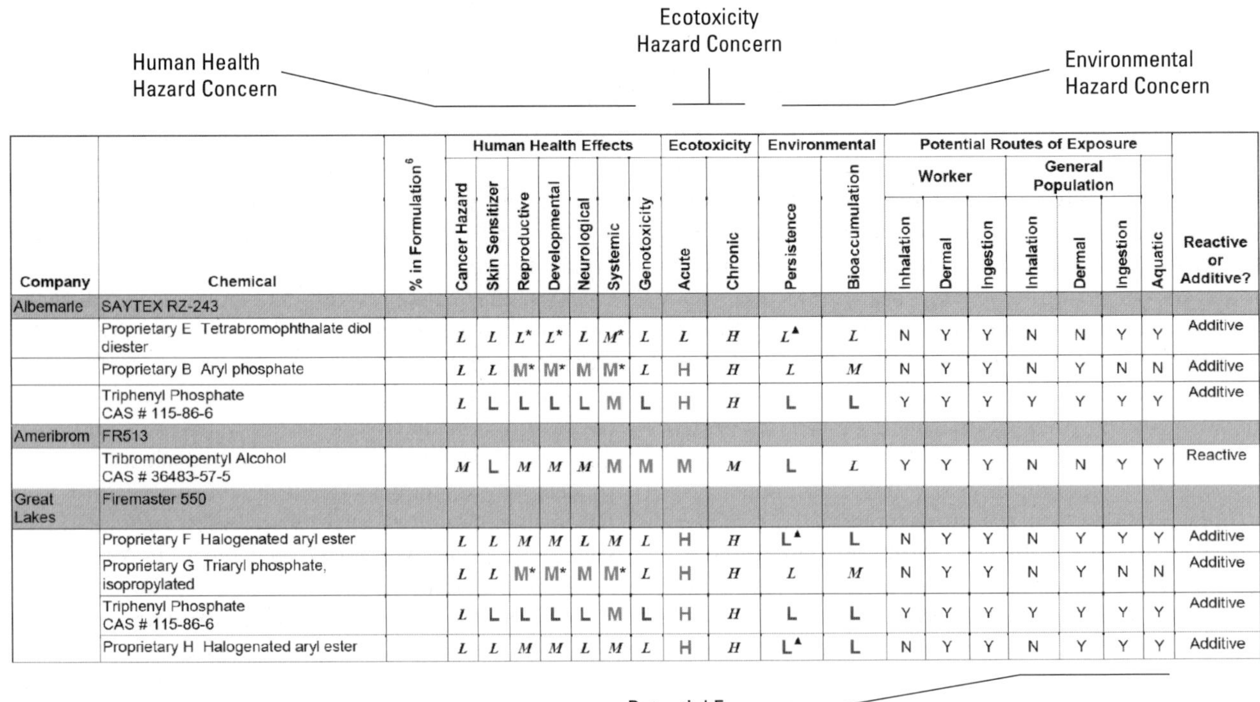

Company	Chemical	% in Formulation[6]	Human Health Effects							Ecotoxicity		Environmental		Potential Routes of Exposure							Reactive or Additive?
														Worker			General Population				
			Cancer Hazard	Skin Sensitizer	Reproductive	Developmental	Neurological	Systemic	Genotoxicity	Acute	Chronic	Persistence	Bioaccumulation	Inhalation	Dermal	Ingestion	Inhalation	Dermal	Ingestion	Aquatic	
Albemarle	SAYTEX RZ-243																				
	Proprietary E Tetrabromophthalate diol diester		_L_	_L_	_L_*	_L_*	_L_	_M_*	_L_	_L_	_H_	_L_▲	_L_	N	Y	Y	N	N	Y	Y	Additive
	Proprietary B Aryl phosphate		_L_	_L_	M*	M*	M	M*	_L_	H	_H_	_L_	_M_	N	Y	Y	N	Y	N	N	Additive
	Triphenyl Phosphate CAS # 115-86-6		_L_	L	L	L	L	M	L	H	_H_	L	L	Y	Y	Y	Y	Y	Y	Y	Additive
Ameribrom	FR513																				
	Tribromoneopentyl Alcohol CAS # 36483-57-5		_M_	L	_M_	_M_	_M_	M	M	M	_M_	L	_L_	Y	Y	Y	N	N	Y	Y	Reactive
Great Lakes	Firemaster 550																				
	Proprietary F Halogenated aryl ester		_L_	_L_	_M_	_M_	_L_	_M_	_L_	H	_H_	L▲	L	N	Y	Y	N	Y	Y	Y	Additive
	Proprietary G Triaryl phosphate, isopropylated		_L_	_L_	M*	M*	M	M*	_L_	H	_H_	_L_	_M_	N	Y	Y	N	Y	N	N	Additive
	Triphenyl Phosphate CAS # 115-86-6		_L_	L	L	L	L	M	L	H	_H_	L	L	Y	Y	Y	Y	Y	Y	Y	Additive
	Proprietary H Halogenated aryl ester		_L_	_L_	_M_	_M_	_L_	_M_	_L_	H	_H_	L▲	L	N	Y	Y	N	Y	Y	Y	Additive

Figure 13.4. Furniture flame retardancy partnership summary table. Symbols: L (low hazard concern), M (moderate hazard concern), H (high hazard concern), *L, M, H* (endpoints assigned using estimated values and professional judgement such as structure activity relationships, * (ongoing studies that may result in a change in this endpoint), and ▲ (persistent degradation products expected).

focusing on potential hazards and exposure routes, as well as the potential for the chemicals to bioaccumulate and persist in the environment. The assessment results were presented at three levels, to meet the needs of a range of audiences and to maximize transparency. At the top level, the partnership created a summary table (Figure 13.4 is an excerpt) for decision-making, which shows qualitative information on each formulation. This summary table indicates the potential hazard concern (high, moderate, low) and its basis (test data or SAR) for each key human health and environmental endpoint, and indicates where there is a potential for exposure, based on physical-chemical properties. The key chemical information that forms the basis for this table is included in summary assessments. These quantitative summaries include toxicity and exposure data from publicly available literature as well as EPA's databases, studies conducted by industry, and the professional judgment of EPA experts. Key to maximizing transparency, especially for proprietary chemicals, is the detailed hazard review for each chemical, including summaries of key studies and an indication of data adequacy for a wide range of environmental, human health, and environmental fate endpoints, as well as physical-chemical properties.

The results from this partnership are now being used by foam manufacturers in choosing alternative flame retardants. The results have also been cited by the Consumer Product Safety Commission as important information to consider in the context of developing a national flammability standard for residential upholstered furniture. Their preliminary conclusion is that foam flame retardants are available that would not pose appreciable health risks. The FFRP Alternatives Assessment methodology was referenced by the NPPTAC, a committee advising EPA under the Federal Advisory Committee Act, as a model for conducting alternatives assessments [45]. The NPPTAC called the partnership's methodology an effective tool for communicating information on chemical alternatives and enabling industry to consider environmental and health impacts along with cost and performance as they make chemical decisions and ensure informed substitution.

DfE's *formulator program* encourages partners to reformulate chemical products to be environmentally safer, cost competitive, and effective. Drawing on expertise from OPPT, the formulator program makes chemical properties and toxicology information available in an understandable format, suggests safer substitutes for ingredients of concern, and emphasizes continuous improvement. The program uses recognition as an incentive to steer industry sectors towards formulations that are safer for human health and the environment [46].

Businesses value the formulator program because it looks at product formulation from their perspective, based on ingredient functionality. The formulator program is flexible and improvement-oriented, and provides the quality information product formulators need to meet the marketplace's ever-increasing demand for greener products.

To begin the partnership process, a company submits for review a complete list of product ingredients with specific chemical identifiers (chemical abstract service registry number, functionality, percent in formula, etc.). DfE then develops a profile on each ingredient based on available physical-chemical property and environmental and human health hazard information, either provided by the submitter or researched in the literature; in the absence of data, the program uses Agency predictive models and tools to estimate key assessment endpoints.

When the profiles for each ingredient are ready, DfE convenes a product review work group, comprised of technical area experts, to review the profiles individually and then as a whole product. It is this workgroup's experience and skill—at assessing chemical hazards, applying predictive tools, and identifying safer substitutes—that distinguishes the DfE program from all other product recognition programs. The workgroup compares the ingredients' characteristics to other chemicals in the same use class (e.g., to other surfactants, solvents), considers possible negative synergies between ingredients, and places the ingredients on a continuum of improvement relative to other similar chemicals (Figure 13.5). In this way, DfE helps formulators select from among the safest chemical in an ingredient class.

If an ingredient shows room for improvement, the workgroup suggests a potential safer substitute, if available. Substitution of some ingredients and product formulation often follows this review. Once the work group agrees that a product's ingredients have a consistently more positive profile than other chemicals in their class, the basis for a partnership and product recognition exists.

DfE and the company form a partnership through a memorandum of understanding, signed by the director of OPPT and a senior company official. In exchange for the company's commitment to formulate a product as described to the Agency, or with even safer ingredients over time, the partnership agreement permits a company to use DfE recognition, including the DfE logo, on

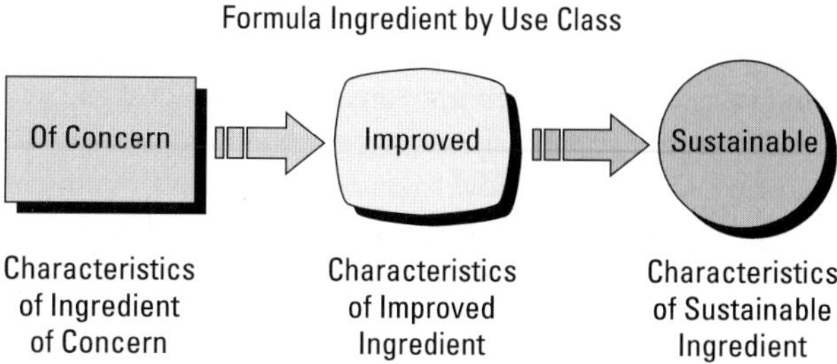

Figure 13.5. Continuum of improvement.

recognized products. On an annual basis, partners inform DfE of their environmental progress, including the quantities of chemicals of concern no longer used in their formulations and other reformulation benefits. By the end of 2005, DfE had partnered with over 30 companies to recognize more than 130 products, primarily within the industrial/institutional cleaning sector.

13.5.4 Sustainable futures

The Sustainable Futures (SF) Initiative encourages voluntary efforts between the EPA and the chemical industry to facilitate increased prescreening activities for industrial chemicals, allowing identification of chemical hazards and risks in early research and development (R&D) stages. SF is an example of an innovative voluntary effort that builds on the chemical risk screening expertise developed by EPA through chemical regulation activities, and promotes technology transfer of that expertise to other chemical stakeholders.

The Agency launched SF in 2002. SF is designed to help industry develop new chemical substances that are both economically and environmentally sustainable. Regulatory relief is offered to participating companies submitting qualifying new chemical substances. SF was based on the successes of the Pollution Prevention (P2) Framework efforts, which combined the new chemical assessment methodologies developed through the new chemicals program. Through the P2 Framework, EPA began working with the chemical industry in 1995 to help transfer the methods to developers of new chemicals and new chemical products. The P2 Framework captured OPPT expertise and made these methodologies available to the public. It encouraged risk screening early in the development process when the potential for cost savings and innovation are the greatest. Industry participants in the P2 Framework outreach efforts investigated the applicability of the P2 Framework risk screening methodologies to their efforts, and were able to show that prescreening chemicals using these tools gave them a competitive advantage through increased costs savings, regulatory certainty, and innovative chemical design.

Through SF, EPA is offering expedited review to qualified submitters of low hazard/ low risk new chemical notices that have been prescreened for hazard and risk concerns. SF participants receive, at no cost, hands-on training in the use of the chemical risk screening methods developed by EPA. The benefits to SF participants include:

- Identification and commercialization of safer chemicals.
- Increased P2 opportunities.
- Increased innovation.
- More focused testing.
- More efficient processes.
- Reduced generation of chemical waste.

Companies participating in SF receive copies of the SF/ P2 Framework risk screening methods, hands-on training in the proper use of the methods, and extensive training materials developed for SF. The screening methods contained in the SF/ P2 Framework efforts include models to predict the following endpoints [47].

Cooperative efforts by industry partners, environmental advocates, and the Agency contribute to the continuing development of improved computerized screening methods. For example, the *PBT Profiler* and the *Analog Identification Methodology* (AIM) are tools created to address specific needs identified by participants in SF and the P2 Framework efforts.

The PBT Profiler (Table 13.2) predicts a chemical's potential persistence, bioaccumulation, and chronic aquatic toxicity values and compares the predictions to EPA's new chemicals program PBT criteria to determine if the substance is a potential PBT of concern [48].

An online computerized tool called the Analog Identification Methodology was developed to help users evaluate the potential health effects of untested chemicals through identification of close chemical analogs that have measured data on the endpoints of concern. AIM identifies chemical analogs through a fragment-based structural similarity approach, and points the user to publicly available databases or other sources of information where experimental data on the related chemical analogs can be found.

13.5.5 Hospitals for a healthy environment

In 1998, the American Hospital Association and the U.S. Environmental Protection Agency signed a landmark agreement that launched Hospitals for a Healthy Environment (H2E) to advance pollution prevention efforts in U.S. hospitals [49].The memorandum of understanding, which is the cornerstone of H2E, sets out three main goals:

- Virtually eliminating mercury-containing waste from the health care waste stream.
- Reducing the overall volume of waste (both regulated and non-regulated waste).
- Identifying hazardous substances for pollution prevention and waste reduction opportunities, including hazardous chemicals and persistent, bioaccumulative, and toxic pollutants.

H2E has developed the following tools to assist its members. These tools include:

- A Web site full of practical tools, information, and resources.
- A peer-to-peer list serve that allows health care professionals to ask technical questions and receive advice and feedback from their peers.
- Monthly free teleconferences for H2E partners and champions where expert speakers address practical solutions to the many environmental challenges faced by today's health care facilities.
- A monthly newsletter, STATGreen, that includes an H2E partner's success story in dealing with a particular environmental challenge, information about the upcoming H2E teleconference, and a variety of other features.
- Model waste minimization, mercury elimination, and other pollution prevention plans.

13.5.6 Partnership results

The OPPT voluntary partnerships yield impressive results. By year 2005, DfE projects touched 200,000 business facilities and two-million workers. The furniture and electronics industries are using DfE information to choose safer alternatives to 19 million pounds of flame retardants annually. By making chemical and toxicological information available in an understandable format and suggesting safer substitutes, the formulator program has been able to reduce an estimated 40 million pounds of chemicals of concern in 2005, through recognizing industry partners that take action to make formulations that are safer for human health and the environment. From 2005 to 2011, DfE estimates cumulative reductions of chemicals of concern to increase from 237 million to 836 million pounds and cumulative cost savings to increase from $914,000 to $406,000,000.

The *voluntary Sustainable Futures* Initiative has been popular with stakeholders. Currently, more than 240 companies are participating in SF, and nearly 500 individuals have taken SF hands-on training. A positive impact has been noticed by the Agency in that the percentage of new chemical notices submitted containing some level of self-assessment (prescreening), including SF model results, has increased from 10% in 2003 to 17% for the first 6 months of 2006. Internal review by EPA staff scientists has shown that prescreened new chemical notices require fewer resource-intensive detailed risk assessments than new chemical notices that have not been prescreened. Stakeholders have conducted more than 75,000 chemical-specific PBT assessments using the PBT Profiler since its release in September 2002. Various stakeholders have told EPA that the PBT Profiler may be the most widely used publicly available chemical screening tool in the world.

As of March 26, 2006, the *H2E Program* has 1,170 partners representing 6,431 facilities: 1,320 hospitals, 3,124 clinics, 1,015 nursing homes and 972 other types of facilities, and 110 champions representing some of the biggest names in the healthcare sector, including Kaiser Permanente, Healthcare Corporation of America, the Veterans Health Administration, Baxter Healthcare, Sodexho, and 7 Group Purchasing Organizations that represent over 90 percent of the medical supplies market. The number of Partners represents a growth of 178 new Partners in 2005, and over 50 new Partners so far in 2006. The culmination of the voluntary partnership program's success has been that in spring of 2006 the H2E became the first EPA partnership program to become a fully independent, non-profit organization. The new organization will continue to receive support from the EPA, the American Hospital Association, American Nurses Association, and Health Care Without Harm, all the while having the ability to raise funds through other avenues, including fees for services provided to the healthcare sector.

13.6 CONCLUSIONS

Within the USEPA the OPPT is responsible for implementing TSCA and the PPA. Within TSCA chemical risk reduction activities have statutory and voluntary dimensions. As such EPA is gatekeeper/guardian for chemical hazard and facilitator of environmental stewardship. These different approaches are summarized below.

13.6.1. Statutory risk reduction

TSCA gives EPA broad statutory authority to protect against unreasonable risks of toxic substances through information-gathering, testing requirements, and controls on the production, import, processing, distribution in commerce, use, and disposal of chemicals. Under these authorities the Agency:

- Compels testing of chemical substances which are already in commerce.
- Requires submission of information on new chemicals (not already in commerce) to the Agency for its determination of whether use of the new chemical may present unreasonable risk of harm.
- Establishes controls on production and/or use of chemicals already in commerce:
- Compels submission to EPA of data showing "substantial risks".

To facilitate these tasks, the Agency maintains an inventory of the approximately 80,000 chemical substances in commerce in the United States.

13.6.2 Statutory/voluntary approaches

EPA's chemical regulatory program has evolved over time to encompass a mix of regulatory and voluntary components. This chapter has presented successful risk reduction approaches that combine statutory or regulatory activities by EPA with voluntary action by manufacturers. Examples include PFOS, PFOA, lead, and the HPV Challenge. These efforts rely and build on the established structure of the Agency's authorities under TSCA, which provides a solid, consistent structure for developing regulations, guarantees a level playing field, and serves as a backstop if necessary. But this also serves as an excellent foundation for partnerships and voluntary actions by progressive companies who wish to innovate or go beyond the TSCA requirements to achieve better or faster results in terms of testing, risk assessment, or risk reduction.

13.6.3 Voluntary environmental stewardship

As explained in the sections of this Chapter, chemical risk management at EPA is accomplished through varying means, under a range of statutory authorities. Sometimes, however, the most cost and time-effective risk management is accomplished through voluntary partnerships with broad-based stakeholder groups that do not depend on the Agency's legal authorities. EPA acts as a convener of stature for these partnerships, bringing together groups that might not otherwise communicate and cooperate. EPA ensures a level playing field across industries and brings unique technical tools and expertise that enhance understanding of potential human health and environmental outcomes.

As shown in the examples above, there are many different incentives for partnerships that are not driven directly by environmental statutes and regulations. Availability of appropriate information can enable companies to take the initiative to shift toward safer alternatives, or to choose safer alternatives as they redesign processes or reformulate products.

In the coming years, EPA's success in protecting human health and the environment from chemical risk depends on its ability to maintain the crucial balance of its roles as gatekeeper/guardian and facilitator of environmental stewardship.

13.6 FURTHER READING

An extensive list of readings regarding USEPA OPPT regulations, new (voluntary) approaches for chemicals management, as well as further readings on chemicals such as PCBs, asbestos, and HPVs, among many others topics, is provided in Annex 13.1 at the end of this chapter.

REFERENCES

1. U.S. Environmental Protection Agency. October, 2006. Pollution Prevention & Toxics (OPPT). Last updated on Monday, October 16th, 2006. USEPA, Washington, D.C. http://www.epa.gov/oppt/
2. U.S. Environmental Protection Agency. October, 2006. Pollution Prevention. Last updated on Wednesday, October 11th, 2006. USEPA, Washington, D.C. http://www.epa.gov/p2/
3. U.S. Environmental Protection Agency. October, 2006. Toxic Substances Control Act | US EPA: Related Internet Sites - New Chemicals Program. Last updated on Tuesday, October 10th, 2006. USEPA, Washington, D.C. http://www.epa.gov/oppt/newchems/pubs/tscasite.htm
4. U.S. Environmental Protection Agency. March, 2006. TSCA – Civil Enforcement - Statute, Regulations & Enforcement. Last updated on Thursday, March 23rd, 2006. USEPA, Washington, D.C. http://www.epa.gov/compliance/civil/tsca/tscaenfstatreq.html
5. U.S. Environmental Protection Agency. October, 2006. TSCA §5 New Chemicals Program. Last updated on Tuesday, October 3rd, 2006. USEPA, Washington, D.C. http://www.epa.gov/opptintr/newchems/
6. U.S. Environmental Protection Agency. October, 2006. Federal Environmental Regulations Section E. Toxic Substances Control Act. Last updated on Tuesday, October 10th, 2006. U.S. USEPA, Washington, D.C. http://www.epa.gov/dfe/pubs/pwb/tech_rep/fedregs/regsecte.htm
7. U.S. Environmental Protection Agency. August, 2006. Exposure Assessment Tools and Models. Last updated on Thursday, August 10th, 2006. USEPA, Washington, D.C. http://www.epa.gov/opptintr/exposure/
8. U.S. Environmental Protection Agency. October, 2006. Exposure Assessment Tools and Models, Estimation Program Interface (EPI) Suite. Last updated on Tuesday, October 3rd, 2006. USEPA, Washington, D.C. http://www.epa.gov/oppt/exposure/pubs/episuite.htm
9. U.S. Environmental Protection Agency. October, 2006. Green Engineering, Software: Exposure Assessment Tools and Models, OncoLogic™. Last updated on Friday, October 6th, 2006. USEPA, Washington, D.C. http://www.epa.gov/oppt/greenengineering/pubs/software.html#exposure_tools
10. U.S. Environmental Protection Agency. June, 2005. Pollution Prevention (P2) Framework. EPA-748-B-04-001. Manual. USEPA, Washington, D.C. pp. 61-63 and pp. 171-175 of the PDF, or pp. 55-57 and pp. 165-169 of the hard copy. http://www.epa.gov/opptintr/newchems/pubs/sustainablefutures.htm
11. U.S. Environmental Protection Agency. September, 2006. New Chemicals Program, Ecological Structure Activity Relationships (ECOSAR). Last updated on Monday, September 18th, 2006. USEPA, Washington, D.C. http://www.epa.gov/oppt/newchems/tools/21ecosar.htm
12. U.S. Environmental Protection Agency. September, 2006. PBT Profiler. Last updated on Thursday, September 21st, 2006. Syracuse Research Corporation under contract to the U.S. Environmental Protection Agency, USEPA, Washington, D.C. http://www.pbtprofiler.net/
13. U.S. Environmental Protection Agency. October, 2006. Exposure Assessment Tools and Models, Exposure and Fate Assessment Screening Tool (E-FAST). Last updated on Tuesday, October 3rd, 2006. USEPA, Washington, D.C. http://www.epa.gov/opptintr/exposure/pubs/efast.htm
14. U.S. Environmental Protection Agency. October, 2006. Exposure Assessment Tools and Models, Chemical Screening Tool For Exposures & Environmental Releases (ChemSTEER). Last updated on Tuesday, October 3rd, 2006. USEPA, Washington, D.C.

http://www.epa.gov/opptintr/exposure/pubs/chemsteer.htm

15. Organization for Economic Cooperation and Development. 1994. *US EPA/EC Joint Project on the Evaluation of (Quantitative) Structure Activity Relationships (QSARS).* OECD, Paris, France. Environment Monographs No. 88. USEPA.
16. U.S. Environmental Protection Agency. September, 2006. New Chemicals Program, TSCA 5(e) Exposure-Based Policy: Criteria. Last updated on Monday, September 18th, 2006. USEPA, Washington, D.C. http://www.epa.gov/oppt/newchems/pubs/expbased.htm
17. U.S. Environmental Protection Agency. March, 2006. TSCA 8(e). Last updated on Friday, March 3rd, 2006. USEPA, Washington, D.C. http://www.epa.gov/opptintr/tsca8e/
18. U.S. Environmental Protection Agency. October, 2006. Prevention, Pesticides and Toxic Substances. Last updated on Wednesday, October 18th, 2006. USEPA, Washington, D.C. http://www.epa.gov/oppts/
19. U.S. Environmental Protection Agency. April, 2006. Polychlorinated Biphenyls (PCBs). Last updated on Tuesday, April 18th, 2006. USEPA, Washington, D.C. http://www.epa.gov/pcb/
20. U.S. Environmental Protection Agency. October, 2006. Asbestos. Last updated on Tuesday, October 17th, 2006. USEPA, Washington, D.C. http://www.epa.gov/asbestos/
21. U.S. Environmental Protection Agency. August, 2006. Lead Programs. Last updated on Tuesday, August 8th, 2006. USEPA, Washington, D.C. http://www.epa.gov/lead/
22. U.S. Environmental Protection Agency. August, 2006. High Production Volume Chemicals Challenge Program. Last updated on Tuesday, August 29th, 2006. USEPA, Washington, D.C. http://www.epa.gov/chemrtk/
23. U.S. Environmental Protection Agency. August, 2006. High Production Volume Chemicals Challenge Program. Last updated on Tuesday, August 29th, 2006. USEPA, Washington, D.C. http://www.epa.gov/hpv/index.htm
24. U.S. Environmental Protection Agency. August, 2006. High Production Volume Chemicals Challenge Program, Information Sources. Last updated on Thursday, August 17th, 2006. USEPA, Washington, D.C. http://www.epa.gov/hpv/pubs/general/guidocs.htm
25. U.S. Environmental Protection Agency. August, 2006. HPV Challenge Sponsoring Organizations. Last updated on Monday, August 21st, 2006. USEPA, Washington, D.C. http://www.epa.gov/chemrtk/pubs/update/spncomp.htm
26. U.S. Environmental Protection Agency. November, 2004. Status and Future Directions of the High Production Volume Challenge Program. EPA-743-R-04-001. Report. USEPA, Washington, D.C.
27. U.S. Environmental Protection Agency. September, 2006. High Production Volume Information System (HPVIS). Last updated on Friday, September 22nd, 2006. USEPA, Washington, D.C. http://www.epa.gov/hpvis/index.html
28. United Nations Economic Commission for Europe. June, 2006. Globally Harmonized System of Classification and Labelling of Chemicals (GHS). Last updated on Monday, June 26th, 2006. UNECE, Geneva, Switzerland. http://www.unece.org/trans/danger/publi/ghs/ghs_welcome_e.html
29. U.S. Environmental Protection Agency. July, 2006. National Pollution Prevention and Toxics Advisory Committee. Last updated on Thursday, July 27th, 2006. USEPA, Washington, D.C. http://www.epa.gov/oppt/npptac/
30. U.S. Environmental Protection Agency. September, 2006. Perfluorooctanoic Acid (PFOA). Last updated on Tuesday, September 5th, 2006. USEPA, Washington, D.C. http://www.epa.gov/oppt/pfoa/
31. U.S. Environmental Protection Agency. July, 2006. 2010/15 PFOA Stewardship Program. Last updated on Tuesday, July 18th, 2006. USEPA, Washington, D.C. http://www.epa.gov/oppt/pfoa/pubs/pfoastewardship.htm
32. U.S. Environmental Protection Agency. June, 2006. Polybrominated diphenylethers (PBDEs). Last updated on Tuesday, June 13th, 2006. USEPA, Washington, D.C. http://www.epa.gov/oppt/pbde/
33. U.S. Environmental Protection Agency. March, 2006. Polybrominated Diphenyl Ethers (PBDEs) Project Plan. USEPA, Washington, D.C. http://www.epa.gov/oppt/pbde/pubs/proj-plan32906a.pdf
34. U.S. Department of Health and Human Services. August, 2005. Public Health Service (PHS), Centers for Disease Control and Prevention (CDC). Preventing Lead Poisoning in Young Children; A Statement by the Centers for Disease Control and Prevention. Statement. USDHHS, Washington, D.C.
35. U.S. Environmental Protection Agency. July, 2004. Integrated Risk Information System; Lead and compounds (inorganic) (CASRN 7439-92B1). Report. USEPA, Washington, D.C.
36. U.S. Environmental Protection Agency. August, 2006. Lead Programs. Last updated on Tuesday, August 8th, 2006. USEPA, Washington, D.C. http://www.epa.gov/lead/
37. U.S. Department of Health and Human Services. May, 2005. Centers for Disease Control and Prevention (CDC), Update: Blood Lead Levels – United States, 1999 to 2002, MMWR 2005; 54 (20); 513-516. Weekly Report.

USDHHS, Washington, D.C.

38. U.S. Environmental Protection Agency. October, 2006. Pollution Prevention (P2) Basic Information. Last updated on Tuesday, October 10th, 2006. USEPA, Washington, D.C. http://www.epa.gov/p2/pubs/basic.htm
39. U.S. Environmental Protection Agency. October, 2006. Pollution Prevention (P2). Last updated on Wednesday, October 11th, 2006. USEPA, Washington, D.C. http://www.epa.gov/p2/
40. U.S. Environmental Protection Agency. October, 2006. Pollution Prevention (P2) Partnerships. Last updated on Tuesday, October 3rd, 2006. USEPA, Washington, D.C. http://www.epa.gov/p2/pubs/partnerships.htm
41. U.S. Environmental Protection Agency. September, 2006. Design for the Environment (DfE), Partnerships for a Cleaner Future. Last updated on Monday, September 18th, 2006. USEPA, Washington, D.C. http://www.epa.gov/dfe/index.htm
42. U.S. Environmental Protection Agency. September, 2006. Design for the Environment (DfE), Furniture Flame Retardancy Partnership. Last updated on Monday, September 18th, 2006. USEPA, Washington, D.C. http://www.epa.gov/dfe/pubs/projects/flameret/
43. U.S. Environmental Protection Agency. September, 2006. Design for the Environment (DfE), Lead-Free Solder Partnership. Last updated on Monday, September 18th, 2006. USEPA, Washington, D.C. http://www.epa.gov/dfe/pubs/projects/solder/index.htm
44. U.S. Environmental Protection Agency. September, 2005. Furniture Flame Retardancy Partnership: Environmental Profiles of Chemical Flame-Retardant Alternatives for Low-Density Polyurethane Foam, Volumes I and II, EPA 742-R-05-002A and EPA 742-R-05-002B. Report. USEPA, Washington, D.C.
45. U.S. Environmental Protection Agency. July, 2006. National Pollution Prevention and Toxics Advisory Committee. Last updated on Thursday, July 27th, 2006. USEPA, Washington, D.C. http://www.epa.gov/oppt/npptac/
46. U.S. Environmental Protection Agency. September, 2006. Design for the Environment (DfE), Formulator Program. Last updated on Monday, September 18th, 2006. USEPA, Washington, D.C. http://www.epa.gov/dfe/pubs/projects/formulat/index.htm
47. U.S. Environmental Protection Agency. October, 2006. New Chemicals Program, Sustainable Futures Initiative. Last updated on Tuesday, October 3rd, 2006. USEPA, Washington, D.C. http://www.epa.gov/opptintr/newchems/pubs/sustainablefutures.htm
48. U.S. Environmental Protection Agency. September, 2006. PBT Profiler. Last updated on Thursday, September 21st, 2006. Syracuse Research Corporation under contract to the USEPA, Washington, D.C. http://www.pbtprofiler.net/
49. U.S. Environmental Protection Agency. October, 2006. Pollution Prevention (P2), Hospitals for a Healthy Environment (H2E). Last updated on Tuesday, October 3rd, 2006. USEPA, Washington, D.C. http://www.epa.gov/p2/pubs/h2e.htm

ANNEX 13.1. Selected references for further reading

TSCA

1. U.S. Environmental Protection Agency. October, 2006. Toxic Substances Control Act-Related Internet Sites. http://www.epa.gov/oppt/newchems/pubs/tscasite.htm
2. U.S. Environmental Protection Agency. March, 1997. Chemistry Assistance Manual for Premanufacture Notification Submitters. EPA 744-R-97-003. Report. USEPA, Washington, D.C. http://www.epa.gov/opptintr/newchems/pubs/chem-pmn/
3. U.S. Environmental Protection Agency. September, 2006. New Chemicals Program: Guidance Materials for PMN Submitters. USEPA. http://www.epa.gov/opptintr/newchems/pubs/guideman.htm

General

4. U.S. Environmental Protection Agency. Overview: Office of Pollution Prevention and Toxics Programs. December 24, 2003. http://www.epa.gov/oppt/pubs/oppt101c2.pdf
5. U.S. Environmental Protection Agency. Appendices to Overview: Office of Pollution Prevention and Toxics Programs. December 24, 2003. http://www.epa.gov/oppt/pubs/oppt101c-append1a.pdf
6. U.S. Environmental Protection Agency. July 11, 2005. EPA Authorities under TSCA. USEPA, Washington, D.C. http://www.epa.gov/oppt/npptac/pubs/tscaauthorities71105.pdf
7. U.S. Environmental Protection Agency. March, 2006. Design for the Environment Partnership Highlights. http://www.epa.gov/opptintr/dfe/pubs/about/highlights1.htm
8. U.S. Environmental Protection Agency. February, 2006. Wanted: Innovative Green Chemistry Technologies Deserving National Recognition. EPA 744-F-06-001. Fact Sheet. USEPA, Washington, D.C.

Structure Activity Relationships (SAR)

9. U. S. Environmental Protection Agency. 1994. USEPA/EC Joint Project on the Evaluation of (Quantitative) Structure Activity Relationships. USEPA, Washington,

DC: Office of Pollution Prevention and Toxics, United States Environmental Protection Agency, EPA Report #EPA-743-R-94-001. Available from National Technical Information Service, U.S. Department of Commerce, 5285 Port Royal Road, Springfield, Virginia 22161, Tel: 703-487-4650, and http://www.epa.gov/oppt/newchems/21ecosar.htm

10. Organization for Economic Co-Operation and Development. 1994. US EPA/EC Joint Project on the Evaluation of (Quantitative) Structure Activity Relationships. OECD Environment Monographs No. 88. OECD/GD(94)28. OECD, Paris, France.

PCBs, Asbestos, Lead

11. U.S. Environmental Protection Agency. April, 2006. Polychlorinated Biphenyls (PCBs). http://www.epa.gov/pcb/
12. U.S. Environmental Protection Agency. August, 2006. Asbestos Information Resources. http://www.epa.gov/oppt/asbestos/pubs/pubs.html
13. U.S. Environmental Protection Agency. August, 2003. Today's Lesson: Asbestos. EPA 745-K93-017. Pamphlet. USEPA, Washington, D.C. http://www.epa.gov/oppt/asbestos/pubs/abcsfinal.pdf
14. U.S. Environmental Protection Agency. June, 2003. Protect Your Family From Lead In Your Home. EPA 747-K-99-001. Booklet. USEPA, USCPSC, USDHUD, Washington, D.C. http://www.epa.gov/lead/pubs/leadpdfe.pdf

HPV Chemicals

15. U.S. Environmental Protection Agency. July, 2000. Voluntary Participation in the HPV Challenge Program. EPA 745-F-98-002b. Fact Sheet. USEPA, Washington, D.C. http://www.epa.gov/chemrtk/pubs/general/hpvvol2.pdf
16. U.S. Environmental Protection Agency. March, 1999. High Production Volume Chemicals Frequently Asked Questions. EPA 745-F-09-002g. Report. USEPA, Washington, D.C. http://www.epa.gov/chemrtk/pubs/general/hpvq&a.pdf
17. U.S. Environmental Protection Agency. November, 2004. Status and Future Directions of the High Production Volume Challenge Program. EPA 743-R-04-001. Report. USEPA, Washington, D.C.
18. U.S. Environmental Protection Agency. September, 2006. Policy Regarding Acceptance of New Commitments to the High Production Volume (HPV) Challenge Program. http://www.epa.gov/chemrtk/pubs/general/hpvpolcy2.htm
19. National Pollution Prevention and Toxics Advisory Committee (NPPTAC)
U.S. Environmental Protection Agency. July, 2006. NPPTAC Recommendations. Last updated on Thursday, July 13th, 2006. USEPA, Washington, D.C. http://www.epa.gov/oppt/npptac/pubs/recommendations.htm

Pollution Prevention (P2)

20. U.S. Environmental Protection Agency. Fall, 1993. Pollution Prevention Incentives for States. EPA 242-F-93-0012. Brochure. USEPA, Washington, D.C. http://www.epa.gov/p2/pubs/ppisbro.pdf
21. U.S. Environmental Protection Agency. July, 2001. Why Should You Care About Preventing Waste? Small Business Guide. EPA 742-E-01-001. Booklet. USEPA, Washington, D.C. http://www.epa.gov/oppt/p2home/pubs/assist/sbg.htm
22. U.S. Environmental Protection Agency. June 15, 1993. P2 Policy Statement: New Directions for Environmental Protection: Carol Browner, EPA Administrator. http://www.epa.gov/oppt/p2home/pubs/p2policy/policy.htm
23. U.S. Environmental Protection Agency. June, 1996. Pollution Prevention Incentives for States (PPIS). EPA 742-S-96-001. Report. USEPA, Washington, D.C.
24. U.S. Environmental Protection Agency. Spring, 1994. Pollution Prevention Incentives for States. EPA 742-K-93-001.
Guidebook. USEPA, Washington, D.C. http://www.epa.gov/p2/pubs/ppispam.pdf

Design for the Environment (DfE)

25. U.S. Environmental Protection Agency. April, 1999. Design for the Environment: EPA and the Navy: A Pollution Prevention Partnership. EPA 744-B-99-001. Article. USEPA, Washington, D.C.
26. U.S. Environmental Protection Agency. August, 2002 (revision). Design for the Environment Program Fact Sheet. EPA 744-F-00-019. Fact Sheet. USEPA, Washington, D.C. http://www.epa.gov/dfe/pubs/tools/dfefactsheet/dfefacts8-02.pdf
27. U.S. Environmental Protection Agency. March, 2001. Design for the Environment: Partnerships for a Cleaner Future. EPA 744-F-00-020. Brochure. USEPA, Washington, D.C. http://www.epa.gov/opptintr/dfe/pubs/tools/DfEBrochure.pdf

Sustainable Futures Initiative / P2 Framework

28. PPG Industries, Inc. and U.S. Environmental Protection Agency, Jean Chun, Vince Nabholz, and Maggie Johnson Wilson. March, 2001. Comparison of Measured Aquatic Toxicity Data with EPA, OPPT SAR Predictions.

Report. PPG Industries, Inc., Pittsburgh, PA and USEPA, Washington, D.C.
http://www.epa.gov/opptintr/newchems/pubs/sustainable/ppg-sar-study-1999-2000.pdf

29. SC Johnson and Syracuse Research Corporation, John A. Weeks, Fred H. Martin, and Jay Tunkel, Ph.D. 2002. Case Study: SC Johnson's Use of the EPA PBT Profiler to Screen SC Johnson's Chemical Inventory, a Joint Study Conducted by SC Johnson and Syracuse Research Corporation. Report. SC Johnson, Racine, WI and SRC, Syracuse, NY. http://www.epa.gov/opptintr/newchems/pubs/sustainable/scjohnson-study-2002.pdf
30. Bayer Corporation. May, 2003. PBT Profiler Case Study: Bayer's Experience In The Use And Application Of The PBT Profiler for Predicting Persistence, Bioconcentration and Aquatic Toxicity of Chemical Substances. Case Study. Bayer, Pittsburgh, PA.
http://www.epa.gov/opptintr/newchems/pubs/sustainable/bayer-study-2003.pdf
31. Burleigh-Flayer, Heather, PPG Industries. May, 2000. Comparison of Environmental Fate Predictions from the PBT Profiler with Data found in the Literature. PPG Industries, Pittsburgh, PA.
http://www.epa.gov/opptintr/newchems/pubs/sustainable/profiler-predictions1.pdf
32. Eaton Aeroquip Inc. October, 2004. PBT Profiler Case Study Eaton Aeroquip Inc.'s Experience In the Use and Application of the PBT Profiler for Predicting Persistence, Bioconcentration, and Toxicity of Chemical Substances. Case Study. Eaton Aeroquip, Inc., Williamsport, MD.
http://www.epa.gov/opptintr/newchems/pubs/sustainable/eaton-aeroquip-study-2004.pdf
33. Votta, Thomas J. and Allen L. White, Ph.D. August, 2000. Design for Competitive Advantage: The Business Benefits of the EPA Pollution Prevention Assessment Framework In New Product Development. Report. Tellus Institute, Boston, MA.
http://www.epa.gov/opptintr/newchems/pubs/sustainable/kodak-tellus1b2c.pdf

Voluntary Programs

34. U.S. Environmental Protection Agency. January, 2000. The Lean and Green Supply Chain: A Practical Guide for Materials Managers and Supply Chain Managers to Reduce Costs and Improve Environmental Performance. EPA 742-R-00-001. Report. USEPA, Washington, D.C. http://www.epa.gov/oppt/library/pubs/archive/acct-archive/pubs/lean.pdf
35. U.S. Environmental Protection Agency. October, 2000. Fact Sheet: Hospitals for a Healthy Environment (H2E), Voluntary Partnership with the American Hospital Association to Reduce Hospital Waste. EPA 742-F-99-016R. Fact Sheet. USEPA, Washington, D.C.
http://www.epa.gov/pbt/pubs/h2efactsht2.pdf

Green Chemistry

36. U.S. Environmental Protection Agency. April, 2005. EPA's Green Chemistry Program Celebrates 10 Years of Results. EPA 744-F-05-011. Fact Sheet. USEPA, Washington, D.C.
37. U.S. Environmental Protection Agency. March, 2002. Green Chemistry Program Fact Sheet. EPA 742-F-02-003. Fact Sheet. USEPA, Washington, D.C.
http://www.epa.gov/oppt/greenchemistry/pubs/docs/general_fact_sheet.pdf
38. U.S. Environmental Protection Agency. May, 2003. Awards Opportunities Fact Sheet. EPA 744-F-02-001. Fact Sheet. USEPA, Washington, D.C.
http://www.epa.gov/oppt/greenchemistry/pubs/docs/awardsfactsheet.pdf
39. U.S. Environmental Protection Agency. May, 2003. Research Grant Opportunities Fact Sheet. EPA 744-F-03-003. Fact Sheet. USEPA, Washington, D.C.
http://www.epa.gov/oppt/greenchemistry/pubs/docs/grantsfactsheet.pdf
40. U.S. Environmental Protection Agency. July, 2005. Presidential Green Chemistry Challenge 2005 Award Recipients. EPA 744-K-05-001. Brochure. USEPA, Washington, D.C.
41. U.S. Environmental Protection Agency. June, 2005. The Presidential Green Chemistry Challenge Awards Program, Nomination Package for 2006 Awards. EPA 744-K-05-002. Report. USEPA, Washington, D.C.
http://www.epa.gov/greenchemistry/pubs/docs/nomination_package_for_2006_awards.pdf

Environmentally Preferable Purchasing

41. U.S. Environmental Protection Agency. April, 2005. Integrating Green Purchasing Into Your Environmental Management System (EMS). EPA 742-R-05-001. Report. USEPA, Washington, D.C. http://www.epa.gov/oppt/epp/pubs/grn-pur/green-pur-ems1a3a.pdf
42. U.S. Environmental Protection Agency. February, 2004. Buying Green Online. EPA 742-R-04-001. Report. USEPA, Washington, D.C. http://www.epa.gov/oppt/epp/pubs/buying_green_online.pdf

14. THE MANAGEMENT OF INDUSTRIAL CHEMICALS IN JAPAN

E. TODA

14.1 INTRODUCTION

Like many other countries, Japan has developed a number of systems for the assessment and management of different types of chemicals. While pharmaceuticals, food additives, pesticides, acutely toxic substances and other specific types of chemicals have a longer history of regulation, the general regulation on industrial chemicals started in the 1970's. It was in the 1970's that industrialized countries were confronted with severe pollution affecting human health and the environment (see also Chapter 1), and Japan was no exception. As a result of pollution caused by PCBs the first regulations in Japan focused on persistent, bioaccumulative and toxic substances. Successive amendments have been made to address less harmful substances. As a result, the chemical management system has gradually shifted from a hazard-based approach (with a focus on the intrinsic properties of chemicals) to an approach where more consideration is given to exposure to chemicals.

The risk assessment of industrial chemicals is at an initial stage in Japan. Risk assessments were conducted for certain environmental pollutants prior to the introduction of emission regulations, but efforts to screen chemicals of concern and to conduct risk assessments started only recently with a pilot project on initial risk assessment by the Ministry of the Environment (then the Environment Agency) from 1997 to 2000.

This chapter describes the current risk assessment and management system for industrial chemicals in Japan, together with some information about the historical background. Emphasis will be placed on the scientific methodology and procedural aspects. Although the legislation and extensive guidance documents are available (in Japanese) through the internet (e.g., the websites of the Ministry of the Environment and the National Institute for Technology Evaluation), most of this information is not available in English. The English websites of Ministries, other government bodies, and reports such as the National Profile on Chemical Management (Inter-ministerial Meeting on IFCS 2003) only provide basic information [1]. This chapter will help provide readers with an understanding of the overall picture of chemicals regulation in Japan compared with other countries.

14.2 CHEMICAL SUBSTANCES CONTROL ACT

The management of industrial chemicals in Japan is based on the "Act Concerning the Evaluation of Chemical Substances and Regulation of their Manufacture, etc." (Act No. 117 of 1973), known as the Chemical Substances Control Act or Law (CSCL). This legislation establishes a notification and evaluation system for new industrial chemicals and regulates the production, import and use of chemicals based on the hazardous properties of the chemicals. The CSCL is aimed at the prevention of damage to human health and ecosystems as a result of environmental pollution caused by these chemicals. The CSCL covers all types of industrial chemicals with the exception of:

- Specific controlled poisonous substances under the Poisonous and Deleterious Substances Control Act (Act No. 303 of 1950).
- Narcotic and stimulant drugs covered by relevant legislation.
- Radioactive substances.

In order to avoid duplication with other legislation, the CSCL states that its provisions do not apply to substances regulated under the following legislation when these other acts provide measures to protect human health and the environment:

- Food, food additives, etc. under the Food Sanitation Act (Act No. 233 of 1947).
- Agricultural chemicals under the Agricultural Chemicals Regulation Act (Act No. 82 of 1942).
- Fertilizers under the Fertilizer Control Act (Act No. 127 of 1950).
- Feed additives under the Act Concerning Safety Assurance and Quality Improvement of Feed (Act No. 35 of 1953).
- Pharmaceuticals, quasi-drugs, cosmetics and medical devices under the Pharmaceutical Affairs Act (Act No. 145 of 1960).

Since the purpose of the CSCL is to prevent adverse effects on human health and ecosystems caused by environmental pollution, regulation from the viewpoint of occupational health and safety, or consumer protection, takes place in the framework of other legislation. The Industrial Health and Safety Act (Act No. 57 of 1972) has its own requirements concerning the evaluation of

C.J. van Leeuwen and T.G. Vermeire (eds.), Risk Assessment of Chemicals: An Introduction, 575–589.

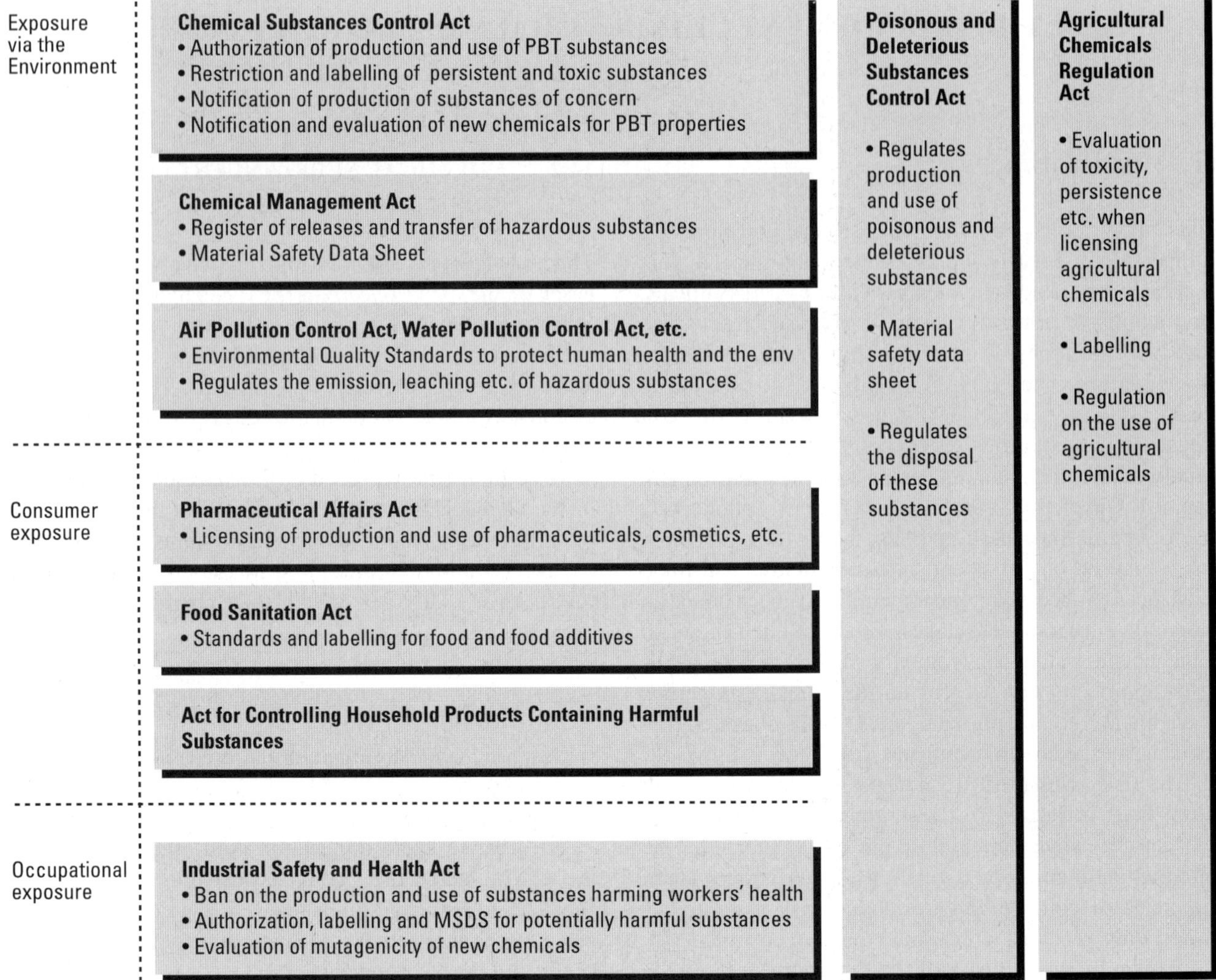

Figure 14.1. Chemical regulations in Japan.

new chemicals. Direct human exposure to chemicals is controlled by such legislation as the Food Sanitation Act or the Control of Household Products Containing Harmful Substances Act (Act No. 112 of 1973).

Figure 14.1 illustrates the relationship between some of these laws related to the management of chemicals. It is important to note that each piece of legislation can be categorized on the basis of exposure route, i.e., exposure via the environment, consumer exposure and occupational exposure.

14.2.1 History of the legislation and amendments to the CSCL

The Chemical Substances Control Act was passed in 1973 in response to widespread public concern about persistent bioaccumulating toxic substances such as PCBs. In the late 1960's, concern about the environmental pollution caused by PCBs spread worldwide, with reports of this very persistent, bioaccumulative and toxic group of substances being detected in fish and birds, and later in other animals and human breast milk. In 1968, a widespread episode of human poisoning in Western Japan (called "Yusho", or "rice bran oil disease") increased concern in Japan about the safety of PCBs. The PCB-contaminated rice bran oil resulted in a severe form of acne called chloracne, fatigue, nausea, and liver disorders as well as an increase in liver cancer mortality. The CSCL was passed further to the widespread concern over PCBs triggered by this serous health threat. The production, import and use of PCBs and other similar chemicals was banned. These persistent, bioaccumulative and chronically toxic substances are referred to as Class I Specified Chemical Substances under the current legislation.

The legislation also introduced the authorization process for the production and import of new industrial chemicals to prevent chemicals with properties similar to PCBs entering the market. In fact, the Japanese pre-marketing evaluation system precedes similar legislation enacted in the United States (Toxic Substance Control Act, 1976) and the European Community (Sixth Amendment to Directive 67/548/EEC, 1979) by a few years.

When the legislation was first enacted, the regulation was purely hazard-based, i.e., production, import and use were either permitted or not, depending on the persistence, bioaccumulation and long-term human toxicity. Since then, the legislation has undergone two major amendments to address issues such as: (1) persistent but non-bioaccumulating chemicals, and (2) hazards to environmental species. These amendments partially introduced a risk-based approach that takes into account the exposure potential of a chemical.

The first major amendment in 1986 was in response to the call for regulatory measures to address persistent and chronically toxic substances that do not have a high bioaccumulative potential. Examples of these are trichloroethylene and tetrachloroethylene. The amendment introduced a system for prior notification of production and import, allocation of production and import quantity (if necessary), product labelling, and guidance for the prevention of pollution by such chemicals (Class II Specified Chemical Substances). The amendment also introduced a system of Designated Chemical Substances which are persistent in the environment and suspected to be hazardous to human health, but lack definitive data on long-term toxicity. The production and import volume of these substances is periodically published based on mandatory reporting by manufacturers and importers. The government has the authority to direct manufactures and importers to investigate hazardous properties when this is deemed necessary in terms of risk.

What prompted the amendment in 2003 was, among other things, the OECD Environmental Policy Review of Japan in 2002 [2]. The OECD recommended that Japan should further improve the effectiveness and efficiency of its chemicals management and extend the scope of its chemical regulation to include ecosystem protection. Three government bodies, namely the Industrial Structure Council (Ministry of Economy, Trade and Industry), the Health Sciences Council (Ministry of Health, Labour and Welfare) and the Central Environment Council (Ministry of the Environment) reviewed the chemical evaluation and regulation system in force and published a joint report in February 2003. In line with this joint report, the government submitted a bill to parliament to amend the CSCL legislation in March, which was passed by Japan's legislator, the Diet, in May 2003. The Diet consists of two houses: the House of Representatives and the House of Councillors.

The 2003 amendment introduced four new features to the Act. Firstly, the consideration of ecological effects was introduced. Requirements for the submission of ecotoxicity data were introduced as a part of the notification and assessment of new chemicals, although such data had been submitted in the past on a case-by-case basis. Controls were introduced for persistent chemicals with known ecotoxicity.

Secondly, the concept of exposure information was introduced. However, chemicals used only as intermediates, chemicals used in closed systems, or chemicals for export only (provided that a pre-import evaluation system for new chemicals is in place in the export country) were exempted from notification. New chemicals manufactured or imported in volumes of up to 10 tonnes per year were also exempted from the requirement to submit toxicity data, if the substances were classified as persistent but not highly bioaccumulative and were not considered to present a significant risk to human health or the environment further to an evaluation based on the information already known.

Thirdly, measures were introduced to deal with chemicals known to be persistent and highly bioaccumulative. Reporting on the actual manufacturing quantities, import and use of such substances became mandatory. If a certain risk potential is identified based on a preliminary evaluation of toxicity data by the government authorities, guidance and advice shall be given to businesses on measures for risk reduction in order to minimize emissions into the environment. After risk reduction measures have been taken, the manufacturers and importers may be directed to investigate long-term toxicity, which will be used for the classification of controlled substances.

Fourthly, a requirement was placed on manufacturers and importers of chemicals to submit all the information they have indicating the hazardous properties of their substances. These amendments entered into force in April 2004. Other minor amendments include:

- 1983 Amendment, which allowed foreign manufacturers to submit a new chemical notification.
- 1999 Amendment, which designated the Ministry

Table 14.1. Types of chemicals regulated under CSCL.

Name	Explanation	Number of substances (as of September 2006)
Class I Specified Chemical Substances	Persistent, bioaccumulative, and hazardous (long-term human toxicity or ecotoxicity to higher predators)	15
Class II Specified Chemical Substances	Persistent, hazardous (long-term human toxicity or ecotoxicity to living organisms), with concern for long-term presence in the environment	23
Type I Monitoring Chemical Substances	Persistent and bioaccumulative, but hazardous properties unknown	22
Type II Monitoring Chemical Substances	Persistent and suspected as hazardous to human health	859
Type III Monitoring Chemical Substances	Persistent and hazardous to living organisms	51

of the Environment as an enforcement authority alongside the Ministry of Health, Labour and Welfare and the Ministry of Economy, Trade and Industry.

14.2.2 Types of chemicals regulated under the CSCL

Having undergone two major amendments, the basic structure of the CSCL has remained unchanged since. The regulation focuses on two main aspects:

1. Regulation of specified chemicals based on information on the potential for adverse effects.
2. Notification of new chemicals and their evaluation based on the requirements of the regulation.

Figure 14.2 illustrates the types of chemicals controlled by the Act and the regulatory procedures involved.

The Act establishes the following five types of "chemicals of concern", as shown in Table 14.1. Chemical substances that do not fall within the categories below are not regulated under this legislation. For substances that break down easily in the environment and where the degradation product meets the conditions given in Table 14.1, the parent substance will be classified according to the properties of the degradation products.

Table 14.2 presents the names of Class I Specified Chemical Substances. These substances are strictly regulated under the CSCL. A licence, which can only be issued to satisfy domestic demand, is required from the Minister of Economy, Trade and Industry for the manufacture and import of these chemicals. However, the use of these chemicals is only permitted for specific purposes and under such conditions as will ensure that environmental pollution does not occur. Therefore, there is little demand for these substances. In practical terms, their production and import are banned. The law also prohibits the import of products and articles containing these substances. When Class I Specified Chemical Substances or products or articles containing them are already on the market, the government can issue an order for them to be recalled. The government can also issue recommendations to control the manufacture, import or use of chemicals suspected as having the properties of a Class I Specified Chemical Substance, in order to prevent environmental pollution before the substances are officially designated as Class I chemicals.

As can be seen from Table 14.2, all the 10 intentionally produced POPs (Persistent Organic Pollutants) under the Stockholm Convention [4] are designated as Class I Specified Chemical Substances. Indeed, the criteria for designating these substances are similar to the criteria in the Stockholm Convention. When new substances are added to the Stockholm Convention in the future, it will be necessary to consider whether or not they should also be added to the list of Class I Specified Chemical Substances.

Table 14.3 provides an overview of Class II Specified Chemical Substances. Since these substances are not bioaccumulative, there is less concern about their risks when exposure levels are sufficiently low. Therefore, rather than totally banning the production, use and import of these chemicals, it was decided to make it mandatory to notify the authorities of the planned manufacture and import quantity. On the basis of this information the government authorities can issue an order to change

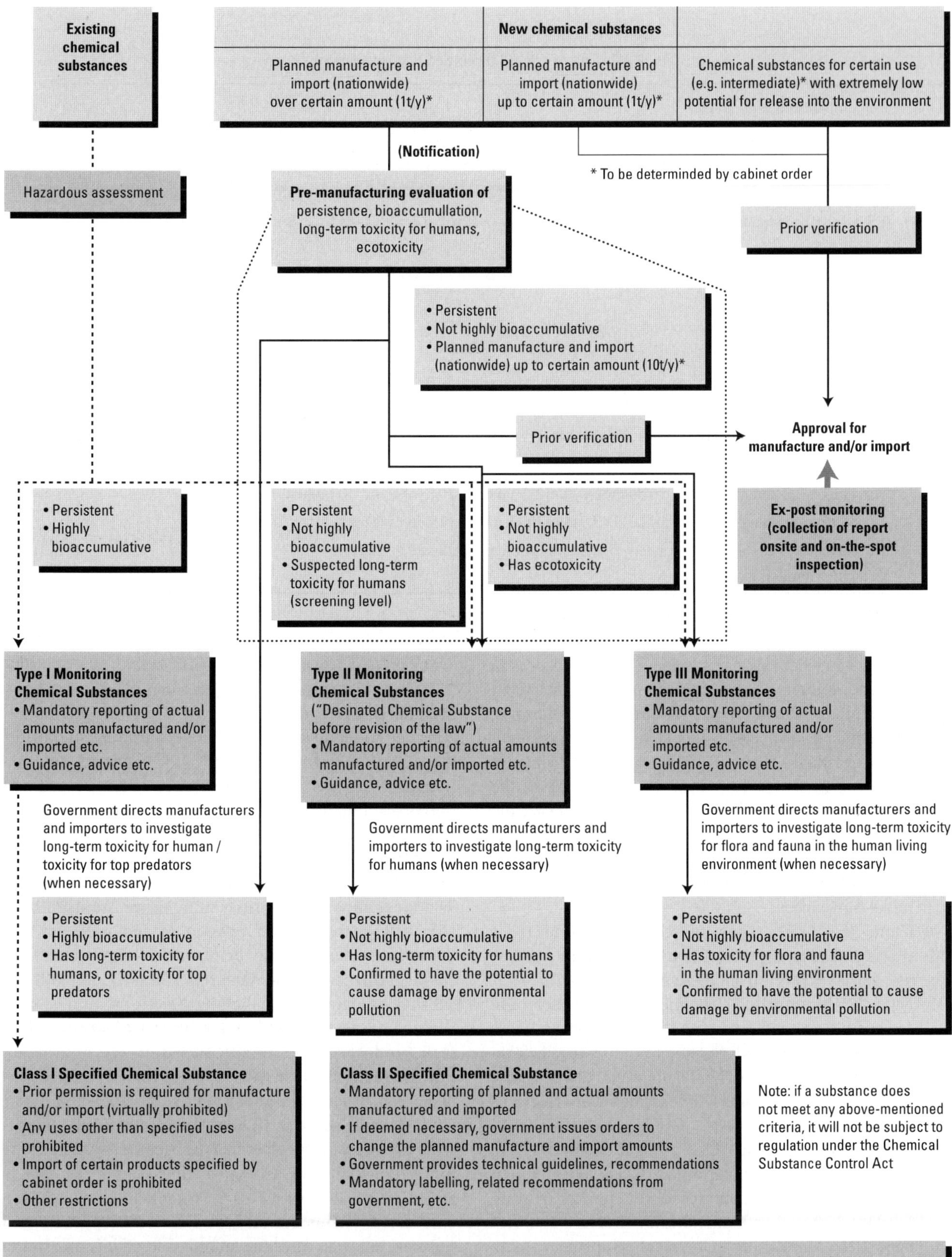

Figure 14.2. New framework for the management of chemicals under the amended Chemical Substances Control legislation [3].

Table 14.2. List of Class I Specified Chemical Substances.

1	Polychlorinated biphenyls
2	Polychlorinated naphthalenes (only those containing more than 2 chlorine atoms in the molecule)
3	Hexachlorobenzene
4	1,2,3,4,10,10-Hexachloro-1,4,4a,5,8,8a-hexahydro-exo-1,4-end-5,8-dimethanonaphthalene (Synonym: Aldrin)
5	1,2,3,4,10,10-Hexachloro-6,7-epoxy-1,4,4a,5,6,7,8,8a-octahydro-exo-1,4-end-5,8-dimethanonaphthalene (Synonym: Dieldrin)
6	1,2,3,4,10,10-Hexachloro-6,7-epoxy-1,4,4a,5,6,7,8,8a-octahydro-end-1,4-end-5,8-dimethyanonaphthalene (Synonym: Endrin)
7	1,1,1-Trichloro-2,2-bis(4-chlorophenyl)ethane (Synonym: DDT)
8	1,2,4,5,6,7,8,8-Octachloro-2,3,3a,4,7,7a-hexahydro-4,7-methano-1H-indene, 1,4,5,6,7,8,8-heptachloro-3a,4,7,7a-tetrahydro-4,7-methano-1H-indene and their analogous compounds (Synonym: Chlordane or Heptachlor)
9	Bis(tributyltin) oxide
10	N,N'-Ditolyl-p-phenylenediamine, N-Tolyl-N'-xylyl-p-phenylenediamine, or N,N'-Dixylyl-p-phenylenediamine
11	2,4,6-tri-tert-butylphenol
12	Polychloro-2,2-dimethyl-3-methylidenebicyclo[2.2.1]heptane (Synonym: Toxaphene)
13	Dodecachloropentacyclo[5.3.0.0(2,6).0(3,9).0(4,8)]decane (Synonym: Mirex)
14	2,2,2-trichloro-1,1-bis(4-chlorophenyl)ethanol (synonym: kelthane or dicofol)
15	hexachlorobuta-1,3-diene

this quantity when deemed necessary. In order to prevent environmental pollution caused by the use of these chemicals, certain types of products containing these chemicals have to be labelled. The government has also published technical guidance on the measures to be taken by the users of these chemicals. As in the case of the Class I Specified Chemical Substances, the government can also issue recommendations to control the production, import and use of such substances before they are officially designated as Class II Specified Chemical Substances.

There are three types of Chemical Substances Monitoring used as a precautionary measure to manage the risk that may result from these chemicals:

1. Type I Monitoring Chemical Substances are designated as candidates for classification as Class I Specified Chemical Substances, identified from existing substances. (The distinction between new and existing substances is explained in Sections 14.2.3 and 14.3.)
2. Type II Monitoring Chemical Substances are candidates for classification as Class II Specified Chemical Substances, identified from new or existing substances, based on human health effects.
3. Type III Monitoring Chemical Substances are candidates for classification as Class II Specified Chemical Substances, identified from new or existing substances, based on ecotoxicological effects.

Manufacturers and importers of these substances are expected to report annually to the government authorities the actual quantity produced and imported, and the intended use for these substances. The government publishes this import and or production information when the total amount exceeds 1 tonne for Type I Monitoring Chemical Substances, and 100 tonnes for Type II and Type III Monitoring Chemical Substances. The government authorities may issue guidance and advice to manufacturers, importers and users on how these substances should be handled. The government can also direct the manufacturers and importers of these substances to investigate their hazardous properties, when deemed necessary based on the information provided on their production and import. Based on the information provided on the hazardous properties, the authorities decide on whether the substances should be classified as Class I or II Specified Chemical Substances or excluded from the Monitoring Chemical Substances system. Some Type III Monitoring Chemical Substances may not be added to the list of Class II Chemical Substances or excluded from the list of Type III Monitoring Chemical Substances, even after the submission of information on ecotoxicity.

Table 14.3. List of Class II Specified Chemical Substances.

1	Trichloroethylene
2	Tetrachloroethylene
3	Carbon tetrachloride
4	Triphenyltin N,N-dimethyldithiocarbamate
5	Triphenyltin fluoride
6	Triphenyltin acetate
7	Triphenyltin chloride
8	Triphenyltin hydroxide
9	Triphenyltin salt of fatty acid (only those containing 9,10 or 11 carbon atoms in the fatty acid)
10	Triphenyltin chloroacetate
11	Tributyltin methacrylate
12	Bis(tributyltin) fumarate
13	Tributyltin fluoride
14	Bis(tributyltin) 2,3-dibromosuccinate
15	Tributyltin acetate
16	Tributyltin laurate
17	Bis(tributyltin) phthalate
18	Copolymer of alkyl acrylate, methyl methacrylate and tributyltin methacrylate (only those containing 8 carbon atoms in alkyl group of alkyl acrylate)
19	Tributyltin sulfamate
20	Bis(tributyltin) maleate
21	Tributyltin chloride
22	Mixture of tributyltin cyclopentanecarboxylate and its analogous compounds (Synonym: Tributyltin naphthenate)
23	Mixture of tributyltin 1,2,3,4,4a,5,6,10,10a-decahydro-7-isopropyl-1,4a-dimethyl-1-phenanthrenecarboxylate and its analogous compounds (Synonym: Tributyltin salt of rosin)

14.2.3 Notification and evaluation of new chemicals

Manufacturers and importers of new chemical substances are required to submit prior notification to the government with relevant information on the properties of the substances. The information submitted is evaluated and the authorities decide on whether or not these substances fall within the criteria of the regulated substances, as explained in the previous section. New substances cannot be produced or imported before this government decision has been made and published. In Japan, a "new chemical substance" means any chemical except the following (see also the exemptions explained in Section 14.2):

- Substances in the Inventory of Existing Chemical Substances, which lists the chemical substances already manufactured, imported or in use in 1973 when the CSCL took effect.
- Substances already designated as Specified Chemical Substances or Monitoring Chemical Substances.
- Substances that have already been evaluated by the government authorities and officially designated as substances that do not need to be classified as Specified Chemical Substances or Monitoring Chemical Substances.

Notification of new chemical substances is exempted in the following cases:

- Substances manufactured or imported for research and development purposes, or for use in experiments.
- Substances manufactured or imported as intermediates for the manufacture of other substances, provided that measures are taken to prevent environmental pollution during the process in which the original substances are used.
- Substances used in closed systems with little or no possibility of release outside the facilities, provided

that measures are taken to prevent environmental pollution throughout the process until their disposal.

- Substances for export only, provided that the destination of the export is identified as a country where a pre-import evaluation system for new chemicals is in place, and that measures are taken to prevent environmental pollution until their export. These countries are specified in a ministerial order, and include Australia, Canada, China, the Republic of Korea, New Zealand, Switzerland, the United States, and the countries where EU legislation applies.
- Substances produced or imported in an amount less than one tonne per year, provided that their manufacture or import is notified to the authorities and the substances are government approved, i.e., there is no safety concern.
- Imported chemicals already notified by foreign manufacturers and approved by the Japanese government.

Manufacturers and importers must first submit a form specifying the name of the substance, its chemical formula (if unknown, the description of the manufacturing process), physical and chemical properties, constituents, intended use, and planned amount of annual production or import over a period of up to three years. Information on the hazardous properties of the substance, including test results, may be submitted at this stage. Within three months (or four months, in the case of notification by foreign companies), a decision is made based on the available information about which of the following six categories the substance falls into. The manufacturer or importer is notified of this.

1. Class I Specified Chemical Substance.
2. Substance is classified as both a Type II and a Type III Monitoring Chemical Substance.
3. Type II Monitoring Chemical Substance (and not Type III).
4. Type III Monitoring Chemical Substance (and not Type II).
5. Substance is neither Class I Specified Chemical Substance nor Type II or Type III Monitoring Chemical Substance (i.e., non-regulated substance).
6. Substance cannot be classified on the basis of the available information.

Based on the submitted data (see Section 14.2.4), the authorities take an official decision on which of the categories the chemical belongs in:

1. When a substance is identified as a Class I Specified Chemical Substance (category 1), the designation is published and production or import of the substance is virtually banned.
2. When a substance is identified as a Type II and/or Type III Monitoring Chemical Substance (categories 1-4), the designation is also published, and the requirements for notification of production/import apply. On the basis of further assessment, with the manufacturer/importer possibly being requested to conduct toxicity testing, these substances may later be designated as Class II Specified Chemical Substances.
3. When none of these criteria apply, the chemical is designated as a non-regulated chemical (category 5). When a substance is classified as a non-regulated substance, the name of the chemical is published after five years, following which other companies may also manufacture or import it without the need for notification. If other companies wish to produce or import the same substance before this publication date, these other companies must follow the same procedure as the first notifier.
4. When the substance is identified as falling into category 6, test data on persistence, bioaccumulation, mammalian toxicity and/or ecotoxicity must be submitted to the authorities.

It is important to note that new chemicals cannot be classified as Type I Monitoring Chemical Substances, since toxicity testing will be required for persistent and bioaccumulating substances. In these cases a decision will be taken on whether these substances should be classified as a Class I Specified Chemical Substance or not. Thus only existing substances are eligible to be designated as Type I Monitoring Substances.

14.2.4 Data requirements and assessment criteria

As explained in the previous section, manufacturers or importers of new chemicals are requested to submit test data unless the authorities can designate the substance as a regulated or non-regulated substance based on the available information. The data requirements are shown in Table 14.4.

As non-persistent chemicals are not controlled under CSCL (as explained in Section 14.2.1), degradation test data are the first to be considered. If the OECD 301C test results (302C may be used when 301C test results show that degradation continues after the test period) show ready biodegradability, the substance will be identified as a non-regulated substance, and further tests will not be required. However, as the regulation also considers the degradation products, test data on degradation products will also be required. Chapter 16 provides an overview of the OECD test guidelines.

Table 14.4. Test data required for a new chemical notification under the CSCL.

Name	Explanation	Production volume	
		1-10 t/y	≥ 10 t/y
Fate properties	Ready biodegradability	x	x
	n-octanol-water partition coefficient or bioconcentration in fish	x[1]	x[1]
Mammalian toxicity	Repeated dose 28-day oral toxicity in mammals		x[1]
	Bacterial reverse mutation test		x[1]
	Chromosome aberration in mammalian cell culture		x[1]
	Mammalian chronic toxicity, toxicity to reproduction and offspring, teratogenicity, carcinogenicity, biotransformation and pharmacological effects		[2]
Ecotoxicity	Algae growth inhibition		x[1]
	Daphnia acute immobilization		x[1]
	Fish acute toxicity		x[1]
	Avian reproduction toxicity, and mammalian toxicity to reproduction and offspring		[2]

[1] If a substance is found to be readily biodegradable, other tests are not required. However, tests for the degradation products are also required.

[2] These tests are needed to assess long-term toxicity to humans or high predators to be able to designate a chemical as a Class I Specified Chemical Substance. Therefore, these tests are only required later.

Bioaccumulation is evaluated based on the fish bioconcentration test (OECD 305) or *n*-octanol-water partition coefficient (OECD 107 or 117). If the bioconcentration factor (BCF) is ≥ 5000, the substance is judged to be highly bioaccumulative. If the BCF is <1000 or the log K_{ow} (*n*-octanol-water partition coefficient) is less than 3.5, the substance is judged not to be highly bioaccumulative. If the BCF lies between 1000 and 5000, bioaccumulation is assessed by taking into account other test data and relevant information.

As explained in the next section, toxicity data submitted are evaluated by experts in order to take a decision on the designation of the substance as a Specified or Monitoring chemical substance. The three Ministries provide guidance on the evaluation of mammalian toxicity data (see also Chapter 6) with a view to classifying the substance as a Type II Monitoring Chemical Substance. For example, a substance with data showing high subchronic toxicity (No Observed Effect Level, NOEL, ≤ 25mg/kg/day) or strong *in vitro* mutagenicity will be classified as a Type II Monitoring Chemical Substance. The same applies to medium subchronic toxicity (NOEL between 25 and 250 mg/kg/day) and weak mutagenicity.

The Ministry also provides guidance for evaluating ecotoxicity data (see also Chapter 7) with a view to classifying the substance as a Type III Monitoring Chemical Substance. Substances with LC50 or EC50 values of less than 1 mg/L in one of the three tests (fish, daphnids or algae), or with fish-LC50 data between 1 and 10 mg/L are classified as Type III Monitoring Chemical Substances. A substance with an EC50 for *Daphnia* between 1 and 10 mg/L may be classified by taking into account other information. A substance with a NOEL of ≤ 0.1mg/L in an algae growth inhibition test, a *Daphnia* reproduction test or a fish early life stage test will also be classified as a Type III Monitoring Chemical Substance.

14.2.5 Procedures for the evaluation of new and existing chemicals

The CSCL stipulates that the government, in designating chemicals as Class I or Class II Specified Chemical Substances or Type I, II or III Monitoring Chemical Substances, should consult with three Councils, i.e., the Drug and Food Sanitation Council (under the Ministry of Health, Labour and Welfare), the Chemicals Council (under the Ministry of Economy, Industry and Trade) and Central Environment Council (under the Ministry of the Environment). These councils consist of experts

Table 14.5. New chemicals notification in Japan, the EU and the US [5].

Name	1996	1997	1998	1999	2000	2001	2002	2003	2004	2005
Japan CSCL	320	325	352	323	373	322	292	362	426	443
EU Directive	441	437	408	430	424	352	398	337	312	174
US TSCA	1472	998	1105	1418	1233	948	1083	869	959	844

in relevant scientific fields and representatives from different sectors of society.

New chemicals notifications are evaluated by a joint subcommittee of these Councils. The subcommittee meets approximately once a month. The number of new chemical notifications per year is given in Table 14.5. The data in Table 14.5 are from the Ministry of Economy, Trade and Industry, the European Chemicals Bureau and the US Environmental Protection Agency (Please note that the EU number for 2005 is still tentative).

Although the meetings of government councils are usually open to public, new chemical notifications are evaluated by the subcommittee in closed sessions in order to protect confidential business information associated with the notification.

The designation to Class I and Class II Specified Chemical Substances is discussed by these councils in open sessions. A recent example is the future addition to Class I Specified Chemical Substances of 2-(2H-1,2,3-Benzotriazol-2-yl)-4,6-di-tert-butylphenol. This chemical, an ultraviolet absorber for plastic resins, is currently designated as a Type I Monitoring Chemical Substance. The toxicity tests required were conducted to consider further regulatory needs. In November 2005, the test results were submitted to the three Councils. On the basis of this information the Councils recommended in January 2006 that this substance be designated as a Class I Specified Chemical Substance. The government implemented this recommendation and issued administrative guidance that this chemical should no longer be produced or used. In July 2006, the Councils recommended measures for products containing this substance. The government plans to proceed with the official designation of this substance further to public participation and WTO notification procedures.

14.3 EXAMINATION OF EXISTING CHEMICALS

When the CSCL was passed by the Diet in1973, the House of Councillors requested the government to conduct an overall review of the safety of existing chemicals, and designate Specified Chemical Substances as necessary on the basis of this review. Further to this parliamentary resolution, environmental monitoring was conducted and hazard information collected by the government authorities. The hazard assessment was also carried out within the framework of the OECD High Production Volume (HPV) Chemicals Programme. It has been recognized, both domestically and internationally, that major resources are required for the assessment of existing chemicals and international collaboration is therefore essential. In 1998, through the International Council of Chemical Associations (ICCA), the global chemical industry announced its intention to work with the OECD to establish a working list of 1000 priority substances for investigation (see Chapter 16). When the CSCL was amended in 2003, a new resolution by the House of Councillors stated that the investigation of the safety of existing chemicals should be done strategically, through co-operation between government and industry, and sharing the work internationally. The following sections outline the achievements so far and future plans for the evaluation of existing chemicals in Japan.

14.3.1 Collection of hazard information – Japan Challenge Program

The "Japan Challenge Program", designed after the HPV Chemicals Challenge Program in the United States, was announced by the Ministry of Health, Labour and Welfare, the Ministry of Economy, Trade and Industry and the Ministry of the Environment in June 2005. It is a joint initiative of the Japanese government and industry, especially manufacturers and importers of chemicals in Japan, set up to facilitate the collection of safety information on HPV chemicals.

Before the 2003 amendment of the CSCL, hazard testing in Japan was primarily done by the government. By the end of 2004, 1455 substances had been tested for degradation and bioconcentration, while 275 chemicals had been tested for mammalian toxicity and 438 for ecotoxicity, respectively. These numbers look modest, considering that the inventory of existing chemicals includes approximately 20,000 substances, but these figures represent a good start considering the overall international cooperation involved in the testing and assessment of existing chemicals.

Under the OECD HPV Programme, Japan has sponsored the assessment of 191 chemicals, of which 109 have already been evaluated, with hazard data generated by government laboratories and manufacturers. However, the progress of the OECD Programme has been slow and some measures were needed to achieve the ambitious goal established in 2005 to finish screening assessments for an additional 1000 chemicals by 2010.

The Japan Challenge Program was established to accelerate the assessment of existing industrial chemicals produced in Japan, and to contribute to the OECD Programme. The target chemicals for the programme were selected from organic substances produced or imported into Japan in amounts of over 1000 tonnes per year. However, to ensure the effective development and application of the category approach, the programme does not exclude low production volume chemicals. To avoid duplication of testing, the programme excludes chemicals for which there are past, ongoing and planned safety information collection activities, such as the OECD HPV and the US Challenge Program. The list is regularly updated with new information on production and assessment activities.

Manufacturers or importers inform the government of their willingness to sponsor information gathering for the listed chemicals. Sponsors gather safety information in line with the data requirements of the OECD Screening Information Data Set. Forming a consortium is recommended to avoid duplication of testing and ensure the efficient use of resources. The use of category approaches in test proposals is encouraged. Safety information collected through this programme will be made available to the public through the Internet. As of May 2006, sponsors have been identified for 78 substances, with the participation of 62 companies and 3 associations. This number represents about half the number of substances for which sponsors are required.

The progress of the Japan Challenge Program is reviewed at the end of each fiscal year. The names of the sponsors and chemicals are published. It is anticipated that the collection of the required information under the current Japan Challenge Program will be accomplished by the end of the Japanese fiscal year 2008. The overall review of progress will be conducted after April 2008.

14.3.2 GHS classification

The Globally Harmonised System for Classification and Labelling of Substances and Mixtures (GHS) was published by the United Nations in 2003 [6]. The United Nations GHS Subcommittee recommended that the GHS be implemented in every country by 2008. The government of Japan started a number of activities to implement the GHS in its chemical management system, starting with the amendment of the Occupational Safety and Health Act in 2006. Another activity in this context is a joint project of the Ministry of Agriculture, Forestry and Fishery, the Ministry of Health, Labour and Welfare, the Ministry of Economy, Trade and Industry, and the Ministry of the Environment, to classify around 1500 chemicals in accordance with the GHS criteria.

The aim of this project is to enable industry to provide material safety data sheets (MSDS) in line with GHS. In Japan, MSDS are required for approximately 1500 chemicals regulated under the Occupational Health and Safety Act, the Poisonous and Deleterious Substances Control Act, and the Chemical Management Act. The four Ministries jointly apply the GHS criteria to these chemicals, using available information. In order to do this, these Ministries have published a manual indicating the priorities for data sources and providing guidance on how to interpret these data for classification purposes. The results from this project are not intended to be compulsory, and industrial users may use their own data to classify chemicals on the basis of their own judgement.

Classification is undertaken by experts from governmental or independent laboratories, and checked by an inter-ministerial committee on the GHS. The results are published on the Internet [7], and comments on individual classifications are invited. As of July 2006, classification results for more than 800 chemicals have been made publicly available.

14.3.3 Environmental monitoring

In preventing environmental pollution by hazardous chemicals, knowledge on the presence of chemicals in the environment provides basic information for policy making. The Ministry of the Environment conducts extensive surveys of the concentration of chemicals in different environmental media. Air pollution monitoring for about 20 priority substances is conducted by hundreds of monitoring stations. About 50 water pollutants have been measured in many waterways. However, data on cross-media pollution by a larger number of chemicals are needed for the assessment and management of risks to human health and the environment.

When the CSCL was enacted in 1973, the then Environment Agency of Japan (now Ministry of the Environment) initiated environmental monitoring. The General Inspection Survey of Chemical Substances for Environmental Safety evaluated the presence and persistence of existing chemical substances in the environment. In 1978 the Environment Agency selected about 2000 priority substances and undertook the First Comprehensive Survey between 1979 and 1988. Substances were selected from the list, analytical methods were developed, and monitoring was conducted. In 1987, the second priority list of about 1000 chemicals was drawn up, and the Second Comprehensive Survey was conducted in the period between 1989 and 2001. During these periods, about 800 substances were measured in the environment.

However, with the enactment of legislation related to chemical management and the progress in international activities, such as the Stockholm Convention on Persistent Organic Pollutants [4] and OECD Programmes [8], it became increasingly clear that a different monitoring policy was needed for chemical assessment and management activities. Therefore, in 2002 the Special Committee for the Assessment of Chemical Substances under the Central Environment Council approved a revised monitoring policy, where substances are selected every year by the Expert Group on Substance Selection under the Central Environment Council, corresponding to the needs of various divisions of government and other organizations, so that the survey results can be utilized to prevent environmental pollution by chemical substances in a more timely manner.

Currently, the environmental survey conducted by the Ministry of the Environment consists of three types of surveys. Firstly, the Initial Environmental Survey aims to confirm whether certain chemicals are found in ambient air, water, sediment and aquatic animals at a detectable level. Substances are selected from the Monitoring Chemical Substances list under the Chemical Substances Control Act, candidate substances for the Pollutant Release and Transfer Register, chemicals for the Initial Environmental Risk Assessment, and from other lists of substances subject to policy. Secondly, the Detailed Environmental Survey is intended to quantify environmental concentration which can then be used for various purposes, including the Initial Environmental Risk Assessment. This survey covers water, sediment, ambient air and biota, and may include food where any special need arises. Thirdly, a Monitoring Survey is conducted with the aim of evaluating the annual trend in chemicals of concern, especially persistent organic pollutants (POPs) under the Stockholm Convention, and Class I and Class II Specified Chemical Substances. In 2005 a biomonitoring programme to monitor POPs in human breast milk and blood was initiated as one of the elements of the monitoring survey. The results of these surveys are made available to the public through the website of the Ministry of the Environment. The most recent report in English contains the results of the survey carried out in 2003 [9].

14.3.4 Initial risk assessments

The Ministry of the Environment conducts Initial Environmental Risk Assessments to screen chemical substances which may pose relatively high environmental risks. The Initial Environmental Risk Assessment process started in 1997, and the results from the pilot project on 39 chemicals were published in 2001. As of December 2006, initial assessments for 116 substances have been published, as well as test data for some other chemicals.

These initial assessments are conducted following a guideline established by the Ministry of the Environment. The guideline is continuously updated and included in the most recent assessment reports [10]. Chemicals are selected every year, taking into consideration the requirements of the relevant chemical management legislation and regulations. Current assessment priorities are substances on the Pollutant Release and Transfer Register (PRTR) and Monitoring Chemical Substances under the CSCL. The assessment results in four possible conclusions: (1) candidate for more detailed assessment, (2) chemicals requiring further information collection, (3) chemicals that do not require further action at this stage, and (4) chemicals that could not be evaluated.

The assessment starts with an exposure assessment. Measured data are collected for environmental media and food. Data from the same monitoring point in the same year are averaged. In view of the need for conservative assessment at this initial stage, maximum concentration data are selected and an evaluation of the reliability of the data carried out. Predicted Environmental Concentrations (PECs) are calculated from these data. If measured data are not available, partition into environmental media estimated from PRTR or other emission data are used as supplementary information. Human exposure is calculated using standard exposure parameters (e.g., respiration volumes of $15m^3$/day/50kg, drinking water consumption of 2L/day/50kg, a soil ingestion (intake) of 0.15g/day/50kg and a food consumption of 2000g/day/50kg).

Table 14.6. Criteria for Initial Environmental Risk Assessment.

Conclusions	Human health – with threshold	Human health – without threshold	Ecological effects
	Margin of exposure	Excessive occurrence at max. exposure	PEC/PNEC
Candidate for more detailed assessment	Less than 10	10^{-5} or more	1 or more
Chemicals requiring further information collection	10– 100	10^{-6} – 10^{-5}	0.1 – 1
Chemicals that do not require further action at this stage	100 or more	Less than 10^{-6}	Less than 0.1
Chemicals that could not be evaluated	Unavailable	Unavailable	Unavailable

The human exposure estimation is compared with health hazard information obtained from the literature and limited testing. Where there is a threshold level for human toxicity, the margin of exposure is calculated by dividing the estimated maximum exposure by the No Observed Adverse Effect Level. When there is no threshold, excessive occurrence of adverse effects at the maximum exposure level is used for the assessment. For ecological assessment, the PEC is compared with the Predicted No Effect Concentration (PNEC). The PNEC is derived from existing test data using the safety factor approach (see Chapters 6 and 7). The criteria for these judgements are summarized in Table 14.6.

14.3.5 Other risk assessments

Substances identified as candidates for further assessment are investigated in terms of whether they may be subject to regulatory measures. One such example is the risk assessment for hazardous air pollutants. Under the Air Pollution Control Act of 1996, 22 substances are listed as priority substances. Environmental Quality Standards for four of these chemicals and guideline values for another four substances have been established so far. In June 2006, draft guideline values for yet another four substances were published for comments. The proposals for these standards and guidelines are accompanied by detailed risk assessments for air.

The government has also launched other risk assessment initiatives. These include the project funded by the New Energy Development Organization and implemented by the National Institute for Technology Evaluation (NITE) and the National Institute of Advanced Industrial Science and Technology (AIST). In this project, initial risk assessments are being conducted by NITE for about 150 chemicals selected from the chemicals on the PRTR or subject to the MSDS requirement under the Chemicals Management Act. AIST is also conducting detailed risk assessments for 16 substances. The results from this project can be downloaded from the NITE and AIST websites.

14.4 FUTURE OF CHEMICALS MANAGEMENT IN JAPAN

This chapter reviewed the development and current status of chemicals regulation in Japan. As described above, the CSCL has evolved to incorporate risk-based approaches. It has also responded to developments in international programmes, such as the OECD Environment, Health and Safety Programme [8] and the Stockholm Convention on POPs [4].

The word "environmental risk" was mentioned for the first time in an official governmental document, in the Basic Environment Plan, first approved by the cabinet in 1993, following the introduction of the Basic Environment Act [11]. The Plan included a brief section stating that chemical management would be pursued to reduce the environmental risk posed by chemicals. Here, "environmental risk" is not limited to ecological effects, but includes the possibility of adverse effects on human health caused by environmental exposure as a result of environmental pollution. How the risk-based approach has been incorporated in the CSCL is explained in the preceding sections.

Another important piece of legislation which

Policy objectives toward 2025

- Scientific risk assessments
- Application of precautionary approach
- Actions by all the stakeholders
- Contribution to international harmonization

Bridging the information gap on hazard and exposure

- Accelerating the investigation of existing chemicals
- Strengthening the environmental monitoring including biomonitoring
- Collecting exposure information such as production/use amount and use categories
- Obtaining mass flow data of major chemicals from production to disposal by 2020.

Policy mix for risk management

- Achieving environmental quality standards and guidelines including hot spots
- Applying best available techniques and best environmental practices
- Best policy mix of regulatory and voluntary approaches

Strengthening risk communication

- Making hazard information available to consumers
- Promoting environmental education

Contribution to global issues with enhanced visibility

- Promoting international chemicals management in line with SAICM
- Leading environmental monitoring and chemical management based on Japanese experience
- Learning from chemical management systems in other countries
- Enhanced contribution to harmonization of chemical assessment & management methods
- Introduce GHS classification and labelling by 2008

Figure 14.3. Policies for reducing the environmental risks of chemicals as presented in the Third Basic Environment Plan of 7 April 2006.

addresses chemical risks, explicitly taking into account exposure is the PRTR system, introduced by the Chemical Management Act in 2000. The enactment of this law was also prompted by the OECD Council Recommendation on Implementing Pollutant Release and Transfer Registers [12]. Currently, factories and other facilities have a duty to report annually the emission quantities of 354 chemicals to air, water and land. The transfer of chemicals to waste is also taken into account. The government estimates the emissions of these chemicals that are not reported by individual facilities, e.g. small businesses and non-point sources. These figures are summarized and published. Reported data can also be obtained on request following the information disclosure procedures. The PRTR data provide a valuable information source for exposure assessments.

The cabinet approved the third Basic Environment Plan in February 2006 [13]. The Plan includes a chapter setting out the future policy for reducing the environmental risks of chemicals. The structure of the chapter is illustrated in Figure 14.3. As this figure shows, the plan addresses major international commitments, such as the Millennium Goal agreed at the World Summit for Sustainable Development in Johannesburg in 2002 to minimize the adverse effects of the production and use of chemicals to human health and the environment by 2020 [14]. Others include the UN recommendation to implement GHS by 2008, and the worldwide commitments expressed in the Strategic Approach to International Chemicals Management (SAICM) [15].

The legislation on chemicals management is subject to a statutory review requirement. The PRTR will be reviewed in 2007, and the CSCL will be reviewed in 2009. In September 2006, an expert group under the Ministry of the Environment published a report on the implementation of the PRTR and future challenges [16]. In December 2006, the Central Environment Council began its deliberations on future environmental chemicals policy, including the review of these two pieces of legislation. This review is expected to address the issues identified in the Third Basic Environment Plan, such as the need for further action on existing chemicals, a shift towards a risk-based approach, a focus on chemicals in products and articles, and international co-operation and harmonization.

14.5 FURTHER READING

1. A provisional English Translation of the CSCL and other chemicals-related legislation are available from http://www.safe.nite.go.jp/english/kasinn/kaiseikasinhou.html. For a full understanding the legal requirements, however, it is essential to refer to the Japanese legal texts. These are available from http://www.env.go.jp/chemi/kagaku/kashinkaisei.html.
2. OECD New Chemicals Programme compares the chemicals regulation systems of OECD member countries. One of the results is OECD (2004) New Chemical Assessment Comparisons and Implications for Work Sharing.

REFERENCES

1. Inter-ministerial Meeting on IFCS. 2003. National Profile on Chemical Management – Japan.
2. Organization for Economic Co-operation and Development. 2002. OECD Environmental Performance Reviews – Japan. OECD, Paris, France
3. Ministry of Economy, Trade and Industry (METI), the Ministry of Health, Labour and Welfare (MHLW) and the Ministry of the Environment (MoE). 2003. Outline of the 2003 Partial Amendment to the CSCL. Available from http://www.nite.go.jp/english/.
4. United Nations Environment Program. 2001. Stockholm Convention on Persistent Organic Pollutants. UN, Nairobi, Kenia.
5. Ministry of Economy, Trade and Industry. 2006. Paper for the Subcommittee on Basic Policy on Chemicals, Industrial Structure Council (In Japanese).
6. United Nations. 2003. Globally Harmonized System of Classification and Labelling of Chemicals (GHS). UN, Geneva, Switzerland
7. Inter-ministerial Meeting on GHS. 2006. Classification of Chemicals Following GHS. (In Japanese). Available from http://www.safe.nite.go.jp/ghs/index.html.
8. Organization for Economic Co-operation and Development. 2005. OECD Environment, Health and Safety Programme. OECD, Paris, France. Available from http://www.oecd.org/ehs/.
9. Ministry of the Environment. 2005. Chemicals in the Environment FY 2005. Available from http://www.env.go.jp/en/.
10. Ministry of the Environment. 2002-2005. Initial Environmental Risk Assessment of Chemicals Volume 1-4 (In Japanese).
11. Government of Japan. 19994. First Basic Environment Plan. Available from http://www.env.go.jp/en/.
12. Organization for Economic Co-operation and Development. 1996. Recommendation of the Council on Implementing Pollutant Release and Transfer Registers (C(96)41/Final). OECD, Paris, France.
13. Government of Japan. 2006. Third Basic Environment Plan. Will be available from http://www.env.go.jp/en/.
14. United Nations. 2002. Report of the World Summit on Sustainable Development. UN, Geneva, Switzerland.
15. SAICM Secretariat. 2006. Strategic Approach to International Chemicals Management. Available from http://www.chem.unep.ch/saicm/.
16. Forum on Chemical Management Act. 2006. Implementation of the Chemical Management Act and Future Challenges (In Japanese).

15. THE ASSESSMENT AND MANAGEMENT OF INDUSTRIAL CHEMICALS IN CANADA

M.E. MEEK AND V.C. ARMSTRONG

15.1 INTRODUCTION

The chemical industry is one of the largest manufacturing sectors in Canada and employs more than 90,000 people; nearly every major global chemical company in the world has production or research and development facilities. In 2003, more than two thousand companies, including 21 of the 25 world's largest manufacturers, had operations in this country. Shipments of chemical products were worth \$ 42 billion (~ € 26.4 billion). The industrial chemicals subsector, which includes companies manufacturing petrochemicals, industrial gases, pigments, other inorganic and organic chemicals, resins, and synthetic fibres, accounted for close to 50% of this amount [1].

In this chapter, emphasis is placed on the progressive, legislated requirements for assessment and control of significant numbers of existing substances under the Canadian Environmental Protection Act (CEPA), which include emissions and by-products associated with chemicals production [2]. This has involved the in-depth assessment of 69 substances (including complex mixtures and groups of substances) identified as priorities under the first CEPA (CEPA-1988) [3] in two mandated five-year timeframes. These assessments were followed by the implementation of risk management measures for a significant proportion that were deemed to present a risk to the environment or human health.

More recently, under CEPA–1999 [4], precedent-setting provisions to systematically identify, in a timely manner, priorities for assessment and management from among the approximately 23,000 existing commercial substances have been introduced. This has necessitated the development of innovative methodology including evolution of the previously linear or sequential steps of risk assessment and risk management to a more iterative approach where the need for, and focus of, potential control options are identified at an early stage of assessment. It has also required development of assessment products that efficiently dedicate resources, investing no more effort than is necessary to set aside a substance as a non-priority or to provide necessary information to permit risk management. Similarities to, and variations from, approaches adopted or contemplated under US and European chemicals control legislation are also outlined.

It is to be stressed that the nature of the actions taken under CEPA-1999 and the associated methodological developments necessitated by the provisions of this Act continue to evolve. Therefore this chapter can provide only an overview of the status of industrial chemicals management as it was at the time of its completion, that is, early in 2007. Additional detail and further developments are and will be available at website references listed in the bibliography.

15.2 LEGISLATIVE BACKGROUND

15.2.1 Federal-provincial regulatory structure

Canada is a federation of ten provinces and three territories, for which responsibilities for matters pertaining to the environment are shared. Indeed, the Supreme Court of Canada has ruled that environmental protection is of such importance that it requires action by governments at all levels. In January, 1998, the provinces, with the exception of Quebec, and the federal government signed an Accord on Environmental Harmonization, which sets the framework for collective goals and action to protect the environment [5,6]. The CEPA, the cornerstone of federal environmental protection, is entirely consistent with the Harmonization Accord and is the tool for implementation of Harmonization Agreements.

International aspects of the assessment and control of toxic chemicals fall under the purview of the federal government; these include responsibilities related to international air and water pollution and participation in international initiatives of, for example, the International Programme on Chemical Safety (IPCS), the United Nations Environment Programme (UNEP) and the Organization for Economic Co-operation and Development (OECD).

CEPA has been structured to avoid duplicating effective measures that have already taken or are proposed to be taken by other federal and provincial departments or ministries. Should an assessment conducted under CEPA indicate the need to take action to protect health or the environment and should such action not be undertaken under other Canadian statutes, the risk management provisions of CEPA can be invoked.

C.J. van Leeuwen and T.G. Vermeire (eds.), Risk Assessment of Chemicals: An Introduction, 591–621.

Any actions that are to be taken under other legislation as a result of initiatives under CEPA must be deemed to be equivalent to those proposed under this Act. In order to coordinate work with the provinces and avoid duplication, especially with respect to the development of regulations, a National Advisory Committee has been established as required under CEPA-1999.

With respect to assessment and control of the environmental impacts of new substances, provisions of other federal statutes must be equivalent those of CEPA in terms of requiring notification prior to import into, or manufacture in, Canada and assessment of potential risks to both the environment and human health. These equivalency provisions have had an impact on the assessment and control of substances that fall under the purview of federal legislation such as the Food and Drugs Act [7], the Feeds Act [8] and the Fertilizers Act [9].

15.2.2 Evolution of the Canadian Environmental Protection Act

In contrast to some of its other health protection statutes such as the Food and Drugs Act which dates back to 1920, Canada's environmental protection laws have been developed relatively recently. The Department of the Environment was created in 1972 and the first federal environmental protection act, the Environmental Contaminants Act (ECA) [10] "an act to protect human health and the environment from substances that contaminate the environment." was promulgated in 1975. This legislation, like its successors, was administered jointly by the Department of Health and the Department of the Environment and was developed, in part, to provide a means to respond domestically to international environmental initiatives such as those being undertaken by the OECD to control polychlorinated biphenyls (PCBs). A number of other substances of concern at that time (e.g., polybrominated biphenyls, polychlorinated terphenyls, mirex, and lead from secondary lead smelters, asbestos and mercury), were subsequently banned or controlled under the ECA.

While the ECA required companies to identify chemicals not previously used as "new" there was no systematic testing or assessment of chemicals for toxic effects prior to their introduction. "New" was defined as previously unused by the company and while relevant quantities were also to be specified, notification was required only after introduction of the substances into commerce. Submission of information and testing by industry could be required only if the Ministers of Health and of the Environment had "reason to believe that a substance is entering/may enter the environment in amounts that are a danger to health or the environment" based on consideration of information that had already been generated or obtained from other sources. While information on a large number of new chemicals was examined, the administrative procedure for effecting control was complex and therefore none was undertaken through the short history of the program.

Proposals from an Environmental Contaminants Act Amendments Consultative Committee [11,12] were considered during the Parliamentary review of the ECA and the Canadian Environmental Protection Act (CEPA), "an act respecting the protection of the environment and of human life and health", came into force in 1988. CEPA-1988, a much more comprehensive piece of environmental protection legislation, not only superseded the Environmental Contaminants Act, but also subsumed other Canadian environmental protection statutes (and their regulations) such as the Clean Air Act, the Ocean Dumping Control Act, part of the Canada Water Act and part of the Department of the Environment Act into a single piece of legislation. One of the salient new features of this Act was embodiment of the notion of pre-import/ pre-manufacture notification and assessment of new substances including biotechnology products with adoption of a minimum data set based on that developed under the Chemicals Programme of the OECD [16]. CEPA-1988 was first passed into law in June, 1988, amended in June, 1989 and was replaced with a new and further expanded Act which received royal assent in September, 1999 (CEPA-1999), following a review of its operation and implementation as required to be undertaken within 5 years of the promulgation of the Act.

15.2.3 The current Canadian Environmental Protection Act (CEPA-1999)

CEPA-1999 is entitled "an act respecting pollution prevention and protection of the environment and health in order to contribute to sustainable development" [4]. New principles to guide the application of this Act are spelled out in its preamble; thus in implementing the various provisions of CEPA, Environment Canada and Health Canada are expected to ensure:

- That consistency between federal government departments and collaboration with other jurisdictions results in effective and integrated approaches, policies and programs to manage the health and environmental risks of toxic substances.
- Recognition that risks from toxic substances are a matter of national concern that transcends geographic

boundaries.

- That the importance of an ecosystem approach is recognised.
- That there is a commitment to implement pollution prevention as a national goal and as the primary mechanism to promote environmental protection.
- That the Government of Canada is able to fulfil its international obligations in respect of the environment.
- Implementation of the precautionary principle.
- Implementation of the "polluter pays" principle.
- That there are public participation, openness and transparency in decision-making and that there are mechanisms available for supporting these goals.

Prevention and management of risks posed by toxic and other harmful substances remain as the principal objectives of the revised CEPA. The provisions for implementing pollution prevention, investigating and assessing substances, controlling toxic substances, and those for fuels, international air and water pollution, motor emissions, nutrients, environmental emergencies, for regulating the effects of the federal government's own operations and waste disposal at sea and the import and export of wastes were added or expanded upon. Recognition of the growing importance of biotechnology led to the creation of a specific section with provisions that parallel those for chemical substances.

CEPA-1999 also provides for the gathering of information for research and the creation of inventories of data and for the development of environmental objectives, guidelines and codes of practice. In addition, new rights were bestowed on Canadians to participate in decisions on environmental matters, including the ability to compel investigation of an alleged contravention of the Act and the possibility of bringing civil action when the government is not enforcing the law. Aboriginal governments have the right to be represented on a National Advisory Committee which must be established as a way of "enabling national action to be carried out taking cooperative action in matters affecting the environment and for the purposes of avoiding duplication in regulatory activity among governments".

The new CEPA contains 343 operative sections and six schedules, and is divided into the parts shown in Table 15.1.

Table 15.1 The operative parts of CEPA-1999.

Part	Title
1	Administration
2	Public participation
3	Information gathering, objectives, guidelines and codes of practice
4	Pollution prevention
5	Controlling toxic substances
6	Animate products of biotechnology
7	Controlling pollution and managing wastes
8	Environmental matters related to emergencies
9	Government operations and federal and aboriginal land
10	Enforcement
11	Miscellaneous matters

15.2.4 Administration of the CEPA

Responsibility for the administration of CEPA is shared between Canada's Department of the Environment and Department of Health. The Minister of the Environment has overall responsibility for the administrative aspects of the Act and for most of its other provisions, notably those for enforcement and compliance. Also in most of the instances where the Minister of Health is named, responsibilities must generally be carried out collaboratively. These joint responsibilities include assessing and controlling toxic substances and assessing the impacts of (international) air pollution, and (international) water pollution. The Minister of Health can act independently in conducting health-related research investigations and other studies, setting environmental objectives, guidelines and codes of practice to protect health, and in establishing advisory committees with respect to these responsibilities. It has also been customary for the Health Department to provide advice to Environment Canada on health related issues arising under parts of the Act in which the Health Minister is not explicitly named, (e.g., with respect to the potential health effects of fuels and vehicle, engine and equipment emissions).

15.2.5 The CEPA definition of environment

The definition of "environment" in CEPA (Box 15.1) is sufficiently broad to encompass the occupational as well as the general environment; however, since the provinces and territories are generally responsible for the health and safety of their workers, assessments of the impacts on health of substances under CEPA have been confined to those on members of the general public.

Box 15.1. CEPA definition of "environment"

"Environment" means the components of the earth and includes:

a. Air, land and water.
b. All layers of the atmosphere.
c. All organic and inorganic matter and living organisms.
d. The interacting natural systems that include components referred to in paragraphs (a) to (c).

Box 15.2. CEPA definition of "substance"

"Substance" means, in part: "any distinguishable kind of organic or inorganic matter, whether animate or inanimate, and includes....any mixture that is a combination of substances...,...any complex mixtures of different molecules that are contained in effluents, emissions or wastes that result from any work, undertaking or activity."

15.3 CEPA'S PROVISIONS FOR TOXIC SUBSTANCES

Canada's environmental protection strategy is based on sustainable development; a key component of this is controlling substances that can be harmful to human health or the environment in order to ensure that the risks are prevented or reduced. CEPA-1999 requires the Minister of the Environment, in carrying responsibilities with respect to Toxic Substances, "…to the extent possible, to cooperate and develop procedures with jurisdictions other than the Government of Canada, (that is other governments in Canada or those of member states of the OECD), to exchange information respecting substances that are specifically prohibited or substantially restricted by, or under, the legislation of those jurisdictions for environmental or health reasons".

Controlling toxic substances is viewed as a two-phase process, risk assessment and risk management. The first of these entails a science-based evaluation to enable decision-making on whether a substance poses a risk to health or the environment; the second phase identifies the most suitable control measures [13]. The Act provides a framework for the identification and control of existing substances and management of those considered to pose a risk to human health and/or the environment. This framework is broad, transparent and evidence-based, taking into account aspects (i.e., exposure and effects) of a substance in relation the potential risk it may pose.

15.3.1 CEPA definitions of substance and toxic

The broad definition of "substance" (Box 15.2) under CEPA encompasses not only discrete (industrial) chemical compounds but also complex mixtures formed naturally or as a result of chemical reactions, emissions and effluents and products of biotechnology. All such substances are therefore candidates for assessment under the legislation. Animate biotechnology products can be whole organisms, or parts, or products of organisms, including those developed through genetic engineering. The definition of "substance" is somewhat more restrictive with respect to "new substances" in that articles, physical mixtures and effluents and emissions are excluded.

The purpose of carrying out an assessment under CEPA is to determine whether a substance is or is not "toxic". The definition of "toxic" (Box 15.3) is a legal one and embodies the notion that the ability of a substance to harm the environment or human health is a function of its release into the environment, the intrinsic toxicity (i.e. toxicity in the traditional sense) and the concentration of the substance to which a person (or other environmental receptor), is exposed. Also, inclusion of the word "may" in the definition with respect to both entry into the environment and the potential danger or harm (i.e., effects) allows the approach to designating "toxic" to be developed in a manner which takes into account uncertainties and is consistent with the generally accepted principles of health risk assessment. Thus, "risk" is considered more precisely as depending on the nature of the possible effects and the likelihood of their occurrence; the probability (that any given effect will occur) in turn is a function of the potency of the toxicant, the susceptibility of the exposed individual, or species, and the level of exposure.

The existence of information that is consistent with the designation of a substance as "toxic" under the Act sets the stage for reviewing options for controlling risks to human health and/or to the environment and, hence, for adding the substance to Schedule I of CEPA (the "List of Toxic Substances").

15.3.2 Provisions for new substances

Under the New Substances Program, companies or individuals wishing to import or manufacture substances that are new to Canada must notify the government of

Box 15.3. CEPA definition of "toxic"

A substance is "toxic" if it is "...entering or may enter the environment in a quantity or concentration or under conditions that:

a. Have or may have an immediate or long-term harmful effect on the environment or its biological diversity.
b. Constitute or may constitute a danger to the environment on which life depends; or
c. Constitute or may constitute a danger in Canada to human life or health.

that intent so that the substances can be assessed for possible effects on the environment and human health; certain information specified in regulations must also be provided [14,15]. The New Substances Notification Regulations for chemicals and polymers first came into force in July, 1994; in October, 2005, these were replaced with amended regulations.

The new substances provisions were a critical component in the introduction of CEPA-1988 since they allowed Canada to meet its obligation to honour the OECD Council Decision [16] concerning the requirement for a Minimum Pre-market Data Set for assessing new chemicals. CEPA allows for the control of a new substance before it is manufactured or imported whenever there is a "suspicion" that the substance is "toxic" under the Act.

In order to distinguish commercial substances that are new to Canada and those already in use, a Domestic Substances List (DSL) [17] was compiled under CEPA-1988; the DSL included some 22,400 substances nominated to Environment Canada that were, between January 1, 1984, and December 31, 1986:

- In Canadian commerce.
- Used for commercial manufacturing purposes or
- Manufactured in, or imported into, Canada in a quantity of 100 kg or more in any calendar year.

Substances on the DSL are referred to as "existing" substances. Under CEPA, an existing substance can also be one that is released as a single substance, an effluent, a mixture or a contaminant in the environment. A Non-Domestic Substances List (N-DSL) [18] was also compiled for substances not on the DSL but believed to be in international commerce, though not in Canada, during the reference period. The N-DSL was based on the 1985 Toxic Substances Control Act Inventory (excluding DSL entries), published by the US EPA, chosen as representative of substances that were in commercial use in an "ecozone" similar to that of Canada over the 1984-1986 reporting period [12]. The N-DSL now comprises more than 58,000 substances and is updated bi-annually. Information requirements for substances which are listed on the N-DSL when notified as new to Canada are reduced. For additional information, see: http://www.ec.gc.ca/substances/nsb/eng/home_e.shtml.

The DSL is amended from time to time to include new substances that have been assessed for their risks to human health and the environment and which are deemed not to require the imposition of conditions; substances for which Significant New Activities (SNAc) provisions have been imposed can also be added (see Section 15.3.7).

Between January, 1987 and the coming into force of the New Substances Notification Regulations (NSNR) for chemicals and polymers (July, 1994), about 4,400 commercial chemicals were imported into, or manufactured in, Canada; CEPA-1988 included transitional provisions for post-market notification of these substances.

Substances that are not on the DSL or the N-DSL cannot be imported into, or manufactured in Canada in quantities greater than those stated in the NSNR (Chemicals and Polymers) until prescribed information has been notified to Environment Canada. These regulations specify the information that must be provided to meet the notification obligations. The main features are [19]:

- Establishment of categories of substances (e.g., chemicals, biochemicals, polymers, biopolymers, and organisms).
- Identification of administrative and other information requirements.
- Specification of conditions, test procedures and laboratory practices to be followed in developing test data.
- Timing of notification before manufacture or import or activity outside the scope of a previously issued SNAc Notice.
- Assessment periods for the submitted information.

The establishment of different categories of substances enables different levels of notification requirements to be established depending on the characteristics of the substance and the quantities in which it is to be imported or manufactured. Thus, substances are first generally categorized by type (i.e., chemicals, polymers, biopolymers or organisms) and, then, each substance type is further separated into notification groups based on factors such as use, volume of manufacture or import use and whether the substance is on the N-DSL. Eight Schedules of information requirements are specified for

Box 15.4. Possible outcomes of the assessment of information

1. The substance is not toxic or capable of becoming toxic.
2. The substance is toxic or capable of becoming toxic.
3. The substance is not toxic or capable of becoming toxic, but a suspicion that a significant new activity in relation to the substance may result in the substance becoming toxic.

chemicals and polymers and one for biochemicals and biopolymers under the NSNR (Chemicals and Polymers) [14]. There are reduced requirements for special category substances, those for research and development, contained site-limited intermediates and contained for export only. There are also reduced requirements for certain polymers that meet the "reduced regulatory requirement" criteria. Additional information may be required for chemicals and polymers released to the aquatic environment in high quantities or to which the public may be significantly exposed. The most comprehensive data package is required for substances that are not on the N-DSL and are to be imported or manufactured in a quantity greater than 10,000 kg/year.

Submitted information is assessed (Box 15.4). If the substance is not suspected to be toxic, the notifier may import or manufacture the substance after the assessment period has expired. Where the substance is suspected of being toxic, or becoming toxic, the government may take measures under the Act to ensure that the substance is handled in ways that will adequately manage these risks. These measures could include imposing conditions under which the substance may be used, prohibiting import or manufacture of the substance or requesting additional information or test results that would enable a determination of whether or not the substance is toxic. If the substance is not suspected to be toxic but could become so by means of a significant new activity, it can be subject to a re-notification through the issue of a Significant New Activity (SNAc) Notice (see Section 15.3.7).

The time periods that Health Canada and Environment Canada have to assess the notified information and to impose any controls prescribed within the NSNR (Chemicals and Polymers) and the NSNR (Organisms) vary depending on the notification requirements and range from 5 to 75 days for chemicals and polymers [14], and 30 to 120 days for organisms [15]. Failure to assess a new substance within the legislated time period automatically permits the manufacture or import of the substance in(to) Canada with no (environmental) restrictions on how it can be used. In such cases, CEPA still provide measures for addressing the substance; even though the time period for assessment has expired and the substance has been added to the DSL.

The New Substances Program is regarded as a first line of defence against the release of harmful substances into the Canadian environment; the notification regulations are seen as an integral part of the federal government's national pollution prevention strategy. Approximately 800 substances new to the Canadian marketplace are assessed annually [20].

15.3.3 Provisions for existing substances

Under Part II of CEPA-1988, a framework for systematically determining the toxicity of substances deemed to be of high priority was implicit in the legislation. Thus, the Ministers of Health and the Environment were required to establish a list of substances (the Priority Substances List) deemed to be of highest concern with respect to health or the environment and to assess the risks of these substances (whether CEPA "toxic"). Ministers were also required to respond (within 90 days) to public nominations for additions to the List. If a report of an assessment was not published within 5 years of the substance being added to the List, establishment of a Board of Review could be requested under the Act. A summary of each assessment was to be published in the *Canada Gazette* along with an indication of whether Ministers intended to recommend the development of regulations to control the substance.

Two lists of priority substances (PSL-1 and PSL-2) were generated prior to the introduction of CEPA-1999. The first priority substances list, published in February 1989, comprised 44 substances. A second list comprising 25 substances was published in December, 1995. Both lists included classes of substances and complex mixtures as well as discrete industrial chemicals. They were developed by panels of experts (Ministers' Expert Advisory Panels) drawn from stakeholders and convened under the authority of the Act. Annexes 15.1 and 15.2 list these priority substances.

As described below (Section 15.4.1), assessment of the health and environmental risks of priority substances entailed a comprehensive and scientifically rigorous approach to decision-making. Examination of the 69 listed priority "substances" resulted in assessment of far more than this number in terms of discrete chemical entities because of the complex nature of some of

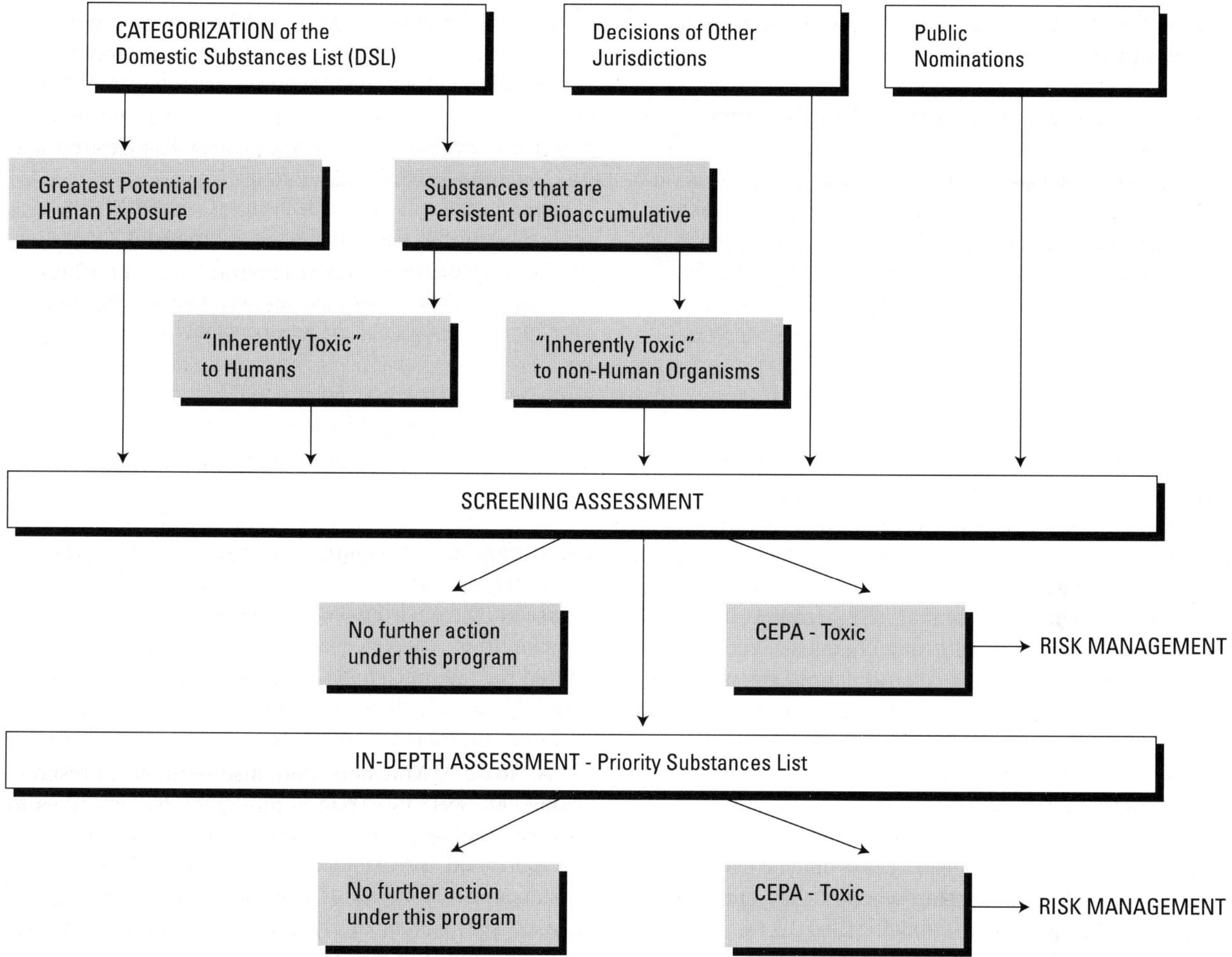

Figure 15.1 Existing Substances Program under CEPA-1999 [4].

the entries (i.e., mixtures and classes). Nevertheless, public expectation to consider the potential health and the environment impacts of all 22,400 or so existing industrial chemicals in Canada was increasing, a trend evident also in other parts of the world (see, for example, Chapter 12). This expectation was reflected in the views of the Parliamentary Committee that reviewed CEPA-88 and by the Commissioner on Environment and Sustainable Development [21]. As a result, significant changes were made to the provisions for existing substances in the renewed Act (CEPA-1999).

15.3.4 Categorization of the Domestic Substances List

CEPA-1999 incorporates a number of requirements to ensure that more existing substances are assessed for health and environmental risks in shorter timeframes, while at the same time retaining the PSL Assessment Program for substances, mixtures or effluents deemed to require a more in-depth assessment. Figure 15.1 depicts the processes for selecting and assessing existing substances. The three principal phases of identification and assessment of priorities for risk management specified under CEPA-1999 are categorization, screening assessment and in-depth (Priority Substances List) assessment.

An internationally leading provision was that all chemical substances on the Domestic Substance List, together with all the living organisms that were subsequently nominated for addition to the DSL, were to be "categorized" by September, 2006 to determine whether they possess certain characteristics that could indicate that they pose a risk to the environment or to

Box 15.5. Substances are identified for further work if they meet the following criteria:

- May present, to individuals in Canada, the greatest potential for exposure; or
- Are persistent and/or bio-accumulative in accordance with regulations (see Section 15.5.1), and
- Are "inherently toxic" to human beings or non-human organisms. Note that in this context the meaning of toxic is that in the generally accepted scientific sense, as determined by laboratory or other studies.

human health. (Box 15.5). Substances added to the original DSL, having undergone assessments under the NSNR, were not examined in the DSL categorization exercise.

This examination of the DSL did not, in itself, entail assessment of potential risks; rather it was a sorting process for identifying chemicals for further consideration. Systematic identification of priorities from among the 22,400 industrial substances on the DSL presented a significant and precedent setting challenge, similar initiatives having not been undertaken in any other country at that time.

Substances identified as priorities from categorization or other selection mechanisms must undergo screening risk assessments to determine whether they are "toxic" or capable of becoming "toxic" (Box 15.6). Another mechanism for triggering an assessment of toxicity under CEPA-99 is the requirement to review decisions made by other jurisdictions to prohibit or substantially restrict a substance for environmental or health reasons. The requirement to establish a list of priority substances, and the mechanism for doing so are retained under CEPA-1999.

The primary objective of screening and in-depth assessments is to determine whether a substance is "CEPA-toxic" as defined under the Act, which may then set the stage for addition of the substance to Schedule 1 (the List of Toxic Substances) of the Act and for reviewing options for controlling risks to human health and/or the environment.

15.3.5 Options for controlling existing substances

A wide range of regulatory instruments can be used under CEPA to control exposure to substances deemed to be toxic with respect to any aspect of their lifecycle, from the research and development stage to manufacture,

Box 15.6. Possible outcomes of a screening level risk assessment, a risk assessment of a priority substance, or a review of a decision made by another jurisdiction:

- No further action (typically if the substance is found not to be toxic).
- A recommendation (to the Federal Cabinet) that the substance be added to the List of Toxic Substances with a view to developing controls and, if applicable, be subject to virtual elimination in order to adequately manage the risks to the environment or to human health.
- The substance is added to the PSL for further review (if the substance is not already on the PSL).

use, storage and transport and, ultimately, disposal (Box 15.7). Regulations can address, for example, the amounts released to the environment and where releases can occur, the conditions of release, quantities manufactured or offered for sale in Canada, quantities imported, countries from, or to, which a substances may be imported or exported, the manner in, and conditions under, which a substance is advertised or offered for sale, how it is to be handled, stored and transported. Provisions also allow for the partial or total prohibition of manufacture, import or export, and for the submission of information on the substance, the conduct of analyses and monitoring and of tests, submission of samples to the government and the maintenance of records. Before any such regulations are made, it must be ascertained that a regulation does not address an aspect already effectively regulated under another Act (see Section 15.2.1).

Controls can also take the form of guidelines, standards, codes of practice, plans and voluntary or non-regulatory initiatives and may include any other measures deemed appropriate based on the known level of risk, available technology, and socio-economic considerations. The Act states that, in developing the regulations or other control options, priority is to be given to pollution prevention actions.

For substances that are "categorized in" and for which subsequent screening assessment indicates that they are "toxic" to human health and/or the environment, addition to the List of Toxic Substances requires that a proposed regulation or other control instrument respecting preventative or control actions in relation to the substances be published in the *Canada Gazette* within two years of the additions. Final regulations or instruments must normally be developed and published

Box 15.7. Risk management tools that can be considered in identifying options for managing toxic substances under CEPA 1999 [22]

- Regulations, pollution prevention plans, environmental emergency plans, administrative agreements, codes of practice, environmental quality objectives or guidelines, release guidelines.
- Voluntary approaches - Environmental Performance Agreements, Memoranda of Understanding.
- Non-CEPA 1999 economic instruments - financial incentives and subsidies, environmental charges and taxes.
- Joint federal/provincial/territorial initiatives - Canada-wide Standards, guidelines, codes of practice.
- Provincial/territorial Acts - regulations, permits, or other processes.
- Other federal Acts - e.g., Fisheries Act, Pest Control Products.

in the *Canada Gazette* within 18 months following the proposal. Figure 15.2 is a schematic representation of the steps involved in developing control measures for toxic substances.

15.3.6 Virtual elimination

When a substance is deemed to be "toxic" under CEPA and also meets certain criteria for persistence and bioaccumulation, is not a naturally occurring radionuclide or naturally occurring inorganic substance, and its presence in the environment results primarily from human activity, the substance is then proposed for virtual elimination under the Act (Box 15.8).

A *Virtual Elimination List* (the "List") specifies the level of quantitation for each substance included in the List. Virtual elimination would generally be achieved through a series of progressive release limits set by regulations and/or other risk management measures.

Box 15.8. Definition of virtual elimination

Virtual elimination is "the ultimate reduction of the quantity or concentration of the substance in the release below the level of quantitation specified by the Ministers in the (virtual elimination) List".

15.3.7 Significant new activities

Provisions for dealing with significant new activities with respect to chemical and biotechnological substances were introduced in CEPA-1999; these provisions address any new activity that results in, or may result in, significantly greater quantities or concentrations of a substance in the environment, or a significantly different manner or circumstances of exposure to a substance. They are intended to provide additional flexibility and refinement in the application of both the new and existing substances provisions by triggering re-notification of the substance under certain circumstances.

The Significant New Activity (SNAc) provisions can be used to require a re-notification of a new substance. A SNAc Notice may be issued defining what constitutes a significant new activity in relation to the substance, by inclusion or exclusion. The criteria under which a notification is required and information requirements are also specified therein. This information is further assessed prior to the commencement of any significant new activity to allow the substance to be imported, manufactured, used or released in ways that would not pose a risk to the environment and/or human life or health.

Significant New Activity Notifications (SNANs) contain all prescribed information specified in the SNAc Notice and must be provided within the prescribed time and prior to a company undertaking the significant new activity. Assessment of the information must be completed within the prescribed assessment period [19].

A new substance subject to a SNAc Notice can be added to the DSL with a SNAc ("S") flag; this allows any individual to manufacture, import, use and release the substance in ways that are not defined as a "new activity" under the terms of the definition of "significant new activity".

For existing substances, if an activity can be reasonably anticipated which could substantially change the exposure and consequently the risk posed to the environment and/or human life or health, an amendment to the DSL can be published in the *Canada Gazette*. This amendment would include publishing a SNAc Notice and placing a SNAc ("S") flag on the substance. This again allows any individual to manufacture, import, use and release the substance in ways that are not defined as a "new activity" under the terms of the SNAc Notice.

15.3.8 Information gathering

Provisions for gathering and generating information required for the assessment or control of existing

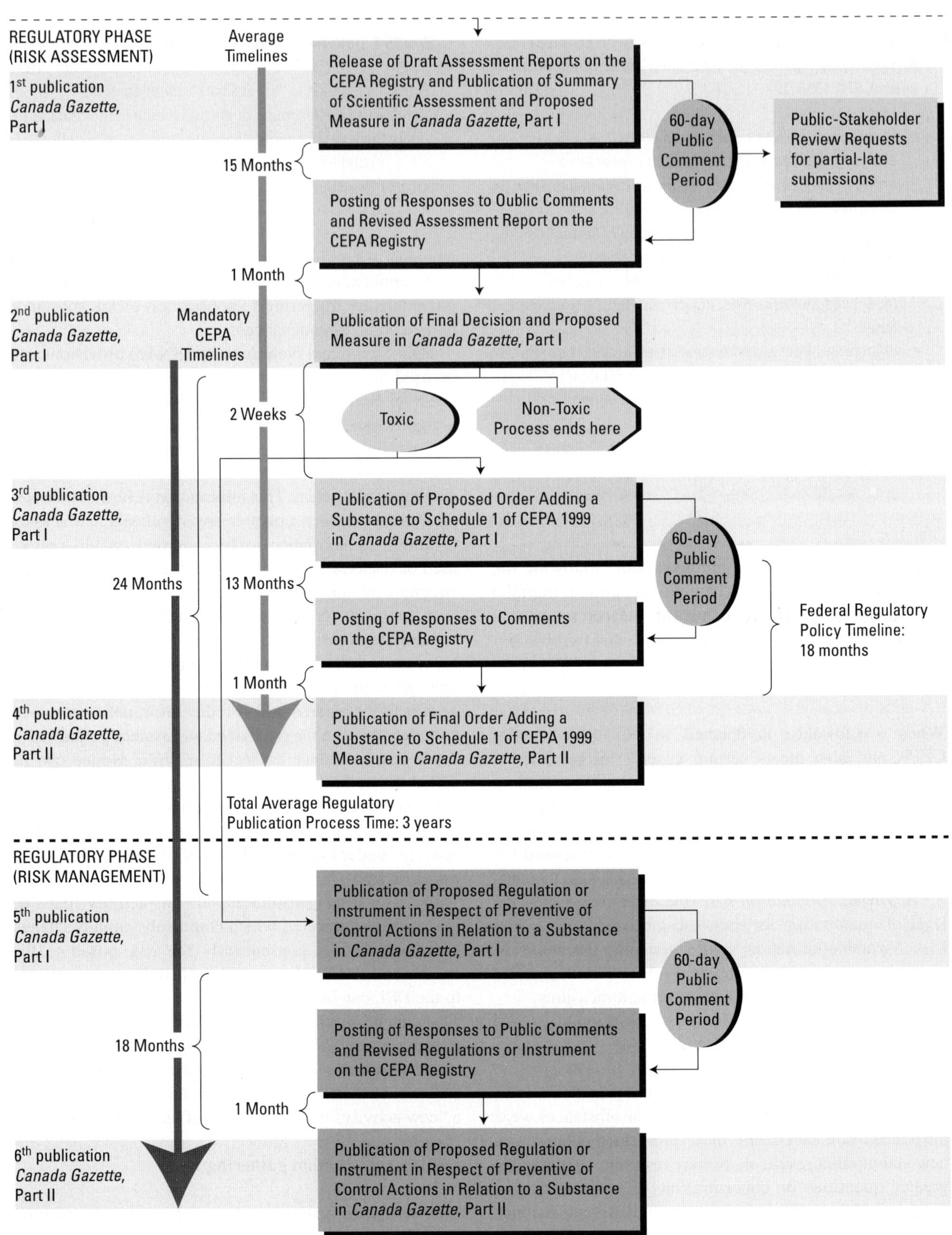

Figure 15.2 Regulatory publication process.

substances under the Toxic Substances provisions of CEPA include ones to ascertain who is using the substance, and to furnish the government with any existing information (e.g., toxicological information, monitoring data, uses, quantities in use) or samples and to conduct toxicological or other tests. Powers to require industry to carry out testing or studies cannot be invoked unless there is "reason to suspect that the substance is toxic or capable of becoming toxic or it has been determined under this Act that the substance is toxic or capable of becoming toxic". Also a user, manufacturer or importer of a substance is required to provide to the government any information that supports the conclusion that the substance is toxic or capable of becoming toxic.

15.3.9 Consultation and communication

The results of an assessment of an existing substance (i.e., screening, PSL) or a review of a decision made by another jurisdiction must be made public by issuing a notice in the *Canada Gazette*. The notice must indicate whether no further action is to be taken or whether the substance is to be added to the Priority Substances List (for further assessment) or to the List of Toxic Substances. A 60-day comment period follows the issuing of these proposals. Provisions exist for objections to be raised if no recommendations are made to add a priority substance to the List of Toxic Substances; establishment of a Board of Review may be requested to review the assessment conclusions. If it is proposed that a substance be added Schedule 1 (the Toxic Substances List), consultation with the public is required through the publication of a Notice in the *Canada Gazette*.

Control instruments are developed through consultations with stakeholders, including industry and industry associations, non-governmental organisations (e.g. environment, health and labour), provincial governments, economists, enforcement officials and legal services. Provincial and territorial governments may be involved in developing and implementing the options. All actions regarding toxic substances should be consistent with the Toxic Substances Management Policy [23] (see also Section 15.6).

15.4 HEALTH ASSESSMENTS UNDER CEPA

This section includes a brief description of the approaches used to implement the key health-related components of the toxic substances provisions of CEPA-1999, with emphasis on novel methodologies developed to address progressive and precedent-setting requirements of the legislative mandate for Existing Substances (see references listed under Section 15.8 for information relevant to assessment of New Chemicals).

Central to the evaluation of Existing Substances are two types of assessments, namely screening and in depth (PSL). Differences and similarities between these two types of health assessments are presented in Annex 15.3.

The provisions of CEPA-1999 for selecting (categorization being the most significant), assessing (screening and in-depth) and managing the risks of existing chemical substances, as depicted in Figure 15.1, are consistent with the principles outlined in Health Canada's "Decision-Making Framework for Identifying, Assessing, and Managing Health Risks" [24].

15.4.1 Comprehensive framework for health risk assessments under CEPA-1999

The objective of the health-related components of DSL categorization is the identification, for additional consideration in screening, of substances that are highest priorities in relation to their potential to cause adverse effects on the general population. Figure 15.3 illustrates the steps (phases) involved in identifying and assessing health priorities in an integrated and iterative framework for priority setting and assessment. To maximize efficiency, the complexity of priority setting, the assessment and associated documentation is tailored to invest only that amount of effort required to identify non-priorities while, at the same time, ensuring that the assessment provides essential support for undertaking risk management of substances where this is deemed to be required.

Phase-1: tools-based priority setting (categorization) and assessment

An element of the categorization mandate relevant to human health was the identification of substances that present the greatest potential for the exposure of the general population of Canada (GPE). Additionally, substances considered inherently toxic to humans (iT-human) and persistent or bioaccumulative, (the criteria for which are specified in regulations under CEPA), were to be identified.

In order to identify true health priorities, however, a risk-based framework encompassing both exposure and hazard for *all substances* was developed, rather than restricting consideration of the criterion "inherently toxic to humans" to the subset of substances considered to be persistent or bioaccumulative. This required multiple stages of increasing complexity, involving development

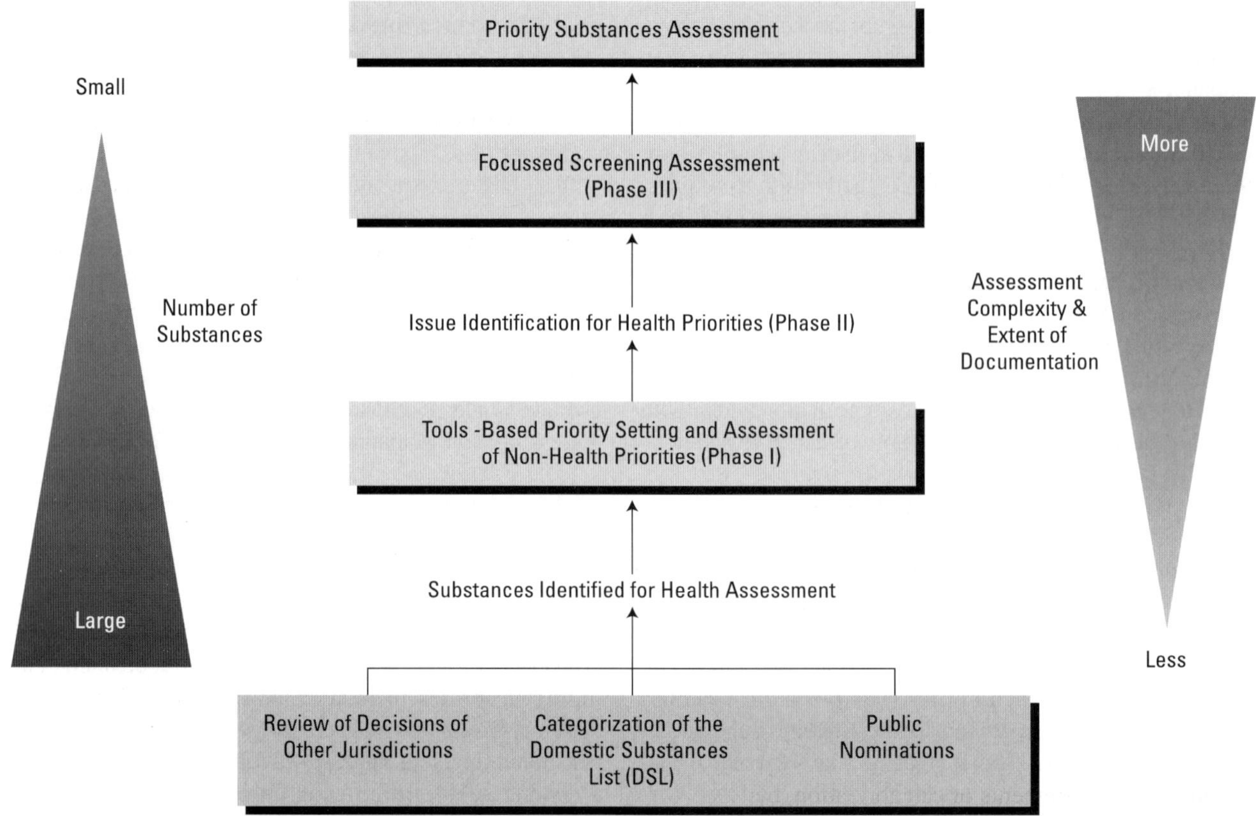

Figure 15.3 Phases in identifying and assessing health priorities.

and application of simple and complex exposure and hazard tools (Box 15.9).

The simple exposure tool (SimET) was developed to accommodate information submitted during the compilation of the DSL and has three lines of evidence:

- Quantity in commerce in Canada.
- Number of companies involved in commercial activities in Canada.
- Weighting by experts of the potential for human exposure based on consideration of various use codes.

Based on collective consideration of these three components, it was possible to relatively rank all substances in relation to their potential for exposure. Based on application of specific criteria for each of the components, all substances on the DSL were grouped into one of three categories, i.e., those presenting "greatest", "intermediate" or "lowest" potential for exposure (GPE, IPE or LPE). The results of relative ranking on this basis indicated that volume is not a good surrogate for exposure with many of the highest volume substances presenting "lowest potential for exposure".

Simple (SimHaz) and complex hazard tools (ComHaz) as well as a complex exposure tool (ComET) were and continue to be developed and implemented within an integrated framework for the health-related components of DSL categorization. The complex tools contribute considerably to predictive methodology for both exposure and hazard, including the development of significant numbers of additional consumer exposure scenarios and a systematic weight-of-evidence approach to take into account data, results of a suite of quantitative structure-activity models and analogue approaches. For additional information on the tools see: http://www.hc-sc.gc.ca/ewh-semt/alt_formats/hecs-sesc/pdf/contaminants/existsub/framework-int-cadre_e.pdf .

The simple exposure and hazard tools were applied to the entire DSL leading to a draft "maximal list" of health priorities, released in October, 2004 [25]. The potential for persistence or bioaccumulation to additionally contribute to exposure for certain subsets of substances, namely, those that are organic, was also taken into account (Figure 15.4).

Box 15.9. Tools for health-related components of DSL categorization

Exposure

- SimET (Relative ranking of all DSL substances based on submitters (S), quantity (Q) and expert ranked use (ERU).
- ComET (Quantitative plausible maximum age-specific estimates of environmental and consumer exposure for individuals based on use scenario (sentinel products), physical/chemical properties & bioavailability).

Hazard [High (H) or Low (L)]

- SimHaz (Identification of high or low hazard compounds by various agencies based on weight of evidence for multiple endpoints).
- ComHaz (Hierarchical approach for multiple endpoints & data sources (e.g. quantitative structure-activity relationships) including weight-of-evidence).

This approach offered a number of advantages and exceeded the requirements of categorization, by:

- Drawing maximally on work completed in other jurisdictions while avoiding continued focus on data-rich compounds.
- Not only identifying substances for screening assessment on the basis of exposure, hazard, and/or risk, but also *prioritizing* them on the basis of potential exposure, hazard, and/or risk to human health.
- Identifying true priorities for both assessment and data generation, since exposure and complex hazard components of the framework were unbiased in relation to data availability.
- Identifying not only those substances that were iT-human for a subset of substances, but all of the approximately 22,400 existing substances based on criteria for weight of evidence of hazard consistent with those for priority substances or screening health assessment.

Implementation of this framework and associated tools has application well beyond simply identifying substances for assessment. These tools enable the efficient prioritization and subsequent screening health assessment of *any substance* considered by the program. Health priorities from categorization have been prioritized by group, based on whether they are exposure, hazard, or risk based, and within groups, based on consideration of their relative potential for exposure. Continued application of the complex tools additionally focused the content of the results of categorization and will contribute to screening assessment for prioritized compounds.

The development of the tools and related products for categorization drew upon considerable prior program experience gained in developing the methodology for conducting in-depth, detailed human health risk assessments of the 69 "priority substances"; most of these assessments were published in the peer-reviewed scientific literature and/or served as the basis for international criteria documents [26].

Substances considered as health priorities based on application of the simple tools are addressed in Phase-II, issue identification.

Phase-II: issue identification

To increase the efficiency in assessment, it is envisaged that screening health assessments for existing substances will incorporate an early stage of issue identification. The objective is to ensure timely and maximum utilization of previously well documented peer reviewed assessments and adequate and accurate focus on more recent information and critical issues. While the process for input and content are still in development, robust senior internal technical review and external peer input would be critical to ensure integrity of the product. Formats for supporting use and toxicity profiles have been developed and draw maximally on available information, based on comprehensive and well documented search strategies and solicitation for submission of relevant information.

This stage provides risk managers and stakeholders with the opportunity to contribute information, for example, in the preparation of use and exposure profiles (i.e., identification of specific end uses and potential for exposure); it also provides early indication of potential focus of the assessment and (possible) subsequent risk management action.

Phase-III: (focussed) screening assessments

The objective of a screening health assessment is, to efficiently consider whether or not a substance poses a risk to human health. To increase efficiency, the focus of the assessment is limited principally to information which is considered most critical with respect to exposure to, and health-related effects of, a substance, in particular the critical aspects identified during Issue Identification. Substances are assessed only to the extent necessary to deem them to be non-priorities, or to provide necessary guidance as a basis for risk management. Depending

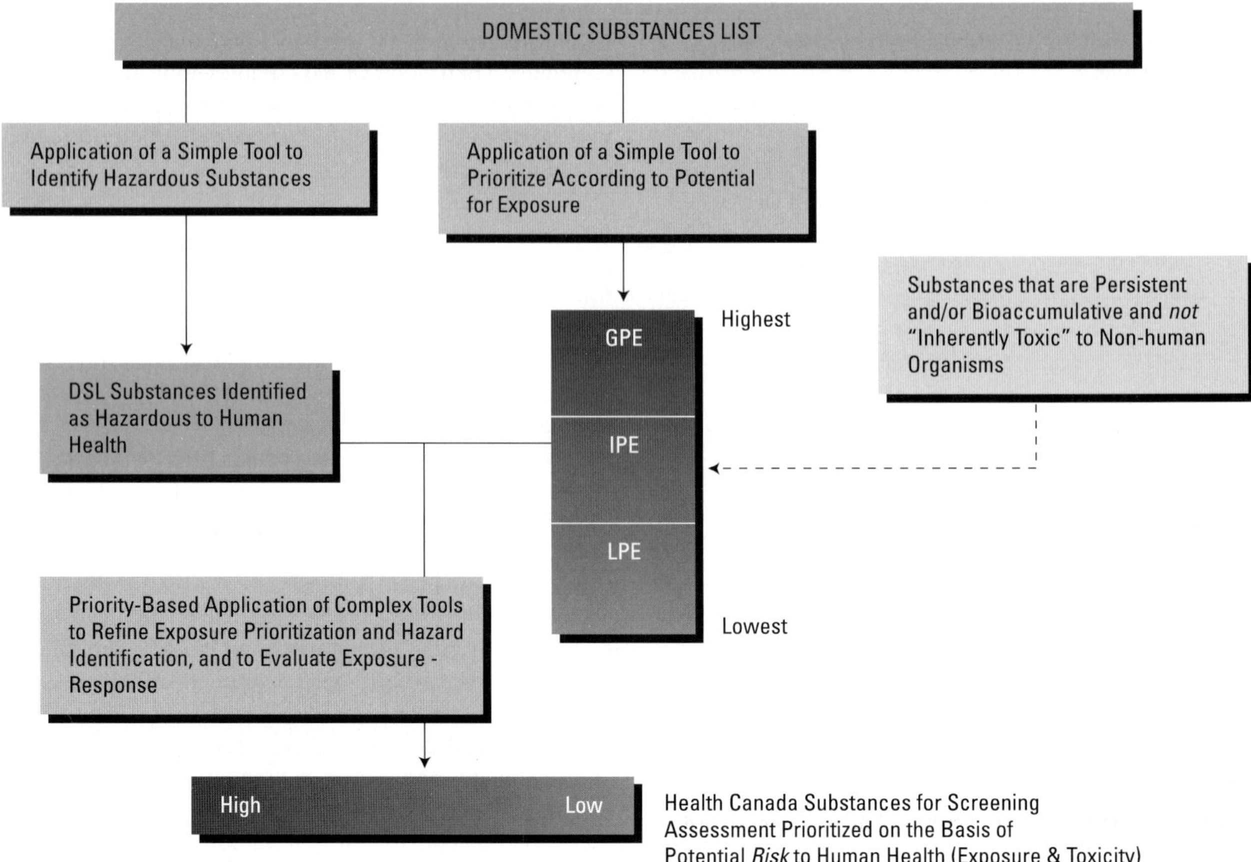

Figure 15.4 Tools-based approach for health-related components of DSL categorization. GPE = greatest potential for exposure, IPE = intermediate potential for exposure and LPE = lowest potential for exposure.

upon complexity of the issues, complexity of process for peer input may increase (e.g., more of the nature of that for priority substances). The objective is to maintain scientific rigour, depending, for example, on the priority for health effects evaluation (based on application of the simple tools) and on the extent of the database, but to vary the degree of detail (and hence the level of effort for the assessment).

Focussed screening health assessments result in the issue of a state of the science report which undergoes internal and external peer review and is posted on the web and/or sent to stakeholders. The state of the science report presents only the technical and scientific information on a substance or a group of substances and serves to provide an early indication of the basis for forthcoming conclusions and recommendations; the conclusion of whether or not a substance is "toxic" under the Act and any proposed Ministerial recommendations are published in the *Canada Gazette* which also serves to link the health and environmental assessments.

With respect to the health of the general public, it is the potential for adverse effects following long-term exposure to the, generally, low environmental levels that is often of importance as a basis for decision-making (that is, to set a substance aside with no further action, add it to the Priority Substances List, or to recommend addition to the List of Toxic Substances). Hazard characterization for both screening and in-depth (PSL) health assessments entail an examination of the effects critical to adults' and children's health, such as potential organ-specific effects or more specialized hazards such as immunotoxicity, neurological/behavioural toxicity, reproductive toxicity, genotoxicity, cancer and developmental effects. Exposure analyses include consideration of all relevant media and are based on six different age groups (an example is provided in Table 15.2).

Decision-making for screening health assessments is based on analysis of a margin of exposure (MOE), that is, comparison of critical effect levels with estimates of exposure taking into account the confidence/uncertainties

Table 15.2. Upper-bounding estimates of daily intake of 1,2-dibromoethane. From [27]. Citations in footnotes are as in [27].

Route of exposure	Estimated intake (μg/kg$_{bw}$ per day) of 1,2-dibromoethane by various age groups						
	0–6 months[1,2,3]		0.5–4 years[4]	5–11 years[5]	12–19 years[6]	20–59 years[7]	60+ years[8]
	formula fed	not formula fed					
Ambient air[9]	0.0050	0.0050	0.011	0.0084	0.0048	0.0041	0.0036
Indoor air[10]	0.0044	0.0044	0.0095	0.0074	0.0042	0.0036	0.0031
Drinking water[11]	0.0043	0.0016	0.0018	0.0014	8.0×10^{-4}	8.0×10^{-4}	9.0×10^{-4}
Food and beverages[12]	0.0043	0.078	0.058	0.037	0.022	0.020	0.016
Soil[13]	1.6×10^{-5}	1.6×10^{-5}	2.6×10^{-5}	8.4×10^{-6}	2.0×10^{-6}	1.7×10^{-6}	1.7×10^{-6}
Total intake	0.014	0.089	0.080	0.054	0.032	0.028	0.024

1 No data were identified on concentrations of 1,2-dibromoethane in breast milk.

2 Assumed to weigh 7.5 kg, to breathe 2.1 m^3 of air per day, to drink 0.8 L of water per day (formula fed) or 0.3 L/day (not formula fed) and to ingest 30 mg of soil per day (EHD, 1998).

3 For exclusively formula-fed infants, intake from water is synonymous with intake from food. The concentration of 1,2-dibromoethane in water used to reconstitute formula was based on data from City of Toronto (1990). No data on concentrations of 1,2-dibromoethane in formula were identified for Canada. Approximately 50% of non-formula-fed infants are introduced to solid foods by 4 months of age, and 90% by 6 months of age (NHW, 1990).

4 Assumed to weigh 15.5 kg, to breathe 9.3 m^3 of air per day, to drink 0.7 L of water per day and to ingest 100 mg of soil per day (EHD, 1998).

5 Assumed to weigh 31.0 kg, to breathe 14.5 m^3 of air per day, to drink 1.1 L of water per day and to ingest 65 mg of soil per day (EHD, 1998).

6 Assumed to weigh 59.4 kg, to breathe 15.8 m^3 of air per day, to drink 1.2 L of water per day and to ingest 30 mg of soil per day (EHD, 1998).

7 Assumed to weigh 70.9 kg, to breathe 16.2 m^3 of air per day, to drink 1.5 L of water per day and to ingest 30 mg of soil per day (EHD, 1998).

8 Assumed to weigh 72.0 kg, to breathe 14.3 m^3 of air per day, to drink 1.6 L of water per day and to ingest 30 mg of soil per day (EHD, 1998).

9 Based on the highest concentration (0.143 μg/m^3) detected for 1,2-dibromoethane in 6766 of 8275 samples of ambient air collected in a national survey across Canada between 1998 and 2002 (Environment Canada, 2002). This survey was selected due to its expansiveness and its currency, which will likely reflect declining use of 1,2-dibromoethane in Canada. Canadians are assumed to spend 3 h per day outdoors (EHD, 1998). Data from which the critical data were selected included Health Canada (2003), Environment Canada (1991, 1992, 1994, 1995 and 2001b), OMEE (1994) and CMHC (1989).

10 In the absence of measured data, the detection limit (0.018 μg/m^3) for a recent indoor air study of 75 homes in Ottawa, Ontario, was used (Health Canada, 2003b). Canadians are assumed to spend 21 h indoors every day (EHD, 1998). Data from which the critical data were selected included Otson (1986), Cal. EPA (1992) and Cohen et al. (1989).

11 In the absence of measured data, the detection limit (0.04 μg/L) from 7 bottled and 27 tap water samples in Toronto, Ontario, was used (City of Toronto, 1990). Data from which the critical data were selected included OME (1988), OMEE (1993) and Golder Associates (1987).

12 In the absence of Canadian monitoring data, detection limits were used in the calculations. A single 1,2-dibromoethane measurement of 13 μg/kg in sweet cucumber pickles in 1995 (U.S. FDA, 2003) was not considered, as the use of detection limits overcompensated its contribution to the overall intake of vegetables in the calculations. In addition, older studies (Gunderson, 1988) in which 1,2-dibromoethane was detected were not used to calculate intake levels, as pesticidal use of 1,2-dibromoethane at that time likely led to levels in food that would not be representative currently. Food: dairy products, fats, fruits, vegetables, cereal products, meat & poultry, fish, eggs, foods primarily sugar, mixed dishes & soups, nuts & seeds, soft drinks & alcohol. Amounts of foods consumed on a daily basis by each age group are described by Health Canada (EHD, 1998).

13 The method detection limit (4.0 ng/g) for soil measurements in urban (59 samples) and rural (102 samples) parklands in Ontario was used to represent the maximum exposure concentration of 1,2-dibromoethane (OMEE, 1993). Data from which the critical data were selected included Golder Associates (1987).

in the available exposure and toxicological databases and other relevant data (e.g., ancillary data on toxicokinetics and/or mode of action). This ensures maximal utilization of available data with several MOEs for potentially critical effects and studies being considered along with associated uncertainties. Delineation of the relative uncertainty and degree of confidence in the exposure and effects databases forms, therefore, a central component of the documentation for screening (and PSL) health assessments. For example, the adequacy of the margin for human health protection takes into account whether exposures are based upon only modelling, measured concentrations of a substance in media important to estimating human exposure (i.e., air, foodstuffs, drinking water, soil, consumer products) or human biomonitoring studies that provide a measure of actual human exposure. It also takes into account the extent of the database as the basis for characterization of hazard and dose-response for all effects particularly those considered critical, including degree of conservatism in the selection of the critical effect. Reliance on MOEs rather than TDIs in screening assessments contributes additionally to efficiency of the process by enabling assessment of larger numbers of substances, drawing maximally on the available database, while minimizing the need for development of exposure-based guidance values for substances that are not considered priorities for further action.

Where relevant and available, toxicokinetic data and weight-of-evidence for hypothesized modes of action and human relevance are taken into account in transparent analytical frameworks [28, 29]. An example of a margin of exposure analysis which appears in the State of the Science report on "perfluorooctane sulphonate (PFOS), its salts and precursors containing the $C_8F_{17}SO_3$ moiety" is presented in Table 15.3.

Decisions on the adequacy of margins take into account the experience gained by conducting screening assessment of large numbers of chemicals and considering the adequacy of various margins for human health protection for chemical substances with a wide variety of datasets. The approach by which these factors are considered in decision-making for screening health assessment has been built upon the experience gained, and are consistent with, decision-making in the health risk assessment of priority substances. Weight-of-evidence for effects for which available data on mode of action indicate that there is a probability of harm at all levels of exposure is also considered in decision-making.

Where it is ascertained from a screening assessment that a more comprehensive analysis of available data (e.g., a complex analysis of exposures from consumer products) and/or the generation of additional data (e.g., on mode of action) is warranted to more fully inform decision-making in order to reach a definitive conclusion, more detailed assessments are undertaken. An option stipulated in CEPA-1999 is to recommend that a substance be added to the PSL. Any recommendation for this action would necessitate defining what needs to be done to further develop the assessment (e.g., request additional information from industry to be able to better assess exposure, examine mode of action) and ascertaining whether such a course would be more advantageous to the assessment outcome.

15.4.2 PSL assessments

In view of the objective of CEPA-1999 to assess much larger numbers of substances more efficiently, the comprehensive and process intensive approach adopted for priority substances will likely be confined to very limited numbers of compounds and/or specific aspects of assessments on specific substances, which warrant a complex process and content.

For substances on the first PSL (PSL-1), chemicals were classified formally into discrete groups with respect to both their potential carcinogenicity and mutagenicity in humans based on clearly defined criteria for weight-of-evidence which took into account the quantity, quality and nature of the results of available toxicological and epidemiological studies [31]. For the assessment of PSL 2 substances (and more recent screening assessments), descriptions of the weight-of-evidence for carcinogenicity are more narrative in nature, in the interest of accommodating increasing availability of data on mode of action. To provide guidance in setting priorities for managing substances considered to present a risk of cancer and/or heritable mutations, exposure was compared with the dose associated with a specified (5%) increase in tumour incidence as a basis for development of a measure of dose-response (i.e., Exposure/Potency Indices, EPIs).

For some priority substances, the critical effect for decision-making was considered to be associated with a mode of action for which where there is a dose or exposure concentration below which adverse health effects are not likely to be observed (i.e., organ specific toxicity and/or cancer associated with same). For these substances, Tolerable Intakes or Tolerable Concentrations (TI or TC) (i.e., the intake or concentration to which it is believed a person can be exposed daily over a lifetime without deleterious effect), were derived by dividing the critical effect level (e.g., Benchmark Dose or Concentration (BMD or BMC) or No or Lowest Observed

Table 15.3. Margins of exposure for PFOS. From [30]. Citations in footnotes are as in [30].

Critical study and effect	PFOS dose metric at critical effect	Metric(s) of human exposure to PFOS	Margin of exposure (critical effect/ human exposure)
Microscopic changes in the liver of rats (m + f) receiving PFOS in the diet for 2 years[1]	Serum PFOS level: 13.9 μg/mL[2]	Mean serum PFOS level in adults in Canada[3]: 0.028 μg/mL	496
		95th percentile of human serum PFOS level in adults in Canada[3]: 0.0631 μg/mL	220
		Mean serum PFOS level in children in the United States[4]: 0.0375 μg/mL	371
		95th percentile of serum PFOS level in children in the United States[4]: 0.097 μg/mL	143
	Liver PFOS level: 40.8 μg/g[5]	Mean[6] liver PFOS level: 0.0188 μg/g	2170[7]
Thymic atrophy (f), reduced serum high-density lipoprotein (m), cholesterol (m), triiodothyronine (m) and total bilirubin (m) in monkeys administered PFOS for 26 weeks[1]	Serum PFOS level: 14.5 μg/mL[8]	Mean serum PFOS level in adults in Canada[3]: 0.028 μg/mL	518
		95th percentile of human serum PFOS level in adults in Canada[3]: 0.0631 μg/mL	230
		Mean serum PFOS level in children in the United States[4]: 0.0375 μg/mL	387
		95th percentile of serum PFOS level in children in the United States[4]: 0.097 μg/mL	149
	Liver PFOS level: 19.8 μg/g[9]	Mean[10] liver PFOS level: 0.0188 μg/g	1053[11]

1 Covenance Laboratories, Inc. (2002a).
2 Average of mean levels in males (7.6 μg/mL) and females (20.2 μg/mL).
3 Kubwabo et al. (2002).
4 3M Medical Department (2002).
5 Average of mean levels in males 26.4 (μg/g) and females (55.1 μg/g).
6 Mean level of PFOS in livers from 30 cadavers (Olsen et al., 2003).
7 Published data on 95th percentile not available; margin of exposure based upon highest level of PFOS in human liver from this study (0.057 μg/g) is 716.
8 Average of mean levels in males (15.8 μg/mL and females (13.2 μg/mL) (week 26).
9 Average of mean levels in males 17.3 (μg/g) and females (22.2 μg/g) (week 27).
10 Mean level of PFOS in livers from 30 cadavers.
11 Published data on 95th percentile not available. Margin of exposure based upon highest level of PFOS in human liver from this study (0.057 μg/g) is 347.

(Adverse) Effect Levels (NO(A)EL or LO(A)EL) by an uncertainty factor. The Benchmark Dose/Concentration is the effective dose/concentration (or lower confidence limit) that produces a specified increase in incidence above control levels. The basis for uncertainty factors is clearly delineated and, where available data permit, replaced by chemical-specific adjustment factors [31].

Details of the application of the above-mentioned approaches are available in the assessment reports for the priority substances, all of which are available at http://www.hc-sc.gc.ca/ewh-semt/contaminants/existsub/eval-prior/index_e.html. See also Annex 15.3 for information on differences and similarities between screening and PSL assessments.

15.4.3 Mixtures of substances

The approach for priority setting and/or assessing mixtures of chemicals depends on the nature of data available [31]. In some instances, the chemical composition of a mixture may be well characterized, levels of exposure of the population known, and detailed toxicological data on the mixture available. More frequently, however, not all components of the mixture are known, exposure data are uncertain and the toxicological data are limited. Thus the approach that can be used will depend on whether data are available:

- For the mixture as a whole.
- Only for components of mixture.
- For similar mixture(s).

15.4.4 Data requirements, information gathering and peer involvement

Search strategies

Search strategies for all aspects of the program are comprehensive and documented in the reports of various stages of priority setting (categorization), and assessment (issue identification, screening and PSL). For substances that are considered as health priorities in screening, in most cases, there is no legislated minimum dataset; however the Screening Information Data Set (SIDS) in the OECD High Production Volume Chemicals Program [32], or equivalent, is considered an appropriate basis to complete the assessments. Mechanisms for gathering information under CEPA are described in Section 15.3.8. Experience acquired in use-profiling from public sources for hundreds of chemicals based on hierarchical, evolving and comprehensive search strategies as part of input for the complex exposure tool provides a consistent and robust basis for understanding the use patterns of the vast majority of chemicals of interest.

Documentation

The conclusions and findings of an assessment, proposed and final regulations and other proposed or final actions under CEPA must ultimately be announced in the *Canada Gazette*. Reports of the outcome of screening assessments, assessments of priority substances and reviews of other jurisdictions' decisions can also be made available "in any other manner that the Minister (of the Environment) considers appropriate". These announcements must state whether the Ministers intend to take no further action respecting the substance, to add a substance to the PSL (unless it is already on the List), or to recommend that the substance be added to the List of Toxic Substances.

The nature and scope of the documentation for the health assessments have been modified with the change in emphasis from assessment of priority substances to categorizing and screening substances on the DSL. Available reports include State of the Science reports for screening assessments and their supporting documentation (including Issue Identifications and use and hazard profiles), PSL assessment reports and supporting documentation and briefer tabulations of output for non-priority tools based assessments, with much more extensive documentation available on the methodology. Considerable efficiency is gained in tailoring the level of documentation to the task at hand, involving no more effort than is necessary to set aside a substance as a non-priority or to provide necessary information to permit risk management (Figure 15.3).

Concise State of the Science reports of the screening health assessments constitute an essential basis for documenting and communicating to the public the scientific basis for the conclusions and decisions required under CEPA. The objective is to produce as concise a document as possible containing only the critical (relevant) information that supports the ultimate conclusion of whether a substance is "toxic", "suspected of being "toxic", or "not considered to be toxic", and the decision/recommendation for any further action; thus the initial focus is on the most critical effects and conservative effect levels and upper-bounding estimates of exposure. State of the Science reports are issued without the Ministerial conclusions and recommendations as a means of alerting stakeholders to the scientific underpinning upon which any recommendations will be based; these conclusions appear subsequently in the *Canada Gazette* Notice along with a synopsis of the technical findings. The State of the Science report and Conclusions in the *Canada Gazette* represent the Screening Assessment Report under CEPA.

Critical information included in screening assessments comprises the identity, production and uses of the substances, sources and levels of human exposure, and health effects. The screening assessment report also outlines the objective of the screening assessment, and delineates the databases which serve as the basis for determining the critical effect levels and upper bounding exposure estimates. For brevity, both hazard (health effects) and exposure (intake) data are tabulated where possible. Screening assessment reports are made available following external peer review and Departmental management approval of their content and of the process followed in their preparation.

More detailed documentation supports the summarized technical data presented in the assessment reports. For the PSL assessment program, the supporting documentation comprised detailed text, multiple tables and a comprehensive reference list; this documentation was designed to present and describe in detail all relevant data needed to demonstrate how the critical exposure and effects were determined. The textual content of the supporting documentation was extensive and required investment of considerable time and resources.

In the interest of meeting objectives to more efficiently assess larger numbers of substances, supporting documentation for the screening assessments is much more issue-focussed and comprises a series of background documents and data tabulations prepared during the course of an assessment rather than integration into a comprehensive criteria document. The extent of this documentation is necessarily dependent upon the tools-based designated priority of the substance and complexity of the issues. For substances designated as priorities for assessment, it includes, as a minimum, issue identifications, exposure and hazard profiles. Tabular summaries of supporting information focussed in critical areas will also be available. These may include survey results, exposure scenarios, robust data summaries for critical studies, framework analyses for weight-of-evidence of specific endpoints (cancer/genotoxicity), and hypothesized modes of action and relevance to humans.

For "tools-based" assessments the results of which indicate non-priorities for additional work, supporting documentation is limited principally to the more extensive documentation on methodology (see http://www.hc-sc.gc.ca/ewh-semt/alt_formats/hecs-sesc/pdf/contaminants/existsub/framework-int-cadre_e.pdf) with some chemical-specific information being available on request (e.g., weight-of-evidence for cancer/genotoxicity based on data, quantitative structure-activity analyses and consideration of analogues). For such substances, State of the Science reports and Gazette Notices report conclusions on significant numbers of substances in tabular format.

Peer involvement

Because of CEPA's requirement to set priorities from thousands of existing chemicals and the associated need to develop novel methodologies and the expectation for rapid review of prioritized substances, the program provides opportunity for incorporation of increasingly complex peer involvement not, only in the assessments of individual or groups of substances, but also in the development of novel methodology for categorization.

Table 15.4. Peer involvement for each stage of product development.

STAGE	Peer Input	Peer Consultation	Peer Review
Problem Formulation	√		
Draft Work Product		√	
Final Draft Work Product			√

Specifically, the program has incorporated to an increasing extent more formal peer input at the earlier stages of development for both methodology and assessments. In addition, the complexity of peer input, consultation, and peer review is greater for the more robust assessments of substances of highest priority and complex issues such as methodology development. This approach maximizes efficiency while maintaining the defensibility of output of the three different levels of priority setting and assessments of increasing complexity within the program (categorization, screening assessment, and full assessment).

The three types of peer involvement, the level and complexity of which increases with the stage of development of documentation and complexity of issues as discussed by Meek et al. [28] and their utilization in various stages of the program are presented in Table 15.4

Full assessments for priority substances generally include early peer input to identify relevant data followed by external panel peer reviews at the end of the process. On the other hand, for screening assessments, there is an early issue identification stage to solicit peer input on identification of relevant data and issues and confirming the focus of the assessment. Since screening assessments are less complex, at a later stage in their development, their peer review is generally restricted to written comments by several external experts. Panel meetings are convened only where there are subsequent outstanding issues. Development of methodology for priority setting and/or assessment of risk often entails all three stages of peer involvement (i.e., input, consultation and review).

15.5 ENVIRONMENTAL ASSESSMENT OF EXISTING SUBSTANCES

15.5.1 Categorization

As indicated in Section 15.3.4, in addition to categorizing all substances on the DSL to identify those that presented "Greatest Potential" for exposure and "inherently toxic" to humans, substances were to be identified for further work if they were persistent (P) and/or bioaccumulative (B) in accordance with regulations and "inherently toxic" (iT) to non-human organisms (i.e., if they were PiTs, BiTs, or PBiTs).

Persistence and bioaccumulation
The CEPA regulations for characterizing persistence and bioaccumulation came into force in March, 2000 [33]. Under these regulations, a substance is considered persistent if its transformation half-life, based on degradation through chemical, biochemical, and photochemical processes, satisfies the criterion in any one environmental medium (Table 15.5). Alternatively, it is considered persistent if it is subject to long-range transport (e.g., transported to remote regions such as the Arctic).

Criteria for bioaccumulation that were applied to organic substances are also presented in Table 15.5. Bioaccumulation factors (BAF) are preferred over bioconcentration factors (BCF); in the absence of BAF or BCF data, the *n*-octanol–water partition coefficient (log K_{ow}) may be used.

Table 15.5. Criteria for persistence and bioaccumulation.

Persistence		Bioaccumulation
Medium	Half-life	
Air	≥ 2 days	BAF ≥ 5000
Water	≥ 6 months	or
Sediment	≥ 1 year	BCF ≥ 5000
Soil	≥ 6 months	or
		log K_{ow} ≥ 5

Inherently toxic to non-human organisms
Inherently toxic to non-human organisms is evaluated based on aquatic (including benthic) toxicity data. This choice reflects the comparative availability of test data in aquatic/pelagic versus terrestrial species and is consistent with required elements in international initiatives such as the OECD SIDS [32]. It also takes into account the absence of recognized standard tests/methods and paucity of data on inhalation toxicity for invertebrates, amphibians, reptiles, or birds. In addition, owing to the relative lack of experimental data, most of the values on which categorization for inherently toxic to non-human organisms was based were modelled. Virtually all decisions were based on external effect concentrations, either median lethal (LC50), effective (EC50) or no observed effect concentrations (NOECs).

The criteria for inherently toxic to non-human organisms are presented in Table 15.6. The thresholds chosen are consistent with those in various European Union (EU) and U.S. Environmental Protection Agency (USEPA) initiatives. Where both acute and chronic effects data are available, for reasons of consistency preference is for application of the acute toxicity values, since most available data are for acute endpoints.

Table 15.6. Criteria for acute and chronic toxicity to aquatic species (algae, invertebrates, fish).

Exposure duration	Criteria
Acute	LC50 (EC50) < 1 mg/L
Chronic	NOEC ≤ 0.1 mg/L

15.5.2 Ecological screening assessments

As with health assessments, the process for preparing ecological screening assessments continues to evolve. Approaches applied vary depending upon the nature of the substance (PBiT, PiT or BiT) and potential for exposure (based on quantities in Canadian commerce).

A typical screening assessment is similar to that for a substance on the PSL, involving consideration of entry, exposure and effects characterization, and risk analysis steps.

The objective of entry characterization is to identify the various uses and sources of the substance in Canada, the quantity of the substance released from each of these sources, and how the substance is released over time, to air, water or soil. Entry characterization includes all phases of the substance's life cycle, from manufacture or importation, through transportation and use, to final disposal. Information gathered during this phase is the first step in determining exposure. If the substance is found to be "toxic" under CEPA, this information is also used to guide the development of risk management options.

In exposure characterization, data from modelling and/or monitoring studies are used to describe the spatial and temporal trends in concentrations of the substance in various environmental media (air, water, sediment, soil) in Canada. At this stage the exposure relevant to each identified receptor organism is typically quantified, i.e., a Predicted Exposure Concentration (PEC) is calculated for each. A PEC is usually selected to reflect a high-end exposure value. However, in some cases exposure may be characterized as a distribution of exposure values.

The aim of effects characterization is to describe the types of impairment that can result when different classes of organisms (e.g., plants, aquatic and terrestrial invertebrates, fish, mammals) are exposed to the substance. Typically a Critical Toxicity Value (CTV), or the lowest concentration of a substance that will cause a certain adverse effect, is identified for each type of receptor organism. A CTV is usually calculated from the results of short-term (acute) and long-term (chronic) laboratory toxicity tests and/or from modelling (e.g., QSAR) estimates. CTVs generally represent low or no effects toxicity values for sensitive organisms. However, in some cases effect levels may be represented by distribution of threshold effects values for an ecologically relevant range of species.

In risk characterization, a weight-of-evidence approach is used to determine the nature of, and where possible and appropriate, the likelihood of adverse effects. A first step usually involves deriving a Predicted No Effect Concentration (PNEC) for each assessment endpoint by dividing the CTV by an assessment factor. The magnitude of the assessment factor varies depending upon the quantity and quality of available effects data, and increases in proportion to uncertainties associated with extrapolating from effects in the laboratory to those anticipated in the field (e.g., the possibility that species found in the wild may be more sensitive than laboratory species; fluctuations in temperature in the field, which may increase susceptibility to effects).

Generally an important line of evidence when evaluating risk is the magnitude of the PEC/PNEC quotient. Quotients are estimated for each assessment endpoint, and values above 1 are typically interpreted to indicate risk. Risk may alternatively be quantified in a probabilistic or semi-probabilistic manner by comparison of the distributions of exposure and/or no effects values. However, decisions about ecological risks may be influenced by additional factors. These include information on the persistence and bioaccumulation potential of the substance, its inherent toxicity, and its distribution in Canadian environmental media. When available, evidence that environmental concentrations are changing with time or, evidence of effects in field situations is also is considered. Evidence that a substance is both persistent and bioaccumulative (as defined in the Persistence and Bioaccumulation regulations of CEPA-1999), when combined with evidence of toxicity and potential for release into the Canadian environment is interpreted to indicate potential to cause ecological harm. This is consistent with the requirements of CEPA-1999, which requires application of a weight-of-evidence approach and the precautionary principle in conducting and interpreting the results of assessments.

A number of substances that met the categorization criteria based on their PiT or BiT status was deemed unlikely to pose an ecological risk because of the very small amounts anticipated to be in commerce in Canada. For these substances, a "rapid screening" assessment process was developed to determine whether they are indeed unlikely to cause harm.

Ecological screening assessments of existing substances under CEPA-1999 are protective in that they typically focus on high-end risks (i.e., actual cases where exposures and sensitivities of receptor organisms are expected to be highest). Plausibly *conservative* assumptions may be made in the face of *uncertainty* resulting from data gaps in such assessments. Stakeholders are provided opportunity to submit additional information before the assessment is finalized to reduce identified uncertainties; where no such information is received, the conclusion (based partly on conservative assumptions) may stand.

When combining several conservative assumptions, steps are generally taken to ensure that the overall amount of conservatism in an assessment is not extreme. However, in certain cases where the harm the substance could cause is judged to be serious or irreversible, and the uncertainties are especially large, a conclusion may incorporate a high amount of conservatism. In such cases, a conclusion that a substance may cause harm could be called *precautionary* in that consequent risk management actions could be viewed as being consistent with the precautionary principle [34].

15.5.3 PSL assessments

From a methodological perspective, priority substance and screening ecological assessments are quite similar. PSL assessments, however, typically involve a more in depth analysis and reporting of available data, and can take up to 5 years to complete. In addition for PSL assessments there is greater opportunity to fill data gaps and increase the realism of the assessment, especially when there is a possibility of concluding that ecological harm is occurring. Thus for PSL assessments, the amount of conservatism associated with a finding that a substance may be causing ecological harm is typically lower.

Environment Canada convenes an Environmental Resource Group (ERG) for each priority substance assessed. The ERG can include experts from industry, academia and members of other government departments or other levels of government who have particular expertise with the substance. The ERG participates in the assessment process by performing tasks ranging from the review of draft documents and assistance in data collection to the development or writing of sections of supporting documents or assessment reports.

15.6 MANAGEMENT OF TOXIC SUBSTANCES

15.6.1 Priority substances

Annexes 15.1 and 15.2 show the chemical entities on the Priority Substances Lists that have been deemed to be toxic with respect to health and/or the environment. By December, 2006, more than 40 had been added to Schedule 1 and, for nearly all of these, measures to control their presence in, or entry into, the environment had been put in place or were under development. The actions taken with respect to the priority substances make up only part of those taken to control the entry of toxic substances into the environment under CEPA. Thus, as a result of this and initiatives taken under other provisions of the Act, some 80 existing substances including classes of substances, complex mixtures, effluents and emissions had been placed on Schedule 1 as of early 2007 [35].

For substances deemed to be "toxic" as a result of a Priority Substances List assessment, a screening assessment, or a review of a decision by another jurisdiction, it is required that a proposed regulation or instrument be published that will lead to *preventive* or *control* actions for managing the substances and hence reductions in, or elimination of, their risks to the environment or human health (Box 15.10). As described in Section 15.3.5, there are time limitations for issuing and finalizing these control instruments.

The "Toxics Management Process" has been set up to ensure that risk management tools are developed with due input from affected parties within timelines set out in the CEPA-1999 [36]. A key aspect of this management process is a Risk Management Strategy document which is prepared by Environment Canada and Health Canada in consultation with the CEPA National Advisory Committee [22] and affected stakeholders. This document sets out how the risks to health and the environment posed by release of each toxic substance are to be addressed.

Box 15.10. CEPA-1999 instruments that can be used to satisfy the requirement for establishing *preventive* or *control* actions [36]:

- Regulations.
- Environmental codes of practice.
- Pollution prevention plans.
- Environmental emergency plans.
- Environmental release guidelines.

Management strategies are not necessarily specific to each "toxic" (priority) substance. Thus, if control of several substances in one industrial sector is required, a sector-specific strategy could be developed. An example, of a broader strategy is that for acrolein and other aldehydes through the following set of initiatives [37]:

- Environmental emergency regulations.
- Off-road compression-ignition engine emission regulations.
- On-road vehicle and engine emission regulations.
- Off-road small spark-ignition engine emission regulation.

The rationale for these engine regulations was that several aldehydes including ones on the second Priority Substances List are emitted from vehicle exhausts thereby leading to efficiencies by regulating them together under the same instrument.

15.6.2 The Chemicals Management Plan

In December 2006 the Government of Canada announced its Chemicals Management Plan (Box 15.11) in which it was sought not just to deal with the industrial chemicals identified as high priorities from categorization but also to strengthen CEPA's integration with other federal chemical regimes administered under statutes such as the *Hazardous Products Act*, the *Food and Drugs Act*, and the *Pest Control Products Act* [38]. With respect to industrial chemicals, the Plan consolidated a number of actions already underway and, significantly, sought, by way of a "Challenge", to increase Industry's role in proactively identifying and managing the risks associated with the those it produces and/or uses. Ultimately, appropriate actions to deal with identified priorities are to be taken by 2020.

Box 15.11. The Chemicals Management Plan includes:

- A "challenge" to industry.
- Prohibitions and virtual elimination.
- Rapid screening of lower risk chemicals.
- Restrictions on re-introduction and new uses.
- Integrating government activities (e.g., relating to pesticides).
- Monitoring, surveillance and research.
- Industry stewardship of chemical substances.

The challenge to industry

Several hundred chemicals were identified through categorization as being priorities for action; these substances were identified as meeting:

- Each of the ecological categorization criteria for persistence (P), bioaccumulation (B) and inherent toxicity to aquatic organisms (iT), *and* were believed to be in commerce in Canada and/or
- The criteria for greatest or intermediate potential for exposure (GPE or IPE) *and* were identified as posing a high hazard to human health (that is showed evidence of carcinogenicity, mutagenicity, developmental toxicity or reproductive toxicity).

Inherent to taking action on these substances under the Chemicals Management Plan is reliance on strong stewardship on the part of Canadian manufacturers, importers and users who were challenged to provide new information on these chemicals in order to:

- Improve, where possible, information for risk assessment (e.g., P, B and iT data).
- Identify industrial best practices in order to set benchmarks for risk management and product stewardship.
- Collect environmental release, exposure, substance and/or product use information.

Industry was also required to provide new information about how it is manages specified chemicals, their use patterns, release and exposure pathways, potential substitution options, analytical methods and the financial implication of eliminating the specified substances.

The absence of information does not preclude action being taken to ensure that human health and the environment are safeguarded; thus if the requisite information is not provided by stakeholders, the federal government will implement controls, as appropriate.

Additional aspects pertaining to industrial chemicals announced under the Plan included the prohibition of two substances, perfluorooctane sulphonate and its salts and polybrominated diphenyl ethers that had been subject to screening assessments and the placement of restrictions on some substances under the SNAc provisions of CEPA (see Section 15.3.7).

In December 2006, Canada began issuing requirements under the SNAc provisions of CEPA-1999 that affect those high-hazard chemical substances not currently in use in Canada. In accordance with these provisions, industry must provide data to the government under the New Substances provisions of CEPA before any of the subject chemicals can be re-introduced into Canada. SNAc provisions are also to be applied to an additional 150 chemical substances that were found to be hazardous to humans. While the current uses of these substances may be adequately managed, this measure is designed to ensure that any new or increased use does not occur without the conduct of an informed assessment and implementation of appropriate controls.

15.7 CONCLUDING REMARKS

Developments in Canada have included the integration of sectoral legislation such as the Environmental Contaminants Act, the Clean Air Act and the Canada Water Act into a single piece of environmental protection legislation, the Canadian Environmental Protection Act. The definitions of "environment" and "toxic" therein require that all aspects of the environment (multi-sectoral) be addressed in assessing the environmental and health risks. Assessment approaches are risk based and management of identified risks can be taken on a "cradle-to-grave" basis. Moreover, the safety net provisions of CEPA require a multi-media approach to health risk assessment and management even though some aspects of management may ultimately be taken under other federal of provincial legislation. While Canada's early environmental legislation addressed both new and existing chemicals, subsequent developments in both of these areas have resulted in increasingly more prescriptive approaches and the need for greater transparency and accountability for the actions taken.

Specifically, since its introduction in the late 1980s, chemicals control legislation in Canada has imposed time limited mandates for multimedia assessment of considerable numbers of existing chemicals and subsequent risk management of those considered to pose a risk to the health of the general population and/or environment. In the mid 1990s, Notification Regulations for New Substances (chemicals and polymers) were introduced which required companies or individuals wishing to import or manufacture substances new to

Canada to provide certain information specified in regulations.

More recently, precedent-setting provisions introduced to systematically identify in a timely manner, priorities for assessment and management from among the approximately 22,400 existing substances has necessitated the development of innovative methodology including evolution of the previously linear or sequential steps of risk assessment and risk management to a more iterative approach where the need for, and focus of, potential control options are identified at an early stage of assessment. It has also required development of assessment products that efficiently dedicate resources, investing no more effort than is necessary to set aside a substance as a non-priority or to provide necessary information to permit risk management.

These provisions lead in many respects similar developments in other countries. It was noted in Chapter 12 that, since the 1960s, the scope of the European Community's chemicals control legislation has expanded with respect to the media of interest (air, water, products) and in terms of consideration of environmental as well as human health effects. These developments have been accompanied by a change in approach from a hazard-based to a risk-based one. This has required the development and refinement of approaches to risk assessment and, in particular, the development of exposure assessment methodologies and risk assessment models. The need to adopt a life-cycle approach to effectively manage the harmful effects of chemicals was also recognised. Recently existing chemicals are increasingly being emphasized since their presence in the environment was viewed as cause for greater concern than for new chemicals (see Chapter 12).

Canada's approach to assessing new chemicals is in line with that followed in the EU in that it requires the up-front provision of prescribed data to permit a meaningful first assessment of the health and environmental risks. The data requirements are based on the data-set developed within the OECD Chemicals Programme but can vary depending on the quantities involved and the nature of the substance. In the US system, initial reporting of test data is not required; rather results from quantitative structure-activity relations can be used to require the generation of experimental data on new chemicals.

One of the most far reaching provisions of the current CEPA-1999 was the categorization of the 22,400 chemicals on Canada's Domestic Substances List. Canada became the first country to have carried out such an analysis of its list of existing commercial chemicals and, by doing so, attempted to respond to the concerns expressed that, under earlier legislation, too few chemical were being addressed. Similar concerns about the slow rate of progress in managing industrial chemicals in the EU are noted in Chapter 12. Initiatives such as those dealing with High Production Volume Chemicals have been undertaken by the US and internationally through the OECD. The results of categorization have identified other types of priorities thereby providing Canada an opportunity to contribute to international efforts to control toxic substances by identifying its own unique strategy for managing industrial chemicals.

In Canada, the responsibility for assessing and developing strategies for managing industrial chemicals has rested primarily on the shoulders of the environment and health departments of the Federal Government. Under the REACH legislation in Europe (see Chapter 12), it is industry's responsibility to assess the risks of existing substances and to develop the means to adequately control any identified risks. Based, in part, on this European approach, fundamental to Canada's Chemicals Management Plan announced in December, 2006 is the "Challenge" made to industry to strengthen its role in proactively identifying and managing risks associated with the chemicals it produces.

A significant difference between CEPA and REACH is the considerable emphasis placed in the latter on assessment and control of occupational exposures to existing substances; occupational exposures are not addressed under CEPA because of the responsibility that the provinces have for occupational health and safety.

The number of substances managed to date as a result of the PSL assessment program compares favourably with actions taken by other jurisdictions. Many of the 69 priority substances on Lists 1 and 2 were deemed to pose a risk to health or the environment and measures to control the entry into the environment of most of these toxic substances have either been implemented or are under development. The challenging provisions of CEPA-1999 to determine and address priorities from among all substances on the DSL are only starting to be implemented (as summarized in Section 15.6.2) and are expected to result in a significant increase in the rate at which existing substances are assessed and, where toxic, managed

Based on the progressive nature of the legislation which has necessitated increased efficiency in assessment and management of industrial chemicals, Canada continues also to contribute extensively to developments in international programs. This includes development of formats and processes for priority setting and assessments and assurance of their integrity through, for example, contribution to development of a robust peer review process.

15.8 FURTHER READING

1. Environment Canada, 2003. Guidance manual for the categorization of organic and inorganic substances on Canada's Domestic Substances List. Determining persistence, bioaccumulation potential, and inherent toxicity to non-human organisms. Existing Substances Branch. Ottawa, Canada.
2. Environment Canada, 2006. Administrative policy and process for conduction ecological risk assessments. Existing Substances Division, draft, March, 2006. Ottawa, Canada.
3. Environment Canada. 2006. Pollution prevention - a federal strategy for action. www.ec.gc.ca/pollution/strategy
4. Environment Canada. 2006. Sustainable development strategy, 2004-2996. http://www.ec.gc.ca/sd-dd_consult/SDS2004/index_e.cfm
5. Environment Canada. 2007. Toxic substances management policy. www.ec.gc.ca/toxics/TSMP
6. Environment Canada. 2007. Understanding the toxics management process. www.ec.gc.ca/toxics/en/index.cfm
7. Health Canada. 2007. Screening health assessments of existing substances. http://www.hc-sc.gc.ca/ewh-semt/contaminants/existsub/screen-eval-prealable/index_e.html
8. Health Canada. 2007. Determination of "toxic" for the purposes of the new substances provisions (chemicals and polymers) under the Canadian Environmental Protection Act - Human health considerations. New Substances Assessment and Control Bureau, Product Safety Programme, Healthy Environments and Consumer Safety Branch. Ottawa, Canada (in press).
9. Health Canada. 2007. Assessment and management of new substances in Canada. http://www.ec.gc.ca/substances/nsb/eng/home_e.shtml
10. Privy Council Office. 2006. A framework for the application of precaution in science-based decision-making about risk. www.pcobcp.gc.ca/default.asp?Language=E&Page=publications&Sub=precaution&doc=precaution_e.htlm
11. Government of Canada. 2005. Guidelines for the notification and testing of new substances: chemicals and polymers, pursuant to section 69 of the *Canadian Environmental Protection Act*, 1999, Version 2005. SBN 0-662-69285-3 Cat. no.: En84-25/2005 EPS M-688. Ottawa, Canada.

ACKNOWLEDGEMENTS

The authors wish to extend their appreciation to Dr. D. Krewski, Professor and Director of the McLaughlin Centre for Population Health Risk Assessment, Institute of Population Health, University of Ottawa for his careful review and helpful comments on this Chapter. Several other persons have provided helpful information or suggestions for its content. In particular the authors wish to acknowledge George Enei, Director, and Pat Doyle of the Existing Chemicals Evaluation Division of Environment Canada, Mark Bonnell, Head, New Substances Evaluation Unit, Environment Canada for reviewing and providing detailed suggestions and additional information for inclusion in the sections addressing the environmental aspects of the Chapter, Jaqueline Sitwell, Director, New Substances Assessment and Management Bureau, Health Canada, for providing information on and reviewing the section on the New Substances Program, Stephen MacDonald, Head, Toxic Substances Management, Health Canada for information and advice on toxics management, Bob Liteplo, Head, Scientific Coordination, Policy and Communication Section, Existing Substances Division, Health Canada for providing figures and diagrams as well as advice on various aspects of Health Canada's CEPA programs, and Eric Millward, Program Officer, Existing Substances Division, Health Canada, for verifying the status of priority substances.

REFERENCES

1. Government of Canada. 2006. Invest in Canada: chemical sector. http://www.investincanada.comen921Industry.html June 23, 2006.
2. Environment Canada. 1999. Canadian Environmental Protection Act, 1999 (CEPA-1999). http://www.ec.gc.ca/CEPARegistry/the_act/
3. Canada. 1994. Canadian Environmental Protection Act (CEPA-1988). Office Consolidation. R.S., 1985, c. 16 (4th Supp.). Minister of Supply and Services Canada, 1994, ISBN 0-61326-0.
4. CEPA (Canadian Environmental Protection Act), 1999. Statutes of Canada 1999. Chapter 33. Act assented to 14 September 1999. Environment Canada, Ottawa, Canada.
5. A Canada-wide accord on environmental harmonization. http://www.ccme.ca/assets/pdf/cws_accord_env_harmonization.pdf
6. Canadian Council of Ministers of the Environment. 1998. Guide to the Canada-wide accord on environmental harmonization. http://www.ccme.ca/assets/pdf/guide_to_accord_e.pdf
7. The Food and Drugs Act (FDA). http://lois.justice.gc.ca/en/F-27/

8. The Feeds Act. http://lois.justice.gc.ca/en/F-9/index.html
9. The Fertilizers Act. http://lois.justice.gc.ca/en/F-10/index.html
10. Government of Canada. 1985. Environmental Contaminants Act, R.S.C. (1985), c. E-12.
11. Environment Canada. 1983. New chemicals workshop, EPS 3-EP-83-4, September, 1983 (ISBN: 0-662-12519-3). Ottawa, Canada.
12. Environmental Contaminants Act Consultative Committee. 1986. Final report of the Environmental Contaminants Act Consultative Committee (ISBN 0-662-15047-3), Ottawa, Canada.
13. Government of Canada. 2007. "Overview of the Existing Substances Program". April 2007 compact disk, Existing Substances Program, Environment Canada, Gatineau, Quebec.
14. Government of Canada. 2005. New substances notification regulations (chemicals and polymers). http://canadagazette.gc.ca/partII/2005/20050921/html/sor247-e.html
15. Government of Canada. 2005. New substances notification regulations (organisms). http://canadagazette.gc.ca/partII/2005/20050921/html/sor248-e.html
16. Organization for Economic Co-operation and Development. 1982. Decision of the Council concerning the minimum pre-marketing set of data in the assessment of chemicals, C(32) 196 (final). OECD, Paris, France.
17. Environment Canada. 2006. The Domestic Substances List. http://www.ec.gc.ca/CEPARegistry/subs_list/Domestic.cfm
18. Environment Canada. 2006. The Non-Domestic Substances List. http://www.ec.gc.ca/CEPARegistry/subs_list/NonDomestic.cfm
19. Government of Canada. 2005. Guidelines for the notification and testing of new substances: chemicals and polymers, pursuant to Section 69 of the Canadian Environmental Protection Act, 1999. Version 2005 (ISBN 0-662-69285-3 Cat. no.: En84-25/2005 EPS M-688). Ottawa, Canada.
20. Government of Canada. Assessing chemicals for risk: protection the health of Canadians and their environment, undated pamphlet.
21. Commissioner for the environment and sustainable development. Understanding the risks from toxic substances: cracks in the foundation of the Federal House. http://www.oag-bvg.gc.ca/domino/reports.nsf/html/c903ce.html
22. Environment Canada. Identifying risk management tools for toxic substances under CEPA 1999. http://www.ec.gc.ca/CEPARegistry/gene_info/fact_02.cfm
23. Environment Canada. 1995. Toxic substances management policy. www.ec.gc.ca/toxics/TSMP
24. Health Canada. 2000. Decision-making framework for identifying, assessing, and managing health risks. http://www.hc-sc.gc.ca/ahc-asc/pubs/hpfb-dgpsa/risk-risques_cp-pc_e.html
25. Health Canada. 2005. Draft "maximal" list of substances prioritized by Health Canada for consideration in screening assessment under CEPA 1999. http://www.hc-sc.gc.ca/ewh-semt/contaminants/existsub/categor/max-list/index_e.html
26. International Programme on Chemical Safety. 2007. Concise International Assessment Documents (CICADs). http://www.who.int/ipcs/publications/cicad/en/index.html
27. Health Canada. 2004. Health state of the science report for 1,2-dibromoethane http://www.hc-sc.gc.ca/ewh-semt/pubs/contaminants/existsub/dibromoethane/index_e.html
28. Meek ME, Patterson J, Strawson J, Liteplo R. 2007. Engaging expert peers in the development of risk assessments. *Risk Analysis* (in press).
29. Seed J, Carney EW, Corley RA, Crofton KM, DeSesso JM, Foster PMD, Kavlock R, Kimmel G, Klaunig J, Meek ME, Preston RJ, Slikker W, Tabacova S, Williams GM, Wiltse J, Zoeller RT, Fenner-Crisp P, Patton, DE. 2005. Overview: Using mode of action and life stage information to evaluate the human relevance of animal toxicity data. *Crit Rev Toxicol* 35:663-672.
30. Final health state of the science report on PFOS. http://www.hc-sc.gc.ca/ewh-semt/pubs/contaminants/existsub/pfos-spfo/intro_e.html#a
31. Health Canada. Human health risk assessment for priority substances. http://www.hc-sc.gc.ca/ewh-semt/pubs/contaminants/existsub/approach/index_e.html.
32. Organization for Economic Co-operation and Development. Manual for investigation of high production volume chemicals. Chapter 2. SIDS, the SIDS plan and the SIDS dossier. http://www.oecd.org/dataoecd/60/43/1947477.pdf
33. Environment Canada. Persistence and bioaccumulation regulations. http://www.ec.gc.ca/CEPARegistry/regulations/DetailReg.cfm?intReg=35
34. United Nations. 1992. Principle 15 of the Rio Declaration on environment and development. United Nations Conference on Environment and Development, Rio de Janeiro, 1992. http://www.unep.org/Documents.multilingual/Default.asp?DocumentID=78&ArticleID=1163&l=en
35. Environment Canada. CEPA Environmental Registry. Current regulations. Toxic Substances List-Updated Schedule 1 as of December 27, 2006. http://www.ec.gc.ca/CEPARegistry/subs_list/Toxicupdate.cfm .

36. Environment Canada. 2007. Understanding the toxics management process. http://www.ec.gc.ca/toxics/tmp/en/index.cfm February 2.
37. MacDonald S. Head, Toxic Substances Section, Health Environments and Consumer Safety Branch, Health Canada. Personal communication, November 2006.
38. The Government of Canada. Challenge for chemical substances that are a high priority for action. http://www.chemicalsubstanceschimiques.gc.ca/challenge-defi/notice-avis_e.html

ANNEX 15.1. STANDING OF SUBSTANCES ON PSL-1

Substance	Conclusion-health[a]	Conclusion-environment[a]	Scheduled
Aniline	Proposed T	NT	No
Arsenic and its compounds	T for inorganic arsenic compounds	T for dissolved and soluble forms of inorganic arsenic	Yes[d]
Benzene	T	NT	Yes
Benzidine	T	NT	Yes
Bis (2-chloroethyl) ether	No conclusion[b]	NT	No
Bis (chloromethyl) ether	T	NT	Yes
Bis (2-ethylhexyl) phthalate	T	No conclusion	Yes
Cadmium and its compounds	T for inorganic cadmium compounds	T for dissolved and soluble forms of inorganic cadmium compounds	Yes[d]
Chlorinated paraffin (CP) waxes	T for short and proposed T for medium and long chain CPs	Proposed T for short, medium and some long chain CPs	No
Chlorinated wastewater effluents	No conclusion	T	Yes
Chlorobenzene	NT	NT	No
Chloromethyl methyl ether	T	NT	Yes
Chromium and its compounds	T for hexavalent chromium compounds; NT for trivalent chromium compounds	T for hexavalent chromium compounds; no conclusion for trivalent compounds	Yes[d]
Creosote-impregnated waste materials	No conclusion	T for creosote-impregnated waste materials from creosote-contaminated sites	Yes[d]
Dibutyl phthalate	NT	NT	No
1,2-Dichlorobenzene	NT	NT	No
1,4-Dichlorobenzene	NT:	NT	No
3,3'-Dichlorobenzidine	T	NT	Yes
1,2-Dichloroethane	T	NT	Yes
Dichloromethane	T	T	Yes
3,5-Dimethylaniline	No conclusion[b]	NT	No
Di-n-octyl phthalate	Proposed NT	NT	No
Effluents from pulp mills using bleaching	No conclusion	T	Yes
Hexachlorobenzene	T	T	Yes
Inorganic fluorides	NT	T	Yes
Methyl methacrylate	NT	NT	No
Methyl tertiary butyl ether	NT	NT	No
Mineral fibres	T for refractory ceramic fibres; NT for other vitreous fibres	No conclusion	Yes[d]

Substance	Conclusion-health[a]	Conclusion-environment[a]	Scheduled
Nickel and its compounds	T for oxidic, sulphidic and soluble inorganic nickel compounds; NT for nickel metal	T for dissolved and soluble inorganic nickel compounds	Yes[d]
Organotin compounds (non-pesticidal uses)	Proposed NT	NT	No
Pentachlorobenzene	NT	T	Yes
Polychlorinated dibenzodioxins	T	T	Yes
Polychlorinated dibenzofurans	T	T	Yes
Polycyclic aromatic hydrocarbons	T for benzo[*a*]pyrene, benzo[*b*]fluoranthene, benzo[*j*]fluoranthene, benzo[*k*]fluoranthene, and indeno[*1,2,3-cd*]	T	Yes
Styrene	NT	NT	No
Tetrachlorobenzenes	NT	T	Yes
1,1,2,2-Tetrachloroethane	Proposed NT	NT	No
Tetrachloroethylene	NT	T	Yes
Toluene	NT	NT	No
Trichlorobenzenes	NT	NT	No
1,1,1-Trichloroethane	Not assessed (see note c)	Not assessed (see note c)	Yes
Trichloroethylene	T	T	Yes
Waste crankcase oils	No conclusion	Proposed T for Use Crankcase oils	No
Xylenes	NT	NT	No

[a] Conclusions relate to those reached under paragraphs 11a (environment) or 11c (health) of CEPA-1988, or under 64a (environment) or 64c (health) of CEPA-1999. Generally, findings under 11b or 64b were NT unless otherwise specified. NT = not toxic; T = toxic.

[b] Not used or imported into Canada; therefore proposed deferring further action pending a submission under CEPA New Substances Notification Regulations.

[c] Following addition of methyl chloroform (1,1,1-trichloroethane) in 1990 to the list of ozone-depleting substances under the Montreal Protocol, this substance was added to the List of Toxic Substances under CEPA; further efforts to assess methyl chloroform as a priority substance was therefore considered to be unwarranted.

[d] The forms found to be toxic are placed on Schedule 1.

ANNEX 15.2. STANDING OF SUBSTANCES ON PSL-2

Substance	Conclusion-health[a]	Conclusion-environment[a]	Scheduled
Acetaldehyde	T	NT (toxic under 64b)[b]	Yes
Acrolein	T	NT	Yes
Acrylonitrile	T	NT	Yes
Aluminum chloride, aluminum nitrate, aluminum sulphate	Assessment period extended pending new data		No
Ammonia in the aquatic environment	No health conclusion included	T	Yes
1,3-Butadiene	T	T	Yes
Butylbenzylphthalate	NT	NT	No

Substance	Conclusion-health[a]	Conclusion-environment[a]	Scheduled
Carbon disulfide	No health conclusion included	NT	No
Chloramines	No health conclusion	T for inorganic chloramines	Yes[c]
Chloroform	NT	NT	No
N,N-Dimethylformamide	NT	NT	No
Ethylene glycol	Assessment period extended pending new data		No
Ethylene oxide	T	NT	Yes
Formaldehyde	T	NT (toxic under 64b)[b]	Yes
Hexachlorobutadiene	NT	T	Yes[g]
2-Methoxy ethanol, 2-ethoxy ethanol, 2-butoxy ethanol	T for 2-Methoxy ethanol and butoxy ethanol; NT for 2-ethoxyethanol	NT	Yes[c]
N-Nitrosodimethylamine	T	NT	Yes
Nonylphenol and its ethoxylates	(see footnote d)	T	Yes
Phenol	NT	NT	No
Releases from primary and secondary copper smelters and copper refineries	T (see footnote e)	T	Yes
Releases from primary and secondary zinc smelters and zinc refineries	T (see footnote e)	T	Yes
Releases of radionuclides from nuclear facilities (impacts on non-human species)	No assessment	T (see footnote h)	No[i]
Respirable particulate matter less than or equal to 10 microns	T (for PM_{10} and, especially, for $PM_{2.5}$)	No assessment	Yes
Road salts	No assessment	T (see footnote f)	Addition proposed
Textile mill effluents	No conclusion	T	No

[a] Conclusions relate to those reached under paragraphs 11a (environment) or 11c (health) of CEPA-1988, or under 64a (environment) or 64c (health) of CEPA-1999. Generally, findings under 11b or 64b were NT unless otherwise specified. NT = not toxic; T = toxic.

[b] Deemed to "..constitute or may constitute a danger to the environment on which life depends….' Paragraph 64 b of CEPA-1999

[c] The species found to be toxic are placed on Schedule 1.

[d] Based on consideration of the MoE between effect levels and reasonable worst-case or bounding estimates of intake by the general population from environmental media, NP and NPEs are not considered a priority for investigation of options to reduce human exposure through control of sources that are addressed under CEPA 1999. However, the relatively low MoE estimated for some products indicates that there is an important need for refinement of this assessment to determine the need for measures to reduce public exposure to these substances in products through other Acts under which they are regulated.

[e] Emissions of metals (largely in the form of particulates) and of sulphur dioxide from copper smelters and refineries and zinc plants are deemed toxic under paragraph 64a of CEPA-1999, and emissions of PM10, of metals (largely in the form of particulates) and sulphur dioxide from copper smelters and refineries and from zinc plants of PM10 are deemed toxic under paragraph 64c of CEPA-1999.

[f] Road salts that contain inorganic chloride salts with or without ferrocyanide salts are toxic under paragraphs

[g] Added to the virtual elimination (VE) list.

[h] Releases of uranium and uranium compounds from uranium mines and mills are deemed to be toxic under paragraph 64a of CEPA-1999.

[i] Controls to be imposed under other federal legislation.

ANNEX 15.3. COMPARISON OF SCREENING AND PSL HEALTH ASSESSMENTS

Issue	Screening assessment	Priority Substances List assessment
Concept	Initial assessment of whether a substance poses a risk to human health	A critical and comprehensive analysis of the risks to human health
Possible Outcomes	There could be no further action on the substance, it could be considered for risk management or it could be considered for more in-depth PSL risk assessment	There could be no further action on the substance or it could be considered for risk management
Information Gathering	Comprehensive information search strategies employed, similar to those for PSL assessments. Greater reliance on other peer-reviewed assessments for identification and assessment of previously reviewed data	Comprehensive information search strategies employed. The search strategies are noted in the PSL assessment reports
Evaluation of Exposure	Focus on upper-bounding estimates of exposure, after consideration of all identified information	Detailed analysis (e.g., probabilistic) of exposure, after consideration of all identified relevant information
Evaluation of Effects	Focus directly on health-related effects, which occur at lowest concentration or dose	Detailed review of all relevant health-related data and full weight of evidence analysis for hazard characterization. This includes weighting of all relevant data, taking into account factors such as consistency, plausibility of observed effects
Hazard Characterization	Initial focus directly on the most conservative effect level associated with the critical health-related effect and/or identification of substances with high intrinsic toxicity to human health	Weight of evidence approach with in-depth evaluation of mode of action (i.e., how a substance induces its toxic effects), toxicokinetics (how the substance is absorbed and distributed within the body), metabolism and exposure–response (e.g., benchmark dose) relationships, where data permit
Approach to Dose–Response Assessment	Margin of exposure approach, i.e., magnitude of the ratio between conservative effect level for effect considered critical and upper-bound estimated (or measured) level of human exposure	Development of tolerable daily intakes/ concentrations, employing default or compound-specific adjustment factors where data permit. Consideration/incorporation of physiologically based pharmacokinetic models or biologically motivated case-specific models, where data permit
Confidence/ Uncertainties in the Assessment	Deals principally with characterization of the extent of the available database that serves as basis for the delineation of the critical data on exposure and effects. Specified in the screening assessment report and supporting working documentation	Deals with characterization of the extent of the available database that serves as basis for the delineation of the critical data on exposure and effects, but primarily with the characterization of specific aspects of dose–response. Specified in the PSL assessment report
Documentation Prepared	Screening health assessment report (published). Supporting working documentation (unpublished). Short amalgamated summary of health and environmental screening assessments published in the *Canada Gazette*	Amalgamated health and environmental risk evaluations published in a PSL assessment report. Supporting documentation (unpublished) for the health components (exposure and effects) assessment. Synopsis of amalgamated health and environmental assessments published in the *Canada Gazette*

Issue	Screening assessment	Priority Substances List assessment
Delineation of Follow-up Actions	When the recommendation is to add the substance to the PSL for more in-depth assessment, the additional work required is clearly delineated in the screening assessment report. When the recommendation is to consider the substance "toxic" under Paragraph 64(c) of CEPA 1999, the appropriate considerations for follow-up and guidance on the priority for the development of options to reduce exposure are provided to risk managers	When the recommendation is to consider the substance "toxic" under Paragraph 64(c) of CEPA 1999, the appropriate considerations for follow-up and guidance on the priority for the development of options to reduce exposure are delineated in the PSL assessment report.
Scientific Review — Internal	Internal review meetings by senior technical staff to consider critical issues and the conclusion of the assessment	Review by senior technical staff.
Scientific Review — External	External review by small number of experts primarily to address adequacy of data coverage and defensibility of the conclusion or to address specific questions on identified critical issues. All reviewers must have declared non–conflict of interest	External review often by convened panels of experts for adequacy of data coverage, defensibility of selection of the critical data, dose–response analysis and exposure assessment. All reviewers must have declared non–conflict of interest
Public Comment	Sixty-day public comment period mandated under CEPA 1999	Sixty-day public comment period mandated under CEPA 1999

16. THE OECD CHEMICALS PROGRAMME

R. DIDERICH

16.1 INTRODUCTION

The Organization for Economic Cooperation and Development (OECD) is an intergovernmental organization. Its principal aim is to promote policies for sustainable economic growth and employment, a rising standard of living, and trade liberalisation. By "sustainable economic growth" the OECD means growth that balances economic, social and environmental considerations. At the time of writing, the OECD groups 30 member countries: Australia, Austria, Belgium, Canada, the Czech Republic, Denmark, Finland, France, Germany, Greece, Hungary, Iceland, Ireland, Italy, Japan, Korea, Luxembourg, Mexico, the Netherlands, New Zealand, Norway, Poland, Portugal, the Slovak Republic, Spain, Sweden, Switzerland, Turkey, the United Kingdom and the United States. The European Commission also takes part in the work of the OECD. The OECD brings together the member countries to discuss and develop both domestic and international policies. With active relationships with some 70 other countries, industry, environmental NGOs and other intergovernmental organizations, it has a global reach (Figure 16.1). It analyses issues, recommends actions, and provides a forum in which countries can compare their experiences, seek answers to common problems, and work to co-ordinate policies [1,2].

The chemicals programme of the OECD was established in 1971. The main objectives of the chemicals programme are to:

- Assist OECD member countries' efforts to protect human health and the environment through improving chemical safety.
- Make chemical control policies more transparent and efficient and save resources for government and industry.
- Prevent unnecessary distortions in the trade of chemicals and chemical products.

Figure 16.1. The global reach of the OECD.

C.J. van Leeuwen and T.G. Vermeire (eds.), Risk Assessment of Chemicals: An Introduction, 623–638.

In the following sections of this chapter, the main activities of the OECD chemicals programme are summarised: the mutual acceptance of data (Section 16.2), the existing chemicals programme (Section 16.3), the new chemicals programme (Section 16.4), the risk assessment programme (Section 16.5), the risk management programme (Section 16.6) and other activities related to environment, health and safety (Section 16.7).

16.2 THE BASIS: MUTUAL ACCEPTANCE OF DATA

As one of the first priorities under the OECD chemicals programme, member countries recognised the need to encourage the generation of high quality test data for chemicals assessment. The guidelines for the testing of chemicals and the principles of good laboratory practice (GLP) were developed at OECD in the late 1970's in the broader context of the concept of mutual acceptance of data (MAD). Both of these instruments for ensuring harmonised data generation and data quality are an integral part of a legally binding 1981 Council Decision on MAD [3]. OECD's 30 member countries agreed to implement the Decision, which states that "data generated in the testing of chemicals in an OECD member country in accordance with OECD test guidelines and OECD principles of good laboratory practice shall be accepted in other member countries for purposes of assessment and other uses relating to the protection of man and the environment." Since 1997 non-member countries can adhere to the Council Decisions on MAD; South Africa, Slovenia and Israel are currently full adherents.

The practical consequence of this Council Decision is that new non-clinical environmental health and safety data for notification or registration of a chemical or chemical product, developed in a member country under these conditions and submitted for fulfilling regulatory requirements in another country, cannot be refused, and thus need not be developed a second time.

The MAD system has allowed OECD countries and adhering non-members to avoid non-tariff trade barriers which can be created by different national regulations while improving protection of human health and the environment. Duplication of expensive safety testing is avoided by the industry and time to market for new chemicals and chemical products is shortened, saving further resources (it has been calculated that the annual saving for governments and industry, resulting from MAD in this respect amounts to approximately $60 million). Another consequence is that fewer animals will be used in testing, thereby promoting animal welfare, which is an issue of concern in member countries. In light of the World Trade Organization (WTO) agreements which require the use of relevant international standards as the basis for national technical regulations, the OECD system has taken on a more global meaning.

The practical implementation of the MAD Council Decision is ensured by the OECD test guidelines programme and the OECD principles of GLP as outlined below (Figure 16.2).

16.2.1 The OECD test guidelines programme

The OECD test guidelines programme provides the supporting structure for developing and updating OECD test guidelines. These are a collection of standard methods used by professionals in governments, industry, academic institutions, and independent laboratories for non-clinical health and environment safety testing of chemical substances. They cover tests for physical-chemical properties, effects on biotic systems (ecotoxicity), environmental fate (degradation and accumulation) and health effects (toxicity). An overview is presented in Table 16.1.

The OECD test guidelines are periodically updated in order to keep pace with progress in science. In addition, new test guidelines are developed and agreed upon, based on specific regulatory needs identified by OECD member countries. OECD-wide networks of national coordinators and national experts, in which about 6000 people are involved, provide the opportunity for input from scientists in government, academia and industry [4]. The expert review process is given in Figure 16.3. The standard project submission forms for test guidelines are evaluated at a national coordinators meeting. Draft test guidelines are circulated to the national coordinators for comments. After consultation of the experts in expert

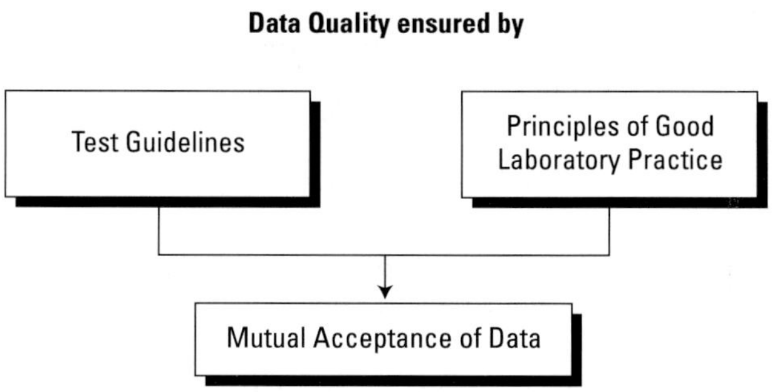

Figure 16.2. The mutual acceptance of data is built on the OECD test guidelines and GLP principles.

Table 16.1. Overview of available OECD test guidelines up to September 2006 [4].

No.	Title	Original Adoption	No. of Updates	Most Recent Version
SECTION 1 - PHYSICAL-CHEMICAL PROPERTIES				
101	UV-VIS absorption spectra	12 May 1981	0	---
102	Melting point/melting range	12 May 1981	1	27 July 1995
103	Boiling point	12 May 1981	1	27 July 1995
104	Vapour pressure	12 May 1981	2	23 March 2006
105	Water solubility	12 May 1981	1	27 July 1995
106	Adsorption/desorption using a batch equilibrium method	12 May 1981	1	21 January 2000
107	Partition coefficient (*n*-octanol-water): shake flask method	12 May 1981	1	27 July 1995
108	Complex formation ability in water	12 May 1981	0	---
109	Density of liquids and solids	12 May 1981	1	27 July 1995
110	Particle size distribution/fibre length and diameter distributions	12 May 1981	0	---
111	Hydrolysis as a function of pH	12 May 1981	2	13 April 2004
112	Dissociation constants in water	12 May 1981	0	---
113	Screening test for thermal stability and stability in air	12 May 1981	0	---
114	Viscosity of liquids	12 May 1981	0	---
115	Surface tension of aqueous solutions	12 May 1981	1	27 July 1995
116	Fat solubility of solid and liquid substances	12 May 1981	0	---
117	Partition coefficient (*n*-octanol-water), HPLC method	30 March 1989	2	13 April 2004
118	Determination of the number-average molecular weight and the molecular weight distribution of polymers using gel permeation chromatography	14 June 1996	0	---
119	Determination of the low molecular weight content of a polymer using gel permeation chromatography	14 June 1996	0	---
120	Solution/extraction behaviours of polymers in water	14 June 1996	1	21 January 2000
121	Estimation of the adsorption coefficient (K_{oc}) on soil and on sewage sludge using high performance liquid chromatography (HPLC)	22 January 2001	0	---
123	Partition coefficient (*n*-octanol-water): slow-stirring method	23 March 2006	0	---
SECTION 2 – EFFECTS ON BIOTIC SYSTEMS				
201	Alga, growth inhibition test	12 May 1981	2	23 March 2006
202	*Daphnia* sp. acute immobilisation test and 14-day reproduction test	12 May 1981	2	13 April 2004
203	Fish, acute toxicity test	12 May 1981	2	17 July 1992
204	Fish, prolonged toxicity test: 14-day study	4 April 1984	0	---
205	Avian dietary toxicity test	4 April 1984	0	---
206	Avian reproduction test	4 April 1984	0	---
207	Earthworm, acute toxicity tests	4 April 1984	0	---
208	Terrestrial plants, growth test	4 April 1984	1	19 July 2006
209	Activated sludge, respiration inhibition test	4 April 1984	0	---
210	Fish, early-life stage toxicity test	17 July 1992	0	---
211	*Daphnia magna* reproduction test	21 September 1998	0	---
212	Fish, short- term toxicity test on embryo and sac-fry stages	21 September 1998	0	---
213	Honeybees, acute oral toxicity test	21 September 1998	0	---
214	Honeybees, acute contact toxicity test	21 September 1998	0	---

No.	Title	Original Adoption	No. of Updates	Most Recent Version
215	Fish, juvenile growth test	21 January 2000	0	---
216	Soil microorganisms: nitrogen transformation test	21 January 2000	0	---
217	Soil microorganisms:carbon transformation test	21 January 2000	0	---
218	Sediment-water chironomid toxicity test using spiked sediment	13 April 2004	0	---
219	Sediment-water chironomid toxicity test using spiked water	13 April 2004	0	---
220	Enchytraeidae reproduction test		13 April 2004	0
221	*Lemna* sp. growth inhibition test	Draft new test guideline	0	---
222	Earthworm reproduction test *(Eisenia fetida/Eisenia andrei)*	13 April 2004	0	---
227	Terrestrial plant test: vegetative vigour test	19 July 2006	0	---
SECTION 3 - DEGRADATION AND ACCUMULATION				
301	Ready biodegradabiility 301A : DOC die-away test 301B : CO_2 evolution test 301C : Modified MITI test (I) 301D : Closed bottle test 301E : Modified OECD screening test 301F : Manometric respirometry test	12 May 1981	1	17 July 1992
302A	Inherent biodegradability: modified SCAS test	12 May 1981	0	---
302B	Inherent biodegradability: Zahn-Wellens/EMPA test	12 May 1981	1	17 July 1992
302C	Inherent biodegradability: modified MITI test (II)	12 May 1981	0	---
303	Simulation test – aerobic sewage treatment a: activated sludge units b: biofilms	12 May 1981	1	22 January 2001
304A	Inherent biodegradability in soil	12 May 1981	0	---
305	Bioconcentration: flow-through fish test	12 May 1981	1	14 June 1996
306	Biodegradability in seawater	17 July 1992	0	---
307	Aerobic and anaerobic transformation in soil	24 April 2002	0	---
308	Aerobic and anaerobic transformation in aquatic sediment systems	24 April 2002	0	---
309	Aerobic mineralization in surface water – simulation biodegradation test	13 April 2004	0	---
310	Ready biodegradability - CO_2 in sealed vessels (headspace test)	Draft new test guideline	0	---
311	Anaerobic biodegradation of organic compounds in digested sludge - method by measurement of gas production	Draft new test guideline	0	---
312	Leaching in soil columns	13 April 2004	0	---
SECTION 4 - HEALTH EFFECTS				
401	Acute oral toxicity	12 May 1981	1	Date of deletion: 20 December 2002
402	Acute dermal toxicity	12 May 1981	1	24 February 1987
403	Acute inhalation toxicity	12 May 1981	1	---
404	Acute dermal irritation/corrosion	12 May 1981	2	24 April 2002

No.	Title	Original Adoption	No. of Updates	Most Recent Version
405	Acute eye irritation/corrosion	12 May 1981	2	24 April 2002
406	Skin sensitisation	12 May 1981	1	17 July 1992
407	Repeated dose 28-day oral toxicity study in rodents	12 May 1981	1	27 July 1995
408	Repeated dose 90-day oral toxicity study in rodents	12 May 1981	1	21 September 1998
409	Repeated dose 90-day oral toxicity study in non-rodents	12 May 1981	1	21 September 1998
410	Repeated dose dermal toxicity:28-day	12 May 1981	0	---
411	Subchronic dermal toxicity: 90-day	12 May 1981	0	---
412	Repeated dose inhalation toxicity: 28/14-day	12 May 1981	0	---
413	Subchronic inhalation toxicity: 90-day	12 May 1981	0	---
414	Prenatal developmental toxicity study	12 May 1981	1	22 January 2001
415	One-generation reproduction toxicity	26 May 1983	0	---
416	Two-generation reproduction toxicity study	26 May 1983	1	22 January 2001
417	Toxicokinetics	4 April 1984	0	---
418	Delayed neurotoxicity of organophosphorus substances following acute exposure	4 April 1984	1	27 July 1995
419	Delayed neurotoxicity of organophosphorus substances: 28-day repeated dose study	4 April 1984	1	27 July 1995
420	Acute oral toxicity – fixed dose procedure	17 July 1992	1	17 December 2001
421	Reproduction/developmental toxicity screening test	27 July 1995	0	---
422	Combined repeated dose toxicity study with the reproduction/developmental toxicity screening test	22 March 1996	0	---
423	Acute oral toxicity – acute toxic class method	22 March 1996	1	17 December 2001
424	Neurotoxicity study in rodents	21 July 1997	0	---
425	Acute oral toxicity: up-and-down procedure	21 September 1998	1	17 December 2001
427	Skin absorption: *in vivo* method	13 Ap		
428	Skin absorption: *in vitro* method	13 April 2004	0	---
429	Skin sensitisation: local lymph node assay	24 April 2002	0	---
430	*In vitro* skin corrosion: transcutaneous electrical resistance test (ter)	13 April 2004	0	---
431	*In vitro* skin corrosion: human skin model test	13 April 2004	0	---
432	*In vitro* 3T3 NRU phototoxicity test	13 April 2004	0	---
435	*In vitro* membrane barier test (method for skin corrosivity)	19 July 2006	0	---
451	Carcinogenicity studies	12 May 1981	0	---
452	Chronic toxicity studies	12 May 1981	0	---
453	Combined chronic toxicity/carcinogenicity studies	12 May 1981	0	---
471	Bacterial reverse mutation test	26 May 1983	1	21 July 1997
472	Genetic toxicology: *Escherichia coli*, reverse assay	26 May 1983	0	Date of deletion: 21 July 1997 (method merged with TG 471)
473	*In vitro* mammalian chromosome aberration test	26 May 1983	1	21 July 1997
474	Mammalian erythrocyte micronucleus test	26 May 1983	1	21 July 1997
475	Mammalian bone marrow chromosome aberration test	4 April 1984	1	21 July 1997
476	In vitro mammalian cell gene mutation test	4 April 1984	1	21 July 1997
477	Genetic toxicology: sex-linked recessive lethal test in *Drosophila melanogaster*	4 April 1984	0	---
478	Genetic toxicology: rodent dominant lethal test	4 April 1984	0	---

No.	Title	Original Adoption	No. of Updates	Most Recent Version
479	Genetic toxicology: *in vitro* sister chromatid exchange assay in mammalian cells	23 October 1986	0	---
480	Genetic toxicology: *Saccharomyces cerevisiae*, gene mutation assay	23 October 1986	0	---
481	Genetic toxicology: *Saccharomyces cerevisiae*, mitotic recombination assay	23 October 1986	0	---
482	Genetic toxicology: DNA damage and repair, unscheduled DNA synthesis in mammalian cells *in vitro*	23 October 1986	0	---
483	Mammalian spermatagonial chromosome aberration test	23 October 1986	1	21 July 1997
484	Genetic toxicology: mouse spot test	23 October 1986	0	---
485	Genetic toxicology: mouse heritable translocation assay	23 October 1986	0	---
486	Unscheduled DNA synthesis (UDS) test with mammalian liver cells *in vivo*	21 July 1997	0	---

meetings and in the validation management group the proposal goes back to the national coordinators. After endorsement by the national coordinators, the OECD secretariat forwards the proposal to the Chemicals Committee and the Environment Policy Committee for approval. Final adoption takes place in the OECD Council.

New and updated test guidelines should improve risk management in countries and/or lead to a further reduction of animal use and improvements in animal welfare. Regulatory acceptance of new and updated test guidelines require that the test guidelines:

- Have been subjected to a transparent and independent peer review process.
- Demonstrate a linkage between the new test and the existing test method or effects in the target species.
- Provide a comparable or better level of protection.
- Be time and cost effective.

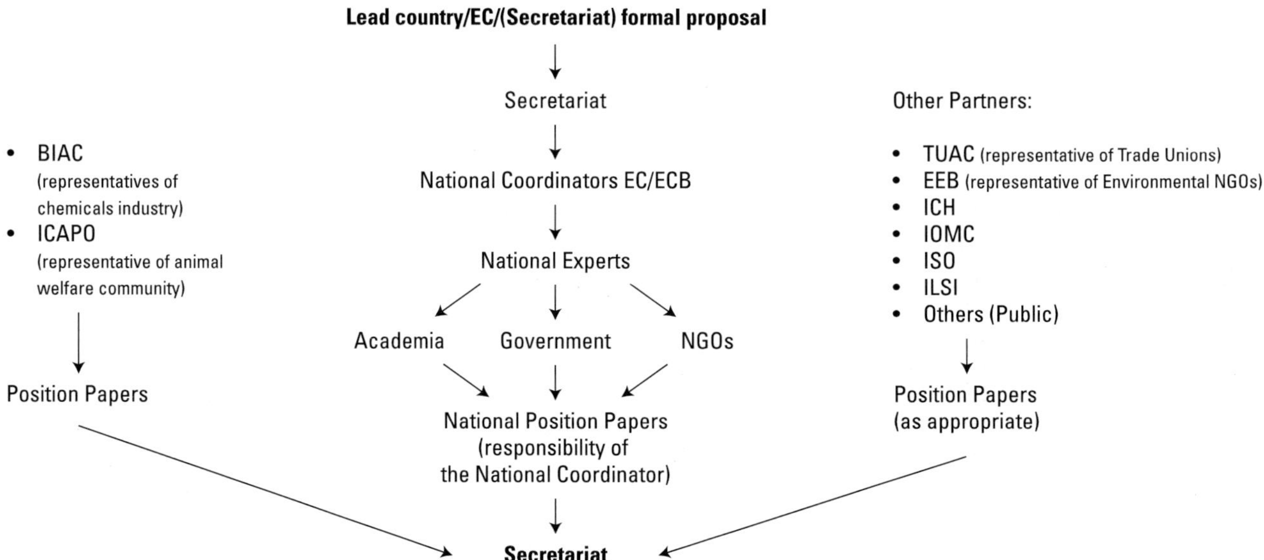

Figure 16.3. The expert review process of the OECD test guidelines programme. Abbreviations: BIAC (Business and Industry Advisory Committee), ECB (European Chemicals Bureau), EEB (European Environment Bureau), ICAPO (International Council on Animal Protection in OECD Programmes), ICHC (International Conference on Harmonization of Technical Requirements for Registration of Pharmaceuticals for Human Use), IOMC (Inter-Organization Programme for the Sound Management of Chemicals), ILSI (International Life Sciences Institute), ISO (International Organization for Standardization), NGOs (Non-Governmental Organizations) and TUAC (Trade Union Advisory Committee).

Box 16.1. OECD conceptual framework for the testing and assessment of endocrine disrupting chemicals [5]

Level 1 Sorting & prioritization based upon existing information	- Physical & chemical properties, e.g., MW, reactivity, volatility, biodegradability, - Human & environmental exposure, e.g., production volume, release, use patterns - Hazard, e.g., available toxicological data	
Level 2 *In vitro* assays providing mechanistic data	- ER, AR, TR receptor binding affinity - Transcriptional activation - Aromatase and steroidogenesis *in vitro* - Aryl hydrocarbon receptor recognition/binding - QSARs	- High Through Put Prescreens - Thyroid function - Fish hepatocyte VTG assay - Others (as appropriate)
Level 3 *In vivo* assays providing data about single endocrine mechanisms and effects	- Uterotrophic assay (estrogenic related) - Hershberger assay (androgenic related) - Non -receptor mediated hormone function - Others (e.g. thyroid)	- Fish VTG (vitellogenin) assay (estrogenic related)
Level 4 *In vivo* assays providing data about multiple endocrine mechanisms and effects	- Enhanced OECD 407 (endpoints based on endocrine mechanisms) - Male and female pubertal assays - Adult intact male assay	- Fish gonadal histopathology assay - Frog metamorphosis assay
Level 5 *In vivo* assays providing data on effects from endocrine & other mechanisms	- 1-Generation assay (TG415 enhanced)[1] - 2-Generation assay (TG416 enhanced)[1] - Reproductive screening test (TG421 enhanced)[1] - Combined 28 day/reproduction screening test (TG 422 enhanced)[1] [1] Potential enhancements will be considered by OECD TG Program	- Partial and full life cycle assays in fish, birds, amphibians & invertebrates (developmental and reproduction)

Note 1: Entering at all levels and exiting at all levels is possible and depends upon the nature of existing information needs for hazard and risk assessment purposes
Note 2: In level 5, ecotoxicology should include endpoints that indicate mechanisms of adverse effects, and potential population damage
Note 3: When a multimodal model covers several of the single endpoint assays, that model would replace the use of those single endpoint assays
Note 4: The assessment of each chemical should be based on a case by case basis, taking into account all available information, bearing in mind the function of the framework levels.
Note 5: The framework should not be considered as all inclusive at the present time. At levels 3, 4 and 5 it includes assays that are either available or for which validation is under way. With respect to the latter, these are provisionally included. Once developed and validated, they will be formally added to the framework.
Note 6: Level 5 should not be considered as including definitive tests only. Tests included at that level are considered to contribute to general hazard and risk assessment.

- Be sufficiently robust (insensitive to minor changes in the protocol).
- Are transferable among properly equipped and staffed laboratories.

Furthermore adequate test data should be provided for chemicals or products representative of the type of chemicals for which the test is proposed and justification (scientific, ethical, economic) should be provided for the new method with respect to existing ones.

Currently the test guidelines programme devotes special attention to endocrine disruption, due to concerns in many countries that certain chemicals act upon hormone systems and diminish the reproductive capacity of wildlife and also possibly have damaging effects in humans. The test guidelines programme is developing and validating a range of test methods to detect endocrine disrupters such as the ER (estrogen receptor), AR (androgen receptor) and TR (thyroid receptor) assays. In addition to developing tests for endocrine disruptors in the human health and environmental fields, OECD has also developed a conceptual framework for endocrine disruptor testing, outlining consecutive steps that could be followed (Box 16.1).

Many of the current OECD test guidelines are based on tests conducted in laboratory animals, with clear guidance to minimize pain and suffering in the animals during testing. The programme has also actively worked towards the development of methods which replace animal tests, or to refine existing tests so that better information is obtained, and/or to reduce the number of animals needed for the testing of chemicals and reduce the pain associated with animal testing. A number of OECD test guidelines are now based on non-animal tests, including skin corrosion, phototoxicity and skin absorption. As new tests which can meet the regulatory safety requirements of the OECD member countries are developed and validated in the programme, it is expected that the range of non-animal test guidelines available will continue to increase.

16.2.2 The OECD principles of GLP

"Harmonisation" means more than using the same standards for laboratory testing and management and having legal instruments on the books which state that data developed under these standards must be accepted. The OECD principles of GLP provide quality assurance concepts concerning the organization of test laboratories and the conditions under which laboratory studies are planned performed, monitored, and reported. This means that the whole system of verification of compliance with the principles of GLP needs to be harmonised among countries, so that they are speaking a common language when they are exchanging information about laboratories and so that they understand and have confidence in the procedures used for monitoring compliance. The principles of GLP were established in the 1981 Council Decision on MAD [3].

After adoption of the GLP principles in 1981, OECD began to concentrate on activities to facilitate internationally harmonised approached to compliance monitoring and assurance, and in 1989, the OECD Council adopted an act on compliance with principles of GLP [6]. This act contains a decision that member countries shall:

1. Establish national procedures for monitoring compliance with GLP principles, based on laboratory inspections and study audits.
2. Designate national compliance monitoring authorities ("GLP inspectors").
3. Require the management of test facilities to issue a declaration, where applicable, that a study was carried out according to GLP principles.

It is not very efficient for countries to carry out GLP inspections abroad to verify compliance with their own national legislation for their own national purposes. With more and more laboratories requesting entrance into national GLP programmes, with more and more countries establishing such programmes, and with more and more areas of testing being done under GLP - for instance, field studies - it is not only not very efficient; it is virtually impossible for national Monitoring Authorities to personally verify the compliance of foreign laboratories with GLP, except in special situations. Therefore, the 1989 Council Act also establishes a framework for international liaison and recognition of compliance assessment by member countries.

The practical information flow between member countries (including adhering non-member countries) and test facilities for GLP compliance monitoring is illustrated in Figure 16.4. In case a regulatory authority in country A would like to receive information on the GLP compliance of a test facility from country B, it can contact the GLP monitoring authority in country A. Through the network of OECD monitoring authorities, country A can then ask for an inspection or study audit of the test facility by the GLP monitoring authority in country B. The OECD working group on GLP, made up of representatives of national GLP compliance monitoring authorities, oversees the programme on GLP and develops common positions on the administration of compliance monitoring. The working group on GLP has developed a number of guidance documents which are published in the OECD series on principles of good laboratory practice and compliance monitoring [7].

16.3 THE OECD EXISTING CHEMICALS PROGRAMME

16.3.1 The OECD high production volume chemicals programme

Through OECD Council Decisions in 1987 and 1990 [8,9] member countries decided to undertake the investigation of high production volume (HPV) chemicals

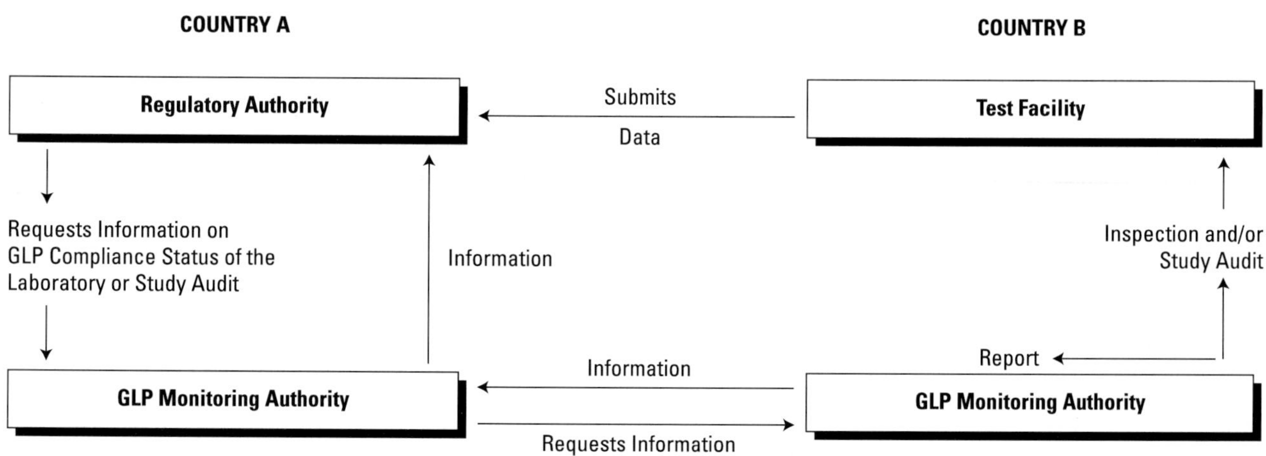

Figure 16.4. Information exchange between member countries for GLP compliance monitoring.

in a co-operative way. These HPV chemicals include all chemicals reported to be produced or imported at levels greater than 1,000 tonnes per year in at least one member country or in the European Union region. The Decision means that member countries will co-operatively:

- Select the chemicals to be investigated.
- Collect characterisation, effects and exposure information from government and public sources and encourage industry to provide information from their files.
- Complete the agreed dossier for the Screening Information Data Set (SIDS) by testing.
- Make an initial assessment of the potential hazard of each chemical investigated.

The OECD list of HPV chemicals serves as the overall priority list from which chemicals are selected for data gathering and initial hazard assessment. The most recent list is that compiled in 2004, which contains 4843 substances [10]. No further priority setting is performed. Member countries, in collaboration with industry, select the substances they intend to assess from this list.

At a minimum, for all sponsored substances, the Screening Information Data Set (SIDS) should be available. The SIDS is a list of data elements similar to those which governments in most OECD countries require from industry before a new chemical can be marketed. It includes information on the chemical's identity, its physical and chemical properties, its behaviour and fate in the environment as well as data on ecotoxicity and toxicity (Table 16.2).

When a full SIDS dossier on a chemical is available, an initial assessment of the information is undertaken and conclusions are drawn on the potential hazard(s) posed by the chemical and recommendations are made on the need for further work. The conclusions present a summary of the hazards of the chemical, written with sufficient detail and clarity as to be informative and to assist countries with classification work and other hazard-based national decision-making; and exposure information to put the hazard information into context. The recommendation, based on these conclusions, can be either that the chemical is currently of low priority for further work or that it is a candidate for further work to clarify its potential risk (for example, that further information is required to clarify concerns identified in the SIDS process, and that post-SIDS testing is recommended).

In the policy bodies of OECD, member countries discuss and agree on any follow-up actions on chemicals for which further work is recommended, and discuss and confirm all conclusions and recommendations made on all chemicals which have undergone SIDS initial assessments. When full SIDS dossiers and initial assessment reports are finalised, the results are made available worldwide through UNEP Chemicals [11].

The chemical industry supports the OECD activities on HPV chemicals because this work avoids duplication of efforts to test chemicals to fulfil various national and regional requirements and international commitments. In 1998 the global chemical industry through the International Council of Chemical Associations (ICCA) announced its intention to work with OECD by using the OECD HPV chemicals list to establish a working list of approximately 1000 substances as priorities for investigation (based on presumed wide dispersive use,

Table 16.2. OECD's Screening Information Data Set (SIDS).

Chemical Identity
- CAS number
- Chemical name

Exposure information
- Production volume (in at least one country)
- Use pattern (in at least one country)
- Sources of exposure

Physical Chemical Data
- Melting point
- Boiling point
- Vapour pressure
- Water solubility
- Octanol-water partition coefficient
- Dissociation constant

Environmental Fate and Pathways
- Photodegradation
- Abiotic degradation (hydrolysis)
- Biodegradation
- Transport and distribution in the environment

Ecotoxicological data
- Acute toxicity to fish
- Acute toxicity to invertebrates (e.g. daphnids)
- Toxicity to algae
- Long-term toxicity to aquatic organisms (necessity determined based on physical chemical properties of the chemical)
- Toxicity to terrestrial organisms (if significant exposure to soil is expected)

Toxicological Data
- Acute toxicity
- Repeated dose toxicity
- Genetic toxicity
  - o Point mutation
  - o Chromosomal aberration
- Reproductive toxicity
  - o Toxicity to fertility
  - o Developmental toxicity

production in two or more global regions or similarity to another chemical meeting either of these criteria). Its members have set a goal to complete SIDS and initial hazard assessments on these chemicals. This initiative is an important source of assessments for consideration in the programme. Member countries work with the chemical industry in order to make the most efficient use of the information compiled through the ICCA initiative in meeting their commitments to investigate a certain proportion of the chemicals on the OECD HPV chemicals list.

16.3.2 Collaboration with national/regional programmes

The aim of the OECD HPV chemicals programme is to elaborate initial hazard assessments for chemicals which can then be used by member countries for their national/regional decision-making (e.g. classification, risk assessments etc.). The OECD programme is therefore designed to achieve the highest possible level of integration with national/regional programmes. Furthermore, many national/regional programmes have been designed to fully use the outcome of the OECD programme. The relationship between the OECD programme and national/regional programmes is illustrated with a few examples below.

REACH

Detailed information on REACH can be found in Chapter 12. While there are many differences in requirements and procedures between REACH and the OECD HPV chemicals programme, reporting requirements between the two programmes are the same, which allows for a high level of integration. In both programmes the reporting requirements are based on the following main concepts:
- Concept of the robust study summary.
- OECD harmonised templates for reporting summary information.
- Template for a hazard assessment report.

Both programmes use the concept of the robust study summary to report study results [12]. A robust study summary reflects the objectives, methods, results and conclusions of a full study report in sufficient detail to allow a technically qualified person to make an independent assessment as to its reliability and completeness - minimising the need to go back to the full study report. In both programmes robust study summaries are provided for all the key studies, i.e. those studies which form the basis for the final hazard assessment.

In 2006, the OECD published harmonised templates for reporting summary information of the results from chemical testing. These templates, which have been prepared for database developers, prescribe the format by which results should be entered into and maintained in databases so that the data can easily be exchanged electronically across countries [13]. The IUCLID database software has implemented the OECD harmonised templates and is the recommended tool for gathering and exchanging robust study summaries in both programmes. Furthermore, the format of the hazard assessment part of the Chemicals Safety Report (CSR) used under REACH and the OECD SIDS Initial Assessment Report (SIAR) [12] are almost identical and the hazard assessment can easily be transferred from one report to the other.

Any assessment work performed for any of the two programmes can therefore feed into or be taken from the other programme and does not have to be rewritten. For HPV chemicals, the data requirements under REACH fulfil the data requirements for the OECD HPV chemicals programme. The hazard assessment part of a REACH registration can therefore be submitted to the OECD programme. Furthermore, after agreement within the OECD HPV chemicals programme, the SIDS dossier and SIAR can be amended (e.g. with an exposure assessment and risk characterisation) to build the documents necessary for a REACH registration.

Furthermore efforts are currently underway to further harmonise across the OECD the use of different tools and methods to fulfil information requirements. The most prominent example is the use of chemical categories (see Chapter 11). A chemical category is a group of chemicals whose physicochemical and toxicological properties are likely to be similar or follow a regular pattern as a result of structural similarity. In the category approach, closely related chemicals are considered as a group, or category, rather than as individual chemicals. Thereby not every chemical needs to be tested for every required endpoint. Rather, the overall data for that category must prove adequate to support a hazard assessment. The overall data set must allow the estimation of the hazard for the untested endpoints. The same guidance document for the development and use of chemical categories is used in both programmes [12], thereby ensuring a harmonised approach to fulfilling information requirements through chemical categories in both programmes and avoiding duplications.

US HPV Challenge Program

Detailed information on the US HPV challenge program can be found in Chapter 13. The cooperation between the US HPV challenge program and the OECD HPV chemicals programme was effective since the beginning of the US HPV challenge program in 1998. As with the REACH, there are there are many differences in requirements and procedures, but the reporting requirements between the two programmes are the same. The concept of robust study summaries was first invented in the context of the US HPV challenge program and subsequently implemented within the OECD programme. Furthermore, the objective of the US programme is to collect robust study summaries for SIDS endpoints for all HPV chemicals on the US market. The assessments performed within the OECD HPV chemicals programme thereby fulfil the requirements of the US HPV challenge programme, which avoids duplications. Manufacturers/ importers in the US can decide whether to submit their datasets and/or assessments to either programme.

Japan HPV challenge program

Detailed information on the Japan HPV challenge programme can be found in Chapter 14. This programme was set up in 2005 and is similar in scope to the US HPV challenge program. Its objective is to collect robust study summaries for SIDS endpoints for all HPV chemicals on the market in Japan. The datasets collected within the US HPV challenge programme and the assessments performed within the OECD HPV chemicals programme both fulfil the requirements of the Japan HPV challenge program. In the first instance, the Japanese programme will therefore focus on gathering data for chemicals which are currently not addressed within either the US or the OECD programme, thereby contributing significantly to the overall objective of assessing the hazards of HPV chemicals.

16.3.3 The OECD project on (Quantitative) Structure-Activity Relationships

(Quantitative) Structure-Activity Relationships or (Q)SARs are methods for estimating properties of a chemical from its molecular structure. The use of (Q)SARs for predicting physicochemical properties and fate is described in detail in Chapter 9. QSARs for the prediction of endpoints of toxicity and ecotoxicity are presented in Chapter 10. The aim of the (Q)SAR project at OECD is to improve the regulatory acceptability and use of (Q)SARs within OECD member countries. As a first step, the OECD principles for the validation, for regulatory purposes, of (Q)SAR models were developed and adopted in 2004 (Box 16.2; see also Chapter 10).

These principles are scientific goals that provide generic baseline guidance for integrating the use of (Q)SAR models into regulatory frameworks. More practically, the principles enable (Q)SAR developers and authorities in member countries to document in a consistent and transparent manner the validation of individual (Q)SAR models and to make these models available for regulatory use by other authorities. The principles only identify the type of information that

Box 16.2. OECD principles for validation, for regulatory purposes, of (Q)SAR models

The agreed OECD principles for the validation, for regulatory purposes, of (Q)SAR models, which are intended to be read in conjunction with the associated explanatory comments, are as follows:

To facilitate the consideration of a (Q)SAR model for regulatory purposes, it should be associated with the following information:

1. a defined endpoint
2. an unambiguous algorithm
3. a defined domain of applicability
4. appropriate measures of goodness-of–fit, robustness and predictivity
5. a mechanistic interpretation, if possible

Notes

1. The intent of Principle 1 (defined endpoint) is to ensure clarity in the endpoint being predicted by a given model, since a given endpoint could be determined by different experimental protocols and under different experimental conditions. It is therefore important to identify the experimental system that is being modeled by the (Q)SAR. Further guidance is being developed regarding the interpretation of "defined endpoint". For example, a no-observed-effect level might be considered to be a defined endpoint in the sense that it is a defined information requirement of a given regulatory guideline, but cannot be regarded as a defined endpoint in the scientific sense of referring to a specific effect within a specific tissue/organ under specified conditions.

2. The intent of Principle 2 (unambiguous algorithm) is to ensure transparency in the model algorithm that generates predictions of an endpoint from information on chemical structure and/or physicochemical properties. It is recognized that, in the case of commercially-developed models, this information is not always made publicly available. However, without this information, the performance of a model cannot be independently established, which is likely to represent a barrier for regulatory acceptance. The issue of reproducibility of the predictions is covered by this Principle, and will be explained further in the guidance material.

3. The need to define an applicability domain (Principle 3) expresses the fact that (Q)SARs are reductionist models which are inevitably associated with limitations in terms of the types of chemical structures, physicochemical properties and mechanisms of action for which the models can generate reliable predictions. Further work is recommended to define what types of information are needed to define (Q)SAR applicability domains, and to develop appropriate methods for obtaining this information.

4. The revised Principle 4 (appropriate measures of goodness-of–fit, robustness and predictivity) includes the intent of the original Setubal Principles 5 and 6. The wording of the principle is intended to simplify the overall set of principles, but not to lose the distinction between the internal performance of a model (as represented by goodness-of-fit and robustness) and the predictivity of a model (as determined by external validation). It is recommended that detailed guidance be developed on the approaches that could be used to provide appropriate measures of internal performance and predictivity. Further work is recommended to determine what constitutes external validation of (Q)SAR models.

5. It is recognised that it is not always possible, from a scientific viewpoint, to provide a mechanistic interpretation of a given (Q)SAR (Principle 5), or that there even be multiple mechanistic interpretations of a given model. The absence of a mechanistic interpretation for a model does not mean that a model is not potentially useful in the regulatory context. The intent of Principle 5 is not to reject models that have no apparent mechanistic basis, but to ensure that some consideration is given to the possibility of a mechanistic association between the descriptors used in a model and the endpoint being predicted, and to ensure that this association is documented.

are considered useful for the regulatory application of (Q)SAR models in a regulatory context. The definition of criteria for determining the national regulatory acceptability of (Q)SAR models are outside of the scope of the OECD project and will be considered by national authorities.

To improve the regulatory acceptance of (Q)SAR models, the OECD experience of member countries with

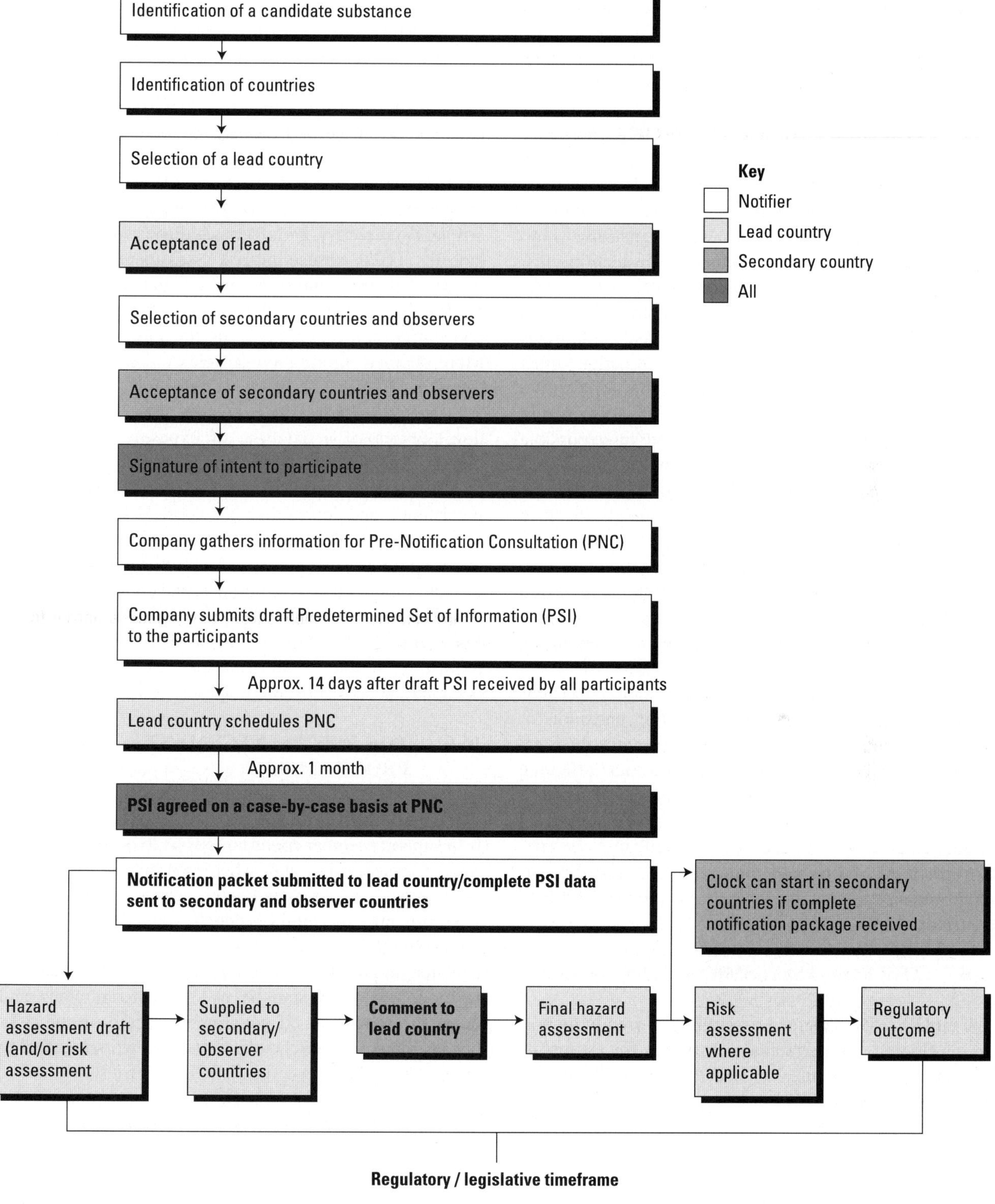

Figure 16.5. Mutual acceptance of notifications: pilot phase of the parallel process.

(Q)SAR applications has been summarised in a report on the regulatory uses and applications in OECD member countries of (Q)SAR models in the assessment of new and existing chemicals [14].

16.4 THE NEW CHEMICALS PROGRAMME

The main aim of the OECD new chemicals programme is to simplify and streamline the access to multiple markets for new chemicals, while ensuring high standards of health and environmental protection [15]. This is achieved through the so-called parallel process, which refers to a process by which a company notifies to multiple jurisdictions in parallel and authorizes participating governments to share information when conducting their reviews (Figure 16.5).

A pilot phase of the parallel process was launched in 2006. Within the parallel process, there are three possible ways in which countries can participate: as lead, as secondary or as observer. The lead country develops the hazard assessment considering the comments received by all participants and the final hazard assessment is issued to the notifier. Secondary countries are those who receive a formal notification but not until the final hazard assessment is complete. The lead country is responsible for reviewing the information submitted, developing the hazard assessment and sharing the hazard assessment with the secondary countries for review. Observer status allows the country to receive information and monitor how assessments are conducted and provide informal input on the hazard assessment, without affecting the timing or content of the process. Jurisdictions participating in the parallel process utilize current evaluation processes to conduct their notification reviews. Nevertheless, throughout this process, jurisdictions retain the sovereign right to make their own risk-based decisions.

16.5 THE RISK ASSESSMENT PROGRAMME

The OECD risk assessment programme currently focuses on developing and harmonising methods for estimating environmental exposures and more specifically on the following four areas [16]:

- Release estimation.
- Exposure modelling.
- Use of monitoring data.
- Reporting of exposure information.

For improving the estimation of releases of chemicals, the OECD is developing Emission Scenario Documents (ESDs) that describe the conditions and parameters for release estimation in specific industry and use categories. An ESD is a document that describes the sources, production processes, pathways and use patterns with the aim of quantifying the emissions (or releases) of a chemical into water, air, soil and/or solid waste. An ESD ideally includes all the following stages: (1) production, (2) formulation, (3) industrial use, (4) professional use, (5) private and consumer use, (6) service life of product/article, (7) recovery, and (8) waste disposal (incineration, landfill). ESDs are used in risk assessment of chemicals to establish the conditions on use and releases of the chemicals, which are the bases for estimating the concentration of chemicals in the environment [17]. More detailed information on release estimation can be found in Chapter 2.

Recent activities on exposure models include the development of a database of exposure models, and the development of a guidance document on the use of multimedia models in the assessment of overall persistence and long-range transport [18]. The OECD activity on environmental monitoring data aims at improving the availability of monitoring data, better design of monitoring programmes, and enhanced dialogue between monitoring and risk assessment communities. Furthermore, the OECD has developed guidance on reporting summary information on environmental, occupational and consumer exposure [19].

16.6 THE RISK MANAGEMENT PROGRAMME

The core objective of OECD's work on risk management is to support member countries' efforts to develop national policies and actions, and, where appropriate, to develop international risk management measures [20]. OECD governments, academia, industry and environmental NGOs work together to identify best practices and new techniques for managing risks, and then develop methodologies that can be used by governments and industry. In addition, if governments agree on the risks posed by a particular chemical, they can work together to take concerted action across OECD countries.

Today, most of OECD's work is focused on developing guidance for risk management that apply to the chemical industry as a whole. This includes: guidance for conducting socioeconomic analysis; guidance on risk communication; listing tools to help companies screen potentially dangerous chemicals before they are manufactured; and promoting the development of environmentally benign chemicals.

16.7 OTHER ACTIVITIES RELATED TO ENVIRONMENT, HEALTH AND SAFETY

Other OECD activities in the environment, health and safety areas are concerned with pesticides [21], biocides [22], chemical accident prevention, preparation and response [23], pollutant release and transfer registers [24] and, harmonisation of regulatory oversight in biotechnology and the safety of novel foods and feeds [25]. These activities are closely connected with the work in the chemicals programme, and are carried out in cooperation with other parts of the OECD and other international organisations.

16.8 FURTHER READING

1. Organization for Economic Cooperation and Development. 2006. OECD Environment Programme 2005 – 2006. OECD, Paris, France.
2. Organization for Economic Cooperation and Development. 2006. OECD's Environment, Health and Safety Programme. Brochure. OECD, Paris, France.
3. Organization for Economic Co-operation and Development. 2001. Environmental outlook for the chemicals industry. OECD, Paris, France.

REFERENCES

1. Organization for Economic Cooperation and Development. 2006. Chemical Safety. OECD, Paris, France. [http//www.oecd.org/ehs].
2. Organization for Economic Cooperation and Development. 2006. OECD's Environment, Health and Safety Programme. Brochure. OECD, Paris, France.
3. Organization for Economic Cooperation and Development. 1981. Decision of the Council concerning the Mutual Acceptance of Data in the Assessment of Chemicals. 12 May 1981 – C(81)30/Final amended on 26 November 1997 – C(97)186/Final. OECD, Paris, France.
4. Organization for Economic Cooperation and Development. 2006. Chemicals Testing Guidelines. OECD, Paris, France [http//www.oecd.org/env/testguidelines].
5. Organization for Economic Cooperation and Development. 2006. Conceptual framework for endocrine disrupter testing and assessment. OECD, Paris, France.
6. Organization for Economic Cooperation and Development. 1989. Decision-Recommendation of the Council on compliance with principles of good laboratory practice. 2 October 1989 - C(89)87/Final amended on 9 March 1995 - C(95)8/Final, OECD, Paris, France.
7. Organization for Economic Cooperation and Development. OECD Series on Principles of Good Laboratory Practice and Compliance Monitoring. OECD, Paris, France.
8. Organization for Economic Cooperation and Development. 1987. Decision-Recommendation of the Council on the systematic investigation of existing chemicals. 26 June 1987 – C(87)90/Final. OECD, Paris, France.
9. Organization for Economic Cooperation and Development. 1990. Decision-Recommendation of the Council on the co-operative investigation and risk reduction of existing chemicals. 31 January 1991 – C(90)163/Final. OECD, Paris, France.
10. Organization for Economic Cooperation and Development. 2004. The 2004 OECD list of high production volume chemicals. OECD, Paris, France.
11. UNEP Chemicals. 2006. Screening Information Data Sets for High Production Volume Chemicals. OECD, Paris, France [http://www.chem.unep.ch/irptc/sids/OECDSIDS/sidspub.html].
12. Organization for Economic Cooperation and Development. 2006. Manual for the Investigation of HPV Chemicals. OECD, Paris, France.
13. Organization for Economic Cooperation and Development. 2006. OECD Harmonised Templates. OECD, Paris, France.
14. Organization for Economic Co-operation and Development. 2006. Report on the regulatory uses and applications in OECD member countries of (quantitative) structure-activity relationship [(Q)SAR] models in the assessment of new and existing chemicals. Environmental Health and Safety Publications. Series on Testing and Assessment 58. OECD, Paris, France.
15. Organization for Economic Cooperation and Development. 2006. The OECD programme on new chemicals. OECD, Paris, France.
16. Organization for Economic Cooperation and Development. 2006. OECD activities on environmental exposure assessment. OECD, Paris, France [http://www.oecd.org/env/riskassessment].
17. Organization for Economic Cooperation and Development. 2006. Emission scenario documents. OECD, Paris, France.
18. Organization for Economic Cooperation and Development. 2004. Guidance document on the use of multimedia models for estimating overall environmental persistence and long-range transport. OECD Series on Testing and Assessment No 45. OECD, Paris, France.
19. Organization for Economic Cooperation and

Development. 2003. Guidance document on reporting summary information on environmental, occupational and consumer exposure. Series on Testing and Assessment No 42. OECD, Paris, France.

20. Organization for Economic Cooperation and Development. 2006. OECD activities on risk management. OECD, Paris, France. [http://www.oecd.org/env/riskmanagement].
21. Organization for Economic Cooperation and Development. 2006. The OECD pesticide programme. OECD, Paris, France [http://www.oecd.org/env/pesticides].
22. Organization for Economic Cooperation and Development. 2006. The OECD Biocides Programme. OECD, Paris, France. [http://www.oecd.org/env/biocides].
23. Organization for Economic Cooperation and Development. 2006. The OECD programme on chemical accidents. OECD, Paris, France. [http://www.oecd.org/env/accidents].
24. Organization for Economic Cooperation and Development. 2006. Pollutant release and transfer registers. OECD, Paris, France. [http://www.oecd.org/env/prtr].
25. Organization for Economic Cooperation and Development. 2006. Biosafety - BioTrack. OECD, Paris, France. [http://www.oecd.org/env/biotrack].

GLOSSARY

C.J. van Leeuwen

(artificial) neural network (ANN) Artificial neural networks are computational models that make predictions by simulating the functioning of human neurons.

abiotic Not associated with living organisms.

abiotic transformation Process in which a substance in the environment is modified by non-biological mechanisms.

absorbed dose Amount of substance absorbed by an organism or organs and tissues of interest.

absorption Process of active or passive transport of a substance across biological membranes or other barriers into an organism. In the case of a mammal, this is usually through the lungs, gastrointestinal tract or skin. Absorption through the gills is an important transport route for many aquatic species.

abundance The degree of plentifulness.

acceptable daily intake Estimated maximum amount of an agent, expressed on a body mass basis, to which an individual in a (sub) population may be exposed daily over its lifetime without appreciable health risk. To calculate the daily intake per person, a standard body mass of 60 kg is used. The ADI is normally used for food additives, while the tolerable daily intake (TDI) is used for contaminants. *See also* reference dose, tolerable daily intake.

acceptable risk This is a risk management term. The acceptability of the risk depends on scientific data, social, economic and political factors, and on the perceived benefits arising from exposure to an agent.

acclimatization (1) Steady-state compensatory adjustments by an organism to the alteration of environmental conditions. Adjustments can be behavioural or physiological/biochemical. (2) An adaptation of organisms to some experimental conditions, including adverse stimuli. (3) Acclimation also refers to the time period prior to the initiation of a toxicity test in which aquatic organisms are maintained in untreated, toxicant-free dilution water with physical and chemical characteristics (e.g., temperature, pH, hardness) similar to those to be used during the toxicity test.

accumulation Successive additions of a substance to a target organism, organ or environmental compartment, resulting in an increasing amount or concentration of the substance in the organism, organ or environment.

accuracy Reflects the agreement between the measured and the "true" value.

acid volatile sulphide (AVS) An extractable reactive pool of solid-phase sulphide that is associated with and available from the mineral surfaces of sediment to bind metals and may render that portion unavailable and non-toxic to biota. Metals associated with the sulphide fraction of suspended matter and sediments in anaerobic environments include zinc, lead, copper, cobalt, nickel, cadmium, arsenic, antimony, mercury, manganese, and molybdenum.

actors in the supply chain In the context of REACH means all manufacturers and/or importers and/or downstream users in a supply chain.

acute Within a short period in relation to the lifespan of the organism, usually ≤ 4 d for fish and ≤ 14 d for rats. It can be used to define either the exposure or the response to an exposure (effect). An acute toxic effect would be induced and observable within a short period of exposure. It can refer to an instantaneous exposure (oral gavage, injection, dermal application, etc.) or continuous exposures ranging from a few minutes to a few days.

acute tests Short-term tests (relative to generation time) usually at high concentrations.

acute toxicity The adverse effect occurring within a short time of exposure (relative to generation time). *See also* chronic and subchronic toxicity.

adaptation (1) Change in an organism, in response to changing conditions of the environment (specifically chemical), which occurs without any irreversible disruption of the given biological system and without exceeding the normal (homeostatic) capacities of its response, and (2) a process by which an organism stabilizes its physiological condition after an environmental change.

added risk Difference between the incidence of an adverse effect in a treated group of organisms or a group of exposed humans and a control group (of the same organisms or the spontaneous incidence in humans).

additive effect An effect which is the result of chemicals acting together and which is the simple sum of the effects of the chemicals acting independently. *See also* antagonistic effect and synergism/potentiation.

additive toxicity The toxicity of a mixture of chemicals which is approximately equivalent to that expected from a simple summation of the known toxicities of the individual chemicals present in the mixture (i.e. algebraic summation of effects).

adenocarcinoma A malignant tumour originating in glandular tissue.

adenoma A tumour, usually benign, occurring in glandular tissue.

ADI *See* acceptable daily intake.

adjuvant In immunology, a substance injected with antigens (usually mixed with them but sometimes given prior to or following the antigen) which non-specifically enhances or modifies the immune response to that antigen.

***ad libitum* feeding** Providing continuous access to food.

adsorption The adhesion of molecules to surfaces of solids.

advection Physical transport or movement of a substance with its medium (air, water, sediment).

adverse effect Change in the morphology, physiology, growth, development, reproduction or life span of an organism, system, or (sub) population that results in an impairment of functional capacity, an impairment of the capacity to compensate for additional stress, or an increase in susceptibility to other influences.

aerobic Requiring molecular oxygen.

aerosol A compound dispersed as minute droplets or particles in a gas which allows widespread environmental distribution, entry to the body via the respiratory tract, and widespread contamination of clothing, skin and eyes.

aetiology The study of the causes or origins of disease.

AFNOR Association Francaise de Normalisation.

AFNOR test Biodegradation test, which monitors the reduction in DOC.

age class A group of organisms of the same age within a population.

age composition The distribution of organisms among the various age classes present in the population. The sum of the number of individuals in all age classes equals the population size.

age distribution The composition of a population in terms of how its abundance is distributed across age classes.

age-specific fecundity The pattern of fecundity of the various age classes.

age-specific mortality The pattern of survival rates of each age class.

Agency In the context of REACH means the European Chemicals Agency as established by this Regulation.

aggregation error The model error resulting from the use of a single set of parameters to represent a collection of distinct entities, such as individuals, in a population.

agonist A chemical with a positive action in the body.

ALARA principle As Low As Reasonably Achievable. This principle is a powerful regulatory tool in that risk managers are expected to do everything possible to reduce risks to a limit which they can justify to their organization and to the regulatory authorities. *See also* precautionary principle (Chapter 1).

algicidal Lethal to algal population.

algicide *See* pesticide.

algistatic Inhibits algal population growth.

alkalinity The acid-neutralizing (i.e. proton-accepting) capacity of water; the quality and quantity of constituents in water which shift the pH towards the alkaline side of neutrality.

alkylating agent A substance which introduces an alkyl radical into a compound in place of a hydrogen atom.

allergy Symptoms or signs occurring in sensitized individuals following exposure to a previously encountered substance (allergen) which would otherwise not cause such symptoms or signs in non-sensitized individuals.

allometry The relationship between growth rates of different parts of an organism or the study of the change in proportions with increase in size.

alloy means a metallic material, homogenous on a macroscopic scale, consisting of two or more elements so combined that they can not be readily separated by mechanical means.

ambient concentration The concentration of a chemical in a medium resulting from the addition of an incremental concentration to a background concentration.

ambient standard *See* environmental quality standard.

anabolism Metabolic processes concerned with synthesis.

anadromy Fish born in freshwater, migrate to the sea for growth and development but return to freshwater for reproduction. *See* catadromy.

anaerobic Not requiring molecular oxygen.

analysis (1) Detailed examination of anything complex, made in order to understand its nature or to determine its essential features, and (2) a formal, usually quantitative, determination of the effects of an action (as in risk analysis and impact analysis).

analysis of extrapolation error A method of risk analysis in which the probability density of an assessment endpoint, with respect to the concentration of a chemical (or other measure of exposure), is estimated by statistical extrapolation from toxicological data and the assessment endpoint.

analysis of variance A method for testing the significance of mean differences in which the total variation in a set of scores is divided into separate parts; a parametric statistical procedure for evaluating the hypotheses about mean differences. In toxicology, a typical application of this method is to test for mean differences when comparing more than two experimental conditions.

aneuploidy Deviation from the normal number of chromosomes excluding exact multiples of the normal haploid complement.

ANOVA *See* analysis of variance.

anoxia Strictly, the total absence of oxygen, but sometimes used to mean decreased oxygen supply in tissues.

antagonist A chemical which diminishes the effect of another chemical; the opposite of an agonist.

antagonistic effect The effect of a chemical in counteracting the effect of another; for example, a situation where exposure to two chemicals together has less effect than the simple sum of their independent effects; such chemicals are antagonists.

anthropogenic Caused or influenced by human activities.

antibody A protein that specifically recognizes and binds to an antigen.

antigen A substance that elicits a specific immune response when introduced into the tissues of an animal.

APHA American Public Health Association.

application factor (AF) A dimensionless value: the chronically toxic threshold concentration of a chemical divided by its acutely toxic concentration. The AF is usually reported as a range and is multiplied by the median lethal concentration of a chemical as determined in a short-term (acute) toxicity test to estimate the expected no effect concentration under chronic exposure.

artefact Finding or product of an experimental or observational technique that is not properly associated with the system under investigation.

article In the context of REACH means an object which during production is given a special shape, surface or design which determines its function to a greater degree than does its chemical composition.

artisol test Tests in which earthworms are exposed to chemicals in amorphous silicagel.

assessment Evaluation or appraisal of an analysis of facts and the inference of possible consequences concerning a particular object or process.

assessment endpoint Qualitative/quantitative expression of a specific factor with which a risk may be associated as determined through an appropriate risk assessment.

assessment factor Numerical adjustment used to extrapolate from experimentally determined (dose-response) relationships to estimate the agent exposure below which an adverse effect is not likely to occur. *See also* safety factor and uncertainty factor.

ASTM American Society for Testing and Materials.

asymptotic threshold concentration The concentration of a chemical at which some percentage of a population of test organisms is in a state of approximate homeostasis for a prolonged period of time (not necessarily absolute). This can be demonstrated as the concentration at which the toxicity curve is approximately asymptotic (parallel) to the time axis. The asymptotic LC50 is the concentration at which acute mortality has essentially ceased and will not change substantially with exposure time. That is, there is no evidence of significantly increasing effects due to a longer period of exposure.

atrophy Wasting away of the body, or an organ or tissue.

aufwuchs Floral/faunal communities attached to submerged surfaces.

autopsy *See* necropsy.

axenic Germ free. An axenic culture is a culture containing only one type of micro-organism or one microbial species.

background concentration The concentration of a chemical in a medium prior to the action under consideration or the concentration that would have occurred in the absence of the prior action.

bactericide *See* pesticide.

baseline toxicity The baseline toxicity is the toxicity corresponding to non-polar narcosis, so-called because no chemicals have been found to display lower toxicity than that due to non-polar narcosis. The baseline is often illustrated as a linear dependence of toxicity on the octanol-water partition coefficient (log K_{ow}). It is normally assumed that baseline toxicity is the minimal (acute) toxic effect that a substance may produce, assuming complete solubility and no volatility.

base pairing The linking of complementary pairs of polynucleotide chains of nucleic acids by means of hydrogen bonds between opposite purine and pyrimidine pairs.

benchmark dose The statistical lower confidence limit of the dose corresponding to a small increase in effect over the background level. Typically, a 1% or 10% response level above the background is selected.

benefit A gain to a population, human health or a company. Expected benefit incorporates an estimate of the probability of achieving a gain.

benign Relating to a growth which does not invade surrounding tissue (nonmalignant).

benthic Living on the bottom of aquatic systems.

bias An error caused by the systematic deviation of an estimate from the true value.

bioaccumulation The net result of multiple physiological processes (ADME: Absorption, Distribution, Metabolism and Excretion) of a substance due to all routes of exposure.

bioaccumulation factor (BAF) A partition coefficient for the distribution of a chemical between an organism exposed through all possible routes and an environmental compartment (air, water, soil or sediment) or food.

bioaccumulation factor, lipid based The quotient of the test chemical substance concentration in the lipid fraction of the test organism and the concentration of the truly dissolved compound in the test water, when the rate of uptake and clearance are equal.

bioactivation Biotransformation of a compound to a more toxic product.

bioassay Test used to evaluate the relative potency of a chemical by comparing its effect on a living organism with the effect of a standard preparation on the same type of organism. Bioassays are frequently used in the pharmaceutical industry to evaluate the potency of vitamins and drugs. Bioassay and toxicity test are not synonymous. *See* toxicity tests.

bioavailability The ability of a substance to interact with the biosystem of an organism. Systemic bioavailability will depend on the chemical or physical reactivity of the substance and its ability to be absorbed through the gastrointestinal tract, respiratory tract or skin. It may be locally bioavailable at all these sites.

biochemical mechanism A chemical reaction or series of reactions, usually enzyme catalyzed, which produces a given physiological effect in a living organism.

biochemical (or biological) oxygen demand (BOD) A measure of the rate at which molecular oxygen is consumed by microorganisms during oxidation or organic matter. The standard test is the 5-day BOD test, in which the amount of dissolved oxygen required for oxidation over a 5-day period is measured. The results are measured in mg of oxygen/L (mg/L).

biocide Non-agricultural pesticides to control the severity and incidence of pests or diseases. Examples are disinfectants and slimicides for the control of algal, fungal or bacterial growth.

biocommunity The biotic part of an ecosystem.

bioconcentration The net result of the uptake, distribution and elimination of a substance due to water-borne exposure of an organism.

bioconcentration factor (BCF) (1) The quotient of the test chemical substance concentration in the test organisms (C_f) and the concentration in the test water (C_w) at steady-state conditions, i.e. when the rate of uptake and clearance are equal, and (2) the ratio between the uptake rate constant (k_1) and depuration constant (k_2), assuming first-order kinetics.

bioconcentration factor, lipid based (BCF) The quotient of the test chemical substance concentration in the lipid fraction of the test organism and the concentration in the test water, when the rate of uptake and clearance are equal.

biodegradation Breakdown of a substance catalyzed by enzymes. *See also* primary biodegradation, ultimate aerobic biodegradation and mineralisation.

biokinetics *See* pharmacokinetics

biological half-life (*t*1/2) The time needed to reduce the concentration of a test chemical in environmental compartments or organisms to half the initial concentration, by transport processes (e.g. diffusive elimination), transformation processes (e.g. biodegradation or metabolism) or growth. $t1/2 = \ln 2/k$ (for first-order kinetics).

biological monitoring Analysis of the amounts of potentially toxic substances or their metabolites present in body tissues and fluids as a means of assessing exposure to these substances and aiding timely action to prevent adverse effects. The term is also used to mean assessment of the biological status of populations and biocommunities at risk in order to protect them and to obtain an early warning of possible hazards to human or environmental health. Also called biomonitoring.

biomagnification Result of the processes of bioconcentration and bioaccumulation by which tissue concentrations of bioaccumulated chemicals increase as the chemical passes up through two or more trophic levels. The term implies an efficient transfer of chemical from food to consumer, so that residue concentrations increase systematically from one trophic level to the next.

biomagnification factor (BMF) Quantitative measure of a chemical's tendency to be taken up through the food. It is obtained in feeding experiments by dividing the concentration of a chemical substance in a living organism by the concentration of the chemical substance in its food at steady-state.

biomarker The use of physiological, biochemical, and histological changes of exposure and/or effects of xenobiotics at the suborganismal and organismal level.

biomarker of effect A measurable biochemical, physiologic, behavioural, or other alteration in an organism that, depending on the magnitude, can be recognized as associated with an established or possible health impairment or disease.

biomarker of exposure The chemical or its metabolite or the product of an interaction between a chemical and some target molecule or cell that is measured in a compartment or in an organism.

biomarker of susceptibility An indicator of an inherent or acquired ability of an organism to respond to the challenge of exposure to a specific chemical substance.

biomonitoring (or biological monitoring) Monitoring methods to better understand the complex relationships between external and internal exposure and, consequently, the potential adverse health and environmental effects. In ambient monitoring living organisms are used as "sensors" in water/sediment quality surveillance and compliance to detect changes in an effluent or water body and to indicate whether aquatic life may be endangered. In health monitoring it is a general term comprising the following subcategories: (a) biological monitoring - biomarkers of exposure such as internal dose or body burden, (b) biochemical effect monitoring - biomarkers of effective dose (also tissue dose), (c) biological effect monitoring - biomarkers of effect, and (d) clinical parameters - biomarkers of disease.

biosphere The part of the earth's surface which supports and is inhabited by living things.

biota-sediment accumulation factor (BSAF) A specific type and form of bioaccumulation factor that is the ratio of lipid-normalized tissue chemical residue to organic carbon-normalized sediment chemical concentration. Note that various other terms for this are used in the literature, including accumulation factor (AF).

biotic indices Use of biota to indicate the quality of the surrounding environment.

biotransformation Enzyme-catalyzed conversion of one xenobiotic compound to another via phase-I and phase-II reactions.

bioturbation Mixing of sediment/soil, by biological action, e.g. burrowing.

blank Is used interchangeably with the term "control".

Blok test Also known as "repetitive die away test". Is a BOD assessment of biodegradation.

BOD *See* biochemical oxygen demand.

body burden Total amount of a chemical present in an organism at a given time.

boundary of exposure scenario Limitations of the applicability of exposure scenario, e.g. substances within a range defined by their properties, maximum duration and frequency of activities, specified target groups, etc. *See also* exposure scenario.

broad exposure scenario An exposure scenario covering more than one activity, process, sector, or substance for which the same risk management measures are sufficient for ensuring adequate control of risks. *See also* exposure scenario.

broodstock Adult fish undergoing physiological changes to produce either eggs or sperm.

BSAF *See* biota-sediment accumulation factor.

bw Abbreviation for body weight.

calcinosis Any of various pathologic conditions characterized by the deposition of calcium salts in tissues.

cancer Disease which results from the development of a malignant tumour and its spread into surrounding tissues. *See* tumour.

carcinogenesis The development of cancer. Any chemical which can cause cancer is said to be carcinogenic.

carcinogenicity Capacity of chemical, physical or biological agents to induce malignant neoplasms.

carrying capacity (K) The maximum number of organisms that can be supported by a given unit of habitat. Often calculated as the long-term average abundance.

catabolism Metabolic processes concerned with degradation.

catadromy Fish born at sea which migrate to freshwater for growth and development but return to the sea for reproduction. *See* anadromy.

catalase A haem-based enzyme which catalyses the decomposition of hydrogen peroxide into oxygen and water. It is found e.g. in peroxisomes in the liver.

CBA *See* cost-benefit analysis.

CEC Commission of the European Communities; the executive body of the European Communities.

cell line A defined population of cells which has been maintained in a culture for an extended period and which has usually undergone a spontaneous process of transformation conferring an unlimited culture lifespan on the cells.

CEN Comité Européen de Normalisation (European Standardization Committee).

CERCLA (US) Comprehensive Environmental Response, Compensation and Liability Act.

chelation The trapping of a multivalent ionic species by ionic bonding to a larger water soluble molecule to make the ion inactive in the biological matrix and to aid excretion.

chemical oxygen demand (COD) When organic materials are not easily degraded by microorganisms, strong oxidizing agents (e.g., potassium permanganate) are used to enhance oxidation. COD is thus measured instead of BOD (*see* BOD). COD values will be larger than BOD values.

chemical safety assessment (CSA) According to REACH this is the tool used by industry to determine which risk management measures and operational conditions are adequate for protecting human health and the environment.

chemical safety report (CSR) According to REACH this is the tool used by industry to document that the risk management measures and the related operational conditions are sufficient for protecting human health and the environment.

cholinesterase inhibitor A substance which inhibits the enzyme cholinesterase and thus prevents transmission of nerve impulses from one nerve cell to another, or to a muscle.

chromosomal aberration An abnormality in chromosome number or structure.

chromosome The heredity-bearing gene carrier in the cell nucleus, composed of DNA and protein.

chronic Extended or longterm exposure to a stressor (conventionally taken to include at least a tenth of the lifespan of a species) or the effects resulting from such exposure. Long-term effects related to changes e.g. in metabolism, growth, reproduction, or ability to survive. Exposure concentrations are usually low. *See also* acute and subchronic toxicity.

chronic toxicity Chronic toxicity refers to the long-term adverse effects on an organism following constant dosing of a toxicant over a significant time period. The endpoint may be lethality, in which case, the dose or concentration causing a 50% effect (LD50 or LC50) may be determined. In addition to lethality, additional information may be gained from a chronic toxicity study, for example organ toxicities and tumour formation. *See also* chronic value and NOEC.

chronic value The geometric mean of the NOEC and LOEC in tests with a chronic exposure.

clastogens Agents which cause chromosomes to break.

clearance The process of losing a chemical substance from a test organism. Also called depuration or elimination.

clone A large number of cells or molecules genetically identical to a single ancestral cell or molecule.

CNS Central Nervous System.

coefficient of variation (CV) The standard deviation of a sample relative to the mean. It can be expressed as a fraction or as a percentage.

cohort A group of individuals, identified by a common characteristic, that are studied over a period of time.

cometabolism Process by which a normally non-biodegradable substance is biodegraded only in the presence of an additional carbon source.

community Collection of populations living together in same place and at same time which, therefore, may interact with one another. *See* ecosystem.

competent authority In the context of EC legislation it means the authority or authorities or bodies established by the Member States to carry out the obligations arising from a Regulation or a Directive.

compliance In accordance with legislative or regulatory requirements.

concentration The quantifiable amount of a chemical in air, water, food, sediment, tissue, or any other medium.

concentration-effect relationship Relationship between the exposure, expressed in concentration, of a given organism, system or (sub) population to an agent in a specific pattern during a given time and the magnitude of a continuously-graded effect to that organism, system or (sub) population. *See also* effect assessment, dose-response relationship.

conductivity A numerical expression of the ability of an aqueous solution to carry an electric current. This ability depends on the concentration of ions in solution, their valence and mobility, and the temperature of the solution. Conductivity is normally reported in the SI unit millisiemens/metre, or as micromho/centimetre (1 mS/m = 10 mmhos/cm).

confidential business information (CBI) Information that a manufacturer/importer or downstream user is not free to disclose, in order not to harm their competitive ability on the market.

congeners (1) Substances whose structure, function or origin are similar to others and may match the same structure-activity relationship (SAR), and (2) refers to species which belong to the same genus.

conjugate A water soluble derivative of a chemical formed by its combination with glucuronic acid, glutathione, sulphate, acetate, glycine, etc.

continuous effect A response that can be measured on a continuum from zero (or even a negative value) to positive values such as growth and reproduction. *See* quantal effect.

continuous flow Tests in which solutions in test vessels are renewed continuously by the constant inflow of a fresh solution, or by a frequent intermittent inflow. *See also* flow-through.

control A treatment in a toxicity test that duplicates all the conditions of the exposure treatments but contains no test material. The control is used to determine the absence of toxicity under basic test conditions (e.g. health of test organisms, quality of dilution water).

control/dilution water The water used for diluting the test substance, or for the control test, or both.

control limit The limiting airborne concentration of potentially toxic substances which is judged to be "reasonably practicable" in the working environment and which must not normally be exceeded.

convection A type of fluid flow like advection but likely to occur in high-energy environments (e.g., streams) where sand ripples may be found and contaminated sediment are not likely to be deposited.

corrosion The process of contact damage due to a destructive agent in tissues.

cost-benefit analysis (CBA) The procedure for determining whether the expected benefits of a proposed action outweigh the expected costs.

covalent binding An irreversible interaction between xenobiotics or their metabolites and macromolecules such as lipids, proteins, nucleic acids.

criterion The level of exposure (concentration and duration) of a contaminant in a particular medium that is thought to result in an acceptably low level of effect on populations, communities, or use of the medium (e.g. water-quality criteria, air-quality criteria).

critical body residue (CBR) The whole-body concentration of a chemical that is associated with a given adverse biological response. This assumes organisms are a single compartment, rather than the multiple compartments that they actually are, but it has considerable utility as a first approximation of dose.

CSA *See* chemical safety assessment.

CSR *See* chemical safety report.

culture A stock of animals or plants raised under well-defined and controlled conditions to produce healthy test organisms. As a verb, it means to carry out this procedure of raising organisms.

cytochrome P-450 A haemprotein involved in phase-I reactions of xenobiotics.

cytogenetics The branch of genetics that correlates the structure and number of chromosomes with heredity and variation.

cytoplasm Cell contents, in which the nucleus, endoplasmic reticulum, mitochondria and other organelles are found.

cytotoxic Causing disturbance to cellular structure or function, often leading to cell death.

damage A loss of inherent quality suffered by an entity. *See also* harm.

dechlorinated water Chlorinated water (usually municipal drinking water) that has been treated to remove chlorine and chlorinated compounds from solution.

degradation Chemicals that are released in the environment are subject to different (biotic and abiotic) degradation processes: biodegradation by microorganisms, photolysis by light, hydrolysis by water, oxidation by different oxidants (for instance, in the atmosphere by hydroxyl and nitrate radicals or by ozone). These degradative processes are usually modelled in terms of the rate constants of the corresponding chemical reactions.

degradation rate constant A first order or pseudo first order kinetic constant, k (d^{-1}), which indicates the rate of degradation processes. For a batch experiment K is estimated from the initial part of the degradation curve obtained after the end of the lag phase.

deionized water Water that has been purified to remove ions from solution by passing it through ion exchange resin columns or a reverse osmosis system.

delayed effects Effects or responses that occur some time after exposure. Carcinogenic effects of chemicals typically have a long latency period; the occurrence of a tumour may take years after the initial exposure.

***de minimis* risk** A risk that is too small to be of real concern to society (negligible risk). Risks below 10^{-5} or 10^{-6} are generally viewed as *de minimis* in the US.

demography The study of populations, especially their age structure and growth rates.

denitrification Reduction of nitrate to nitrite, nitrous oxide or dinitrogen (N_2) catalyzed by facultative aerobic bacteria under anaerobic conditions.

density dependence A change in the influence of any factor (a density-dependent factor) that influences population growth as population density changes.

deoxyribonucleic acid (DNA) The constituent of chromosomes which stores the hereditary information of an organism in the form of a sequence of nitrogenous bases. Much of this information relates to the synthesis of proteins.

depuration The loss of a substance from an organism due to elimination and degradation. The rate of depuration is expressed by its half-life or the time needed to eliminate 50% of the substance in a non-contaminated medium. This term is often referred to as the depuration time (DT50).

derived characteristics Properties of a chemical that are defined by, dependent upon, or are approximations of fundamental properties and the prevailing environmental conditions.

dermal corrosion Defined in OECD test guideline 404 as the production of irreversible damage to skin; namely, visible necrosis through the epidermis and into the dermis, following the application of a test substance for up to four hours. Corrosive reactions are typified by ulcers, bleeding, bloody scabs, and, by the end of observation at 14 days, by discolouration due to blanching of the skin, complete areas of alopecia, and scars.

dermal irritation. Defined in OECD test guideline 404 as the production of reversible damage of the skin following the application of a test substance for up to 4 hours.

detergent A cleaning or wetting agent which possesses both polar and non-polar terminals or surfaces allowing interaction with non-polar molecules making them miscible with a polar solvent.

determinants of exposure Main parameters required for estimating exposure of humans or the environment. Such parameters may pertain to the exposure scenario (process conditions), worker and consumer exposure conditions, environmental conditions or properties of substances and preparations. A list of "determinants of exposure" is used to systematically check the life-cycle of the chemical for relevant exposure and which information is needed to perform the exposure assessment.

deterministic analysis An analysis in which all population and environmental parameters are assumed to be constant and accurately specified.

deterministic model A mathematical model which is fully specified and does not include a stochastic component.

detoxification (1) A process which renders a toxic molecule less toxic by biotransformation, removal, or the masking of active functional groups, and (2) the treatment of patients suffering from poisoning in order to reduce the probability or severity of harmful effects.

detritus Organic debris from decomposing plants and animals.

detrivorous Organisms living on detritus.

diffuse source Emission sources which are not point sources.

diffusion (nonbiological) Nonadvective transport due to migration and mixing of dissolved suspended solutes (including particulates) in natural waters in response to concentration gradients. Diffusion can be at the molecular level, due to Brownian motion producing random movements of the solute molecules (molecular diffusion), or it can be movements of solutes (including particles) due to turbulent eddies, velocity shear, or bioturbation (turbulent diffusion). The two types of diffusion result in mixing and dispersal of dissolved and bound chemicals.

direct toxicity Toxicity that results from and is readily attributable to the toxic agents(s) acting more or less directly at the sites of toxic action in and/or on the exposed organisms that are exhibiting the adverse biological response in question.

disaster (1) An act of nature or an act of man which is or threatens to be of sufficient severity and magnitude to warrant emergency assistance, and (2) a disruption of the human environment which the affected community cannot absorb and manage with its own resources.

dissolved organic carbon (DOC) The fraction of the organic carbon pool that is dissolved in water and that passes trough a 0.45 μm glass fibre filter. DOC quantifies the chemically reactive fraction and is an accurate measure of the simple and complex organic molecules making up the dissolved organic load. The majority of the DOC is humic substances.

dissolved organic matter (DOM) It is analogous to DOC (*see* DOC), but it refers to the entire organic pool that is dissolved in water.

dispersant A chemical substance which reduces the surface tension between water and a hydrophobic substance (e.g. oil), thereby facilitating the dispersal of the hydrophobic substance throughout the water as an emulsion.

distilled water Water that has been passed through a distillation apparatus to remove impurities.

distribution Dispersal of a xenobiotic and its derivatives throughout an organism or environmental matrix, including tissue binding and localization.

distributor In the context of REACH means any natural or legal person established within the Community, including a retailer, who only stores and places on the market a substance, on its own or in a preparation, for third parties.

diversity index Measure of richness of biota (number of taxa) and, usually, the evenness of their distribution in communities.

DOC *See* dissolved organic carbon.

DOM *See* dissolved organic matter.

domain of applicability The applicability domain (AD) of a (Q)SAR model is the response and chemical structure space in which the model makes predictions with a given reliability.

dose Total amount of an agent administered to, taken up or absorbed by an organism, system or (sub) population. *See also* external dose and internal dose.

dose-effect relationship Relationship between the total amount of an agent administered to, taken up or absorbed by an organism, system or (sub) population and the magnitude of a continuously-graded effect to that organism, system or (sub) population. *See also* effect assessment, dose-response relationship, concentration-effect relationship.

dose-related effect Any effect to an organism, system or (sub) population as a result of the quantity of an agent administered to, taken up or absorbed by that organism, system or (sub) population.

dose-response Relationship between the amount of an agent administered to, taken up or absorbed by an organism, system or (sub) population and the change developed in that organism, system or (sub) population in reaction to the agent. *See also* dose-effect relationship, dose-effect assessment, effect assessment, concentration-effect relationship.

dose-response assessment Analysis of the relationship between the total amount of an agent administered to, taken up or absorbed by an organism, system or (sub) population and the change developed in that organism, system or (sub) population in reaction to that agent, and inferences derived from such an analysis with respect to the entire population. Dose-response assessment is the second of four steps in risk assessment. *See also* hazard characterization, dose-effects relationship, effect assessment, dose-response relationship, concentration effect relationship.

dose-response curve Graphical presentation of a dose-response relationship.

dose-response relationship Relationship between the total amount of an agent administered to, taken up or absorbed by an organism, system or (sub) population and the changes developed in that organism, system or (sub) population in reaction to the agent. *See also* dose-effect relationship, effect assessment, concentration effect relationship.

downstream user In the context of REACH means any natural or legal person established within the Community, other than the manufacturer or the importer, who uses a substance, either on its own or in a preparation, in the course of his industrial or professional activities. A distributor or a consumer is not a downstream user. A re-importer exempted pursuant to Article 2(4)(c) of REACH shall be regarded as a downstream user; EC European Communities.

ECETOC European Centre for Ecotoxicology and Toxicology of Chemicals.

ecosystem Collection of populations (microorganisms, plants and animals) that occur in the same place at the same time and can, therefore, potentially interact with each other as well as their physical and chemical environments and thus form a functional entity.

ecotoxicology The study of the toxic effects of chemical and physical agents in living organisms, especially on populations and communities within defined ecosystems; it includes the transfer pathways of these agents and their interaction with the environment.

ECx Effective concentration; the concentration which affects X% of a testpopulation after a specified exposure time. The EC50 usually relates to effects other than lethality (e.g. immobilization, growth rate, equilibrium, developmental abnormality or deformity) in 50% of the test organisms. The effect concentration may refer to other percentages such as 10%, 70%, e.g. EC10 and EC70 etc.

ED50 Dose that affects a designated criterion (e.g. behavioural trait) of 50% of the population observed. Also known as median effect concentration/dose. *See also* ECx, LC(D)50 and IC(D)50.

edaphic Pertaining to the soil.

eddy diffusion Irregularity in the diffusion of solute molecules which occurs in a porous chromatographic support. The phenomenon is due to the fact that (a) the pathlengths of some solute molecules are either shorter or longer than those of most of the molecules, and (b) the rate of solvent flow varies in different regions of the porous support.

EEC *See* estimated (or expected) environmental concentration.

effect Change in the state or dynamics of an organism, system or (sub) population caused by the exposure to an agent.

effect assessment Combination of analysis and inference of possible consequences of the exposure to a particular agent based on knowledge of the dose-effect relationship associated with that agent in a specific target organism, system or (sub) population.

effluent A complex waste material (e.g., liquid industrial discharge or sewage) that may be discharged into the environment.

EINECS European Inventory of Existing Commercial Chemical Substances: a list of all chemicals either separately or as components in preparations supplied to a person in an EC Member State at any time between 1 January 1971 and 18 September 1981.

electrophilicity Electrophilicity is the molecular or substructural property of having an attraction for electrons or negative charge. Molecular electrophilicity is often described by the molecular orbital properties the energy of the lowest unoccupied molecular orbital (E_{LUMO}) and superdelocalisabilities.

elimination The combined process of metabolism and excretion which results in the removal of a compound from an organism.

elimination rate constant (k_e) The first-order one-compartment constant to describe the overall clearance of a chemical substance from an organism.

ELS Early Life Stage.

elutriate An aqueous solution obtained after adding water to a solid substance (e.g. sediment, tailings, drilling mud, or dredging spoil), shaking the mixture, then centrifuging or filtering it or decanting the supernatant.

embryo The undeveloped young animal, before it is born or hatches from the egg. *See also* foetus.

emission Release of a substance from a source, including discharges into the wider environment.

emission standard A quantitative limit on the emission or discharge of a potentially toxic substance from a particular source. The simplest example is a uniform emission standard where the same limit is placed on all emissions of a particular contaminant. *See also* limit value.

emulsifier A chemical substance that aids the fine mixing (in the form of small droplets) with water of an otherwise hydrophobic substance. *See also* dispersant.

endocrine Pertaining to hormones or glands that secrete hormones directly into the blood stream.

endogenous Arising within or derived from the organism.

endoplasmic reticulum A complex pattern of membranes that permeates the cytoplasmic matrix of cells and in which biotransformation reactions of the mono-oxygenase enzyme systems occur. May be isolated as microsomes following cell fractionation procedures.

endpoint In toxicity testing and evaluation it is the adverse response in question that is measured. Endpoints vary with the level of biological organization being examined but include changes in biochemical markers or enzyme activities, mortality or survival, growth, reproduction, primary production, and changes in structure (and abundance) and function in a community. Endpoints are used in toxicity tests as criteria for effects.

endpoint assessment A quantitative or quantifiable expression of the environmental value considered to be at risk in a risk analysis. Examples include a 25% or greater reduction in gamefish biomass or local extinction of an avian species.

enthalpy Heat content or the thermodynamic function H in $H = E+PV$, where E is the internal energy of the system, P is the pressure exerted on the system and V is the volume of the system.

entropy (1) A thermodynamic quantity that changes in a reversible process by an amount equal to the heat absorbed or emitted divided by the thermodynamic temperature, (2) function that is a measure of that part of the system that cannot perform useful work, and (3) the degree of randomness or disorder of a system.

environmental availability The portion of the total chemical material present in all or part of the environment that is actually involved in particular processes and is subject to all physical, chemical, and biological modifying influences. It defines the total amount of material potentially available to organisms.

environmental bioavailability The ratio of the uptake clearance divided by the rate at which an organism encounters a given contaminant in a medium (e.g., water, food) being processed by the organism. This is a measure of an organism's extraction efficiency, via respiratory, dietary, and surface absorption processes, from the environmentally available portion of a material.

environmental compartments Subdivision of the environment which may be considered as separate boxes, and which are in contact with each other. A simple model would separate the environment into air, water, and soil, with biota, sediment (bottom and suspended), layering of water bodies and many refinements being allowed if data to support their inclusion are available. *See also* Chapters 3 and 4.

environmental fate Destiny of a chemical or biological pollutant after release into the natural environment.

environmental impact assessment A type of assessment that attempts to reveal the consequences of proposed actions as an aid to decisionmaking.

environmental quality objective (EQO) The quality to be aimed for in a particular aspect of the environment. For example, the quality of water in a river should be such that coarse fish can maintain healthy populations. *See also* quality objective.

environmental quality standard (EQS) The concentration of a potentially toxic substance which is permitted in an environmental component, usually air (air quality standard) or water, over a defined period. Synonym: ambient standard. *See also* limit value.

environmental risk analysis Determination of the probability of adverse effects on humans and other biota resulting from an environmental hazard (a chemical, physical, or biological agent occurring in or mediated by the environment).

environmental transport The movement of contaminants from their point of release through the various compartments to locations where exposure is assumed to occur.

enzyme A protein which acts as a highly selective catalyst permitting reactions to take place rapidly in living cells under certain physiological conditions.

enzyme induction *De novo* synthesis of an enzyme or activation of an existing enzyme.

enzyme inhibition A process leading to the reduced activity of an enzyme.

enzymic (or enzymatic) process A chemical reaction or series of reactions catalyzed by an enzyme or enzymes.

EPA Environmental Protection Agency.

epibenthic Living on the bed of aquatic systems.

epidemiology The study of the incidence, distribution and causes of disease, or the statistical study of categories of persons and the patterns of diseases from which they suffer in order to determine the events or circumstances causing these diseases.

epifauna Living on the surface of sediment in aquatic systems. *See also* infauna.

epigenetic changes Changes in an organism brought about by alterations in the action of genes. Epigenetic transformation refers to those processes which cause normal cells to become tumour cells without any mutations occurring. *See also* mutation, transformation and tumour.

episodic Discontinuous effect, e.g. due to accidental spill or periodic stormwater discharges from sewers.

EPPO European and Mediterranean Plant Protection Organization.

equilibrium In a thermodynamic sense this indicates that both a steady state of flux and an equivalence in chemical activity have been reached in compartments or phases separated by a membrane or boundary across which the chemical fluxes occur. *See also* steady-state.

equilibrium partitioning (EqP) An approach for estimating the fate of chemicals (primarily organics) in the aquatic environment that is based on the assumption than a steady-state can be achieved, and usually is achieved, between the activity of chemicals (usually approximated as concentration) in the various component phases-water, sediment, organisms. The EqP approach is often exploited, for interpretation and extrapolation purposes, by normalizing chemical concentrations based on the lipid content of the aquatic organisms and the organic carbon content of the sediments. These normalized BSAF values are considered to be independent of particular sediments and species.

ESIS European chemical Substances Information System.

estimated (or expected) environmental concentration (EEC) The concentration of a material estimated to be likely to occur in environmental waters to which aquatic organisms are exposed as a result of planned manufacture, use, and disposal.

ET50 The time it takes for a toxicant to affect 50% of a population with respect to a specific criterion.

EU European Union.

eukaryote An organism (e.g. plant or animal) whose cells contain a membrane-bound nucleus and other membranous organelles. *See also* prokaryote.

European Inventory of Existing Commercial Chemical Substances *See* EINECS.

EUSES European Union System for the Evaluation of Substances.

eutrophic Nutrient rich (aquatic) system with a high or excessive rate of biological production. *See also* oligotrophic.

eutrophication A complex series of interrelated changes in the chemical and biological status of a water body, most often manifested as a depletion of the oxygen content caused by the decay of organic matter resulting from a high level of primary productivity and typically caused by enhanced nutrient input.

excretion Removal of a substance or its metabolites from an organism by the discharge of biological material, including urine, faeces, expired air, mucus, milk, eggs, and perspiration.

existing chemicals Chemicals listed in the EINECS (EC legislation). *See also* EINECS.

exogenous Resulting from events or derived from materials external to an organism. *See also* endogenous.

expected environmental concentration The calculated concentration of a chemical in a particular medium at a particular location at a particular time.

expert judgement Opinion of an authoritative person on a particular subject.

expert system Any formalised system, not necessarily computer-based, which enables a user to obtain rational predictions about the properties or activities of chemicals. All expert systems for the prediction of chemical properties or activities are built upon experimental data representing one or more effects of chemicals in biological systems (the database), and/or rules derived from such data (the rulebase).

exponential growth The growth of cells, organisms or populations in which the number/mass increases exponentially and growth at any time is proportional to the number/mass present.

exposure Concentration or amount of a particular agent that reaches a target organism, system or (sub) population in a specific frequency for a defined duration.

exposure assessment Evaluation of the exposure of an organism, system or (sub) population to an agent (and its derivates). Exposure assessment is the third step in the process of risk assessment. In the context of REACH two tiers can be distinguished: (a) Tier 1 exposure estimation. This is the first exposure estimation step leading to risk characterisation with the purpose of assessing whether risks are adequately controlled. In the first tier of the exposure estimation, simpler assessment steps, model interfaces and models are used to identify potential risks. The main goal of a tier 1 model is to allow simplified (automated) use of models, thus increasing the efficiency of the risk assessment process. (b) Tier 2 exposure estimation. This is the second, more refined exposure estimation step leading to risk characterisation with the purpose of assessing whether risks are adequately controlled. Tier 2 exposure models may introduce additional model parameters, a more complicated interface or more complicated model.

exposure model A conceptual or mathematical representation of the exposure process.

exposure pathway The course an agent takes from the source to the target.

exposure period The time of continuous contact between an agent and a target.

exposure route The way an agent enters a target after contact (e.g., by ingestion, inhalation, or dermal absorption).

exposure scenario (1) According to the definition of OECD and IPCS an exposure scenario is a set of conditions or assumptions about sources, exposure pathways, amount or concentrations of agent(s) involved, and exposed organism, system or (sub)population (i.e. numbers, characteristics, habits) used to aid in the evaluation and quantification of exposure(s) in a given situation. (2) In the context of REACH exposure scenario means the set of conditions, including operational conditions and risk management measures, that describe how the substance is manufactured or used during its life-cycle and how the manufacturer or importer controls, or recommends downstream users to control, exposures of humans and the environment. These exposure scenarios may cover one specific process or use or several processes or uses as appropriate. *See also* boundary of exposure scenario, broad exposure scenario, final exposure scenario and iteration.

external dose Amount of a chemical that is inhaled, ingested, or comes in dermal contact and is available for systemic absorption. External dose is usually expressed in units of mg of chemical per kg body weight per day (mg/kg/d) *See also* internal dose.

extinction probability The probability that a population will become extinct within a specified interval of time.

extrapolation An estimation of a numerical value of an empirical (measured) function at a point outside the range of data used to calibrate the function or the use of data derived from observations to estimate values for unobserved entities or conditions.

extrapolation factor A quantity used in effect and exposure assessments to adjust estimated exposures or concentrations/doses for uncertainties, to make corrections in the data, or to improve safety.

eye corrosion Defined in OECD test guideline 405 as the production of tissue damage in the eye, or serious physical decay of vision, following application of a test substance to the anterior surface of the eye, which is not fully reversible within 21 days of application.

eye irritation Defined in OECD test guideline 405 as the production of changes in the eye following application of a test substance to the anterior surface of the eye, which are fully reversible within 21 days of application.

FAO Food and Agriculture Organization (of the United Nations).

fate Pattern of distribution of an agent, its derivates or metabolites in an organism, system, compartment or (sub) population of concern as a result of transport, partitioning, transformation or degradation.

FDA (US) Food and Drug Administration.

fecundity (1) Ability to produce offspring frequently and in large numbers, and (2) in demography, the physiological ability to reproduce.

FIFRA (US) Federal Insecticide, Fungicide and Rodenticide Act.

final exposure scenario The final exposure scenario in the context of REACH is the outcome of the "tentative" exposure scenario and the iterative exposure estimate and risk characterisation in the chemical safety assessment process. It specifies the operational conditions and risk management measures required for adequate control of risks.

first-order process A chemical process in which the rate of reaction is directly proportional to the amount of chemical present.

first-order reaction A chemical reaction in which the velocity of the reaction is proportional to the concentration of one reactant.

first-pass effect Biotransformation of a xenobiotic before it reaches the systemic circulation. The biotransformation of an intestinally absorbed xenobiotic chemical by the liver is referred to as a hepatic first-pass effect.

fitness When used in a Darwinian sense, refers to capacity to reproduce and survive.

flocculation The formation of a light, loose precipitate (i.e. a floc) from a solution.

flow-through Tests in which solutions in test vessels are renewed continuously by the constant inflow of a fresh solution, or by a frequent intermittent inflow. *See also* continuous flow.

foci A small group of cells occurring e.g. in the liver, which are distinguishable in appearance or histochemically from the surrounding tissue. They are indicative of a lesion in its early stage, which may lead to the formation of neoplastic nodules or hepatocellular carcinomas.

foetus (fetus) The young of mammals when fully developed in the womb. In human beings, this stage is reached after about three months of pregnancy. Prior to this, the developing mammal is in the embryo stage.

Freundlich adsorption isotherm An empirical equation that describes the adsorption of a contaminant to soil. The equation is as follows: $x/m = K_f C_e^{1/n}$, where x/m is the concentration of the contaminant in soil (mg/kg), C_e is the contaminant concentration in the aqueous phase at equilibrium (mg/L), K_f is the equilibrium constant (the Freundlich adsorption constant) and 1/n is the contaminant-specific exponent. *See also* Chapter 3.

fugacity (1) The tendency for a substance to transfer from one environmental medium to another. (2) Analogous to chemical potential as it pertains to the tendency of a chemical to escape from a phase (e.g. from water).

full study report In the context of REACH means a complete and comprehensive description of the activity performed to generate the information. This covers the complete scientific paper as published in the literature describing the study performed or the full report prepared by the test house describing the study performed.

functional group Organic compounds can be thought of as consisting of a relatively unreactive backbone and one or more functional groups. The functional group is an atom, or a group of atoms, which has similar chemical properties whenever it occurs in different compounds. It defines the characteristic physical and chemical properties of families of organic compounds.

fungicide *See* pesticide.

gametes The eggs or sperm obtained from mature adult animals.

gastrointestinal Pertaining to or communicating with the stomach and intestine.

GEMs Genetically Engineered Microorganisms.

gene A part of the DNA molecule which directs the synthesis of a specific polypeptide chain.

generation time The average length of time between the birth of parents and the birth of offspring.

genetic toxicology The study of chemicals which can produce harmful hereditary changes in the genetic information carried by living organisms, in the form of deoxyribonucleic acid (DNA).

genome Chromosomal DNA information.

genomics Techniques available to identify the DNA sequence of the genome.

genotoxicity Ability to cause damage to genetic material or an adverse effect in the genome, e.g. mutation, chromosomal damage, etc. that may lead to a cancer. *See also* carcinogenicity and mutagenicity.

genotype The genetic constitution of an organism. *See also* phenotype.

Gibbs free energy That component of the total energy of a system that can do work under conditions of constant temperature and pressure. The Gibbs free energy (G) is expressed by the thermodynamic function $G = H\text{-}TS$, where H is the enthalpy, T is the absolute temperature and S is the entropy.

good laboratory practice (GLP) Fundamental rules incorporated in national regulations concerned with the process of effective organization and the conditions under which laboratory studies are properly planned, performed, monitored, recorded and reported.

growth An increase in size or weight as a result of proliferation of new tissues.

guidance value Value, such as concentration in air or water, which is derived after allocation of the reference dose among the different possible media (routes) of exposure. The aim of the guidance is to provide quantitative information from risk assessment to risk managers to enable them to make decision. *See also* reference dose.

half-life The half-life (commonly denoted as $t_{1/2}$) is the time interval that corresponds to a concentration decrease by a factor 2. Environmental half-life data generally reflect the rate of disappearance of a chemical from a medium, without identifying the mechanism of chemical loss. For example, loss of water may be due to a combination of evaporation, biodegradation and photolysis. If the elimination rate involves transport and transformation processes that follow first-order kinetics, the half-life time (d) and elimination rate constant k (d^{-1}) are related by the equation $t_{1/2} = -\ln 2/k = 0.693/k$.

Hansch analysis Hansch analysis is the investigation of the quantitative relationship between the biological activity of a series of compounds and their physicochemical substituent or global parameters representing hydrophobic, electronic, steric, and other effects, using a multiple regression method.

haploid The condition in which a cell con-tains only one set of chromosomes.

hardness The concentration of all cations in water that will react with a sodium soap to precipitate an insoluble residue. In general, hardness is a measure of the concentration of calcium and magnesium ions in water and is frequently expressed as mg/L calcium carbonate or equivalent.

harm (1) A loss to a species or individual as a result of damage. (2) A function of the concentration to which an organism is exposed and the time of exposure.

hazard Inherent property of an agent or situation having the potential to cause adverse effects when an organism, system or (sub) population is exposed to that agent. *See also* risk.

hazard assessment A process designed to determine the possible adverse effects of an agent or situation to which an organism, system or (sub) population could be exposed. The process includes hazard identification and hazard characterization. The process focuses on the hazard in contrast to risk assessment where exposure assessment is a distinct additional step.

hazard characterization The qualitatitive and, wherever possible, quantitative description of the inherent properties of an agent or situation having the potential to cause adverse effects. This should, where possible, include a dose-response assessment and its attendant uncertainties. Hazard characterization is the second stage in the process of hazard assessment, and the second step in risk assessment. *See also* dose-effect relationship, effect assessment, dose-response relationship, concentration–effect relationship.

hazard identification The identification of the type and nature of adverse effects that an agent has as inherent capacity to cause in an organism, system or (sub) population. Hazard identification is the first stage in hazard assessment and the first in the process of risk assessment.

hazard quotient The PEC/PNEC ratio, i.e. the definition of environmental and/or health risks by combining the results of the exposure assessment (PECs) with the results of the effect assessment (PNECs or NOAEL). Although there is a clear difference between hazard and risk, hazard and risk quotients are often used synonymously.

HCp (HC5) Hazardous concentration for p% (5%) of the species, derived by means of a statistical extrapolation procedure. *See* Chapter 7.

Henry's law constant The Henry's law constant (H) is an air-water partition coefficient that expresses the tendency of a chemical to volatilise from an aqueous medium. It can be determined by measurement of the solute concentrations in both phases. Due to the difficulty of accurate analytical determination, the H constant is mainly calculated as the ratio of vapour pressure to solubility. *See also* Chapters 3 and 9.

Henry's law Inchemistry. Henry's law is one of the gaslaws, formulated by William Henry. It states that: at a constant temperature, the amount of a given gas dissolved in a given type and volume of liquid is directly proportional to the partial pressure of that gas in equilibrium with that liquid.

hepatocytes Liver cells.

hepatotoxicity Toxicity to the liver.

herbicides *See* pesticides.

histology The study of the anatomy of tissues and their cellular structure.

histopathology The study of adverse changes in the structure of tissues, usually using a microscope.

humic acid Humic substance that is insoluble at acidic pHs.

homeostasis The tendency of an organism to maintain physiological and psychological stability.

hormesis An improvement in the state or performance of an organism in response to low levels of exposure to a chemical that is toxic at higher exposure levels and is not a nutrient.

HPVC High Production Volume Chemical. *See also* EINECS and Chapters 12, 13 and 16.

hybridisation Formation of a double strand from two different, more or less complementary single nucleic acid strands.

Hydra Genus of small freshwater hydrozoan coelenterates. It is a ubiquitous genus with few species.

hydrophilic Describes the character of a molecule or atomic group which has an affinity for water.

hydrophilicity Hydrophilicity refers to the affinity of a molecule or substitutent for a polar solvent (especially water) or for polar groups. It represents the tendency of a molecule to be solvated by water.

hydrophobic Describes the character of a molecule or atomic group which has a tendency to repel water.

hydrophobicity Hydrophobicity refers to the association of non-polar groups or molecules in an aqueous environment, which arises from the tendency of water to exclude non-polar molecules. It is related to lipophilicity. It represents the tendency of a molecule to partition between a polar and a non-polar phase, and is therefore often measured by a partition coefficient between a polar and non-polar phase (usually, but not always, octanol and water).It is often highly related to biological activity due to its strong relationship with the transport and distribution of a molecule, particularly through phospholipid membranes.

hydrosphere Water above, on or in the earth's crust, including oceans, seas, lakes, groundwater and atmospheric moisture.

hyper Prefix meaning above or excessive.

hypoxic (1) Abnormally low in oxygen content or tension, or (2) deficiency of oxygen.

IC(D)50 Concentration that induces a 50% inhibition of a designated process in an exposed population. Also known as median inhibitory concentration/dose. *See also* EC(D)50 and LC(D)50.

ICp The inhibiting concentration to produce a (specified) percentage effect. It represents a point estimate of the concentration of test substance that causes a designated percent impairment in a quantitative biological function such as growth, reproduction or respiration. For example, an IC25 could be the concentration estimated to cause a 25% reduction in growth of fish, relative to the control.

identified use In the context of REACH means a use of a substance on its own or in a preparation, or a use of a preparation, that is intended by an actor in the supply chain, including his own use, or that is made known to him in writing by an immediate downstream user.

idiosyncrasy Specific (and usually unexplained) reaction of an individual to e.g. a chemical exposure to which most other individuals do not react at all. Example: some people react to their very first aspirin with a potentially fatal shock. General allergic reactions do not fall into this category.

immune response Selective reaction by the body to substances that are foreign to it or that the immune system identifies as foreign, as shown by the production of antibodies and antibody-bearing cells or by a cell-mediated hypersensitivity reaction.

immunotoxic Poisonous to the immune system.

immunotoxicology The science that deals with the immunotoxic effects of chemicals.

impermeable The extent to which the membrane, skin or exoskeleton prevents the passage of molecules (e.g., water, ions, proteins, fats or toxicants).

import In the context of REACH means the physical introduction into the customs territory of the Community.

importer In the context of REACH means any natural or legal person established within the Community who is responsible for import.

incipient LC50 The concentration of a chemical which is lethal to 50% of the test organisms as a result of exposure for periods long enough for acute lethal action to cease. The asymptote (part of the toxicity curve parallel to the time axis) of the toxicity curve approximately indicates the value of the incipient LC50.

incremental unit risk estimate For an air pollutant, this is the additional lifetime cancer risk occurring in a hypothetical population in which all individuals are exposed continuously from birth throughout their lifetimes to a concentration of 1 $\mu g/m^3$ of the pollutant in the air they breathe.

indicator A characteristic of the environment, e.g. a species, that provides evidence of the occurrence or magnitude of exposure or effects. Formal expressions of the results of measuring an indicator are referred to as measurement endpoints. Abundance, yield, and age/weight ratios are indicators of population production. A low cholinesterase level is an indicator of exposure to cholinesterase-inhibiting pesticides.

indicator species A species that is surveyed or sampled for analysis because it is believed to represent the biotic community, some functional or taxonomic group, or some population that cannot be readily sampled or surveyed.

indirect toxicity Adverse effects or toxicity that results from the agent(s) acting on and producing changes in the chemical, physical, and/or biological environment external to the organisms under study (e.g., decrease in food for predatory species due to direct toxicity from a chemical to prey may produce adverse effects in the predator species due to starvation rather than inducing any direct chemical toxicity in predator organisms).

individual risk Probability that an individual person will experience an adverse effect.

induction Increase in the rate of synthesis of an enzyme in response to the action of an inducer (substance that causes induction) or environmental conditions, often the substrate of the induced enzyme or a structurally similar substance that is not metabolized.

infauna Lives in the sediment of aquatic systems. *See also* epifauna.

inhibition concentration (IC) A point estimate of the chemical concentration that would cause a given percent reduction (e.g., IC25) in a nonlethal biological measurement of the test organisms, such as reproduction or growth.

initiating event The specific action that results in a risk being incurred.

initiation The ability of an agent to induce a change in tissue which leads to the induction of tumours after a second agent, called a promotor, is repeatedly administered to the tissue. *See also* promotor.

initiator An agent which starts the process of tumour development, usually by acting on the genetic material.

insecticides *See* pesticides.

interindividual variability Biological variation between people.

intake The process by which an agent crosses an outer exposure surface of a target without passing an absorption barrier, i.e. through ingestion or inhalation. *See* dose.

intermediate In the context of REACH means a substance that is manufactured for and consumed in or used for chemical processing in order to be transformed into another substance (hereinafter referred to as "synthesis"). REACH distinguishes 3 types of intermediates: **1. Non-isolated intermediate** means an intermediate that during synthesis is not intentionally removed (except for sampling) from the equipment in which the synthesis takes place. Such equipment includes the reaction vessel, its ancillary equipment, and any equipment through which the substance(s) pass(es) during a continuous flow or batch process as well as the pipework for transfer from one vessel to another for the purpose of the next reaction step, but it excludes tanks or other vessels in which the substance(s) are stored after the manufacture. **2. On-site isolated intermediate** means an intermediate not meeting the criteria of a non-isolated intermediate and where the manufacture of the intermediate and the synthesis of (an)other substance(s) from that intermediate take place on the same site, operated by one more legal entities. **3. Transported isolated intermediate** means an intermediate not meeting the criteria of a non-isolated intermediate and transported between or supplied to other sites.

internal dose In exposure assessment, the amount of a substance penetrating the absorption barriers (e.g. skin, lung tissue, gastrointestinal tract) of an organism through either physical or biological processes. *See also* external dose.

interstitial water The water in sediment or soil that surrounds the solid particles. The amount of interstitial water is calculated and expressed as the percentage ratio of the weight of water in the sediment to the weight of the wet sediment.

intra-individual variability Biological variation within people.

in vitro In glass, referring to studies in the laboratory usually involving isolated organs, tissues, cells or biochemical systems.

in vivo Within the living organism.

IPCS International Programme on Chemical Safety.

IRPTC International Register of Potentially Toxic Chemicals.

ischaemia Local deficiency in the blood supply and hence oxygen to an organ or tissue due to constriction or obstruction of the blood vessels.

ISO International Organization for Standardization.

IT50 Time required for a toxicant to inhibit a specified process in 50% of the observed population. Also known as median inhibitory time. *See also* ET50 and LT50.

iteration In the context of risk assessment it is a cycle in the exposure assessment to change assessment elements: hazard information, risk management measures, defaults of the exposure assessment, conditions of use, etc. As a result, the exposure scenario can be adapted to reflect the additional information and insight gained. *See also* exposure scenario.

i.v. Abbreviation for intravenous (administration).

joint action Two or more chemicals exerting their effects simultaneously.

karyotoxicity Disruption of chromosomal structure.

knockout animals Genetically engineered animals in which one or more genes, usually present and active in the normal animal, are absent or inactive.

lag phase The time from the start of a test until adaptation of the degrading micro-organisms is achieved and the biodegradation degree of a chemical substance or organic matter has increased to a detectable level (e.g. 10% of the maximum theoretical biodegradation, or lower, dependent on the accuracy of the measuring technique).

Langmuir adsorption isotherm An equation that describes the adsorption of a gas onto a solid which takes the same mathematical form as the Michaelis-Menten equation.

larva A recently hatched fish or other organism that has different physical characteristics than those seen in the adult.

LC(D)50 The median lethal concentration/dose (i.e., the concentration/dose of substance that is estimated to be lethal to 50% of the test organisms). The LC50 and its 95% confidence limits are usually derived by statistical analysis of mortalities in several test concentrations, following a fixed period of exposure. The duration of exposure must be specified (e.g. 96-h LC50). *See also* median lethal concentration/dose.

LDn The dose of a toxicant lethal to n% of a test population.

leachate Water or wastewater that has percolated through a column of soil or solid waste in the environment.

lentic Non-flowing or still water; e.g. lakes, ponds.

lesion A pathological disturbance such as an injury, infection or a tumour.

lethal Causing death by direct action. Death of fish is often defined as the cessation of all visible signs of movement or other activity.

lethal body burden (LBB) The body residue of chemical that is associated with mortality in short-term exposures.

life-cycle Series of stages, from a given point in one generation to same point in next generation, e.g. egg-larva-adult-egg (hyphenated when used as an adjective, e.g. life-cycle strategy). *See also* life-cycle study and life history.

life-cycle study A chronic (or full chronic) study in which all the significant life stages of an organism are exposed to a test material. Generally, a life-cycle test involves the entire reproductive cycle of the organism.

life history Sometimes considered synonymous with life cycle, but some see it as segment of life cycle, e.g. egg to adult not egg to egg (hyphen used in the same way as with life cycle).

ligand A small organic molecule bound to a macromolecule in a stable, but not covalent, bond. It is used in toxicology, particularly to describe molecules being transported by blood proteins or molecules binding to the haem iron of haem proteins. The nature of ligand binding and the strength of the bonds involved are important.

limit test (screening) A test in which organisms are exposed to a maximal agreed upon substance concentration and a control to determine the potential toxicity of a toxicant.

limit value The limit at or below which Member States of the European Communities must set their environmental quality standards and emission standards. These limits are set by Community Directives.

linear free energy relationship (LFER) An LFER is an empirical relationship in which numerical parameters are associated with small perturbations in a parent molecule and are subsequently correlated with the change in the free energy of a certain reaction of the parent molecule versus the perturbed molecules. A classical example is the well-known Hammett equation, in which sigma values are constructed for specific substituents on a benzoic acid parent molecule. The linear correlation is then between the (summed) sigma values for a set of compounds and the pK_a (free energy of acid dissociation) of those compounds.

lipophilic (1) Having an affinity for fat and high lipid solubility, and (2) a physicochemical property which describes a partitioning equilibrium of solute molecules between water and an immiscible organic solvent, which favours the latter. Correlates with bioaccumulation.

lipophilicity Lipophilicity refers to the affinity of a molecule or of a substituent for a lipophilic environment.

liver nodule A small node, or aggregation of cells in the liver.

loading Ratio of animal bio-mass to the volume of test solution in an exposure chamber.

LOEC(L) Lowest Observed Effect Concentration (Level). The lowest concentration of a material used in a toxicity test that has a statistically significant adverse effect on the exposed population of test organisms compared with the controls. When derived from a life cycle or partial life-cycle test, it is numerically the same as the upper limit of the MATC. The LOEC is generally reserved for sublethal effects but can also be used for mortality, which might sometimes be the most sensitive effect observed. *See also* NOEC.

logistic curve A function, often applied to growth curves, fitting the general equation: $y = k/(1 + e^{a+bt})$, where t represents time, y the bw or population size and b is greater than 0. In the logistic equation the percentage rate of increase decreases linearly as size increases. The resulting curve continually rises, slowly at first, more rapidly in the middle phase and slowly again near the end of growth. *See also* Chapter 7.

logit transformation A transformation that relates the response to a given concentration or dose of a toxicant to the response in the absence of the toxicant, using the following formula: logit = log $[B/(B_0\text{-}B)]$ where B is the toxicant and B_0 the response in the absence of toxicant. Usually the logit function is plotted against the log of the concentration of the toxicant, to yield a linear relationship.

lognormal distribution A positively skewed distribution of a random variable which, when subjected to a logarithmic transformation, tends to take the shape of a normal distribution.

lordosis An anteroposterior curvature of the spine, generally in the lumbar region. Also called hollow back or saddle back.

lotic Flowing water, such as rivers and streams.

LT50 Time taken for 50% of the observed population to die. Also known as median lethal time (MLT). *See also* ET50 and IT50.

lux A unit of illumination based on units per m^2. One lux = 0.0929 foot candles and one foot candle = 10.76 lux.

lysimeter A laboratory column of selected representative soil or a protected monolith of undisturbed field soil with facilities for sampling and monitoring the movement of water and chemicals.

MAC *See* Maximum Allowable Concentration.

macrocosm Large multi-species test system. *See also* microcosm.

macrophages A large phagocytic cell found in connective tissues, especially in areas of inflammation.

macroscopic (gross) pathology The study of tissue changes which are visible to the naked eye.

malignancy A cancerous growth. A mass of cells showing both uncontrolled growth and the tendency to invade and destroy surrounding tissues.

malignant *See* tumour and malignancy.

manufacturer In the context of REACH means any natural or legal person established within the Community who manufactures a substance within the Community.

manufacturing In the context of REACH means production or extraction of substances in the natural state.

margin of exposure Ratio of the no observed adverse effect level (NOAEL) for the critical effect to the theoretical, predicted or estimated dose or concentration. *See also* margin of safety.

margin of safety For some experts the margin of safety has the same meaning as the margin of exposure, while for other, the margin of safety means the margin between the reference dose and the actual exposure dose or concentration. *See also* margin of exposure.

mass balance equation An equation that expresses the total mass of a chemical in terms of all the various forms and concentrations in different environmental compartments (including biota) in which it occurs. *See also* Chapters 3 and 7.

MATC *See* Maximum Acceptable Toxicant Concentration.

maximum acceptable toxicant concentration (MATC) The hypothetical toxic threshold concentration lying in a range bounded at the lower end by the highest tested concentration having no observed effect (NOEC) and at higher end by the lowest tested concentration having a significant toxic effect (LOEC) in a life cycle (full chronic) or partial life cycle (partial chronic) test. This may be represented as NOEC < MATC < LOEC. The MATC may be calculated as the geometric mean of the LOEC and NOEC. Calculation of an MATC requires quantitative life cycle toxicity data on the effects of a material on survival, growth, and reproduction.

maximum allowable concentration (MAC) Regulatory value defining the concentration which if inhaled daily (for workers: 8 h/d over a working week of 40 h, for the general population: 24 h) does not appear capable of causing appreciable harm in the light of present knowledge. *See also* threshold limit value.

measurement endpoint Measurable (ecological) characteristic that is related to the valued characteristic chosen as an assessment point.

measurement error Error that results from inaccuracy and imprecision in the measurement of parameter values.

median effective concentration (EC50) The concentration of the material in water to which test organisms are exposed that is estimated to be effective in producing some sublethal response in 50% of the test organisms. The EC50 is usually expressed as a time dependent value (e.g., 24-h or 96-h EC50). The sublethal response elicited from the test organisms as a result of exposure to the material must be clearly defined. For example, test organisms may be immobilized, lose equilibrium, or undergo physiological or behavioural changes.

median effective dose (ED50) The exposure dose of material estimated to be effective in producing some sublethal response in 50% of the test organisms. It is appropriately used with test animals such as rats, mice, dogs, but it is rarely applicable to aquatic organisms because it indicates the quantitative of a material introduced directly into the body by injection or ingestion rather than concentration of the material in water in which aquatic organisms are exposed during toxicity tests.

median effective time (ET50) The time required for half of the organisms in a toxicity test to exhibit a given nonlethal response end point at a given exposure concentration.

median lethal concentration (LC50) The concentration of material in air, water, soil or sediment to which test organisms are exposed which is estimated to be lethal to 50% of the test organisms. The LC50 is usually expressed as a time-dependent value (e.g. 24-h or 96-h LC50; the concentration estimated to be lethal to 50% of the test organisms after 24 or 96 h of exposure). The LC50 may be derived by *observation* (i.e. 50% of the test organisms can be seen to be dead at a given test concentration), by *interpolation* (i.e. more than 50% of the test organisms died at one test concentration and fewer than 50% of the test organisms died at a lower test concentration, and the LC50 is estimated by interpolation between these two data points), or by *calculation* (i.e. the LC50 is statistically derived by analysis of mortality data from a series of test concentrations).

median lethal dose (LD50) The dose of material that is estimated to be lethal to 50% of the test organisms. It is suitable for use with test animals such as rats, mice and dogs, but it is rarely applicable to aquatic organisms because it indicates the quantity of a material introduced directly into the body by injection or ingestion rather than the concentration of the material in water to which aquatic organisms are exposed during toxicity tests. The LD50 has often been used to classify and compare toxicity between chemicals but its value for this purpose is doubtful. One commonly used classification of this type is:

median tolerance limit (TL_m or TL50) The concentration of material in air, water, sediment or soil at which 50% of the test organ-isms survive after a specified time of exposure. The TL50 (equivalent to the TL_m) is usually expressed as a time-dependent value (e.g. 24-h or 96-h TL50; the estimated concentration at which 50% of test organisms survive after 24 or 96 h of expo-sure). Unlike lethal concentration and lethal dose, the term tolerance limit is applicable in designating the level of any measurable lethal condition (e.g., extremes in pH, temperature, dissolved oxygen). TL_m and TL50 have been replaced by median lethal concentration (LC50) and median effective concentration (EC50).

medulla The central portion of an organ or tissue, such as the medulla of the mammalian kidney or a plant thallus; bone marrow and pith (adjective: medullary).

meiofauna Animals living in interstices of soil or sediment of aquatic systems

mesocosm *See* microcosm.

mesothelioma A malignant tumour of the mesothelium of the pleura, pericardium or peritoneum, arising as a result of the presence of asbestos or caused by exposure to mining or smelting processes.

meta-analysis The process of using statistical methods to combine the results of different studies. In the biomedical sciences, the systematic evaluation of a problem using information (commonly in the form of statistical tables and other data) from a number of independent studies. A common application is the pooling of results from a number of small randomized controlled trials, none in itself large enough to demonstrate statistically significant differences, but, capable of doing so in aggregate. Meta-analysis has a qualitative component, i.e. application of predetermined quality criteria (e.g. completeness of data, absence of bias) and a quantitative component, i.e. integration of numerical information. Meta-analysis includes overview and data pooling aspects, but implies more than either of these processes. Meta-analysis carries the risk of several biases reinforcing each other.

metabolic activation The biotransformation of relatively inert chemicals to biologically reactive metabolites.

metabonomics Techniques available to identify the presence and concentrations of metabolites in a biological sample.

MFO *See* mixed function oxidase.

microcosm Artificial multi-species test system that simulates major characteristics of the natural environment for the purposes of ecotoxicological effects and risk assessment. Such systems are normally terrestrial or aquatic and may contain plants, animals (vertebrates and invertebrates) and micro-organisms. The terms mesocosm and macrocosm are used to refer to larger and more complex systems than microcosms but the distinction is often not clearly defined. *See* Chapter 7.

Microtox A test involving the "luminous" marine bacterium *Photobacterium phosphoreum*. Changes in light output are taken as indications of stress.

migration (population) The movement of an individual or group into or out of a new population or geographical region.

mineralisation The breakdown of a chemical substance or organic matter by micro-organisms in the presence of oxygen to carbon dioxide, water and mineral salts of any other elements present.

minimum significant difference (MSD) The difference between groups (in tests with e.g. salmonoid fish, the difference in average weights or average mortality) that would have to occur before it could be concluded that there was a significant difference between the groups. The MSD is provided by Dunnett's multiple-range test, a standard statistical procedure.

MIT *See* IT50.

mixed function oxidase An enzyme that catalyzes reactions between an organic compound and molecular oxygen in which one atom of the oxygen molecule is incorporated into the organic compound and one atom of the oxygen molecule is reduced to water. Involved in the metabolism of many natural and xenobiotic compounds giving both unreactive products and products of different or increased toxicity from that of the parent compound (phase-I reactions). *See* Chapter 3.

mixing zone An area where an effluent discharge undergoes initial dilution and is extended to cover the secondary mixing in the ambient water body. A mixing zone is an allocated impact zone where water quality criteria can be exceeded as long as acutely toxic conditions are prevented.

model A formal representation of some component of the world or a mathematical function with parameters which can be adjusted so that the function closely describes a set of empirical data. A *mathematical* or *mechanistic* model is usually based on biological, chemical or physical mechanisms, and its parameters have real world interpretations. By contrast, *statistical* or *empirical* models are curve-fitted to data where the mathematical function used is selected for its numerical properties. Extrapolation from mechanistic models (e.g. pharmacokinetic equations) usually carries higher confidence than extrapolation using empirical models (e.g. the logistic extrapolation models). A model that can describe the temporal change of a system variable under the influence of an arbitrary "external force" is called a *dynamic* model. To turn a *mass balance* model into a dynamic model, theories are needed to relate the internal processes to the state of the system, expressed e.g. in terms of concentrations. The elements required to build dynamic models are called process models.

model error The element of uncertainty associated with the discrepancy between the model and the real world.

mole The SI (Système International) metric unit for reporting amounts of chemicals whose relative molecular mass is known. One mole of a chemical is Avogadro's number (6.023×10^{23}) of molecules of that chemical. Metric mass-based units can be converted to molar units by dividing the former by the grammolecular weight of the chemical in question. Molarity refers to the number of moles of chemical per litre of solution and is denoted as M.

molecular descriptor A molecular descriptor is a structural or physicochemical property of a molecule, or part of a molecule, which characterises a specific aspect of a molecule and is used as an independent variable in a QSAR.

molecular orbital Like atomic orbitals for single atoms, molecular orbitals are the energy levels in a molecule that can be occupied by (pairs of) electrons. Besides an energy level, these molecular orbitals have a specific spatial arrangement (more accurately, a specific spatial distribution of the electron density), and can thus be viewed as relatively localised to a certain part of a molecule. For example the Lowest Unoccupied Molecular Orbital (LUMO), is one of the most important energy levels in reactions where electrons or electron pairs are accepted, has both an energy level, which dictates whether a reaction is feasible or not, and a "localisation" which dictates whether it is accessible, and if so, where the reaction centre is. In most quantum-chemical formulations, molecular orbitals are constructed as minimum-energy, self-consistent linear combinations of atomic orbitals, with the atomic orbitals taken from the constituent atoms.

monitoring Long-term, standardised measurement, evaluation, and reporting of specified properties of the environment, in order to define the current state of the environment, and to establish environmental trends. Surveys and surveillance are both used to achieve this objective.

monitoring test A test designed to be applied on a routine basis, with some degree of control, to ensure that the quality of an environmental compartment, biological endpoint or effluent has not exceeded some prescribed criteria range. In a biomonitoring test, organisms are used as "sensors" to detect changes in the quality of water or effluent. A monitoring test implies generation of information, on a continuous or other regular basis.

monomer Means a substance which is capable of forming covalent bonds with a sequence of additional like or unlike molecules under the conditions of the relevant polymer-forming reaction used for the particular process.

mono-oxygenase *See* mixed function oxidase.

Monte Carlo simulation A technique used to obtain information about the propagation of uncertainty in mathematical simulation models. It is an iterative process involving the random selection of model parameter values from specified frequency distributions, simulation of the system, and output of predicted values. The distribution of the output values can be used to determine the probability of occurrence of any particular value, given the uncertainty in the parameters.

multigeneration study A toxicity test in which at least three generations of the test organism are exposed to the chemical being assessed. Exposure is usually continuous.

multivariate analysis Multivariate analysis is the analysis of multi-dimensional data matrices by using statistical methods. Such data matrices can involve multiple dependent and/or independent variables.

mutagenesis Introduction of hereditary changes (mutations) in the genotype of a cell as a consequence of genetic alterations or the loss of genes or chromosomes (or parts of them). Any chemical that causes mutations is said to be mutagenic. Some mutagenic chemicals are also carcinogenic. *See also* carcinogenesis and transformation.

nanoscale Having one or more dimensions of the order of 100 nm or less.

nanoscience The study of phenomena and manipulation of materials at atomic, molecular and macromolecular scales, where properties differ significantly from those at a larger scale.

nanotechnology The design, characterization, production and application of structures, devices and systems by controlling shape and size at the nanoscale.

nanomaterial Material with one or more external dimensions, or an internal structure, which could exhibit novel characteristics compared to the same material without nanoscale features.

nanoparticle Particle with two or more dimensions at the nanoscale.

nanocomposite Composite in which at least one of the phases has at least one dimension on the nanoscale.

nanostructured Having a structure at the nanoscale.

narcosis Narcosis is a non-specific mode of toxic action, normally associated with a reduction in central nervous system activity (and hence anaesthesia) and ultimately death. The effect is brought about by non-reactive chemicals and is thought to result from an accumulation of the toxicant in cell membranes, diminishing their functionality. The narcotic effect is reversible, so that an organism will recover when the toxicant is removed. Narcotic effects are strongly associated with molecular hydrophobicity and hence good relationships have been found between the acute toxicity of narcotics and log K_{ow}. Within the narcotic mode of toxic action, a number of mechanisms are often been distinguished. These mechanisms, which are especially apparent in environmental species, include non-polar narcosis, polar narcosis, amine narcosis, ester narcosis, as well as other narcotic mechanisms that have yet to be determined.

necropsy Examination of the organs and body tissues of a dead animal to determine the cause of death or any pathological condition.

necrosis Cell death or death of areas of tissue, usually indicating that the affected tissue is surrounded by healthy tissue. Necrosis may be due to chemical agents acting locally, or secondary to physiological insult, infection or loss of circulation.

nematocide *See* pesticide.

neonate A newly-born or newly-hatched individual (e.g. first instar daphnid, < 24 h old).

neoplasm A genetically altered, relatively autonomous growth of tissue. A neoplasm is composed of abnormal cells, the growth of which is more rapid than that of other tissues and is not coordinated with the growth of other tissues.

nephrotoxity Toxicity to the kidney.

neurotoxic Any toxic effect on any aspect of the central or peripheral nervous system. Such changes can be expressed as functional changes (such as behavioral or neurological abnormalities) or as neurochemical, biochemical, physiological or morphological changes.

new chemicals In the EC, those produced since 1981 and not listed on the EINECS.

NIMBY principle Public acceptance of necessary provisions (e.g. waste incinerators) provided they do not affect the individual's quality of life (NIMBY = not in my backyard).

NMR Nuclear Magnetic Resonance, a technique to identify atoms in a sample by measuring the signal given off by the relaxation of e.g. protons previously aligned in a strong magnetic field.

no observed adverse effect level (NOAEL) *See* no observed effect concentration.

no observed effect concentration (NOEC) The highest concentration of a material in a toxicity test that has no statistically significant adverse effects on the exposed population of test organisms compared with the controls. When derived from life cycle or partial life test, it is numerically the same as the lower limit of the MATC. Also called no observed effect level (NOEL) or no observed adverse effect level (NOAEL).

no observed effect level (NOEL) *See* no observed effect concentration.

NO(A)EL No observed (adverse) effect level. *See* N(O)EC.

N(O)EC No (observed) effect concentration. The highest concentration of a test substance to which organisms are exposed, that does not cause any observed and statistically significant adverse effects on the organism compared with the controls. For example, the NOEC might be the highest tested concentration at which an observed variable, such as growth, did not differ significantly from growth in the control. The NOEC customarily refers to sublethal effects, and to the most sensitive effect unless otherwise specified. NEL, NOAEL, NEC and NOEC are equivalent terms.

non-genotoxic carcinogen A substance that causes cancer, not primarily damaging the genetic material, but by mechanisms that stimulate cell proliferation, thus increasing the chances for natural mutations to be produced, and/or selection of specific cell populations that may derange in a later stage.

non-target organisms Those organisms which are not the intended targets of a particular use of a pesticide.

normal distribution The classical statistical bell-shaped distribution which is symmetric and parametrically simple in that it can be fully characterized by two parameters: its mean and variance. A normal distribution is observed in situations where many independent additive effects influence the values of the variates.

not chemically modified substance in the context of REACH means a substance whose chemical structure remains unchanged, even if it has undergone a chemical process or treatment, or a physical mineralogical transformation, for instance to remove impurities.

notified substance According to REACH a notified substance for which a notification has been submitted and which could be placed on the market in accordance with Directive 67/548/EEC. *See also* new chemical.

nucleophilicity Nucleophilicity refers to the molecular or substructural property of having a repulsion for electrons or an attraction for positive charge. Molecular nucleophilicity is often described by the energy of the highest occupied molecular orbital (E_{HOMO}) and by superdelocalisabilities.

nucleotide In this case the basic building block of DNA and RNA: a base/sugar/phosphate complex, these nucleotides from a codon, coding for one amino acid.

occupational hygiene An applied science concerned with the recognition, evaluation and control of chemical, physical and biological factors arising in or from the workplace which may affect the health or well-being of those at work or in the community.

octanol-water partition coefficient (K_{ow}) The ratio of a chemical's solubility in n-octanol and water at equilibrium; also expressed as P. The logarithm of K_{ow} or P (i.e. log K_{ow} or log P) is used as an indication of a chemical's propensity for bioconcentration by aquatic organisms or skin permeability (*see also* Chapters 9 and 10).

ocular Relating to the eye.

OECD Organization for Economic Co-operation and Development.

oligotrophic Nutrient poor (aquatic) system. *See also* eutrophic.

oncogene A retroviral gene that causes transformation of the infected mammalian cell. Oncogenes are slightly changed equivalents of normal cellular genes called protooncogenes. The viral version is designated by the prefix v, the cellular version by the prefix c.

one-hit model Dose-response model of the form $P(d)$=1-exp(-bd) where $P(d)$ is the probability of cancer death from a continuous dose rate (d) and b is a constant. The one-hit odel is based on the concept that a tumour can be induced after a single susceptible target or receptor has been exposed to a single effective unit dose of an agent.

operational conditions In the context of REACH these are conditions under which a manufacturing or use process takes place.

organelle A structure with a specialized function which forms part of a cell.

orifice An opening or aperture.

palate The partition separating the nasal and oral cavities.

parameter uncertainty The element of uncertainty associated with estimating model parameters. It may arise from measurement or extrapolation.

parameterise The allocation of values to the variables.

PARCOM Paris Commission.

parenchymal cell A cell of the functional tissue of a gland or an organ.

parthenogenesis Process by which eggs develop without fertilization.

particulate organic carbon (POC) The fraction of the organic carbon pool that is not dissolved in water, but is retained on a 0.45 μm glass fibre filter. POC is identical to suspended organic carbon (SOC) and is composed of plant and animal organic carbon and organic coatings on silt and clay.

particulate organic matter (POM) it is analogous to POC (*see* POC) but it refers to the entire organic pool that is not dissolved in water.

partition coefficient A partition coefficient is the ratio of the concentrations of a substance between two phases when the heterogeneous system of two phases is in equilibrium. In QSAR analysis, the octanol-water partition coefficient (log K_{ow}) is often used as a descriptor of hydrophobicity, where K = $[chemical]_{octanol}$ / $[chemical]_{water}$, by using a logarithmic relationship, K becomes an additive property, Log K_{ow} = log $[chemical]_{octanol}$ - $[chemical]_{water}$. *See* Chapters 3, 4, 9 and 10.

parts per billion (ppb) One unit of chemical (usually expressed as a mass) per 1,000,000,000 (10^9) units of the medium (e.g. water) or organism (e.g. tissue) in which it is found. For water, the ratio commonly used is micrograms of chemical per litre of water, 1 μg/L= 1 ppb; for tissues, 1 μg/kg = 1 ng/g = 1 ppb.

parts per million (ppm) One unit of chemical (usually expressed as a mass) per 1,000,000 (10^6) units of the medium (e.g. water) or organism (e.g. tissues) in which it occurs. For water, the ratio commonly used is milligrams of chemical per litre of water, 1 mg/L = 1 ppm; for tissues, 1 mg/kg = 1 μg/g = 1 ppm.

parts per thousand (ppt) One unit of chemical (usually expressed as a mass) per 1000 (10^3) units of the medium (e.g. water) or organism (e.g. tissues) in which it occurs. For water, the ratio commonly used is grams of chemical per litre of water, 1 g/L = 1 ppt; for tissues, 1 g/kg = 1 ppt. This ratio is also used to express the salinity of seawater, where the grams of salt per litre of water is denoted by the symbol ppt. Full-strength seawater is approximately 35 ppt.

parts per trillion (pptr) One unit of chemical (usually expressed as a mass) per 1,000,000,000,000 (10^{12}) units of the medium (e.g. water) or organism (e.g. tissues) in which it is found. The ratio commonly used is nanograms of chemical per litre of water, 1 ng/L = 1 pptr; for tissues, 1 ng/kg = 1 pptr.

PBPK Physiologically-based pharmacokinetic model.

PCA Principal component analysis. A multivariate technique to derive a set of orthogonal parameters (principal components) from a large number of properties.

PEC Predicted environmental concentration. The estimated concentration of a chemical in a particular medium at a particular location at a particular time. The PEC can be based on either measured or calculated data. *See* Chapters 1, 4 and 12.

perceived risk *See* risk perception.

percentiles Divides frequency distribution into 100 equal portions. Hence the 95 percentile is the value that 95% of the population does not exceed.

permissible exposure limit (PEL) *See* threshold limit value (TLV).

peroxisome A cytoplasmic organelle present in animal and plant cells, which contains catalase and other peroxidase oxidative enzymes.

persistence Attribute of a substance which describes the length of time that the substance remains in a particular environment before it is physically removed or chemically or biologically transformed.

pesticide A chemical used in agriculture and in other non-agricultural areas, to control the severity and incidence of pests and diseases which would otherwise reduce agricultural yields or hinder other processes. Pesticides are used to control bacteria, fungi, algae, higher plants, nematodes, mollusca, mites and ticks, insects, rodents (e.g. mice and rats) and other organisms. This generic term is also used to cover: bactericides, fungicides, algicides, herbicides, nematocides, molluscicides, acaricides, insecticides and rodenticides.

pH The negative logarithm of the activity of hydrogen ions in gram equivalents per litre. The pH value expresses the degree or intensity of both acidic and alkaline reactions on a scale from 0 to 14, with 7 representing neutral, numbers less than 7 signifying increasingly acidic reactions, and numbers greater than 7 indicating increasingly basic or alkaline reactions.

phagocytosis The ingestion of microorganisms, cells, and foreign particles by phagocytes; hence phagocytic macrophages.

pharmacodynamics Process of interaction of pharmacologically active substances with target sites, and the biochemical and physiological consequences leading to therapeutic or adverse effects. Also known as toxicodynamics, but this term strictly refers to the study of substances other than drugs.

pharmacokinetics Process of uptake of drugs by the body, the biotransformation they undergo, the distribution of the drugs and their metabolites in the tissues, and the elimination of the drugs and their metabolites from the body. Both the amounts and the concentrations of the drugs and their metabolites are studied. The term has essentially the same meaning as toxicokinetics and biokinetics, but this term strictly refers to the study of substances other than drugs.

phase-I reactions Enzymic modification of a xenobiotic by oxidation, reduction, hydrolysis, hydration, dehydrochlorination or other reactions catalyzed by enzymes of the cytosol of the endoplasmatic reticulum (microsomal enzymes) or other cell organelles. *See also* MFO.

phase-II reactions Binding of a substance or its metabolites from a phase-I reaction with endogenous molecules (conjugation) to create more watersoluble derivatives which can be excreted in the urine or bile. *See* Chapter 3.

phase-in substance In the context of REACH means a substance which meets at least one of the following criteria: (a) it is listed in the European Inventory of Existing Commercial Chemical Substances (EINECS); (b) it was manufactured in the Community, or in the countries acceding to the European Union on 1 January 1995 or on 1 May 2004, but not placed on the market by the manufacturer or importer, at least once in the 15 years before the entry into force of this regulation, provided the manufacturer or importer has documentary evidence of this; (c) it was placed on the market in the Community, or in the countries acceding to the European Union on 1 January 1995 or on 1 May 2004 before entry into force of this Regulation by the manufacturer or importer and was considered as having been notified in accordance with the first indent of Article 8(1) of Directive 67/548/EEC but does not meet the definition of a polymer set out in this Regulation, provided the manufacturer or importer has documentary evidence of this.

phenology Life history.

phenotype Total of observable features of an organism, as the result of interaction between the genetic material (genotype) and the environment.

photodegradation Any breakdown reaction of a chemical that is initiated by sunlight (ultraviolet light), or more accurately, by the influence of a high-energy photon. This can be either by direct photodegradation, in which the photon photolysis or ionises the relevant molecule itself, which then reacts with other species in its vicinity, or by indirect photodegradation, in which the relevant molecule reacts with ions or radicals created by photolysis of other species.

photoperiod The duration of illumination and darkness over a 24-h day.

placing on the market In the context of REACH means supplying or making available, whether in return for payment or free of charge, to a third party. Import shall be deemed to be placing on the market.

PLS Partial least square analysis. A multivariate technique to relate *Y* values for a series of objects to a set of *X* variables for the same objects.

PMN Premanufacture notification. Regulation for new chemicals as required by the Toxic Substances Control Act in the US. *See also* Chapter 13.

PNEC Predicted no effect concentration: environmental concentration which is regarded as a level below which the balance of probability is that an unacceptable effect will not occur. *See* Chapters 1, 7 and 12.

p.o. Abbreviation for oral administration *(per os)*.

point source Emission source(s), either single or multiple, which can be quantified by means of location and the amount of substance emitted per source and emission unit (e.g. amount per time unit).

pollutant A potentially harmful agent occurring in the environment, products or at the workplace as a result of human activities.

pollution Release to the environment of a chemical, physical, or biological agent that has the potential to damage the health of human or other organisms.

polymer In the context of REACH means a substance consisting of molecules characterised by the sequence of one or more types of monomer units. Such molecules must be distributed over a range of molecular weights wherein differences in the molecular weight are primarily attributable to differences in the number of monomer units. A polymer comprises the following: a) a simple weight majority of molecules containing at least three monomer units which are covalently bound to at least one other monomer unit or other reactant; b) less than a simple weight majority of molecules of the same molecular weight. In the context of this definition a "monomer unit" means the reacted form of a monomer substance in a polymer.

polymerase chain reaction Technique enabling a rapid multiplication of selected parts of a DNA and RNA strand.

polymorphism In this context, the existence of inter-individual differences in DNA sequences coding for one specific gene. The effects of such differences may vary dramatically, ranging from no effect at all to the building of inactive proteins, or not even building the protein.

POM Particulate organic matter. *See also* TOC.

population A group of interacting and, typically, interbreeding organisms (sharing genes) of the same species.

population biomass The total mass or weight of organisms in a population, given by the sum of the masses (or weights) of all the individual members of the population.

population growth rate The relative increase in the population per unit of time.

population size The total number of organisms in a population.

pore water (or interstitial water) Water found in spaces between particles of soil or sediment.

porous pot test Biodegradation test that simulates the continuous activated sludge (sewage treatment) system.

potentiation The effect of a chemical which enhances the toxicity of another chemical. *See also* synergism.

power of a test The power of a statistical test is the probability of rejecting the zero-hypothesis when it is false and the alternative hypothesis is correct.

precision A measure of the degree of agreement among individual results obtained from the same or identical specimens with the same method and by the same analyst and laboratory.

precautionary principle The general principle by which all that can reasonably be expected is done to prevent unnecessary risks. *See also* ALARA and Chapter 1.

precipitation (1) The formation of a solid (i.e. precipitate) from a solution and (2) rain, snow, etc. formed by condensation of water vapour in air.

predicted environmental concentration (PEC) The concentration of a chemical in the environment, calculated on the basis of available information on certain of its properties, its use and discharge patterns and the quantities involved.

predicted no effect concentration (PNEC) *See* PNEL.

predicted no effect level (PNEL) The maximum level (dose or concentration) which on the basis of current knowledge is likely to be tolerated by an organism without producing any adverse effect.

predictive risk assessment A risk assessment performed for a proposed future action, such as the use of a new chemical or the release of a new effluent.

predictivity The predictivity (or predictive capacity/ability) of a model is a measure of its ability to make reliable predictions for chemical structures not included in the training set of the model.

preliminary test *See* screening test.

preparation means a mixture or solution composed of two or more substances.

primary biodegradation The structural change (transformation) of a chemical substance by micro-organisms resulting in the loss of chemical identity.

probability A quantitative statement about the likelihood of a specific outcome. Probability values can range from 0 to 1.0.

probit A probit, or probability unit, is obtained by modifying the standard variate of the standardized normal distribtion by the addition of a constant value of 5 (to avoid negative numbers). Converting a cumulative percent response to probits followed by plotting it against concentration or dose can provide useful information about the distribution of the response and estimates of the L(E)D50 or L(E)C50 values. This transformation is used for the analysis of dose-response data.

probit/log transform The probability unit obtained from the standardized normal distribution plotted against the logarithm of the concentration or dose of a substance when a quantal or graded response has been measured. A linear plot provides evidence that the distribution is lognormal. Estimates of the L(E)C50 and L(E)D50, as well as the standard deviation for the distribution, can then be made.

producer of an article In the context of REACH: means any natural or legal person who makes or assembles an article within the Community.

product and process orientated research and development (PPORD) In the context of REACH means any scientific development related to product development, the further development of a substance, on its own, in preparations or in articles in the course of which pilot plant or production trials are used to develop the production process and/or to test the fields of application of the substance.

product use category Preparations or articles used for the same purpose (technical function, service, convenience …) and hence likely to be used under comparable conditions for which the same exposure scenario may then be applicable. *See also* exposure scenario.

prokaryote Simple unicellular organism, primarily bacteria and cyanobacteria, that have no nuclei to contain their genetic material. They have a few subcellular structures. *See also* eukaryote.

proliferation Multiplication, i.e. an increase by frequent and repeated reproduction or growth by cell division.

promoter In carcinogenesis this is an agent which enhances tumour growth caused by a chemical after exposure to an initiator. *See also* initiation/initiator.

protein binding The process by which drugs and toxins are bound to proteins other than the receptor in the plasma or, less commonly, intracellularly. The bound fraction is inactive but in equilibrium with the free fraction in the cell or plasma.

proteomics Techniques available to identify the proteins in a biological sample.

public health impact assessment Applying risk assessment to a specific target population. The size of the population needs to be known. The end product will be a quantitative statement about the number of people affected in this specific target population.

pulmonary alveoli Minute thin-walled air sacs, surrounded by bloodvessels. Found in the lungs of vertebrates.

QSAR *See* quantitative structure-activity relationship.

quality assurance (QA) A programme organized and designed to provide accurate and precise results. Included are selection of proper technical methods, tests, or laboratory procedures; sample collection and preservation; selection of limits; evaluation of data; quality control; and qualifications and training of personnel.

quality control (QC) Specific actions required to provide information for the quality assurance program. Included are standardization, calibration, replicates, and control and check samples suitable for statistical estimates of confidence of the data.

quality criteria Quality guidelines based on the evaluation of scientific data.

quality guidelines Numerical limits or text statements established to support and maintain designated uses of the environment or to protect human health.

quality objectives Numerical limits or narrative statements established to protect and maintain human health or designated uses of the environment at a particular site.

quality standards Fixed upper limits for exposure to certain chemicals recognized under law by one or more levels of government. Well-known examples include the air, water and soil quality standards, as well as threshold limit values for air pollutans in the workplace.

quantal effect Discontinuous response such as death or survival or the presence/absence of a behavioural response. *See* continuous effect.

quantitative structure-activity relationship (QSAR) The relationship between the physical and/or chemical properties of substances and their ability to cause a particular effect, enter into certain reactions, etc. *See also* Chapters 9-11.

quantitative structure-activity relationship (QSAR) A quantitative structure-activity relationship is a quantitative relationship between a biological activity (e.g. toxicity) and one or more descriptors that are used to predict the activity. *See also* Chapters 9-11.

quantitative structure-property relationship (QSPR) A quantitative structure-property relationship is quantitative relationship between a physicochemical property or environmental parameter (e.g. a partition coefficient) and one or more descriptors that are used to predict the property *See also* Chapters 9 and 10.

quantum-chemical parameters In principle, anything that can be derived from the results (mostly the density and eigenvector/eigenvalue matrix) of a quantum-chemical calculation for any atom or molecule. This includes basic parameters, such as the total energy or heat of formation of a molecule, overall molecular parameters, such as the electronegativity, hardness or dipole moment (or even the molecular surface area or volume, based on certain cut-off values for electron densities), orbital-specific parameters, such as the energy level of the highest occupied molecular orbital (HOMO); which is the same as the ionisation potential) or lowest unoccupied molecular orbital (LUMO); also the electron affinity), or atom-based parameters, such as partial charge, superdelocalisability, self-polarisability, etc.

quotient method Calculation of the quotient of the measured or predicted environmental concentration (PEC) of a contaminant and the predicted no effect level (PNEL), used as an expression of hazard or risk. Higher quotients constitute greater evidence of a hazard or a greater risk. *See also* hazard quotient.

range-finding test *See* screening test.

REACH Registration, Evaluation, Authorisation and restriction of CHemicals. *See* Chapter 12.

reasonable worst case Reasonably unfavourable but not unrealistic situation. Combining the most adverse environmental circumstances and worst-case release parameters necessarily results in an unrealistic overall worst-case estimation, which is extremely unlikely to occur.

receiving water Surface water (e.g. in a stream, river, or lake) that has received a discharged waste, or is about to receive such a waste (e.g. just upstream or up-current from the discharge point).

recipient of an article In the context of REACH means an industrial or professional user being supplied with an article but does not include consumers.

recipient of a substance or a preparation In the context of REACH means a downstream user or a distributor being supplied with a substance or a preparation.

recommended limit A maximum concentration of a potentially toxic substance which is expected to be safe. Such limits often have no statutory basis, in which event any control or statutory limit should not be exceeded.

reconstituted water De-ionized or glass-distilled water to which reagent-grade chemicals have been added. The resultant synthetic fresh water will be free from contaminants and have the desired pH and hardness characteristics.

reference compound Standard substance whose known toxicological, ecotoxicological or physicochemical properties can be used to check the results of a test.

reference dose An estimate of the daily exposure dose that is likely to be without deleterious effect even if continued exposure occurs over the lifetime. *See also* acceptable daily intake.

reference environment A generalized description of the environment into which contaminants will be released and in which organisms will be exposed. Reference environments are used when there is no specific site at risk.

reference site A relatively unpolluted site used for comparison with polluted sites in environmental monitoring studies, often incorrectly referred to as a control site.

registrant In the context of REACH means the manufacturer or the importer or the producer or importer of an article submitting a registration for a substance.

registrant's own use In the context of REACH means an industrial or professional use by the registrant.

regression analysis A statistical procedure for determining the constants and coefficients in regression equations from an analysis of observed data for two or more variables. *See also* regression coefficient.

regression coefficient A parameter which describes the rate of change of a dependent variable relative to an independent variable; any coefficient in a regression equation, such as the parameters a and b in the linear regression equation $y = a+bx$. *See also* regression analysis.

remediation Concerned with correction and clean-up of chemically contaminated sites.

renal Associated with the kidneys.

replicate A single test unit, such as a container or aquarium, containing a prescribed number of organisms exposed to one concentration or dose of the test compound. An aquatic toxicity test comprising five test concentrations and a control, with three replicates, would require 18 aquaria. For each concentration or control, there would be three aquaria or replicates. A replicate is an independent test unit, thus, any transfer of organisms or solutions from one replicate to another would invalidate the test.

reproducibility Measure of the extent to which different laboratories obtain the same result with the same reference test compound.

reproductive toxicology The study of the adverse effects of chemicals on the embryo, foetus, neonate and prepubertal animal and the adult reproductive and neuroendocrine systems.

resistance time The period of time that an organism is able to live beyond the incipient lethal level.

response Change developed in the state or dynamics of an organism, system or (sub) population in reaction to exposure to an agent.

restriction In the context of REACH means any condition for or prohibition of the manufacture, use or placing on the market.

retrospective risk assessment A risk assessment performed for hazards that began in the past and may have ongoing effects, e.g. waste disposal sites and oil spills.

rhizosphere Zone of soil immediately surrounding the roots.

ribonucleic acid (RNA) A generic term for a group of nucleotide molecules, similar in composition to deoxyribonucleic acid (DNA), which perform a number of functions in programming the genetic code in cells. There are several types of RNA, e.g. messenger RNA, ribosomal RNA, transfer RNA.

ring test (1) A conjoint test conducted under strictly standardized and uniformly applied conditions to assess the precision and accuracy with which different laboratories can determine the toxicity of a chemical or effluent, and (2) a test designed to measure statistically the reproducibility of a test method, or to compare the results obtained from the use of different test methods.

risk The probability of an adverse effect in an organism, system or (sub) population caused under specified circumstances by exposure to an agent.

risk analysis (1) A process for controlling situations where an organism, system or (sub) population could be exposed to a hazard. The risk analysis process consists of three three components: risk assessment, risk management and risk communication, and (2) sometimes used as an equivalent term to risk assessment, especially in older literature.

risk assessment A process intended to calculate or estimate the risk to a given target organism, system or (sub) population, including the identification of attendant uncertainties, following exposure to a particular agent, taking into account the inherent characterizations of the agent of concern as well as the characterization of the specific target system. The risk assessment process includes four steps: hazard identification, hazard characterization (related terms: dose-response assessment and effect assessment), exposure assessment, and risk characterization. It is the first component in a risk analysis process.

risk-benefit analysis The next step after risk classification. It is the process of drawing up a balance sheet of the respective risks and benefits of a proposed risk-reducing action. It is a multidisciplinary task in which the risk manager has to consider not only the risk assessment but also other important aspects such as technical feasibility, economic factors, social and cultural factors as well as legislative and political factors. *See also* Chapter 1.

risk characterization The qualitative and, wherever possible, quantitative determination, including attendant uncertainties, of the probability of occurrence of known and potential adverse effects of an agent in a given organism, system or (sub) population, under defined exposure conditions. Risk characterization is the fourth step in the risk assessment process.

risk classification The valuation (or weighting) of risks in order to decide whether risk reduction is required. It is a complex process of determining the significance or value of the identified hazards and estimated risks to those concerned with or affected by the decision. It therefore includes the study of risk perception and the balancing of perceived risks and perceived benefits. *See also* risk evaluation.

risk communication Interactive exchange of information about health or environmental risks among risk assessors, managers, news media, interested groups and the general public.

risk estimation Quantification of the probability, including attendant uncertainties, that specific adverse effects will occur in an organism, system or (sub) population due to actual or predicted exposure.

risk evaluation Establishment of a qualitative or quantitative relationship between risks and benefits of exposure to an agent, involving the complex process of determining the significance of the identified hazards and estimated risks to the system concerned or affected by the exposure, as well as the significance of the benefits brought about by the agent. It is an element of risk management. Risk evaluation is synonymous with risk-benefit evaluation. *See also* risk classification.

risk management Decision-making process involving considerations of political, social, economic, and technical factors with relevant risk assessment information relating to a hazard so as to develop, analyse, and compare regulatory and non-regulatory options and to select and implement appropriate regulatory response to that hazard.

risk management measures (RMMs) Measures in the control strategy for a substance that reduce the emission and exposure to a substance, thereby reducing the risk to human health or the environment.

risk monitoring Process following up the decisions and actions within risk management in order to ascertain that risk containment or reduction with respect to a particular hazard is assured. Risk monitoring is an element of risk management.

risk perception An integral part of risk evaluation. The subjective perception of the gravity or importance of the risk based on the individual's knowledge of different risks and the moral and political judgement attached to them and their importance.

risk quotient A comparison of exposure with effects, i.e. the PEC/PNEC ratio. This risk quotient is often used to express the risk posed by a particular chemical. *See also* hazard quotient.

risk reduction Taking measures to protect man or the environment against the risks identified.

RMM Risk Management Measure.

robustness A measure of intralaboratory day-to-day variation induced by small changes in procedure.

robust study summary In the context of REACH and the OECD Chemicals Program a robust study summary is a detailed summary of the objectives, methods, results and conclusions of a full study report providing sufficient information to make an independent assessment of the study minimising the need to consult the full study report.

rodenticide *See* pesticide.

round-robin test Synonym for ring test.

ruggedness A method's reproducibility under the influence of variation in analyst, instrumentation, day of testing, and laboratory.

run-off The portion of the precipitate on the land that ultimately reaches streams and, eventually, the sea.

safe concentration Concentration of material to which prolonged exposure will cause no adverse effect.

safety Practical certainty that adverse effects will not result from exposure to an agent under defined circumstances. It is the reciprocal of risk.

safety factor Composite (reductive) factor by which an observed or estimated no-observed-adverse-effect (NOAEL) is divided to arrive at a criterion or standard that is considered safe or without appreciable risk. *See also* assessment factor and uncertainty factor.

safety (toxicological) Defined as a high probability that adverse effects will not result from exposure to a substance under specific conditions of quantity and manner of use.

salinity The total amount of salts, in g, dissolved in 1 kg of water. It is determined after all carbonates have been converted to oxides, all bromide and iodide have been replaced by chloride, and all organic matter has been oxidized. Salinity can also be measured directly using a salinity/conductivity meter or other means. It is usually reported in g/kg or parts per thousand.

SAM Standardized aquatic microcosm.

saprophyte An organism that obtains its nutrients from dead and decaying matter.

SARA (US) Superfund Amendment and Reauthorization Act.

satellite groups In toxicity testing, organisms or groups of organisms treated in a similar fashion for special additional studies.

SCAS test Biodegradation test to monitor the decay of DOC.

scientific research and development In the context of REACH means any scientific experimentation, analysis or chemical research carried out under controlled conditions in a volume less than 1 tonne per year.

screening test (preliminary test or range-finding test) (1) A test conducted to estimate the concentrations to be used for a definitive test, and (2) a short-term test used early in a testing programme to evaluate the potential of a chemical (or other substance) to produce a given adverse effect (e.g. mortality).

SDS Safety Data Sheet

secondary poisoning The product of biomagnification and toxicity.

semistatic Exposure system in which the test volume is renewed at intervals during the study.

sensitization Immune process whereby individuals become hypersensitive to substances, pollen or other agents which then induce a potentially harmful allergy when they are subsequently exposed to the sensitizing material (allergen).

sister chromatid exchange A reciprocal exchange of DNA between the two DNA molecules of a replicating chromosome.

site In the context of REACH means a single location, in which, if there is more than one manufacturer of (a) substance(s), certain infrastructure and facilities are shared.

small and medium enterprise (SME) In the context of REACH means small and medium-sized enterprises as defined in the Commission Recommendation of 6 May 2003 concerning the definition of micro, small and medium-sized enterprises.

sorption Term used instead of adsorption or absorption, when it is difficult to discriminate experimentally between these processes.

source term An estimate of the total amount released, or the temporal pattern of the rate of release of a pollutant from a source.

spawning The release of eggs or sperm from mature adult fish, or refers to behaviour related to the readiness of mature adult fish to release gametes.

speciation Determination of the exact chemical form or compound in which an element occurs in a sample, for example whether arsenic occurs in the form of trivalent or pentavalent ions or as part of an organic molecule, and the quantitative distribution of the different chemical forms that may coexist.

specific exposure scenario An exposure scenario covering one or a few specific uses or activities for which more specific restrictions of activities or more specific risk management measures are needed, compared to a broad exposure scenario. *See also* exposure scenario.

specificity The ability to identify and quantify the target analyte in the presence of chemically similar interfering compounds.

stable age distribution The abundance of relative age classes which a population approaches if it is allowed to grow exponentially.

standard An environmental quality standard is the limiting concentration of a chemical (or degree of intensity of some other adverse condition, e.g. pH) which is permitted in an environmental compartment (soil, effluent or waterway). Standards are established for regulatory purposes and are determined on the basis of a judgement of the criteria involved. The standard is dependent on the use (e.g. drinking water or agricultural water for irrigation). Standards are derived from criteria, often by applying safety factors (e.g. quality standards for air, water and soil).

static Exposure system in which the test volume is not renewed during the study.

static renewal Describes a toxicity test in which test solutions are renewed (replaced) periodically, usually at the beginning of each 24-h period. Synonymous terms are batch replacement, renewed static, renewal, static replacement and semi-static.

statistically significant effects Effects (responses) in the exposed population that are different from those in the controls at a statistical probability level of $p < 0.05$. Biological endpoints that are important for the survival, growth, behaviour and perpetuation of a species are selected as criteria. Endpoints differ depending on the type of toxicity test to be conducted and the species used. The statistical approach also depends on the type of toxicity test conducted.

steady-state The non-equilibrium state of a system in which matter flows in and out at equal rates so that all of the components remain at constant concentrations (dynamic equilibrium). In a chemical reaction, a component is in a steady-state if the rate at which the component is being synthesized (produced) is equal to the rate at which it is being degraded (used). In multimedia exposure models and bioaccumulation models it is the state at which the competing rates of input/uptake and output/elimination are equal. An apparent steady-state is reached when the concentration of a chemical remains essentially constant over time. Bioconcentration factors are usually measured at steady-state. *See also* equilibrium.

stochastic Due to, pertaining to or arising from chance and, hence, involving probability and obeying the laws of probability. The term stochastic indicates that the occurrence of effects so named, would be random. This means that, even for an individual, there is no threshold of dose below which the effect will not occur and the chance of experiencing the effect increases with increasing dose. Hereditary effects and cancer induced by radiation are considered to be stochastic effects.

stochastic analysis An analysis in which one or more parameters is represented by statistical distribution rather than a constant.

stochasticity Randomness determining or influencing a process. Variability in parameters or in models containing such parameters resulting from the inherent variability of the system described.

stochastic model A mathematical model founded on the properties of probability so that a given input produces a range of possible outcomes which are due to random effects.

stock solution A concentrated aqueous solution of the substance to be tested. Measured volumes of a stock solution are added to dilution water to prepare the required strengths of test solutions.

stoichiometry The quantitative relationship between the elements in a compound or between the reactants and the products in a chemical reaction.

STP Sewage Treatment Plant.

stress The proximate (or immediate) cause of an adverse effect on an organism or system.

structural alert A structural alert is a molecular (sub)structure associated with the presence of a biological activity.

structure-activity relationship (SAR) The correlation between molecular structure and biological/chemical/physicochemical activity. It is usually applied to the observation of the effect that the systematic structural modification of a particular chemical entity has on a defined biological, chemical or physicochemical endpoint. *See also* QSAR and Chapters 9-11.

study summary In the context of REACH means a summary of the objectives, methods, results and conclusions of a full study report providing sufficient information to make an assessment of the relevance of the study for hazard assessment.

Sturm test Biodegradation test to measure CO_2 production.

stygobiont Organism which lives only in groundwater.

stygophile Organism which lives in groundwater and in surface water.

subacute *See* subchronic.

subchronic Short-term tests that give an indication of long-term effects, often by focusing on critical (or sensitive) stages. Sometimes referred to as subacute but, in the light of this definition, this would seem to be misleading. The period of exposure usually does not exceed 10% of the life span.

sublethal Below the concentration that causes immediate death. Exposure to sublethal concentrations of a material may produce less obvious effects on the behaviour, biochemical and/or physiological function, and histology of organisms.

substance In the context of REACH means a chemical element and its compounds in the natural state or obtained by any manufacturing process, including any additive necessary to preserve its stability and any impurity deriving from the process used, but excluding any solvent which may be separated without affecting the stability of the substance or changing its composition.

substances which occur in nature In the context of REACH means a naturally occurring substance as such, unprocessed or processed only by manual, mechanical or gravitational means; by dissolution in water, by flotation, by extraction with water, by steam distillation or by heating solely to remove water, or which is extracted from air by any means.

substructure A substructure is an atom, or group of adjacently connected atoms, in a molecule.

supplier of an article In the context of REACH means any producer or importer of an article, distributor or other actor in the supply chain placing an article on the market.

supplier of a substance or a preparation In the context of REACH means any manufacturer, importer, downstream user or distributor placing on the market a substance, on its own or in a preparation, or a preparation.

surfactant A surface-active subtance (e.g. a detergent) which reduces surface tension and facilitates dispersion of substances in water. *See also* detergent.

surrogate A test organism, or population that is cultured under laboratory conditions to serve as a substitute in toxicity testing for indigenous organisms, communities or populations.

surveillance Measurement of environmental or health characteristics over an extended period of time to determine status or trends in some aspect of environmental quality or human health.

survival time The time interval between initial exposure of an organism to a harmful chemical and death.

susceptibility The condition of organism or other ecological system lacking the ability to resist a particular disease, infection or intoxication. It is inversely proportional to the magnitude of the exposure required to cause the response.

synergism A phenomenon in which the toxicity of a mixture of chemicals is greater than that which would be expected from the total toxicity of the individual chemicals present in the mixture.

targeting of exposure assessment The targeting phase of the exposure assessment refines the scope of the exposure scenario: which target groups (worker, consumer, environment) are exposed and to what degree. Initial information on exposure is reviewed to identify relevant exposure routes and exposure conditions. The targeting phase consists of the first two steps in the development of exposure scenarios: Step 1: Identification of uses; Step 2: Specification of manufacturing or use conditions. It ends with the description of the tentative exposure scenario. *See* Chapters 2, 4 and 12.

TDI *See* tolerable daily intake.

tentative exposure scenario The tentative exposure scenario forms the starting point for the exposure estimate and risk characterisation. A tentative exposure scenario is a set of assumptions (using the determinants of exposure) on how a process is conducted and which risk management measures that are used or should be implemented. *See* exposure scenario and Chapters 2, 5 and 12.

teratogen Agent which, when administered prenatally to the mother, induces permanent structural malformations or defects in the offspring.

teratogenesis The potential or capacity of a substance to cause defects in embryonic and foetal development.

terrestrial Relating to land, as distinct from water or air.

test material A chemical, formulation, effluent, sludge, or other agent or substance under investigation in a toxicity test.

test solution or test treatment Medium containing the material to be tested to which the test organisms will be exposed. Different test solutions contain different concentrations of the test material.

threshold Dose or exposure concentration of an agent below that a stated effect is not observed or expected to occur.

threshold-effect concentration (TEC) The concentration calculated as the geometric mean of NOEC and LOEC. Chronic value or subchronic value are alternative terms that may be appropriate depending on the duration of exposure in the test. The TEC is equivalent to the (maximum acceptable toxicant concentration (MATC) used in other countries.

threshold limit value (TLV) Concentration in air of a substance to which it is believed that most workers can be exposed daily without adverse effect (the threshold between safe and dangerous concentrations). These values are established (and revised annually) by the American Conference of Governmental Industrial Hygienists and are time-weighted concentrations for a 7-h or 8-h working day and a 40-h working week. For most substances the value may be exceeded to a certain extent, provided there are compensating periods of exposure below the value during the working day (or in some cases, the week). For a few substances (mainly those that produce a rapid response) the limit is given as a ceiling concentration (maximum permissible concentration, designated by "C") that should never be exceeded. *See also* maximum allowable concentration.

tiered testing strategy Sets out a structured approach to assessing the fate and effects of substances, where tests in higher tiers may be required depending upon the results of tests at earlier stages (i.e. lower tiers). Under a tiered structure, for example, data requirements for effects testing might progress from acute to chronic laboratory studies to field studies.

time-independent (TI) test An acute toxicity test with no predetermined temporal endpoint. This type of test, sometimes referred to as a "threshold" or "incipient" lethality test, is allowed to continue until acute toxicity (mortality or a defined sublethal effect) has ceased or nearly ceased and the toxicity curve (plot of effect against time of exposure) indicates a threshold or incipient concentration. With most test materials, this point is reached within 7 to 10 d, but it may not be reached within 21 d. Practical or economic reasons may dictate that the test has to be stopped at this point and a test be designed for a longer period of time.

time-weighted average concentration (TWA) The concentration of a substance to which a person is exposed in the ambient air, averaged over a period, usually 8 h. For example, if a person is exposed to 0.1 mg/m^3 for 6 h and 0.2 mg/m^3 for 2 h, the 8 h TWA will be $(0.1 \times 6 + 0.2 \times 2) / 8 = 0.125$ mg/m^3.

TLV *See* threshold limit value.

TOC Total organic carbon, often expressed as kg OC/kg solid. The organic matter content of soil and sediment is often determined by measurement of organic carbon. Typically, about half of all natural organic matter consists of carbon (OC ≈ 0.6 x OM).

tolerable daily intake (TDI) Analogous to an acceptable daily intake. The term tolerable is used for agents which are not deliberately added such as contaminants in food.

tolerable intake Estimated maximum amount of an agent, expressed on a body mass basis, to which each individual in a (sub) population may be exposed over a specific period without appreciable risk.

tolerance The ability to experience exposure to potentially harmful amounts of a substance without showing an adverse effect.

topical Pertaining to a particular (skin) area, e.g. a topical effect, that involves only the area to which the causative substance has been applied.

total organic carbon (TOC) The sum of dissolved organic carbon (DOC) and particulate organic carbon (POC) or suspended organic carbon (SOC).

total organic matter (TOM) The sum of dissolved organic matter (DOM) and particulate organic matter (POM) or suspended organic matter (SOM).

toxic Able to cause injury to living organisms as a result of physicochemical interaction.

toxic endpoint A toxic endpoint is a measure of the deleterious effect to an organism following exposure to a chemical. A large number of toxic endpoints are used in regulatory assessments of chemicals. These include lethality, generation of tumours (carcinogenicity), immunological responses, organ effects, development and fertility effects. It is the purpose of a toxicity test to determine whether a chemical has the potential to exhibit the toxic effect of interest, and in some cases, to determine relative potency. In QSAR analysis, it is important to develop models for individual toxic endpoints, and different methods may be required for different endpoints.

toxicant An agent or material capable of producing an adverse response (effect) in a biological system, seriously injuring structure and/or function or producing death.

toxicity Inherent property of an agent to cause an adverse biological effect.

toxicity curve The curve obtained by plotting the median survival times of a group of test organisms against the concentration on a logarithmic scale.

toxicity equivalency factor (TEF) Factor used in risk assessment to estimate the toxicity of a complex mixture, most commonly a mixture of chlorinated dibenzo-*p*-dioxins, furans and biphenyls: in this case, TEF is based on relative toxicity to 2,3,7,8-tetra-chlorodibenzo-*p*-dioxin.

toxicity identification evaluation (TIE) Describes a systematic pre-treatment sample (e.g. pH change, filtration, or aeration) followed by tests for toxicity. This evaluation is used to identify the agent(s) primarily responsible for lethal or sublethal toxicity in a complex mixture.

toxicity test Determination of the effect of a substance on a group of selected organisms under defined conditions. A toxicity test usually measures either the proportion of organisms affected *(quantal)*, or the degree of effect shown *(graded or quantitative)*, after exposure to specific levels of a stimulus (concentration or dose, or mixture of chemicals).

toxicodynamics *See* pharmacodynamics.

toxicokinetics *See* pharmacokinetics.

toxic unit The strength of a chemical (measured in some unit) expressed as a fraction or proportion of its lethal threshold concentration (measured in the same unit). The strength may be calculated as follows: toxic unit = actual concentration of chemical in solution / LC50. If this number is greater than 1.0, more than half of a group of organisms will be killed by the chemical. If it is less than 1.0, more than half the organisms will not be killed. 1.0 toxic unit = the incipient LC50.

Toxiguard Biomonitoring system comprising a submerged bed of continually developing microorganisms.

toxin Natural poison; a toxic organic substance produced by a living organism.

transcription Formation of mRNA, complementary to a string of DNA.

transcriptomics Techniques available to identify the mRNA from actively transcribed genes.

transgenic animals Genetically engineered animals carrying genes from a different species.

triggers/trigger values are criteria applied to results from tests (for fate or effects) which would prompt further studies, e.g. moving to the next tier.

TSCA (US) Toxic Substances Control Act.

tumour (neoplasm) Growth of tissue forming an abnormal mass. Cells of a benign tumour will not pread and cause cancer. Cells of a malignant tumour can spread through the body and cause cancer.

turbidity The extent to which the clarity of water has been reduced by the presence of suspended or other matter that causes light to be scattered and absorbed rather than transmitted (in straight lines) through the sample. It is generally expressed in terms of Nephelometric Turbidity Units.

ultimate aerobic biodegradation The breakdown of a chemical substance by micro-organisms in the presence of oxygen to carbon dioxide, water and mineral salts of any other elements present (mineralisation) and the production of new biomass and organic microbial biosysnthesis products.

ultimate median tolerance limit The concentration of a chemical at which acute toxicity ceases. Also called the incipient lethal level, lethal threshold concentration and asymptotic LC50.

UN United Nations.

UNCED UN Conference on Environment and Development (held in Rio de Janeiro (Brazil) in 1992).

uncertainty Imperfect knowledge concerning the present or future state of an organism, system or (sub) population under consideration.

uncertainty factor Reductive factor by which an observed or estimated no observed adverse effect level (NOAEL) is divided to arrive at the criterion or standard that is considered safe or without appreciable risk. *See also* assessment factor and safety factor.

upstream water Surface water (e.g. in a stream, river or lake) which is not influenced by the effluent (or other test substance), because it is removed from the source in a direction against the current or sufficiently far across the current.

uptake The process of sorbing a test chemical substance into or onto the test organ-isms.

uptake rate constant The first-order one-compartment constant to describe the uptake of a chemical substance by an organism from water.

use In the context of REACH means any processing, formulation, consumption, storage, keeping, treatment, filling into containers, transfer from one container to another, mixing, production of an article or any other utilisation.

use and exposure category In the context of REACH means an exposure scenario covering a wide range of processes or uses where the processes or uses are communicated, as a minimum, in terms of the brief general description of use.

USES Uniform System for the Evaluation of Substances. *See* Chapter 12.

validation Process by which the reliability and relevance of a particular approach, method, process or assessment is established for a defined purpose. Different parties define "Reliability" as establishing the reproducibility of outcome of the approach, method, process or assessment over time. "Relevance" is defined as establishing the meaningfulness and usefulness of the approach, method, process or assessment for a defined purpose.

van der Waals forces weak mutual attractions between molecules which can contribute to bonding between atoms.

verification Comparison of predicted with measured values, and the testing of assumptions and the internal logic of the model. This includes: (1) scientific verification that the model includes all major and salient processes, (2) the processes are formulated correctly, and (3) the model suitably describes observed phenomena for the use intended.

wastewater A general term that includes effluents, leachates and elutriates.

weight composition The distribution of organisms among the various weight classes present in the population. The sum of individual weights over all weight classes equals the population biomass. *See also* population biomass.

WHO World Health Organization.

WWTP Waste water treatment plant.

xenobiotic A man-made chemical or material not produced in nature and not normally considered a constituent component of a specified biological system. This term is usually applied to manufactured chemicals.

xenobiotic metabolism The chemical transformation of compounds foreign to an organism by various enzymes present in that organism. *See also* biotransformation and xenobiotic.

year Per year in the context of REACH means per calendar year, unless stated otherwise, for phase-in substances that have been imported or manufactured for at least three consecutive years, quantities per year shall be calculated on the basis of the average production or import volumes for the three preceding calendar years.

Zahn-Wellens test Biodegradation test to monitor the decay of DOC.

REFERENCES

1. Calow P. ed. 1993. *Handbook of Ecotoxicology*. Blackwell Scientific Publications, London, UK.
2. Duffus JH. 1993. Glossary for chem-ists of terms used in toxicology, IUPAC. *Pure and Appl Chem* 65:2003-2122.
3. Hodgson E, Mailman RB, Chambers JE. 1988. *Macmillan Dictionary of Toxicology.* Macmillan Press Ltd, London, UK.
4. Last JM. 1988. *A Dictionary of Epidemiology*. Oxford University Press, Oxford, UK.
5. Rand GM. 1995. *Fundamentals of Aquatic Toxicology*. 2nd ed. CRC Press, Washington, DC.
6. Richardson ML. 1990. *Risk Assessment of Chemicals in the Environment*. Royal Society of Chemistry, Cambridge, UK.
7. Stenesh J. 1989. *Dictionary of Biochemistry and Molecular Biology*. Wiley, New York, NY.
8. Suter GW. 1993. *Ecological Risk Assessment*. Lewis Publ, Chelsea, MI.
9. Organization for Economic Co-operation and Development. 2003. Description of selected key generic terms used in chemical hazard/risk assessment. Joint project with IPCS on the harmonization of hazard/risk assessment terminology. OECD Environment, Health and Safety Publications. Series on Testing and Assessment 44. OECD, Paris, France.
10. Commission of the European Communities. 2006. Regulation (EC) No 1907/2006 of the European Parliament and of the Council of 18 December 2006 concerning the Registration, Evaluation, Authorisation and Restriction of Chemicals (REACH), establishing a European Chemicals Agency, amending Directive 1999/45/EC and repealing Council Regulation (EEC) No 793/93 and Commission Regulation (EC) No 1488/94 as well as Council Directive 76/769/EEC and Commission Directives 91/155/EEC, 93/67/EEC, 93/105/EC and 2000/21/EC. *Off J Eur Union*, L 396/1 of 30.12.2006.
11. Commission of the European Communities. 2005. Opinion on the appropriateness of existing methodologies to assess the risks associated with engineered and adventitious products of nanotechnologies. Scientific Committee on Emerging and Newly Identified Health Risks (SCENIHR), Brussels, Belgium.
12. European Chemicals Bureau. 2005. Technical guidance document on preparing the chemical safety report under REACH. REACH Implementation Project 3.2. Report prepared by CEFIC, RIVM, the Federal Institute for Risk Assessment (BfR), Federal Institute for Occupational Safety and Health (BAuA), Ökopol, DHI Water & Environment and TNO Chemistry. European Commission, Joint Research Centre, Ispra, Italy (http://ecb.jrc.it/REACH/).
13. Organization for Economic Co-operation and Development. 2006. Guidance document on the validation of quantitative structure-activity relationships-(Q)SARs. Appendix 1. Glossary of (Q)SAR terminology. OECD, Paris.
14. European Centre for Ecotoxicology and Toxicology of Chemicals. 2001. Genomics, transcript profiling, proteomics and metabonomics (GTPM). An introduction. ECETOC Document No 42. ECETOC, Brussels, Belgium.
15. European Centre for Ecotoxicology and Toxicology of Chemicals. 2004.. ECETOC Technical Reports 90, 93 and 97. ECETOC, Brussels, Belgium.
16. National Research Council. 2006. *Human Biomonitoring for Environmental Toxicants*. National Academic Press, Washington, DC.
17. Zartarian V, Bahadori, T, McKone T. 2005. Adoption of an official ISEA glossary. J Exp Anal Environ Epidemiol 15:1-5.

INDEX